Medicinal Orchids of Asia

Eng Soon Teoh

Medicinal Orchids
of Asia

 Springer

Eng Soon Teoh, M.D., F.R.C.O.G., F.A.C.S.
Singapore
Singapore

ISBN 978-3-319-79598-0   ISBN 978-3-319-24274-3   (eBook)
DOI 10.1007/978-3-319-24274-3

Printed on acid-free paper

This Springer imprint is published by Springer Nature
The registered company is Springer International Publishing AG Switzerland

Dedicated to my wife Teoh Phaik Khuan

Preface

Rebecca Northern's *Home Orchid Growing* provided recreational reading when I was studying for my professional examination in London in 1968. I continue to enjoy reading about orchids and growing the plants. Noticing that orchids are now receiving attention as possible sources of medicinal compounds, I decided to study this area in depth because, having been privileged to participate in laboratory and clinical research for many years, I could employ my medical background to examine the extensive research that has been conducted on the use of orchids as medicine. There is maximum, continuous usage of orchids as medicine in Asia, for the longest period, and because Asia is my home, and also the area of much of my travel, the coverage of medicinal orchid species will be restricted to this region.

Before the advent of scientific medicine, all civilizations relied on plants and other natural materials to cope with pain, trauma and disease. Dominance and sex being important elements of Man's animal heritage, plants with alleged invigorating or aphrodisiac properties were eagerly sought after. Growth hormone, testosterone, sidenafil and tadalafil are only modern substitutes for much the same need.

Herbal medicine is practised worldwide, and not in Asia alone. In many continents, orchids have been used to nourish or to heal. *Salep* derived from terrestrial Mediterranean orchids and extracts of North American *Cypripediums* featured in European and American pharmacopoeia well into the later nineteenth century. Nevertheless, using orchids as medicine is most entrenched in China, which has a well-documented and the longest history of continuous usage, as well as the longest list of orchids in medicinal use. Worldwide, there are over 25,000 species in approximately 850 genera but only 2 % have a published medicinal usage. It is estimated that approximately 1200 species of orchids belonging to 174 genera occur in China. More than 250 species (or over 20 % of total species) in approximately 100 genera are already recorded in Chinese herbals as having medicinal usage. Four orchids are classical medicinal herbs whose origins can be traced to antiquity. They are still commonly employed in the Orient, from China to Korea, Japan and Southeast Asia, and by Chinese communities in far-flung countries. Chinese herbal soups occasionally contain an orchid.

Orchids also feature in Ayurvedic medicine. They constitute half of the eight items in the famed *Asthvargha* tonic. Nearly two dozen Indian medicinal names are attached to *Dendrobium plicatile*, many of which denote "life". It is one of two candidates that fulfil the description of a mysterious

Himalayan herb that restored life to the *Ramayana* hero, Lakshmana. Truckloads of this orchid are imported into India from neighbouring countries. Tribes in India and the surrounding countries employ orchids in various ways to treat trauma and disease.

Having experienced a shortage of medicines during World War II, after the war scientists embarked on an extensive search for alkaloids in plants. There was some difficulty in studying alkaloids in orchids because rarity or small plant size sometimes made it difficult to collect sufficient materials for the identification of chemical compounds. Nevertheless, this did not deter the Swedish team led by Bjorn Luning and Kurt Leander or the Australian team of Slaytor and Walker from undertaking alkaloid screening programmes. Together, by 1994 they had published the data on 1750 orchid species. Japanese, Chinese and Korean scientists were more interested in studying orchid species employed in traditional Chinese, Japanese and Korean medicine. There are extensive publications on phytochemicals from *Gastrodia elata*, *Dendrobium* spp. and *Bletilla striata*.

Forced to rely on its own resources to promote the health of its people when the "Bamboo Curtain" descended over the Peoples' Republic of China, the country undertook an extensive in-depth study of ethno-medicine which hitherto had only been provincial or tribal knowledge. In the process, hundreds of orchid species were incorporated into an expanded *Ben Cao* (*Materia Medica*). By unveiling the reproductive biology and pharmacology of important medicinal species, scientists in Guizhou and Yunnan managed to promote the cultivation of such species on an industrial scale and to manufacture specific compounds to meet with demand.

As a consequence of the remarkable economic and scientific advances made by China during the last three decades, traditional Chinese medicine is witnessing an interphase between tradition and science, with science dominating the transition. Concurrently, there has been an explosion of publications, including many on orchids. Several traditional medicinal orchids have since been shown to contain compounds that exert a wide range of pharmacological effects. Some compounds are unique in their actions, some are capable of protecting brain cells, while a few act against malaria parasites or liver flukes. Orchid extracts may protect the skin or liver. There are compounds that kill cancer cells, while others reverse cancer cell resistance to chemotherapeutic agents. *Bletilla striata* starch is employed for embolization to treat inoperable liver cancer and is being developed to assist in the delivery of drugs. Such compounds are being actively studied to determine whether they can be turned into potent drugs.

There are many studies on ethno-medicine in India, but until the advent of the internet knowledge of the subject was mainly confined to the nation. In Calcutta, Majumder and his team of collaborators isolated a long list of compounds from numerous Himalayan orchids. Korean scientists have also investigated the pharmacology of their orchids. However, it is difficult to extract these data because they are either not available in the English language or they are published in journals which are not read by doctors, nor by orchid enthusiasts with no interest in pharmacology. Research materials originating in China and Japan are mostly recorded in journals which employ their native languages and such data may remain unnoticed by Western readers for decades. I am fortunate to have been helped by medical colleagues who have mastered the Chinese language. They provided me with the necessary support that has enabled me to research the Chinese literature.

This volume thus sets out to compile a list of the medicinal orchids employed in Asia, briefly describe their identification, distribution, habitat and flowering season, how they are used and their pharmacology. Medicinal orchid species are described individually, grouped under genera. An overview sums up and discusses the research findings in each genus. Clinical information is occasionally included when it was necessary to explain the significance of the research data. Findings that hold promise of possible therapeutic application are highlighted.

The Introduction gives an overview of herbal medicine, its role, the extent of its usage and the risks involved. Since many Western readers may be unfamiliar with Asian medicinal tradition, the various traditions are described, followed by a discussion of the way herbs are collected and prepared. A brief survey of plant metabolites is intended as an introduction for the general reader to the types of bioactive chemical compounds isolated from orchids in which scientists are interested. Currently, the effects of medicinal orchid species are being proven by laboratory research on tissues or laboratory animals, but there is a paucity of data on human subjects. Nevertheless, this is a beginning. A placebo-controlled randomized clinical trial being the gold standard to prove efficacy of any new drug, a chapter is included to emphasize that it would be difficult, today, to endorse a herbal remedy without such experimental proof. A short final chapter explains the difference between anecdotal evidence and data which are statistically significant.

The use of orchids as aphrodisiacs in India, and the high Chinese demand for some medicinal orchids, worry conservationists who are concerned about the status of wild orchids in India (especially in the Himalayan region), China and continental Southeast Asia. Conservation, the preservation of biodiversity and problems in the enforcement are discussed in detail.

I wish to stress that this book was not written to endorse the use of orchids as medicinal herbs. I do not practise Chinese herbal medicine. Scientific research on medicinal orchids, although extensive and exciting, is still at an early stage, and it is inappropriate to draw clinical inferences or conclusions from in vitro studies and animal experiments. Much remains to be done to see that proper clinical trials are conducted to substantiate or disprove the claims for specific therapeutic uses of individual orchids or their compounds. I have kept an open mind on the subject, and I have tried to be impartial in my assessments of published articles. The claims that are made in the quoted publications are reported as they appear and no attempt was made to check their validity. On the issue of tribal and folk medicine, it seems unlikely that the authors actually witnessed the orchids they described being used to treat disease and the results, as ethnomedicinal botany is basically constituted from statements of folk practitioners. This being so, the portion of the present book which deals with herbal usage should be viewed as just a historical record and not as recommendations for use.

Scientists recognize that, when dealing with herbal products, there is the problem of dosage standardisation. Pharmacological potency of biological products vary from batch to batch and the potency of a herb cannot be accurately determined by weight. Only highly purified, single compounds can be standardised by weight. Recommendations for the use of chromatographic analyses to be employed for quality control have been made, but this is not universally practised. Furthermore, since chromatograms of batches will certainly vary, every batch requires bioassay and that is hardly possible to put into practice. There is also the problem of the proper identification of species. The best way to handle this last problem is for medicinal orchids to be commercially cultivated. Clonal propagation might bring herbs closer to a uniform standard, albeit the environment will produce differences.

Readers will notice that a good percentage of the references are made up of material published in Chinese. For their translation, I wish to thank my colleagues and friends, Ong Siew Chey, MD (Chicago), FACS, and Wu Dong Yin, MD (Shanghai), who devoted a great deal of time to this effort. I am grateful to Wibowo Sutjahto, MD, who translated the Dutch texts for me, and especially to Joseph Arditti, who spent many days going over the text with me, for his advice on how botanical data should be presented and for access to his treasure house of references. The Library of the Singapore Botanic Gardens and the National Library of Singapore were valuable sources of reference materials, and I thank their staff for their kind assistance. I also wish to record my thanks to Hong Hai, PhD, for reading through my text and making valuable suggestions. For colour photographs, I am indebted to Bhaktar Bhadhar Raskoti for his numerous images of Himalayan orchids, the Plant Photo Bank of China and Luo Yibo, PhD, for photographs of Chinese orchids, as well as Henry Oakley, Peter O'Byrne, Nima Gyeltshen, Liu Ming, Hubert Kurzweil, Tim Yam, Sathish Kumar, Suranjan Fernando, Jagdeep Varma and Ang Wee Foong for additional photographs of medicinal orchids. I thank the Orchid Society of Southeast Asia for giving me the privilege to photograph orchid plants exhibited at their meetings and shows, and my many friends in Singapore, Thailand, India, Bhutan and China who helped to provide access to orchid species: in particular, Mak Chin On, John Elliott, Rapee Sagarik, Apichart Jitnuyanond, Michael Ooi, Robert Ang, Udai C. Pradhan and Ganesh Pradhan, Ngawang Gyeltshen, Nima Gyeltshen and Rajendra Yonzone.

To complement the short descriptions of the plants, black and white drawings are featured to illustrate the general appearance of the plants in various genera. These were originally planned to be line drawings borrowed from old, classic publications. As it turned out, it was not possible to provide a comprehensive coverage through this avenue. However, I did manage to borrow some excellent line drawings from Abraham and Vatsala's *Introduction to Orchids* and I wish to thank the Director of Jawaharlal Nehru Tropical Garden and the Botanical Research Institute, Parlode, for permission to reproduce the drawings. For the remaining black and white illustrations, I am grateful to the Plant Illustrations Organization and the Missouri Botanical

Gardens for access to their vast collection of old publications whose material is now in the public domain. I also wish to thank the other libraries for the materials which are reproduced with their named source.

Singapore   Eng Soon Teoh, MD
January 2016

Contents

Part I

General Information

Introduction

Once, when the author expressed his concern over the breadth and depth of knowledge required of him as a doctor staying abreast with science, a learned colleague commented that, whereas it was difficult to pass a medical college examination, the actual practice of medicine was relatively easy. The reason stated was that, given rest and time, the body usually heals itself—a phenomenon that was well understood by ancient physicians. Hippocrates referred to this healing power of the body as "the vital force", almost an equivalent to the *qi* of the Yellow Emperor, father of Chinese medicine.

The body's capacity to heal itself supports many forms of complementary and alternative medical practices, with placebo effectively relieving pain and anxiety in up to a third of instances. Although complementary and alternative medicine (CAM) is no match for modern scientific medicine, which is now armed with sophisticated modalities for diagnosis and effective remedies for acute conditions, CAM is on the rise, even in developed countries, because of the growth of their elderly populations causing a concomitant increase in degenerative lesions which are accompanied by discomfort, pain, handicap, depression and dementia, conditions for which modern medicine is still lacking fully effective, safe and affordable remedies. Medical science is still struggling to achieve a better understanding of senescence so that it can either slow down the process or temper its

consequences. Meanwhile, effective treatments for sustained pain relief, loss of flexibility, strength, drive, mobility and confidence are all lacking. Consequently, many people turn to CAM for symptom relief, sometimes even with the overly optimistic hope of a cure. The Center for Disease Control of the United States of America (USCDC) reported that in 2004 slightly over a third of Americans had used at least one form of CAM in the previous year, expending billions of dollars from their own pockets (such expenses not being covered by medical insurance). In their search for new compounds to help people cope with old age and its attendant illnesses, scientists have looked again at orchids, and here indeed there seem to be compounds that might prove to be useful (Institute of Medicine (US), Committee on the Use of Complementary and Alternative Medicine in the United States 2005).

Before the nineteenth century, all forms of traditional medicine were pre-scientific, be they Western, Indian, Chinese or tribal (French 2003). They had their own systems of classifying disease, reasoning which stood apart from logic, and treatments that often baffle modern science. Nevertheless, many ancient observations on the history and prognosis of several diseases were accurate, and within each tradition numerous remarkable cures were achieved (Huard and Wong 1968; Horner 1982). Each tradition had its own array of medicinal plants alleged to possess properties relevant to the illness for which

© Springer International Publishing Switzerland 2016
E.S. Teoh, *Medicinal Orchids of Asia*, DOI 10.1007/978-3-319-24274-3_1

they were intended. During the past 100 years or so, advances in chemical, biochemical, pharmacologic and pharmaceutical research have enabled scientists to identify the key constituents of potent medicinal plants and poisons. Numerous compounds have been isolated, studied, purified, modified and improved upon to give us many of the therapeutic weapons that we use today against fever, pain, hypertension, infections and cancer.

The concerted search for new remedies from plants during the period of colonial exploration and expansion led to the discovery of such useful drugs as morphine (from Poppy) for pain relief; ephedrine (from the Chinese *ma hung*, or *Ephedra*) for treating asthma; quinine (from the Indonesian *Cinchona* bark) for malaria; reserpine (from the Indian herb, *Rauwolfia serpentina*) to lower blood pressure; curare (from the Amazonian arrow poison) which paralyses muscles so that safe surgery can be performed; and vincristine (from the Madagascan periwinkle), a cytotoxic agent that is effective against some forms of cancer. There are many others. The search continues and new compounds are constantly being added to the list. For instance, artemisinin (*qinghaosu*) from *Artemisia annua* for chloroquine-resistant cerebral malaria, and taxol from the Pacific Yew, which destroys ovarian and breast cancers, were only added to the medical armamentarium during the last 30 years.

Nevertheless, much of the ancient *Materia Medica* was non-specific, and often did not work. Proficient ancient healers were adept at separating minor discomforts and disabilities from serious illnesses, and non-fatal illnesses from fatal conditions. Reputable physicians did not arrive at diagnosis by guesswork. They honed their skills at clinical diagnosis by long, painstaking study, and by making careful observations on the background and living conditions of the patient, the history of the illness, its symptoms and signs and the patient's personality. This holistic approach determined the remedy that the physician recommended.

Famous physicians in China such as Hua Tuo (died 208 CE) of the Three Kingdoms Period and Sun Simiao (581–682 CE) of the Tang Dynasty recognised the greater prevalence of disease among the underprivileged, and they devoted their lives to the treatment of this group. Hua Tuo and Sun Simiao declined imperial appointments and often gave away their earnings to poor patients whose "illness" was principally lack of food. Sun Simiao exploited the Chinese concern for progeny to emphasize rest and good nutrition for pregnant or lactating women. Pregnancy was a special time when an ancient daughter-in-law might be treated as someone who was extra-special in the family (Teoh and Lam 1985). Sun Simiao's ideas on a healthy lifestyle and proper nutrition are incorporated in his *Qian Jin Yao Fang* (*A Thousand Golden Remedies*) and *Supplement to A Thousand Golden Remedies*, two works that summed up Chinese medical achievements up to the seventh century (Xie and Huang undated). Herbs were not considered to be a first-line treatment: they would be used only when simple measures did not work or if the illness worsened or was serious. Sun Simiao stressed that there was a proper time for the gathering of individual herbs which should thereafter be properly handled and processed; otherwise, 50–60 % of patients would experience no improvement when the herbs were administered. In other words, he appreciated that there is an appreciable placebo effect of perhaps 50 % with any treatment. Sun Simiao is revered as a god of medicine and is commonly shown in control of both *yin* (riding a tiger) and *yang* (holding a dragon) influences (Figs. 1.1 and 1.2).

Such physicians showed great wisdom, for, in the absence of really potent remedies such as antibacterial, antihypertensive and cardiotrophic drugs, and denied the ability to perform safe surgery for most conditions, the best part of a physician's remedy was the provision of an optimal setting for the body to heal itself. Hans Agren (1975), in his assessment of Chinese traditional medicine (TCM), reiterated that a good doctor often cures his patient without the aid of drugs with verified pharmacological properties. What in effect constitutes a good doctor is, among other things, trustworthiness and the power to fulfil the patient's needs for comfort and hope (Figs. 1.3 and 1.4).

Fig. 1.1 Sun Simiao depicted as a god of medicine at White Cloud Temple, Beijing. A Chinese Hippocrates, he placed great stress on medical ethics, described 4500 prescriptions, the proper gathering and handling of herbs and reverence for all forms of life (Photo: E.S. Teoh)

Fig. 1.2 Portrait of Sun Simiao on Chinese postage stamp

Fig. 1.3 White Cloud Temple in Beijing dating from the Yuan Dynasty (1271–1368) is dedicated to famous physicians (or gods of medicine). Main shrine in the background houses images of the Yellow Emperor and Shen Nong. A five-clawed dragon decorating the incense urn in the foreground denotes the imperial status of the shrine's occupants (Photo: E.S. Teoh)

Historical records state that ancient Indian and Chinese physicians gave a prognosis right at the start, and directed a patient to seek the expertise of other practitioners if they were unable to cope with the problem at hand. Nevertheless, if the physician was practising in an isolated area, there was no recourse to higher talent. He was required to rely on his own skills, and his own, few, medicinal resources.

The account of the life of Jivaka Komarabhacca who lived in Rajagriha (in modern Bihar) around 600 BC illustrates how a resourceful Indian physician dealt with medicinal herbs. Jivaka was the royal physician of King Bimbisara, and his duties involved looking after

Fig. 1.4 Shrine at Beijing's White Cloud Temple dedicated to Sun Si Miao also houses an image of Hua Tuo. This reverence for famous physicians characterizes the high regard accorded to Traditional Chinese Medicine in Chinese culture (Photo: E.S. Teoh)

the Buddha and the order of monks. In his youth, he went to study under a famous Master in the city of Taxila (now in Pakistan). After 7 years of study, Jivaka enquired from his Master when his studies would be complete. The Master gave him a basket and asked him to go round the city to collect all the plants with unknown medicinal properties so that the Master could explain their usage to round off his knowledge. At nightfall, Jivaka came back empty-handed. He told the Master that within one *yojana* (13 km) of Taxila, he did not find a single plant whose usage he did not comprehend. Thereupon, the Master announced that his training was complete and he gave Jivaka some money to help with his journey home.

Jivaka proved to be an excellent physician and surgeon. The record of the cures that he achieved showed that his skills far surpassed those of the leading physicians throughout all of northern India (Horner 1982).

On the question of the herbs, young Jivaka understood that it was not always possible to find the plants that contained the specific medicinal properties required to treat the disease on hand. However, as a physician, he could not tell his patient that he did not have the correct remedy, so he would need to use a placebo. For that purpose, almost any non-poisonous plant could be used. In the event, should someone be on hand to record the identity of the placebo plant used by such a distinguished physician, it is conceivable that plant might eventually be incorporated into the local *Materia Medica*.

In the High Chinese Medical Tradition, the *Ben Cao* (*Materia Medica*) forms a major component of the physician's therapeutics. It lists the officially recognised herbs (*Guan Yao*, official medicines). Chinese physicians additionally employ physical methods to treat disease, such as acupuncture, manipulation, moxibation (applying external heat) and cupping. Shamanistic therapists employ apotropaics (incantations, amulets, charms) as well as local herbs. They belong to the Small or Low Tradition that is also practised by lay men and lay women. In the old days, grandmothers in Indonesia and Malaysia (mine among them) would utter a mantra over the lesion when they applied herbal poultices on painful insect and snake bites. From my childhood recollection, oftentimes the patient was instantly relieved. Within this Low Tradition, there is a folkloric *Materia Medica* which is derived by extracting popular remedies from classical pharmacopeias and adding to them simple folk remedies based on locally available plants, referred to by Chinese writers as "uncured or wild herbs" (Hu 1971).

Orchids as Medicine or Food

Apart from three main drugs which happen to be orchids in the *Shen Nong Ban Cao Jing* or the *Materia Medica of Shen Nong*, namely *Tianma* (*Gastrodia elata*) (Fig. 1.5), *Shihu* (various species of lithophytic *Dendrobium*) (Fig. 1.6) and *Baiji* (*Bletillia striata*) (Fig. 1.7), the vast majority of orchids used as medicinal herbs in China

Fig. 1.5 *Tianma* (tubers of *Gastrodia elata*) offered in a herbal shop in Lijiang, Yunnan Province, China (Photo: E.S. Teoh)

Fig. 1.6 *Dendrobium catenatum* (syn. *Dendrobium officinale*) is one of two *Dendrobium* species mentioned as *shihu* in *Shen Nong Bencao Jing* (first century CE) (Photo: E.S. Teoh)

Fig. 1.7 *Baiji* are tubers of *Bletilla striata* (as *Cymbidium hyacinthinum* Sm. in Curtis Botanical Magazine, vol. 36: t. 1492 (1812) drawn by S.T. Edwards. Courtesy of Missouri Botanical Gardens, St. Louis, USA

fall within this Low Tradition. The same is probably the case with the use of orchids in Indian Ayurvedic and Unnani Medicine (Akarsh 2004; Rao and Sridhar 2007; Trivedi et al. 1980; Dagar and Dagar 2003; Sivakumar et al. 2003; Dash et al. 2008). Medicinal usage of native orchids in Thailand, Malaysia, Indonesia, the Philippines and the rest of tropical Asia is similarly the provenance of provincial herbalists (Burkill and Haniff 1930; Gimlette and Thomson 1939; Beekman 2002; Chuakul 2002).

It is estimated that about 25–50 % of medications used today are derived directly from plants or are modified forms of secondary metabolites of plants. Many pharmaceutical firms and research institutes devote a sizable portion of their resources to the investigation of medicinal plants and related species, even though the search is seldom fruitful. Nevertheless, since only a tiny fraction of the botanical world has been examined by chemists and pharmacologists (Kong et al. 2003), there is much scope for new discoveries.

Apart from being decorative plants grown primarily for the sake of their flowers, orchids have other uses. Worldwide, there are 25,000–30,000 species of orchids, and about 600 species (2 %) have recorded medical usage. China is an exception: approximately 20 % of its native orchids enjoy medicinal usage.

Four orchid species have been used in China as herbal remedies since the dawn of history. All four orchids are remarkable in their ecological characteristics. *Dendrobium catenatum* (syn. *D. officinale*) (Fig. 1.6) and *D. moniliforme*, or *shihu* in TCM terminology, were consumed as tonics. In what is possibly the oldest Chinese account of an orchid habitat, Tao Hongjing's *Ming Yi Bie Lu* (*Additional Records of Famous Physicians*) published around 520, reported that *shihu* grew on rocks by the waterside, in the valleys of Lu An (Chen and Tang 1982). This saxicolous (lithophytic) characteristic endowed *shihu* with tonic potential. Exotic *Chi Jian* (red arrow), now universally referred to as *Tianma* (*Gastrodia elata*), was a parasitic orchid which lived underground, only sending out its arrow-shaped red inflorescence when it sought to cross-fertilize. *Tianma* enjoyed a reputation for an ability to promote recovery from strokes and other neurological disorders. *Baiji* (*Bletilla striata*) flourished in extensive colonies by producing potent phytoalexins which kept other plants at bay. It was used to treat tuberculosis and bronchiectasis, two very common and incurable conditions in impoverished ancient China, and as a remedy for chapped skin and bleeding sores. These three orchidaceous herbs are still widely consumed, their usage extending to Korea, Japan and Southeast Asia, countries that long ago experienced an influx of Chinese culture (Han et al. 1998; Zhao et al. 2006). Today,

the list of orchid species used in Chinese remedies run into the hundreds, with dozens of saxicolous and epiphytic *Dendrobium* species now classified as *shihu*.

Orchids have also seen medicinal usage for centuries in India, Southeast Asia, the Middle East, Africa, Europe, the Americas and in Oceania. Vanilla comes from an orchid with the same name. Aztecs used *tlilxochitl* (the original name for vanilla) to flavour *choklatl* (de la Cruz and Bandianus 1552; Ossenbach 2009), a delicacy which was introduced by the Spaniards into Europe in the sixteenth century. It was used as appetiser, digestive, tonic and aphrodisiac (Fig. 1.8). Over time, vanilla was put into perfume, food and drinks (Ecott 2004). To meet the demand for vanilla, use was made of slave labour to grow and hand-pollinate *Vanilla planifolia* in Reunion and other islands in the Indian Ocean, far removed from the orchid's homeland in Central America, but in climates similar to the original (Arditti et al. 2009). The chemical

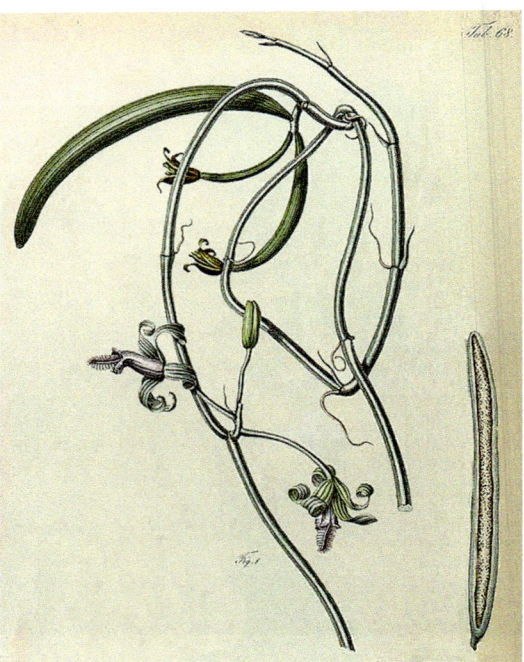

Fig. 1.8 *Vanilla planifolia* Jacks. ex Andrews. Adapted from Blume C.L., *Rumphia*, vol. 1: t.68, Fig. 2 (1835). Courtesy of Missouri Botanical Gardens, St. Louis, USA

constituents of the vanilla spice have been extensively studied. Artificial vanillin is now available and made by hydrolysing lignin.

Vainilla chica (little vanilla) and *vainilla en arbol* (vanilla on a tree) refer to the tall South American slipper orchid, *Selenipedium chica*, the seed capsules of which are sometimes used as a substitute, when vanilla is not available, by the Indians living in the mountains of Panama (Fig. 1.9). In Brazil, fruit capsules of the native species, *Leptotes bicolor*, have been used to flavour ice cream (Hawkes 1943) (Fig. 1.10). However, worldwide, vanilla is a dominant flavour in the dessert.

Faham tea, derived from the leaves of *Jumella fragrans* (syn. *Angraecum fragrans*), an epiphytic orchid endemic to Reunion and Mauritius, was introduced to Paris just before 1866 (Fig. 1.11). It was intended to replace Chinese tea which had not been well received in France "owing to the wakefulness resulting from its

Fig. 1.10 *Leptotes bicolor* Lindl. (as *Leptotes Serrulato* Lindl.). From: Lindley J., *Sertum Orchidaceum*: t.11 (1838). Courtesy of Missouri Botanical Gardens, St. Louis, USA

use"; Faham did not have this effect. The manufacturer's blurb mentioned that Faham possessed a "most agreeable perfume; after being drunk it leaves a lasting fragrance in the mouth, and in a closed room the fragrance of it can be recognized long after." Spirits, especially rum, in small amounts enhanced its aroma. An English writer commented that the perfume from the teapot was very agreeable; it was an indisputable novelty and, should Faham come to general use, the teapot would serve the additional function of a perfume vaporiser (Jackson 1866).

Faham tea failed and is no longer in fashion. This was possibly due to cost; it was not possible to obtain large amounts of leaves from the *Jumella* plants without unduly stressing them. However, Faham still enjoys a role as a native remedy for childhood eczema and diarrhoea. It possesses antispasmodic properties. The volatile

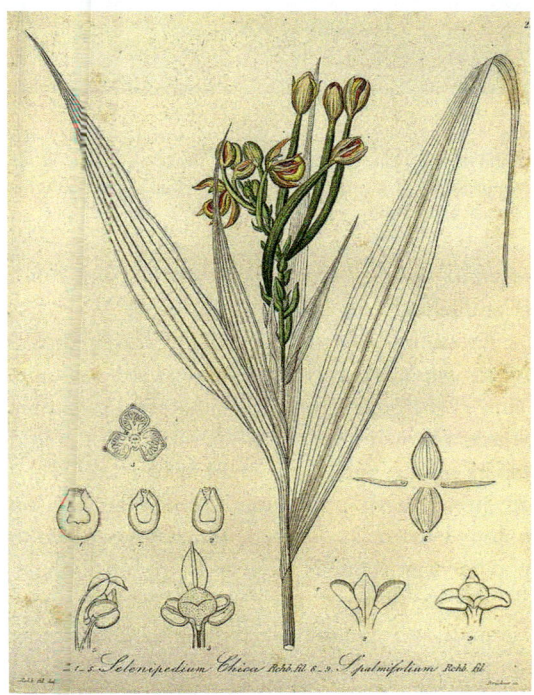

Fig. 1.9 *Selenipedium chica* Rchb. f. From: Reichenbach, H.G., Arnott G.A.W., *Xenia Orchidacea*, vol. 1: t. 2 (1900). Courtesy of Missouri Botanical Gardens, St. Louis, USA

Fig. **1.11** *Jumellea fragrans* (Thouars.) Schltr. (as *Angraecum fragrans* Thouars.) *Curtis Botanical Magazine*, vol. 117 (ser. 3, vol. 47): t. 7161 (1891) drawn by Matilda Smith. Courtesy of Missouri Botanical Gardens, St. Louis, USA

Fig. 1.12 *Dendrobium chrysotoxum* Lindl. is easily distinguished from other yellow flowered, deciduous *Dendrobium* by its fat, ribbed, hard pseudobulbs (Photo: E.S. Teoh)

constituents of faham are made up of coumarin and 99 minor components which include kaurenes and phytadienes (Sing et al. 1999).

Currently, a new tea prepared from dried *Dendrobium chrysotoxum* flowers is popular in Xishuangbana, Yunnan Province in China (Fig. 1.12). Leaves of *D. salaccense* Lindl. (syn. *Grastidium salaccense* Bl.) impart a delicate aroma to rice if added during cooking (Burkill 1936). The orchid is not widely distributed nor cultivated for domestic usage, and thus the practice is not in vogue. Common *pandan* leaves are used instead for that purpose. In Fiji, *Dendrobium tokai* (Fig. 1.13) and *Calanthe hololeuca* are used as tonics, and *Oberonia equitans* (syn. *O. glandulosa*) and *Taeniophyllum*

fasciola (syn. *T. parhamiae*) for pain relief (Cambie and Ash 1994).

Australian aborigines resort to eating orchid tubers when there is a desperate food shortage. This is different from the situation in Zambia where, following urbanisation, millions of orchid tubers are brought to the markets for sale. They are mixed with groundnuts to make an African polony known as *chikanda*, which has become so popular that there is even a recipe on Google. Such is the demand that Zambia's orchid supply is rapidly being depleted, and the tubers are now sourced from the Southern Highlands of Tanzania, Angola, the Democratic Republic of Congo and Malawi. The main species in the trade belong to three genera, *Disa*, *Satyrium* and *Habenaria* (Veldman et al. 2014).

Fig. 1.13 *Dendrobium tokai* Rchb. f., a native of Papua New Guinea and some South Pacific Islands. From: Seeman, B, *Flora Vitiensis,* vol. 2: t. 92 (1873) (drawn by W. H. Fitch). Courtesy of Missouri Botanical Gardens, St. Louis, USA

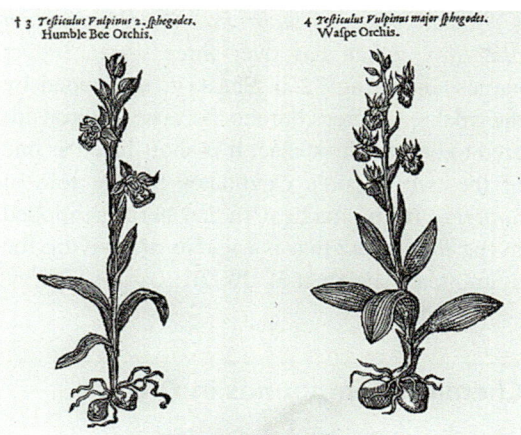

Fig. 1.14 *Ophrys* from Gerarde's *Herball* (1597)

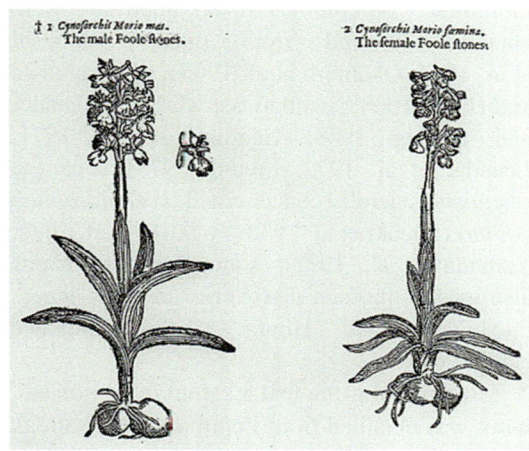

Fig. 1.15 *Orchis* from Gerarde's *Herball* (1597)

Orchids as Aphrodisiacs

In Europe, the Middle East and India, tubers of various terrestrial orchids which resemble a pair of testicles of uneven size were considered to be potent aphrodisiacs. For almost 2000 years, *Ophrys* and *Orchis* tubers (Figs. 1.14 and 1.15) were ground up to make a drink called *salep* (*saalep*). Lacking these orchids, Indians substituted *Dactylorhiza hatagirea* and various species of *Eulophia* (Dymock 1885; Chopra 1930; Caius 1936; Duggal 1972; Trivedi et al. 1980; Jalal et al. 2008). *Salep* is still available in Turkey where, *inter alia*, it is made into ice cream. Many *salep* orchids became so rare that the Turkish government had to ban the export of *salep*. Nowadays, *salep* usually consists of flour with artificial orchid flavour.

Asthavarga is an Indian tonic which also substitutes as a local aphrodisiac. It is a cocktail of eight herbs, which currently include up to four orchid species, namely: *Malaxis muscifera* (Ayurvedic name: *Jivaka*); *Crepidium acuminatum* (syn. *M. acuminata*) (*Rishbhaka*); *Habenaria intermedia* (*Riddhi*); and *Platanthera edgeworthii* (syn. *H. edgeworthii* (*Vriddhi*) (Singh and Duggal 2009; Dhayani et al. 2011). Nevertheless, Sanskrit works on which this preparation is based carry only casual descriptions of the plants involved, and there are insufficient details for plant taxonomists to be absolutely certain about the identities of the orchids (Van Steenis 1948).

Flickingeria fimbriata (*Dendrobium plicatile*), which has over three dozen Indian names, including 32 in Sanskrit, is shipped by the truckload across borders because it is considered to be an aphrodisiac. It is short-listed as one of the two possible candidates for the role of *Sanjeevani*, the magical Indian herb mentioned in the *Ramayana* that is capable of reviving the dying (Ganeshaiah et al. 2009)!

Chemical Compounds in Orchids

During the 1960s, Japanese and Swedish researchers began to re-examine orchids for alkaloids because many useful drugs such as morphine, quinine and ergometrine were alkaloids derived from plants (Inubushi et al. 1964; Okamoto et al. 1966a, b; Nishikawa and Hirata 1968; Nishikawa et al. 1969; Leander and Luning 1968; Hedman et al. 1971; Leander et al. 1973; Luning 1974). Japanese scientists generally concentrated on *Dendrobium* (*shihu*) (Suzuki et al. 1932; Inubushi et al. 1972; Yamada et al. 1972); some Japanese teams also made important discoveries in other genera (Nishikawa and Hirata 1968; Nishikawa et al. 1969).

Morphine was the first alkaloid to be isolated, but it was obtained from Poppy and not from an orchid. The use of opium for pain relief is ancient, as it was mentioned by Theophrastus (371–287 BC) and Dioscorides (c. 40–90 CE) (Figs. 1.16 and 1.17). However, crude opium was unreliable and dangerous because sometimes the standard dose did not work, and on other occasions the same amount was lethal. To solve this problem, in 1803, Friedrich Wilheim Serturner (1783–1841), a chemist's assistant with no scientific education and working under primitive conditions in the Westphalian town of Pederborn, tried using solvents to see what he could extract from crude opium (Fig. 1.18). Crystals appeared when he poured liquid ammonia over the raw opium. By washing and further treatment with other solvents, Serturner managed

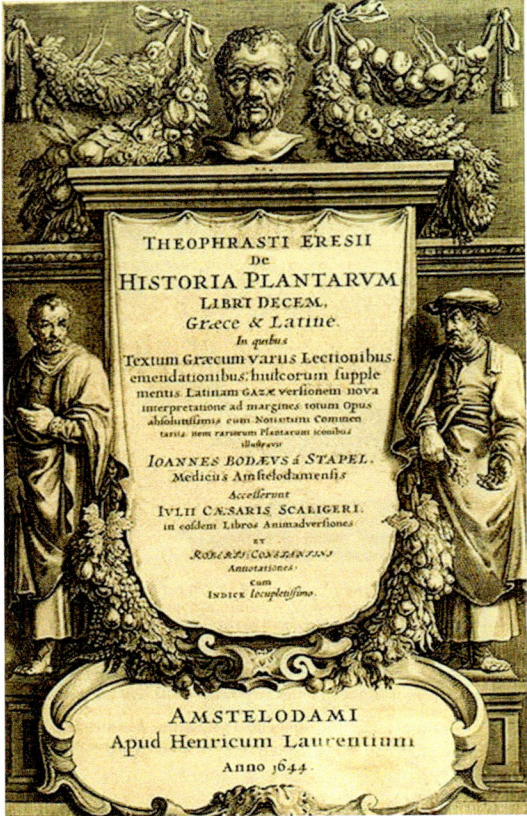

Fig. 1.16 Frontispiece of a seventeenth century edition of *Historia Plantarum* by Theophrastus (371–287 BC)

to isolate a white crystalline residue. He called it the "somniferous principle of opium". In the evenings after work, Serturner fed the crystals to mice and dogs in increasing amounts to observe the effects, ending only when the quantities were sufficient to kill the animals. Serturner thus established the two critical values now required for every drug that is in use: the minimum effective dose, and the lethal dose. He then experimented on himself and his friends. After 14 years of study, he concluded that the white powder from opium was capable of relieving pain and, because of the dreams associated with its use, he changed its name to morphium after the Grecian Morpheus, son of Somnus and god of dreams. The term was later changed to "morphia" and "morphine". Saturner's discovery

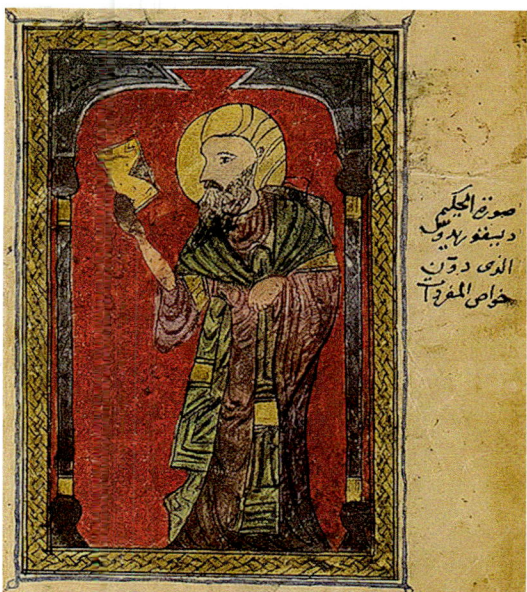

اور الحكيم
ديسقو لريدس
الذي دون
خواص العقار

Fig. 1.17 Dioscorides (40–90 CE) exerted tremendous influence on the therapeutics of Europe and the Middle East up to the end of the eighteenth century

Fig. 1.18 Friedrich Wilhelm Serturner (1783–1841) made the first extraction of a pure compound from raw herbal materials. He isolated morphine from opium

of every compound in a plant. Most importantly, in clinical practice, dosage could be defined by weight of pure substance rather than by weight of raw material. Dose standardisation is unreliable for raw biological products.

Using similar methods, Johann Buchner in Munich isolated salicin from willow bark, an ancient remedy from pain which had been in use from the time of Hippocrates. In 1897, Felix Hoffmann produced acetylsalicylic acid (or aspirin) from salicin and showed that this derivative had fewer side effects than the original compound. J. R. Vane in 1971 discovered that acetylsalicylic acid worked by suppressing prostaglandins, thromboxanes and cyclo-oxygenase. The discovery opened the door for the development of a whole new class of non-steroidal anti-inflammatory drugs. In 1982, Vane was awarded the Nobel Prize for Medicine.

Ergometrine is an alkaloid from *Claviceps purpura*, the fungus which causes miscarriages in cattle. It contracts the uterus and is used to prevent bleeding after childbirth. It has saved thousands of lives. During the twentieth century, separation of organic compounds has been much simplified by chromatography, with its numerous refinements. These processes facilitated the study of the constituents of orchids.

In 1995, it was reported that a phenanthrene, denbinobin, isolated from *Dendrobium nobile* (Fig. 1.19), exerted cytotoxic effects on several strains of human cancers in test-tube culture (A549 human lung cancer, SK-OV-3 human ovarian adenocarcinoma and HL-60 human promyelocyte leukaemia), but no clinical studies followed (Lee et al. 1995). Now, it has been shown, again in vitro, that denbinobin, diminishes the levels of expression of the decoy receptor-3 and acts synergistically with Fas-ligand to induce apoptosis (programmed cell death) in a human pancreatic cancer cell line (Magwere 2009). Whether denbinobin can be developed into an adjuvant to treat drug-resistant cancers is the $64 million question.

Other cytotoxic agents have been identified in *Dendrobium* (Ho and Chen 2003; Zhang et al. 2008; Chen et al. 2008) and several

was hailed by numerous scientific bodies in Europe, and he was conferred doctorates by the universities of Jena, Marburg, Berlin, St. Petersburg, Batavia, Paris and Lisbon (Krikorian 1967).

The importance of Serturner's work is that it demonstrated the possibility of isolating the active principle or principles from a medicinal plant simply by making use of their differential solubility. This allows a scientist to study the actions of each compound in isolation and to discover the synergistic and antagonistic actions

Fig. 1.19 *Dendrobium nobile* Lindl. From: Lindley, J, *Sertum Orchidaceum,* t. 3 (1838) (drawing by Miss Drake). Courtesy of Missouri Botanical Gardens, St. Louis, USA

Fig. 1.20 *Dienia ophrydis* (J. Koenig) Seidenf. (syn. *Malaxis latifolia* Sm.) (From: Teoh Eng Soon, *Asian Orchids,* Singapore: Times Books Intern., 1980)

other orchid genera. There are claims of a hepatoprotective effect of *Anoectochilus formosanus* extracts; neuroprotective effects of *Gastrodia elata*, *Dactylorhiza viriis* (syn. *Coeloglossum viride*) and *Dendrobium chrysotoxum*; and antiviral activities in *Cymbidium* and *Epipactis helleborine* (see Chap. 8). Malaxin, first isolated from *Dienia ophrydis* (syn. *Malaxis latifolia*) (Fig. 1.20), is used for treating uncomplicated falciparium malaria which is resistant to quinine and chloroquine.

This explosion of laboratory research on the pharmacology of orchids in China, Taiwan and Japan in the past two decades is a welcome development. One can expect the constant arrival of new information given the rapid advancement of science in China and her wealth of medicinal orchids. Nevertheless, we lament the absence of clinical trials which are crucial to evidence gathering. There is no substitute for well-designed and well-conducted clinical trials to prove the efficacy and safety of any treatment of disease.

Are Orchids the Real Source of the Chemical Constituents?

Many of the chemicals isolated from orchids come from the plants themselves, but in the case of the parasitic orchid, *Gastrodia elata*, several potent chemicals have been isolated from culture media supporting pure cultures of its symbiotic fungus, *Amillaria mellea* (Xiong and Huang 1998; Ng et al. 2007). This suggests that the fungus, not the orchid, is the true site for the production of such substances. Research teams are already setting out to determine whether this may also be the case with autotrophic orchids, but the issue is complex. Scientists at the Chinese Academy of Medical Science in Beijing who studied fungal species associated with 10 species of *Dendrobium* encountered extensive biodiversity in the endophytic fungi. They uncovered 37 genera and 60 species in 401 fungal endophytic cultures. Furthermore, considerable variability occurred in host fungus specificity among the orchids (Chen et al. 2011).

Orchids as Talismans

Although the full extent of ethnobotany will not be explored in the present work, a few orchids which are used as charms (apotropaics) are described because this may be regarded as a form of psychotherapy. This is certainly the case in Africa (Chinsamy et al. 2011). On this point, it is interesting to note that the Chinese word for orchid, *lan* (originally restricted to *Cymbidium*), was derived from a word which means 'to ward off'. In his famous pharmacopoeia published in 1596, Li Shihzhen mentioned using 'orchids to ward off'. Hu (1971) explained that the ancients wore *Cymbidium* flowers in their hair when they climbed mountains to ward off evil spirits who might cause them to fall ill (Fig. 1.21). The floral scent masked the human odours which would otherwise attract the spirits.

The *Shih Qing* (*Book of Songs*) which preceded Confucius (sixth century BC) describes the practice:

The waters of the Chen and Wei have begun to swell.
  Young men and women wear lan in their hair.
  Travelling together they reach the headwaters of the Wei.
  Tender, adorable, happy, and zestful,
  This young man and his mate.
  They amuse each other: play tricks mutually.
  Presently, he offers her a peony.

Fig. 1.21 *Cymbium ensifolium* (L.) Sw. (Photo: E.S. Teoh)

References

Agren H (1975) A new approach to Chinese traditional medicine. Am J Chin Med 3(3):207–212

Akarsh M (2004) Newsletter of ENVIS NODE on Indian Medicinal Plants 1(2), June

Arditti J, Rao AN, Nair N (2009) Hand pollination of vanilla: how many discoverers? In: Kull T, Arditti J, Wong SM (eds) Orchid biology, vol 10. Springer, Dordrecht

Beekman EM (trans ed) (2002) Rumphius orchids. Orchid texts from the Ambonese Herbal by Georgius Everhardus Rumphius. Yale University Press, New Haven

Burkill IH, Haniff M (1930) Malay village medicine. Gard Bull Straits Settl 6:165–321

Burkill IH (1936) A dictionary of economic products of the Malay Peninsula, vol II. Crown Agents for the Colonies, London (with contributions by Birtwistle W, Foxworthy FW, Scrivenor JB, Watson IG)

de la Cruz M, Bandianus J (1552) Codex de la Cruz-Bandianus (The Bandanus Manuscript)

Caius JF (1936) The medicinal and poisonous plants of India. J Bombay Nat Hist Soc 38(4):791–799

Cambie RC, Ash J (1994) Fijian medicinal plants. CSIRO, East Melbourne

Chen J, Hu KX, Hou XQ, Guo SX (2011) Endophytic fungi assemblages from 10 dendrobium plants (Orchidaceae). World J Microbiol Biotechnol 27 (5):1009–1016

Chen TH, Pan SL, Guh JH, Chen CC, Huang YT, Pai HC, Teng CM (2008) Denbinobin induces apoptosis by apoptosis-inducing factor releasing and DNA damage in human colorectal cancer HCT-116 cells. Naunyn Schmiedebergs Arch Pharmacol 378(5):447–457

Chen SC, Tang T (1982) A general review of the orchid flora of China. In: Arditti J (ed) Orchid biology. Reviews and perspectives, vol 2. Cornell University Press, Ithaca

Chinsamy M, Finnie J, van Staden J (2011) The ethnobotany of South African medicinal orchids. S Afr J Bot 77:2–9

Chopra RN (1930) The indigenous drugs of India, Calcutta: the art press. Republished as Chopra's Indigenous Plants of India, 2nd edn. Academic, Kolkata (1986)

Chuakul W (2002) Ethnomedical uses of Thai Orchidaceous plants. Mohidol Univ J Pharm Sci 29(3–4):41–45

Dagar HS, Dagar JC (2003) Plants used in ethnomedicine by the Nicobarese of Islands in Bay of Bengal, India. In: Singh V, Jain AP (eds) Ethnobotany and medicinal plants of India and Nepal. Scientific, Jodhpur, pp 773–784

Dash PK, Sahoo S, Bal S (2008) Ethnobotanical studies on orchids of Niyamgiri Hill Ranges, Orissa. India Ethnobot Leaft 12:70–78

Dhayani A, Nautiyal BP, Nautiyal MC (2011) Importance of Astavarga plants in traditional systems of medicine in Garhwal, Indian Himalaya. Int J Biodivers Sci Ecosyst Serv Manage 6(1–2):13–19

Duggal SC (1972) Orchids in human affairs (a review). Pharm Biol 11(2):1727–1734

Dymock W (1885) The vegetable materia medica of Western India. Education Society's Press, Bombay

Ecott T (2004) Vanilla. Travels in search of a luscious substance. Michael Josep, London

French R (2003) Medicine before science. Cambridge University Press, Cambridge

Ganeshaiah KN, Vasudeva R, Shaanker RU (2009) In search of Sanjeevani. Curr Sci 97(4):484–489

Gimlette JD, Thomson HW (1939) A dictionary of Malayan medicine. Oxford University Press, London

Han BH, Suh Y, Chi HJ (1998) Medicinal plants of the Republic of Korea. WHO Regional Publications. Pacific series no. 21. WHO, Manila

Hawkes AD (1943) Economic importance of the Orchidaceae. Am Orchid Soc Bull 11:412–415

Hedman K, Leander K, Lunin B (1971) Studies on orchidaceae alkaloids. N-isopentenyl derivatives of dendroxine and 6-hydroxydendroxine from

Dendrobium fredricksianum Lindl. and Dendrobium hildebrandii Rolfe. Acta Chem Scand 25 (3):1142–1144

Ho CK, Chen CC (2003) Moscatilin from the orchid Dendrobium loddigesii is a potential anticancer agent. Cancer Invest 21(5):729–736

Horner IB (ed) (1982) Book of discipline, vol I–IV. Pali Text Society, London

Hu SY (1971) Orchids in the life and culture of the Chinese people. Chuing Chi J 10:1–26

Huard P, Wong M (1968) Chinese medicine (trans: Fielding B). World University Library, London

Institute of Medicine (US), Committee on the Use of Complementary and Alternative Medicine by the American Public (2005) Complementary and alternative medicine in the United States. National Academies Press (US), Washington, DC

Inubushi Y, Tsuda Y, Konita T, Matsumoto S (1964) Shihunine. A new phthalide pyrrolidine alkaloid. Chem Pharm Bull 12:749–750

Inubushi Y, Kikuchi T, Ibuka T et al (1972) Total synthesis of the alkaloid dendrobine. J Chem Soc Chem Commun 1972:1252–1253

Jackson JR (1866) Orchid Tea. The Gardeners' Chronicle and Agricultural Gazette for 1866, p 315

Jalal JS, Kumar P, Pangtey YPS (2008) Ethnomedicinal orchids of Uttarakhand, Western Himalayas. Ethnobot Leaft 12:1227–1230

Kong JM, Goh NK, Chia LS, Chia TF (2003) Recent advances in traditional plant drugs and orchids. Acta Pharmacol Sin 24(1):7–21

Krikorian AD (1967) The story of opium. Cornell Plant 23 (2):19–24

Leander K, Luning B (1968) Studies on Orchidaceae alkaloids. 8. An imidazolium salt from Dendrobium anosmum Lindl. and Dendrobium parishii Rchb.f. Tetrahedron Lett 8:905–908

Leander K, Luning B, Westin L (1973) Studies on orchidaceae alkaloids. XXXIV. The absolute configuration of cryptostyline I, II, and III, three 1-Phenyl-1,2,3,4-tetrahydroisoquinolines from Cryptostylis fulva Schlr. Acta Chem Scand 27:710

Lee YH, Park JD, Baek NI, Kim SI, Ahn BZ (1995) In vitro and in vivo anti-tumoral phenanthrenes from the aerial parts of Dendrobium nobile. Planta Med 61 (2):178–180

Luning B (1974) Alkaloids of the Orchidaceae. In: Withner CL (ed) The orchids. Scientific studies. Wiley, New York

Magwere T (2009) Escaping immune surveillance in cancer: is denbinobin the panacea? Br J Pharmacol 157 (7):1172–1174

Ng LT, Wu SJ, Tsai JY et al (2007) Antioxidant activities of cultured Armillaria mellea. Appl Biochem Microbiol 43(4):444–448

Nishikawa K, Hirata Y (1968) Chemotaxonomical alkaloid studies III. Further studies on liparis alkaloids. Tetrahedron Lett 9:6289–6291

Nishikawa K, Miyamura M, Hirata Y (1969) Chemotaxonomical alkaloid studies. Structures of liparis alkaloids. Tetrahedron 25(13):2723–2741

Okamoto T, Natsume M, Onaka T, Uchimaru F, Shimizo M (1966a) The structure of dendroxine. The third alkaloid from *Dendrobium nobile*. Chem Pharm Bull (Tokyo) 14(6):672–675

Okamoto T, Natsume M, Onaka T, Uchimaru F, Shimizo M (1966b) The structure of dendramine (6-oxydendrobine) and 6-oxydendroxine. The fourth and fifth alkaloid from *Dendrobium nobile*. Chem Pharm Bull (Tokyo) 14(6):676–680

Ossenbach C (2009) Orchids and orchidology in Central America. 500 years of history. Lankesteriana 9 (1–2):1–268

Rao TA, Sridhar S (2007) Wild orchids in Karnataka. A pictorial compendium. Institute of Natural Resources Conservation, Education, Research and Training (INCERT), Bangalore

Sing AS, Smadja J, Brevard H et al (1999) Volatile constituents of Faham [Jumella fragrans (Thou.) Schltr.]. J Agric Food Chem 40:642–646

Singh A, Duggal S (2009) Medicinal orchids: an overview. Ethnobot Leaft 13:351–363

Sivakumar A, Subramanian MS, Karunakaran M, Burkanudeen A (2003) Ethnobotany of Poliyars of Anailamai Hills, Tamil Nadu. In: Singh V, Jain AP (eds) Ethnobotany and medicinal plants of India and Nepal. Jodhpur, Scientific, pp 679–685

Suzuki H, Keimatsu I, Ito M (1932) Alkaloid of the Chinese drug "Chin-Shih-Hu". II. Dendrobine. J Pharm Soc Jpn 52:1049–1060

Teoh ES, Lam D (1985) Pregnancy. Times Books International, Singapore

Trivedi VP, Dixit RS, Lal VK (1980) Orchids in the drug markets of Bareilly, Kanpur and Nearby Districts. Nagarjun (Calcutta) 23(8):157–163

Veldman S, Otieno J, van Andel T et al (2014) Efforts urged to tackle thriving illegal orchid trade in Tanzania and Zambia for chikanda production. Traffic Bull 26(2):47–50

Van Steenis CGGJ (ed) (1948) Flora Malesiana, vol 4. -Noordhoff-Kolff NV, Batavia

Xie ZF, Huang XK (eds) (undated) Dictionary of traditional Chinese medicine. The Commercial Press, Hong Kong

Xiong J, Huang J (1998) The A1- and non A1-effects of N6-(5-hydroxy—2-pyridyl)-methyl-adenosine on rat vas deferens. Yao Xue Xue Bao 33:175–179

Yamada et al (1972) J Am Chem Soc 94:8279

Zhang CF, Wang M, Wang L et al (2008) Chemical constituents of Dendrobium gratiosissimum and their cytotoxic activities. Indian J Chem 47(6):952–956

Zhao ZZ, Yang ZJ, Iida O (2006) Supply and cultivation of medicinal plants in Japan. In: Zheng HC (ed) 2003: Edible and medicinal plants of China. Shanghai Lexicographic Publishing House, Shanghai

Traditional Chinese Medicine, Korean Traditional Herbal Medicine, and Japanese Kanpo Medicine

2

Origins

According to the Chinese *Book of Origins*, the nation was founded by Fuxi who was assisted by two capable ministers. All three of them authored classic works. Fuxi was the author of the *Yi Jing* (*The Book of Change*), which divided phenomena into *ying* (negative, female) and *yang* (positive, male) and stressed their balance. The *ying–yang* concept continues to dominate Chinese thinking in many aspects of life, not least in health and disease. Table 2.1 provides some idea of how this concept is used to describe nature and to manage disease.

The first minister's name was Huangdi, or Yellow Emperor, which possibly implied that he also looked after civil affairs. He wrote *The Yellow Emperor's Classic of Internal Medicine*, a book on the maintenance of health and the treatment of disease. The other minister, Shen Nong, taught people to till the land and cultivate grain (Fig. 2.1). To discover which plants were edible and could serve as food or medicine, Shen Nong tasted a hundred herbs daily, and through this process he discovered that a few orchidaceous plants could be used to cure illness. He described *Chi Jian* (red arrow) or *Tianma* (*Gastrodia elata*), *Shihu* (*Dendrobium catenatum* and *D. moniliforme*), and *Baiji* (*Bletilla striata*), and explained their medicinal applications.

Tradition maintains that the first Chinese *Materia Medica*, renowned as the *Sheng Nong Ben Cao*, was written over 4600 years ago (on bamboo slabs), but it was formally published as an official document only in the first century, during the Eastern Han period. This edition, known as *Shen Nong Ben Cao Jing* (*The Materia Medica of Shen Nung*, alternatively referred to as *The Divine Husbandman's Classic of Materia Medica*), described 365 different medicinal substances, 237 of which were botanical, and the rest animal, mineral or from other sources (Hou 1977; Unschuld 1977). As scholars have questioned the existence of Chinese medical publications before the first century, it might be pertinent to point out that, when Li Ssu recommended the infamous burning of books during the Qin Dynasty (221–210 BC), he mentioned that "writings on medicine, divination, agriculture and forestry should be spared from destruction". Ssu-ma Ch'ien concludes his account with the words: "An imperial decree granted approval" (Bodde 1986). Thus, writings on medicine, agriculture and forestry already existed long before the *Current Era*, although there is no extant copy of the *Shen Nong Ben Cao* dating from this period.

Other early manuscripts discovered in archeological sites dating to the Neolithic and Han Periods (206–220 BC), such as the *Mawangdui Medical Manuscripts* (dated second century BC) attest that the central themes of interest to the ancients were individual self-preservation and the continuation of the race.

Medical dissertations focused on prolonging life, and on promoting sexual prowess or fecundity. However, the prescriptions in the *Mawangdui Medical Manuscripts* were the private property of a noble family and not chapters of an official pharmacopoeia, and they did not mention any orchid (Harper 1998). Nevertheless, this omission suggests that, whereas *salep*, a drink made from the tubers of Mediterranean terrestrial orchids, was valued as an aphrodisiac in Europe and the Middle East, orchids were not used as an aphrodisiac in ancient Orient.

Sheng Nong demonstrated uncanny wisdom. He listed five "divine" and principal crops, four of which were cereals (barley, wheat, millet and rice), while the fifth one was soya bean. Barley has the shortest growing period and is the only cereal crop which can be grown in areas like Xizang (Tibet), which enjoys a brief summer. Wheat and millet are temperate crops, and millet with a short growing period is tolerant of drought. Rice is suited to the warmer, wetter southern provinces. These four cereals thus provided staples for the entire country. *Glycine soja*, the wild soya, has been domesticated for at least 4500 years (Kiple 2007). Today, we know that humans need eight essential amino-acids, and that rice contains only seven. The missing amino acid is found in soya bean. Had soya not been included in a rice diet would there be 1.4 billion Chinese today?

Over the next 1500 years, the Chinese *Materia Medica* was revised, expanded, and re-edited under different titles (Hou 1977; Leslie 1977). The *Zhenghe Bencao* (printed in 1249) contains over 1000 items. Physicians of the Yuan Period found this too unwieldy and preferred a *Materia Medica* that discussed the use of

Table 2.1 Manifestations of *yin* and *yang*

Yin	Yang
Female	Male
Night	Day
Clear	Opaque
Winter	Summer
Cold	Hot
Passive	Active
Steady	Agitated
Inwards	Outwards
Downwards	Upwards
Deflexion	Thrust
Interior, abdomen	Exterior, back
Lower part of body	Upper portion of body
Bones, sinews	Skin, hair
Blood	*qi*
Deficient	Excessive

 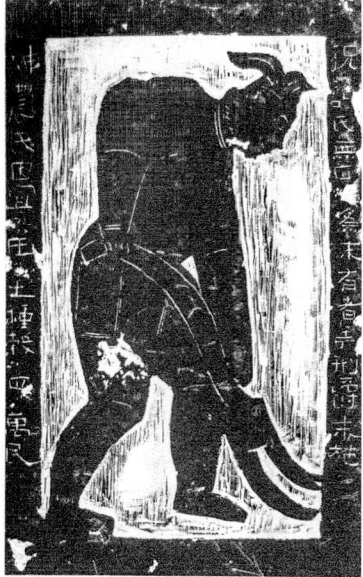

Fig. 2.1 Fanciful depictions of Shen Nong (*left*) as the original herbalist and (*right*) as a farmer. He is credited with the introduction of agriculture in China

around 100 drugs (Lee 1955). The famous, posthumous, monumental pharmacopoeia, *Bencao Gangmu (Pen Ts'ao Kung Mu)* of the Ming Dynasty physician and pharmacologist Li Shizhen (1518–1593), published 3 years after the author's death, described 1892 medical substances, of which 1094 were botanical and 444 zoological. Within its 52 volumes, 1160 line drawings were included to assist in the identification of the items (Wu 2005).

What inspired Li Shizhen to undertake this massive compilation was his observation that, in the preceding centuries, much confusion had arisen regarding the identity and usage of numerous medicinal compounds. New drugs had been introduced, some imported from distant lands. Sometimes, different drugs shared the same name. Li Shizhen took 40 years to complete his study. He travelled around the important drug-producing provinces of the time (Hunan, Jiangsu, Jiangxi and Anhui) collecting specimens for study (Fig. 2.2).

Li's tome is not merely a *Materia Medica*, it is also a great work on taxonomy, because he made careful observations on hundreds of plants and animals, classifying plants according to species, noting the varietal differences within species and the synonyms of plant names. He paid attention to the morphology and distribution of the species. Li Shihzhen described stable variations and mutations in the lotus *(Nelumbo nucifera)* (Figs. 2.3 and 2.4), poultry and goldfish *(Carassius auratus)*, the last a mutated form of the crucian carp. His descriptions recognised hereditary influences, familial lines and artificial selection. He was a naturalist in the class of Linneaus and Darwin. Charles Darwin acknowledged the contributions of Li Shizhen when he wrote *Variation of Animals and Plants under Domestication*. Darwin referred to the *Ben Cao Kang Mu* as "a Chinese Encyclopaedia published in 1596" and noted that it described several breeds of fowl "including what we should now call jumpers or creepers, and likewise fowls with black feathers, bones and flesh." Goldfish were "introduced into Europe only two or three centuries ago, but kept in confinement from an ancient period in China." Of interest to modern

Fig. 2.2 Li Shizhen (1518–1593) is a popular subject for porcelain portraits (Photo: E.S. Teoh)

Fig. 2.3 Single petal pink lotus: the primary form (From: Teoh Eng Soon: *LOTUS Photographs and Chinese Poems*. Singapore: Editions Dideire Millet, 2015)

medical science is Li Shizhen's description of the extraction of steroidal hormones from urine, which preceded the manufacture of *Premarin*® (equine oestrogens from pregnant mares' urine) by some 370 years.

Fig. 2.4 "A Thousand Petal Lotus". In this peloric form, the number of petals have greatly increased with the stamens taking on the form of small petals. Li Sizhen described such mutations in his *Bencao Guangmu* (Photo: E.S. Teoh)

Li Shizhen understood that some substances were heat-stable while others were destroyed by heat. Artemisinin, effective in the treatment of chloroquine-resistant falciparum malaria, was isolated from *Qinghaosu* in 1971. In 1596, Li Shihzhen wrote that, to treat ague, employ "a handful of *Artemisia*, crushed and taken with 2 l of water". The therapeutic agent needed low-temperature extraction in a large volume of water because it was heat-labile (Huang 2010).

Revision and expansion of the *Chinese Pharmacopoeia* has continued to the present time. During the Qing Dynasty (1644–1911), two new *Materia Medica* were published. *Zhi Wu Ming Shih Tu Kao* edited by Wu Qijun and published in 1848 described 1714 herbs in 38 volumes. After the founding of the People's Republic of China, an extensive systematic study was made of provincial and folk remedies, and this was collated by the Jiangsu College of New Medicine in the monumental *Zhongyao Da Cidian* (Encyclopaedia of Traditional Chinese Medicinal Substances) in 1977. It contained 5767 entries. The contemporary *Zhongyao Da Cidian* (1986) which is semi-bilingual and *Zhonghua Bencao* (2000) contain over 6000 items.

Purists who practice traditional Chinese medicine (TCM) distinguish between the herbs mentioned in the *Shen Nong Bencao Jing* and those that are mentioned in the current *Materia Medica*

Fig.   2.5 *Dendrobium   moniliforme*   (L.) SW. [as *Dendrobium japonicum* (Blume) Lindl.]— From: *Curtis Botanical Magazine*, vol. 90 [ser. 3, vol. 20]: t.5482 (1864). Painting by W.H. Fitch. Courtesy of Missouri Botanical Gardens, St. Louis, Missouri. This is one of the two original species that constitute the herb, *Shihu* (growing on rocks) in the seminal *Shen Nong's Bencao Jing*

(*Zhongyao Da Cidian* and *Zhonghua Bencao*). Only those mentioned in the *Shen Nong Bencao Jing* are accepted by the purists to be indisputable TCM, whereas later additions would be considered under Chinese herbal medicine (CHM) (Figs. 2.5 and 2.6).

In his overview of the development of CHM and the *Bencao*, Joseph P. Hou (1977) stated that, over the centuries, more than 50 different and revised *Ben Cao* (*Materia Medica*) which included compendiums on dietary therapy were published officially and unofficially. The complete collection of traditional *Chinese Herbals*

Fig. 2.6 *Cymbidium ensifolium. Lan* is not included in the *Shen Nong Bencao Jing* but it is described in Li Shizhen's *Bencao Guangmu*. This gives it a respectable status in Chinese herbal medicine. Line drawing from *Bencao Guangmu*

spans 800 volumes and possibly represents "the world's richest knowledge of natural remedies".

Using Chinese Herbs

Traditional Chinese medicine proposes that, when a person is healthy, all the functions in the body are in equilibrium. Imbalance of this natural life force (*qi*) results in disease. TCM aims to restore the body to its ideal state, one that is also in harmony with nature. Food and herbs are important aspects of management, in addition to practising ideal behaviour such as exercise, rest, calmness, relaxation and meditation. Whatever its cause, the common final pathway of an illness is an imbalance in the body: too

much *yang* (positive, hot, aggressive, lustful) or too little (weakness, cold, resigned, lack of libido) or insufficient *yin* cause imbalances in the various organ systems such as kidneys, spleen, liver, heart, etc. Confusion often arises when translating the use of Chinese herbs because the organ systems as designated in herbal texts do not refer to the organs themselves, but to what TCM considers as the functions embodied by the name of the organ. For instance, everyone who understands Chinese knows that "heart" also means "mind". "Kidney" refers not only to functions of organs arising from the uro-genital ridge, i.e renal and sexual functions, but also to growth and ageing. However, the "liver meridian" does not refer to the liver at all; instead, it denotes the central nervous system. Thus, *Tianma* (*Gastrodia elata*) is said to "nourish the *yin* fluid, calm the liver and extinguish wind". It is used to treat all forms of *internal wind*, manifested by "wind" stroke, seizures, aphasia, slurred speech, blurred vision, numbness, tingling of the extremities, headache, vertigo and dizziness. In addition to the regulation of blood and circulation, "spleen" embraces the functions of the alimentary tract. TCM has its own classification of syndromes (Hong 2014). They do not always correspond to those of modern medicine.

TCM recognises that there may be different causes for a symptom, and its treatment will vary according to the cause. For instance, headache could be due to weak *qi* or excessive liver *yang*, and TCM's treatment of the headache would differ accordingly. Although a painkiller is the modern, short-term, and ready remedy for a headache, practitioners of Western medicine would also look for the cause of persistent headache which might be migraine or something far more serious, such as a brain infection, intracranial haemorrhage or a tumour.

Chinese herbs are hardly ever used as single items. A prescription always contains multiple items which are decocted together to produce an herbal potion. Liniments and balms also consist of several compounds. The concept is that when herbs are used together they act synergistically. Chinese herbs are said to have a mild effect, and they function by assisting the body to regain its balance (*ying–yang* balance) and to heighten its immunity. Afterwards, the body heals itself. Where the effect of an herb is too mild, the addition of other herbs may increase its potency. Sometimes, the additional herb reduces the toxicity of the principal herb.

When *Shih Hu* (*Dendrobium*) is used to treat "tidal fever or night sweats due to kidney deficiency and *yin* deficiency", *Bai Wei* (*Cyananchum atratum*) and *Zhimu* (*Anemarrhena asphodelioides*) are included in the preparation of the decoction (Wu et al. 1999). The two herbs are listed among the 50 most important herbs in the ancient *Shen Nong Ben Cao Jing* (c. first century), and by themselves they have the ability to clear heat and reduce fever. *Bai Wei* is also a diuretic.

Some people complain of a dry throat but, perhaps from habit, they are reluctant to drink adequate amounts of liquids. This condition of "hypopsia" is attributed to *yin* deficiency of the liver and kidney in TCM. The *Shi Hu Guang Pill* prescribed for the condition consists of *Shi Hu* (*Dendrobium*) together with *Juhua* (*Chrysanthemum* flower) and *Guo Qizi* (*Lycium barbatum* or *L. chinensis*, Chinese wolfberry) (Wu et al. 1999). The latter two are popular ingredients of Chinese thirst quenchers (Zhongyao Da Cidan 1986; Wu 1994) .

Some herbal combinations employing orchids are supported by biochemical data. Among its numerous applications, *Tianma* (*Gastrodia elata*) is used to treat rheumatic pain. In this application, the decoction calls for the addition of *Qinjiao* and *Qianghuo* (Wu et al. 1999). *Qinjiao* (*Gentiana macrophylla*, Large leaf gentian root) is an antiprostaglandin and would therefore relieve inflammation and pain. *Qinghuo* stimulates the release of ACTH from the pituitary thereby increasing the output of anti-inflammatory corticosteroids. Together, they should have an ability to relieve joint pains.

Of 12 prescriptions in the ancient *Inner Classic of Huangdi*, only 4 consisted of a single item and 3 contained four substances. Contemporary prescriptions generally contain 6–12 substances. However, the dangers of poly-pharmacy have long been well understood. In his *Bencao Gangmu*, Li Shihzhen pointed out that some combinations produce mutual antagonism (*xing*

lu). When two substances oppose each other, they cancel out their positive effects. In another situation, mutually incompatible substances may produce harmful side effects that neither possesses in the absence of the other.

When several herbs are properly combined into a single prescription, there is usually a dominant compound which determines the direction of the treatment. This is the principal medicine, the so-called king (sovereign, lord) medicine (*jun, yaowang*). Only select herbs are suited for such important functions. *Gastrodia elata* (*Tianma*), various species of *Dendrobium* (*Shihu*) and *Anoectochilus formosanus* (*Jinxian Lan*) are three examples of *yaowang* from the *Orchidaceae* (Hew et al. 1997). Some herbs, known as the ministers or associates (*chen*), assist or promote the action of the king medicine. In a third category are the assistants or adjutants (*zuo*), which are used to moderate the harshness or toxicity of the prescription, and/or to relieve symptoms associated with the condition. *Bletilla striata* (*Baiji*) is a member of this class. Finally, there are the so-called messengers or envoys (*shi*) which exert a harmonious influence, guiding the herbs into a specific channel (Bensky et al. 2004). It is not necessary for a prescription to contain all four components; sometimes, a single herb may play multiple roles.

Isolating compounds from orchids and other herbs, and studying them individually for their pharmacological properties, mark a paradigm shift away from the central hypothesis of TCM. It is revolutionary and is an attempt to discover new drugs for treating disease; hopefully, it may transform ancient remedies into scientific cures.

Herbal Names

Herbal names may denote their:

1. Appearance

For example, the beautiful white flowers of *Habenaria dentata* resulted in this herb's numerous popular names which describe the appearance of its delicate flowers: *Emaoyufeng Hua* (Feather Jade Phoenix Flower), *Baifeng Lan* (White

Phoenix Orchid), *Dalucao* (Large Heron Grass), *Yufeng Lan* (Jade Phoenix), *Dongpuyufeng Lan* (Dongpu Flaked Teeth Heron Orchid), *Dongfubaifeng Lan* [Dongpu White Phoenix (White Heron) Orchid], *Chipianlu Lan*, and *Emaoyufenghua* (Goose-Feather Jade Blossom); and *Chi Jian* (Crimson arrow) alluding to the appearance of an inflorescence of saprophytic *Gastrodia elata* which resembles an arrow stuck in the ground. Even more descriptive is *Bai Long Pi* (White Dragon Skin) which describes the crinkled surface of the medicinal tubers of *Gastrodia elata*, or *Wugong Lan* (Centipede Orchid) which describes the appearance of the small, segmented, green rhizome of *Cleisostoma scolopendrifolium*. Many species of *Goodyera* are used in TCM; the name for the genus is *Banye Lan* (Spotted Leaf Orchid). *Congye Lan* (Onion Leaf Orchid) and *Jiuye Lan* (Garlic Leaf Orchid), the two names for *Microtis unifolia*, aptly describe the appearance of the rod-shaped leaf of the species.

2. Colour

Gastrodia elata was originally called *Chi Jian* (Crimson Arrow) from the appearance of its flowers. *Habenaria dentata* is known as *Baifeng Lan* (White Phoenix Orchid). The jewel orchid, *Ludisia discolor*, is called *Xueye Lan* (Blood Leaf Orchid).

3. Scent

Bulbophyllum ambrosia whose flowers are fragrant is known as *Xiangshidou Lan* (Fragrant Stone Bean Orchid).

4. Taste

Orchid preparations do not carry any words that describe their taste, but many Chinese herbal preparations do. The sweet taste of glycerine is denoted in *Gan Cao* (*Radix Glycyrrhizae*), which means 'Sweet Herb', while *Suan Zao Ren* (*Ziziphi sp.nosae*) is 'Sour Date Seed'. Flavours

hint at the usefulness of a herb, and herbs which share a common flavour are likely to have similar functions. There are five flavours: sour, bitter, sweet, pungent and salty, with a sixth category, bland (Wu 2005).

Sour covers astringent, and herbs with this flavour are used to treat deficiency syndromes or diarrhoea. Sweet herbs are tonics used in the treatment of deficiency syndromes. *Tianma* (*Gastrodia elata*) is sweet and slightly bitter, and has the function of treating *yin* deficiency. Bitterness implies draining and drying functions, and bitter herbs are used to clear heat, purge fire, treat constipation, resolve dampness, or lower rebellious *qi*. They preserve *yin* (Wu 2005). Alkaloids are bitter, and for this reason a bitter herb may exert a powerful effect.

Pungent substances have the function of dispersing. They promote *qi* and blood circulation. A famous herb with a pungent property is *Herba Ephedrae* (*Ma Huang*) which contains ephidrene, a compound related to adrenalin. It is a bronchodilator but it also raises the pulse and affects the blood pressure. *Ma Huang* is not an orchid. Salty compounds soften and soothe sore throats. Many Chinese resort to salted plums when they have a sore throat.

Although the flavour of a herb serves as a guideline to its efficacy, this is not an absolute criterion for its function. In many cases, other characteristics are more important (Wu 2005).

5. Derivation (Entire Plant, Leaves, Stems, Roots, Flower, Seed, Stamens, Etc.)

Spiranthes sinensis goes by the name *Shoucao* (Tassel Grass). It is the first orchid mentioned in Chinese literature (Chen and Tang 1982), and the entire plant ("grass") is decocted to make a preparation for strengthening the kidneys. *Baiji* (White Root) designates the part of *Bletilla striata* that is considered medicinal. The rule is not absolute. *Yangersuan* (Sheep's Ear Garlic) is the common name for all species of *Liparis* the pseudobulbs of which do indeed resemble garlic, but generally the entire plant is used as medicine.

6. Habitat

Dendrobium species used in TCM are collectively known as *Shihu* (living on rocks) which originally referred to the two lithophytic species found in the northern limits of *Dendrobium* distribution around the Huanghe, namely *D. moniliforme* and *D. catenatum (syn. D. officinale)*. *Shanlanhua* (Mountain Orchid Flower) describes the occurrence of *Cymbidium ensifolium* in the hills and mountains. Sometimes, a common name combines habitat and appearance, e.g. *Shantao* (mountain peach) for *Pholidota.*

7. Source (A Particular Mountain, Province or Country)

Alishanxiaozhu Lan (A Li Mountain Small Pillar Orchid) is the name given to *Malaxis monophyllos*, designating the Taiwanese mountain where it was first discovered and named, although the orchid is not restricted to Taiwan. It enjoys a much wider distribution from Yunnan northwards through Sichuan to Xinjiang, and eastwards across the northern Chinese provinces to Mongolia.

Yunnan Yangercao (Yunnan Sheep Ear Herb) is the name given to *Liparis yunnanensis* which is used to treat pneumonia. The herb comes from Yunnan as well as from the provinces of Tibet, Sichuan, Guangxi and Guangdong. *Xizang Yangersuan* (Tibetan Sheep Ear Garlic) is the name for *L. tschangii* whose distribution is in Tibet and adjacent Yunnan Provinces. The pseudobulbs are used as a haemostatic. There is also *Guangdongshidou Lan* (Guangdong Stone Bean Orchid) which is *Bulbophyllum kwangtungense*. It is used to treat coughs, reduce "heat", clear "wind" and for convulsions in children. *Dianmiangeju Lan* (Yunnan Province and Myanmar Separate Distance Orchid) for *Cleisostoma williamsonii* is used in treating pulmonary tuberculosis, viral encephalitis, stroke, polio and backache, as well as indigestion in children.

Riben Chun Lan (Japanese Spring Orchid) is the name given to *Cymbidium goeringii*.

Kanchajianiaochao Lan (Kamchatka Bird's Nest Orchid) is an adaptation from the scientific name of *Neottia camtschatea*.

8. History

Shancigu (Kind Mountain Lady) and *Maocigu* (Kind Furry Lady) which are common names for *Cremastra wallachii* are probably derived from legend. The orchid is used to treat impotence, fever, snake bites and poisoning in general. A paste from it is used to treat surface abscesses. Some herbs which are not orchids are named after the people who discovered them. For instance, *Qian Niu Zi* (Walk Cow Seed) for *Pharbitidis semen* recounts how the herb was accidentally discovered by an old man who herded his cows.

9. Preparation

There is no orchid name based on the preparation of a herb. *Dendrobium* stems undergo various treatments, as will be described in a later chapter, and the freshly harvested stems may be distinguished from the final product in the apothecary by prefixes, but the herbal name, *Shihu*, is unchanged.

10. Function

Huaqi Lan (Clearing Gas Orchid) is *Cymbidium faberi*. Many spin doctors would be proud were they the originators of such Chinese herbal names as *Yang Qi Shi* (which denotes *Actinolitum*, not an orchid), literally *Yang Lifting Rocks*. It tonifies the *yang* and is said to be so powerful that it renders the penis strong enough to lift rocks! In Ayurveda, its usage as an aphrodisiac possibly earned the name *Jiwanti* for *Flickingeria nodosa*.

Jiegucao (Bone Setting Herb), the medicinal name for *Holcoglossum amesianum* (syn. *Vanda amesiana*), identifies its usage to treat fractures. The orchid has other non-specific names like

Diao lan (Hanging Orchid) and *Wanda Lan* (Ten Thousand Generation Orchid).

Jiejinsixiantao (Node Stem Rock Immortal Peach) is the name for *Pholidota articulata*. The reference to rock denotes its habitat. *Pholidota* is substituted for *shihu*, sometimes when the latter is not available, but more often as an adulterant. The part of the name which says 'Immortal Peach' is intended to advertise its medicinal efficacy in prolonging life.

In other medicinal traditions, for example in Malaysia, *Acriopsis javanica*, which is used to treat earache and fever, is known as *Anggerek darat* (River Bank Orchid) as well as *Sakat ubat kepialu* (medicinal epiphyte for severe fever), whereas in Indonesia it goes by the Sundanese name *kip'engpeng*, alluding to the clicking sounds in the ear of someone suffering from tinnitus (van den Brink 1937).

11. Legendary Origin

Tianma (Heaven's fibre), which replaced *Chi Jian* as the common name for *Gastrodia elata* during the Song Dynasty (960–1279), was probably derived from popular legends on the origin of the herb which supposedly arose as a divine gift for people afflicted with nerve disorders and stroke. *Xicaren Zhijia Lan* (Immortal's finger nail orchid) referring to the flowers of *Aerides* with their pointed spurs is fancifully legendary (Hu 1965).

Specificity of Herbal Names

Most of the time, herbal names refer to a single botanical species, but this is not always the case. *Shihu*, for instance, originally referred to just two species of lithophytic *Dendrobium*, but now, in addition to adulterants, it is applied to more than 25 species of *Dendrobium* and the list keeps expanding. To narrow the identification, prefixes which describe their appearance or source are applied to *shihu*. The popular *Guoshangye* used in Tuja and Miao folk medicine may consist of *Pholidota yunnanensis*, *Bulbophyllum andersonii*,

B. odoratissimum, or *B. kwangtungense* (Qu, Qin and Yang, 2006). *Shancigu* usually refers to *Cremastra wallachii* but *Pleione bulbocodiodes* (Franch.) Rolfe and *Pleione yunnansis* Rolfe are common substitutes. Lack of precision is undesirable when using a medicinal product, but this may partly be the fault of the practitioner lacking detailed knowledge of the product that he prescribes.

Classification of Herbs

Chinese herbs are sometimes designated as superior, medium or inferior. This classification is based on the *Shen Nong Ban Cao* which divided medicinal herbs into three classes: superior, common and inferior.

Superior herbs are said to be highly efficacious, with few or mild side effects. They can be taken over a long period. Two of the three classical orchidaceous herbs (*shihu* and *tianma*), ginseng and dried lychee qualify as superior herbs. In the old, traditional Chinese family, a senior member would be held as an expert, but his or her expertise may be based merely on hearsay. Thus, it was important to identify those "medicines" or tonics which were safe, to distinguish them from others which require some sort of medical supervision.

Medium herbs are those which have a therapeutic effect but they carry some serious side effects, and thus require professional supervision. Unfortunately, such supervision is seldom provided because some of these herbs are well known, although their side effects are not. Many common herbs fall into this category, including *Dong quai* (*Angelica sinensis*) which is extremely popular among women and consumed for a variety of menstrual disorders. Common herbs are said to be effective for a few disorders, and they can be slightly toxic. We know today that many Chinese herbs contain salicylates (aspirin) which have antiplatelet activity and are anticoagulants even in very small amounts. Such herbs are hazardous when used in combination with warfarin and other blood thinners, and they can be dangerous if they are taken before surgery. *Angelica* also has contradictory muscle contracting or relaxing effects on the uterus depending on the duration of boiling. It is dangerous when consumed by pregnant women close to term or after delivery.

Anoectochilus formosanus, used in Taiwan for treating liver disease, is regarded as a king medicine, but it is best classified under the category of a medium herb because excessive dosage causes liver damage whereas low dosages protect the liver from carbon tetrachloride poisoning in rats. It also shows some oestrogenic and antitumour activity, and here again its dosage and interaction with other compounds like oestrogen would determine how it behaves in a living body.

Inferior herbs are toxic, and apothecaries were forbidden to store or sell such herbs during the Yuan Dynasty (1260–1368) because the Mongol overlords feared being poisoned. *Fu Zi* (aconite) is an example. Inferior drugs have a role only for specific illnesses. Caution is advised because they are toxic and may be harmful. Hou (1977) lists the orchid *Baiji (Bletilla striata)* in this class although *Baiji* appears to be quite harmless.

Li Shihzhen (1518–1593) disliked this oversimplification. He preferred to describe each drug in detail, and it would be left to the knowledgeable practitioner to decide on the requirements of each patient. In his *Ban Cao*, each item is headed by its general name, followed by other common, popular or local names, habitat and characteristics, processing and preservation, properties, principal uses and the origin and development of the medicine. Finally, there is a collection of prescriptions (Li 1578). In the present work, we have tried to follow this approach wherever it is possible to do so.

Today, we know that toxicity is always related to dosage. Ginseng is described as a superior herb, and many people take it to mean that the herb can be used for a wide variety of illnesses because they think it is absolutely non-toxic. However, when consumed in excess in the USA, ginseng results in the well-known *ginseng abuse syndrome* (GAS) which is characterised by anxiety states, agitation, chronic insomnia, skin rash, hypertension, fluid retention, loose stools, etc. Taken together, the effects of GAS mimic those of corticosteroid poisoning (Siegel 1976). Ginseng binds to oestrogen and progesterone receptors in the muscle cells of the human uterus (Punnonen and Lukola 1980).

Several species of *Dendrobium* which are used as *Shihu* (classified as a superior herb) contain powerful cytotoxic compounds. Therefore, theoretically, *Shihu* could be poisonous if taken in large amounts. On the other hand, if someone is harbouring a malignant tumour, continuous consumption of *Shihu* might render that person's tumour insensitive to the cytotoxic contents of the *Shihu*. When a tumour has acquired the ability to bypass apoptotic pathways, it becomes more difficult or even impossible to treat that person's cancer by chemotherapy.

Foreign Influences

Chinese traditional medicine (TCM) is not exclusive. It is pragmatic. It has adapted foreign inputs whenever it saw these as being useful to the understanding of disease or improvement of treatment. For instance, TCM practitioners in China now employ ultrasound and X-rays for diagnosis. During the Tang Dynasty, Wei Tzu Tsay was conferred the title of King of Medicine by the Emperor. He is worshipped in the temple of the Gods of Medicine in Beijing in the company of great Chinese physicians of the past like Shen Nong (c. 4600 BC), Huangdi (c. 4600 BC), Hua Tuo (110–208) (Fig. 2.7) and Sun Simiao (581–682) (Figs. 1.1 and 1.2). He is an Indian who arrived in China in 733 with gourds full of herbs and began to treat patients without charge (Tao 1940). During the Jin (1127–1234) and Yuan (1234–1368) Dynasties, China absorbed elements of Arabic (more correctly, Persian) and European medicine. A Moslem Institute of Medicine under the direction of Arabs (or Persians) was set up in Beijing at the end of the thirteenth century to cater for the needs of Arabs, Persians and Europeans who had arrived at the capital with the occupying Mongols (Lee 1955).

Fig. 2.7 Hua Tuo (120–218 C.E.), portrait in a Chinese postage stamp. A physician-surgeon who lived during the Three Kingdoms Period, Hua Tuo is not known for introducing new herbs. He performed amputation, employed anaesthesia for abdominal surgery. He was executed when he proposed to perform neurosurgery on the de facto ruler of China, Cao Cao

The Swedish psychiatrist, Hans Agren (1975), employed the term pre-scientific to describe TCM, placing it on a par with pre-scientific Western medicine which was the mode prior to the seventeenth century. Indeed, up to that time, the Western pharmacopoeia was not more effective in curing disease. During the period of colonial expansion, which extended from the seventeenth to the mid-twentieth century, a concerted search was made to discover new medicines from folk remedies and poisons used in various parts of the world. This led to the discovery of a number of useful drugs, many of which are still in use today, either in their original form or as analogs (see Chap. 1). Thus, one should not dismiss pre-scientific traditional cures out of hand, but only after proper investigation. In the Far East, herbs are being subjected to scrutiny by modern science employing the latest technology.

Today, everyone is eagerly waiting for TCM to boldly take a quantum leap and embrace the approach of double-blind, randomised control clinical trials to confirm the efficacy of its treatments. While anecdotes mark a beginning, proof of efficacy must not be only anecdotal. It should be based on double-blind comparisons with placebos showing significant statistical differences between the two (see Chap. 9).

Nevertheless, probability and double-blind randomised control trials of treatment can still mislead us, so the more trials there are the better. To take the initial step, traditional practitioners working in hospitals or those who see large numbers of a particular illness should collect accurate observational data on the patients that they treat. Quantitative observational data on herbal remedies are not readily available in the English literature.

Korean Traditional Herbal Medicine

Traditional Korean medicine (TKM) is closely related to the theory of TCM due to the ancient links between the two countries. Fragments of Chinese medical classics reached Korea as early as 200 CE, during which period Korean scholars used Chinese to record history and medical treatment. The Korean monk, Kwan Rok (*Kwanroku* Japanese), taught TCM practice in Japan in 602 (Huard and Wong 1968).

The oldest TKM book in existence today is King Sejong's 85-volume *Hyang Yak Jip Sung Bang* compiled in 1433. Twelve years later, King Sejang (1397–1450) headed a team of scholars to compile the 266-volume *Eni Beng Yoo Chin*, an enormous medical encyclopaedia. It contains the classical TCM prescriptions from the Han to the Ming Dynasties, a period of 1700 years, but it is not easily accessible because the only extant copy is in the Japanese Imperial Household (Huard and Wong 1968). Contemporary Korean herbal medicine is most strongly influenced by three later works, the *Dong Eni Bo Gan* compiled by Huh Ju (1546–1615), the *Dong Eui Soo See*

Bon Won compiled by Jae Ma Lee (1836–1900) and the *Bay Yah Hap Pyon*.

The number of herbal items in Korea exceeds 1000, but the number of medicinal herbs listed in the *Korean Pharmacopoeia* is 514. Since 1987, the Korean Medical Insurance System, which is subscribed to by 5.6 % of the Korean population, covers the use of TKM, and this has undoubtedly encouraged herbal usage. The domestic consumption of herbs has risen more than tenfold since 1990, and it is estimated that over 20 % of the population of Korea resort to TKM. *Tianma* (*Gastrodia elata*) is a traditional herb in Korea for the treatment of neurological disorders such as scotodinia, paralysis and epilepsy (Han et al. 2014).

The Korean Food and Drugs Administration (Korean FDA) controls the quality and safety of herbal materials. Good Manufacturing Practice (GMP) Guidelines are followed in the manufacture of 68 herbal extracts, but Shi et al. (2008) noted that the safety of herbal materials is not highlighted. Few research studies documented their toxicity when herbs were used to treat diseases of the liver, kidneys, urinary tract, cardiovascular system and skin. Commenting on this, Shi et al. (2008) stressed the need to accurately assess the safety, quality and cost effectiveness of herbal materials in Korea.

Ethnic Traditional Medicine in Japan

Kanpo (Chinese Style) Medicine

Ethnic traditional medicine in Japan was originally based on TCM introduced during the Nara Period (710–784) by the Buddhist monk, Gangin (Jianzhen, 687–763), who had been invited by the Emperor to establish a proper Buddhist Order in Japan. Skilled craftsmen, scholars and herbalists who accompanied him introduced to Japan the art of making *tofu* (bean curd) and *miso* (fermented soya-beans), two major items in Japanese cooking. When the Emperor Shomu died in 756, Empress Komyo presented Todaiji with a stock of 60 medicines and other gifts. The *Japanese Pharmacopoeia* contains some

150 prescriptions of medicinal herbs which are said to carry few side effects.

Doctors graduating from medical schools teaching Western medicine are allowed to prescribe *kanpo* medicine in Japan, although their curriculum does not provide extensive coverage of herbal medicine. The Japanese Public Health Insurance System covers treatment involving the use of *kanpo* medicine (Fukiji et al. 2008). Many young Japanese doctors, unlike their older predecessors, readily embrace the practice of *kanpo* medicine. Over 80 % of the members of the over 9000-strong Japan Society for Oriental Medicine are medical doctors (Zhao et al. 2006). Nevertheless, there are still many doctors who limit their practice to Western medicine because they do not trust the efficacy and safety of herbal medicine in Japan (Motoo 2008).

The *Japanese Pharmacopoeia* (14th edition) recorded 193 crude drugs, whereas there are 390 crude drugs circulating in Japan. In the official list, there is only one orchid, *Gastrodia elata*. Sixty-seven medicinal herbs with a total value of 1.2 trillion yen (approximately US$ 1 billion) are cultivated in Japan which established its own Good Agricultural Practice (J-GAP) in 2005. *Gastrodia elata* is not cultivated (Zhao et al. 2006). This is odd considering that a Japanese scientist was the first person to demonstrate a symbiotic relationship between the *Gastrodia elata* and *Amillaria mellea* (Kusano 1911).

Although *Dendrobium moniliforme* is not included in the contemporary *Japanese Pharmacopoeia*, it was introduced as a medicinal herb into Japan during the eighth century. The Edo Period (Tokugawa Period, 1603–1868) saw an interest in the cultivation of orchids (*Cymbidium, Calanthe, Dendrobium, Neofinetia*) as ornamental plants. *D. moniliforme* then acquired its second name, *Chouseiran*, on account of its long-lasting blooms.

Considerable misunderstanding surrounds the role of herbal medicine in cancer prevention and treatment. Many compounds extracted from plants show an anticancer effect in vitro but not in vivo, or else the effective dose is too close to the lethal dose. *Kanpo* medicine has not proven to be effective in cancer prevention and nor is it effective in reducing cancer growth. The principal roles of medicinal plants are centred on their ability to relieve the gastro-intestinal effects of chemotherapy such as nausea, vomiting and diarrhoea. They also have a role in relieving malaise, insomnia, depression and pain. Japan is one of the few Asian countries that has been involved in pharmacological and pharmaceutical research on medicinal plants for over a century. Such research is very much alive today, and medicinal orchids have not been neglected.

References

Agren H (1975) A new approach to Chinese traditional medicine. Am J Chin Med 3(3):207–212

Bensky D, Clayey S, Stoger E, Gambie A (2004) Herbal medicine materia medica, 3rd edn. Eastland Press, Seattle, WA

Bodde D (1986) The state and empire of Ch'in. In: Twitchett D, Loewe M (eds) The Cambridge history of China Vol. 1 The Ch'in and Han Empires (221 BC – AD 220). Cambridge University Press, Cambridge

Chen SC, Tang T (1982) A general review of the Orchid Flora of China. In: Arditti J (ed) Orchid biology: reviews and perspectives II. Cornell University Press, Itcaca and London

Fukiji K, Nakamura M, Taniguchi K (2008) Japanese botanical medicine in liver and gastrointestinal health. Risks and benefits. In: Watson RR, Preedy VR (eds) Botanical medicine in clinical practice. CABI, Cambridge, MA, pp 99–104

Han YJ, Je JH, Kim SH et al (2014) Gastrodia elata shows neuroprotective effects via activation of P13K signaling against oxidative glutamate toxicity in HT22 cells. Am J Chin Med 42(4):1007–1019

Harper DJ (1998) Early Chinese medical; literature. The Mawangdui medical manuscripts. Kegan Paul Intern, London

Hew CS, Arditti J, Lin WS (1997) Three orchids used as herbal medicines in China: an attempt to reconcile Chinese and Western pharmacology. In: Arditti J, Pridgeon AM (eds) Orchid biology: reviews and perspectives VII. Kluwer Academic, Dordrecht, pp 213–283

Hong H (2014) The theory of Chinese medicine. A modern interpretation. World Scientific, Singapore

Hou JP (1977) The development of Chinese herbal medicine and the Pen-ts'ao. In: Comparative medicine east and west, Vol. V (2). Institute for Advanced Research in Asian Science and Medicine, New York, pp 117–122.

Hu SY (1965) Whence the Chinese generic names of orchids. Am Orchid Soc Bull 34:518–521

Huang KC (2010) The pharmacology of Chinese herbs, 2nd edn. CRC Press, London

Huard P, Wong M (1968) Chinese medicine. Translated from French by B. Fielding. World University Library, London

Kiple KF (2007) A moveable feast. Ten millennia of food globalization. Cambridge University Press, Cambridge

Kusano S (1911) Gastrodia elata and its symbiotic association with *Armillaria mellea*. J Coll Agric Imp Univ Tokyo IV(I):1–73

Lee T (1955) Chinese medicine during the Chin (1127-1234) and Yuan (1234-1368) Eras. Chin Med J 73:241–253

Leslie C (1977) Asian medical systems: a comparative study. University of California Press, Berkeley

Li SZ (1578) Ben Cao Gang Mu (reprinted 1977 by People's Health Publishing Co., Beijing)

Motoo Y (2008) Japanese herbal medicine in the western style modern medical system. In: Watson RR, Preedy VR (eds) Botanical medicine in clinical practice. CABI, Cambridge, MA, pp 105–111

Punnonen R, Lukola A (1980) Oestrogen-like effect of ginseng. Br Med J 281:1110

Qu XY, Qin SY, Yang DQ, Li QS, Peng FS (2006) Study on resource and varieties of Guoshangye. Zhongguo Zhong Yao Za Zhi 31(2):110–4

Shi A, Kim YK, Lee LM (2008) Korean traditional herbal materials, herbal formulas in health promotion. Benefits and safety. In: Watson RR, Preedy VR (eds) Botanical medicine in clinical practice. CABI, Cambridge, MA, pp 112–120

Siegel RK (1976) Ginseng abuse syndrome. Problems with the Panacea. JAMA 241(15):1614–1615

Tao L (1940) Ten celebrated physicians and their temple. Chin Med J 58:267–274

Unschuld PU (1977) The development of medical-pharmaceutical thought in China. In: Comparative medicine east and west Vol. V(2). Institute for Advanced Research in Asian Science and Medicine, New York, pp 109–115

Van den Brink RCB (1937) Synopsis of the vernacular names and the economic uses of indigenous orchids of Java. Blumea 29 VI:38–51

Wu XR (1994) A concise edition of medicinal plants in China. Guangdong Higher Education Publication House, Guangdong (in Chinese)

Wu JN (2005) An illustrated Chinese Materia medica. University Press, Oxford

Wu KJ, Lu G et al (eds) (1999) Traditional Chinese materia medica. Wuhan University Press, Wuhan

Zhao ZH, Yang ZJ, Iida O (2006) Supply and cultivation of medicinal plants in Japan. In: PC Leung, H Fong, CC Xue (eds) Annals of traditional Chinese medicine Vol 2. Current review of Chinese medicine. Quality control of herbs and herbal medicine. Singapore: World Scientific, pp 59–73

Zhongyao Da Cidian (1986) Compiled by Nanjing College of New Medicine. Science & Technology Publishing, Shanghai

Indian Traditional Medicine and Other Asian Traditions

Before the arrival of Western medicine, Indian medicine consisted of five traditions, two of which were imported:

1. *Ayurvedic* medicine
2. *Siddha* medicine
3. *Unani tibb*
4. Tibetan medicine
5. Tribal medicine

These traditions are still robust today despite the popularity of Western medicine, which dominates in the cities. Approximately 80 % of Indians still rely on *Ayurveda* or *Siddha* medicine, and the pharmacopoeia of *Ayurveda* is continually expanding through the absorption of tribal cures.

Ayurveda

Ayurveda translates as "the knowledge of life" (Thatte 1995). Its ancient exponent, Kasyapa, who wrote the *Kasyapasamhita*, said that "by this, longevity is achieved". According to tradition, the god Brahma possessed special knowledge of curative principles, procedures and substances which he transmitted to Indra and Atreya. Mankind received the knowledge through Atreya (Varier 2005). *Ayurveda* embraces philosophical and spiritual issues in addition to purely medical conditions. It provides knowledge of nutrition and medicines, an understanding of their properties and actions, and whether such will be wholesome or unwholesome to a person. Vedic physicians in ancient times were said to be skilled in the use of medicinal plants, but the *Rigveda*, which presented most of the knowledge, did not offer it in a systematic order. That had to await the compilation of the *Charakasamhita* (compiled in the third century B.C., with multiple later editions) and the *Susrutasamnita* (compiled in the third century but attributed to Susruta of Varanasi, who lived in the sixth century B.C. This treatise on surgery mentioned 700 herbs). The Buddhist *Vinaya Pitaka* (originating in the sixth century B.C.) described illnesses cured by Jivaka, as well as self-taught remedies administered by monks that did not work, even for simple conditions like headache, which intermittently afflicted the monk, Pindola. It was axiomatic then, as it still is today, that, of the four legs (*catuspada*) essential for effective treatment (physician, medicine, caregiver and patient), the most important member was the physician. Nevertheless, both the *Carakasamhita* and the *Vinaya* cautioned against pretentious physicians who feigned knowledge and skill in order to exploit the patient in every possible way. During the Buddhist Period (mid-sixth century B.C. to mid-sixth century C.E.), there was a rule that "all physicians who treat their patients wrongly shall pay a fine." In practice, the law was probably quite lenient even

© Springer International Publishing Switzerland 2016
E.S. Teoh, *Medicinal Orchids of Asia*, DOI 10.1007/978-3-319-24274-3_3

when physicians resorted to the use of adulterated drugs (Chopra 1933). The Dhauli Edict of Asoka encouraged the establishment of gardens for the cultivation of medicinal herbs (Figs. 3.1 and 3.2).

Ayurveda continued to evolve until the end of the Gupta Period (mid-sixth century), but it went into decline during the period of civil war that followed (Varier 2005). The Muslim invasion led by Bakhtuyar Khilji in 1193 and the burning of the great library at Nalanda caused the greater part of the original knowledge to be lost.

There are three principal doctrines in *Ayurveda*:

1. *Panchamahabhuta Siddhanta* (Philosophy and Fundamentals)
2. *Tridosa Siddhanta* (Form and Function)
3. *Rasagunaviryavipaka Siddhanta*, or *Dravyaguna* (Pharmacodynamics)

The first, *Panchamahabhuta Siddhanta*, provides the basic Vedic concept of existence. The entire universe, both living and non-living, is constituted by the elements *akasa, vayu, agni, jala* and *prithvi* (Nanal 1995).

Coming down to the physical body, it is seen to possess three essential components referred to collectively as *tridosa*, namely *dosa, dhatu* and *mala*. *Dosa* controls physiological functions, *dhatu* provides the structure, and *mala* regulates the excrements. When the three components are in balance, the individual is sound. When they are askew, a person's health breaks down.

The third component of *Ayurveda*, its pharmaco-dynamics, comes into play when the

Fig. 3.1 Dr. Teoh Eng Soon and his wife, Teoh Phaik Khuan at the Ashokan rock edict in Dhauli, Odisha, India which is marked by a stone elephant. The Second Rock Edict reads as follows: *Everywhere in the empire of Beloved of the Gods, the King Piyadassi* (referring to Emperor Ashoka, r. circa 269 – 232 B.C.E.), *and even in lands on its frontiers, those of the Colas, Pandyas, Satyaputras, Keralaputras, Ceylon, and of the Greek king Antiochus and his neighbours, everywhere, two medical services of Beloved of the Gods, King Piyadassi have been provided. These consist of the medical care for both man and animals. Medicinal herbs useful either to humans or animals have been brought and planted, wherever they did not grow: also roots and fruit have been brought and planted, wherever they did not grow. Wells have been dug and trees planted along the roads for the use of men and beasts*

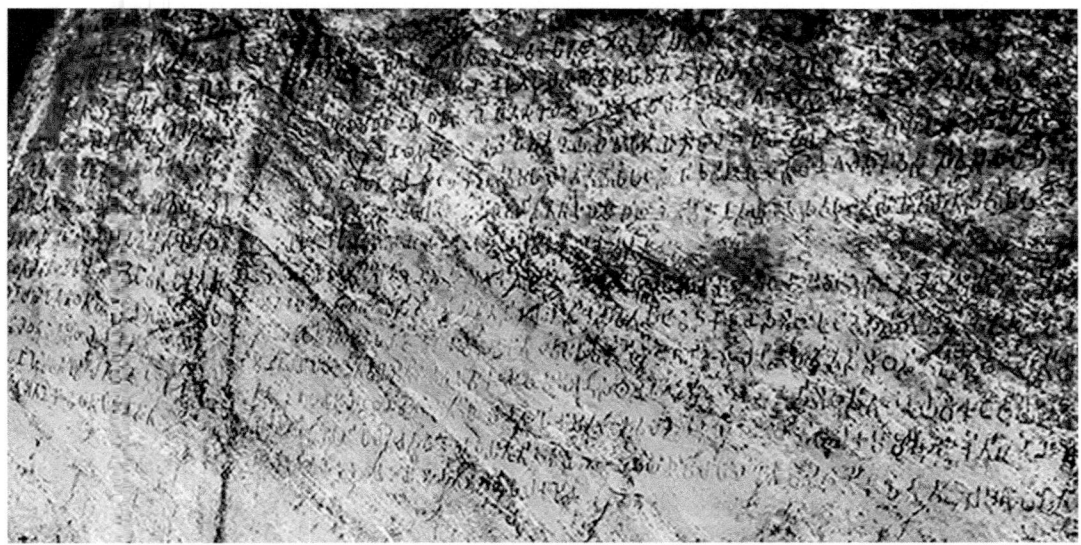

Fig. 3.2 The Asokan Edict at Dhauli, Odisha is inscribed on a massive rock in Brahmi script using Maghadi Prakrit. Photo shows a segment of the Edict which is protected by a glass encasing [Photo: E.S. Teoh]

body becomes unsound. Ayurvedic practitioners pay attention to:

1. The complaints and the illness
2. The appropriate drug or substance
3. Its properties *(guna)*. There are 20 variations, e.g. light or heavy, which refer to subjective experience and not to absolute weight (Table 3.1).
4. The taste *(rasa)*
5. Its metabolism *(vipaka)*. This refers to the ultimate "taste" after digestion by the body's *agni* (fire). The taste may be sweet, sour or pungent, but *Vipaka* describes the action of the drug rather than its ordinary taste.
6. Its potency *(virya)* and whether it remains in the body *(adhivasa)*.
7. *Prabhava* which refers to specific potency, e.g. a specific action on the heart or some other particular organ. *Prabhava* focuses on a specific effect, while *virya* discusses the overall effect.
8. *Karma* referring to the result following the administration of the drug. It is broken down into
   (a) *adhikarma*—site of action

Table 3.1 The properties *(guna)* or characteristics of drugs according to *Ayurvedic* classification (Sharma 1995)

Heavy	Light
Cold	Hot
Unctuous	Rough
Dull	Sharp
Immobile	Dynamic
Soft	Hard
Non-slimy	Slimy
Smooth	Rough
Fine, minute	Bulky, gross
Solid	Liquid

   (b) *upaya*—method of administration, mode of action
   (c) *kala*—time of administration
   (d) *phala*—the final effect.

Drug incompatibility *(vairodhika)* is another important aspect of *Ayurveda*, as it is with Chinese traditional medicine (TCM). They also deal with poisons and antidotes.

In 1931, Caius and Mkhasar tested 333 traditional herbal remedies which included two orchids (*Vanda tessellata* and *Flickingeria fimbriata*) used in the treatment of snake bites. Their experiments were conducted on animals at the Haffkine Institute in Bombay. Without exception, none of the remedies showed any preventive, antidotal or therapeutic effect. Hundreds of remedies (including the two orchids) for scorpion bites were similarly ineffective (Chopra 1933). Caius and Mkhasar conducted their tests by injecting dogs with cobra venom and daboja venom which were probably too poisonous for the herbs to counter.

Among the *Ayurvedic* herbs that Western medicine has considered useful, the best known is *Sarpagandhe* (*Rauwolfia serpentina*) which is used to treat hypertension. Reserpine extracted from this herb was a mainstay of hypertensive treatment for several decades during the mid-twentieth century. It has since been replaced by newer drugs that target specific loci of blood pressure control. *Ayurveda* has joined in the search by showing that *Pseudathria viscida* and *Triumfetta rhomboidea* possess angiotensin receptor blocker (ARB) properties and are nearly comparable to the ARB, captopril (Hansen et al. 1995).

While much of Ayurveda's interest parallels those of mainstream medicine in disciplines like internal medicine, general surgery, ophthalmology, otorhinolaryngology, toxicology, psychiatry and paediatrics, two areas find parallels only in oriental medicine, albeit there is an awakening of interest among entrepreneurial Western physicians. These are the disciplines of *Rasayana* (translated broadly as rejuvenation) and *Vajikarna* ("the sexual power of a horse", which also covers fertility). Similar interest has been displayed in the ancient Chinese herbal dated to the second century BC found at Mawangdui in Changsha, China (Harper 1998). The so-called tonics or health-promoting, life-prolonging elixirs are dealt with in *Rasayana* which focuses on people above 40 years of age. Some of them include dietary supplements and orchids, but *Rasayana* treatment, in its very preliminary stages, involves systems of cleansing with emetics, laxatives, enemas, nasal drops and venesection. In the *Vinaya Texts*, it is recorded that, when the Buddha complained of a tummy upset, his physician Jivaka made him inhale the fragrance of water-lilies. The Buddha felt better after purging several times. Tibetans scholars think that the Buddha was offered *haritaki*, (*Terminalia chebula*), which is used in *Rasayana* as an emetic (Puri 2003). More worrying is the common usage of poisonous herbs for rejuvenation or "longevity", such as aconite, and red sulfide of mercury, which element has also sent many Chinese emperors seeking immortality to an early grave.

Aphrodisiacs occupy a central place in *Ayurveda*. A entire chapter in the ancient *Atherveda* (allegedly compiled in 1000 BC) is devoted to the subject, and nearly all subsequent Indian treatises on medicine discussed plants with aphrodisiac properties (Puri 1970b). *Vajikarna* are aphrodisiac preparations that may be used as an alternative to *Rasayana*. According to the central theory, *shukra*, which is a vital fluid and the essence of all bodily functions, provides the basis of beauty, strength, power and fame. A person whose *shukra* is not depleted does not require *vajikarna* treatment, but if it is depleted, then an aphrodisiac or *rasayana* needs to be prescribed (Puri 2003). This belief finds a parallel in contemporary Western practice supposedly dealing with "anti-ageing", which recommends the use of testosterone or other male hormones to increase libido and various drugs for erectile dysfunction. The reliance on *salep* prepared from a large range of terrestrial orchids for aphrodisiac was a Grecian heritage transmitted to India and Central Asia, even reaching Mongolia. *Eulophia* species constitute the main source of *salep* in India. However, the full range of Indian *salep* includes *Dactylorhiza hatagirea* (*Orchis latifolia*), *Orchis mascula*, *Eulophia hormusjii* (*Salam lahori* is the local name), *E. campestris*, *E. nuda*, *Habenaria commelinifolia*, *H. edgeworthii*, *Cymbidium aloifolium*, *Pholidota articulata*, *Pectalis suzannae*, *Pogonia gammiena*, *Herminium monophyllum*, *H. lanceum*, *Satyrium nepalensi*, *Goodyera repens*, *Zeuxine strateumatica*, and even *Dendrobium chrysanthum*, *D. lindleyi* and *Flickingeria fimbriata* (= *D. plicatile*) (Pandey et al. 2003). Such usage has caused several species to become rare. The Nepal Gazette, 2001, banned the collection, use, sale and distribution transportation and export of *Dactylorhiza hatagirea* (local name, *Panchaule*). In Nepal, only two medicinal plants are considered to be as precious and endangered (Rajbhandary and Ranjitkar 2006).

The Problem of Correct Identification

During the classical Indian period (Vedic Period to the Gupta Dynasty), Indian herbal tradition required direct apprenticeship with a Master. Many classes were conducted in the forest, and repeated field trips ensured that the herbs collected were correctly identified. With the appearance of written texts, students paid more attention to the texts than to field work, and over time misconceptions arose. Descriptions of herbs were not sufficiently detailed in the texts because a proper botanical system had yet to evolve, so lacking first-hand field experience, it became impossible to ensure the correct identification of every herb, especially if the herb was rare or difficult to obtain. Adding to the confusion was the long list of names given to some herbs, and the fact that dissemination of knowledge was generally verbal. Over time, the identities of rare herbs became confused. The situation was worsened when herbalists suggested using substitutes to treat their rich and powerful patients.

Medical books from the medieval Indian period (seventh to eleventh century) mention that *Ashtavarga* is a rejuvenator, tonic and nutrient constituted by eight medicinal plants. The Sanskrit names for these eight plants are *Meda*, *Mahameda*, *Kakoli*, *Ksheer Kaoli*, *Riddhi*, *Vriddhi*, *Rishbhaka*, and *Jeevak* (*Jivaka*). *Dhanwantri Nighantu* specified the characteristics of each herb. *Jeevak* is sweet, cooling, cures *pitta*, *vata*, fire, fever and tuberculosis and aggravates *kapha*. It increases seminal output. *Rishbhak* is also sweet, suppresses *pitta*, *vata*, tuberculosis and fever, and is an aphrodisiac. *Riddha* and *Vriddhi* are also sweet, and pacify *pitta* and *vata* while increasing *kapha*, thereby producing a balance. They also increase semen output. With this emphasis on health

restoration, rejuvenation and sexual enhancement, over-collection of plants on the list was unavoidable, and this resulted in a rapid disappearance of these plants from their natural habitats. Substitutes were then offered, and soon the very identities of the plants became uncertain in the confusion.

After independence, Indian scholars began taking an interest in herbal medicine and several groups of scientists conducted proper botanical studies and field surveys to identify the constituent plants in *Ashtavarga* and other herbal remedies. Indian experts currently agree on the correct identities of the eight constituent herbs in *Ashtavarga*. They all occur in the Himalayas at elevations of 1200–4000 m and belong to three families. Four of the herbs are orchids, whose names as given are now, without exception, revised in the new nomenclature.

Jivaka and *Rishbhaka* were from the Himalayas and possessed garlic-like bulbs. Thus, it is surmised that *Jivaka* or *Jeevak* is *Malaxis acuminata* D. Don (= *Crepidium acuminatum*) and *Rishbhaka* is *Microstylis muscifera* Ridl. (= *Malaxis muscifera*). *Riddhi* and *Vriddhi* are species of *Habenaria*, possibly *H. edgeworthii* Hook f. ex Collett (= *Platanthera edgeworthii*), and *H. intermedia* D. Don (= *H. arietina*) respectively (Nadkarni 1954, 1976; Puri 1970a; Yadkin et al. 2003; Singh and Duggal 2009). Jalal et al. (2008) assigned *Riddhi* and *Vriddhi* to the two orchid species in the reverse order, which is not surprising given that the local names for the two *Habenaria* species are identical [namely, *Lakshmi*, *Mangala*, *Rathanga*, *Rishisrista*, *Saravajanpriya*, *Siddhi*, *Sukha*, *Vasu* and *Yuga* (Singh 2006)]. Thus, of the eight herbs, four are geophytic orchids. Today, due to the scarcity of these orchids, they have been substituted by other herbs: for instance, *Pueraria tuberosa* is recommended in *Bhavparkasha* as an acceptable substitute for the original *Jeevak* and *Rishbhaka*. The Indian yam (*Dioscorea bulbifera*) is a cheaper substitute for *Riddhi* and *Vriddhi* (Puri 1970a; Yadkin et al. 2003).

A recent survey conducted in Garwal in the Indian Himalayas found that *Ashtavarga* is very much alive in the folk medicine of the region. The survey involved interviewing 92 people (84 men and 8 women) who were traditional *vaidyas*, local healers and elderly persons. Of the eight plants used together, four were orchids [*Habenaria arietina* Hook f. (syn. *H. intermedia*); *Platanthera edgeworthii* (Hook f. ex Collett) R.K. Gupta (syn. *H. edgeworthii*); *Crepidium acuminatum* (D. Don.) Szlach. (syn. *Malaxis acuminata*) and *M. muscifera* (Lindl.) Kuntze.] and the remaining four belonged to *Liliaceae* (*Polygonatum cirrhifolium*, *P. ventricillatum*, *Fritillaria roylei* and *Lilium polyphylum*). *Ashtavarga* was used to treat sexual problems, physical disability, respiratory problems, pain, fever and urinary complaints and to slow the process of ageing (which included, in particular, the loss of libido). The predominant usage was for sexual problems that encompassed impotence, poor semen production and poor reproductive performance. Sometimes, the plants were used singly, and on occasion the particular herb may have an usage additional to its role in *Ashtavarga*. For instance, *Habenaria edgeworthii* and *Malaxis muscifera* were also used as a galactagogue (Dhayani et al. 2011).

Whereas the *Ashtavarga* concept has been revived in some medical circles under the catchy term "anti-ageing", the modern movement relies on hormones, not on herbs, with or without aesthetic surgery. It is also claimed that herbal therapy improves the immune system, an area which finds some research but no conclusive findings in experimental work with orchids. Russian scientists coined a new term, *adaptogens*, for this new class of herbs, but use of the term is confined to alternative medicine.

Siddha Medicine *(Siddha vaidyam)*

Siddhas are sages. In India, it is a term reserved for holy men with divine or intuitive knowledge. *Siddha* medicine is practised in the Tamil-speaking areas of southern India and as such it is perhaps the older form of native Indian medicine in the Deccan. *Ayurveda* medicine came from northern India. Siddha medicine is contemporaneous with Egyptian, Greek, Mesopotamian and Chinese medicine, and it was already flourishing in the sixth to seventh century BC, at the commencement of the Classical Age of India. Nevertheless, prior to its systemisation, *Siddha vaidyam* was generally referred to as *vattu vaidyam* (folk medicine), *pattichonna vaidyam* (grannies' medicine), *cintamani chikista* (treatment with a "precious jewel") or *angali chikitsa* (treatment with crude drugs) (Vaidyar 1995). Its pharmacology does not differ much from *Ayurveda* but the *Siddha* system also encompasses astrology, philosophy and yoga.

The origin of *Siddha* medicine is attributed to the god Shiva. Early *Siddha* texts are said to be the work of Agasthiar, its founder-physician and 18 other siddhas who maintained that an intimate relationship existed between man and nature, and that the soul and the mind are as much part of his constitution as the five elements of earth, water, heat, air and ether. Astral influences, disturbances of the soul and bad psychological states (such as passions, evil desires, disturbed thoughts or morbid imagination), spiritual causes, poisons and disturbances of the three principal humours can individually, or in combination, develop into illness. The *Materia Medica* of the *Siddhas* exists as separate literature from the Ayurvedic literature and is written in Tamil and in translation (Thirunavukkarasu 1995). With 2000 drugs on record, only 300 are in contemporary common usage (Vaidyar 1995). *Habenaria* and *Platanthera* species are used in Siddha medicine to prepare tonics or to treat lapses of consciousness. *P. edgeworthii* (syn. *H. acuminata*) is extensively used in Karnataka as a tonic (Rao 2004).

Tribals living in Tamil Nadu on the southeastern part of the Indian Peninsula have scarce knowledge of the medicinal usage of orchids. *Nervilia biflora* (Poliyar: *Oarilai thamarai*) is the only orchid among 54 plants used medicinally by the Poliyars of the Anaimalai Hills of Tamil Nadu. Leaf paste is used for skin diseases (Sivakumar et al. 2003). A second tribe, the Natti Vaidyas of Kolli Hills in Tamil Nadu, reported using 71 medicinal plants but only

1 orchid; their men consume tubers of *Habenaria longicoriculata* (Tamil: *Kozhikilangu*) when they develop scrotal swelling (Subramani and Goraya 2003). Valaiyans of Vellimalai Hills in Tamil Nadu mentioned only *Cymbidium aloifolium* among the 84 angiospermic plant species that they used as medicine. Leaves of *Panaipulluruvi* (*Cymbidium aloifolium*) are roasted over a fire and the juice expressed from them is instilled into the external ear to relieve earache (Ganesan and Kesavan 2003). Valaiyans living in the Alakgakoil Hills (Reserved Forest) and Piramalai Hills (Reserved Forest) of Tamil Nadu also make use of *Cymbidium aloifolium* (local name: *Panaipulluruvi*) in the same manner (Ganesan et al. 2007 Sandhya et al. 2006), but the Valaiyan community living in Karundamailai Hills of Tamil Nadu does not include an orchid among its numerous medicinal herbs (Kottaimuthu, 2008), and nor do other tribals living in Tamil Nadu (Muthu et al. 2006; Darairaj and Kanaraj 2013; Senthikumar et al. 2013; Sivasankari et al. 2013). One hazards a guess that this might have something to do with a dearth of orchids among Siddha medicinal herbs and a propensity in the past for secrecy among healers.

Unnani tibb

Unni tibb (Greek medicine) is the third tradition in India. It draws its lineage from Hippocrates (c. 460–377 BC) and Persian physicians such as Rhazes (858–925) and Avicenna (980–1037) with additional influences from the medicine of the Middle East, India and China. According to the theory of humours developed by Hippocrates, a person's health is influenced by the four humours, or fluids, which constitute the body (blood, phlegm, yellow bile and black bile), balance being the key to good health. Some of these ideas possibly seeped into India when remnants of Alexander's armies established the kingdom of Bactria in 300 BC.

Persian *Unani* physicians followed the tide of Islam which swept over northern India in the eleventh century. In their new environment, they studied the native drugs and added useful ones to their armamentarium. Thus, the *Unani* medicine derived from this period is actually quite different from the old tradition of Hippocrates and Galen. In the early twentieth century, Masihul Mulk Hakim Ajamal Khan (1864–1927) was the driving force behind *Unani* medicine and he tried to encourage further research and development in its practices. More recently, the Indian government has established separate central research institutes for *Ayurveda*, *Sidda* and *Unani* medicine to exploit the full potential of India's natural resources. The Central Council for Research into *Unani* medicine has started a programme for the systemic survey of forest areas to collect medicinal plants of interest in the *Unani* system, to extend the knowledge of the entire Indian flora and to identify every plant that may have a potential medicinal value. While more emphasis is being placed on the procurement of genuine *Unani* herbs, up to 1995, 600 folk claims for different ailments have also been collected (Siddiqui 1995).

Salep prepared from the tubers of various geophytic (terrestrial) orchids is much valued in *Unnani* medicine as a remedy for impotency and nervous disorders. *Salep*, or more correctly *salem/salam* as it is called in India, goes by different names depending on the appearance of the tubers: *salam punja* for the palm-like variety; *salam labsuinia* for the garlic type; *salam mishri* if translucent and globular like candied sugar; or *salam badshah* (emperor *sclam*) if large and globular (Puri 2003). *Habenaria* and *Platanthera* species are also used in Unnani medicine to prepare tonics or to treat lapses of consciousness (Rao 2004).

Tibetan Medicine

Tibetans constitute a significant portion of the population of Ladakh, Srinagar and Kashmir in northern India. Tibetan medicine is practiced in this region and at remote Dolpo in NW Nepal.

Tibetan medicine or the *Amchi* system is practised by isolated Tibetan communities residing in Ladakh. *Amchis* are local healers who acquire their skill through apprenticeship. The *Four Tantras* which form the basis of Tibetan medicine set up their concepts in a systemic order, but at the same time they incorporate legends of a religious nature. The system makes use of herbs, hot water spring baths, minerals, moxibation, venepuncture and incantations (*mantras*) to treat various ailments. At least three important herbs are said to have arisen from the tears shed by the three principal Bodhisattvas who embody the compassion, knowledge, wisdom, power and blessings of the Buddha (Dolma Khangkar 1986).

In a survey conducted in Zarskar, a remote community of 8500 Tibetans occupying 43.1 km^2 of harsh scrubland, Kaul, Sharma and Singh (1989) managed to identify 45 crude drugs prepared from herbs. There was not a single orchid among the herbs. A more recent listing of 60 plants used in Tibetan medicine also did not include an orchid. In China, *guoshanye* which contains three orchid species is used in Tibetan medicine, but not much is known about its applications. In Dolpo, the culturally Tibetan region of NW Nepal, two orchids, *Cypripedium himalaicum* (Amchi *khu juk pa*) and *Dactylorhiza hatagirea* (Amchi *wangpo lagpa*) are employed by Amchi physicians practising Tibetan medicine (Lama et al. 2001).

Indian Tribal Medicine

Like the Tibetans, other ethnic tribes living in remote areas of India have their own concepts of illness which do not entirely agree with the major classical traditions. Every tribe has its own herbal lore that it guards jealously. Choice of herbs is generally decided by availability. The tribes of Rajasthan used 30 different herbs to treat snake bites and 10 for scorpion stings. These herbs do not overlap (Joshi 1995). In a study of one sedentary and three nomadic tribal communities in the Kashmir Himalayas conducted in 1985–1987, Kaul and Gaur (1995) recorded 957 examples of home remedies which contained a total of 154 items used to treat 1 or several of 55 illnesses. Among these 154 herbs, there was only a single orchid. *Dactylorhiza*

hatagirha (*Orchis latifolia*; tribal name *salimpanja*) which was used in 8 remedies out of the 957 (or 0.8 %), including high fever and dysentery.

In contrast, the primitive Dongria Kandha tribe in the Niyamgiri hills of southwest Orissa employs 20 species of orchids (including 4 terrestrials) to treat 33 different diseases (Dash et al. 2008). Further south along the Bay of Bengal, Reddy et al. (2005) found that 23 orchid species were regarded as medicinal by various ethnic groups in the Eastern Ghats of Andra Pradesh.

Common complaints require ready remedies; thus, the commoner the complaint, the more plant species are available for its treatment. If this premise holds true, one can get an idea of the usage of herbal remedies by Lepchas residing in their Reserve in Northern Sikkim by examining the chart compiled by Pradhan and Badola (2008). This shows that 36 species are used to treat stomach-related disorders, 23 species for trauma, 19 species for skin disorders, another 19 for respiratory disorders and 17 for fever. Heart disease, cancer, dementia and other degenerative diseases related to ageing did not feature among the common complaints.

Tribal identity influences the choice of medicinal plants. Different tribes living in Nepal do not share a common knowledge on the use of plants as food and medicine. When narrowed to the usage of orchids, the Nepali ranked highest. They made use of 17 genera of orchids (*Brachycorythis, Calanthe, Coelogyne Cymbidium, Cypripedium, Dactylorhiza, Dendrobium, Gymadenia, Habenaria, Luisia, Malaxis, Plantathera, Pleione, Rhynchostylis, Satyrium, Thunia* and *Vanda*) and left out only *Epipactis* and *Pholidota* which were used by other tribes. The Tamang made good use of plants but used only two genera of orchids, namely *Coelogyne* and *Satyrium*. The Chepang were familiar with *Coelogyne, Epipactis* and *Satyrium*, while Sherpas used *Dactylorhiza* and *Gymadenia*. Surprisingly, the much travelled Newari, whose artistic works dating from several hundred to over a thousand years ago can be seen in such far-flung regions as Xinjiang and Gansu Provinces in China, in Mongolia, and in

Myanmar only made use of *Cymbidium*. Ten Nepalese tribes were totally unfamiliar with the use of orchids as medicinal herbs (Manandhar and Manandhar 2002). Why there should be this ethnic variation remains a mystery. It may be partially explained by the approach used during the surveys and the chance meeting with someone really knowledgeable about medicinal plants. An approach used to study medicinal orchid usage by six aboriginal native tribes living in the Nicobar Islands, together numbering 26,000, resorted to interviewing the following knowledgeable people: elderly men and women with experience in the use of herbs, Hakeems (*Hinlona*) Voo-doos (*kamasoons*), witch doctors (*tamiloos*) and "other individuals who had knowledge of plants and of their traditional life". Such people together only volunteered five medicinal orchids (Dagar and Dagar 2003).

Comment on Indian Tradition Medicine

The Ayurvedic tradition claims to be 3000 years old. Its original seat of learning was Takshashila (Taxila) in present-day Pakistan, and through Jivaka Kumarabhacca and others it was introduced into Bihar (Rajghr, Pataliputra, Nalanda) around 600 BC. As a northern tradition, its *Materia Medica* was necessarily based on the flora of northern India.

Siddha medicine is a southern tradition. Its herbs consisted mainly of plants from tropical India. When north and south India became one country, the traditions merged. Dissimilar plants were then used to treat the same illness. They shared a common name, *Rasna* (an alternate spelling is *raasnaa*), a herb with anti-inflammatory, analgesic and laxative properties, which is the root of the orchid, *Vanda tessellata* (*V. roxburghii*) in the east. The root of *Pluchea lanceolata* (not an orchid) and the rhizome of *Alpina galangal* (also not an orchid) are used for the same purpose in the north and the south, respectively. *Jeevanti*, a tonic, restorative, aphrodisiac, and an important component of the acclaimed *Asthavarga*, is the orchid *Flickingeria fimbriata* (= *Dendrobium plicatile*) in the

east, whereas two non-orchidaceous plants, *Leptadenia reticulata* in the north and *Holostemma annulare* in the south, are used as *Jeevanti* (Sarin 1996).

Until fairly recently, attempts to screen Indian plants for biological activity generally omitted orchids, possibly because the chemists preferred to work on plants which were readily available in bulk (Dhar et al. 1968, 1973; Bhakuni et al 1969, 1971). In 1982, Majumder and his team in Calcutta began isolating secondary metabolites from Himalayan orchid species (Majumder et al. 1982a, b; Majumder, Datta, Sarkar, and Chakraborti, 1982a), and this team has been extremely productive. Nevertheless, the medicinal and poisonous orchids of India are spread over at least 15 genera [*Acampe, Cymbidium, Dendrobium, Eulophia, Flickingeria (Desmotrichum), Habenaria, Hetaeria, Luisia, Oberonia, Orchis, Rhynchostylis, Saccolabium, Vanda, Vanilla* and *Zeuxine*] (Caius 1936), and not all have been screened. Indian publications on medicinal plants, like their Chinese versions, are often short in their listing of medicinal orchids. The list of 1636 medicinal plants from Mysore contains only 7 orchids (*Eulophia campestris, Flickngeria fimbriata, Habenaria grandiflora, Orchis latifolia, Saccolabium papillosum, Vanda tessellata* and *Vanilla walkeriana*), giving their Latin and Kannada names but without mentioning their usage (Narasimhachar 1952).

Between 2006 and 2007, medicinal usage of wild orchids in the Chittagong Hill Tract, Sylhet district, and the Dhaka, Sundarban and Khulna divisions of Bangladesh was investigated by Musharof Hossain who interviewed 40 "Kobiraj", the local practitioners of native herbal medicine. These practitioners were asked to identify the plants by their vernacular names, specify the parts to be used and their properties, and advise on their usage, preparation, dosage and administration. Unfortunately, the nature of a herbal preparation was often a guarded secret, considered essential for safeguarding their professional interest. For any single plant, at least three Kobiraj were interviewed to determine whether there was consensus. Altogether, 29 indigenous orchids were identified, distributed in 10 terrestrial and 10 epiphytic genera. Their usage were varied but together they extended over the treatment of paralysis, asthma, chest pain, abdominal disorders, joint pain, muscle ache, earache, fractures, snake and scorpion bites, swellings, fever, tuberculosis, malaria, syphilis, constipation and diarrhoea, jaundice, menstrual disorders, sexual dysfunction, skin disease and worm infestation (Musharof Hossain 2009).

Arab Medicine

Genghis Khan conquered the Eurasian steppes and almost the whole of the Middle East in the mid-thirteenth century, and his grandson Kubulai Khan captured China in 1279. The Yuan (Mongol) Dynasty (1280–1368) that followed lasted almost 89 years. The Mongol hordes were ruthless, annihilating entire cities and civilisations (e.g. the Xixia) which resisted. Under Kubulai Khan, the tax-based census of China's population fell from 100 million in the period of Southern Song to 59 million at the end of the Yuan Dynasty (Fitzgerald 1935). However, the lives of people who were useful to the Mongols, like physicians, artisans, soothsayers, shamans and priests, were spared. Mongols depended on cavalry for speed of conquest. Dislocations and fractures were common among the horsemen.

Arab and Persian physicians taken to China started a Moslem Institute of Medicine in Beijing and translated Arab books on surgery and medicine into Chinese. Arab medicine was favourably received by the Chinese physicians because of an apparent similarity in the classification of disease. Arab medicine which was influenced by Greek medicine propounded that there were four bodily fluids whose dysfunction produced disease. Chinese medical theory originally attributed illness to an imbalance of the six vapours but, in the fourteenth century, these six vapours were reduced to four, namely wind, moisture, cold and heat.

Many herbs of Middle Eastern origin were incorporated into the Chinese pharmacopoeia. Glaring omissions are the much touted aphrodisiacs derived from *Dactylorhiza*, *Orchis*, *Ophrys* and other terrestrial orchids. This is surprising given the inordinate ancient Chinese interest in aphrodisiacs and sex attested by the *Mawangdui Medical Manuscripts* from the second century BC and the classic novels of the Ming Dynasty (1368–1644), such as *Golden Lotus*, *Water Margin*, and *Journey to the West*. Perhaps this was due to Mongol law which established adultery as a capital offence.

The Yuan Dynasty (1280–1368) was an important period for scientific exchange between China and the Middle East, although it was not the first time that such an exchange occurred. The Silk Road had been opened by Zhang Jian (fl. c. 125 BC) during the reign of Han Emperor Wudi (r. 140–87 BC) (Twitchett and Loewe 1986). By the third century, maritime relations had been established between Guangdong (Canton) and the Persian Gulf, and during the Tang Dynasty 30 Arab emissaries visited the court in Changan. At least two pharmacopoeias compiled during the Tang Dynasty mentioned drugs imported from the Middle East, Persia, Sogdiana and India. Persian alfalfa to feed the noble horses from Ferghana was not the only herb brought back to China during the second century BC. There was myrrh (Arabic *murr*, Chinese *moyo*), fenugreek (Arabic *hulba*, Chinese *huluba*), opium (Arabic *afiun*, Chinese *afujong*, *apian*) and theriac (Arabic *tiryaq*, Chinese *ti-yih-chia*), whose Chinese names clearly indicate their Arab origin, and later came almonds, basil, carrots, castor oil, coriander, cumin, dates, flax, gall nut, grapes, henna, indigo, jasmine, lettuce, locust beans, manna, myrobolan, narcissus, *nux vomica*, onions, pepper, pistachio, pomegranates, saffron, sesame, shallots, spinach, sugar-cane, walnuts, water-melon and many others. Items taken from China to the west included apricot, bamboo, China rose, cinnamon, *coptis tecta*, kaolin, peaches, rhubarb and sarsaparilla (China root). Al-Razi (Rhazes, 864–925) is credited with identifying small-pox, but it had already been described in China several centuries earlier by Ko Hung (281–340). Scholars discovering similarities between the *Canon of Medicine of Ibn Sina* (Avicenna, 980–1037) and the *Ben*

Cao Gang-mu also speculate that the noted Arab herbalist could have expanded his list of drugs by not limiting himself to local sources (Needham 1954; Huard and Wong 1968).

Arab medicine has not been well studied by Western scholars, and many prejudices undoubtedly exist. Currently, there is no training programme on Arab traditional medicine, and Arab medicine is not officially offered as an alternative form of treatment. Nevertheless, in the Arab-Islamic world, rural folk rely on traditional self-administered remedies and consult with traditional healers who may be spiritual healers, homeopaths or herbalists. Around 200–250 native plants in the region have a role as medicinal herbs (Saad et al. 2008).

Arab physicians at the beginning of the current era based their practice on the theories and methods of Hippocrates and Galen. Herbal remedies copied Dioscorides (40–90 CE). The most notable Arab physicians were Al Razi (known in the west as Rhazes) and Ibn Sina (Avicenna to Europeans).

Abu Bakr al Razi was born in Rayy (located near Tehran) in 865 and died in 924. He was a lute player who decided to study medicine in Baghdad after sustaining an eye injury. He challenged the theories of Galen (129–199), in particular the concept that good health depended on a balance among the four kinds of body fluids or humours (blood, phlegm, yellow bile and black bile), and that a sick person needed to take special diets supplemented with herbs, or even undergo invasive treatments like bloodletting (venesection) to restore the balance. Al-Razi emphasised that theories must be proven by experimentation. On one occasion, he allowed one group of meningitis patients to be treated by bloodletting, leaving the second group untreated. In the event, the group that received the Galenic treatment improved while the other group did not! [This showed that well-intentioned but poorly designed experiments can go wrong and its result can be confounding (see Chap. 24)]. Al-Razi recognised that when the body needed to mobilise its defences it generally did so by raising its temperature. He authored several treatises on different topics, a

23-volume encyclopedia on medicine, and two books entitled *Why People Prefer Quacks and Charlatans to Skilled Physicians* and *On the Fact That Even The Most Skillful of Physicians Cannot Heal All Diseases* (Masood 2009). Al-Razi was familiar with *Satyrion* which "is hot, increases sperm production and cures gout". It is strange that this sceptic did not question the claims of Dioscorides. Al-Razi described *Asahasafra* which possibly corresponded to *Orchis morio* or *Dactylorhiza maculata* (Jacquet 1994).

Ibn Sina (Avicenna), the other renowned physician and herbalist was born near Bukhara in Uzbekistan in 980. He died in Hamadan, Persia, in 1037. A child prodigy, Ibn-Sina was already an acknowledged physician at the age of 16. Before that, he had memorised the entire Koran! His writings bear evidence of his faith (Hameed 1983) but this did not stop him from realizing that all physical phenomena had to have known causes, a fact that could not be violated. He dismissed the possibility of miracles and resurrection. Ibn-Sina challenged Galen's theory that the eye emitted energy which allowed it to see. Nevertheless, he subscribed to the Greek theory of the four humours.

In his treatise *On the Treatment of Cardiac Conditions with Herbs* (*al adwiyat al-Qalbia*), Ibn Sina related the four humours to the emotions. Thus, notwithstanding this scientific approach which Al-Razi, Ibn-Sina and their contemporaries tried to promote, the theory of the four humours remained in Arab medicine and was transmitted to China during the Yuan Dynasty.

When he succeeded in treating the Samanid ruler, Ibn-Sina gained access to the royal library at Bukhara. That privilege greatly widened his knowledge. (Similarly, China's great Ming pharmacologist Li Shizhen also had access to the imperial medical library in Beijing for a short period.) Ibn-Sina was a renaissance man before that term was invented. He was a prolific writer, authoring 270 books on a great variety of topics, including astronomy, physics, mathematics, poetry and medicine. His main claim to fame arose from his two great books, *Kitab ash-Shifa*

(*Book of Healing*) and *al-Qann fi al-Tibb* (*The Canon of Medicine*). Translated into Latin, the latter was a standard text in Europe for six centuries until the arrival of germ theory. In the *Canon*, he stated that nerves had sensory and motor functions, that tuberculosis was contagious and that diseases could spread through water and soil. The *Canon* contained a list of 760 drugs, the majority of which were herbal. Its pertinence to our discussion lies only in its use of certain terrestrial orchids as aphrodisiacs. In *The Canon*, Ibn Sina mentioned three orchids: *Alisma sive Damasonia* (used as an antidote for coughs and asthma), *Satyrio chasi altaleb* and *Testiclus chasi alchelb*. Ibn Sina suggested that the orchid tubers could be used as aphrodisiac, appetite stimulant, for mucus production and to promote recovery from strokes (Sezik 1967).

Ibn Bairtar (c. 1197–97), the botanist (El Achachb), who had travelled widely in the Mediterranean before settling as director of gardens in Damascus, gave orchids several names some of which are close to the ones in use today, *khossa-el kelb* (canine testicle), *Chusa elthalab* and *Qatel akhihi* (fratricide) and *Efibakthis* (Jacquet 1994). The *Flora of the Kingdom of Saudi Arabia Illustrated Volume III* described only 9 orchid species divided among 6 genera, *Holotrix*, *Epipa ctis*, *Orchis*, *Nonatea*, *Zeuxine* and *Eulophia* (Chaudhary 2001).

Three orchids, probably *Orchis* species or *Dactylorhiza*, are also mentioned in the writings of Ibn Mansour who lived in Persia during the tenth century. Ibn Mansour used the term *Abu Beiden* to describe an orchid introduced from the East, possibly India, which cured nerves, and which was transcribed as *Buzeiden* during the sixteenth century and became *Digiti citrini*, referring to its hand-like shape (Jacquet 1994). The orchid is probably a species of *Dactylorhiza* which was mentioned by Shakespeare in *Hamlet*. The Indian citrus which has the shape of a human hand is the chebulic myrobalan (*Terminalia chebula*), a symbol of Bhaishajiaguru (Medicine Buddha) in Tibet. Under the name *Haritaki*, it has been used in Ayurvedic medicine for over 2000 years. It is used for diarrhoea and other

abdominal complaints, not to treat nerve disorders. It is not an orchid.

Under the *Doctrine of Signatures*, many species of *Dactylorhiza*, *Orchis*, *Orphrys* and *Eulophia* whose tubers resemble a pair of testicles were regarded as sex stimulants from India and the Middle East to Europe. The drug was called *Khus yatu's salab* (fox's testicle) or *Khus yaty'l klab* (dog's testicle); thus, the English names of the orchids: dog stones, fox stones, hare stones, and goat stones. Their alleged properties originated from the writings of Dioscorides whose *Herbal* the Arabs adopted, and *salab* (*salep*) was promoted by herbalists and physicians of the Middle Ages. Other Arab terms for *salab* were *Chafi alkes*, *Chafi alchels*, *Chafi attraleb*, *Safi alchaleb* and *Tartarichi*. Ibn Baitar, also known as El Achchab (c. 1197–1248), referred to such orchids as *Qatel akhihi* (fratricide), a misunderstanding from the belief that the new bulb caused the demise of the older bulb (Jacquet 1994). In the nineteenth century, Arabs still called several orchids *El-hay* (the dead) and *El Meit* (the living), referring to their paired tubers (Jacquet 1994). *Sahlab* was usually prescribed in the fresh state as the tubers had to be plump. Shrivelled tubers were discarded. After processing, these orchid tubers lose their mildly bitter taste and peculiar odour. They were sold in the form of a powder and made into a drink known as *sahlab*, rendered *salep* in the English language. They were also known as *satyrion*. The two names were used interchangeably (Lawler 1984). *Salep* has no medicinal powers but, because of its ability to congeal into a jelly when mixed with water, it came to be regarded as highly nutritious; again, merely a belief, not a fact. Nevertheless, as recently as the end of the nineteenth century, the annual import of *salep* from Smyrna into England was reported to be 642,500 kg (Fluckiger and Hanbury 1879).

In his extensive review of *salep*, the distinguished Australian orchidologist and biochemist, Leonard J. Lawler, stated that its medicinal usage was so extensive as to make it a panacea whose role in herbal medicine even exceeded that of *Panax ginseng* (see *Orchis*). The common factor running through the various conditions treatable by *salep* was the need for an easily digested sustenance, and *salep* appeared to be well fitted for that role. Many European and one Japanese pharmacopoeia, including a 1924 edition of the British *Martindale*, were referenced; however, Lawler did not include an Arab *Materia Medica* in his list. Its specific applications in Arabic medicine are unclear (Lawler 1984).

The extent of *salep's* usage in the Arab countries, in India, Turkey, and continental Europe at that time is unknown, but from the various published descriptions, it was considerable. Turkey and Greece supplied Europe and the Middle East with *salep*. India received its main supply from Afghanistan but also made use of native species of *Eulophia*, and, additionally, *Habenaria commelinifolia*, *H. pectinata*, *Zeuxine strateumatica*, and *Cymbidium aloifolium* (Lawler 1984). However, a Japanese expedition from Kyoto University to Afghanistan and Pakistan reported that it found only five species of orchids in Afghanistan [*Cephalanthera longifolia*, *Epipactis helleborine* subsp *helleborine* (syn. *E. latifolia* L.), *E. royleana* Lindl., *E. veratrifolia* Boiss & Hohen and *Dactylorhiza incarnata* subsp. *incarnata* (syn. *Orchis latifolia* L.) (Kitamura 1960, 1964)]; hence, it is quite likely that the *salep* used in Afghanistan was imported from Iran or Turkey; furthermore, the country merely served for trans-shipment of *salep* from these sources to Pakistan/India.

In Arabia, *Aceras anthropophora* R. Br. was used as a stimulant and diaphoretic (Dragendorff 1898). A drink prepared from the bulbs of *Anacamptis*, *Ophrys*, *Orchis* and *Spiranthes* available in Turkish markets was used as a home remedy for colds. The drink was sometimes sold by street vendors. The tuber of *Orchis mascula* was used as a tonic and nutrient in infantile diarrhoea in Iraq, and also as a demulcent, astringent and nervine (Al-Rawi and Chakrararty 1988, quoted by Lawler 1984). Orchids are not mentioned in the list of medicinal plants in current usage for the treatment of diabetes, liver disease, psoriasis, atopic dermatitis,

cancer and sexual impairment in the Arab countries (Saad et al. 2008).

In the above discussion, I have used the term Arab Medicine because the sources that I used employed this term; however, it should be noted that the greatest "Arab" physicians, Avicenna and Al Razes, were Persians.

Iranian Medicine

Historically, Persia was linked more closely to Europe than to central or eastern Asia. Its floristic composition is similar to that of southern and central Europe. Ancient Iranian medicine was much influenced by Greek medicine and the *Materia Medica* of Pedanius Dioscorides (40–90 CE) compiled around 70 CE which influenced European medicine until the nineteenth century. The tenth century was a flourishing period in the Middle East and Persia with freedom of movement throughout the Islamic states. Medical practice was influenced by physicians who were regarded by both Persians and Arabs as their own. The famous Persian physician, Avicenna [Ibn Sina, (full name: Abu Ali al-Husayn Ibn Abd Allah Ibn Sina) 980–1037], also wrote extensively on herbs and shared many of the beliefs of Dioscorides. Avicenna followed Dioscorides in stating that orchids were aphrodisiacs.

Available literature on Iranian interest in medicinal orchids still focuses on *salep*, which is constituted in Iran by species of *Anacamptis*, *Dactylorhiza*, *Himantoglossum*, *Ophrys*, *Orchis*, *Platanthera*, *Serapias*, *Spiranthes* and *Steveniella* (Voth 1973; Lawler 1984; Ghorbani et al. 2014).

Iran is not endowed with an abundance of orchids. The highland plateau that forms a major part of Iran is a semi-desert which is almost totally devoid of orchid species. There are only 46 species and subspecies of orchids in Iran, distributed in the northern and western mountain regions. Tubers are present in 30 species and subspecies. The tubers of 16 species or subspecies are collected for sale in the herb markets or exported; 11 identical species are collected in Turkey for *salep*. Fresh tubers are washed, soaked in water to achieve maximum weight and then packed into 50- or 100-kg bags for shipment to middlemen for further processing before shipment abroad, mainly to Turkey. It is estimated that 5.5–6.1 million orchid plants are destroyed annually by this trade which started in 2006 (Ghorbani et al. 2014).

Residents in some areas where such orchids are harvested do not use orchids as medicine. However, in the cities, *salep* still enjoys a role in Iranian folk or traditional medicine. *Attari* (herbal) practitioners feed malnourished children with *salep*. They also recommended it as a stimulant, an expectorant for breathing difficulties and for painful joints and bone strengthening. *Salep* is sometimes used as a starter in the preparation of yogurt (Ghorbani et al. 2014).

Malay and Indonesian Medicine

Traditional Malay medicine was practised by *bomohs* (magic practitioners who also treated illness) and by knowledgeable matriarchs, such knowledge being acquired through vertical transmission in the family line. The *bomoh* generally preferred to use exotic drugs to enhance his reputation. He was usually fond of polypharmacy and might prescribe rare items that were only found in the jungle. The matriarch, on the other hand, resorted to the use of plants found in her kitchen garden or others growing in the neighbourhood, and on preserved spices. The 20 orchids which were used in Malay medicine (Ridley 1906; Burkhill and Haniff 1930; Burkill 1935) were familiar village items. Their usage was probably common knowledge.

The manuscript of a *Medical Book of Malayan Medicine* translated into English by Ismael Moonshee in Penang was found in the possession of the Pharmaceutical Society of Great Britain in 1928. After revision by I.D. Gimlette and I.H. Burkill (1930), it was published in the *Gardens Bulletin of the Straits Settlements*. It gives us some clues on the origin of the Malay tradition. Plant names were polyglot, with input

from Arabic, Tamil, Javanese, Sundanese and Balinese. This suggests that many of the medical practices were derived from Arabic and Indian traditions. Certain practices were common to Malay, Indian, and Thai medicine, for instance, the use of *Eria pannea* Lindl. (= *Mycaranthes pannea*) in a medicinal bath to treat ague (Ridley 1894; Trivedi et al. 1980; Chuakul 2002). The old Malayan medicine was also similar to that existing in Sumatra. Juices were extracted from plants and consumed, instilled to treat earache, or applied to snake and scorpion bites and boils. Plants were pounded to make poultices. Leaves were heated before application as counter-irritants. Spells were chanted. And the patient was instructed on prohibitions (*pantang*, Malay). The *bomoh*, being Muslim, did not employ alcohol in his prescriptions. Therefore, alcoholic tinctures were unknown (Gimlette and Burkill 1930; Gimlette and Thomson 1939).

At the time when medicinal orchids were in vogue, Malays did not distinguish between species as do botanists, and they often substituted one species of *Dendrobium* for another. For instance, whereas the pigeon orchid was used for poultices in the western part of Peninsular Malaysia, in the eastern part of the peninsula, *Dendrobium purpureum* was used. *D. calcaratum*, *D. planibulbe*, *D. pumilum* and *D. subulatum*, all covered by the common name *anggrek*, were also used in a similar manner (Burkill 1935). In treating earache, juice squeezed from a heated pseudobulb was used, and the orchid chosen would be one that was readily available in the village, such as *Bulbophyllum vaginatum* or *Dendrobium crumenatum*.

It should be mentioned that traditional Malay medicine is now of only historical interest because Malaysia is a rich, progressive country. Today, most villagers will consult a qualified doctor for an illness or an injury.

Thai Herbal Medicine

Edward F Anderson, reporting on his study of the ethnobotany of the hill tribes in northern

Thailand, stated that tribal practitioners were trained informally through long apprenticeships which started in childhood. They often become proficient at a young age and they have not ceased to practice. However, although many tribal herbalists still collect their materials themselves from the forest, when they are too old they will send trusted apprentices. Some herbalists guard their knowledge by mashing plant materials to render them unrecognisable, and most do not agree on the usage of individual plants, nor on the identities of the correct herbs to be used for a particular ailment. Plant usage varies within a tribe, and from one tribe to another. Their knowledge is fast disappearing because young members of the tribe are reluctant to undergo herbalist training, and herb gathering has been made difficult through deforestation (Anderson 1993).

Wongsatit Chuakul of Mahidol University in Bangkok is perhaps the only person who has done a proper study of orchids used in Thai herbal medicine. Chuakul surveyed the usage of medicinal plants in 9 provinces by interviewing 10 herbalists, collecting the plant specimens and confirming their identity by making careful taxonomic comparisons with specimens in Thai National Herbaria. Forty-two medicinal Thai orchids from 25 genera were identified. Their usage was based on statements of herbalists from the provinces of Sukhothai, Phitsanulok, Maha Sarakham, Ubon Ratchathani, Chaiyaphum, Yasothon, Surin, Krabi and Yala. No herbalist was familiar with herbal usage of every Thai medicinal orchid, and for individual species, the number of herbalists that identified its usage varied from one to six, and most of the time only two or three herbalists could do so (Chuakul 2002). A similar methodology was recently used to study ethnomedicinal usage of orchids in India and elsewhere. Chuakul's approach should be taken as the standard for all studies on tribal or native medicine.

Analysis of Chuakul's report indicates that the medicinal usage of orchids is not as extensive in Thailand as the total list would suggest.

References

Al-Rawi A, Chakrararty HL (1988) Medicinal plants of Iraq, 2nd edn. Ministry of Agriculture and Irrigation, Baghdad

Anderson EF (1993) Plants and people of the golden triangle. Ethnobotany of the hill tribes of Northern Thailand. Dioscorides Press, Portland

Bhakuni DS, Dhar D et al (1969) Screening of Indian plants for biological activity: part II. Indian J Exp Biol 7:250–262

Bhakuni DS, Dhar ML, Dhar MM et al (1971) Screening of Indian plants for biological activity: Part III. Indian J Exp Biol 9:91–102

Burkhill IH, Haniff M (1930) Malay village medicine. Gard Bull Straits Settlem 6:165–321

Burkill IH (1935) (1966 reprint, 2nd ed., with contributions by Birtwistle W, Foxworthy FW, Scrivenor JB, Watson IG). A dictionary of economic products of the Malay Peninsula, Vol II. Crown Agents for the Colonies, London. Ministry of Agriculture & Co-operatives, Kuala Lumpur

Caius JF (1936) The medicinal and poisonous plants of India. J Bombay Nat Hist Soc 38(4):791–799

Chaudhary SA (ed) (2001) Flora of the kingdom of Saudi Arabia illustrated, vol III. Ministry of Agriculture and Water Research Centre Natural Herbarium, Riyadh

Chopra RN (1933) The indigenous drugs of India, Calcutta: The Art Press. Republished as Chopra's Indigenous Plants of India, 2nd ed. Academic, Kolkata (1986)

Chuakul W (2002) Ethnomedical uses of Thai orchidaceous plants. Mahidol Univ J Pharm Sci 29(3–4): 41–45

Dagar HS, Dagar JC (2003) Plants used in ethnomedicine by the Nicobarese of Islands in Bay of Bengal, India. In: Singh V, Jain AP (eds) Ethnobotany and medicinal plants of India and Nepal. Scientific Publishers (India), Jodhpur, pp 773–784

Darairaj P, Kanaraj M (2013) Traditional medicinal plant resources of Southern Pachchmalais in Trichirapalli of Tamil Nadu, India. Implication of traditional knowledge in health care systems. Int J Res Hum Arts Lit 1 (6):39–46

Dash PK, Sahon S, Bal S (2008) Ethnobotanical studies on orchids of Niyamgiri hill ranges, Orissa, India. Ethnobot Leaflets 12:70–78

Dhar ML, Dhar MM, Dhawan BN et al (1968) Screening of Indian plants for biological activity: part 1. Indian J Exp Biol 6:232–247

Dhar ML, Dhar MM, Dhawan BN et al (1973) Screening of Indian plants for biological activity: part 1V. Indian J Exp Biol 11:43–54

Dhayani A, Nautiyal BP, Nautiyal MC (2011) Importance of Astavarga plants in traditional systems of medicine in Garhwal, Indian Himalaya. Int J Biodiv Sci Ecosyst Serv Manag 6(1–2):13–19

Dolma Khangkar L (1986) Lectures on Tibetan medicine. Compiled and edited by K. Dhondup. Library of Tibetan Works and Archives, New Delhi

Dragendorff JGN (1898) Die heilpflanzen der verschiedenen Volker und Zeiten lhre Anwendung wesentlichen Bestandtheile und Geschichte: ein Handbuch fur Arzte, Apotheker, Botniker und Droguisten. Reprinted by Nabu Press, Charleston, South Carolina, 2011

Fitzgerald CP (1935) China. A short cultural history. Praeger, New York

Fluckiger F, Hanbury D (1879) Pharmacographia.: a history of drugs. Macmillan, London

Ganesan S, Kesavan L (2003) Ethnomedicinal plants used by the ethnic group Valaiyans of Vellimalai Hills (Reserved Forest), Tamil Nadu, India. J Econ Taxon Bot 27(3):754–760

Ganesan S, Pandi NR, Banumathy (2007) Ethnomedicinal survey of Alagarkoil Hills (reserve forest), Tamil Nadu, India. eJ Indian Med 1:1–18

Ghorbani A, Gravendeel B, Naghibi F, de Booer H (2014) Wild orchid tuber collection in Iran: a wake-up call for conservation. Biodivers Conserv. doi:10.1007/s10531-014-0746-y

Gimlette JD, Burkill IH (1930) The medical book of Malay medicine. Gard Bull Straits Settlem 6:323–474

Gimlette JD, Thomson HW (1939) A dictionary of Malayan medicine. Oxford University Press, London

Hameed HA (ed.) (1983) Avicenna's tract on cardiac drugs and essays on arab cardiotherapy. Institute of History of Medicine and Medical Research, Indian Institute of Health and Tibbi (Medical) Research Karachi, Pakistan, New Delhi. Hamdad Foundation Press, Karachi

Hansen K, Nyman U, Smitt UW, Adsersen A, Gudiksen L, Rajasekharan S, Pushpangadan P (1995) In vitro screening of Indian medicinal plants for anti-hypertensive effect based on inhibition of angiotensin converting enzyme (ACE). In: Pushpangadan P, Nyman U, George V (eds) Glimpses of Indian ethnopharmacology. Tropical Botanic Garden and Research Institute, Thiruvananthapuram, pp 263–273

Harper DJ (1998) Early Chinese medical literature. The Mawangdui medical manuscripts. The Sir Henry Wellcome Asian series. Kegan Paul International, London

Huard P, Wong M (1968) Chinese medicine. Translated from French by B. Fielding. World University Library, London

Jacquet P (1994) History of orchids in Europe, from antiquity to the 17th century. In: Arditti J (ed) Orchid biology reviews and perspectives VI. Wiley, New York

Jalal JS, Kumar P, Pangtey YPS (2008) Ethnomedicinal orchids of Uttarakhand. Western Himalaya Ethnobot Leaflets 12:1227–1230

Joshi P (1995) Ethnomedicine of Tribal Rajasthan – an overview. In: Pushpangadan P, Nyman U, George V (eds) Glimpses of Indian ethnopharmacology. TBGRI, Thiruvananthapuram, pp 147–162

Kaul MK, Gaur RD (1995) Characteristics of ethnopharmacological resources in Kashmir Himalaya. In: Pushpangadan P, Nyman U, George V (eds) Glimpses of Indian ethnopharmacology. TBGRI, Thiruvananthapuram, pp 185–209

Kaul MK, Sharma PK, DSingh V (1989) Ethnobotanic studies of Northwest and Trans-Himalaya VI: contributions to the ehnobotany of Bani-Basholi region J & K state. Bull Bot Surv India 31(1-4):89–94

Kitamura S (1960) Flora of Afghanistan. Committee of Kyoto University Expedition Karakoram and Hindo Kush. Kyoto University, Japan

Kitamura S (1964) Plants of West Pakistan and Afghanistan. Kyoto University, Japan

Kottaimuthu R (2008) Ethnobotany of the Valaiyans of Karandamalai, Dindigul District, Tamil Nadu, India. Ethnobot Leaflets 12:195–203

Lama YC, Ghimire SK, Ameeruddy-Thomas Y et al (2001) Medicinal plants of Dolpo. Amchis' knowledge and conservation. WWF Nepal Program, Kathmandu

Lawler LJ (1984) Ethnobotany of the orchidaceae. In: Arditti J (ed) Orchid biology reviews and perspectives, vol 3. Cornell University Press, Ithaca

Majumder PL, Datta N, Sarkar AK, Chakraborti J (1982a) Flavidin, a novel 9,10dihydrophenanthrene derivative of the orchids Coelogyne flavida, Pholidota articulata and Otochilus fusca. J Nat Prod 45(6):730–732

Majumder PL, Laha S, Datta N (1982b) Coelonin, a 9.10-dihydrophenanthrene from the orchids Coelogyne ochracea and Coelogyne elata. Phytochemmistry 21:478–480

Manandhar NP, Manandhar S (2002) Plants and people of Nepal. Timber, Portland

Masood E (2009) Science and Islam. Icon Books, London

Musharof Hossain M (2009) Traditional therapeutic uses of some Indigenous Orchids of Bangladesh. Med Aromat Plant Sci Biotechnol 3:100–106

Muthu C, Ayyanar M, Raja N, Ignacimuthu S (2006) Medicinal plants used by traditional healers in Kancheepuram District of Tamil Nadu, India. J Ethnobiol Ethnomed 2:43

Nadkarni AK (1954) Dr. K.M. Nadkarni's Indian Materia Medica, vol 2, 3rd edn. Popular Book Depot, Bombay

Nadkarni N (1976) Indian natural medica, vol 1. Pop Prakashan, Bombay

Nanal VM (1995) Fundamental principles of ayurveda and their application in the study of Parnabeeja (Kalanchoe pinnata). In: Pushpangadan P, Nyman U, George V (eds) Glimpses of Indian Ethnopharmacology. Tropical Botanic Garden and Research Institute, Thiruvananthapuram

Narasimhachar SG (1952) Latin and Kannada names of indigenous and medicinal plants of Mysore. Government Press, Bangalore

Needham J (1954) Science and civilization in China. Cambridge University Press, Cambridge

Pandey NK, Joshi GC, Mudaiya RK et al (2005) Management and conservation of medicinal orchids of Kumaon and Garhwal Himalaya. In: Singh V, Jain

AP (eds) Ethnobotany and medicinal plants of India and Nepal. Scientific Publishers (India), Jodhpur, pp 114–118

Pradhan BK, Badola HK (2008) Ethnomedicinal plant use by Lepcha tribe of Dongzu valley, bordering Khangchendzonga Biosphere Reserve, in North Sikkim, India. J Ethnobiol Ethnomed 4:22–39

Puri HS (1970a) Indian medicinal plants used in elixirs and tonics. Quart J Crude Drug Res 10:1555–1566

Puri HS (1970b) Vegetable aphrodisiacs of India. Quart J Crude Drug Res 11:1742–1748

Puri HS (2003) Rasayana. Ayurvedic herbs for longevity and rejuvenation. Taylor and Francis, London

Rajbhandary S, Ranjitkar S (2006) Herbal drugs and pharmacognosy. Monographs on commercially important medicinal plants of Nepal. Ethnobotanical Society of Nepal (ESON), Katmandu

Rao AN (2004) Medicinal orchid wealth of Arunachal Pradesh. Newsl ENVIS Node Indian Med Plants 1 (2):1–7

Reddy KN, Subha Raju GV, Sudhakar Reddy Ch, Raju VS (2005) Ethnobotany of Certain Orchids of Eastern Ghats of Andra Pradesh. EPTRI-ENVIS Newsl II(3):5–9

Ridley H (1894) The Orchidaceae and Apostasiaseae of the Malay Peninsula. J Linn Soc 32:335–338

Ridley HN (1906) Malay drugs. Agric Bull Straits Settlem Fed Malay States 5, 245 and 277

Saad B, Azaizeh H, Said O (2008) Arab herbal medicine. In: Watson RR, Preedy VR (eds) Botanical medicine in clinical practice. CAB International, Wallingford, pp 31–39

Sandhya B, Thomas S, Isabel W, Shenbarathai R (2006) Ethnomedicinal plants used by the Valaiyan community of Piranmalai Hills (Reserved Forest), Tamil Nadu, India—A pilot study. Afr J Tradit Complement Altern Med 3(1):101–114

Sarin YK (1996) Illustrated manual of herbal drugs used in Ayurveda. Indian Council of Medical Research, New Delhi

Senthikumar K, Aravindan V, Rajendran A (2013) Ethnobotanical survey of medicinal plants used by Malayali tribes in Yercaud Hills of Eastern Ghats. J Nat Res 13(2):118–132

Sezik E (1967) Salep (Doctoral Thesis). Istanbul Universitesi Eczacilik Fakultesinde

Sharma PV (1995) Some aspects of the basic concepts of ayurvedic pharmacology. In: Pushpangadan P, Nyman U, George V (eds) Glimpses of Indian Ethnopharmacology. TBGRI, Thiruvananthapuram, pp 1–9

Siddiqui MK (1995) State of Unani Tibb (Unani Medicine) in India. In: Pushpangadan P, Nyman U, George V (eds) Glimpses of Indian ethnopharmacology. TBGRI, Thiruvananthapuram, pp 85–98

Singh AP (2006) Asthavarga – rare medicinal plants. Ethnobot Leaflets 10:104–108

Singh A, Duggal S (2009) Medicinal orchids: an overview. Ethnobot Leaflets 13:351–363

Sivakumar A, Subramanian MS, Karunakaran M, Burkanudeen A (2003) Ethnobotany of Poliyars of Anaimalai Hills, Tamil Nadu. J Econ Taxon Bot 27 (3):679–685

Sivasankari B, Pitchaimani S, Anandharaj M (2013) A study of traditional medicinal plants of Uthapuram, Madurai District, Tamil Nadu, South India. Asian Pac J Trop Biomed 3(12):975–979

Subramani SP, Goraya GS (2003) Some folklore medicinal plants of Kolli Hills. Record of a Natti Vaidyas Sammelan. J Econ Taxon Bot 27(3):665–678

Thatte U (1995) Medicine Sans barriers. In: Pushpangadan P, Nyman U, George V (eds) Glimpses of Indian ethnopharmacology. TBGRI, Thiruvananthapuram, pp 129–136

Thirunavukkarasu S (1995) Theoretical and conceptual foundations of Siddha medicine. In: Pushpangadan P, Nyman U, George V (eds) Glimpses of Indian ethnopharmacology. TBGRI, Thiruvananthapuram, pp 71–76

Trivedi VP, Dixit RS, Lal VK (1980) Orchids in the drug markets of Bareilly, Kanpur and Nearby Districts. Nagarjun (Calcutta) 23(8):157–163

Twitchett DS, Loewe M (eds) (1986) The Cambridge History of China. Vol. 1. The Ching and Han Empires 221 B.C. – A.D. 220. Cambridge University Press, Cambridge

Vaidyar KE (1995) The role of Kuppameni (Acalypha indica) in Sidha Vaidya. In: Pushpangadan P, Nyman U, George V (eds) Glimpses of Indian ethnopharmacology. Tropical Botanic Garden and Research Institute, Thiruvananthapuram, pp 99–100

Varier NVK (2005) History of ayurveda. Arya Vaidya Sala, Kottakkal

Voth W (1973) Salep im turkischen Speiseeis. Orchidee 24:29–32

Yadkin KC, Sharma A, Tiwari RK, Shanka T, Trivedi VP (2003) Standardization and Quality Control of Rshbhak (Microstylis wallachii Lindl). In: National symposium on emerging trends in Indian medicinal plants, Lucknow, p 74

Herbal medicine generally makes use of orchids in their raw state, but, in traditional Chinese medicine (TCM) practice, such herbs are cleaned, processed distributed, and stored for extended periods. Dried herbs may be used for many years after their preparation. Proper processing is therefore an important aspect of herbal medicine.

The objectives in processing herbs are to (Zhang 2008):

1. Remove contaminants.
2. Prevent spoilage, prolong shelf life, and minimise the loss of active components.
3. Make them presentable.
4. Enhance their therapeutic actions.
5. Reduce side effects or toxicity.
6. Increase surface area and permit optimum extraction.
7. Modify taste and smell to make them more acceptable.
8. Change their property and efficacy to suit a patient's specific requirements.
9. Broaden their application.

Washing and Cleaning

The initial cleaning process for all herbs involves washing to remove dirt, debris, odours, weeds, other plants, and parts that are spoilt or unwanted. One may need to brush or scrape to obtain a clean surface. If one looks at *Shihu* or *Tianma*, the processed final products are very different from the wild *Dendrobium* or the newly harvested rhizomes of *Gastrodia*. The final products are yellow or white and are very presentable (Figs. 4.1 and 4.2). Sulfur fumes are used to bleach *baichi* (Hu 1971).

Sometimes, repeat washing is required to remove contaminants or to eliminate odours. Soaking softens plant tissues and occasionally it is a prerequisite to slicing. Dried *Tianma*, for instance, is as hard as wood and it is difficult to slice through it.

The *Shui fei* (flying in water) technique is used to obtain the finest powder. In this process, after thorough grinding, the product is mixed with a large amount of water and stirred. Smallest particles will float while the larger heavier ones sink. The suspension is collected to recover a fine powder. The sediment undergoes further blending until it is reduced to a size that floats. This process is used for preparing ophthalmic products.

During processing, adjuvants may be added to improve the therapeutic effect. Common adjuvants are alcohol, vinegar, ginger extract and honey.

© Springer International Publishing Switzerland 2016   51
E.S. Teoh, *Medicinal Orchids of Asia*, DOI 10.1007/978-3-319-24274-3_4

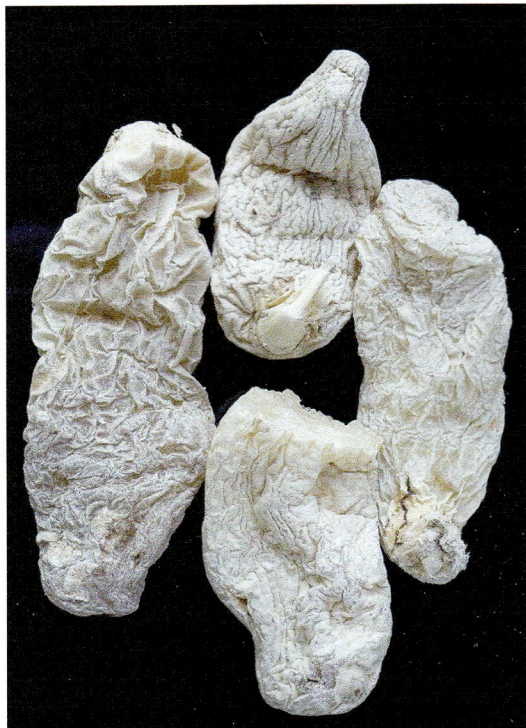

Fig. 4.1 *Tianma* (tubers of *Gastrodia elata* Bl.) bought from a supermarket in Guiyang, Guizhou Province, China. Guizhou produces the best-quality *Tianma* (Photo: E.S. Teoh)

Heat Treatment

Further processing of medicinal herbs involves steaming, boiling, simmering, quenching, roasting and dry roasting; boiling in honey, alcohol, vinegar, saline or ginger; and calcining. Steaming and boiling are used to reduce toxicity, e.g. aconite is extremely toxic and needs to be processed with *Gan cao* or *Radix glycyrrhizae* and *Hei dou* or soya beans to minimise its toxicity. Sometimes, this enhances the therapeutic effect. Simmering is used for extraction of active constituents which are then added to pills, pastes and powders. Duration of heat exposure may also affect the effect of a herb. For instance, decoction of *Dong Qui* (*Angelica sinensis*) exerts a contractile effect on the uterus when it is decocted for a short time. It relaxes the uterus when the herb is boiled for a long period. The anticoagulant effect of *Dong Qui* is retained regardless of how long it is heated.

Fermentation

Except for *Vanilla*, orchids are not fermented.

Examples

Handling of the three classical herbs derived from orchids are described to illustrate the processes involved in preparation of herbal remedies.

1. *Tianma* (*Gastrodia elata* Bl.)
   After harvesting, *Tianma* tubers are washed free of soil and grouped according to size. They are boiled in water in batches, the timing being determined by the mass of the individual tubers. Those weighing:
   1. More than 150 g, 10–15 min
   2. 100–150 g, 7–10 min
   3. Under 100 g, 5–8 min
   After boiling, tubers are exposed to sulfur fumes for 30 min and then dried over a low fire at 50–60 °C until they are about 80 % dry. Each tuber is then hand-flattened, and dried at 70 °C to complete desiccation.
   Tianma is usually sold as whole tubers which may be: (1) large, white and flawless if obtained from a pharmaceutical source, or (2) yellow, of uniform size, and (3) yellow or grey and of variable size, if obtained from a village herb seller. We managed to locate only one mention of a pill being prepared exclusively from *Tianma*. The other prescriptions contained dried *Tianma* mixed with other herbs (Table 4.1). A Wei family formula used for treating fretfulness in children is in the form of a pill prepared by grinding 25 g of *Tianma* with 50 g of *Buthus martensii*, 25 g of *Ansaemi japonicum*, and *Bombyx batvyticatus* into a fine powder and subsequently heating the powder in wine to make into pills. The famous *Po Chai* pills from Hong Kong,

Fig. 4.2 *Tianma* (tubers of *Gastrodia elata* Bl. being sold in Lijiang, Yunnan Province, China). Although these tubers are small, they have the same contents as the larger, more attractive tubers (Photo: E.S. Teoh)

which are a popular Southeast Asian treatment for diarrhoea and over-indulgence, contain 6 % of *Tianma* together with *Rosa Bankiae*, *Pochyma Cocos*, *Chrysanthemum sinense*, *Mentha arvensis*, *Cinnabar*, *Angelica aromala*, *Oryza sativa*, tangerine peel and *Lophanthus rugosus*.

2. *Baiji* (*Bletilla striata* Rchb. f.)

In the preparation of *Baiji*, tubers are harvested from September to March, 3–4 years after planting. The tubers are soaked in water for an hour, and washed clean of soil. Root strands are removed. After steaming, they are dried in the sun or by heating over fire until the surface is dry, hard, and free of glue. Next, the tubers are fumigated overnight with sulfur, dried again, and checked for any remnant root strands which are removed. Its final appearance should be clean, smooth and white, with just a tinge of yellow. *Baiji* is also available in powder form. It is combined with other herbs to make into pills.

3. *Shihu* (*Dendrobium moniliforme*, *D. catenatum*, *D. nobile*, etc.)

Shihu is prepared from the matured stems of several *Dendrobium* species. The stems may be used fresh, in which case the roots are removed, and the stems are separated and set aside, or divided into 8-cm lengths and set aside. They are stored in cool wet sand or in a bamboo basket, but must not be allowed to become frozen. Orchids tend to turn into mush

if frozen for longer than 4 h. In spring, newly collected *Dendrobium* are commonly spread over slate or a rock and periodically dampened with water.

For long-term storage, the stems are washed in water and the roots and membraneous streaks are removed from the stems. They are dried in the sun or over a slow fire. *Huangcao Shihu* (Golden Herb *D. nobile*) and *Jinchai Shihu* (Golden Hairpin *Dendrobium* which consists of *D. nobile* and *D. linawianum*) are sold in the market in this form.

Huohu from Guangxi Province refers to *Dendrobium* which have been collected in April or May, cleaned and washed, and then dipped in boiling water before the stems are dried by rubbing (Hu 1971).

Segments of the stems of *D. moniliforme* with roots removed are washed clean and divided into 4- or 8-cm lengths. Cut stems are placed in a large iron container which is heated over a small fire to soften the stems. They are then twisted into a spiral form and allowed to cool. Twisting may be repeated until the stems are dry. These are called *Er Huan Shihu* (Earring *Shihu*).

Another processing method involves soaking the stems in alcohol overnight, followed by drying in the sun, steaming, and drying over a low fire. The drying process of *D. moniliforme* has a profound influence on the

Table 4.1 Prescriptions containing *Shihu* (*Zhongyao Da Cidian*   1986)

1. Indication: For fever with sweats and dark coated tongue
Composition
Fresh *Shihu* 15 g
Forsythia supensa (莲翘) 15 g
Trichosanthes kinlowii (天花粉) 10 g
Rehmannia glutinosa (生地) 20 g
Linope graminifoia (麦冬) 20 g
Preparation
Boil with water
(Source: *Treatise of Current Diseases*)
2. Indication: For "gastric fire" surging and causing depression and fear
Composition
Shihu 50 g
Scrophularia oldhami (玄参) 10 g
Preparation
Boil with water
(Source: *Record of Evidence*)
3. Indication: For thirst after recovery from illness
Composition
Fresh *Shihu* (*Dendrobium candidum*) 9 g
Linope graminifoia 9 g
Schizandra chinensis (五味子) 9 g
Preparation
Boil with water and drink as tea
4. Indication: For lung heat and dry cough
Composition
Fresh *Shihu* (*Dendrobium candidum*) 9 g
Juice of *Enbotrya japonica* (枇杷) 9 g
Glycyrrhiza globra (甘草) 3 g
Playtcodon grandiflorum (桔梗) 3 g
Preparation
Boil with water
(Source: *Record of Herbs of Zhejiang Province*)
5. Indication: For poor night vision
Composition
Shihu 50 g
Epimedium macranthum (仙灵脾) 50 g
Atractylodes japonium (苍术) 25 g
Preparation
Ground to fine powder. Take 15 g, twice a day
(Source: *Sheng Ji General Record*)
6. Indication: For weak *yin qi*, backache, urinary frequency, spontaneous ejaculation, wet and pruritic scrotal skin
Composition
Equal portions of *Shihu, Herpestis monniera* (巴戟天) Tenodera
Sinensis (桑螵蛸), *Euonymus chinensis* (杜仲)
Preparation
Make mixture in pill form. Take 10 pills with wine, twice a day
(Source: *Record of Tested Prescriptions*)

concentrations of polysaccharides and alkaloids in different parts of the herb. Ideally, the drying process should involve drying over fire at high temperature followed by desiccation at 60 °C (Chen et al. 2001).

Dried *Tianma* and *Shihu* need to be soaked in water until they regain 80 % moisture. They are then cut into thinner slices or shorter lengths. This permits better extraction of their contents.

Contemporary Methods for Extracting Chemicals from Medicinal Herbs

Traditional methods for extracting medicinal herbs involve slicing, pressing, pounding in a mortar, adding cold or hot water (which yields an infusion, e.g. in the preparation of teas and dyes), boiling (to produce a decoction), or soaking in alcohol, oil (ghee is widely used in Ayurveda), grease or wax. Often, several herbs are extracted or blended together to achieve a synergistic action.

Modern Methods Involve:

1. Rupturing Entire Plant Tissues
   Rupture of cell walls results in the release and activation of numerous enzymes which break down complex substrates into simpler compounds. Proteases digest proteins, lipases and fats, etc. Such enzymatic activity is undesirable as they may destroy the active component of a medicinal herb. Catabolism may be inhibited by lowering the temperature, or by the addition of specific inhibitors and buffers. Liquid nitrogen freezes tissues without producing any chemical damage. Adding nitrogen to a mortar is a simple way to protect the desired compounds during the extraction process which involves pounding the herb into a fine powder.
2. Aqueous Extraction
   Aqueous extractions are used to extract proteins, carbohydrates and pigments. The soluble compounds are separated from

particulate matter by centrifugation and filtration. They then undergo separation and filtration to recover individual compounds of interest. Finally, to obtain a solid extract, the sample is lyophilised to dryness.
3. Organic Solvent Extraction
   Many organic solvents are used to extract (non-polar) compounds which are insoluble in water. Liquid-phase solvents include methanol, ethanol, acetone, chloroform, diethyl ether and methylene dichloride or a combination of solvents. When dealing with living plant material which needs to be kept alive, detergents, alcohols (methanol, ethanol and long-chain alcohols) and DMSO (dimethyl sulfoxide, an additive used in cryopreservation) may be used to extract organic compounds.
   The above methods are also used to obtain material for research.
4. Physical Methods: Ultrasound, Micro-wave
   In addition to employing any or several of the above separation methods, ultrasound extraction and micro-wave extraction are used in industrial separation processes which demand greater efficiency and the ability to cope with large quantities of raw materials. Such extraction processes are used to recover tartaric and malic acids from grape seeds, taxanes from *Taxus* spp., artemesinin from *Artemesia annua*, and ginsenosides from ginseng (*Panax ginseng*). Supercritical fluid extraction is another method, which is used to extract hops (*Humulus lupulus*) for beer production and to decaffeinate green coffee beans (Cseke et al 2006).
5. Chromatographic Separation
   Chromatography is an indispensable tool for the separation of complex compounds. Numerous refinements of chromatography exist today (for a review, see Cseke et al. 2006). Column chromatography has been most widely used for the isolation of chemical compounds from medicinal orchids. Choice of the solvent used in these systems is dependent on the interest of the investigators, and which family of compounds they wish to investigate.

Dosage of a Medicinal Compound

Medicines which have a useful therapeutic effect also possess side effects and toxicity. Therefore, modern medicines are generally classified as poisons. Sometimes, there is even a lethal dose determined by experiments on animals. An amount equivalent to half the lethal dose is designated as the LD50 (i.e. 50 % of the lethal dose) for that particular compound. For a medicine to be considered safe, it should have minimal side effects, and the difference between the therapeutic dose and the LD50 should be a 100-fold or greater. One can look at this mathematically, and the relationship between dose and efficacy/LD50 would then take the form of a sigmoid curve. The LD50 of a drug is based on animal experiments and does not guarantee absolute safety in humans.

A good, safe medicine would have (1) a wide effective range, (2) beginning at the low end of the ascending portion of the curve, (3) a toxic range which is far above the therapeutic range, and (4) no lethal dose, or one which is a considerable distance from the toxic dose and a great distance from the minimum effective dose.

This curve can also be applied to herbal remedies. It is claimed that most herbs have a wide therapeutic range and that they are safe, thereby allowing for the use of both poor-quality and high-quality herbs with the same name. If the herb is of poor quality (i.e. it possesses a small amount of the active ingredient), it is used in larger amounts. If the herb is of high quality, a small amount will suffice. The medicinal dosage of an herb is the amount that allows it to be effective, also equivalent to the amount in common usage. Chinese herbs are currently measured in grams, and the Chinese National Bureau of Measurement stipulates the conversion for the old prescriptions (Zhang 2008):

1 liang (16 radix) = 30 g
1 qian = 3 g
1 fen = 0.3 g
1 li = 0.03 g

Several factors are taken into consideration when prescribing the dosage, such as age, weight, constitution, severity of the illness, stage of disease, occupation and environment, eating habits, season and climate, the flavour of the herb, whether it is used singly or in combination, whether it is the principal herb or a supporting item, and whether it is a fresh herb or has been preserved.

In the commercial market today, the fundamental claim is that most herbs possess a wide therapeutic range, thus eliminating the necessity for determining proper dosage through bio-assay. This is an excuse for sloppy manufacture and a disregard for strict quality control. When Steven C. Schachter, professor of neurology at Harvard Medical School, decided to test the effect of Hyperzine A on epileptic patients, he first assayed the actual content of Hyperzine A in products that were available on the US market. The marketing label claimed that each capsule contained 50 mg of the herb, but the assays showed that the actual contents varied from 30 to 120 mg per capsule, and only one firm met the requirements for use in a clinical trial (Schachter 2010).

Expiry Date

Li Shihzhen emphasised the need to ensure that herbal medicines are properly stored. He appreciated that over time there could be deterioration and loss of active principles (Lu 1976). Deterioration would be accelerated by improper storage. Like any biological product, herbs have a shelf-life, but what this is for each item is generally unstated.

Administration of a Herbal Product

Apart from oral intake and direct application on the skin, the other routes that have been classically described include sublingual, inhalation, trans-nasal (all three used by both TCM and

Ayurvedic Medicine), per rectal and per vaginal. The vaginal route is dangerous because it avoids a first pass through the liver, and the blood levels could be 10 times higher than if the same amount is taken by mouth. The fatal overdoses of prohibited stimulant drugs are known to result from their vaginal application. Currently, some TCM products are administered by an intravenous or subcutaneous route. The choice of administration is determined by the desired effect. In one study described later in this book, *Gastrodia elata* extract was administered through the nasal route, allegedly to better cross the blood–brain barrier.

Decoction

Decoction is the commonest way to prepare a herbal prescription filled at a Chinese apothecary. The herbalist generally instructs the buyer on the method of decoction because this determines the effect of the herb. Herbs are decocted in earthenware or glass; iron, copper and aluminium utensils are avoided because the minerals may react with the herbal chemicals. Soaking prior to decoction is desirable. It improves the extraction of the essential compounds, reduces the decoction time, lessens the blockage of vascular channels in the plant by coagulated starch or protein and prevents destruction of heat-labile chemicals. Herbs may be wrapped in gauze before they are decocted. The amount of heat (strong fire or slow fire), the duration of decoction, sometimes the sequence of addition of the various components, separate decoction and subsequent mixing are carefully detailed. Squeezing the herb after decoction may increase the yield. A prescription can be decocted twice or three times; thrice to make the best use of the herbs and avoid wastage (Zhang 2008). Prescriptions may be decocted several times before they lose their potency. The commonest example is the famous Fujian tea, *Ti Kuanyin* (*Tieguanyin*) which can be decocted at least a dozen times before the tea leaves need to be thrown away (Heiss and Heiss 2007), making it very popular with restaurants.

Herbs are commonly decocted together, but on occasion they may need to be boiled sequentially because decoction time affects the property and clinical application of some herbs (Zhang 2008). Table 4.1 shows how several herbs could be decocted together, using *shihu* (medicinal *Dendrobium*) as an example.

These are some details that ought to be documented for tribal remedies because none of the reports that we have read have made any mention of the method of decoction.

Timing and Dietary Avoidance

These are two additional considerations. Herbs are generally administered inbetween meals. TCM recommends an antihelmintic herb to be taken in the morning on an empty stomach so that it can rapidly reach the intestines and kill the worms. The preprandial timing is preferred for medicines that act on the bowel. Herbs are taken warm. Uncooked, cold, oily, greasy, fishy and stimulant foods are best avoided (Zhang 2008).

References

Chen YL, Zhang M, Hua YF, He GQ (2001) Studies on polysaccharide alkaloids and minerals from *Dendrobium moniliforme* (L.) Sw. Zhongguo Zhong Yao Za Zhi 26(10):709–710

Cseke CJ, Kaufman PB, Warber S et al (eds) (2006) Natural products from plants, 2nd edn. CRC Press, Boca Raton, pp 264–317

Heiss ML, Heiss RJ (2007) The story of tea: a cultural history and drinking guide. Ten Speed Press, Berkeley

Hu SY (1971) Orchids in the life and culture of the Chinese people. Chuing Chi J 10:1–26

Lu GD (1976) China's greatest naturalist: a brief biography of Li Shih-Chen. Am J Chin Med 4(3):209–218

Schachter SC (2010) Botanical therapies for epilepsy. Oral Presentation. Recent Development in Chinese Herbal Medicine, Nanyang Technology University, 2010

Zhang TM (ed) (2008) Chinese Materia Medica. The five-year national planned textbooks for higher education. National Planned University Textbooks for International Traditional Chinese Medicine Education. Higher Education Press, Beijing

Zhongyao Da Cidian (1986) Compiled by Nanjing College of New Medicine. Science & Technology Publishing, Shanghai

Secondary metabolites are substances manufactured by plants that make them competitive in their own environment. These small molecules exert a wide range of effects on the plant itself and on other living organisms. They induce flowering, fruit set and abscission, maintain perennial growth or signal deciduous behaviour. They act as antimicrobials and perform the role of attractants or, conversely, as repellents. Over 50,000 secondary metabolites have been discovered in the plant kingdom. Medicinal herbs and many modern medicines rely on secondary plant metabolites for their actions.

The search for new secondary products in plants with the hope of discovering new products or, even better, new approaches for the treatment of disease is an on-going process involving academic and pharmaceutical institutions. At one time, another reason was the hope that understanding the distribution of natural products would assist in the classification of plants (Lawler 1986a, b). This secondary reason is no longer important today because plant classification is being increasingly approached by comparing DNA sequences.

Phytoalexins

In their natural environment, orchids are naturally exposed to many micro-organisms, and in response to such microbial challenge they produce phytoalexins, which are low-molecular-weight compounds that confer resistance against such organisms (Letcher and Nhamo 1975; Stoessl and Arditti 1984). The main phytoalexins of orchids are 9,10-dihydrophenanthrenes. Orchinol was the first phytoalexin to be isolated, from *Orchis militaris* infected with *Rhizoctonia repens* (Boller et al. 1957). Loroglossol, an isomer of orchinol, was next isolated from *Loroglossum hircinum* infected with *Rhizoctonia versicolor* (Hardegger et al. 1963), followed by hircinol. More than 40 dihydrophenanthrene phytoalexins have been isolated from orchids, and many, including the three original phytoalexins, have been synthesised (Stoessl and Arditti 1984). Feeding experiments using radioactive L-phenylalanineas as a precursor demonstrated that the biosynthetic sequence for production of 9,10-dihydrophenanthrenes starts with L-phenylalanine and passes through *m*-coumaric acid, dihydro-*m*-coumaric acid, and 3,3′,5-trihydroxybibenzyl (Fritzememeier and Kindl 1983). From their biosynthesis, dihydrophenanthrenes can be classified as stilbenoids because they are derivatives of dihydrostilbenes or bibenzyls (Reinecke and Kindl 1994). Resveratrol is the best publicised stilbenoid. Present on the skin of grapes and playing a role in warding off attack by fungi and bacteria on the fruit, it is alleged to have many beneficial effects on plants and animals and even cytotoxic activities, but its global benefit is controversial and has never been replicated in humans.

Ordinarily, phytoalexins are present only in minute amounts in healthy orchids. When attacked by pathogenic fungi, the orchid responds by an intense activation of genes encoding phytoalexin enzymes, but this response is transient (Reinecke and Kindl 1994a, 1994b). Phytoalexin concentrations decline markedly when symbiosis is established between the orchid and the mycorrhiza. Nevertheless, upon destruction of the mycorrhiza, phytoalexin production increases in proportion to the amount of fungal material present (Gehlert and Kindl 1991). In young, sterile plants of *Phalaenopsis*, bibenzyls and their oxidative products, the 9,10-dihydrophenanthrene phytoalexins, are not present. Following infection with fungi, such as *Botrytis cinerea* and *Rhizoctonia* spp., there is a greater than 100-fold increase in bibenzyl synthase activity. This is the key enzyme for the formation of phytoalexins (Reinecke and Kindl 1994b). Since mycoheterotrophic plants are unable to photosynthesise, they are totally dependent on their mycorrhiza for carbon supplies, and therefore they need to be able to defend themselves against microbes and herbivores (Roy et al. 2013). Over 50 chemical substances have been isolated from *Gastrodia elata*, and it would not be surprising if similarly large numbers of phytochemicals are also found in other highly successful, large, mycoheterotrohic orchids (e.g. various species of *Gastrodia, Galeola, Cephalantera, Corallorhiza*, and *Cymbidium micorhizon*).

Orchinol and loroglossol inhibit spore germination of *Phytophora infestans* at 0.000006 M concentration and disrupt vegetative growth of newly germinated *Monilinia fructicola* (Ward et al. 1975). Phytoalexins are bacteristatic and fungistatic, while being neither bactericidal nor fungicidal. In this respect, it is interesting to note that, in Nepal, pertaining to skin lesions, native medicine only makes use of orchids for minor conditions like wounds (employing *Coelogyne, Dactylorhiza, Gymadenia, Rhynchostylis* and *Vanda*), pimples (*Dendrobium*), boils (*Coelogyne, Cymbidium, Dendrobium, Pholidota* and *Vanda*) or as a demulcent (*Dactylorhiza*); orchids are not used for sores or carbuncles (Manandhar and Manandhar 2002).

Phytoalexins are also produced by a large number of plants consumed by humans, but generally they are in such small amounts that they would not cause problems unless the vegetable in question is consumed excessively. Garden peas contain pisatin and green bean phaseolin, both of which will lyse bovine red blood cells, the former at a concentration of 200 ppm (parts per million), the latter at 17.5 ppm. Carrots contain myristicin which is insecticidal and in humans produces cerebral excitation. However, a 70-kg man would need to consume 5 kg of carrots to experience hallucinations. Damaged sweet potato is toxic to cattle and humans due to elevated levels of ipomeamarone which damages the liver and lungs. Blighted white potato is known to have caused poisoning and deaths in humans due to the presence of two glycoalkaloids, alpha-sloanine and alpha-chaconine. Good agricultural practice reduces the amount of phytoalexins in agricultural crops and is also important from the consumer acceptance standpoint (Surak 1978).

Hydrocarbons

These are the simplest compounds. They contain only hydrogen and carbon. They occur as straight chains (aliphatic hydrocarbons) or with ring forms, and form the basic skeleton of more complex molecules. A carbon atom is capable of binding to four hydrogen atoms, and when fewer hydrogen atoms are present relative to the carbon, the hydrocarbon is said to be unsaturated. Such compounds carry double or triple bonds. Marsh gas, methane (CH_4), is a saturated hydrocarbon, and the four bonds of carbon are all attached to hydrogen. The waxy coat on leaves and fruits contain many saturated hydrocarbons which are insoluble in water. They prevent water sticking on the surface of leaves and fruit. Olive oil also contains a number of saturated hydrocarbons.

Another gas, ethylene ($H_2C{=}CH_2$), and an example of an unsaturated hydrocarbon, is a plant hormone. It is released by apples and by fading flowers of *Papilionanthe* and their hybrids. Ethylene causes ripening of fruit, abscission of leaves and fading of adjacent flowers, especially in an enclosed space which prevents dissipation of the gas.

When a hydroxyl group (−OH), hydrogen and oxygen, is attached to a hydrocarbon, the latter becomes an alcohol, for drinking, or ethanol, C_2H_5OH.

Terpenes (Terpenoids and Steroids)

Terpenes are important plant metabolites. They include substances like floral fragrances, which serve as insect attractants, pine oil, growth inhibitors, the two plant hormones, gibberelic acid and abscisic acid, and some which are insecticidal. The 30,000 terpenes that have been identified share one common characteristic: they all possess repeating five-carbon isoprene units (a five-carbon ring, Fig. 5.1).

The number of five-carbon isoprene units determines their classification into:

1. Hemiterpenes (single isoprene unit)
2. Monoterpenes (two isoprene units)
3. Sesquiterpenes (three isoprene units)
4. Diterpenes (four isoprene units)
5. Sesterterpenes (five isoprene units)
6. Triterpenes (six isoprene units)
7. Carotenoids (eight isoprene units).

Although their structures were first elucidated in the nineteenth century, terpene-based essential oils, found in frankincense, for instance, have Biblical usage. Monoterpenes such as linalool are major components of the scent produced by orchids (Kaiser 1993). The modern antimalarial, artemesinin, a sequiterpene, comes from the Chinese medicinal plant *Quinhao* (*Artemesia annua*) which had been in use as a fever medicine for over two millennia. It was mentioned in the *52 Remedies* recovered from the Mawangdui Tomb dating from the Han Dynasty (206 BC–221) located in Henan Province (Harper 1998). It has the empiric formula C15H22O5 and is chemically known as 3R,5aS,6-R,8aS,9R,12S,12aR)-Octahydro-3,6,9-trimethy-l-3,12-epoxy-12H-pyranol[4,3-j]-1,2-benzodioxepin-10(3H)-one. Artemesinin is effective against the dangerous chloroquine-resistant falciparium malaria which sometimes involves the brain (Anonymous 1979). Another life-saving terpene is placitaxol (a diterpene with a very complex molecular structure) effective against ovarian, breast, colon, non-small cell lung cancer and malignant melanoma. It has the empiric formula $C_{47}H_{51}NO_{14}$ and is known as 5beta,20-epoxy-1,2alpha,4,7beta,10beta,13alpha-hexhydroxyt-ax-11-en-9-one 4,10 diacetate 2-benzoate 13-ester with (2R,3S)-N-benzoyl-3-phenylisoserine (Evangelista 1995).

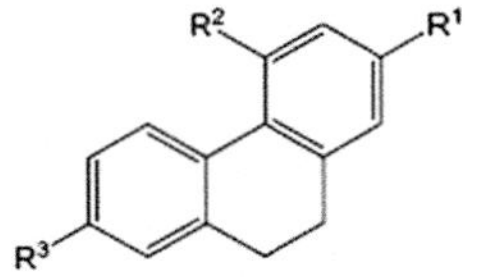

Fig. 5.1 Orchinol is the first phytoalexin to be isolated, from *Orchis militaris* infected with *Rhizoctonia repens*, followed by loroglossal (Hardegger et al. 1963) and hircinol (Fisch et al. 1973). Analogues of orchinol like coelonin and fusianthridin occur in other orchid species

R1 = OMe, R2= OMe, R3 = OH

Orchinol

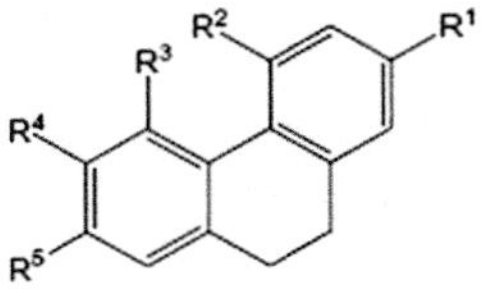

R1 = OMe, R2 = OMe, R3 = OH, R4 = H, R5 = H

Loroglossal

R1 = OH, R2 = OMe, R3 = OH, R4 = H, R5 = H

Hirsinol

Terpenoids (diterpenoids, sesquiterpenoids, triterpenoids) and lignoids also possess antiviral activities, and at least 22 have been shown to inhibit corona-viruses, including the dangerous SARS-Corona Virus which created such havoc in the Far East in 2007. Betulinic acid and savinin are competitive inhibitors of a protease (an enzyme which breaks down proteins) produced by the SARS-CoV 3CL virus. Terpenes in orchids are therefore a topic of great interest to researchers.

Triterpenes and Steroids

Tetracyclic triterpenes (compounds) and steroids have similar structures, but are biosynthesised through different pathways. The plant steroids contain three six-membered and one five-membered rings. Such steroids exert profound physiological effects on animals. Some are employed as an oestrogen substitute in menopausal women. Cardiac glycosides consisting of a sugar molecule bound to a steroid, such as digitalis (digitoxin) from foxglove (*Digitalis purpurea,* not an orchid), are used to treat cardiac insufficiency. Steroidal saponins are important precursors for the manufacture of steroid drugs ranging from anti-inflammatory agents to sex hormones such as androgens, oestrogens, progestogens and oral contraceptives. Triterpene saponins have antitussive (cough preventing), expectorant, analgesic, anti-inflammatory and cytotoxic effects. Liquorice, which is used in the treatment of coughs, is one example. The ginsenocides from ginseng are another. All saponins are surfactants, and when mixed with water and shaken, they form a foamy solution. Many saponins are haemolytic (they rupture red blood cells). They are toxic to cold-blooded animals like fish (de Padua et al. 1999).

Stilbenoids and Bibenzyls

Bibenzyl is a hydrocarbon whose basic structure consists of two benzene rings attached to ethane. They occur commonly in plants. Bibenzyls in

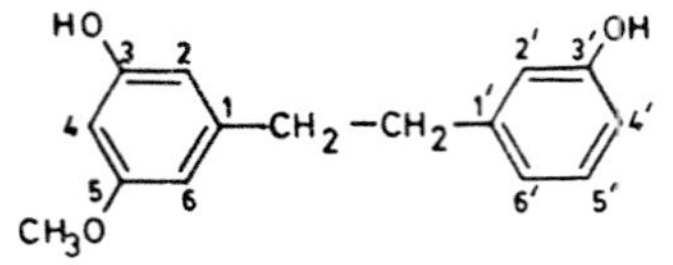

Fig. 5.2   Batatasin III, a common bibenzyl in orchids

orchids are synthesised from dihydro-*p*-coumaric acid and acetate or malonate (Fritzemeier and Kindl 1983; Friederich et al. 1998). Gigantol and batatasin III are the two commonest bibenzyls occurring in orchids (Chen et al. 2008) which have cytotoxic activity. Gigantol (from *Dendrobium draconis*) inhibits migration of non-small cell lung cancer in vitro (Charoenrungruang et al. 2014). Erianin, a bibenzyl which occurs naturally in *Dendrobium chrysotoxum*, is often employed as an antipyretic and analgesic in traditional Chinese medicine (Su et al. 2011). Erianin possesses antiangiogenic properties (Gong et al. 2004a, b); furthermore, it induces apoptosis in human leukaemia HL-60 cells and hepatocarcinoma (HCC) Huh7 cells, in vitro (Li et al. 2001; Su et al. 2011). Should it be applicable as an antitumour agent, that would make the drug more valuable. Erianin was successfully synthesised in 2008 (Zou et al. 2008).

Tamoxifen and diethylstilbenoids are examples of synthetic stilbenoids which have been used to treat hormone-dependent breast cancer, and clomiphene is a synthetic stilbenoid that is used for ovulation induction Fig. 5.2.

Phenanthrenes

Phenanthrene, C14H10, is an angular polynuclear hydrocarbon which is related to certain alkaloids like morphine, and figure in the structure of steroids. It is postulated that they are formed through the oxidative coupling of the aromatic rings in stilbene or diterpenoid precursors. Many phenanthrenes occur in higher plants, particularly orchids, in such medicinal genera as *Bletilla, Bulbophyllum, Dendrobium, Coelogyne, Cymbidium, Eria* and *Flickingeria*.

Table 5.1 Properties of phenanthrenes present in medicinal orchid species

Pharmacological action	Orchid species
Anti-allergic	Gymadenia conopsea
Anti-inflammatory	Dendrobium moniliforme
Antimicrobial	Bletilla striata
	Cypripedium macranthos
Anti-oxidant	Pholidota yunnanensis
Antithrombotic	Dendrobium loddigesii
	Dendrobium xantholeucum (syn. Ephemerantha lonchophylla)
Cytotoxic	Bulbophyllum kwangtungense
	Cremastra appendiculata
	Dendrobium catenatum
	Dendrobium nobile
	Dendrobium chrysanthum
	Dendrobium thrysiflorum

There is on-going interest in natural phenanthrenes because some of them have been shown to be cytotoxic against specific human cancer cell lines, while other possess anti-allergic, antimicrobial, anti-inflammatory, anti-oxidant, antiplatelet (antithrombotic) and spasmolytic properties (Kovacs et al. 2008). Examples of such laboratory-demonstrated pharmacological actions found in various species are shown in Table 5.1.

Antitumour effects are probably the most important property of phenanthrenes to be investigated. Monomeric phenanthrenes, generally the commonest, in Cremastra appendiculata were ineffective in all tested cancer cell lines, whereas its biphenanthrenes and triphenanthrene displayed antitumour activity (Xue et al. 2006; Kovacs et al. 2008). Denbinobin, a phenanthroquinone, and lusianthridin, a dihydroxymethoxy phenanthrene from Dendrobium nobile, exhibit cytotoxic effects in vitro and in vivo, with denbinobin being more potent. A free phenolic hydroxyl group appears to be essential for the inhibitory activity (Lee et al. 1995; Kovacs et al. 2008).

Phenanthrenes from orchids are classified into three main groups: monophenanthrenes, diphenenthrenes and triphenenthrenes. There are 210 compounds in the first group, the monophenenthrenes, of which almost half are only hydroxyl- and/or methoxy-substituted. Almost all the remainder are 8,10-dihydro- or dehydro derivatives (Kovacs et al. 2008). Glycosides are rare, but three were discovered in Bletilla striata (Yamaki et al. 1993) and one in Dendrobium chrysanthum (Ye et al. 2003). A unique monophenanthrene with a spirolactone ring was also isolated from Bletilla striata; it was named blespirol (Yamaki et al. 1993). An additional monophenanthrene with a spironolactone ring was isolated from Dendrobium chrysanthum and named dendrochrysanene (Yang et al. 2006). Phenanthraquinones form another group of monomeric phenanthrenes and have been isolated from Spiranthes sinensis and Cremastra appendiculata (Tezuka et al. 1990; Xue et al. 2006). Bibenzyl derivatives of phenanthrenes were discovered in Pleione bulbocodioides (Bai et al. 1996) and Pholidota yunnanensis (Guo et al. 2006).

Diphenthrenes are less common. They have been isolated from Agrostophyllum callosum and A. khasiyanum (Majumder and Sabzabadi 1988), Bletilla striata (Honda and Yamaki 1989, 2000; Bai et al. 1991), B. formosana (Lin et al. 2005), Bulbophyllum reptens (Majumder et al. 1999), B. maculosum (Cirrhopetalum maculosum) (Majumder et al. 1990) B. vaginatum (Leong and Harrison 2004), Cremastra appendiculata (Xue et al. 2005), Dendrobium plicatile (Honda and Yamaki 2000), D. thyrsiflorum (Zhang et al. 2005), Eria flava (Majumder and Banerjee 1988), Eulophia nuda (Tuchinda et al. 1988) Gymadenia conopsea (Matsuda et al. 2004), Pleione bulbodioides (Bai et al. 1996) and Pholidota yunnanensis (Guo et al. 2006). The single orchidaceous triphenenthrene was isolated from the tubers of Cremastra appendiculata (Xue et al. 2006). Their phytochemistry and pharmacology have been well reviewed by Kovacs et al. (2008).

Some of the bioactive compounds may originate in the endophytic fungi associated with the orchid. Ten endophytic fungi from Dendrobium devonianum and 11 from D. thyrsiflorum exhibited antimicrobial activity against at least one species of bacteria or fungus among the six pathogenic microbes that were tested (*Escherichia coli, Bacillus subtilis, Streptococcus aureus, Candida albicans, Cryptococcus neoformans* and *Aspergillus fumigatus*). Antibacterial activity of *Epicoccum nigrum* from *D. thyrsiflorum* was stronger than ampicillin. *Fusarium* from the two *Dendrobium* species was effective against both bacteria and fungus (Xing et al. 2011). These findings suggest that tribal usage of orchids to treat infection may be based on experience of beneficial effects.

Alkaloids

The term alkaloid is used as a name for plant-derived compounds, containing one or more nitrogen atoms, usually in a heterocyclic ring (an amine functional group), and which have a marked effect on animals, including humans. They are optically active. Like proteins, they are derived from amino acids, but they differ in being alkaline. The term has an Arabic origin. Soda ash is known as *al qali* in Arabic. Alkaloids are bitter to taste. Among their functions, they are thought to play a role in germination and in protecting plants from predators, in particular herbivores and microbes. They are present in around 20 % of higher plants. Sometimes, they are also present in animals, for instance in the skin of some species of frogs.

Many alkaloids act on the nervous system. Poppy was employed in the Middle East over 3000 years ago, and coffee drinking originated in Ethiopia. Poppy is narcotic, caffeine and nicotine are stimulants, while cocaine is an anaesthetic, and scopolamine induces "twilight sleep." Codeine, which is more commonly employed by doctors to suppress severe coughing, is also present in the latex of the poppy capsule, and structurally very similar to morphine. Codeine is now a controlled drug. Aminophylline is a bronchodilator, while papaverine is a vasodilator which had a role in treating erectile dysfunction before the discovery of Viagra. Reserpine which lowers blood pressure is an ancient Indian remedy derived from *Rauwolfia serpentina,* now totally replaced by a wide range of more potent and reliable antihypertensives. Many alkaloid stimulants (e.g. morphine, cocaine, nicotine) are addictive. Improperly applied, some stimulants and sedatives are deadly. Strychnine is used as a rat poison. In 339 BC, the Greek philosopher, Socrates, was killed by being forced to drink hemlock which contains the alkaloid, coniine.

Taxol which has a diterpenoid core possesses an alkaloid side chain. It is an indispensable component in the chemotherapeutic cocktail employed in the treatment of ovarian and breast cancer. Vincristine and vinblastine are two alkaloids derived from the periwinkle, *Catharanthus roseus* (not an orchid), and are also cytotoxic agents but their use is limited to late-stage cancers because of their high toxicity. Camptothecin, a quinoline alkaloid obtained from the Chinese 'tree of joy' (*Camptotheca accuminata*), is used for treating advanced ovarian cancer that is resistant to taxol. Many synthetic compounds are derived from natural plant materials, and some of these are safer to use although they retain some toxicity along with the beneficial properties. Codeine derived from morphine is one of these. Sometimes, the derivative is more potent and far more dangerous, like heroine, which is also derived from the hydrolysis of morphine.

A shortage of quinine and several medicinal alkaloids during World War II precipitated by the interruption of supplies provided the impetus for governments, the pharmaceutical industry and scientists to undertake extensive screening of plants for alkaloids during the 1950s and 1960s (Lawler 1986a, b).

Alkaloids being so important in the pharmaceutical industry, it is not surprising that they were among the first secondary metabolites to be studied in orchids (Suzuki et al. 1932; Chen and Chen 1935; Yamamura and Hirata 1964; Inubushi et al. 1964; Luning 1964, 1967, 1974, 1975, 1980; Nishikawa and Hirata 1967, 1968; Brandange and Granelli 1973; Slaytor 1977; Lawler 1984), but many species that were screened did not contain appreciable amounts of the such metabolites. In 1974, Luning reported that 2044 species of orchids from 281 genera had been screened for alkaloids. Over half (numbering 30, or 53.6 %) of the 56 medicinal orchid

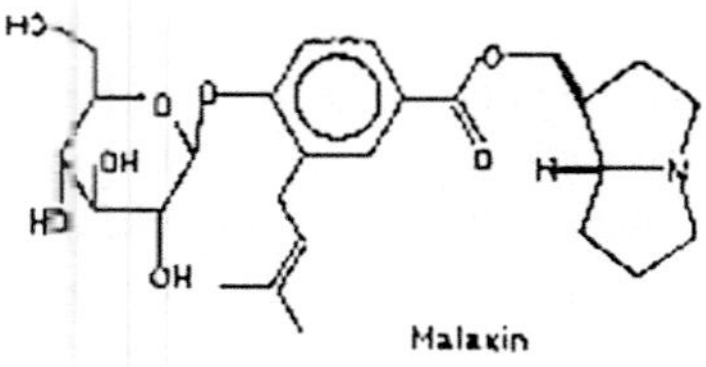

Fig. 5.3 Malaxin is a useful anti-malarial alkaloid

genera from Asia that were screened contained species that tested positive for alkaloids, albeit not all their species were medicinal. Only 14.6 % of all orchid species tested gave a positive test for alkaloids (i.e. present in amounts of 0.1 % or more). Genera that contained the largest number of alkaloid-positive species were *Liparis* Fig. 5.3 (with 28 species), *Dendrobium* (24), *Phalaenopsis* (19), *Malaxis* (18) and *Bulbophyllum* (9) (Table 5.2). Alkaloid-rich species were found in only four genera, *LIparis, Malaxis, Oberonia* and *Bulbophyllum*, when 314 orchid species in Bougainville, Papua New Guinea, were screened for alkaloids (Lawler and Slaytor 1969). (It should be noted that *Liparis* and *Malaxis* species are related. In the recent taxonomic revision, many species have been reassigned to different genera.) There were no appreciable amounts of alkaloids in 29 genera that had medicinal species (Table 5.3). However, single species of plants are not homogenous in their chemical content and individual plants of species that tested negative in past studies may actually contain undiscovered alkaloids. For instance, 8 out of 10 Himalayan *Coelogyne* species (*Coelogyne cristata, Coelogyne elata, Coelogyne flavida, Coelogyne nitida, Coelogyne ovalis* and *Coelogyne virescens* (=*Coelogyne brachyptera* Rchb. f) tested negative for alkaloids when they were screened by Luning (1964), but ten (different) species of *Coelogyne* from Bougainville, Papua New Guinea, were found to contain small amounts of alkaloids (Lawler and Slaytor 1969). Most of these alkaloid-rich genera occur in India and Southeast Asia. Only 5–10 % of their species have been screened, so there is much opportunity for good work to be done.

Table 5.2 Alkaloid-positive medicinal orchid genera from Asia

Genus	Number positive	% positive	Number tested
Anoectochilus	1		11
Arachnis	1	50	2
Bulbophyllum	9	6.5	138
Calanthe	2	7.7	26
Coelogyne	2	7.7	26
Corymborkis	1	25	4
Cymbidium	2	5.4	37
Cyrtochis	1	16.7	6
Dendrobium	24	8.3	384
Eria	14	18.2	77
Eulophia	2	15.4	13
Gastrochilus	1	50	2
Gastrodia	1	50	2
Goodyera	1	9.1	11
Habenaria	2	16.7	12
Liparis	28	41.8	67
Malaxis	18	36.7	49
Malleola	1	100	1
Nervilia	4	33	12
Oberonia	5	17.2	29
Paphiopedilum	1	4.3	23
Phalaenopsis	19	50	38
Plocoglottis	2	28.6	7
Renanthera	1	20	5
Cleisostoma (as *Sarcanthus*)	2	25	8
Vanda	3	13.6	22
Zeuxine	1		205
Total	149	14.6	1015

Reference: Luning 1974
Note: *Doritis* and *Kingiella* are now in *Phalaenopsis*. *Cirrhopetalum* is in *Bulbophyllum*. *Sarcanthus* are *Cleisostoma*. *Eria* species are not assigned contemporary nomenclature because of insufficient data in the original.

Orchid alkaloids commonly fall into two main classes: alkaloids of the pyrrolizidine type and (2) alkaloids of the dendrobine type (Fig. 5.4). *Dendrobium* is the genus richest in alkaloids, but their most important alkaloids are pyrrolizidine compounds, not the dendrobine type (Hausen 1984). Bibenzyl alkaloids have been identified in many orchid species. A picrotoxinin-type alkaloid has recently been isolated from the *Dendrobium*, *D*. Snowflake "Red Star" (Morita et al. 2000).

Dendrobine, the first alkaloid discovered in *Chin Shih Hu* (*Dendrobium nobile*) was isolated

Table 5.3 Alkaloid-negative medicinal orchid genera from Asia

Genera	Number tested negative
Arundina	1
Bletilla	1
Bronheadia	2
Cephalanthera	1
Cremastra	1
Cypripedium	4
Dactylorhiza	3
Epipactis	3
Geodorum	3
Grammatophyllum	5
Gymnadenia	1
Hetaeria	5
Luisia	8
Neottia	1
Nephelaphyllum	3
Orchis	4
Ornithochilus	2
Pelantheria	2
Phaius	13
Pholidota	14
Platanthera	1
Pleione	4
Polystachya	15
Rhynchostylis	2
Robiquetia	3
Satyrium	2
Spathoglottis	8
Spiranthes	7
Vanilla	5
Total	124
Total number of species in alkaloid-positive genera	1015
Total number of species in medicinal genera reported	1139

Reference: Luning (1974)
Note: *Doritis* and *Kingiella* are now in *Phalaenopsis*. *Cirrhopetalum* is in *Bulbophyllum*. *Eria* species are not assigned contemporary nomenclature because of insufficient data in the original

by Suzuki, Keimatsu and Ito in Japan in 1932, and pharmacological action was reported by Chen and Chen in 1935. It is the major alkaloid in *D. nobile*, and was subsequently found to be also present in *D. linawianum* (Suzuki et al. 1932, 1934). Another 14 alkaloids related to dendrobine are present in *Dendrobium* species. These include nobiline or nobilonine (Yamamura and Hirata 1964; Onaka et al. 1965), dendramine (6-hydroxydendromine),

dendrine, dendroxine, 4-hydroxydendroxine and 6-hydroxydendoxine in *D. nobile* (Inubushi and Nakano 1965; Inubushi et al. 1966; Okamoto et al. 1966a, b); 2-hydroxydendrobine in *D. finlayanum* (Granelli et al. 1970); 6-hydroxynobilonine in *D. hildebrandii* (Elander and Leander 1971); and the isopentenyl derivatives of dendroxine and 6-hydroxydendroxine in *D. hildebrandii* and *D. friedricksianum* (Hedman et al. 1971).

More than 30 alkaloids have now been isolated from the genus *Dendrobium*. Although *Dendrobium* is the genus richest in alkaloids, only 8.33 % of the 384 species tested were found to have an alkaloid content which amounted to 0.1 % or greater (Luning 1974). A more recent tally discovered alkaloids to be present in appreciable amounts in 42 species of *Dendrobium*, particularly those of the northern clade, among which are species included within *shihu* (such as *D. nobile*, *D. liniawanium*, *D. findlayanum*, *D. moniliforme*, *D. hildebrandii*, *D. friedricksianum*, *D. wardianum*, *D. crepidatum*, *D. aphyllum*, *D. chrysanthum*, *D. lohohense*, *D. primulum*, *D. parishii* and *D. anosmum*) (Zhang et al. 2003; Liu et al. 2007).

Malaxin is the first of the pyrrolizidine-based alkaloids to be elucidated. Present in *Malaxis* and *Liparis*, malaxin was first isolated from *M. congesta* by Luning and Leander in 1967 (Luning 1974) and subsequently discovered in *L. bicallosa* and *L. hachijoensis* (Nishikawa and Hirata 1968; Nishikawa et al. 1967). Malaxin is dihydroartemesinin, an ester of laburnine (an alkaloid present in *Trudelia cristata*, *Vanda hindsii* and *V. helvola*) and malaxinic acid. Following its synthesis in 1969 (Tanino et al. 1969), it is now employed in Korea and several African countries for the treatment of uncomplicated falciparium malaria (Jackson et al. 2006; Anonymous, undated, http://www.act.watch.info.). Treatment failures with artemisine-based therapies have been reported, but these may be due to suboptimal dosage caused by poor pharmaceutical practice (Green et al. 2001; Jackson et al. 2006). Fakes have also been reported.

More complex pyrrolizidine alkaloids are present in several genera of monopodial orchids like *Vanda*, *Vandopsis Phalaenopsis*, and *Doritis*

Fig. 5.4 Structure of benzopyran and cyanidin-3-glycoside

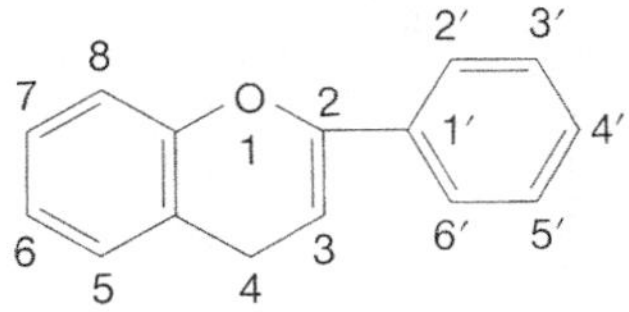

benzopyran

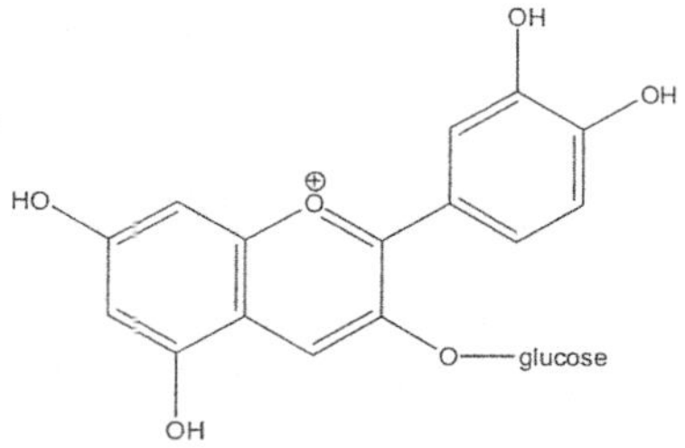

cyanidin-3-glycoside

(Slaytor 1977). Shihunine isolated from *Dendrobium lohohense* is an early example of a phthalide-pyrrolidine alkaloid (Inubushi et al. 1964).

Phenols

Probably the largest group of secondary metabolites, phenols range from simple compounds with a single aromatic ring to complex compounds which are polymers like tannins and lignins. They include coumarins, quinones, napthoquinones and anthraquinones, all flavonoids which give odour, scent and colour to plants. Some of the compounds have physiological effects on animals. Vanillin is widely used to flavour food. It is a simple phenol. The ubiquitous salicylic acid is a precursor of aspirin.

Denbinobin, a 1,4-phenanthrenequinone first isolated from *Ephermerantha lonchophylla* (*Flickingeria xantholeuca*), and subsequently found to be present in *Dendrobium nobile* and *D. candidum* (Lin et al. 2001; Li et al. 2010; Yang et al. 2011), has been found to inhibit HIV-1 replication through an NF-kappaB-dependent pathway (Sanchez-Duffhunes et al. 2008). In vitro studies also show that denbinobine causes apoptosis of numerous human cancer cell lines (leukaemia; breast, lung, colorectal and stomach cancers) (Huang et al. 2005; Kuo et al. 2008, 2009; Chen et al. 2008; Sanchez-Duffhues et al. 2009; Chen et al. 2011; Song et al. 2012). Additionally, it may suppress tumour growth by blocking angiogenesis (Tsai et al. 2011) and prevent invasion or spread of

breast and stomach cancers (Chen et al. 2011; Song et al. 2012). By causing selective apoptosis in hepatic stellate cells but not in normal hepatic cells, denbinobin exerts an antifibrotic effect on the liver and may thus be a useful starting point for developing compounds to protect the liver against cirrhosis (Yang et al. 2011). Denbinobin has been synthesised (Kraus and Zhang 2002; Wang et al. 2005), and so more studies and clinical testing should be forthcoming. This is probably the most promising phenanthrene or phenol that has been isolated from orchids.

Flavonoids

Flavonoids, phenols and tannins are aromatics. By that it is meant that their chemical structure contains a cyclic carbon (aromatic) ring instead of being merely straight or branched chains. The double bonds of the benzene ring effectively absorb ultraviolet radiation, and modifications to the ring move the absorbance towards longer wavelengths. Through their absorbance of various wavelengths of visible light, flavonoids give rise to the various colour pigments in plants (Lee 2007).

Flavonoids are constituted by a large family of compounds, estimated as exceeding 10,000. Their structural diversity results from various modification reactions, an important example being O-methylation regulated by a wide range of *O*-methyl transferases. The biological activities of a flavonoid and its O-methylated derivative are dissimilar (Kim et al. 2010). Flavonoids are commonly recommended because of their antioxidant activity. Quercetin, the most abundant dietary flavonoid (it is also present in *Dendrobium catenatum*), is a potent antioxidant with anti-allergic and anti-inflammatory properties. Some flavanoids are antibacterial, antiviral (against the common cold sore virus), anti-allergic, anti-inflammatory, antiplatelet and antineoplastic (Liu 2011).

Anthocyanins, which impart yellow, red, mauve, pink, magenta and purple colouring to flowers, play an important role in insect mimicry. Colour in many orchids is often decided by two genes. In *Spathoglottis plicata*, the presence of both the dominant gene for pink colour (P) and the dominant gene for pale pink (T, which results in a flowers with a tinge of colour) results in flowers of deep purple. The presence of a single dominant gene results in either pink or tinge, and two recessive genes produce white (Storey 1950, 1958).

The parent compound of anthocyanins is 2-phenylbenzopyran. Few studies have been conducted on orchid anthocyanins. The bulk of these have been focused on orchids from the New World, and comments on anthocyanin pigments in Asian orchids are sometimes speculative. Hybrids of the *Vanda-Aranda-Renanthera* group of cultivated orchids contain a single anthocyanin which is cyanidin-based and only present in the flowers. Cyanidin-3-glycoside was chemically identified in flowers of *Cymbidium finlaysonianum*, *Grammatophyllum speciosum*, and *Pogonia japonica*, and cyanidin-3-rutinoside in the *Dendrobium* hybrid, *Dendrobium* Caesar (Arditti and Fish 1977).

Flavonoids have a wide range of pharmacological activities that include anti-oxidant, anti-microbial, anti-inflammatory, antimutagenic, antitumour, antidiabetic vaso-relaxant, immunomodulatory and both oestrogenic and anti-oestrogenic activities (Lin et al. 2014). Anti-oxidant activity is exhibited by floral anthocyanins extracted from a hybrid between *Papillionanthe teres* and *P. hookeriana* (*Vanda Miss Joaquim*) (Junka et al. (2012). Coumarin class compounds which exhibit anticoagulant or antiplatelet activities are phenylpropanoids (with three carbon side chains attached to a phenol).

Podophyllotoxin is a lignan used to treat warts. Etoposide and related anticancer drugs are derived from podophyllotoxin. Unfortunately, these derivatives are extremely toxic, and ordinarily they would only be employed as a last resort.

Among the flavonoids are phyto-oestrogens: quercetin which possesses anti-oxidant activity, and genistein and galangin which show some antibacterial activity. The potency of these compounds is weak and much work needs to be done to enhance their therapeutic value. Nevertheless, in their present state, they may have a role in tribal medicine.

Bulbophyllum odoratissimum is employed to treat respiratory infections and injuries in China, and this usage may have some justification because the orchid contains chrysin, a flavanoid with anti-inflammatory and pain-relieving properties. Chrysin suppresses lipopolysaccharide- induced cyclooxygenase-2-expression (COX2 expression) through the inhibition of nuclear factor for IL-6 (NF-IL6) DNA-binding activity (Woo et al. 2005). In the health supplement trade, chrysin is promoted as an aromatase inhibitor on the basis of in vitro testing; and from this it is inferred that it may encourage muscle development and possibly enhance libido. However, in vivo studies found that orally-administered chrysin did not alter steroid levels in humans nor in experimental animals (Saarinen et al. 2001). In nature, the commonest source of chrysin is the blue passion flower, *Passiflora caerulea*.

The other flavonoid present in *Bulbophyllum odoratissimum* is pinobanksin, subsequently also isolated from sunflower honey. It exhibits anti-oxidant activity against low density lipoproteins (LDL) (Oridrias et al. 1997). Oxidation of LDL is thought to contribute to atherosclerosis. When vitamin E was found to possess anti-oxidant activity against LDL, many studies for atherosclerosis prevention included prophylactic vitamin E supplementation. The intervention studies failed to show any benefit (Upston et al. 1999), the reason being that, under different conditions, vitamin E can be either pro- or anti-oxidant (Thomas et al. 1997).

Flavonoids are abundant in the plant kingdom. Orchids being relatively rare and smaller plants would seldom be a choice to supply a source for their isolation.

Polysaccharides

Bioactive polysaccharides or carbohydrates with beta 1–3, 1–4 or 1–6 branch-chains from herbs are widely promoted in TCM and Kanpo medicine as tonics and anticancer agents. They are principally derived from fungi but some are also present in other herbs, such as aloe, cinnamon, gingers, ginseng and *Gallang*. Polysaccharides in orchids are attracting scientific attention in China and Japan, the work still being restricted to the classic traditional herbs like *shihu* and *baiji* (Diao et al. 2008; Hua et al. 2004; Hsieh et al. 2008; Luo et al. 2008, 2010; Sun et al. 2005; Tagaki et al. 1983; Wang et al. 2006, 2010; Wu et al. 2010; Yamaki et al. 1985; Zhao et al. 2007). They exhibit immuno-modulatory activity in vitro. Other actions include an antimicrobial action against *Streptococcus mutans*, induction of cell differentiation, inhibition of angiogenesis and an antimetastatic effect. Polysaccharides vary greatly in their efficacy; their greater complexity in the branch chains and higher molecular weight are directly related to higher bioactivity.

They are usually administered in conjunction with conventional chemotherapy and radiotherapy. Lack of standardisation and a paucity of acceptable controlled trial data restrict their acceptance as adjunctive therapy.

The Orchids

The secondary metabolites of many orchids have been studied. They are discussed in the concluding 'OVERVIEW' of the various orchid genera which have been used as medicinal orchids in Asia. These compounds include alkaloids, terpenes, stilbenoids, bibenzyls, phenanthrenes, coumarins and flavonoids. Polysaccharides of orchids with medicinal properties are being intensively studied in China.

Genetic transformation is currently being studied as a tool to improve orchids of horticultural value (Sanjaya and Chan 2007). When the process is mastered, it could be employed to improve medicinal orchids or to extend their range of pharmaceutically important compounds.

Comment

To qualify for testing in a clinical situation, a compound must be effective at extremely low dosage (indicated by IC50), be non-lethal or with a lethal dose (LD50) much below 1 % of the minimum effective dose, and possess few serious side effects or none at all. Animal experiments are essential before human trials. The compound's structure must be known and, preferably, synthesis of it achieved. How it acts is explained at the molecular level. Exceptions may be made for anticancer agents; many of them elicit serious side effects which have to be carefully monitored. New compounds should always be introduced via clinical trial studies and their efficacy proven beyond doubt before they are approved for clinical use. There should be a system for voluntary notification of side effects.

References

Anonymous (1979) Editorial: Qinghaosu Project. Further Progress along the road to integrating Chinese and Western Medicine. J New Med Pharmacol, 10–11

Arditti J, Fish MH (1977) Anthocyanins of the Orchidaceae: distribution, heredity, functions, synthesis, and localization. In: Arditti J (ed) Orchid biology reviews and Perspectives, I. Cornell University Press, Ithaca and London

Bai L, Kato T, Inoue K, Yamaki M (1991) Blestrianol A, B and C, Biphenanthrenes from Bletilla striata. Phytochemistry 30(8):2733–2735

Bai L, Yamaki M, Tagaki S (1996) Stilbenoids from Pleione bulbocodioides. Phytochemistry 42(3): 853–856

Boller AH, Corrodi F, Gaumann E et al (1957) Uber induzierte Abwehrstoffe bei Orchideen Pt. 1. Helv Chim Acta 40:1062–1066

Brandange S, Granelli I (1973) Studies on Orchidaceae Alkaloids. XXXVI. Alkaloids from some Vanda and Vandopsis species. Acta Chem Scand 73(3): 1096–1097

Charoenrungruang S, Chanvorachote P, Sritularak BC, Pongrakhananon V (2014) Gigantol, a Bibenzyl from *Dendrobium draconis*, inhibits the migratory behavior of non-small cell lung cancer cells. J Nat Prod 77(6): 1359–1366

Chen KK, Chen AL (1935) The alkaloid of Chin-shih-hu. J Biol Chem 111:653–658

Chen TH, Pan SL, Guh JH, Chen CC, Huang YT, Pai HC, Teng CM (2008) Denbinobin induces apoptosis by apoptosis-inducing factor releasing and DNA damage in human colorectal cancer HCT-116 cells. Naunyn Schmiedebergs Arch Pharmacol 378(5):447–57

Chen PH, Peng CY, Pai HC et al (2011) Denbinobin suppresses breast cancer metastasis through the inhibition of Src-mediated signaling pathways. J Nutr Biochem 22(8):732–740

de Padua LS, Bunyapraphatsara N, Lemmens RHMJ (1999) Medicinal and poisonous plants. PROSEA. Nordic J Bot 19(5):612

Diao H, Li X, Chen J et al (2008) Bletilla striata polysaccharide stimulates inducible nitric oxide synthase and proinflammatory cytokine expression in macrophages. J Biosci Bioeng 105(2):85–9

Elander M, Leander K (1971) Studies on Orchidaceae alkaloids. XXI. 6-hydroxynobiline, a new alkaloid from *Dendrobium hildebrandii* Rolfe. Acta Chem Scand 25:717–720

Evangelista LF (managing ed.) (1995) MIMS Annual. p. 1082

Fisch MH, Flick BH, Arditti J (1973) Structure and fungal activitiy of hircinol, loroglossol and orchinol. Phytochemistry 12:437–441

Friederich S, Maier UH, Deus-Neumann BD et al (1998) Biosynthesis of cyclic bis(bibenzyls) in *Marchanta polymorpha*. Phytochemistry 50(4):589–598

Fritzememeier KH, Kindl H (1983) 9,10-dihydrophenanthrenes as phytoalexins of Orchidaceaea. Biosynthetic studies in vivo and in vitro proving the route from L-phenylalanine to dihydro-m- coumaric acid, dihydrostilbene and dihydro-phenanthrenes. Eur J Biochem 133:545–550

Gehlert R, Kindl H (1991) Induced formation of dihydrophenanthrenes and bibenzyl synthase upon destruction of orchid mycorrhiza. Phytochemistry 30:457–460

Gong Y, Fan Y, Liu L et al (2004a) Erianin induces a JNK/SAPK-dependent metabolic inhibition in human umbilical vein endothelial cells. In Vivo 18(2): 223–238

Gong YQ, Fan Y, Wu DZ et al (2004b) In vivo and in vitro evaluation of erianin, a novel anti- angiogenic agent. Eur J Cancer 40(10):1554–1565

Granelli I, Leander K, Luning B (1970) Studies on orchidaceae alkaloids. XVI. A new alkaloid, 2-hydroxydendrobine, from *Dendrobium findlayanum* Par. Ex Rchb. f. Acta Chem Scand 24(4):1209–12

Guo XY, Wang J, Wang NL, Kitanaka S, Liu HW, Yao XS (2006) New stilbenoids from *Pholidota yunnanensis*

and their inhibitory effect on nitric oxide production. Chem Pharm Bull (Tokyo) 54(1):21–5

Hardegger E, Schellenbaum M, Corrodi H (1963) Uber onduzierte Abwehrstoffe bei Orchideen. Part 2. Helv Chim Acta 46:1171–1180

Harper DJ (1998) Early Chinese medical literature. The Mawangdui medical manuscripts. Kegan Paul Intern, London

Hausen BM (1964) Toxic and allergic orchids. In: Arditti J (ed) Orchid biology reviews and perspectives, III (1984). Cornell University Press, Ithaca and London, pp 261–282

Hedman K, Leander K, Lunin B (1971) Studies on orchidaceae alkaloids. XXV. N-isopenttenyl derivaties of dendroxine and 6-hydroxydendroxine from *Dendrobium fredricksianum* Lindl. and *Dendrobium hildebrandii* Rolfe. Acta Chem Scand 25(3): 1142–4

Honda C, Yamaki M (2000) Phenanthrenes from Dendrobium plicatile. Phytochemistry 53(8):987–990

Hsieh YSY, Chien C, Liao SKS et al (2008) Structure and bioactivity of the polysaccharides in medicinal plant *Dendrobium huoshannense*. Bioorg Med Chem 16 (11):6054–68

Hua YF, Zhang M, Fu CX, Chen ZH, Chan GY (2004) Structural characterization of a 2-0-acetylglucomannan from *Dendrobium officinale* stem. Carbohydr Res 339(13):2219–24

Huang YC, Guh JH, Teng CM (2005) Denbinobin-mediated anticancer effect in human K562 leukaemia cells: role in tubulin polymerization and Bcr-Abi activity. J Biomed Sci 12(1):113–121

Inubushi Y, Nakano J (1965) Structure of dendrine. Tetrahedron Lett 31(Aug):2723–2728

Inubushi Y, Tsuda Y, Konita T, Matsumoto S (1964) Shihunine. A new phthalide pyrrolidine alkaloid. Chem Pharm Bull 12:749–750

Inubushi Y, Tsuda Y, Katarao E (1966) The structure of dendramine. Chem Pharm Bull Tokyo 14:668

Jackson Y, Chappuis F, Loutan L, Taylor W (2006) Malaria treatment failures after artemesinin-based therapy in three expatriates: could improved manufacturer information help to decrease the risk of treatment failure. Malaria J 5:81

Junka N, Kantayanarat S, Buanong M, Wongs-Aree C (2012) Characterization of floral anthocyanins and their antioxidant activity in Vanda hybrid (*V teres* × *V hookeriana*). J Food Agric Environ 10(2):221–226

Kaiser (1993) The scent of orchids: olfactory and chemical investigations. Editiones Roche, Basel

Kim BG, Sung SH, Chong YH et al (2010) Plant flavonoid O-Methyltransferases: substrate specificity and application. J Plant Biol 53(5):321–329

Kovacs A, Vasas A, Hohmann J (2008) Natural phenanthrenes and their biological activity. Phytochemistry 69:1084–1110

Kraus GA, Zhang N (2002) A direct synthesis of denbinobin. Tetrahedron Lett 43(52):9597–9599

Kuo CT, Hsu MJ, Chen BC, Chen CC, Teng CM, Pan SL, Lin CH (2008) Denbinobin induces apoptosis in human lung adenocarcinoma cells via Akt inactivation, Bad activation, and mitochondrial dysfunction. Toxicol Lett 177(1):48–58

Kuo CT, Chen BC, Yu CC et al (2009) Apoptosis signal-regulating kinase-1 mediates denbinobin-induced apoptosis in human lung adenocarcinoma cells. J Biochem Sc 16:43

Lawler LJ (1984) Ethnobotany of the orchidaceae. In: Arditti J (ed) Orchid biology reviews & perspectives III. Cornell University Press, Ithaca

Lawler LJ (1986) Orchid ethnobotany in the Asean Area. In: Rao AN (ed): Proc 5th Asean Orchid Congress. Singapore, Parks & Recreation Department, Ministry of National Development, pp. 42–45

Lawler LJ (1986b) Alkaloids in orchids. In: Rao AN (ed) Proc 5th Asean Orchid Congress. Parks & Recreation Department, Ministry of National Development, Singapore, pp 28–30

Lawler LJ, Slaytor M (1969) The distribution of alkaloids in orchids from the Territory of Papua New Guinea. Proc Linnean Soc NSW 94:419–421

Lee D (2007) Nature's palette. The science of plant color. University of Chicago Press, Chicago & London

Lee YH, Park JD, Baek NI, Kim SI, Ahn BZ (1995) In vitro and in vivo antitumoral phenanthrenes from the aerial parts of Dendrobium nobile. Planta Med 61(2): 178–80

Leong YW, Harrison LJ (2004) A biphenanthrene and a phenanthro(4,3 beta)furan from the orchid Bulbophyllum vaginatum. J Nat Prod 67:1601–1603

Letcher RM, Nhamo LRM (1975) Structure of orchinol, loroglossol and hircinol. J Chem So Perkin Trans 1:1263–1265

Li YM, Wang HY, Liu GQ (2001) Erianin induces apopstosis in human leukemia HL-60 cells. Acta Pharmacol Sin 22(11):1018–1012

Li Y, Wang CL, Wang YJ et al (2010) Chemical constituents of Dendrobium candidum. Zhongguo Zhong Yao Za Zhi 35(13):1715–1719

Lin TH, Chang SJ, Chen CC, Wang JP, Tsao LT (2001) Two phenanthraquinones from Dendrobium moniliforme. J Nat Prod 64(8):1084–6

Lin YL, Chen WP, Macabalang AD (2005) Dihydrophenanthrenes from Bletilla formosana. Chem Pharm Bull (Tokyo) 53(9):1111–1113

Lin LG, Liu QY, Ye Y (2014) Naturally occurring homoisoflavonoids and their pharmacological activities. Planta Med 80:1053–1066

Liu HW (2011) Identification, analysis, bioassay, and pharmaceutical and clinical studies. In: Liu WJ (ed) Traditional herbal medicine research methods. Wiley, Hoboken, New Jersey

Liu WH, Hua YF, Zhang ZJ (2007) Moniline, a new alkaloid from Dendrobium moniliforme. J Chem Res 2007(6):317–8

Luning B (1964) Studies on the Orchidaceae alkaloids. 1. Screening of species for alkaloids. I. Acta Chem Scand 18:1507–1516

Luning B (1967) Studies on the Orchidaceae alkaloids IV. Screening of the species for alkaloids 2. Phytochemistry 6:857–861

Luning B (1974) Alkaloids of the Orchidaceae. In: Withner CL (ed) The orchids: scientific studies. Wiley, New York, pp 349–382

Luning B (1975) Hunting orchids for chemistry. In: Senghas SK (ed) Proceedings, 8th World Orchid Conference, Frankfurt, 538–9

Luning B (1980) Alkaloids of the Orchidaceae. In: Sukshom Kashemsanta MR (ed.) Proceedings of 9th World Orchid Conference, Bangkok

Luo JP, Deng YY, Zha XQ (2008) Mechanism of polysaccharides from Dendrobium huoshanense on streptozotocin-induced diabetic cataract. Pharmaceut Biol 46(4):243–9

Luo AX, He XJ, Zhou SD et al (2010) Purification, composition analysis and antioxidant activity of the polysaccharides from Dendrobium nobile Lindl. Carbohydr Polym 79(4):1014–1019

Manandhar NP, Manandhar S (2002) Plants and people of Nepal. Timber, Portland

Majumder PL, Banerjee S (1988) Structure of Flavanthrin, the first dimeric 9,10-dihydrophenanthrene derivative from the orchid, Eria flava. Tetrahedron 44(23):7303–7308

Majumder PL, Pal A, Joardar M (1990) Cirrhopetalanthrin, a dimeric phenanthrene derivative from the orchid Cirrhopetalum maculosum. Phytochemistry 29(1):271–274

Majumder PL, Pal S, Majumder S (1999) Dimeric phenanthrenes from the orchid Bulbophyllum reptans. Phytochemistry 50:891–897

Majumder PL, Sabzabadi E (1988) Agrostophyllin, a naturally occurring phenanthropyran derivative from Agrostophyllum khasiyanum. Phytochemistry 27(6): 1899–1901

Matsuda H, Morikawa T, Xie H, Yoshikawa M (2004) Antiallergic phenanthrenes and stilbenes from the tubers of Gymnadenia conopsea. Planta Med 70(9): 847–55

Morita H, Fujiwara M, Yoshida N, Kobayashi J (2000) New Picrotoxin-type and Dendrobine-type Sesquiterpenoids from Dendrobium Snowflake 'Red Star'. Tetrahedron 56(32):5801–5805

Nishikawa K, Hirata Y (1967) Chemotaxonomical alkaloid studies. I Structure of nervosine. Tetrahedron Lett 27:2591–2596

Nishikawa K, Hirata Y (1968) Chemotaxonomical alkaloid studies. III. Further studies on Liparis alkaloids. Tetrahedron Lett 9:6289–6291

Nishikawa K, Miyamura M, Hirata Y (1967) Chemotaxonomical studies structures of Liparis alkaloids. Tetrahedron 25(13):2723–2741

Okamoto T, Natsume M, Onaka T, Uchimaru F, Shimizo M (1966a) The structure of dendroxine. The third alkaloid from *Dendrobium nobile*. Chem Pharm Bull (Tokyo) 14(6):672–5

Okamoto T, Natsume M, Onaka T, Uchimaru F, Shimizo M (1966b) The structure of dendramine (6-oxydendrobine) and 6-oxydendroxine. The fourth and fifth alkaloid from *Dendrobium nobile*. Chem Pharm Bull (Tokyo) 14(6):676–80

Onaka TS, Kamata T, Maeda Y et al (1965) The structure of nobilonine. The second alkaloid from *Dendrobium nobile*. Chem Pharm Bull 13:745–747

Oridrias K, Stasko A, Hromadova M et al (1997) Pinobanksin inhibits peroxidation of low density lipo-protein and it has electron donor properties reducing alpha-tocopherol radicals. Pharmazie 52(7):566–567

Reinecke T, Kindl H (1994) Inducible enzymes of the 9,10 dihydro-phenanthrene pathway. Sterile orchid plants responding to fungal infection. Mol Plant Microbiol Interact 7(4):449–454

Reinecke T, Kindl H (1994b) Characterization of Bi-benzyl synthase catalyzing the biosynthesis of phyto-alexins of orchids. Phytochemistry 35(1): 63–66

Roy M, Gonneau C, Rocheteau A et al (2013) Why do mixotrophic plants stay green? A comparison between green and achlorophyllous orchid individuals in situ. Ecol Monogr 83(1):95–118

Saarinen N, Joshi SC, Ahotupa M, Li XD et al (2001) No evidence for in vivo activity of aromatase inhibiting flavonoids. J Steroid Biochem Mol Biol 78(3): 231–239

Sanchez-Duffhunes G, Calzado MA, de Venuesa AG et al (2008) Denbinobin, a naturally occurring 1,4-phenthrenequinone, inhibits HIV-1 replication through an NF- kappaB-dependent pathway. Biochem Pharmacol 76(10):1240–1250

Sanchez-Duffhues G, Calzado MA, de Venuesa AG et al (2009) Denbinobin inhibits nuclear factor-kB and induced apoptosis via reactive oxygen species generation in human leukaemic cells. Biochem Pharm 77(8): 1401

Sanjaya, Chan MT (2007) Genetic transformation as a tool for improvement of orchids. In: Chen WH, Chen HH (eds) Orchid biotechnology. World Scientific, New Jersey

Slaytor MB (1977) The distribution and chemistry of alkaloids in the orchidaceae. In: Arditti JA (ed) Orchid biology reviews and perspectives, vol 1. Cornell University Press, Ithaca and London

Song JI, Kang YJ, Yong HY et al (2012) Denbinobin, a phenanthrene from *Dendrobium nobile*, inhibits invasion and induces apoptosis in SNU-484 human gastric cancer cells. Oncol Rep 27(3):813–818

Stoessl A, Arditti J (1984) Orchid phytoalexins. In: Arditti J (ed) Orchid biology, reviews and perspectives, III. Cornell University Press, Ithaca & London

Storey WB (1950) Genetics in flower colour in Spathoglottis cross. Pac Orchid Soc Bull 8(4):1–5

Storey WB (1958) Additional observations on the genetics of flower colour in Spathoglottis. Pac Orchid Soc Bull 11:17–25

Su P, Wang J, An JX et al (2011) Inhibitory effect of erianin on Hepatocellular carcinoma (HCC) Huh7 Cells. Chin J Appl Environ Biol 17(5):662–665

Sun J, Wang C, Zhang J (2005) Effect of polysaccharides from *Bletilla striata* on the adhesion of human umbilical venous endothelial cells. Zhong Yao Cai 28(11): 1006–8

Surak JG (1978) Phytoalexins and human health—a review. Proc FL State Hort Soc 91:256–258

Suzuki H, Keimatsu I, Ito M (1932) Alkaloid of the Chinese drug "Chin-Shih-Hu". II. Dendrobine. J Pharm Soc Japan 52:1049–1060

Suzuki H, Keimatsu I, Ito M (1934) Alkaloids of the Chinese drug "Chin-Shih-Hu". III. Dendrobine. J Pharm Soc Japan 54:802–819

Tagaki S, Yamaki M, Inoue K (1983) Antimicrobial agents from *Bletilla striata*. Phytochemistry 22: 1011–1015

Tanino H, Inoue S, Nishikawa K, Hirata Y (1969) Synthesis of tetra-acetyl malaxin and kuramerine. Tatrahedron 25(15):3033–3037

Tezuka Y, Ji L, Hirano H et al (1990) Studies on the constituents of orchidaceous plants IX. Constituents of *Spiranthes sinensis* (Pers.) *Amesvar amoena* (*M. Bieberson*) Hara. (2) Structures of spiranthesol, spiranthoquinone, spiranthol-C and spirasineol-B new isopentenyldihydrophenanthrene. Chem Pharm Bull 38:629–635

Thomas SR, Neuzil J, Stocker R (1997) Inhibition of LDL oxidation by ubiquinol-10. A proactive mechanism for co-enzyme Q in atherogenesis? Mol Aspects Med 18(Suppl):85–103

Tsai AC, Pan SL, Lai CY et al (2011) The inhibition of angiogenesis and tumor growth by denbinobin is associated with the blocking of insulin-like growth factor-1 receptor signaling. J Nutr Biochem 22 (2011):625–633

Tuchinda P, Udchachon J, Khumtaveeporn K et al (1988) Phenanthrenes of Eulophia nuda. Phytochemistry 27:3267–7

Upston JM, Terentis AC, Stocker R (1999) Tocopherol-mediated peroxidation of lipoproteins: implications for vitamin E as a potential antiatherogenic supplement. FASEB J 13(9):977–94

Wang YC, Lin CH, Chen CM, Liou JP (2005) A concise synthesis of denbinobin. Tetrahedron Lett 46(47): 8103–8104

Wang CM, Sun J, Luo Y et al (2006) A polysaccharide isolated from the medicinal herb *Bletilla striata* induces endothelial cells proliferation and vascular growth factor expression in vitro. Biotechnol Lett 28(8):539–43

Wang JH, Luo JP, Yang XF, Zha XQ (2010) Structural analysis of a rhamnoarabinogalactan from the stems of *Dendrobium nobile* Lindl. Food Chem 122(3): 572–576

Ward EWB, Unwin CH, Stoessl (1975) Loroglossol: an orchid phytoalexin. Phytopathology 65(5):632–633

Woo KJ, Jeong YJ, Inoue H et al (2005) Chrysin suppresses lipopolysaccharide-induced cyclooxygenase-2-expression through the inhibition of nuclear factor for IL-6 (NF- IL6) DNA-binding activity. FEBS Lett 579(3):705–711

Wu XG, Xin M, Chen H et al (2010) Novel mucoadhesive polysaccharide isolated from *Bletilla striata* improves the intraocular penetration and efficacy of levofloxacin in the topical treatment of experimental bacterial keratitis. J Pharm Pharmacol 62(9):1152–1157

Xing YM, Chen J, Cui JL et al (2011) Antimicrobial activity and biodiversity of endophytic fungi in *Dendrobium devonianum* and *Dendrobium thyrsiflorum* from Vietnam. Curr Microbiol 62(4): 1218–1224

Xue Z, Li S, Wang S, Wang Y, Yang Y, Shi J, He L (2006) Mono-, Bi-, and triphenanthrenes from the tubers of *Cremastra appendiculata*. J Nat Prod 69(6):907–13

Xue Z, Li S, Wang SJ, Yang YC, He DX, Ran GL, Kong LZ, Shi JG (2005) Studies on the chemical constituents from the corm of Cremastra appendiculata. Zhongguo Zhong Yao Za Zhi 30(7):511–3

Yamaki M, Bai L, Inoue K, Tagaki S (1989) Biphenanthrenes from *Bletilla striata*. Phytochemistry 28(12):3503–3505

Yamaki M, Bai L, Kato et al (1993) Blespirol, a phenanthrene with a spirolactone ring from *Bletilla striata*. Phytochemistry 33(6):1497–1498

Yamamura S, Hirata Y (1964) Structures of nobiline and dendrobine. Tetrahedron Lett 5:79–87

Yang L, Qin LH, Bligh SW, Bashall A, CF Z, Zhang M, Wang ZT, Xu LS (2006) A new phenanthrene with a spirolactone from *Dendrobium chrysanthum* and its anti-inflammatory activities. Bioorg Med Chem 14(10):3496–3501

Yang H, Lee PJ, Jeong EJ et al (2011) Selective apoptosis in hepatic stellate cells mediates the antifibrotic effect of phenanthrenes from *Dendrobium nobile*. Phytother Res 26(7):974–980

Ye QH, Zhao WM, Qin GW (2003) New flourenone and phenanthrene derivatives from *Dendrobium chrysanthum*. Nat Prod Res 17(3):201–5

Zhang GN, Bi ZM, Wang ZT, Xu LS, Xu GJ (2003) Advances in studies on chemical constituents from plants of *Dendrobium* Sw. Chin Trad Herbal Drugs 34:S5–S8 (Appendix)

Zhang GN, Zhong LY, Bligh SW, Guo YL, Zhang CF, Zhang M, Wang ZT, Xu LS (2005) Bi-cyclic and bi-tricyclic compounds from Dendrobium thyrsiflorum. Phytochemistry 66(10):1113–1120

Zhao Y, Son YO, Kim SS, Jang YS, Lee JC (2007) Antioxidant and anti-hyperglycaemic activity of polysaccharide isolated from *Dendrobium chrysotoxum* Lindl. J Biochem Mol Biol 40(5):670–7

Zou YX, Xiao CF, Zhong RQ et al (2008) Synthesis of combretastatin and erianin. J Chem Res 2008(6): 354–356

Discovery, Testing and Improving the Production of Herbs and New Drugs

6

It is reported that, once, when Li Shizhen (1518–1593) was out searching for medicinal plants, he came across an injured snake that was bleeding. As he watched, the snake slithered away. A short while later, Li saw the injured snake coiling and rubbing itself against a bush, and to his surprise it stopped bleeding. Not long afterwards, its skin began to heal. Li Shihzhen concluded that the herb had healing properties, and he recommended its use in trauma (Li 1578).

A medical scientist would carry his investigation a bit further. He would get a snake handler to trap a number of snakes, then separate them into two groups and retest the effect of the herb on the injuries of one group, leaving identical injuries of the second group untreated in order to compare the results. Afterwards, the test would be repeated on several species of mammals before being tested on humans. At this stage, the advice of a statistician would be sought to determine the numbers of test animals required to obtain a statistically valid observation of efficacy.

If the results are sufficiently spectacular to induce a pharmacological examination, the chemical constituents of the herb would be isolated in pure, crystalline form by extraction and chromatographic separation and their properties studied. Should the experiments yield promising results, a large amount of the herb would be gathered and column chromatography used to obtain the effective substance in a quantity that would permit its chemical identification.

Nowadays, numerous methods would permit its rapid and accurate identification. Knowledge of its molecular structure would allow the scientist to determine whether the substance was new to science or already in the pharmaceutical armamentarium. A new substance might then be synthesised.

Before it can be marketed, the compound has to go through several stages of laboratory testing on a cellular level to determine its mode of action, then testing on animals to confirm its efficacy and toxicity before going on to human testing. Tests on human subjects require research approval by research committees, which make use of independent referees to evaluate the soundness of a study and the qualifications of the investigators. Next, an institutional review board examines the ethical aspects of the research. Sound statistical advice, careful planning and proper supervision are essential to ensure that the trial is well conducted and that the observations and conclusions will be valid. Proof of this lies in the publication of a research paper in a reputable, peer reviewed, international journal and an absence of valid negative comments. Large, international, multicentre trials are required to convince drug regulatory bodies of major nations to include the drug on their approved list of medications.

Post-marketing surveillance for side effects continues for many years is necessary to convince doctors and regulatory bodies that the drug is safe for the public.

© Springer International Publishing Switzerland 2016
E.S. Teoh, *Medicinal Orchids of Asia*, DOI 10.1007/978-3-319-24274-3_6

A modern success story is the discovery of penicillin through the astuteness of Alexander Fleming who made an appropriate deduction when he observed the accidental contamination of a bacterial culture in 1928. Today, one is no longer dependent on mould for penicillin, as the compound has been synthesised. Its molecular structure has been modified to provide for ease of delivery (it can be taken by mouth instead of having to be injected), for longer intervals between administration, and, either in modified form or in combination with other agents, a broader spectrum against the various types of bacteria.

Screening Plant Extracts for Possible Medicinal Properties

The search for secondary metabolites in orchids started with the screening of plant extracts for alkaloids. Alkaloids can be suspected if a plant extract is bitter to taste, but their presence needs to be confirmed by a simple procedure which employs the Dragendorff reaction, a colour test performed on the extracts. The first report was published more than a century ago (Boorsma 1902) but it was not until the 1960s that scientists in Japan, Sweden and Australia (Luning 1967, 1974; Lawler and Slaytor 1969) decided to undertake full-scale mass-screening of various orchid genera. Their work is not complete, but the approach today would be different because of advancements in screening technology which come within the ambit of metabolomics.

Mass Screening for Cytotoxic Compounds with Anticancer Potential

Many cytotoxic agents employed in the treatment of cancer destroy cancer cells by activating apoptosis or programmed cell death. Caspase-3 is activated at the final stage of a cascade of intracellular events which result in apoptosis. By developing a fluorescence resonance energy transfer (FRET)-based caspase sensor cell line (Hela C3), Kathy Qian Luo and her team at Nanyang Technological University in Singapore and Simon Han of the Chinese University of Hongkong, among others, achieved a high throughput screening platform that can simultaneously screen very large numbers of herbal extracts for cytotoxic potential (Luo 2010; Han 2010).

The technology employs trays with multiple rows of wells each containing the sensor cell line. Following incubation with the test material, the fluorescent reagents are added. If a cytotoxic agent is present, the cells show a blue fluorescence; if absent, they are green.

Once the anticancer potential has been identified, an extract will undergo purification procedures. The structure of the effective compounds will be defined by high-performance liquid chromatography-mass spectroscopy or nuclear magnetic resonance spectroscopy. MTT assays (another colour-based assay checking on cell survival following exposure to a suspected cytotoxic agent) and other investigations will be performed to confirm its cytotoxic action. Through this approach Xu and his colleagues managed to isolate oblongfolia C, a polyprenylated bensoylphorogucinol from *Garcinia yunnanensis*, a plant related to mangosteen (*Garcinia mangostana*) (Xu et al. 2008; Luo 2010). The process offers great potential for the mass screening of orchids for potential anticancer properties.

Avenues and Goals in Ethnobotany

Today, there appears to be considerable interest in examining orchids for their pharmacological, pharmaceutical, fragrance and pesticide potentials. This comes within the ambit for worldwide search for new drugs. From past experience, we know that:

1. Sometimes, this is achieved by using a drug in an unmodified form, unpurified or purified but without modifying the molecular structure of its basic components. (If we take a cytotoxic agent, for instance, vincristine is an unmodified product, and perhaps for that reason, it is

very toxic.) Most Chinese medicinal products are presented in this way, with improvements by rendering them in powder form or in capsules, but chemically the herb is unmodified. *Tianma* is a herb that may find acceptance in the Western pharmacopoeia because there is still an ongoing search to find better and longer acting remedies to treat parkinsonism and other nervous disorders, but that would require its efficacy and safety to be proven by randomised clinical trials.

2. Sometimes, a chemical constituent may provide the starting molecules for the synthesis of potent compounds, although in their original form they lack pharmacological properties. A case in point is the Mexican yam which is the starting point for the synthesis of oral contraceptives and other oestrogenic compounds.

3. Sometimes, a herbal constituent may provide new modes of pharmacological actions. There has been a suggestion that denbinobine and a few other orchid-related compounds may help to overcome the drug resistance of some cancers, but the findings are very preliminary and it will require a lot more research to determine whether this is really true, and then to improve the compound.

4. Sometimes, it provides an addition to the established armamentarium. However, the idea that a preparation that is "natural" is safer than a synthetic compound is not valid. Some of the most potent poisons are natural compounds.

Enhancing the Metabolite Content of Herbs Through Biology

Medicinally active compounds from plants belong to four groups of secondary metabolites, i.e. alkaloids, terpenes, aromatics (phenolics) and polysaccharides. Various enzymes and hormones are involved in their synthesis, which involves the secondary metabolic pathway separate from photosynthesis, but dependent on the latter for energy and carbon supply. Enhancement of the desired secondary metabolites can be achieved by efficient management in the cultivation, timing of harvest, use of plant hormones, stress exposure, in vitro culture and genetic manipulation.

Farmers know that fruits are sweeter if the plants receive an adequate supply of potassium which plays an essential role in translocation. The use of fertilisers almost doubles the yield of psoralen and xanthotoxin in *Ammi majus* and *Psoralea corylifolia* (Abdin 2007; Aberoumard 2009). Calcium is essential for healthy roots and critical for the production of peanuts (groundnuts). Enhanced calcium supply to root-based medicinal herbs might possibly enhance their yield of medicinal compounds. Fertiliser usage is only possible if the plants are cultivated. In the wild, soil types determine quality. To indicate the quality of their herbs apothecaries qualify them according to their source. This has been done for many orchids.

The age of a plant, and the age of maturation, flowering and senescence, have a significant bearing on its content of secondary metabolites. The content of artemesinin in *Artemesia annua* rises in a straight line to reach a peak at 130 days, after which there is a rapid progressive decline to values reached at 50 days. Production of artemesic acid in the same species escalate after 80 days and peaks at 100, well before the peak of artemisinin. The level of arteannuin B shows a sharp rise from day 50 to day 130, reaching a high peak on the same day as artemisinin. Its decline thereafter is not as rapid as the other two compounds (Abdin et al. 2001). Similar studies on medicinal orchids have not been published, but undoubtedly they would be studied in this manner in the cultivated varieties. More well established is the timing of harvest. Orchids and several medicinal herbs are traditionally harvested just prior to flowering, or towards the end of the growing season, because it is thought, or has been shown, that these plants have the desired potency if harvested at this time. That is not to say that they must have the maximal metabolic content at this time. Xanthotoxin yield at different phonological stages of *Ammi majus*, for instance, reaches maximum value post-flowering, not before. Indeed, the yield is over 16 times higher than the content in pre-flowering plants (Abdin 2007). Thus, every medicinal orchid has to be individually studied. This would be in the tradition of Li Shizhen.

Studies conducted on 11 species of *shihu* confirmed that ecological factors played an important role in the chemical components of several notable species of medicinal *Dendrobium*, namely: (1) polysaccharide content of *D. officinale* varied with soil type; (2) dendrobine content of *D. nobile* correlated with annual rainfall; and (3) erianin content of *D. chrysotoxum* was affected by ambient temperature. Zhejiang Province was most suitable for *D. officinale* whereas Guizhou was favoured for *D. nobile* and Yunnan was the best province for *D. chrysotoxum* (Li et al. 2013).

The flowering period of individual species is stated when the orchids are described (in Chap. 7, but flowering time varies depending on location and variation in climate conditions from year to year). In southern India, Abraham and Vatsala (1981) noted that there are two peak periods of flowering for orchids. The first peak occurs after the period of heavy rains when terrestrial orchids like *Habenaria*, *Liparis* and *Malaxis* come into bloom after a period of rapid growth. The rains also induce flowering in epiphytic species that require a dry resting period before they can bloom, species like *Rhynchostylis retusa* and *Aerides crispum*. The second peak involves epiphytic species which require the stimuli of stronger sunlight and higher temperatures to bloom. *Acampe praemorsa*, *Dendrobium crepidatum*, *Eulophia spectabilis*, various *Luisia* species and *Vanda testacea* bloom in April to early summer. There is a third group of southern Indian orchids which respond to cool nights and they bloom in January and February. Examples are *Zeuxine*, *Anoectochilus* and *Goodyera*.

The application of growth hormones IAA and GA separately increases the accumulation of artemisinin under laboratory conditions, and a modest additional increase results when both hormones are applied together. The increase in artemisin is associated with an increased in the level of HMGCo-A reductase, the enzyme which diverts carbon flux away from the primary photosynthetic pathway towards the secondary metabolite pathway (Abdin 2007). Salt and lead stress applied at the bolting stage also greatly increased artemesinin yield from the leaves of *Artemesia annua* (Abdin 2007). The use of plant hormones has been proposed to induce flowering in orchids, but its usage for enhanced medicinal production has yet to be announced.

Clonal selection has been used to select plants possessing the highest medicinal content for propagation. Classification according to geographic region is the simplest form of genetic selection. Plants collected from the best locale may then undergo a secondary screening step to pick out those with the highest yield. Genetic transformation which introduces genes that encode key rate-limiting enzymes, which control the secondary metabolic pathways, offer the best hope for enhancing medicinal products from plants. The production of artemesinin in genetically transformed plants is far higher than in the wild types. The incorporation of isopentyl transferase increased the yield by 70 % (Sa et al. 2001), while the incorporation of farnesyl diphosphate synthase increased the yield by 300 % (Chen et al. 2000). Tobacco plants genetically manipulated to overexpress HMGCo-A reductase enzyme from the *Hevea brazilensis* gene overproduced sterols by 600 % (Schaller et al. 1995). Chinese scientists have begun their study of the genetic constitution of *Shihu Dendrobium* in an effort to produce superior plants that will be capable of yielding a high content of medicinally desirable compounds.

If an orchid is truly discovered to produce an important disease-curing metabolite, this would not result in the overcollection and disappearance of the species from the wild. Rather, scientists will identify the most productive clones from specimens collected in the wild and genetically modify the orchid so that it will yield a large amount of the desired metabolite. Afterwards, the orchid will be grown in tissue culture, supplied with optimal nutrients and hormones and given optimum light. The metabolites will probably be extracted from protocorms, not from mature plants. Finally, one could envisage the metabolite being further modified either chemically in the laboratory or by the use of microorganisms or plant cell suspensions to produce an even more efficacious medicine (Fig. 6.1).

Fig. 6.1 *Vanda coerulea* Griff. Ex Lindl. New compounds isolated from this beautiful orchid that protect skin fibroblast cells from ultraviolet light damage would have a place in cosmetic preparations

Combating Insect Pests and Red Spiders

Insect pests of all kinds cause damage to an estimated 50–60 % of medicinal and aromatic plants in India (Sarmma et al. 2008). Red spider mites are an even greater problem with orchids. How these pests are handled, and the nature of the pesticides if any are used, greatly impact on the safety of the final medicinal, flavouring or cosmetic product. A biological approach using mites to fight red spider mites has been used with some success.

Combating Viral Disease in Orchids

Viral diseases are a bane to orchid growers. Infected plants grow poorly and they produce fewer flowers which exhibit colour breaks and other abnormalities. All orchids are susceptible to viral infection, but *Cymbidium* is a genus which has been most widely studied. Thus, attempts to control viral disease by genetic manipulation have focused on *Cymbidium*. However, the results of the early attempts were not encouraging (Rubino et al. 1993; Lupo et al. 1994). No correlation was found between

the level of expression of the integrated *Cymbidium Ringspot Tombusvirus* gene and the level of resistance to the challenging virus (Lupo et al. 1994).

Monoclonal antibodies to *Cymbidium* mosaic virus (CyMV) raised in mouse myeloma cells fused with spleen cells immunised with CMV particles has greatly simplified the detection of CMV in infected plants using antigen-coated plate ELISA (ACP-ELISA) (Meng et al. 2007). Transgenic tobacco plants were produced by introducing the coat protein gene of *Cymbidium Ringspot Virus* (*CyRSV*) into normal tobacco plants (*Nicotiana benthamiana*). When these transgenic plants were challenged with *CRSV*, they were immune to infection if the inoculums were 0.05 mcg/ml or lower, but they could not be protected if the viron concentrations were 0.5 mcg or higher. They were also not protected against synthesised genomic RNA of *CRSV* (Rubino et al. 1993).

A Korean team has reported success with *Cymbidium Mosaic Virus* (*CyMV*). Using CyMV-Ca isolated from a naturally infected *Cattleya*, they worked out its nucleotide sequence at the 3′terminal region, and found that it contained an open reading frame which coded for the viral coat protein (CP) and three other ORFs (triple gene block or movement protein) of the virus. This CP gene encodes a large polypeptide chain of 220 amino acids (one could compare it with human growth hormone which has 188 amino acids), and a molecular mass of 23,760 Da (Lim et al. 1999). With such a large size, it is possible to raise antibodies to this CP gene in mammals and not use the entire virus, but the workers were not interested in such a project. Instead, they deduced the CP sequence and used it to make a construct of the *CyMV-Ca* CP gene in the antisense orientation in the plant expression vector pMBP-1. This was transferred to tobacco (*Nicotonia occidentalis*), a propagation host of CMV, via *Agrobacterium tumefaciens*-mediated transformation. When the T1 progeny of the transgenic *Nicotiana* were inoculated with *CyMV*, they were shown to be highly resistant to the *CyMV* infection (Lim et al. 1999). Chang et al. (2005) obtained transgenic *Dendrobium* moderately resistant to *CyMV* through particle bombardment of protocorms with the *CyMV*-capsid protein gene, which they had synthesised and sequenced. This is therefore an approach to obtain virus-resistant plants of desirable and/or endangered medicinal orchids.

Gene stacking would take this one step further; by that we mean stacking genes of multiple pathogens which cause severe damage to orchids. Chan et al. (2005) of Taiwan National University have stacked the genes of *Cymbidium Mosaic Virus* (*CyMV*) and *Erwina carotovora* (a bacterium which causes soft rot) in a *Phalaenopsis* orchid, and proved that the transgenic orchid had dual resistance to the plant pathogens. They also employed *Agrobacterium tumefaciens* for transfection, and the gene integration and expression in the transformed *Phalaenopsis* lines were confirmed by Southern blot and northern blot analysis.

The advantage of such genetic manipulation is that disease resistance would be passed on to their progeny: mericlones, selfings and hybrids. Genetic manipulation which confers resistance to pesticides is a different matter, and should not be pursued in respect of medicinal orchids.

Genetic Manipulation to Produce New Drugs

For over a decade, infertile women have benefited from the use of recombinant pure human follicular-stimulating hormone (hFSH) and recombinant pure human-luteinising hormone (hLH) to produce super-ovulation for the purpose of in vitro fertilisation (IVF). These hormones were produced by Serono which also market recombinant human growth hormone (hGH) which is employed to hasten growth in short children before they reach puberty. This technology can be employed to produce pharmaceutically important compounds from orchids, and remove the reliance on actual orchid plants.

Several key genes have been introduced into *Artemisin annua* L to encourage higher yields of the valuable antimalarial drug, artemisin, than is obtainable from normal plants (Liu et al. 2011; Tang et al. 2014; Wang et al. 2014). Studies are being undertaken to create transgenic orchid plants using *Agrobacterium* (Nan et al. 1998; Yu et al. 2001; Semiarti et al. 2007, 2009, 2010, 2015; Belrmino and Mii 2000; Chi and Mii 2011) or particle bombardment (Chia et al. 2001; da Silva et al. 2011), but these are principally directed towards producing plants with resistance to cold or disease and improving flower quality (Obsuwan et al. 2003). Transgenic *Dendrobium* have been produced in Singapore and Hawaii (Chia et al. 1994); the challenge is to produce transgenic *Dendrobium* that will boost the supply of *Shihu* and stop the plunder of plants from the wild.

A patent has already been filed with the United States Patent and Trademark Office Patents (Patent No. US 0780007869) for transgenenic plants and plant tissues including plant cells which contain a DNSA construct encoding *Gastrodia* Antifungal Protein (GAFP), also known as gastrodianin, present in *G. elata*. Plants involved in the patent range from herbaceous to woody plants and fruit trees, the idea being to provide a disease-resistant rootstock (Schnabel et al. 2010).

References

Abdin MZ (2007) Enhancing bioactive molecules in medicinal plants. In: Zhu YZ, Tan BKH, Bay BH, Liu CH (eds) Natural products. Essential resources for human survival. World Scientific, Singapore

Abdin MZ, Israr M, Srivastava PS et al (2001) In vitro production of artemisinin, a novel antimalarial compound from Artemisia annua. J Med Arom Plant Sci 22/4A, 23/1A:378–384

Aberoumard A (2009) Preliminary assessment of nutritional value of plant based diets in relation to mineral nutrition. Int J Food Sci Nutr 60(Suppl 4):155–162

Abraham A, Vatsala P (1981) Introduction to orchids, with illustrations and descriptions of 150 South Indian orchids. TPGRI, Trivandrum

Belrmino MM, Mii M (2000) Agrobacterium-mediated genetic transformation of a phalaenopsis orchid. Plant Cell Rep 19:435–442

Boorsma WG (1902) Pharmakologische Mitteilungen I. Bull l'INst Botan Buitenzorg 14:1–39

Chan YL, Lin KH, Sanjaya et al (2005) Gene stacking in phalaenopsis orchid enhances dual tolerance to pathogen attack. Transgenic Res 14(3):279–288

Chang C, Chen YC, Hsu YH et al (2005) Transgenic resistance to cymbidium mosaic virus in dendrobium expressing the viral capsid protein gene. Transgenic Res 14(1):41–46

Chen DH, Li HC, Li GF (2000) Expression of a chimeric famesyladiphosphate synthase gene in *Artimisia annua* L. transgenic plants via *Agrobacterium tumefaciens*-mediated transformation. Plant Sci 155:179–185

Chi DP, Mii M (2011) Dwarf phalaenopsis plant produced by over-expression of giberrelin2-oxidase gene. In: Proceedings of the NIOC Nagoya, Japan, 11 Mar 2011

Chia TF, Chan YS, Chua NH (1994) The firefly luciferase gene as a non-invasive reporter for dendrobium transformation. Plant J 6(3):441–446

Chia TF, Lim AYH, Luan Y, Ng I (2001) Transgenic dendrobium (orchid). Biotech Agri For 48:95–106

da Silva JAT, Chin DP, Pham TV, Mii M (2011) Transgenic orchids. Sci Hortic 4:673–680

Han SQB (2010) Quick characterization of apoptosis inducers from natural products by a drug discovery platform composed of high-speed counter-current chromatography and the fluorescence-based caspase-3 biosensor detection. In: Conference abstracts on: recent developments in Chinese herbal medicine. Nanyang Technological University, 2010, p 24

Lawler LJ, Slaytor M (1969) The distribution of alkaloids in orchids from the territory of Papua New Guinea. Proc Linnean Soc NSW 94:419–421

Li SZ (1578) Ben Cao Gung Mu (reprinted 1977 by People's Health Publishing Co.,Beijing)

Li WT, Huang LF, Du J, Chen SL (2013) Relationship between dendrobium quality and ecological factors based on partial least square regression. Ying Yong Sheng Tai Xue Bao 24(10):2787–2792

Lim SH, Ko MK, Lee SJ et al (1999) Cymbidium mosaic virus coat protein gene in antisense confers resistance to transgenic *Nicotiana occidentalis*. Mol Cells 9 (60):503–608

Liu BY, Wang H, Duet ZG et al (2011) Metabolic engineering of artemesinin biosynthesis in *Artemesia annua* L. Plant Cell Rep 30:689–694

Luning B (1967) Studies on the Orchidaceae alkaloids IV. Screening of the species for alkaloids 2. Phytochemistry 6:857–861

Luning B (1974) Alkaloids of the Orchidaceae. In: Wittner CL (ed) The orchids: scientific studies. Wiley, New York

Luo KC (2010) Cellular, molecular and animal studies for the anti-cancer effect of oblongifolin C discovered from Garcinia yunnanensis by bioassay-guided screening. In: Hew CS (ed) Recent development in Chinese Herbal Medicine Abstracts. Nanayang Technological University, Singapore, p 34

Lupo F, Rubino L, Russo M (1994) Immuno-detection of the 33K/92K polymerase proteins in cymbidium ringspot virus-infected and in transgenic plant tissue extracts. Arch Virol 138(1–2):135–142

Meng CM, Wu JK, Xie L et al (2007) Production of monoclonal antibodies to cymbidium mosaic virus and application in orchids virus detection. Wei Sheng Wu Xue Bao 47(5):928–931

Nan GL, Kado CI, Kuehnle AR (1998) Transgenic dendrobium orchid tissue through agrobacterium-mediated transformation. Malayan Orchid Rev 32:93–96

Obsuwan K, Borth W, Hu J et al (2003) Development of transgenic dendrobium orchid resistant to CyMV. In: American society of plant pathology annual meeting, Abstract #732

Rubino L, Apriotti G, Lupo R, Russo M (1993) Resistance to cymbidium ringspot tombusvirus infection in transgenic *Nicotiana benthamiana* plants expressing the virus coat protein gene. Plant Mol Biol 21 (4):665–672

Sa G, Mi M, He-chun Y et al (2001) Effects of IPT gene expression on the physiological and chemical characteristics of Artemesia annua L. Plant Sci 160 (4):691–698

Sarmma S, Senthikumar N, Das SK (2008) Insect pests of medicinal and aromatic plants and their management: an overview. Indian Forester 134(1):105–118

Schaller H, Grausem F, Benveniste B et al (1995) Expression of the Hevea brazilensis (H.B.K) Mull. Arg.3-Hydroxy-3-Methyl glutaryl Coenzyme A Reductase 1 in Tobacco Results in Sterol Overproduction. Plant Physiol 109(3):761–770

Schnabel G, Scorza R, Lauyne DR (2010) Increases resistance of plants to pathogens from multiple higher order phylogentic lineages. Patent No: US 07807869. Official Gazette of the United States Patent and Trademark Office Paterts OCT 5 2010

Semiarti E, Indrianto A, Pirwantoro A et al (2007) Agrobacterium-mediated transformation of the wild species *Phalaenopsis amabilis*. Plant Biotechnol 24 (3):265–272

Semiarti E, Indrianto A, Suyono EA et al (2010) Genetic transformation of the Indonesian black orchid (*Coelogyne pandurata* Lindley) through *Agrobacterium tumefaciens* for micropropagation. In: Proceedings of NIOC 2010, Nagoya Dome, Japan, 16–20, 2010

Semiarti E, Indriarto A, Pirwantoro A et al (2015) Agrobacterium mediated transformation of Indonesian orchids for micropropagation. www.intechopen.com

Semiarti E, Purwantoro A, Dwiyani R et al (2009) Perbandingan karakter morfologi dan molekuler Vanda tricolor Lindl var sauvis forma Merapi dan Vanda tricolor Lindl. var auvis forma Bali. UIN Maulana Malik Ibrahim, Yogyakarta

Tang K, She Q, Yan T, Fu X (2014) Transgenic approach to increase artemisinin content in *Artemisia annua* L. Plant Cell Rep 33(4):605–615

Wang YX, Long SP, Zeng LX et al (2014) Enhancement of artemisinin biosynthesis in transgenic *Artemisia annua* L. by overexpressed HDR and ADS genes. Yao Xue Xue Bao 49(9):1346–1352

Xu G, Feng C, Zhou Y et al (2008) Bioassay and ultra-performance liquid chromatography/mass spectrometry guided isolation of apoptosis-inducing benzophenones and xanthone from the pericarp of Garcinia yunnanensis Hu. J Agri Food Chem 56 (23):11144–11150

Yu H, Yang SH, Goh CJ (2001) Agrobacterium-mediated transformation of a dendrobium orchid with the class 1 knox gene DOH1. Plant Cell Rep 20:301–305

Medicinal Orchids in Asia by Genus and Species
Introduction to Part II

This part presents the medicinal plants belonging to the Family Orchidaceae which are employed as herbal medicine in the Asian continent. For ease of identification, they are grouped together by genus. In some genera, the medicinal species have similar usage, but in the majority their applications vary. Names of individual plants are based on the Kew Monocot List:

http://apps.kew.org/wcsp/prepareChecklist.do;jsessionid=44C110337630 E14F3148698099593A15?checklist=selected_families%40%403461212201 01919117.

Such names are in bold print. Synonyms preferred by authors of *Herbals, Materia Medica* and articles on medicinal usage of orchids in contemporary journals and magazines are included for ease of cross-reference. Local and medicinal names are additional means of identification. Chinese names follow those in the *Flora of China* Vol. 25 Orchidaceae (2009). In the case of Indian orchids, the presence of numerous names in Sanskrit may suggest an old *Ayurvedic* usage for the orchid plant. Variations in spelling are commonly encountered with Indian and Thai local names of orchids because pronunciations vary with district and dialect.

Synonymous Latinised names for the various orchid species are not included unless they happen to have been previously employed in publications relating to medicinal usage or phytochemistry. However, all local names for the species are retained because of the Asian focus of this volume.

Genus: *Acampe* Lindl.

The name *Acampe* originates from Greek *akampes* (rigid), which refers to its overall character pertaining to all aerial parts of the plant. This name recalls *Thalia Maravara*, or the 'rigid air flower', the name mentioned by van Rheede in the first published description of an Asian orchid by a European, in 1703 (van Rheede 1693–1703). Rumphius, who lived in Maluka, described orchids much earlier but his book was published after 1703 (Beckman 2002).

Acampe is a robust, monopodial epiphyte with coriaceous leaves and rigid flowers on a short raceme (Fig. 7.1). Tight clustering of the flowers and a tendency to cup are dominant traits transmitted to its hybrids. It is not popular among orchid growers as a breeding parent. It is distributed in India, Sri Lanka, southern China and Southeast Asia. There are about ten species in *Acampe*, several with numerous names. Three *Acampe* species are used medicinally.

Acampe carinata (Griff.) Panigrahi

Indian Name: *Kano Kato*.
Thai name: *Phaya mue lung*

Description: A robust, monopodial epiphyte with coriaceous leaves, 8.5–20 by 0.6–2 cm sheathing at the base. Inflorescence is branching, bearing 5–12 small, yellowish flowers with transverse brown bars, 6 mm across. The lip is white, with fine purple spots (Vaddhanaphuti 2001). It flowers in December in peninsular India (Santapau and Kapadia 1966) and in November to December in Thailand (Vaddhanaphuti 2005). *Acampe carinata* occurs throughout Thailand, and is also found in Myanmar, Sikkim, Mumbai, the Western Ghats and Sri Lanka.

Herbal Usage: The entire plant is used in rural Thailand as a tonic to strengthen the body (Chuakul 2002). Root paste is applied externally on scorpion and snake bites in the eastern peninsular Indian state of Orissa. Here, leaf paste is consumed with a clove of garlic daily for 7 days to obtain relief from chest or epigastric pain (Dash et al. 2008). At Uttarakhand in Western Himalaya, *A. carinata* is used to treat rheumatism, sciatica and nerve pain (Jalal et al. 2008)

Acampe multiflora (Lindl.) Lindl. (see **Acampe rigida Hunt**)
Acampe papillosa Lindl. [see **Acampe praemorsa** (Roxb.) Blatt. & McCain]

Acampe praemorsa (Roxb.) Blatt. & McCain

syn. *Acampe papillosa* Lindl., *Acampe wightiana* (Lindl. ex Wight) Lindl.

Fig. 7.1 *Acampe praemorsa* (Roxb.) Blatt & McCain. Reproduced with permission from *Introductions to Orchids* by Abraham and Vatsala, Parlode, Thiruvananthapuram: Tropical Botanic Garden and Research Centre (TBGRI), 1981

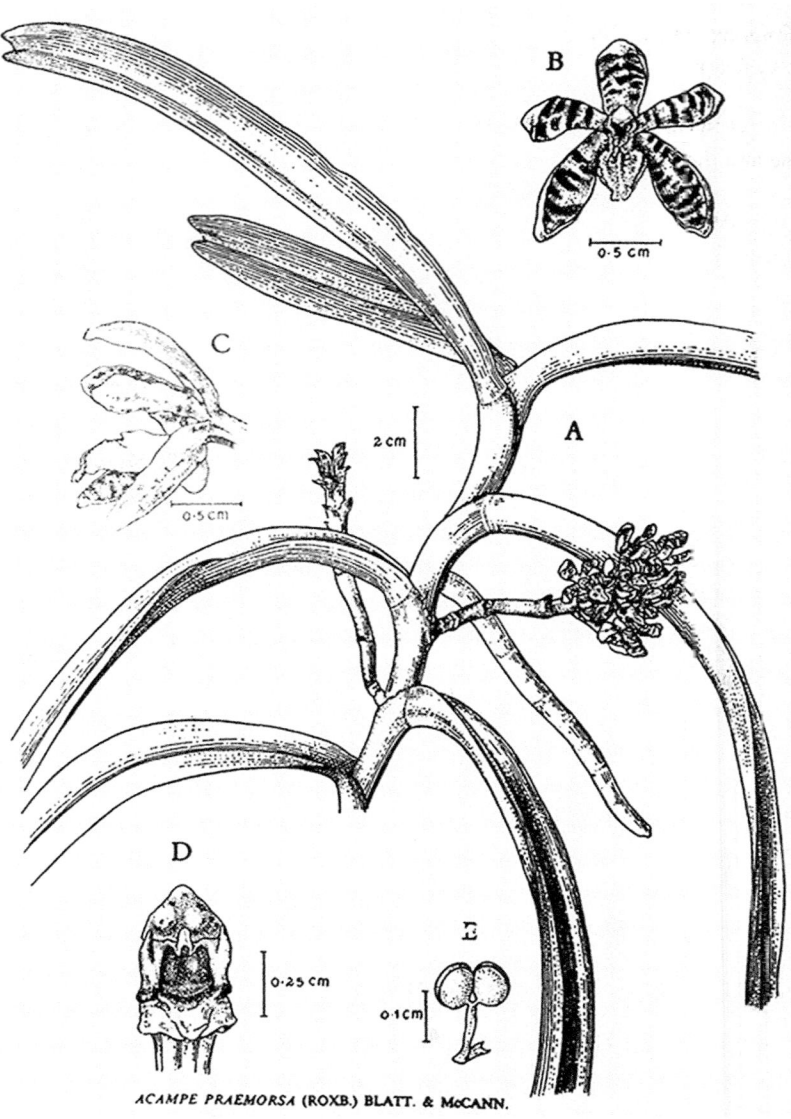

ACAMPE PRAEMORSA (ROXB.) BLATT. & McCANN.

Indian names: *Marabale* in the Canarese dialect, *Maravasha, Khanbher, Nakul, Rasna* (Marathi), *Taliyamaravada* (Malayanam), *Rasna* (Nakuli), *Kano-kato* (Orissa), *Gandhata* (Sanskrit, Malayanam)

Nepali names: *Parajivi* (name is not specific and is widely applied to epiphytic orchids)

Chinese name: *Duanxucui Lan* (short crispy orchid)

Myanmar name: *Mee ma long pan*

Thai name: *Chang saraphi noi*

Description: *A. praemorsa* is a large, robust, monopodial epiphyte with a stout stem, up to 30 cm by 1–1.5 cm in diameter. Leaves are distichous, thick, coriaceous, channeled, 10–30 by 2–3 cm, which appear to be bitten off at the tip (*praemose*). The plant produces several short inflorescences simultaneously, each 3–4 cm long, bearing a crowded cluster of 8–12 long-lasting, fragrant flowers that are yellow, spotted, or barred with crimson. The lip is white, caruncled, and sparsely speckled with magenta to dark brown (Fig. 7.2).

Fig. 7.2 *Acampe praemorsa* (Roxb.) Blatt. & McCain [Photo: Bhaktar B. Raskoti]

Different Indian authors have indicated diverse flowering periods. It was reported to flower in April to August in Bombay (Santapau and Kapadia 1966), December to January in Karnataka and Kerala, also in western peninsular India (Rao and Sridhar 2007), May in Nilgiris (Joseph 1982), March to June in southern India (Abraham and Vatsala 1981), June to October in peninsular India (Misra 2007) and September to November in Sri Lanka (Jayaweera 1981). Flowering season is November to January in Thailand (Vaddhanaphuti 2005) and December to January in China (Chen and Wood 2009a).

A. praemorsa is the commonest orchid in peninsular India, distributed from Bengal to both the Eastern Ghats (Tamil Naidu and Andra Pradesh) and the Western Ghats (Kerala and Karnataka). It grows at sea level to 500 m (Abraham and Vatsala 1981). It is also found in Nepal, Bhutan, Sikkim, Assam, Myanmar, Thailand, China (in Yunnan and Hainan), Vietnam, Laos and Sri Lanka on tree trunks in low-lying forests up to an altitude of 700 m. It is often referred to as *A. papillosa* in northern India because the northern variety was thought to be a separate species, although Indian herbalists believed that medicinally there was no difference between northern and southern plants. In the Bombay state of India, *A. praemorsa* occurs in abundance, in epiphytic masses on *Mangifera indica*,

Syzygium spp. and *Terminalia* spp. (Santapau and Kapadia 1966). In Sri Lanka, *A. praemorsa* is epiphytic on roadside rain trees (*Albizzia saman* syn. *Samanea*) at low elevations (Jayaweera 1981). This is an interesting phenomenon because *A. saman* is not a native Sri Lankan tree: it was introduced from Madagascar.

Herbal Usage: *Rasna*, a decoction of the roots of *A. praemorsa,* is a bitter tonic that is considered to be a specific remedy for rheumatism in India (van Rheede 1693; Caius 1936; Trivedi et al. 1980; Rao 2004; Rao and Sridhar 2007). Its usage also extends to the treatment of sciatica, neuralgia, syphilis and uterine disorders in the country. It is sold as *rasna* (Duggal 1971; Rao 2004), or as a substitute for *Vanda tessellata*. However, it has been reported to be therapeutically inert. It is a substitute for sarsaparilla (Caius 1936; Trivedi et al. 1980).

The primitive Dongria Kandha tribe that resides in the Niyamgiri Hills of southeastern Orissa consume a tablespoon of a paste prepared from the roots of *A. praemorsa* and *Asparagus racemosus* (not an orchid) on an empty stomach, twice daily for 15 days, when they suffer from arthritis (Dash et al. 2008). To the south, in Andra Pradesh on the Eastern Ghats, the Koya tribe uses the pulverised plant, mixed with egg white and calcium (sic; presumably referring to the lime that is included in *serai*), to produce a paste for application on fractured limbs to promote healing (Akarsh 2004). In Nepal, the powdered root of *A. praemorsa* (syn. *A. papillosa*) is used to treat rheumatism or to produce a cooling effect (Subedi et al. 2013).

Phytochemistry: *A. praemorsa* contains flavidinin and the phenanthropyran named praemorsin (Anuradha and Prakash 1994a, b).

Acampe rigida Hunt

syn. *Acampe multiflora* (Lindl.) Lindl.

Chinese names: *Duohuacui Lan* (many-flowered, rigid or crisp orchid), *Jiawandailan* (fake 10,000-generation or *Vanda* orchid), *Taiwanhouchun Lan* (Taiwan thick-lipped orchid), *Changyejiawandai Lan* (long-leaved,

Fig. 7.3 *Acampe rigida* Hunt [Photo: Bhaktar B Raskoti]

fake *Vanda* orchid), banana orchid (in Hong Kong); in Taiwan, *Jiao Lan* (fake 10,000-generation or *Vanda* orchid), *pa chio lan*
Chinese medicinal name: *Heishanzhe*
Thai names: *Chaang sarapee, Ueang sarapi, Ueang jed poi*

Description: The stem is short, stout, usually unbranched and entirely covered by leaf bases. Leaves are distichous, 15–40 by 3.5–4 cm, fleshy and leathery. Inflorescence is branched, carrying a crowded cluster of fragrant, orange-yellow flowers marked with crimson bars. Flowers do not open widely (Fig. 7.3). It flowers in August on the Chinese mainland (Chen et al. 1999), in Hong Kong, August to September (Wu et al. 2001), in Taiwan, August to October (Lin 1977), in Thailand, October to January (Vaddhanaphuti 1997) or November to February (Nanakorn and Watthana 2008), in India, June to August (Misra 2007), and in Sri Lanka, September and October (Jayaweera 1981).

This is a common tropical epiphyte widely distributed from East Africa across tropical Asia, generally in lowland forest. In China, it is found in Guangxi, Yunnan and Hainan, on trees or shady cliffs, at the edge of forests and on trees or rocks in Hong Kong and Lantau Island. It is predominantly saxicolous in Taiwan, proliferating into huge clumps (Wu et al. 2002).

Herbal Usage: The Chinese herb *Heishanzhe* (*A. rigida*) is obtained from Guangdong, Guangxi and Yunnan (Wu 1994). Chinese medicinal texts state that its roots and leaves relax muscles and joints, promote blood circulation and relieve pain. *Heishanzhe* is used to treat traumatic injuries and fractures. Leaves are harvested in summer or autumn, and either used fresh or dried and cut into sections for storage. A decoction is prepared with 6–15 g of *Heishanzhe* and consumed. The taste is acrid but "neutral in nature" (Wu 1994; *Zhonghua Bencao*, 2000; Ou et al. 2003). In Laos, leaves were used in making mats (Vidal 1963). In Thailand, the entire plant is used as a tonic to strengthen the body (Chuakul 2002).

Phytochemistry: 4-hydroxybenzoic acid, 4-hydroxybenaldehyde and 4-methoxymethyl phenol were isolated from Acampe rigida (Cakova 2013).

Acampe wightiana (Lindl. ex Wight) Lindl. [see **Acampe praemorsa (Roxb.) Blatt. & McCain**]

Overview
Kano-kato refers to both *A. carinata* and *A. praemorsa* in Orissa, but their usages are different. Root paste of *A. carinata* is used to treat scorpion or snake bites, and a leaf paste is used for pain in the chest or abdomen. A primitive tribe in Orissa uses the root of *A. praemorsa* to treat arthritis (Dash et al. 2008). Elsewhere in India, the principal usage of *A. praemorsa* is to treat rheumatism (van Rheede 1693; Caius 1936; Trivedi et al. 1980; Rao 2004; Rao and Sridhar 2007).

Aqueous extracts of *A. praemorsa* and *A. ochracea* showed inhibitory activity against antibiotic-susceptible, penicillin-resistant and kanamycin-resistant strains of *Escherichia coli* (Chowdhury et al. 2013). However, skin, ear or other infections that might be caused by *E. coli* are not being treated with preparations containing *A. praemorsa*. Medicinal usage of *A. ochracea* has not been reported.

Cyanobacteria are ubiquitious on the aerial but not on the substrate roots of *A. praemorsa* (syn. *A. papillosa*); on the other hand, *Cyanobacteria* are present in the substrate roots of *Dendrobium moschatum* (Tsavkelova et al. 2003). They are phototrophic organisms that produce auxins, and they are capable of stimulating root growth in other plants (Tsavakelova et al. 2005). *Cyanobacteria* are commonly present in freshwater lakes and reservoirs. Under suitable conditions, they can dominate the phytoplankton and cause nuisance blooms. There are anecdotal reports of allergic skin or gastrointestinal reactions to cyanobacteria. Other effects are headache, fever, myalgia, vertigo, blistering of the mouth and pneumonia. Some species of *Cyanobacteria* produce poisons, neurotoxins and saxitoxins that damage the nervous system (van Apeldoorn et al. 2007; Araoz et al. 2009), and also hepatotoxins, which damage the liver (Zurawell et al. 2005). Human fatality has been reported from ingesting cyanobacteria growing in a golf-course pond (Stewart et al. 2006). However, apart from possible skin allergy, it is unlikely that contact with orchid *Cyanobacteria* will cause any serious problem.

Flavidirin and praemorsin are present in *Acampe praemorsa*; 4-hydroxybenzoic acid, 4-hydroxybenzaldehyde and 4-methoxymethyl phenol are present in *Acampe rigida*. Their medicinal roles, is any, have not been demonstrated. *Acampe* species are not under threat.

Genus: *Acriopsis* Reinw. ex Bl.

Acriopsis is a genus of small, sympodial epiphytes with short, thick rhizomes, ovoid pseudobulbs bearing 2–4 lanceolate, glabrous leaves. Inflorescence is terminal, loosely many-flowered but only a few flowers open at any one time. Flowers are insignificant but they have the unusual characteristic of fused lateral sepals, whereas the column and the lip are joined to form a long slim tube. The name is derived from Greek *acris* (locust) and *opsis* (resembling).

There are nine species in the genus distributed from Sikkim and Assam in India across Southeast Asia to the Solomon Islands and Queensland, Australia. They are common in lowland forests and on roadside trees throughout Southeast Asia.

Fig. 7.4 *Acriopsis liliifolia* (J. Konig) Seidenf [Photo: E.S. Teoh]

Acriopsis liliifolia (J. Konig) Seidenf.

syn. *Acriopsis javanica* Reinw. ex. Blume

Malay Names: *Anggerek darat* (river bank orchid), *Sakat Ubat Kepialu* (medicinal epiphyte for severe fever), *Pemolek*

Description: Pseudobulbs are clustered, ovoid, 2.5–5 cm long, up to 1 cm in diameter, with 2–3 narrow, thin leaves, up to 20 by 1.2 cm near the top. Inflorescence is arching, branched, up to 40 cm long with many well-spaced flowers which face in different directions. The small pink flowers resemble insects in flight with outstretched wings. This appearance is brought about because the flower is tetramerous, the perianth consisting of two very narrow petals that stretch out horizontally and an erect dorsal sepal and two narrow lateral sepals that are fused along their length and arranged vertically, thus resembling the body of an insect. The white lip lies anterior to the petals and sepals (Fig. 7.4). Peak blooming season is March to May but it can flower throughout the year. Ants often build gardens around its pseudobulbs. It is thought that lipids on the seed coats of the orchid attract ants that assist in their dispersal. Such plants are called myrmecochores (Benzing and Clements 1991). *Acriopsis liliifolia* is a small, common, lowland, epiphytic orchid that is widely distributed from Sikkim, Myanmar (in Tenasserim) and Thailand, through Malaysia, Indonesia, the Philippines and New Guinea to the northern tip of Queensland and the Solomon Islands (Seidenfaden and Wood 1992).

Herbal Usage: A decoction of the leaves and roots was used as an antipyretic in Malaya (Ridley 1907; Burkill 1935). Alvins, who collected the information around Malacca between 1884 and 1888, reported that the decoction was taken for any prolonged or severe fever which the Malays called *kepialu* (Burkill 1935). A similar usage was subsequently reported from India (Duggal 1971). In Malacca, *A javanica* was used to treat headaches, whereas in Indonesia, juice from the pseudobulbs was dropped into the ear to cure earache or tinnitus, and pulverised pseudobulb was plastered on the head or abdomen to treat fever and hypertension (van den Brink 1937). Roots are used for treating rheumatism in the Western Ghats in India (Rao 2004).

Overview

Employment of *A. liliifolia* to treat fever by Malays and Indonesians during the nineteenth and the first half of the twentieth century most probably originated from a similar usage in India. Dropping juice of heated orchid pseudobulbs into the ears to treat earache was similarly a common practice from India to Malaya and Indonesia.

A. liliifolia lacks horticultural value and is not endangered. There is no chemical or pharmacological information on *Acriopsis*. It would be interesting to investigate whether this common orchid possesses any antimicrobial activity against microbes (viruses, bacteria or plasmodium).

Genus: *Aerides* Lour.

Chinese name: *Zhijia Lan*

Aerides is a genus of attractive, monopodial, epiphytic orchids with elongated, pendulous

stems that produce many offshoots near the base, thus forming large clumps when well established. Leaves are tough, duplicate and arranged in two rows. Old leaves turn reddish-brown at the base where they sheath the stems. Inflorescence is lateral, arching or pendulous, and many-flowered. Flowers are medium-sized with widespread petals and sepals, a lip with three lobes, and a prominent, hooked spur. There are some 20 species distributed from Sri Lanka and India eastwards through the Asian tropics.

The generic name, 'children of the air', is derived from Greek *aer* (air) and *eides* (resembling), referring to its epiphytic nature and the way such orchids are cultivated. Joao de Loureiro, a Jesuit missionary who described several important orchid species in his *Flora Cochinensis*, coined this name when he saw *Aerides* flowering in wooden hanging baskets in Annam (Fig. 7.5).

Aerides crispa Lindl.

Description: *A. crispa* is a large, tough, robust epiphyte. Stem is stout, erect, reaching up to 1.7 m in length, 1–2 cm in diameter, and of a dull purple or brownish-violet, with spreading leaves, 12–20 by 4–6 cm that are widely separated from one another. Leaves are thickly coriaceous, oblong, with two unequal lobes at the apex and sheathing at the base. Young leaves are typically covered with purple spots. Inflorescence is up to 35 cm long, drooping, branching, loosely many-flowered. Flowers are 5 cm across, white, tinged with rose-purple at the tips of the sepals and petals. Lip is large, fringed with a large patch of bright cerise over the mid-lobe. Flowers smell of pineapple. In southern India, it flowers from May to June (Santapau and Kapadia 1966; Abraham and Vatsala 1981), but at Nilgris, April or May to November (Joseph 1982); in Myanmar, June and July, the flowers lasting for 2–3 weeks (Grant 1895; Christensen 1993). *A. crispa* is distributed in Indian Himalaya, the Western Ghats and in Myanmar. It has become rare in the Western Ghats because of overcollection on account of its showy

Fig. 7.5 *Aerides crispa.*
From: Wight, R, Icones
Plantarum Indiae
Orientalis, vol. 5 (1):
t. 1677bis (1346). Drawing
by Govindoo. Courtesy of
Missouri Botanical
Gardens, St Louis, USA

flowers (Santapau and Kapadia 1966). Introduced into cultivation in the west over 200 years ago, *A. crispa* was the most popular species until the discovery of *A. lawrenceae* which has now totally eclipsed the former species as the top horticultural species (Cootes 2001).

Herbal Usage: Ear-drops, prepared by boiling the pulverised plant in neem oil, are instilled 2–3 drops at a time into the ear every night to treat earache in the Western Ghats (Rao 2004).

Phytochemistry: *A. crispa* contains aeridin, a bactericidal phenanthropyran. (Anuradha and Prakash 1998; Singh and Duggal 2009). Nevertheless, the contribution of aeridin to the management of earache is undetermined.

Aerides falcata Lindl. & Paxton

Chinese name: *Zhijia Lan*
Thai names: *Ueang Kulaab Krapao Perd*

Description: A showy, monopodial, epiphytic orchid which forms large clumps on trees, with stems that may reach 1.6 m in length. When it blooms, *A. falcata* produces numerous sprays of extremely fragrant, white flowers, about 30 to a spray,. *Falcata* (*falcate* means sickle-shaped) describes the side lobes of the lip that are stretched out, distinguishing this species from that other equally fragrant species, *A. odorata*, whose lip is folded over the column. *A. falcata* is distributed throughout Thailand, Indochina and Myanmar (Tenasserim), but not further south. The cultivated plant blooms well in the lowlands. Flowering season is April to June (Kamemoto and Sagarik 1975; Vaddhanaphuti 2001).

Herbal Usage: In Vietnam, it is fed to weak infants as a tonic. Its seeds are sprinkled on boils and other skin disorders to help heal the lesions (Lawler 1984).

Fig. 7.6 *Aerides multiflora* Roxb [Photo: Bhaktar B. Raskoti]

Aerides multiflora Roxb.

Thai name: *Uang Kulap Malai Daeng*

Description: This is a beautiful, robust, epiphytic species of horticultural importance. Stem is 25–30 cm tall; leaves oblong, distichous, deeply channeled, 12–34 by 1.3–3.5 cm. Inflorescence is long, compact and carries up to 50 well-arranged, purplish flowers. Petals and sepals are white, spotted with purple near the base, and flushed with purple at the tip. Lip is purple (Fig. 7.6). Flowering season is March to June in Bhutan (Gurung 2005), April to May in Thailand (Kamemoto and Sagarik 1975; Vaddhanaphuti 1997) and May to July in Nepal (Raskoti 2009). It is widely distributed from the Himalayan foothills through Nepal, Bhutan, Sikkim and Assam to Myanmar and Thailand.

Herbal Usage: *A. multiflora* is used to treat wounds in India (Rao 2004). In Nepal, leaf paste is also applied to cuts and wounds (Pant and Raskoti 2013), whereas powdered leaf constitutes a tonic (Subedi et al. 2013). The tubers exhibit an antibacterial effect in vitro (Singh and Duggal 2009). What needs to be demonstrated is that the leaves possess similar antimicrobial effects.

Aerides odorata Lour.

Common name: Fragrant *Aerides*
Chinese name: *Xianghuazhijia Lan* (fragrant flowered *Zhijia* orchid)
Indonesian names : *Angkrek Lilin, Lau Bintang* in Kalimantan
Thai name: *Ueang Kulaab Krapao Pid*
Indian name: *Hameri* in Orissa

Description: This is a widespread, variable species of *Aerides* which grows into a magnificent clump if it is well anchored in the crotch of a tree, and especially if it receives direct sunlight for half a day and is located near water. The unusual brownish coloration at the stems and leaf bases of *Aerides* distinguishes it from strap leaf *Vanda* when the *Aerides* is not in bloom.

Stems are droopy, stout, up to 1.8 m tall, freely branching. Leaves are thick, leathery, unequally bilobed at the tips, 15–20 by 2.5–4.6 cm. Inflorescences are numerous, appearing simultaneously, racemose, nodding, 15–30 cm long with 20–30 fragrant flowers that open widely, and are white to pink, tipped or spotted with purple. Spur is greenish-yellow (Fig. 7.7). Flowering period in China is May (Chen and Wood 2009a, b, c, d, e, f), in Nepal,

Fig. 7.7 *Aerides odorata* Lour [Photo: E.S. Teoh]

and nose (Rao 2004). Vietnamese herbalists believe that, if seeds are sprinkled over the lesions, they help to heal boils and other skin disorders (Lawler 1984). Hill tribes in Orissa combine the fresh root of *A. odorata* with root powder from *Saraca asoca*, bark from *Azadirachta indica* and common salt to prepare an oral medicine for painful swollen joints. They also use juice from the leaves to treat tuberculosis (Dash et al. 2008). In Nepal, a poultice prepared from the leaves is applied over cuts and wounds (Pant and Raskoti 2013).

Overview

Aqueous extract of *A. odorata* exhibits inhibitory activity against antibiotic-sensitive, penicillin-resistant and kanamycin-resistant strains of *Escherichia coli*, common organisms in stools, on skin and in superficial infections. Phytoalexins such as aeridin possess antimicrobial effects. These findings lend support for the principal usage of the various medicinal species of *Aerides,* which is to prevent and treat local infections (wounds, boils, other skin disorders and earache). Nevertheless, effectiveness in treatment of infections would depend on the potency of the associated phytoalexin and how it is delivered.

May to July (Raskoti 2009), in Bhutan, March to June (Gurung 2006) and in Singapore, August to September Flowers last for 2 weeks and are easily recognised by the funnel-shaped lip which extends into a horn-like spur.

A. *odorata* occurs in southern China, Nepal, India, Myanmar, Thailand, Indochina, Malaysia, Indonesia and the Philippines from sea level to 2000 m. According to Sagarik and Kamemoto (1975), plants in northern Thailand, and presumably those in China, are tetraploid whereas those from the south are diploid. Tetraploid plants have erect, twisted stems with shorter, thicker, sturdier leaves. Floral scape is similarly more erect, but the waxy, fragrant flowers are similar. Flowers are white to mauve or lavender, extremely fragrant and produced in abundance.

Herbal Usage: Fallen fruits of *A. odorata* are used to heal wounds in India. Juice extracted from the leaves is used to treat boils in the ear

An oral Indian preparation for treating painful, swollen joints contains four herbal products, one of which is *A. odorata* (Dash et al. 2008). Any of the four herbs might possibly contain a salicylate (the basis of Aspirin®). Orchids have been shown to have a salicylic acid-related defence mechanism that helps them to respond to viral invasion (Lu et al. 2012).

There are no data on the phytochemistry on medicinal *Aerides* species, but some data have been published on non-medicinal *A. rosea* Lodd. ex Lindl & Paxton. In addition to gigantol, imbricatin, methoxycoelonin and coelonin, it contains five minor constituents, namely a phenanthropyran, two phenenthrenes and two dihydrophenenthrene derivatives. The two newly described phenanthrene derivatives are aerosanthrene (5-methoxyphenenthrene-2,3,7-triol) and aerosin (3-methoxy-9,10-dihydro-2,5,7-phenanthrenetriol) (Cakova et al. (2015).

Five species of *Aerides* occur in China, but there is no record of any being used medicinally.

Genus: *Agrostophyllum* Bl.

Chinese name: *Heye Lan*

Agrostophyllum is a genus of epiphytic orchids with clustered, erect stems, with a single thin, flat leaf at each internode. Flowers are small, resupinate, numerous, white or yellow and self-pollinating. There are with 40–50 species distributed from the Seychelles across tropical Asia to the Pacific. There is one species in China, *A. callosum* Rchb. f., but it is not used as medicine (Li et al. 2000).

The generic name comes from Greek, *agrostis* (grass) and *phyllon* (leaf), alluding to the leaves of many species in this genus. *Agrostophyllum* has no horticultural value and is almost unknown in cultivation (Yong 1990).

Agrostophyllum bicuspidatum J.J. Smith. [see **Agrostophyllum stipulatum ssp. bicuspidatum (J.J.Sm.) Schuit.**]

Agrostophyllum stipulatum ssp. *bicuspidatum* (J.J.Sm.) Schuit.

syn. *Agrostophyllum bicuspidatum* J.J. Smith

Description: Stems are close to one another, 15–40 cm in length, with oblong leaves 4 by 1 cm. Inflorescence is apical, carrying a head of single-flowered spikes. Flowers are 0.6–1 cm across, white or a pale yellow, and they open widely. Petals are very narrow, 3 mm long, the ends curving backwards. Sepals are 3–4 mm long, broad, the upper erect and concave, the lateral ones forming a broad mentum. Lip is sac-shaped at its base (Comber 2001).

A. stipulatum ssp. *bicuspidatum* is found in lowland forests at 300 m in Sumatra, Java, Sarawak and Sulawesi, peninsular Malaysia and southern and upper northeastern Thailand.

Medicinal Usage: The Kalabit in Sarawak wear parts of the orchid as talismans to protect against curses (Christensen 2002).

Phytochemistry: No phytochemical investigation has been conducted on *A. stipulatum* ssp. *bicuspidatum* (J.J.Sm.) Schuit., but terpenoids, stilbenoids and derivatives have been isolated from Indian species in the genus: i.e. a naturally occurring phenanthropyran derivative, agostrophyllin from *A. khasiyanum* (Majumder and Sabzabadi 1988); two stilbenoids (agrostophyllol and isoagrostophyllol) and two diasteromeric 9,10-dihydrophenanthropyran derivatives from *A. callosum* Rchb. f. (Majumder et al. 1995, 1996; Majumder et al. 1998; Majumder et al. 1999); and two terpenoids (agrostophyllinol and agrostophylline) from *A. brevipes* Ridley and *A. callosum* Rchb. f. (Majumder et al. 2003).

Overview

Talismans and charms are very much a part of native medicine. They can be viewed as items of health promotion or preventive medicine; on the other hand, they may also be used to treat illnesses that are considered by native healers to be caused by spirits. This is not the only orchid used in this manner in Southeast Asia (see *Dendrobium crumenatum*). When such treatment did not work, people accepted that the evil spirit taking possession of the patient was too powerful to be put off by the talisman.

Several bio-active compounds (three stilbenoids and several dimeric phenanthrenes) have been isolated from two Indian species, *A. callosum* and *A. khasiyanum* (Majumder et al. 1996; Majumder et al. 1998), but these two orchids are not used medicinally. Nevertheless, with around 100 species in *Agrostophyllum*, several of which are large plants, this genus appears to be a good subject for phytochemical research. From a medicinal perspective, stilbenoids, phenanthrenes, alkaloids and other phytochemicals are important because many have been found to possess antimicrobial, antiprotozoal, antihelminthic, anti-inflammatory, antiplatelet and spasmolytic properties, are cytotoxic against specific human cancer cell lines, or are capable of protecting tissues against toxic damage by chemical compounds (Kovacs et al. 2007). In the *Maxillariinae* species of

Fig. 7.8 Compounds
isolated from
Agrostophyllum

Fig. 7.8 Compounds isolated from *Agrostophyllum*

R = H; Agrostophyllin
R = OMe; Agrostophylloxidin

R = H; Agrostophyllidin
R = Me; Callosinin

Agrostophyllol (5-OH *axial*)
Isoagrostophyllol (5-OH *equatorial*)

R = H; Agrostophyllone
R = Me; Agrostophylloxin

Agrostophyllanthrol (5-OH *equatorial*)
Isoagrostophyllanthrol (5-OH *axial*)

R_1 = Me, R_2 = H; Agrostonin
R_1 = H, R_2 = Me; Agrostonidin

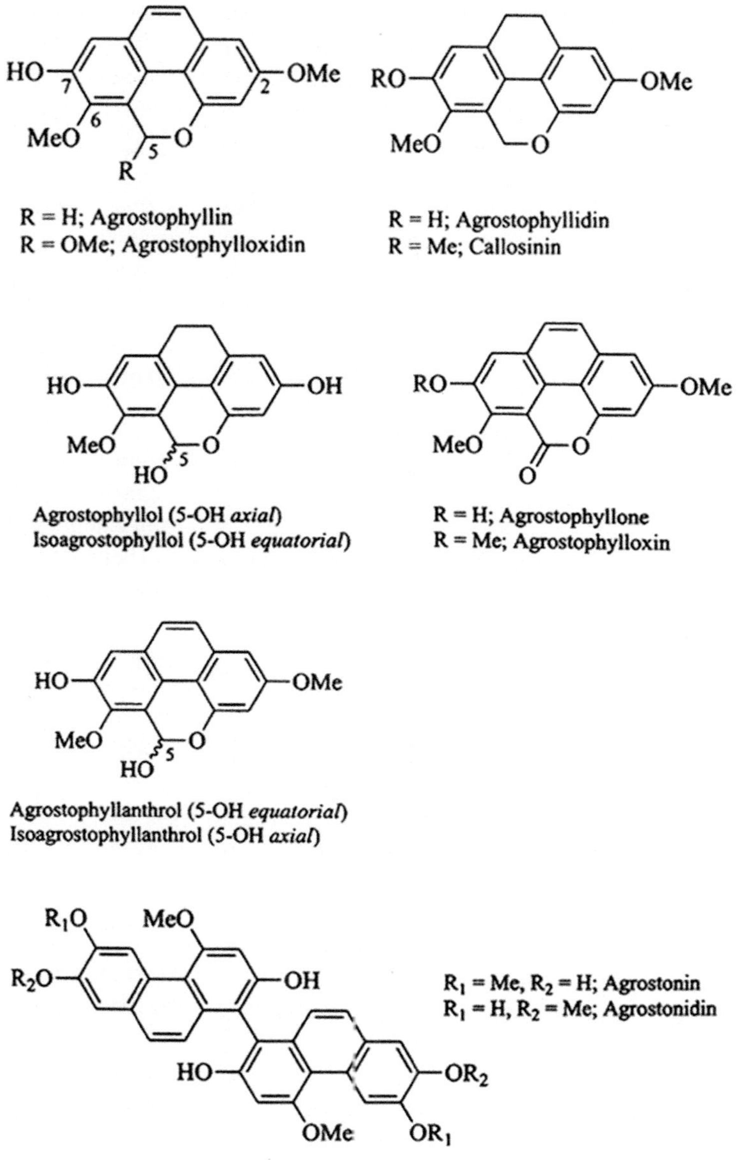

South American orchids, triterpenoids are the major compounds present in the labellar secretions that constitute the reward for bees attracted to the non-fragrant flowers (Fig. 7.8) (Flach et al. 2004).

Genus: *Amitostigma* Schltr.

Chinese name: *Wuzhu Lan* (no pillar orchid)

Amitostigma are small, montane, terrestrial orchids of the Himalayas, China and Japan with flowers that resemble those of *Habenaria*. There is one species in Thailand and one in Vietnam (Schuiteman and de Vogel 2000). Many species are found on wet mossy rocks or in humus-covered soil in forests and meadows or on hill slopes and cliffs. Plants are small, with spheroid, subterranean tubers and short stems that bear one or two ellipsoid, glabrous leaves, which ensheath the stem at the base. Flowers are also small, resupinate, trilobed, and borne on a slim, tall, erect inflorescence. Many Chinese species have pink to purple flowers. The exceptional species is *Amitostigma simplex* with its large, yellow flowers (Chen et al. 1999).

The generic name is derived from three Greek words, *a* (not) *mitos* (thread) and *stigma* (stigma).

Amitostigma chinense (Rolfe) Schltr. [see *A. gracile* (Blume) Schltr.]

Amitostigma gracile (Blume) Schltr.

syn. *Amitostigma chinense* (Rolfe) Schltr.

Chinese names: *Xitingwuzhu Lan* (slim standing no pillar orchid), *Xiewuzhu Lan* (slim standing no pillar orchid), *Huawuzhu Lan* (no pillar/ column orchid)
Chinese medicinal name: *Duyeyizhiqiang*

Description: Tubers are ovoid, globose, bearing a single, oblong-elliptic, membranous leaf, 3.5 by 1.5 cm, which sheathes the slim erect stem at the base. Inflorescence is terminal, lax and carries 5–12 small light pink to purple flowers which resemble *Habenaria*. Mid-lobe of the lip is shaped like a butterfly. Flowering season is June and July (Chen et al. 2009a, b).

Rolfe discovered the species in 1909 at 3800 Steps Pass at 899 m, a small herb growing on moss-laden hill slopes. Subsequently, von Schlechter (1919) reported that it was present in Fujian and Jiangsu. *A. gracile* is distributed from Guangxi northwards in eastern China through Guizhou, Hunan, Jiangxi, northern Fujian, Zhejiang, Anhui, Hubei, Sichuan, Shaanxi, Henan, Jiangsu, Shandong, Hebei and Liaoning to Korea, Japan and Taiwan. It grows on damp, rocky soils in forests, valleys and crevices at 200–3000 m (Chen et al. 2009a, b).

Herbal Usage: Although *A. gracile* and *A. chinense* are regarded by botanists as a single species, in *A Concise Edition of Medicinal Plants in China*, Wu Xiu Ren (1994) described them as two separate species despite their usage being similar. Separate origins or differences in

Table 7.1 Herbal Usage of *Amitostigma gracile* (*Duyeyizhiqiang*)

1. Indications: Detoxification, relief of swelling and haemostasis   Boil whole plants 30–60 g. for consumption. For external application, grind fresh stems and roots.
2. Indication: Venomous snake bite   Grind roots and stems and mix with rice water for application
3. Indication: External injuries, haematemesis   Prepare decoction with fresh whole plants, 30–90 g, for consumption
4. For dysmenorrhea and metrorrhagia   Prepare decoction with 9–15 g of dried herb

vegetative form might have led herbalists to distinguish between the two. The herb is obtained from Zhejiang Province (Wu 1994).

Their alleged properties are that they are cool, antitoxic and with an ability to reduce swellings and arrest bleeding. The whole plant together with its roots is used to treat snake bites, as an antidote for traumatic injury, or to treat dysmenorrhea and menstrual irregularities (*Zhonghua Bencao*, 2000). Several prescriptions for the use of *A. gracile* (*Duyeyizhiqiang*), and shown in Table 7.1, were originally from the *Zhejiang Commonly Used Folk Herbs* (*Zhongyao Da Cidian*, 1986).

Amitostigma pinguicula (Rchb.f. & S.Moore) Schltr.

Chinese names: *Dahuawuzhu Lan* (big flower no pillar orchid)

Description: *A. pinguicula* is a larger plant, though still small, with ovoid tubers 10–15 mm in diameter. Leaf is single, linear to narrowly elliptic, 1.5–8 by 0.6–1.2 cm. Flower is rose red to purplish-red (Fig. 7.9). It is found on rocky soils in forests, valleys and moist grasslands at 200–400 m in northeast Zhejiang. It flowers from April to May (Chen et al. 2009a, b).

An endemic species, it grows on rocky soils in wet grasslands, forests and valleys at 200–400 m in northeast Zhejiang (Chen et al. 2009a, b).

Herbal Usage: The Chinese herbal name, *Duyeyizhiqiang*, also refers to this species of

Fig. 7.9 *Amitostigma pinguicula* (Rchb.f. & S.Moore) Schltr. [Photo: Courtesy of Plant Photo Bank of China]

Fig. 7.10 *Am.tostigma simplex* Tang & F.T. Wang [Photo: Liu Ming]

Amitostigma despite the fact that the flowers of *A. pinguicula* and *A. gracile* are quite different (*Zhonghua Bencao*, 2000). The entire plant is used in preparing medicine. It is used for detoxification and used to reduce noxious swellings, in the treatment of trauma and snake bites, as an antidote for poisons and to treat haemetemesis. The medicinal plant is cultivated in Fujian, Zhejiang, Hubei and Sichuan (Wu 1994; *Zhonghua Bencao*, 2000).

Amitostigma simplex **Tang & F.T. Wang**

Chinese name: *Huanghuawuzhu Lan* (yellow flower no pillar orchid)

Description: The yellow-flowered *A. simplex* has the biggest flower with a tri-lobed lip that is 1–1.6 cm in length and 2.5 cm wide. A small plant, its tubers are ovoid, 4–5 mm in diameter (Fig. 7.10). The single leaf is linear to oblong-elliptic, 1.5–4 by 0.3–0.6 cm. It is endemic to China where it occurs on grassy slopes above 2300–4400 m in western Sichuan and Southwestern Yunnan. It flowers in July (Chen et al. 1999, 2009a, b). This plant is on the 2006 IUCN Red List of Threatened Species.

Herbal Usage: Similar to *Amistostigma pinguicula*.

Overview

In describing Item 7057 in his *Concise Edition of Medicinal Plants in China*, Wu Xiu Ren used the scientific name, *Amitostigma pinguicula*. He listed two Chinese names: (1) *Dahuawuzhu Lan* which means 'big flower no-pillar orchid or 'big flower *Amitostigma*' the local name for the red to purple *A. pinguicula*; and (2) *Huanghuawuzhu Lan* (yellow flower no pillar orchid) which refers to yellow-flowered *A. simplex*. In fact, the latter species has the largest flower in the genus. The sources of the medicinal herb are stated as: Zhejiang, Fujian, Hebei and Sichuan provinces. *A. pinguicula* is only found in northeast Zhejiang, on rocky soils in forests, moist grasslands and valleys at 200–400 m. *A. simplex* is found in grassy slopes at 2300–4400 m in western Sichuan and northwest Yunnan (Chen et al. 2009a, b). Item 7057 therefore consists of at least two species of *Amitostigma*. Nevertheless, since the medicinal usage of the two or three species of the two rare, endemic mountain orchids is identical, their correct botanical identification is not crucial. In the recent *Zhonghua Bencao* (2000), the medicinal name *Duyeyizhiqing* covers two species, namely *A. gracile* and *A. pinguicula*, but the *Herbal* makes no mention of *A. simplex*. Since *A. pinguicola* is endemic and present only in a small area in northeast Zhejiang, collection of the herb for medicinal usage is not sustainable. Thus, medicinal *A. pinguicola* is cultivated in Fujian, Zhejiang, Hubei and Sichuan (Wu 1994). Although *A. simplex* has similar medicinal usages, it is not cultivated because its native habitat is the Gaoligongshan Mountains at 2300–4400 m and a similar environment is difficult to replicate, whereas *A. pinguicola* occurs at 200–400 m in a subtropical, coastal province. Amitostigma being endemic in continental China, its medicinal usage is only described in the Chinese mainland.

The search for records of pharmacological investigation was unsuccessful.

Genus: *Anacamptis* Rich.

The genus derives its name from Greek *anakamptein* (to bend back), possibly referring to the shape of the slender spur at the base of the lip or to its reflexed pollinia (Alrich and Higgens 2008). Plants have the habit of *Orchis*, each with two subterranean, globose tubers that resemble testicles (Fig. 7.11). The plant produces a rosette of leaves in autumn which lasts through winter and spring, then senesces with fruit set so that the plant is leafless during the summer heat (Neiland 2001). A dozen species are distributed in montane meadows and grasslands in the northern Iran, the Middle East and southern and central Europe. Tubers of *Anacamptis* are harvested to make *salep*, once thought to be an aphrodisiac and super-nutrient throughout Europe.

Anacamptis coriophora R.M.Bateman, Pridgeon & M.W. Chase

Common name: Bug orchid

Description: Tubers are paired, globose and sessile. Plants are generally 20–40 (up to 60)

cm tall. Stem is gabrous, sheathed with scale leaves below and bearing a few lanceolate leaves, without spots, 10 cm long, arranged in a rosette. Inflorescence is densely many-flowered (20–40). Bracts are longer than the ovaries, with a green centre bordered in white. Flowers are small, green to deep wine-red. Tepals are joined to form a hood above the column. Lip is shaped like a broad trident which is white and spotted or splashed with red patches centrally, and purple on the distal halves of the lobes. Flowering season is for 2 months in summer, from early May to August, depending on region.

Commonly known as the bug orchid because flowers of many strains possess an unpleasant smell, this is a widespread European species which is distributed from the British Isles across continental Europe to the Middle East and northern Iran. *A. coriophora* occurs in pine forests, dry meadows, dunes and river banks from sea level to 1500 m but it may be found at elevations of up to 2000 m (Neiland 2001).

Phytochemistry: When challenged with *Rhizoctonia repens, A. coriophora* produces *p*-hydroxybenzyl alcohol, an aglycone of gastrodin. This compound has been isolated from many terrestrial and mycoheterotrophic orchid species (Stoessl and Arditti 1984). In rats, gastrodin and *p*-hydroxybenzyl alcohol facilate memory consolidation and retrieval but not its acquisition (Hsieh et al. 1997). By suppressing dopiaminergenic and serotobergic activites, *p*-hydroxybenzyl alcohol improves learning when rats are evaluated with an avoidance test (Wu et al. 1996). Anti-oxidant-related gene expression on the rat brain induced by exposure to *p*-hydroxybenzyl alcohol may explain the compound's ablity to reduce focal ischaemic brain injury in rats (Yu et al. 2005; Kam et al. 2011). This stroke-protective effect which is unique to 4-hydroxybenzyl alcohol is thought to involve the induction of protein disulfide isomerase because it is blocked by bacitracin (Descamps et al. 2009). At doses of up to 200 mg/kg, *p*-hydroxybenzylalcohol was found to be devoid of neurotoxicity (Descamps et al. 2009).

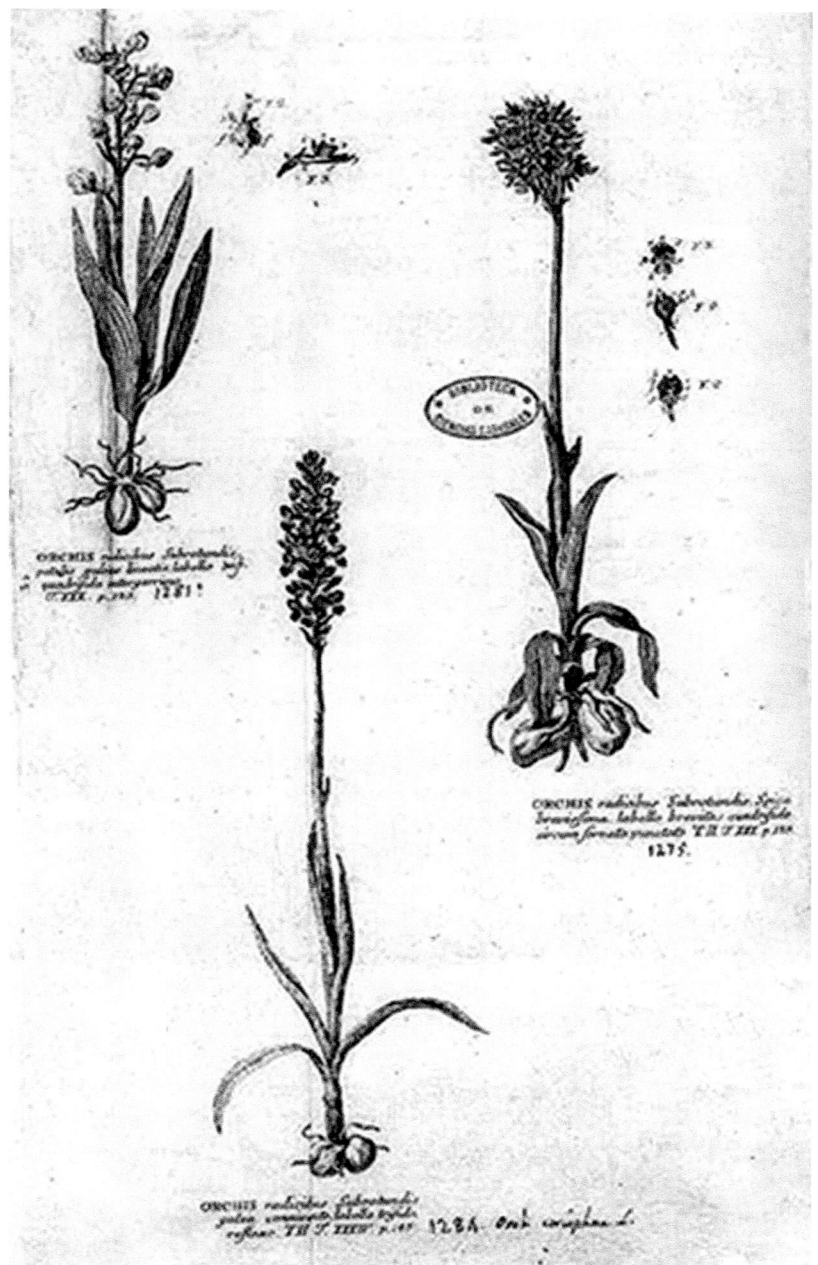

Fig. 7.11 *Anacamptis coriophora* (L.) R.M. Bateman, Pridgeon & M.W. Chase. From: Haller, A von, *Historia stirpium indigenaarum Helvetiae inchoata* vol. 2: t. 34 (1768). Courtesy of Real Jardin Botanico, Madrid, Spain

p-Hydroxybenzyl alcohol may also have a role in skin whitening because it inhibits tyrosinase, the enzyme that catalyses the formation of the skin pigment, melanin (Liu et al. 2007). In co-cultures of melanocytes and keratinocytes (the major component of skin cells), the effect of hydroxybenzyl alcohols (HBAs) was even more evident. HBAs exhibited low toxicity in vitro (Liu et al. 2008). In the cosmetic industry, tyrosine inhibitors are regularly used in the preparation of skin whiteners.

Herbal usage: Tubers of *A. coriophora* are harvested in Iran for use as *salep* (Ghorbani et al. 2014a, b).

Anacamptis laxiflora (Lam.) R.M. Bateman, Prigeon & M.W. Chase

syn. *Orchis laxiflora* Lam.

Common name: Jersey Orchid
Indian name: *Salep misri* (Bengal), *shala misriri* (Madras Presidency in 1933)

Description: Plant is slender, erect, 60 cm tall with a basal rosette of 3–8 narrow leaves, 10–15 by 1–2 cm, and two ovoid tubers that resemble testicles, the current developing tuber being larger than the previous year's. Ten to twelve pink to purple flowers, are loosely arranged on an erect inflorescence that is 6–25 cm tall; hence, *laxiflora*. Lip has a central streak of pure white extending from the base to the apex. It flowers around May (Pridgeon et al. 2001; Wood and Ramsey 2004). Rare cases of triploidy are present in the species (Pridgeon et al. 2001); such plants would be floriferous but sterile.

Plants produce a rosette of leaves in autumn that remain green and functional throughout winter and spring. Leaves become senescent when the plant flowers. Following fruit set, tubers develop and these allow the plant to live through the summer drought. When conditions are favourable, more than one new tuber develops from buds located at the base of the aerial stem. Vegetative reproduction is important in *Anacamptis* because fruit set is low. However, *A. laxiflora* in the Mediterranean is a weed species and it quickly invades abandoned fields (Pridgeon et al. 2001). It is found in fens and marshy, calcareous meadows with slightly acidic to alkaline soils which are permanently wet, predominantly in the Meditarranean–Atlantic region, *A. laxiflora* also occurs in the Middle East, Iran and Afghanistan. Apparently, in Iran and Afghanistan, grazing goats and sheep avoid this orchid (Lawler 1984).

Herbal Usage: In India, the tubers were used as an expectorant, an astringent and as nourishment (Chopra 1933).

Fig. 7.12 *Anacamptis morio* (L.) R.M. Bateman, Pridgeon & M.W. Chase. [Photo: Henry Oakley]

Anacamptis morio ssp. *picta* (Loisel.) Jacquet & Scaooat

Description: Tubers are paired, globular and sessile. Stem is erect and glabrous, 20–40 cm tall. Leaves are basal, lanceolate, unspotted. Inflorescence is laxly 6- to 25-flowered. Dorsal sepal and petals form a helmet whereas lateral sepals, which are green or purple, "wing" above the helmet. Lip is 3-lobed. Lateral lobes are larger than the mid-lobe and they fold backwards. Central lobe is red-purple to pink and carries a central spur which is convex at the apex. The centre of the mid-lobe is white with purple spots. Subspecies *picta* is smaller than the type species and has fewer flowers, which are either purple or white (Fig. 7.12). Flowering season is February to April, or May to June, depending on location.

A. morio ssp. *picta* (syn. *A. picta*) occurs in dry grassland that is wet in winter and dry in summer, in meadows and pastures, and in calcareous soil that is either neutral or alkaline. It is a common orchid in the Catabrian Mountains of

Spain (Heiningen 2014) and is distributed across Europe to the Middle East and Iran.

Herbal usage: Tubers are harvested in Iran, at the eastern edge of its distribution, for use as *salep* (Ghorbani et al. 2014b). In Europe, *Anacamptis* species are protected.

Anacamptis palustris (Jacq.) R.M. Bateman, A.M. Pridgeon, M.W. Chase

Common name: Bog orchid

Description: A slender plant with two globose tubers; stem is erect, 30–60 cm tall. Leaves are 3–5, narrow, lanceolate, unspotted, drawn up, held vertically, and gutted near at the lower portion to ensheath the stem. Bracts are leaf-like and reddish. Inflorescence is erect, 7.5–25 cm, and laxly several-flowered. Flowers are large, pink to purple. Lip has 3 lobes, 16 mm wide. Spur is convex, white with three vertical stripes. *A. palustris* flowers from May to July. It is distributed predominantly in western Europe but is also found in Greece, the Aegen, Turkey and Iran.

Herbal usage: Tubers are harvested in Iran for use as *salep* (Ghorbani et al. 2014b).

Anacamptis pyramidalis (L.) A. Rich

Description: This is a variable species. Tubers are paired, ovoid, 1.2–5.2 (mean 2.2) long and 0.6–4.0 (mean 1.5) cm in diameter. Plants are terrestrial, 20–65 cm tall, stem erect, with 2–14, narrow, lanceolate, basal leaves, the largest being 6.5–23 by 0.6–2.2 cm, that senescence at flowering. Inflorescence is densely many-flowered, with numerous small, lanceolate, subtending bracts, 4–18 mm long. Raceme is initially pyramidal, gradually becoming ovoid as the flowers open. Flowers are pale to bright red or carmine, sometimes pink, rarely white, 1 cm long. Lip is darker than the other floral segments. Lip is flat, trilobed, all lobes of equal size, oblong side lobes well spread out at an

Fig. 7.13 *Anacamptis pyramidalis* (L.) A. Rich. [Photo: Henry Oakley]

angle of 75° to the vertical (Fig. 7.13). Flowering season is June to July. Butterflies and moths pollinate the flowers. The species is distributed in south and central Europe to Turkey and northern Iran, occurring in sunny meadows or bushy slopes, from lowlands to the foothills, on slightly acidic, neutral or calcareous soil (Hoskovrc 2007; Sevgi et al. 2012).

Herbal usage: Used as *salep* in Iran and Turkey (Ghorbani et al. 2014a). Best grades of *salep* should have a mucin content greater than 40 % and an ash content which is lower than 5 %. *A. pyramidalis* has a mucin content of 44.72 % and an ash content of 1.72 % (Sezik 1967), which places it among the top six most marketable orchids for *salep*.

Tubers are harvested by collectors of *salep* from the Golestan Province of Iran. The main collection areas are the woodlands between Maraveh Tappeh and Golidagh, especially around the village of Aq-Eman in eastern Golestan. Total annual trade in *salep* from Maraveh Tappeh (not restricted to *A. pyramidalis*) in 2013 amounted to 3500 kg of fresh tubers (Ghorbani et al. 2014b).

Phytochemistry: When challenged with *Rhizoctonia repens*, *A. pyramidalis* produces two phytoalexins, orchinol and *p*-hydroxybenzyl alcohol (Veitch and Grayer 2003a, b). Orchinol is bacteriostatic and fungistatic but, being neither bactericidal nor fungicidal, it does not have much pharmaceutical application. *p*-hydroxybenzyl alcohol is neuroprotective and may also have a role as a skin whitener (see *A. coriophora*). Extracts of flowers and above-ground parts of *A. pryamidalis* exhibit anti-oxidant and scavenging capacities in vitro (Stajner et al. 2010).

Overview

Following the classic *Doctrine of Signatures,* which once determined the role of herbs, the spherical tubers of *A. pyramidalis*, *A. laxiflora*, *A. palustris* and many terrestrial Mediterranean orchids which resembled testicles were reputed to possess aphrodisiacal properties. Powder prepared from dried tubers was used to prepare *salep*, a drink that was alleged to boost one's libido and sexual performance. *Salep* drinking boomed during the heyday of the Ottoman Empire when even sultans took to eating halva made with *salep*, but its reputation dates from a much earlier period, the result of an anecdote spun by Theophrastus (371–287 BC) that on one occasion it caused a man to have 70 consecutive acts of coitus (Wedeck 1961). The belief became widespread with the inclusion of orchids as aphrodisiacs in the *Herball of Dioscorides* (Diocorides, 40–90 CE), a text used by Western physicians well into the 19th century. The famous Persian physician Avicenna (Ibn Sina, 980–1037) promoted the *Herball of Dioscorides* and his influence on neighbouring India further reinforced the belief. Unnani (Greek) medicine in India holds a similar egard for *salep*. However, native Indian *salep* is generally constituted with pseudobulbs of *Eulophia*.

Although also patently untrue, it was once believed that *salep* contained the greatest amount of nourishment in the smallest bulk (Culpeper 1653). A small amount of *salep* in a large volume of warm water converted into a jelly-like substance which was believed to be superior to rice. To protect against famine at sea, it was proposed that *salep* should constitute part of a ship's provision at all times (Hooper and Akerly 1829; Hooper 1937). *Salep* was seldom used in the United States, "except in the composition of Castillon powders, a nutritive and bland article of diet for invalids" (Griffith 1847). Orchid tubers (not *salep*) are fed to weak children cut off from other supplies (Hedley 1888), and eaten by Australian aborigines during periods of privation (Low 1987). Indian *salep* made with *A. laxiflora* is still advocated as a source of nourishment. *Salep* contains mainly mucilage. Starch, reducing sugar, nitrogen (0.92 %), moisture and ash are present in small amounts (Sezik 1990).

Salep imported from Turkey would certainly contain *A. pyramidalis* and *A. laxiflora*. *A. pyramidalis* was shown to synthesise orchinol and *p*-hydroxybenzyl alcohol when incubated with *Rhizoctonia repens* (Stoessl and Arditti 1984). This probably also holds true for *A. laxiflora*, although that has not been purposefully demonstrated. Anti-oxidant enzymes (superoxide dismutase, catalase, peroxidase, etc.) with scavenging activity were demonstrated in the flowers and above-ground parts of *A. pyramidalis* (Stajner et al. 2010). However, their presence does not lend support to the alleged aphrodisiac property of *salep* because they have not been shown to exert such action in humans nor in animals.

At one time, Nepal exported about 5 tons of *A. laxiflora* (syn. *Orchis latifolia*) tubers annually and a considerable quantity would have made its way into India. Several million plants would need to be harvested to obtain this amount of tubers. It is claimed that the orchid is replanted, but experts think that the following year's harvest probably comes from the smaller bulbs which were originally ignored (Lawler 1984) because the plump and fleshy (daughter) tubers that are collected are those that would have produced the following year's crop. Shrivelled (mother) tubers with flowering or senescent plants cannot give rise to another plant. Although wild populations have declined due to drainage of wetlands and modern farming practices (Pettersson 1976), currently, members of the genus are not threatened in Europe (Wood and

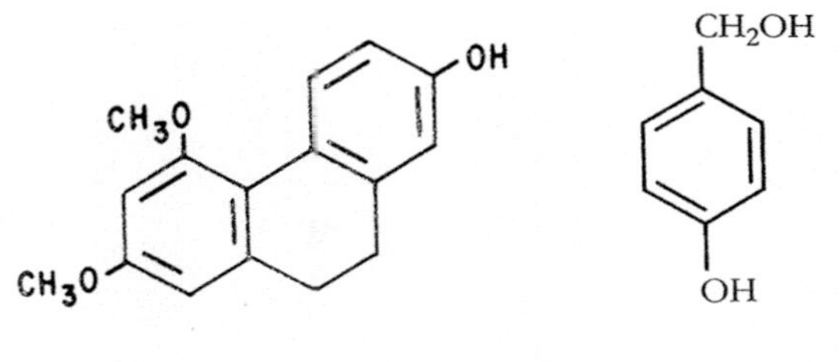

Fig. 7.14 Formulae of Orcinol and *p*-hydroxybenzyl alcohol

Orchinol   *p*-hydroxybenzyl alcohol

Ramsey 2004). Their existence is only precarious in Turkey and Iran (Ghorbani et al. 2014b).

Harriet J. Muir at Kew managed to raise seedlings of *A. laxiflora* through symbiotic germination using mycorhiza obtained from roots of *Orchis morio* and *Dactylorhiza fuchsia*, and also with *Ceratobrasidium corrigerum* from the Commonwealth Mycological Institute. She used this method to provide seedlings for reintroduction into the field (Fig. 7.14) (Muir 1987).

Genus: *Anaphora* Gagnep.

The species mentioned below is the sole species in this genus which belongs to the subtribe *Liparidinae*. It is now classified as *Dienia* (Alrich and Higgins 2008). In Greek the word *anaphora* means 'a carrying back'. It refers to the lip which is adnate (united) to half the length of the column (Schultes and Pease 1963).

Anaphora liparioides Gagn. [see *Dienia ophrydis* (J. Konig) Ormerod &Seiden.]

Genus: *Anoectochilus* Blume

Chinese name: *Jinxian Lan* (gold thread orchid)
Chinese medicinal name: *Jianxianlan* (referring to *Anoectochilus formosanus* and *A. roxburghii*)

A genus of terrestrial orchids, *Anoectochilus* has beautiful, soft, velvety, ovate-lanceolate (oval but terminally pointed) foliage decorated with a network of fine, yellow-orange or "golden" veins. They belong to the Jewel Orchids, so-called because as a group they possess distinctive velvety foliage with attractively coloured veins and/or multi-coloured blotches. Leaves are arranged in a spiral fashion near the apex of the soft, fleshy, succulent stem.

The generic name, *Anoectochilus*, is derived from Greek, *anoektos* (open) and *cheilos* (lip). There are around 40 species in the genus, distributed from Sri Lanka and India eastward across southern China, the Ryukyu Islands and Southeast Asia to the Pacific islands.

Anoectochilus formosanus **Hayata**

Chinese names: *Jinxian Lan* (gold thread orchid), similar to its generic name; *Benshanshisong* (mountain stone pine), *Jinqianzicao* (golden currency notes baby grass), *Shucan Lian* (tree and grass lotus), *Yaowang* (King of Medicine); *Yaofu* (strong medicine), *Wusen*, Taiwan jewel orchid. In Taiwanese (Hokien dialect): *Kim soa lian* (gold thread lotus), *Kim chi a chha* (gold streaked herb), *Oa ke chahau* (black herb)

Description: *A. formosanus* is a terrestrial herb with creeping stems which produce leaves as they bend upwards towards the light. Leaves are ovate, pointed at the tip, 3–4 cm long and 2–3 cm broad, dark green, greenish-purple on the underside, with white venation. A terminal inflorescence bears few white flowers with an interesting, fimbriate lip terminating in two lobes (Fig. 7.15). *A. formosanus* is found throughout Taiwan in primeval forests or in bamboo stands at 500–1500 m, and in the Ryukyu Islands. It flowers in October and November (Liu and Su 1977).

Herbal Usage: Herbs are obtained from Taiwan and Fujian Province. The entire plant is used in

Fig. 7.15 *Anoectochilus formosanus* Hayata [Photo: E.S. Teoh]

Table 7.2 Chemical compounds isolated from *Anoectochilus roxburghii* and *A. formosanus*

A. roxburghii (He et al. 2006; Cai et al. 2008)
Quercetin-7-o-beta-D-(6″-o-(trans-feruloyl))-glucopyranoside (compound 1)
8-C-*p*-hydroxybenzylquercetin (compound 2)
Isorhamnetin-7-0-beta-D-glucopyranoside (compound 3)
Isorhamnetin-3-0-beta-D-glycopyranoside (compound 4)
Kaempferol-3-0-beta-D-glucopyranoside (compound 5)
Kaempferol-7-0-beta-D-glucopyranoside (compound 6)
5-hydroxy-3′,4′,7-trimethoxyflavonol-3-0-beta-D-rutinoside (compound 7)
Isorhamnetin-3-0-beta-D-rutinoside (compound 8)
Beta-D-glucopyranosyl-(3R)-hydroxybutanolide (I),
Stearic acid (II)
Palmatic acid (III), betasitosterol (IV)
Succinic acid (V),
p-hydroxybenzaldehye VI)
daucosterol (VII)
methyl 4 beta glucopyranosyl-hutanoate (VIII)
p-hydroxy cinnamic acid (IX)
0-hydroxy phenol (X)
A. formosanus (Du et al., 2008)
(3R)-3-(beta-D-glucopyranosyloxy) butanolide (kinsenoside; 1)
(3R)-3-(beta-glucopyranosyloxy)-4-hydroxybutanoic acid (2)
2-((beta-D-glucopyranosyloxy)methyl)-5hydroxymethylfuran (3)
Isopropyl-beta-D-glucopyranoside (4)
®-3,4-dihydroxybutanoic acid gamma-lactone (5)
4-(beta-D-glucopyranosyloxy) benzyl alcohol (6)
(6R,9S)-9-(beta-D-glucopyranosyloxy)megastigma-4,7-dien-3-one (7),
(3r)-3-(beta-D-glucopyranosyloxy)-4-hydroxybutanolide (8)

Reference: *Zhongyao Da Cidian* (1986) and *Zhonghua Bencao*, (2000) and *Zhonghua Bencao*, (2000)

Chinese medicine for cooling the blood, to smooth the liver, as an antipyretic and for detoxification. It is used to treat tuberculous patients who suffer from haemoptysis, and also to treat diabetes, bronchitis, kidney and bladder infections, cramps in children, snake bites and stomach ache (Liu 1952 quoted by Perry and Metzger 1980; Ou et al. 2003). The entire plant is used for treating pain at the waist and knee, numbness, haemetemesis, nocturnal emission, nephritis, vaginal discharge and convulsions affecting children (Wu 1994; *Zhonghua Bencao* 2000)

Phytochemistry: At Peking Union Medical College, He Chun-Nian and his colleagues managed to isolate eight compounds by ethanolic extraction of entire plants of *A. formosanus*. The flavonoid glucoside, quecetin-7-O-beta-D (6″-O-(trans-feruloyl) glucopyranodie, was shown to be a potent anti-oxidant while the remaining compounds possessed weak activity. The remaining seven compounds are: 8-C-*p*-hydroxybenzylquercetin; isorhamnetin-7-O-beta-D-glucopyranoside; isorhamnetin-3-O-beta-D-glucopyranoside; kaempferol-3-O-beta-D-glucopyranoside; kaempferol-7-O-bet-Dguco pyranoside; 5-hydroxy-3′,4′,7-trimethoxyflavo nol-3-O-beta-D-rutinoside; and isorhaemnetin-3-O-beta-D-rutinoside (He et al. 2005). Ten compounds were isolated at the Guangdong Pharmaceutical University by JY Cai and his colleagues (Cai et al. 2008). Scientists at Seiwa Pharmaceuticals in Tokyo found that *A. formosanus* grown in the wild and propagated by tissue culture both contained ten compounds including kinsenoside (Table 7.2) (Du et al. 2008). Kinsenoside administered to rats fed a high fat diet significantly reduced their body and liver weights (Du et al. 2001). The hepato-protective property of *A. formosanus* is conferred by Kinsenoside (Wu et al. 2007).

Anoectochilus koshunensis Hayata

Local name: *Gaoxiong Jinxian Lan* (Gaoxiong golden thread orchid or, in Taiwanese spelling, Kao-hsiung jewel orchid).

Other common names: *Hengchunjinxianlian* (Hengchun golden thread lotus); *Jinxian Lan* (golden thread orchid). In Taiwanese (Hokien) dialect: *Ko hiong kim soa lian; Heng chhun kim soa lian*

Description: Stems are rigid but soft and succulent, 20 cm long bearing 4–5 ovate, dark green leaves with the typical reticulate pattern, and 5–6 white flowers terminally. Flower is usually non-resupinate, i.e. lip is dorsal. Nearly halfway along its length, the lip splits to produce a pair of divergent blades. Flowering season is from July to October. This jewel orchid is found in broad-leaved forests below 1500 m in the central and southern parts of Taiwan and the Ryukyu Islands (Liu and Su 1977).

Herbal Usage: The whole plant "cools the blood, smoothes the liver", is antipyretic and removes toxins. It is used to treat haemoptysis resulting from tuberculosis, diabetes, bronchitis, nephritis, cystitis, infant convulsions and snake bites (Ou et al. 2003).

Phytochemistry: Five sterols, including a new one with a non-conventional side chain [26-methylstigmasta-5,22,25, (27)-trien-3 beta-ol, together with a megastigmane glucoside and 2'-deoxyadenosine, was isolated from the whole plant of *A. koshunensis* by Ito et al. (1994). Kinsenoside is also present (Du et al. 2001).

Anoectochilus regalis Bl. [see ***Anoectochilus roxburghii*** **(Wall.) Lindl.**]

Anoectochilus reinwardtii Blume

Description: Plants are up to 20 cm tall with 4–6 round leaves, 5 by 3.5 cm, of a dark velvety green

Fig. 7.16 *Anoectochilus reinwardtii* Blume [Photo: Peter O'Byrne]

with a reticulation of pink to red veins. Inflorescence is 15 cm tall, pubescent, bearing 2–4 flowers near the apex. Flowers are 2 cm across. Sepals are lanceolate, reddish and hairy on the outside, lighter and smooth on the inside. Petals are white and of the same length as the sepals. Lip is white, extending beyond the sepals, and divides into two claws, each lined with seven teeth on their lateral margins (Fig. 7.16). *A. reinwardtii* is found in Sumatra, Java, Borneo and Maluku at 1400–1700 m (Comber 2001).

Herbal Usage: The Iban and Kelabit tribes of Borneo use the orchid to treat infertility. It is believed to possess magical properties: supposedly infertile woman would conceive if leaves of a single plant are placed under their sleeping mat. From the appearance of the leaves, some people claimed that they could even predict the sex of the child (Christensen 2002)!

Anoectochilus roxburghii **(Wall.) Lindl.**

Synonym: *Anoectochilus regalis* Bl.

Sri Lankan name: *Wanna rajah* ("that which glistens in the woods". In this instance, *rajah* has the meaning "to shine" rather than "king"—Cooray 1940)

Fig. 7.17 *Anoectochilus roxburghii.* Adapted from a water-colour painting by A.J. Wendel in Blume C.L., *Collection des Orchidees les plus remarquables de l'archipel Indien et du Japan,* t. 12b, fig. analysis (1858). Courtesy of plantillustrations.org

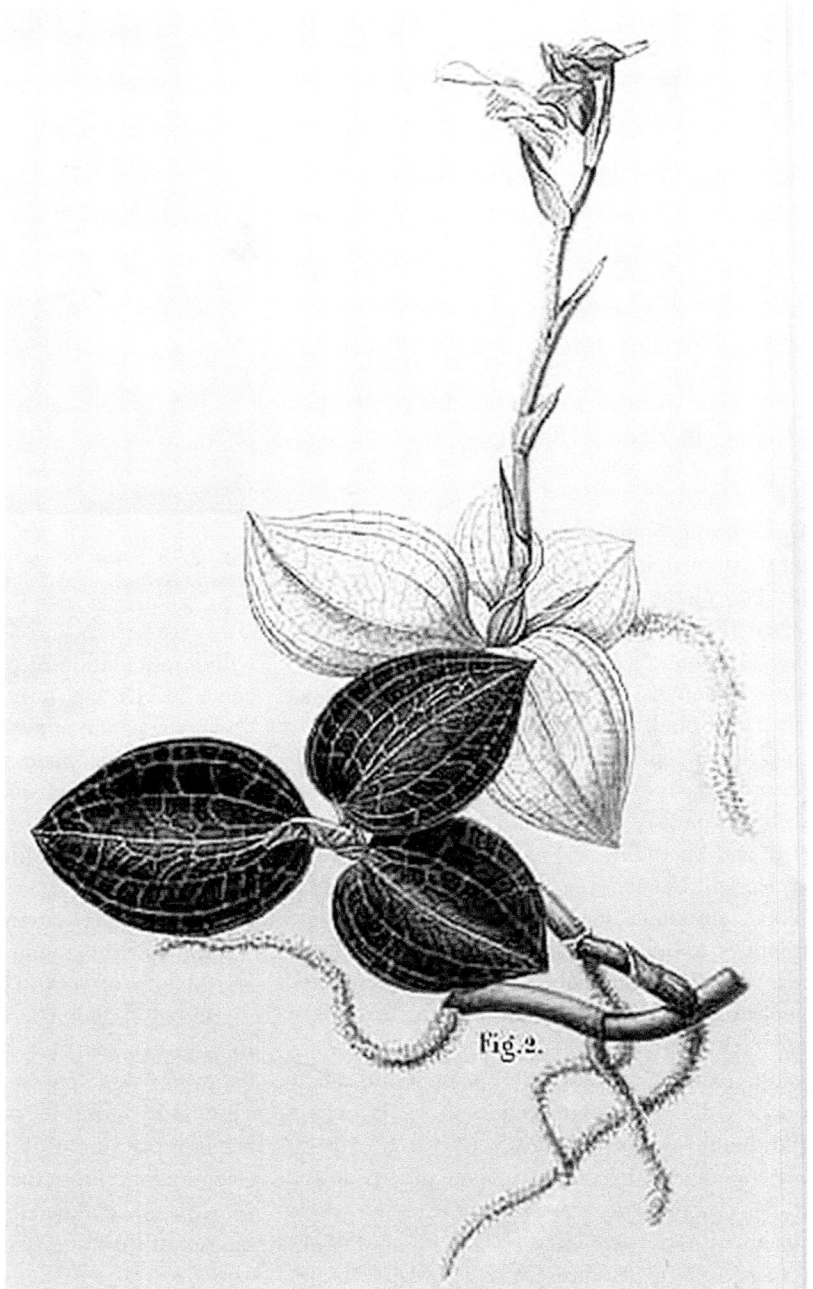

Taiwanese herbal name: *Yaowang* (King medicine)

Description: Plants are up to 30 cm tall; leaves few, 2.5–4 by 1.7–2.5 cm, of a dark velvety-green or purplish-red with a complex network of golden veins. Inflorescence is up to 27 cm tall, pubescent, with a few white flowers, 1.8 cm across. Sepals are glandular, pubescent, ovate and acute, with a central vein. Petals are narrow and pointed. Lip is 1.1 cm long, clawed, the margins of the claw carrying 8 filiform lobes on each side (Fig. 7.17).

Fig. 7.18 *Anoectochilus roxburghii* (Wall.) LIndl., from a herbal kitchen in Xiamen, Fujian Province, China [Photo: E.S. Teoh]

It flowers in December to January and May to September in Sri Lanka (Jayaweera 1981), in October to December in Bhutan (Gurong 2006) and August to December in China (Chen et al. 2009c).

This small, terrestrial, jewel orchid thrives in humus rich soil on sparsely wooded slopes at 600–800 m in continental East Asia, from China to Vietnam, Laos and Myanmar to the Himalayan foothills and Sri Lanka. It also occurs in Japan. *A.roxburghii* grows in humid primary forests in crevices or in rich humus that is constantly dampened by mists and splash, and along steep water courses at 300–1800 m in India (Sathish Kumar and Manilal 1994).

Herbal Usage: *A. roxburghii* is not included among the medicinal herbs in the popular Chinese classic *Herbals*. Nevertheless, it is a medicinal herb that enjoys widespread provincial usage in Taiwan and Fujian (Fig. 7.18). Herbalists use it to treat hepatitis, splenic disorders, hypertension, cancer, tuberculosis, impotence, fever,

snake bites and even slow development in children (Lin et al. 2000). In the southern Chinese province of Fujian, *A. roxburghii*, and the cultivated, imported species, *A. formosanus*, are a panacea for numerous ailments that include pleurodynia and other pulmonary conditions, liver disease, hypertension, paediatric malnutrition and snake bites. The Fujian Institute of Traditional Medicine likewise refers to it as "King Medicine" (He et al. 2006).

Stems and leaves of *A. roxburghii* (syn. *A. regalis*) are included in certain medicinal oils in India, but there is no mention of how it is being used (Cooray 1940; Rao 2004). Nevertheless, it has a long history of Indian medicinal usage (Chopra 1930).

Phytochemistry: Eighteen compounds have been isolated from *A. roxburghii*. *Zhonghua Bencau* Kinsenoside, a glycoside whose antiglycaemic activity has been demonstrated in streptozotoin-induced, diabetic rats is the major constituent of *A. roxburghii* (Zhang et al. 2007a). Numerous compounds were isolated by Cai and his colleagues: beta-D-glucopyranosyl-3R)-hydroxybutanolide, stearic acid, palmitic acid, beta-sitosterol, succinic acid, *p*-hydroxybenzylaldehyde, daucosterol, methyl 4-beta-D-glucopyranosyl-hutanoate, *p*-hydroxycinnamic acid and 0-hydroxy phenol. Ferulic acid, quercetin, daucosterol and cirsilineol in addition to *p*-hydroxybenzylaldehyde were present in the chloroform fraction (Cai et al. 2008). In the same year, two novel sorghumol triterpenoid acyl esters, a new alkaloid (anoecochine) and a known triterpenoid (soghumol), were isolated from *A. roxburghii* by Han et al. (2008).

Overview

Four species of *Anoectochilus* are used in herbal medicine. *A. roxburghii* enjoys a wide distribution, but its medicinal usage is mainly limited to southern China where its application for a broad spectrum of conditions has earned it the title of *Yaowang* or "King of Medicines". In Nepal, plants are used to treat tuberculosis (Pant and Raskoti 2013). *A. roxburghii* is present in some Indian medicinal oils (Cooray 1940; Rao 2004). *A. formosanus* and *A. koshunensis* are endemic to Taiwan and the Ryukyu Islands. The former species has a high reputation in Taiwan where it enjoys a broader range of medicinal uses than the latter although their applications overlap significantly. *A. formosanus* has been researched extensively in Taiwan (see below). *A. regalis* found in southern India and Sri Lanka is incorporated in local medicinal oils. Both H. N. Ridley and the Dutch botanist A. H. Berkhout, writing in the early years of the twentieth century, observed that Chinese in Malaya cultivated *Anoectochilus* species to use as medicine, but they could not discover the details of its usage (Burkill 1935). *Zhonghua Becao* (2000), the authoritative Chinese *Materia Medica*, stated that the medicinal name *JInxianian* refers to both *A. roxburghii* and *A. formosanus* whose vegetative forms resemble one another, and their usage is similar. In Indonesia, it was also used to treat tuberculosis (Van Steenis quoted by Perry and Metzger 1980) and this probably held true for Malaya during that period.

The cultivated "*Anoectochilus*" of Ridley and the *Anoctochilus* (sic) of Perry and Metzger (1980) is likely to be *Ludisia discolor* (Ker Gawler) A. Rich which is native to Malaysia and Indonesia, popularly cultivated in the region and still being offered as medicinal plants whereas *A. formosanus* only occurs in Nansei-shoto and Taiwan and the distribution of *A. roxburghii* does not extend south of the major Thai–Indochina land mass. Although *A. reinwardtii* Bl. occurs in Malaysia and Indonesia, it is not commonly cultivated.

A magico-medicinal usage is of anthropological interest: the Kelabit in Sarawak place leaves of *A. reinwardtii* Bl. (*udo anak* in Kelabit) under the sleeping mat to boost the fecundity of a woman who has not conceived (Christensen 2002).

Commercial Cultivation

A. formosanus, the golden-striped lotus, a beautiful, terrestrial, jewel orchid, is used by herbalists who use it to treat disorders of the liver and spleen, cancer, fever, hypertension, tuberculosis, impotence, snake bites and even slow development in children (Lin et al. 2000). The herb is so popular that it can fetch a price of US$100 per kg fresh weight (JETRO TTPP 2008). In the southern Chinese province of Fujian, the related and far more widely distributed *A. roxburghii*, the so-called "King of Medicines" (*yaowang*), is a panacea for numerous ailments (He et al. 2006). Fujian supplies *A. formosanus* to the herbal market although the species is not native to mainland China but is a cultivated herb.

Attempts have been made to promote the cultivation of these two orchid species as commercial crops because of the high commercial value of the herbs. Plants are harvested when they reach a height of 10–15 cm. Seedlings can be bought and raised by those who aspire to grow their own herbs, as tea packs, syrup and in dried form (Anonymous 2008).

Such widespread interest in this herb has naturally led to studies to facilitate its propagation through asymbiotic and symbiotic seed germination, synchronous flowering to promote pollination, propagation through the culture of shoot tips and nodal explants and improved methods of cultivation (Wu 1997; Gao and Guo 2001; Tsay 2002; Chou and Chang 2004; Tang and Guo 2004; Shiau et al. 2006). When *A. formosanus* is raised in pot culture, 10 weeks symbiotic co-culture with the addition of F-23 fungus (belonging to the genus *Mycena*, but not identified by species) resulted in gains in shoot height (16.6 %), shoot dry weight (31.3 %), leaf numbers (22.5 %) and, more importantly, increased contents of kinsenosides (by 85.5 %), isorhamnetin-3-0-beta-rutinoside (by 226.1 %) and isorhamnetin-3-0-bta-D-glucopyranoside (by 196.0 %). This suggests that the use of endophytes for other medicinal orchid species should also be investigated (Zhang et al. 2013). In 1997, *Mycena anoectochila* was isolated from mycorrhizal roots of *A. roxburghii* from Xishuanbanna (Guo et al. 1997).

Light intensity affects growth, photosynthetic capability and total flavonoid accumulation of *Anoectochilus* plants. A good cultivation system should employ optimum light intensity; too much light may be almost as bad as too little

(Ma et al. 2010). The possibility of accurately identifying *Anoectochilus* species based on rDNA ITS sequences is also being studied (Gao et al. 2009).

Pharmacological Studies

Most of the research published or abstracted in English comes from Taiwan and focuses on *A. formosanus*. The study group led by C.C. Lin of Kaoshing Medical College in Taiwan started their research in 1993, and they have been the most active in this field. They have also conducted studies on other medicinal plants. Their findings are summarised and discussed below. Recently, two renowned Chinese medical schools have also published papers on *A. roxburghii* (He et al. 2005; Wu et al. 2007; Cai et al. 2008; Han et al. 2008).

Anti-inflammatory Effect of *A. formosanus*

Extracts of whole plants of *A. formosanus* administered to rats reduced paw oedema chemically induced by administration of carrageenan. This anti-inflammatory effect was a delayed phenomenon, apparent only after 4 h (Lin et al. 1993). Kinsenoside, the major active compound from *A. formosanus*, inhibited inflammatory reaction in peritoneal macrophages and protected mice from endotoxin shock (Hsiao et al. 2011). Aqueous extract of *A. formosanus* fed to rats suppressed allergic asthma triggered by ovalbumin by modulating cytokine production and T-cell subpopulations (Hsieh et al. 2010).

Protection Against Liver Damage

Toxic damage to the liver is reflected by sharp increases in two liver enzymes in the blood, SGPT (serum glutamate-pyruvate transaminase), also known as ALT (alanine aminotransferase), and SGOT (serum glutamate-oxaloacetate transaminase), also known as AST (aspartate aminotransferase). Feeding rats with an aqueous extract of *A. formosanus* (hereafter called AFE) reduces the extent of their liver damage when these animals were exposed to carbon tetrachloride, the dry-cleaning solvent that is extremely toxic to the liver. The steep increase in liver enzymes following exposure to carbon tetrachloride is blunted, and histological evidence of liver damage is less when rats are fed AFE (Lin et al. 1993).

Such hepato-protective effects demonstrable through animal experiments are not uniquely confined to AFE. Lin and his colleagues reported similar anti-inflammatory and hepato-protective effects of an aqueous extracts of *Solanum alatum* (Lin et al. 1995), leaves of *Alstonia scholaris* (Lin et al. 1996) and *Terminalia catappa* (Lin et al. 1997a, 1998), root wood of *Cudrania cochinchinensis* (Lin et al. 1999), burdock (*Arctium lappa*) (Lin et al. 2002b), aqueous extracts of Chinese yam *(Dioscorea alata)*, which is commonly used in Chinese medicine (Lee et al. 2002a, b), aqueous root extracts and chloroform leaf extracts of *Limonium sinensis* (Chuang et al 2003), and various folk (herbal) medicines such as *xiao-chai-hu-tang* (Yen et al. 1991), *thang-kau-tin* (Lin et al. 1992b), *mu-mien* (Lin et al. 1992a), *ban-zhi-lian* (Lin et al. 1997b), *tao-shang-tsao* (Lu et al. 2000), *simo yin, guizhi fuling wan, xieqing wan, sini san* (Lin et al 2001) and *peh-hue-juwa-chi-cao* (Lin et al. 2002a). Researchers elsewhere have found similar hepato-protective properties in extracts of *Artemisia asiatica* (Ryu et al. 1998), *Trichilia roka* (Germano et al. 2001) and fruit juice of *Aronia melanocarpa* (Valcheva-Kuzmanova et al. 2004). These herbs are not orchids, but three species of the heterotrophic orchid, *Goodyera*, namely *G. schlectendaliana*, *G. matsumurana* and *G. discolor*, also displayed hepato-protective properties (Du et al. 2000) (see *Goodyera*). The cholesterol-lowering agent, simvastatin, reduces liver damage caused by carbon tetrachloride in rats, whereas alcohol aggravated the damage (Okovityi et al. 2007). Finally, there is a long list of other plants which reduce liver damage produced by other hepato-toxins such as acetoaminophen and beta-D-galactoseamine (Lin et al. 1992b, 1994, 1995, 1997a, b, 1998, 1999, 2002a; Lin, Huang, and Lin 2000; Lee et al. 2002a, b).

Nevertheless, work on the hepato-protective effect of AFE continues. More recently, in 2008, Fang and his colleagues demonstrated that surrogate markers of liver damage from carbon tetrachloride, such as plasma glutamate-pyruvate transaminase (GPT) and hepatic levels of hydroxyproline and malondialdehye, were significantly lower in rats receiving AFE compared with controls. Hydroxyproline is an important component of collagen, and its elevation is an indicator of tissue breakdown. Malondialdehyde is formed by degradation of polyunsaturated fatty acids and its elevation is an indicator of oxidative stress. Therefore, low levels of the liver enzyme GPT, hydroxyproline and malondialdehye can be seen as evidence of liver protection. Treatment with AFE increased expression of methionine adenosyltransferease 1A essential for liver repair; and decreased the expression of collagen (alpha 1) and transforming growth factor-beta 1, reflecting suppression of inflammation (Fang et al. 2008)

AFE protected rats against fibrous injury induced by intraperitoneal injections of dimethylnitrosamine (DMN). When portions of the liver were subsequently removed, AFE encouraged cell proliferation and liver regeneration in the residual organ accompanied by increase in liver weight. Such liver regeneration did not occur if portions of the liver were removed in normal rats, not damaged by DMN (Shih et al. 2004).

Oral administration of AFE reduced thioacetamide (TAA)-induced liver fibrosis in mice. Animals fed with AFE had significantly reduced plasma alanine aminotransferase (ALT) activity, lowered liver weights, reduced hepatic hydroxyproline, and there was less fibrosis on histological examination of the liver. AFE treatment reduced mRNA expression of collagen (alpha I), lipopolysaccharide binding protein, CD14, TLR4 and TNF receptor 1. The investigators led by Lin Wen-Chuan from the China Medical University in Taichung, Taiwan,

to postulate that the reduced TAA-induced liver fibrosis in mice fed AFE was probably mediated through inhibition of hepatic Kupffer cell activation (Wu et al. 2010).

The hepatoprotective effect is dose-dependent, and one could overdo a good thing. Aqueous extracts of AF at 300–500 mg/kg enhanced recovery from liver injury caused by acetoaminophen in male Wister albino mice. Methanol extracts at 100–300 mg/kg were similarly hepato-protective. However, alcoholic extract at 500 mg/kg resulted in serious injury (Lin et al. 2000). Organic solvent extractions are generally used to obtain pure forms of the active constituents that are more potent. However, it appears that the old approach of decocting a medicinal herb is generally safer for the layman (Lin et al. 2000).

Antidiabetic and Lipid Lowering Properties

Aqueous extracts of *A. formosanus* (AFE) and *A. roxburghii* both exhibited anti-oxidant and sugar-lowering properties in rats rendered diabetic by treatment with streptozoin. Following forced feeding with AFE (2 g/kg) for 21 days, rats had lower fasting blood glucose, lower serum fructosamine, lower triglycerides and lower total cholesterol levels than rats in the control group (Du et al. 2001; Shih et al. 2003). Kinsenoside is responsible for the antiglycaemic activity of *A. roxburghii* (Zhang et al. 2007a, b). Du's Japanese team found that AFE from cultured *A. formosanus* lowered triglyceride levels in the liver and blood and reduced the deposition of adipose tissue (Du et al. 2003). AFE delayed the oxidation of human LDL (low-density lipoprotein, the harmful cholesterol) by scavenging biological oxidants species like superoxide anion and hydroxyl radicals, and thus shows promise as an agent for preventing atherosclerosis (Shih et al. 2002, 2003). Kinsenone appears to be the strong anti-oxidant. Flavonoid glycosides and their derivatives present in *A. formosanus* also possessed anti-oxidant activity (Wang et al. 2002).

AFE lowered the plasma triglyceride (Du et al. 2008). Aqueous AFE increased HDL in the blood and vitamin E levels in the liver and kidneys of alloxan-induced diabetic mice (Cui et al. 2013). The protective effect is most likely conferred by kinsenoside (Shiau et al. 2006). This compound also exhibits antihyperglycaemic and antihyperliposis effects in rats (Zhang et al. 2007a; Du et al. 2001). Kinsenocide obtained by extraction and purification of AF reduced body weight, liver size and decreased the liver triglyceride level in rats (Wang et al. 2002).

The effect of daily 450-mg doses of *A. formosanus* (AF) on their serum lipids was tested on a small group of 66 human volunteers who were divided into four groups according to their blood chemistry: 14 had high triglyceride alone; 11 had high cholesterol alone; 5 had both high triglyceride and high cholesterol; and 36 were healthy individuals who served as controls. After 6–12 months, AF significantly decreased the serum levels of cholesterol, LDL cholesterol and very-low-density lipoprotein (VLDL) in all volunteers. These results suggest that some constituents of AF might function like a statin (Du et al. 2007), but the numbers are too small to justify usage of AF to lower blood lipids.

A simple method to demonstrate oxidative stress in vitro is to add hydrogen peroxide to cultured cells. This causes DNA in the cells to disintegrate and the cells to undergo apoptosis (programmed cell death). Using HL-60 cells and the hydrogen peroxide model, Wang and his team at Taipei Medical University showed that oxidative stress damage was prevented in a concentration-dependent manner by AFE (Wang et al. 2005).

Improved Endurance

Amphetamines (e.g. 'Speed' and 'Ecstasy') are drugs that improve endurance capacity and exercise performance, but they have dangerous side effects and are banned by sports organisations and responsible governments. Swimming endurance in rats (swimming to exhaustion) was used to demonstrate the psycho-stimulant effect of such drugs (Estler and Gabrys 1979). When green tea was tested, it enhanced fatty acid usage which was associated with an improved endurance of 8–24 % (Murase et al. 2005). Similar results were observed in mice fed *A. formosanus* extract (Ikeuchi et al. 2005) or a Korean medicinal preparation of *Rubus acoreanus* Miquel (Jung et al. 2007).

Antitumour Activity

Taiwanese inventors filed a United States patent in 2002 for a product comprising alpha-amyrin trans-*p*-hydroxy cinamate and isorhametin obtained by stepwise extraction of *A. formosanus* for the chemoprevention and treatment of human cancer. They showed that the product inhibited the growth of three types of tumour cells in vitro, namely mouse B16 melanoma, MCF-7 human breast cancer and HepG2 human liver cancer (United States Patent 7033617; Yang et al. 2004). Subsequently, the team demonstrated the inhibitory effect of AFE on CT-26 murine (sheep) colon cancer cells which were implanted on the skin of BALB/c mice (Shyur et al. 2004).

Shyur and her team at the Institute of Bio Agricultural Sciences in Taipei prepared a bioactivity-guided ethyl acetate-partitioned fraction of *A. formosanus*. When tested by apoptosis (programmed cell death) induction of cultured MCF-7 human breast cancer cells, it showed an enhanced antitumour activity over aqueous AFE. They also explored the apoptotic signalling pathway of the MCF cells exposed to the ethylacetate extract of *A. formosanus* (Tseng et al. 2006). These antitumour studies are still at an early stage and more testing will have to be performed before human studies are attempted.

Immunomodulating Activities

Finding that AFE activated phagocytosis and stimulated interferon-gamma production in lymph node cells, Lin and Hsieh (2005) postulated that *A. formosanus* might have a role in boosting defence against infection. However, animal and clinical studies based on this postulate are not available.

A. *formosanus* produces a polysaccharide known as Type II Arabinogalactan (AGAF). When AGAf was fed to mice treated with an old anticancer agent, 5-fluorouracil, it reduced the leucopaenia resulting from the anticancer treatment (Yang et al. 2013).

Prevention of Osteopaenia

Bone mass reaches a maximum in humans around the age of 30–32, after which there is an inexorable decline. This contributes to the dramatic rise in the incidence of fractures in old age. In women, an additional drop in universal bone mineral density and loss of collagen from spongy bone in the spine occur at menopause, when the ovaries cease to produce oestrogen and progesterone (Teoh and Teoh 1991). Removal of the ovaries in mammals eliminates the all-important source of oestrogen and progesterone and replicates the effect of menopause. In an experiment conducted on rats whose ovaries were removed, AFE and the potent human oestrogen, 17-beta oestradiol, both prevented bone loss and shrinkage of the pituitary gland. Elevation of circulating alkaline phosphatase, which is present when there is excessive bone loss, was also suppressed by AFE feeding (Shih et al. 2001).

Bone remodelling relies on a balance between the activities of osteoblasts which build up bone and osteoclasts which remove calcium from bone. Excessive osteoclastic activity erodes bone strength. AFE blocks the formation of osteoclasts without suppressing the formation of osteoclast progenitor cells from bone marrow stem cells (Masuda et al. 2008). Kinsenoside is the active compound that suppresses the formation of osteoclasts and bone loss associated with removal of the ovaries (Hsiao et al. 2013).

Oestradiol preserved the vagina and uterus in the experimental animals but A. *formosanus* did not (Shih et al. 2001). In other words, A. *formosanus* behaved like a selective estrogen-receptor modulator (SERM). In their search for an oestrogen-like substance which preserves bone while not increasing the risk of breast or uterine cancer, the pharmaceutical industry came up with SERMs, examples of which are tamoxifen, tibolone and raloxifene. Tamoxifen and raloxifene have been shown to reduce the risk of breast cancer recurrence, while tamoxifen enjoys worldwide usage in the long-term management of oestrogen receptor-positive breast cancer (Johnston and Howell 2002). It might be worthwhile to study the effect of A. *formosanus* on normal and malignant breast tissues.

Foetal Lung Maturation

Dexamethasone is used to accelerate foetal lung maturation when it becomes necessary to deliver babies prematurely (Crowley 1995; RCOG Guideline No. 7 1996). In 2004, a team of pediatricians from the Taipei Medical University Hospital led by C.M. Chen tested the effects of the "King Medicine" *(A. formosanus)* on foetal lung maturation in rodents and compared its effect with that achieved with dexamethasone. They fed pregnant rats with AFE for 7 days, from Days 12 to 18 and delivered them by caesarean section on Day 19. Another group of rats received an intraperitoneal injection of dexamethasone on Day 18 and were sectioned on Day 19. The control group received intraperitoneal saline. AFE treatment and dexamethasone both increased growth hormone levels in the pregnant rats and saturated phosphatidylcholine (surfactant) levels in the foetal lung tissue. A surfactant is required to allow the lungs of mammalian newborn to expand at birth. Lungs of the newborn rats treated with AFE and dexamethasone showed histological evidence of accelerated lung maturation (Chen et al. 2004).

The normal duration of human pregnancy is 38 weeks from the date of fertilisation, and protection of premature human babies from respiratory distress is achieved 36–48 h after the administration of high doses of dexamethasone. Thus, the exposure of pregnant rats to

Fig. 7.19 Structure of some compounds present in *Anoectochilus*

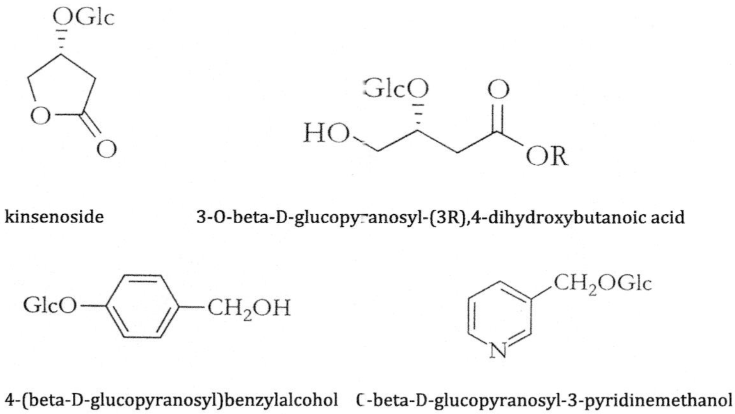

kinsenoside

3-O-beta-D-glucopyranosyl-(3R),4-dihydroxybutanoic acid

4-(beta-D-glucopyranosyl)benzylalcohol

C-beta-D-glucopyranosyl-3-pyridinemethanol

AFE in the experiment is inordinately prolonged and cannot be said to be comparable to that of dexamethasone, since the latter acts much more rapidly in the human. It would be logical to prove the efficacy and safety of AFE in veterinary medicine before any test is conducted on humans.

Effect on Amnesia

A small study on rats by researchers in Taiwan showed that rats fed with extracts of AF had better memory retention of an unpleasant experience when they were treated with scopolamine to induce amnesia. The anti-amnesic effect was boosted by the simultaneous administration of neostigmine, a drug which is known to enhance cholinergic neural transmission (Cheng et al. 2003).

Comment

Such a wide range of possible beneficial effects of a common herbal remedy with considerable safety ought to be properly investigated by randomised controlled clinical trials on human subjects. To date, the report on the lipid-lowering effect of kinsenocide is the sole published human data (Du et al. 2007), but the study sample is too small for a definitive conclusion to be made. Many modern drugs that show great promise either in the laboratory or with surrogate markers in humans fail to demonstrate real benefits when

they are subjected to randomised clinical trials. Therefore, until large human trials validate the beneficial effects of *A. formosanus*, the extent of the benefits of this orchid, if any, is still subject to speculation.

Considerable overlap in the medicinal effects of *A. formosanus* and *A. roxburghii* is likely because both contain kinsenocide, the principal pharmacologic constituent (Fig. 7.19).

Genus: *Anthogonium* Lindl.

Chinese name: *Tongban Lan* (barrel petal orchid)

This is a small genus of terrestrial or occasionally saxicolous herbs with a single species in China (Figs. 7.20–7.22). The generic name is derived from Greek, *anthos* (flower) and *gonia* (angle). It probably refers to "the curious angle at which the tubular flower is joined to the pedicellate ovary" (Schultes and Pease 1963).

Anthogonium gracile Wall ex Lindl.

Anthogonium griffithii Rchb. f.

Chinese name: *Tongban Lan* (barrel petal orchid)
Chinese medicinal name: *Honghuaxiaodusuan*
Thai name: *Wan phrao*

Description: *A. gracile* is a small, slender, terrestrial or occasionally saxicolous, sympodial

Fig. 7.20 *Anthogonium
gracile* Wall ex Lindl.
From: *Annals of the Royal
Botanic Gardens,
Calcutta*, vol. 8 (3): t.134,
(1891) Drawing by
R. Pantling. Courtesy of
Missouri Botanical
gardens, St. Louis, USA

orchid, with subterranean corm-like pseudobulbs the size of a hazel nut or walnut. Stem is slender, 14–40 cm tall. Leaves are 1–3, narrow, petioled, pleated, 15–30 by 0.8–2.5 cm and deciduous (Fig. 7.21). Inflorescence is erect, simple or branched, reaching or exceeding the tip of the leaves, with a few greenish-white or pink, oddly shaped, non-resupinate flowers, loosely arranged and opening in succession (Grant 1895). Lip is purple (Fig. 7.22). It occurs at mid- to high elevation in deciduous forests and scrub from Sri Lanka and the eastern Himalayas to the southern Chinese provinces of Xizang, Yunnan, Guizhou and Guangxi, at 1200–2300 m, and in the northern parts of Myanmar, Thailand, Laos and Vietnam. It flowers in October in Thailand (Vaddhanaphuti 2005), from July to November in China (Chen and Wood 2009a) and July to September in India (Misra 2007).

Fig. 7.21 *Anthogonium gracile*, plant [Photo: E.S. Teoh]

Phytochemistry: No alkaloid was detected in *Anthogogium gracile* (Luning 1967), but the plant produces Batatasin III. (Veeraju, et al. 1989).

Herbal Usage: *Honghuaxiaodusuan* is obtained from Guangxi, Yunnan and Xizang. In Chinese Herbal Medicine, the orchid is used to treat menstrual disorders and to prevent pain (Wu 1994; *Zhonghua Bencao*, 2000).

Overview

Pharmacological information on *Anthogonium* is lacking. Batatsin II is a phytoalexin with antifungal properties.

Genus: *Apostasia*, Blume

Chinese name: *Ni Lan*

Apostasia is a genus of seven primitive terrestrial orchids with erect stems and narrow, lanceolate leaves and inconspicuous flowers

Fig. 7.22 *Anthogonium gracile* Wall ex Lindl. [Photo: Bhaktar B. Raskoti]

(Fig. 7.23). Petals and sepals and sometimes even the lip are similar in shape and open widely. It is not resupinate. Flowers carry two anthers which have not evolved into pollinia. However, there is a primitive column which carries two or three stamens (Ridley 1894). If someone was describing a hypothetical ancestor of the modern orchid, this flower would fit their description (Fernier 2006). There are six species in Southeast Asia (Comber 2001), none with horticultural interest (Kocyan 2010).

Apostasis is a Greek word which means 'divorced or desertion', i.e. the tribe and genus are separated from the usual more advanced orchids in which the anthers are replaced by pollinia.

Apostasia nuda R. Br.

Malay names: *Si sarsar bulang, Si marsari sari, Duhut bane-bane, Poko pulumpus bedak, Dudulu ingap, Kniching pelandok*

Description: Plant is grass-like, 20–60 cm tall, with linear-lanceolate, green leaves, 15–30 cm by 0.6–1.2 cm, spirally arranged around the

Fig. 7.23 *Apostasia nuda.*
Adapted from a water-
colour painting by C.M
Curtis in Wallich N:
*Plantae Asiatica
Rariores*, vol. 1: t. 85
(1830). Courtesy of
Missouri Botanical
Gardens, St. Louis, USA

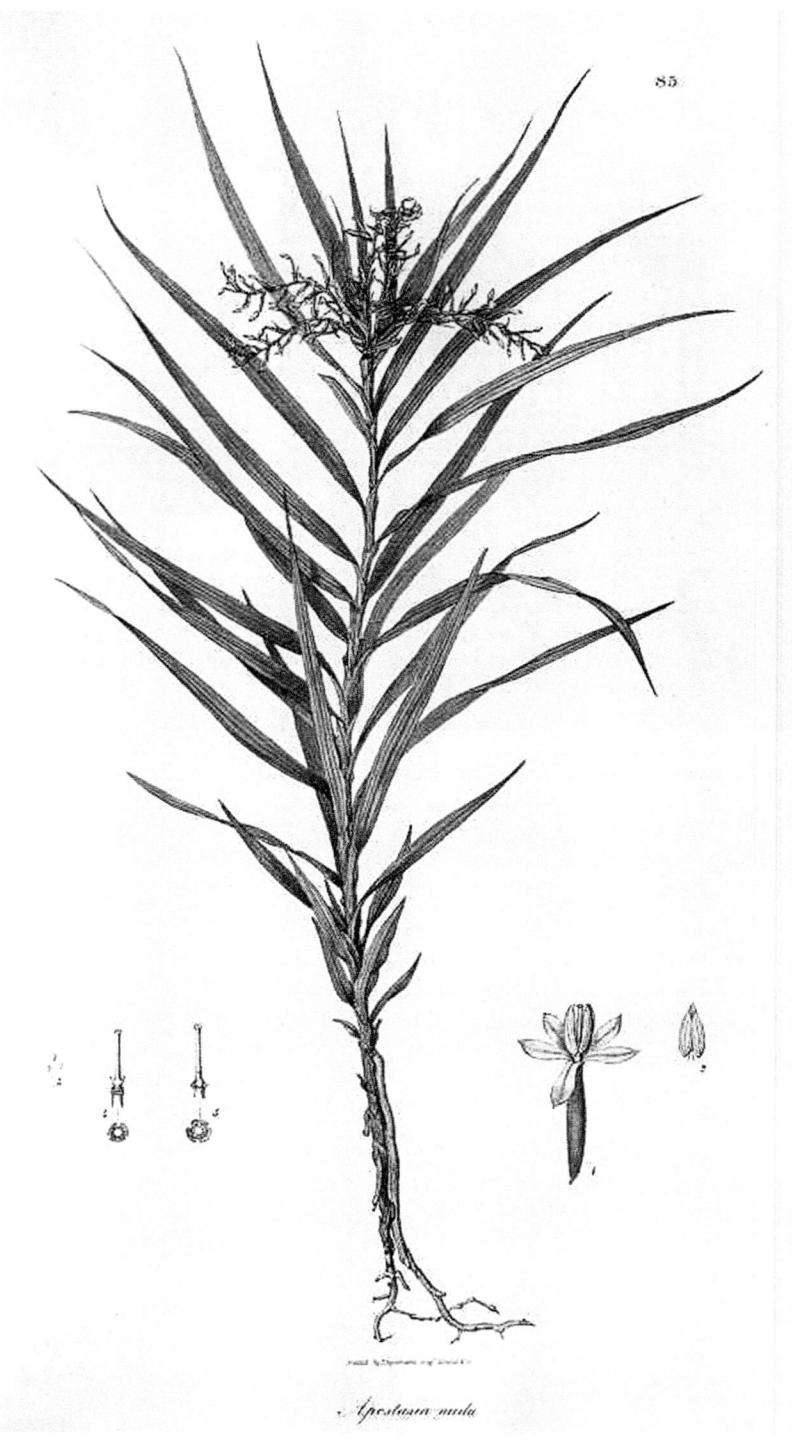

stem. Aerial roots arise from lower parts of the stem, 2–3 mm in diameter. Inflorescence is terminal, emerging horizontally and becoming pendulous as it matures, branching, the base of the branches covered with numerous, narrow, overlapping bracts. Raceme is pendulous.

Fig. 7.24 *Apostasia nuda* R. Br. [Photo: Tim Yam]

Fig. 7.25 *Apostasia wallichii* R. Br. [Photo: Peter O'Byrne]

Flowers are 15–20, yellow or white, 5–8 mm across, loosely arranged. Tepals and lip are of nearly equal size (4 by 0.5–1 mm) and appearance, lanceolate, and rolled backwards. Ovary is prominent, 8 mm long, lengthening to 1.2 cm in fruit (Fig. 7.24). It flowers from May to June (Seidenfaden and Wood 1992; Nanakorn and Wathana 2008). *Apostasia nuda* is distributed in Myanmar, southern Thailand, Malaysia and Indonesia in lowland forests at 100–1300 m, in shade. It is fairly common in lowland dipterocarp forests in Malaysia but is not easily recognised as an orchid (de Vogel 1969; Seidenfaden and Wood 1992; Beaman et al 2001).

Herbal Usage: Roots were boiled and made into poultices to treat diarrhoea in Malaysia, and an infusion of the fruit was a local remedy for sore eyes (Burkill 1935). Root was also used to treat dog bite (de Vogel 1969).

Apostasia wallichii R. Br.

Chinese name: *Jianyeni Lan*
Thai name: *Tan khamoi, Ma thon lak*
Indonesian name: *Djukut mayang kasintu* (grass like tail feather of jungle cockerel)
Malay name: *Hanching fatimah, Kenching fatimah*

Description: Plants are 40 cm tall bearing a rosette of narrowly lanceolate leaves 30 by 1.2 cm. Inflorescence is 8 cm long, with a few, long, arching branches not densely covered with bracts at their base. Flowers are 5–25, yellow. Petals, sepals and lip are identical in form, not recurved, 5.5 mm long (Fig. 7.25). Ovary is 1.2–1.5 cm long, lengthening to 2.5 cm in fruit (Seidenfaden and Wood 1992). A widespread, variable, lowland, terrestrial species, *A. wallichii* is distributed from Nepal and Assam to Sri Lanka and through Southeast Asia to New Guinea and Australia. It is also found in Sri Lanka, and at 1000 m in Hainan and southwest Yunnan in China. It grows on the forest floor in lowland forests at 250–1200 m in the tropics (de Vogel 1969). Flowering is not seasonal.

Herbal Usage: The root is used as a tonic in Thailand (Chuakul 2002). It was used as an antidiabetic agent in Malaya (de Vogel 1969).

Overview

As a primitive genus, it would be interesting to compare its constituents with those of more advanced genera. Unfortunately, phytochemical and pharmacological information on *Apostasia* are lacking.

Fig. 7.26 *Appendicula cornuta* Blume [Photo: E.S. Teoh]

Genus: *Appendicula* Blume

Appendicula is a genus with 163 species, distributed from India and China throughout Southeast Asia to the Pacific. Plants are epiphytic or saxicolous, rarely terrestrial, small or large, with stems that are erect or pendulous, simple or branched, and enclosed by permanent leaf sheaths. Leaves are arranged in two ranks, regular, flat and usually held obliquely to the stems. Leaves are twisted at the base so that the blades all lie in one plane. Inflorescences are terminal and axillary, generally short with one or a few, small, white or greenish flowers.

Appendicula cornuta Blume

syn. *Dendrobium bifarium* Lindl.

Description: Stems are up to 30 cm in length with short internodes ensheathed by two ranks of oblong leaves that are rounded at the tip, 5 by 1.1 cm. Flower is solitary, single, white or slightly greenish, 1.2 cm across; lip flat, large, without side lobes, white, and marked by several, central, warty

ridges (Fig. 7.26). *A. cornuta* is the commonest species of *Appendicula* and the most widely distributed. It is a fairly common, small lowland orchid in Sikkim, southern China, Myanmar, Thailand, Indochina, Malaysia and Indonesia.

Phytochemistry: Alkaloid was detected in *A. cornuta* (Luning 1967; Chen and Gale 2009).

Usage: Juice extracted from the stems of *A. cornuta* (syn. *Dendrobium bifarium*) was used in Maluku as a medicine for whitlow (Dragendorff 1898, quoted by Lawler 1984).

Overview

Being a fairly common and widely distributed lowland species in Southeast Asia, it is not surprising that it was used in Indonesian folk medicine.

Genus: *Arachnis* Blume

Chinese name: *Zhizhu Lan*

The Greek word *arachne* (spider, scorpion) is well known and the generic name describes the overall shape of the tough, fleshy flower which looks like a scorpion (Fig. 7.27). Members of the genus are commonly known as Scorpion Orchids. Spider Orchid is an alternative name. *Arachnis* is a robust, sun-loving, monopodial, tropical epiphyte that often has its roots scrambling over the ground or over rocks. Leaves are leathery, strap-shaped, ensheathing the stem and arranged in two straight alternating rows. Inflorescence is simple or branched, straight or arching, and carries several well-spaced flowers. *Arachnis* is distributed in Peninsular Thailand, Malaysia, Singapore and Indonesia. Hybrids between *Arachnis* and other monopodial orchids travel well, and they were used to initiate the transcontinental cut flower orchid industry. Intergeneric hybridisation involving *Arachnis* is extensive.

Arachnis flos-aeris (L.) Rchb. f.

syn. *Arachnis moschifera*

Iban name: *Wi buntak*

Fig. 7.27 *Arachnis flos-aeries*, from: Smith J.J., Die Orchideen von Java Figurenatlas, t. 437 (1905–1914). Drawing by M. Kromohardjo. Courtesy of the Swiss Orchid Foundation

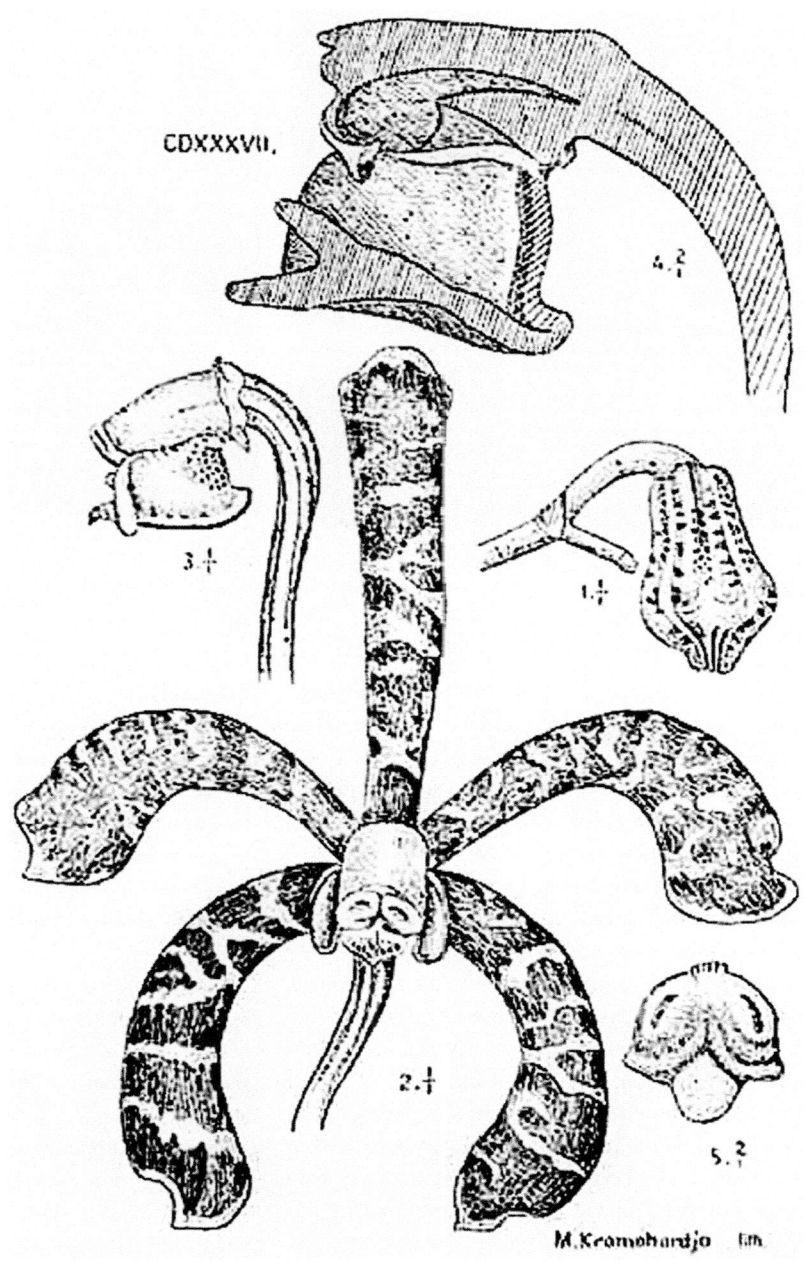

Thai name: *Ueang maeng mum*

Description: *A. flos-aeries* is found frequently on limestone in Perak and Pahang in Peninsular Malaysia, Peninsular Thailand, Sumatra and Borneo. The variety *insignis* has dark maroon petals and sepals. Although this species is not known to occur in Singapore, C.E. Carr managed to collect its natural hybrid, *A. maingayi*, which was growing on coastal mangrove in Pulau Seletar. *A. maingayi* is a natural hybrid between *A. flos-aeries* and *A. hookeriana* (Holttum 1964).

Plants are large and rambling, over 2 m in length, with leaves 15–20 cm long and 5 cm

Fig. 7.28 *Arachnis flos-aeris* (L.) Rchb. f [Photo: E.S. Teoh]

Fig. 7.29 *Arachnis flos-aeris* (L.). Rchb. f. var *Insig* [Photo: E.S. Teoh]

across. Inflorescence is up to 150 cm long, branching, bearing many widely spaced, large flowers 10 cm tall and 8 cm across. Sepals and petals are narrow, pale yellow-green marked with broad, irregular bars of purplish-brown (Fig. 7.28). Their strong musk scent is a lead when tracking down the species in the forest. The species was formerly known as *A. moschifera* because of the scent.

There are two natural varieties in addition to the type. *A. flos-aeris* var. *insignis,* the black scorpion orchid, occurs in Sumatra. Sepals and petals are dark maroon, almost black. Lip is white (Fig. 7.29). When not in bloom, the plant can be recognised by the purple flush on its young leaves (Comber 2001; Teoh 2005). Flowers of *A. flos-aeris* var. *gracilis* are four-fifths the size of the type, and curvature of lateral sepals and petals is more pronounced. Native to the mangrove swamps of Selangor and Negri Sembilan in Peninsular Malaysia, this variety has a strong, unpleasant scent (Teoh 2005). *A. flos-aeris* blooms twice a year, in May and November. In Thailand, flowering season is July (Vaddhanaphuti 2005). The variety *gracilis* blooms only in June.

Herbal Usage: Ibans of Sarawak apply sap from the orchid plant onto the painful site to relieve toothache (Christensen 2002).

Overview

Although *Arachnis* is widespread in Malaysia and Indonesia, it is not widely used as a medicinal genus and pharmacological information on the genus is not available.

Genus: *Arundina* Blume

Chinese name: *Zhuye Lan* (bamboo leaf orchid).

In Taiwanese (Hokien) dialect: *chiau a hoe*

Arundina is a grass-like, terrestrial orchid which resembles bamboo in vegetative form, and *Cattleya* in the appearance of its flowers. It is an attractive, sun-loving, free-flowering genus which enjoys a wide distribution from India, Nepal and Sri Lanka across southern China and Southeast Asia to the Ryukyu Islands and Tahiti. Flowers are variable in colour, size and fullness of their form, but experts who are very familiar with *Arundina* have opined that the differences do not constitute sufficient criteria to constitute several species; they all belong to a single variable species (Holttum 1964; Seidenfaden and Wood 1992). Nevertheless, a second dwarf species, *A. graminifolia* var. *revoluta* A.L. Lamb and C.L. Chan, which is native to Indochina and

Fig. 7.30 *Arundina graminifolia* (D. Don) Hochr. (as *Arundina bambusifolia* Lindl.) adapted from a colour painting in Warner R., Williams B.S., *The Orchid Album* vol. 1884: t. 139 (1884) (painting by J.N. Fitch). Courtesy of Missouri Botanic Gardens, St. Louis, USA

ARUNDINA BAMBUSÆFOLIA.

Borneo, was recently described and recorded as a distinct species, *A. caespitosa* Aver.

Arundina is a familiar orchid in Southeast Asia. It has the ability to establish itself in secondary scrubland from sea level to moderate elevations and may be unaffected by logging. *Arundo* is Latin for 'reed'. The generic name describes the stems of the orchid (Fig. 7.30).

Arundina graminifolia (D. Don) Hochr.

Common name: Bamboo Orchid
Chinese names: *Zhuye Lan* (bamboo leaf orchid);
   Changgan Lan (long stem orchid) *Shiyu Lan*
(jade stone orchid); *Hu Lian* (lake lotus);
   Caojiang (ginger grass); *Dayeliaodiaozhu*
   (big leaf bamboo); in Taiwan: bird orchid
Indonesian name: *Anggerik Bamb;* in Sundanese:
   Handjuwang Sapu
Malaysian name: *Phanyar* among the aboriginal
   Jakuns of Johor
Myanmar name: *Wah thitkhw*
Thai name: *Ueang Pai*
Vietnamese name: *Lan say*

Description: Stems and leaves are reed-like, or rather like diminutive bamboo. Flowers are borne singly or two at a time, successively on a slender erect stem. They are medium-sized to large,

Fig. 7.31 *Arundina graminifolia* (D. Don Hochr) [Photo: E.S. Teoh]

looking a bit like *Cattleya* (Fig. 7.31). Petals and sepals are white to deep pink. Lip is deep purple with a yellow throat. There is a dwarf species with paler flowers. Jim Comber discovered an *alba* variety in North Sumatra at 1550 m which is also present in Borneo (Comber 2001; Beaman et al. 2001). It flowers throughout the year in Malaysia, Singapore and Indonesia (van Steenis 1958). In Thailand, the flowering season extends from August to March (Vaddhanaphuti 2001), but in the Kachin and Shan states of adjacent Myanmar, the flowering season is reversed, March to July (University of Myanmar Department of Botany, 2004). It flowers from July to December in Hong Kong (Wu et al. 2001). An extremely common orchid in open scrub, lowland forests, and in the foothills throughout southern China and Southeast Asia, this is an elegant orchid for lowland gardens in the tropics. However, lowland plants usually have smaller and less colourful flowers than plants from the highlands.

Herbal Usage: In Indian traditional medicine, applications of the scrapings of the bulbous stem are used to heal cracks on the skin. It is thought to have antibacterial properties (Singh and Duggal 2009).

The Hong Kong Chinese Medical Research Institute commented that *A. graminifolia* (syn. *A. chinensis* Blume) which flowers locally from July to December could be collected throughout

the year, and the whole plant or pseudobulb was either used fresh or divided into small pieces and sun-dried. The orchid was used for a variety of apparently unrelated conditions ranging from hepatitis and jaundice to urinary tract infections, oedema, rheumatic pain, trauma and snake bites. To treat a snake bite, the following prescription was recommended: (1) decoction of *Arundina*, 10–15 g in water, to be drunk; (2) additionally, a poultice made from mashed fresh bulbs to be applied to the wound. The herb is bitter, neutral, anti-inflammatory and diuretic (Li 1988).

Dispelling "heat" and toxicity, antirheumatism, anti-inflammation, diuresis, rheumatism with waist and thigh pain, gastric pain, urethritis, leg oedema and food poisoning are reasons for using the herb in Yunnan. In Guangxi, a decoction prepared by boiling 9–15 g of the whole plant of *A. graminifolia* is used for pain relief, the treatment of bruises, oedema, abdominal pain, intestinal parasitic infestation, jaundice, pulmonary tuberculosis, mental illness, rheumatism and bleeding from knife wounds (*Zhongyao Da Cidian*, 1986).

Phytochemistry: *Arundina* tested negative for alkaloids with the Dragendorf reaction (Lüning 1974, Luning 1975). However, it contains numerous stilbenoids, namely arundin and its analogues: arundinin, isoarudinin-I, isoarundinin II; lusianthridin, flavanthrin; flavidin; and batatasin III. Arundinol (a triterpenoid) and *p*-hydroxybenzyldehyde are also present. The first four compounds had not previously been isolated from other orchids (Majumder and Ghosal 1991, 1993, 1994).

Recently, one new benzyldihydrophenanthrene has been isolated from *A. graminifolia*. Named arundinaol, it is 7-hydroxy-1-(*p*-hydroxybenzyl)-2,4-dimethoxy-9,10-dihydrophenanthrene (Liu et al. 2004a, b, 2005a). Five phenanthrene constituents were separately reported by Liu et al. (2005b), namely: orchinol or 7-hydroxy-2,4-dimethoxy-9,10-dihydrophenanthrene; 4,7-dihydroxy-2-methoxy-9,10-dihydrophenanthrene; 2,7-dihydroxy-4-methoxy-9,10-dihydrophenanthrene; 7-hydroxy-2-methoxyphenanthrene-1,4-dione or densiflorol B; and 7-hydroxy-2-methoxy-9,10-dihydrophenanthrene-1,4-dione. The pharmacologic actions of these compounds have not been reported.

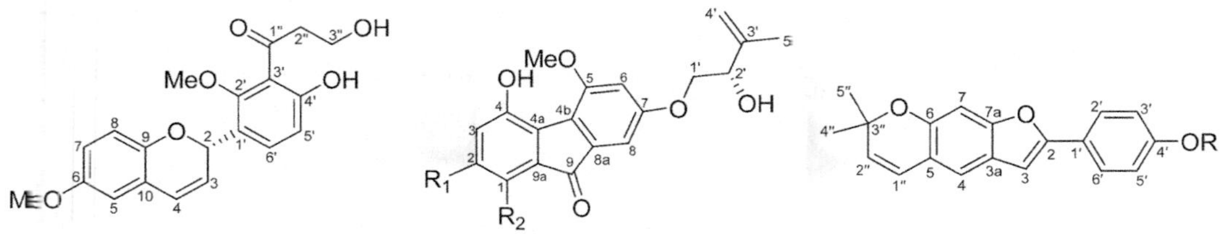

1   **2** R$_1$= H, R$_2$ = OH; **3** R$_1$= OH, R$_2$ = H   **4** R= H; **5** R= Me

Fig 7.32 Molecular structure of graminiphenols C (1), D (2) and (F). Graminiphenols C and F inhibit Tobacco Mosaic Virus; graminiphenol D shows anti HIV-1 Activity (From: Hu, Zhou, Huang, et al., *J Nat Prod*, 76: 292–296, 2013)

Subsequently, an additional six compounds were isolated from pseudobulbs of *A. graminifolia*, namely: 2,7-dihydroxy-1-(*p*-hydroxylbenzyl)-4-methoxy-9,10-dihydrophenanthrene; 4,7-dihydroxy-1-)*p*-hydroxybenzyl)-2-methoxy-9,10-dihydroxyphenanthrene; 3,3'-dihydroxy-5-methoxybibenzyl; (2E)-2-propenoic acid-3(4-hydroxy-3-methoxyphenyl)-tetracosyl ester; (2E)-2-propenoic acid-3-(4-hydroxy-3-methoxyphenyl) pentacosyl ester; and pentadecyl acid. Antitumour activity was exhibited by the first three compounds, maximal in compound 3 which possesses an open bibenzyl ring (Liu et al. 2012).

Graminiphenols A and B, and four known phenols [9'-dehydroxy-vladinol F, vladinol F, 9-O-beta-D-xylopyranoside-vladinol F and 4 9-dihydroxy-4',7-epoxy-8'9-dinor-8,5'-neolign a-7'-oic acid] were recently isolated from air-dried, powdered whole plant of *A. graminifolia* by Gao et al. (2013). Only the last compound exhibited stronger inhibition of tobacco mosaic virus than the control, whereas Graminiphenols B showed moderate and Graminophenol A only exhibited weak activity against the virus (Gao et al. 2013). Five additional new phenolic compounds, graminiphenols C–G together with eight known phenols (graminiphenol B, moracin M, 2,4,7-trihydroxy-5-methoxy-9*H*-flouren-9-one, candenatenin A, catechin, kaempferol and quercetin), were recently isolated from *A. graminifolia* at the Key Laboratory of Chemistry in Ethnic Medicinal Resources, Yunnan University of Nationalities in Kunming, China (Fig. 7.32). Graminiphenols B, D and E exhibited anti-HIV-

1 activity with therapeutic index above 100:1. Graminiphenol C4 and G possessed anti-TMV (tobacco mosaic virus) activity and they may have a role in protecting the orchid from the virus (Hu et al. 2013a). Continuing their investigations on the species, the team then isolated eight new c-4-alkylated deoxybenzoins, three new diphenylethylenes and five known diphenylethylenes from the plant. Two of the alkylated deoxybenzoins and the three new diphenylethylenes showed significant cytotoxicity against human prostate cancer (PC3 cell line), human alveolar basal epithelial cell (A549 cell line) and human neuroblastoma (SHSYSY cell line) (Hu et al. 2013b).

Overview

Residents of tropical Asia are familiar with *A. graminifolia* because this is a common, lowland to low montane, terrestrial orchid throughout the region, It is surprising that it has medicinal usage only in Yunnan and Hong Kong. Its usage in the latter province is curious because the orchid is not used medicinally in the adjacent coastal provinces of Guangdong and Fujian. The orchid contains several compounds with pharmacological potential. Animal and human studies need to be conducted to determine whether they are useful as anti-HIV agents.

A. graminifolia is a popular garden plant but its collection for local cultivation has not placed this tough, widespread species under threat. In fact, the orchid has become naturalised in Hawaii, Venezuela and Trinidad (Recart et al. 2014), It was one of three orchid species that were among the earliest plants to reappear in Krakatoa after the massive volcanic eruption in 1883 (Goldman 2005).

References

Abraham A, Vatsala P (1981) Introduction to Orchids, with illustrations and descriptions of 150 South Indian Orchids. TPGRI, Trivandrum

Akarsh (2004) Newsletter of ENVIS NODE on Indian Medicinal Plants 1(2): June 2004)

Alrich P, Higgins W (2008) The Marie Selby Botanical Gardens illustrated dictionary of orchid genera. Comstock Books, Carson City, NV

Anonymous (Advert) 2008 Window on Taiwan, Puli Township

Anuradha V, Prakash NS (1994a) Revised structure of flavidinin from Acampe praemorsa. Phytochemistry 35:273–274

Anuradha V, Prakash NS (1994b) Praemorsin, a new phenanthropyran from Acampae praemorsa. Phytochemistry 37:909–910

Anuradha V, Prakash NS (1998) A phenanthropyran from Aerides crispum. Phytochemistry 48(1):185–186

Araoz R, Molgo J, Tandeau Marsac NT (2009) Neurotoxic cyanobacterial toxins. Toxicon 56:813–28

Beaman TE, Wood JJ, Beaman RS, Beaman JH (2001) Orchids of Sarawak. Natural History Publications, Kota Kinabalu

Beckman EM (2002) Rumphius Orchids. Orchid texts from the Ambonese Herbal by Georgius Everhardus Rumphius. Yale University Press, New Haven

Benzing DH, Clements MA (1991) Dispersal of the Orchid Dendrobium insigne by the Ant Tridomyrmex cordatus in PNG. Biotropica 23(4, Part B):604–607

Burkill IH (1935) (1966 reprint, 2nd ed., with contributions by Birtwistle W, Foxworthy FW, Scrivenor JB, Watson IG) A dictionary of economic products of the Malay Peninsula, Vol. II. London, Crown Agents for the Colonies. Kuala Lumpur: Ministry of Agriculture & Co-operatives

Cai JY, Gong LM, Zhang YH et al (2008) Studies on chemical constituents from Anoectochilus roxburghii. Zhong Yao Cai 31(3):370–372

Caius JF (1936) The medicinal and poisonous plants of India. J Bombay Nat History Soc 38(4):791–799

Cakova V (2013) Contribution a l'etude phytochimique d'orchidees tropicales: identification des constituants d'Aerides rosea et d'Acampe rigida. Techniques analytiques et preparatives appliquees a Vanda coerulea et Vanda teres. Doctoral thesis, Universite de Strasbourg

Cakova V, Urbain A, Antheaume C et al (2015) Identification of phenanthrene derivatives in Aerides rosea (Orchidaceae)using the combined systems of HPLC-ESI-HRMS/MS and HPLC-DAD-S-SPE-UV-NMR. Phytochem Anal 26(1):34–39

Chen CM, Wang LF, Cheng KT, Hsu HH, Gau B, Su B (2004) Effects of Anoectochilus formosanus Hayata extract and glucocorticoid on lung maturation in preterm rats. Phytomedicine 11(6):509–515

Chen SC, Tsi ZH, Luo YB (1999) Native Orchids of China in Colour. Science Press, Beijing

Chen XQ, Gale SW (2009) Arundina Blume. In: Chen XQ, Zj L, Zhu GH et al (eds) Flora of China—Orchidaceae. Science Press, Beijing

Chen XQ, Gale SW, Cribb PJ (2009a) Amitostigma Schltr. In: Chen XQ, Zj L, Zhu GH et al (eds) Flora of China—Orchidaceae. Science Press, Beijing

Chen XQ, Gale SW, Cribb PJ (2009b) Apostasia Bl. In: Chen XQ, Zj L, Zhu GH et al (eds) Flora of China—Orchidaceae. Science Press, Beijing

Chen XQ, Gale SW, Cribb PJ, Ormerod P (2009c) Anoectochilus Blume. In: Chen XQ, Zj L, Zhu GH et al (eds) Flora of China—Orchidaceae. Science Press, Beijing, p 79

Chen XQ, Wood JJ (2009a) Acampe Lindley. In: Chen XQ, Zj L, Zhu GH et al (eds) Flora of China—Orchidaceae. Science Press, Beijing

Chen XQ, Wood JJ (2009b) Aerides Loureiro. In: Chen XQ, Zj L, Zhu GH et al (eds) Flora of China—Orchidaceae. Science Press, Beijing

Chen XQ, Wood JJ (2009c) Agrostophyllum Blume. In: Chen XQ, Zj L, Zhu GH et al (eds) Flora of China—Orchidaceae. Science Press, Beijing

Chen XQ, Wood JJ (2009d) Anthogonium Wall ex Lindl. In: Chen XQ, Zj L, Zhu GH et al (eds) Flora of China—Orchidaceae. Science Press, Beijing

Chen XQ, Wood JJ (2009e) Appendicula Blume. In: Chen XQ, Zj L, Zhu GH et al (eds) Flora of China—Orchidaceae. Science Press, Beijing

Chen XQ, Wood JJ (2009f) Arachnis Bl. In: Chen XQ, Zj L, Zhu GH et al (eds) Flora of China—Orchidaceae. Science Press, Beijing

Cheng HY, Lin WC, Kiang FM, Wu LY, Peng WH (2003) Anoectochilus formosanus attenuates amnesia induced by scopolamine in rats. J Chin Med 14 (4):235–245

Chou LC, Chang DCN (2004) Asymbiotic and symbiotic seed germination of Anoectochilus formosanus and Haemaria discolor and their F1 hybrids. Bot Bull Acad Sin 45:143–151

Chopra RN (1933) The indigenous drugs of India. The Art Press, Calcutta

Chowdhury A, Paul P, Nath D, Bhattacharjee MK (2013) Antimicrobial efficacy of orchid extracts as potential inhibitors of antibiotic resistant strains of *Escherichia coli*. Asian J Pharm Clin Res 6(3):108–111

Chuakul W (2002) Ethnomedical uses of Thai Orchidaceous plants. Mohidol Univ J Pharm Sci 29 (3-4):41–45

Chuang SS, Lin CC, Lin J, Yu KH, Hsu YF, Yen MH (2003) The hepatoprotective effects of Limonium sinense against carbon tetrachloride and beta-D-galactosamine intoxication in mice. Phytother Res 17 (7):784–791

Christensen E (1993) Aerides. Am Orchid Soc Bull 62 (6):594–609

Christensen H (2002) Ethnobotany of the Iban and the Kelabit. Forest Department Sarawak; NEP Con Denmark; and Aarhus: University of Aarhus

Comber JB (2001) Orchids of Sumatra. Natural History Publications (Borneo) and Singapore Botanic Gardens, Kota Kinabalu

Cooray DA (1940) Orchids in oriental literature. Orchids Zelandica 7:73–80

Cootes J (2001) Orchids of the Philippines. Marshall Cavendish, Singapore

Crowley P (1995) Antenatal corticosteroid therapy: a meta-analysis of the randomized trials, 1972–1994. Am Obstet Gynecol 173:322–335

Cui SC, Yu J, Zhang XH et al (2013) Antihyperglycemic and antioxidant activity of water extract from Anoectochilus roxburghii in experimental diabetes. Exp Toxicol Pathol 65(5):485–488

Culpeper N (1653) Complete herbal: consisting of a comprehensive description of nearly all herbs with their medicinal properties and directions for compounding the medicines extracted from them

Dash PK, Sahoo S, Bal S (2008) Ethnobotanical studies on orchids of Niyamgiri Hill Ranges, Orissa, India. Ethnobot Leaflet 12:70–78

Descamps E, Petrault-Laprais M, Maurois P et al (2009) Experimental stroke protection induced by 4-hydroxybenzyl alcohol is cancelled by bacitracin. Nerosci Res 64(2):137–142

Du XM, Irino N, Furusho N, Hiyashi J, Shoyama Y (2008) Pharmacologically active compounds in the Anoectochilus and Goodyera species. Nat Med (Tokyo) 62(2):132–148

Du XM, Sun N, Tamura T, Mohri A, Sugiura M, Toshizawa T, Irino N, Hayashi J, Shoyama Y (2001) Higher yielding isolation of kinsenoside in Anoectochilus and its antihyperliposis effect. Biol Pharm Bull 24(1):65–69

Du XM, Sun NY, Chen Y, Irino N, Shoyama Y (2000) Hepatoprotective aliphatic glycosides from three Goodyera species. Biol Pharm Bull 23(6):731–734

Du XM, Sun NY, Hayashi J, Chen Y, Sugiura M, Shoyama Y (2003) Hepatoproptective and antihyperliposis activities of in vitro cultured Anoectochilus formosanus. Phytother Res 17 (1):30–33

Du XM, Sun NY, Furusho N, Hayashi J, Shoyama Y (2007) Effect of in-vitro cultured Anoectochilus formosanus on lipid metabolism in clinical uses. Am J Chin Med 35(5):735–741

Duggal SC (1971) Orchids in human affairs (A review). Pharm Biol 11(2):1727–1734

Estler CJ, Gabrys MC (1979) Swimming capacity of mice after prolonged treatment with psychostimulants, II. Effect of methamphetamine on swimming performance and availability of metabolic substrates. Psychopharmacology (Berl) 60(2):173–176

Fang HL, Wu JB, Lin WL, Ho HY, Lin WC (2008) Further studies on the hepato-protective effects of Anoectochilus formosanus. Phytother Res 22(3):291–6)

Ferrier H (2006) Cypripediums in China. Part I: A history of Cypripedium in China. Orchids 75(10):764–771

Flach A, Dondon RC, Singer RB et al (2004) The chemistry of pollination in selected Brazilian Maxillariinae orchids: floral rewards and fragrances. J Chem Ecol 30 (5):1045–1056

Gao C, Zhang FS, Zhang J et al (2009) Identification of Anoectochilus based on rDNA ITS sequences alignment and SELDI-TOF-MS. Int J Biol Sci 5 (7):727–735

Gao WW, Guo SX (2001) Effects of endophytic fungi hyphae and their metabolites on the growth of Dendrobium candidum and Anoectochilus roxburghii. Zhongguo Yi Xue Ke Xue Yuan Xue Bao 23 (6):556–559

Gao XM, Yang LY, She YQ et al (2013) Phenolic compounds fro Arundina graminifolia and the anti-Tobacco Mosaic Virus activity. Bull Kor Chem Soc 33(7):2447–2449

Germano MP, D'Angelo V, Sanoogo R, Morabito A, Pergolizzi S, De Pasquaale R (2001) Hepatoprotective activity of Trichillia roka on carbon tetrachloride-induced liver damage in rats. J Pharm Pharmacol 53 (11):1569–1574

Ghorbani A, Gravendeel B, Zarre S, de Booer H (2014a) Illegal wild collection and international trade in CITES-listed terrestrial orchid tubers in Iran. Traffic Bull 26(2):52–58

Ghorbani A, Gravendeel B, Naghibi F, de Booer H (2014b) Wild orchid tuber collection in Iran: a wake-up call for conservation. Biodivers Conserv 23:2749. doi:10.1007/s10531-014-0746-y

Goldman D (2005) Arundina, ecology. In: Pridgeon AC, Cribb PJ, Chase MW, Rasmussen FN (eds.) Genera Orchidacearum, Vol 4 Epidendroideae (Part One). Oxford, University Press

Grant B (1895) The Orchids of Burma. Hanthawaddy Press, Rangoon

Griffith RE (1847) Medical Botany of descriptions of the more important plants used in medicine, with their history, properties and mode of administration. Lea and Blanchard, Philadelphia

Guo SX, Fan L, Cao WQ et al (1997) Mycena anoectochila nov. isolated from mycorrhizal roots of Anoectochilus roxburghii from Xishuanbanna. Mycologia 89(6):952–954

Gurong DB (2006) An illustrated guide to the orchids of Bhutan. DSB Publications, Thimphu

Gurung DB (2005) An illustrated guide to the orchids of Bhutan. DSB Publications, Thimpu

Han MH, Yang YW, Jin YP (2008) Novel terpenoid acyl esters and alkaloids from Anoectochilus roxburghii. Ohytochem Anal 19(5):438–443

He CN, Wang CL, Guo SX, Yang JS, Xiao PG (2005) Study on chemical constituents in herbs of Anoectochilus roxburghii II. Zhongguo Zhong Yao Za Zhi 30(10):761–763 (in Chinese)

He CN, Wang CL, Guo SX, Yang JS, Xiao PG (2006) A novel flavonoid glucoside from Anoectochilus roxburghii (Wall.) Lindl. J Integr Plant Biol 48 (3):359–363

Hedley C (1888) Uses of Queensland plants. Proc R Soc Queensland 5:10–13

Heiningen MV (2014) Overview of the wild orchids that occur in the Cantabrian mountains. http://natuur-cantabrisch.blogspot.com/

Holttum RE (1964) Orchids of Malaya, 3rd edn. Government Printers, Singapore

Hooper D (1937) Useful drugs of Iran and Iraq. Chicago, Field Museum of Natural History IX (3)

Hooper R, Akerly S (1829) Lexicon medium or medical dictionary. 4th American edition. Collins and Hannay, New York, p 137

Hoskovrc L (2007) Anacamptis pyramidlis (L.) Rich. – Rrudhiavek pyramidal/Red pyramidal. Botany.cz. 15.7.2007

Hsiao HB, Lin H, Wu JB, Lin WC (2013) Kinsenoside prevents ovariectomy-induced bone loss and suppresses osteoclastogenesis by regulating classical NF-kB pathways. Osteoporos Int 24(5):1663–1676

Hsiao HB, Wu JB, Lin H, Lin WC (2011) Kinsenoside isolated from Anoectochilus ormosanus suppresses LPS-stimulated inflammatory reactions in macrophages and endotoxin shock in mice. Shock 35 (2):184–190

Hsieh MT, Wu CR, Chen CF (1997) Gastrodin and p-hydroxybenzyl alcohol facilitate memory consolidation and retrieval, but not acquisition, on the passive avoidance task in rats. J Ethnopharmacol 56:45–54

Hsieh CC, Hsiao HB, Lin WC (2010) A standardized aqueous extract of Anectochilus formosanus modulated airway hyperresponsiveness in an OVA-inhaled murine model. Phytomed (Jena) 17(8-9):557–562

Hu QF, Zhou B, Huang JM (2013a) Antiviral phenolic compounds from Arundina graminifolia. J Nat Prod 76 (2):292–296

Hu QF, Zhou B, Ye YQ et al (2013b) Cytotoxic deoxybenzoins and diphenylethylenes from Arundina graminifolia. J Nat Prod 76(10):1854–1859

Ikeuchi M, Yamaguchi K, Nishimura T, Yazawa K (2005) Effects of Anoectochilus formosanus on endurance capacity in mice. J Nutr Sci Vitaminol (Tokyo) 51 (1):40–44

Ito A, Yasumoto K, Kasai R, Yamasaki K (1994) A sterol with an unusual side chain from Anoectochilus koshunensis. Phytochemistry 36(6):1465–1467

Jalal JS, Kumar P, Pangtey YPS (2008) Ethnomedicinal Orchids of Uttarakhand. Western Himalayas Ethnobot Leaflets 12:1227–1230

Jayaweera DMA (1981) A revised handbook of the flora of Ceylon, vol II. A.A. Balkema, Rotterdam

JETRO TTPP 8/9/2008: *Anoectochilus formosanus*

Johnston SRD, Howell A (2002) Endocrine treatment of advanced breast cancer: selective estrogen–receptor modulators (SERMS). In: Miller WR, Ingle JN (eds) Endocrine therapy in breast cancer. Marcel Dekker, New York

Joseph J (1982) Orchids of Nilgiris. Records of the Botanical Survey of India, vol XXII. Botanical Survey of India (Department of Environment), Howrah

Jung KA, Han D, Kwon EK, Lee CH, Kim YE (2007) Antifatigue effect of Rubus coreanus Miquel extract in mice. J Med Food 10(4):689–693

Kam KY, Yu SJ, Jeong N et al (2011) p-Hydroxybenzyl alcohol prevents brain injury and behavioral impairment by activating Nrf2, PDI and neurotrophic factor genes in a rat model of brain ischaemia. Mol Cell 31(3):209–215

Kamemoto H, Sagarik R (1975) Beautiful Thai orchid species. Orchid Society of Thailand, Bangkok

Kocyan A (2010) Apostasioideae—the least known orchid sub-family. Malesian Orchid J 5:125–138

Kovacs A, Vasas A, Hohmann J (2007) Natural phenanthrenes and their biological activity. Phytochemistry 69:1084–1110

Lawler LJ (1984) Ethnobotany of the orchidaceae. In: Arditti J (ed) Orchid biology reviews & perspectives 3. Cornell University Press, Ithaca

Lee SC, Tsai CC, Chen JC, Lin CC, Hu ML, Lu S (2002a) The evaluation of reno- and hepato- protective effects of huai-shan-yao (Rhizome Dioscoreae). Am J Chin Med 30(4):609–616

Lee SC, Tsai CC, Chen JC, Lin JG, Lin CC, Hu ML, Lu S (2002b) Effects of "Chinese yam" on hepato-nephrotoxicity of acetaminophen in rats. Acta Pharmacol Sin 23(6):503–508

Li H, Guo HJ, Dao ZL (eds) (2000) Flora of Gaoligong Mountains. Science Press, Kunming

Li NH (ed) (1988) Chinese medicinal herbs of Hong Kong, vol 2. Hong Kong Chinese Medicinal Research Institute, Hong Kong

Lin CC, Chen SY, Lin JM, Chiu HF (1992a) The pharmacological and pathological studies on Taiwan folk medicine (VIII): the anti-inflammatory and liver protective effects of "mu-mien". Am J Chin Med 20 (2):136–146

Lin CC, Chen YL, Lin JM, Ujiie T (1997a) Evaluation of the antioxidant and hepatoprotective activity of *Terminalia catappa*. Am J Chin Med 25 (2):153–161

Lin CC, Hsu YF, Lin TC, Hsu FL, Hsu HY (1998) Antioxidant and hepatoprotective activity of punicalagin and punicalin on carbon tetrachloride-induced liver damage in rats. J Pharm Pharmacol 50(7):789–794

Lin CC, Huang PC, Lin JM (2000) Antioxidant and hepatoprotective effects of Anoectochilus formosanus and Gynostemma pentaphyllum. Am J Chin Med 28 (1):87–96

Lin CC, Lee HY, Chang CH, Yang JJ (1999) The anti-inflammatory and hepatoprotective effects of fractions from *Cudrania cochinchinensis* var. gerontogea. Am J Chin Med 27(2):227–239

Lin CC, Lin JM, Chiu HF (1992b) Studies on folk medicine "thang-kau-tin" from Taiwan. (I). The anti-inflammatory and liver-protective effect. Am J Chin Med 20(1):37–50

Lin CC, Lin WC, Yang SR, Shieh DE (1995) Anti-inflammatory and hepatoprotective effects of Solanum alatum. Am J Chin Med 23(1):65–69

Lin CC, Ng LT, Yang JJ, Hsu YF (2002a) Anti-inflammatory and hepatoprotective activity of peh-hue-juwa-chi-cao in male rats. Am J Chin Med 30(2–3):225–234

Lin CC, Shieh DE, Yen MH (1997b) Hepatoprotective effect of fractions of Ban-zhi-lian on experimental liver injuries in rats. J Ethnopharmacol 56(3):193–200

Lin JM, Lin CC, Chiu HF, Yang JJ, Lee SG (1993) Evaluation of the anti-inflammatory and liver-protective effects of Anoectochilus formosanus, Ganoderma lucidum and Gynostemma pentaphyllum in rats. Am J Chin Med 21(1):59–69

Lin KJ, Chen JC, Tsauer W, Lin CC, Lin JG, Tsai CC (2001) Prophylactic effect of four prescriptions of traditional Chinese medicine on alpha-naphthylisothiocyanate and carbon tetrachloride induced toxicity in rats. Acta Pharmacol Sin 22 (12):1159–1167

Lin SC, Lin CC, Lin YH, Shyuu SJ (1994) Hepatoprotective effects of Taiwan folk medicine: Wedelia chinensis on three hepatotoxin-induced hepatotoxicity. Am J Chin Med 22(2):155–168

Lin SC, Lin CH, Lin CC, Lin YH, Chen CF, Chen IC, Wang LY (2002b) Hepatoprotective effects of Arctium lappa Linne on liver injuries induced by chronic ethanol consumption and potentiated by carbon tetrachloride. J Biomed Sci 9(5):401–409

Lin SC, Lin CC, Lin YH, Supriyatna S, Pan SL (1996) The protective effect of Alstonia scholaris R. Br On hepatotoxin-induced acute liver damage. Am J Chin Med 24(2):153–164

Lin TP (1977) Native orchids of Taiwan, Vol. 2, pp. 166–7, colour photo 98

Lin WC, Hsieh CC (2005) Commercial application of Anoectochilus formosanus: immunomodulating activities. Int J Appl Sci Eng 3(3):175–178

Lin M, Lv H, Ding Y (2012) Antitumoral bibenzyl derivatives from tuber of Arundina graminifolia. Zhongguo Zhong YaoZa Zhi 37(1):66–70

Lin MF, Ding Y, Zhang DM (2005a) Phenanthrene constituents from rhizome of Arundina graminifolia. Zhongguo Zhong Yao Zazhi 30(5):353–356

Lin MF, Han Y, Xing DM et al (2004a) A new stilbenoid from Arundina graminifolia. J Asian Nat Prod Res 6 (3):229–232

Lin MF, Han Y, Xing DM et al (2004b) Chemical constituents from the rhizome of Arundina graminifolia. Zhongguo Zhong Yao Zazhi 29(2):147–149

Lin MF, Han Y, Xing DM, Wang W, Xu LZ, Du LJ, Ding Y (2005b) One new benzyldihydrophenanthreene from Arundina graminifolia. J Asian Nat Prod Res 7 (5):767–770

Liu SH, Pan IH, Im C (2007) Inhibitory effect of p-hydroxybenzyl alcohol on tyrosinase activity and melanogenesis. Biol Pharm Bull 30(6):1135–1139

Liu SH, Chu IM, Pan IH (2008) Effects of hydroxybenzyl alcohols on melanogenesis in melanocyte-keratinocyte co-culture and monolayer culture of melanocytes. J Enzyme Inhib Med Chem 23(4):526–534

Liu TS, Su HJ (1977) Flora of Taiwan. Taiwan National University, Taipei

Low T (1987) Australian wild foods. Ground orchids—Salute to Saloop. Aust Nat Hist 22(5):202–203

Lu KL, Chang YS, Ho LK et al (2000) The evaluation of the therapeutic effect of tao- tsang –tsao on alpha-napthylisothiocyanate and carbon tetrachloride-induced acute liver damage in rats. Am J Chin Med 28(3-4):361–370

Lu HC, Hsieh MH, Chen CE et al (2012) A high-through-put virus-induced gene-silencing vector for screening transcription factors in virus-induced plant defense response in orchid. Mol Plant Microbe Interact 25 (6):738–746

Luning B (1967) Studies on Orchidaceae alkaloids. IV. Screening of species for alkaloids 2. Phytochemistry 6:857–861

Lüning B (1974) Alkaloids of the Orchidaceae. In: Withner CL (ed) The orchids: scientific studies. John Wiley & Songs, New York

Luning B (1975) Hunting orchids for chemistry. In: Senghas SK (ed) Proceedings, 8th World Orchid Conference, Frankfurt, 538–9

Ma ZQ, Li SS, Zhang MJ et al (2010) Light Intensity affects growth, photosynthetic capability, and total flavonoid accumulation of Anoectochilus plants. Hortsci 45(6):863–867

Majumder PL, Banerjee S, Maiti DC, Sen S (1995) Stilbenoids from the orchids Agrostophyllum callosum and Coelogyne flaccida. Phytochemistry 39:649–653

Majumder PL, Banerjee S, Sen S (1996) Three stilbenoids from the orchid Agrostophyllum callosum. Phytochemistry 42:847–852

Majumder PL, Banderjee S, Lahari S et al (1998) Dimeric phenanthrenes from two Agrostophyllum species. Phytochemistry 47:855–860

Majumder PL, Ghosal S (1991) Arundinol, a new triterpene from the orchid Arundina babbusifolia. J Indian Chem Soc 68:88–91

Majumder PL, Ghosal S (1993) Two stilbenoids from the orchid Arundina bambusifolia. Phytochemistry 32:439–444

Majumder PL, Ghosal S (1994) Two stilbenoids from the orchid Arundina bambusifolia. Phytochemistry 35 (1):205–208

Majumder PL, Majunder S, Sen S (2003) Triperoenoids from the orchids Agrostophyllum brevipesand Agrostophyllum callosum. Phytochemistry 62:591

Majumder PL, Sabzabadi E (1988) Agrostophyllin, a naturally occurring phenanthropyran derivative from Agrostophyllum khasiyanum. Phytochemistry 27 (6):1899–1901

Majumder PL, Sen S, Banerjee S (1999) Agrostophyllol and isoagrostophyllol, two novel disteromeric 9,10-dihydrophenanthropyran derivatives from the orchid Agrostrophyllum callosum. Tertahedron 55 (21):13

Masuda K, Ikeuchi M, Koyama T, Yamaguchi K, Woo JT, Nishimura T, Yazawa K (2008) Suppressive effects of Anoectochilus formosanus extract on osteoclast function in vitro and bone resorption in vivo. J Bone Miner Metab 26(2):123–129

Misra S (2007) Orchids of India A Glimpse. Bishen Singh Mahendra Pal Singh, Delhi

Muir HJ (1987) Symbiotic micro propagation of Orchis laxiflora. Orchid Rev 95:27–29

Murase T, Haramizu S, Shimotoyadome A, Nagasawa A, Tokimitsu I (2005) Green tea extract improves endurance capacity and increases muscle lipid oxidation in mice. Am J Physiol Regul Integr Comp Physiol 288 (3):R708–R715

Nanakorn W, Watthana S (2008) Queen Sirikit Botanic Garden (Thai Native Orchids 1 and 2). Wanida Press, Chiang Mai

Neiland MRM (2001) Anacamptis, ecology. In: Pridgeon AM, Cribb PJ, Chase MW (eds.) Genera Orchidacearum Vol. 2 Orchidoideae (Part 1). Oxford, Oxford University Press

Okovityi SV, Arkad'eva AV, Bezborodkina NN, Sakuta GA, Iaroslavtsev MI, Shulenin SN, Kudriavtsev BN (2007) New protective effect of simvastatin in rats with experimental steatohepatitis. Eksp Klin Farmakol 70(3):43–5 (in Russian) (English Abstract in PubMed)

Ou JC, Hsieh WC, Lin IH, Chang YS, Chen IS (eds) (2003) The catalogue of medicinal plant resources in Taiwan. Department of Health, Executive Yuan, Taipei

Pant B, Raskoti BB (2013) Medicinal orchids of Nepal. Kathmandu, Himalayan Map House (P) Ltd

Perry LM, Metzger J (1980) Medicinal plants of East and Southeast Asia: attributed properties and uses. MIT Press, Cambridge, MA

Pettersson B (1976) Orchids and men in European landscapes. In: Senghas K (ed) Proceedings of 8th World Orchid Conference. Deutsch Orchideen Gesell-schaft, Heidelberg, pp 80–83

Pridgeon AM, Cribb PJ, Chase MW, Rasmussen FN (eds) (2001) Genera Orchidacearum, Vol. 2. Orchidoideae (Part 1). Oxford University Press, Oxford, pp 251–255

Rao AN (2004) Medicinal orchid wealth of Arunachal Pradesh. Newslett ENVIS NODE Indian Med Plant 1 (2):1–7

Rao TA, Sridhar S (2007) Wild Orchids in Karnataka. A pictorial compendium. Institute of Natural Resources Conservation, Education, Research and Training (INCERT), Bangalore

Raskoti BB (2009) The Orchids of Nepal. Bhakta Bahadur Raskoti and Rita Ale, Kathmandu

RCOG Guideline No. 7 (1996) Antenatal corticosteroids to prevent respiratory distress syndrome

Recart W, Ackerman JD, Falcon W, Hernandez P (2014) Here and there and everywhere: the invasion experience differs among islands for the bamboo orchid. www.ibparticipation.org/pdf/wilnelia_Phase3_Poster. pdf

Ridley H (1894) The Orchidaceae and Apostasiaseae of the Malay Peninsula. J Linn Soc 32:335–338

Ridley H (1907) Materials for a flora of the Malay Peninsula, vol 1. Methodist Publishing House, Singapore

Ryu BK, Ahn BO, Oh TY, Kim SH, Kim WB, Lee EB (1998) Studies on protective effect of DA-9601, Artemisia asiatica extract, on acetaminophen and CCl4-induced liver damage in rats. Arch Pharm Res 21 (5):508–513

Santapau H, Kapadia Z (1966) The orchids of Bombay. Government of India Press, Calcutta

Sathish Kumar C, Manilal KS (1994) A catalogue of Indian orchids. Bishen Singh Mahendra Pal Singh, Hehra Dun

von Schlechter R (1919) Orchidelogiae Sino-Japonicae Prodromus Eine Kritische Besprechung der Orchideen Ost-Asiens. Berlin, Verlag Des Repertoruins, pp. 92–93, 99. 50, 125–135

Schuiteman A, de Vogel EF (2000) Cac Ci Ho Lan (Orchidaceae) Cua Thai Lan, Lao, Campuchia Va Viet Nam. Orchid Genera of Thailand Laos, Cambodia and Vietnam. (Vietnamese-English edition). Leiden, National Herbarium Nederland

Schultes RE, Pease AS (1963) Generic names of Orchids. Their origin and meaning. Academic, New York & London

Seidenfaden G, Wood JJ (1992) The orchids of peninsular Malaysia and Singapore. Olsen & Olsen, Fredensborg

Sevgi E, Attandag E, Kara O et al (2012) Studies in the morphology, anatomy and ecology of *Anacamptis pyramidalis* (L.) L. C. M. Richard (Ochidaceae) in Turkey. AGRIS Records, FAO

Sezik E (1967) Turkiye'nin Salepgilleri Ticari Salep Cesitleri ve Ozellikle Mugla Salebi Uzerinde Arastirmalar. Doctoral Thesis. Istanbul Universitesi Eczacihk Fakultesinde (In Turkish. Summary in English)

Sezik E (1990) Turkiye'nin orkideleri. Bilim ve Teknik 269:5–8 (quoted by Ericisli, Esitken, 2002)

Shiau YJ, Hsia CN, Tsay HS (2006) In vitro study, conservation and utilization of medicinal plant Anoectochilus formosanus in Taiwan. ISHS Acta Horticulturae 764: XXVII International Horticultural Congress—IH 2006: International Symposium on Plant Biotechnology: From Bench to Commercialization

Shih CC, Wu YW, Lin WC (2001) Ameliorative effects of Anoectochilus formosanus extract on osteopaenia in ovariectomised rats. J Ethnopharmacol 77 (2-3):233–238

Shih CC, Wu YW, Lin WC (2002) Antihyperglycaemis and anti-oxidant properties of Anoectochilus formosanus in diabetic rats. Clin Exp Pharmacol Physiol 29(8):684–688

Shih CC, Wu YW, Lin WC (2003) Scavenging of reactive oxygen species and inhibition of the oxidation of low density lipoprotein by aqueous extraction of Anoectochilus formosanus. Am J Chin Med 31 (1):25–36

Shih CC, Wu YW, Hsieh CC, Lin WC (2004) Effect of Anoectochilus formosanus on fibrosis and regeneration of the liver in rats. Clin Exp Pharmacol Physiol 31 (9):620–625

Shyur LF, Chen CH, Lo CP, Wang SY, Kang PL, Sun SJ, Chang CA, Tzeng CM, Yang NS (2004) Induction of apoptosis in MCF-7 human breast cancer cells by phytochemicals from Anoectochilus formosanus. J Biomed Sci 11(6):928–939

Singh A, Duggal S (2009) Medicinal orchids: an overview. Ethnobot Leaflets 13:351–363

Stajner D, Popovic B, Kapo A et al (2010) Antioxidant and Scavenging Capacity of Anacamptis pyramidalis—Pyrimidal Orchid from Vojvodina. Phytother Res 24(5):759–763

Stoessl LJ, Arditti J (1984) Orchid Phytoalexins. In: Arditti J (ed) Orchid biology reviews and perspective, vol III. Comstock Publishing Associates, Cornell University Press, Ithaca

van Steenis CGGJ (1958) Magic plants of the Dayak. Sarawak Museum J 11:432–436

Stewart I, Webb PM, Schluter PJ, Shaw GR (2006) Recreational and occupational field exposure to freshwater cyanobacteria—a review of anecdotal and case reports, epidemiological studies and the challenges for epidemiologic assessment. Environ Health 5:6

Subedi A, Kunwar B, Choi Y et al (2013) Collection and trade of wild-harvested orchids in Nepal. J Ethnobiol Ethnomed 9:64–73

Tang MJ, Guo SX (2004) Effect of endophytic fungi on the culture and four enzyme activities of Anoectochilus formosanus. Zhongguo Zhong Yao Za Zhi 29(6):517–520

Teoh ES (2005) Orchids of Asia, 3rd edn. Marshall Cavendish, Singapore

Teoh ES, Teoh K (1991) Over 45 feeling fabulous. Menopause and the hormone replacement controversy. Times Editions, Singapore

Trivedi VP, Dixit RS, Lal VK (1980) Orchids in the drug markets of Bareilly, Kanpur and nearby districts. Nagarjun (Calcutta) 23(8):157–163

Tsavakelova A, Cherdyntseva TA, Netrusov AI (2005) Auxin production by bacteria associated with orchids. Mikrobiologica 74(1):55–62

Tsavkelova EA, Lobakova ES, Kolomeitseva GL et al (2003) Localization of associative cyanobacteria on the roots of epiphytic orchids. Mikrobiologicia 72 (1):99–104

Tsay HS (2002) Tissue culture of Anoectochilus formosanus. In: Proceedings of the 16th World Orchid Conference, Kuala Lumpur, p. 438

Tseng CC, Shang HF, Wang LF, Su B, Hsu CC, Kao HY, Cheng KT (2006) Antitumor and immunostimulating effects of Anoectochilus formosanus Hayata. Phytomedicine 13(5):355–370

United States Patent 7033617 filed by Shyur LF, Yang NS, Kang PL, Sun SJ, Wang SY (2002). Use of Anoectochilus formosanus plant extracts and their derived fractions as herbal medicines or nutraceutical supplements for chemoprevention or treatment of human malignancies. University of Yangon Department of Botany (undated, possibly 2004): Myanmar Native Orchids

Vaddhanaphuti N (1997) A field guide to the wild orchids of Thailand. Silkworm Books, Chiang Mai

Vaddhanaphuti N (2001) A field guide to the wild orchids of Thailand, 3rd edn. Silkworm Books, Chiang Mai

Vaddhanaphuti N (2005) A field guide to the wild orchids of Thailand, Fourth and expanded edn. Silkworm Books, Chiang Mai

Valcheva-Kuzmanova S, Borisova P, Galunska B, Krasnalaaiev I, Belcheva A (2004) Hepatoprotective effect of the natural fruit juice from Aronia melanocarpa on carbon tetrachloride-induced acute liver damage in rats. Exp Toxicol Pathol 56 (3):195–201

van Apeldoorn ME, van Egmond HP, Speijers GJ, Bakker GJ (2007) Toxins of cyonobacteria. Mol Nutr Food Res 51(1):7–60

van den Brink RCB (1937) Synopsis of the vernacular names and the economic use of the indigenous orchids of Java. Blumea Suppl 1:38–51

van Rheede HA (1693) Hortus Indicus Malabaricus, vol 12. Dutch East India Company, Kerala

Veitch NC, Grayer B (2003a) Phytochemistry of Anacamptis. In: Pridgeon AM, Cribb PJ, Chase MW, Rasmussen FN (eds) Genera Orchidacearum, vol 3, Orchidoideae (Part Two) Vanilloideae. Oxford University Press, Oxford

Veitch NC, Grayer B (2003b) Phytochemistry of Anoectochilus. In: AM Pridgeon, PJ Cribb, MW Chase, FN Rasmussen (eds) Genera Orchidacearum, Vol. 3. Orchidoideae (Part Two) Vanilloideae. Oxford, University Press

Veeraju P, Rao P, Rao NSJ et al (1989) Bibenzyls and phenanthreroids of some species of Orchidaceae. Phytochemistry 28:3031–3034

Vidal J (1963) Les plantes utiles Du Laos. Cryptograms—Gymnospermes—Monocotyledones. Museum National d'Historie Naturelle, Paris

de Vogel EF (1969) Monograph of the tribe Apostasieae (Orchidaceae). Blumea 17:313–350

Wang LF, Lin CM, Shih CM, Chen HJ, Su B, Tseng CC, Gau BB, Cheng KT (2005) Prevention of cellular oxidative damage by an aqueous extract of Anoectochilus formosanus. Ann NY Acad Sci 1042:379–386

Wang SY, Kuo YH, Chang HN, Pl K, Tsay HS, Lin KF, Yang NS, Shyur LF (2002) Profiling and characterization antioxidant activities in Anoectochilus formosanus Hayata. J Agric Food Chem 50 (7):1859–1865

Wedeck HE (1961) Dictionary of Aphrodisiacs. Philosophical Library, New York, p 216

Wood J, Ramsey M (2004) Plate 482. Anacamptis laxiflora Orchidaceae. Curtis's Botanical Magazine 21(1):26–33

Wu CR, Hsieh MT, Liao J (1996) p-Hydroxybenzyl alcohol attenuates learning deficits in the inhibitory avoidance task: involvement of serotonergic and dopaminergic systems. Clin J Physiol 39:265–273

Wu TL, Hu QM, Xia NH, Lai PCC, Yip KL (2001) Check list of Hong Kong plants 2001. Agriculture, Fisheries and Conservation Department Bulletin 1, Hong kong

Wu JB, Lin WL, Hsieh CC, Ho HY, Tsay HS, Lin WC (2007) The hepatoprotective activity of kinsenoside from Anoectochilus formosanus. Phytother Res 21 (1):58–61

Wu JB, Chuang HR, Yang LC, Lin WC (2010) A standardized extract of Anoectochilus formosanus ameliorated thioacetamide-induced liver fibrosis in mice. The role of Kupffer cells. Biosci Biotechnol Biochem 74(4):781–787

Wu K (1997) Quick propagation and immediate test on Anoectochilus formosanus. Zhong Yao Cai 20 (12):595–597 (in Chinese)

Wu TL, Hu QM, Xia NH, Lai PCC, Yip KL (2002) Check list of Hong Kong plants 2001. Hongkong, Agriculture, Fisheries and Conservation Department Bulletin 1 (Revised)

Wu XR (1994) A concise edition of medicinal plants in China. Guangdong Higher Education Publication House, Guangdong (in Chinese)

Yang LC, Lu TJ, Lin WC (2013) A Type II Arabinogaactan from Anoectochilus formosanus for G-CSF production in macrophages and leukopenia improvement in CT-28 bearing Mie treated with 5-Fluorouravil. Evid Based Complement Alternat Med

Yang NSD, Shyur LF, Chen CH, Wang SY, Tzeng CM (2004) Medicinal herb extract and a single-compound drug confer similar complex pharmacogenomic activities in mcf-7 cells. J Biomed Sci 11(3):418–422

Yen MH, Lin CC, Chuang CH, Liu SY (1991) Evaluation of root quality of Bupleurum species by TLC scanner and the liver protective effects of xiao-chai-hu-tang" prepared using three different Bupleurum species. J Ethnopharmacol 34(2-3):155–165

Yong HS (1990) Orchid portraits. Tropical Press Sdn Bhd, Kuala Lumpur

Yu SJ, Kim JR, Lee CK et al (2005) Gastrodia elata Blume and an active component, p-hydroxybenzyl alcohol reduce focal ischaemic brain injury through antioxidant related gene expression. Biol Pharm Bull 28:1016–1020

Zhang FS, Lu YL, Zhao Y, Guo SX (2013) Promoting role of an endophyte on the growth and contents of kinsenosides and flavonoids of Anoectochilus formosanus Hayata, a rare and threatened orchidaceous plant. J Zhejiang Univ Sc B 14(9):785–792

Zhang Y, Cai J, Ruan H et al (2007a) Antiglycemic activity of kinsenoside, a high yielding constituent of Anoectochilus roxburghii in streptozoin diabetic rats. J Ethnopharmacol 114(2):141–145

Zhang YH, Cai JY, Pi HF, Wu JZ (2007b) Antihyperglycaemic activity of kinsenoside, a high yielding constituent from Anoectochilus roxburghii in steptozoin treated rats. J Ethnopharmacol 114 (2):141–145

Zhongyao Da Cidian (1986 reprint of 1997): compiled by the Jiangsu College of New Medicine in 1977

Zhonghua Bencao vol. 8 (2000) edited by Hu XM, Zhang WK, Zhu QZ, et al. Shanghai, Shanghai Scientific and Technical Press, 1999

Zurawell RW, Chen H, Burke JM, Prepas EE (2005) Hepatotoxic cyanobacteria: a review of the biological importance of microcystins in freshwater environments. J Toxicol Environ Health B Crit Rev 8(1):1–37

Genus: *Bletilla* Rchb. f.

Chinese name: *Baiji*

Bletilla are terrestrial orchids with only seven species in the genus, distributed from China to Japan, Myanmar and Vietnam. They are small to medium-sized sympodial plants with subterranean, tuberous rhizomes and short stems that bear a few plicate leaves sheathing at the base; they are deciduous. Inflorescence is terminal and loose with several flowers. Flowers are medium-sized, star-shaped and showy, with plain coloured tepals and a lip that is marked by yellow-coloured keels. *Bletilla* are found in open shrubbery at low elevations. In cultivation, they thrive on sandy soil boosted with compost and are reported to be hardy, given a temperate climate (Weaver and Stoutamire 2008).

One species, *B. striata* (*baiji*), has been in continuous medicinal usage in China for over 2000 years (Fig. 8.1). The generic names (*Bletia* and *Bletilla*) honour Don Luis Blet, a Spanish apothecary who lived in the eighteenth century.

Bletilla foliosa (King & Pantl.) Tang & F.T. Wang

Syn. *Bletilla sinensis* (Rolfe) Schltr.; *Bletilla chinensis* Schltr.

Chinese name: *Xiaobaiji* (small *Baiji*; small white root)

Vietnamese name: *Bach cap*; *Hoa lan tia*

Description: Plants are 15–20 cm tall with a subglobose rhizome, 1–1.5 cm in diameter; leaves are 2–3 elliptic-lanceolate, 5–12 by 0.8–3 cm, ensheathing the short stem at the base. Inflorescence is up to 15 cm tall with 1–3 pale-purple, nodding flowers, 2 cm across. Tepals are linear-lanceolate, white, tinged with purple. The lip is white, fimbriate and purple towards the tip. Flowering season is May to June. The species is found in Yunnan, Myanmar and Thailand (Chen et al. 2009). In Thailand, it flowers in January to February (Vaddhanaphuti 2005).

Medicinal Usage: Identical to *baiji* (see *B. striata*). Herbs are obtained from Hubei, Sichuan and Yunnan (Wu 1994).

Bletilla formosana (Hayata) Schltr.

Chinese names: *Taiwanbaiji* (Taiwan *Baiji*), hyacinth orchid, Chinese ground orchid, white rhizome orchid, *Xiao Baiji* (Small *Baiji*)
Japanese name: *Shi-ran* (purple orchid)

Description: This species is similar to *B. striata*, differing from it only in the narrower leaves and smaller flowers, white flushed with pink and of various intensities but never rose-purple, with lips of various colours. Flowers do not open widely. Plants are of variable size, from 12 to

© Springer International Publishing Switzerland 2016
E.S. Teoh, *Medicinal Orchids of Asia*, DOI 10.1007/978-3-319-24274-3_8

Fig. 8.1 *Bletilla striata* (Thunb.) Rchb.f. (as *Bletia gebina* Lindl.). From: Edward's Botanical Register, vol 33: t. 60 (1847) (Original drawing by S.A. Drake in colour with black and white). Courtesy of Missouri Botanical Gardens, St. Louis, USA

60 cm tall. A perennial herb found in sunny locations on rocks from the coast to an elevation of 1800 m in Taiwan, it also occurs in southern Gansu, southern Shaanxi, Sichuan, Yunnan, Jiangxi and Guangxi at 1200–3100 m in open country. Flowering season is March to June, but it may extend to October (Hsu 1994).

The variety *kotoensis* (Hay) T.P. Lim [syn. *B. kotoensis* (Hay) Schltr.], whose flowers are white except for the yellow keels and reddish-brown spots on the lip, occurs only on Lanyu (Orchid) Island.

Herbal Usage: *Xiao Baiji* (*B. formosana*) is obtained from Shaanxi, Sichuan, Yunnan, Guizhou, Guangxi and Taiwan (Wu 1994). Traditional Chinese medicine (TCM) uses the stems to strengthen the lungs, stop bleeding and reduce

swelling. They are used to treat patients suffering from tuberculous cough, bronchiectasis, bleeding peptic ulcers and nose-bleed (Ou et al. 2003). In India, scrapings of the stem are applied to treat cracks on the heel (Rao 2004).

Phytochemistry: Working with the whole plant of *B. formosana*, Lin et al. (2005) managed to isolate 12 dihydrophenanthrenes including blestriarene B, which is also present in *B. striata* and has been shown to have antimicrobial effects on two pathogenic bacteria: *Staphylococcus aureus*, a common cause of skin infections, and *Streptococcus mutans*, which causes dental decay (Yamaki et al. 1989).

Bletia hyacinthina R. Br. (see **Bletilla striata Rchb. f.**)

Bletilla ochracea Schltr.

Chinese name: *Huanghua Baiji* (yellow flower white mucilaginous root)

Description: A terrestrial orchid up to 50 cm tall with leaves that are longer but narrower than those of *B. striata*, it bears a long terminal inflorescence carrying 3–8 yellowish flowers that are sometimes tinged a purplish-brown dorsally, 3.5 cm across. The lip is white, streaked with brown and purple, and carries five longitudinal lamellae on its midlobe (Fig. 8.2). The flowers and plant resemble *Spathoglottis*. Flowering season is June to July (Yang et al. 1998). The species is saxicolous or terrestrial, in grassy locations in open forests and along valleys at 900–2400 m, with a distribution involving southern Shaanxi, southern Gansu, Hubei, Hunan, Guangxi, Sichuan, Guizhou and Yunnan and Vietnam (Chen et al. 2009).

Herbal Usage: Identical to *baiji* (see *B. striata*). *Huanghua Baiji* (*B. ochracea*) comes from Gansu, Shanxi, Sichuan, Hunan, Hubei, Yunnan, Guizhou and Guangxi.

Phytochemistry: Phenanthrenes with antimicrobial activity against three Gram-positive bacteria (*Staphylococcus aureus*, *S. epidermidis* and *Bacillus subtilis*) are present in *Bletilla ochracea* (Yang et al. 2012).

Fig. 8.2 *Bletilla ochracea* Schltr. (Photo: Stan Shebs. Through Courtesy of Wikimedia Commons)

Fig. 8.3 *Bletilla striata* Rchb. f. (Photo: Courtesy of Plant Photo Bank of China)

Bletilla sinensis (Rolfe) Schltr. [see **Bletilla foliosa (King & Pantl.) Tang & F.T. Wang**]

Bletilla striata **Rchb. f.**

Chinese name: *Baiji* (white mucilaginous root, white chicken), *Gangen* (sweet root), *Baigen* (white root), *Baijiertou* (white hen's head/top), *Shantianji* (mountain frog) *Lian Ji Cao. Bak-kup* in Hong Kong, Taiwanese (Hokien) dialect: *Peh kiu* (white ginger)

Japanese: *Shiran* (purple orchid, a name also applied to *B. formosana*)

Korean name: *Jaran*

Vietnamese names: *Bach cap*; *Hua lan tia*

Description: *B. striata* is a terrestrial plant varying in size from 15 to 50 cm in height, with an irregular, compressed, ovoid, white pseudobulb and a short stem carrying four or five oblong, lanceolate leaves, 8–30 by 4–5 cm. Inflorescence is terminal, bearing 3–8 purple flowers with an attractive lip marked with purple longitudinal striations over a white and yellow background (Fig. 8.3). There is an *alba* form. Flowering season is April to June (Chen et al. 2009). The plant is terrestrial or lithophytic, growing on grassy slopes or scrub at 1100–3200 m. It is distributed from Sichuan, Gansu and Shaanxi in the west to Guangdong and Jiangsu in the east of China, and in Korea and Japan. To meet demand, it is also grown in Guizhou, Yunnan, Jiangxi and Guangxi. Guizhou produces the largest quantity of top-quality *Baiji*.

Herbal Usage: Herbs are obtained from Huadong (in Guangdong), Huanan (in eastern Heilongjiang), Sichuan, Yunnan, Guangxi, Guizhou, Shanxi and Gansu provinces (Wu 1994). Li Shizhen (1578) stated that the herb grows in many places on stony mountains and in the valleys, emerging in March or April. Herbs are collected from February to August. The taste of the tubers of *B. striata* is stated as bitter, sweet and acerbic; its nature is slightly cold (Fig. 8.4). According to TCM, it benefits the lungs, liver and stomach meridians. For consumption, the tubers are cut into thin slices or crushed into a fine powder after cleaning and drying. Slices are irregular in shape. The cut surface is translucent, white, with a suggestion of veins. They are brittle and somewhat sticky. There is a faint odour. Slices become gluey when they are chewed. *Baiji* is stored in dry containers in well-ventilated stores.

Baiji powder is slightly yellowish, odourless and bitter. When mixed with water it turns gluey.

Fig. 8.4 *Baiji (Bletilla striata),* an illustration from Li Shizhen, *Bencao Gangmu* (1578)

The powder is stored in tightly sealed containers to protect against moisture.

In TCM, *baiji* is used principally as a haemostatic. The effects of the medicine are haemostatic, reduces swelling and promotes regeneration of muscle and other tissues (*Zhongyao Da Cidian* 1986; *Zhonghua Bencao* 2000; Wu et al. 1999; Ou et al. 2003; Anonymous 1989, 2004). Tubers are used in the treatment of swelling and haemorrhage, for instance, when patients with tuberculosis cough up blood, bleeding occurring in bronchiectasis, gastric bleeds, bleeding from trauma or burns, bleeding pustules, bleeding ulcers, fissure-in-ano and skin fissures of extremities caused by exposure to cold. They are used to treat sores and pustules, and dry and chapping skin (Wang and Wang 2006) or scaled skin (Anonymous 2004).

For oral decoctions, 3–10 g of *baiji* is used to prepare an oral powder of 1.5–3 g. It is combined with different herbs that are varied according to the condition for which it is used (Wu et al. 1999). *Baiji* is incompatible with *Radix aconite* and allied drugs, and they should not be used together (Guo 1996).

In Vietnam, *B. striata* is made into an emollient for the treatment of burns. Pseudobulbs are prescribed for tuberculosis, haemoptysis and other pulmonary diseases. Experiments conducted using an aqueous extract of the orchid by Nguyen van Doung (1993) showed that it produced a significant reduction in bleeding time.

Perry and Metzger (1980) reported that the rhizomes were collected from August to November with a non-metallic tool. In the case of planted crops, harvesting takes place in September and October, 3–4 years after planting. The tubers are crushed, or if dried, they are powdered, and mixed with oil to produce an ointment that is then applied to burns and scalds, on swellings and cracked skin. The extract of the tubers is also an insecticide. Sometimes, *baiji* is used as glue (Chen and Tang 1982). Essential oils, mucilage and glycogen have been isolated from the tubers.

Fourteen examples of TCM prescriptions which contain *baiji* are illustrated in Table 8.1. The usual dosage is 3–10 g in decoction for oral consumption, 1.5–3 g as oral powder, and adequate amounts for external use (Wu et al. 1999).

Phytochemistry: Investigations into the constituents of *B. striata* were initiated in Japan during the 1980s by the team of Tagaki, Yamaki and Inoue, of the Mukogawa Women's University in Nishinomiya, Hyogo, and over the years, this team have continued to discover new compounds in the orchid. Five antimicrobial agents consisting of three new bibenzyls and two new dihydrophenanthrenes were isolated from the tubers of *B. striata* in 1983. The structures of the bibenzyls were determined to be 4,7-dihydroxy-1-*p*-hydroxybenzyl-2-methoxy-9,10-dihydrophenanthrene, 3,3′-dihydroxy-2′6′-bis (p-hydroxybenzyl)-5-methoxybibenzyl and 2,6-bis(p-hydroxybenzyl)-3′5-dimethoxy-3-hydroxybibenzyl. The two dihydrophenanthrenes were 3,3′-dihydroxy-5-methoxy-2,5′6-tris (p-hydroxybenzyl) bibenzyl and 4,7-dihydroxy-2-methoxy-9,10-dihydrophenanthrene. Four known

Table 8.1 Manufactured products and prescriptions incorporating *baiji* (References: Items 1–4: *Zhongyao Da Cidian. Chinese Medical Encyclopaedia Vol. 1.* Shanghai Science and Technology Publishers, 1986. Items 5–9: KJ Wu and G Wu, chief eds. Wuhan: *Traditional Chinese Materia Medica*, Wuhan University Press, 1999; Items 10–13 *Zhonghua Bencao*, 2000)

---

1. *Baiji San* (白及散): *Baiji* powder 800 g mixed with sugarcane sugar powder 200 g to form a greyish-white powder with a sweet and slightly bitter taste. It has the effect of stopping bleeding from lung. It is used to treat longstanding coughs with haemoptysis. It is taken by mouth 15 g twice a day.
Source: *Standard Drugs of Hunan Province (1982)*

2. *Ning Fei San* (宁肺散): *Baiji* 200 g, *Pseudoginseng* (三七) 25 g, and "cat claw grass" (猫爪草) 100 g mixed in powder form to make 1,000 tablets, each 0.3 g. It can stop bleeding and cause tissue swelling to subside and healing to take place. Take by mouth, 8 tablets, three times a day. Dosage is halved for children.
Source: *Standard Drugs of Guizhou Province (1983)*

3. *Zhi Xue San* (止血散): *Baiji* 2,000 g, *Rhizoma cyperi royundi* (香附) 500 g, *Galla rhoic* (五倍子) 500 g, mixed, crushed into powder and strained, and packed in packets of 3 g each. It is used to stop bleeding of various causes from the digestive tract. Take by mouth 1 packet, 3 times a day before each meal.
Source: *Clinical Pharmaceutics of Da District (1985)*

4. *Du Sheng Wan* (独勝丸): *Baiji* 500 g, *Fritillariae cirrhosae* (川贝母) 200 g, *Fructus germinatus hordei vulgaris* (麦牙) 100 g, mixed and crushed into powder and strained. Add honey 1.2 parts to 1 part of powder. Concentrate the mixture over fire and convert it to pills, 7.5 g per pill. It can stop coughing and bleeding from lung, make tissue swelling subside and induce healing to take place. It can be used to treat tuberculous cavities in lung. Take one pill by mouth twice a day.
Source: *Criteria of Pharmaceutics of Liaoning Province (1982)*

5. *Baiji Pipa Pill.* For dry coughs and haemoptysis due to deficiency of lung *yin*. Baiji with Ejiaozhu (gelatin beads), Shengdi (*Rehmannia*) Pipaye (loquat leaf), etc. Amounts not stated.

6. *Wu Ji Powder.* For haemetemesis. *Baiji* with Haipiaoxiao (Cuttlefish bone, predominantly calcium carbonate) and Shigao (Gypsum, calcium sulfate). Amounts not stated

7. *Nei Xiao Powder.* For early sores. Baiji with *Jinyinhua* (*Lonicera japonica*, Japanese honeysuckle) and *Zhaojiaoci* (thorn of *Gleditsia sinensis*, Chinese honeylocust). Function of *Zhaojioci* is to extract pus and it has some antibacterial function. *Jinyinhua* contains compounds that may have a role in stopping bleeding and wound healing. Amounts are not stated.

8. *Misc.* For cracked skin on extremities. *Baiji* powder mixed with sesame oil for application.

9. *Misc.* Protracted sores. Apply powdered *baiji* directly.

10. For pharyngitis, bleeding from the trachea or broncheictasis
Baiji 1 liang, Pipaye (*Eriobotryae folium*) 5 chien, Lotus rhizome 5 chien.
Ground the 3 items into powder, mix and make into pills. A 1.5-cm-diameter pill is consumed daily

11. For tuberculosis and haemoptysis from any cause
Powder made with Cuttlefish "bone" 180 g, Baiji 180 g, and San Qi (Panax notoginseng) 180 g.
9 g of the powder is consumed 3 times daily.

12. For gastric and Intestinal Bleeding
Baiji and Bloodwort root (*Sanguisorba officinalis*) in equal amounts, toasted and made into powder, 3 g of which is consumed at any one time.

13. Burns
Apply *Baiji* powder suspended in oil

---

compounds were isolated in their pure forms from the acidic fraction of the *B. striata* extract and these were identified as *p*-hydroxybenzoic acid, protocatechuic acid, *p*-hydroxybenzaldehyde and cinnamic acid. The compounds are predominantly active against Gram-positive bacteria (*Bacillus subtilis*, *B. cereus* var. *mycoides*, *Nocardia gardneri* and *Staphylococcus aureus*), but they are ineffective against Gram-negative bacteria, in vitro. They are weakly active against certain fungi (*Candida albicans* or thrush and *Trichophyton mentagrophytes*). The introduction of the *p*-hydroxybenzyl group enhances their activity whereas the introduction of a methoxyl group decreases it (Tagaki et al. 1983). The narrowness of their antimicrobial spectrum limits their usefulness. Nevertheless, on the basis of their bacteriostatic action against *Staphylococcus aureus*, baiji probably has an active role in the healing of sores and pustules, bleeding ulcers, fissure-in-ano, skin fissures of extremities following exposure to cold, dry or chapping skin, burns and sore throat. Phenanthrenes in *B. ochracea* are similarly active against three Gram-positive bacteria, namely, *Staphylococcus aureus, S. epidermidis* and *Bacillus subtilis* (Yang et al. 2012).

Continuing their studies, Takagi, Yamaki and Inoue found three new biphenanthrenes (blestriarene A–C) together with the known compounds batatasin III and 3'-0-methyl batatasin III in *B. striata*. These compounds possessed antimicrobial activity against *S. aureus* and *S. mutans*. Of the five compounds, blestriarene B was the most effective, with minimum inhibitory concentrations (MIC) ranging from 6.25 to 25 mcg per ml in various tests (Yamaki et al. 1989).

S. mutans ferments glucose, thereby producing sufficient acid to cause decalcification of enamel and dental caries (Hare 1956; Duchin and van Houte 1978). This action is blocked by casein phosphopeptide–amorphous calcium phosphate (CCP-ACP) nano-complexes prepared from bovine milk. CCP-ACP binds to the surface of the bacteria, stabilises the bioavailable calcium and phosphate ions at the tooth surface, and promotes remineralisation of the enamel subsurface. In the presence of fluoride, it stimulates the formation of flourapatite deep in the subsurface lesion. CCP-ACP and fluoride are superior to fluoride alone in preventing dental caries (Reynolds, Cai, Cochrane et al. 2008). Prior to the introduction of routine fluoridation of tap water, dental caries was a serious problem in all societies, but there are no reports that orchids have ever been used in the promotion of dental health. Innovative Japanese companies have incorporated CCP-ACP into dental *mousse*, and perhaps *baiji*, or the purified blestriarene B might also find its way into toothpaste.

Over the next 4 years, Yamaki's team isolated numerous compounds from the tubers of *B. striata* and elucidated their structures: three biphenanthrenes, blestrianol A, B and C (Bai et al. 1991); three benzylphenanthrenes (Yamaki et al. 1992); two benzylphenanthrene ethers, blestrin C and D (Yamaki et al. 1992); three dihydrophenanthropyran derivatives, bletilols A, B and C (Yamaki et al. 1993a); four phenanthrene glucosides (Yamaki et al. 1993a); and a

phenanthrene with a spirolactone ring named blespirol (Yamaki et al. 1993c) (Figs. 8.5 and 8.6).

A polysaccharide obtained from tubers of *B. striata* was shown to contain four parts of D-mannose to one part of D-glucose (Ohtsuki 1937; Tomoda et al. 1973). In a recent study, a glucomannan of molecular weight 20 kDa isolated from *B. striata* had a mannose:glucose ratio of 3.5:1 (Zhang et al. 2014). Mucilage in other orchid tubers were later shown to also be glucomannans, albeit the ratios of their hexoses vary. For instance, in *Cremastra variabilis*, the ratio of D-mannose to D-glucose is 3:1 (Ernst and Rodriguez 1984). *B. striata* also contains a galactoglucomannan with some immunological activity. The compound contains mannose, glucose and galactose in the compound in a molar ratio of 9.4:2.6:1.0 (Peng et al. 2014).

B. striata polysaccharide (BSP) enhances the expression of inducible nitric oxide synthase, tumour necrosis factor alpha (TNF-alpha) and interleukin-1 beta but has no effect on interferon gamma (Diao et al. 2008). The pro-inflammatory factors may enhance the body's defence against infection. A mucoadhesive polysaccharide from *Bletilla striata* improves the intraocular penetration and therapeutic efficiency of levofloxacin in experimental bacterial keratitis (Wu, Xin, Chen et al. 2010). It may be a useful component to add in ophthalmic preparations.

A water-soluble glucomannan designated *B. striata* polysaccharide b (BSPb) inhibited the proliferation of human mesangial cells in vitro at a concentration of 20 mcg/ml when incubated with 5 ng/ml of TGF-beta-1, whereas lower concentrations had no inhibitory effect. This antifibrosis effect may contribute to the response of wounds to "Weiping burn regereation ointment" which has been used in China for over 10 years (Wang et al. 2014).

Polysaccharide from *B. striata* stimulated the proliferation of human umbilical cord vascular endothelial cells by up to 156 %, and the expression of endothelial growth factor by 147 % (Wang et al. 2006). These processes play important roles in wound healing. BSPb promoted the adhesion of human umbilical venous endothelial cells in vitro, albeit not in a dose-dependent manner (Sun et al. 2005).

Fig. 3.5 Phenanthrenes isolated from *Bletilla striata*

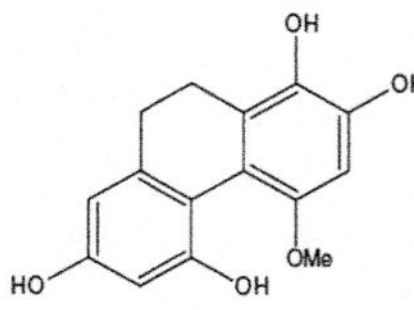

4-methoxy-9,10dihydrophenethrene-1,2,7-triol

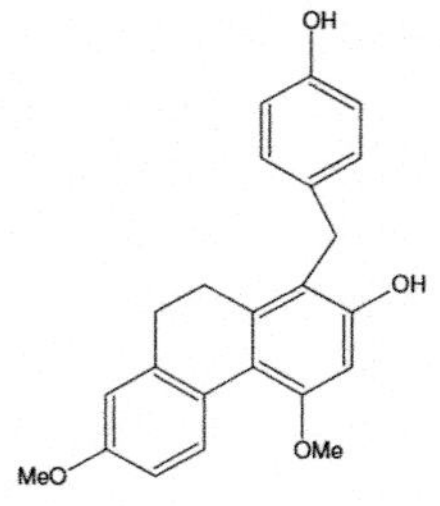

1-(4-hydroxybenzyl)-4,7-dimethoxy-

9,10dihydrophenanthrene-2-ol

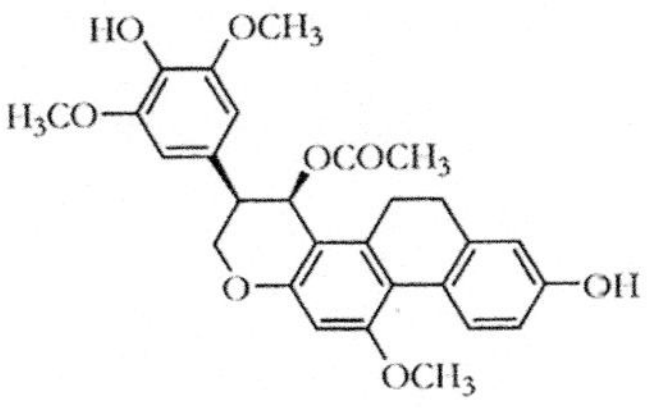

bletilol A,

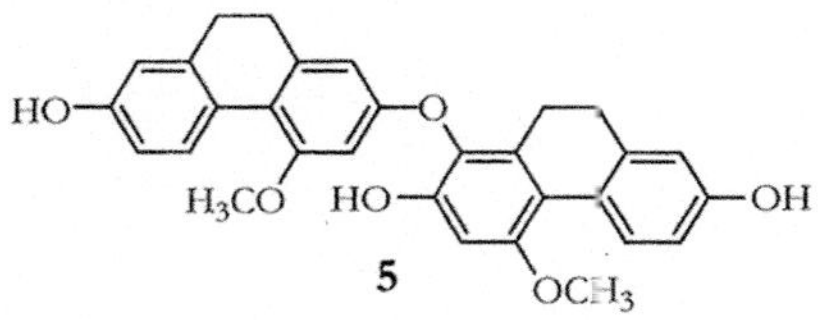

blestrine A

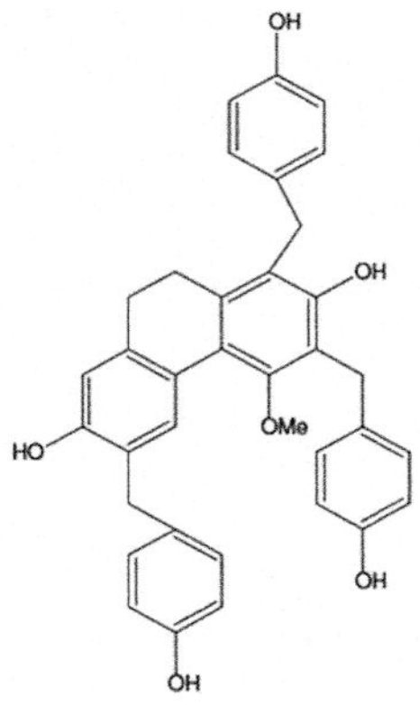

1,3,6-tri(4-hydroxybenzyl)-4-

methoxydihydrophenanthrene-2,7-diol

An interesting finding on *Bletilla striata* polysaccharide (BSP) is that when the nematode species, *Caenorhabdtis elegans* was treated with BSP at a dose of 50 mcg/ml, it became more active and its mean life-span was increased from 19 to 22 days (Zhang, Lv, Li, et al. 2015). Studies need to be conducted on higher animals.

Spectral analysis showed that blestriarene B is 4,4′-dimethoxy-9,10-dihydro[1,1′-biphenanthrene]-2,2′7,7′-tetrol (Yamaki et al. 1989). *B. striata* also yielded stilbenoids (Bai et al. 1993), biphenanthrenes, benzylphenanthrenes (Yamaki et al. 1990), biphenanthrene ethers, bis(dihydrophenanthrene) ethers, a phenanthrene with a spirolactone ring, phenanthrenes and bibenzyls with anthroquinones, dihydrophenanthropyrans,

phenanthrene glucosides, and polysaccharides (Bai et al. 1990, 1991; Yamaki et al. 1991, 1993a, b; Guo 1996). The stilbenoids were 2,4,7,-trimethoxyphenanthrene, 2,4,7,-trimethoxy-9,10-dihydrophenanthrene, 2.3.4.7-tetramethoxyphenanthrene, 3,5-dimethoxy bibenzyl, and 3,3′, 5-trimethoxybibenzyl (Yamaki et al. 1991). The biphenanthrenes were blestriarenes A, B, C, batatasin III, and 3′-O-methylbatatatsin (Yamaki et al. 1989), while the bis(dihydrophenanthrene) ethers were named blestrin A and B (Bai et al. 1990). The anthroquinone was physcion (Yamaki et al. 1991). Physcion is the principal compound in rhubarb and has antifungal and antitumour activities. Physcion plays an important role in the plants' defence against fungi. Blespirol is the phen-

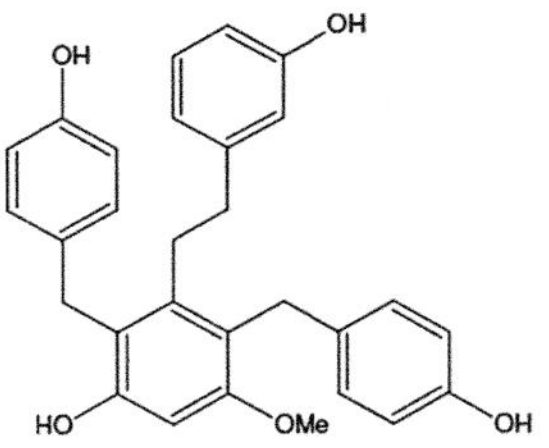

3,3′-dihydroxy-2′6′-bis(p-hydroxybibenzyl)-

5methoxybibenzyl

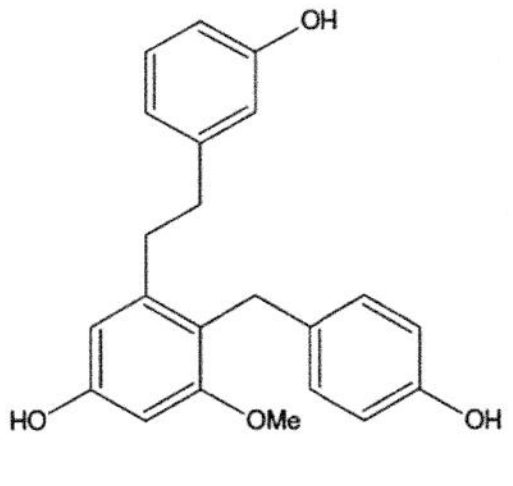

3,5-dihydroxy-2′(p-hydroxybenzyl)-

3methoxybibenzyl

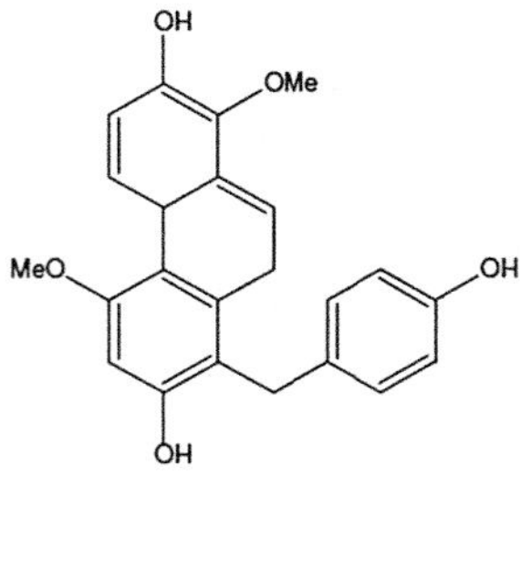

1-(p-hydroxybenzyl)-4,8-
dimethoxyphenanthrene

Fig. 8.6 Phenanthrenes from *Bletilla striata* exhibiting antitumour properties in vitro

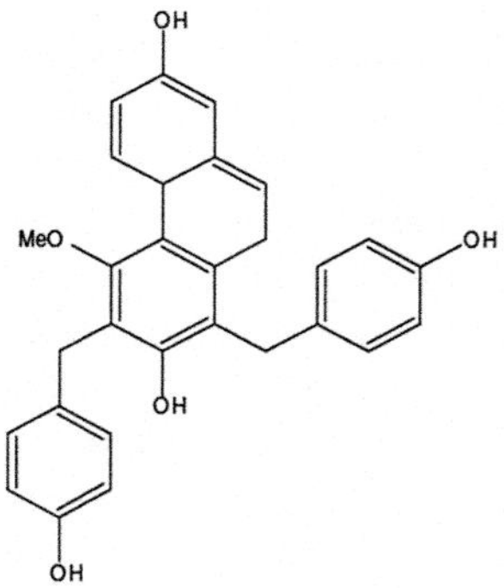

2,7dihydroxy-1,3-bis(p-hydroxybenzyl)

4-methoxy-9,10,dihydrophenanthrene

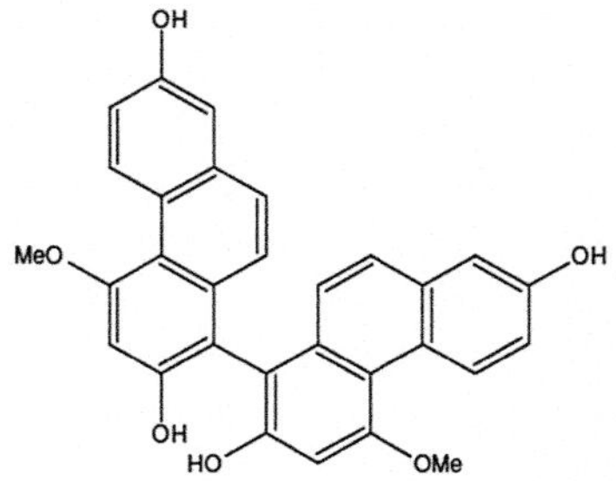

blestriarene B

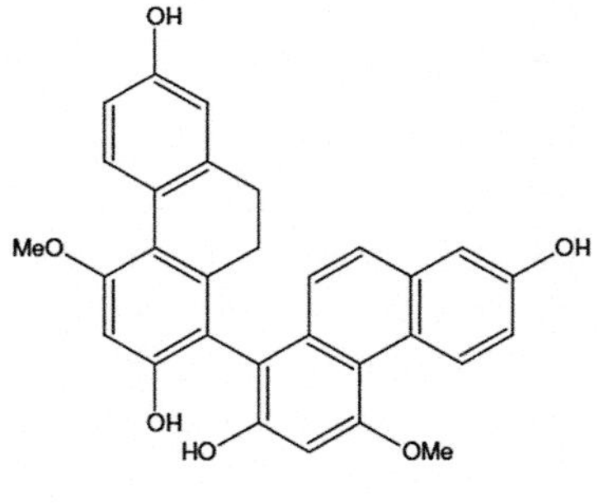

blestriarene

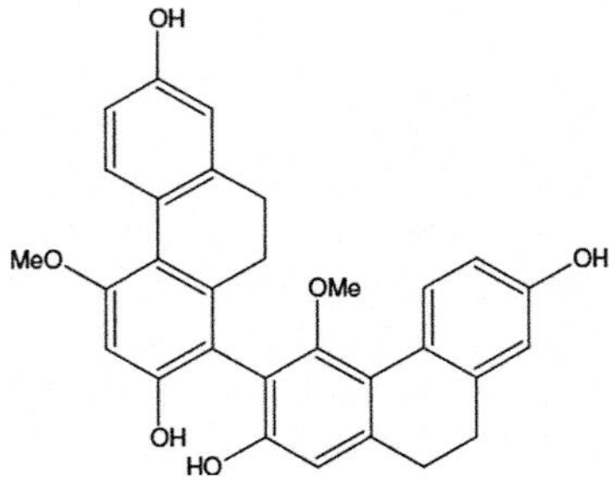

blestrianol A

Fig. 8.6 (continued)

anthrene with a spirolactone ring (Yamaki et al. 1993a).

The isolation of four phenanthrene glucosides from *B. striata* was the first report of the isolation of stilbenoid glucosides from orchids (Yamaki et al. 1993a). This finding is significant because water-soluble, polar compounds are more easily transported and delivered to the site of a microbial invasion. On the basis of spectral analysis, the four phenanthrene glucosides were determined to be:

2,7-dihydroxy-4-methoxy-phenanthrene-2-*O*-glucoside;

2,7-dihydroxy-4-methoxy-phenanthrene-2,7-*O*-diglucoside;

3,7-dihydroxy-2,4-dimethoxyphenanthrene-3-*O*-glucoside;

and 2,7-dihydroxy-1-(4'-hydroxybenzyl)-4-methoxy-9,10-dihydrophenanthrene-4'-*o*-glucoside (Yamaki et al. 1993a).

A new 2-(2-methylpropyl)butanedioic acid derivative, bletillin A (1), and a new bibenzyl, bletillin B (2), together with 17 known compounds were isolated from the tubers of *B. ochracea* by Yang et al. (2012). The two new compounds did not exhibit antimicrobial properties.

An early attempt at the Institute of Drug Inspection in Shenzhen to study the chemical constituents of *B. striata* managed to isolate and identify hexacosanoic alcohol 3-(4-hydroxy-3-methoxybenzol)-trans-acryliceylenate, physcion and cyclobalanol, the first being a new compound (Wang et al. 2001). After they moved to Shanghai, the same team isolated three more compounds from the roots of *B. striata*: 5-hydroxy-4-(*p*-hydroxybenzyl)-3'-3-dimethoxybibenzyl, schizandrin and 4,4''-dimethoxy-(1,1'-biphenanthrene)-2,2',7,7'-tetrol. The first compound was a new bibenzyl derivative (Han, Wang, Gu and Zhang 2002).

Formation of bibenzyls and 9,10-dihydrophenanthrenes in orchids is triggered by fungal infection and physical or chemical injury to the plant tissues. In an elegant experiment on sliced pieces of pseudobulbs of *B. striata*, Reinecke and Kindl (1994) from Philippe

Universitat in Marburg, Germany, showed that slicing and, in particular, exposure to fungi (*Botrytis* and *Rhizotonia*) caused a sharp rise in bibenzyl synthase activity in slices of pseudobulbs peaking 4 h after induction. Published studies on the isolation of bibenzyls and phenanthrenes from *Bletilla* have not stated that any attempt was made to enhance production and higher yields of such compounds from fresh tubers of the orchid.

Eight stilbenoids were isolated from the tubers of *B. striata* by the Japanese team of Morita et al. (2005). Bisbenzyls 4 and 5 inhibited the polymerisation of tubulin at IC50 of 10 μM, while bisbenzyl 4 potentiated the cytotoxicity of SN-38 in BCRP-transduced K562 (K562/BRCP) cells. The researchers concluded that the interesting feature of the antimitotic activity of the stilbenoid was its ability to counter breast cancer-resistant protein-mediated drug resistance. Recently, resensitisation of cancer cells to apoptosis has also been reported with dendronobin (see *Dendrobium*.)

The ability of a stilbenoid in *B. striata* to counter cancer drug resistance and re-sensitise cancer cells to programmed cell death (apoptosis) is a new promising area for the drug industry to investigate, and hopefully it will lead to novel new therapies.

The content of militarine, cinnamic acid, 1,8-bi(4-hydroxybenzyl)-4-methoxyphenanthrene-2,7-diol, and 4,7-dihydroxy-1-*p*-hydroxybenzyl-2-methoxy-9,10-dihydrophenanthrene in *B. striata* raised though micropropagation was higher throughout the vegetative phase than during the flowering phase. Ideal harvest time for *B. striata* in Taiwan was September to October (Wu, Chen, Lay, 2010).

Overview

Although we started off describing four species of medicinal *Bletilla*, the classic *baiji* is *B. striata*. The others are substitutes that are used in a similar manner when *B. striata* is not available. The four species which have a similar function are distributed in many provinces in central China, and it is not surprising that, given its geography and climate, the salubrious province of Guizhou produces the best quality *baiji*. Traditional Chinese medicinal tracts mentioned that the main effects of *baiji* are haemostasis, abatement of tissue swelling, and promotion of healing. It is used for treatment of haemoptysis, haematemesis, epistaxis, rectal bleeding, bleeding of external wounds, carbuncles, burns, frostbite, skin fissures of the hands and feet, and fissure-in-ano. A contemporary adaptation of this usage is the Chuangyiling dressing designed by the Orthopaedics Department of Union Hospital in Wuhan. Traditional medicines are mixed into a scaffold of gelatin and *Bletilla* gum to produce a porous sponge. The dressing has met the requirements for safety (absence of acute toxicity, skin irritation, sensitisation and cytotoxicity) set by the Ministry of Health of China (Peng et al. 2005).

Numerous antimicrobial compounds that are predominantly active against Gram-positive bacteria (such as *Staphylococcus aureus*, which is a common cause of skin and wound infections) have been isolated from *B. striata* and *B. ochracea*. These laboratory studies support the use of *baiji* as a wound dressing (Tagaki et al. 1983; Yamaki et al. 1989; Luo et al. 2010). A natural polysaccharide (BSP) hydrogel prepared by oxidation and cross-linking methods was shown to promote excellent wound healing in a full-thickness trauma mouse model. Eleven days after surgery, the wound area in the treated group was reduced by 67 to 80.5 % compared to the untreated controls. Infiltrating inflammatory cells and tumour necrosis factor (TNF), the latter a marker of tissue damage, were attenuated while epidermal growth factor was increased (Luo et al. 2010). The advantages in using *B. striata* polysaccharide are its good biocompatibility, controllable properties and abundance (Gong et al. 2009), which also means that it is inexpensive.

Baiji is one of the several components of *Xiao Wei Yan Powder* (XWYP) that is used for treating intestinal metaplasia (IM) and atypical hyperplasia (AP) of the gastric mucosa (Liu et al. 1992). *Baiji* has also been used with sepium as well as on its own to treat mucosal damage of the bowel and bleeding peptic ulcers (Anonymous 1959). Its

antibiotic activity against *Streptococcus mutans* (Tagaki et al. 1983; Yamaki et al. 1989) may possibly assist gastric mucosal repair.

As a Gene Carrier

Gene therapy is an important new modality in the treatment of disease. The absence of or damage to the p53 gene predisposes to many forms of cancer including cancers of the breast and liver (hepatocarcinoma). In a rabbit model, transarterial adenovirus (Ad)-p53 gene therapy has been found to reduce tumour growth of transplanted liver cancer (Gu et al. 2007).

A novel gene carrier has been developed from bioactive glucomannan, a polysaccharide isolated from *B. striata* (BSP) after modification with N,N'-carbonyldiimidazole (CDI)/ethylenediamine to confer it with nucleic acid-binding affinity. Cationised BSP (cBSP) bound to DNA to form nano-scaled compact, stable complexes and promoted the transfection of oligodeoxynucleotide (ODN). These cBSP had high affinity to macrophages. In the experiment, they successfully inhibited the expression of TNF-alpha. The team of scientists lead by L Dong at the State Key laboratory of Pharmaceutical Biotechnology at Nanjing University expects cBSP to be capable of conveying antisense nucleotides (e.g. oligodeoxynucleotide and small interference RNA) for anti-inflammatory therapy (Dong et al. 2009). Cholesteryl modification of *B. striata* glucomannan rendered it more amphiphilic, and this macromolecule could self-assemble into nanoparticles in aqueous solution (Zhang et al. 2014).

Carrier for Cytotoxic Agent to Tumours

Conjugated *B. striata* polysaccharide is a promising avenue for the delivery of cytotoxic agents to tumours based on the following observations. Tumour-associated macrophages (TAM) are large white blood cells that reside in the tumour environment. Unlike normal macrophages which are scavenger cells, TAM promote the growth of blood vessels in the tumour site, tumour spread and evasion of host immune response by the tumour. *B. striata* polysaccharide (BSP) is a branching glucomannan with repeating units of mannose in its main chain, and this structure gives it high binding affinity to macrophage mannose receptors which are abundant in TAMs. Targeted delivery of oligonucleotide into macrophages by cationic BSP inhibited expression of tumour necrosis factor alpha (Dong et al. 2009). Alendronate possesses in vitro macrophage-inhibiting activity. At Nanjing University School of Life Sciences, scientists developed and tested an alendronate–BSP conjugate (ALN-BSP) both in vitro and in vivo. BSP-ALN's spheroidal shape and nanoparticle size (10–100 nm) expedited cell uptake. ALN-BSP preferentially accumulated in macrophages and induced apoptosis. Treatment with ALN-BSP eliminated TAMs from subcutaneously transplanted cancers (S180 sarcoma) in a mouse model. ALN-BSP inhibited angiogenesis and suppressed tumour progression without altering the systemic immune response. These effects were not seen when unconjugated alendronate or saline was administered (Zhan et al. 2014).

Further experiments are being designed to evaluate the systemic safety and therapeutic efficacy of the conjugate in various tumour models (Zhan et al. 2014). Mannan-modified solid lipid DNA has been shown to be capable of targeted gene delivery to the lungs of rats in an in vivo evaluation experiment (Yu et al. 2010), but ALN-BSP has not been tested in a similar manner. Unconjugated alendronate is an established drug for the prevention and treatment of osteoporosis but it has numerous side effects.

An Embolising Agent

Animal experiments and clinical application of *B. striata* as an embolising agent were introduced in China before 1985 (Feng 1985). Similar experiments have been repeated in Europe in the past decade (Qian et al. 2003a; Maataoui et al. 2005a, b). Microspheres containing mitomycin were highly effective in reducing liver cancer growth in rats (Qian et al. 2003a). Injecting suspensions of *B. striata* into the intrahepatic portal veins of dogs and rabbits invariably resulted in portal hypertension, and this has been used as a model for studying the dynamics of portal hypertension and the means

to relieve it in order to prevent or to treat cirrhosis of the liver (Wang 1993; Luo et al. 1996; Qian et al. 1998).

Subsequently, embolisation was also used to treat unresectable liver tumours. The rationale for this therapeutic modality is as follows. Primary and secondary liver cancers are notoriously difficult to treat without sacrificing too much healthy liver, unless they are small, solitary and conveniently excised. Systematic chemotherapy is generally ineffective or at best it produces a poor response. Intra-arterial therapy is better, but still improves survival by only a few months. The drugs enter the blood stream almost immediately after passing through the liver, and at high circulating levels they are toxic to healthy tissues. When degradable starch micro-particles are co-administered intra-arterially to the liver, they achieve a transient to permanent vascular obstruction, and the simultaneously administered chemotherapeutic agent is lodged in the target area for prolonged periods. Circulating drug concentration as measured by a concentration–time curve (area under the curve, or AUC) is significantly reduced, thus reducing generalised toxic side effects. Studies have shown that there is a selectively higher uptake of the drugs by tumours as compared to healthy liver tissue when TACE (transarterial chemoembolisation) is used (Hakansson et al. 1997). Co-administration of Spherex® (a pharmaceutical grade of starch microparticles) and carboplatin increased tumour concentration of carboplatin by a factor of 47 in an experimental study using liver tumour-bearing rabbits (Pohlen et al. 2000).

Addition of intra-arterial *Bletilla* to TACE (transarterial chemoembolisation) with 0.1 mg mitomycin and 0.1 ml lipiodol resulted in a 31.3 % non-significant reduction in tumour growth in rats implanted with solid MORRIS hepatoma 3924A. Rodents that simultaneously underwent hepatic artery ligation had a far greater, and significant, 83.2 % reduction in the rate of tumour growth (Qian et al. 2003b).

In experiments on New Zealand white rabbits using MRI and Chinese ink casting to view the effects, arterial chemoembolisation with 5FU-*Bletilla* microspheres showed a slightly, but not significantly, better reduction in blood flow through implanted liver tumours than lipicdiol mixed with mitomycin C. Subsequently, microvessels appeared around the tumours in the lipicdiol group but there were none in the *Bletilla* group. Zhao et al. (2004), who conducted the experiment at the Sixth Affiliated Hospital of Shanghai Jiaotong University, concluded that TACE with *Bletilla* microspheres may enhance its antitumour effect by inhibiting angiogenesis. Lyophilised powder of fresh *Gekko chinensis* (GCLP) also inhibited H22 hepatocarcinoma angiogenesis in rats causing tumour cell apoptosis, and down-regulation of VEGF and bFGF protein expression (Song et al. 2006).

In the contemporary medical setting in China where hospitals treat cancer patients with both modern chemotherapy and herbal medicine, *baiji* has found a unique use for the embolisation of unresectable cancer metastases in the liver because of its physical properties. *Baiji* coagulates in the blood stream and blocks the blood supply to a specific site when it is introduced into the appropriate artery. Although amylases digest starch in the body and *baiji* is biodegradable, embolisation with *Bletilla* was found to produce extensive and permanent vascular obstruction and was superior to gelfoam powder which produced only a temporary occlusion (Feng et al. 1995, 1996, 2003; Zheng et al. 1996; Qian et al. 2005). In trials conducted at Tongji Medical University in Wuhan, 56 patients with hepatic carcinoma were treated with *Bletilla* embolisation and 50 by conventional gelfoam embolisation. Embolisation with *baiji* resulted in permanent vascular obstruction, marked shrinkage of tumour size, decrease in serum alphafetoprotein (a marker of liver cancer), with late development of only a few collaterals allowing for extended treatment intervals, averaging 7 months. One-, two- and three-year survival of patients treated by embolisation with *baiji* was 81.9 %, 44.9 % and 33.6 %, respectively, compared to 48.9 %, 31.1 % and 16.0 % with gelfoam. The median survival time with *baiji* embolisation was 19.8 months (Zheng et al. 1996, 1998; Feng et al. 1996).

Baiji is not the only starch-based embolisation agent used in medicine. An alternative source of biodegradable starch-microsphere is Spherex® (Pharmacia), a 45-μm, biodegradable microsphere suspension prepared by X-linking partially hydrolysed potato starch. It is broken down by serum amylase and has an in vitro half-life of 20–35 min. It has been used to deliver mitomycin-C, doxorubicin, cisplatin nitrosoureas and 5FU, and has been extensively studied (Teder et al. 1988; Teder and Johansson 1993; Taguchi 1995; Johansson 1996; Carr et al. 1997; Hakansson et al. 1997; Pollen, Berger, Binnenhe et al. 2000). In a study conducted at the Department of Liver Transplant at the University of Pittsburgh Medical School in the USA, Carr and his co-workers used Spherex® to deliver doxorubicin 30 mg/m^2 plus escalating doses of cisplatin up to 100 mg/m^2 into the hepatic artery until there was slowing or reversal of blood flow. The treatments were repeated every 4–6 weeks. Of the 35 patients who could be evaluated, 22 (63 %) showed objective tumour response, with 2 showing complete response. Four patients had reversal of tumour-induced portal vein thrombosis. The complications include one death from drug toxicity, another death of uncertain cause, two cases of hepatitis, one of pancreatitis, two with shortness of breath and hypotension and four of hepatic artery thrombosis. Six patients (17.1 %) were alive after 2 years, and 10 more (45.7 %) were still alive after 1 year (Carr et al. 1997). A Phase II study of transarterial embolisation in 50 consecutive European patients with Stage 1 and Stage 2 primary liver cancer produce a favourable response in 81 % of patients with three deaths shortly after the procedure, two from tumour progression and one from liver failure. One-year survival was 65 % and 2-year survival was 38 % (Bruix et al. 1994). These results are not superior to those obtained in Wuhan with TACE using *baiji*.

This improved survival may not be due to entirely to the vascular obstruction. In vitro studies showed that *Bletilla* colloid suppressed ECV-304 endothelial cell growth in a dose-dependent manner (Feng et al. 2003). Transarterial neoadjuvant chemoembolisation has been proposed as a bridging therapy for patients awaiting liver transplant, and in a small series of 67 patients it was shown to reduce the incidence of post-transplant recurrence of disease (Tsochatzis et al. 2013).

A retrospective study of TACE (transarterial chemoembolisation) in 165 patients with unresectable hepatocellular carcinoma conducted at Changhai Hospital in Shanghai between January 2002 and December 2007 found that the addition of orally administered *Jiedufang* (JDF) granules extended median overall survival from 5.9 months to 9.2 months. One-, two- and three-year survivals were 41.2 %, 18.4 % and 9.6 %, respectively, very respectable figures for unresectable hepatocellular carcinoma (primary liver cancer), which is generally rapidly fatal. The number of patients in the TACE + JDF group was 80 and the number receiving TACE alone was 85 (Yu et al. 2009a, b). Their results are almost similar to those of the earlier Pittsburgh study that used Spherex® to deliver doxyrubicin and cisplatin (Carr et al. 1997). *Jiedufang* consists of four traditional herbal medicines: root of *Actinidia valvata*, root of *Salvia chinensis*, bulb of *Cremastra appendiculata* (a medicinal orchid) and gizzard membrane of *Gallus g. domesticus* in a proportion of 1:1:4:0.4 (Yu et al. 2009a).

Embolisation with *B. striata* may be useful in palliative treatment of craniofacial tumours and certain vascular diseases, but this observation is based on a very small series of only seven patients treated at the Union Hospital, Tongji Medical University in Wuhan (Du et al. 1998). The authors have not reported a larger series in the following decade.

At the First Affiliated Hospital of the Sun Yat Sen University in Guangzhou, an attempt has been made to use TACE as a pre-operative adjunct to limb salvage surgery for treating osteosarcoma. From January 1998 to December 2003, 32 patients with osteosarcoma were treated with various embolic materials: (1) adriblastina gelatin microspheres, (2) anhydrous alcohol, (3) common *Bletilla* tuber, and (4) gelatin sponge particles. Following treatment, there was a decrease in the alkaline phosphatase level (a marker of tumour activity), and when the tumours were resected, 85.5 % showed large areas of necrosis. No serious complications followed the use of transcatheter arterial chemoembolisation (TACE). There was no difference in the results following the use of the first three embolic materials but gelatin sponge particles were inferior. The survival rates at 1, 2 and 5 years were 95.5 %, 72 % and 42 %, respectively, with the longest follow-up of 86 months. Dr. Chu and colleagues stated that the best time to operate was 10–14 days after TACE (Chu et al. 2007). A larger series comparing adriblastina microspheres and *Bletilla* would be interesting because it would show whether the addition of a cytotoxic agent made a difference.

In vitro studies on Hep-G2 primary liver cancer cells did not show any difference in the rate of proliferation, VEGF levels or apoptosis rate between cultures treated with *B. striata* colloid and untreated cultures. On the other hand, ECV-304 human endothelial cells in tissue culture responded dramatically to *Bletilla* colloid in a dose-dependent manner, with inhibition rates of 57.6 % at 0.5 mcg/ml, 66.7 % at 1 mcg/ml, 86.4 % at 2 mcg/ml, 87.5 % at 4 mcg/ml and 94.8 % at 8 mcg/ml. The microvascular density (MVD) determined by counting Factor VIII-positive endothelial cells was significantly lower in the 5FU–*Bletilla* group than in the 5FU normal saline or the poppy seed oil-treated group. There was no difference in expression of VEGF and b-FGF. In an earlier study on rats, the scientists found that there was no difference in the endothelial cells of transplanted Walker 256 hepatoma following treatment with normal saline, 5FU, or 5FU with lipidiol (Li et al. 2003). Feng and their colleagues (2003) concluded that *Bletilla* colloid inhibits angiogenesis following TACE, possibly by blocking the binding of VEGF to its receptor. Terpenoids from *B. striata* inhibit vascular endothelial cells through apoptosis (Liu et al. 2008). In a surprising discovery, Wang and colleagues (2006) from Nanjing University found that a polysaccharide from *B. striata* (Thunb.) Rchb. f., which they had isolated, purified and characterised, induced proliferation of human umbilical vascular endothelial cells by 56 % and the expression of VEGF by 47 % after 72 h. The opposing actions of the different constituents of *B. striata* appear to support the TCM notion that the action of an herb is dependent on the collective action of its constituents and not on any single substance. A new glucomannan from *B. striata* protected against renal fibrosis in vitro (Wang et al. 2014).

Liver Flukes

Clonorchiasis is a disease caused by eating raw freshwater fish or snails that carry the encysted cercaria of the liver fluke, *Clonorchis sinensis* (Choi 1984). In the intestines, the parasites excyst and the worms multiply in the intestinal and biliary tract, engulfing biliary epithelial tissues and blood cells. Infestation may pass unnoticed, but a blood count will show a rise in eosinophils. Sometimes, it may cause abdominal discomfort often located over the region of the gall bladder. Eggs of the liver fluke are shed with the faeces and the presence of eggs in the stool confirms the diagnosis (Wang, Zhang, Cui, et al. 2004). Clonorchiasis leads to cholecystitis (infection of the gall bladder), bilary adenomatous hyperplasia, bile duct obstruction, cholangiofibrosis, cirrhosis of the liver and an increased risk of cholangiocarcinoma. The magnitude of the risk may be gleaned by the statistics from Korea. The country has the highest liver cancer incidence in the world, and 20 % of liver cancers in the Pusan area are due to cholangiocarcinoma (Shin et al. 1996). It is estimated that about seven million people living in East Asian countries are infected by the fluke (Wang, Zhang, Cui et al. 2004). Praziquentil developed by Bayer in the mid-1970s is the only drug used to treat the worm infestations; it reduces the worm burden by 50–95 %. Side effects of the drug result from poisons released by dead flukes, and are therefore related to the intensity of the infestation. In 1982, the Korean team led by J.K. Rhee of Joenbug National University reported that they had isolated an antihelminthic substance from the tubers of *B. striata* growing on Gangweon-do in the Korean peninsula. The substance was extracted with ethyl ether and further isolated by serial passage through various organic solvents and column chromatography. It killed the excysted metacercaria within 14 min and adult *Clonorchis* within 128 min of exposure (Rhee et al. 1982). However, no additional data have since appeared reporting on the further development of the *Bletilla* extract. Two artemisinins, OZ78 and tribendimidine, have meanwhile all been shown to be effective in killing the worms in the rat (Keiser et al. 2009), and malaxin (dihydoartemisinin) might perhaps also be effective but it has not been tested.

Propagation and Preservation of *B. striata*

Medicinal species of *Bletilla* are widely distributed in the provinces south of the Qingling Mountains, albeit commercial *baiji* production is centered in Guizhou. Hubei, Hunan, Honan, Zhejiang and Shanxi are other provinces where *Bletilla* is also cultivated. The harvesting season starts in August and ends in November. Besides being used as medicine, *baiji* is also used as a glue (Chen and Tang 1982).

Research directed at finding the best means of propagating and growing *B. striata* in vitro are ongoing (Zeng et al. 2004; Zhang et al. 2009). Immature seeds collected 2 and 4 months after pollination, precultured on New Dogashima medium supplemented with 0.3 M sucrose, then cryopreserved by vitrification, showed no decrease in germination rate and developed into normal plantlets after thawing. When assessed for viability by staining with 2,3,5-triphenyltetrazolium chloride, 2-month-old premature seeds had a slightly higher survival rate of 92 % compared with 81 % for the 4-month-old premature seeds (Hirano et al. 2005).

Propagation, Cultivation and Chemical Studies on *B. formosana*

In Taiwan, *B. formosana* is the medicinal herb replacing *B. striata* as *baiji*. Seven compounds, militarin, formoside, gymnoside IX, benzyl alcohol, transcoumaric acid methyl ester, coelonin and batatasin III, have been isolated from mature pseudobulbs of cultivated plants, proving that these were comparable in medicinal value to plants collected from the wild.

Bletia hyacinthina R. Br.

In 1929, A. Hooper recorded *B. hyacinthina* as a medicinal product in Malaya (Peninsular Malaysia), and this information has been repeated in numerous later publications. Its inclusion here is intended to clarify the mystery of a "Malayan medicinal plant" which does not thrive in the tropics.

B. hyacinthina is a synonym for *Bletilla striata* which does not occur in the Malay Archipelago. In the past, the generic name *Bletia* was inappropriately applied to some species of *Calanthe* and *Arundina*. Holttum (1953) mentioned *B. verecunda* introduced to Malaya from Central America which had the appearance of *Spathoglottis plicata*, although its flowers were not quite as fine as those of the latter.

However, Hooper was not referring to *S. plicata*. Hooper (1929) observed that this '*Bletia hyacinthina*' formed a Malay remedy for dysentery, haemorrhoids and ague. It was also applied as a demulcent (a soothing medication) over the abdomen of children with dyspepsia. He was probably quoting from F. Porter Smith's translation of the *Chinese Materia Medica: Vegetable Kingdom,* revised by Rev. G. A. Stuart, MD in 1911, which stated: "*Bletia Hyacinthina* (sic) –(Pai-chi, 935). . . . . . is considered demulcent, and is used in the diseases of children, especially those of a dyspeptic character, as well as in dysentery, haemorrhoids and ague. It has much repute in the treatment of burns, wounds and other injuries, and also in various kinds of skin diseases" (Stuart 1911).

Genus: *Brachycorythis* Lindl.

Chinese name: *Baoye Lan*

Brachycorythis is a genus of over 30 predominantly terrestrial orchids distributed in Africa and tropical Asia. The name is derived from the Greek *brachy* (short) and *korys* (helmet), referring to the flower. Tubers are fusiform or ellipsoid. Stems are erect and leafy, leaves numerous and overlapping.

Fig. 8.7 *Brachycorythis obcordata* (Lindl.) Summerh. (Photo: Bhaktar B. Raskoti)

Brachycorythis obcordata (Lindl.) Summerh.

Nepali name: *Gangdol*
Common name: *Heart shaped* Brachycorythis

Description: *B. obcordata* is a small to medium-sized terrestrial herb with paired globose to oblong tubers. Stem is 20 cm tall bearing numerous, sessile, lanceolate leaves, 2–4 by 0.8–1.8 cm, which ensheath the stem. Pale purple flowers, 1.2 cm long, are borne singly at the axils of leaves. Sepals and petals are small, lanceolate, approximately 5 by 1.5 mm. Lip is an inverted heart-shape, obscurely three-lobed and spurred, the spurs marked with deep purple (Fig. 8.7). *B. obcordata* is distributed in India, Bhutan, Bangladesh and Myanmar at 900–2300 m on sunny, rocky slopes (Pearce and Cribb 2002; Gurung 2006) with a wider altitude range of 600–2600 m in Nepal (Raskoti 2009). It flowers June to August in Bangladesh (Hoque and Huda 2003); July to August in Nepal (Raskoti 2009); July to September in Bhutan (Gurong 2006).

Herbal Usage: The root is an astringent, expectorant, antidiarrhoeal, with use as a tonic in Nepal. Boiled pseudobulbs are eaten as food. Leaves and shoots are also cooked and eaten as a vegetable (Manandhar and Manandhar 2002; Pant and Raskoti 2013).

Overview

Medicinal usage of this genus appears to be localised to Nepal where a single species is used. There, the species is under threat from excessive cattle grazing and overexploitation as vegetable and as a medicinal plant (Raskoti 2009). In the IUCN Nepal National Register of Medicinal Plants (Shrestha 2000), *B. obcordata* is one of four orchids on the list of plants slated for protection, the others being *Dactylorhiza hatagirea*, *Dendrobium fimbriata* and *Otochilus porrectus*. In Africa, *Brachycorythis* is not used medicinally, but *B. ovata* enjoys cultural tribal usage as a charm (Chinsamy et al. 2011). There are no pharmacological data on this genus.

Genus: *Bromheadia* Lindl.

Bromheadia are terrestrial or epiphytic, sympodial orchids with long, thin, tough stems carrying many leaves. Those with leaves that are flattened horizontally look rather like bamboo or grass. The other group has leaves that are compressed laterally and looking rather like orchids of the section *Aporum* in *Dendrobium*. Both types have ephemeral flowers that open in the morning and fade by midday. There are 17 species distributed from Sri Lanka across Southeast Asia to Australia. Two Malaysian species, *B. brevifolia* and *B. ruprestris*, occur in the highlands. *B. finlaysoniana* is a Malaysian lowland species (Fig. 8.8). The generic name honours Edward French Bromhead, an English botanist.

Fig. 8.8 *Bromheadia finlaysoniana* (Lindl.) Miq. (as *Bronheadia palustris* Lindl.) From: Wight, R., *Icones Plantarum Indiae Orientalis*, vol 5(1): t. 1740 (1846). Line drawing by Govindoo. Courtesy of Missouri Botanic Garden, St. Louis, USA

Bromheadia finlaysoniana (Lindl.) Miq.

Malaysian names: *Seraman* in Kelabit, *Wi buntak** (Iban), *Busak paya* (Malay)

Description: A common, lowland, sun-loving species, it often presents in a large clump on peaty and sandy soils throughout the Malay Archipelago, extending towards southern Thailand and Indochina. The wiry stems are up to a metre tall, with stiff leaves 10 by 2.5 cm. The fragrant, white flowers are usually borne singly or up to four on an inflorescence according to O'Byrne (2001), but they only last a day; in fact, only the morning hours (Fig. 8.9). Yong (1990) observed that on some days the flowers are plentiful whereas on other days there are no flowers.

Fig. 8.9   *Bromheadia finlaysoniana* (Lindl.) Miq. (Photo: E.S.Teoh)

This phenomenon was explained as follows: when flower buds reach 12 mm in length, their growth abruptly slows down and this allows younger buds to catch up. An unusually cool day accelerates their development and 7 days later the orchid flowers gregariously. Bud development varies from 19 to 34 days with an average of 24 (Holttum 1949). *B. finlaysonniana* was once a popular garden plant in Southeast Asia, but it is now totally eclipsed by the flood of long-blooming, sun-loving orchid hybrids (Teoh 2005, 2011).

Herbal Usage: I.H. Burkill (1935), quoting Alvins, reported that, in Malacca, a decoction of the roots was consumed for rheumatism. In Peninsular Malaysia, flower stalks are chewed for the juice which is thought to be effective for treating asthma. In Sarawak, it was used to treat body aches and tired muscles (Go and Hamzah 2008). Ibans use the sap to treat toothache. Kelabit eat the flowers, cooked or raw, simply as a vegetable (Christensen 2002).

Overview

There are no chemical or pharmacological data on this genus.

- Note: *Wi buntak* also refers to *Arachnis flos-aeris*

Genus: *Bulbophyllum* Thouars

Chinese name: *Shi duo Lan*

An enormous genus comprising of over 2000, or perhaps even 2700, species (Siegerist 2001), *Bulbophyllum* is distributed throughout the tropics and subtropical regions, with the greatest diversity in Southeast Asia. Most species are epiphytic, but some are saxicolous, and there are a few terrestrial species. The generic name is derived from the unique appearance of round to ovate pseudobulbs which bear a single fleshy leaf, from Greek *bulbo* (bulb) and *phyla* (leaf). Inflorescence arises from the base of the plant and carries a single flower or a raceme. A hinged lip is characteristic. *Bulbophyllums* are very easy to grow and they flower readily, usually following a drop in temperature occasioned by a storm. Collectors are currently seeking new species. In the past, the *Bulbophyllum* was better known through its notorious members that stank of carrion!

Medicinal usage have been reported for only 23 species.

Bulbophyllum affine Wall ex Lindl.

Chinese name: *Chichunshidou Lan* (red lip stone bean orchid), *Gaoshifodou Lan* (Gaoshifo bean orchid), *Wenxing Lan* (stripe star orchid)

Description: This is a common epiphyte in lowland and low montane forests, from sea level to 2000 m, widely distributed from the Ryukyu Islands of Japan through China, Indochina, Thailand and Myanmar to the foothills of the Himalayas. On Lantau Island in Hong Kong, it grows on rocks along streams (Wu et al. 2002). Rhizome is stout and densely rooting, bearing pseudobulbs spaced 5 cm apart. Pseudobulbs are cylindrical, 1–2 cm

Fig. 8.10 *Bulbophyllum affine* Wall ex Lindl. (Photo: Bhaktar B. Raskoti)

long, and bear single, fleshy leaves, 6–15 by 1–2 cm. Flowers are small, star-shaped, pale yellow, and marked with linear reddish-brown stripes. Lip is orange (Fig. 8.10). In Thailand, it flowers in April to June (Vaddhaphuti 2005), elsewhere somewhat later; for example, in China and Nepal, May to July (Chen et al. 1999; Raskoti 2009), and in Bhutan and India, from June to July (Gurong 2006; Bose and Bhattacharjee 1980).

Herbal Usage: In Taiwan, the entire plant is used as a tonic, antipyretic, to reduce phlegm, and to stop bleeding (Ou et al. 2003).

Bulbophyllum ambrosia (Hance) Schltr.

Chinese name: *Fangxuianshidou Lan* (fragrant stone bean orchid), *Xiangshidou Lan* (fragrant stone bean orchid)
Chinese medicinal name: *Fangzhucao*

Description: This epiphytic species occurs in southern China, Hong Kong (where it is often sold as *B. watsonianum*) and Vietnam (Siegerist 2001). Its 3.5-cm-tall, yellowish-orange-coloured pseudobulbs are spaced 5–8 cm apart, and bear solitary leaves, up to 13 cm long. Fragrant flowers are borne singly, 2 cm across, white with crimson longitudinal stripes on the sepals that are longer

and wider than the petals. Dorsal sepal is ovate, Lateral sepals are 2–2.5 cm long, conjoined and recurved. Petals are triangular. Flowering period is February to May (Chen et al. 1999).

Herbal Usage: Herbs are obtained from Fujian, Guangdong, Guangxi and Yunnan and may be harvested throughout the year (Wu 1994; *Zhongyao Da Cidian* 1986). After cleaning, they are used fresh or steamed and dried. Taste is dry, bland and cold in nature. Chinese herbal medicine (CHM) uses a decoction of the plant, prepared by boiling 6–15 g of the dried herb (or 12–30 g of the fresh plant) to treat hepatitis, coughs and heat in the lungs (*Zhongyao Da Cidian* 1986; *Zhongyao Bencao* 2000; Wu 1994),

Bulbophyllum andersonii (Hook. f.) J.J. Smith

Syn. *Cirrhopetalum andersonii* Hook. f., *Cirrhopetalum henryi* Rolfe

Chinese names: *Shumaojuanban Lan* (comb hat roll petal orchid), *Chenhongjuanban Lan* (orange color roll petal orchid)
Chinese medicinal name: *Yipicao*

Description: This pretty *Cirrhopetalum* is a member of the Section *Recurvae*. However, the term *Cirrhopetalum* is no longer in use at the Royal Horticultural Society and the orchid should be referred to as *B. andersonii*. It is native to Guangxi, SW Sichuan, Guizhou and Yunnan in China, and in Vietnam, Myanmar, Sikkim and India, and is epiphytic or saxicolous at 400–2000 m.

Pseudobulbs are narrowly ovoid, 2–5 cm long, carrying a single leaf, 7–21 cm long. Umbel carries numerous flowers, commonly six, which are white to greenish-white with purplish-red-spotted veins. Sepals are 2–2.5 cm long, apex elliptic. Lip is deep purple accentuated by a yellow central band. In China, flowering period is February to October (Chen et al. 1999); in Sikkim, September to October (Pearce and Cribb 2002).

Herbal Usage: The entire plant is used for expelling wind and removing dampness, improving blood flow, stopping coughs and clearing retention of food (Wu 1994). A decoction of 6–15 g of the plant is used to treat rheumatism. For feminine weakness, the prescription calls for a soup to be prepared with 6–15 g of the herb and chicken meat or pork (*Zhongyao Da Cidian* 1986).

Phytochemistry: *B. andersonii* contains a novel phenanthrene derivative named cirrhopetalin (7-hydroxy-4-methoxy-2,3-methylene-dioxyphenanthrene; two bibenzyl derivatives, cirrhopetalidin (2′,3-dihydroxy-3′-methoxy-4,5-methylenedioxybibenzyl) and cirrhopetalinin (3,3′-dihydroxy-4,5-methylenedioxybibenzyl) and batatasin-III (3,3′-dihydroxy-5-methoxybibenzyl); two stilbenoids, respectively, a cirrhopetalanthridin (4,7-dihydroxy-2,3-methylenedioxy-9,10-dihydro-phenanthrene), and cirrhopetalidinin (3,2′-dihydroxy-5′-methoxy-4,5-methylene-dioxybibenzyl); and gigantol (Majumder and Basak 1990, 1991a, b).

Bulbophyllum calodictyon Schltr. [see ***Bulbophyllum griffithii* (Lindl.) Rchb. f.**]

Bulbophyllum careyanum (Hook.) Sprengel

Description: Pseudobulb is conical with a single elliptic leaf at its apex, measuring 8–12 by 2–4 cm. Inflorescence arises from the base of the pseudobulb. Rachis is densely many-flowered, giving the appearance of scales tightly layered over one another. Flowers are 1 cm across, yellowish-brown, foul smelling. Bracts are lanceolate (Fig. 8.11). It flowers in September to November (Raskoti 2009). This epiphytic *Bulbophyllum* is distributed in Nepal, Bhutan, Myanmar, Thailand and Vietnam. In Nepal, it occurs at 600–2100 m, on *Shorea robusta* (sal) and *Schima* (Pant and Raskoti 2013).

Herbal Usage: Poultices made with the pseudobulbs and leaves are used to treat burns on the skin (Pant and Raskoti 2013).

Bulbophyllum cariniflorum Rchb. f.

Syn. *Bulbophyllum densiflorum* Rolfe

Fig. 8.11 *Bulbophyllum careyanum* (Hook.) Sprengel (Photo: Bhaktar B. Raskoti)

Fig. 8.12 *Bulbophyllum cariniflorum* Rchb. f. (Photo: Bhaktar B. Raskoti)

Chinese name: *Jianyeshiduo Lan*
Indian name: *Sumura*

Description: Unlike most *Bulbophyllum*, each ovoid pseudobulb, which are 1–3 by 1.5–2 cm in this species, carries two oblong, leathery leaves, 12–15 by 2.7–4 cm, and deciduous. Raceme is many-flowered but the flowers are not so tightly packed as in the preceding species. Flowers are yellow, thickly textured, small (under 1 cm in diameter), not opening widely (Fig. 8.12). *B. cariniflorum* is saxicolous in mixed forests at 2100–2200 m in southern Xizang. It is also found in northern Thailand, northeast India, Bhutan and Nepal. It grows in moist habitats, and between May and June, following the arrival of the monsoon rains, the young, under-developed pseudobulb bears a lateral inflorescence (Misra 2007) and flowers appear in July in Tibet (Chen and Vermeulen 2009) and in July to August in Bhutan and Nepal (Gurong 2006; Raskoti 2009).

Herbal Usage: A paste made from the dried roots of *B. cariniflorum,* black pepper and cow's milk is taken for several days to induce abortion during the first trimester in the districts of Mondanala and Sutanguni in the Niyamgiri Hill Ranges of Orissa, India (Dash et al. 2008).

Bulbophyllum congestum Rolfe [see ***Bulbophyllum odoratissimum* (Sm.) Lindl. ex Wall**]

Bulbophyllum cylindraceum Wall. ex Lindl.

Local Name: *Dabaoshidou Lan* (large bud stone bean orchid)

Description: This unusual *Bulbophyllum* has very small, almost spherical pseudobulbs, 3–5 mm in diameter bearing a terminal, petiolated, leathery, elliptical leaf, 15–25 cm long and 2–4 cm wide. . Rhizome is very stout, 6 mm in diameter, with pseudobulbs spaced 5–7 mm apart. Scape arises from base of pseudobulb. Flowers are 8 mm across, dark purple, densely arranged on a nodding

Fig. 8.13 *Bulbophyllum cylindraceum* Wall. ex Lindl. (Photo: Bhaktar B. Raskoti)

raceme in a catkin-like manner on an erect inflorescence (Fig. 8.13). In China, it flowers from October to January (Chen and Vermeulen 2009), in October to January in Bhutan (Gurong 2006) and November in Nepal (Raskoti 2009). The species is distributed from west to southeast Yunnan at 1400–2400 m (Chen and Vermeulen 2009), thence to Sikkim Himalaya and Khasia Hills at 2000–2300 m (Bose and Bhattacharjee 1980). It is epiphytic and lithophytic.

Herbal Usage: Herb is obtained from Sichuan and Yunnan. Entire plant is used to treat painful joints and numbness (Wu 1994).

Bulbophyllum flabellum-veneris (J. Koenig) Aver.

Syn. *Bulbophyllum lepidum* (Blume) J.J. Sm., *Cirrhopetalum lepidum* (Blume) Schltr.

Thai names: *Phet phra in, Sa mai, Khon dam phi*

Description: *B. flabellum-veneris* is widely distributed in Southeast Asia but not further east than Borneo. It is found in the lowlands up to 900 m, and is popular with orchid growers because of the ease of cultivation. Pseudobulbs are ovoid, 1.5 cm tall, and spaced 2 cm apart on a creeping rhizome. The thick leaves are lanceolate, 16 by 3 cm, with a blunt apex. Scape is thin,

20 cm long and carries 7–10 flowers spread around its tip in a semi-circle. The lateral sepals are 2.5 cm long, 4 mm wide, and fused along their medial margin to form a petaloid structure which is cream-coloured, blushing into deep crimson towards the base. Dorsal sepal is yellow, hairy over the margin, and it forms a hood over the column. Petals are 5–6 mm long, creamy, and covered with hair-like bristles on their lateral margins. The hinged lip is well adapted for insect pollination. It is olive-green in colour. *B. flabellum-veneris* flowers throughout the year with a peak during the rainy seasons which are determined by monsoons and the orientation of mountain barriers (Teoh 2005).

Herbal Usage: Despite its small size, in Thailand, the pseudobulb is widely used to treat oedema whereas the entire plant is used to treat liver dysfunction (Chuakul 2002).

Bulbophyllum griffithi (Lindl.) Rchb. f.

Syn. *Bulbophyllum calodictyon* Schltr., *B. chitoense* S.S. Ying, *Sarcopodium griffithi* Lindl.

Chinese name: *Duanchishiduo Lan* (stone bean orchid), *Xiaolushidaolan* (small green stone bean orchid)
Chinese medicinal name: *Shichuanlian*

Description: This is an "understory epiphyte" in forests at 1000–1700 m in central Taiwan, southeast Yunnan, Vietnam, northeast India, Bhutan and Nepal (Chen and Vermeulen 2009). Pseudobulbs are ovoid up to 3 cm long. Leaves are solitary, broad, lanceolate, sessile, 2.2–12 by 1–2.5 cm. Scape is very short. Flower is single, moderately open, resupinate, 2.5 cm across, yellow with reddish-brown spots. It flowers in February, August and October to November in China (Chen and Vermeulen 2009), and in August at the Khasia Hills and in Sikkim in northeast India at 3000 m (Bose and Bhattacharjee 1980).

Herbal Usage: Herb is obtained from Yunnan Province (Wu 1994). Pseudobulbs are used for treating pulmonary condition to relieve coughs, pain and reduce inflammation. Main usage is to treat chronic coughs, bronchitis and sore throat. They are also applied to fractures, infected breasts, abscesses and all types of sores (Wu 1994). Plants are harvested in summer and autumn; pseudobulbs are used fresh or dried. Character is dry, bland and neutral in nature. In Yunnan, 30–60 g of fresh pseudobulbs are boiled for drinking, or 6 g of powder are made into a paste with an appropriate amount of honey and consumed. To heal breast abscesses or infected traumatic injuries, a paste made with fresh pseudobulbs is applied (*Zhongyao Bencao* 2000).

Bulbophyllum inconspicuum Maxim

Japanese name: *Mugiran* (Wheat Orchid)
Chinese medicinal names: *Maihu* (Wheat epiphyte), *Guoshanye* (leaf over fruits) *Yiguayu* (one hanging fish); *Yangnaicao* (goat's milk herb); *Yaquezui* (bird mouth), *Linzhijiao*; *Shiyangmei* (rock berry); *Wangniantao* (10,000 peach); *Zishangye* (leaf on paper); *Guazilian* (seed lotus); *Shilongshiwei* (stone dragon and stone tail); *Qixiantao* (seven immortal peach); *Shixiantao* (stone immortal peach); *Xiaokuozi Lan* (small button orchid); *Shiwenchong* (rock mosquito) *Huangdouxian* (soyabean whip) *Loushanglou* (building over building) *Shilianzi* (stone lotus seeds) *Genshangzi* (seeds above the roots); *shi yu* (stone bean).

Description: This tiny, evergreen *Bulbophyllum* well deserves its name. It grows in clusters on the trunks of old trees at low altitude in the southern half of Japan and in central China. Pseudobulbs are tiny, flattened, oval, with crenate surface, and rather remote on a slender rhizome. Leaves are oblong 1.35 cm by 6–8 mm. The olive-green, solitary flowers (sometimes in pairs) with thinly

Table 8.2 Herbal Remedies employing *Bulbophyllum inconspicuum* [Sources: *Zhongyao Da Cidian* (1986) and *Zhonghua Bencao* (2000)]

1. Indication: fever and coughs
Boil *B. inconspicuum* 6 g with *Aralia chinensis* (Herbal name: *Ci Lau Bao*) 9 g and consume
(*Quizhou Folk Herbs*)

2. Indication: hundred-day cough (whooping cough, pertussis)
Boil *B. inconspicuum* 30 g, with *Coptis chinense* 3 g, and honey 15 g, and consume
(*Quizhou Folk Herbs*)

3. Indication: bronchiectasis: Boil *B. inconspicuum* 30 g, with *Stenoloma chusana* 15–30 g to produce a decoction. One dose per day.
(*Jiangxi Medicinal Herbs*)

4. Indication: coughs associated with tuberculosis
Cook *B. inconspicuum* 15 g, with *Bai Zhe Er* 15 g, and pork 500 g. Consume meat and soup in one or two portions. Repeat twice.
(*Quizhou Folk Herbs*)

5. Indication: Tuberculosis
Make a decoction for consumption with *B. inconspicuum* 30 g, *Ardisia japonica* 30 g. and one fresh lotus leaf.
(*Hubei Journal of Traditional Medicine*)

6. Indication: Damage of *yin* by febrile illness and polydipsia
Prepare decoction with *B. inconspicuum* 15 g, bamboo leaves 9 g, reed rhizome 12 g, *Scutellaria baicalensis* 4.5 g.
(*Anhui Medicinal Herbs*)

7. Indication: Painful, swollen joints
Prepare a soup with *B. inconspicuum* 60 g, Japanese honeysuckle stem 30 g, 1 pig's trotter, 200 ml yellow millet wine, water in appropriate amount: braise.
(*Hubei Journal of Traditional Chinese Medicine*)

8. Indication: children's frightfulness, coughs with hoarse voice,
Cook fresh *Mai Hu* (*B. inconspicuum*) 45–60 g, one pig's pancreas, and rock sugar, and consume
(*Eastern Fujian Herbs*)

9. Indication: external injuries
Boil fresh *Mai Hu* and consume
(*Records of Hunan Medicine*)

10. Indication: facial acne
Mix a handful of *Mai Hu* with rock sugar and apply externally
(*Eastern Fujian Herbs*)

11. Indication: skin ulcers
Boil *Mai Hu* 30 g and consume with red sugar
(*Handbook of Jiangxi Herbs*)

12. Indication: menstrual irregularities
Fry fresh *Mai Hu* 30 g, *Yue Ji Hua* 15 g, with an egg and consume
(*Fujian Chinese Herbs*)

13. Indication: Toothache
Boil fresh *Mai Hu* 30–60 g. Add some salt and consume.
(*Fujian Chinese Herbs*)

membraneous bracts appear in May to June (Ohwi 1965).

(Note: In the recent *Flora of China 25—Orchidaceae*, Chen and Vermeulen (2009) did not list *B. inconspicuum* as occurring in Mainland China. Instead, they described a *B. brevipedunculatum* T.C. Hsu and S.W. Chung which has an appearance which is similar to *B. inconspicuum*, the former species occurring in Taiwan. Thus, the identity of the Chinese herb which numerous Chinese Pharmacopoeia described extensively remains a mystery. Nevertheless, we also describe it here because many authoritative Chinese medicinal works mention the orchid. There is a long list of prescriptions which include *B. inconspicuum*.)

Herbal Usage: *B. inconspicuum* is collected in summer and autumn, washed and dried for future use. Entire plant of the tiny orchid is used to treat tuberculosis in Sichuan Province (Anonymous 1974) and coughs in Quizhou and Fujian. It is also used for fretfulness in children, acne, skin ulcers, menstrual irregularities and external injury. Eight examples of prescriptions are listed in Table 8.2 compiled from *Zhongyao Da Cidian* (1986) and *Zhonghua Bencao* (2000).

Bulbophyllum kwangtungense Schltr.

Medicinal name: *Guangdongshidou   Lan* (Guangdong Province stone bean orchid)

Description: A miniature *Bulbophyllum* with cylindric pseudobulbs, 1–2.5 cm tall and 2–5 mm in diameter spaced 2–7 cm apart with a terminal, oblong, leathery leaf 2.5–4.7 by 0.5–1.4 cm. Inflorescence bears an umbel of 2–4, occasionally up to 7, small yellow flowers with sepals of 8–10 by 1–1.3 mm. Flowering season is May to August. It thrives on rocks in forests from Zhejiang, Hubei and Hunan to the southern Chinese provinces of Fujian, Guangdong, Guangxi, Jiangxi, Guizhou and Yunnan (Chen and Vermeulen 2009). In Hong Kong and Lantau Islands, it is reported to be epiphytic (Wu, Hu, Xia et al. 2002).

Herbal Usage: The entire plant is used to treat coughs, reduce "heat", clear "wind", and convulsions in children (Wu 1994). Harvested in summer or autumn, pseudobulbs obtained from Guangdong, Guangxi and Zhejiang may be used fresh or made into powder after drying and steaming. Taste is dry, bland and cold in nature. It releases heat, nourishes *yin* (see Chap. 2) and relieves swelling. It is used to treat sore throat, coughs, rheumatism and arthritic pain, traumatic injuries and mastitis. For external use, pound pseudobulbs and apply to affected parts. The Chinese *Materia Medica* advises that should be used with caution by people who have a "cold body" (*Zhonghua Bencao* 2000).

Phytochemistry: *B. kwantungense* Schltr. contains cumulatin, densiflorol A, plicatol B, and three dihydrodibenzoxepins, namely 7,8-dihydro-5-hydroxy-12,13- methylenedioxy-11-methoxyldibenz[B,F]oxepin, 7,8-dihydro-3-hydroxy-12,13-methylenedioxy-11-methoxyldibenz[B,F]oxepin, (1-3) and the known compound Densiflorol A. The three dihydrobenzoxepins exhibited antitumor activities against HeLa and K562 human tumor cell lines (Wu et al. 2006, 2008; Wu, Chen, Lay 2010; Wu, Chen, He, Pan 2011). Another new dibenz [b,f]oxepin was found to be produced as a stress metabolite from the leaves and stems of *B. kwangtungense* Schltr., in response to abiotic stress induced by exposure to copper chloride (Chen et al. 2011).

Bulbophyllum laxiflorum (Blume) Lind.

Syn. *Bulbophyllum radiatum* Lindl.

Chinese names: *Fusheshidou Lan* (radiating stone bean orchid); *Yashe Lan* (Duck tongue orchid)

Chinese medicinal names: *Shizao* (Stone date); *Shiduo* (stone bean); *Yanduo* (stone bean); *Jinduo* (golden date); *Shimi* (Stone rice); Duyiyanzhu (Single leaf cliff pearl)

Description: This is a small *Bulbophyllum* belonging to the Section *Corymbosia* which has small flowers (here white or yellowish) with thin segments, and one leaf to each pseudobulb. Pseudobulbs are 2 cm tall, the leaf 5 cm in length. The species is found in southeastern China, Myanmar, Thailand and Peninsular Malaysia, in the lowlands.

Herbal Usage: Herb is obtained from Gansu, Shaanxi and Guangxi (Wu 1994). *B. laxiflorum* enriches the *Yin*, and benefits the lungs (clears phlegm, stops haemoptysis) and stomach (improves appetite, helps digestion, relieves dry throat) and speeds recovery from trauma and fractures (Wu 1994). It is also used to treat rheumatism, high fever and epilepsy (*Zhongyao Da Cidian* 1986).

Bulbophyllum leopardinum (Wall.) Lindl. ex Wall.

Chinese name: *Duantingshiduo Lan*

Description: Pseudobulbs are cylindrical, close or sparse, with single oblong-elliptic leaf, 5–7.5 by 2–3 5 cm. Inflorescence is basal, bearing a single flower, sometimes two. Flowers 3 cm across, cream to pale green, with brown, red or purple spots. Lip is crimson. It is distributed in Nepal, Bhutan, India, China (southern Xizang and Yunnan) Laos and northern Vietnam. Epiphytic or saxicolous, it occurs in humid forests, in Nepal at 1500–2000 m and in China at 1300–3000 m. Flowering season is July in Nepal, and April to August, then October in China (Pant and Raskoti 2013, Chen and Vermeulen 2009).

Herbal Usage: Poultice made with the pseudobulbs is used in Nepal to treat burns on the skin (Pant and Raskoti 2013).

Phytochemistry: In one of their early studies on the chemical constituents of Himalayan orchids, Majumder's team isolated the phenenthrine, bulbophyllanthrin, from *B. leopardinum* (Majumder et al. 1985).

Bulbophyllum levinei Schltr.

Chinese name: *Chibanshidou Lan* (tooth petal stone bean orchid)
Chinese medicinal name: *Duyeyanzhu*

Description: A small *Bulbophyllum*, pseudobulbs are bottle-shaped, up to 1 cm long and 2–4 mm in diameter, with a terminal, lanceolate leaf, 3–4 cm long. Inflorescence carries a raceme of 2–6 white flowers, tinged with purple. Flowering season is May to August. It grows on tree trunks and rocks from Zhejiang and Hunan to the southern Chinese provinces of Fujian, Guangdong, Jiangxi and Yunnan at 800 m and also in Vietnam (Chen and Vermeulen 2009).

Herbal Usage: Herbs are obtained from Zhejiang, Guangdong, Yunnan and Xizang (Wu 1994). Entire plant is used to treat swellings. It "benefits *yin*, reduces heat", and is thus used to treat sore throat, tonsillitis, mouth ulcers, fever and thirst (Wu 1994). For decoction, standard practice is to boil 6–15 g of dried herb or 30–60 g of fresh herb. For external use, the herb is pulverised to make a paste. Another method of making the poultice, mentioned in the *Zhejiang Journal of Medicinal Plants Research*, is to combine 9–15 g each of *B. levinei* whole plant, radish (*dingluobo*), *didancao*

Fig. 8.14 *Bulbophyllum lobbii* Lindl. (Photo: E.S. Teoh)

Fig. 8.15 *Bulbophyllum siamensis* (=*Bulbophyllum lobbii*, a very variable species) (Photo: E.S. Teoh)

(*Elephantopus scaber*, Asteraceae) and *chuanxinlian* (*Andrographis paniculata*, Acanaceae) (Zhonghua Bencao 2005). The additional herbs are commonly used to treat swellings and superficial infections.

Bulbophyllum lobbii Lindl.

Syn. *Bulbophyllum siamense* Rchb. f.

Thai name: *lin fa* in Ubon Rachthani

Description: *B. siamense* is now considered conspecific with the epiphytic *B. lobbii*. Flowers are solitary, 5–6 cm across, yellow with thin, longitudinal brown stripes across all floral segments. Petals and sepals are shaped like blades. Dorsal sepal is upright and pointed, lower sepals are bow-legged and pointed. Petals are horizontal and curve downwards towards the tip (Figs. 8.14 and 8.15). Flowering season in Thailand is variously given as December to April (Nanakorn and Watthana 2008) and May to July (Vaddhanaphuti 2001). The species is found at mid-elevations between 500 and 1300 m in evergreen forests.

Herbal Usage: In Thailand, leaves of *B. siamense* (Fig. 8.15) are used to treat burns (Chuakul 2002).

Bulbophyllum neilgherrense Wight

Description: Plant is epiphytic with compressed, smooth, ovoid to conical pseudobulbs, 1.5–2.5 by 1.0–3.0 cm, sparsely spaced along a creeping rhizome. Leaves are oblong-obtuse, coracious, 4.3–10.5 by 1.5–2.6 cm. Inflorescence is up to 8.5 cm long, shorter than the leaves, drooping, and many-flowered. Flowers are dull yellow to purplish-green and appear in February (Joseph 1982) (Fig. 8.16). An endemic species, *B. neilgherrense* is distributed in the southern states of India, in Malabar and Nilgiri Hills.

Herbal usage: Pseudobulbs are used as a tonic for rejuvenation (Das 2004; Jonathan and Raju 2005).

Bulbophyllum odoratissimum (Sm.) Lindl. ex Wall.

Chinese names: *Mihuashidou Lan* (small flowered stone bean orchid), *Xiaohaoshi Ganlan* (small stone olive), *Shimi* (stone rice); Chinese name: *Mitoushidou Lan* (dense head stone bean orchid)
Medicinal name: *Guoshangye* (leaves on fruit); *Xiaoguoshangye* (small *Guoshangye*); *Shicuanlian* (rock string lotus)
Myanmar name: *Thazin hmwe*

Description: Pseudobulbs are spaced 4–8 cm apart, and are 2.5–5 cm tall. Leaves are oblong, 4–5 cm by 0.8–1 cm. One or two inflorescences arise from the base of the pseudobulb, each 10–14 cm long, bearing 10–20, fragrant, white flowers arranged in a globose head or umbel (Fig. 8.17). Some flowers are tipped with yellow on the sepals and have a reddish-brown lip. The species is epiphytic or saxicolous at 200–2300 m, and is distributed from Fujian and Guangdong westwards to Guangxi, Sichuan, Yunnan and Tibet, then to Nepal, Bhutan, Sikkim and southwards to Laos, Vietnam, Thailand and Myanmar (Kachin and Shan states). In China, it flowers from April to August (Chen et al. 1999); Myanmar, April to June (Tanaka et al. 2003); Bhutan, May to September (Gurong 2006); Nepal, April to August (Raskoti 2009); and in Thailand, May to July (Nanakorn and Watthana 2008).

Phytochemistry: A phenanthraquinone, bulbophyllanthrone, and two flavonoids, chrysin and pinobanksin, were isolated from *B. odoratissimum* by Majumder and Sen (1991). Shenyang Pharmaceutical University scientists managed to isolate and subsequently synthesise two natural dihydrostilbenes from *B. odoratissimum*, namely, 3,[2-7-methoxy-benzo (d)(1,3)dioxol-5-yl)ethyl]phenol with an overall yield of 28 % and 6-(3-hydroxy-phenethyl)benzo (d)(1,3)dioxol-4-ol with an overall yield of 20 %. They also obtained 9 analogues. Several compounds possessed antitumour activity (Zhang et al. 2007a). Two new dimeric phenanthrenes, bulbophythrins A and B, isolated from the orchid

Fig. 8.16 *Bulbophyllum neilgherrense* Wight. Reproduced with permission from *Introductions to Orchids* by Abraham and Vatsala, Parlode, Thiruvananthapuram: Tropical Botanic Garden and Research Centre (TBGRI), 1981

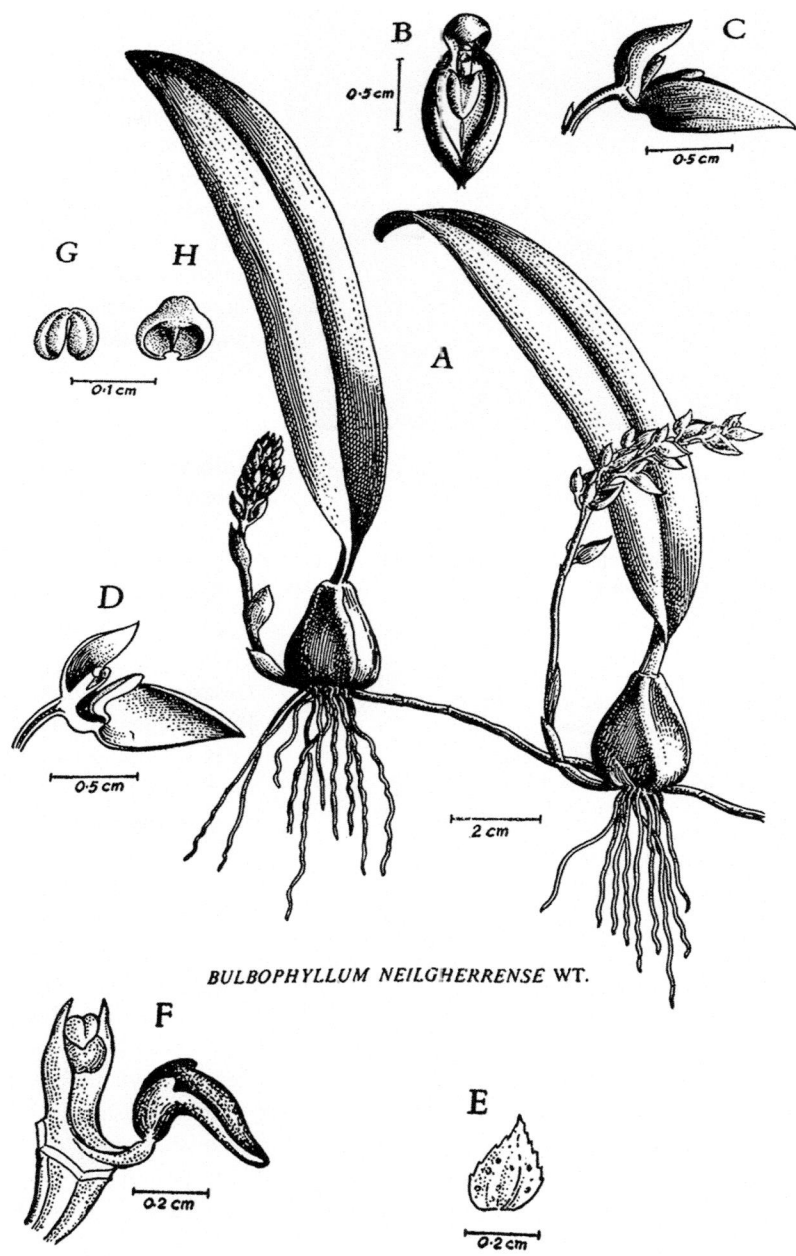

BULBOPHYLLUM NEILGHERRENSE WT.

showed cytotoxic activity against human leukaemia cell lines K562 and HL-60, human lung adenocarcinoma A549, human hepatoma BEL-7402 and human stomach cancer SGC-7901 (Xu et al. 2009). Pseudobulbs also contain four glycosides, namely, bulbophyllinoside [3-hydroxyphenethyl alcohol 4-0-(6'-0-beta-apiofurannosyl)-beta-D-glucopyanoside], syringin, 3-methoxy-phenethyl alcohol 4-O-beta-D-glucopynanoside and 3, 5-dimethoxy-phenethyl alcohol 4-O-alpha-D-glucopynanoside (Liu et al. 2006). Two new dihydrostilbenes, 5-(2-benzo[1,3]dioxole-5-ylethyl)-6-methoxy

Fig. 8.17 *Bulbophyllum odoratissimum* (Sm.) Lindl. ex Wall. (Photo: Bhaktar B. Raskoti)

Table 8.3 Chinese herbal prescriptions that employs *Bulbophyllum odoratissimum* (*Zhongyao Da Cidian* 1986)

1. Indication: fractures
Mix *Guoshangye*[a] (e.g. *B. odoratissimum*) powder with wine to form a glue and apply externally. Change daily (*Yunnan Selected Chinese Herbs*)

2. Indication: pulmonary tuberculosis
Boil *Guoshangye* 30 g, *Xiao Baiji* (*Bletilla striata* which is another orchid) 30 g, and *Qi Xing Cao* 15 g with red sugar and consume
(*Kunming Commonly Used Folk Herbs*)

3. Indication: pain caused by hernia
Boil *Guoshangye* 30 g, *Jiangwei Cao* 3 g, and *Xiaonanmuxiang* 6 g, with red sugar; and consume
(*Kunming Commonly Used Folk Herbs*)

[a]*Guoshangye* is a tuja and Miao folk medicine which makes use of *B. andersonii, B. kamgtimgemse* and *B. odoratissimum* or *Pholidota* species. It does not appear to designate a single specific herb or orchid

benzo[1,3]dioxole-4-ol and 5-(2-benzo [1,3] dioxole-5-ylethyl)benzo[1,3]dioxole-4,7-diol, from *B. odoratissimum* Lindl. showing significant cytotoxicity toward human cancer cell lines have been synthesised (Zhang et al. 2007b).

Herbal Usage:

B. odoratissimum is obtained from Fujian, Guangdong, Guangxi, Sichuan, Yunnan and Xizang Provinces. *B. odoratissimum* var. *odoratissimum* is obtained from Yunnan Province. The herb is collected year round. The entire plant is used. For storage, it is washed and dried, or washed, steamed and dried. Chinese herbals state that *Guoshangye* benefits the lungs, clears phlegm, stops the coughing of blood in tuberculosis, and relieves bronchitis, sore throat and joint pain, and assists in the healing of fractures, knife wounds and other traumatic injuries (*Zhongyao Da Cidian* 1986; Wu 1994). Examples of Chinese herbal prescriptions using *B. odoratissimum* are shown in Table 8.3. In Nepal, it is also used to treat tuberculosis and fractures (Pant and Raskoti 2013).

Bulbophyllum pectenveneris (Gagnep) Seidenf.

Chinese name: *Huanghuajuanban Lan* (yellow flower folding petal orchid), *Huanghuashitou Lan* (yellow flower stone bean orchid)

Description: This is a small *Bulbophyllum*. Pseudobulbs are ovoid, 5–10 mm in diameter. Leaf is oblong-lanceolate, 1–6 by 0.7–1.8 cm, thick and leathery. Inflorescence is 5–12 mm long. Umbel carries 3–9 flowers distinguished by their slender, yellow sepals that are slightly tinged with brown. Flowering period is April to September (Chen and Vermeulen 2009). *B. pectenveneris* is distributed from Hubei and Anhui Provinces in eastern China southwards to Fujian, Taiwan, Hong Kong, Guangxi, Hainan and across to Laos and Vietnam, growing on trees in lowland forests up to 1000 m.

Herbal Usage: The whole plant is believed to have the ability to reduce stasis of blood, improve circulation and stop pain in the joints, muscles and bones (Ou et al. 2003).

Bulbophyllum pectinatum Finet.

Description: Epiphytic or saxicolous. Pseudobulbs are distant, ellipsoid to ovoid, 2–3.5 by 1–2 cm. Leaf is elliptic, 6–12 by 2.4–4 cm. Inflorescence is erect with a single flower, 2.5–3.5 cm across. Flower form is full, sepal and petals yellowish-green with densely spotted maroon veins. Dorsal sepal is broad, deltoid, 2 cm by 1 cm. Lateral sepals measure 2 by 1.5 cm. Lip is white and irregularly spotted with red on its upper surface. There is an *alba* form. Flowering period is April to May in China (Chen et al. 1999). This small *Bulbophyllum* species enjoys a wide distribution in Taiwan, Yunnan, Vietnam, northern Thailand, Myanmar and northeast India at 100–2500 m.

Herbal Usage: Plants are obtained from Hubei, Hunan, Guangxi and Yunnan. The herb is used with the intention to sooth the lungs, stop coughs and for pain relief. Whole plant is used in the treatment of tuberculous cough, asthma, sprains and fractures (Wu 1994).

Bulbophyllum radiatum Lindl. [see *Bulbophyllum laxiflorum* (Blume) Lindl.]

Bulbophyllum reptans (Lindl.) Lindl. ex Wall.

Chinese name: *Fushengshidou Lan* (conceal life stone bean orchid); *Baihuashidou Lan; Shidou Lan*
Chinese medicinal name: *Shilianzi*

Description: *B. reptens* has small, ovoid-conical pseudobulbs, 0.4–1 cm diameter, spaced 5–9 cm apart, thus forming sparse colonies. Leaves are lanceolate and narrow, up to 15 cm long. Flowers are a pale yellow with a deeper orange-coloured lip, with 3–9 flowers on an erect raceme. The Indian variety is spotted with purple. Flowering season is October to December in northeast India (Garhwal, Sikkim, Naga and Khasia Hills) (Bose and Bhattacharjee 1980), possibly flowering only when there is rain, but in China it flowers from January to October (Chen et al. 1999). It is epiphytic or saxicolous at 1000–2800 m, and is distributed from Hainan and Guangxi to Yunnan, southeast Xizang, Vietnam, Myanmar, Sikkim, Bhutan, Nepal and India (Chen et al. 1999).

Phytochemistry: Numerous phenanthrenes have been isolated from *Bulbophyllym reptens*,

including coelonin, flavanthrin, confusarin, gynopusinblestriarene A (flavanthrin), cirrhopetalanthrin, reptanthrin and isoreptanthrin (Majumder et al. 1999).

Herbal Usage: Similar to *B. laxiflorum* (syn. *B. radiatum*). It "enriches *yin*, and benefits the lungs (clears phlegm, stops haemoptysis) and stomach (improves appetite, helps digestion, relieves dry throat) and speeds recovery from soft tissue trauma and fractures". Herbs are obtained from Guangdong, Yunnan and Xizang (Wu 1994).

Bulbophyllum retusiusculum Rchb. f.

Chinese name: Yellow Comb Orchid in Taiwan
Thai name (in Ubon Rachthani): *Ma tak khok*

Description: This is small but striking "*Cirrhopetalum*" that produces 5–8, sometimes brilliant, orange-coloured flowers arranged in an umbel-like raceme, the lateral sepals 14 by 2.2–3.5 mm, yellowish-brown or orange marked with purplish veins at the base. Pseudobulbs are ovoid, 1–2 cm by 10.8–1 cm, spaced about 1.5 cm apart. Flowers appear from September to December. The species enjoys a wide distribution from Hunan, Taiwan, Hainan, Sichuan and Yunnan southwards to Myanmar, Thailand, Indochina and Peninsular Malaysia, and westwards to Nepal, Sikkim and Bhutan. It is found growing on trees in sparse woods at the edge of forests, at 500–2000 m. In India, it is one of the very few epiphytic orchids that grow in scrubby *Betula,* conifer and rhododendron–juniper scrub (Misra 2007).

Herbal Usage: Thai herbalists use the whole plant to make a tonic (Chuakul 2002).

Bulbophyllum rufinum Reichb. f.

Thai names: *Ma tak khok, Ueang kip ma ya, hang krarok, om nin* (in Ubon Rachthani)

Description: *B. rufinum* is found in broad-leaved, evergreen forests in north, northeast, east and southern Thailand, Myanmar, Indochina and southern Yunnan. Pseudobulbs are globose or ovoid, 2.5–4 by 1.5–2 cm, well spaced on a stout rhizome. Leaves are oblong-elliptic, 13–14 by 3 cm. The 8-cm-long, arching, green or purple rachis bears 20–40 yellow flowers with purple striping, 0.3 cm wide and 0.8 cm long, that are only partially open, and sparsely arranged. Lip is a dark yellow. In Thailand, it flowers in October (Vaddhanaphuti 2005; Nanakorn and Watthana 2008) and in November in China (Chen and Vermeulen 2009).

Herbal Usage: In Thailand, the entire plant is used as a tonic and to treat asthma (Chuakul 2002).

Bulbophyllum sterile (Lam.) Suresh

Syn. *Bulbophyllum nilgherrense* Wight
Cirrhopetalum neilgherrense (Wight) Wight, *B. rosemarianum,* Sattish et al.; *B. kaitense* Rchb. Description: A common epiphyte in southern India, it occurs in lowlands up to 900 m. The four-cornered, 3 by 2 cm, green, smooth pseudobulbs are well spaced along a slender creeping rhizome, and they bear a single leaf 10–15 cm long by 2–3 cm wide. The arching rachis carries many small, dull yellow or purple flowers with a deep purple lip. Petals are minute. Flowers are not well expanded. They may be tightly or loosely arranged on the rachis. Flowering season is February (Joseph 1982; Abraham and Vatsala 1981).

Phytochemistry: In vitro testing showed that leaf and pseudobulb extracts of *B. sterile* inhibited the growth of five strains of bacteria, namely *Escherichia coli, Staphylococcus aureus, Bacillus pumilus, Pseudomonas aeroginosa* and *P. putida.* Leaf extracts were more effective against the first three organisms, while pseudobulb extracts were more effective against the last two. However, their antibacterial activity was weak compared to the old antibiotic, strep-tomycin. No attempt was made to determine the nature of the antimicrobial compound in this study (Priya and Krishnaveni 2005).

Herbal Usage: Tubers of this orchid are eaten for good health by the Valmikis tribe of

Visakhapatnam district in Andhra Pradesh (Reddy et al. 2005).

Bulbophyllum umbellatum Lindl.

Syn. *Cirrhopetalum maculosum* (Lindl) Rchb. f.

Description: Plant is medium-sized, with ovoid-conical, ridged pseudobulbs, solitary ellip-tic, thick leaves of up to 15 cm length, and a scape of 7 cm. Umbels carry up to 5 small, yellow-green flowers, 1.2 cm across. It is distributed in the western Himalayas (Siegerist 2001) and Thailand. It flowers in June in Thailand (Vaddhanaphuti 2005). Indian writers refer to this orchid as *Cirrhopetalum maculosum.*

Phytochemistry: Cirrhopetalanthrin, a dimeric phenanthrene derivative, has been isolated from *C. maculosum* (= *B. umbellatum* Lindl.) (Majumder, Pal and Joardar 1990).

Herbal Usage: *C. maculosum* (= *B. umbellatum* Lindl.) was used as *swarna jivanti* in Ayurvedic preparations (Duggal 1972). The roots of the orchid are said to promote longevity (Vij 1995).

Bulbophyllum vaginatum (Lindl.) Rchb. f.

Syn. *Cirrhopetalum vaginatum* Lindl.

Malay name: *Magrah batu* (stone orchid)

Description: *B. vaginatum* is a common lowland orchid which forms large colonies on the branches and trunks of mature trees. In Singapore, their favourite perch seems to be the popular imported species, *Samanea samaan* (commonly known as the Rain Tree), planted to provide shade at the roadside. In Malaysia, it is also found on old *Pterocarpus indicus* (Angsana), *Eugenia grandis* (Jambu Laut) and other trees. As with the pigeon orchid, flowering is triggered by a sudden drop in temperature generally brought on by a rainstorm. *B. vaginatum* presents an impressive sight with its abundance of creamy-yellow flowers, each umbel with pairs of long tapered sepals, medusa-like, blowing in the wind (Figs. 8.18

Fig. 8.18 *Bulbophyllum vaginatum* (Lindl.) Rchb. f. (Photo: E.S. Teoh)

and 23.13). Once split off from *Bulbophyllum* and given a generic name *Cirrhopetalum* which appropriately describes the hair-like petals of some of its members (e.g. *B. medusae*), it has recently been returned to the genus *Bulbophyllum*.

Phytochemistry: At the National University of Singapore, Leong, Harrison and co-workers isolated a total of 21 phenanthrenes, dihydrophenanthrenes, bibenzyls, phenanthrofuran and other aromatic compounds from the locally prevalent *B. vaginatum* (Leong et al. 1997, 1999; Leong and Harrison 2004). Most of the compounds are also present in other orchids but four have not been isolated from other orchid species. A unique compound of *B. vaginatum* is the complex molecule, phenanthro(4,3-b)furan (Leong and Harrison 2004).

Herbal Usage: In Malaya, juice from the roasted pseudobulb of this orchid was dropped into the ear to treat earache (Alvins quoted by Burkill 1935).

Overview

Before the founding of the People's Republic of China in 1949, *Bulbophyllum* was not included in the *Chinese Materia Medica*. *B. inconspicuum* was described in *Barefoot Doctors' Manual*, the first *Herbal* in English to list a *Bulbophyllum* (Anonymous 1974). In Taiwan, *B. affinis* and *B. pectenveneris* are included among the medicinal plant resources (Ou et al. 2003). By 1994, Prof. Wu Xiu Ren in his magnificent *A Concise Edition of Medicinal Plants in China* was able to identify 11 *Bulbophyllum* species used as medicinal remedies in China. A single species, *B. neilgherrense* (syn. *B. rosemarianum*), features in Ayervedic medicine. The ubiquitous *B. vaginatum* (syn. *Cirrhopetalum vaginatum*) is used to treat earache in Malaysia. *Sakat bawang* is the common Malay name for *Bulbophyllum* in general and despite other conspicuous *Bulbophyllum* species being common on village trees (Holttum 1964), the *Sakat bawang* of Malay medicine referred specifically to *B. vaginatum* (Burkill 1935). Formerly, the generic name *Cirrhopetalum* was applied to *Bulbophyllums* with flowers that are arranged in a fan shape and possessing long sepals. Hence the name cirrhopetalins for numerous compounds isolated from *Bulbophyllum*.

Going through the list of traditional medicinal uses for *Bulbophyllum*, one is struck by its usage to treat common, non-life threatening conditions. In China and Nepal, *B. odoratissimum* is used to treat respiratory infections and fractures. The authoritative *Zhongyao Da Cidian* mentions five prescriptions for the use of *B. inconspicuum* to treat various conditions associated with inflammation of the respiratory system (throat, bronchus and lungs) (Prescriptions 1–5 in Table 8.2). In each instance, the orchid is not the sole item, but is always used in combination with at least another herb that is a more traditional remedy for inflammation, coughs, sore throat, and bronchitis, such herbs being *Aralia chinensis*, *Coptis chinensis*, *Stenoloma chusana*, *Ardisia japonica* or *Sophora subprostata*.

B. andersonii, B. odoratissimum and *B. kwangtungense* and *Pholidota* sp. are used in Tuja and Miao folk medicine under the name *Guoshangye* (Qu et al. 2006). These are used in the treatment of fracture, tuberculosis and pain from hernias. *B. affine* has a moderate bactericidal activity against *Staphyloccocus aureus* (a common cause of skin infections) but none against *Bacillus subtilis, Klebsiella pneumonia, Escherichia coli* or *Vibrio cholera* (Marasini and Joshi 2012), but it is not used in Nepal where the research was conducted, whereas *B. careyanum* and *B. leopardinum* are used in the treatment of burns (Raskoti 2009), but antimicrobial testing has not been reported for the latter species. *B. sterile* exhibits weak antibacterial activity against five infectious bacterial species (Priya and Krishnaveni 2005); however, this orchid is not used by Indian tribes to treat infection.

In the Ayeyarwady Delta and Thaninthayi Region of Myanmar, women prepare a hair tonic and shampoo by mixing ground pseudobulbs of various species of *Bulbophyllum* with pulverised bark, seeds and fruit (species not identified) to make a sticky fluid used when washing their hair. This is reputed to cure dandruff, promote hair growth and improve hair colour (Kurweil and Lwin 2014).

The applications described above would not merit a second look at *Bulbophyllums* from researchers who have their mind set on finding multi-billion dollar medicinal cures. Nevertheless, many species of *Bulbophyllum* are common and easy to find, and they produce large colonies on trees so there is an abundance of material to work on.

In 1974, Luning reported that he screened 134 species of *Bulbophyllum* for alkaloids and discovered that just 9 species contained significant amounts. However, *Bulbophyllums* are rich in phenanthrenes and stilbenoids (Majumder et al. 1985, 1997; Leong and Harrison 2004), and lignans was discovered in *B. triste* (Majumder et al. 1994). Majumder and his team isolated 18 different chemical compounds from

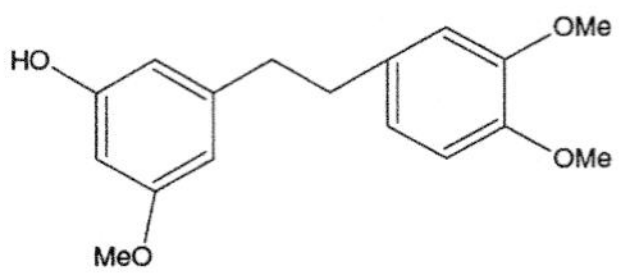

cumulatin

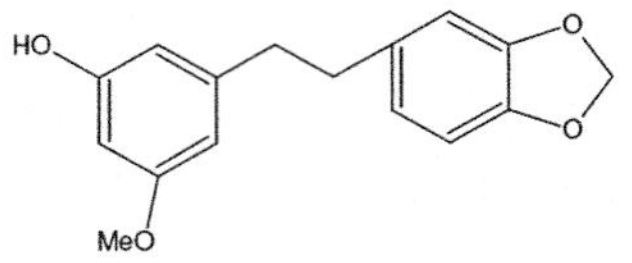

densiflorol

Fig. 8.19 Bibenzyls from *Bulbophyllum*

six species of *Bulbophyllum* during a 13-year period from 1985 to 1997, but only three of these species, *B. andersonii, B. odoratissimum* and *B. umbellatum*, were medicinal (Majumder and Pal 1985, 1993; Majumder et al. 1985, 1987, 1997; Majumder and Banerjee 1988, 1989; Majumder and Basak 1990, 1991a, b; Majumder et al. 1990; Majumder and Sen 1991) (Fig. 8.19). The Indian scientists indicated that the phenenthrenes from *Bulbophyllum* should be investigated for antitumour activity (Majumder, Pal and Joardar, 1990), but they did not proceed to do the testing themselves. *B. umbellatum* is used in the preparation of an Ayurvedic tonic, *swarna jivanti*, which is purported to promote vitality and longevity (Duggal 1972; Vij 1995). The Valmikis tribe in Andhra Pradesh also believes that eating the tubers of *B. sterile* (*B. neilgherrence*) helps to maintain good health (Reddy et al. 2005).

A 50 % ethanolic extract of *B. gymnopus* Hook f. (one of the species from which chemicals were obtained) failed to show any antimicrobial or anti-inflammatory activity, and it had no effect on respiration, the cardiovascular system or the central nervous system in experimental animals. It caused diuresis in rats and caused contraction of an isolated strip guinea pig's small intestine (Bhakuni et al. 1990). The correct structure of gymopusin, a phenanthrenediol from *B. gymnopus*, was elucidated and the compound synthesised by Hughes and Sargent (1989).

Fig. 8.20 Phenanthrenes from *Bulbophyllum kwangtungense* displaying cytotoxic activity in-vitro

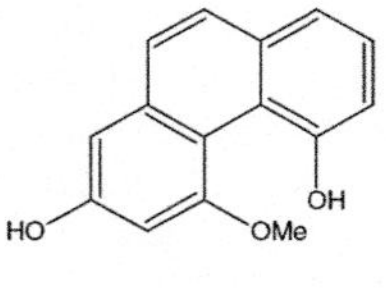

plicatol

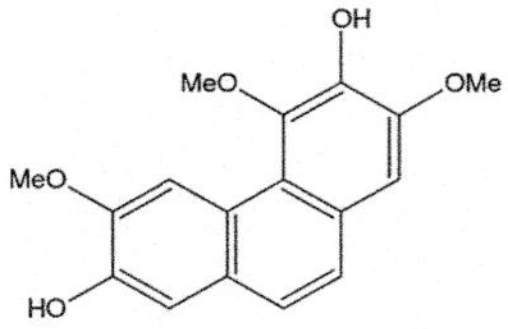

3,7-dihydroxy-2,4,6-trimethoxyphenanthrene*f*

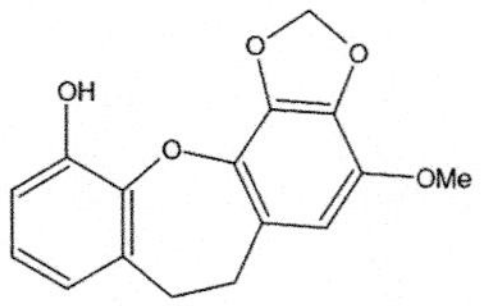

7,8-dihydro-5-hydroxy-12,13-methylenedioxy-11-methoxydibenz

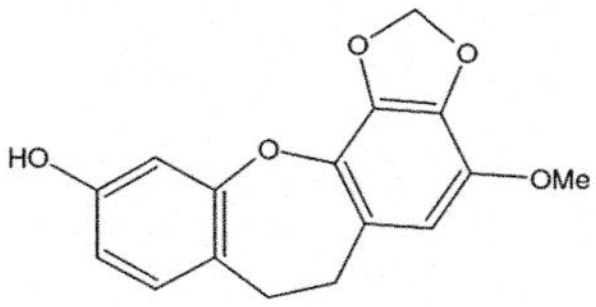

7,8-dihydro-4-hydroxy-12,13-methylenedioxy-

11-methoxydibenz (*bf*)oxepin

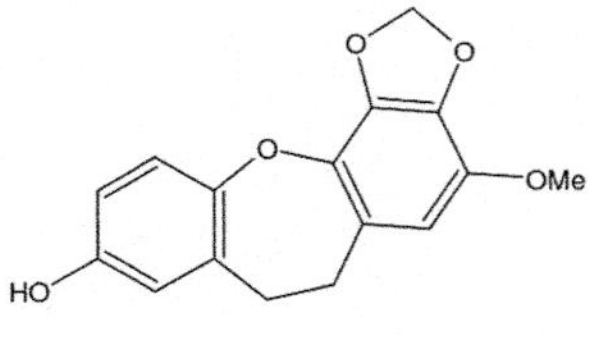

7,8-dihydro-3-hydroxy-12,13-methylenedioxy-

11-methoxydibenz (*bf*)oxepin

Medical interest in *Bulbophyllum* has been re-activated with the discovery of three new compounds that showed antitumour activity on human cancer cell lines in the test tube (Wu et al. 2006; Xu et al. 2009) (Fig. 8.20). Animal experiments and human trials are awaited before any comment can be made on these substances.

Several Chinese herbs enhance the transformation of lymphocytes by the Epstein–Barr (EB) virus and the virus early antigen

(EA) induction in the Raji cell system (Zeng et al. 1983; Furukawa et al. 1986; Hu 1985). There is some concern that they may have a role in the induction of nasopharyngeal carcinoma (cancer of the nose) and other malignant diseases. *B. inconspicuum* has been studied and found to have no enhancing effect on the antigen induction by the EB virus (Hu 1985).

B. vinaceum produces an array of phenylpropanoids which attract male fruit flies to effect pollination. They are building blocks for the production of pheromones that the male flies employ to attract the females. Tan et al. (2006) working in Penang, Malaysia, identified the major floral volatile components as methyl eugenol, transconiferyl alcohol, 2-allyl-4,5-dimethoxyphenol, and trans-3,4-dimethoxy-cinnamyl acetate. The minor components were eugenol, euasarone, trans-3,4-dimethoxy cinnamyl alcohol, and cis-conferyl alcohol. The highest concentration of these chemicals was present in the osmophores (odour glands) located on the upper surface of the lip (Teixeira et al. 2004; Tan and Nishida 2005; Tan et al. 2006).

B. apertum and *B. cheiri* emit a raspberry fragrance to attract *Bactricera* fruit flies which use the methyl eugenol for its defence (Tan et al. 2002; Tan and Nishida 2005). Methyl eugenol metabolites (transconiferyl alcohol and 2-allyl-4,5-dimethoxyphenol) excreted in the rectal glands of the orchid fruit fly are poisonous to the Malaysian spiny lizard, *Gekko monarchus*, that feeds on insects (Tan et al. 2002; Wee and Tan 2001). Some *Bulbophyllum* species, like *B. beccarii*, attract blue-bottle flies by producing the stench of rotting meat (Burkill 1935; Teoh 2011). However, none of these floral products seem to interest scientists and medical men involved in finding medicinal uses for *Bulbophyllum* species.

References

Abraham A, Vatsala P (1981) Introduction to Orchids, with illustrations and descriptions of 150 South Indian Orchids. TPGRI, Trivandrum

Anonymous (1959) Preliminary observations on the treatment of haemorrhage of digestive ulcers with sepium and Bletilla hycinthina. Zhonghua Nei Ke Za Zhi 7(1): 9

Anonymous (1974) Barefoot doctors' manual. National Institutes of Health, Bethesda

Anonymous (1989) Medicinal plants of China. Selection of 150 commonly used species. WHO pacific series no. 2. Regional Office of the Pacific, Manila

Anonymous (2004) The new century Chinese-english dictionary of traditional Chinese medicine. People's Military Medical Press, Beijing

Bai L, Yamaki M, Inoue K, Takagi S (1990) Blestrin A and B, bis(dihydrophenanthrene) ethers from Bletilla striata. Phytochemistry 29(4):1259–1260

Bai L, Kato T, Inoue K, Yamaki M (1991) Blestrianol A, B and C, biphenanthrenes from Bletilla striata. Phytochemistry 30(8):2733–2735

Bai L, Kato T, Inoue K, Yamaki M, Tagaki S (1993) Stilbenoids from Bletilla striata. Phytochemistry 33 (6):1481–1483

Bhakuni DS, Goel AK, Goel AK et al (1990) Screening of Indian plants for biological activity. Part XIV. Indian J Exp Biol 28:619–632

Bose TK, Bhattacharjee SK (1980) Orchids of India. Naya Prokash, Calcutta

Bruix J, Castells A, Montanya X et al (1994) Phase II study of transarterial embolization in European patients with hepatocellular carcinoma: need for controlled trials. Hepatology 20(3):643–650

Burkill IH (1935) (1966 reprint, 2nd edn, with contributions by Birtwistle W, Foxworthy FW, Scrivenor JB, Watson IG): A dictionary of economic products of the Malay Peninsula, vol II. Crown Agents for the Colonies, London. Ministry of Agriculture & Co-operatives, Kuala Lumpur

Carr BI, Zajko A, Bron K, et al (1997) *Phase II study of Spherex)degradable starch micospheres) injected onto the hepatic artery in conjunction with doxyrubicin and cisplatin in the treatment of advanced stage hepatocellular carcinoma: interim analysis.* Semin Oncol 24(Suppl 6):S6-97–S6-99

Chen SC, Tang T (1982) A general review of the orchid flora of China. In: Arditti J (ed) Orchid biology. Reviews and perspectives II. Cornell, Itcaca

Chen XQ, Vermeulen JJ (2009) Bulbophyllum. In: Chen XQ, Liu ZJ, Zhu GH et al (eds) Flora of China—Orchidaceae. Science Press, Beijing

Chen SC, Tsi ZH, Luo YB (1999) Native orchids of China in colour. Science Press, Beijing

Chen XQ, Gale SW, Cribb PJ (2009) Bletilla Rchb. f. In: Chen XQ, Zj L, Zhu GH et al (eds) Flora of China - Orchidaceae. Science Press, Beijing

Chen J, Zhang H, Chen L, Wu B (2011) New stress metabolite from Bulbophyllum kwangtungense. Nat Prod Commun 6(1):53–54

Chinsamy M, Finnie JF, Van Staden J (2011) The ethnobotany of South African medicinal orchids. South Afr J Bot 77(1):2–9

Choi DW (1984) Clonorchis sinensis: life cycle, intermediate hosts, transmission to man and geographical distribution in Korea. Arzneimittelforschung 34(9B): 1145–1151

Christensen H (2002) Ethnobotany of the Iban and the Kelabit. Forest Department Sarawak; NEP Con Denmark; and University of Aarhus, AarhCus

Chu JP, Chen W, Li JP et al (2007) Clinicopathologic features and results of transcatheter arterial chemoembolization for oesteosarcoma. Cardiovasc Intervent Radiol 30(2):201–206

Chuasul W (2002) Ethnomedical uses of Thai orchidaceous plants. Mohidol Univ J Pharm Sci 29(3-4): 41–45

Das SP (2004) Indian orchids in indigenous medicine system. In: Brirto SJ (ed) Orchids diversity and conservation. The Rapinat Herbarium, St John's College, Tiruchirappalli

Dash PK, Sahoo S, Bal S (2008) Ethnobotanical studies on orchids of Niyamgiri Hill Ranges, Orissa, India. Ethnobot Leaflets 12:70–78

Diao H, Li X, Chen J et al (2008) Bletilla striata polysaccharide stimulates inducible nitric oxide synthase and proinflammatory cytokine expression in macrophages. J Biosci Bioeng 105(2):85–89

Dong L, Xia S, Luo Y et al (2009) Targeting delivery oligonucleotide into macrophages by a cationic polysaccharide from Bletilla striata successfully inhibited the expression of TNF-alpha. J Control Release 134 (3):214–220

Du D, Feng G, Liang H (1998) Embolotherapy of craniofacial tumors and vascular diseases: assessing the effect of different embolic agents. Lin Chuang Er Bi Yan Hou Ke Za Zhi 12(6):266–268

Duchin S, van Houte J (1978) Relationship of Streptococcus mutans and lactobacillus to incipient smooth surface dental caries in man. Arch Oral Biol 23:779–786

Duggal SC (1972) Orchids in human affairs (a review). Pharm Biol 11(2):1727–1734

Duong NV (1993) Medicinal plants of Vietnam, Cambodia and Laos. World Health Organization, Manila

Ernst R, Rodriguez E (1984) Carbohydrates of the Orchidaceae. In: Arditti J (ed) Orchid biology. Reviews and perspectives III. Cornell University Press, Ithaca

Feng GS (1985) Animal experiment and clinical application of Bletilla striata as an embolizing agent. Zhonghua Fang She Xue Za Zhi 19(4):193–196

Feng XS, Qiu FZ, Xu Z (1995) Experimental studies of embolization of different hepatotropic blood vessels using Bletilla striata in dogs. J Tongji Med Univ 15(1): 45–49

Feng G, Kramann B, Zheng C, Zhou R (1996) Comparative study of the long-term effect of permanent embolization of hepatic artery with Bletilla striata in patients with primary liver cancer. J Tongji Med Univ 16(2):111–116

Feng GS, Li X, Zheng CS et al (2003) Mechanism of inhibition of tumor angiogenesis by Bletilla colloid: an experimental study. Zhonghua Yi Xue Za Zhi 83 (5):412–416

Furukawa M, Komori T, Ishiguro H, Umeda R (1986) Epstein-Barr virus early antigen induction in nasopharyngeal hybrid cells by Chinese medicinal herbs. Auris Nasus Larynx 13(2):101–105

Go R, Hamzah KA (2008) Orchids of Peat Swamp Forests in Peninsular Malaysia. Kuala Lumpur: Peat Swamp Forest Project, UNDP/GEF Funded (MAL/99/G321) Ministry of Natural Resources & Environment

Gong YH, Wang CM, Lai RC et al (2009) An improved polysaccharide hydrogel: modified gellan gum for long term cartilage regeneration in vitro. J Mater Chem 19:1968–1977

Gu T, Li CX, Feng Y et al (2007) Trans-arterial gene therapy for hepatocellular carcinoma in a rabbit model. World J Gastroenterol 13(14):2113–2117

Guo JX (1996) Bletilla striata (Thunb.) Reichb. f. In: Kimura et al (eds) International collation of traditional and folk medicine. World Scientific, Singapore, vol 1, p 205

Gurong DB (2006) An illustrated guide to the orchids of Bhutan. DSB Publications, Thimphu

Hakansson L, Hakansson A, Morales O et al (1997) Spherex (degradable starch microspheres) chemoocclusion – enhancement of tumor drug concentration and therapeutic efficacy: an overview. Semin Oncol 24(2 Suppl 6) S6-100–S6-109

Hare R (1956) An outline of bacteriology and immunity. Longmans Green & Co., London

Han GX, Wang LX, Gu ZB, Zhang WD (2002) A new bibenzyl derivative from Bletilla striata. Yao Xue Xue Bao 37(3):194–195

Hirano T, Godo T, Mii M, Ishikawa K (2005) Cryopreservation of immature seeds of Bletilla striata by vitrification. Plant Cell Rep 23:534–539

Holttum RE. Flora of Malaya I. Orchids of Malaya, 2nd edn. Government Printers, Singapore

Holttum RE (1949) Gregarious flowering of the terrestrial orchid, Bromheadia finlaysonniana. Gard Bull Singapore 12(2):295–302

Holttum RE (1964) Orchids of Malaya, 3rd edn. Government Printers, Singapore

Hooper D (1929) On Chinese medicine: drugs of Chinese pharmacies in Malaya. Gard Bull Straits Settl 6:1–163

Hoque MM, Huda MK (2008) Brachycorythis obcordata (Lindl.) Summerh. (Orchidaceae): a new angiospermic record for Bangladesh. Bangladesh J Bot 37(2):199–201

Hsu HJ (1994) Orchidaceae. In: TC Huang (editor in chief) Flora of Taiwan: flora of Taiwan, vol 5. Editorial Committee of the Flora of Taiwan, Taiwan National University, Taipei

Hughes AB, Sargent MV (1989) Structure and synthesis of gymnopusin, a novel phenanthrenediol from the orchid, Bulbophyllum gymnopus. J Chem Soc Perkin Trans 1:1787

Hu YL (1985) Enhanced transformation of human lymphocytes by Chinese herbs. Zhonghua Zhong Liu Za Zhi 7(6):417–419

Johansson CJ (1996) Pharacokinetic rationale for chemotherapeutic drugs combined with intra-arterial degradable starch microspheres (Spherex). Clin Pharmacokinet 31(3):231–240

Jonathan KH, Raju AJS (2005) Terrestrial and epiphytic orchids of Eastern Ghats. EPTRI – ENVIS Newsl 11(3):2–4

Joseph J (1982) Orchids of Nilgiris. Records of the botanical survey of India, vol XXII. Botanical Survey of India (Department of Environment), Howrah

Keiser J, Xiao SH, Smith TA, Utzinger J (2009) Combination chemotherapy against Clonorchis sinensis: experiments with artemether, artesunate, OZ78, praziquantel, and tribendimidine in a rat model. Antimicrobiol Agents Chemother 53(9):3770–3776

Kurweil H, Lwin S (2014) A Guide to orchids of Myanmar. Natural History Publicatios (Borneo), Kota Kinabalu

Leong YW, Harrison LJ (2004) A biphenanthrene and a phenanthro(4,3 beta)furan from the orchid Bulbophyllum vaginatum. J Nat Prod 67:1601–1603

Leong YW, Kang CC, Harrison LJ, Powell AD (1997) Phenanthrenes, dihydrophenanthrenes and bibenzyls from the orchid *Bulbophyllum vaginatum*. Phytochemistry 44:157–165

Leong YW, Harrison LJ, Powell AD (1999) Phenanthrene and other aromatic constituents of *Bulbophyllum vaginatum*. Phytochemistry 50:1237–1241

Li Shizhen (1578) *Bencao Gangmu*. Anonymous

Li NH (ed) (1988b) Chinese medicinal herbs of Hong Kong, vol 4. Hong Kong Chinese Medical Research Institute, HK, p 200

Li X, Feng GS, Zheng GS et al (2003) Influence of transarterial chemoembolization on angiogenesis and expression of vascular endothelial growth factor and basic fibroblast growth factor in rats with Walker-256 transplanted hepatoma: an experimental study. World J Gastroenterol 9(11):2445–2449

Lin YL, Chen WP, Macabalang AD (2005) Dihydrophenanthrenes from *Bletilla formosana*. Chem Pharm Bull 53:1111–1113

Liu XR, Han WQ, Sun DR (1992) Treatment of intestinal metaplasia and atypical hyperplasia of gastric mucosa with Xio Wei Yan powder. Zhongguo Zhong Xi YiJie He Za Zhi 12(10):602–603

Liu DL, Pang FG, Zhang X et al (2006) Water-soluble phenolic glycosides from the whole plant of Bulbophyllum odoratissimum. Yao Xue Xue Bao 41 (8):738–741

Liu MZ, Tang JZ, Zhang JS et al (2008) Angiogenesis inhibition in vascular endothelial cells by terpenoid compounds from Bletilla striata is via apoptosis pathway. Fen Zi Xi Bao Sheng Wu Xue Bao 41(5): 383–392

Luo B, Li J, Ma J (1996) Color Doppler in evaluating the effects of octreotide on portal hemodynamics. Zhonghua Yi Xue Za Zhi 76(3):200–202

Luo Y, Diao HJ, Xia SH et al (2010) A physiologically active polysaccharide hydrogel promotes wound healing. J Biomed Mater Res 94A(1):193–204

Maataoui A, Qian J, Mack MG et al (2005a) Liver metastases in rats: chemoembolization combined with interstitial laser ablation for treatment. Radiology 237(2):479–484

Maataoui A, Qian J, Vossoughi D et al (2005b) Transarterial chemoembolization alone and in combination with other therapies: a comparative study in an animal HCC model. Eur Radiol 15(1):127–133

Majumder PL, Banerjee S (1988) A ring-B oxygenated phenanthrene derivative from the orchid *Bulbophyllum gymnopus*. Phytochemistry 27(1):245–248

Majumder PL, Banerjee S (1989) Revised structure of gymnopusin, a ring-B oxygenated phenanthrene derivative isolated from the orchid *Bulbophyllum gymnopus*. Indian J Chem 28B:1085–1088

Majumder PL, Basak M (1990) Cirrhopetalin, a phenanthrene derivative from *Cirrhopetalum andersonii*. Phytochemistry 29:1002–1004

Majumder PL, Basak M (1991a) Two bibenzyl derivatives from the orchid *Cirrhopetalum andersonii*. Phytochemistry 30:321–324

Majumder PL, Basak M (1991b) Two stibenoids from the orchid *Cirrhopetalum andersonii*. Phytochemistry 30:3429–3432

Majumder PL, Pal A (1985) 24-Methylene cycloartanyl p-hydroxycinnamate from the orchid *Cirrhopetalum elatum*. Phytochemistry 24:2120–2122

Majumder PL, Pal A (1993) Cumulatin and tristin, two bibenzyl derivatives from the orchids Dendrobium cumulatum and Bulbophyllum triste. Phytochemistry 32:1561–1565

Majumder PL, Sen RC (1991) Bulbophyllanthrone, a phenanthraquinone from Bulbophyllum odoratissimum. Phytochemistry 30:2092–2094

Majumder PL, Kar A, Shoolery JN (1985) Bulbophyllanthrin, a phenanthrene of the orchid Bulbophyllum leopardinum. Phytochemistry 26:1127–1129

Majumder PL, Pal A (nee Kundu), Lahiri S (1987) Structure of pholidotin, a new triterpene from orchids Pholidota rubra and *Cirrhopetalum elatum*. Indian J Chem 26B:297–300

Majumder PL, Pal A, Joardar M (1990) Cirrhopetalanthrin, a dimeric phenanthrene derivative from the orchid *Cirrhopetalum maculosum*. Phytochemistry 29(1):271–274

Majumder PL, Lahiri S, Pal S (1994) Occurance of lignans in the Orchidaceae plants Luisia volcris and Bulbophyllum triste. J Indian Chem Soc 71: 645–647

Majumder PL, Roychowdhury M, Chakraborty S (1997) Bibenzyl derivatives from the orchid *Bulbophyllum protractum*. Phytochemistry 44:167–172

Majumder PL, Pal A, Majumder S (1999) Dimeric phenanthrenes from the orchid *Bulbophyllum reptans*. Phytochemistry 50:891–897

Manandhar NP, Manandhar S (2002) Plants and people of Nepal. Timber, Portland

Marasini R, Joshi S (2012) Antibacterial and antifungal activities of medicinal orchids growing in Nepal. J Nepal Chem Soc 29:104–109

Misra S (2007) Orchids of India A Glimpse. Bishen Singh Mahendra Pal Singh, Delhi

Morita H, Koyama K, Sugimoto Y, Kobayashi J (2005) Antimitotic activity and reversal of breast cancer resistance protein-mediated drug resistance by stilbencids from Bletilla striata. Bioorg Med Chem Lett 15(4):1051–1054

Nanakorn W, Watthana S (2008) Queen Sirikit Botanic Garden (Thai Native Orchids 1 and 2). Wanida Press, Chiang mai

O'Byrne P (2001) A to Z of South East Asian orchid species. Orchid Society of South East Asia, Singapore

Ohtsuki T (1937) Untersuchung uber das Bletillamannan aus den Knollen von Bletilla striata. Acta Phytochim 10:29–41

Ou JC, Hsieh WC et al (eds) (2003) The catalogue of medicinal plant resources in Taiwan. Department of Health, Executive Yuan, Taipei

Ohwi J (ed) (1965) Flora of Japan. English edition: Meyer FG, Walker EH (eds). Smithsonian Inst, Washington

Pant B, Raskoti BB (2013) Medicinal orchids of Nepal. Himalayan Map House (P) Ltd, Kathmandu

Pearce NR, Cribb PJ (2002) The orchids of Bhutan. Royal Botanic Gardens, Edinburgh and Royal Government of Bhutan, Thimpu

Peng R, Zheng Q, Hao J et al (2005) Biological evaluation of ChuangYuLing dressing – a multifunctional medicine carrying biomaterial. J Huazhong Univ Sci Technol Med Sci 25(1):72–74

Peng Q, Li M, Xue F et al (2014) Structure and immunobiological activity of a new polysaccharide from Bletilla striata. Carbo Poly 107:119–123

Perry LM, Metzger J (1980) Medicinal plants of East and Southeast Asia: attributed properties and uses. MIT Press, Cambridge, MA

Pohlen U, Berger G, Binnenhei M et al (2000) Increased carboplatin concentration in liver tumors through temporary flow retardation with starch microspheres (Spherex) and gelatin powder (Gelfoam): an experimental study in liver tumor- bearing rabbits. J Surg Res 92(2):165–170

Priya K, Krishnaveni C (2005) Antibacterial effect of Bulbophyllum neilgherrense Wt. Orchidaceae): an in vitro study. Anc Sci Life 25(2):50–52

Qian J, Feng G, Liang H (1998) Action of DDPH in the interventional treatment of portal hypertension induced by liver cirrhosis in rabbits. J Tongji Med Univ 18(2):108–112

Qian J, Truebenbach J, Graepier F et al (2003a) Application of poly-lactide-co-glycolide- microspheres in the transarterial chemoembolization in an animal model of hepatocellular carcinoma. World J Gastroenterol 9(1):94–98

Qian J, Vossoughi D, Woitaschek D et al (2003b) Combined transarterial chemoembolization and arterial administration of Bletilla striata in treatment of liver tumor in rats. World Gastroenterol 9(12):2676–2680

Qian J, Vossoughi D, Oppermann E et al (2005) Experimental study in the effects of transarterial immuno-chemoembolization in hepatocellular carcinoma. J Huazhong Univ Sci Technol Med Sci 25(3):329–331

Qu XY, Qin SY, Yang DQ, Li QS, Peng FS (2006) Study on resource and varieties of Guoshangye. Zhongguo Zhong Yao Za Zhi 31(2):110–114

Rao AN (2004) Medicinal orchid wealth of Arunachal Pradesh. Newsl ENVIS NODE Indian Med Plants 1 (2):1–7

Raskoti BB (2009) The Orchids of Nepal. Bhakta Bahadur Raskoti and Rita Ale, Kathmandu

Reddy KN, Reddy CS, Raju VS (2005) Ethno-orchidology of orchids of Eastern Ghats of Andra Pradesh. EPRTI Newsl 11(3)

Reinecke T, Kindl H (1994) Characterization of Bibenzyl synthase catalyzing the biosynthesis of phytoalexins of orchids. Phytochemistry 35(1):63–66

Reynolds EC, Cai F, Cochrane NJ et al (2008) Fluoride and casein phosphopeptide-amorphous calcium phosphate. J Dent Res 87(4):344–348

Rhee JK, Kim PG, Baek BK et al (1982) Isolation of anthelminthic substance on Clonorchis sinensis from tuber of Bletilla striata. Kisaengchunghak Chapchi 20(2):142–146

Shin HR, Lee CU, Park HJ et al (1996) Hepatitis B and C virus, Clonorchis sinensis for the risk of liver cancer: a case-control study in Pusan, Korea. Int J Epidemiol 25(5):933–940

Shrestha R (2000) Some medicinal orchids of Nepal. The Himalayan plants, can they save us? In: Watanabe T, Takano A, Bista MS, Saiju HK (eds) Proceedings of Nepal-Japan joint symposium on conservation and utilization of Himalayan medicinal resources. SCDHMR, Japan

Siegerist ES (2001) Bulbophyllums and their allies. Timber Press, Portland

Song P, Wang XM, Xie S (2006) Experimental study on mechanisms of lyophilized powder of fresh Gekko chinensis in inhibiting H22 hepatocarcinoma angiogenesis. Zhongguo Zhong Xi Yi Jie He Za Zhi 26(1): 558–562

Stuart GA (1911) Chinese Materia Medica. Vegetable Kingdom. (A revision of a work by F. Porter Smith.) American Presbyterian Mission Press, Shanghai

Sun J, Wang C, Zhang J (2005) Effect of polysaccharides from Bletilla striata on the adhesion of human umbilical venous endothelial cells. Zhong Yao cai 28(11): 1006–1008

Tagaki S, Yamaki M, Inoue K (1983) Antimicrobial agents from Bletilla striata. Phytochemistry 22:1011–1015

Taguchi T (1995) Liver tumor targeting of drugs: Spherex, a vascular occlusive agent. Gan To Kagaku Ryoho 22(7):969–976

Tan KH, Nishida R (2005) Synomone or kairomone? – Bulbophyllum apertum flower releases raspberry ketone to attract Bactrocera fruit flies. J Chem Ecol 31(3):497–507

Tan KH, Nishida R, Toong C (2002) Floral synomone of a wild orchid, Bulbophyllum cheii, lures Bactrocera fruit flies for pollination. J Chem Ecol 28(6): 1161–1172

Tan KH, Tan LT, Nishida R (2006) Floral phenylpropanoid cocktail and architecture of *Bulbophyllum vinaceum* orchid in attracting fruit flies for pollination. J Chem Ecol 32(11):2429–2441

Tanaka Y, Htun N, Yee TT (2003) Wild orchids in Myanmar, vol 1–3. The Foundation of Agricultural Development and Education, Bangkok

Teder H, Johansson CJ (1993) The effect of different dosages of degradable starch microspheres (Spherex ®)on the distribution of doxorubicin regionally administered to the rat. Anticancer Res 13(6):2161–2164

Teder H, Nilsson M, Aronsen KF et al (1988) Influence of degradable starch microspheres (Spherex) on the retention of pertechnetate in a solitary rat liver tumor. Eur J Surg Oncol 14(4):327–333

Teixeira Sde P, Borba EL, Semir J (2004) Lip anatomy and its implications for the pollination mechanism of Bulbophyllum species (Orchidaceae). Ann Bot (Lond) 93(5):499–505

Teoh ES (2005) Orchids of Asia, 3rd edn. Marshall Cavendish, Singapore

Teoh ES (2011) Find paper in WOC proceedings or MOR

Tomoda M, Nakatsuka S, Tamai M et al (1973) Isolation and characterization of a mucous polysaccharide, Bletilla-glucomannan from Bletilla striata tubers. Chem Pharm Bull 21:2667–2671

Tsochatzis E, Garcovich M, Marell L et al (2013) Transarterial embolization as neo- adjuvant therapy pretransplantation in patients with hepatocarcinoma. Liver Int 33(6):944–949

Vaddhanaphuti N (2001) A field guide to the wild orchids of Thailand, 3rd edn. Silkworm Books, Chiang Mai

Vaddhanaphuti N (2005) A field guide to the wild orchids of Thailand, fourth and expanded edition. Silkworm Books, Chiang Mai

Vij SP (1995) Orchid genetic diversity in India: conservation and commercialization. Proceedings of the 5th Asia pacific orchid conference and show, Fukuoka, pp 20–39

Wang WM (1993) Experimental observation of high portal venous resistance in dogs. Zhonghua Yi Xue Za Zhi 73(6):349–351

Wang and Wang (2006) Color illustrations of Chinese materia medica, 3rd edn (in Chinese). Fujian Science & Technology Publishing House, Foozhou

Wang KX, Zhang RB, Cui YB et al (2004) Clinical and epidemiological data of patients with clonorchiasis. World J Gastroenterol 10(3):446–448

Wang LX, Han GX, Shu Y et al (2001) Studies on chemical constituents of Bletilla striata (Thunb.) Reichb. f. Zhongguo Zhong Yao Za Zhi 26(10): 690–692

Wang CM, Sun J, Luo Y et al (2006) A polysaccharide isolated from the medicinal herb Bletilla striata induces endothelial cells proliferation and vascular growth factor expression in vitro. Biotechnol Lett 28(8):539–543

Wang Y, Liu D, Chen SJ et al (2014) A new glucomannan from Bletilla striata: structural and anti-fibrosis effect. Fitoterapia 92(2014):72–78

Weaver RE Jr, Stoutamire WP (2008) Bletilla species. Hardy orchids with mass appeal. Orchids 77(5): 358–365

Wee SL, Tan KH (2001) Allomonal and hepatotoxic effects following eugenol consumption in Bactrocera papayae male against Gekko monarchus. J Chem Ecol 27(5):953–964

Wu XR (1994) A concise edition of medicinal plants in China. Guangdong Higher Education Publication House, Guangdong (in Chinese)

Wu KJ, Lu G et al (eds) (1999) Traditional Chinese Materia Medica. Wuhan University Press, Wuhan

Wu TL, Hu QM, Xia NH, Lai PCC, Yip KL (2002) Check List of Hong Kong Plants 2001. Agriculture, Fisheries and Conservation Department Bulletin 1 (Revised), Hongkong

Wu B, He S, Pan YJ (2006) New dihydrodibenzoxepins from *Bulbophyllum kwantungense*. Planta Med 72: 1244–1247

Wu B, He S, Pan YJ (2008) New dihydrodibenzoxepins from *Bulbophyllum kwangtungse*. Planta Med 72(13): 1244–1247

Wu TY, Chen CC, Lay HL (2010a) Study on the components and antioxidant activity of Bletilla plant in Taiwan. J Food Drug Anal 18(4):279–289

Wu XG, Xin M, Chen H et al (2010b) Novel mucoadhesive polysaccharide isolated from Bletilla striata improves the intraocular penetration and efficacy of levofloxacin in the topical treatment of experimental bacterial keratitis. J Pharm Pharmacol 62(9): 1152–1157

Wu B, Chen JB, He S, Pan YJ (2011) Oxepine and Bibenzyl compounds from Bulbophyllum kwantungense. Chem J Chinese Univ 29(2):305–308

Xu JJ, Yu H, Qing C et al (2009) Two new biphenanthrenes with cytotoxic activity from Bulbophyllum odoratissimum. Fitotherapia 80 (7):381–384

Yamaki M, Bai L, Inoue K, Tagaki S (1989) Biphenanthrenes from Bletilla striata. Phytochemistry 28(12):3503–3505

Yamaki M, Bai L, Inoue K, Tagaki S (1990) Benzylphenanthrenes from Bletilla striata. Phytochemistry 29(7):2285–2287

Yamaki M, Kato T, Bai L, Inoue K, Takagi S (1991) Methylated stilbenoids from Bletilla striata. Phytochemistry 30(8):2759–2760

Yamaki M, Bai L, Kato T et al (1992) Bisphenanthrene ethers from Bletilla striata. Phytochem 31(11): 3985–3987

Yamaki M, Bai L, Kato T et al (1993a) Three dihydrophenanthropyrans from Bletilla striata. Phytochemistry 32(2):427–430

Yamaki M, Bai L, Kato T et al (1993b) Blespirol, a phenanthrene with a spirolactone ring from Bletilla striata. Phytochemistry 33(6):1497–1498

Yamaki M, Kato T, Bai L, Inoue K, Takagi S (1993c) Phenanthrene glucosides from Bletilla striata. Phytochemistry 34(2):5535–5537

Yang ZH, Zhang QT, Feng ZZ, Lang KY, Li H. English edition translated by ZR Xiong (1998) Orchids. China Esperanto Press, Beijing

Yang X, Tang C, Zhao P et al (2012) Antimicrobial constituents from the tubers of Bletilla ochracea. Planta Med 78(6):606–610

Yong HS (1990) Orchid Portraits. Tropical Press Sdn Bhd, Kuala Lumpur

Yu Y, Lang QB, Chen Z et al (2009a) Prognostic analysis of transarterial chemoembolization combined with a traditional herbal medicine formula for treatment of unresectable hepatocellular carcinoma. Chinese Med J 122(17):1990–1995

Yu Y, Lang QB, Chen Z et al (2009b) The efficacy for unresectable hepatocellular carcinoma may be improved by trancatheter arterial chemoembolization in combination with a traditional herbal medicine formula: a retrospective study. Cancer 115(22): 5132–5138

Yu WY, Liu CX, Liu Y et al (2010) Mannan-modified solid lipid nanoparticles for targeted gene delivery to alveolar macrophages. Pharm Res 27(8): 1584–1596

Zeng Y, Zhong JM, Mo YK, Miao XC (1983) Epstein-Barr virus early antigen induction in Raj cells by Chinese medicinal herbs. Intervirology 19(4):201–249

Zeng S, Huang X, Chen Z et al (2004) Asepsis sowing and tissue culture of Bletilla striata. Zhong Yao Cai 27(9): 625–627

Zhan XD, Jia LX, Niu YM et al (2014) Targeted depletion of tumour associated macrophages by an alendronate-glucomannan conjugate for cancer chemotherapy. Biomaterials 35:10046–10057

Zhang WG, Zhao R, Ren J et al (2007a) Synthesis and anti-proliferative in-vitro activity of two natural dihydrostilbenes and their analogues. Arch Pharm (Weinheim) 340(5):244–250

Zhang WG, Lin JG, Niu ZY et al (2007b) Total synthesis of two new dihydrostilbenes from Bulbophyllum odoratissimum. J Asian Nat Prod Res 9(1):23–28

Zhang Y, Li B, Li SF (2009) Germination and seedling morphogenesis of Bletilla striata under different culture media. Xibei Zhiwu Xuebao 29 (8):1584–9

Zhang MS, Sun L, Zhao WC et al (2014) Cholesteryl-modification of a glucomannan from Bletilla striata and its hydrogel properties. Molecules 19(7): 9089–9100

Zhao JG, Feng GS, Kong XQ et al (2004) Changes of tumor microcirculation after transcatheter arterial chemoembolization: first pass perfusion MR Imaging and Chinese ink casting in a rabbit model. World J Gastroenterol 10(10):1415–1420

Zheng C, Feng G, Zhou R (1996) New use of Bletilla striata as embolizing agent in the intervention treatment of hepatic carcinoma. Zhonghua Zhong Gu Za Zhi 18:305–307

Zheng C, Feng G, Liang H (1998) Bletilla striata as a vascular embolizing agent in interventional treatment of primary hepatic carcinoma. Chin Med J 111: 10060–11063

Zhang Y, Lv T, Li M et al (2015) Anti-aging effect of polysaccharide from Bletilla striata on nematode Caenorhabditis elegans. Pharmacogn Mag 11 (43):449–454

Zhonghua Bencao vol. 8 (2000) edited by Hu XM, Zhang WK, Zhu QZ, et al. Shanghai Science and Technology Publication, Shanghai

Zhongyao Da Cidian (1986 reprint) Compiled by Najing College of New Medicine in 1977

Genus: *Calanthe* Brown

Chinese name: *Xiaji Lan* (prawn spine orchid)
Japanese name: *Ebine*

Calanthe are sympodial orchids with short stems and several plicate, elliptical leaves that are spirally arranged, or arranged in two rows, ensheathing the stem. Inflorescences arise at the side or from the base and carry many showy flowers on a short raceme. Approximately 100 species of *Calanthe* are distributed across tropical Asia to the Pacific Islands, tropical and southern Africa. A single species occurs in Central America. They are mainly terrestrial, with a few epiphytic members and occur at low to high elevations. Some species are deciduous, others evergreen. The plant and its flowers turn bluish when they are bruised.

Calanthe is the first orchid species to be artificially hybridised by man. Many species have attractive flowers. Indeed, the name, *Calanthe*, was derived from two Greek words, *kalos* (beautiful) and *anthe* (bloom), meaning 'beautiful flower'. The shape of the lip is an important criterion for distinguishing among species.

Calanthe alismifolia Lindl.

Calanthe nigropuncticulata Fukuyama

Chinese names: *Zexiexiaji Lan* (glossy prawn spine orchid), *Xidiangenjie Lan* (small spots

segmented root orchid); in Taiwan: black-spotted *Calanthe*; white flower *Calanthe*
Chinese medicinal name: *Zongyeqi*

Description: Plant is 21–40 cm tall with thick fibrous roots. Pseudobulb is conical-ovoid 15–2 by 1 cm in diameter. Leaves are 3–5 in number, elliptic, and with a short petiole. Flowers are white, tinged with violet, of good form, 1.5 to 2 cm across, arranged in two rows. Individual flowers are closely arranged near the top of the 30-cm-tall scape, with many unopened buds above. Petals and sepals are broadly oval. Lip is deeply trilobed, the side lobes linear and the mid-lobe fanning out towards the apex and deeply cleft at the tip. Several yellow, warty calli are present at the base of the lip (Fig. 9.1). Flowering period is June and July. *C. alismifolia* thrives in dense forests at 800–1700 m. It is distributed from southern Japan and Taiwan to Hubei, Hunan, Sichuan, Yunnan and Xizang in China to northeast India, Sikkim and Nepal (Pearce and Cribb 2002).

Herbal Usage: Plants are harvested in summer and autumn, washed clean and sun-dried for future use. The herb is acrid in taste, slightly bitter and considered to be cool in nature. In Traditional Chinese Medicine (TCM), the whole plant is antipyretic and it detoxifies, removes gas and humidity, reduces stasis of blood, reduces swellings, improves blood circulation, and heals ulcers, scrofula, haematuria and traumatic injuries, the last being its principal

© Springer International Publishing Switzerland 2016
E.S. Teoh, *Medicinal Orchids of Asia*, DOI 10.1007/978-3-319-24274-3_9

Fig. 9.1 *Calanthe alismifolia* Lindl. (Photo: E.S. Teoh)

usage. For consumption, decoction is made with 6–12 g of the whole dried plant (Wu 1994; *Zhongyao Bencao* 2000; Ou et al. 2003).

Calanthe alpina Hook. f. ex Lindl.

Syn. *Calanthe fimbriata* Franch.

Chinese name: *Liusuxiaji Lan* (tassels prawn spine orchid)

Medicinal names: *Mayaqi* (horse teeth seven)—the name is shared with *Calanthe davidii; Jiuxilian* (nine son lotus), *Daxiancao* (large divine herb)

Description: Plant is 30–50 cm tall. Pseudobulbs are small, 7 mm in diameter, conical, and terminating in a short pseudo-stem which is ensheathed by three or four thin, elliptic leaves with undulating margins. Leaves are 11–26 by 3–6 cm. Inflorescence is axillary, erect, 3–12 cm tall. Up to a dozen, nodding, pink flowers are carried on the erect inflorescence. Flowers are 3–4 cm across and are darker-coloured on the dorsal surface than on the under-surface of the petals and sepals. Lip is white or pale yellow veined with purplish-red (Chen et al. 1999; Perner and Luo 2007). Flowering period is June to September.

C. alpina is widespread in northern Sichuan in mixed broad-leaved, evergreen forest at 2200–2450 m but it is not common (Rathore 1983; Perner and Luo 2007). The species is distributed in southern Shaanxi, southern Gansu, Sichuan, west Yunnan and southern Xizang to Sikkim, Taiwan and Japan. It occurs in montane forests or on grassy slopes at 1500–3500 m (Chen et al. 1999).

Herbal Usage: The herb is collected in summer from Hebei, Shanxi, Hunan, Hubei, Guizhou, Yunnan and Sichuan (Wu 1994). Roots and stem are used to remove "heat" and toxins, relieve pain, and dispel "wind" or to hasten the disappearance of ecchymosis. *C. alpina* is prescribed for stomach ulcer, acute distension of the stomach, hepatitis, scrofula, toothache, sore throat, common colds, painful joints, fatique, snake bite, and traumatic and chest injuries (*Zhongyao Da Cidian* 1986; Wu 1994; *Zhonghua Bencao* 2000). It can be used in three ways: (1) by itself in decoction, using 15–30 g fresh herb; (2) to treat chronic pharyngitis, also in decoction, *Mayaqi* (*C. fimbriata/C. alpina*) 30 g together with *Ba zhao long* 60 g; and (3) as a paste for external application, just grinding a suitable amount of *Mayaqi*. These prescriptions were originally published in *Shaanxi Chinese Herbs* (*Zhongyao Da Cidian* 1986).

(Illustration: see Flora of China Orchids, p. 136, Fig. 28/29)

Calanthe brevicornu Lindl.

Syn. *Calanthe lamellosa* Rolfe

Chinese name: *Shenchunxiaji Lan* (kidney lips prawn spine orchid)

Description: This is a pretty *Calanthe*. Pseudobulb is conical, 2 cm thick, bearing 3 or 4 elliptic, plicate leaves, 10 by 5–11 cm sheathing a pseudo-stem, 5–8 cm in length. Scape is up to 30 cm long with many widely-spaced flowers of white to yellowish-green (brick red in the Indian varieties), 2–3 cm across, well expanded. Lip is trilobed and carries prominent, symmetrical kidney-shaped red patches on its large mid-lobe. It flowers from May to June. The species is distributed from northeast India, Bhutan and Nepal to Xizang, Yunnan, Sichuan, Hubei and Guangxi in China (Chen et al. 1999). At Huanglong in Sichuan Province, it occurs at 1800–2300 m in open scrub. Flowers in some clones smell of cinnamon (Perner and Luo 2007).

Usage: CHM employs the root to counter 'heat', promote diuresis, arrest bleeding, reduce swelling, and to treat nephritis or the presence of blood in the urine. It is usedused to promote expulsion of an incompletely delivered placenta, recovery after a stillbirth, or to stop abdominal pain caused by 'poor air flow' (Wu 1994). There is no mention of the herb in *Zhongyao Bencao* (2000).

Fig. 9.2 *Calanthe cardioglossa* Schltr. (Photo: E.S. Teoh)

Calanthe cardioglossa Schltr.

Thai names: *Ueang namton, Uang liam*

Description: This terrestrial orchid is 7–15 cm tall. Pseudobulbs are shaped like small flasks. Leaves are elliptic, 18–25 by 7–10 cm, deciduous. Inflorescence is erect, 20–40 cm with 10–15 flowers but usually only 2 are open at a time. Flowers are 1.2–1.5 cm across, ranging from white to light lavender to deep rose, changing to orange or yellow before dropping. The trilobular lip is prominent, carries a thin, long spur, and is about half the size of the flower. Side lobes are marked with deep purple blotches and striping (Fig. 9.2). Flowering season is November to February (Vaddhanaphuti 2005; Nanakorn and Watthana 2008). The species is distributed in northern, northeastern, eastern and southern Thailand and in Laos and Vietnam.

Herbal Usage: The stem is used as a tonic in Thailand (Chuakul 2002).

Calanthe ceciliae Rchb. f.

Malay name: *Sebueh*

Description: *C. ceciliae* is a large *Calanthe*, 25–40 cm tall. Leaves are elliptic, plicate, up to 40 by 15 cm. Inflorescence is erect, arising from the base of the stem, 1 m tall, 3 cm wide, with

few, mauve to pale violet flowers that turn a brownish-orange or apricot as they age (Fig. 9.3). Its distinctive feature is an upward pointing, slim, white spur (Rhodehamel 2005). It flowers from August to October in Thailand (Nanakorn and Watthana 2008), The species is distributed in Assam, Thailand, Peninsular Malaysia Sumatra and Java at 500–2000 m (Comber 2001). In lowland forests, it sometimes occurs on limestone (Yong 1990). The white variety is cleistogamous (Handoyo 2010).

Herbal Usage: Burkill (1935) quoting K. Heyne (1927) reported that in Sumatra the flowers were used as a poultice to relieve the pain of ulcers.

Calanthe davidii Franch.

Chinese name: *Jianyexiaji Lan* (sword leaf prawn spine orchid) *Changyegenjie Lan*
Chinese medicinal name: *Mayaqi* (the name is shared with *Calanthe alpina*)

Description: Plant is 30–50 cm tall with 6–10 long and narrow, ensiform or lorate, membranous leaves, 20–60 by 1–4 cm. Inflorescence is 60–80 cm tall but may reach a height of 120 cm (Perner and Luo 2007). Flowers are numerous, crowded, randomly orientated, pale green or

Fig. 9.3 *Calanthe ceciliae* Rchb. f. (as *Calanthe wrayi* Hook. f.) From: Hooker W.J., Hooker, J.D., *Icones Plsntarum*, vol 22: t. 2114 (1894). Drawing by M. Smith. Courtesy of Missouri Botanical Gardens, St. Louis, USA

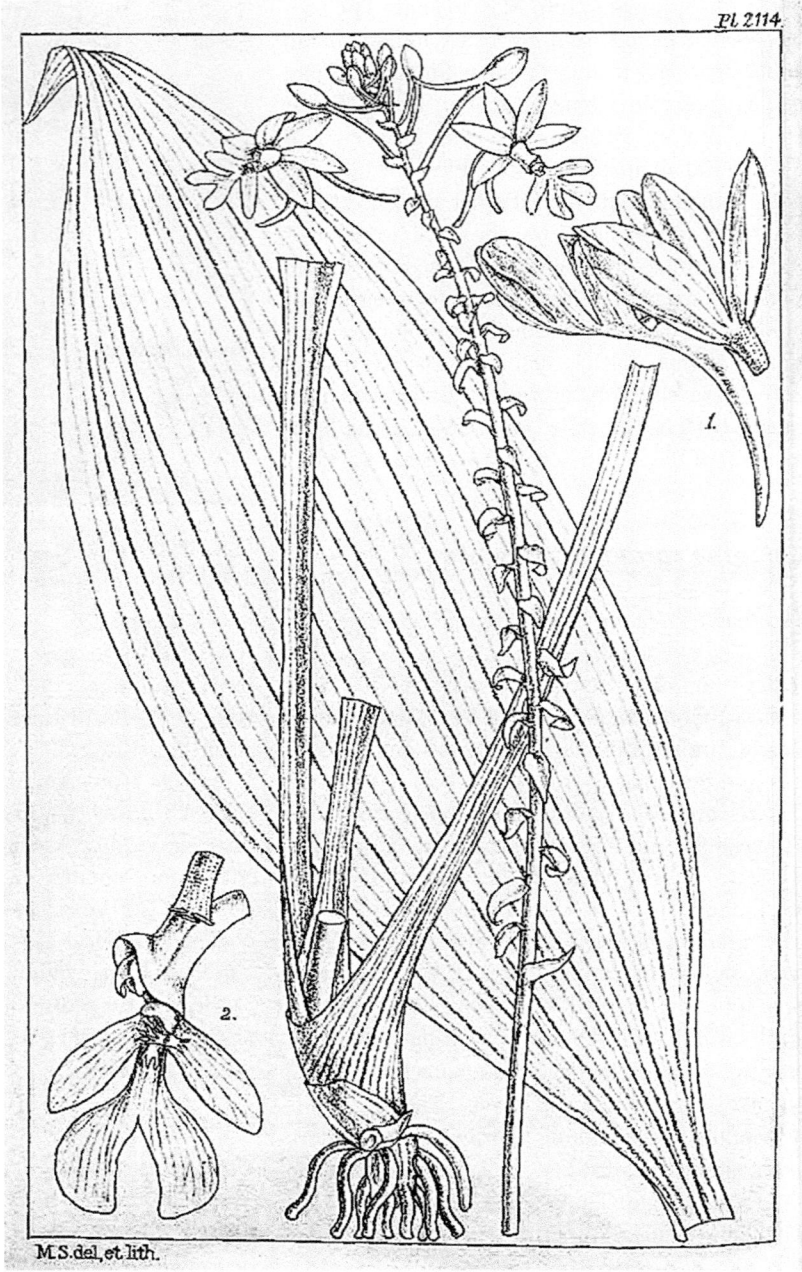

white with reflexed petals and sepals. Lip is white, three-lobed, and quite variable in shape and size. *C. davidii* usually flowers in June and July (Chen et al. 2009a).

  C. davidii is distributed from India across southern China to Taiwan and southern Japan, occurring in forests at 1200–2000 m. However,

the last time it was collected in India was in July 1899 (Rathore 1983). It is probably seriously endangered or extinct in the Himalayas. Perner (2007) found it growing and flowering in between *Cypripedium wardii* at 1620 m in pine forest in Sichuan in late May. Plants were most luxuriant at the grass borders.

Herbal Usage: Herb (*Mayaqi*) refers to both *C. alpina* and *C. davidii* although the vegetative appearance of the two species is dissimilar. They are generally supplied by the same provinces, Hebei, Shanxi, Hunan, Hubei Guizhou, Yunnan and Sichuan. Guangxi also supplies *C. alipina* but does not produce *C. davidii*. Stems and roots of *C. davidii* are used in the same manner and for the same conditions as those of *C. alpina* in Chinese Herbal Medicine (Wu 1994; *Zhongyao Bencao* 2000).

Calanthe densiflora Lindl.

Chinese name: *Zhuyegenjie Lan* (bamboo leaf segmented root orchid), *Mihuaxiaji Lan* (prawn spine flower orchid)

Description: A terrestrial herb found in shaded hardwood forests. Rhizomes are long, terete, about 8 mm in diameter, from which arise 7–9 cm tall shoots bearing 2–3 lanceolate or narrowly elliptic leaves of 20–40 by 2.3–6.5 cm, and 5-ribbed. Scape is 20 cm tall, arising from the rhizome adjacent to a mature shoot, and carries a crowded head of yellow flowers at the apex, each about 1.5 cm across. Lip is trilobed: side lobes ovate, middle lobe large, oblong and with two large keels at the base. The species is found in the southern Himalayas, China, Indochina and Japan: in Taiwan, below 1500 m throughout the island. It is the last *Calanthe* to flower in Taiwan, flowering from October to December (Lin 1975).

Herbal Usage: The *Taiwanese Chinese Herbal* states that the whole plant improves blood circulation, and reduces stasis of blood and swellings. It removes gas and humidity and is usedused to treat rheumatism, backache, pain affecting the lower limbs, running sores and traumatic injuries (Lin et al. 2003).

Calanthe discolor Lindl.

Chinese names: *Xiaji Lan* (prawn spine orchid)
Chinese medicinal names: *Jiuzilianhuancao* (nine united sons flowering herb)—this name also refers to *Calanthe tricarinata; zhu chuan zhu* (string of beads); *ye baiji* (night white chicken); *Roulainhuan* (meat in circles); *Jiujiechong* (nine segment bug); *Yichuanniuzi* (string of buttons).

Description: *C. discolor* is a robust *Calanthe*, 40–65 cm tall, with 3 elliptic-oblong leaves, 13–25 by 3–9 cm, not deciduous. Pseudobulbs are small, 1 cm in diameter. Inflorescence is axillary, erect, 20–30 cm tall, densely pubescent bearing 10 round, nodding, purplish-brown to pale maroon flowers with a white lip. Flowers may be clustered or spaced out. They are 2 cm across and appear from April to May. *C. discolor* grows in considerable profusion on the forest floor, preferring sloping terrain, at 700–1500 m in an arc in southeastern China from Guizhou to Guangdong, Fujian and Zhejiang to Japan. It is widespread throughout most of the Japanese islands (Japan Calanthean Society 1987).

Phytochemistry: Calanthoside (a novel indole, *S,O*-bisdesmoside), glucoindican, calaliukiuenoside, calaphenanthrenol, tryptanthrin, indirubin, isatin and indicant were obtained by methanolic extraction of *C. discolor* and *C. liukiuensis*. The first four compounds improved blood flow through the skin and promoted hair growth (Yoshikawa et al. 1998).

Acid hydrolysis of calanthoside yielded indirubin and isatin, whereas enzymatic hydrolysis with beta-glucosidase furnished tryptanthrin and small amounts of indirubin and isatin. Judging from their relative concentrations in fresh and dried plants of the two *Calanthe* species, the investigators postulate that calanthoside may be a common, genuine glycoside of tryptanthrin, indirubin and isatin in the plants (Yoshikawa et al. 1998).

Herbal Usage: Herb is obtained from Huadong (in Guangdong Province). It is usedused to dissolve extravasated blood and improve circulation (Chen and Tang 1982). Entire plant, roots and stem are used to improve blood flow, and to heal abscesses, scrofula, rheumatism, bone pain and traumatic injuries (Wu 1994). It is also used to treat skin ulcers and haemorrhoids (Table 9.1) (*Zhongyao Da*

Table 9.1 Chinese herbal prescriptions employing *Jiuzilianhuancao* (*Calanthe discolor* or *C. tricarinata*. The latter species is similar to *C. discolor* but it has fewer and slightly bigger flowers) (*Zhongyao Da Cidian* 1986; *Zhonghua Bencao* 2000)

1. Indication: skin ulcers,
(a) Mix *Jiuzilianhuacao* with vinegar and apply three times a day.
(b) Cook *C. discolor* 15 g with meat and consume. Additionally, mix *C. discolor* 6 g with chives 3 g for external application.
2. Indication: hemorrhoids and prolapsed piles
Mix powdered *C. discolor* 15 g in vegetable oil for application.
(*Quizhou Herbs*)
3. For swelling and pain in the throat, rheumatism, trauma and hepatitis
Use 9 g *Jiuzilianhuancao* in decoction

Cidian 1986). The herb is collected in spring or summer when the plant finishes flowering.

Calanthe fimbriata Franch. [see **Calanthe alpina (Hook. f. ex Lindl.)**]

Calanthe graciliflora Hayata

Syn. *Calanthe hamata* Hand. Mazz

Chinese names: *Goujuxiaji Lan* (splayed hooks prawn spine orchid), *Xiyegenjie Lan* (fine leaved, segmented root orchid), *Xihuagenjie Lan* (fine flowered, segmented root orchid), *Zhihuagenjie Lan* (brocade flower segmented root orchid), *Goujuxiaji Lan* (hooked prawn spine orchid): in Taiwan: slender flower *Calanthe*
Chinese medicinal name: *Silima*

Description: A robust terrestrial orchid, *C. gracifolia* has tufted, ovoid pseudobulbs, 2 cm in diameter, which bears 3–4 large, lustrous, lanceolate leaves with undulating edges that taper towards the base, 30–50 by 4–7 cm. Inflorescence is slender, 45–60 cm in length, nodding, with 9–20 flowers, 2 cm across, also nodding, and loosely arranged: the petals and sepals are pale maroon dorsally, yellowish ventrally. Lip is white and flat. It flowers in June and July (Chen et al. 2009a).

It is found in shady, moist locations in forests and along ravines at 600–1500 m in southern China from Yunnan and Sichuan in the west to Zhejiang, Guangdong, Hong Kong and Taiwan in the east (Chen et al. 1999). In Taiwan, it is found in broad-leaf forests at 1000–1500 m (Su 1985).

Herbal Usage: In Taiwan, the entire plant of *C. graciliflora* (syn. *C. hamata*) is usedused to relieve fever and for detoxification. It boosts *yin* elements, benefits the lungs, improves blood flow, reduces stasis of blood, detumescence, and stops pain and coughing (Ou et al. 2003).

On the mainland, the entire plant of *C. hamata* is used to treat rheumatism, bone pain, traumatic injuries. Herb is obtained from Hunan Province (Wu 1994). Several prescriptions are provided in *Zhonghua Bencao* (2000).

Calanthe hamata Hand. Mazz. (see **C. graciliflora**)
Calanthe lamellosa Rolfe (see **C. brevicornu**)

Calanthe mannii Hook. f.

Local Name: *Xihuaxiaji Lan* (Small Flowers Prawn Spine Orchid)

Description: Plants are 18–35 cm tall with small, conical pseudobulbs and 4–5 thin, narrowly elliptic, 5-veined, plicate leaves, 18–35 by 3–4.5 cm. Inflorescence is axillary. Rachis is erect, many-flowered, the flowers closely arranged all round. Sepals and petals are brownish; lip yellow, trilobed, mid-lobe splaying distally into 2 lobules which are rounded at their apices (Jin et al. 2009). Flowering season is May in China (Chen et al. 2009a); April to June in Bhutan (Gurong 2006); and April to May in Nepal (Raskoti 2009). *C. manii* is found in the Himalayas, Nepal, Myanmar, southern China, Vietnam and Kyushu Island in Japan in dense, broad-leaved evergreen forests at 600–2600 m (Gurong 2006; Chen et al. 2009a; Raskoti 2009).

Herbal Usage: Herb is obtained from Sichuan, Yunnan and Guizhou. In Chinese herbal medicine, *C. manii* is used for stomach heat, scrofula and abscess (Wu 1994).

Calanthe masuca (D. Don) Lindl.

Syn. *Calanthe sylvatica* (Thou.) Lindl.

Chinese names: *Changjuxiaji Lan* (long spur prawn spine orchid), *Zihuaxiaji Lan* (purple flower prawn spine orchid), *Shankala, Shanzhizhu* (mountain spider)
Myanmar Name: *Thazin gyi myo kywe*
Nepali Name: *Pakha phul*

Description: The "species is widely known as *C. masuca* in mainland Asia" (Seidenfaden and Wood 1992), but numerous taxonomists working in Asia named the species as *C. sylvatica* (Thou.) Lindl. (Seidenfaden and Wood 1992; Chen et al. 1999; Matthew 1995; Comber 2001; Gurong 2006; Raskoti 2009). However, according to the Kew Monocot List, the Asian species is *C. masuca* (D. Don) Lindl. *C. sylvatica* is distributed in Africa and in some islands in the south of the Indian Ocean but it does not occur in mainland Asia. For the sake of consistency and easy reference, the present volume follows the accepted names given in the Kew Monocot List, so the name of this species is *C. masuca*.

At one time, this pretty, moderate-sized, montane, pink *Calanthe* was a popular orchid in Europe. It has the distinction of being one of the parents of the first hybrid orchid to be bred by man.

Pseudobulbs are short and stout with annual scars and oval-elliptical leaves, 20–40 by 10 cm. Inflorescence arises from the leaf axil, erect, 45–75 cm tall, and crowded with 10–15 flowers,

2.5–5 cm across, of a delicate mauve or pale rose outlined by a border of purple and accentuated by a solid-coloured purple lip (Fig. 9.4). This is a highly variable species (Liu and Su 1978). Plants from the Chinese mainland are larger, with leaves that are 50 cm long and 15 cm wide (Chen et al. 1999). Flowers are 3–5 cm wide, pale purple to white, and open in succession resulting in a long flowering period. Nepalese plants have flowers of deep violet, 3 cm across (Raskoti 2009) (Fig. 9.5). Flowers in Bhutan are 2–4 cm across, violet to purple. In Thailand, flowers are 2.5–3 cm across. It flowers in July in Thailand (Vaddhanaphuti 2001), from May to July in the Kachin State of Myanmar (University of Myanmar Department of Botany 2004), followed by August to September in India and Bhutan (Bose and Bhattacharjee 1980; Gurong 2006), September in Nepal (Raskoti 2009), and all the time in Kerala (Abraham and Vatsala 1981). The Sri Lankan *C. purpurea*, which differs from the Indian glabrous-leaved *C. masuca* only by having downy leaves, flowers in February, July and August (Jayaweera 1981). In China, *C. sylvatica* enjoys a long flowering period which extends from April to September (Chen et al. 1999).

C. masuca enjoys a wide distribution from Japan across southern China (Taiwan, Hunan, Guangxi, Guangdong, Hong Kong, southern Yunnan and southeast Xizang), Bhutan, Nepal, Sikkim, India, Sri Lanka to Madagascar and South Africa. Its northern-most distribution is in a few scattered southern islands at the southern tip of Japan and on Mikurajima Island just below 34°N latitude and south of Tokyo (Japan Calanthean Society 1987). In the south, it is found in Thailand, Malaysia and Indonesia. It occurs at an elevation of 800–2000 m in shaded, moist locations, in broad-leaved low montane forests.

Herbal Usage: Herb is obtained from Fujian, Jiangxi, Yunnan, Hunan, Guangxi, Guangdong and Xizang. The entire plant is used as an anodyne. It also reduces swellings, removes toxins and repairs wounded tissues. It is usedused in the treatment of abscesses especially if foreign bodies in the body are not surgically removed

Fig. 9.4 *Calanthe masuca*
(D. Don) Lindl.
Reproduced with
permission from
*Introductions to
Orchids* by Abraham and
Vatsala, Parlode,
Thiruvananthapuram:
Tropical Botanic Garden
and Research Centre
(TBGRI), 1981

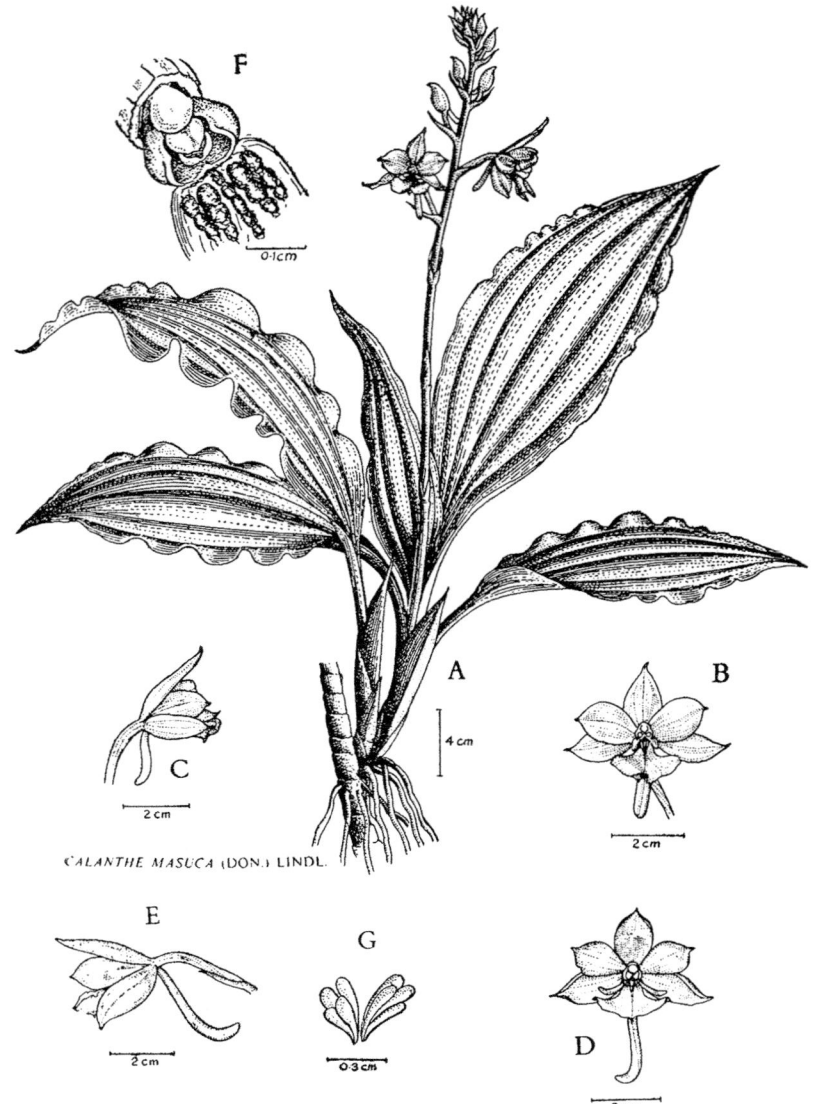

(Wu 1994). In Nepal, a paste made from the stem is applied to treat dislocated bones (Manandhar and Manandhar 2002). The flowers are used to arrest epistaxis (Rao 2004; Singh and Duggal), Pseudobulbs are also used to treat nose bleeds (Baral and Kurmi 2006).

Calanthe plantaginea Lindl.

Description: Plants are 40–65 cm tall with coni-cal pseudobulbs 1.5–2 cm in diameter, covered with 4 sheaths. Leaves, 2–4, are elliptic-lanceolate, plicate, 10–28 by 4–12 cm, with pseudopetiole that forms a pseudostem 20 cm tall. Inflorescence arises from apex of pseudobulb, 30–40 cm with many-flowerd rachis. Flowers 12–24, are scented, white to lilac, 3–4 cm across, facing all directions. Sepals and petals are narrow, elliptic to ovate-laceolate. Lip is trilobed, side lobes erect, mid-lobe splayed into three tongues on its distal half and carrying a faint orange flare at its midpoint (Fig. 9.6). Flowering season is March to April in China

Fig. 9.5 *Calanthe masuca* (D. Don) Lindl. (syn. *Calanthe sylvatica* (Thou.) Lindl. (Photo: E.S. Teoh)

Fig. 9.6 *Calanthe plantaginea* Lindl. (Photo: E.S. Teoh)

(Chen et al. 2009a), January to June in Bhutan (Gurong 2006) and April in Nepal (Raskoti 2009). *C. plantaginea* is distributed from Yunnan and Tibet to Sikkhim, Bhutan, Nepal and Kashmir, occurring in broad-leaved, temperate forests at 1500–2500 m (Chen et al. 2009a; Pant and Raskoti 2013) and between 100 and 300 m in Bhutan (Gurong 2006).

Herbal Usage: Harvested tubers are washed, sun-dried and rendered into powder form. In Nepal, the powder is mixed with milk is consumed as a tonic and aphrodisiac (Pant and Raskoti 2013).

Calanthe puberula Lindl.

Syn. *Calanthe similis* Schltr.

Chinese names: *Lianexiaji Lan* (sickle lip orchid), *Fanjuangenjie Lan* (counter folding root segment orchid), *Juanegenjie Lan* (folding calyx root segment orchid), *Lianyexiaji Lan* (sickle leaf prawn spine orchid): *Jiaxiaji Lan* (fake prawn spine orchid), *Xiangsixiaji Lan* (similar prawn spine orchid) *Zigenjie Lan* (purple root orchid)

Chinese medicinal name: *Lianexiaji Lan* (sickle lip orchid)

Description: *C. puberula* is a beautiful, small to medium-sized, pink *Calanthe* with lovely foliage and attractive pale pink flowers. Raceme bears up to a dozen flowers which are 2–3 cm across. Dorsal sepal is broad, triangular, and curves forward. Petals and lateral sepals are filiform, undulate and bend forward. The inverted trident-shaped lip is distinctive (Fig. 9.7). Flowering period is May to August at Gaolingongshan in

Fig. 9.7 *Calanthe puberula* Lindl. (Photo: E.S. Teoh)

Yunnan (Jin et al. 2009), July and August else-where (Chen et al. 1999; Pearce and Cribb 2002; Misra 2007).

C. puberula grows in mixed broad-leaved evergreen forests at 1200–2500 m in western Yunnan and adjacent southeastern Tibet; in Sikkim and northeast India at 2000 m; in Taiwan at 1300–2500 m; and in Vietnam. It is widely distributed in Japan with a northern limit at Okushiri Island (Japan Calanthean Society 1987).

Herbal Usage: Herb is obtained from Yunnan, Guangxi and Guandong. The whole plant is usedused in Chinese Herbal Medicine (CHM) to treat scrofula, and sores that itch (Wu 1994). It is antipyretic and detoxifies. Used for running sores, it improves blood flow and stops pain. It is used to treat ulcers, scrofula, mange, scarlet fever, amenorrhoea, trauma and dysentery in Taiwan (Ou et al. 2003). *C. puberla* Lindl. and *C. similes* Schltr. are mentioned as separate species by Wu (1994), but botanically they are not different and the first name has precedence.

Calanthe similis Schltr. (see **C. puberula Lindl.**)
Calanthe sylvatica (Thou.) Lindl. [see **Calanthe masuca (D. Don) Lindl.**]

Calanthe tricarinata Lindl.

Chinese name: *Sanlengxiaji Lan* (triangular prism prawn spine orchid)
Chinese medicinal name: *Jiuzilianhuancao* (nine united sons flowering herb) This name also refers to *Calanthe discolor; Roulianhuan*

Description: Stem of this attractive, evergreen *Calanthe* is sheathed in bracts and leaves in its lower half and bears 2–3 elliptic leaves 30 by 7 cm. Flowers are 2.5 cm in diameter, opening widely, and are loosely arranged in the lower part of the raceme. Sepals are greenish, petals white. Lip is rose purple with a white border; large, oblong and trilobed with barely any space between the large middle lobe and the two smaller lateral lobes. It flowers in April to June in Indian Himalaya, Bhutan and at

Gaoligongshan in western Yunnan (Bose and Bhattacharjee 1980; Japan Calanthean Society 1987; Pearce and Cribb 2002; Jin et al. 2009). *C. tricarinata* grows in the shade on the forest floor at an altitude of 2000 m in Pakistan, Kashmir, Sikkim, western China and Japan, being widespread in the Japanese islands.

Herbal Usage: Herb is obtained from Shanxi, Hubei, Sichuan, Yunnan, Guangxi, Guizhou and Xizang. In China, the root is used to stimulate blood circulation, relax muscles and joints, remove wind and stop bleeding. It is usedused in the treatment of stomachache, arthritis, lumbar muscle degeneration and traumatic injuries. Its use is contra-indicated during pregnancy (Wu 1994; *Zhonghua Bencao* 2000). Leaf paste is used to treat wounds and eczema in Nepal. In that country, leaves and pseudobulbs are valued as aphrodisiacs (Pant and Raskoti 2013). In Uttarakhand, West Himalaya, roots and leaves are usedused to treat jaundice and typhoid (Maikhuri et al. 2014).

Calanthe triplicata (Willimet) Ames

Calanthe veratrifolia R. Br. ex Ker Gawl.

Chinese names: *Sanzhexiaji Lan* (three layered shrimp's spine), *Baihe Lan* (white crane orchid), *Shishangjiao* (leaf on the stone), *Roulianhuan* (meaty chain of rings); *Paiwan* (put in order and bend in a stream), embossed banana leaf orchid
Chinese medicinal name: *Shishangjiao* (leaf on the stone)
Japanese name: *Tsuru Ran* (crane orchid)
Thai name: *Ueang Kao Tog*
Indonesian names: *Lau Bawang* in Kalimantan Barat; *Angkrek Popotjongan, Ahan Malona* (Amboin); *Bunga Tiga Lapis* (Maluku); *Guru ni Hambing* (Batak Toba) *Lumbu Hutan* (Sumatra and Timor); *Seugeundeu* (Gajo Singkut in Batak Karo) *Anggrek bayi tidur* (Sulawesi)

Description: Plant without its inflorescence is about a foot tall. Leaves are 4–6, ovate-lanceolate, plicate, undulate, furrowed, petiolate and dark green, 30–60 by 5–12 cm. Inflorescence is carried well above the tips of leaves and bears up to 50 beautiful white flowers which are well arranged around the scape. Lip has four lobes and is marked by a vertical, linear, yellow or reddish, wart-like callus (Fig. 9.8). Flowers open successively over an extended period, with up to 12 simultaneously open.

Flowering period is April to September in Singapore–Malaysia (O'Byrne 2001) but the author's plants in Singapore are still putting out new inflorescences in October, March to September in the Shan state of Myanmar, April to May on the Chinese mainland (Chen et al. 1999), March to July on Hong Kong and Lantau Islands (Wu et al. 2001), June in Thailand (Vaddhanaphuti 2001), April to June at the Western Ghats in southern India (Bose and Bhattacharjee 1980), May to July in southern India (Abraham and Vatsala 1981; Joseph 1982) or May to October (Misra 2007), and April to May and July to December in Bhutan (Pearce and Cribb 2002).

Phytochemistry: Leaves of *C. triplicata* produce indigo when bruised. Alkaloid is present (Luning 1967).

Fig. 9.8 *Calanthe triplicata* (Willimet) Ames (Photo: E.S. Teoh)

Herbal Usage: Chinese herbalists in Taiwan use the root to treat rheumatism, backache and traumatic injuries including fractures (Ou et al. 2003). The whole plant is a diuretic.

Ananda Rao and Sridhar (2007) reported that in Karnataka the roots are used for diarrhoea and toothache. In Arunachal Pradesh, the roots are an ingredient in a remedy for swollen hands, and, in a separate combination, usedused for treating diarrhea. Various parts of the plant are usedused to treat toothache. Pseudobulbs are a masticatory for a variety of gastro-intestinal disorders while flowers are used to relieve toothache (Rao 2004). Root extract is usedused to treat diarrhoea and toothache (Das 2004).

Rumphius (late seventeenth century) who lived on the island of Amboin (Sulawesi) observed that the plant was "quite sharp" and cautioned regarding its use. Initially, the taste of the roots is insipid, but suddenly it becomes quite sharp, "like some *Gentiana*, burning the mouth, so that one's lips will swell, one's throat gets hoarse, and one even feels this sharpness in the leaves, wherein it differs from all *Angreks*". The roots were used together with nutmeg, cloves and two types of ginger "rubbed together and tied to" the swollen hands. "The natives have such tough mouths, that they dare to take these sharp roots internally, and chew it along with *pinang* (betel-nut), nutmeg and ginger, against persistent diarrhoea caused by cold or raw dampness" (quotations from Beekman's translation, 2002) (Rumphius 1741–1750). In Sumatra, the flowers are used to relieve pain from caries (Heyne (1927). In Sulawesi, the rhizome is a cure for toothache (Yuzammi and Hidayat 2002).

Calanthe vestita Wall ex Lindl.

Thai name: *Khao Malila*
Myanmar name: *Thazin gyi ahphyu*

Description: *C. vestita* is a terrestrial orchid with above-ground pseudobulbs that are broadly ellipsoid, 7 by 2.5 cm. Leaves arise near apex of pseudobulb, 4 in number, lanceolate-elliptic, glabrous, plicate petiolated, 40 by 5 cm, and deciduous. Inflorescence is up to 50 cm tall, arching, pale green, covered with short white hairs, and it carries up to a dozen attractive, white to pink flowers, that are marked with a blotch of yellow at the throat. Sepals and petals are lanceolate, spread out, sepals slightly larger than the petals, 2.5 cm long and hairy whereas petals are glabrous (Fig. 9.9). Flowering period is October to February in Thailand (Vaddhanaphuti 2005) and Myanmar (Grant 1895); November to December in the Western Ghats (Abraham and Vatsala 1981); elsewhere, it is after the dry season. Several cultivars were described during the nineteenth century following the introduction of this species into Europe (Grant 1895). This popular *Calanthe* is distributed from Assam to Papua New Guinea. It is found in the limestone areas of Southeast Asia that experience a distinct dry season.

Phytochemistry: Leaves of *C. vestita* contain flavone C-glycosides (Williams 1979).

Herbal Usage: In Vietnam, crushed bulbs are rubbed over aching bones of people suffering from rheumatism (Petelot quoted by Perry). Six bacterial strains belonging to the genera *Athrobacter, Bacillus, Mycobacterium* and *Pseudomonas* have been isolated from the

Fig. 9.9 *Calanthe vestita* Wall ex Lindl. (Photo: E.S. Teoh)

underground roots of *C. vestita* var. *rubrooculata* (Tsavkelova et al. 2001), but whether this impacts on its medicinal usage is unknown.

Overview
There is discordance between the traditional medical usage and the modern scientific/pharmaceutical interest in this genus of orchids.

In Chinese herbal medicine, 14 *Calanthe* species provide a remedy for a variety of conditions, in particular swellings of different aetiology (abscess, trauma, arthritis), and pain (painful joints, pain at the extremities, toothache, pharyngitis, pain from stomach ulcers or abdominal distension, snake bites and trauma). *Calanthe* reduces stasis of blood, improves blood circulation and detoxifies. A number of species (*C. alismaefolia, C. davidii, C. gracilifolia, C. fimbriata, C. lamellose* and *C. puberula*) are said to be antipyretic. Several *Calanthe* are used for treating wounds and infected skin (*C. alismaefolia, C. densiflora, C. discolor, C. manii, C. mascula, C. puberula* and *C. similes*) (Wu 1994; Lin et al. 2003). Among the ways of using the orchid, wine fortified by the roots of some *Calanthe* species was reportedly usedused in China to treat traumatic injuries and internal bleeding (Hu and Cheo, quoted by Perry and Metzger 1980). Interestingly, use of *C. tricarinata* is contra-indicated during pregnancy (Wu 1994).

C. vestita which is common in the limestone areas of Southeast Asia is usedused in Vietnam to treat rheumatism. This usage is probably derived from a Chinese tradition set during the Chinese Tang Dynasty's (618–907) suzerain over Vietnam (Annam).

Another common Southeast Asian species, *C. triplicata* (*C. veratrifolia*) has similar uses in Ayervedic medicine and in Indonesia. It is used to treat rheumatism, backache and trauma. More unique is its alleged ability to correct intractable diarrhoea. In Amboin, the root of *C. triplicata* is a component of a remedy for swollen hands (Rumphius, late seventeenth cent.). In Sumatra, its flowers are used to relieve pain from dental caries (Heyne, quoted by Perry and Metzger 1980).

Contemporary research completely ignores the traditional usage of *Calanthe* and focuses on two areas: (1) a hair-restoring property, and (2) a possible anticancer agent. Both areas have tremendous economic potential but there is no report of any relevant drug trial. Calanthoside, glucoindican, calaliukiuenoside and calaphenanthrenol present in *C. discolor* and *C. liukieuensis* (= *C. lyroglossa* var. *lyroglossa*) improved blood flow through the skin and promoted hair growth (Yoshikawa et al. 1998).

Calanquinone A from *C. arisanensis* exhibited potent antitumour activity against lung (A549), prostate (PC-3 and DU145), colon (HCT-8), breast (MCF7), nasopharyngeal (KB) and vincristine-resistant nasopharyngeal (KB-VIN) cancer cell lines. Most exciting was the finding that this compound showed an improved drug resistance profile compared to paclitaxel. (The latter is a highly effective cytotoxic agent from the Pacific yew that has saved so many women suffering from ovarian cancer.) Lee and his co-workers have managed the total synthesis of Calaquionone A (Lee et al. 2008). Calanquinone A induces s-phase arrest and apoptosis of glioblastoma (brain tumour) cell types A172, T98 and U87 by decreasing cellular glutathione. Glioblastoma is resistent to radiotherapy and chemotherapy, so it would be helpful if this action of calantquinone can be translated into antiglioblastoma therapy (Liu et al. 2014a).

C. arisanensis also contains other calanquinones (B and C). Four new 9,10-dihydrophenanthrenes, calanhydroquinones A, B, C and calanphenanthrene A, and several known compounds are also present in *C. arisanensis*. Calanquinones B and C, and calanhydroquinones A, B and C, all showed cytotoxic activity against several human cancer cell lines. Calanquinone B exhibited the highest potency (EC(50) < 0.5 mg/mL) against seven cancer cell lines (human lung A549, prostate PC-3 and DU145, colon HCT-8, breast MCF-7, nasopharyngeal KB and vincristine-resistant nasopharyngeal KBVIN cancer, with the greatest activity against breast cancer MCF-7 cells [EC (50) <0.02 µg/mL] (Lee et al. 2009). Alkaloids

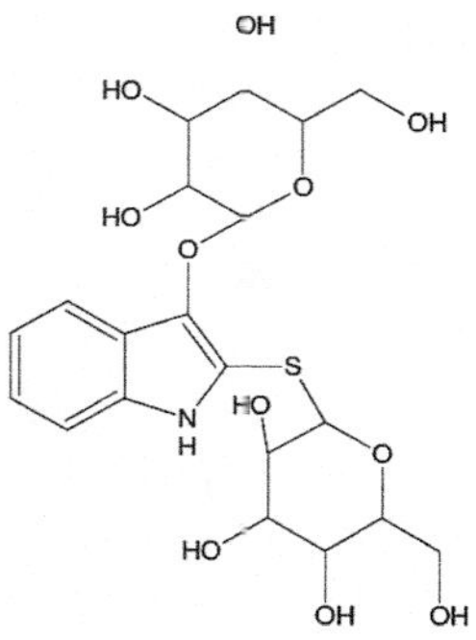

calanthoside

Fig. 9.10 Structure of calanthoside. The compound promotes blood flow through the skin and promotes hair growth

are present in *C. triplicata, C. vestita* and several other species. Indoles are present in *C. triplicata* and flavone C-glycoside in *C. vestita* (Williams 1979; Veitch and Grayer 2007a).

Germination of mature seeds of *C. tricarinata* in asymbiotic culture has been achieved using a "New Dogashima" medium supplemented with napthelene acetic acid (NAA) and benzyladenine (BA) (Godo et al. 2010). The process can be used for mass propagation to protect the wild species which is valued as an aphrodisiac in Nepal (Pant and Raskoti 2013) (Fig. 9.10).

Genus: *Callostylis* Blume

Chinese name: *mei zhu lan*

A small genus with only five or six species distributed from the Himalayan region to China and southwards to Myanmar, Thailand, Indochina, Malaysia and Indonesia, these are epiphytic orchids that were once included under *Eria*. Pseudobulbs are bulbous or terete, with base loosely covered by dry sheaths, leafy on the upper part, and are well spaced along a stout, creeping rhizome. Leaves are leathery, 2–5, arising at or near the apex of the stem. Inflorescence is axillary, with many cream-coloured or yellow flowers. Sepals are covered with brown hairs abaxially.

Callostylis bambusifolia (Lindl.) S.C. Chen & J.J. Wood

Syn. *Eria bambusifolia* Lindl.

Indian name: *Mundabai*

Description: This is a tall epiphyte with tufted stem, 20–70 (occasionally 90) cm in height and 3–7 mm in diameter with approximately 10 long, lanceolate leaves near the top, measuring 10–22 by 1–3 cm. Inflorescences are axillary, pendulous, appearing a few at a time, with 5–7 flowers each. Flowers are 1.5–2 cm across, white, striped with purple, and covered with short brown hairs abaxially (Fig. 9.11). Flowering period is November to December. The species is distributed from Vietnam, Thailand, Myanmar and southern Yunnan to northeast India down to Orissa state. It grows on trees in sparse woods at 900–1200 m (Chen et al. 1999; Jin et al. 2009).

Herbal Usage: The entire plant is used to treat stomach upsets in India. A plant of *Callostylis bambusifolia* (syn. *Eria bambusifolia*) and another of *Aegle marmelos* (not an orchid) are

Fig. 9.11 *Callostylis bambusifolia* S.C. Chen & J.J. Wood. From: *Annals of the Royal Botanic Gardens, Calcutta.* Vol. 8(3): t. 163 (1891). Original drawing by R. Pantling in colour with black and white

ERIA BAMBUSIFOLIA, Lindl.

separately burnt to ashes in earthen pots, and thereafter their ash is mixed in a 1:1 ratio. Half a tablespoon is administered on an empty stomach twice a day for one week to treat hyperacidity and stomach upsets. This is the practice of the Dongria Kandha hill tribe in the southwest of Orissa State in India (Dash et al. 2008).

Genus: *Cephalanthera* Rich.

Chinese name: *Tourui Lan*

Cephalanthera is a genus made up by a dozen species of robust, terrestrial orchids generally inhabiting the temperate regions of Eurasia to the Himalayas and North Africa. There is one species in Taiwan (Tang and Su 1978) and another in Laos (Schuiteman and de Vogel 2000). Plants have underground rhizomes and erect stems sheathed with ovate-lanceolate leaves. Inflorescences are terminal and carry several small, resupinate, white, red, or green flowers. Lips are trilobed and do not possess spurs. In some species, the flowers do not open widely (Chen et al. 2009b).

The generic name is derived from the Greek *kephale* (head) and refers to the fanciful impression that its anther is held high like a head. *Cephalanthera* are not present in orchid collections.

Cephalanthera erecta (Thunb.) Blume

Chinese name: *Yin Lan* (silver orchid)
Chinese medicinal name: *Yin Lan* (silver orchid)

Description: *C. erecta* is native to the Eastern Himalaya, China and Japan, growing amidst grasses and low shrubbery. Tubers are paired, irregular and underground. Stems are slender, 30–40 cm tall with 3–4 lanceolate leaves near the apex, 3–6 cm long, plicate and pointed. The Huanglong variety is small, generally only 15 cm tall. Inflorescence is short with up to 8 flowers, loosely arranged. Flowers are 1–1.2 cm long, green or white, barely opening, and appear in April to June (Hawkes 1965; Perner and Luo 2007; Jin et al. 2009).

This species resembles *C. longifolia* but the plant is smaller. Chin-shaped spur is a prominent characteristic of the flowers (Perner and Luo 2007).

Herbal Usage: Herb is obtained from Shanxi, Hubei, Zhejiang, Jiangxi, Guangdong and Sichuan. Plant is used to treat fever, thirst, urinary infection It is diuretic (Wu 1994).

Cephalanthera falcata (Thunb.) Lindl.

Chinese names: *Jin Lan* (gold orchid), *Lianyetourui Lan* (pistal above sickle leaf orchid)
Chinese medicinal name: *Jin Lan* (gold orchid)

Description: *C. falcata* is a lowland terrestrial orchid of the temperate zone. Its slender stems arise from slender, creeping rhizomes with numerous roots. Stem is 25–35 cm tall, with 4–6 broadly elliptic or lanceolate, plicate, pointed leaves, 5–8 cm long. Inflorescence is up to 15 cm long, loosely 9–18 flowered. Flowers are 1.6 cm long, a clear golden yellow in Japan, or white or green flushed with white, and are fragrant. The golden Japanese variety is quite handsome (Kanda 1977).

The species is found in Yunnan Province in China, and in Korea and Japan. It flowers in April and May at Gaoligongshan in Yunnan (Jin et al. 2009), appearing a bit later, from May to July, further north (Kanda 1977). Accelerated growth observed when the orchid exists in tripartate symbiosis with Telephoraceae fungi, and *Quercus serrata* in pot culture in Sapporo suggests that under inclement conditions *C. falcata* may become mycoheterotrophic or even purely mixotrophic (Yagame and Yamato 2013).

Herbal Usage: Herb is obtained from Hubei, Hunan, Guangdong, Guangxi, Yunnan and Sichuan. The entire plant is antiheat, and relieves fever. It is used to treat sore throat and toothache (Wu 1994).

Cephalanthera longifolia (L.) Fritsch.

Chinese names: *Changyetourui Lan* (pistal above the long leaf orchid), *Tourui Lan*

Description: A terrestrial inhabiting forests or grassy slopes and shrubberies, *C. longifolia* has an erect stem 20–47 cm tall with 4–7 lanceolate, plicate leaves, 4–13 by 0.5–2.5 cm (Fig. 9.12). Tap root is long and sparsely branched with a few short side roots (Rasmussen 1995). Inflorescence is terminal, 1.5–6 cm long bearing 2–15 white flowers that do not open widely, about 1.5 cm across (Fig. 9.13). Flowering period is May to June. The species is found in Central China from Shanxi to Xizang (at 2300–3000 m) and is widespread in Northern India, Central Asia, Europe and North Africa (Chen et al. 1999).

Phytochemistry: Alkaloids, quercetin and kaempferol-*O*-glycosides are present in *C. longifolia* (Luning 1967; Williams 1979). *Loroglossin* is present in *C. damasonium* (syn. *C. grandiflora*) and *C. rubra* (Veitch and Grayer 2007b) (Fig. 9.14).

Fig. 9.12 *Cephalanthera longifolia* (as *Cephalanthera acuminata* Wall. ex Lindl.). From: Jacquemont, V., *Voyage dans l'Inde pendant les annees* 1828 a 1832, vol. 4(3): t. 164 (1844). Courtesy of Missouri Botanic Gardens, St. Louis, USA

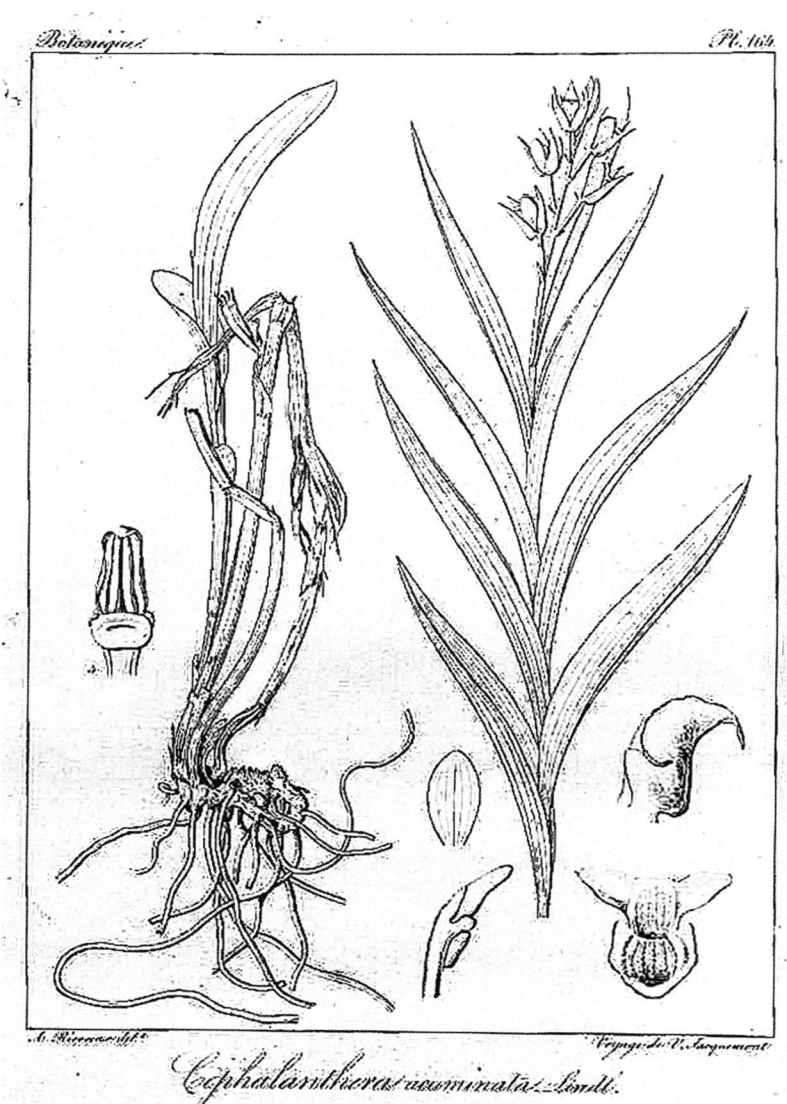

Fig. 9.13 *Cephalanthera longifolia* (L.) Fritsch. (Photo: E.S. Teoh)

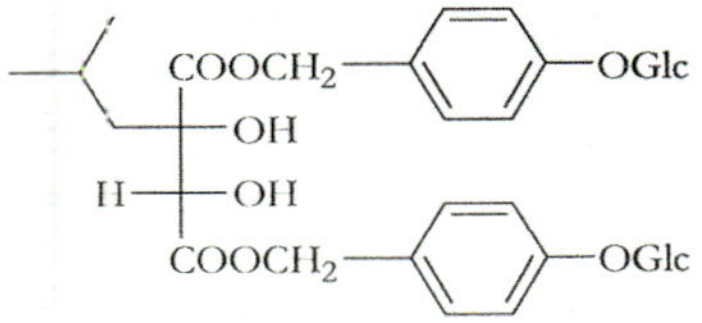

COOCH$_2$——⬡——OGlc
——OH
H——OH
COOCH$_2$——⬡——OGlc

loroglossin

Fig. 9.14 Loroglossin, a phytoalexin produced by *Cephalanthera damasonium*

Herbal Usage: Roots and stems are used for nocturesis and enuresis in TCM (Wu 1994). In Arunachal Pradesh, roots and rhizomes are valued as tonic (Duggal 1971; Rao 2004).

Overview

Significant amounts of alkaloid were not detected when a single species of *Cephalanthera* was screened by Luning's Swedish team (Luning 1974a, b). The newly discovered saprophytic species might be more promising as a source of secondary metabolites (Chen and Lang 1986). Antibiotics might be present in the mixotrophic *C. falcata* which is used to treat toothache and sore throat.

C. longifolia is on the red list of endangered orchid species (Duffy et al. 2009) and attempts to germinate the seeds and propagate *C. longifolia* were uniformly unsuccessful (Rasmussen 1995). However, Yamazaki and Miyoshi (2006) succeeded in the asymbiotic germination of *C. falcata* which is reported to be endangered in Japan, using green pod culture 70 days from pollination. They found that after this date the viability of the seeds declined and minimal germination was seen in seeds harvested 100 days post-pollination or later (Yamazaki and Miyoshi 2006).

Recently, it was shown that light intensity is the decisive factor influencing autotrophic versus heterotrophic behaviour in adult *Cephalanthera* plants. When the light intensity is low, for instance during winter, the orchid may switch to strong dependency on fungi for its carbon nutrition (Preiss et al. 2010). *C. falcata* photosynthesises but simultaneously obtains carbon supplies from mycorrhiza (Thelephraceae fungi growing on oak), i.e. it is mixotrophic (Yagame and Yamato 2013). *C. longifolia* obtains 33 % of its carbon and 86 % of its nitrogen from Thelephraceae fungi (Abadie et al. 2006). Tripartate symbiosis should be considered when attempts are being made to conserve terrestrial orchids growing in deep shade or in temperate regions.

Genus: *Changnienia* Chien

Chinese name: *Duhua Lan* (solitary flower orchid)

Changnienia is a recently discovered, monotypic, terrestrial genus which belongs to the subtribe Calypsoeae. It is endemic in China and enjoys a wide distribution in the central provinces at elevations of 400–1500 m. Pseudobulbs are subellipsoid or ovoid, 1.5 cm, cloaked with numerous membranous sheaths. Leaf is apical, solitary, broadly ovate-elliptic to broadly elliptic. Flowers are pink; however, if the soil is alkaline (pH 4.5–5.0), they are blue.

The fleshy, corm-like, subterranean pseudobulb sends up a single elliptic leaf in September. Inflorescence appears in November but the solitary flower does not open until March or April and fruit is set in May to June. There is considerable variation within the species, allegedly an adaptive mechanism for pollination (Sun et al. 2005). Bumblebees of the species *Bombus trifaciatus* Smithi pollinate the orchid. Visits by other species of insects have also been observed in Shenggongjia, but their visits do not result in pollination (Sun et al. 2003).

The species is named for Chang-nien Chen, a botanical collector of the early twentieth century who worked for the Academia Sinica (now renamed the Chinese Academy of Sciences) in Nanjing.

Changnienia amoena Chien

Chinese names: *Duhua Lan* (solitary flower orchid); elder blue*
Chinese medicinal name: *Changnian Lan*

Description: This terrestrial orchid is found in humus-rich soil in shady spots along ravines at 700–1800 m in southern Shaanxi, Jiangsu, Anhui, Zhejiang, Jiangxi, Hubei, Hunan and Sichuan Provinces in China. Pseudobulbs are subterranean, corm-like, fleshy, ovoid, pale yellowish-white, 1.5–2.5 cm long and 1–2 cm in diameter. It bears a single, terminal broadly elliptic, undulate leaf at the end of a 3.5–8 cm petiole thrust above the humus and ground cover. Leaf is green on its upper surface, reddish-purple on the under-surface, 6.5–11.5 by 5–8 cm. Scape is terminal, 10–17 cm long and carries a single pink flower which is 5–6 cm across. Dorsal sepal and lateral petals are close together and form a hood over the lip. Lateral sepals are linear and well extended at a 160° angle. Lip has three lobes. Side lobes are erect and form a hood over the column; mid-lobe is broad and irregularly undulate, extending backwards into an iron-shaped spur. Lip is white with pink spots on the three keels in the throat and at the edges; spur is long and pointed. Flowers appear in April (Chen et al. 1999).

Herbal Usage: Herb is obtained from Zhejiang, Jiangsu, Hunan and Sichuan. The whole plant together with its roots is regarded as antiheat and antitoxic. It cools the blood. It is useddused in the treatment of coughs, blood-streaked sputum, sores and furuncles (Wu 1994). To treat bloody phlegm, a decoction is prepared with 15–30 g of dried or 60–90 g of fresh herb, then sweetened with white sugar and drunk day and night before meals.

To treat sores, a poultice is made by mixing fresh pulverised plant with salt. The dressing is changed at least every day (*Zhonghua Bencao* 2000).

Overview
Only discovered and botanically published in 1935, it is remarkable that it already has a herbal use. The reason being that Chinese herbalists have long recognised *C. amoena* as a distinctive species but botanists were not aware of its existence. A similar explanation exists for many medicinal orchid species long known to tribal people and much later "discovered" and described in journals.

There is no pharmacological information on the species.
Genus: *Cirrhopetalum*
This genus is now included under ***Bulbophyllum***

Cirrhopetalum andersonii Hook. f. (see ***Bulbophyllum andersonii* Hook f.**)
Cirrhopetalum vaginatum Lindl. (see ***Bulbophyllum vaginatum* (Lindl.) Rchb. f.**)

Genus: *Cleisostoma* Blume

Chinese name: *Geju* Lan

Cleisostoma is a large genus of small to medium-sized epiphytic, monopodial orchids with some 90 members. It enjoys a widespread distribution from Sri Lanka and India through China to Japan and across Southeast Asia to Papua New Guinea and the Pacific Islands. Many species formerly classified as *Sarcanthus* are now in *Cleisostoma*, whereas numerous others formerly in *Cleisostoma* have been moved to other genera

(Seidenfaden and Wood 1992). The epicenter for the genus is Thailand. They are rarely cultivated as garden plants.

Stems may be short or long, erect or pendant, with leaves that are terete, semi-terete or flat. Flowers are small with spreading petals and sepals of equal size. Lip is trilobed, with a spur and the conspicuous callus that distinguishes the species. The name *Cleisostoma* is constituted by two Greek words, *kleistos* (closed) and *stoma* (mouth), referring to the calli blocking the entrance to the spur thus producing a narrowed mouth.

Cleisostoma birmanicum (Schltr.) Garay

Syn. *Sarcanthus ophioglossa* Guillaumin

Laotian name: *Ka dam phi*

Description: This is a stout, monopodial epiphyte with an 8- to 9-cm stem carrying several thick leaves 15 by 1.5 cm. Inflorescences are lateral, paniculate, extending beyond the tips of the

Fig. 9.15 *Cleisostoma birmanicum* (Schltr.) Garay (Photo: E.S. Teoh)

leaves, with 10–12 reddish-maroon flowers with yellow lips whose mid-lobes extend downwards into pairs of thread-like tails (Fig. 9.15). Flowering period is April to May. It occurs in Hainan, Indochina, Thailand and Myanmar (Chen et al. 1999)

Herbal Usage: The orchid was used to treat orchitis in Laos (Vidal 1963).

Cleisostoma flagelliforme (Rolfe ex Downie) Garay (see **Cleisostoma fuerstenbergianum Kraenzl.**)

Cleisostoma fuerstenbergianum Kraenzl.

Syn. *Cleisostoma flagelliforme* (Rolfe ex Downie) Garay

Chinese name: *Changyegeju Lan* (long leaf separate distance orchid)
Thai names: *Kloi nam thai* in Ubon Ratchathani; *kang pla*

Description: Stems are terete, branching, pendulous, green, 40–60 cm long. Leaves are terete, slim, curved, acute, 12–15 by 0.8 cm. A 12- to 15-cm-long, pendulous inflorescence is produced 180° degrees from (opposite) a leaf. Flowers are 1 cm across, 5–18 in number, well spaced, and have dark brown petals and sepals. Lip is white, turning yellow with age (Fig. 9.16). It flowers in May to June in China (Chen and Wood 2009) and in Chiang Mai, Thailand, August to September (Vaddhanaphuti 2005), and November to February (Nanakorn and Watthana 2008). This is a handsome, steady epiphyte growing on trees in broad-leaved evergreen forests at 700–2000 m in Yunnan, Guizhou, Hainan, Indochina and Thailand

Herbal Usage: Herb is obtained from Hainan and Yunnan. The whole plant is used as a remedy for heat and toxins, sore throat and tonsillitis in China (Wu 1994). Leaves are used to treat diabetes in Thailand (Chuakul 2002).

Cleisostoma hongkongense (Rolfe) Garay [see **C. williamsoni (Rchb .f.) Garay**]

Fig. 9.16 *Cleisostoma fuerstenbergianum* Kraenzl. (Photo: Courtesy of Plant Photo Bank of China)

Cleisostoma paniculatum (Ker-Gawl) Garay

Local names: big centipede orchid; tiger stripes; Taiwan centipede; purple stripes

Description: Its local name is curious, seeing that this species has small yellow flowers only 7 mm across. Plant is 20 cm tall, intermediate in size among medicinal Chinese species of *Cleisostoma*. The slim, erect stem is sometimes branched and carries numerous flat, linear, oblong leaves, 10–25 by 0.8–2 cm. Sepals and petals are yellow-green with brown lines. Lip is yellow with a brownish tinge at the edge of the side lobes. Inflorescence is axillary, branching with 10–15 flowers. Flowering season is June. The species is distributed from Taiwan southwards to Fujian, southern Guangdong, Hong Kong, Guangxi, Hainan, Jiangxi, Sichuan and Xizang; also Vietnam, northeastern and central Thailand and northeast India (Vaddhanaphuti 2001).

Herbal Usage: The entire plant is used to boost *yin*, treat coughs and to strengthen the lungs in Taiwan (Lin et al. 2003).

Cleisostoma scolopendrifolium (Makino) Garay [see **Pelantheria scolopendrifolium (Makino) Aver.**]

Cleisostoma tenuifolium (L.) Garay

Common name: delicate leafed *Cleisostoma*
Old Malabarese name: *Mau Tsjerou Maravara, Ambo keli; Kolli Tsjerou Mava-maravara, Abo-tia*

Description: A rather rare, miniature epiphyte that bears a small, short cluster of brownish flowers with a pale purple lip. The brown on the tepals are overlaid on a green background. Flowers are 5 mm across (Fig. 9.17). It is distributed in southern India, Sri Lanka and Thailand, from sea level to 300 m. Flowering season is August to October in southern India and Sri Lanka (Jayaweera 1981) and December in Thailand (Vaddhanaphuti 2005). In Sri Lanka, it is sometimes found in association with three other orchid species, namely *Bulbophyllum thwaitesii* Rchb. f., *Cymbidium aloifolium* (L.) Sw., *Pholidota pallida* Lindl. and several non-orchidaceous epiphytes (Jayaweera 1981). As this is the only species of *Cleisostoma* that occurs on the Malabar Coast, the drawing of a *Cleisostoma* without flowers and labeled as *Kolli Tsjerou Mava-maravara* or *Abo-tia* in van Rheede's *Hortus Indicus Malabaricus* (1703) may be identified as *C. tenuifolium* (van Rheede 1703).

Herbal Usage: The whole plant was used in western peninsular India to treat kidney disorders, leucorrhoea, gonorrhoea and scalds (van Rheede 1703). Made into a poultice, *Kolli Tsjerou Mava-maravara* was used to reduce pain and swelling of abscesses and to promote their rupture. Plant was also blended in vinegar and administered to expel kidney stones, treat dysuria, gonorrhoea, other forms of white vaginal discharge and heavy menstrual loss (Van Rheede 1703).

Cleisostoma williamsonii (Reichb. f.) Garay

Syn. *Cleisostoma hongkongense* (Rolfe) Garay

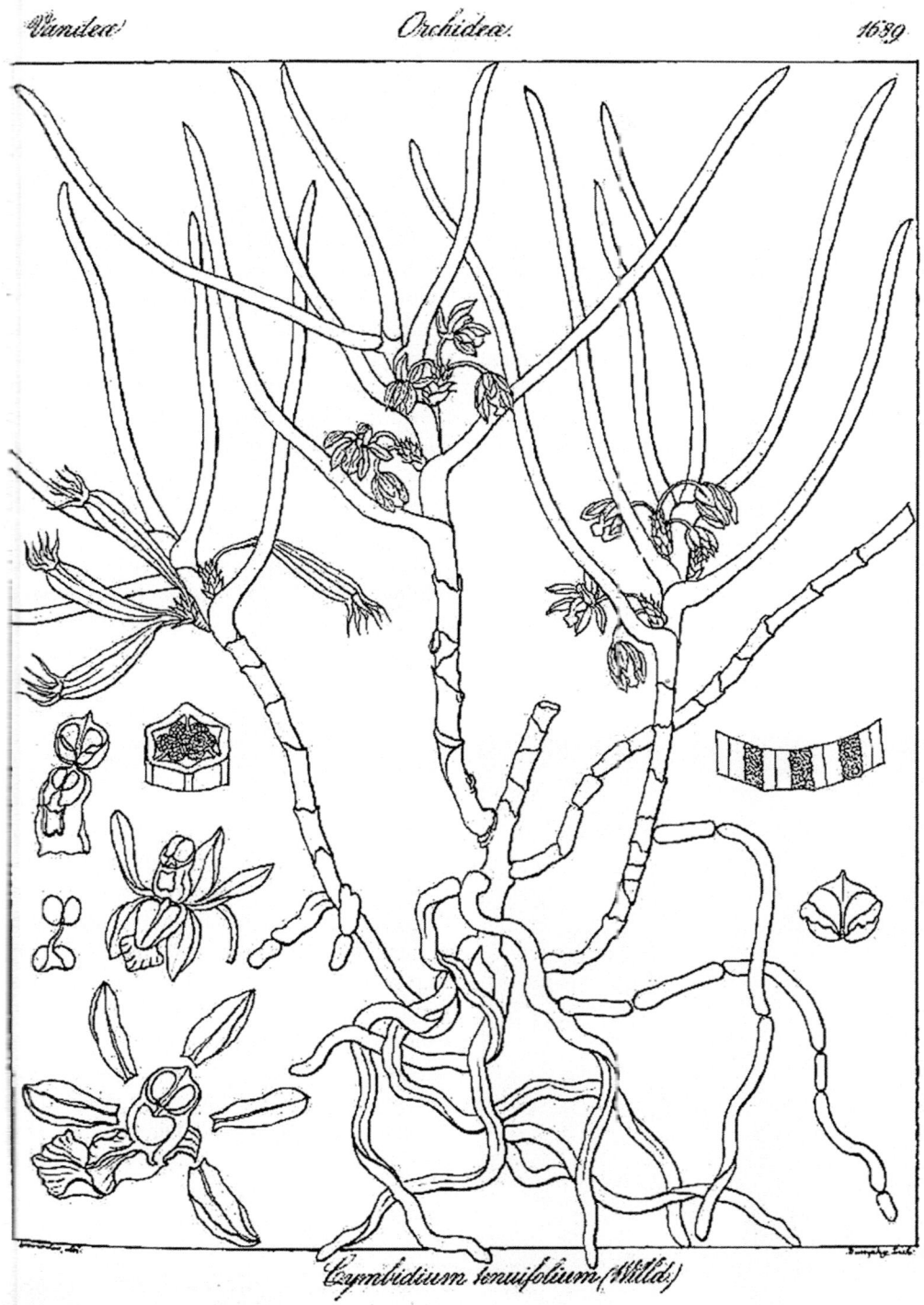

Vandea   *Orchidea.*   1689

Cymbidium tenuifolium (Willd.)

Fig. 9.17 *Cleisostoma tenuifolium* (L.) Garay. From: Wight R., *Icones Plantcrum Indiae Orientalis*, vol. 5(1): t. 1689 (1846). Drawing by Govindoo. Courtesy of Missouri Botanical Gardens, St. Louis, USA

Common Name: *Dianmiangeju Lan* (Yunnan -
    Myanmar separate distance orchid)
Chinese medicinal name: *Longjiaocao*

Description: Stems are fleshy, 30–70 cm long,
3 mm thick, with slim, terete leaves up to 18 cm
long, 2.5 cm in thickness. Inflorescence is usually
simple, with 20–25, well-spaced flowers of pale
olive green with a contrasting purple lip
(Fig. 9.18). Flowering season of *C. williamsonii*
is from April to June in China (Chen et al. 1999),
July to August in the Shan state of Myanmar
(Grant 1895) and September in Thailand
(Vaddhanaphuti 2005). Seidenfaden (1985) listed
a dozen synonyms for the species which enjoys a
distribution from Bhutan to Hong Kong and
Hainan, and southwards through Thailand, northern Peninsular Malaysia (Perlis and Langkawi) to
Sumatra, Java and Kalimantan. In China, it is
epiphytic or saxicolous in forests and along
valleys at 300–2000 m in Yunnan, Guizhou,
Guangxi, Guangdong, Hong Kong and Hainan.

Herbal Usage: The Chinese herb is obtained
from Guangdong, Guangxi and Yunnan. Plants
can be harvested throughout the year, washed
and sun-dried. Taste is mildly sweet and sour,
and it is neutral in nature. The whole plant is used
to improve blood circulation, relax muscles and
joints, clear phlegm and stop coughs. It is
usedused in treating pulmonary tuberculosis,
viral encephalitis, stroke, polio, backache and
indigestion in children (Wu 1994).

The herb may be harvested at any time during
the year and the entire plant is sun-dried for
future use. It is mildly sweet and sour in flavour,
neutral in nature. It simulates circulation, relaxes
muscles and joints, stops coughing and is an
expectorant. A Guangxi Materia Medica mentioned that it was usedused during epidemics of
encephalitis B and to treat patients with tuberculosis or paralysis resulting from stroke or polio-
myelitis, and malnourished children. A simple
decoction is prepared by boiling 9–15 g of the
herb. Alternatively, a soup may be prepared by
boiling 9–15 g of *Lionjiaocao* (*C. williamsonii*),
9 g of *Zanthoxylum bungeanum* Maxim. (Chinese prickly ash, Sichuan pepper) root bark and
lean pork. The preparation can alternatively be
turned into a stew (*Zhaoyao Bencao* 2000).

Overview

In *A Concise Edition of Medicinal Plants in
China*, Wu Xiu Ren listed *C. hongkongense* as
a separate species from *C. williamsonii* and the
two are given different Chinese names;
Honghuageju Lan (red flower separate distance
orchid) for *C. hongkongense* and *Dianmiangeju
Lan* for *C. williamsonii* with *Longjiaocao* as the
medicinal name for the latter. Botanically, however, these two species are identical. There may
be varietal differences between the two, but both
herbs are reported to be obtained from
Guangdong, Guangxi and Yunnan. Furthermore,
the fact that the 'two' *Cleisostoma* are similar in
their usage shows that Chinese herbalists do recognise that they are one and the same species.

Although a widespread genus, *Cleisostoma*
does not appear to have a medicinal use outside
China, apart from some limited usage in the
region of the Western Ghats during the early
seventeenth century (van Rheede 1703). In the

Fig. 9.18 *Cleisostoma williamsonii* (Reichb. f.) Garay
(Photo: Peter O'Byrne)

latter region, it was not mentioned again when Cains compiled his description of medicinal plants in the region of Bombay (Mumbai) in 1935. Most Chinese medicinal books make no mention of the orchid, and it is only in Wu's (1994) extensive listing that some species are listed and their usage described. There is much overlap and confusion over taxonomic identification but the different species appear to have a similar usage in a wide spectrum of unrelated illnesses.

The earliest search for medicinally active compounds focused on alkaloids (Luning 1974a, b, 1980; Slaytor 1977), those bitter compounds produced by plants, the most famous one of which is possibly quinine. Five *Cleisostoma* species were screened for alkaloids and none were found to have at least 0.1 % alkaloid content which was designated as the diagnostic criterion for classification as an alkaloid-accumulating species. However, several non-medicinal species (*C. appendiculata, C. discolor, C. racemiferum, C. subulatum*) contained small amount (0.01–0.1 % dry weight) of alkaloid (Luning 1974a, b).

Coelogiossum viride Hartm. [see **Dactylorhiza viridis (Linn.) R.M. Bateman, Pridgeon and M.W. Chase**]

Genus: *Coelogyne* Lindl.

Chinese name: *Beimu Lan* (pearl shell orchid)

The genus *Coelogyne* is constituted by more than 200 epiphytic species, of which about 80 are in cultivation. They are distributed from northern India and southern China across Southeast Asia to the Pacific, in lowland and montane forests. Introduced by John Lindley in 1822, the name is derived from Greek, *koilos* (hollow) and *gyne* (female), possibly referring to the deep stigmatic cavity. Many species bear pendulous inflorescences with flowers that open simultaneously resulting in spectacular displays, but in some species single flowers open in succession on a short inflorescence. Flowers of most species

Fig. 9.19 *Coelogyne occultata* Hook. f. From Engler, H. G.A., *Das Pflanzenreich, Orchidaceae—Monoandreae—Coelogyninae*, vol. 50 (II.B.7.): (Hef 32), p. 58, Fig. 19, 1907. Courtesy of University of Toronto Library, Canada

short-lived, usually lasting less than a week. Pseudobulbs are prominent and carry one or two plicate leaves which do not sheath the pseudobulb (Fig. 9.19).

Six Chinese species and two Thai species are used for medicinal purposes, while *C. ovalis* is used in India as *Jeevanti* (a substance which promotes life). Eight species from Nepal are medicinal (Pant and Raskoti 2013).

Table 9.2 Prescriptions employing *Feng Lan* (*Coelogyne barbata* Lindl. ex Griff.) Reference: *Zhonghua Bencao* 2000

Preparation: for decoction, use 15–30 g Feng Lan (*C. barbata*)
For external use: an appropriate amount, pulverise and apply
1. Indications: Cough with "lung heatiness"   Decoction with Plantain 30 g
2. Indication: sore throat   Decoction with Plantain 30 g, Prunella 15 g
3. Indication: Pain associated with hernia or scrotal swelling   Decoction with Plantain 30 g, Tangerine seed 15 g
4. Indication: Bruises and Sprains   Prepare paste with *C. barbata* and Plantain, and apply fresh to wounded part

5. Indication: Chapped hands and feet.
     Prepare paste with equal amounts of *C. barbata* and Plantain and apply to wounds.
     (Source: Xizang Chinese Materia Medica)

Coelogyne barbata Lindl. ex Griff.

Local name: *Xuchunbeimu Lan* (beard and lip pearl shell orchid), *Ranmaobeimu Lan*
Chinese medicinal name: *Fengian*

Description: Plants are epiphytic or lithophytic. Pseudobulbs are clustered, pale green, almost round, up to 10 cm in diameter, and carry 2 leathery, narrowly lanceolate, stalked leaves; up to 45 by 6 cm. Inflorescence is arching with few, crowded, white, musk-scented flowers, 5–7.5 cm across. Lip is white and bears three deep sepia-brown, fringed crests at the centre: it is fringed around the distal third with similarly coloured projections. Flowering season is autumn to winter. The species is distributed in Nepal, Bhutan, the Khasia Hills in Bangladesh, Myanmar and Gaoligongshan in Yunnan (Hawkes 1965; Jin et al. 2009).

Herbal Usage: Herb obtained from Sichuan and Yunnan may be harvested without regard to season. After collection, plant is washed and dried, then further sun-dried for storage. It is sweet in taste and cool in nature. In TCM, the whole plant is valued for its ability to counter 'heat', relieve thirst, and stop coughs and lessen pain. It is used to treat sore throat, pain at hernias, swelling of the scrotum, chappy extremities, traumatic injuries and

'lung-heat' (Wu 1994; *Zhonghua Bencao* 2000). Some prescriptions on the usage of the herb are shown in Table 9.2 (*Zhonghua Bencao* 2000).

Coelogyne corymbosa Lindl.

Pleione corymbosa (Lindl) Kuntze

Chinese name: *Yanbanbeimu Lan* (eye spotted pearl shell orchid), *Beimu Lan* (pearl shell orchid), *Zhixueguo* (haemostatic fruit); *Shibajiao* (stone palm leaves); *Duiyeguo* (fruit with a pair of leaves); *Xiaoluji* (small green Chinese elder)
Chinese medicinal name: *Beimu Lan* (pearl shell orchid); *Guoshangye* (leaves above the fruit)
Newari name: *Tuyu kenbu swan*

Description: *C. corymbosa* is a small, pretty, epiphytic or terrestrial orchid with clustered pseudobulbs, 1–4.5 cm in length, each with a pair of oblong, coriaceous leaves of 4.5–15 cm length. Raceme is curved or drooping and bears two to four white or greenish-tinged flowers with a distinctively decorated lip. Tepals are lanceolate, concave and pointed. There are four large yellow patches (or "eyes") bordered with orange-red on the upper half of the pointed lip, a feature that earned the orchid its local name (Fig. 9.20). In China it flowers from May to July (Chen et al. 1999). In Bhutan, it flowers from February to June (Gurong 2006). Flowering season is shorter in Nepal, March to May (Raskoti 2009).

C. corymbosa is distributed from Yunnan to southern Tibet into Myanmar, Sikkim, Bhutan, Nepal and northern India. It is epiphytic on the tree trunks at the edge of forests and wet cliffs at 1300–3100 m.

Herbal Usage: Herb is obtained from Yunnan and Xizang. It may be collected at any time of the year. Pseudobulbs or entire plant are used to treat fractures and soft tissue injuries. The herb is usedused as a haemostatic and to relieve pain. It reduces heat, stops coughs, and is taken for coughs, flu and bronchitis. Four prescriptions are reproduced in Table 9.3 (*Zhonghua Da Cidian* 1986; *Zhonghua Bencao* 2000; Wu 1994).

Fig. 9.20 *Coelogyne corymbosa* Lindl. (Photo: E.S. Teoh)

Table 9.3 Four prescriptions employing *Coelogyne corymbosa* (*Zhongyao Da Cidian* 1986; *Zhonghua Bencao* 2000)

1. Indication: bronchitis, flu
   Use entire plant 15–30 g in decoction
      (Source: Yunnan Selected Chinese Herbs)
2. Indication: soft tissue injuries
   Use sheaths from base of pseudobulbs. Apply as powder or paste externally.
      (Source: Yunnan Selected Chinese Herbs)
3. Indication: fractures
   Grind *Coelogyne corymbosa* 100 g with Pteris multifidapoir 1 g
   Apply to site of fracture after reduction and splinting.
   Then add Man Shan Xiang powder to wound directly and apply another layer of the mixture.
   Change medicine daily or on alternate days
      (Source: Quan Zhan Selected Chapters)
4. Indication: bleeding from external wounds
   Apply powdered, or a paste of grounded, fresh *Coelogyme corymbosa* to the wound
      (Source: Wen Shan Chinese Herbs)

In India and Nepal, paste made with pseudobulbs is applied on the forehead to relieve headaches (Manandhar and Manandhar 2002; Baral and Kurmi 2006; Pant and Raskoti 2013). Juice of pseudobulbs is applied on wounds for pain relief and to treat burns (Das 2004; Baral and Kurmi 2006; Pant and Raskoti 2013).

Fig. 9.21 *Coelogyne cristata* Lindl. (Photo: Bhaktar B. Raskoti)

Coelogyne cristata Lindl.

Coelogyne speciosissmum D. Don

Chinese name: *Beimu Lan* (pearl shell orchid). Note that this name does not distinguish it from the preceding species.
Indian name: *Hadjojen* (bone joiner)
Nepali names: *ban maiser, jhyanpate* in Chepang dialect; *chandi gabha* (Nepali), *syabal* (Tamang)

Description: *C. cristata* is an epiphytic or saxicolous orchid. Pseudobulbs are oblong, 2.5–4 by 1–1.7 cm, spaced 1.5–3 cm apart, and each carries two sessile, lanceolate leaves. Leaves are linear-lanceolate, 10–17 by 0.7–1.9 cm. Inflorescence is 8–12 cm long with 2–10 flowers. Flowers are white, large and fragrant, sepals and petals with undulating borders that fold backwards in parts (Fig. 9.21). Flowering period is February and March in India, May in China.

Distributed throughout Nepal, Bhutan and northern India at 1000–2000 m (Manandhar and Manandhar 2002), Bangladesh and Myanmar, it is found on large rocks in southern Tibet at 1700–1800 m (Chen and Clayton 2009).

Phytochemistry: Ethanolic extract of *C. cristata* is strongly bacteriostatic against *Staphylococcus aureus* and moderately against *Escherichia coli* (Marasini and Joshi 2012), a property which supports its usage in animal husbandry. Coelogin and coeloginin, and two novel 9,10-dihydrophenanthrene derivatives, coeloginanthridin and coeloginanthrin, were isolated from air-dried, finely-ground whole plant of *C. cristata* by Majumder's group (Majumder et al. 1982a, 2001). The four compounds possess the biological activities of phytoalexins and endogenous plant growth regulators. Sensitivity testing of soil and other bacteria to individual phytoalexins should be performed to determine which compounds, if any, could be reasonable remedies for superficial infections, but such data are not available.

Ethanolic extract of *C. cristata* also restored trabecular bone without producing uterine changes when fed to mice rendered oestrogen-deficient by oophorectomy. Coelogin promoted surrogate markers of osteoblastic differentiation and activity in vitro (elevated alkaline phosphatase; increased calcium nodule formation). Together, they support the notion that the folk tradition of using *C. cristata* to treat fractured bones in the Kumaon region of Uttarakhand may have a rational basis. Perhaps there might be a role for compounds present in *C. critsta* for managing post-menopausal osteoporosis (Sharma et al. 2014).

Herbal Usage: *Hadjogen* (Indian, bone joiner; *C. cristata*) is used in the Himalaya to treat fractured bones in animals (Jaiswal et al. 2004). It is used to treat dysentery and diarrhoea in Myanmar (Naing et al. 2010) whereas in Nepal an infusion of pseudobulbs is used to correct constipation. Nepalese also use this orchid as an aphrodisiac (Pant and Raskoti 2013). Fresh juice or paste made with *Coelogune cristata* is consumed to relieve headache, fever and indigestion (Subedi et al. 2013). Juice squeezed from the

pseudobulbs is applied to boils and to wounds on the hooves of animals (Manandhar and Manandhar 2002).

Coelogyne elata Lindl. [see **Coelogyne stricta (D. Don) Schltr.**]

Coelogyne fimbriata Lindl.

Chinese name: *Liusubeimu Lan* (tassels pearl shell orchid)
Myanmar name: *Ngwe hnin phyu myo kywe*

Description: Epiphytic on trees growing in the edge of forests, or lithophytic in shady spots in ravines, at 500–2300 m in China, *C. fimbriata* is a single-flowered or sometimes two-flowered *Coelogyne* with pale yellow flowers. Pseudobulbs are ovoid-ellipsoid, 2.5–6 by 0.8–1.5 cm with 2 leaves at the apex. Leaves are oblong-elliptic, 6–14 by 1.2–2.4 cm. Flowers are 3 cm across, yellow with a distinctive lip. Lip is large, with fimbriate margins, and heavily marked with brown (Fig. 9.22). Some racemes bear two flowers which open in succession. It flowers from March to October in Hong Kong and August to October on the mainland (Chen and Clayton 2009); in Bhutan, flowering season

Fig. 9.22 *Coelogyne fimbriata* Lindl. (Photo: E.S. Teoh)

is June to November (Gurong 2006). The species is distributed across southern China from Hainan, southern Jiangxi and Guangdong, Hong Kong, across Yunnan and Xizang to India, and south-wards through Indochina and Thailand to Malaysia, Sumatra and Kalimantan.

Herbal Usage: Herb is obtained from Hainan, Guangdong, Yunnan and Xizang. In Chinese herbal medicine, the whole plant is used to reduce heat (Wu 1994).

Coelogyne flaccida Lindl.

Chinese names: *Lilinbeimu Lan* (chestnut scales pearl shell orchid), *Guishangye* (the leaf above fruits)
Chinese medicinal name: *Jidatui*
Nepali name: *Thur gava*

Description: An impressive, epiphytic or saxicolous *Coelogyne*, plants in bloom carry a long spray of pendulous, pale, straw-coloured flowers. Pseudobulbs are conical to ovoid-cylindrical, spaced 2–3 cm apart, 6–12 cm long and 1.5–3 m broad bearing two coriaceous, lanceolate leaves, 13–19 by 3–4.5 cm. Inflorescences arise from young, leafless pseudobulbs reaching a length of 20 cm and bear 8–12 loosely arranged, pale-coloured flowers with yellow to brownish markings on the lip. Flowers are 3–5 cm across (Fig. 9.23). It flowers in March in China (Chen et al. 1999); April to May in South India (Abraham and Vatsala 1981); April in Nepal (Raskoti 2009); February to May in Bhutan (Gurong 2006). *C. flaccida* occurs at 1600 m in Guangxi, Guizhou, Yunnan, Laos, Myanmar, Sikkim and Nepal.

Phytochemistry: Phenanthrenes and stilbenoids have been isolated from this species: flaccidin in 1988 (Majumder and Maiti 1988), and soflaccidin and isooxoflaccidin in 1991 (Majumbder and Maiti 1991). Callosin originally isolated from the orchid, *Agrostophyllum callosum,* was discovered in *C. flaccida* in 1995 (Majumder et al. 1995; Kovacs et al. 2007).

Herbal Usage: Herb is collected from Guizhou and Yunnan. Known as *Guoshangye,*

Fig. 9.23 *Coelogyne flaccida* (Photo: E.S. Teoh)

C. flaccida is a popular medicine among the minority tribes in both provinces. In China, the whole plant is used to clear heat, counter dryness, promote the production of body fluids, and to clear phlegm and stop coughs (Wu 1994; *Zhonghua Benco* 2000). Pseudobulbs are made into a paste in Nepal and applied to the forehead to treat headache, while the juice treats indigestion (Manandhar and Manandhar 2002). The paste is also used for boils (Baral and kurmi 2006).

Coelogyne flavida Hook. f. (see ***Coelogyne prolifera* Lindl.**)

Coelogyne fuscescens Lindl.

Chinese name: *Hechunbeimu Lan*
Thai names: *Sing to, phaya rat, phao hin*

Description: Pseudobulbs are clustered, narrowly sub-oblong, 2–3 cm long and 5–7 mm in diameter, with two elliptic leaves, 12 by 1.5–2 cm (Chen and Clayton 2009). In Bhutan, *C. fuscescens* Lindl. var. *fuscescens* is much larger. Pseudobulbs are 8–14.5 by 1–3.2 cm (Pearce and Cribb 2002; Gurong 2006). Inflorescence carries only one or two pale yellow flowers which are 5 cm across, with golden brown edges at the sidelobes of the lip and three golden brown keels on the mid-lobe. It flowers in December in Thailand (Vaddhanaphuti 2001), October to November in Nepal (Raskoti 2009), June in Yunnan (Chen and Clayton 2009). This epiphytic or saxicolous species is found in northern and northeastern Thailand, Laos, Vietnam, Myanmar, southern Yunnan Sikkim, Bhutan and Nepal at 1300 m. It is lithophytic in Yunnan.

Herbal Usage: The whole plant is a Thai aphrodisiac. When the plant was shown to ten Thai herbalists, three stated that it was an aphrodisiac, but seven did not. Stems are also used to treat burns and otitis media (Chuakul 2002). In Nepal, abdominal pain is treated with juice extracted from pseudobulbs or a poultice made with it (Baral and Kurmi 2006; Pant and Raskoti 2013).

Coelogyne leucantha W.W. Sm.

Chinese name: *Baihuabeimu Lan* (white flower pearl shell orchid)

Description: Pseudobulbs are ovoid-oblong, 1.5–5 cm long and 8–15 mm in diameter, spaced 1–2 cm apart. Two leaves at the apex are oblong-lanceolate, 5–15 by 1.1–3 cm wide, with a long, narrow petiole. Inflorescence is apical, on matured pseudobulbs, erect, 15–20 cm tall, with 3–11 slightly droopy flowers on raceme. Flowers are 3–5 cm across, not fully opened, white with yellow blotch on the lip. Petals are filiform (Chen et al. 1999). Flowering season is May to July in China (Chen and Clayton 2009).

C. leucantha is epiphytic or saxicolous in broad-leaved evergreen forests below 2500 m in south and northwest Yunnan, southwest Sichuan, Myanmar and Vietnam. In the Gaoligongshan area, it is found on so many trees that, from May to June when the *Coelogyne* is in bloom, the trees are beautifully garlanded with their white flowers (Yang et al. 1998).

Herbal Usage: Herb is obtained from Yunnan. Pseudobulbs and sometimes the entire plant is used to lessen heat, stop coughs, improve blood flow, reduce pain, promote the union of fractured bones and repair torn tendons (Wu 1994).

Coelogyne nitida (Wall ex D. Don) Lindl.

Syn. *Coelogyne ochracea* Lindl.

Chinese name: *Mijingbeimu Lan*
Nepali names: Silver Orchid in English, *bhyan pat* (Chepang), *Salida, Sanit* (Gurung), *Chandi gabha, para phul* (Nepali)

Description: An small epiphytic orchid with oblong pseudobulbs, 1.5–3 cm long and 1–1.5 cm in diameter, carrying a pair of lanceolate, leathery leaves, 7 by 1.5 cm, and white to pale yellow flowers with a dark brown centre on lax racemes (Fig. 9.24). It flowers in March in China (Chen and Clayton 2009); May in northern Thailand (Vaddhanaphuti 2005); January to June in Bhutan (Gurong 2006); April to June in Nepal (Raskoti 2009). It is found in northern India, Bangladesh, Nepal, Bhutan and Myanmar, the adjacent part of Yunnan at 1300–2400 m, Indochina and Thailand.

Phytochemistry: Ochrolide, a phenanthropyrone, and Ochrone A, a novel 9.10-dihydro-1,4-phenanthraquinone together with coelonin are present in *C. ochracea* (= *C. nitida*) (Bhaskar et al. 1989, 1991). Ochrolic, a monomeric phenanthrene derivative and a

Fig. 9.24 *Coelogyne nitida* (Wall ex D. Don) Lindl. (Photo: E.S. Teoh)

precursor to phenanthropyrones, was isolated from *C. nitida* (Anuradha et al. 1994).

Herbal Usage: Juice of the pseudobulb is recommended for stomach ache in Nepal (Manandhar and Manandhar 2002; Baral and Kurmi 2006).

Coelogyne occultata Hook. f.

Pleione occulta (Hook f.) Kuntze

Chinese name: *Luanyebeimu Lan* (ovate leaf pearl shell orchid)
Chinese medicinal names: *Luanyebeimu Lan* (ovate leaf pearl shell orchid); *Youguashihu* (squashed epiphyte)

Description: Plant has small pseudobulbs 1.5–5 cm by 0.5–1.5 cm with two ovate, coriaceous leaves at the apex, 1.5–6 by 1–2.5 cm. Flowers are relatively large, 5–6 cm across; white with two large yellow eyes connected by a transverse band on the proximal half of the lip whose side lobes are streaked with brown veins. It grows on tree trunks or cliffs at 1300–3000 m in Yunnan, Tibet, Myanmar, Sikkim and Bhutan. Flowering season is June to August (Chen et al. 1999; Chen and Clayton 2009).

Usage: Herb is obtained from Yunnan and Xizang. It nourishes *yin*, protects the kidney, nourishes the stomach and promotes the production of body fluids. Plant is used to treat hot flushes, fever, nocturnal emission, backache, anorexia and gastritis (Wu 1994; *Zhonghua Bencao* 2000). Decoction is prepared with 6–9 g of the herb for consumption to promote *yin*, relieve thirst and dry throat, or to treat tuberculosis, night sweats, chronic gastritis, lack of gastric acid, anorexia, nocturnal emission, waist pain, fatigue and haemorrhoids (*Zhongyao Da Cidian* 1986). It helps digestion (Chen and Tang 1982).

Coelogyne ochracea Lindl. [see **Coelogyne nitida (Wall ex D. Don) Lindl.**]

Coelogyne ovalis Lindl.

Chinese name: *Changlinbeimu Lan*
Indian name: *Jeevanti*

Description: *C. ovalis* is a few-flowered, epiphytic or saxicolous species. Psuedobulbs are fusiform, 4–9 by 1.5–2 cm. Leaves are elliptic-oblong, 9–17 by 2.5–4 cm. Inflorescence arises from apex of pseudobulb, 1–5 flowered. Flowers are 3–4 cm across, with linear petals, pale buff brown sepals and a lip that is marked with dark brown and is hirsute at the edges (Fig. 9.25). It flowers from September to November in Yunnan (Yang et al. 1998) and January in Thailand (Vaddhanaphuti 2005); October to December in South India (Abraham and Vatsala 1981), April to July in Bhutan (Gurong 2006) and September to December in Nepal (Raskoti 2009). *C. ovalis* is distributed across the western Himalaya to Tibet, Yunnan, Myanmar, Thailand and Vietnam at 1750–2000 m, and in Nilgiris and Mysore in South India.

Phytochemistry: 2,7-dihydroxy-3,4,6-trimethoxy-9,10-dihydrophenanthrene, coelogin, coeloginin, favidin, flavidinin, batatasin III, imbricatin, beta-sitosterol and its glycoside and a new bibenzyl compound, 3′-*o*-methylbatatasin III, are present in *C. ovalis* (Majumder and laha 1981; Majumder et al. 1982c; Majumder 1984; Sachdev and Kulshreshtha 1986). An alcoholic extract of the orchid pseudobulbs which contained flavidin and coelogin showed spasmolytic activity. Flavidin produced 50 % and 90 % inhibition of barium chloride-induced spasm of the guinea pig ileum at 1.0 and 2.0 mcg/ml doses. Coelogin showed 50 and 51 % activity at 0.5 and 1.0 mcg/ml (Sachdev and Kulshreshtha 1986).

Herbal Usage: It is known as *Jeevanti* which means 'promoting life', and in this respect it is usedused as a tonic. In Nepal, pseudobulbs are regarded as aphrodisiacs (Pant and Raskoti 2013), hence the name *Jeevanti*. However, *Jeevanti* may also refer to other popular "aphrodisiac" orchids, *Flickingeria fugax* (= *Dendrobium*

Fig. 9.25 *Coelogyne ovalis* Lindl. Reproduced with permission from *Introductions to Orchids* by Abraham and Vatsala, Parlode, Thiruvananthapuram: Tropical Botanic Garden and Research Centre (TBGRI), 1981

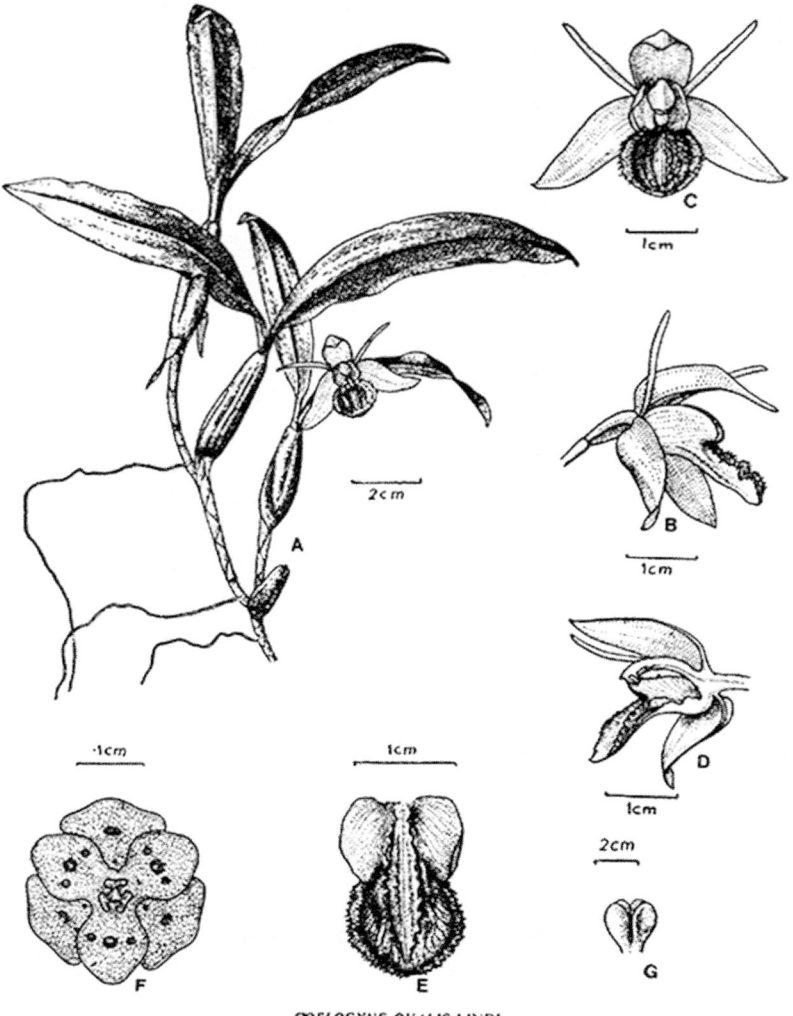

COELOGYNE OVALIS LINDL

fugax) and *Flickingeria fimbriata* (= *Dendrobium plicatile*) in eastern India, or to *Leptadenia reticulata* in northern India and *Holostemma annulare* in southern India, these last two being plants that are not orchids (Sarin 1995). Baral and Kurmi (2006) reported that paste made with pseudobulbs is used as aphrodisiac!

The entire plant of *C. ovalis* is also usedused in western and southern India to treat coughs, urine infections and eye disorders (Rao 2004).

Coelogyne prolifera Lindl.

Syn. *Coelogyne flavida* Hook. f.

Chinese name: *Huanglubeimu Lan*
Nepali names: *Liso* in Gurong dialect, *Thurgaujo* (Nepali)

Description: An epiphytic orchid, its pseudobulbs are spaced 2.5–4 cm apart and measure 2.2.–3.7 long and 1 cm in diameter. Leaves are paired, oblong to lanceolate, 8–13 by 1.6–2.1 cm, with a 2–2.5 cm long petiole. Inflorescence is borne apically on leafy pseudobulbs, 10–15 cm long with 4–6 greenish-white to yellow green flowers, 1 cm in diameter. Sepals are oblong, petals linear. Flowering period is June in China (Chen et al. 1999; Jin et al. 2009), January in Thailand (Vaddhanaphuti 2005), April to June in Nepal (Raskoti 2009), March to July in Bhutan (Pearce and Cribb 2002; Gurong 2006) and May to June in South India (Abraham and Vatsala 1981). *C. prolifera* is found on rocks and trees at 1100–2200 m in west and southern Yunnan, Myanmar, Thailand, Laos, Bangladesh, Sikkim, Bhutan and Nepal.

Phytochemistry: Flavidin, a novel 9,10-dihydrophenanthrene derivative, has been isolated from *C. flavida* Hook. f. (syn. *C. prolifera* Lindl.). This phytoalexin also occurs in two other Himalayan orchids, namely *Pholidota articulata* and *Otochilus fuscus* (Majumder et al. 1982b). Four related compounds, flavidinin and oxoflavidinin (Majumder and Datta 1982), a 9,10-dihydrophenanthropyran named flaccidin (Majumder and Maiti 1988) and imbricatin are also present in the orchid. The last compound had earlier been isolated from *Pholidota imbricata*: hence its name (Majumder and Sarkar 1982).

Herbal Usage: Paste made with pseudobulbs of *C. prolifera* is rubbed on the back to relieve backache (Manandhar and Manandhar 2002) and to treat boils in Nepal (Pant and Raskoti 2013).

Coelogyne punctulata Lindl.

Description: Psedobulbs are contiguous on a stout, rigid rhizome. They are oblong, 2.5–4 by 0.7–1.3 cm, bright yellow when dried. There are 2 leaves at the apex and papery sheaths at the base. Leaf is lanceolate, 8–14 by 1.3–2.5 cm, petiolated. Inflorescence is 8–15 cm long, carrying 2–4 white flowers 4 cm across. Lip is trilobed and bears a central white keel. A bright yellow patch outlined with a thin, orange rim is present on either side of the keel and on the medial aspect of the side lobes. Flowering season is November in China (Chen and Clayton 2009). *C. punctulata* is distributed from central Himalaya to Bangladesh, Myanmar, Thailand, southeastern China (SE Xizang and West Yunnan), Thailand and Vietnam. It is epiphytic or lithophytic in forests at 100–2900 m.

Usage: Pseudobulbs are dried and made into powder for use to treat wounds and burns in northern India (Das 2004). It is usedused to treat dry coughs and bleeding resulting from trauma in Vietnam (Hung 2014).

Coelogyne stricta (D. Don) Schltr.

Coelogyne elata Lindl.

Indian name: *Harjojan*

Description: *C. elata* is an epiphyte with cylindrical to narrowly ovoid pseudobulbs, 15 by 2.5–6.5 cm, carrying 2 lanceolate, leathery leaves marked by prominent veins, 18–30 by 4–7 cm. Inflorescence is erect, up to 60 cm long with 4–10 flowers. Flowers are fragrant, 2.5–6 cm across, white, with a forked yellow central band on the lip. It flowers in March to June in Bhutan (Pearce and Cribb 2002; Gurong 2006), rarely also in October to November (Gurong 2006) and in Nepal, April to June (Raskoti 2009). *C. stricta* is found between 1100 to 2000 m in Yunnan, Myanmar and Indochina (Chen and Clayton 2009), 1400 and 2000 m in Nepal (Raskoti) and over a wide range of elevations, from 500–3300 m, in Bhutan (Pearce and Cribb 2002; Gurong 2006).

Phytochemistry: A 9,10-dihydrophenanthropyrone was isolated from *C. elata* (correct name: *C. stricta*) and *C. nitida* (Majumder et al. 1982c), and a 9,10-dihydrophenanthrene named coelonin was obtained from *C. elata* (= *C. stricta*) (Majumder and Datta 1984). The phytoalexins exhibit bacterostatic and fungistatic activities (Marasini and Joshi 2012).

Usage: In northeast India, it is used to promote healing of bones and is applied externally to fractured limbs (Trivedi, Dixit and Lal 1980). Poultice made with pseudobulbs is applied to relieve headache and fever in Nepal (Baral and Kurmi 2006; Pant and Raskoti 2013).

Coelogyne trinervis Lindl.

Thai name: *Ueang mak*

Description: Pseudobulbs are ovoid, 9 cm long, yellow green, with two narrow leaves, 40 by 3.5 cm that taper to a stalk towards the base. There are 5–6 white to creamy, fragrant flowers, 4–6 cm across, on the inflorescence. Tepals are narrow, 2.2 cm by 2.5 mm. Lip is marked with brown lines and three keels on the mid-lobe. Flowering period is November (Seidenfaden and Wood 1992; Vaddhanaphuti 2001). *C. trinervis* is found throughout Thailand, in Assam, Myanmar, Indochina, Peninsular Malaysia, Java and Maluku at 700–1000 m (Handoyo 2010).

Herbal Usage: In Thailand, the tuber is used to treat fractures and sprains (Chuakul 2002).

Overview

Coelogyne is a huge genus with a wide distribution, and it is surprising that only the few species in China, Nepal and Thailand, at the periphery of its distribution, should find medicinal usage, whereas in Malesia, where the genus has the most number of species, there is not much medicinal application. *C. asperata* is sacred in some parts of Indonesian Borneo (Kalimantan) and here it was believed that the abundance of the rice harvest could be predicted by seasonal profusion of its flowers (Lawler 1986).

C. cristata is used to treat dysentery and diarrhoea in Myanmar whereas infusion of its pseudobulbs is used to treat constipation or indigestion in Nepal (Pant and Raskoti 2013; Subedi et al. 2013). This apparent paradox suggests that the pseudobulbs of the orchid may contain heat-labile and heat-stable compounds with opposing actions, as is the well-known case with *Angelica sinensis*. *C. cristata* contains coelogin (a 9,10-dihydrophenanthrene derivative) which has been shown to reduce intestinal spasms (Sachdev and Kulshreshtha 1986). The observation supports the Burmese herbal usage of the orchid.

C. cristata is sometimes usedused to treat headache and fever in Nepal (Subedi et al. 2013). Three species of *Coelogyne* are regarded as aphrodisiacs: *C. cristata* and *C. ovalis* in Nepal (Pant and Raskoti 2013) and *C. fuscescens* Lindl. var. *brunnea* Lindl. in Thailand (Chuakul 2002). This usage is probably not widespread because the three species are still not endangered.

Majumder's group in India have been most active in the investigating the chemical constituents of *Coelogyne*. In 1982, they reported the isolation of coelonin and coeloginin, two 9,10-dihydrophenanthrenes from *C. ochracea* (= *C. nitida*) and *C. elata* [= *C. stricta* (D. Don) Schltr.] known in Chinese as *Shuangzhebeimu Lan*] (Majumder et al. 1982c). *C. nitida* had not been reported as a medicinal plant at that time, and the isolation of coelonin in both *C. nitida* and *C. stricta* demonstrates the value of examining many species when searching for pharmacologically active compounds within a genus. This is standard practice.

Uniflorin, a steroidal ester, was isolated from *C. uniflora* [= *Panisea uniflora* (Lindl.) Lindl], another Himalayan orchid without medicinal usage (Majumder and Pal 1985, 1990) Later, his group found four phenanthrene derivatives, coelogin, coeloginin, coeloginanthridin and coeloginanthrin, in *C. cristata* (Majumder et al. 2001). The last four compounds possess the biological activities of phytoalexins and endogenous plant growth regulators (Fig. 9.26).

Micropropagation of *C. cristata* with the intent of conserving this medicinal plant for Myanmar was achieved through the assistance of a scientific team led by Ki Byung Lim in Korea (Naing et al. 2010).

Fig. 9.26 Phenanthrenes and stilbenoids from *Coelogyne*

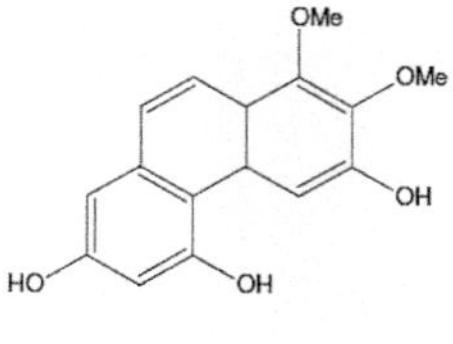

Coeloginanthrin

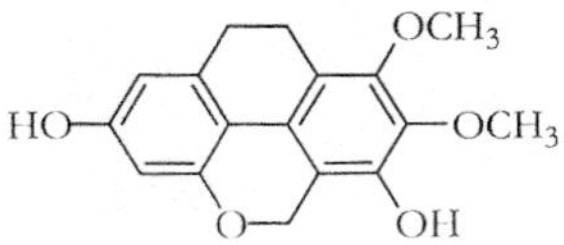

coelogin

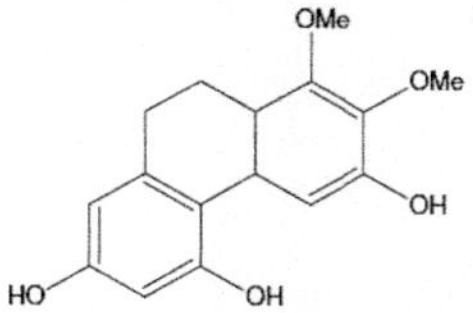

Coeloginanthridin

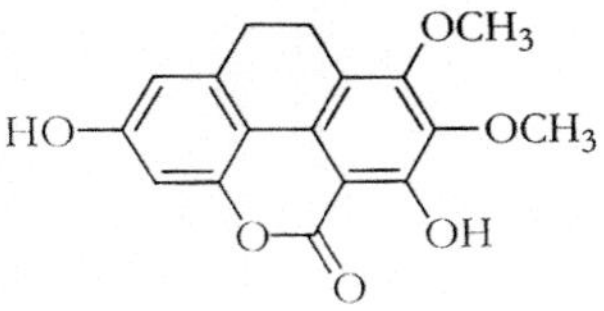

coeloginin

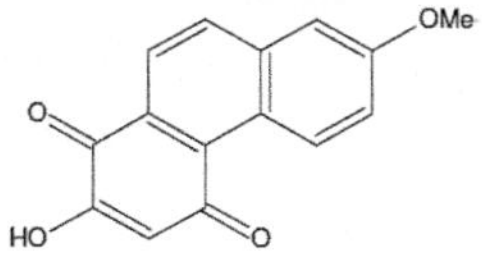

Ochrone A

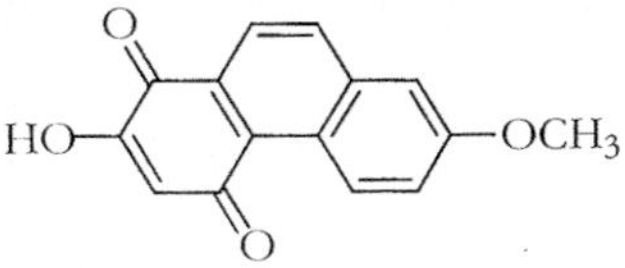

ochrone B

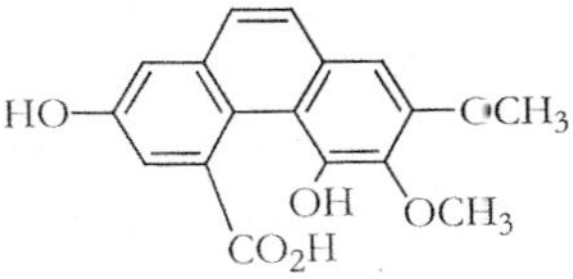

ochrolic acid

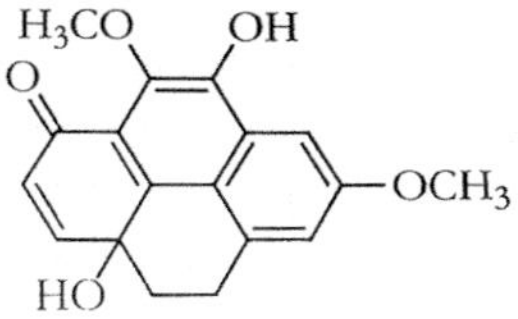

ochrolone

Genus: *Conchidium* Griff.

Conchidium are dwarfed, epiphytic herbs that were until fairly recently listed under *Eria*. There are ten species distributed from the Himalayas to southern China, Japan, and continental Southeast Asia excluding Peninsular Malaysia. Plants are epiphytic or saxicolous, often forming mats on tree trunks or rocks (caespitose). Pseudobulbs have a single internode and bear 1–4 obvate, lanceolate leaves at the apex. Inflorescence is terminal with a few or a single white, pale green or yellow flower.

Conchidium muscicola (Lindl.) Rausch.]

Syn. *Eria muscicola* (Lindl.) Lindl.

Sanskrit Name: *Jivanti*

Description: A dwarf epiphyte with oval, flattened pseudobulbs, 1 by 0.5 cm, crowded together on a creeping rootstalk. It carries 2–3, oval to lanceolate, dark green leaves on the side, never at the top. Leaves are petioled and measure 0.4–2 by 0.3–0.7 cm. A zig-zag rachis carries 3–6 tiny, greenish-white flowers that are 2–3 mm in diameter (Fig. 9.27). Flowering period is July to

Fig. 9.27 *Conchidium muscicola* (Lindl.) Rausch. (Photo: Bhaktar B. Raskoti)

November with maximum flowering in November. Common near watercourses up to 2000 m in Sri Lanka, its distribution extends northward into east and northeast India, Nepal, Bhutan and Myanmar (Jayaweera 1981; Karthikeyan et al. 1989; Gurong 2006), the Andaman Islands, Thailand, Laos and Vietnam.

Herbal Usage: Lawler (1984) reported that the Sanskrit name for this orchid is *Jivanti*, a name that is more commonly applied to *Flickingeria fimbriata* (*Dendrobium plicatile*). The Sanskrit word *Jiva* means 'life' and the term *Jivanti* is used for many herbs which are considered to be powerful tonics possessing rejuvenating and life-prolonging properties. They also act as aphrodisiacs. For comparison, some members of the contemporary medical fraternity assign a similar role to testosterone. Apart from this usage, the pseudobulbs of *Eria muscicola* are usedused in India to treat diseases of the heart and lungs, disorders of the nervous system, eye, ear and skin, facial tumours, fever and rabies (Hoernle, quoted by Lawler 1984). Usage in Nepal is fairly similar: it is usedused to treat heart, lungs and psychiatric disorders (Baral and Kurmi 2006).

Genus: *Corymborkis* Thouars.

Chinese name: *Guanhua Lan*

The generic name is derived from the clustering of the flowers which appear like a ready-made bouquet – Greek *corymbos* (cluster of flowers) and *orchis* (orchid). Flowers are short-lived. *Corymborkis* are tall, sympodial orchids, usually exceeding a metre, with short rhizomes and stout, broad-leafed stems. Leaves are plicate, glabrous, persistent, sheathing at the base. Flowers are numerous, medium-sized, arranged in a panicle on a lateral inflorescence, and all species have white flowers (Fig. 9.28). The five species in this genus are widely distributed in the lowlands throughout the tropics. They are rarely cultivated because their flowers are ephemeral.

Corymborkis veratrifolia (Reinw.) Blume

Chinese name: *Guanhua Lan*
Malaysian name: *Kayu Hok* in aboriginal Semang
Description: *C. veratrifolia* is a beautiful, shade-loving, tall, terrestrial orchid. Stems are erect, unbranched, 60–300 cm tall, sheathed by large, eliptic, lanceolate, plicate, dark green leaves up to 45 by 15 cm. Inflorescenes are axillary, numerous, branching, up to 17 cm long, and many-flowered. Flowers are white, 5 cm long and 3 cm across, facing all directions (Fig. 9.29). They do not open widely. The species occurs in the lowlands and up to 1000–1300 m, the mountain variety bearing smaller flowers (Seidenfaden and Wood 1992). It is distributed from the Himalayas to Myanmar (Andaman Islands), southern China (southern Yunnan, southwest Guangxi and Taiwan), the Ryukyu Islands, Thailand, Malaysia, Indonesia, northern Australia and the Pacific Islands (Seidenfaden and Wood 1992; Comber 2001: Chen et al. 1999). It has also been found in the Western Ghats but is rare there, and in Sri Lanka (Jayaweera 1981).

Fig. 3.28 *Corymborchis veratrifolia* Blume. Reproduced with permission from *Introductions to Orchids* by Abraham and Vatsala, Parlode, Thiruvananthapuram: Tropical Botanic Garden and Research Centre (TBGRI), 1981

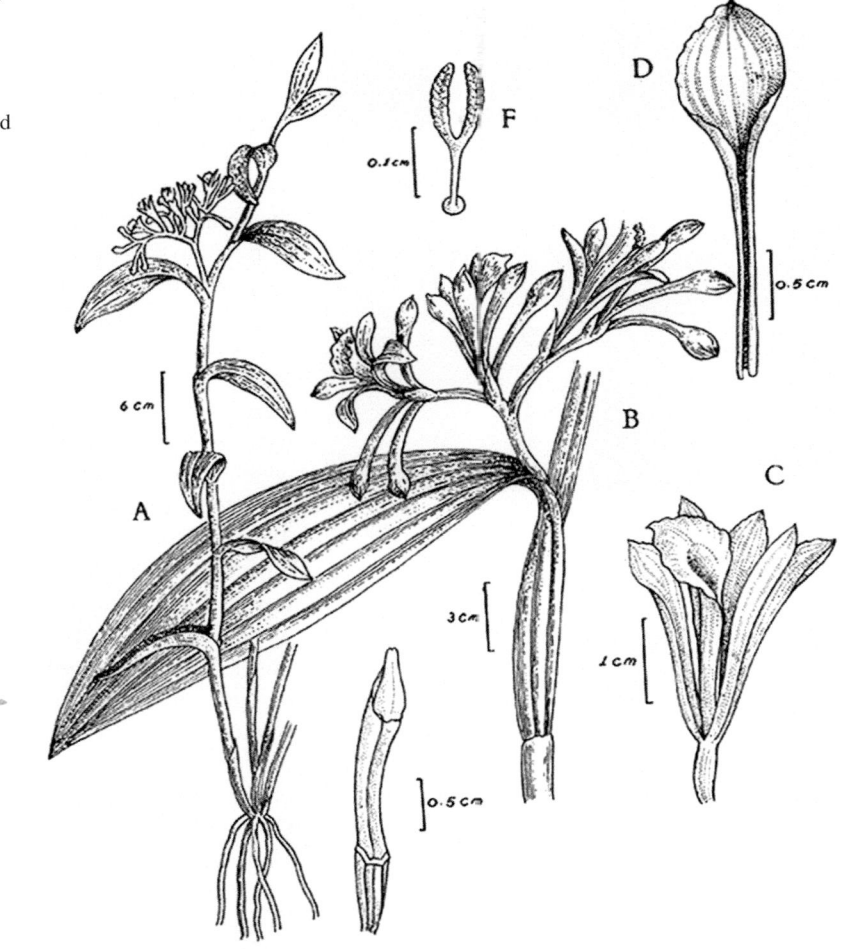

CORYMBORCHIS VERATRIFOLIA BL.

Phytochemistry: Alkaloid is present in *Corymborkhis veratrifolia* (Lawler and Slaytor 1967).

Herbal Usage: In 1906, Ridley received a specimen of *Corymbokis veratrifolia* with a note from Dr. J.D. Gimlette who was the British Resident Physician in Kelantan, the most northeastern state of the Malay Peninsula. The note read "Collect the green leaves; bruise them in quantity; administer the juice either alone or with fine scrapings of *Akar Bertak* (which is not an orchid). It will cause vomiting. Use for ague *(Demum kura)*, especially in children. No water to be mixed with juice. It is customary to cultivate a plant for the occasion" (Ridley 1906;

Gimlette and Thomson 1939). In India, juice freshly extracted from the leaves is usedused as an emetic (Rao 2007). It is usedused to treat cuts on the feet in the British Solomon Islands (Henderson and Hancock 1988).

Overview
The instruction given by Gimlette recalls Li Shizhen's prescription for *Artemesia*. Since *C. veratrifolia* is also usedused to treat ague (malaria or some other illness characterised by fever and rigors, perhaps dengue), it might be worthwhile to test the orchid against *Plasmodium falciparum*. It should be remembered that artemisinin (Chinese *Qinghaosu*) is heat-labile

Fig. 9.29 *Corymborkis veratrifolia* (Reinw.) Blume. (Photo: E.S. Teoh)

and any active ingredient in *C. veratrifolia* may also be destroyed by heat. There are currently no published pharmacological data on this orchid.

Genus: *Cremastra* Lindl.

Chinese name: *Dujuan Lan* (Azalea orchid)

Cremastra are sympodial terrestrial orchids with creeping rhizomes and partially subterranean, tuberous pseudobulbs that bear single large long-petioled, plicate, lanceolate leaves. The distinctive feature is the tall, erect, many-flowered inflorescence that bears many elongated, droopy, partially opened flowers and looking much like a floral standard (Fig. 9.30). There are only a handful of species found in open montane forests from Nepal, Bhutan, across China, Thailand and Indochina to Japan.

The generic name is derived from Greek *kremastra* (flower stalk) which is the conspicuous feature of the genus.

Cremastra appendiculata (D. Don) Makino

Syn. *Crematra variabilis* (Blume) Nakai; *Cymbidium wallichiana* Lindl.

Chinese name: *Mabian Lan* (horse whip orchid), *Dujuan Lan* (Azalea orchid); *Shancigu* (kind mountain lady), *Maocigu* (kind furry lady), *Sandangu* (three layer hoop)

Japanese: *Sai-hai ran* (purple orchid standard), *Sanjiko*

Korean: *Sanjago, Yaknancho*

Medicinal names: The Chinese *Shancigu* also refers to *Pleione bulbocoides*. It is *Sanjiko* in Japanese, and *Sanjago* in Korean (Kimura et al. 2001). Their similarity denotes a ancient common origin.

Description: *C. appendiculata* is a terrestrial herb with tuberous, clustered pseudobulbs, each of which bears a single, large, plicated, three-ribbed, long-petioled, elliptical leaf, 20–30 by 4–6 cm. Floral scape arises from the side of the tuber and carries a dozen floppy, scented, tubular flowers each up to 4 cm long rather like lilies, which do not open widely, together looking rather like a standard. Flowers are yellow to orange with a white lip. Lip and petals are spotted with bluish-violet (Fig. 9.31). Flowering period is May and June. It is a variable species distributed from the Himalayas across most of China south of the Yellow River, Thailand, Vietnam and Japan, in forests at 300–2900 m.

C. appendiculata is endangered because of habitat disturbance. At the Guizhou Biotechnology Institute, in vitro methods are being developed for mass propagation of the orchid from seed and meristems (Mao et al. 2007).

Phytochemistry: 5,7-dihydroxy-3-(3-hydroxy-4-methioxybenzyl)-6-methoxychroman-4-one is the homoisoflavanone isolated from the pseudo bulbs of *C. appendiculata* by Shim et al. (2004) of Korea (Fig. 9.32). It inhibits basic fibroblast growth factor (bFGF)-induced, in vitro and in vivo angiogenesis of the chorioallantoic membrane of the chick embryo, without demonstrating any toxicity. It also inhibits inflammatory and allergic response in mast cells, and ultraviolet beam-induced skin inflammation (Hur and Kim 2009a, b) by reducing cyclooxidagenase-2-expression and NF-kappa B nuclear localisation (Hur et al. 2010). *C. appediculata* extract up-regulates tyrosinase activity in vitro (Yan et al. 2002). Tyrosine is the enzyme that promotes melanin

Fig. 9.30 *Cremastra appendiculata* (D. Don) Makino (as *Cremastra wallichiana* Lindl.). From: *Annals of the Royal Botanic Gardens, Calcutta,* vol. 8(3): t. 246 (1891) Drawing by R. Pantling. Courtesy of Missouri Botanic Garden, St. Louis, USA

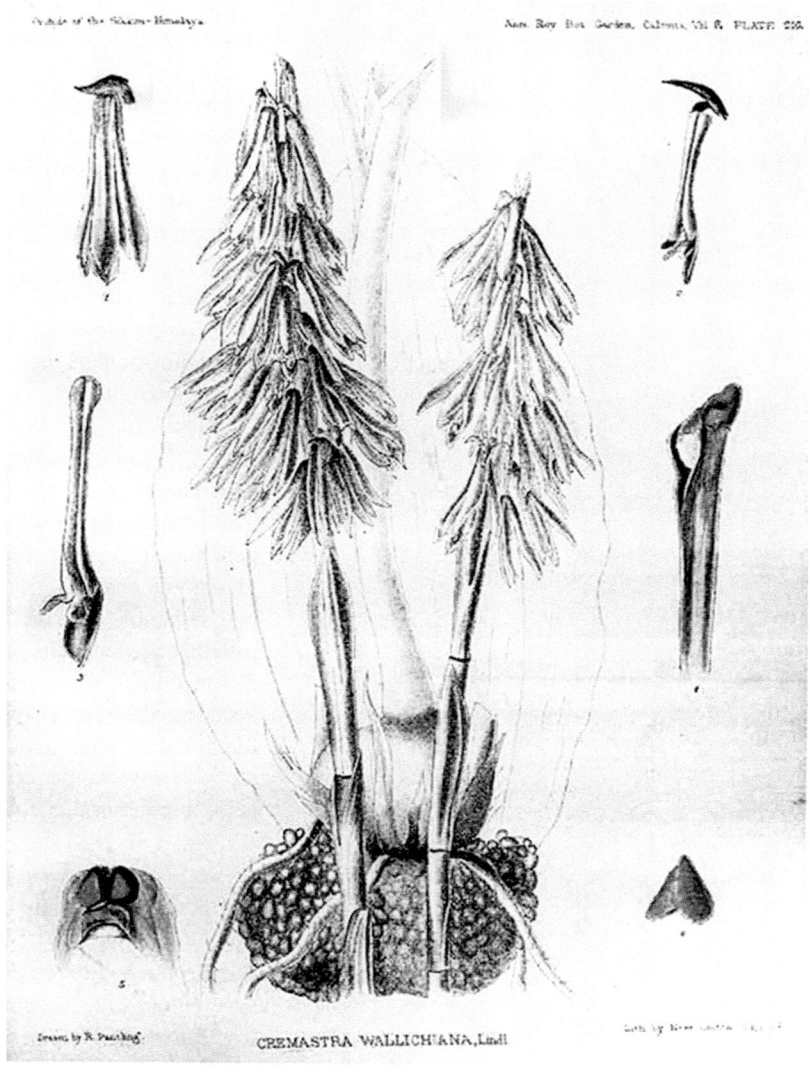

formation, darkening of skin and UV protection. Xue et al. (2005) from the Institute of Materia Medica in Beijing's Chinese Academy of Medical Sciences isolated and identified six compounds from the "corm" of *C. appendiculata*: isohircinol, flavanthrinin, *p*-hydroxyphenylethyl alcohol, 3,4-dihydroxyphenylethyl alcohol, daucosterol and beta-sitosterol. A few months later, they reported the isolation of eight compounds from the "tubers" (pseudobulbs) of *C. appendiculata*: cirrhopetalanthrin, 7-hydroxy-4-methoxyphenyl-1-0-beta-D-glucoside, 4-(2-hydroxyethyl)-2-methoxy phenyl-1-0-beta-glucopyranoside, tyrosol 8-0-beta-D-glucopyranoside, vanilloloside, *p*-hydroxybenzyl aldehyde, sucrose and adeniosine (Fig. 9.33). Except for cirrhopetalanthrin which showed non-selective moderate cytotoxicity, none of the other compounds showed any cytotoxicity against human colon cancer (HCT-8), human hepatoma (Bel77402), human stomach cancer (BGC-823), human lung adenocarcinoma (A549), human breast cancer (MCF-7) and human ovarian cancer (A2780) cell lines (Xia et al. 2005). Next, employing ethanol as a solvent, they recovered six new phenanthrene derivatives from the pseudobulbs of the orchid which consisted of three monophenanthrenes, two biphenanthrenes and one triphenanthrene. The compounds were

Fig. 9.31 *Cremastra appendiculata* (D. Don.) Makino (Photo: Liu Ming)

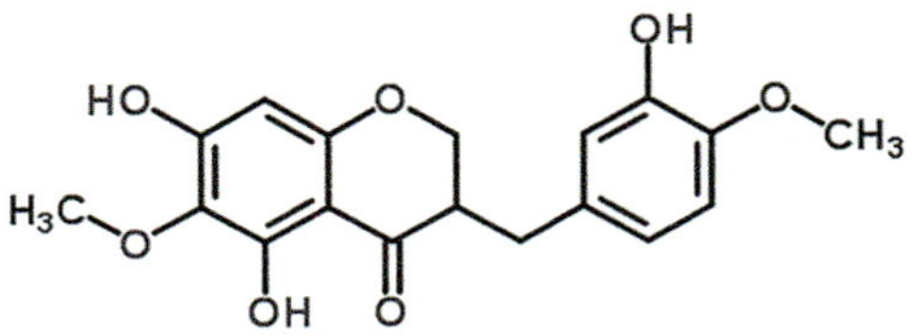

Fig. 9.32 Chemical structure of Cremastranone, a homoisoflavanone with anti-angiogenesis properties isolated from *Cremastra appendiculata*

screened for possible cytotoxicity but they tested negative (Xue et al. 2006). They also managed to isolate two new terpenoids, cadinane sesquiterpene and ent-kaurane diterpene diglycoside, together with a known triterpene with 32 carbon atoms. The triterpene with 32 carbon showed selective cytotoxicity against human breast cancer cell line (MCF-7) in vitro but not against other human cancer cell lines, while the two terpenoids tested negative throughout (Li et al. 2008). A new pyrrolizidine alkaloid, cremastrine, was isolated from the pseudobulbs by Ikeda et al. (2005) at Mitsubishi Pharma Corporation in Osaka (Fig. 9.34). Seven compounds isolated with silica gel, reverse-phase silica gel, and Sephadex column

chromatography were identified by Liu et al. (2008) as 5-methoxybibenzyl-3,3'-di-*O*-beta-D-glucopyranoside, militarine, loroglossin, protocatechuic acid, succinic acid, gastrodin and daucosterol. Compound 1 is new and the others were isolated from *C. appendiculata* for the first time. At least one of these compounds has a neuroprotective effect, but the effect of extracts of *C. appendiculata* on the nervous system has not been fully studied. Subsequently, another 14 compounds were isolated from petroleum ether and ethyl acetate extracts of *Shancigu*, namely: 4,4'-dimethoxy-9,9', 10,10-'-tetrahydro-(1,1'-biphenanthrene)2,2', 7,7'-tetrol; 4,4', 7,7'-tetrahydroxy-2,2'dimethoxy-1,1'-biphenanthrene; 3,5-dihydroxy-2,4-dimethoxyphenanthrene; physcion; chrysophanol; emodin; genkwanin; quercetin; quercetin3'-*O*-beta-D-glucopyranoside; 3-methoxy-4-hydroxyphenylethanol; syringic acid; vanillin; and p-hydroxybenzaldehyde (Liu et al. 2014b). Seven compounds were isolated from an ethyl acetate extract of *Creamstra appendiculata* and identified as fumaric acid, dimethylhexyl phthalate, L-pyroglutamic acid, 2-furoic acid, vanillic acid, *p*-coumaric acid and protocatechuic acid (Zhang et al. 2011).

Recently, an additional 11 new and 23 known compounds were isolated from *C. appendiculata*. They include 20 phenanthrene or 9,10 dihydrophenanthrene derivatives, five bibenzyls, seven glucosides, adenosine and gastrodin. When tested for cytotoxic activity, only one compound showed moderate activity against A549 tumour cell line (Wang et al. 2013).

In the hexosan present in tubers of *C. variabilis* [= *C. appendiculata* var. *variabilis* (Bl.)I.D. Lund], the ratio of D-mannose to D-glucose is 3:1 (Ernst and Rodriguez 1984).

Herbal Usage: *C. wallichiana* (= *C. appendiculata*) was first listed as a medicinal herb in Chen Can Qi's *Ben Cao Shi Yi* (*Omissions from the Medica Medica*) compiled around 720 during the Tang Dynasty. Stem was used to treat impotence, tuberculosis, fever, frostbite, snake bites and poisoning in general. It was also usedused to treat abscesses and swellings. Paste made with the pseudobulb was spread over a boil to heal it. Pseudobulbs are

Fig. 9.33 Bibenzyls from *Cremastra appendiculata*. Cirrhopetalanthrin possesses moderate cytotoxic activity

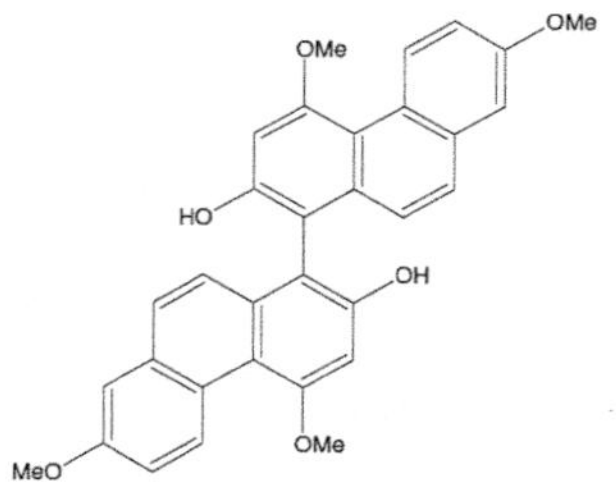

2,7,2',7',2'-pentahydroxy-4,4'4",7"-tetramethoxy-

1,8,1'1"-triphenanthrene

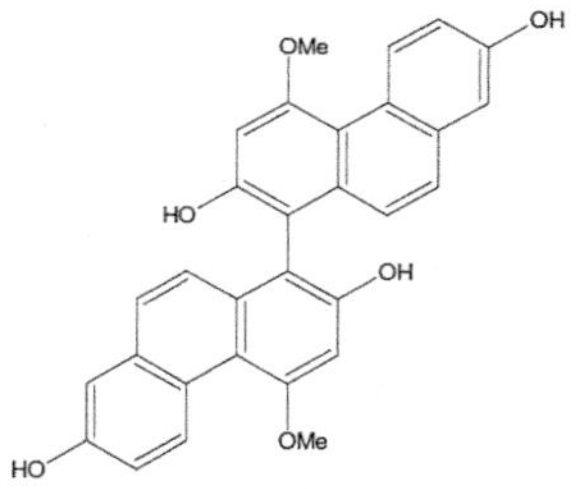

cirrhopetalanthin

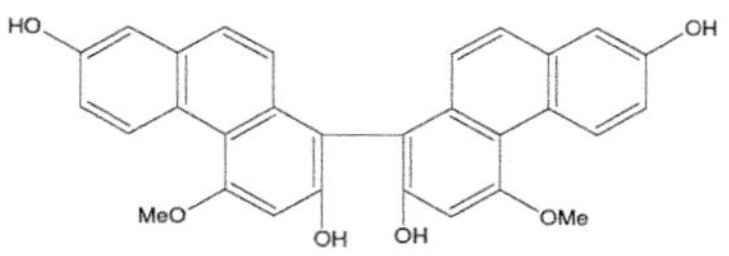

cirrhopetalanthrin

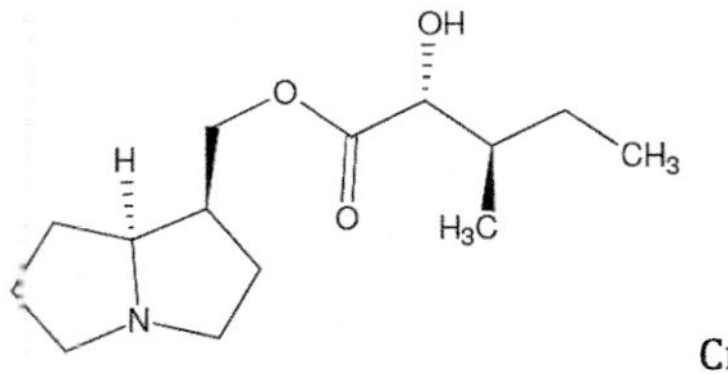

Cremastrine

Fig. 9.34 Alkaloids from *Cremastra appendiculata*

harvested in May and June, detached from the leaves and roots, washed clean, cut into slices, and sun-dried before use. Several Chinese herbal prescriptions are listed in Table 9.4 (*Zhongyao Da Cidian* 1986; *Zhonghua Bencao* 2000). In Japan, the Ainu chew on a pseudobulb of *C. appendiculata* to relieve a toothache. They also use it to treat snake bites and insect bites (Lawler 1984).

Overview

Neovascularisation (an overgrowth of new blood vessels) in the eye is the commonest cause of blindness. It occurs in premature retinopathy, diabetic retinopathy, age-related macular degeneration and sickle cell anaemia. Jeong Hun Kim and his colleagues found that, in a rat model, the compound, a homoisoflavanone, extracted from *C. appendiculata* significantly reduces retinal neovascularisation. The scientists proposed that the compound might be useful for the treatment of vaso-proliferative retinopathies (Kim, Kim, Kim et al. 2007; Kim et al. 2008). A synthetic isomer of this homoisoflvanone code-named SH-11052 exhibits antiproliferative activity against human umbilical vein endothelial cells and human retinal microvascular endothelial cells. Although it did not induce apoptosis, it might be able to complement existing anti-angiogenic drugs usedused in the treatment of neovascular eye diseases (Basavarajappa et al. 2014). An ethanol extract of *Dendrobium chrysotoxum* was also found to be capable of alleviating retinal angiogenesis in

Table 9.4 Prescriptions employing *Shancigu* (*Cremastra wallachii* or *Pleione bulbocodioides* or *Pleione yunnansis*) (Source: *Zhongyao Da Cidian* 1986; *Zhonghua Bencao* 2000)

1. For reduction of swelling, dissolution of phlegm, detoxification, carbuncle, tuberculous lymphadenitis, throat numbness, swelling and pain, snake bites, and (?) rabies:
   Wen Ha (a species of frog?) 90 g,
   Cremastra. variabilis (Shan Ci Gu) 60 g,
   Moschus moschiferns 900 mg,
   Qian Jin Zi 30 g,
   Euphorbia Perkinensis 45 g.
   Cook with glutinous rice and make 40 tablets.
   Take one tablet each time.
   Original Source: Essentials of External Diseases

2. For carbuncle, jaundice:
   Grind Cremastra. variabilis with roots and Can Er Cao,
   Mix and take with wine, 9 g each time.
   Original Source: Qiankun Sheng Yi

3. For ulcers, sores, scrofula, snake bite
   Shancigu 9–15 g in decoction for oral consumption and also applied to affected part

4. For malignant sores and jaundice
   (a) Shancigu with roots
   Xanthium sibiricum (Siberian cocklebur)
   Pulverise. Mix the two ingredients with wine; filter. Filtrate is the medication.
   (b) Render into powder Shancigu. Add 9 g to wine for consumption

5. For cracked skin
   Pulverise the sheath of the stems and apply to affected part

6. Cough
   Decoction made with 9–15 g of Shancigu

7. To treat cancer of the oesophagus
   Shancigu 9 g
   Cloves 9 g
   Diospyros kaki (persimmon) 5
   Boil and drink.

streptozotocin-induced diabetic rats (Gong et al. 2014), but the identity of the compounds with this property was not defined.

Homoisoflavonoids exhibit a broad range of bioactivities that include antimicrobial, antimutagenic, anti-oxidant, immunomodulatory, antidiabetic, cytotoxic, anti-angiogenesis, vasorelaxant, anti-inflammatory and anti-allergic effects (Lin, Liu, Ye 2014; Lee et al. 2014a, b; Basavarajappa et al. 2015). Therefore, there is still much about *C. appendiculata* that could be explored. Perhaps homoisoflavanone would also find a use in the treatment of tumours. There is a

single case report of a 74-year-old patient with metastatic bladder cancer who refused chemotherapy and was treated with oral and nebuliser Korean herbal therapy which included *C. appendiculata* tubers. Serial X-rays showed diminution of the multiple metastatic nodules in the lungs and his symptoms disappeared. The herbal remedy is complex and, besides *C. appendiculata*, it contained *Cordyceps militaris*, *Panax ginseng radix*, *Commiphora myrrha*, *Calculus bovis*, margarita, *Boswellia carteri*, *Panax notoginseng* radix: the nebuliser solution was made with wild ginseng and *Cordyceps sinensis* distillate (Lee, Kim, Seong, et al. 2014). The Korean team is also studying other compounds from herbs with similar properties, for instance decursin extracted from roots of the non-orchidaceous plant, *Angelica gigas* Nakai (Kim, Kim, Lee, et al. 2009).

The anti-angiogenic homoisoflavanone, cremastranone, has now been synthesised. This synthetic compound was shown to inhibit proliferation, migration and tube formation of human retinal microvascular endothelial cells (Basavarajappa et al. 2014; Lee et al. 2014a).

In Japan, *C. appendiculata* is mixotrophic. Plants usually occur on the heavily shaded forest floor which rather limits their capacity for photosynthesis despite the presence of green leaves. However, the cortical cells of its underground rhizomes are heavily colonised by fungi (*Coprinellus*, Psathyrellaceae) which supply the orchid with additional carbon (Yagame et al. 2013). Orchids which associate with saprobic mycobionts like *Gastrodia elata* have been shown to contain neuroprotective compounds. Gastrodin has been isolated from *C. appendiculata*, but such therapeutic possibilities of the orchid for neuroprotection have not been explored.

Pleione bulbocodioides (Franch.) Rolfe and *Pleione yunnansis* Rolfe are substitutes for *C. wallachii* when the term *Shancigu* is used (*Zhonghua Bencao* 2000; Bensky, Clavey, Stoger, Gamble 2004). Japanese and Korean medicinal names for *C. appendiculata* are derived from the Chinese. *Shancigu* entered the *Chinese Pharmacopoeia* around 720 during the height of the Tang Dynasty (618–907), a period which saw an active transfer of Chinese learning

Fig. 9.35 *Crepidium acuminatum* (D. Don) Szlach. (as *Malaxis acuminata* D. Don). Reproduced with permission from *Introductions to Orchids* by Abraham and Vatsala, Parlode, Thiruvananthapuram: Tropical Botanic Garden and Research Centre (TBGRI), 1981

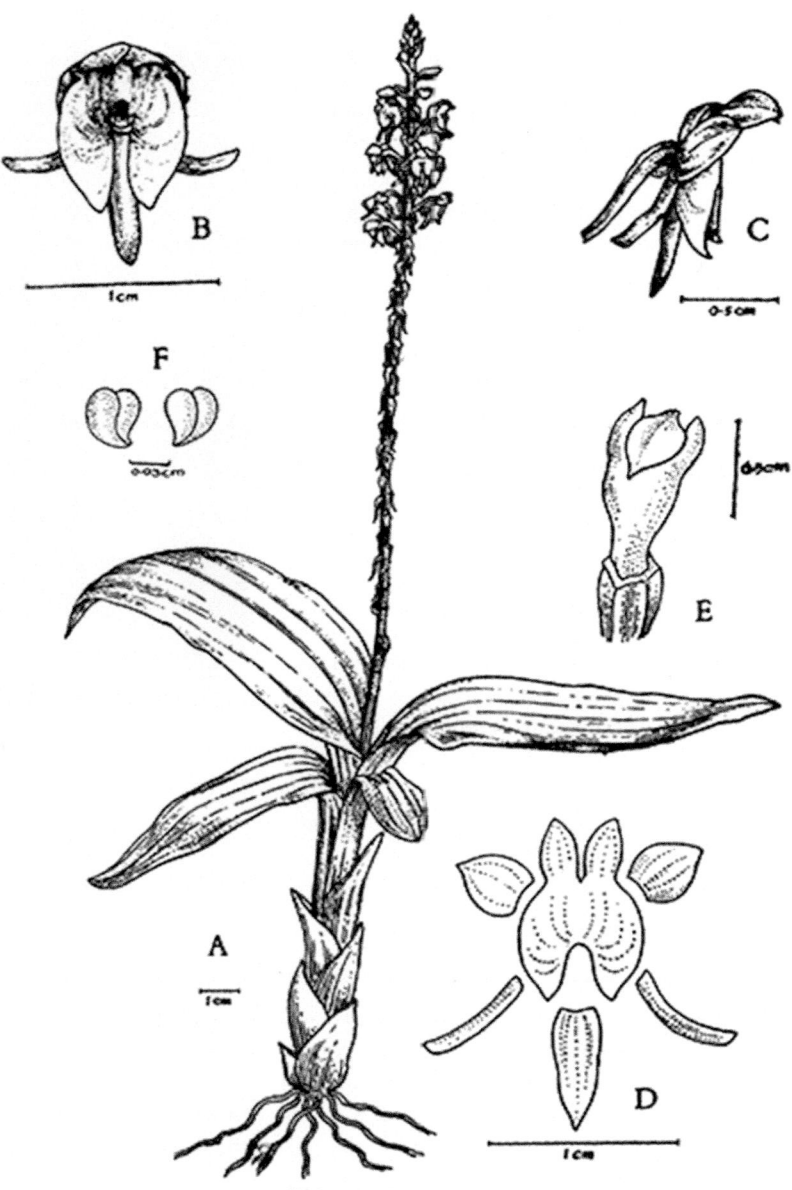

MALAXIS ACUMINATA D. DON.

and culture to Korea and thence to Japan. The similar sounding Korean and Japanese names, *Sanjaco* and *Sanjiko*, respectively, for the medicine reflect the timing of *C. appendiculata's* entry into Korean and Japanese herbal medicine.

Genus: *Crepidium*

Syn. Genus: *Seidenfia* Szlach.

Crepidium (syn. *Seidenfia* Szlach) is an Indo-Sri Lankan genus with 280 species of terrestrial

herbs with hairy roots that were generally classified under *Malaxis* or *Liparis*. Six species are
present in Peninsular India, and there is one in
Seychelles. The genus was named by Blume and
revived by Szlachetko.

Stems are cylindric to pseudobulbous, leaves
2 or several, petiolated, sheathing the stem, plicate, membraneous or fleshy. Inflorescence is
erect with persistent floral bracts. Flowers usually non-resupinate, green to yellow and purple,
with an erect, relatively large, prominent, flat lip
(Chen and Wood 2009) (Fig. 9.35).

Crepidium acuminatum (D. Don.) Szlach.

Syn. *Malaxis acuminata* D. Don., *Microstylis
wallichii* Lindl.

Indian names: *Jeevak* in Hindi, *Jivak* (Tamil),
   Jivakam (Malayalam), *Jivakam* (Tekugu,
   Jivakamu (Kannada), *Jivaka* (Sankrit):
   Lahsunia (vernacular name in Kumaun
   Himalaya)
Ayurvedic names: *Jivak, Rishvak, Rishbhaka,
   Bandhura, Dhira, Durdhara, Gopati,
   Indraksa, Kakuda, Matrika, Visani, Vrisa,
   Vrisnabha*

Description: *C. acuminatum* is a variable, robust,
terrestrial herb. Stem is succulent, 10 cm tall
without inflorescence (Abraham and Vatsala
1981); shoot with inflorescence is 16–27 cm tall
(Joseph 1982), with round pseudobulbs, and 1–4
unequal, ovate sessile leaves, 3–14 by 1.3–4 cm.
Inflorescence is 10–25 cm long, with pale, yellow to green, or pink to dull purple flowers
(Fig. 9.36). In India, it flowers from July to
September depending on location (Abraham
and Vatsala 1981; Joseph 1982; Matthew 1995).

The preceding description fits the Indian variety. In Thailand, leaves are lanceolate-elliptic,
thin, plicate, 5–9 by 2–3 cm, 4–6 leaves per
plant. Thai flowers are 1 cm across (Nankorn
and Watthana 2008). The orchid is also found at
Gaoligongshan in western Yunnan where it
blooms from May to July (Jin et al. 2009).

Fig. 9.36 *Crepidium acuminatum* (D. Don) Szlach.
(Photo: E.S. Teoh)

C. acuminatum is widely distributed from the
southern Himalayas to Myanmar, Thailand,
southern China (Xizang, Yunnan, Guizhou,
Guangdong and Taiwan), Indochina and the
Philippines to Australia at 300–2100 m. It is
found mostly in pine or oak forests in the
Himalayas (Jain 2003). Considered a medicinal
plant and a protected species which is seriously
threatened in India, it was discovered growing in
the Haat Kali sacred grove in Uttarakhand in
Central Himalaya (Singh 2010), and at
1800–2300 m in Garhwal (Dhayani et al. 2011).

Herbal Usage: *C. acuminatum* (syn. *Malaxis
acuminata*) is one of eight component of
ashtavarga (Dhayani et al. 2011). It is one of
several herbs that could be considered as *Jivak*,
another being *Pueraria tuberosa* (Indian name:
kudzu) which is not an orchid (Puri 1970a). In
Ayurvedic classification, it is sweet in taste (as a
matter of fact, it is slightly bitter), cold in
potency, pacifies *vata* and aggravates *kapha*. It

is cooling, thus causing fever to abate, and promotes sperm formation. It is administered to men whose wives are unable to conceive. Pseudobulbs are used to treat bleeding disorders, fever, tuberculosis and a sensation of heat, emaciation, dysentery, rheumatism and insect bites (Pushpa et al. 2001). Sometimes, they are substituted with *Pueraria tuberosa* (Singh and Duggal 2009). Pseudobulbs of *Crepdium acuminatum* (syn. *Microstylis wallichi*) were considered to be simultaneously a tonic and an aphrodisiac (Duggal 1971). Dried pseudobulbs of *C. acuminatum* are incorporated into the Ayurvedic tonic "*Chyavanprash*", a popular herbal preparation for promoting health and preventing illness (Lawler 1984; Bhattacharjee 1998; Cheruvathur et al. 2010). It is a diuretic in addition to being a tonic (Pandey et al. 2003). In Bangladesh, it is used as a tonic to treat tuberculosis (Musharof Hossain 2009).

Jeevak or *Jivak* (*C. acuminatum*) features in the following formulations: *Astavargha churna*, *Chyanprash rasayan*, *Chitrakadi taila*, *Vachadi taila*, *Mahakalyan ghrita*, *Mahamayura ghrita*, *Manapadma taila*, *JIvaniya ghrita*, *Vajkaran ghrita*, *Brahini gutika* and *Himvana agada*. *Malaxis cylindrostachya* (Lindl.) Kuntze and *Malaxis mackinnoni* (Duthie) Ames are sometimes usedused when *C. acuminatum* is not available. Other substitutes are *Pueraria tuberosa* (*Vidara kand*), *Centaurea behen* (*Safed behmen*), *Centaurium roxburghii* (D. Don) Druce (or *Lal behmen*) and *Tinospora cordifolia* (*Guruchi*). The last four herbs are not orchids (Chinmay et al. 2011; Balakrishna et al. 2012).

In Ayurvedic practice, to prepare the tonic for increasing sperm production and improving the reproductive tissues, 1 g of powdered *C. acuminatum* pseudobulb is mixed with the powdered *Malaxis monophyllos* (syn. *Malaxis muscifera*) pseudobulb, *Lilium polyphyllum* bulb, *Fritillaria roylei* bulb and *Asparagus racemossus*. This is consumed in the morning (Dhayani et al. 2011). *Crepdium acuminatum* has become rare in Kamaun Himalaya due to overexploitation (Jain 2003).

Crepidium resupinatum (G. Forst.) Szlach.

Syn. *Seidenfia rheedii* (Sw.) Szlach. (see *Liparis rheedii* Sw.); *Seidenfia versicolor* Marg. & Szlach.

Description: This is a variable, terrestrial herb with stems 8–21 cm long, 1.5–2 cm in diameter, with pseudobulbs along its length supporting 2–3 sessile, thin, lanceolate, plicate leaves 6–10 by 3–5.5 cm, and 7 veined. Inflorescence is erect and carries numerous small greenish-yellow to orange or purple flowers, 3.4 mm across. It continues to lengthen and produces new flowers which open successively over a long period. Flowers are non-resupinate. Lip is large, semicircular, with dentate margin, the teeth long and pronounced in some varieties, and barely visible in others (Abraham and Vatsala 1981). Plants are found in shaded locations between 400 and 1800 m (Jayaweera 1981).

The colour of the plant and flowers is influenced by light intensity: pure green in bright light, deep purple in the shade, and yellowish inbetween. It is because of this variation in colour that Lindley gave the species the epithet *vesicolor* (Santapau and Kapadia 1966). It was formerly referred to as *Microstylis vesicolor*, then as *Seidenfia rheedii*, and now as *C. resupinatum*. The species occurs in southern India (Karnataka, Kerala and Tamil Nadu) and in Sri Lanka.

Phytochemistry: Pseudobulb of *C. acuminatum* contains an alkaloid, glycosides, flavonoids, beta sitosterol, piperitone, 0-methylbatatasin, 1,8-cineole, citronelal, eugenol, glucose, rhamnose, coline, limonene, *p*-cymene and ceryl alcohol (Pushpa et al. 2001; Balakrishna et al. 2012).

Herbal Usage: In the western part of the Indian peninsula, a potion made with the plant is used to treat fever, biliousness and infantile epilepsy (Delgardo, quoted by Lawler 1984).

Overview

C. acuminatum is an ingredient of the popular Indian rejuvenating tonic, *Asthavarga* in Uttarakhand in the western Himalayas and in many other parts of the country (Jalal et al. 2008). On account of its popularity as an ingredient in such Ayurvedic preparations and its rapid disappearance from its natural Indian habitat, Cheruvathur et al. (2010) undertook to propagate *C. acuminatum* in tissue culture by inducing adventitious shoots in cultured internodal explants. Meanwhile, Deb and Temjensangba (2006) succeeded with in vitro immature seed germination of another threatened terrestrial Indian *Crepidium* species, *C. khasianum* (Hook f.) Szlach. [syn. *Malaxis khasiana* (Hook f.) Kuntz.]. The plantlets showed 65 % survival under field conditions. Although this is not a medicinal species, it would appear that the medicinal *C. acuminatum* could also be seed-germinated.

Pseudobulb extracts of *C. acuminatum* contain polyphenols which possess anti-oxidant activity. This has been successfully exploited for the green synthesis of gold nanoparticles which will have applications in nanobiodiagnostics, pharmaceuticals, catalysis, and other applications of nanoscience (Gopal et al. 2014). However, they neither explain nor support the herbal usage of *Crepidum acuminatum* as a tonic and aphrodisiac.

An alkaloid, grandifoline, isolated from *C. grandifolium* (Schltr.) Szlach. (syn. *Malaxis grandifolia* Schltr.) is a glycosidic derivative of nervogenic acid esterified with laburnine (Lindstrom et al. 1971). Grandiflorine has also been isolated from *Delphinium geyeri* (low larkspur). It is closely related to the neurotoxin methyllycaconitine, and it has comparable neurotoxicity in mouse bioassays, whereas its synthetic monoacetate is significantly less toxic (Manners et al. 1998). Grandiflorine is one of several alkaloids in low larkspurs which are sometimes fatally ingested by cattle in the western USA (Gardner and Pfister 2009).

Genus: *Cymbidium* Sw.

Chinese name: *Lan* (orchid)

Cymbidiums are epiphytic or terrestrial orchids with extremely short rhizomes and pseudobulbs which carry many long, often arching, lanceolate, duplicate leaves which ensheath the pseudobulb at their base. Inflorescence arises laterally and carries several to numerous, showy, medium-sized to large flowers. *Cymbidium* is distributed in tropical East Asia from India eastwards to China, Japan and Southeast Asia in lowland and montane forests. Its hybrids play an important role in the cut flower industry but they are not grown extensively in Southeast Asia because the large, showy types require cool temperatures to initiate flowering. Approximately 68 species have been described, with 49 occurring in China. The generic name is derived from Greek *kymbos* (boat–shaped cup), alluding to the lip of the flower.

Cymbidium aloifolium (L.) Sw.

Syn. *Cymbidium pendulum* (Roxb) Sw.

Chinese name: *Wenban Lan* (stripe petal orchid), *Yingyediao Lan* (stiff leaf hanging *Cymbidium*), *Chuihuadiao Lan* (pendulant flower *Cymbidium*), *Diao Lan* (hanging *Cymbidium*), *Dabi Lan* (lean-on-the-wall *Cymbidium*)
Chinese medicinal name: *Yingyediao Lan* (stiff leaf hanging *Cymbidium*)
Thai name: *Ka Re Ka Ron*
Vietnamese Name: *Kim bien*
Laotian names: *Lung khao, Huan so pet, Kin loum, Khi mot top*
Indian name: *Supurn* in Orissa State, boat orchid; *panaipulluruvi* (Valaiyans in Tamil Nadu)
Myanmar name: *Thit tet lin nay*
Nepalese name: *Harjor* in Tharu
English name: boat orchid

Description: Pseudobulbs are small, slightly flattened, bearing 4–5 rigid, thick, coriaceous leaves,

Fig. 9.37 *Cymbidium aloifolium* (L.) Sw. (Photo: E.S. Teoh)

Fig. 9.38 *Cymbidium aloifolium* (L.) Sw. Palm tree trunks are a favourite perch for thick-leaved *Cymbidium* plants; here in the grounds of a Buddhist temple in Chiang Mai, Thailand (Photo: E.S. Teoh)

40–90 cm in length and 4–5 cm wide. Scape is lateral, pendulous, bearing 15–35 (in Yunnan, 25–48) well-spaced, lightly scented flowers 3–4 cm across, of pale yellow or buff with broad maroon central striping (Fig. 9.37). Flowering period does not vary greatly throughout its distribution: March to April in Trivandrum, Kerala State in southern India (Abraham and Vatsala 1981), in May in Mumbai only slightly north in the adjacent state the west Deccan (Santapau and Kapadia 1966), April to May in Nilgiris, Tamil Nadu (Joseph 1982), March and April in Sri Lanka (Jayaweera 1981), March to June in Myanmar (Grant 1895), March to May in Thailand (Vaddhanaphuti 2005) and April to May in China (Chen et al. 1999).

This tough, epiphytic *Cymbidium* occurs as clumps on trees in sparse forests and on cliffs in ravines at 100–1100 m across northeast India, southern India, Sri Lanka, southern China, Myanmar and the Andaman Islands, Thailand, Vietnam, Peninsular Malaysia, Sumatra and Java (Chen and Tsi 1998; Comber 2001). It has been reported as very common on *Borassus flabellifer* L. in northern Sri Lanka, but also occurs on many host plants which include *Albizzia falcata, Wormia triquetra, Artocarpus nobilis, Artocarpus heterophyllus, Cassia nodosa, Samarea saman, Eugenia* sp., *Garcinia* sp., *Mangifera indica* and *Terminalia arjuna* (Jayaweera 1981). In Thailand and Peninsular Malaysia, it is epiphytic on palms (Fig. 9.38).

Santapau and Kapadia (1966) highlighted the confusion between this species and *C. pendulum* and with *C. bicolor. C. aloifolium* is sometimes confused with *C. paucifolium* in China, but the latter is distinguishable by its shorter, broader leaves and few flowers (usually 6–11), and also with *C. mannii*, but in the latter the leaves are thinner and the two lamellae of the lip are not broken in the middle (Liu et al. 2006). There is also confusion between this species and *C. finlaysonianum* Lindl., a common lowland orchid in Malaysia and Indonesia (Du Puy and Cribb 2007). The latter has pendulous scapes which commonly reach 90–100 cm in length with widely spaced flowers.

Phytochemistry: *C. aloifolium* contains several phenanthrenes: aloifol I and II, coelonin and 6-methoxycoelonin (Juneja et al. 1987), cymbinodin A (Barua et al. 1990), cymbinodin B (Ghosh et al. 1992), a novel polyoxygenated phenanthrene derivative designated pendulin, and a 3,7-dihydroxy-2,4,8-trimethoxyphenanthrene named denthyrsinin, the last which had earlier been isolated from *Eulophia nuda* and *Dendrobium thyrsiflorum* (Majumder and Sen 1991). An ethanolic extract of *C. aloifolium* leaves produced an anti-inflammatory and analgesic effect in mice (Howlader et al. 2011). It would be good to know which of any of the six phenanthrenes isolated so far have anti-infammatory, analgesic or haemostatic effects.

Pendulin, a polyoxygenated phenanthrene derivative, was isolated from *C. pendulum* (Majumder and Sen 1991). Unfortunately, the publication did not permit specific identification of the species because the orchid name might refer to any of the following: *C. pendulum* (Roxb. Sw. [= *C. aloifolium* (L.) Sw.); *C. pendulum* var. *atropurpureum* Lindl. [= *C. atropurpureum* (Lindl.) Rolfe]; *C. pendulum* var. *brevilabre* Lindl [= *C. finlaysonianum* Lindl.] or *C. pendulum* var. *purpureum* W. Watson [= *C. crassifolium* Herb.].

Herbal Usage: In Indian traditional medicine, juice is extracted from the whole plant by pounding it with ginger and a small amount of water is usedused to induce vomiting and diarrhoea (Caius 1936), or to cure chronic illness, weakness of the eyes, vertigo and paralysis (Lawler 1984). Reddy et al. (2005) who researched the region of the Eastern Ghats found that aboriginal Konda reddis of East Godavari district used the aerial roots of the orchid to make a paste for treating cracks on the feet, whereas aboriginal Koyas of Khammam district usedused a similar preparation for setting fractures. On the other side of the Deccan in the Uttara Kannada district, roots of *C. aloifolium* are added to tubers of a common terrestrial orchid, *Zeuxine strataeumatica*, to prepare a tonic (Rao 2004). Tribal residents at Kudremukh National Park in Karnataka use the mucilage extracted from the orchid leaves to stop bleeding from leech bites, as the sap is said to promote blood coagulation. Powdered pseudobulb is made into a drink to cause vomiting and purging by the same tribes (Rao 2007). *C. aloifolium* also forms an ingredient of medicinal oil that is usedused to treat both benign and malignant tumours. To treat paralysis, the Dongria Kandha tribe in Southwestern Orissa uses a twice-daily dose of a mixture of cow's milk with powdered root of the orchid, ginger and black pepper for a period of 2 months (Dash et al. 2008). Santapau and Kapadia (1966) reported that the leaf sap had styptic properties. This useful medicinal property caused the plant to be collected to such an extent that it disappeared from some areas. Crushed leaves are used to stop bleeding from leech bites (Rao and Sridhar 2007). It is sometimes used as a vegetable aphrodisiac or as *salep* in India (Puri 1970b, c). Valaiyans living in the Vellimalai Hills of Tamil Nadu heat the leaves over a fire and administer the hot juice into the ear to relieve earache (Ganesan and Kesaavan 2003).

C. aloifolium is a constituent of a Sri Lankan oily embrocation usedused in the treatment of tumours (Soysa, quoted by Lawler 1984). In Indochina, a decoction of the plant is used as a medicinal bath for sickly children, or to treat women suffering from irregular menstruation (Petelot, quoted by Perry and Metzger 1980). Pseudobulbs are also used to treat cuts, sores and burns in Luang Prabang (Spire 1907, quoted by Vidal 1963). In Thailand, the leaves are used to treat ear infection while the root is usedused for kidney disorders (Chuakul 2002). In Myanmar, pseudobulbs are used to treat earache, stomach ache and dysentery, whereas leaves are usedused for fractures (Kurzweil and Lwin 2014).

In Nepal, a country which has imbibed many Indian traditions, the plant is used as an emetic, purgative and demulcent. It is also usedused in the form of a paste to treat dislocated bones (albeit many other orchids are also usedused for such purposes; see *Calanthe masuca*) (Manandhar and Manandhar 2002; Baral and Kurmi 2006; Pant and Raskoti 2013).

Chinese herbalists use the whole plant and its seeds to improve the condition of the lungs, stop

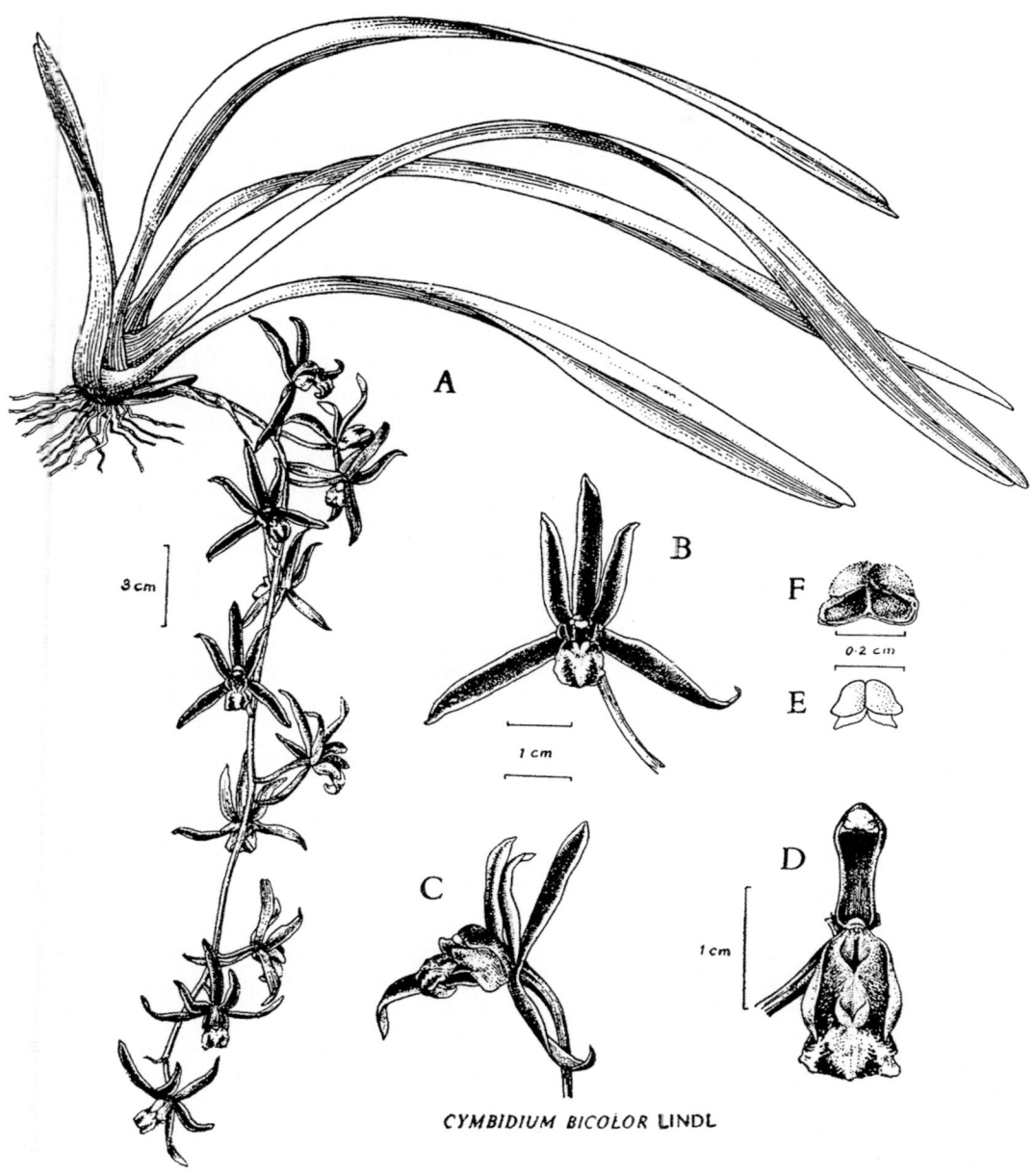

CYMBIDIUM BICOLOR LINDL

Fig. 9.39 *Cymbidium bicolor* Lindl. Reproduced with permission from *Introductions to Orchids* by Abraham and Vatsala, Parlode, Thiruvananthapuram: Tropical Botanic Garden and Research Centre (TBGRI), 1981

coughs, establish regular menstruation, and to treat haemetemesis, discharge and bleeding from injuries. The medicinal plants are collected from Guangdong, Guangxi, Guizhou and Yunnan (Wu 1994).

Cymbidium bicolor Lindl.

Sri Lankan name: *Visa Dhooli* (Poison Dust); *Beyudhuru* (not specific; also used for *Pholidota imbricata*)

Description: *C. bicolor* is a hardy, strap leaf, low-land, epiphytic *Cymbidium*. Plant resembles *C. finlaysonianum* vegetatively but leaves of this species, measuring 45 by 1.5 cm, are smaller than those of *C. aloifolium* (Abraham and Vatsala 1981; Seidenfaden and Wood 1992). Inflorescence is pendulous, up to 25 cm, arises from the base of the pseudobulb and carries numerous flowers, spaced 2–3 cm apart. Flowers also resemble *C. finlaysonianum* but are less full and are marked by a broad, central, deep purple stripe along the length of the sepals and petals. Edges of sepals and petals are pale green or buff. Lip is yellow spotted with purple. Column is dark purple at the back, yellow with purple spots in front (Seidenfaden and Wood 1992) (Fig. 9.39 and Fig. 9.40). It is distributed from India, Sri Lanka, the Andaman and Nicobar Islands to Sumatra, Malaysia, Kalimantan, Philippines and Sulawesi.

Herbal Usage: Leaves are used for treating fractures in southern China (Wu 1994).

Cymbidium crassifolium **Herb.**

Syn. *Cymbidium mannii* Rchb.

Chinese   name: *Rouye   Lan*   (tender   leaf *Cymbidium*)

Fig. 9.40 *Cymbidium bicolor* Lindl. (Photo: Bhaktar B. Raskoti)

Description: This epiphytic orchid has thick, duplicate, coriaceous leaves 20–90 cm by 1–3 cm. sheathing at the base. Inflorescence is lateral, pendulous with 10–20 flowers, 3–4.5 cm across, well spaced on the raceme. Sepals and petals are narrow, whitish or a pale yellow to brown, with a broad central streak of crimson. Petals are not well extended. Lip is trilobed, crimson with two longitudinal, yellow-coloured lamellae. It flowers in March and April in China, January to May in Thailand (Vaddhanaphuti 2005) and April to May elsewhere. The species is distributed from Yunnan eastwards across Guizhou and Guangxi to Guangdong and Hainan, southwards to Myanmar, Thailand, Indochina and Indonesia, and westwards to Nepal, Bhutan and Bangladesh, from 100 to 1600 m. It thrives in sunny locations, on trees in forests or in thickets. The medicinal plant is collected from Guangxi and Yunnan.

Herbal Usage: Leaves of *C. bicolor* are used for treating fractures (Wu 1994).

Cymbidium devonianum **Paxton**

Chinese name: *Fu Lan*
Nepali name: *Thir gava*
Vietnamese name: *Gam ngu sac*

Description: A striking, handsome, epiphytic or saxicolous *Cymbidium* of moderate size, up to 30 cm tall, with inflorescences of up to 45 cm with numerous (20–40), closely well–arranged, star-shaped, flat flowers, 3.5 cm across of variable colour, reddish-yellow, olive green to pale brown, sometimes speckled or streaked with red. Lip is purple with darker blotches on the side lobes. Leaves are suberect, oblong to oblanceolate, coriaceous, 22–27 by 3–4 cm, 2 to 4 arising from a short subcylindrical pseudobulb 1.5–2.5 by 1 cm diameter (Fig. 9.41). It flowers from March to June, depending on location. The species is distributed in Nepal, Bhutan and Bangladesh, southeast Yunnan, Myanmar, Thailand and Vietnam at 1500–1600 m, in exposed locations, on mossy rocks and trees.

Herbal Usage: In Nepal, a paste of the root is applied on boils. Plant is decocted until the liquid

Fig. 9.41 *Cymbidium devonianum* Paxton (Photo: E.S. Teoh)

Fig. 9.42 *Cymbidium elegans* Lindl. (Photo: Bhaktar B. Raskoti)

volume is reduced to half, salt is added and the decoction is consumed in small amounts three times a day for coughs and colds. Proportions are not stated (Manandhar and Manandhar 2002).

Cymbidium elegans Lindl. var. *elegans*

Syn. *Cymbidium longifolium* D. Don.

Chinese name: *Suocao Lan*

Description: Plants are epiphytic or saxicolous. Pseudobulbs are ovoid, laterally compressed, 4–9 by 2–3 cm, enclosed by persistent leaf bases. Leaves are numerous, linear, 45–80 by 1–1.8 cm. Inflorescence usually pendent or arching with pendulous rachis, arising from lower portion of pseudobulb, 40–50 cm long. Rachis carries over 20–35 nodding, bell-shaped, cream-coloured flowers. Sepals and petals are lanceolate 3–4 cm long. Lip is oblong-lanceolate, 3–4 cm long, trilobed; in some plants, spotted

with red (Fig. 9.42). Flowering season is October to December in China, September to November in Nepal, and September to December in Bhutan (Chen et al. 1999; Raskoti 2009; Gurong 2006). It occurs in forests at 1700–2800 m in Sichuan, Xizang and Yunnan in China (Liu et al. 1969), at 1000–2500 m in Bhutan (Gurong 2006) and at 1500–2500 m in Nepal (Raskoti 2009). In the last country, it is localised and threatened by deforestation and overexploitation as an ornamental plant (Raskoti 2009).

This species varies from the type in having more numerous flowers on the rachis. Lamellae on the lip are without any appendages (Chen and Cribb 2009).

Herbal Usage: *Salep* made with the plant is usedused as demulcent or emetic in India (Das 2004; Jalal et al. 2010). Fresh juice extracted from the leaves is usedused to arrest bleeding, especially from deep wounds (Baral and Kurmi 2006).

Cymbidium ensifolium (L.) Sw.

Chinese names: *Lan* (orchid*), Guo Lan* (National
   Orchid), *Gog Lan* (Nation's Orchid), *Jian Lan*
   (Jian orchid), *Dajing Lan* (large, lush orchid),
   Jinbaolisuxin Lan (golden centered, quietly
   elegant orchid), *Suxin Lan* (quietly elegant
   heart orchid), *Guanlanhua* (official orchid
   flower), *Lancao* (orchid herb, orchid grass),
   Shanlanhua (mountain orchid flower),
   Kienlan (Fukien or Fujian Orchid), four season orchid, rock orchid, etc.
Medicinal names: *Jian lan hua* (famous orchid
   flower); *Jian lan gen*; *Qiu Lan* (autumn *Cymbidium*); *Ba Yue Lan* (Eighth month *Cymbidium*); *Guan Lan* (official *Cymbidium*)
Thai name: *Chu lan*

Description: This is the popular, fragrant *Cymbidium* which many oriental scholars identified
with Confucius when he likened the company of
good friends to a room full of fragrant orchids. It
is frequently featured in Chinese paintings of the
orchid that suggest grace and contentment. A
standard advice to aspiring Chinese artists is to
*"Paint Cymbidium ensifolium when you are
happy; bamboo when you are angry."*

C. *ensifolium* has numerous Chinese names
(not including those of prized cultivated
varieties). It is referred to as the National Orchid
(*Guo Lan*) in Taiwan (Ou et al. 2003). However,
it is not China's national flower (*Guo hua*)—that
is the peony (*Paeonia*).

Pseudobulbs are cylindrical, up to 2.5 cm in
length with 2–6 narrow leaves 30–60 cm by
1.5 cm, erect to suberect, curving in the middle.
Scape is lateral 20–30 cm in length with 3–9
fragrant flowers of variable coloration, usually
beige to pale yellow-green, marked with purple.
Lip is three-lobed, often marked with red to
purple and the central lobe curling backwards
(Fig. 9.43). Flowering period is variable, but
commonly June to October. There are over a
hundred named varieties of this popular, widespread, floriferous species which is easy to cultivate (Fung 1999), with numerous peloric forms.

Fig. 9.43 *Cymbidium ensifolium* (L.) Sw. (Photo:
E.S. Teoh)

The species is widely distributed throughout
subtropical Asia, China south of the Yangzi,
Japan, Indochina, Thailand, Malaysia,
Philippines and Sri Lanka at 500–1800 m. It
favours sparsely wooded, grassy slopes and
open, hardwood forests which are not too humid.

Herbal Usage: The herb is collected in autumn.
TCM states that its taste is acrid; neutral. For
internal use, it is decocted or made into a tea. The
entire plant is used to reinforce the body fluids,
nourish blood (*yin*), 'smooth the lungs', reduce
phlegm and stop coughing. It also stops pain, especially dysmenorrhoea, and corrects amenorrhoea,
leucorrhoea (vaginal discharge), giddiness,
coughing and haemoptysis in tuberculosis. It

helps to heal burns. It is diuretic. Flowers brighten the eyes if consumed over a long period, and relieve coughs, chest pain, glaucoma and cataract. Decoction of the root was formerly used in China to treat gonorrhoea (*Zhongyao Da Cidian* 1986; Wu 1994). To treat bronchitis, it is recommended that two leaves be boiled and the decoction consumed. For chronic coughs, juice is extracted from 30 g of fresh root and mixed with rock sugar for a dose of medicinal syrup. For leucorrhoea, 30–60 g of fresh root is cooked with lean pork and served as a soup. The plant is freshly mashed up to prepare a poultice for application to boils and abscesses (Li 1994). A decoction made with the roots and rhizome of *C. ensifolium* was mixed with rice wine (fermented glutinous rice) and eaten as a remedy for stomach ache (Hu 1971). Regular consumption of a tea made with *C. ensifolium* is a method used used to correct stagnation of qi. The orchid is also usedused for the induction of labour (*Zhonghua Bencao* 2000) Examples of Fujian and Sichuan prescriptions employing *C. ensifolium* are illustrated in Table 9.5. Given the place of these two provinces in the history of Late Tang (ninth century CE), they may date from that period.

Table 9.5 Three Fujian and Two Sichuan Prescriptions employing *Cymbidium ensifolium* (Source: *Zhongyao Da Cidian* 1986; *Zhonghua Bencao* 2000)

1. Indication: chronic cough,
   Prepare decoction with 14 flowers of Jian Lan Hua (Xiamen New Treatment and Selected Chapters of Chinese Herbs)

2. Indication: tuberculosis with cough and haemoptysis
   Squeeze juice from fresh Jian Lan Gen and cook with rock sugar.
   Take 15–24 g each time.
   (Quanzhou Herbs)

3. Indication: hematuria or dysuria
   Boil Jian Lan Gen 45 g, onion 3–5 bulbs, and take with brown sugar
   (Quanzhou Herbs)

4. Indication: leucorrhea
   Cook Jian Lan Gen, Tian Dong, Lilium brownii, Bai Jie Ou with chicken
   (Records of Sichuan Chinese Medicine)

5. Indication: feminine 'dryness'
   Cook Jian Lan Gen, Bai Jie Ou, Shi Zhu Gen, and Polygonatum chinense with pork
   (Records of Sichuan Chinese Medicine)

In Indochina, the flowers were used as an ophthalmic wash, leaves as a diuretic, and roots for chest ailments (Petelot, quoted by Perry and Metzger 1980). Decoction of the flowers has a similar usage in Indonesia (Usher 1971), and to treat sore eyes in India (Das 2004). Also in India, rhizomes are boiled and the extract is consumed to treat gonorrhoea (Das 2004).

Cymbidium faberi Rolfe

Chinese names: *Jiuhua Lan* (nine flower orchid), *Yijingjiuhua* (nine splendour flower), *Tubaibu* (wild hundred steps); *Taiwanyijingjiuhua* (Taiwan Jiuhua blossom), *Hui Lan* (pure heart orchid), *Changye Lan* (long leaf orchid), *Huaqi Lan* (clearing gas orchid) . In Taiwan: multi-flowered orchid
Medicinal name: *hua qi lan*

Description: *C. faberi* is a large terrestrial herb with inconspicuous pseudobulbs and 6–10 linear, grass-like leaves, 60–90 cm long and 8–12 mm wide with serrated margins. Raceme carries 12–18 loosely arranged, fragrant flowers of pale green or yellow tinged with light purple, with purplish-red patches on the lip (Liu et al. 2006). Some varieties are very fragrant. Flowering period is February to May. In cultivation, it likes dampness but a well-drained medium. *C. faberi* occurs south of the Yellow River in China, in Taiwan, and in Nepal, Bhutan and northeast India, in sunny grassland or sparse forests at 700–3000 m, often in association with *Miscanthus* spp. (a perennial grass) or the Bhutan white pine, *Pinus bhutanica* (Gurong 2006).

Herbal Usage: Herb is collected in autumn. After washing, leaves and roots are removed from the pseudobulbs which are sun-dried for storage. The root is bitter, sweet, mild and slightly poisonous. *C. faberi* is used for the relief of headache or coughs, and to destroy insects, worms and lice. A decoction is taken for headache, while 6 g in decoction is consumed with white wine once a day to relieve coughs. To clear the bowel of *ascaris* (round worms), 500 g of *C. faberi* pseudobulb is added to wheat powder and made into buns. These are consumed over 3 days (*Zhongyao Da Cidian* 1986).

Cymbidium finlaysonianum Lindl.

Malay name: *Sepuleh*
Thai name: *Ka Re ka Ron Pak Pet*

Description: A common lowland orchid in Southeast Asia, it is tolerant of strong sunlight, often growing as large clumps on trees or rocks near the roadside. Pseudobulbs are short and carry thick, leathery, strap leaves that measure 75 by 4 cm. The pendulous, metre-long inflorescence bears 20–24, well-spaced, yellow to chocolate-coloured flowers that are streaked with red. Lip is white with purple on the side lobes and it carries a crescent-shaped patch on its curled tip (Fig. 9.44). Flowering season is May.

Fig. 9.44 *Cymbidium finlaysonianum* Lindl. (Photo: E.S. Teoh)

Phytochemistry: 7-*O*-glycosides of vitexin and isovitexin was identified from *C. finlaysonianum* (Williams 1979).

Herbal Usage: Burkhill and Haniff (1930) reported that the Malay medicine men used it to remove bewitchment in Telok Anson, in the northwest of Peninsular Malaysia. At that time, malevolent spirits were thought to be the cause of numerous serious illnesses. Such employment of *C. finlaysonianum* and other orchids in the Malay magical approach to treating illness is indicated by their common Malay name, *sepuleh* which ranslates as "restorative", i.e. restoring to health.

Cymbidium flaccidum Schltr. (see: **Cymbidium crassifolium Herb.**)

Cymbidium floribundum Lindl.

Syn. *Cymbidium floribundum* Lindl. var. *pumilum* (Rolfe) Y.S. Wu et S.C. Chen; *Cymbidium pumilum* Rolfe

Local name: *Duohua Lan* (many flowered *Cymbidium*)

Description: Pseudobulbs are ovoid and a little flattened, 2.5–3.5 cm long, carrying 5 or 6 thin, coriaceous leaves, 50 by 0.8–1.8 cm. Inflorescence is suberect, lateral, with numerous flowers on the raceme. Flowers are well arranged and displayed, 3–4 cm across, of variable coloration, reddish-brown to green or brownish-grey with a white lip that is spotted with red. Flowering season is April to August. A large clump is very handsome when it produces numerous sprays of reddish flowers (Liu et al. 1969, 2006).

This epiphytic, occasionally terrestrial or saxicolous *Cymbidium* is widely distributed through central and southern China (Zhejiang, Jiangxi, Fujian, Taiwan, Guangdong, Guangxi, Hunan, Guizhou, Hubei, Sichuan, Yunnan and Xizang) at 100–3300 m. Plants are found in forests, at the edge of forests or on sunny cliffs and along ravines, and very rarely on rocky soil.

Herbal Usage: Herb is obtained from Huadong (Gouangdong Province), Huanan

(Heilongjiang Province) and Tibet. Entire plant is used in the same manner as *C. ensifolium* (*Zhongyao Da Cidian* 1986; Wu 1994).

Japanese honey bees (workers, drones, queens and absconding bees) are attracted by fragrances emitted by *C. floribundum* which resemble compounds present in their mandibular glands. These are a mixture of 3-hydroxy octanoic acid and 10-hydroxy (E)-2-decenoic acid (Sugahara et al. 2013). Shiseido markets a perfume that contains the scent of *Cymbidium*. The French perfume *Diorissimo*® attracts *Euglossine* bees in South America (Pijl and Dodson 1966) but we are not able to determine whether Shiseido's perfume attracts Japanese honey bees.

Cymbidium floribundum Lindl. var. *pumilum* (Rolfe) Y.S. Wu et S.C. Chen (see ***Cymbidium floribundum Lindl.***)

This variety is given different species status in the *Chinese Materia Medica* (2000), but it is not separated from *C. floribundum* Lindl. in *Flora of China* (Liu et al. 1969). Its medicinal usage is similar to that of *C. floribundum* (*Zhongyao Da Cidian* 1986; Wu 1994).

Cymbidium goeringii (Rchb. f) Rchb. f.

Local names: *Chun Lan* (spring orchid), *Riben Chun Lan* (Japanese spring orchid), *Diaolanhua* (hanging orchid flowers); *Cao Lan* (grass orchid); *Shan Lan* (mountain orchid); *Shuangfeiyan* (twin flying sparrow)

Japanese: *Hokuro* (black seeds/age spots); *Jiji-baba* (grandpa and grandma)

Description: This is a terrestrial orchid with an inflorescence that carries a single fragrant flower, rarely two. Pseudobulbs are small, ovoid 1–2.5 by 1–1.5 cm. and enclosed by leaf bases. Leaves are 4–7, slender, lorate, 20–40 cm by 0.5–1 cm, slightly serrated on the margins. Inflorescence arises near the base of the pseudobulb, is obliquely erect, short, 2–5 cm, and usually 1-, occasionally 2-flowered. Typical varieties are green or a straw-coloured, with a white lip stippled with maroon spots (Fig. 9.45). It flowers from January to March.

Fig. 9.45 *Cymbidium goeringii* (Rchb. f.) Rchb.f. ((Photo: E.S. Teoh)

Widely distributed throughout most of China (excluding the very northern provinces in Manchuria, Inner Mongolia, Xinjiang and Xizang), it is also found in Bhutan (at 500–3000 m with *Pinus bhutanica, P. roxburghii, Quercus lanata* and in mixed broad-leaved forests), in India, Korea and Japan (Liu and Nakayama 2007). It prefers stony habitats, shrubby slopes or sparse forests, at 300–2200 m. The southern regions of Korea represent its northernmost distribution, and here the clumps face south where they are exposed to wind speeds of 3 m/s. The swift wind cools the leaves in summer. In winter, wind blowing from the south maintains leaf temperature above minus 6 °C. Elsewhere in Korea, these conditions are not met and *Cymbidium goeringii* can barely survive in the wild (Cho and Beyoung 1995).

Phytochemistry: Cymbidine A, a monomeric peptidoglycan-related compound isolated from *C. goeringii*, possesses diuretic and hypotensive activities (Watanabe et al. 2007). Gigantol isolated from whole plants of *C. goeringii* by Won et al. (2006) exhibits inhibitory effects of LPS-induced nitric oxide and prostaglandin E2 (PGE2) production in macrophages. It is a potent inhibitor of tumour necrosis factor-alpha (TNF-alpha), interleukin-1beta (IL-1beta) and interleukin-6(IL-6) release, and it influences mRNA expression of these cytokines in a dose-dependent manner. These effects are produced through its ability to block nuclear factor kappa B (NF-kappaB) activation (Won et al. 2006).

Three new diketopiperazines were recently isolated from the fungus *Chaetomonium cochliodes* 88194 recovered from *C. goeringii* collected from Xinning in Hunan Province, China in 2008. Of the three compounds, only chaetocochin G arrested cell proliferation and induced apoptosis of MCF-7 human breast cancer cells in vitro. The other two compounds did not exhibit cytotoxicity (Wang et al. 2015).

C. goeringii is admired for its faint floral fragrance. This is constituted by a blend of methyl-cis(z)-dehydrojasmonate, (E) neroldol, 1,2,4-trimethoxybenzene, 1,2,3,5-tetramethoxybenzene and other jasmonates (Kaiser 1993).

Herbal Usage: According to TCM, roots improve blood flow, cool the blood and detoxify. The herb is used to treat traumatic injuries, bleeding from such injuries, and fractures, clear heat in the lungs, and relieve coughs and sore throat, stop the production of blood streaked phlegm, and treat haematuria and rabies. The entire plant is used to treat fever, large round worm infestation (ascariasis), abdominal colic associated with worm infestation, poor health, weak kidneys, dizziness, backache, sweating and piles (*Zhongyao Da Cidian* 1986).

Cymbidium hookerianum Rchb. f.

Chinese name: *Hutou Lan*
Chinese medicinal name: *Hutou Lan*

Description: *C. hookerianum* is an attractive, saxicolous or epiphytic *Cymbidium* with strongly fragrant (Chinese variety only slightly fragrant) flowers of apple green with a contrasting cream-coloured lip spotted or blotched with maroon, 5–7 cm across. The inflorescence is arching, 30–40 cm long, and carries 6–15 flowers. The species forms large clumps each pseudobulb carrying 4–6, occasionally 8 narrow, long, flat, arching sometimes slightly twisted, dark green

Fig. 9.46 *Cymbidium hookerianum* Rchb.f. (Photo: E.S. Teoh)

leaves measuring 35–60 by 1.4–2.3 cm (Fig. 9.46). It is found on trees or steep banks in dense forests and evergreen oak forests at 1660–2330 m in Bhutan (Pearce and Cribb 2002), in eastern Nepal, northeast India and southwest China (Yunnan and the adjacent portions of its neighbouring states only), at 1100–2700 m) (Chen and Tsi 1998; Chen et al. 1999). It flowers from February to May in Bhutan; January to April in China.

Herbal Usage: Seeds are applied on cuts and injuries as a haemostatic in India (Rao 2004). The Chinese herb which consists of the whole plant obtained from Yunnan is used to treat fractures and traumatic soft tissue injuries (Wu 1994).

Cymbidium iridioides D. Don

Syn. *Cymbidium giganteum* Wall ex Lindl.

Chinese name: *Huang chan Lan*

Description: Plant is epiphytic or saxicolous. Pseudobulbs are ovoid, with 6–10 linear-lanceolate leaves 70–90 by 2–4 cm, pointed at the tips. Inflorescence is suberect, raceme laxly many-flowered. Flowers are reddish-brown, 7.5 cm across, lasting for months on the plant. Flowering season is September to December in Nepal (Raskoti 2009), August to December in China (Liu et al. 1969). *C. iridioides* is distributed in a narrow band from central and eastern Nepal (at 1500–2800 m) to northern Vietnam across Bhutan, Sikkim, Myanmar, SE Xizang, SW Sichuan, NW and SE Yunnan and SW Guizhou at 900–2800 m (Liu et al. 1969).

Phytochemistry: *C. iridioides* contains a triterpene glucoside, cymbidoside (Dahmen and Leander 1978) and a taraxerane triterpenoid, taraxerone, gigantol and sitosterol (Juneja et al. 1985). Taraxerone was inactive against leukaemia, or renal and ovarian cancer cell lines (Pub Chem CID 392170).

Herbal Usage: In Nepal, juice from the leaves of *C. iridioides* is used as a haemostatic on wounds (Baral and Kurmi 2006; Pant and Raskoti 2013). In the Khasi Hills in India, leaf juice is usedused to stop bleeding from wounds and for diarrhoea (Jalal et al. 2010).

Cymbidium kanran Makino

Local Name: *Han lan* (frigid *Cymbidium* orchid), winter orchid, *Cao Lan* (grass orchid)

Description: This olive-green, scented *Cymbidium* is characterised by very narrow petals and sepals and very long (40–85 cm), thin (1–1.8 cm broad), elegant leaves. Pseudobulbs are ovoid, 2–4 by 1–1.5 cm, enclosed by leaf bases. There are several colour variants, from light apple green to dark olive green, some with striping, and brown (Liu et al. 2006) (Fig. 9.47). It grows in rocky but moist soils along ravines or forests with light shade at 400–2400 m in southern China to Anhui and Zhejiang, and also in Taiwan (in mountainous regions at 800–1500 m in broad-leaved forests near the ridges on the southeast slopes), Korea and Japan. It flowers from August to January but mostly in December and January.

Fig. 9.47 *Cymbidium kanran* Makino (Photo: E.S. Teoh)

This magnificent species enjoys much popularity among Chinese growers on account of its elegant form and strong fragrance.

(line drawing, Chen and Tsi 1998)

Herbal Usage: Herb is obtained from Huadong, Huanan and Yunnan. Chinese herbalists employ the entire plant to "purify the heart", smooth the lungs, or to stop coughs and asthma. Roots are used for treating gastroenteritis and ascariasis (infestation of large intestinal round worms) (Wu 1994).

Cymbidium lancifolium Hook.

Chinese names: *Soushan Hu* (searching mountain tiger), *Zhupo Lan* (bamboo and pine

Cymbidium orchid): *Tuer Lan* (rabbit ear orchid); *Diqingmei* (green floor plum); *Xuli Cao* (Through-the-ages herb, everlasting Herb). In Taiwan: white bamboo-leaf orchid. Indonesian name : *Ki Adjag* in Sunda

Description: *C. lancifolium* is a small, terrestrial or saxicolous, and sometimes epiphytic, species found at low to moderate elevations (at 300–2200 m) throughout China south of the Yangzi and in Southeast Asia. It also occurs in southern Japan, south-eastern Tibet, Bhutan and northeast India. It is found in open forests, broad-leaved forests, bamboo forests, on the edge of forests and on humus-rich rocks along valleys.

Pseudobulbs are fusiform, naked, 5–15 by 0.5 to 1.0 cm, and carry 2–4 leaves on its apex. Leaves are oblong-elliptic to oblong-lanceolate, 6–35 by 2–5.5 cm. When not in bloom, it may be identified by the minute teeth along the upper margins of the leaves. The lateral, erect, short scape bears 2–8 flowers, each 4 cm across, with narrow, pointed petals and sepals of white to pale green, and with purple-maroon markings on the lip (Fig. 9.48). It

flowers in April in Hong Kong and from May to August on the Chinese mainland. In Southeast Asia, its flowers may be seen throughout the year, with peak flowering from August to December after the period of heavy rains. This is possibly the least attractive species in *Cymbidium*.

Phytochemistry: Saponins are present in *C. lancifolium*. Alkaloids are absent (Boorsma 1902).

Herbal Usage: In China, the entire plant is used to relieve rheumatism, improve blood flow, and to treat traumatic injuries (Wu 1994).

Cymbidium longifolium D. Don (see **Cymbidium elegans Lindl. var. elegans**)
Cymbidium mannii Rchb. f. (see: **Cymbidium crassifolium Herb.**)
Cymbidium pendulum (Roxb) Sw. [see **Cymbidium aloifolium (Linn) Sw.**]
Cymbidium pumilum Rolfe (see **Cymbidium floribudnum** Lindl.)

Cymbidium macrorhizon Lindl.

Description: *C. macrorhizon* is a holomycotrophic species which has no green leaves. It lives under-ground and can only be detected when it sends out flowering inflorescences. Rhizomes are subterra-nean, fleshy, white, slightly hairy, 5–10 by 0.3–0.7 cm, branched and with short roots of up to 1 cm length. Inflorescence is usually apical, 11–18 cm, peduncle purplish-red at the base, green above, with 2–5 flowers, 3–4 cm across. Floral segments white or yellow with purplish-red markings. It flowers in June to August (Fig. 9.49). It occurs along riversides, at forest margins or on open grassy slopes from southern China, Vietnam and Thailand to Myanmar and the southern Himalayas up to Pakistan (Liu et al. 1969).

Herbal Usage: In northern India, rhizomes of *C. macrorhizon* are usedused as diaphoretic and febrifuge, and also to treat boils and rheumatism (Lawler 1984; Oudhia 2013).

Fig. 9.48 *Cymbidium lancifolium* Hook. (Photo: E.S. Teoh)

Fig. 9.49 *Cymbidium macrorhizon* Lindl. (Photo: Hubert Kurzweil)

Cymbidium sinense (Jacks.) Willd.

Syn. *Cymbidium chinense* Heynh.

Chinese names: *Baisui Lan* (New Year Greeting Orchid, Pay a New Year's Call Orchid), *Baosui Lan* (Congratulations for the New Year), *Chun Lan* (spring orchid), *Mo Lan* (dark orchid)

Description: This is a large terrestrial species with ovoid pseudobulbs, 2.5–6 by 1.5–2.5 cm that carry 3–4 lustrous, dark green, coriaceous, linear, lanceolate leaves, 60–90 by 2–3.5 cm. The tall, erect inflorescence bears 10–20 or more deep purple or purplish-brown flowers, that are strongly fragrant. They smell like violets (Nakamura, Tokuda and Omata 1990). Sepals and petals are narrow and pointed. Lip is cream, striped with red on the side lobes, and splashed with a border of red on the mid-lobe. It grows in shady, moist, well-drained soil in forests and along ravines at 300–2000 m. Its distribution extends from India to Myanmar, northern Thailand, Vietnam and east China to the Ryukyu Islands of Japan. In China, it

has a long flowering period that extends from October to March (Chen et al. 1999) or to May (Chen, Liu, Chu, et al. 2009) with a peak around Chinese New Year, but in Thailand the chartreuse-coloured variety flowers in August (Vaddhanaphuti 2001) and the dark purplish-brown variety flowers from October to December (Nanakorn and Watthana 2008).

The typical Chinese variety is known as the var. *sinense*. *C. sinense* var. *albo-juncundissimum* (Hayata) Masamune, the Ink Orchid, is native to Hong Kong, occurring at Tai Mo Shan and Sunset Peak (Wu et al. 2002). The var. *haematodes* which has a more southerly distribution which extends to Sri Lanka, Indonesia and Papua New Guinea has broader leaves and scapes that are longer than the leaves. The floral morphology is similar, with red or green forms. Chinese and Japanese collectors admire *Cymbidium* for the shape of its leaves, and they are fascinated by the dark green leaves which carry fine yellow or white lines. Such Golden Thread Orchids are worth a king's ransom. Peloric floral forms also occur.

Herbal Usage: The whole plant or just the roots may be usedused. Roots are collected in autumn and sun-dried for storage. They purify the heart and lungs, and stop coughs and asthma. A decoction prepared with 30 g of the herb is usedused to treat dry coughs (Li 1994).

Cymbidium wilsonii (Rolfe ex De Cock) Rolfe

Chinese names: *Duanyechutou Lan* (short leaf tiger head orchid), *Diannanhutou Lan*

Description: This epiphytic *Cymbidium* has ovoid, flattened pseudobulbs 4–6 by 2.5–3.5 cm. Leaves are slim, pointed, 70–100 by 1.3–3.2 cm. Inflorescence is 25–70 cm long and arching, with 5–15 chartreuse flowers, 9–10 cm across. Petals and sepals are narrow and pointed. Lip is cream, with chestnut stripes on the side lobes and purplish blotches along the edges of the mid-lobe. The lip becomes purplish-red after pollination. Flowering season is

February to April. It is found in southern Yunnan and Vietnam at 2000 m (Liu et al. 2006).

Herbal Usage: Herb is collected from Yunnan. Roots used to treat weak lungs, coughs, bronchitis, tonsillitis and body ache (Wu 1994).

Overview

C. goeringii is the first orchid to receive the attention of mankind. It was mentioned in the *Book of Odes* (collated in the sixth century BC) under its ancient Chinese name "Wild Grass". Confucius likened the character of gentlemen to the nature of *Cymbidium*—simple in its needs, modest, discreet, Spartan, resilient, reclusive and disdainful of honours, noble. When Kubulai Khan conquered China and instituted the most horrendous genocide, Zhen Xuxiao (1250–1300) expressed his grief by painting uprooted *C. goeringii*. This painting is now in the National Palace Museum in Taipei, but there are similar paintings in the collection of the Osaka Municipal Museum. It is one of the four seasonal flowers, and represents spring (Teoh 1982). Although many writers make the mistake of linking *C. ensifolium* with spring, that species flowers from June to October. *Chun Lan* (Spring Orchid) is the Chinese name for *C. goeringii*.

Cymbidiums have been cultivated as house plants in the Far East for more than 2000 years. In the course of time, new varieties have appeared, possessing variegated leaves, multiple flowers and various colour forms, such as narrow central purple streaks along the dorsal sepal and petals, or an orange-coloured flower (the variety *Takahime* or Martial Princess in Japan). Every one of the popular cultivated species (*C. goeringii, C. ensifolium, C. kanran, C. faberi, C. tortisepalum, C. sinense, C. serratum, C. cyperifolium, C. hookerianum, C. floribundum)* have produced mutations and there are several books in the Chinese language dedicated to describing such mutations. However, in so far as the aesthetic appreciation of *Cymbidium* emphasises simplicity and gracefulness of form and delicacy of fragrance, this should immediately disqualify most of the mutants.

Hybrids of *Cymbidium* enjoy a favoured position in the orchid cut flower trade because of the flower's thick texture and longevity. Classified into four coloured groups, pink, yellow, green and white, their pigmentation is strongest in the lip, and even white forms carry a trace of pigment there. Six anthocyanins were isolated from 8 *Cymbidium* hybrids of four colours by Wang et al. 2014 working in New Zealand. The anthocyanins are: cyanidin-3-*O*-glucoside; cyanidin-3-*O*-rutinoside; peonidin-3-*O*-glucoside; peonidin-3-*O*-rutinoside; cyaniding-3-*O* (malonyl)-glucoside; and peonidin-3-*O* (malonyl)-glucoside. Anthocyanin, which is responsible for the lip coloration, is also present in high concentration in the tepals of pink flowers, but only trace to small amounts occur in flowers of other hues. Chlorophyll and its green breakdown products, pheophytin *a* and *b* responsible for the colour of green flowers, are also present in small amounts in yellow flowers. Yellow flowers carry beta-carotenoid and some additionally contain lutein. Three flavonols, glucosides of kaempferol, quercetin and isohamnetin, are present in all flowers. They are co-pigments that determine the final colour in association with anthocyanins (Wang et al. 2014).

Shiseido offers a perfume, *Tentatrice,* which carries the scent of *Cymbidium.* Ten aromatic glycosides were isolated from fresh flowers of the hybrid *Cymbidium* Great Flower 'Marie Laurencin', and two of these were new discoveries, namely marylaurensinosides D and E (Yoshikawa et al. 2014). Methyl cis(Z) dehydrojasmonate and related compounds are responsible for the fragrance of *C. goeringii* (Kaiser 1993), the orchid whose pleasant scent was usedused as a simile for friendship.

When a *Cymbidium* is used medicinally in TCM, the following names apply:

Lanhua refers to *C. ensifolium, C. goeringii, C. faberi, C. floribundum* and *C. kanran;*

Lanhuaye and *Lanhuagen* refer to *C. ensifolium, C. kanran* and *C. floribundum* var. *pumilum;*

Huaqilan and *Huishi* refer to *C. faberi;*

Fig. 9.50 Some compounds isolated from *Cymbidium*

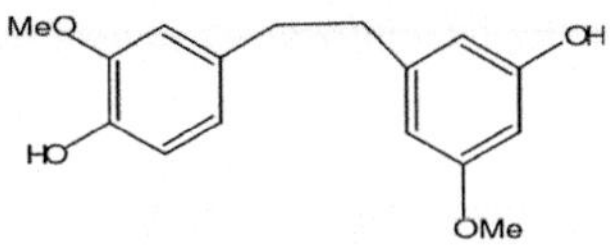

gigantol

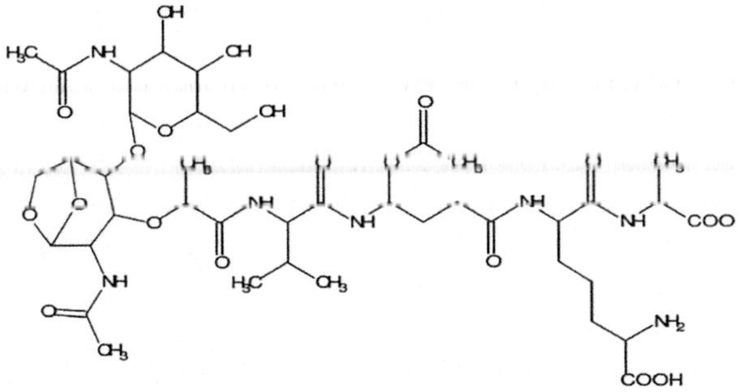

cymbidine A

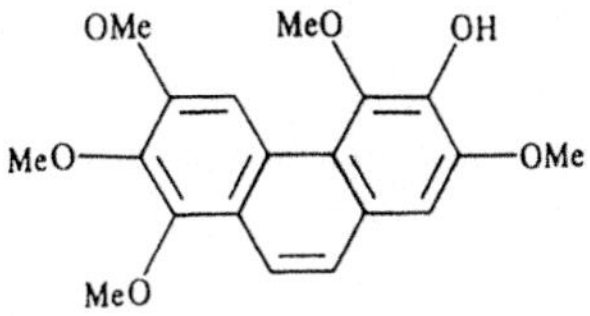

pendulin

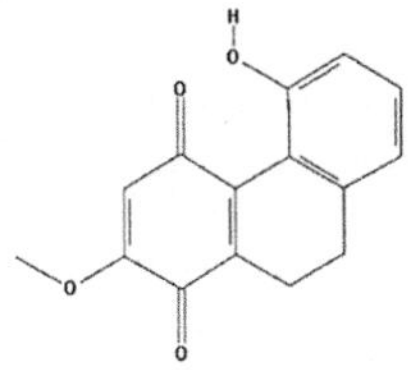

ephemeranthroquinone B

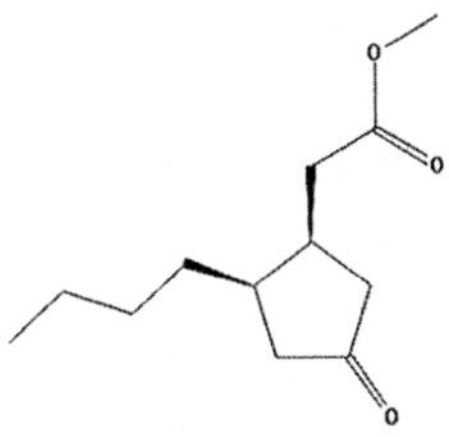

methyl cis dihydrojasmonate

Niujiaosanqi refers to *C. floribundum;*
Hutoulan refers to *C. hookerianum;*
Yingyediaolan refers to *C. pendulum.*

Although coughs and tummy upsets are common conditions for which various *Cymbidium* species are a traditional medicinal remedy, there is no actual extensive usage of *Cymbidium* in TCM. This is probably due to the ready availability of many alternative remedies that have a longer history of usage, are less fancy, cheaper, and perhaps more efficacious. Four strains of mycorrhiza isolated from wild strains of *C. goeringii* showed strong inhibitory activity when tested against two pathogenic bacteria, *Escherichia coli* and *Sarcina lutea* (Min et al. 2012), but presently, mycorrhiza have not been usedused as probiotics.

Whereas *C. aloifolium* with its more southerly distribution enjoys many medicinal usages in the Indian sub-continent and Myanmar, it is totally ignored in TCM despite the orchid being present in several Chinese provinces (Yunnan, Guizhou, Guangxi, Guangdong and Hong Kong). Two Australian *Cymbidium* species are usedused by Queensland aborigines. Delicate children are reared on the readily available *C. canaliculatum* when accidents deny alternative supplies. The Australian aboriginal name for this useful plant is *"dampy-ampy".* White Australians call it "native arrowroot" (Hedley 1888). Seeds of *C. madidum* are usedused as an oral contraceptive. The aborigines also chew on bulbs of any *Cymbidium* when afflicted by dysentery (Lawler and Slaytor 1970).

Ephemeranthroquinone B derived from a *Cymbidium* hybrid, *Cymbidium* Great Flower Marie Laurencin, inhibits *Bacillus subtilis* and has moderate cytoroxicity on lung promyeolcytic leukaemia (HL60) cells in vitro (Yoshikawa et al. 2012). Two new phenanthrenes, marylaurencinols C and D, and a new phenylpropanoid, ephermeranthoquinone, were isolated from *Cymbidium* Great Flower "Marylaurencin" and tested for antimicrobial activity against *Bacillus subtilis, Klebsiella pneumonia* and *Grichophyton rubrum* (Yoshikawa et al. 2014a).

Gigantol present in *C. goeringii* also occurs in *Dendrobium draconis, Dendrobium nobile,*

Dendrobium densiflorum and in several medicinal orchids from Meso-America; there, it was demonstrated to possess anti-inflammatory, analgesic, phytotoxic and spasmolytic activities (Fig. 9.50). Several analogues of gigantol have been synthesised with a view to developing new antitumour, anaesthetic, antidepressant, antipsychotic and spasmolytic agents (Reyes-Ramirez et al. 2011). Gigantol isolated from *Dendrobium nobile* was shown to possess antimutagenic activity (Miyazawa et al. 1997). Gigantol induces apoptosis in non-small cell lung cancer (H460) in vitro (Charoenrungruang et al. 2014).

An interesting group of compounds in *Cymbidium* are the lectins, a conceptionally new class of antivirals which bind to N-linked oligosaccharide elements of enveloped viruses (van de Meer et al. 2007). Test tube experiments showed that mannose-specific lectins from a *Cymbidium* hybrid and *Epipactis helleborine* prevent human immunodeficiency (AIDS) viruses (HIV-1 and HIV-2) and cytomegalovirus (CMV) from reproducing themselves. The 50 % effective concentration of the *Cymbidium* hybrid agglutinin (CA) and *Epipactus helliborine* agglutinin for HIV ranged from 0.04 to 0.08 mcg/ml, which is about 3 orders of magnitude below their toxic threshold (Balzarini et al. 1992). This suggests that they would not be poisonous when administered at the proper therapeutic dosage. However, they would need to be tested on laboratory animals and later on human volunteers before they can be medicine.

Large-scale National Cancer Institute (NCI) screening of 20,000 plant extracts for possible anti-HIV activity showed that approximately 5 % of organic plant extracts tested positive (Cragg and Boyd 1996). Some agents which show promise in vitro may not be usable because either they are not bio-available by oral administration, are only effective in high/near-toxic concentrations, or they possess serious side effects.

Nevertheless, the team from the Rega Institute for Medical Research at Ghent in Belgium and their associates have repeatedly stated that the properties of the plant lectins, which include *Cymbidium* agglutinin, among others, should be taken into consideration in the eventual choice of moving microbiocide candidate drugs into the clinical setting (Balzarini et al. 2004; Turville et al. 2005; Balzarini 2007; Balzarini et al. 2007; Pollicita et al. 2008; Auwerx et al. 2009).

Perhaps more importantly, *Cymbidium* agglutinin (CA) and a number of plant lectins strongly inhibited coronaviruses (transmissible gastroenteritis virus, infectious bronchitis virus, feline coronaviruses serotypes I and II, mouse hepatitis virus), arteriviruses (equine arteritis virus and procine respiratory and reproductive syndrome virus) and torovirus (equine Berne virus) (van de Meer et al. 2007). Scientists at Utrecht University in the Netherlands usedused three antiviral tests based on different evaluation principles to study the plant lectins: (1) cell viability (MTT-based colorimetric assay); (2) the number of infected cells (immunoperoxidase assay); and (3) the amount of viral protein expression (luciferase-based assay). These findings are important because there are no antivirals to combat infection with the Nidovirales (the Order grouping which includes the toroviruses, arteriviruses, roniviruses and coronaviruses). A coronavirus was the cause of the sudden SARS outbreak that caused numerous deaths and brought havoc to the Far East in 2002–2003. In their review, World Health Organization (WHO) experts on SARS from the Centre for Disease Control (CDC) concluded that, after they had examined 54 SARS treatment studies, "it was not possible to determine whether treatments benefited patients during the SARS outbreak. Some may have been harmful" (Stockman et al. 2006). Finding an effective antiviral agent is important. Can *Cymbidium* lectins provide an answer?

Cymibidium macrorhizon, being a holomycotrophic geophyte, should contain some interesting compounds. Unfortunately, published data have not appeared. There are altogether 170 species of holomycotrophic orchids within Asia (including eastern Russia) and the western Pacific (Campbell 2014) which can provide enormous opportunities for new discoveries. Chaetocochin, a diketopiperazine produced by *Chaetomium cochliodes*, isolated from *C. goeringii*, was recently shown to possess cytotoxic activity againt breast cancer cells in vitro (Wang et al. 2015).

Epiphytes absorb nutrients from the atmosphere. *C. aloifolium* growing in the environment of Kaiga, on the southwest coast of India where two nuclear power reactors were being constructed in 2001, contained elevated levels of the radionuclide 137 V. Higher fallout occurred when it rained (Karunakara et al. 2001). *Cymbidium* and other orchids may therefore be useful for minitoring the effects of environmental pollution: epiphytes for atmospheric contamination of radioactive elements, terrestrials for soil pollution of heavy metals and radioactive compounds. The findings further suggest that medicinal herbs harvested in contaminated localities might contain undesirable compounds.

Genus: *Cypripedium* Linn.

Chinese name: *Shao Lan*

The generic name is derived from *Cyprus* and *pedium* (Latin, slipper), the slipper of Cyprus. In Greek mythology, Cyprus (alternatively referred to as Paphos; hence *Paphipedilum*) is the island home of Aphrodite who is most beautiful among Greek goddesses.

Plants are small to large, herbaceous, generally terrestrial, sometimes saxicolous, and rarely epiphytic. Stems are unbranched, obscure or erect if long, and leafy. Roots are numerous, thick and fibrous. Inflorescence is terminal: rachis hairy, glandular or glabrous with single or several flowers. Flowers are showy. Dorsal sepal forms a hood over the lip. Petals are free, spreading or droopy. Lip is shaped like a pouch or slipper. Pollen is powdery or viscid (Chen and Cribb 2009). An ancient genus, *Cypripedium* is distributed predominantly in temperate Eurasia and North America, with China as the centre of biodiversity (Fig. 9.51).

Fig. 9.51 *Cypripedium cordigerum* D. Don From: Jacquemont, V., *Voyage dans l'Inde pendant les annees 1828a 1832*, vol. 4 (3): t. 166 (1844). Courtesy of Missouri Botanical Garden, St. Louis, USA

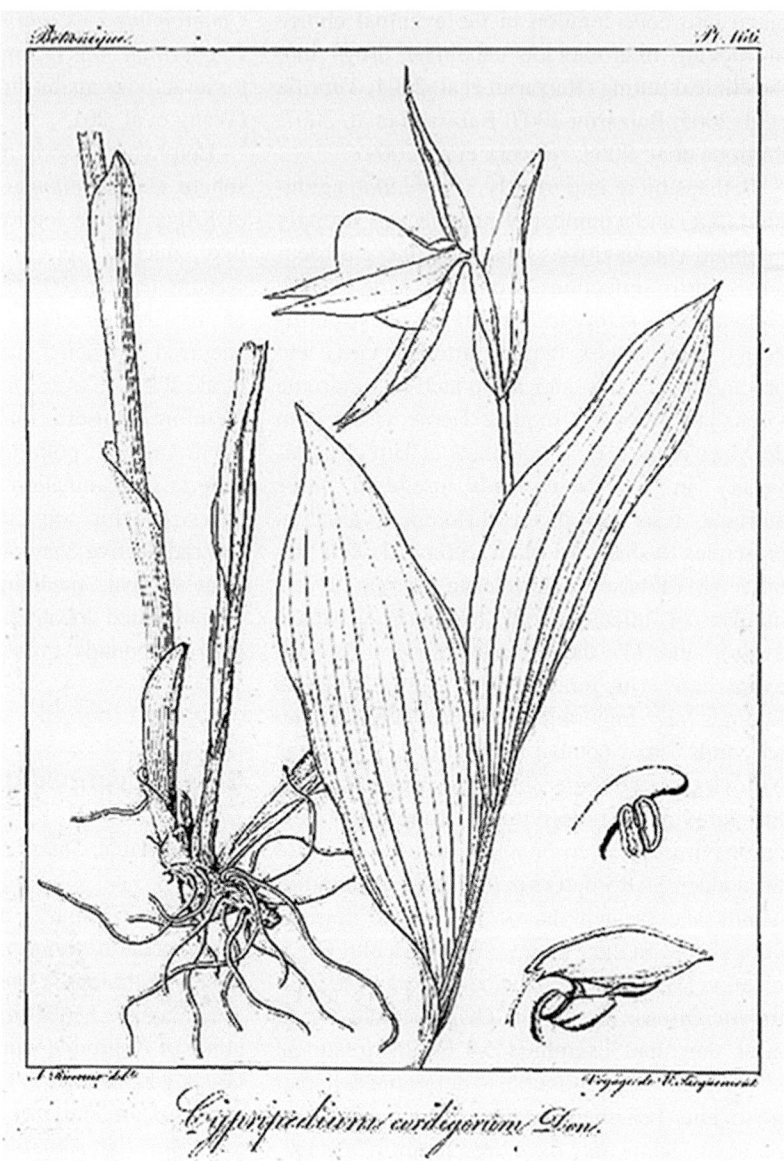

Cypripedium cordigerum D. Don

Chinese name: *Baichunshao Lan*
Nepali Name: *Jibri*

Description: A large terrestrial herb, plant is 50–80 cm tall. Stem is stout, sheathed by elliptic-lanceolate plicate leaves, 7–15 by 5–10 cm. Flower is solitary, 10 cm across. Petals are droopy. Dorsal sepal and petals greenish-yellow, lip white (Fig. 9.52). It is in flower in May to August in Bhutan (Gurong 2006), July and August in Nepal (Raskoti 2009), June to August in Tibet (Chen et al. 1999).

Cypripedium cordigerum occurs in grassland, meadows and pine forests at 2800–3800 m in central and western Nepal, southern Tibet, Bhutan, Kashmir and Pakistan (Chen and Cribb 2009; Pant and Raskoti 2013).

Herbal Usage: Roots are used as a tonic in Nepal (Pant and Raskoti 2013). Young leaves of *Cypripedium cordigerum* D. Don are cooked and

Fig. 9.52 *Cypripedium cordigerum* D. Don (Photo: E.S. Teoh)

eaten as a vegetable (Manandhar and Manandhar 2002).

Cypripedium corrugatum Franch. (see **Cypripedium tibeticum King ex Rolfe**)

Cypripedium debile Reichb. f.

Chinese names: *Duiyeshao Lan* (two leafed spoon orchid), *Xiaoxipuxie Lan* (small Xipu shoe orchid), *Shuangyeshao Lan* (two leaf spoon orchid), *Erye Lan* (two leaf orchid)

Description: A small terrestrial herb, it is probably the least impressive *Cypripedium*. Stem is thin, 8–15 cm tall. Leaves 2, directly opposite each other, small, 2.5–4 by 1–2 cm. A slim, 3-cm-long scape arises from the apex and curls downwards from the weight of the single, 2-cm, pale green flower of typical *Cypripedium* form. Lip is marked with maroon stripes. Flowering season is May to July. It grows on the forest floor in humus or litter-rich soil at 2000–3000 m, and has a restrictive distribution, being found only in southern Gansu, west and northeast Sichuan,

western Hubei, at Mount Shenmachen in Taiwan and in Japan (Chen et al. 1999).

Herbal Usage: The entire plant is used in Taiwan. It improves blood flow, reduces swelling, relieves pain and is diuretic (Ou et al. 2003).

Cypripedium elegans, Rchb. f.

Chinese name: *Yazhishao Lan*

Description: *Cypripedium elegans* is a small plant, 10 cm tall with an erect stem that is densely covered with fine, white hairs. Plant bears 2 sub-opposite, sessile, ovate, apple-green, spreading leaves, 4–5 cm by 3–3.5 cm, laxly pubescent on both surfaces, and marked by 3 prominent veins underneath. Inflorescence is erect, 3–4 cm tall, with a single small flower, 1.5–2 cm long by 0.6–1.0 cm across. Petals and sepals are yellow-green, striped with maroon. Lip is white to yellowish-green and striped with crimson. Flowering period is June to July. It is found in humus-rich habitats on the edge of forests or in thickets at 3600–3700 m in north-west Yunnan and south to southeast Xizang, Nepal, Bhutan, Sikkim and northeast India (Chen et al. 1999)

Herbal Usage: In India, the root is used to treat disorders of the nervous system (Vij 1995). It is used as a nerve tonic and also in hysteria, spasms, fits, insanity, and epilepsy, and in rheumatism in the Himalayan region (Duggal 1971; Das 2004; Baral and Kurmi 2006; Jalal et al. 2010; Pant and Raskoti 2013).

Cypripedium fasciolatum Franch.

Chinese name: *Dayezhuo Lan* (big leaf spoon orchid)
Chinese medicinal name: *Wugongqi*

Description: Plant is robust 35–40 cm tall with a few, well-spaced, cordate, plicated leaves that measure 15–20 by 6–12 cm. Flowers are the largest among the Chinese *Cypripediums*, 10–12 cm across, of a cream to yellow colour, and marked by longitudinal reddish stripes over the synsepalum, petals and pouch. They have a sweet scent when newly open, but the smell becomes pungent when they age (Perner and Luo 2007). Flowering period is May to July. *Cypripedium fasciolatum* is endemic to China and has a small distribution in Sichuan, Hebei and Shanxi, growing in grassland and scrub or in forests at 1650–2100 m.

Herbal Usage: Herb is collected from Sichuan, Hebei and Shanxi. Roots and stems are diuretic and they are usedused to reduce swelling, improve blood flow, clear phlegm and stop pain, in particular joint pains. It is often used to treat generalised oedema, swelling of the lower extremities, fractures and other traumatic injuries (Wu 1994).

Cypripedium formosanum Hayata

Chinese name: *Taiwanshao Lan* (Taiwan spoon orchid), *Taiwanpuxie Lan* (Taiwan ordinary shoe orchid), *Yidianhong* (One Spot of Red), *Taiwanjiapuxie Lan* (Taiwan ordinary shoe orchid)

Description: Plant is 30–40 cm tall. Stem is erect, clothed at the base by the sheath of paired fan-shaped, plicate, finely hirsute, membranous leaves, 8–11 cm wide, and resembling a fan palm. Single, pale pink flowers finely spotted with red, and pubescent towards the base and peduncle, about 8 cm across, are borne on a 10-cm-long peduncle decorated by a few leafy bracts. Lip is pink, dotted with red spots. Flowering season is April to May (Chen and Cribb 2009). *Cypripedium formosanum* is found in forests of the Central Range of mountains in Taiwan (formerly known as Formosa) at 2500–3000 m, in moist sunny locations (Su 1985). It is an endemic, terrestrial species.

Herbal Usage: The entire plant improves blood flow, regulates the menses, expels gas, stops pain and relieves itching. The root and stem also expel gas, improve blood flow, and

they are used to treat malaria, snake bites, traumatic injury and rheumatism (Lin et al. 2003).

Cypripedium franchetii Rolfe

Chinese name: *Maozhuo Lan* (hairy spoon orchid), *Maoshao Lan*
Chinese medicinal name: *Wugongqi*

Description: A pink to purplish-red *Cypripedium* endemic to China, plant is 20–35 cm tall with elliptical leaves 10–16 by 4–6.5 cm, lightly pubescent over the veins on both surfaces. The single flower on the inflorescence is 9–10 cm across, with prominent deep purplish veins on both surfaces of the sepals and petals but lighter markings on the pouch. It flowers in May to July (Chen et al. 1999).

Cypripedium franchetii occurs in humid, humus-rich soil on shrubby slopes and sparse woods at 1500–3700 m in Sichuan and southern Gansu eastwards to southern Shaanxi, southern Shanxi, western Henan and western Hubei (Chen and Cribb 2009).

Herbal Usage: Herb is collected from Shaanxi, Shanxi, Gansu, Henan, Hubei and Sichuan (Wu 1994). It is credited with the ability to regulate the flow of vital energy (*qi*) to eliminate obstruction to its flow, and improve blood circulation (Chen and Tang 1982). It is used to stop coughs, relief pain, "wind stagnation", chest and epigastric pain (Wu 1994). Roots and stems are used in the same manner as *Cypripedium fasciolatum* (*Zhonghua Bencao* 2000).

Cypripedium guttatum Sw.

Chinese name: *Zidianshao Lan*
Chinese medicinal name: *Banhuashaolan*

Description: Plant is 15–25 cm tall, with slender creeping rhizome and an erect, pubescent, glandular stem bearing several sheaths at the base. Leaves, are 2–3, usually two, elliptic or ovate-lanceolate, 5–12 by 2.5–4.5 cm, green, turning black when dried. The small pink flowers are

2.5 cm tall, borne single on a pubescent inflorescence. They appear from May to July (Chen et al. 1999). *Cypripedium guttatum* enjoys the greatest distribution among *Cypripediums* being present in eastern Europe, northern Asia through Japan and North America. It is a small herb which is distributed throughout northern China (except Xinjiang) and southwest China except Guizhou, in forests, scrub and marshes at 1000–4100 m (Yang, Zhang, Feng, et al. 1993). It can be found in both open and shady habitats over a wide geographical range because of a greater efficiency in photochemical utilisation of absorbed light energy and a lower non-photochemical dissipation of excess light energy, but does best with 45 % sunlight (Zhang, Hu, Xu, et al. 2007).

Herbal Usage: The roots and leaves of this orchid have been used in eastern Russia and Siberia to treat epilepsy (Gmelin 1747, Dragendorff 1898, and Hawkes 1944, all quoted by Lawler 1984)

Cypripedium henryi Rolfe

Chinese names: *Luhuazhuo Lan* (green flower spoon orchid);

Chinese medicinal name: *Longshejian* (dragon tongue sword); *Jinlongqi* (gold dragon seven)

Description: Stem is erect, pubescent, 30–60 cm tall, carrying 4–5 broadly elliptical leaves near its tip, 10–18 cm by 6–8 cm. Inflorescence is terminal, 30–70 cm tall, with 2–4 yellowish-green flowers, 6–7 cm across, with narrow petals and sepals and a smooth surfaced, pouched lip. Flowers have the scent of spicy honey and attract a small black wasp, *Lasioglossum sauterum* (Perner and Luo 2007). Flowers appear in April and May.

Cypripedium henryi is an elegant, endemic, terrestrial herb found in open scrub in the Chinese highlands. It thrives in damp humus-rich soil in fairly open places at 800–2800 m from northwest Yunnan towards the northeast to Sichuan, southern Shaanxi, Guizhou and western Hubei.

Herbal Usage: Herb is collected from Shanxi, Gansu, Hubei and from China's southwest (Wu 1994). Roots are used to improve *qi* and blood circulation, reduce swelling and pain, "cold in the stomach", pain around the waist and thighs and pain resulting from injury. A decoction is prepared by boiling 6–9 g of the orchid roots. A standard dose contains 0.3–0.6 g of the roots (*Zhongyao Da Cidian* 1986)

Cypripedium himalaicum Rolfe

Chinese name: *Xiaezhuo Lan* (narrow calyx spoon orchid), *Gaoshanshao Lan*

Nepali name: *Khujukpa*

Description: *Cypripedium himalaicum* is 30–45 cm tall, with uneven broadly elliptic leaves, 5–10 by 2.5–4 cm, and fragrant flowers with lips which are 2–2.5 cm across. Base colour of frontal aspect of the synsepalum is white or greenish-yellow densely marked with purplish-brown, longitudinal stripes; the dorsal surface is a dull purple and striped. Lip is pink to maroon marked with darker stripes (Fig. 9.53). Flowering season is June to August in Nepal, June to July in China. It is widely distributed

Fig. 9.53 *Cypripedium himalaicum* Rolfe (Photo: Nima Gyeltshen)

from the western Himalayas through Nepal, Bhutan, Sikkim, Tibet and northern Burma at rather high altitudes: 3000–4800 m in Nepal, and 3600–4000 m in China (Raskoti 2009; Chen and Cribb 2009). It is usually found in association with dwarf scrub (particularly *Rhododendron*) but has also been discovered growing on grass-clad limestone in association with *C. tibeticum* (Pradhan 1975).

Herbal Usage: *Cypripedium himalaicum* is used in CHM to treat female infertility, hernia and pain at the waist in women (Wu 1994). In Nepal, juice extracted from fresh plants or a drink made with dried plants are used (Subedi et al. 2013). In Nepal, it is used to treat difficulty in passing urine, urinary stones, heart and lung disease and coughs (Baral and Kurmi 2006; Pant and Raskoti 2013).

Fig. 9.54 *Cypripedium japonicum* Thunb. (Photo: Courtesy of Plant Photo Bank of China)

Cypripedium japonicum Thunb.

Chinese name: *Shanmaishao Lan*
Chinese medicinal name: *Shanziqi*
Japanese name: *Kumagiso*

Description: Plant is 35–55 cm tall, erect with 2–3 sub-opposite leaves and several sheaths below. Leaf is broadly fan-shaped with radiating veins, 10–16 by 10–20 cm, glabrous and slightly hirsuite. Inflorescence is terminal, single-flowered, the flower pendulous, 9–10 cm across. Sepals and petals are lanceolate, greenish-yellow spotted with purple. Lip is yellowish-green to white or pale pink, with red spots and veins. Pouch is shaped like a teardrop or egg (Fig. 9.54). Flowering season is April to May. It is found in southern Shaanxi, southern Gansu, Sichuan, Hubei, Guizhou, Hunan, Jiangxi and Anhui, and also in Japan and Korea, in humus-rich soils in forests, thickets, forest margins and shaded slopes at 1000–2000 m (Chen et al. 1999).

Herbal Usage: The herb is said to dispel wind, remove toxins, moderate *qi*, improve blood circulation and relieve pain. It is used in the treatment of tertian malaria, menstrual irregularities,

Table 9.6 Chinese Herbal Prescriptions employing *Cypripedium japonicum* (*Zhongyao Da Cidian* 1986; *Zhonghua Bencao* 2000)

1. Indication: Tertian malaria Pulverizse 1.5 g of roots for consumption with cold water 1 h before onset of symptoms. Caution: Abstain from warm wine and rice for half a day.
2. Indication: pruritic rash Boil whole plants to clean the skin
3. Indication: swellings of unknown etiology Grind whole plants and mix with vinegar for application.

4. Indication: snake bite
(a) Fresh roots of *Cypipedium japonicum* 9–12 g
    Goodyera sp. (Banye Lan) 6 g
    Jinbuhuagen, fresh roots, 60–90 g
    Boil the three ingredients and add to warm wine. Consume three times a day
(b) For application, 60–90 g of *Cypripedium japonicum* mixed with wine and rendered into a paste.

5. Indication: Low backache *Cypripedium japonicum* 6 g decocted and added to wine

pain arising from physical injuries and pruritus (*Zhongyao Da Cidian* 1986; Wu 1994). Prescriptions for its use are mentioned in *Shaanxi Chinese Herbs* (Table 9.6).

Cypripedium macranthos Sw.

Chinese names: *Qilaixipuxie Lan* (big flowered spoon orchid), *Dahuashao Lan* (big flowered spoon orchid), *Dakoudaihua* (large pocket orchid)

Chinese medicinal name: *Dunshengcao, Wugongqi*

Japanese name: *Atsumoriso*

Fig. 9.55 *Cypripedium macranthos* Sw. (as *Cypripedium speciosum* Rolfe). From: *Curtis Botanical Magazine* vol. 137 (ser. 4, vol. 7): t. 8386 (1911) original, a colour drawing by M. Smith. Courtesy of Missouri Botanical Garden, St. Louis, USA

Description: Plant is 25–50 cm tall, carrying alternate elliptic or elliptic-ovate, plicate, pubescent leaves, 15 by 6–8 cm, and a terminal inflorescence with a single flower. The 8-cm-broad, deeply pouched, well-rounded lip reminiscent of the ideal Song porcelain shape is surrounded by the striped, dorsal sepals and petals which curl around it. Holger Perner (2007) found populations of exclusively white forms near Vladivostok. Elsewhere, the colour of the flowers varies from a light pink to deep red or purple. It flowers in June and July (Fig. 9.55). Natural hybrids with *C. calceolus* are common; the resultant natural hybrid, *Cypripedium* × *ventricosum* has been described by many authors and so far bestowed with 14 different names. In Japan, the paler forms are known as var. *speciosum* (Rolfe) Koidzumi, of which the white form is known as var. *speciosum* (Rolfe) Koidzumi f. *albiflorum* (Makino) Ohwi. The yellow form, var. *rebuense* (Kudo) Miyabe et Kudo, occurs only in Rebun Island northeast of Hokkaido.

A large, beautiful, northern Lady Slipper Orchid, *C. macranthos*, occurs in large populations in alpine meadows and scrub land across northern Asia from eastern Russia to the Kamchatka Peninsula, northeastern China (Heilongjiang, Jilin, Liaoning, Nei Mongolia, Hebei and Shandong), Korea and Japan, at 500–2400 m, overlapping with the same distribution as *C. calceolus*, but, unlike the latter species, it does not extend to North America.

A minature *C. macranthos* is present in Taiwan, *C. macranthos* var. *taiwanianum* (Masamune) F. Maekawa, (= *C. segawai* Masamune) or *Ch-lai Cypripedium*. This plant is 20 cm tall with leaves 6 by 3 cm, and flowers 6 cm across. It occurs in the Central Mountains at 3000 m, on ridges and rocky cliffs in the company of dwarf junipers (Perner 2007).

Phytochemistry: Lusianthrin and chrysin were isolated from seedlings of *C. macranthos* that had developed shoots (Fig. 9.56). Lusianthrin was present in minute amounts in aseptic protocorms but levels increased dramatically when protocorms were challenged with *Rhizoctonia* species, its natural symbiont (Shimura et al. 2007, 2009). Chrysin was not present in the infected protocorms. This suggests that lusianthrin maintains a symbiotic balance during germination whereas chrysin protects the adult plant (Shimura et al. 2007). Lusianthrin and chrysin are antifungal compounds.

Herbal Usage: Chinese Herbals state that roots and stem promote diuresis, reduce swelling, clear ecchymosis, expel gas, stop pain and improve blood flow. The orchid is used to treat generalised oedema, swelling of the lower limbs, oliguria, leucorrhoea, gonorrhoea, rheumatism, traumatic injuries, dysentery and illness resulting from overwork (*Zhongyao Da Cidian* 1986; Wu 1994; Ou et al. 2003). Pulverised, dried flowers are used to stop bleeding from wounds. For oral consumption, the decoction is prepared with 6–9 g of the orchid plant and consumed with, or without, wine (*Zhongyao Da Cidian* 1986).

Cypripedium margaritaceum Franch.

Chinese names: *Banyeshao Lan*;
Chinese medicinal name: *Lanhuashuangyecao*

Description: This short, Chinese *Cypripedium* has a 10-cm, erect stem wrapped by two tubular sheaths and carrying two rounded ovate dark green leaves which are heavily spotted with blackish-purple on the superior surface. The leaves are held horizontally across the forest floor. A single flower, 3–4 cm across, appears

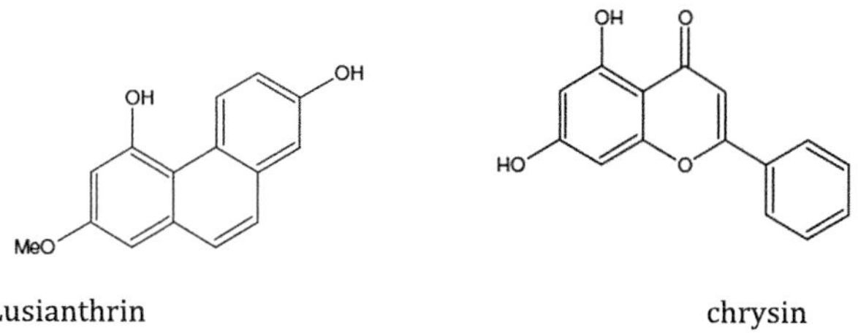

Fig. 9.56 Lusianthrin and chrysin, two phytoalexins from *Cypripedium macranthos* with antifungal properties

Lusianthrin

chrysin

terminally from May to July. Sepals are yellow with red spots and stripes; the petals and pouch white and similarly marked. The dorsal surfaces of the tepals are hairy while the anterior surface of the pouch is warty.

An endemic species, *C. margaritaceum*'s distribution is limited to grassy slopes at 2500–3600 m in southwest Sichuan and northwest Yunnan (Chen and Cribb 2009). It grows in deep shade (Puy and Cribb 1991).

Herbal Usage: *C. margaritaceum* is used when there is a need to nourish the liver and kidneys. It moderates *qi* (vital energy) and blood, promotes diuresis and relieves oedema, and improves blurred vision or night blindness. A commonly used folk herb in Yunnan, decoction is prepared by boiling 9–15 g of the herb (*Zhongyao Da Cidian* 1986).

Cypripedium tibeticum King ex Rolfe

Cypripedium corrugatum Franch.

Chinese name: *Xizang Zhuolan* (Tibetan scoop orchid); *Zhoushao Lan* (crepe spoon orchid)

Fig. 9.57 *Cypripedium tibeticum* King ex Rolfe (Photo: Nima Gyeltshen)

Chinese medicinal name: *Wugongqi; Zhoushao Lan* (crepe spoon orchid)

Description: *C. tibeticum* bears large, nodding, maroon flowers up to 12 cm across with a prominent pouch with a white rim at the mouth. Stem is erect, 15–30 cm tall, carrying three elliptical leaves, 8–16 cm by 3–9 cm, and a terminal inflorescence that has a single flower and a bract-like leaf above the bloom. Flowers are crimson with livid crimson stripes over a light crimson base (Fig. 9.57). Those from Yunnan differ from those of the typical *C. tibeticum* in having brownish-red stripes over a white base, and this feature is important to Chinese herbalists who find a different usage for the Yunnan variety. In Indian Himalaya, *C. tibeticum* rarely exceeds 20 cm in height, and this short stature is a feature which distinguishes it from other *Cypripedium* species when the plant is not in bloom. It flowers from May to August.

C. tibeticum occurs in open forests and on grassy or rocky slopes and at 2300–4200 m from southern Gansu to western Sichuan, Tibet, Sikkim and Bhutan. It is numerous in open scrub, on humus pockets in the central and lower valley of Huanglong, the famous scenic spot in Sichuan (Perner 2002).

Herbal Usage: Herb is collected from Yunnan. Roots of *C. tibeticum* are thought to be anti-inflammatory and capable of preventing pain. They are used to increase urine output, relieve painful swellings, or to improve blood circulation and to treat menstrual disorders. Roots of *C. tibeticum*, 6–9 g in decoction, is consumed in Tibet to treat rheumatism, leg oedema, external injuries, gonorrhoea and leucorrhoea (*Zhongyao Da Cidian* 1986; Wu 1994). With the low boiling point of water at the high altitude of Tibet which can be as low as 56 °C, extraction of the ingredients of any plant would not be as efficient as in the lowlands. This could affect the potency of a decoction.

Phytochemistry: Cypritibetquinone A and B are two new phenanthraquinones isolated from *C. tibetium*. They are 7-hydroxy-2-methoxy-1,4-phenthraquinone and 7-hydroxy-2,10-

dimethoxy-11-4-phenanthraquinone, respectively (Liu et al. 2005).

Overview

Cypripediums are now attracting attraction from taxonomists, gardeners and biochemists. At the Institute of Medicinal Plant Development in Beijing and the Peking Union Medical College (PUMC), Liu et al. (2005) managed to isolate two new phenanthraquinones, cypritibetquinones A and B, from *C. tibeticum*. Cypripediquinone-A isolated from *C. macranthum* in 2000 (Ju et al. 2000) has been synthesised by oxidative coupling using MoC15 (Trosien and Waldvogel 2012), but its medicinal usage, if any, has not been described.

Cypripedium has been used medicinally in western medicine in Europe and in North America. Many *Cypripedium* species are distributed in North America, and American Indians used roots of *C.s* as sedatives and antispasmodics, as well as to treat hysteria and chorea (Griffith 1847). Cherokees prepared root tea with roots of *C. acaule* or *C. calceolus* to treat worms, stomach ache, flu and neuralgia (Hamel and Chiltoskey 1975). *Fluidextractum Cypripedii, U.S.P.* or extract of *C. parvifolium* and *C. pubescens* (sic) was official in the *United States Pharmacopeia* and included in the *British Pharmaceutical Codex* just a hundred years ago. The extract was used to treat nerve disorders and sometimes used as an aphrodisiac. It is interesting to note that, in India, there is a similar usage for *C. elegans*, the root of which is used to treat such disorders of the nervous system as hysteria, spasms, fits, madness and epilepsy (Duggal 1971).

Contact dermatitis occurred in a high percentage of subjects who came into contact with the glandular hairs of *C. spectabile* and *C. pubescens* (MacDougal 1894), but this has not been reported with the Asian species. Contact dermitis following contact with *Cypripedium*s is caused by exposure to quinines and oxalate. Taken internally, overdose of *Cypripedium* induces hallucination (Wilson 2007).

The employment of *C. japonicum* by Chinese herbalists to treat bouts of malaria that recurs every other day (tertian malaria) is interesting and bears further investigation. There are three types of tertian malaria, each caused by a different parasite. Benign tertian malaria produces bouts of high fever with chills and rigours occurring every other day, and is caused by *Plasmodium vivax*. As its name denotes, it only weakens the patient and results in anaemia but is not fatal. Numerous antimalarial drugs like quinine, chloroquine and doxycycline work well for benign tertian malaria. Malignant tertian malaria is quite a different matter. The causative parasite, *Plasmodium falciparum*, does not provoke a high fever response but is capable of infecting the brain, and when it does the disease is often fatal (Hunter 1956). Common antimalarial drugs do not work on *falciparum* malaria but the parasite is killed by artemesinin, which is derived from an ancient Chinese herbal remedy for malaria, *Artemesia annua*. Thus, it would be important to know which type of malaria responds to *C. japonicum*, but that information is not available.

Genus: *Cyrtosia* Blume

Chinese name: *Rou guo lan*

Cyrtosia is a small genus of achlorophyllous, mycotrophic orchids constituted by five Asian species, three of which are found in China. Rhizomes are stout, simple or branched, with fleshy roots and scales at the nodes. Raceme is hairy and many-flowered. Flowers do not open fully and sepals are hairy on their dorsal surface (Chen and Cribb 2009).

C. septentrionalis has the largest seeds among orchids: they weigh 22 mcg each, which is ten times the weight of *Goodyera repens* seed that weigh only 2 mcg each (Rasmussen 1995). Seeds are winged.

Cyrtosia septentrionalis (Rchb. f.) Garay

Syn. *Galeola septentrionalis* Rchb. f.

Chinese name*: Xue hong rou guo lan*
Chinese medicinal name: *Shanshanhu*
Japanese name: *Tsuchi-akebi, Dutuusoo*

Description: This is a large, cool-growing species found in southwest Anhui, and western Henan, Hunan and Zhejiang in China, spreading to Korea, Japan and the Ryukyu Islands. It grows in forests at 1000–1300 m. Its persistent, stout

Fig. 9.58 *Cyrtosia septentrionalis* (Rchb. f.) Garay (Photo: Courtesy of Wikimedia Commons)

1–2 cm diameter, horizontal rhizome occurs 0.5 m below the soil surface and possesses a perennial root system, which reaches upwards to a depth of 5–15 cm below the surface. Each of the 20 main roots is about 1 m long when they are young but eventually they may reach up to 5 m. They produce short secondary roots, many of which establish permanent symbiotic relationships with mycorrhiza, *Armillaria mellea* (Honey Mushroom) or *Armillaria tabescens*, on which the orchid is dependent for its carbon supply (Rasmussen 1995). Inflorescence is up to 90 cm tall with peach-coloured flowers that carry a yellow lip (Fig. 9.58). Flowers appear in late spring and summer (May to July). Fruit is banana red. The aerial portion of the orchid is also achlorophyllous.

In the shaded and sparse understory of forests where *C. septentrionalis* occurs, insect pollinators are limited and the orchid has adapted by developing an effective self-pollinating system. Fruit set following autogamous and xenogamous pollinations were both recorded in central Japan (Suetsugu 2013).

Phytochemistry: Several glycosides were isolated from *C. septentrionalis* (Rchb. f.) Garay in Japan by Inoue et al. (1984). Subsequently, a Chinese team reported the isolation of eight phenolic derivatives from *C. faberi* Rolfe, a related species that occurs in China,

Fig. 9.59 Two phenolic compounds from *Cyrtosia*

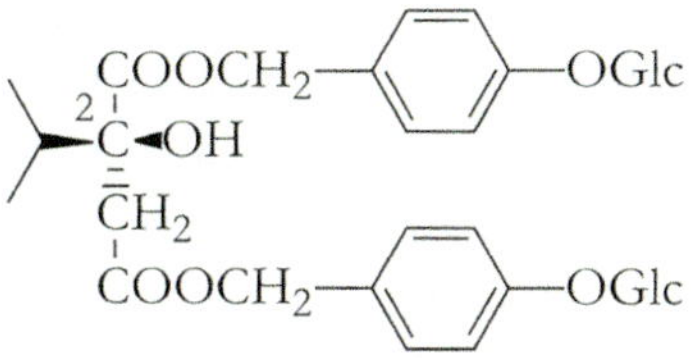

Bis[4(beta-D-glucopyranosyloxy)-benzyl](S)-(-)-2-isopropylmalate

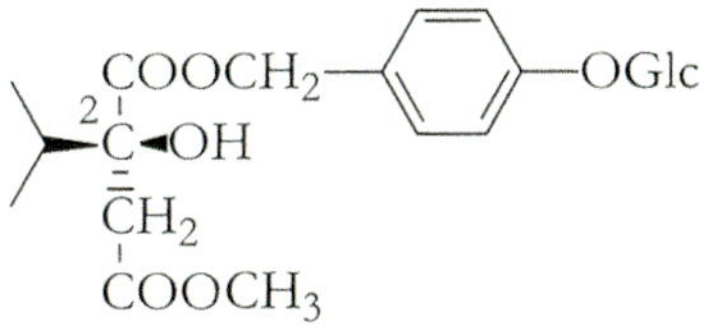

Bis[4(beta-D-glucopyranosyloxy)-benzyl](S)-(-)-2-isopropylmalate methyl ether

Nepal and Sumatra (Li et al. 1993a, b) (Fig. 9.59). The eight phenolic derivatives were:

2,4-bis(4-hydroxybenzyl) phenol

Bis[4(beta-D-glucopyranosyloxy)-benzyl](S)-(-)-2-isopropylmalate

Bis[4-(beta-D-glucopyranosyloxy)benzyl](S)-(-)-2-sec-butylmalate

4-hydroxybenzylaldehyde

4,4′-dihydroxydiphenyl-methane

4-hydroxybenzylalcohol

4-(beta-D-glucopyranosyloxy)benzyl alcohol or gastrodin

5-methoxy-3-(2-[phenyl-E-ethenyl)-2,4-bis (4-hydroxybenzyl)phenol

Herbal Usage: Indications for use of this parasitic orchid mentioned in the *Zhonghua Bencao* are to use the root, 30 g in decoction, to treat stiffness or spasm of the muscles; a paste made with the whole plant and vegetable oil for sores and fungal infection of the skin with ulceration; and fruit with liquorice and prepared in decoction for treating gonorrhea. Another method of preparing the paste for external application to sores, used in Zhejiang, is to fry the plant to dryness and render it as powder before mixing with vegetable oil (*Zhonghua Bencao* 2000). Decoction of the root of *Galeola septentrionalis* was formerly used in Japan to treat gonorrhoea. Ash produced by burning the plant was used as a hair tonic for diseases of the scalp (Lawler 1984).

Overview

In its natural environment in Japan, *C. septentrionalis* was observed not to be capable of attracting pollinators despite its striking flowers, but the species manages to set fruit through autogamy (Suetsugu 2012). It is also not particular about its mycorrhizal association, albeit it is commonly associated with *Armillaria mellea* (Merckx 2012). Germination occurs in sawdust-based medium containing one of four fungal species, *Armillaria gallica, Armillaria mellea,* subsp. *Nipponica, Armillaria tabescens* and *Xylobolus annosus*; germination even occurred in the absence of direct seed–mycobiont contact (Umata et al. 2012). *C. septentrionalis* is therefore quite different from *Gastrodia elata* in

its requirements for germination: *Mycena osmundicola* supports germination of *Gastrodia elata* whereas *Armillaria mellea* is required for its subsequent growth.

It is interesting to note that *Cyrtosia* (*Galeola*) *septentrionalis* has a permanent relationship with either Honey Mushroom (*Armillaria mellea*) or *Armillaria tabescens,* and that gastrodin is present in the tubers of *C. septentrionalis. Armillaria mellea* is responsible for many of the medicinal compounds derived from *Gastrodia elata* (Chinese herbal name: *Tianma*) which is so highly regarded by Chinese herbalists who use it to treat nerve disorders, in particular to promote recovery from stroke (see *Gastrodia elata* in this Chapter). It would not be surprising if either or both of these two achlorophyllous *Cyrtosia* species should share some similar pharmacological properties with *Tianma*. Nevertheless, presently, *C. septentrionalis* is not used in TCM to treat disorders of the nervous system nor hypertension.

Meanwhile, eight phenolic derivatives have been isolated from *C. faberi* Rolfe (reported as *Galeola faberi*) (Li et al. 1993a, b) (Fig. 9.58). Several compounds found in *C. faberi* were also present in *Gastrodia elata,* e.g. gastrodin, 4-hydroxybenzaldehye, 4-hydroxybenzylalcohol (Yang et al. 2007), which is not unexpected since both orchids parasitise on *Armillaria mellea* (Baumgartner et al. 2011). It would not be surprising if similar compounds are present in *C. septentrionalis.*

References

Abadie JC, Puttsepp U, Gebauer G (2006) *Cephalanthera longifolia* (Neottieae, Orchidaceae) is mixotrophic: a comparative study between green and non-green photosynthetic individuals. Can J Bot 84(9):1462–1477

Abraham A, Vatsala P (1981) Introduction to Orchids, with illustrations and descriptions of 150 South Indian Orchids. TPGRI, Trivandrum

Anuradha V, Prakasa Rao NS, Bhaskar MU (1994) Ochrolic acid, a precursor to phenanthropyrones from *Coelogyne ochracea.* Phytochemistry 36:1515–1517

Auwerx J, Francois KO, Vanstreels E et al (2009) Capture and transmission of HIV-1 by the C-type lectin L-SIGN (DC-SIGNR) is inhibited by carbohydrate-binding agents and polyanions. Antivir Res 83 (1):61–70

Balakrishna A, Srivastava A, Mishra R et al (2012) Astavarga plants – threatened medicinal herbs of the North-west Himalaya. Int J Med Arom Plants 2 (4):661–676

Balzarini J (2007) Carbohydrate-binding agents: a potential future cornerstone for the chemotherapy of enveloped viruses? Antivir Chem Chemother 18(1):1–11

Balzarini J, Neyts J, Schols D, Hosoya M, Van Damme E, Peomans W, De Clerq E (1992) The mannose-specific plant lectins from Cymbidium hybrid and Epipactis helleborine and the (N-acetylglucosamine)n-specific plant lectin from Urtica dioica are potent and elective inhibitors of human immunodeficiency virus and cytomegalovirus replication in vitro. Antiviral Res 18 (2):191–207

Balzarini J, van Laethem K, Jatse S et al (2004) Profile of resistance of human immunodeficiency virus to mannose-specific plant lectins. J Virol 78 (19):10617–10627

Balzarini J, Van Herrewege Y, Vermeire K et al (2007) Carbohydrate-binding agents efficiently prevent dendritic cell-specific intercellular adhesion molecule-3-grabbing nonintegrin (DC-SIGN)-directed HIV-1 transmission to T lymphocytes. Mol Pharmacol 71(1):3–11

Baral SR, Kurmi PP (2006) A compendium of medicinal plants in Nepal. Mrs Rachana Sharma and IUCN, Kathmandu

Barua AK, Ghosh BB, Ray S, Patra A (1990) Cymbinobin-A, a phenanthraquinone from Cymbidium aloifolium. Phytochemistry 29:3046–3047

Basavarajappa HD, Lee B, Fei X et al (2014) Synthesis and mechanistic studies of a novel homoisoflavanone inhibitor of endothelial cell growth. PLoS One 9(4), e95694. doi:10.1371/journal.pone.0095694

Basavarajappa HD, Lee B, Lee H et al (2015) Synthesis and biological evaluation of novel homoisoflavonoids for retinal neovascularization. J Med Chem 58 (12):5015–5027

Baumgartner K, Coetzee MP, Hoffmeister D (2011) Secrets of the subterranean pathosystem of Armillaria. Mol Plant Pathol 12(6):515–534

Bensky D, Clavey S, Stoger E, Gambie A (2004) Herbal medicine materia medica, 3rd edn. Eastland Press, Seattle, WA

Bhaskar MU, Rao LJM, Rao NSP, Rao PRM (1989) Ochrolide, a phenanthropyrone from Coelogyne ochracea. Phytochemistry 28(12):3545–3546

Bhaskar MU, Rao LJM, Rao NSP, Rao PRM (1991) Ochrone A, a novel 9.10-dihydro-1,4-phenanthraquinone from Coelogyne ochracea. J Nat Prod 54:386–389

Bhattacharjee SK (1998) Handbook of medicinal plants. Pointer Publishers, Jaipur

Boorsma WG (1902) Pharmakologische Mitteilungen I. Bull de l'Institut Botanique de Buitenzorg 14:1–39

Bose TK, Bhattacharjee SK (1980) Orchids of India. Naya Prokash, Calcutta

Burkhill IH, Haniff M (1930) Malay village medicine. Gard Bull Straits Settl 6:165–321

Burkill IH (1935) (1966 reprint, 2nd edn, with contributions by Birtwistle W, Foxworthy FW, Scrivenor JB, Watson IG). A dictionary of economic products of the Malay Peninsula, vol II. Crown Agents for the Colonies, London. Ministry of Agriculture & Co operatives, Kuala Lumpur

Campbell F (2014) A summary of holomycotrophic orchids. MIOS J 15(4):6–17

Caius JF (1936) The medicinal and poisonous plants of India. J Bombay Nat Hist Soc 38(4):791–799

Charoenrungruang S, Chanvorachote P, Sritularak B, Pongrakhananon V (2014) Gigantol, a bibenzyl from Dendrobium draconis. Inhibits the migratory behavior of non-small cell lung cancer cells. J Nat Prod 77 (6):1359–1366

Chen XQ, Clayton D (2009) Coelogyne. In: Chen XQ, Zj L, Zhu GH et al (eds) Flora of China – Orchidaceae. Science Press, Beijing, pp 315–325

Chen XQ, Cribb PJ (2009) Cyrtosia. In: Chen XQ, Zj L, Zhu GH et al (eds) Flora of China – Orchidaceae. Science Press, Beijing, pp 168–169

Chen SC, Lang KY (1986) Cephalanthera calcarata, a new saprophytic orchid from China. Acta Bot Yunnanica 8:271–274

Chen SC, Tang T (1982) A general review of the orchid flora of China. In: Arditti J (ed) Orchid biology. Reviews and perspectives II. Cornell University Press, Itcaca

Chen SC, Tsi ZH (1998) The orchids of China. Native cymbidiums of China. Wanhai Books, Beijing

Chen SC, Tsi ZH, Luo YD (1999) Native orchids of China in colour. Science Press, Beijing

Chen XQ, Wood JJ (2009) Cleisostoma blume. In: Chen XQ, Zj L, Zhu GH et al (eds) Flora of China – Orchidaceae. Science, Beijing, pp 458–463

Chen XQ, Cribb PJ, Gale SW (2009a) *Calanthe* R. Brown In: Chen XQ, Liu Zj, Zhu GH, et al (eds) Flora of China - Orchidaceae. Science Press, Beijing, pp 292–309

Chen XQ, Gale SW, Cribb PJ, Ormerod P (2009b) Cephalanthera. In: Chen XQ, Liu ZJ, Zhu GH et al (eds) Flora of China – Orchidaceae. Science, Beijing, pp 174–175

Chen XQ, Zj L, Zhu GH et al (eds) (2009) Flora of China – Orchidaceae. Science, Beijing

Cheruvathur MK, Abraham J, Mani B, Thomas TD (2010) Adventitious shoot induction from cultured internodal explants of Malaxis acuminata D. Don, a valuable terrestrial medicinal orchid. Plant Cell Tissue Organ Cult 101(2):163–170

Chinmay R, Suman K, Bishnupriya D et al (2011) Phyto-Pharmacognostical studies of two endangered species of Malaxis (Jeevak and rishibhak). Pharmacogn J 3 (26):77–85

Cho KH, Beyoung HK (1995) Significance of some environmental factors in the natural distribution of Cymbidium goeringii growing wild in Korea and the consideration for cultural practice. Proceedings of 5th Asia Pacific Orchid Conference and Show, Fukuoka, Japan

Chuakul W (2002) Ethnomedical uses of thai orchidaceous plants. Mohidol Univ J Pharm Soc 29(3-4):41–45

Comber JB (2001) Orchids of Sumatra. Natural History Publications (Borneo), Kota Kinabalu

Cragg GM, Boyd MR (1996) Drug development at the National Cancer Institute. In: Balick MJ, Elisabetsky E, Laird SA (eds) Medical resources of the tropical forest: biodiversity and its importance. Columbia University Press, New York

Dahmen J, Leander K (1978) Amotin and amoenin, two sesquiterpenes of the picrotoxane group from Dendrobium amoenum. Phytochemistry 17 (11):1949–1952

Das SP (2004) Indian orchids in indigenous medicine system. In: Brirto SJ (ed) Orchids diversity and conservation. The Rapinat Herbarium, St John's College, Tiruchirappalli

Dash PK, Sahoo S, Bal S (2008) Ethnobotanical studies on orchids of Niyamgiri Hill ranges, Orissa, India. Ethnobot Leaflets 12:70–78

Deb CR, Temjensangba (2006) In vitro propagation of threatened terrestrial orchid, Malaxis khasiana Soland ex Swartz through immature seed culture. Indian J Exp Biol 44(9):762–766

Dhayani A, Nautiyal BP, Nautiyal MC (2011) Importance of Astavarga plants in traditional systems of medicine in Garhwal, Indian Himalaya. Int J Biodiv Sci Ecosyst Serv Manage 6(1–2):13–19

Du Puy D, Cribb PJ (2007) The Genus Cymbidium. Kew Publishing, Royal Botanic Gardens, Kew

Duffy KJ, Kingston NE, Sayers BA et al (2009) Inferring national and regional declines of rare orchid species with probabilistic models. Conserv Biol 23 (1):184–185

Duggal SC (1971) Orchids in human affairs (a review). Pharm Biol 11(2):1727–1734

Ernst R, Rodriguez E (1984) Carbohydrates of the Orchidaceae. In: Arditti J (ed) Orchid biology. Reviews and perspectives III. Cornell University Press, Ithaca

Fung TK (1999) Cymbidium ensifolium. Proceedings of the 16th World Orchid Conference, pp 314–319

Ganesan S, Kesaavan L (2003) Ethnomedicinal plants used by the ethnic group Valaiyans of Vellimalai Hills (Reserve Forest), Tamil Nadu, India. In: Singh V, Jain AP (eds) Ethnobotany and medicinal Plants of India and Nepal. Scientific Publishers (India), Jodhpur, pp 754–760

Gardner DR, Pfister JA (2009) HPLC-MS analysis of toxic norditerpenoid alkaloids: refinement of toxicity assessment of low larkspur (Delphinium spp.). Phytochem Anal 20(2):104–113

Ghosh BB, Ray S, Bhattacharyya P et al (1992) Cymbinodin B, a phenanthraquinone from Cymbidium aloifolium. Indian J Chem 31B:557–558

Gimlette JD, Thomson HW (1939) A dictionary of Malayan medicine. Oxford University Press, London

Godo T, Komori M, Nakaoki E et al (2010) Germination of mature seeds of Calanthe tricarinata Lindl., and endangered terrestrial orchid, by asymbiotic culture in vitro. Vitro Cell Dev Biol Plant 46(3):323–328

Gong CY, Yu ZY, Lu B et al (2014) Ethanol extract of Dendrobium chrysotoxum Lindl. ameliorates diabetic retinopathy and its mechanism. Vascul Pharmacol 62 (3):134–142

Gopal BB, Shankar DS, Kumar PS (2014) Study of antioxidant property of the pseudobulb Extract of Crepidium acuminatum (Jeevak) and its use in the green synthesis of gold nanoparticles. Int J Res Chem Environ 4(3):133–138

Grant B (1895) The Orchids of Burma. Hanthawaddy Press, Rangoon

Griffith RE (1847) Medical botany of descriptions of the more important plants used in medicine, with their history, properties and mode of administration. Lea and Blanchard, Philadelphia

Gurong DB (2006) An illustrated guide to the orchids of Bhutan. DSB Publications, Thimphu

Hamel PB, Chiltoskey MU (1975) Cherokee plants – their uses. A 400 year history. Self published.

Handoyo F (2010) Orchids of Indonesia, vol 1. Indonesian Orchid Society, Jakarta

Hawkes AD (1965) Encyclopaedia of Cultivated Orchids. Faber & Faber, London

Hedley C (1888) Uses of Queensland plants. Proc R Soc Queensland 5:10–13

Henderson CP, Hancock IR (1988) A guide to the useful plants of Solomon Islands. Research Department, University of Agriculture and Lands, Honiara, Solomon Is

Heyne K (1927) De nuttige planten van Nederlandsche Indie, vol 1, pp 508–513. Uitgave van het Departement van Landbouw, Nijverheid & Handel in Nederlandsche-Indie

Howlader MA, Alam M, Ahmed Kh T et al (2011) Antinociceptive and anti-inflammatory activity in an ethanolic extract of Cymbidium aloifolium (L.). Pak J Biol Sci 14(19):909–911

Hu SY (1971) Orchids in the life and culture of the Chinese people. Chung Chi J 10:1–26

Hung T (2014) Lan Thanh Dam – Coelogyne: được tinh (Thanh Lan – Coelogyne: pharmacology). www.hoalanvietnam.org

Hunter D (ed) (1956) Price's textbook of the practice of medicine, 9th edn. Oxford University Press, London

Hur S, Kim T (2009a) Homoisoflavanone, an extract from Cremastra appendiculata Makino, inhibits inflammatory and allergic response in mast cell. J Invest Dermatol 129(Suppl 1):S125

Hur S, Kim T (2009b) Homoisoflavanone, an extract from Cremastra appendiculata Makino, inhibits UVB-induced skin inflammation. J Invest Dermatol 129(Suppl 1):S23

Hur SG, Lee YS, Yoo H et al (2010) Homoisoflavanone inhibits UVB-induced skin inflammation through reduced cyclooxygenase-2-expression and NF-kappa B nuclear localization. J Dermatol Sci 59 (3):163–169

Ikeda Y, Nonaka H, Furumai T, Igarashi Y (2005) Cremastrine, a pyrrolizidine alkaloid rom Cremastra appendiculata. J Nat Prod 68(4):572–573

Inoue S, Wakai A, Konishi T, Kiyosawa S, Sawada T (1984) Studies on Galeola septentrionalis Reichb. fil. I. Isolation and structure of the constituents of "Dotmusoo". Yakugaka Zasshi 104(1):42–49

Jain SP (2003) An inventory of threatened medicinal and aromatic plants of northwestern India. In: Singh V, Jain AP (eds) Ethnobotany and medicinal Plants of India and Nepal. Scientific Publishers (India), Jodhpur, pp 908–913

Jaiswal S, Singh SV, Singh B, Singh HN (2004) Plant used for tissue healing of animals. Nat Prod Rad 3:284–292

Jalal JS, Kumar P, Pangtey YPS (2008) Ethnomedicinal Orchids of Uttarakhand. Western Himalayas Ethnobot Leaflets 12:1227–1230

Jalal JS, Kumar P, Tewari L, Pangtey YPS (2010) Orchids: uses in traditional medicine in India. In: National Seminars on Medicinal Plants of Himalayas. Regional Research Institute, Himalaya, India, Tarikat

Japan Calanthean Society (1987) Brochure, 12th World Orchid Conference, Tokyo

Jayaweera DMA (1981) A revised handbook of the flora of Ceylon, vol II. A.A. Balkema, Rotterdam

Jiangsu College of New Medicine (1986) Zhongyao Da Cidian (Encyclopaedia of Chinese medicine. First class award winner, Vol 1 and II). Shanghai Science and Technology Publishers

Jin XH, Zhao XD, Shi XC (2009) Native orchids from Gaoligongshan mountains China. Science Press, Beijing

Joseph J (1982) Orchids of Nilgiris. Records of the botanical survey of India, Vol XXII, 1982. Botanical Survey of India (Department of Environment), Howrah

Ju JH, Yang JS, Li J, Xiao PG (2000) Cypripediquinone A, a new phenanthraquinone from Cypripedium macranthum (Orchidaceae). Chinese Chem Lett 11(1):37

Juneja RK, Sharma SC, Tandon JS (1985) A substituted 1,2-diarylethane from *Cymbidium giganteum*. Phytochemistry 24:321–324

Juneja RK, Sharma SC, Tandon JS (1987) Two substituted bibenzyls and a dihydrophenanthrene from *Cymbidium aloifolium*. Phytochemistry 26:1123–1125

Kaiser R (1993) The scent of orchids: olfactory and chemical investigations. Editiones Roche, Basel

Kanda K (1977) The native orchids of Japan. Seibundo-Shinkosha, Tokyo

Karthikeyan S, Jain SK, Nayar MP, Sanjappa M (1989) Florae Indicae Enumeratio: Monocotyledonae. Botanical Survey of India, Calcutta

Karunakara N, Somashekarappa HM, Narayana Y et al (2001) 137C concentration in the environment of Kaiga of southwest coast of India. Health Phys 81 (2):148–155

Kim JH, Kim KH, Kim JH et al (2007) Homoisoflavanone inhibits retinal neovascularization through cell arrest with decrease of cdc2 expression. Biochem Biophys Res Commun 362(4):848–852

Kim JH, Kim JH, Yu SH et al (2008) Inhibition of choroidal neovascularization by homoisoflavanone, a new angiogenesis inhibitor. Mol Vis 14:556–561

Kim JH, Kim JH, Lee YM et al (2009) Decursin inhibits retinal neovascularization via suppression of VEGFR-2 activation. Mol Vis 15:1868–1875

Kimura T, But PPH, Guo JX, Sung CK (eds) (2001) International collation of traditional and folk medicine, Northeast Asia Part 1. World Scientific, Singapore, New Jersey, London & Hong Kong

Kovacs A, Vasas A, Hohmann J (2007) Natural phenanthrenes and their biological activity. Phytochemistry 69:1084–1110

Kurzweil H, Lwin S (2014) A guide to orchids of Myanmar. Natural History Publications (Borneo), Kota Kinabalu

Lawler LJ (1984) Ethnobotany of the orchidaceae. In: Arditti J (ed) Orchid biology reviews and perspectives, vol 3. Cornell University Press, Ithaca

Lawler LJ (1986) Orchid ethnobotany in the ASEAN area. In: Rao AN (ed) Proceedings of the 5th ASEAN orchid congress. Parks & Recreation Department, Ministry of National Development, Singapore, pp 42–45

Lawler LJ, Slaytor M (1969) The distribution of alkaloids in New South Wales and Queensland Orchidaceae. Phytochemistry 8:1959–1962

Lawler LJ, Slaytor M (1970) Uses of Australian orchids by aborigines and early settlers. Med J Aust 2:1259–1261

Lee CL, Nakagawa-Goto K, Yu D, Liu YN, Bastow KF, Morris-Natschke SL, Chang FR, Wu YC, Lee KH (2008) Cytotoxic calanquinone A from Calanthe arisanensis and its first total synthesis. Bioorg Med Chem Lett 18(15):4275–4277

Lee CL, Chang FR, Yen MH et al (2009) Cytotoxic phenanthrenequinones and 9.10 dihydrophenanthroquinones from Calanthe arisanensis. J Nat Prod 72 (2):210–213

Lee B, Basavarajappa HD, Sulaiman RS et al (2014a) The first synthesis of the antiangiogenic homoisoflavanone, cremastranone. Org Biomol Chem 12(39):7673–7677

Lee YS, Hur S, Kim TY (2014b) Homoisoflavanone prevents mast cell activation and allergic responses by inhibition of Syk signaling pathway. Allergy 69 (4):453–462

Lee DH, Kim SS, Seong S et al (2014) A case of metastatic bladder cancer in both lungs treated with Korean medicine therapy alone. Case Rep Oncol 7 (2):534–540

Li YM, Zhou ZL, Hong YF (1993a) Studies on the phenolic derivatives from Galeola faberi Rolfe. Yao Xue Xue Bao 28(10):766–771

Li YM, Zhou ZL, Hong YF (1993b) New phenolic derivatives from Galeola faberi. Planta Med 59 (4):363–365

Li L (1994) Chinese medicinal herbs of Hong Kong, vol 6, 198. Hong Kong Chinese Medical Research Institute, Hong Kong

Li S, Xue Z, Wang SJ, Yang YC, Shi JG (2008) Terpenoids from the tuber of Cremasttra appendiculata. J Asian Nat Prod Res 10(7–8):685–691

Lin TP (1858) Calanthe Dominii (Hybrida): Gardeners' Chronicle

Lin TP (1975) Native orchids of Taiwan, vol 1. Southern Materials Center Inc., Taipei

Lin IH, Chang YS, Chen IS, Hsieh WC (2003) The catalogue of medicinal plant resources in Taiwan. China Press, Kuala Lumpur, p 153

Lin LG, Liu QY, Ye Y (2014) Naturally occurring homoisoflavonoids and their pharmacological activities. Planta Med 80(13):1053–1066

Lindstrom B, Luning B, Sirirala-Hansen K (1971) Studies on the Orchidaceae alkaloids XXVI. A new glycosidic alkaloid from Malaxis grandifolia Schltr. Acta Chem Scand 25:1900–1903

Liu YFP, Nakayama H (2007) *Cymbidium goeringii* in Japan. AOS Orchids 76:450–456

Liu ZJ, Chen XQ, Cribb PJ (1969) Cymbidium Swartz. In: Chen XQ, Zj L, Zhu GH et al (eds) Flora of China - Orchidaceae. Science Press, Beijing, pp 260–280

Liu TS, Su HJ (1978) Flora of Taiwan, vol 5. National University of Taiwan, Taipei

Liu D, Ju JH, Zou ZJ et al (2005) Isolation and structure determination of cypritibetquinone A and B, two new phenanthraquinones from Cypripedium tibeticum. Yao Xue Xue Bao 40(3):255–257

Liu ZJ, Chen SC, Ru ZZ, Chen LJ (2006) The genus cymbidium in China. www.sciencep.com

Liu J, Yu ZB, Ye YH, Zhou YW (2008) Chemical constituents from the tuber of Cremastra appendiculata. Yaoxue Xuebao 43(2):181–184

Liu FL, Hsu JL, Lee YJ et al (2014a) Calanquinone A induces anti-glioblastoma activity hrough glutathione-involved DNA damage and AMPK activation. Eur J Pharmacol 73:90–101

Liu L, Ye J, Li P, Tu PF (2014b) Chemical constituents from tubers of Cremastra appendiculata. Zhongguo Zhong Yao Za Zhi 39(2):250–253

Luning B (1967) Studies on the Orchidaceae alkaloids IV. Screening of species for alkaloids 2. Phytochemistry 6:857–861

Luning B (1974a) Studies on Orchidaceae alkaloids. IV. Screening of species for alkaloids 2. Phytochemistry 6:857–861

Luning B (1974b) Alkaloids of the Orchidaceae. In: Withner CL (ed) The orchid. Scientific studies. Wiley, New York, pp 347–382

Luning B (1980) Alkaloids of the Orchidaceae. In: Sukshom Kashemsanta MR (ed) Proceedings of the 9th World Orchid Conference, Bangkok

MacDougal DT (1894) Poisonous influence of various species of cypripedium. Med Plants Nat Hist Bull 9:450–451

Maikhuri RK, Phodani PC, Rao KS et al (2014) Ethnobiology and traditional knowledge of medicinal plants in health care system. In: Uniyal PL, Chamda BP,

Semwal DP (eds) The plant wealth of Uttarakhand. Jagdamba Publishing Co., New Delhi

Majumbder PL, Maiti DC (1991) Isoflaccidinin and Isooxoflaccidin, stilbenoids from Coelogyne flaccida. Phytochemistry 30(3):971–974

Majumder PL, Datta N (1982) Structures of flavidinin and oxoflavidinin, two new modified 9,10-dihydrophenanthrenes of the orchid, Coelogyne flavida. Indian J Chem 21B:534–536

Majumder PL, Datta N (1984) Structure of Oxoflavidin, a 9,10-dihydrophenanthropyrone from Coelogyne elata. Phytochemistry 23:671–673

Majumder PL, Laha S (1981) Occurrence of 2,7-dihydroxy-3,4,6-trimethoxy-9,10-dihydrophenenthrene in Coelogyne ovalis, a high altitude Himalayan orchid: application of C-13 NMR (nuclear magnetic resonance) spectroscopy in structure elucidation. J Indian Chem Soc 58:928–929

Majumder PL (1984) Structure of Oxoflavidin, a 9,10-dihydrophenanthropyrone from *Coelogyne elata*. Phytochemistry 23:671–673

Majumder PL, Maiti DC (1988) Flaccidin, a 9,10-dihydrophenanthropyran derivative from the orchid, Coelogyne flaccida. Phytochemistry 27(3):899–901

Majumder PL, Pal S (nee Roy) (1985) A steroidal ester from *Coelogyne uniflora*. Phytochemistry 29(8):2717–2720

Majumder PL, Pal (Nee Roy) S (1990) A steroidal ester from Coelogyne uniflora. Phytochemistry 29(8):2717–2720

Majumder PL, Sarkar AK (1982) Imbricatin, a new modified 9,10-dihydrophenanthrene derivative of the orchid Pholidota imbricata. Indian J Chem 21B:829–831

Majumder PL, Sen RC (1991) Pendulin, a Polyoxygenated phenanthrene derivative from the orchid, Cymbidium pendulum. Phytochemistry 30(7):2432–2434

Majumder PL, Bandyopadhyay D, Joardar S (1982a) Coelogin and Coeloginin: Two novel 9,10-dihydrophenanthrene derivatives from the orchid Coelogyne cristata. J Chem Soc I:1131–1136

Majumder PL, Datta N, Sarkar AK, Chakraborti J (1982b) Flavidin, a novel 9,10dihydrophenanthrene derivative of the orchids Coelogyne flavida, Pholidota articulata and Otochilus fusca. J Nat Prod 45(6):730–732

Majumder PL, Laha S, Datta N (1982c) Coelonin, a 9.10-dihydrophenanthrene from the orchids Coelogyne ochracea and Coelogyne elata. Phytochemistry 21:478–480

Majumder PL, Banerjee S, Maiti DC, Sen S (1995) Stilbenoids from the orchids Agrostophyllum callosum and Coelogyne flaccida. Phytochemistry 39:649–653

Majumder PL, Sen S, Majumder S (2001) Phenanthrene derivaties from the orchid Coelogyne cristata. Phytochemistry 58(4):581–586

Manandhar NP, Manandhar S (2002) Plants and people of Nepal. Timber, Portland

Manners GD, Panter KE, Pfister JA et al (1998) The characterization and structure-activity evaluation of toxic nordipterpenoid alkaloids from two Delphinium species. J Nat Prod 61(9):1086–1089

Mao TF, Liu T, Liu ZY, Zhu GS, Huang YH (2007) Rapid propagation of Cremastra appendiculata in vitro. Zhong Yao Cai 30(9):1057–1059

Marasini R, Joshi S (2012) Antibacterial and antifungal activities of medicinal orchids growing in Nepal. J Nepal Chem Soc 29:104–109

Matthew KM (1995) An excursion flora of Central Tamilnadu, India. A.A. Balkaema, Rotterdam

Merckx V (2012) Mycoheterotrophy: the biology of plants living on fungi. Springer (eBook)

Min CC, Wang XJ, Liu WB (2012) Preliminary study on isolation of endophytic fungi from Cymbidium goeringii and its antimicrobial activity. Xibei Zhiwu Xuebao 32(3):596–599

Misra S (2007) Orchids of India a glimpse. Bishen Singh Mahendra Pal Singh, Delhi

Miyazawa M, Shimammura H, Nakamura S, Kameoka H (1997) Antimutagenic activity of gigantol from Dendrobium nobile. J Agric Food Chem 45(8):2849–2853

Musharof Hossain M (2009) Traditional therapeutic uses of some indigenous orchids of Bangladesh. Med Arom Plant Sci Biotechnol 3:100–106

Naing AH, Myint KT, Hwang YJ et al (2010) Micropropagation and conservation of the wild medicinal orchid, coelogyne cristata. Hort Environ Biotechnol 51(2):109–114

Nakamura S, Tokado K, Onata A (1990) Japan prize fragrance competition. Am Orchid Soc Bull 59:1031–1036

Nanakorn W, Watthana S (2008) Queen Sirikit Botanic Garden (Thai Native Orchids 1 and 2). Wanida Press, Chiang mai

O'Byrne P (2001) A to Z of South East Asian Orchid Species. Orchid Society of South East Asia, Singapore

Ou JC, Hsieh WC, Lin IH, Chang YS, Chen IS (eds) (2003) The catalogue of medicinal plant resources in Taiwan. Department of Health, Executive Yuan, Taipei

Oudhia P (2013) Medicinal orchid Cymbidium macrorhizon Lindl. based herbal formulations used in Rheumatism in Indian traditional healing: Pankaj Oudhia's Ethnobotanical Surveys 1990–2012

Pandey NK, Joshi GC, Mudaiya RK et al (2003) Management and conservation of medicinal orchids of Kumaon and Garhwal Himalaya. In: Singh V, Jain AP (eds) Ethnobotany and medicinal plants of India and Nepal. Scientific Publishers (India), Jodhpur, pp 114–118

Pant B, Raskoti BB (2013) Medicinal orchids of Nepal. Himalayan Map House (P) Ltd, Kathmandu

Pearce NR, Cribb PJ (2002) The orchids of Bhutan. Royal Botanic Gardens/Royal Government of Bhutan, Edinburgh/Thimpu

Perner H (2002) Orchids and eco-tourism: the world natural heritage and biosphere reserve, Huanglong. Proceedings of the 17th World Orchid Conference, Shah Alam, Malaysia, pp 158–164

Perner H (2007) Cypripediums in China Part IV. Orchids 76(4):291–292

Perner H, Luo Y (2007) Orchids of Huanglong. Huanglong National Park, Sichuan Province, China

Perry LM, Metzger J (1980) Medicinal plants of east and southeast asia: attributed properties and uses. MIT Press, Cambridge, MA

Pollicita M, Schols D, Aquaro S et al (2008) Carbohydrate-binding agents (CBAs) inhibit HIV-1 infection in human primary monocyte-derived macrophages (MDMs) and efficiently prevent MDM-directed viral capture and subsequent transmission to CD4+ T lymphocytes. Virology 370(2):3822–3891

Pradhan UC (1975) The Himalayan Cypripediums. In: Senghas K (ed) Proceedings of the 8th world orchid conference, Frankfurt, 10–17 Apr 1975, German Orchid Society, Frankfurt am Main, 199–204

Preiss K, Adam IKU, Gabauer G (2010) Irradiance governs exploitation of fungi: fine-tuning of carbon gain by two partially myco-heterotrophic orchids. Proc R Soc Biol Sc Ser B 277(1686):1333–1336

Puri HS (1970a) Indian medicinal plants used in elixirs and tonics. Quart J Crude Drug Res 10:1555–1566

Puri HS (1970b) Vegetable Aphrodisiacs of India. Quart J Crude Drug Res 11:1742–1748

Puri HS (1970c) Salep – the drug from Orchids. Am Orchid Soc Bull 39(1):723

Pushpa S, Nipunar M, Pankaj G et al (2001) Malaxis acuminata. A review. Int J Res Ayurveda Pharm 2(2):422–425

Puy DD, Cribb PJ (1991) The genus Cymbidium. Batsford, Dunfermline

Rao AN (2004) Medicinal orchid wealth of Arunachal Pradesh. Newslett ENVISNODE Indian Med Plants 1(2):1–7

Rao TA (2007) Ethno botanical data on wild orchids of medicinal value as practised by tribals at Kudremukh National Park in Karnataka. Orchid Newslett 2(2):1–7

Rao TA, Sridhar S (2007) Wild Orchids in Karnataka. A pictorial compendium. Institute of Natural Resources Conservation, Education, Research and Training (INCERT), Bangalore

Raskoti BB (2009) The orchids of Nepal. Bhakta Bahadur Raskoti and Rita Ale, Kathmandu

Rasmussen HN (1995) Terrestrial orchids from seed to mycotrophic plant. Cambridge University Press, Cambridge

Rathore SR (1983) Endemic and rare Species of Calanthe R. Br. (Orchidaceeae) in India. In: Jian SK, Rao RR (eds) An assessment of threatened plants of India. Botanical Survey of India, Delhi

Reddy KN, Raju GV, Reddy CS, Raju VS (2005) Ethno-orchidology of orchids of Eastern Ghats of Andra Pradesh. EPRTI Newsl 11(3)

Reyes-Ramirez A, Leyte-Lugo M, Figueroa M et al (2011) Synthesis, biological evaluation, and docking studies of gigantol analogs aas calmodulin inhibitors. Eur Med Chem 46:2699–2708

Rhodehamel WA (2005) Calanthe ceciliae. Orchids 74(12):900–902

Ridley HN (1906) Malay Drugs. Agricultural Bull. Straits Settlements and Fed Malay Staes 5, 245 and 277

Rumphius GE (1741–1750, posthumus). *Amboinsch Kruidboek* (*The Amboinese Herbal, Vol. 1–6*). Translated and annotated into English, with an introduction by EM Beekman (2011). Yale University Press, New Haven and London

Sachdev K, Kulshreshtha DK (1986) Phenolic constituents of Coelogyne ovalis. Phytochemistry 25 (2):499–502

Santapau H, Kapadia Z (1966) The Orchids of Bombay. Government of India Press, Calcutta

Sarin YK (1995) Ethnopharmacological perspectives of ayurvedic drugs having controversial botanical identity. In: Pushpangadan P, Nyman U, George V (eds) Glimpses of Indian ethnopharmacology. Tropical Botanic Garden and Research Institute, Thiruvananthapuram, pp 179–184

Schuiteman A, de Vogel EF (2000) Cac Ci Ho Lan (Orchidaceae) Cua Thai Lan, Lao, Campuchia Va Viet Nam. Orchid Genera of Thailand Laos, Cambodia and Vietnam. (Vietnamese-English edition). Leiden: National Herbarium Nederland

Seidenfaden G (1985) Contributions to the Orchid Flora of Thailand. XII. Dendrobium Sw. Opera Botanica 83, Copenhagen

Seidenfaden G, Wood JJ (1992) The orchids of Peninsular Malaysia and Singapore. Olsen & Olsen, Fredensborg

Sharma C, Mansoori MN, Dixit M et al (2014) Ethannolic extract of Coelogyne criststa Lindley (Orchidaceae) and its compound coelogin promote osteoprotective activity in ovariectomized oestrogen deficient mice. Phytomedicine 21(12):1702–1707

Shim JS, Kim JH, Lee JY et al (2004) Anti-angiogenic activity of a homoisoflavanone from Cremastra appendiculata. Planta Med 70(2):171–173

Shimura H, Matsuura M, Takada N, Koda Y (2007) An antifungal compound involved in symbiotic germination of Cypripediumacranthos var. ruenenese (Orchidaceae). Phytochemistry 68(10):1442–1447

Shimura H, Sadamoto M, Matsuura M et al (2009) Characterization of mycorrhizal fungi isolated from the threatened Chprpedium macranthos in a northern island of Japan: two phylogenically distinct fungi associated with the orchid. Mycorrhiza 19 (8):525–534

Singh H (2010) Haat kali sacred grove, central himalaya, Utttarakhand. Curr Sci 98:290

Singh A, Duggal S (2009) Medicinal orchids: an overview. Ethnobot Leaflets 13:351–363

Slaytor MB (1977) The distribution and chemistry of alkaloids in the Orchidaceae. In: Arditti JA (ed) Orchids biology reviews and perspectives, vol 1. Cornell University Press, Ithaca and London

Stockman LJ, Bellamy R, Garner P (2006) SARS: systematic review of treatment effects. PLoS Med 3(9), e343

Su, HJ (1985) Native orchids of Taiwan. Revised Third Ed. Harvest Farm Magazine, Taipei

Subedi A, Kunwar B, Choy Y et al (2013) Collection and trade of wild harvested orchids in Nepal. J Ethnobiol Ethnomed 9:64–73

Suetsugu K (2012) Autogamous fruit set in mycoheterotrohic orchid Cyrtosia septentrionalis. Olant Syst Evol 299(3):481–486

Sugahara MI, Kazunari I, Nishimura Y, Sakamoto F (2013) Oriental Orchid (Cymbidium floribundum) attracts the Japanese honey bee (Apis cerana japoinica) with a mixture of 3-hydroxyoctanoic acid and 10-hydroxy-(E)-2-decanoic acid. Zool Sci 30 (4):99–104

Suetsugu K (2013) Autogamous fruit set in a mycoheterotrophic orchid *Cyrtosia septentrionalis*. Plant Syst Evol 299:481–486

Sun HQ, Liu YB, Song GE (2003) A preliminary on pollination of an endangered orchid, Changnienia amoena in Shennongjia. Acta Bot Sin 45(9):1019–1023

Sun HQ, Li A, Ban W, Zheng XM, Ge S (2005) Morphological variation and its adaptive significance for changnienia amoena, an endangered orchid. Biodiv Sci 13(05):376–386

Tang SL, Su HJ (1978) Flora of Taiwan, vol 5. Department of Botany, Taiwan National University, Taipei

Teoh ES (1982) A joy forever. Vanda Miss Joaquim. Singapore's National Flower. Singapore: Times (reprinted 2008. Singapore: Marshall Cavendish)

Trosien S, Waldvogel SR (2012) Synthesis of Highly funtionalized 9,10-phenenthrenequinones by oxidative coupling using MoC15. OrganLett 14(12):2976–2979

Tsavkelova EA, Cherdyntseva TA, Lobakova ES et al (2001) Microbiota of the orchid rhizoplane. Mikrobiologiia 70(4):567–573

Turville SG, Vermeire K, Balzarini J, Schols D (2005) Sugar binding proteins potently inhibit dendriitic cell human immunodeficiency virus type-1 (HIV-1) infection and dendritic –cell directed HIV-1 transfer. J Virol 79(21):13519–13527

Umata H, Ota Y, Yamada M et al (2012) Germination of the fully-mycoheterotrophic orchid Cyrtosia septentrionalis is characterized by low fungal specificity and does not require direct seed-mycobiont contact. Mycoscience 54(5):343–352

University of Yangon Department of Botany (undated, possibly 2004). Myanmar Native Orchids

Usher G (1971) A dictionary of plants used by man. Constable, London

Vaddhanaphuti N (2001) A field guide to the wild orchids of Thailand, 3rd edn. Silkworm Books, Chiang Mai

Vaddhanaphuti N (2005). A field guide to the wild orchids of Thailand, Fourth and Expanded Edition. Silkworm Books, Chiang Mai

van de Meer FJ, de Haan CA, Schuurman NM et al (2007) Antiviral activity of carbohydrate agents against Nidovirales in cell cultures. Antiviral Res 76(1):21–29

van der Pijl L, Dodson CH (1966) Orchid flowers and their pollination. University of Miami Press, Coral Gables

van Rheede HA (1703) Hortus Indicus Malabaricus, vol 12. Dutch East India Company, Kerala

Veitch N, Grayer R (2007a) Calanthe. Phytochemistry. In: Pridgeon AM, Cribb PJ, Chase MW (eds) Genera Orchidacearum, vol 4, Epidendroideae (Part One). University Press, Oxford, pp 126–127

Veitch N, Grayor R (2007b) Cephalanthera. Phytochemistry. In: Pridgeon AM, Cribb PJ, Chase MW (eds) Genera Orchidacearum, vol 4, Epidendroideae (Part Two). University Press, Oxford, pp 500–501

Vidal J (1963) Les plantes utiles Du Laos. Cryptograms – Gymnospermes – Monocotyledones. Museum National d'Historie Naturelle, Paris

Vij SP (1995) Orchid genetic diversity in india: conservation and commercialization. In: Proceedings of the 5th Asia Pacific orchid conference and Show, Fukuoka, pp 20–39

Wang Y, Guan SH, Meng YH et al (2013) Phenanthrenes, 9,10-dihydrophenanthrenes, bibenzyls with their derivatives, and malate or tartrate benzyl ester glucosides from tubers of Cremastra appendiculata. Phytochemistry 94:268–276

Wang L, Albert NW, Zhang HB et al (2014) Temporal and spatial regulation of anthocyanin biosynthesis provide diverse flower colour intensities and patterning in Cymbidium orchid. Planta 240(5):983–1002

Wang FQ, Tong QY, Ma HR (2015) Indole diketopiperazines from endophytic Chaeomium sp 88194 induce breast cancer cell apoptotic death. Sci Reports 5: Article 9294: 1–9

Watanabe K, Tanaka R, Sakurai H et al (2007) Structure of Cymbidine A, a monomeric peptidoglycan-related compound with hypotensive and diuretic activities, isolated from a higher plant, Cymbidium goeringii (Orchidaceae). Chem Pharm Bull (Tokyo) 55 (5):780–783

Williams CA (1979) The leaf flavonoids of the Orchidaceae. Phytochemistry 18:803–813

Wilson MF (2007) Medicinal plant fact sheet. Cypripedium: Lady slipper orchids. A collaboration of the IUCN Medicinal Plant Specialist Group, PCA Medicinal Plant Working Group, and the North American Pollinator Protection Campaign. Arlington, Virginia. http://www.pollinator.org/Resources/Cypripedium.draft.pdf

Won JH, Kim JY, Yun KJ et al (2006) Gigantol isolated from the whole plants of Cymbidium goeringii inhibits the LPS-induced iNOS and COX-2 expression via NF-kappaB inactivation in RAW 264.7 macrophages cells. Planta Med 72(13):1181–1187

Wu XR (1994) A concise edition of medicinal plants in China. Guangdong Higher Education Publication House, Guangdong (in Chinese)

Wu TL, Hu QM, Xia NH, Lai PCC, Yip KL (2001) Check list of Hong Kong plants 2001. Agriculture, Fisheries and Conservation Department, Hong Kong

Wu TL, Hu QM, Xia NH, Lai PCC, Yip KL (2002) Check List of Hong Kong Plants 2001. Agriculture, Fisheries and Conservation Department Bulletin 1 (Revised), Hongkong

Xia WB, Xue Z, Li S, Wang SJ, Yang YC, He DX, Ran GL, Kong LZ, Shi JG (2005) Chemical constituents from the tuber of Cremastra appendiculata. Zhongguo Zhong Yao Za Zhi 30(23):1827–1830

Xue Z, Li S, Wang SJ, Yang YC, He DX, Ran GL, Kong LZ, Shi JG (2005) Studies on the chemical constituents from the corm of Cremastra appendiculata. Zhongguo Zhong Yao Za Zhi 30(7):511–513

Xue Z, Li S, Wang S, Wang Y, Yang Y, Shi J, He L (2006) Mono-, Bi-, and triphenanthrenes from the tubers of Cremastra appendiculata. J Nat Prod 69(6):907–913

Yagame T, Yamato M (2013) Mycoheteroetrophic growth of Cephalanthera falcata (Orchidaceaea) in tripartite symbiosis with Thelephoraceae fungi and Quercus serrata (Fagaceae) in pot condition. J Plant Res 126(2):215–222

Yagame T, Funabike N, Nagasawa E et al (2013) Identification and symbiotic ability of Psathyrellaceae fungi isolated from a photosynthetic orchid, Cremastra appendiculata (Orchidaceae). Am J Bot 100 (9):1823–1830

Yamazaki J, Miyoshi K (2006) In vitro asymbiotic germination of immature seed and formation of protocorm by Cephalanthera falcata (Orchidaceae). Ann Bot (Lond) 98(6):1197–1206

Yan J, Li C, Chen S et al (2002) The effects of twenty one traditional Chinese medicines on tyrosinase. Zhang Yao Cai 25(10):724–726

Yang ZH, Zhang QT, Feng ZZ, Lang KY, Li H (1998) English edition translated by ZR Xiong (ed.) Orchids. Beijing: China Esperanto Press

Yang XD, Zhu J, Yang R et al (2007) Phenolic constituents from the rhizomes of Gastrodia elata. Nat Prod Res 21:180–186

Yong HS (1990) Orchid Portraits. Tropical Press Sdn Bhd, Kuala Lumpur

Yang ZH, Zhang QT, Feng ZZ, Lang KY, Li H (1993) English edition: Orchids (trans: Xiong ZR). China Esperanto, Beijing

Yoshikawa M, Murakami T, Kishi A, Sakurama T, Matsuda H, Nomura M, Matsuda H, Kubo M (1998) Novel indole S, O-bisdesmoside, calanthoside, the precursor glycoside of tryptanthrin, indirubin and insatin, with increasing skin blood flow promoting effects, from two Calanthe species (Orchidaceae). Chem Pharm Bull (Tokyo) 46(5):886–888

Yoshikawa K, Ito T, Iseki K et al (2012) Phenanthrene derivatives from cymbidium great flower marie laurencin and their biological activities. J Nat Prod 75(4):605–609

Yoshikawa K, Baba C, Iseki K et al (2014a) Phenanthrene and phenylpropanoid constituets from the roots of Cymbidium Great Flower "Marylaurencin" and their antimicrobial activity. J Nat Med 68(4):743–747

Yoshikawa K, Okahuji M, Iseki K et al (2014b) Two novel aromatic glucosides, marylaurencinosides D and E from the fresh flowers of Cymbidium Great Flower 'Marlaurencin". J Nat Med 68 (2):455–458

Yuzammi and Hidayat S (eds) (2002) The unique, endemic and rare Flora of Sulawesi. Centre for Plant Conservation, Bogor Botanic Gardens, Institute of Indonesian Sciences, Bogor

Zhang Y, Huang B, Zhao Z, Zhou Y (2011) Study on the chemical constituents from the ethyl acetate tracts of

Cremastra appendiculata. Zhong Yao Cai 34 (12):1882–1883

Zhang SB, Hu H, Xu K et al (2007) Flexible and reversible responses to different irradiance levels during photosynthetic acclimation of *Cypripedium guttatum*. Plant Physiol 164(5):611–620

Zhonghua Bencao (eds) (2000) Health Department and National Chinese Management Office. Shanghai Science and Technology, Shanghai

Zhongyao Da Cidian (1986) Jiangsu New Medical College (eds). Shanghai Science and Technology, Shanghai

Genus: *Dactylorhizia,* Necker

Chinese name: *Zhanglie Lan*

The generic name *Dactylorhiza* is derived from the Greek *daktylos* (finger) and *rhiza* root); it refers to the digitate subterranean portion of the plant (Fig. 10.1). *Panch anula,* the Nepali and Gurung name for this orchid, carries a similar meaning: *panch* meaning five, and *anula* finger, referring to the root which is shaped like a hand. However, *Panch anula* also applies to another orchid, *Gymnadenia orchidis,* whose roots form the shape of a hand (Manandhar and Manandhar 2002).

There are about 30–40 species in this beautiful, terrestrial, deciduous genus, *Dactylorhiza,* which is widely distributed across the temperate regions of Asia, Europe and Alaska, in wet to dry, low or high elevations, in grasslands, marshes and bogs, in full sunlight, often forming large colonies (Alrich and Higgins 2008; Karel et al. 2011).

Dactylorhiza hatagirea (D. Don) Soo

syn. *Orchis latifolia* Lindl.

Common name: Marsh Orchid
Chinese names: *Kuanyehongmen Lan* (broad leaf red door orchid), *Hongmen Lan* (red door orchid), *Mengguhongmen Lan* (Mongolian red door orchid), *Zhanglie Lan*

Chinese medicinal name: *Hongmen Lan* (red door orchid)
Indian names: *Munjataka* in Ayurveda, *Panja, Salampanja, Salep, Salap* (Hindi), *Salap* (Sanskrit), *Salimpanja* (Kashmir Himalaya), *Hathazari* (Uttarakhand)
Nepali names: *Hathejadi* is the common name; *Lob, Panchaunle* (Gurung), *Panchaunle* (Nepali), *Ongu lakpa* (Sherpa), *dbang-po-lag* (Tibetan), *Wonglak* (Amchi), *Lovha* (Kham), *Airalu* (Sanskrit)
Spanish name: *Palma Christi*

Description: A temperate, alpine orchid, *D. hatagirea* is found on grassy slopes, damp meadows and marshes between 2500 and 4000 m. It shares characteristics with two common Mediterranean genera, *Orchis* and *Anacamptis*. Plants are 30–90 cm tall, herbaceous, with paired tubers and numerous leaves near the base. Leaves are lanceolate or oblong, blotched with purple, progressively smaller towards the top. Inflorescence is up to 15 cm long, densely many-flowered. Flowers are resupinate, purple to a flesh pink, also yellow and white, and arranged around the rachis much like a hyacinth (Fig. 10.2). The flowers appear from May to July.

D. hatagirea is widely distributed in western temperate Himalayas (Kashmir, Arunachal Pradesh, Nepal, Bhutan, Pakistan), China (Heilongjiang, Gansu, Xinjiang, Sichuan and

© Springer International Publishing Switzerland 2016
E.S. Teoh, *Medicinal Orchids of Asia*, DOI 10.1007/978-3-319-24274-3_10

Fig. 10.1 *Dactylorhiza viridis* R.M. Bateman, Pridgeon & M.W.Chase (as *Platanthera viridis*). From: Thome O. M., *Flora von Deutschland Osterreich und der Schweiz, Tafleln*, vol. 1: t.145, Fig.B (1885)

Fig. 10.2 *Dactylorhiza hatagirea* (D. Don) Soo [PHOTO: Bhaktar B.Raskoti]

Tibet), Afghanistan, Central Asia, Europe (to the British Isles) and North Africa.

Phytochemistry: *D. hatagirea* contains a bitter principal and volatile oil but alkaloids are not present in the genera *Dactylorhiza* and *Orchis* (Luning 1974a, b).

Herbal Usage: Rhizomes of *D. hatagirea* are a source of *salep* (Fig. 10.3). They are used as expectorant, astringent and nutrient (Chopra 1933; Caius 1936); also as a tonic (Vij 1995). Tubers are eaten raw for their alleged aphrodisiac properties in India (Sood et al. 2005). *D. hatagirea* is used in the treatment of diabetes, chronic diarrhoea, dysentery, coughs, hoarseness of voice, paralysis, impotence, and during convalescence and to correct malnutrition (Chopra 1933; Shrestha 2000; Das 2004; Singh and Duggal 2009). In Kashmir Himalaya, root extract is a common home remedy for fever, and leaf extract is used for dysentery

(Kaul and Gaur 1995). *Hathazari*, tubers of *D. hatagirea*, are used to treat cuts, wounds and fever in Uttarakhand in western Himalaya (Li et al. 2014), and a paste made with tubers is applied on fractures in some parts of India (Das 2004).

A study of an area which comprised the Dolpa, Humla, Jumla and Mustang districts in the alpine zone of Nepal found that *D. hatagirea* was one of the four commonest medicinal plants. In this region, juice from the rhizome was used for cuts, wounds and gastritis (Kunwar et al. 2006). Manandhar and Manandhar (2002) reported that in Nepal the roots were used in medicine as an expectorant, astringent and demulcent, and root powder was sprinkled on wounds to arrest bleeding. A decoction of the root was taken for stomach ache. It was considered to be highly nutritious! Consumption of *Panchaunle* as a farinaceous food or aphrodisiac (Shrestha 2000)

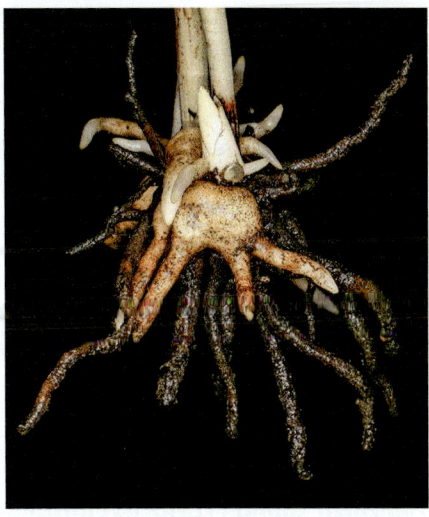

Fig. 10.3 Tubers of *Dactylorhiza hatagirea* [PHOTO: Courtesy of Krish Dulai through Wikimedia Commons]

Fig. 10.4 Tuber and roots of *Dactylorhiza incarnata* (L.) Soo showing new growth emerging from the young multi-digitate tuber. [PHOTO: Henry Oakley]

must lead to the harvest of countless tubers. Although the Nepali government has banned the collection, sale, distribution, transport and export of *D. hatagirea* under the Forests Act, 1993, the local plant population is endangered because the collection of its roots for trade and of leaves for local consumption continues (Manandhar and Manandhar 2002). Tubers of *D. hatagirea*, considered an aphrodisiac in Bhutan, are consumed to strengthen the body and promote longevity (Wangchuk 2009).

Hongmeng Lan (*D. hatagirea*) is obtained from Inner Mongolia, Heilongjiang, Gansu, Xinjiang, Sichuan and Tibet. In Chinese herbal medicine, the entire plant is used as a cardiac stimulant. It is alleged to benefit the kidney, stomach and spleen. It relieves thirst, improves appetite and is used to treat irregular menstruation, anaemia and dizziness (Wu 1994).

Dactylorhiza incarnata (L.) Soo

Common name: Early marsh orchid

Description: An extremely widespread and successful, temperate, terrestrial, tuberous species, *D. incaranata* is a variable species. Tubers are palmate with 3–4 lobes (Fig. 10.4). Stem is 25–35 cm tall, slim, 6 mm in diameter in

Gotland. Leaves are basal, 3–5, oblong-lanceolate, 12 by 2 cm all but one ensheathing the stem. They are evenly green but some populations have leaves that are purple-spotted. Inflorescence is 4–7.5 m tall, erect, densely many-flowered. Flowers are crimson to purple. Lip has three lobes. Flowering season is June to July (Hendren and Nordstrom 2009).

D. incaranata grows in bogs and wet meadows from Ireland and Sweden across Europe to the Middle East, Iran and Xinjiang (China).

Usage: Tubers are collected for *salep* in Iran but the main consumer is Turkey (Ghorbani et al. 2014).

Dactylorhiza romana (Schltr.) Soo *subsp. georgica*

Description: Tubers are large 2 by 1 cm. Plants are 15–40 cm tall. Base of stem is covered by 2–3 large, brownish sheaths followed by 4–11 linear-lanceolate leaves, 8.3–10.3 by 7.5–8.1 cm. Bracts are herbaceous, exceeding flowers in number. Flowers are 5–10 on the rachis, small, violet purple, dark red or yellow, very rarely white. Sepals are oblong 13 by 5 mm. Lateral sepals are reflexed. Petals are obliquely ovate, 10 by 4–7 mm. Lip is ovate, stretches forwards, trilobed, 15 mm long. Spur is 10–25 mm. long. The subspecies occurs at 700–2100 m in abandoned agricultural lands in Turkey, the Caucasus and Iran (Attundag et al. 2012).

Herbal Usage: Tubers are collected in Turkey and Iran for use as *salep* (Attundag et al. 2012; Ghorbani et al. 2014). In Turkey, it is also used as a tonic, as nourishment (used in ice-cream and drinks) and to treat constipation.

Phytochemistry: Tubers of *D. romana* contain mainly mucilage (61.05 %), sugar (4.5 %), water (11.1 %) and traces of starch and ash after burning (Sezik 1967).

Dactylorhiza salina (Turez exLindl.) Soo

Mongolian name: *Martsnii tsakhiram*
Tibetan name: *Ban lag*

Description: This is a small, cold-growing, perennial herb. Roots are palmate. Stem is 10–30 cm tall, glabrous with 4–10 oval-lanceolate, thick leaves, 4–10 by 1–3 cm. Inflorescence is terminal, rachis 3–12 cm with numerous pink to purplish flowers facing all directions. Flowering season is late spring. The species is distributed from the Caucasus to Amur at 3600 m. In Mongolia, it is found in damp, swampy alkaline meadows in almost all natural zones (Purevsuren and Tuya 2013).

Herbal usage: Root tubers are reported as sweet and astringent. They are employed to treat oedema and inflammation, to give strength and enhance life. There is a popular demand for *D. salina* in Mongolia and it is an ingredient in several traditional Mongolian prescriptions (Bayarsukh 2004; Purevsuren and Tuya 2013).

Dactyloriza umbrosa (Kar and Kir.) Nevski

Common name: Dark *Dactylorhiza*
Chinese name: *Yin Sheng Zhang Lie Lan*

Description: This is a robust species which thrives in temperate and sub-artic regions of Asia and looks rather like lilies before inflorescences appear. Tubers are palmate, 3- to 5-lobed. Plant is 15–45 cm tall. Stem is thick, hollow, and bears 2–3 tuberous sheaths at the base. Leaves are 4–9, erect, narrowly lanceolate, spotted with purple on both sides or unspotted, 7–13 by 1–5 cm. Inflorescence is 5–25 cm tall, erect, densely many-flowered. Flowers are dark purple. Lip is shaped like a kite (rhomboid), and bracts are shorter than the flowers. Flowering season is late spring to early summer (May to July). In China, *D. umbrosa* occurs at 600–4000 m in Xinjiang. It is also distributed in Pakistan, Afghanistan and northern Iran (Chen et al. 2009).

Usage: It is used to prepare *salep* and is being harvested in Iran (Ghorbani et al. 2014).

Dactylorhiza viridis (Linn.) R.M. Bateman, Pridgeon and M.W. Chase

syn. *Coeloglossum viride* Hartm.

Chinese names: *Woshe Lan* (nest tongue orchid), *Luhua-woshe Lan (green flower nest tongue orchid, Nanhulinyuzi Lan* (South Lake orchid), *Aoshe Lan* (nest tongue orchid), *Shoushen* (hand ginseng), *Aoshezhanglie Lan*
Tibetan medicinal name: *Wangla*
Alaskan name: frog orchid

Description: Plant is large, 10–30 cm tall, with a single thick, fleshy stem, sheathed at the bottom and carrying several lanceolate leaves, 3–5 cm by 5–17 mm, and large bracts. Inflorescence is terminal, erect, with numerous green flowers, 1–2 cm in diameter, that are tinged with red. There are several underground, lobed tubers which are 2–4 cm long. It flowers from May to August (Liu and Su 1978). The Huanglong Mountains have a strain of the species which varies from the type and this was been given the name *Coeloglossum viride* f. *virescens* (Muehlenberg) Perner & Y.B. Luo. It grows at 2200–3600 m and occurs in two coloured forms. At high altitude, flowers are green; at lower altitudes, flowers are reddish-brown. Flowering occurs in June (Perner and Luo 2007).

A variable species, *D. viridis* (syn. *Coeloglossum viride*) is an alpine orchid of the sub-polar region and the high mountains of Europe, eastern Asia, the Himalayan region and Alaska. In Bhutan, it is found in open grassland in upper montane *Betulag/Rhodendron/Lonicera* forest at 3500–4500 m (Pearce and Cribb 2002). In Taiwan, it is limited to the alpine tundra of Nanhutashan (Liu and Su 1978). It survives in intense cold and ice, In Alaska, the temperature may drop to minus 50 °C in winter, with severe desiccation during long winter months, yet the orchid survives.

Phytochemistry: Two phytoalexins, orchinol and *p*-hydroxybenzyl alcohol, were isolated from *D. viridis* following incubation with *Rhizoctonia repens* (Nuesch 1963) but they do not appear to play a role in the medicinal applications of the orchid. Orchinol is fungistatic. It is bacteriostatic at low concentration (in vitro inhibitory concentration 50 mcg/ml) against *Staphylococcus aureus,* but only at high concentration (invitro inhibitory concentration 500 mcg/ml) against *Escherichia coli* (Urech et al. 1963).

Twenty-four compounds have been isolated from the rhizome of *D. viride* (*Coeloglossum viride* var. *bracteosum*) by Huang and his team at the Institute of Materi Medica in Beijing. Initially, they reported isolating eight compounds from the orchid, namely: 4-(4-hydroxyphenyl) methoxybezenemethanol, 4,4′-dihydroxydiphenyl methane, 4,4′-dihydroxybenzyl ether, gastrodin, 4-hydroxy benzenemethanol, 4-hydroxybenz aldehyde, beta-sitosterol and beta-daucosterol. The first compound was new to the chemical world (Huang et al. 2002a). The following month, in a separate publication, they reported that ethanolic extraction of the rhizome of *C. oeloglossum viride* yielded eight compounds: Dactylorhin B, loroglossin, dactylorhin A, militarine, coelovirin A, gastrodin, thymidine and quercetin-3,7-di-*O*-beta-glucopyranoside. The fifth item, coelovirin A, is a new compound (Huang et al. 2002b).

Over the next 2 years, Huang and his co-investigators (Huang et al. 2004) isolated seven new compounds (Coelovirin A–G) and 17 known compounds from the rhizome of *C. viride*, namely:

1-4-beta-D-glucopyranosyloxybenzyl)-(2R,3S)-2-isobutyltatrate (1),

4-(4-beta-glucopyranosyloxybenzyl)-(2R,3S)-2-isobutyltatrate (2),

1-(4-beta-D-glucopyranosyloxybenzyl)-(2R,3S)-2-beta-D-glucopyranosyl-2-isobutyltartrrate (3),

4-(4-beta-D-glucopyranosyloxybenzyl)-(2R,3S)-2-beta-D-glucopyranosyl-2-isobutyltartrrate (4),

(2R,3S)-2-beta-Dglucopyranosyl-2-isobutyl-tartaric acid (5),

bis(4-beta-D-glucopyranosyloxybenzyl)-(2R,3S)-2-(beta-D-glucopyranosyl-(1→4)-beta- D glucopyranosyl)-2-isobutyltartrate (6)

bis(4-beta-D-glucopyranosyloxybenzyl)-(2R)-2-(beta-D-glucopyranosyl-(1→4)-beta-D-glucopyranosyl)-2-isobutylmalate (7):

4-hydroxybenzaldehyde,

4-hydroxybenzyl alcohol,

4,4′-dihydroxydibenzyl ether,

4,4′-dihydroxydiphenylmethane,

4-(4-hydroxybenzyloxy)benzyl alcohol,

Gastrodin,

Quercetin-3-7-diglucoside,

Thymidine, loroglossin,

Militarine,

Dactylorhin A,

Dactylorhin B, beta-sitosterol,

Daucosterol.

The functional groups of the compounds suggest that *D. viride* extracts may have a neuro-protective effect. To study this role, Zhang and his colleagues at the Peking Union Medical College in Beijing induced senescent ageing with memory impairment in mice by consecutive intra-peritoneal injections of D-galactose and sodium nitrite over a 60-day period. (D-galactose produces a metabolic disturbance whereas sodium nitrite causes ischaemia and hypoxia in many organs including the brain.) From Day 47 to Day 60, groups consisting of 10 treated mice were fed with 2.5, 5, or 10 mg *D. viride* rhizome extract per kilogram per day; another group of ten treated mice received piracetin (90 mg/kg/day). The untreated (no D-galactose and no sodium nitrite) control mice and the treated control mice were fed with saline. A water maze test was used to evaluate learning and memory function in the six groups of mice. Subsequently, the activity of various enzymes and the content of MDA in the rodents' brains were measured. The aged rats fared poorly in the water maze test compared with the untreated control group. Piracetin and *D. viride* extract treatment improved their performance (Zhang and Zhang 2005a, b; Zhang et al. 2006c).

In the aged mice, activities of superoxide dismutase (SOD Na+K(+)-ATPase, and Ca^{2+} Mg(2+)-ATPase decreased while the malondialdehyde (MDA) and the activities of monoamine oxidase-A (MAO-A) and mono-amine oxidase-B (MAO-B) were increased, indicating brain tissue damage. *D. viridis* extract at all tested dosages, both high and low, ameliorated these changes (Zhang and Zhang 2005a, b).

Tau 2 is classically increased in the brain of patients with Alzheimer's disease. Histochemical staining revealed that the hippocampal area of the brain of rats treated solely with D-galactose and sodium nitrite had an increase of Tau 2. The brain of the animals who additionally received *D. viridis* extracts did not. Additionally, the amount of NT-3 and Bcl-2 was reduced in the brain of senescent rats, but it was not reduced in the animals that received CE. *Coeloglossum* extract also prevented the increase in Bax-and capase-3-positive cells in the treated mice (Zhang et al. 2006d). The neuro-protective effect of the *D. viridis* extract is at least partly attributable to the presence of Dactylorhin B which has been shown to reduce toxic effects of beta-amyloid fragment (25–35) on neuron cells and isolated rat brain mitochondria (Zhang et al. 2006c, e), beta-amyloid fragments in the brain being associated with Alzheimer's disease. Gastrodin may also play a protective role. *Gastrodia elata* extract, and two of its pure components, gastrodin and 4-hydroxybenzyl alcohol, inhibited the expression of C/EBP homologous protein (CHOP), a pro-apoptotic protein (i.e. a protein which hastens natural cell death), and protected cultured rodent brain cells against beta-amyloid (Lee et al. 2012a). *Coeloglossum viride*

(=*D. viridis*) extract attenuated learning and memory deficits, motor functional disability and brain cell loss when fed to rats (5 mg/kg) who either (1) underwent transient middle cerebral artery occlusion to effect focal cerebral ischaemia or (2) four-vessel occlusion to effect transient global forebrain ischaemia (Ma et al. 2008).

A host of other ethnomedicines possessing characteristic functional group is being investigated for possible a neuroprotective effect, with the following testing positive: *facteur thymique serique* (Zhao et al. 2013), puerparin (Xu and Zhao 2002), the root of *Cynanchum auriculatum* (Zhang et al. 2007a), *Liriope platyphylla* (Jiang et al. 2007), catalpol from fresh roots of *Rehmannia*, the last a popular Chinese herbal remedy (Zhang et al. 2008c). Additionally, there is a synthetic compound which is a cyclic squamosamide analogue, Compound FLZ (Fang and Liu 2007), as well as the natural female hormone, oestradiol (Nielsen et al. 2006).

Herbal Usage: *Woshe lan* enriches blood and air, encourages salivation and stops thirst. It is used in the treatment of weak lungs, asthma, weight loss, weak kidneys, discharge, enuresis, haemorrhage and failure of lactation. The Tibetan medicine *wangla* is derived from the rhizome of *D. viride*.

Overview

The alleged aphrodisiac property of *Dactylorhizia* runs parallel to the old Mediterranean *Theory of the Signatures* wherein the shape of a plant discloses its usage. In the case of *Dactylorhiza* (*Orchis*) it is the testicular form of its paired tubers.

Aphrodisiacal properties of *D. hatagirea* were studied in Wistar strain male albino rats by Thakur and Dixit (2007) who fed the animals with 200 mg of lyophilised aqueous extracts of the orchid tubers. The investigators compared the anatomy and behaviour of these rats with those given 0.5 mg/kg of testosterone propionate by injection twice a week, and a control group that did not receive treatment. Rats fed the orchid had a 2.5-fold increase in attraction to females compared with the testosterone-treated rats. The sexual behaviour of the treated animals were similar but far superior to that of the untreated rats. Body weight of the animals increased, suggesting an anabolic effect which might be due to increased endogenous male hormone production because an aqueous extract would not contain steroids.

D. hatagirea contains mucilage, starch, glucoside, loroglossin, albumen, volatile oil and ash, together with five dactylorhins A–E and two dactyloses A and B (Kizu et al. 1999), none of which excite the senses. Nevertheless, today in Singapore and other parts of southern Asia, one may find the orchid among the constituents of health supplements which are marketed as revitalising, rejuvenating, or aphrodisiac. *Dactylorhiza* as *salep* is also considered to be an aphrodisiac in Iran, Turkey and the Colchis Forest area of Georgia in the Caucasus (Ghorbani et al. 2014; Averjanova et al. 2014).

From a medical standpoint, the neuroprotective effects of *D. viridis* is more interesting, *Wangla* is a potential candidate for the treatment of vascular dementia (Ma et al. 2008). However, all compounds have only been tested on rodents. Continued investigation of *D. viridis* extract in a clinical setting appears to be worthwhile, but, so far, no report has been published.

Data on the ecology of *Dactylorhiza* are derived from studies on the genus in Europe but basic principles should still apply to *Dactylorhiza* species occurring in Iran and the Himalayas. *Dactylorhiza* is a very successful, cold-growing, temperate, generally montane genus because members adapt well to a wide range of soils and produce lots of seeds. However, they only thrive in full sun, or require at least semi-shade, and populations have declined when the requirements for bright light coupled with low temperatures are not met. *Dactylorhiza* reproduces via seed production. Vegetative reproduction, in *D. virides*, for instance, is almost non-existent. Plants reach maturity and flower in one to several years. Flowering is achieved only when plants reach a substantial size with 10 or more leaves per plant. A Dutch study reported that only half the population of *D. viridis* flowered each year. Fruit set is highly variable among populations, from 10 to 39 %, and seed count varies from 2000 to 5000 per capsule. High seed output enables *Dactylorhiza* to recolonise

habitats from which they have disappeared and to establish populations in new habitats (Neiland 2001). Therefore, conservation of *Dactylorhiza* which is so highly valued in Himalaya should always ensure adequate seed recruitment and control grazing to sustain the wild populations. Recently, rapid in vitro protocorm multiplication of *D. hatagirea* (D. Don) Soo has been achieved (Warghat et al. 2014). This holds promise for large-scale propagation of the much desired orchid. Some *Dactylorhiza* hybrids are very vigorous (Cribb 2001). This may offer an approach to increase commercial productivity.

Genus: *Dendrobium* Sw

Chinese name: *Shihu* (living on rocks)

Dendrobium is a huge family constituted by approximately 900 species (Fig. 10.5). There are an estimated 74 species of *Dendrobium* in China, of which approximately 40 % are purported to have medicinal usage. Several excellent monographs on *Dendrobium* have been published in the last two decades (Baker and Baker 1996; Cribb 1986; Lavarack et al. 2000; etc.)

The term *Dendrobium* is derived from Greek, *dendros* (tree) and *bios* (life), which refer to the habitat of most members of this ubiquitous Asian epiphyte. It is thus surprising that the Chinese name for *Dendrobium* is *shihu* with a homonym that means 'living on rocks'. This term is appropriate for the two species mentioned in the seminal *Shen Nong Ben Cao Jing* (*The Materia Medica of Shen Nong*) compiled at the beginning of the current era, during the Han Dynasty. *D. catenatum* (*D.. officinale*) and *D. moniliforme* (syn. *D. candidum*) are saxicolous. The *Ming Yi Pieh Lu* (*Extra Accounts of Renowned Drugs*) compiled by Tao Hung-Ching (552–536) who also produced an revised edition of the *Shen Nong Bencao Jing* during the sixth century, stated that "shihu grew on rocks by the waterside in the valleys of Lu An" (Chen and Tang 1982). The medicinal usage of *Dendrobium* is possibly related to a Chinese equivalent of the

Fig. 10.5 *Dendrobium nobile* Lindl. From: *Annals of the Royal Botanic Gardens, Calcutta,* vol 8 (2): t. 71 (1891). [Drawing by R. Pantling] Courtesy of Missouri Botanical Gardens, St. Louis, U.S.A

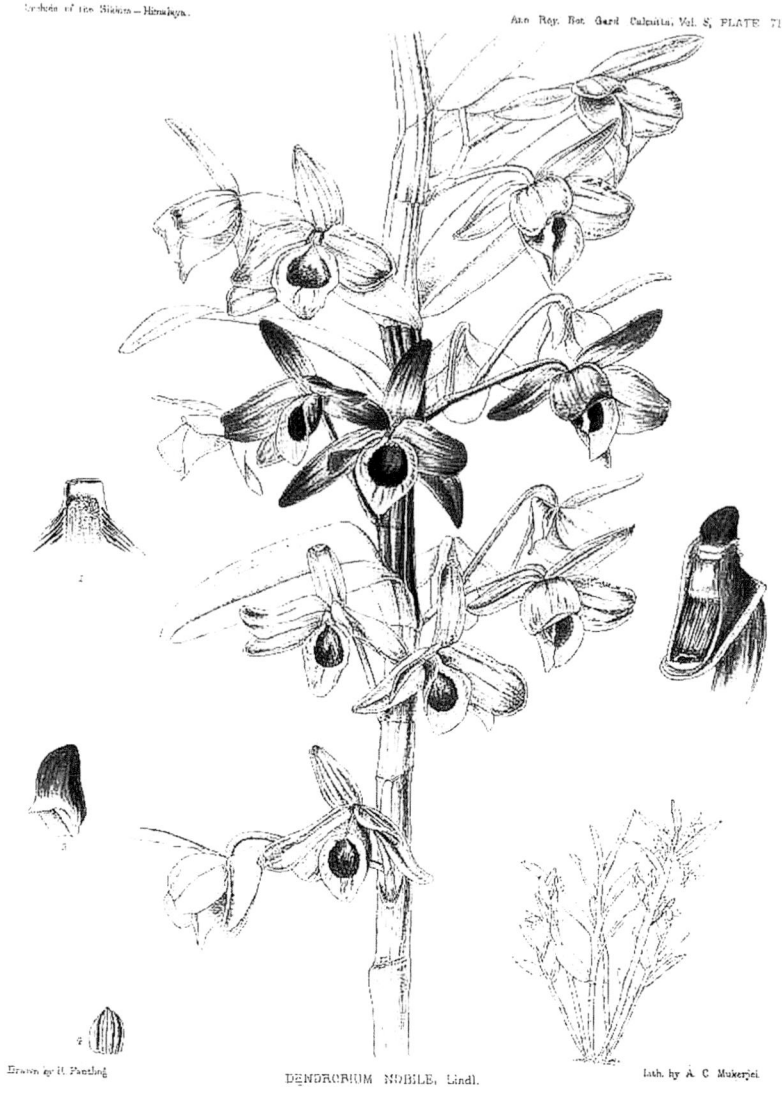

DENDROBIUM NOBILE, Lindl.

Doctrine of Signatures which states that the form of a plant determines its properties. As *Dendrobium* lives on rocks, it must be extremely robust and resilient, and thus such a plant would strengthen the body (Hu 1970, 1971). *Shihu* is principally used as a tonic.

Golden, pink and white flowers predominate among the medicinal *Dendrobium*, and an untrained observer may find difficulty in distinguishing some species from others. The confusion is compounded when the plant is not in bloom. To cope with the twin problems of substitution and contamination, it was proposed that morphological details be supplemented with histological characteristics and TLC spectra to facilitate proper identification (Zhao et al. 1998). This is now superceded by the various PCR procedures which allow for a far more precise identification (Lau et al. 2001; Xu et al. 2001, 2006; Qian et al. 2008). While PCR elicits correct species identification, by itself, correct identification does not ensure quality of the product. Factors like source of supply (Bai et al. 2007), age of the plant, season of harvest (Table D-2, Ding et al. 1998), portion of the stem (Chen et al. 2001), processing (Wu et al. 2007) and storage all affect the final quality (Bai et al. 2007; Ding et al. 1998). Capillary electrophoretic fingerprinting has also been proposed as a means of identification (it has 15 characteristic peaks). *D. candidum* can be classified according to source and quality by this process (Zha et al. 2009).

Chinese people in many parts of the world resort to traditional Chinese medicine (TCM) for cures and for tonics to improve their quality of life. *Shihu* is a popular tonic which has a long history of use outside China (Hooper 1929).

Dendrobium species are also used in Ayurvedic and other medicinal traditions. They are included in the list which follows. Five *Dendrobium* species are used in Thai native medicines: they are *D. cumulatum, D, draconis, D. indivisum, D. leonis* and *D. trigonopus* (Chuakul 2002). We have allowed two spellings for the Thai word used to describe *Dendrobium* (*Uang* or *Ueang*) because the names are derived from different sources, and possibly both spellings are used in Thailand (Seidenfaden and Smitinand 1960; Vaddhanaphuti 2001). The same problem occurs in the transliteration of Indian names (e.g. *rasna* and *raasnna*).

In the Malay Archipelago, *Dendobium crumenatum* is used to treat earache (Lawler 1986). *D. crumenatum* was the sole orchid listed among 194 species by the Administration Department of the Japanese Army in Singapore in their first compilation of useful plants produced in July 1944 (*Compilation of Medicinal Plants in the Malay District*). Burkill (1935) observed that the Malays do not distinguish between species in the manner of botanists and may use dissimilar species for the same medicinal purposes. For instance, whereas *D. crumenatum* was the common *Dendrobium* used for poulticing, sometimes in East Malaysia, *D. purpureum* was used instead; in West Malaysia, *D. planibulbe* might be used in its place. In Perak, *D. subulatum* was an acceptable substitute.

De Waldemann (1892) observed a high alkaloid presence in *D. nobile* when undertaking the first investigations on orchid alkaloids. But it was not until 40 years later that Suzuki and co-workers (1932) managed to obtain a crystalline alkaloid from '*Chin Shih Hu*', a herbal preparation derived from dried stems of *D. nobile*. Subsequently, alkaloids were isolated from other species of *Dendrobium* (e.g. *D. linawianum, D. findlayanum, D. fredricksianum, D. hildebrandii*) (Leander and Luning 1968; Hedman et al. 1971; Elander et al. 1973; Luning 1974a, b). Another 40 years passed before the correct structure of dendrobine was defined. Inubushi et al. (1972) and Yamada et al. (1972) independently succeeded in synthesising the alkaloid. Plant chemistry has since progressed by leaps and bounds, and much is known about alkaloids and other steroidal compounds produced by orchids. Much of the work still centres on *Dendrobium*. Since alkaloids are also present in hybrids of medicinal *Dendrobium*, the use of hybrid *Dendrobium* opens a vast resource for the preparation of *shihu* (Morita et al. 2000).

Perhaps the most exciting discoveries on *shihu* centres on the reports that certain compounds from *Dendrobium* may help kill cancer cells when they have become resistant to conventional cancer drugs. Combined treatment with dendrinobin (a well-studied, naturally occurring phenathroquinone present in various *Dendrobium* species) and another anticancer agent (Fas-ligand) has a synergistic cytotoxic effect in human pancreatic adenocarcinoma cells. The hope is that dendronobin can be developed as an adjuvant for use in combination therapies aimed at killing cancer cells which evade the host's immune surveillance by relying on decoy receptors. That would be a major advance in chemotherapy (Magwere 2009; Yang et al. 2009). Another promising compound is erianin which is present in *D. chrysotoxum*. It halts the division of cancer cells and causes them to undergo programmed cell death (apoptosis) in a manner that is similar to that of placitaxel, and also interferes with their blood supply (anti-angiogenesis). However, such usage for cancer treatment is far removed from the original intent for consuming *shihu*.

Dendrobium acinaciforme Roxb

D. spatella Rchb.f, *D. banaense* Gagn.,

Chinese name: *Jianye Shihu* (sword leaf *Dendrobium*)
Thai name: *Uang Takhap*

Fig. 10.6 *Dendrobium acinaciforme* Roxb. [PHOTO: E.S. Teoh]

Description: Pseudobulbs are 15–30 cm long, slender, with 8–12 narrow, pointed leaves 4–5 cm long. Flowers are small (up to 1 cm wide) and are borne singly on the leafless nodes, at the distal portion of the stem. Colour is variable, from white to yellow and pink, with a yellow patch in the centre of the lip (Fig. 10.6). Flowering period is June to July in Assam, India (Nath and Das 2013) and August in China and Thailand (Chen et al. 1999b; Vaddhanaphuti 2005).

A common *Dendrobium* in the Chiangmai area of northern Thailand, *D. acinaciforme*, occurs at 800–2100 m in Thailand, Indochina, Myanmar, northeast India and south China (from Yunnan to Hong Kong and Lantau Island), Malay Peninsula and Malaku.

Herbal Usage: The entire plant is used in TCM as a tonic during the recuperation period of an illness to eliminate fever, thirst, lassitude and malaise. This is a standard application of *shihu*.

Dendrobium aduncum Lindl.

D. scorianum W. W. Smith, *D. faulhaberianum* Schltr.

Chinese names: *Gouzhuangshihu* (noble hook *Dendrobium*), *Huangcaoshihu* (yellow herbal *Dendrobium*); *Dahuangcao* (large yellow herbal *Dendrobium*); *Honglancao* (red orchid herb); *Jishengcao* (parasitic herb)

Description: *D. aduncum* has a pendulous habit of growth and is epiphytic and saxicolous. Pseudobulbs are 30–60 cm long, slim, of uniform diameter, with new branching, side internodes appearing on mature stems. There are 4–6 elliptic-lanceolate leaves, 7–10.5 by 1.5–3.5 cm per cane. Inflorescence is 7 cm long with 3–5 fragrant, long-lasting flowers, 2.5–3.5 cm across. Petals and sepals are semi-transparent and pink or white. Lip is white, pointed. Its colour highlights the two red blotches at the top of the column. It flowers in May and June in China, Assam (India) and Thailand (Chen et al. 1999b; Nath and Das 2013; Vaddhanaphuti 2005) or June to August in Hong Kong (Wu et al. 2002).

D. aduncum is widely distributed from the Himalayan foothills to Kwangtung Province and Hong Kong, Lantau and Hainan Islands in southern China, southwards into Myanmar, northern Thailand and Vietnam at 600–1300 m.

Phytochemistry: The sesquiterpene, aduncin, was isolated from *D. aduncum*. It is related to picrotoxinin (Gawell and Leander 1976), and is also present in *D. moniliforme* (Li et al. 2010).

Herbal Usage: as *Shihu*. The stem is nourishing and promotes vitality. Antipyretic, it also benefits the stomach and stops thirst. It is commonly used to help patients who are weak because of illness, and those who have a poor appetite (Wu 1994; Ou et al. 2003).

D. alpestre Royle (see **Dendrobium monticola** P.F. Hunt & Summerh.)

Dendrobium amoenum Wall ex Lindl.

Description: *D. amoenum* is an epiphytic, sun-loving species. Stems are slender, pendulous, 35–70 cm long and 1–2 cm in diameter; leaves oblong-lanceolate, 6.5–10 by 1–1.6 cm,

Fig. 10.7 *Dendrobium amoenum* Wall ex Lindl. [PHOTO: E.S. Teoh]

pointed at the tips. Inflorescence arises from leaf-less stems. Flowers single, 2 cm across, white with blotch of cerise or purple at the tips of the petals and lip. There is a flush of yellow-green at the throat of the lip (Fig. 10.7). Flowering season is May in Nepal (Raskoti 2009). The species is distributed from Himalaya to Myanmar. In Central and eastern Nepal it occurs at 600–1500 m (Raskoti 2009).

Phytochemistry: Two sesquiterpenes of the picrotaxane group, amotin and amoenin, were isolated from *D. amoenum* (Dahmen and Leander 1978). Two bibenzyl derivatives, amoenylin and isoamoenylin, are also present in the orchid, together with moscatilin, amotin, amoenin and flaccidin (Majumder et al. 1999). Isoamoenylin has moderate anti-oxidative and weak antibacterial activities. It has been successfully synthesised from 3,4,5-trimethoxybenzaldehyde (Venkateswarlu et al. 2002). *D. amoenum* also contains a 9,10-dihydrophenanthropyran, amoenumin (Veeraju et al. 1989), amoenylinin, moscatilin, batatasin III, 3,4'-dihydroxy-5-methoxybibenzyl, confusarin, 2,7-dihydroxy-3,4,6-trimethoxy-phenanthrene, imbricatin and flaccidin (Majumder and Bandyopadhyay (2010).

Herbal usage: Stems are used as a tonic in Nepal (Pant and Raskoti 2013). Fresh paste

prepared with pseudobulbs of *D. amoenum* is applied to treat burns and dislocated bones (Subedi et al. 2013; Pant and Raskoti 2013).

Dendrobium amplum Lindl.

syn. *Epigeneium amplum* (Lindl.) Summerh.

Chinese names: *Kuanyehouchun Lan* (broad leaved, thick lipped orchid); *Guoshangye* (leaves above fruit)
Thai names: *Kra chiang, Si thiang*

Description: An epiphyte with creeping rhizome and remote, brownish, ovoid pseudobulbs (separated from one another by 3–14 cm), square in cross-section, 2–5 by 0.7–2 cm, it bears 2 leaves at apex, 6–22 long and 5.5 cm wide. Inflorescence is terminal with a single, light brown, fragrant flower spotted with dark brown. Flower is 7–8 cm across. Lip is dark reddish-brown (Fig. 10.8). Flowering season is November in China (Chen et al. 1999b), and October to January (Nanakorn and Watthana 2008) in Thailand, with a possible peak in November (Vaddhanaphuti 2001).

Epigeneium amplum is found on tree trunks and rocks in mountain forests and along ravines at 1000–1900 m from Guangxi, Guizhou and northern Thailand and Vietnam to Yunnan, Myanmar, Tibet, Sikkim, Nepal, Bhutan and India (Chen et al. 1999b). This is the commonest species of *Epigeneium*.

Herbal Usage: The whole plant is used in Chinese herbal medicine to enrich *yin*, stop coughs and improve blood flow (Wu 1994).

Dendrobium aphyllum (Roxb.) C.E.C. Fisch,

syn. *D. macrostachyum* Lindl.

Thai names: *Uean sai, Ueang sai long laeng, Ueang yumai, Ueang khai nao, Ueang saimai, Ueang lawng laeng*

Fig. 10.8 *Dendrobium amplum* Lindl. [PHOTO: Bhaktar B. Raskoti]

Fig. 10.9 *Dendrobium aphyllum* (Roxb.) C.E.C.Fisch [PHOTO: E.S. Teoh]

Description: Pseudobulbs are slender, terete, pendulous, 60–160 cm, with many short internodes and numerous thin, elliptic-ovate leaves, 6–8 by 2–3 cm, that last for nearly a year. Inflorescences arise from nodes at the distal 60 % of deciduous pseudobulbs and bear 1–3 slightly fragrant flowers which are white to pink mauve, 3–6 cm across, very numerous on the stem. Flowers last 3 weeks. Petals and sepals are narrow and translucent. Lip is hirsute, funnel-shaped, of variable colour, but usually white or cream with purplish veins at the base (Fig. 10.9). A distinct dry season is essential to induce flowering which occurs in spring in China and Thailand (March to April) (Chen et al. 1999b; Vaddhanaphuti 2005; Nanakorn and Watthana 2008), lasting to May in Nepal and Peninsular Malaysia. *D. aphyllum* flowers in summer, July and August, in Upper Assam, India (Nath and Das 2013).

The species is widespread in continental Southeast Asia, spreading to southwest China, Sikkim and Nepal, at 150–1800 m. Usually an epiphyte, in deforested areas it is saxicolous. It

has also been found growing on trees in mangrove swamps (Go and Hamzah 2008).

Phytochemistry: Isoamoenylin, a dihydrostilbene, has been extracted from *D. amoenum* (= *D. aphyllum*) and synthesised. It has anti-oxidative and mild bactericidal properties (Venkateswarlu et al. 2002). The Nanjing group at the Department of Pharmacognosy at China Pharmaceutic University reported the isolation of a total 19 compounds from *D. aphyllum* and specified their structures but not their pharmacological action (Zhang et al. 2006a; Shao et al. 2008). The compounds are gigantol, batatasin, tristin, 3,5,4′-trihydroxylbibenzyl, 3,5-dimethoxyl-4,4′-dihydroxylbibenzyl, moscatin, 2,4,7-trihydroxyl-9,10-dihydroxyphenanthrene, hircinol, 2-(4-hydroxypheny;)ethyl-beta-D-glucopyranoside, salidroside, p-hydroxybenzylacetic acid, 4′-meyhoxyl-tricin, tricin, 7,3′5′-tri-O-methyl-tricetin, syringic acid, (+)-syring-aresinol, D-allitol, sucrose and icariside (Zhang et al. 2008; Shao et al. 2008). A new phenanthrene, aphyllone A, four new bibenzyl derivatives, aphyllone B, aphyllals C and D, together with nine known compounds were isolated from *D. aphyllum* by Yang et al. (2015a). Aphyllone B exhibited significant DPPH radical-scavenging activity of 87.97 at 100 mcg/ml.

Herbal Usage: It is used as *shihu* (Wu 1994). Stems are used to prepare a tonic in Nepal (Pant and Raskoti 2013). Eardrops prepared with juice of young pseudobulbs of *D. macrostachyum* (this name is a synonym for *D. aphyllum*) are used by the Valmikis tribe of Visakhapatnam district in Andhra Pradesh to treat earache (Reddy et al. 2005a). Dried stems of *D. aphyllum* resemble stems of *D. devonianum*. Both species are used as *shihu*.

Dendrobium appendiculatum (Blume) Lindl.

syn. *Flickingeria bifida* A. Hawkes; *Ephemeranta bifida* (Ridley) Hunt et Summerh.

Chinese name: *Youzhua shihu* (Claw *Dendrobium*)

Description: Rhizome is stout, creeping, branching, a shiny olive, 5–7.5 mm in diameter, with internodes 0.75–1 cm long. Stems are 4–10 cm apart. Pseudobulbs are slim, fusiform, 5 cm long. Leaves arise from the top of the pseudobulb, broadly ovate, leathery, 17 by 7 cm. Inflorescences are both abaxial and adaxial. Flowers 2.5 cm across, opening widely, white with purple blotches at the base of the sepals and petals. Lip trilobed, 2.8 cm long, mid-lobe long and narrow, splitting into two long, narrow lobules distally. It is distributed in Peninsular Thailand, Peninsular Malaysia, Sumatra and Java (Seidenfaden and Wood 1992; Comber 2001).

Identification of this plant is somewhat uncertain because Wu (1994) who listed three species of *Ephemerantha* (*Ephemerantha bifida, Ephemerantha fimbriata* and *Ephemerantha. lonchophylla,* now all returned to *Dendrobium*) did not give a description of this species but mentioned that it occurs in Guangxi, Guizhuo and Yunnan Provinces. *Flora of China* (2009)

does not describe an *Ephemerantha* that occurs in all three provinces, but *Ephemerantha* (syn. *Flickingeria*) *fimbriata* (correct name: *D. plicatile*) occurs in Guangxi, Yunnan and Hainan (Zhu et al. 2009).

Usage: The entire plant is used in TCM to enrich *yin*. It benefits the lungs and stops coughs. It is used to treat troublesome tuberculous cough and asthma (Wu 1994).

Dendrobium bellatulum Rolfe

Thai name: *Uang Sae Mon*

Description: Plants are found in deciduous forests at 900–2100 m. Stems are 2–5 cm long, bearing two to four thick leaves, 1.5–4 cm in length. Inflorescence is short, terminal or sub-terminal, and bears one or two creamy white flowers, 3.8 cm across. Lip is large, cadmium orange; throat a deep red. It flowers in January to March or April in Thailand (Vaddhanaphuti 2005; Nanakorn and Watthana 2008), April to June in Yunnan (Chen et al. 1999b). This small, epiphytic species is widespread throughout Myanmar, Thailand, Laos and Vietnam, extending to Sikkim and southern Yunnan.

Herbal Usage: It is used as *shihu* (Zhongyao Da Ci Dian 1986; Wu 1994).

Dendrobium bifarium Lindl. (see ***Appendicula cornuta*** Blume)

Dendrobium blumei Lindl.

Description: Plant resembles *D. crumenatum*, with solitary white flowers that carry red marking at tips of segments. Flowering season is June (Vaddhanaphuti 2005).

It is found in Peninsular Thailand, Peninsular Malaysia (in Perak), Java, Borneo and the Philippines.

Herbal Usage: A poultice made from the pounded leaves and pseudobulbs of the orchid is applied on itching skin and eruptions for relief (Burkhill and Haniff 1930).

Dendrobium candidum Wall. ex Lindl. [see **Dendrobium moniliforme** (L.) Sw.]

Dendrobium capillipes Rchb. f.

Chinese name: *Duanbang Shihu*

Description: *D. capillipes* is a small to medium-sized epiphyte. Pseudobulbs are ellipsoid, 8–15 by 1.5 cm at its widest diameter, fleshy, with few internodes and numerous longitudinal ridges. Leaves are located near the apex of the stem, 2–4, oblong, narrow, 10–12 by 1–1.5 cm (for Chinese plants; 3–5 cm by 1–1.5 cm for plants in Thailand), their base ensheathing top internode of the stem. Inflorescences arise from leafless stems, 12–15 cm long, suberect, laxy 2–5 flowered. Flowers are up to 2–3 cm across, spreading, golden yellow with a rounded lip that is of a deeper golden hue at the base. Lip of flowers in Chinese plants carry crimson stripes (Zhu et al. 2009; Nanakorn and Watthana 2008). Flowering season is February to April in Thailand (Nanakorn and Watthana 2008).

D. capillipes. is distributed from Nepal and Northeast India to Yunnan, Myanmar, Thailand, Laos and Cambodia

Phytochemistry: Scoparone (6,7-dimethoxycoumarine) which is present in *D. capillipes* was shown to have vasodilatory effects on the rat aorta. It opposed the constrictive effect of adrenalin and serotonin but not that of potassium chloride. It boosted the output of 6-keto-prostagladin F1-alpha (Chen et al. 1991; Huang et al. 1992).

Phytohaemagglutinin stimulates white blood cells to undergo rapid replication and the process is used in the laboratory for chromosomal analysis. Scoparone suppressed the response of human mononuclear cells (lymphocytes, a type pf white blood cells) to phytohaemagglutinin and is, therefore, thought to be immunosuppressive (Chen et al. 1991). Scoparone also stimulates dopamine biosynthesis by PC12 cells which are nerve cells from a pheochromocytoma (tumour) of the rat

adrenal medulla (Yang et al. 2009). Dopamine is an important chemical messenger of the brain. Loss of dopamine-secreting cells in the brain is thought to be a cause of Parkinson's disease.

An alkaloid, hygrine, is present in *D. capillipes*. A flavanol glycoside, namely quercetin-3-*O*-alpha-ʟ-rhamnopyranosyl-1(1–2)-betaxylopyranoside together with two other flavanol glycosides and five bibenzyls were isolated from stems of *D. capillipes* (Bao et al. 2001). Some bibenzyls exhibited cytotoxicity against KB, NCI-H187 and MCF-7 cancer cells but the gycosidic flavonoids were inactive (Phechrmeekha et al. 2012).

Herbal Usage: It is used as *shihu* (Bao et al. 2001).

Dendrobium cariniferum Rchb. f.

Common name: Keel carrying *Dendrobium*
Thai names: *Ueang sae dong, Ueang ngoen daeng, Ueang kachok, Ueang tueng*
Myanmar name: *Mahar deiwi*

Description: Pseudobulbs are cylindric to fusiform, yellow when dry, 10–28 cm long, 1.5 cm in diameter, Leaves in two ranks, leathery, oblong, rounded at the tip, slightly and unequally bilobed, 11 by 1.5–4 cm, densely black hirsute where they enseath the stem.

Inflorescences are subterminal, short. Flowers are scented, white, borne singly or in pairs. The common name of this small *Dendrobium* describes the appearance of the lip which has six papillary ridges lining a throat that is stained with orange (Fig. 10.10). Flowers last for up to 2 months on the plant. Flowering season is May to June in China (Chen et al. 1999b); February to April in Myanmar and Thailand (Vaddhanaphuti 2005; Tanaka et al. 2003). *D. cariniferum* occurs in Xixuangbana in Yunnan Province, the Himalayas, the Shan state of Myanmar, northern Thailand and Indochina at 450–1800 m, growing on old dwarf trees in mixed and coniferous forests.

Herbal Usage: as *shihu* (Wu 1994)

Fig. 10.10 *Dendrobium cariniferum* Rchb. f. [PHOTO: E.S. Teoh]

Fig. 10.11 *Dendrobium catenatum* Lindl. (syn. *Dendrobium officinale* K. Kimura et Migo [PHOTO: Courtesy of Plant Photo Bank of China]

Dendrobium catenatum Lindl.

syn. *Dendrobium officinale* K. Kimura et Migo; *Dendrobium huoshanense* G.Z. Tang and S.J. Cheng; *Dendrobium tosaense*

Chinese Medicinal Name: *Shihu*

A widespread Chinese species, *D. catenatum* is one of the two original species constituting the medicinal herb, *shihu*. Nearly all Chinese medicinal texts refer to this *shihu* as *D. officinale* because it is one of the two classic species of *shihu* mentioned in the seminal *Chinese Materia Medica,* the ancient *Shen Nong's Bencao Jing*. Nevertheless, the correct name as stated in the Kew Monocot List is *D. catenatum* Lindl.

Plants collected from rocks at 500 m around Huoshan in Anhui Province were given the name *D. huoshanense* G.Z. Tang and S.J. Cheng in 1984. Only two phytochemical papers referred to the orchid by this name.

Description: Stems are erect, terete, 9–35 cm long, slender (2–4 mm thick), with 3–5 leaves on the upper nodes. They are deciduous. Roots are very fine compared with other *Dendrobium* species. Inflorescences arise from the nodes in the distal portion of leafless stems, 2–4 cm long, with

2–3 flowers. Flowers are 1.5 cm across, star shaped, sepals and petals white or yellowish green; lip pointed, with a reddish blotch, giving off a mild or strong fragrance (Fig. 10.11). They appear from March to June and last a fortnight. In its natural environment in Hainan, the seed setting rate was very poor (0.31 %) and the species is regarded as endangered in that region due to its overcollection in the past (He et al. 2009).

D. catenatum occurs in Anhui, Zhejiang, Fujian, Guangxi, Yunnan and Sichuan, on tree trunks and on rocks in sparse woods in limestone areas up to 1600 m. Its distribution extends to northern Myanmar, Sikkim and Nepal and southern Japan. In Anhui Province, a study found that the distribution of wild *D. catenatum* is patchy, the conditions are harsh, and the wild resources are diminishing so rapidly as to require protection with simultaneous development of alternative supply through cultivation (Jin et al. 2013).

Phytochemistry: A total of 25 genes involved in alkaloid backbone biosynthesis were identified through analysis of the *D. officinale* (*D. catenatum*) transcriptome (Guo et al. 2013) but actual alkaloids have not been isolated and identified. Two new bibenzyl derivatives, dendrocandin T and dendrocandin U together with eight known bibenzyls were isolated from pseudobulbs of *D. catenatum* (Yang et al. 2015b). A few of the bibenzyls promoted neurite development (Yang et al. 2015b). Naringenin content in *D. catenatum* was highest at year 3 of growth, with different levels for different sources (Zhou et al. 2013).

Chinese researchers were mainly interested in recovering polysaccharides from the plant (Fan et al. 2005; Zhang and Liao 2005; Li et al. 2006) employing tube culture to obtain more tissue for polysaccharide recovery and even resorting to sound wave stimulation to promote the synthesis of the desired compounds (Li et al. 2006). A method for obtaining a stable supply of active polysaccharides has been developed using protocorm-like bodies of *D. huoshanense* in long-term cultures (Zha and Luo 2008). In the case of fully grown plants, the traditional method of boiling uncrushed pseudobulbs before drying did not yield as much polysaccharide as drying those pseudobulbs which have been twisted at a temperature of 80 °C (Li et al. 2013a).

A heteropolysaccharide extracted from dried stems of *D. officinale* Kimura and Migo (= *D. catenatum*) was identified as a 2-*O*-acetyl glucomannan composed of mannose, glucose and arabinose in a 40.2:2.8:4.1 molar ratios. Its structure was also elucidated (Hua et al. 2004). A recent study showed that there are additional monosaccharide constituents in the polysaccharides of *D. hushanense* (= *D. catenatum*), namely, rhamnose, fucose, arabinose, xylose, galacturonic acid, glucuronic acid, 4-0-methylglucuronic acid and 2-0-acetylmannose. Homoglaturonan (HGA) and galactomannan and a smaller proportion of rhamnogalacturonan(RG) are present in the petin fractions of leaf and stem. The alkali extractable fractions are mainly glucuronoarabinoxylans (GAXs), fucosylated xyloglucans (XGs) and glucomannan. Mucilage polysaccharide extracted from the leaf and stem are constituted by 2- and 3-*O*-acetyl glucomannan. Whereas mucilage polysaccharide stimulate cellular immunity and haemopoeitic growth factors GM-GSF and G-CSF, the deacetylated mucilage obtained with alkaline treatment is devoid of such activity (Wang et al. 2015).

D. catenatum polysaccharides induce aquaporin-5 translocation to the apical membrane of epithelial cells of the salivary glands located under the jaw. Aquaporins are water channel proteins and by this action, *D. catenatum* promotes salivary secretion. The process is effected by activation of M3 muscarinic receptors and induction of extracellular calcium influx (Lin et al. 2015).

Experiments in mice showed that *Dendrobium officinale* and its polysaccharides fed to the animals enhanced cellular immunity. Although one study found that humoral immunity was enhanced by feeding the plant but not by feeding its polysaccharides (Liu et al. 2011b) several studies report otherwise. *D. catenatum* polysaccharides (DCP) suppressed progressive lymphocyte infiltration and apoptosis in experiments on mice (Lin et al. 2011) and DCP inhibited TNF-alpha-induced apoptosis in a human salivary gland cell line (A-253) (Xiang et al. 2013). Mucilage polysaccharide extracted from the leaves and stems of *D. catenatum* (reported as *D. huoshanense*) activated murine splenocytes to produce cytokines and haemopoietic growth factors (Hsieh et al. 2008). "Crude" (sic) polysaccharides of *D. huoshanense* at 100–200 mcg/ml stimulated tumour necrosis factor-alpha release from peritoneal macrophages; at concentrations of 100–800 mcg/ml, it stimulated interferon gamma release from murine splenocytes. Authors of the study conclude that this reflects an ability of the *D. hushanense* polysaccharide to induce an immune response (in mammals) (Zha et al. 2007).

Enzymatic fingerprints of *D. officinale* by polysaccharide analysis using carbohydrate gel electrophoresis (PACE) showed that herbs from Anhui, Fujian, Guangdong, Guangxi and Yunnan provinces produced distinct patterns enabling them to be identified, but the fingerprints of herbs from Jiangxi, Hunan and Zhejiang provinces were similar. PACE might be a method to identify and control the quality of *shihu* (Zha et al. 2012). Similar findings were reported for monosaccharide composition of polysaccharides in *D. catenatum* from different germplasms (Yuan et al. 2011a).

Timing of harvest had strong influence on polysaccharide content. Plants harvested in May had 25 % higher content of polysaccharide than plants harvested in February (Zhang et al. 2013b).

Four 6,8-Di-C-glycosyl flavonoids and seven known compounds (malic acid, dimethyl malate, N-phenylacetamide, isopentylbutyrate, salicylic acid, shikimic acid and isoshaftocide) were isolated from *D. huoshanense* (= *D. catenatum*) by Chang et al. (2010) in Taiwan's Academia Sinica and a patent on the mucopolysaccharides has been lodged. *Huoshan shihu* (*Dendribium catenatum*, syn. *D. huoshanense*) polysaccharide possessed anti-oxidant activity and was hepato-protective in mice exposed to carbon tetrachloride. Pretreatment with *D. huoshanense* polysaccharide decreased production of malondialdehyde and restored activities of super-oxide dismutase, catalase and glutathione perox-idase and glutathione in the livers of the carbon tetrachloride treated mice (Tian et al. 2013). *D. huoshanense* extract also reduced damage to human nerve cells (human neuroblastoma SH-SY5Y cells) by hydrogen peroxide in-vitro (Suen et al. 2013)

Quercetin isolated from in-vitro propagated *D. tosaense* (correct name: *D. catenatum*) showed significant anti-oxidant activity by DPPH radical anti-oxidate assay. In the same experiment the anti oxidants isolated from *D. moniliforme* were alkyl ferulates (Lo et al. 2004a). In mice injected with ovalbu-min with 2,4,6-trinitro-1-chlorobenzene to pro-voke skin lesions that resembled atopic dermatitis, administration of an ethyl acetate extract of *D. tosaense* (= *D. catenatum*) suppressed cytokine profiles, anti-OVA igE levels, mast cell infiltration and degranulation (Wu et al. 2013).

Cold pretreatment at 10 °C for 1–2 weeks significantly enhanced the conversion of protocorm-like bodies (PLB) into shoots, and if followed by the use of 20 μ M kinetin and 10 g/l maltose in ½ M/S medium it could produce over 1300 shoots from 1 g of PLB after 3 months of culture. This would greatly enhance the supply of the endangered, medicinally desirable variety, *D. houshanense* (Luo et al. 2009b).

Herbal Usage: This is one of the primary species of *Dendrobium* that originally consti-tuted *shihu*. Chinese physicians attribute numer-ous properties to *D. catenatum* (syn. *D. officinale*, *D. tosaense*) namly, it enhances the body's immunity, removes fatigue, promotes digestion, increases salivary secretion, prevents hyperglycaemia, lowers blood pressure protects against liver damage, is an anti-oxidant and has antitumour activity (Lu et al. 2013). The claims are verified by randomized clinical trials.

A pilot, uncontrolled study evaluating the effects of orally administered extract of *D. huoshanense* was conducted aata the Taipei Veterans General Hospital on 27 children aged 4–18 years who suffered from moderate to severe, recalcitrant atopic dermatitis showed that polysaccharides from *D. huoshanese* improved the symptoms and concomitantly there was a reduction in the levels of some allergy related cytokines. The lesions improved initially and the children slept better between weeks 0–4, but did not change during the second month. Itching also decreased between weeks 0 and 4 but increased thereafter. No serious adverse effects occurred. Laboratory testing showed falls in levels of IL-5, IL-13, IFN-gamma and TGF-beta 1, but there was no change in the level of IL-10 (Wu et al. 2011).

Dendrobium ceraia Lindl. (see **Dendrobium crumenatum, Sw.**)

Dendrobium chrysanthum Lindl.

Dendrobium paxtonii Lindl.

Chinese name: *Banchunshihu* (spotted lip noble *Dendrobium*), *Suhuashihu* (bouquet *Dendrobium*) *Dahuangcao* (large yellow herb); *Mabiancao* (horse whip herb); *Shuidabang* (water trashing stick); *Jin Lan* (golden orchid).

Indian name: *Mera leikham* in Manipati dialect

Thai names: *Ueang sai moragole; Ueang thian, Bai morakot, Ueang kham sai, Ueang pu loei, Ueang sai morakot*

Fig. 10.12 *Dendrobium chrysanthum* Lindl. [PHOTO: E.S. Teoh]

Description: Pseudobulbs are club-shaped or spindle-shaped, 10–30 cm long and 3 cm in diameter. Inflorescence is pendulous, with 3–5, golden yellow flowers each 5 cm across from the upper axils. Petals are obovate, retuse, fleshy and broader than the sepals (Dalstrom 2009). Flowers are fragrant and may last 2–3 weeks (Fig. 10.12). Flowering occurs at the end of the dry season, which is May to June in Thailand (Vaddhanaphuti 2005) and September to October in China (Chen et al. 1999b).

D. chrysanthum is an attractive, popularly cultivated orchid species which is distributed from the tropical foothills of the Himalayas across Myanmar to Yunnan, northern and upper northeast Thailand and Indochina where it occurs at an elevation of 1000–1200 m (Grant 1895). Although the plant prefers a slightly cool environment, it will flower in Singapore.

Phytochemistry: Five pyrrolidine alkaloids are produced by *D. chrysanthum*, namely, hygrine, *cis*- and *trans*-dendrochrysine, the latter two unique to this species (Luning and Leander 1965); *trans* and *cis*-dendrochrysanines (Yang et al. 2005b). The last two compounds did not show any immunomodulatory activity in mice (Yang et al. 2005b). Ye et al. (2003) managed

to isolate two new flourene derivatives, denchrysans A and B and a new phenanthrene-diglycoside, denchryside A from the medicinal herb. The following year they isolated another new neolignan glucoside, denchryside B along with three known lignans from the herb (Ye et al. 2004). Yang et al. (2006a) isolated a novel phenanthrene derivative with a spirolactone ring, dendrochrysanene which they found could suppress the mRNA level of TNF-alpha, IL8, IL110 and iNOS in murine peritoneal macrophages.

Herbal Usage: In TCM the properties of this orchid are defined as follows: 'The pseudobulb benefits the stomach, reinforces the *yin* element, reduces fever, and eliminates thirst and anorexia. It is a tonic' (Wu 1994; Anonymous 2004; Xu et al. 2006). There is sufficient pharmaceutical interest in this species as a variety of *shihu* for molecular identification to be proposed to separate it from morphologically allied species and establish its identity (Ding et al. 2002a; Zhang et al. 2005b).

Dendrobium chrysotoxum Lindl.

Dendrobium sauvissimum Rchb. f

Thai names: *Uang Khan, Ueang kham*
Vietnamese name: *Kim diep*
Myanmar names: *Shwe tu, Mout khan war*

Description: *D. chrysotoxum* is a spectacular, golden *Dendrobium* with bright yellow sepals and petals and a round, fringed, hirsute, orange coloured lip. Flowers are fragrant, 4–5 cm in diameter, well arranged, and numerous (20 or more) on an arching or pendent inflorescence (Fig. 10.13). Flowering period is February to June (Chen et al. 1999b; O'Byrne 2001; Vaddhanaphuti 2005; Nath and Das 2013), depending on locale. Stems are clustered, fusiform and many angled, 12–30 cm long with whitish membraneous sheaths, and bearing 2–4 leaves near the apex. Leaves are coriaceous, lanceolate, 10–15 by 2–3.5 cm. *D. chrysotoxum* is distributed across Vietnam and Laos to Thailand, southern Yunnan, Myanmar and Sikkim,

Fig. 10.13 *Dendrobium chrysotoxum* Lindl. [PHOTO: E.S. Teoh]

epiphytic or lithophytic in sparse woods, at 500–1700 m. It does reasonably well in the tropical lowlands.

Phytochemistry: Gong et al. (2006) from the China Pharmaceutical University in Shanghai used column chromatography to isolate ten compounds which they identified as syringare sinol, 5 alpha, 8 alpha-epidioxy-24(R)-methycholesta-6,22-dien-3beta-ol, trans-3-(4-hydroxy-3-methoxyphenyl)-acrylic acid octacosyl ester, defusing, 3,4-dihydroxy benzoic acid, 3,4-dimethoxy-benzoic acid, vanillic acid, 3,4-dimethoxy-benzoic acid methyl ester, 3,5-dibromo-2-aminobenzaldehyde and heptadecanoic acid 2,3-dihydroxy-propyl ester. The therapeutic roles of the compounds were not defined. Two new fluorenones were isolated from the species, namely 2,4,7-trihydroxy-5-methoxy-9-fluorenone and 2,4,7-trihydroxy-1,5-dimethoxy-9-fluorenone (Yang et al. 2004). Their pharmacological activity was not described.

Erianin, a bibenzyl from *D. chrysotoxum* has potent anti-oxidative action in lipid peroxidation and hemolysis assays (Ng et al. 2000). It induces apoptosis in human leukemia HL-60

cells (Li et al. 2001). Erianin retards the growth of xenografted human hepatoma Bel 7402 and melanoma A 375. It induces significant vascular shutdown within 4 h following administration of 100 mg/kg and shows similar anti-angiogenic activities in vitro. Furthermore, it prevents fibroblast-growth factor-induced neovascularization in the chick embryo, inhibits proliferation of human umbilical vein endothelial cells, disrupts endothelial tube formation, and abolished cell migration across collagen and adhesion to fibronectin. Erianin depolymerizes both F-actin and beta-tubulin in proliferating endothelial cells (Gong et al. 2004b). In human umbilical vein endothelial cells, erianin decreases glucose consumption, lactate production and intracellular ATP. These effects are blocked by pretreatment with JNK/SAPK inhibitor SP600 125 which suggests JNK/SAPK as the mechanism involved in erianin's antitumour and anti-angiogenesis actions (Gong et al. 2004a).

Twenty-three new isoerianin derivatives were studied for cytotoxic activity and shows promise as an anticancer agent. At micromolar concentration, it inhibits tubulin polymerization, G2/M phase cell-cycle arrest in H1299 and K562 cancer cells, and induces apoptosis (Massaoudi et al. 2011). ZJU-6 is another derivative of erianin that shows potent antitubulin polymerization and anti-angiogenesis activities (Lam et al. 2011). They are promising anticancer agents. Other teams are studying compounds with similar activities that are not obtained from orchids or modified from natural compounds in orchids (Rasolofonjatovo et al. 2012).

Chrysotoxine, a bibenzyl from *D. chrysotoxum* significantly attenuates 6-hydroxydopamine induced and MPP+ induced apoptosis in the human neuroblastoma cell-line SH-SY5Y in a dose-dependent manner. It has no effect on rotenone neurotoxicity on the cells (Song et al. 2010, 2012b). Chrysotoxine has also been isolated from *D. pulchellum* and shown to facilitate anoikis and proliferation of lung cancer cells (Chanvorachote et al. 2013).

At Chonbuk National University in Chonju, Korea, Zhao et al. (2007) studied the anti-oxidant and antiglycaemic role of DCLP (*D. chrysotoxum* Lindley Polysaccharide), a polysaccharide with a molecular weight of 150 kDa isolated from the stems of *D. chrysotoxum* Lindl. In-vitro and in-vivo experiments demonstrated that DCLP possessed anti-oxidative activity. It offered significant protection against glucose-oxidase mediated cytotoxicity in Jurkat cells in-vitro. DCLP reduced blood glucose levels in alloxan-induced diabetic rats. It also enhanced immune response in rats. An ethanolic extract of *D. chrysotoxum* administered to streptozoin-induced diabetic rats at 30–300 mg/kg dosage decreased retinal angiogenesis and inhibits the expression of vascular endothelial growth factors (VEGF and VEGFR2) and several aother pro-angiogenic factors (Gong et al. 2014).

Two polysaccharides from pseudobulbs of *Dendroboium chrysotoxum*, DCPP-1a and DCPPII, exhibited antiproliferative activity on human lung cancer cell line SPC-A-1 (Sun et al. 2013).

Herbal Usage: It is used as *shihu* in Taiwan (Ou et al. 2003). *D. chrysotoxum* flowers are being sold in China for making tea.

Dendrobium clavatum Wall. ex Lindl. (see **D. denneanum Kerr**.)

Den crepidatum Lindl. & Paxton,

Chinese name: *Meigui Shihu* (rose *Dendrobium*)
Thai name: *Uang sai nam khieo,*

Description: Plant is epiphytic and of variable size, with canes 5–45 cm in length, with 5–9 linear lanceolate leaves 5–15 cm long. Inflorescences appear near the apex of deciduous stems and carry 1–4 waxy, white flowers 2.5–4.5 cm across, with a lilac tint at the tips of the tepals. Lip is heart shaped, slightly cupped, white to cream with a yellow or orange stain near the base. Flowers are mildly fragrant and variable in size, form and colour (Fig. 10.14). Flowering season is April (Vaddhanaphuti 2005). They last for 8–21 days.

Widespread in India from the southern peninsula to the Himalayan region, it is distributed

Fig. 10.14 *Dendrobium crepidatum* Lindl. & Paxton. [PHOTO: E.S. Teoh]

across Nepal, Bhutan, Sikkim, Myanmar and Yunnan, and Guizhou to Thailand, Laos and Vietnam, at 600–2100 m.

Phytochemistry: Three alkaloids, crepidine, crepidamine and dendrocrepine were isolated from *D. crepidatum* (Elander et al. 1973).

Herbal Usage: In TCM, stems are used to enrich *yin*, benefit the stomach, and clear dry or itchy throat (Wu 1994). Stems are used to treat fractures in Nepal (Pant and Raskoti 2013). The orchid is among 199 medicinal plants used by the Hani ethnicity in the Naban River Watershed National Nature Reserve in Yunnan, China. The orchid is commonly cultivated in home gardens by the Hani whereas they usually collect other medicinal plants from the forest (Ghorbani et al. 2011)

Dendrobium crispulum Kim ei Migo [see **D. moniliforme (L) Sw.**]

Dendrobium crumenatum Sw.

Dendrobium caninum Merrill, *D. kwashotense* Hayata, *D. ceraia* Lindl.

Common names: pigeon orchid, dove orchid
Malay: *bunga angin* (wind orchid)*; Daun sepuleh tulang* when it is used as a protective charm (bone restoring leaf)
Indonesian: *Anggerik Merpati* (dove orchid)*; Anggerik Bawang* (onion orchid); *Bunga Angin* (wind flower);

Fig. 10.15 *Dendrobium crumenatum* Sw. [PHOTO: E.S. Teoh]

Thai: *Wai tamoi; Bua klang hoa; Sae phra in; Thiam ling; Dawk mai wai, Ueang Mali*
Vietnamese: *Tuyel mai*
Indian name: *Jivanti*

Description: Pseudobulb is bulbous at the bottom, this portion comprising 4 internodes, 3–4 cm by 1 cm whereas the distal portion is wiry, 30–50 cm long. Leaves are borne only on the thin, upper portion; they are oblong-lanceolate, thick, 9 by 2 cm. *D. crumenatum* is a one-day wonder. Flowering is triggered by a sudden fall in night temperature, usually brought on by a heavy thunderstorm. All the trees in the vicinity flower simultaneously in great abundance 11–12 days after the storm, but the flowers only last the day. However, a clump will flower 6–11 times a year (Burkill 1920). Flowers are fragrant, white with with a patch of yellow keel on the mid-lobe of the lip. They are 3 cm across, petals and sepals well extended (Fig. 10.15). Plantlets develop on the axils of old stems and inflorescence after flowering.

The beautiful, fragrant, white Pigeon Orchid is found growing on trees in exposed places all over Malaysia and is extremely common on mature roadside trees in Singapore, even right in the heart of the city. This lowland orchid (growing at 0–500 m) enjoys a large distribution

extending from India and southern China through the Malay Archipelago to the Philippines.

Phytochemistry: Alkaloids are present in *D. crumenatum* (Chopra 1933).

Herbal Usage: Juice from the crushed pseudobulbs, fresh, boiled or roasted, was dropped into the ear to treat pain caused by small abscesses, boils or other intractable swellings in the external ear (*bunting telinga,* Malay) in Malaysia and earache in Indonesia (Ridley 1906, 1907). Another method, observed by Drs. Gimlette and Thomson in Kelantan in 1919, called for the orchid stem to be stuffed with onion and *jintan manis* seeds, then roasted in hot ashes before the juice was squeezed into the affected ear (Gimlette and Thomson 1939). A poultice made from the residue was then applied around the external ear. This practice might be of Indian origin. Currently, juice from the tender growing tips of *D. herbaceum* Lindl., or of *D. macrostachyum* Lindl. [= *D. aphyllum* (Roxb.) Fischer], is similarly used as ear drops for treating earache by the Valmikis of Visakhapatnam in Andra Pradesh, while the Nukadoras use only the juice of *D. herbaceum* to treat earache (Reddy et al. 2005b). Poultices made with leaves are used on boils and pimples in India (Das 2004).

Malays in Perak used poultices prepared from the pounded leaves of epiphytic orchids to relieve headache in the early twentieth century (Burkhill and Haniff 1930). During the Japanese Occupation of Singapore and Malaya, Koriba and Watanabe (1944) wrote an *Illustrated Useful Plants in Malaya. First Compilation: Medicinal Plants* in which they described 194 plants. There was only one orchid, *Dendobium crumenatum*, Sandrasagaran et al. (2014) in their list. The Pigeon Orchid was also used to treat nervous ailments and cholera in the region (Lawler 1986). In old Malaya, another usage was to form a *besom* for scattering rice in ceremonies to ward off spirits from homes (Gimlette and Thomson 1939) or to induce the return of beneficent spirits (Burkill 1935) (see also *Plocoglottis lowii*). In Vietnam, crushed bulbs were rubbed over aching bones of people suffering from rheumatism (Petelot 1953).

D. crumenatum is used as an antiseptic and applied as a poultice to boils and pimples in Uttar

Pradesh, India. Under the name *Jivanti*, it is also used for affections of the brain and nerves. Patients suffering from cholera are fed a conserve of the flowers and leaves (Trivedi et al. 1980). Burkill figured that the Dutch physician Bontius, who visited Java during the seventeenth century, was referring to *D. crumenatum* when he stated that "the Malays thought nothing to be its equal for affections of the brain and nerves and that a conserve of the flowers and leaves was used for cholera" (Burkill 1935).

Dendrobium crystallinum Rchb. f.

Chinese name: *Hainanjinmao Shihu* (Hainan Province crystal hat *Dendrobium*)
Thai Names: *Ueang sai sam si, Ueang nang fawn*
Myanmar Name: *Setkhu pan*

Fig. 10.16 *Dendrobium crystallinum* Rchb. f. [PHOTO: E.S. Teoh]

Description: *D. crystallinum* is widespread from northern India to Yunnan, Myanmar, Thailand and Indochina into Hainan Island at 500–1700 m in exposed locations. Plants tend to form large clumps. Canes are up to 45 cm long, slender, carrying 2–4 linear, lanceolate leaves 15 by 1.5–2.5 cm. Inflorescences are generally located near the tip of deciduous canes and bear 1–3 flowers. The fragrance of this *Dendrobium* is enhanced by its habit of flowering gregariously, all the flowers opening simultaneously with hundreds of flowers on a single plant. Flowers are 4–5 cm across, white with a magenta blotch at the tip of the petals and sepals. Lip is white or yellow with a white border, orange at the base and blotched with magenta or amethyst at the apex (Fig. 10.16). Flowering period is May in Yunnan and from April to July in Thailand (Chen et al. 1999b; Vaddhanaphuti 2005).

Phytochemistry: Nine compounds were iolated from pseudobulbs of *D. crystallinium* in 2008, namely, 4,4′-dihydroxy-3,5dimethoxy-bibenzyl, gigantol, naringenin, *p*-hydroxybenzoic acid, *n*-tetracosyl *trans-p*-coumarate, *n*-octacosyl *trans-p*-coumarate, *n*-hexacosyl transferulate, stigmasterol and daucosterol (Wang et al. 2008). The following year, the group reported the

recovery of five new compounds from the stems of *D. crystallinum*, namely, dencryol A, dencryol B, crystalltone, crystallinin and 3-hydroxy-2-methoxy-5,6-dimethylbenzoic acid, together with six known compounds, namely, dendronobilin B, syringic acid, apigenin, isoviolanthin, 6‴-glucosyl-vitexin and palmarumycin JC2(Wang et al. 2009b). Nine compounds have been isolated from the species and identified as: 4,4′-dihydroxy-3,5-dimethoxybi-bibenzyl, gigantol, naringenin, *p*-hydroxybenzoic acid, *n*-tetracosyl *trans-p*-cou-marate, n-octacosy *trans-p*-coumarate, *n*-hexacosyl *trans*-ferulate, stigmasterol and daucosterol (Wang et al. 2008).

Herbal Usage: Entire plant is used as medicine. It has a long history of usage as *shihu* and goes by the label, *zhong huan cha* (medium ring hairpin).

Dendrobium cumulatum Lindl.

Thai names: *Thian phaya in, thian thong, Uang sai si dok*

Description: Stems are laterally flattened, thin, 40 cm long; leaves oblong-lanceolate 5–10 by

2–3.5 cm, borne on the distal half of the stem and deciduous. Inflorescence arises from lateral nodes of leafless canes with 4–8 flowers bunched together on the short inflorescence. Flowers are light rose-purple, 3 cm across with a white, spatulate lip. Flowering season is June in Nepal (Raskoti 2009), July in Thailand (Vaddhanaphuti 2005) and May to August in Upper Assam, India (Nath and Das 2013). An epiphyte with a deciduous habit, *D. cumulatum* is found Nepal, Bhutan, northeast India, Myanmar, Thailand, Indochina and Borneo at 300–1500 m.

Phytochemistry: Two bibenzyl derivatives, cumulatin and tristin, were isolated from the orchids *D. cumulatum* and *Bulbophyllum triste* (Majumder and Pal 1993).

Herbal Usage: Stems are used to treat asthma (Chuakul 2002).

Dendrobium dalhousieanum Wall. [see **D. pulchellum Roxb. ex. Lindl.**]

Dendrobium denneanum Kerr

syn. *Dendrobium aurianticum* Rchb

Chinese name: *Ma pien Shihu* (horse whip *Dendrobium*) (Note: *Ma Pien Shihu* is a collective name which includes several species.)

Description: This is a large *Dendrobium* with pencil-thin, erect, terete canes, 40–75 cm in length, fattening towards the base. Leaves are coraceous, narrow, pointed, notched at the tips, and measure 9–11 by 1.7–2.5 cm. Inflorescences arise from nodes near the apex of the pseudobulbs. They are 8 cm long and bear 2–8 flowers. *D. flaviflorum* in Taiwan generally has two flowers on each inflorescence. Flowers are 4–8 cm across, fragrant, a golden yellow or orange (Fig. 10.17). Seidenfaden (1985) commented that *D. clavatum* carried a deep purple blotch on the lip, but *D. auriantiacum* did not. Chen et al. (1999b) observed that D. *deneanum* differed from *D. auriantiacum* (or *D. chryseum*) by having a purple red patch on the lip and stout stems with broad leaves. In China, it flowers in April to May (Chen et al. 1999b;

Fig. 10.17 *Dendrobium denneanum* Kerr [PHOTO: E.S. Teoh]

Vaddhanaphuti 2005). Elsewhere, it flowers several times a year, 7–14 days after a sudden 5 °C drop in temperature. Flowers are short-lived.

Variation in vegetal forms and colour of the flowers in *D. denneanum* partly explain the existence of numerous names for the species. Taxonomists have disagreed as to whether they should all be lumped into one species, but for the present the varieties are lumped together into *D. denneanum*. It is widely distributed in Myanmar, Thailand, Laos, Vietnam and southern China to Sikkim, Bhutan and northern India at 600–2000 m. In China, the plants are found as an epiphyte on trees in forests in Yunnan, Guizhou, Guangxi, Hainan and Taiwan.

Phytochemistry: Three 2-glucosyloxycinnamic acid derivatives with anti-oxidant properties have been isolated from the stems of *D. aurantiacum* var. *denneanum* by Yang et al. (2007b, c) at the Shanghai University School of Traditional Chinese Medicine. They are: *cis*-melliotoside, *trans*-melilotoside and dihydromelilotoside. Three polysaccharide fractions were recovered by column chromatographic separation of a crude extract of *D. denneanum*. Their monosaccharide components were made up of arabinose, mannose, glucose and galactose, with a preponderance of glucose. In one fraction, xylose was also present. The polysaccharides exhibited anti-

oxidant and immune-modulatory activities on in vitro testing (Fan et al. 2009, 2010). Recently, ten compounds were isolated from chloroform and n-butanol fractions of dried stems of *D. denneanum*, namely, coumarin, moscatin, thymidine, trans-syringin, dihydrosyringin, 2,4,6-tromethoxyphenol-1-*O*-beta-D-glucopyranoside, (+)-syringaresinol-*O*-beta-D-glucopyranoside, 3-hydroxy-1-(4-hydroxy-3,5-domethoxyphenyl)-2-[4-(3-hydroxy-1-(E)-propenyl)2,6-dimethoxy-phenoxyl]propyl-7-*O*-beta-D-glucopyranoside, picra-quassioside C and citrusin B (Pan et al. 2013a, b). Li et al. (2014) isolated three new neolignan glycosides together with four known analogs from stems of *D. aurantiacum* var. *denneanum* (= *D. denneanum*). The three new compounds are (−)-(8R,7′E)-4-hydroxy-3,3′,5,5′-tetramethoxy-8,4′-oxyneolign-7′-ene-9,9′-diol 4,9-bis-*O*-beta-D-glucopyranoside, (−)-(8S, 7′E)-4-hydroxy-3,3′,5,5′-tetramethoxy-8,4′-oxyneolign-7′-ene-9,9′-diol 4,9-bis-*O*-beta-D-glucopyranoside and (−)-(8R, 7′E)-4-hydroxy-3,3′,5,5′,9′-pentamethoxy-8,4′-oxyneolign-7′-ene-9-ol 4,9-bis-*O*-beta-D-glucopyranoside.

Herbal Usage: *D. denneanum* is used as a tonic and used in the same manner as *D. chrysanthum* (see above).

Dendrobium densiflorum Lindl.

Dendrobium clavatum Roxb.

Thai names: *Ueang Mon Kai Liam, Uang Min Khai Luang.*
Vietnamese name: *Thy-tien*
Myanmar name: *Ta khun lone shwe*
Nepali name: *Sungabha, Sungava*

Description: This spectacular species with up to 50 cream-coloured flowers, 2–4 cm across, well-arranged but held close together on the inflorescence is distributed in the Vietnam, northern and upper northeast Thailand, Myanmar, Sikkim, Nepal and Bhutan; also, in China, it is found in Hainan, Guangdong, Guangxi and southeast Tibet at 400–1000 m. Flowers are bright butter yellow.

Fig. 10.18 *Dendrobium densiflorum* Lindl. in Kalimpong, West Bengal, India [PHOTO: E.S. Teoh]

Lip is round, pubescent and orange-yellow (Fig. 10.18). It flowers in April to June in China and Sikkim, July to September in the Kachin, Kayin and Shan states of Myanmar and August to September in northern Thailand, the flowers lasting 4–6 days.

Phytochemistry: *D. densiflorum* produces denisfloroside, a 2-(beta-D-glucopyranosyloxy)-4,5-methoxy-transcinnamic acid (Dahmen et al. 1975) which is also present in *D. chrysotoxum, D. denneanum* and *D. thyrsiflorium* (Pridgeon et al. 2014). Zheng et al. (2000) isolated a new coumarin compound, dihydroayapin, and seven known compounds from the stems of *D. densiflorum*. Fan et al. (2001) isolated a unique phenanthrenedione along with 16 known compounds, including lusianthriidin, denthyrsin, densiflorin, densiflorol B, cypripedin, gigantol, moscatilin, moscatin, naringenin, tristin, homoeriodictyol, scopoletin, ayapin, dengibsin, scoparone, oleanolic acid and beta-sitosterol. Five compounds, gigantol, moscatalin, homoeridictyol, scoparone and scopoletin, possessed antiplatelet properties in vitro. Densiflorol B, which is also present in *D. venustum* (not a medicinal orchid), showed strong antimalarial activity and a high selectivity index (Sukphan et al. 2014). Given the rapidly rising incidence of artemisin-resistant falciparum malaria in Cambodia and Thailand, there is an urgent need to discover new and effective antimalarial agents.

Herbal Usage: In India, leaves of *D. densiflorum* are ground into a paste with salt and applied on fractures to help set bones. Pulp of the pseudobulbs is used to remove pimples and boils in Nepal (Manandhar and Manandhar 2002).

Dendrobium denudans D. Don

Description: Plant is small, epiphytic or saxicolous, occurring at 800 - 2200 m in the Himalayas and Thailand. Stems are cylindrical, tapering towards the apex, with thin, lanceolate leaves, 5 - 10 by 1.2 - 2 cm. Inflorescence is apical, pendulous, bearing numerous droopy, spidery, cream-coloured flowers, 2 cm long. Lip is striped with purple. Flowering season is August to September (Raskoti, 2009).

Herbal Usage: Pseudobulbs are eaten raw to treat fever and body ache in Darjeeling Himalaya. They are also employed to make a narcotic preparation (Mao, 2006; Yonzone 2013).

Dendrobium devonianum Paxton

D. pulchellum Lindl.

Chinese name: *Chiban Shihu* (teeth pedal *Dendrobium*)
Thai names: *Miang; Ueang sai man pra in, Ueang sai pha kang, Ueang sai luat*

Description: Its distribution extends from the southern Himalayas to eastern Tibet, Myanmar, Yunnan, Sichuan, Guizhou, Guangxi, Taiwan. Thailand, Laos and northern Vietnam, at 1000–2000 m, but occasionally it is found at

550 m. Canes of this large *Dendrobium* are 90–150 cm long, slender, branching, pendulous, bearing numerous pale green leaves, 10 cm in length and deciduous. Inflorescence is 2.5 cm long, arising from nodes in the distal two-thirds of deciduous canes, and bearing 1–3 creamy-white flowers tinged with pink or lavender darkening to pink at the tips, and 4–8 cm across. Petals are lined with a row of hairs at the margin. Lip is open, heart-shaped, pubescent, with a fimbriate margin, marked by a central dark pink to lavender blotch fringed with white, and with purple lines at the throat. Plants from China and Thailand have two yellow patches on the lip replacing the dark pink blotch (Fig. 10.19). Flowers appear in April or May and last a fortnight.

Phytochemistry: *D. devonianum* from Vietnam yielded 30 endophytic fungi, with *Fusarium* being the dominant species. It exhibited bactericidal and fungicidal activities against pathogenic bacteria and fungi (Xing et al. 2011), but specific compounds were not identified. Two out of nine compounds isolated from *D. devonianum*, *N-trans-p*-feruloyl tyramine and 4-hydroxy-3,5-dimethoxybenzoic acid possessed some anti-oxidant activity whereas the

Fig. 10.19 *Dendrobium devonianum* Paxton in Thailand. [PHOTO: E.S. Teoh]

rest did not. The remaining seven compounds are 2,3,4,9-tetrahydro-1H-pyrido[3,4-b]indole-3-carboxylic acid, 2′-deoxythymidine, adenosine, *N-trans-p*-coumaroyl tyramine, *N-trans-p*-ferulyol tyramine, 3-methoxy-4-hydroxyben-zaldehyde and 3,4-dihydroxybenzoic acid methyl ester (Zhang et al. 2013a). The orchid also contained a unique flavonol glycoside, 5-hydroxy-3-methixy-flavone-7-*O*-[beta-ᴅ-apiosyl-(1–6)]-beta-ᴅ-glucoside (Sun et al. 2014).

Herbal Usage: As for *Shihu*. In Nepal, pulp of the pseudobulb is applied to boils and pimples to encourage healing (Baral and Kurmi 2006; Pant and Raskoti 2013).

Dendrobium discolor Lindl.

Dendrobium undulatum R. Br., *D. elobatum* Rupp., *D. fuscum* Fitzg.

Common Australian name: golden orchid
Papuan names: Rigo twist, Moresby gold, Bensbach yellow.

Description: Pseudobulbs vary in length from 0.5 to 5 m and 1–8 cm in diameter. Leaves are leathery, ovate 5–20 cm long, arranged in two ranks along the distal two-thirds of the stem. Inflorescence is arching, many-flowered. Flowers are yellow, brownish-yellow to chocolate-brown, 3.5–5 cm across. Sepals and petals are twisted, petals with undulating edges (Fig. 10.20). Flowers last for 2 months on the plants. It flowers throughout the year with a peak period in late winter and early spring in Australia (Lavarack et al. 2000).

A member of the *Spatulata* Section of *Dendrobium*, *D. discolor* is a hardy, sun-loving, variable, lowland species from Sulawesi, southern New Guinea, the islands of the Torres Straits and the northeastern tip of Australia.

Herbal Usage: In Mackay, Queensland, aborigines prepared a poultice from young canes of *D. discolor* and used it to draw a boil. A linament for ringworm was prepared with the juice of old canes (Lawler and Slaytor 1970). We

Fig.   10.20 *Dendrobium   discolor*   Lindl.   var. *Bromfieldie*   [PHOTO: E.S. Teoh]

do not know whether aborigines in Papua New Guinea use this orchid medicinally.

Dendrobium draconis Rchb.f.

Thai names: *Ueang ngoen, ueang ngum*
Myanmar Name: *Kein na ri*

Description: This white *Dendrobium* has stems that are 15–45 cm long, finely hirsute, bearing many dark green leaves 6–10 cm long. Short inflorescences are borne at the upper nodes of the stem and carry 2–5 white flowers, 5–6.5 cm across, with orange-red lines at the throat of the lip (Fig. 10.21). Flowers are fragrant and appear in March and April, maximally at the peak of the hot, dry season during Thiugyan (the Buddhist Water Festival) in the Shan state of Myanmar (Tanaka et al. 2003). The Burmese name is probably derived from Sanskrit *kinnari*, a Buddhist mythical half-bird, half-human creature with a beautiful voice, referring to the delicate outspread-winged form of the white flowers; *kinnari* is female for *kinnara*.

A very common species in Thailand, *D. draconis* is distributed from northeast India to Myanmar, Thailand and Indochina.

Fig. 10.21 *Dendrobium draconis* Rchb. f. [PHOTO: E.S. Teoh]

Phytochemistry: Gigantol, a common bibenzyl, was isolated from *D. draconis*. It is cytotoxic to several cancer cell lines and in the present report it suppresses migratory behaviour of non-small cell lung cancer by suppressing filopodia formation (Charoenrungruang et al. 2014). Another team from Thailand earlier reported the isolation of new phenanthroquinone from the orchid, namely 5-methoxy-7-hydroxy-9,10-dhydro-1,4-phenanthroquinone, together with four known stilbenoids, hircinol, gigantol, batatasin III and 7-methoxy-9,10-dihydrophenanthrene-2,4,5-triol. The stilbene derivatives exhibited appreciable anti-oxidant activity (Sritularak et al. 2011).

Herbal Usage: The stem is used as an antipyretic and haematimic in Thailand (Chuakul 2002).

Dendrobium eriiflorum **Griff.**

Description: An epiphytic species with erect stems covered with sheaths bearing linear-lanceolate leaves, 3.6–6.2 by 0.5–1 cm, its pendulous inflorescence with half-open flowers resemble those of *Eria*, hence the common name, Eria-like Dendrobium. Inflorescence is axillary, arching or pendent, flowers numerous, greenish-white, laxly arranged. Lip trilobed, purple. It flowers in September to October in Nepal where it occurs at 1500–2100 m (Raskoti 2009). *D. eriiflorum* is distributed from the eastern Himalaya to Myanmar, Thailand and Java.

Herbal Usage: Dried powdered pseudobulb is used as a tonic, and paste is used to treat fractures and dislocations in Nepal (Subedi et al. 2013; Pant and Raskoti 2013). *D. eriiflorum* was included among 21 samples of *Herba Dendrobii* collected from various herbal sources by Takamiya et al. (2011), but it is not included in standard *Chinese Materia Medica* as *shihu*.

Dendrobium falconeri **Hook.f.**

Dendrobium erythroglossum Hayata

Chinese names: *Xinzhushihu* (Xinzhu noble *Dendrobium*), *Honglishihu* (red crane *Dendrobium*); *Chuanzhushihu* (string *Dendrobium*), and in Taiwan: red oriole *Dendrobium*

Taiwanese name: *Xin Zhu Shi Hu* (new bamboo *Dendrobium*)

Thai Names: *Ueang sai wisut, Rot rueang saeng, Ueang mieng, ueang ya phaet.*

Description: A magnificent *Dendrobium* with pendulous, terete, long, slender, branched and knotted stems, 60–120 cm in length, bearing 2–5 small linear leaves of 4–6 cm by 3–4 mm at the growing tips. Flowers are large, 11–12 cm diameter, bright white, stained a brilliant crimson on the tips of the petals, sepals and lip. They are produced singly over leafless stems after the dry season. Lip is large, and carries an orange disc on a white background; it is stained a dark purple at the throat, and marked by a deep purple at the tip. Flowers appear in May to June in China. It occurs on tree trunks in forests at 800–1800 m from northeast India, Bhutan, Myanmar, Thailand, Vietnam and southern China (Yunnan, Guangxi, Hunan and Taiwan). Seidenfaden and Smitinand (1959) commented that their plants failed to survive when brought to Bangkok.

Phytochemistry: Dendrofalconerol-A, a pure bisbibenzyl isolated from pseudobulbs of

D. falconeri at concentrations of 0.5–5 μmol/L, significantly reduced protein levels of migrating human lung cancer H460 cells in a dose-dependent manner. Expression of migration-related integrins such as integrins beta-1 and alpha-4 were significantly reduced, and epithelial transition into mesenchyme was suppressed (Pengpaeng et al. 2015a). Dendrofalconerol-A also sensitises anoikis and decreased caveolin-1, a protein associated with tumour aggressiveness (Pengpaeng et al. 2015b). Anoikis is a form of programmed cell death which occurs when normal cells detach from their surroundings. However, many cancer cells escape anoikis and this allows the cancer to spread or metastasise. It has been proposed that rendering cancer cells susceptible to anoikis can be a new therapeutic approach for managing cancers

Usage: TCM states that stem nourishes the *yin* elements, benefits the stomach, stops thirst and relieves the feeling of heat, dry mouth and dry throat. It is used to treat people recovering from illness or who are suffering from anorexia (Wu 1994; Bencao 2000).

Dendrobium faulhaberianum Schlect. [see **Dendrobium aduncum Lindl.**]

Chinese name: *Guo Shihu* (hook *Dendrobium*), *Huangcao* (golden herb)

Dendrobium fargesii Finet.

syn. *Epigeneium fargesii* (Finet) Gagnep

Chinese names: *Danyehouchun Lan* (single leaf, thick-lipped orchid), *Maihu* (wheat Dendrobium); *Guoshangye* (leaf above fruit); *Shiduo* (stone bean); *Danyeshizao* (single leaf stone date); *Shi Lan* (stone olive)
Taiwanese name: *Lian Zhu Lan* (chain of pearls orchid), *Xiao Pan Long* (coiled dragon) *San Xing Shi Hu* (three stars *Dendrobium*)

Description: *Epigeneium fargesii* is a small epiphyte with a single ovate leaf and pseudobulbs that are spaced not as far apart as in the former species. Inflorescence carries a single flower with narrow pointed petals and sepals. Lateral sepals curve backwards. Lip is trilobed and shaped rather like a gourd in outline; the middle lobe is oval and extends laterally almost to the border of the lateral lobes.

It is saxicolous along valleys and epiphytic in forests at 400–2400 m in Sichuan, Yunnan, Guizhou, Guangdong, Fujian, Anhui, Guangxi, Hubei, Hunan, Jiangxi and Taiwan, and in northern Vietnam, Thailand and Bhutan Zhu (Ji and Li 2009).

Herbal Usage: The entire plant is used for clearing heat and moistening dryness. It clears phlegm and stops coughs. It is used to treat tuberculous cough, bronchitis, pneumonia, diphtheria, sore throat, gastritis, knife wounds and night sweats (Wu 1994).

Dendrobium fimbriatum Hook. f.

syn. *Dendrobium normale* Fale.

Chinese names: *Liusushihu* (tasseled stone orchid), *Mabianshihu* (Mabian stone orchid)
Indian names: fringed lip *Dendrobium*
Thai name: *Ueang waew mayura, Ueang kaam ta daam*

Description: *D. fimbriatum* is an attractive species with up to a metre-long, thin pseudobulbs, bearing terminal inflorescence when deciduous, carrying 6–12 golden-coloured flowers with a purple blotch at the base of the fringed lip (Fig. 10.22). Chinese flowers are reportedly only 3–4.5 cm across while Grant (1895) reported that the flowers in Myanmar were 5–7.5 cm in diameter. Its distribution extends in a belt that stretches across northern India [Garhwal Himalaya and Uttar Pradesh up to 2100 m. (Bose and Bhattacharjee 1980)], Nepal, Bhutan, Sikkim, Myanmar, Northern Thailand, Vietnam, and the Chinese provinces of Yunnan, Guizhou and Guangxi. *D. fimbriatum* grows on trees in dense forests or on cliffs at 600–1700 m. Flowering period is March in China, and April and May in

Fig. 10.22 *Dendrobium fimbriatum* Hook. f. [PHOTO: E.S. Teoh]

Thailand, Myanmar and Assam (India), the flowers lasting a week (Tanaka et al. 2003; Vaddhanaphuti 2005; Nath and Das 2013).

Phytochemistry: Eight compounds were isolated from *D. fimbriatum*, namely, fimbriatone, confusarin, crepidatin, physcion, rhein, ayapin, scopolinmethyl ether and n-octacostyl ferulate (Bi et al. 2003).

Herbal Usage: In TCM, *D. fimbriatum* is used as *shihu* to improve eyesight (Anonymous 2004. This particular *shihu* polysaccharide is said to enhance T-cell and macrophage immunity, and it possesses anti-oxidant activities, enhances superoxide dismutase (SOD) and reduces lipid peroxidation (Xu et al. 2006).

The whole plant is used to treat liver disorders and nervous debility in Nepal (Baral and Kurmi 2006; Pant and Raskoti 2013). In Garhwal Himalaya and Uttar Pradesh, entire plant of *D. fimbriatum* (syn. *D. normale* Fale.) is used as an aphrodisiac (Bhattacharjee 1998, quoted by Sood et al. 2005).

Seven bibenzyl dimers, fimbriadimer-bibenzyls A–G, together with a new dihydrophenanthrene derivative, (S)-2,4,5,9-tetrahydroxy-9,10-dihydrophenanthrene, and 13 known compounds were isolated from the pseudobulbs of *D. fimbriatum* during a bioassay-guided

chemical investigation. The bibenzys exhibited cytotoxic activity against five human cancer cell lines, with cytotoxicity decreasing in parallel with decreasing oxygen-containing groups (Xu et al. 2014a).

Dendrobium flexicaule Z.H. Tsi. S.C. Sun & L.G. Xu

Chinese name: *Qujing Shihu*

Description: Pseudobulbs are clustered, short (6–11 cm), with 2–4 lanceolate leaves, 3 cm long. Inflorescences arise from the nodes near the apex of deciduous stems and bear 1–2 yellowish green flowers 5–6 cm across. Lip is bordered with purple. Blooming season is May to July (Zhu et al. 2009). In the Fuliu mountain area, a team of scientists and growers are attempting to provide a sustainable source of this rare *shihu* by growing *D. flexicaule* in a manner that simulates the natural habitat of the orchid (Zhang et al. 1999a). *D. flexicaule* is a rare, endemic, saxicolous *Dendrobium* found only in southwest Sichuan, Henan, northwest Hubei and east-central Hunan provinces (on the sacred Taoist mountain, Heng Shan) of China at 1200–2000 m (Zhu et al. 2009).

Usage: The plant qualifies for *shihu* since it is a *Dendrobium* which is exclusively saxicolous. Discovered only in 1986, efforts are now being made to cultivate this "rare medicinal plant" by simulating its habitat at the Xixia Forestry Bureau in Henan Province (Zhang et al. 1999a).

Dendrobium fugax Rchb.f.

syn. *Flickingeria fugax* (Rchb. F.) Seidenf.

Description. Rhizome is profusely branching forming large clumps. Pseudobulb slim, 5 cm long; leaves linear-elliptic 7–20 by 1.7–3 cm. Inflorescence short, rising from apex of pseudobulb, 1- to 2-flowered. Flowers are variable in size, 1–1.5 cm long (Raskoti 2009), 3–3.5 cm across, white to yellow and turning purple with age, and ephermeral. Lip is trilobed,

Fig. 10.23 *Dendrobium fugax* Rchb. f. [syn. *Flickingeria fugax* (Rchb. f.) Seidenf. [PHOTO: Bhaktar B. Raskoti]

Fig. 10.24 *Dendrobium gratiosissimum* Rchb. f. [PHOTO: E.S. Teoh]

mid-lobe bifurcated at tip, with undulating edges and spotted with purple at the base (Fig. 10.23). In some strains, lip is finely spotted with purple right down to the tips of the mid-lip. Flowering season is May to June in Nepal (Raskoti 2009), May to August in Bhutan (Gurong 2006), March to October in Assam, India. It is distributed in Nepal, Bhutan, northeast India, Myanmar, Thailand and Vietnam; in broad-leaved forests at 800–1650 m in Bhutan (Gurong 2006).

Herbal Usage: The whole plant is used to make a tonic or stimulant (Subedi et al. 2013; Pant and Raskoti 2013). *D. fugax* (syn. *Flickingeria fugax*) is frequently confused with *D. plicatile* (syn. *Flickingeria fimbriata*) and in India both herbs are used in the same manner (see *D. plicatile*).

Dendrobium gratiosissimum Rchb. f.

Thai name: *King Dam*
Vietnamese names: *Hoa thao, huang thao, Long tu*

Description: This montane *Dendrobium* is found at 1200–1700 m as an epiphyte on tree trunks in southern Yunnan, and from Laos and Vietnam to

northern Thailand, Myanmar and northeast India. Stems are 30–90 cm long, pendulous, terete and elongate, with numerous swollen nodes. Leaves are deciduous, 7–10 by 1–1.5 cm. Short inflorescences appear at the distal nodes on leafless stems in February in Thailand (Vaddhanaphuti 2005) or April in Yunnan (Chen et al. 1999b), and bear 1–3 pink flowers, 4.5–5.5 cm across in the Chinese variety. Flowers are considerably smaller in the Thai variety, being 3.7 cm across. Petals are white to pink, stained rose at the tips. Lip is broad, sub-obicular, with a large, central orange patch which is densely but finely papillose. It is stained a deep rose purple at the inferior border (Fig. 10.24).

Phytochemistry: Nine compounds have been isolated from *D. gratiosissimum* by Wang et al. (2007b). They are: 3,5,4′-trihydroxybibenzyl (1), 3,4′-dihydroxy-5-methoxybibenzyl (2), 3,4-dihydroxy-4″,5-dimethoxybibenzyl (3), apigenin (4), p-hydroxybenzaldehyde (5), defuscin (6), n-octacostyl ferulate (7), beta-sitosterol (8) and daucosterol.

Herbal Usage: It is used as *shihu* in Indochina (Petelot quoted by Perry and Metzger 1980; Doung 1993).

Fig. 10.25 Dendrobium hancockii Rolfe [PHOTO: E.S. Teoh]

Dendrobium hancockii Rolfe

Dendrobium odiosum Finet.

Common Name: *Xiyeshihu* (slim leaf *Dendrobium*)

Description: Stems are erect, terete, slim, branched and knotted (almost grass-like), bearing a cluster of narrow, green leaves 3–10 cm long and 3 cm wide, near the apex. Flowers are pale to golden yellow, borne singly or in pairs, laterally from the internodes of leafless stems (Fig. 10.25). It flowers in May and June (Chen et al. 1999b; Zhu et al. 2009).

Among the Chinese *Dendrobium* species, *D. hancockii* has the northern-most distribution. It occurs in southern Gansu and Shaanxi, Sichuan, Henan, Hubei, Hunan, Guangxi, Guizhu and Yunnan Provinces of Central China, growing on rocks and trees in forests at 200–1500 m; also in Myanmar and Vietnam.

Herbal Usage: Entire plant is served in pork porridge to treat ordinary coughs and asthma (Wu 1994).

Dendrobium henryi Schltr.

Chinese name: *Shuhua shihu* (sparse flower *Dendrobium*)

Description: Pseudobulbs are suberect, 30–80 cm long, 5–8 cm thick, with numerous nodes and many lanceolate leaves, 6–10 cm in length. Inflorescences arise from the nodes near the centre of leafy stems and bear one or two scented, golden–yellow, thin-textured flowers, 4–5 cm across, not widely open. Flowering season is June to September in China (Chen et al. 1999b), and September in Thailand (Vaddhanaphuti 2005).

D. henryi has a limited distribution extending from southern Hunan, Guangxi, southern Yunnan and southwest Guizhou in China to northern Vietnam and northern Thailand, at 600–1700 m. Plants are are found on tree trunks in mountain forests or on shaded, moist rocks in valleys. The variety *D. daoensis* was found in the Tam-dao Mountain in the Tonkin region of Vietnam.

Herbal Usage: Same as *shihu*.

Dendrobium herbaceum Lindl.

Chinese name: grass *Dendrobium*
Common Indian name: *Sasanga* in Orissa

Description: This species has the smallest flowers in the genus. Stems are pendulous, up to a metre in length, branched and bearing linear, lanceolate, grass-like leaves. Racemes are terminal, extremely short and few-flowered. Flowers are greenish-white, with tepals well spread out. Lip is white. Flowering season is July to September, in the south later and sometimes to December (Santapau and Kapadia 1966). *D. herbaceum* occurs in southern India from the Western Ghats eastwards to Orissa, Bangladesh and the Andaman Islands.

Phytochemistry: Flavonoids, sugars, cyanogenic glycosides and tannins have been isolated from its leaves, but their structures and function have not been elucidated.

Herbal Usage: In India, leaves of *D. hebaceum* are pounded with an equal amount (by weight) of the young shoot of *Andrographis paniculata* into a paste which is applied twice a day for a week on syphilitic ulcers. After 30 min, the paste is removed and the infected part is washed with a leaf decoction of *Azadirachia indica* (Dash et al. 2008).

Although the disease will persist and eventually affect the heart, brain and other organs in the body, syphilitic ulcers are painless and they disappear a week after their onset, giving a false impression that the disease has been "cured" by the treatment.

Dendrobium hercoglossum Rchb. f.

Callista annamensis Kraenz., *C. hercoglossa* (Rchb.) Kuntze, *C. vexans* (Dammer) Kraenz., *Dendrobium poilanei* Guillaumin, *D. vexans* Dammer, *D. wangii* C.L.Tso

Common name: Rampart Lip *Dendrobium*
Thai name: *Ueang dawk makhua*

Description: Pseudobulbs are 20–35 cm long, narrow at the base and swollen at the apex, carrying 4–6 linear, pointed leaves 5–10 cm in length. Inflorescences are 4 cm long, borne at nodes near the apex of newly matured pseudobulbs and carry 2–8 slightly fragrant, flat, waxy, pink, flowers with a white lip. Some flowers are white, apart from the anther cap which is crimson (Fig. 10.26). It flowers in February to May in Thailand (Vaddhanaphuti 2005; Nanakorn and Watthana 2008).

D. *hercoglossum* is distributed throughout southern Chinese provinces of Hunan, Jiangxi, Guangxi, Guangdong, Hong Kong, Hainan, Guizhou and Hunan at 600–1300 m and in Indochina, Thailand, Malaysia and the Philippines. (Zhu et al. 2009)

Herbal Usage: used as *shihu* (Zhongyao Da Cidian, 1986; Bencao 2000)

Dendrobium heterocarpum Wall. ex Lindl.

Chinese name: *Jiandaochun Shihu*

Description: Stems are ascending, thick, fleshy, swollen at the nodes, 5–27 cm long and 1–1.5 cm in diameter; golden-yellow when fresh, sulfhuryellow tinged with dirty black when dry (Zhu

Fig. 10.26 *Dendrobium   hercoglossum* Rchb. f. [PHOTO: E.S. Teoh]

Fig. 10.27 *Dendrobium heterocarpum* Wall. [PHOTO: E.S. Teoh]

et al. (2009). Leaves are oblong-lanceolate, 7–12, by 1.2–2 cm, coriaceous, but papery at the base which ensheaths the stem. Inflorescence arises from nodes of leafless stems, bearing 1–4 flowers. Flowers white to creamy-yellow, scented, 5 cm across. Lip is marked by reddish-purple stripes (Fig. 10.27). Flowering season is March to April in Nepal. It is distributed from Bhutan, central Nepal (at 1000–1600 m) southern and western Yunnan (at 1600–1800 m) to Thailand, Indo-China, Malaysia, Indonesia and the Philippines (Raskoti 2009; Zhu et al. 2009).

Herbal Usage: Paste of stems mixed with wheat flour is applied on fractures and dislocations (Subedi et al. 2013; Pant and Raskoti 2013).

Dendrobium hookerianum Lindl.

Chinese Name: *Jiner Shihu* (gold earring *Dendrobium*)

Description: This is one of the largest *Dendrobium* in terms of pseudobulb length. The yellow, pendulous stems continue to elongate for years, reaching 100–245 cm in length, and their papery, dark green, lanceolate, unusually long (18–30 cm) leaves may last for years. Inflorescence is 15 cm long, slender, droopy, apical or sub-apical, with 2–4 flowered clusters of fragrant, gold-yellow, blossoms along its length. Flowers measure up to 10 cm across, and are among the largest in the *Dendrobium* genus. Lip is fringed and marked by a purple blotch in the centre. Flowering season is July to September (Jin et al. 2009).

This lithophytic/epiphytic *Dendrobium* occurs in central Nepal, northeastern India, Bangladesh, Myanmar, Yunnan and Xizang, at 1000–2000 m. In China, it has been found growing on laval rocks.

Herbal Usage: The pseudobulbs are used to treat malaria and high fever (Wu 1994).

Dendrobium huoshanense G.Z. Tang and S.J. Cheng (see **Dendrobium catenatum, Lindl.**)

Dendrobium hymenanthum, Rchb.f.

syn. *Dendrobium quadrangulare* C.S.P. Parish & Rchb. f.

Description: Pseudobulbs are club-shaped, four-angled, numerous and tightly clustered. They are up to 10 cm long, bearing single, terminal, small, white flowers, 1–2 cm across. It flowers gregariously several times a year in response to a sudden drop in temperature occasioned by heavy rain or a tropical storm. C.E. Carr was quoted by Holttum (1949) to have observed that *D. quadrangulare* and *D. pumilum* flower on different days after the temperature drop, *D. quadragulare* first and *D. pumilum* several days later. *D. quadrangulare* is a minute species that occurs in Myanmar (Tenasserim), Peninsular Thailand, Indochina, Malaysia, Kalimantan and the Philippines.

Herbal Usage: Used to treat dropsy in Peninsular Malaysia (Ridley 1906).

Dendrobium indivisum (Blume) Miq.

Thai Name: *Kang pla*

Description: A tiny *Dendrobium* belong to the Section *Aporum*, stem is flattened, up to 30 cm long, sheathed throughout by thick, flattened leaves 2 cm wide and 2.5 cm long, arranged in alternate rows facing each other. Flowers appear individually or in twos, greenish-yellow with purple stripes 0.8 cm across, in July (Fig. 10.28). The species is found in Myanmar,

Fig. 10.28 *Dendrobium indivisum* (Blume) Miq. [PHOTO: E.S. Teoh]

Thailand, Laos, Vietnam, Malaysia and Indonesia, in exposed localities. Although this is a free-flowering species, it has no horticultural value owing to the insignificant size of its flowers.

Herbal Usage: The whole plant is used to treat headache in Thailand (Chuakul 2002).

Dendrobium jenkinsii **Wall. ex Lindl.**

Thai name: *Ueang phung noi*
Myanmar name: *Yadanar shwe ket lay*

Description: Plant has tiny, dark green, compressed, four-angled pseudobulbs, 1–2.5 cm tall, form large clusters. Leaf is apical 1–3 by 0.5–0.8 cm. Inflorescence is short and carries 1–3, golden-yellow flowers with darker, pubescent lips, 2–2.5 cm across (Fig. 10.29). Flowering season is February and March in Myanmar (Tanaka et al. 2003), March in Thailand (Vaddhanaphuti 2005), April to June in Bhutan (Gurong 2006). This miniature species is distributed from the eastern Himalayas to Myanmar, Thailand, southern Yunnan, Laos and Vietnam.

Herbal Usage: As *shihu*.

Dendrobium leonis **(Lindl.) Rchb.f.**

Thai Names: *Uang takhap yai, kang pla* (note: *kang* pla is also the name for *D. indivisum*)

Description: *D. leonis* belongs to the Section *Aporum* which has very characteristic flattened stems that are completely covered by laterally flattened, thick, fleshy, overlapping leaves. This is a very small *Dendrobium* found in the lowlands throughout Thailand, Malaysia, Indochina and Indonesia. It flowers throughout the year. Flowers are 1 cm across, yellow, borne singly, and smell of vanilla (Fig. 10.30).

Herbal Usage: The entire plant is used to treat headache in Thailand (Chuakul 2002).

Dendrobium linawianum **Rchb. f.**

Chinese names: *Yinshihu* (oriental cherry *Dendrobium*), *Jinshihu* (gold *Dendrobium*), *Lishishihu* (Mr. Lee's *Dendrobium*), *Juchunshihu* (lip *Dendrobium*)

Description: Stems are tufted, erect, slightly flattened, 30–40 cm in length with oblong leaves

Fig. 10.29 *Dendrobium jenkinsii* Wall. ex Lindl. [PHOTO: E.S. Teoh]

Fig. 10.30 *Dendrobium leonis* (Lindl.) Rchb.f. [PHOTO: E.S. Teoh]

4–7 cm by 2–2.5 cm, a bright green on their upper surface and pale on the lower surface. Short subterminal and lateral inflorescences carry two or three large showy flowers, 5 cm in diameter, with tepals that are white purple at the apex fading into white at the base. Lip is white with rim of bright purple. Flowering period is spring and the flowers last two weeks (Lavarack et al. 2000).

Dendrobium linawianum is a small to medium-sized species with a confined distribution being found only in Wulai in broad-leaved forests, at low altitude, in the mountainous north of the Taiwanese island (Liu and Su 1978), and in the southern Guangxi Province in continental China at 400–1500 m (Zhu et al. 2009).

Phytochemistry: *D. linawianum* extract exhibited stronger anti-oxidant activity than *D. moniliform* in a DPPH assay (Lo et al. 2004b).

Herbal Usage: The stem nourishes the *yin* elements, benefits the stomach, stops thirst, and removes the feeling of heat, dry mouth, weakness, poor health, night sweats and joint pain.

Dendrobium lindleyi Steub.

syn. *Dendrobium aggregatum* Roxb.

Description: Plants are epiphytic, with short pseudobulbs 6–10 cm long bearing an oblong-elliptic leaf 8–10 by 3 cm.. Inflorescence is apical paniculate, carrying 12–15 golden yellow, thin-textured, fragrant flowers, 3 cm across resembling flowers of *D. jenkinsii* (Fig. 10.31). Flowering season is February to April in Thailand (Nanakorn and Watthana 2008). Flowers last a week. *D. lindleyi* is distributed in deciduous and evergreen forests at 300–1000 (in - Thailand, up to 1500 m) from northeast India to Bhutan, Myanmar, southern China, Thailand, Laos and Vietnam.

Herbal Usage: as *shihu*.

Dendrobium linearifolium Teijsm.& Binn.

Description: Plants resemble *D. crumenatum* but leaves are linear. Pseudobulbs are yellow or

Fig. 10.31 *Dendrobium lindleyi* Steub. [PHOTO: E.S. Teoh]

reddish, 3.5 by 1.5 cm in diameter, 1–3 cm above the base, and carry wiry stems up to 70 cm long, and often branched. Internodes are 3 cm lower down, shortening to 2 cm distally. Leaves are linear 6 by 0.6 cm borne along entire length of the stem. Flowers appear on every node, white with crimson streaks, 2. 5 by 1.7 cm. Lip is entire and shaped like a kite. Flowering is gregarious. *D. linearifolium* is a mountain species, occurring at 800–2000 m distributed from Bali to Java and Sumatra, in bright light (Comber 2001).

Herbal usage: Pseudobulbs are pounded and squeezed to extract the juice which is instilled into the ear for the relief of earache by the Sepang community in Bali (Astuti et al. 2000).

Dendrobium loddigesii Rolfe

Chinese names: *Fenhuashihu* (pink flower *Dendrobium*), *Meihuashihu (*beautiful flowered *Dendrobium*); *Xiaohuangcao* (small yellow herb)*; huanchahu* (ring-like hairpin)

Description: Stems are pendulous, terete, slender, with numerous nodes; the leaves measure 2–5 by 1–1.5 cm and are numerous. Fragrant flowers are borne singly, predominantly on leaf-less stems, profusely in April (Chen et al. 1999b), and they last for 3 weeks. They are 4–5 cm in diameter, of a pale rose-pink with a yellow blotch on the disc of the fringed lip (Fig. 10.32). Some strains have white flowers.

D. loddigesii is a small, pink, deciduous *Dendrobium* with a straggly habit. It grows on trees or rocks at 400–1500 m in Guangdong, Guangxi, Guizhou, Yunnan and Hong Kong.

Phytochemistry: *D. loddigesii* has been studied from a chemical perspective for some 20 years. In 1991, Li and his team from the China Pharmaceutical University in Nanjing reported the isolation of two alkaloids, shihunidine and shihunine, and a phenol, dendrophenol (4,4′-dihydroxy-3,3′,5-trimethox-ybibenzyl), from fresh stems of *D. loddigesii* (Li et al. 1991). Shihunidine and shihunine were shown to be inhibitors of Na^+, K^+-ATPase in the rat kidney. Chen et al. (1994) isolated moscatilin and moscatin from the stems of *D. loddigesii* and found that the moscatilin possessed strong antiplatelet activity. Ho and Chen (2003) from the Taipei Veterans General Hospital also isolated moscatilin from the stems of *D. loddigesii*. They discovered that, in addition to its antiplatelet action, it had potent cytotoxic

Fig. 10.32 *Dendrobium loddigesii* Rolfe [PHOTO: E.S. Teoh]

effect on cancers arising from the placenta, stomach and lung but not on tumours arising from the liver. This seems to be related to its ability to induce G2-phase arrest but, surprisingly, it had no detectable inhibitory effect on cyclin B-cdc-2 kinase activity. Moscatilin induces apoptosis in human colorectal cancer cells in HCT-116 cell cultures and xenograft models via tubulin depolymerization and DNA damage stress (Chen et al. 2008). Mosacatilin from *D. loddigesii* also suppressed growth of human oesophageal cancer cells of both histological cell types, adenocarcinoma and squamous cell carcinoma (Chen et al. 2013a).

Recently, Ito et al. (2010) reported the isolation, from an 80 % ethanolic extract of *D. loddigesii*, of two new phenanthrenes, loddigesiinols A and B, two new stilbenes, loddigesiinols C and D, and various known compounds, i.e. 5 phenanthrenes, 3 stilbenes, 2 lignans and 3 sterols (moscatin, 5-hydroxy-2,4-dimethoxyphenanthrene, lusianthridin, rotundatin, hircinol,moscatalin, gigantol, batatasin III, stigmasterol, beta-sitosterol, sitostenone, dehydrovomifoliol, pinoresinol and medioresinol). Several phenenathrenes, including the new loddigesiinols A and B, showed significant inhibitory activity on nitric oxide (NO) production. Loddigesiinol D had a weak inhibitory activity but loddigesiinol C had none.

Four polyphenols, lodigesiinols G–J, and crepidatuol B were isolated from the stems of *D. loddigesii*. All five compounds are strong inhibitors of alpha-glucosidase, considerably stronger than trans-resveratrol (Lu et al. 2014b) (line drawing: Chen & Tsi 1998).

Herbal Usage: According to TCM, the stem benefits the stomach, produces saliva, nourishes the *yin* elements, and it is antipyretic. It is used to treat thirst, anorexia, dry vomiting, weak bodies and "muddy eyes" (Ou et al. 2003; Anonymous 2004). Chinese herbal physicians are now using it to treat Type 2 (late-onset) diabetes (Lu et al. 2014b).

Polysaccharide content of *D. loddigesii* harvested from different habitats varies. A study from five regions found that samples from Xingyi and Anlong in Guizhou Province had the highest content, 22.62 and 20.96, respectively. The authors suggest that, for commercial purposes, *D. loddigesii* should be cultivated in these two districts of Guizhou (Bai et al. 2007). Twisting the stems of *D. loddigesii* that have been scaled by boiling water, followed by drying over fire, retains the highest amount of polysaccharide in the herb (Wu et al. 2007).

Dendrobium lohohense Tang & F.T.Wang

Chinese name: *Luohe Shihu* (Luo River *Dendrobium*)
Japanese herbal name: *Chukanso*

Description: Pseudobulbs are slender, 20 cm long, sometimes branched with oblong-elliptic leaves, 3.5–6 cm long, and pointed at the tip. Short inflorescences arise from apical or sub-apical nodes, and carry single yellow blooms, 2.5 cm across. Flowers appear in December.

D. lohohense is a saxicolous *Dendrobium* once thought to be confined to Guangxi Province at 1150 m. It has now been discovered in mountain valleys and on forest margins at 1000–1500 m in Chongqing, N Guangdong, W Guizhou, W Hubei, Hunnan and SE Yunnan (Zhu et al. 2009).

Phytochemistry: *D. lohohense* produces a phthalide-pyrrolidine alkaloid, shihunine, which is related to another alkaloid, pierardine, both of which are found in *D. aphyllum* (syn. *D. pierardii*) (Inubushi et al. 1964). Shihunine is also present in the South American hallucinogenic vine, *yage* (*Banisteriopsis caapi* Spruce ex Griseb., Malpighiaceae), which is used by shamans of the Amazonian Basin in their rituals. However, the hallucinogenic effect of *yage* is due to other alkaloids in the vine and not to shihunine (McKenna 2014).

Shihunine rapidly and almost totally converts into betaine, and it is likely that shihunine is an intermediary for betaine in the living orchid. Betaine is found in dietary supplements. Physicians sometimes prescribe it, together with vitamins B6, B12 and folic acid, for people who have high levels of homocysteine in their blood,

because this increases the risk of coronary heart disease and stroke. However, this does not mean that consuming *D. lohoense* will help to prevent cardiovascular disease.

Herbal Usage: Plants are collected and used in Chinese medicine, as *shihu* (Wu 1994).

Dendrobium longicornu Wall ex Lindl.

Dendrobium bulleyi Rolfe, *D. flexuosum* Griffith, *D. hirsutum* Griffith.

Chinese name: *Changju Shihu*
Nepali name: *Bawar, Kause*

Description: Stems are suberect, terete, thin, 15–60 cm long, with many lanceolate leaves. Leaves are covered with fine black hairs on both sides and this hirsutism extends to the sheaths which cloak the stems. Inflorescence is borne on leafy stems, 1–3 flowered. Flowers are 5 cm long, not well expanded, white with orange streaks on the lip, and fragrant. Flowering season is September to November.

Usually epiphytic on mossy trees in forests, but also occurring as lithophytes on rocks along valleys, the species is distributed in Guangxi, Yunnan and southeastern Tibet in China, as well as in Nepal, Bhutan Sikkim and northeast India, Myanmar and Vietnam, all at 1200–2500 m.

Phytochemistry: Five new and 14 known compounds have been isolated from the stems of *D. longicornu* at the State Key Laboratory of Phytochemistry and Plants Resources in Kunming. The new compounds are a bibenzyl, two phenanthrenes and a lignin glycoside, namely longiconuol A, 4-[2-(3-hydroxyphenol-1-methoxyethyl]-2,6-dimethoxyphenol, 5-hydroxy-7-methoxy-9,10dihydrophenanthrene-1,4-dione; 7-methoxy-9,10-dihydro-phenanthrene-2,4,65-triol, and erythro-1-(4-O-beta-D-gluco-pyranosyl-3-methoxyphenyl)-2-[4-(3-hydroxypropyl)-2,6-dimethoxyphenoxy]-1,3-propanediol (Hu et al. 2008c).

Herbal Usage: The Chinese *Materia Medica 1999* includes this species as *shihu* (Zhang 1999).

In Nepal, juice of the plant mixed with lukewarm water is used to bathe children afflicted with fever. Boiled root is fed to livestock to rid them of coughing (Manandhar and Manandhar 2002; Baral and Kurmi 2006;). Juice of stems is consumed to treat fever (Subedi et al. 2013).

Dendrobium macraei Lindl.

Desmotrichum fimbriatum Bl., *Flickingeria fimbriata* (Bl) Hawkes

Chinese name: *Liusujin Shihu*

Indian names: *Jibai, Jibanti* in Bengal, *Jivanti, Radarudi, Wajhanti* (Gujerati), *Jiban, Joivanti, Sag* (Hindi), *Jivanti* (Marathi), *Bhadra, Jiva, Jivabhadra, Jivada, Jivani, Jivaniya, Jivanti, Jivapatri, Jivapushpi, Jivavardhini, Jivarisha, Jivdatri, Jivya, Kanjika, Kshurajiva, Madhushvasa, Madhusrava, Mangalya, Mrigaratika, Payaswini, Praanada, Putrabhadra, Ratangi, Shakashreshtha, Shashashimbika, Shringati, Srava, Sukhankari, Supringala, Yashaskari, Yashasya* (Sanskrit)

Sri Lankan names: *Jeevaniya* (meaning: supporting life) *Saaka shreshtha* (best of herbs); *Jata Makuta.*

Description: *D. macraei* is a large epiphyte with a creeping rhizome up to a metre long and branching, ending in fusiform pseudobulbs of 5–6 cm length. Leaves are 10–20 cm long. Pseudobulb produces 2–3 white flowers, up to 3 cm across with a sprinkling of red on the side lobes of the lip. It is common in Sikkim at 2200–2600 m, but also occurs in Nepal, China, Thailand, Vietnam Malaysia, Indonesia, Sri Lanka and Papua New Guinea. It flowers in March in Thailand (Vaddhanaphuti 2005),

Herbal Usage: The long list of Indian names is due to the popularity of the plant as a sweet preparation, the much-valued '*Halwa*'. Hailed as a stimulant and tonic, it is taken for debility caused by seminal loss in India and Nepal (Chopra et al. 1958; Suwal 1970). In the markets of Bombay, it is known as *ruttun-purush* and sold at a high price. Apart from its use as a tonic, the plant is used as a remedy for disorders of the bile, blood and phlegm. The fruit is an aphrodisiac (Caius 1936).

The plant is also used as a counter-poison. It is a constituent of the local remedy for snake or scorpion bites and used as a demulcent (Caius 1936; Chopra et al. 1958). However, Mhaskar and Caius found it to be worthless in this respect (Caius 1936). In Nepal, the pseudobulb is used to treat asthma, bronchitis, sore throat, fever, biliousness, diseases of the eye and blood and sexual dysfunction (Baral and Kurmi 2006; Pant and Raskoti 2013). In Nepalese folk (?Ayurvedic) medicine, the plant is sweet and cooling, an alternative (i.e. it is a herb that is capable of reestablishing the healthy functions of the body) (Baral and Kurmi 2006). In Sri Lanka, the whole plant of *D. macraei* is an ingredient in medicinal oils used in massage for treating paralytic lesions (Cooray 1940).

Dendrobium moniliforme (L.) Sw.

syn. *Dendrobium candidum* Wallich ex Lindl., *Dendrobium wilsonii* Rolfe

Chinese name: *Shihu* (Noble Orchid), *Shilan* (stone orchid), *Xijingshihu* (thin-stemmed *Dendrobium*), *Xiaoshihu* (small *Dendrobium*), *Xiaohuancao* (small whorled herb), *Jizhua Lan* (chicken claw orchid), Jingchacao *Tongpi Lan* (copper orchid), *Tongpishihu* (copper *Dendrobium*)

Medicinal names: *Shihu, Huan cao* (whorled herb), (gold hairpin grass), *Erhuancao* (earring herb), *Xicao* (slender herb), *Huangcao* (yellow herb), *Xihuancao* (slender whorled herb), *Xiaojingcha* (small gold hairpin)

Taiwanese names: *Bai Shi Hu* (white *Dendrobium*), *Shi Hu* (medicinal *Dendrobium*), *Jie Gu Cao* (fixing fracture *Dendrobium*)

Japanese name: *Sekkoku*

Korean name: *Seok gok*

Description: *D. moniliforme* is generally considered to be one of the original plants denoted by the term, *shihu* (Fig. 10.33). It is an epiphyte

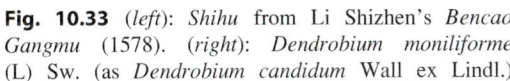

Fig. 10.33 (*left*): *Shihu* from Li Shizhen's *Bencao Gangmu* (1578). (*right*): *Dendrobium moniliforme* (L) Sw. (as *Dendrobium candidum* Wall ex Lindl.) from: *Annals of the Royal Botanic Gardens, Calcutta*, vol. 5 (1): t. 19 (1891). [Drawing by L. Singh]. Courtesy of Missouri Botanical Gardens, St. Louis, USA

or lithophyte with usually erect, sometimes pendulous stems that are 30–40 cm long, terete, slender, 3–5 mm thick and thinning near the base, with numerous nodes, and sheathing near the base. Leaves are lanceolate, 4–6 by 7–9 cm, pointed at the tip. Two white flowers, 2–4 cm across and with narrow tepals, are borne on deciduous stems (Fig. 10.34). Flowering period is April and May. Epiphytic on tree trunks in forests or on rocks in ravines at 600–3000 m, *D. moniliforme* enjoys an extensive distribution across China south of the Huanghe, to Japan and Korea in the east and northeast India in the west. Its distribution in China exceeds that of all other species of *Dendrobium* and one wonders whether this is the result of nature or through human intervention.

D. *moniliforme* enjoys the northernmost distribution among epiphytic species of orchids in

Fig. 10.34 *Dendrobium moniliforme* (L.) Sw. [PHOTO: Courtesy of Plant Photo Bank of China]

Asia. In Japan it is found in sheltered places near the sea just north of Tokyo (Koopowitz 2001). It was developed into a highly desirable ornamental plant in Japan during the early part of the nineteenth century. Originally, the selection focused on variegated leaves, but the present trend is for more flowers, tolerance to cold, response to vernalisation, peak flowering at Christmas and shorter plant size. Tetraploid hybrids that have bigger flowers and higher alkaloid content have also been produced by crossing with *D. nobile* (Karasawa and Shimai 2002). Tetraploidy may have contributed to *Dendrobium* Snowflakes' ability to produce more alkaloids than its parent species (*D. moniliforme* × *D. nobile*).

A Japanese clone, *D. moniliforme* 'Aobabue', is reputed to have a "gorgeous, sophisticated fragrance resembling that of *Camellia japonica*" (Nakamura et al. 1990). In the past, Japanese royalty perfumed their clothing with flowers of *D. moniliforme*.

Phytochemistry: The most important compound to be isolated from *D. moniliforme* is denbinobin (Lin et al. 2001), which was first discovered in *D. nobile* (Lee et al. 1995). Denbinobin has unique antitumour activity and is also effective against the AIDs virus (see *D. nobile*) (Li et al. 2001). Denbinobin has been synthesised (Krohn et al. 2001). The second promising compound is moniliformedequinone which exerts potent cytotoxic activity against hormone-resistant prostate cancer cells in vitro. It induces cellular glutathione depletion leading to DNA damage response and cell-cycle arrest at S-phase (i.e. prevents cell multiplication) and a caspase cascade that culminates in apoptosis (Hsu et al. 2014). Synthesis of moniliformedequinone (which is 2,6-dimethoxy-1,4,5,8-phenanthrenetetrone) was achieved in 2011. It also inhibits breast cancer cells in vitro (Thangraj et al. 2011). Moniliformin is another phenanthrene present in *D. moniliforme*. Nevertheless, Chinese researchers were mainly interested in recovering polysaccharides from the plant (Fan et al. 2005; Zhang and Liao 2005; Li et al. 2006),

employing tube culture to obtain more tissue for polysaccharide recovery and even resorting to sound wave stimulation to promote the synthesis of the desired compounds (Li et al. 2006). Two teams in Chongqing did not find any difference in polysaccharide content between the wild and the cultured *D. candidum* (Fan et al. 2005; Li et al. 2006). A heteropolysaccharide obtained by aqueous extraction of dried stems of *D. officinale* (= *D. catenatum*) was found to be a 2-0-acetyl-glucomannan, composed of annose, glucose and arabinose in 40.2:8.4:1 molar ratio (Hua et al. 2004).

Recently, six phenanthrenes and five bibenzyl derivatives were isolated from *D. candidum* (Li et al. 2008, 2009a, b, c). The six phenenthrenes were: 2,3,4,7-tetramethoxyphenenthrene; nakaharain, 2,5-dihydroxy-3,4-dimethoxyphenanthrene, confusarin, nudol and bulbophyllanthrin (Li et al. 2009a). Twenty compounds were isolated and identified as 3,4'-dihydroxy-5-methoxybibenzyl, dihydroresveratrol, dendromoniliside E, denbinobin(4),2,4,7-trihydroxy-9, 10-dihydrophenanthrene, aduncin, (−)-loliolide; adenosine, uridine, guanosine, sucrose, 5-hydroxymethyl-furaldehyde, *n*-octacostyl ferulate, defuscin, *n*-triacontyl *cis-p*-coumarate, daucosterol, beta-sitosterol, hexadecanoic acid, hentriacontane and heptadecanoic acid. Their structures were elucidated on the basis of spectroscopic data and physicochemical properties. All of the compounds were isolated from this plant for the first time (Li et al. 2010).

Using leucocyte count, weight of immune organs, carbon granule clearance and lymphocyte transformation in mice treated with the cytotoxic agent cyclophosphamide, Gao and colleagues from the Shanghai University of Traditional Chinese Medicine (TCM) found that medicinal materials from both wild and cultured *D. candidum* improved immunological function to the same degree. The maximum tolerated dose was 227 times the effective clinical dose (Gao et al. 2002). *D. candidum* (=*D. moniliforme*) extracts at a dosage of 400 mg/kg protected mice against liver damage caused by carbon tetrachloride (Li et al. 2014a). However, no comparison was made with the dozens of other herbal products that also protect against hepatic damage.

D. moniliforme extract (DME) also protects the kidneys. When administered orally to mice provided with a high fat diet, DME decreased blood sugar and cholesterol levels. There was also lower serum creatinine, less accumulation of kidney fats and renal collagen-IV deposition in the kidneys compared with mice treated with metformin (Lee et al. 2012b).

Methanolic extract of *Mycena dendrobii*, a mycorrhiza isolated from roots of *D. candidum* (*D. moniliforme*), produced analgesic effects when tested on mice (Wang et al. 2001). *Mycena dendrobii* stimulates germination of *Gastrodia elata* seeds as well as seeds of *D. densiflorum* (Guo et al. 1999).

Different batches of *D. candidum* (= *D. moniliforme*) were analysed for their constituents by Ding et al. (1998) who wanted to determine the best time for harvesting newly planted orchid. Their calculated therapeutic index implied that the orchid is currently used in TCM to strengthen immunity, and/or to cure throat and eye disorders. On this basis, the best time for harvest was autumn of the third year from planting if all indices, yield and planting costs are taken together. However, if one wishes to use the orchid to treat throat and eye disease, the best time for harvest is actually the fourth year; and for strengthening immunity, either the first or the third year (Ding et al. 1998).

Traditional Chinese Medicinal Usage: *D. moniliforme* is classic *shihu* (Li 1578). It is the most expensive variety. In his list of 124 important medicinal plants in Chosen, Korea, Ishidoya (1925) included only a single orchid, *D. moniliforme*. The pseudobulb is used to correct body heat, haemoptysis, thirst, malaise, anorexia and to improve the body fluids. Japanese and Korean herbalists use it to treat night sweats, weakness, anorexia, lumbago and impotence (Sung 2002). It inhibits the release of histamine (Hirai) and aldose reductase (Shin et al. 2010). *D. moniliforme* contains the

sequiterpene alkaloid, dendrobine, as well as triterpenes and sterols (Han et al. 1981).

Dendrobium monticola P.F. Hunt & Summerh.

Indian names: *Jiwanti, Bhotia, Lahsan*

Description: A tiny *Dendrobium* with stumpy, clustered pseudobulbs, it bears 2–4 thin, long, lanceolate leaves near the apex. Apical and axillary inflorescences arise from leafy and leafless pseudobulbs. They bear 3–8 greenish-white flowers, 1.5 cm across. Sepals and petals are pointed and concave forming a hood over the lip, with sepals broader than the narrow petals. Lip is trilobed, white with linear, thin, brown striping on the inner surface of the lateral lobes; margin of the mid-lobe is serrated (Fig. 10.35). The species is distributed from the eastern Himalayas to Laos and Vietnam at 1500–2300 m. It grows on rocks at 1700–2200 m in southwest Guangxi Province in

Fig. 10.35 *Dendrobium monticola* P.F. Hunt & Summe [PHOTO: E.S.Teoh]

China (Zhu et al. 2009). It flowers in November in northern Thailand (Vaddhanaphuti 2005).

Usage: This is one of several *Dendrobium* species used as an emollient or in the form of a poultice for pimples, boils and other skin lesions in India and Nepal (Duggal 1972; Das 2004; Baral and Kurmi 2006; Pant and Raskoti 2013). Sold in the drug markets of Kanpur in the Indian northeastern state of Uttar Pradesh as a nerve tonic and antiphlogistic, it is also used for external application in rheumatism (Trivedi et al. 1980).

Dendrobium moschatum (Buch.–Ham.) Sw.

Thai name: *Ueang Champa*

Description: *D. moschatum* is a robust, deciduous, epiphytic Himalayan *Dendrobium* with a wide distribution from the foothills of northeast India, Bhutan and Nepal across Myanmar and Thailand to Laos, Vietnam and Yunnan Province in China. It grows at 300–2000 m in deciduous forests. The slender, pendulous pseudobulbs may attain a length of 2 m, with lanceolate leaves, 8–15 cm in length, in two rows along its entire length. Leaves are shed after 2 years, and the pendulous inflorescences arise from the nodes of leafless stems. Inflorescence carries 7–15 cream-coloured flowers with dark brown at the hirsute throat (Fig. 10.36). Flowering season is late spring, the flowers only lasting a few days (Lavarack et al. 2000); it flowers in May to June in Assam, India (Nath and Das 2013).

Phytochemistry: Moscatin, a new phenanthrene derivative, and moscatilin, a bibenzyl derivative, were isolated from the orchid *D. moschatum* (sic) by Majumder and Sen (1987a, b). Moscatin (syn. Dendrophenol) was subsequently recovered from *D. denneanum* and moscatilin from *D. amoenum* (Majumder et al. 1999), *D. nobile* (Miyazawa et al. 1999), *D. loddigesii* (Ho and Chen 2003), *D. polyanthum* (Hu et al. 2009) and

Fig. 10.36 *Dendrobium moschatum* (Buch.–Ham.) Sw. [PHOTO: E.S.Teoh]

D. pulchellum (Kowitdamrong 2014). Moscatilin has cytotoxic effects on placenta, oesophagus, stomach, colon and lung cancer cells but not those of the liver (Miyazawa et al. 1999; Ho, and Chen 2003; Chen et al. 2008a, 2013a). It suppresses tumour angiogenesis and tumour growth (Tsai et al. 2010). Moscatilin inhibits migration and metastasis of human breast cancer cells (MDA-MB-231) in an animal model (Pai et al. 2013). Studies with human liver cancer cells showed that markers associated with cell metastases were inhibited by moscatilin (Wang et al. 2011). Moscatilin in non-toxic concentrations may be able to inhibit human non-small cell lung cancer cells as suggested by its ability to attenuate reactive oxygen species and down-regulation of focal adhesion kinase (Kowitdamrong 2014).

Herbal Usage: Juice from the leaves is squeezed into a painful ear in Meghalaya state (Rao 1981). Numerous epiphytic orchids, including *D. crumenatum*, in Southeast Asia, are used in this manner, possibly reflecting the influence of a strong Indian culture drift in the past. Pseudobulbs are used to treat fractures and dislocations (Pant and Raskoti 2013).

Dendrobium nobile Lindl.

Dendrobium lindleyanum Griff., *D. coerulescens* Lindl., *D. formosanum* Rchb. f.

Chinese names: *Jinchashihu* (gold hairpin *Dendrobium*), *Meihuashihu* (pretty flowered *Dendrobium*), *Chunshihu* (spring *Dendrobium*), *Yunnanshihu* (Yunnan *Dendrobium*), *Diaolanhua* (hanging orchid), *Bianjincha* (flat golden hairpin), *Xiaohuangcao* (small yellow herb); *Dahuangcao* (big yellow herb—its name in Guizhou Province); *shek huk* in Hong Kong.

Japanese name: *Koki-sekkoku*

Korean name: *Go gwi seok gok*

Vietnamese names: *Thach hoc, Kim thoa thach hoc, Hoang thao, kep thao, hoang thao cang ga, phi diep kep, co vang sao, se kep*

Thai names: *Ueang Gao Giew, Ueang Khao kiu*

Description: Deciduous stems bear several inflorescences near their apices, each carrying 2–4 white, pink, purple, or pink suffused with rosy purple flowers that are 5.5–8 cm across. Lip is purple or white with a purple blotch in the throat (Chen et al. 1999b; Vaddhanaphuti 2001). Several colour forms occur in Sikkim (Figs 10.37, 10.38, 10.39, and 10.40). Benjamin Samuel Williams (1824–1890), the famous English nurseryman who authored the classic *Orchid Growers Manual*, described a variety *nobilis* which measured 10 cm across, undoubtedly the largest flowered form of the species (Grant 1895). Flowering period is March to May (Chen et al. 1999b; Vaddhanaphuti 2005; Nath and Das 2013.

D. *nobile* is an attractive, deciduous epiphyte that is distributed from Taiwan and Hainan across the southern Chinese provinces to Laos, north and upper northeast Thailand, Myanmar, Sikkim, Bhutan, Nepal and northern India. It grows on trees in sparse forests at 500–1700 m.

Phytochemistry: Within the genus, no species has been more extensively investigated by a chemical approach than *D. nobile*, one reason

Fig. 10.37 *Dendrobium nobile* Lindl. perched on a tree in Sikkim at 1700 m, exposed to strong sunlight. [PHOTO: E.S.Teoh]

Fig. 10.38 *Dendrobium nobile* Lindl., typical form, flowering in a private garden in Darjeeling, West Bengal, India. [PHOTO: E.S. Teoh]

Fig. 10.39 *Dendrobium nobile* Lindl., a polyploidy form [PHOTO: E.S. Teoh]

Fig. 10.40 *Dendrobium nobile* Lindl., one of numerous colour variants of the species encountered in eastern Himalaya [PHOTO: E.S.Teoh]

being that it is the commonest variety of *shihu*. The first crystalline alkaloid, dendrobine, was isolated by Suzuki from *D. nobile* in 1932 (Suzuki et al. 1932, 1934). Subsequently, 24 more alkaloids have been isolated from this species: (*n*-methyl dendrobine, 3-hydroxy-2-oxo-dendrobine, dendrobine-*N*-oxide, dendrobane, dendramine, dendroxine, 8-hydroxy-

dendroxine, 6-hydroxy-dendroxine, *n*-isopenteryl-6-hydroxy-dendroxine, dendrine, nobiline, and dendronobiline A–N) (Tadamasa et al. 1965; Okamoto et al. 1966a, b; Hedman et al. 1971; Zhang et al. 2008b). It is also the source of antifibrotic phenanthrenes (Yang et al. 2007a); cytotoxic phenanthrenes (4,7-dihydroxy-2-methoxy-9,10-dihydro phenanthrene of lusianthridin, and denbinobin) (Lee et al. 1995); bibenzyls (moscatilin, and nobilins A–E) (Miyazawa et al. 1999; Zhang et al. 2006f, 2007b; Hu et al. 2008a, 2008b, 2008c); coumarins, immunomodulatory sesquiterpene glycosides (Zhao et al. 2001; Ye et al. 2002; Ye and Zhao 2002; Shu et al. 2004); and phenols (denbinobin) (Lee et al. 1995).

Dendroflorin triggers embryonic lung fibroblasts in the G1 phase to enter the S phase and thereby it benefits cell proliferation. It supports ROS degradation. Could it have anti-senescent properties (Jin et al. 2008)? Nevertheless, cell proliferation has its own problems, from tumour growth to vascular blockage.

D. nobile and *D. moniliforme* contain the largest number of sesqueterpenoids. Dendrobine has a slight pain relieving and antipyretic action but it is weaker than standard over-the counter (OTC) painkillers and fever remedies. It raises blood sugar levels, and in large doses it diminishes cardiac activity, lowers blood pressure, suppresses respiration, inhibits rabbit intestinal contractions and contracts isolated guinea pig uterus. In white mice and rats, the minimum lethal dose is 20 mg/kg. Death is preceded by convulsions (Chen and Chen 1935a). In animal experiments, sodium amytal promoted detoxification of dendrobine (Chen and Rose 1936). Dendrobine reversibly blocked the presynaptic inhibition of the ventral spinal cord of the frog, an effect similar to that of strychnine which is a structurally related compound (Kudo et al. 1983). A likely precursor of dendrobine is nobilonine, the second most abundant alkaloid in *D. nobile* (Yamamura and Hirata 1964), The structures of dendrobine and nobilonine were elucidated by Onaka and his colleagues (Onaka et al. 1964, 1965), while the structure of dendrobine was also studied by Inubishi et al. (1964). Dendrobine was synthesised in 1972 (Inubushi et al. 1972); recently, alternative approaches for its synthesis were developed by Kreis and Carreira (2012).

A few oxygen derivatives of dendrobine such as dendramine (6-hydroxydendrobine), dendrobine-N-oxide, *N*-methyldendrobine N-isopentenyldendrobine and the carbomethoxy-methylene derivative, dendrine, were also isolated from *D. nobile* (Okamoto et al. 1966a, b; Hedman 1972; Inubishi and Nakano 1965; Slaytor 1977). The configuration of dendrine has been determined and the compound success-fully synthesised (Granelli and Leander 1970).

Another group of derivates from dendrobine present in *D. nobile* comprises dendroxine (Okamoto et al. 1966a, b) and its derivatives, 4-hydroxydendroxine, 6 hydroxydendroxine *N*-isopentenyldendroxine and *N*-isopentenyl 6-hydroxydendroxine (Okamoto et al. 1966a, b, 1972; Hedman 1972).

Cyclocopacamphane, cadinene, emmotin and muurolene are found mainly in *D. nobile*. The isolation of new picrotoxinin-type and dendrobine-type sesquiterpenoids from the hybrid *Dendrobium* Snowflake 'Red Star' opens a new direction to employ *Dendrobium* for the production of novel alkaloids, and possibly some of them might find a medicinal use. The new alkaloids are called flakinins A and B and mubironines A, B and C, respectively (Morita et al. 2000). *Dendrobium* Snowflake is a tetra-ploid hybrid that has 75 % *D. nobile* and 25 % *D. moniliforme* in its makeup (Table D-4). It was registered with the Royal Horticultural Society in 1904 and is not a new hybrid.

Two bibenzyl derivatives in *D. nobile*, nobilin D and E, have a stronger inhibitory effect on nitrous oxide production than resveratrol and higher antioxidant activity than Vitamin C (Zhang et al. 2007b). *D. nobile* alkaloids administered to rats reduced memory impairment resulting from injection of lipopolysaccharide into their brain (Li et al. 2011). Treatment with *Dendrobium* alkaloids attenuated rat cortical neuronal damage following oxygen-glucose dep-rivation, the evidence shown by increased cell viability, lesser histological damage and reduced apoptosis (Wang et al. 2010c).

Denbinobin, a phenanthrene from *D. nobile*, inhibits invasion and induces apoptosis in SNU-484 human gastric cancer cells (Song et al. 2012a),; inhibits nuclear factor-kappaB and induces apoptosis in human leukaemic cells (Sanchez-Duffhues et al. 2009). It induces apo-ptosis in human lung adenocarcinoma cells via Akt inactivation, Bad activation and mitochon-drial dysfunction (Kuo et al. 2008). In an in vivo xenograft model, denbinobin was shown to suppress lung adenocarcinoma A549 growth and microvessel formation. It selectively inhibits insulin-like growth factor-1 (IGF-1) in endothelial cells of human umbilical cord. Suppression of IGF $=-1$ is possibly the action that suppresses angiogenesis (Tsai et al. 2011). In human colorectal cancer, denbinobin induces apoptosis by apoptosis-inducing factor releasing and DNA damage (Chen et al. 2008b). It inhibits HIV-1 replication through an NF-kappaB-dependent pathway (Sanchez-Duffhunes et al. 2008).

Moscatilin has potential antimutagenic activ-ity (Miyazawa et al. 1999). Ephemeranthol A is a potent inhibitor of nitric oxide and pro-inflammatory cytokine production; another phenanthrene, dehydroorchinol, also possesses anti-inflammatory properties. Pharmacological actions of the new phenanthrene, 1,5,7-trimethoxyphenanthrene-2-ol, have not been fully investigated but it did not exhibit any anti-inflammatory property (Kim et al. 2014).

Herbal Usage: *D. nobile* is currently the dom-inant species of *shihu* (Anonymous 2004). Chi-nese herbal texts state that the pseudobulb benefits the stomach, encourages secretion of body liquids, and reinforces the *yin* element. It is used to alleviate dehydration, thirst and poor vision and to hasten recovery after an illness. When patients with chronic superficial gastritis had their stomachs perfused with 20 g of *D. nobile*, Chen et al. (1995) observed a signifi-cant increase in serum gastrin concentration and acid secretion without any change in plasma somatostatin concentration.

The medicinal preparation is known as *Sek koku* in Japan where it is used to treat fever, loss of appetite with nausea and lumbago. It is mainly used to treat night sweats in Korea. The Korean name is *Seok gok* (Kimura et al. 2002).

The World Health Organization (WHO), in its publication, *Medicinal Plants of China. A Selection of 150 Commonly Used Species,* only described *D. nobile* as *shihu*. It mentioned that other species were used but did not list them by name. Thirst, dryness of the mouth, malaise and fever during convalescence are the main reasons for taking *shihu* which is consumed at a dose of 6–13 g daily, or 15–30 g if fresh pseudobulbs are used (Anonymous 1989).

The WHO Manual on the *Medicinal Plants in Viet Nam* (Institute of Materia Medica, Hanoi, 1990) made a somewhat different statement about the usage of *D. nobile* in Vietnam:

"The entire plant is applied as a tonic in pulmonary tuberculosis, general debility, flatulence, dyspepsia, reduced salivation, parched and thirsty mouth, night sweats, fever and anorexia. It is likewise effective for sexual impotence, amblyopia, arthralgia, myasthenia, lumbago and pain in the extremities. It is prescribed in a daily dose of 8–16 mg in the form of a decoction, powder or pills." It is one of only two species used as *shihu* in Vietnam (Doung 1993).

Seeds of *D. nobile* are applied on fresh wounds to hasten their healing (Das 2004). Rao and Sridhar (2007) reported that eye diseases in Karnataka were treated with *D. nobile*, but they did not explain how the orchid was used in the treatment. On the other side of the Deccan, Jonathan and Raju (2005) reported that the freshly dried stems of *D. nobile* were used in a preparation that promotes longevity and also as an aphrodisiac and analgesic. Stems are used as tonic, stomachic, pectoral antiphlogistic and tonic in Nepal; also to correct throat dryness and thirst (Baral and Kurmi 2006).

Dendrobium nodosum Dalzell.

syn. *Flickingeria nodosa* (Dalzell) Seidenf.

Indian names: *Purushratna* (Kannada) *Jiwanti* (Ayurveda)

Description: Typical of *Flickingeria*, this epiphyte has a creeping rhizome stems with well-spaced, slender, erect or pendulous, branching pseudobulbs, up to 30 cm in length, ending in a single leaf. Small, solitary, white flowers, 1 cm across, with maroon spots appear from June to October; they are short-lived. The species is found in southern India and Sri Lanka at moderate elevations.

Herbal Usage: In Uttara Kanada, Karnataka Province, Rao and Sridhar (2007) report that a sweet-tasting 'halva' is prepared from the orchid. The *halva* is used as an expectorant to treat asthma, bronchitis, *"tridosha"*, throat infections, and as an astringent for bilousness or to purify the blood. Cold extract of pseudoblubs of *Flickingeria nodosa* (= *D. nodosum*) exhibited antimicrobial activity against *Staphylococcus aureus* (a common cause of superficial infections) and *Staphylococcus citreus*, also a cause of skin infections. Warm extract of pseudobulbs shows antifungal activity against *Trichophyton mentagrophytes*, an organism that causes skin inflammation and infection of scalp, face, body, feet ("athlete's foot") and hair, commonly in rural workers (Nagananda and Satishchandra (2013). Matured fruits of *D. nodosum* are used as an aphrodisiac (Rao and Sridhar 2007).

Dendrobium normale Fale. (see ***Dendrobium fimbriatum* Hook**.)

Dendrobium officinale K. Kimura et Migo (see ***Dendrobium catenatum* Lindl**.)

Dendrobium ovatum (L.) Kraenzl.

Indian Names: *Anantali Maravara* (van Rheede); *Maravar* along the Malabar Coast and Uttar Pradesh; *Nagli* (Marathi)

Description: Its slender stems are tufted, smooth, 30–50 cm in length, when leafless bearing 3–6 cream-coloured flowers on short, apical or axillary racemes. Santapau and Kapadia (1966) observed the species growing on *Tectonia grandis, Salmalia malbrarica, Erythrina*

Fig. 10.41 *Dendrobium ovatum* (L.) Kraenzl. on a mango tree in a private garden in Kerala, Peninsular India. [PHOTO: E.S. Teoh]

suberosa, Mangifera indica and even on palm trees (*Areca* sp.) (Fig. 10.41). Leaves appear in June to September; flowers from September to January in Bengal (Santapau and Kapadia 1966) and in March to May in western Peninsular India. *D. ovatum* is found in Bengal, around Mumbai, in the Western Ghats and in southern India in open deciduous forests.

Herbal Usage: In 1703, van Rheede reported that *D. ovatum* was used for all sorts of chest pain, and especially to relieve stomach ache (van Rheede 1703). In Mumbai, India, it is still used as an emollient, and the juice of the entire plant is prescribed for stomach ache. It acts as a laxative (Caius 1936). In the adjacent state of Karnataka, tribals at the Kudremukh National Park still use the orchid for the same indications (Rao 2007). To treat constipation and stomach ache, in the Western Ghats, juice of *D. ovatum* is obtained by crushing the stems "by hand" (Rao 2004). In northeastern Uttar Pradesh it is reputed to be a tonic, stomachic, pectoral and antiplogistic. There, it is used to treat rheumatism. There are traces of alkaloid in the leaves (Trivedi et al. 1980).

Dendrobium pachyphyllum (Kuntze.) Bakh. F.

syn. *Dendrobium pumilum* Roxb.

Malay Name: *Sakat kalumbai*
Thai name: *Uang song bai*

Description: *D. pumilum* is a tiny, single-flowered *Dendrobium* with stems less than 2–5 cm in height forming dense mats on fruit trees in the Malaysian lowlands. Flowers are white or cream with purple veins, 1–1.2 cm across (Fig. 10.42). *Alba* forms occur. It flowers gregariously several days after a sudden temperature drop. It flowers more frequently than *D. crumenatum* which also flowers after a temperature drop (Comber 2001). In Thailand, peak flowering occurs in August and September (Vaddhanaphuti 2001). Previously, it was a common orchid in orchard trees in Malaya and Indonesia, extending to Vietnam, Thailand, Myanmar and northeastern India.

Herbal Usage: Burkhill and Haniff (1930) reported that Alvins said that a decoction of its

Fig. 10.42 *Dendrobium pachyphyllum* (Kuntze.) Bakh. F. [PHOTO: E.S. Teoh]

roots was used to treat dropsy. This was also reported by Caius (1936).

Dendrobium. parishii Rchb. f.

Local Name: *Xishishihu* (named after the classic southern beauty Xishi),

Other Common Names: *Zibanshihu* (Purple feet *Dendrobium*)

Thai Names: *Ueang Kraang Sai Sun, Ueang Sai Nam Kraang, Ueang Nam Khrang, Ueang attakrit, Ueang lathakrit, Ueang khrang.*

Myanmar Name: *Khayang yaung twin pyin*

Description: This striking, rose-purple, fragrant deciduous *Dendrobium* has a small distribution extending from Indochina into northern Thailand into Yunnan, Myanmar, Bangladesh and Arunachal Pradesh. Epiphytic in deciduous forests from 250 to 1500 m, it requires a distinct dry season to initiate flowering. Pseudobulbs are pendulous, cylindrical, up to 60 cm long, 1 cm in diameter, and carry several short inflorescences at the distal internodes, each with two or three rose-purple flowers, 6 cm across, from March to May (Vaddhanaphuti 2005). Lip is circular, downy, with two purple blotches on either side of the throat, accentuated by a white rim (Fig. 10.43). Its scent has been compared with that of rhubarb. It is very free-flowering.

Fig. 10.43 *Dendrobium. parishii* Rchb. f. [PHOTO: E.S. Teoh]

Phytochemistry: An imidazolium salt was isolated from *D. parishii*, and its structure confirmed by synthesis (Leander and Luning 1968). Anosmine is an imidazole alkaloid from *D. parishii* which has also been synthesised. It is 1(1,2,3,4,6,7,8,9-octahydrodipyrido[1,2,-alpha-1′,2′c]imidazole-10-iumbromide (Hemsccheldt and Spenser 1991)

Usage: Pseudobulb benefits the *yin* element, is antipyretic, stops thirst, and encourages the secretion of body fluids (Ou et al. 2003).

Dendrobium planibulbe Lindl.

Dendrobium tubiferum Hook.f., *Callista tubifera* (Hook. f) O. Kuntze., *Aporum planibulbe* (Lindl.) S. Rauschert.

Description: *D. planibulbe* belongs to the Section *Rhopalanthe* whose typical member is the well-known, roadside pigeon orchid, *D. crumenatum*. Like the latter species, its 40-cm stem is slender and the distal, flower-bearing half is leafless, while the basal internode is pseudobulbous. Pseudobulb measures 3–4 cm in length and is four-cornered. Leaves, placed 2 cm apart, are also similar to those of *D. crumenatum*, but smaller, 3.5–5 by 1 cm. Flowers are white marked with pink stripes, and the trilobed lip carries a yellow callus in the centre. Sepals and the mid-lobe of the lip are fringed with minute hairs. Flowers appear several days after a drop in temperature. They are borne singly on the nodes, are around 2 cm across, not opening widely.

D. planibulbe is found in Peninsular Thailand, only in Perak and Trengganu in Peninsular Malaysia, in Sumatra, Borneo and the Philippines (Seidenfaden and Wood 1992).

Herbal Usage: Pounded into a poultice, it was used to treat sores and infected wounds in Peninsular Malaysia (Burkhill and Haniff 1930; Burkill 1935).

Dendrobium plicatile Lindl.

syn. Flickingeria fimbriata (Blume) A.D. Hawkes; *Dendrobium macraei* Lindl.,

D. plicatile Lindl., *Desmotrichum fimbriatum* Bl., *Ephemerantha macraei* (Lindl.)Hunt et Summerh., *Flickingeria macraei* (Lindl.) Seidenf.

Indian Names: *Jibai, Jibanti* in Bengal, *Jivanti, Radarudi, Wajhanti* (Gujerati), *Jiban, Joivanti, Sag* (Hindi), *Jivanti* (Marathi), *Bhadra, Jiva, Jivabhadra, Jivada, Jivani, Jivaniya, Jivanti, Jivapauri, Jivapushpi, Jivavardhini, Jivarisha, Jivdatri, Jivya, Kanjika, Kshurajiva, Madhushvasa, Madhusrava, Mangalya, Mrigaratika, Payaswini, Praanada, Putrabhadra, Ratangi, Shakashreshtha, Shashashimbika, Shringati, Srava, Sukhankari, Supringala, Yashaskari, Yashasya* (Sanskrit), *Saka* (in Orissa), *pourusha rathna* or *purusha ratana* (Kannada), *Swarn* (Uttar Pradesh)

Ayurvedic name: *Jeewanti* in *Sarangadhara Sanghita*.

Sri Lankan name: *Jata makuta*

Chinese name: *Liusu Jin Shihu* (Tassels gold *Dendrobium*)

Chinese medicinal name: *Youguashihu*

Description: In contrast to the vast array of Indian names, the description of this species is vague and scanty. I have stayed with the version given by Grant in *Orchids of Burma*, 1895, due to lack of a more definitive description. It is a large plant with pendulous stems up to a metre long and branching, ending in fusiform, grooved, yellow pseudobulbs of 6 cm length. Leaves are apical, 10–20 cm long. Pseudobulb produces 2–3 white or creamy flowers, up to 3 cm across, with a sprinkling of red on the side lobes of the lip.

D. plicatile is widely distributed from the Himalayas, India and China to Sri Lanka, Southeast Asia and Papua New Guinea. In Sikkim, it occurs at 2200–2600 m on trees or rocks, The Indian species is thought to be synonymous with *Desmotrichum fimbriatum* Blume or *Flickingeria fimbriata* which is found in Sri Lanka at the lower elevation of 1000 m (Cooray 1940), and in Indonesia and Papua New Guinea. In Peninsular Malaysia, it is found in many localities in the lowlands and at mid-elevations in the hills, (Seidenfaden and Wood 1992). It is most common in areas without a long dry season and often covers tree trunks which overhang the sea or rivers (Go and Hamzah 2008). In China, it is present in Hainan, Guangdong, Guangxi and Guizhou.

The long list of Indian names is due to the popularity of the plant as a sweet preparation, the much-valued '*Halwa*'. Hailed as a stimulant and tonic, It is taken for debility due to seminal loss. In the markets of Bombay, it is known as *ruttun-purush* and sold at a high price. But it is not an attractive plant. It has creepy rhizomes with long, branching pseudobulbs, bearing insignificant, white flowers that barely last a day.

Phytochemistry: Plant contains an alkaloid (Chopra 1933). Three phenanthrenes were isolated from *Ephemerantha fimbriata* (= *D. plicatile*), namely fimbriol A, fimbriol B and ephermeranthol-C (Tezuka et al. 1993). Five more phenenthrenes were isolated by Yamaki and Honda in 1996, namely, lusianthrin, erianthridin, ephemernthroquinone, a dimethoxydihydrophenanthrene (related to nudol and showing high cytotoxicity, smooth muscle relaxing effect on the intestine and blood vessels) and a dihydrophenethrene dimer, An ethanol extract was found to exhibit strong scavenging activity against superoxide anion radicals in vitro (Osugi et al. 1999). Three phenanthrenes have been isolated from the stems of *D. plicatile*, namely, 2,5-dihydroxy-4,9,10-trimethoxyphenenthrene or plicatol A, 2,5-dihydroxy-4-methoxyphenanthrene or plicatol B, moscatin and 5,9-trihydroxy-4-methoxy-9,10-dihydrophenanthrene or plicatol C, already named rotindatin (Honda and Yamaki 2000). Plicatol B has anticoagulant activity (Kovacs et al. 2007).

Employing bioassay-guided fractionation of an ethanolic extract, Chen and his colleagues isolated two new degraded diterpenoids that possess a rare 15,16-dinor-ent-primarane skeleton, named flickinflimilins A and B, a new ent-pimarane type diterpenoid glycoside and 4 known steroids The new compounds all showed antioxidant activity and inhibited tumour necrosis factor-alpha production by mouse macrophages (Chen et al. 2014).

Herbal Usage: Sanskrit writers described the plant as cold, mucilaginous and light. Besides being a tonic, the orchid is used as a remedy for disorders of the bile, blood and phlegm. It is commonly used in decoctions with other plants possessing similar properties. The fruit is an aphrodisiac. The plant is sometimes used in isolation as a stimulant and tonic, the latter to treat debility associated with seminal loss (Caius 1936). It is used in *Rasayana* therapy and sold as *Jibanti* in West Bengal (Chakrabarty et al. 2001). In Uttar Pradesh and far-off Karnataka, it is used to treat asthma, bronchitis, sore throat, stomach ache, biliousness and fever (Trivedi et al. 1980; Rao 2007).

In the *Sushruta*, it is a constituent of the local remedy for snake and scorpion bites, but when Mhaskar and Caius tested it on rats, they found that it was found to be worthless as a counterpoison for both snake and scorpion venom (Caius 1936).

Hill tribes in Orissa used *F. macraei* in the following manner: one spoonful of root paste is consumed with 1 g of powdered black pepper on an empty stomach for 21 days to treat skin allergies. It may also be applied directly on eczematous lesions (Dash et al. 2008). It is also used to treat general debility (Singh and Duggal 2009).

In Sri Lanka, ancient writers called it *Jeevaniya* (supporting life) and *saaka shreshtha* (best of herbs). A decoction of the plant is said to restore normal function to all three humours (*tridosha*) when the *tridosha* is deranged. The whole plant is used in the distillation of medicinal oils for external application to massage paralysed or painful limbs (Cooray 1940). In Nepal, paste prepared with the entire plant is used as a stimulant, restorative, demulcent, and to treat snake bites (Subedi et al. 2013).

D. plicatile participates in the large family of nourishing tonic herbal medicines used in China and Japan. It is a form of *Shihu* in TCM (Wu 1994) and widely used as a substitute for *D. moniliforme* to treat pneumonia, tuberculosis, asthma and pleurisy (Bencao 2000; Chen et al. 2014a).

Dendrobium polyanthum Wall ex Lind.

syn. *Dendrobium primulinum* Lindl.

Chinese name: *Xibianshihu* (fine petal *Dendrobium*), *Baochunhuangshilan* (harbinger of spring yellow *Dendrobium*)
Thai names: *Ueang Sai Naam Pueng, Sai lueang, Sai prasat,*
Myanmar name: *Thinn kyu kyu*

Description: A deciduous *Dendrobium*, *D. polyanthum* (syn. *D. primulinum*), has pendulous, thin pseudobulbs 15–40 cm long, bearing single, fragrant, pale pink flowers, 5–6 cm across, with a contrasting sulfur-yellow lip. Petals are narrow but the oval, somewhat kidney-shaped lip is large, pubescent, fringed, and sometimes outlined in white. It enjoys a distribution from southern China to the Himalayan foothills in the west, Laos and Vietnam in the east and southwards into Myanmar, Thailand and the Malay Peninsula, occurring in deciduous forests from 300 to 1700 m. A distinct dry season is mandatory to initiate flowering. Flowering season is from February to April (early to mid-spring).

Phytochemistry: It contains alkaloids of the indolizidine type, and a high content of hygraine (Luning 1980). *D. polyanthum* produces 5,7-dimethyloctahydroindolizine (Leander and Luning 1968) and its isomer dendroprimine, which is (5R,7S,9R)-5,7-dimethyloctahydroindolizine (Leander and Luning 1968; Biomqvist et al. 1972). A new tetrahydroanthracene, 3,6,9-trihydroxy-3,4-dihydroanthracen-1(2H)-one, together with six phenolics (moscatilin, gigantol, batatasin, moscatin, 9,10-dihydromoscatin, 10-dihydrophenanthrene-2,4,7-triol), and a sesquiterpenoid, corchoionoside C, as well as two sterols, β-sitosterol and daucosterol, were isolated from the stems of *D. polyanthum*. The new tetrahydroanthracene was assessed for cytotoxic activity against two human tumour cell lines (A549 and HL-60) but did not show any (Hu et al. 2009).

Herbal Usage: Whole plant is used to treat burns, eczema and paralysis (Wu 1994).

Dendrobium primulinum Lindl. (see **Dendrobium polyanthum Wall ex Lind.)**

Dendrobium pulchellum Roxb. ex Lindl.

Thai names: *Uang Chang Nao, Uang Kham Ta Kwai*
Vietnamese names: *Hoang thao, Po len, Co anh*

Description: This attractive *Dendrobium* is distributed from the eastern Himalaya to Bangladesh, Myanmar, Thailand Indochina and Peninsular Malaysia from 200 to 2000 m. Pseudobulbs are 100–200 cm but may reach up to 2 m in length, with leaf sheath that are carry purple stripes. Old stems are purplish. Leaves are oblong, 10 cm, carried on the upper half of the pseudobulb and deciduous after a few years. Inflorescence is pendulous, 15–20 cm long, with several large, creamy-yellow flowers with pink veins and a purplish lip with two prominent, maroon blotches. Flowers are 7.5–10 cm across, seven to ten on an inflorescence (Seidenfaden and Wood 1992; Lavarack et al. 2000) (Fig. 10.44). Flowering period is February to March (Vaddhanaphuti 2001).

Fig. 10.44 *Dendrobium pulchellum* Roxb. Ex Lindl. [PHOTO: E.S. Teoh]

Phytochemistry: Four bibenzyls, namely chrysotobibenzyl, chrysotoxine, crepidatin and mosatilin, were isolated from *D. pulchellum* and shown to facilitate anoikis (cell death resulting from failure to attach to extracellular matrix) and inhibit growth of lung cancer cells (Chanvorachote et al. 2013).

Herbal Usage: Petelot reported that Chinese living in Indochina at that time used this orchid and *D. gratiosissimum* Rchb. to make medicine, presumably using them as *shihu* (Perry and Metzger 1980). It is still being used as *shihu* in Vietnam (Doung 1993).

Dendrobium pumilum Roxb. [see **Dendrobium pachyphyllum (Kuntze.) Bakh. f.**]

Dendrobium purpureum Roxb.

Malay name: *Angrec cassomba*

Description: Stems are pendulous, 50–100 cm long, with lanceolate, dark green leaves, 9–13 by 1.4–2.5 cm. Inflorescences are short (2.5 cm), borne at several nodes on leafless stems, Flowers are arranged into a ball around the rachis. They barely open and they last a few weeks on the plant. Colour ranges from white through pink to light purple with green at the tips (Fig. 10.45). *D. purpureum* is a large, deciduous, lowland epiphyte found in steamy forests up to 800 m in the northern Sulawesi, Amboin, Moluccas, Banda Island and Aru Islands of Indonesia, in New Guinea, Bougainville, Vanuatu and Fiji. Rumphius (1627–1702) found them growing on clove trees, seaside *Casuarina* (*Casuarina equisetifolia*) and the banyan (*Ficus benjamina*) in Amboin and they flowered from October to November (Rumphius, second half seventeenth century) (Fig. 10.46).

The subspecies *D. candidulum* is white. It flowers from June to October corresponding to the driest period of the year.

Herbal Usage: The heated pulp of crushed stems of the orchid is used as a poultice for whitlow in Amboin (Sulawesi) (Rumphius, second half seventeenth century).

Fig. 10.45 *Dendrobium purpureum* Roxb. [from: Teoh Eng Soon, Orchids of Adia, 3rd ed. Singapore; Marshall Cavendish, 2005]

Dendrobium quadrangulare C.S.P. Parish & Rchb. f. (see ***Dendrobium hymenanthum*, Rchb.f.)**

Dendrobium sinense **T. Tang & F.T. Wang**

Chinese name: *Hua shihu* (Chinese *Dendrobium*)

Description: Pseudobulbs are clustered, 10 cm long and 3–4 mm thick, non-branching. Distal nodes carry four elliptical leaves which are 4–5 by 1.2 cm and rounded at the apex. Flowers are large, white, with a broad vermillion streak at the throat (Zhu et al. 2009). They are pollinated by wasps. This is an endangered species.

D. *sinense* belongs to the Section *Formosae*. It grows on the trunk of trees at 1000 m in montane woodland on Hainan Island and is endemic.

Phytochemistry: A new phenanthre-nequinone, denbinobin B, with antibacterial activity against *Staphylococcus aureus*, and three known phenanthrenes were isolated from whole plants of *D. sinense* (Chen et al. 2013b). Subsequently, from the same species, the team isolated four new bibenzyl derivatives,

dendrosinens A–D, together with 12 known phenolic compounds, namely, 3,4,30-trimethoxy-5,40-dihydroxybibenzyl, aloifolI, dihydroxy-3,4-dimethoxybibenzyl, longicornuol A, trigonopol A, coniferyl *p*-coumarate, sinapyl *p*-coumarate, coniferyl aldehyde, syringaldehyde, 3-hydroxy-1-(4-hydroxy-3,5-dimethoxyphenyl)-1-propanonc, tcctochrysin and syringaresinol. Against human gastric (SGC-7901), human hepatoma (BEL-7402), and chronic myelogenous leukaemia (K562) cell lines in vitro, four compounds, 3,4,30-trimethoxy-5,40-dihydroxybibenzyl, aloifolI, dihydroxy-3,4-dimethoxybibenzyl, and longicornuol A, showed only slight cytotoxicity when compared with placitaxil. The remainder had none (Chen et al. 2014b).

Herbal Usage: The entire plant is used as *shihu* (Wu 1994).

Dendrobium strongylanthum **Rchb. f.**

Chinese name: *Shuchun Shihu*

Description: This is an unusual, epiphytic herb, droopy and densely flowered Stems are erect, fleshy, fusiform, 3–27 cm and 0.4–1 cm in diameter, enclosed by leaf sheaths on current stems; old leafless stems are yellow. Leaves are alternate, distichous, thin, oblong, 4–10 by 1.7 cm, with pointed tips that are unequally bilobed. Inflorescences are terminal or subterminal, suberect or droopy, densely many-flowered. Flowers are yellowish-green, with dark purple at the base. Sepals and petals are lanceolate, lateral sepals twice the length of the petals. Lip is trilobed; margins of side lobes lined by comb-shaped teeth, mid-lobe margin wrinkled, marked with red streaks and crested. Anther cap is red. Flowers do not open widely. They are 1.4 cm across. The species is epiphytic on tree trunks in evergreen forests at 1000–2100 m in Hainan, Yunnan, Myanmar, northern Thailand and Vietnam (Zhu et al. 2009). It flowers from October to November in Thailand (Nanakorn and Watthana 2008).

Herbal Usage: as *shihu* (Bao, Shun, Chen, et al. 2001)

Fig. 10.46 *Dendrobium purpureum* Roxb. [as *Angraecum purpureum II sylvestre*]. from: Rumphius, G.E., *Herbarium Amboinense* vol. 6: p. 109, t. 50 (1750). *Dendrobium purpureum* is the plant on the right. Courtesy of Missouri Botanical Gardens, St. Louis, USA

Dendrobium subulatum (Blume) Lindl.

Malay name: *Anggerek* (orchid)

Description: A small lowland epiphyte with thin long stems, spreading curved leaves up to 1.5 cm long, and equally small, cream to brownish flowers, 8 mm in diameter. It is distributed in Thailand, Malaysia, Sumatra, Java and Kalimantan (Handoyo 2010).

Herbal Usage: Leaves used as a poultice to relieve headache in the Malay Peninsula. Although this species, *D. subulatum* (Bl) Lindl., was identified by Burkhill and Haniff (1930)

from the specimen supplied by the native medicine-man, the term used by their provider, *anggerek* (orchid), implied that, in practice, one would not be particular about the species of orchid one uses to treat earache. It is not unusual for different plants to share the same name in traditional medicine. This happens in China as well as in India. For instance, *Shancigu* usually refers to *Cremastra wallachii* but it is also applied to *Pleione bulbocodiodes* and *Pleione yunnanensis* because all these several species of orchids serve the same medicinal purpose. It is the efficacy of an herb for a specific treatment that determines its naming not its botanical identity.

Dendrobium thyrsiflorum B.S. Williams

Chinese herbal name: *Huangcao shihu*
Vietnamese name: *Thuy tien vang*
Thai names: *Uang mon khai liam, Uang mon khai luang*

Description: *D. thyrsiflorum* is a beautiful prolific *Dendrobium* which produces long clusters each with 30–50 well-arranged, yellow flowers with an orange lip in spring. The International Orchid Commission notes that *D. thyrsiflorum* Rchb. f. is synonymous with *D. densiflorum* Wall., but the Commission accepts *D. thyrsiflorum* as a separate species in its *Register of Hybrids*. There are subtle differences in the two "species", *D. thyrsiflorum* having larger flowers (3–4 cm vs. 3 cm) which are more tightly arranged; the white petals and sepals are broader with wavy margins, the lip is a darker orange, and the flowers appear in spring (February to April) while those of *D. densiflorum* appear in autumn (August to September) (Vaddhanaphuti 2001). Flowers are fragrant (Figs. 10.47 and 10.48). *D. thyrsiflorum* is found at 1100–1800 m (400–900 m for *D. densiflorum*) in northeast India, Myanmar, Yunnan, Thailand and Indochina. PCR technology will possibly determine whether the two species are different.

Phytochemistry: In one of the earliest pharmacologic studies on *Dendrobium,* Wrigley (1960) described the extraction of ayapin,

Fig. 10.47 *Dendrobium thyrsiflorum* B.S.Williams, pale pink form [PHOTO: E.S. Teoh]

Fig. 10.48 *Dendrobium thyrsiflorum* B.S.William. This is a common colour form. [PHOTO: E.S. Teoh]

scopoletin and 6,7-dimethoxycoumarin from *D. thyrsiflorum*. It is suggested that measurement of these three compounds by HPLC (high-performance liquid chromatography) would be a suitable method for assessing the quality of *Huangcao shihu* (*D. thrysiflorum*) derived from different sources (Zhang et al. 2006b). Zhang and Zhang (2005a) isolated one bicyclic and two bi-tricyclic derivatives of coumarin-benzofuran, phenanthrene-phenanthrene and phenanthrene-phenanthraquinone along with seven other known compounds from the stems of *D. thyrsiflorum*. Four of the compounds, denthyrsin, denthyrsinol, denthyrsinone and denthyrsinin, showed significant cytotoxic activities against Hela, K-562 and MCF-7 cell lines (Table D-3 from Zhang et al. 2003a). Employing laser scanning confocal microscopy, Zheng et al. (2005a, b) found that coumarins were located mainly in vascular bundles, on the walls of outer fibre cells, with maximum concentration towards the apex of the stem. Coumarins are also present in leaves and root, reaching their maximum concentrations during the peak of flowering, less during the vegetative phase, and plunging to the lowest level during the fruit set period (Zheng et al. 2005a).

A unique study from Vietnam investigated the antimicrobial activity and biodiversity of endophytic fungi in local plants *D. thyrsiflorum* and *D. devonianum* From *D. thrysiflorum*, 23 endophytic fungi were isolated, and from *D. devonianum*, 30. Eleven fungi in *D. thyrsiflorum* were bacteriostatic/bactericidal of fungistatic/fungicidal against at least one pathogenic organism from the six tested (*Escherichia coli, Bacillius subtilis, Staphylococcus aureus, Candida albicans Cryptococcus neoformans* and *Aspergillus fumigatus*). The bactericidal property of *Epicoccum nigrum* (from *D. thyrsiflorum*) was even stronger than that of ampicillin. *Fusarium* present in both species of *Dendrobium* was effective against both bacteria and fungi. This opens a new approach for obtaining antibiotics and fungicides (Xing et al. 2011).

Herbal Usage: *Huangcao shihu* is used to nourish *yin* and remove heat. It benefits the stomach and promotes production of body fluids. Using *D. thrysiflorum*, specific microsatellite markers for identification, commercial samples of *Huangcao shihu* (involving *D. thrysiflorum*) sold in Nanjing were found to originate predominantly from Laos and Myanmar (Yuan et al. 2011b). This is not a new phenomenon: *Dendrobium* species were "exported in large bales to Vietnam and Hong Kong. Various species were gathered in large amounts for export to China" (Vidal 1963).

Dendrobium tosaense Makino (see D. catenatum Lindl.)

Chinese name: *Huanhua Shihu* (golden flower *Dendrobium*)
Taiwanese names: *Huang Hua Shi Hu* (yellow flower *Dendrobium*), *Huang Shi Hu* (yellow *Dendrobium*)

In *Flora of China*, Zhu et al. (2009) considered *D. tosaense* to be synonymous with *D. catenatum* (*D. officinale*) which has historical precedence having been proposed by Lindley in 1830. Chinese herbalists and medical researchers refer to their herb (possibly determined by source) as *D. tosaense* (Tang and Cheng 1984; Lo et al. 2004a; Hu et al. 2008b). Their phytochemical studies are discussed under *D. catenatum*.

Dendrobium transparens Wall ex Lindl.

Nepali name: *Parajivi, Thuur*

Description: Stems are terete bearing linear-lanceolate leaves 7–10 by 1–1.5 cm, pointed at their tips. Inflorescence arises from leafless stems at the nodes, is short, and carries 2–3 flowers, white flushed with pink. Lip is ovate, elliptic, with a purple blotch at the throat (Fig. 10.49).

D. transparens is distributed from eastern Himalaya to Yunnan and Myanmar. It is epiphytic on tree trunks in sunlit forests at 700–1500 m in Nepal. It flowers in April to May in Upper Assam (Nath and Das 2013), and May and June in Nepal (Raskoti 2009).

Herbal Usage: Plant is used to treat fractures and dislocated bones in Nepal (Subedi et al. 2011; Pant and Raskoti 2013). Although it is not listed among the various Chinese pharmacopoeia consulted for this publication,

Fig. 10.49 *Dendrobium transparens* Wall ex Lindl. [PHOTO: E.S. Teoh]

D. transparens was present among the 21 samples of *Herba Dendrobii* collected from several herbal markets by Takamiya et al. (2011). In that study, *D. catenatum* was regarded as the most important and commonest, but other unconventional species included as *Herba Dendrobii* were *D. cucullatum*, *D. denudans*, *D. eriiflorum*, *D. lituiflorum*, and *D. regium*.

Dendrobium trigonopus Rchb. f.

Thai names: *Uang kham pak kai, Uueang kham liam*

Description: Pseudobulbs are suberect, clavate and fusiform, 5–11 cm long with 3–4 oblong, thick, coriaceous leaves 7–10 by 3 cm. Inflorescences are short, terminal or subterminal, and carry 2–3 bright, golden-yellow flowers 3–5 cm across. Lip has a green patch at the centre and brown stripes. Flowering season is February to March in Thailand, and March to April in Yunnan (Seidenfaden and Smitinand 1960; Chen et al. 1999b; Vaddhanaphuti 2001).

D. *trigonopus* is distributed in northern Thailand, Myanmar, southern Yunnan and Laos at 1100–1600 m, epiphytic on trees in sparse woods.

Phytochemistry: Two novel bibenzyls, trigonopols A and B, and seven known compounds [gigantol, tristin, moscatilin, hircinol, narigenin, 3-(4-hydroxy-3-methoxyphenyl)-2-propen-1-ol and (−)-syringresinol] were isolated from the stems of the orchid by the well-known group in Kunming. Trigonopol A showed antiplatelet aggregation activity in vitro (Hu et al. 2008a).

Herbal Usage: The stem is used to treat fever and anaemia in Thailand (Chuakul 2002). *D. trigonopus* is not a medicinal plant in China.

Dendrobium umbellatum Rchb. f.

Murut name: *Tingasu*

Description: Plants are dwarf epiphytes, 3–4 cm tall. Pseudobulbs are close-set, slender, terete, with brownish sheath at its lower third. Leaf is solitary elliptic, apical. A solitary, cream-coloured flower is borne on a short raceme at the apex of the pseudobulb. Sepals are ovate, lateral sepals larger than dorsal sepal, petals are linear. Lip is trilobed, side lobes tiny, thin and pointed; mid-lobes discoid, large, glabrous and white. It flowers intermittently throughout the year. The species is distributed from Borneo to Papua New Guinea.

Herbal Usage: Leaves are boiled and drunk as tea to treat stomachache in Sabah (Fasihuddin and Hamsah 1992).

Dendrobium wangii Tso (see **D. hercoglossum Rchb.f.**)
Dendrobium wilsonii Rolfe [see **Dendrobium moniliforme (L.) Sw.**]

Dendrobium williamsonii J. Day & Rchb. f.

Description: Pseudobulbs are cylindric, suberect, 20 by 0.4–1 cm in diameter, unbranched, with several nodes. Leaves are several, coriaceous, oblong to oblong-lanceolate, 5–11 by 1.5–2 cm, borne at the top of the stems, sheaths covered with dense, black hairs. Inflorescence is subterminal, usually single-flowered. Flowers are ivory-white to pale yellow with a large orange to blood-red patch on the lip. Tepals are lanceolate, 2.7–3 by 1.5 cm. Flowering season is April to May (Chen et al. 1999b). *D. williamsonii* is distributed from northeast India to Myanmar, Thailand and Vietnam, and Yunnan, Guangxi and Hainan in China occurring in forests at 1000 m (Chen et al. 1999b; Zhu et al. 2009).

Herbal Usage: On Hainan island, *D. williamsonii* is used as *shihu* (Xu et al. 2005) by the Li minority (Song, personal communication, 2015).

Dendrobium xantholeucum Rchb. f.

syn. *Flickingeria xantholeuca* A. Hawkes; *Ephemerantha lonchophylla* (Hook.f.) P.F.Hunt & Summerh.

Chinese name: *Jiye Jin Shihu* (sharp leaf gold *Dendrobium*)

Description: The present official name for this orchid is *D. xantholeucum* Rchb. f. (though fairly recently, *Flickingeria xantholeuca* Hawkes) but *Ephermerantha lonchophylla* is the name which is most commonly used in ethno-botanical and biochemical texts (Lee et al. 1995; Ma et al. 1998; Chen et al. 1999a, b, 2000, 2008a; Yang et al. 2005a; Chua and Koh 2006; Kuo et al. 2008).

Rhizome is short, bearing numerous flattened, branching stems close to one another. Pseudobulbs are 5.5 cm long and 1 cm diameter. Leaves are lanceolate, pointed, 4–8 by 1–2.2 cm. The short inflorescence arises from the sub-apical node and carries a single white or pale yellow flower with very narrow petals. Lip is 1.2 by 0.95 cm. Mid-lobe of the lip is creamy, deeply notched, and marked by ridges. *D. xantholeucum* is a small epiphyte, distributed throughout Thailand, Malaysia and Indonesia and southern Taiwan from sea level to 1000 m. In China, it is found in Guangxi, Guizhou and Yunnan.

Phytochemistry: Various groups of investigators who worked on *D. xantholeucum* knew this orchid as *Ephemerantha lonchophylla* and, to date, all of them published their results giving *Ephemerantha lonchophylla* as the name of their herb.

Ephemeranthone, denbinobin and 3-methylgigantol were isolated from this orchid by Chen et al. (1999a). They studied the anti-oxidant principles of the compounds and found that only ephemeranthone could inhibit oxidation of human low-density lipoprotein (LDL) in vitro. Subsequently, the group also isolated 3,7-dihydroxy-2,4 dimethoxyphenanthrene and crianthridin from the orchid and found that these two compounds and methylgigantol had anticoagulant properties (Chen et al. 2000). Erianthridin was isolated previously from the orchid together with lusianthridin and ephemeranthol-B by Wang et al. (1997).

Two primarne diterpenoids, lonchophylloids A and B, isolated from the orchid stems exhibited an ability to sensitise cells that expressed multi-drug resistance phenotype to the anticancer drug, doxorubicin (Ng et al. 1998). This holds much promise for application to cancer chemotherapy. Denbinobin, which is also present in *D. nobile* and *D. moniliforme*, is cytotoxic against several human cancer cell types, lung, breast, colon, etc. (Kuo et al. 2008; Chen et al. 2008b). This phenanthroquinone derivative inhibits the formation of blood vessels that support tumour growth (Tsai et al. 2011), and it prevents spread of the tumour to adjacent or distant parts of the body (Chen et al. 2011).

Herbal Usage: The entire plant is used as *shihu*.

Overview

D. moniliforme and *D. catenatum* (syn. *D. officinale*) are two plants that have enjoyed medicinal usage in China for 5000 years. They were the original *shihu* which grew on rocks in *Shen Nong's Bencao Jing*, the classic Chinese Pharmacopoeia that was published at the beginning of the current era. The saxicolous nature of the plants denoted robustness and resilience, and, going by the *Doctrine of Signatures*, these plants were capable of strengthening the body (Hu 1971). Additional lithophytic species of *Dendrobium* occuring at the cradles of Chinese civilization along the Huanghe and Yangzi and in the eastern provinces (Table D1) were listed as *shihu* during the Tang Dynasty (618–907) (Chen and Tang 1982). At this stage, it was noted that *Dendrobium* species also thrived as epiphytes.

Dendrobium species used as Shihu

With the acquisition of territories in the south and the possibility of safe travel accorded by a strong central government, numerous epiphytic, deciduous *Dendrobium* species were encountered and added to *shihu*. In 1936, Koiti Kimura identified 20 different species of *Dendrobium* being sold as *shi-hu* in Chinese markets, and with Hisao Migo he described a species new to science, *D. crispulum*, picked up from the drug market (Kimura and Migo 1936). During a study on raw plants labelled as *Huoshan Shihu*, Tang and Cheng (1984) identified *D. catenatum* (syn. *D. officinale*), *D. tosaense*, and what they thought was a new, undescribed species which they named *D. huoshanense*. However, recent taxonomical studies concluded that *D. tosaense* and *D. huoshanse* are synonymous with *D. catenatum* G.Z. Tang et S.J. Cheng, the last named being a variable species. The herbalists were correct!

Records of China Imperial Customs in 1884 and 1888 revealed that an enormous amount of *Dendrobium* collected from various provinces passed through the numerous ports along the Yangzi and on the China coast. They were given different Chinese names (*chin cha, huang tsao, hsien tuo, hsien hu tou, hsien shihu, mu hu, mu hu pi, huay tsan, ya tuo, huan chai*) by the exporters, but these were invariably re-interpreted by the English officers as *D. ceraia* (an old name for *D. crumenatum*, a species that does not occur in China) (Braun 1888; Hart 1884)! By the Ming Dynasty (1368–1644), demand already exceeded supply. *D. nobile* was accepted by Li Shizhen (1518–1593) as *shihu,* and this led to a flood of other *Dendrobium* species, notwithstanding the fact that many of these were not lithophytic. When Hooper examined the herbs from Chinese medicine shops in the Straits Settlements (Singapore, Penang and Malacca) in 1929, he

recorded only one orchid, *D. nobile*, which went by the local (principally Cantonese) names of *kin sak fook* (gold hairpin), *chen kin cha* (genuine golden herb), *chak hook (shihu),* and *ma pien* (horse whip). He stated that this herb was exported fresh or in a dry state from Hankow and Canton, and that there was a four-fold range in price depending on the quality of the herb (Hooper 1929). Hooper was probably unaware that at least four species of *Dendrobium* were being sold as *ma pien*, and this should not have included *D. nobile* which is classified as *Huangcao shihu*. However, from the 74 species and two varieties in China, 32–35 species carry the designation '*Huangcao Shihu*' (Golden Herb *Dendrobium*) in the herbal medicine market (Bao, Shun, Chen, et al. 2001; Xu et al. 2006). The name *Huangcao* describes the yellow or golden appearance of the pseudobulbs of these species when they are old or dried, not the colour of the flowers (Figs. 10.50 and 10.51). Other orchids such as *Pholidota* which were both lithophytic and epiphytic were initially adulterants in shipments of *shihu*. Subsequently, they were accepted as substitutes (Bao and Shun 1999). The distribution of *Dendrobium* species in the various provinces in China is shown in Table 10.1.

The Catalogue of Medicinal Plant Resources in Taiwan, lists nine species of *Dendrobium* as *shihu*, namely *D. aduncum, D. chrysotoxum, D. falconeri, D. linawianum, D. loddigesii, D. moniliforme, D. nobile, D. parishii* and *D. primulinum* (Ou et al. 2003). The list is limited to the species with pharmacological data because there are almost twice as many distinct species of *Dendrobium* in Taiwan (Tang and Su 1978). It excludes *D. tosaense* Makino (now regarded as a synonym for *D. catenatum*) which occurs in Taiwan and, as noted above, later discovered among *Huoshan Shihu* in Shanghai (Tang and Cheng 1984).

The quality of *shihu* (by that one possibly implies similarity with the Shen Nong original) is sometimes defined by the area from which it was collected. Some idea of the likely species may also be gleaned from the prefixes of the herb

Fig. 10.50 *Shihu* in a Wholesale Herb Market in Guangzhou, Guangdong Province, China, 2013. [PHOTO: E.S. Teoh]

Fig. 10.51 Common presentations of *Shihu* in herbal outlets. (*left*): ear ring form; (*right*): stick form. [PHOTOS: E.S. Teoh]

Table 10.1 Distribution of *Dendrobium* species in China based on data culled from Chen and Tsi (1998)

Anhui Province

D. moniliforme

D. officinale (= *D. catenatum*)

Fujian Province

D. moniliforme

D. officinale (= *D. catenatum*)

Gansu Province

D. moniliforme

Guangdong

D. aduncum

D. densiflorum

D. lindleyi (not found in Yunnan: but see *D. jenkinsii*)

D. loddigesii

D. moniliforme

Guangxi

D. aduncum

D aphyllum

D. chrysanthum

D. denneanum

D. devonianum

D. falconeri

D. fimbriatum

D. gibsonii

D. guangxiense

D. hancockii

D. henryi

D. lindleyi

D. lituiflorum

D. loddigesii

D. longicornu

D. moniliforme

D. nobile

D. officinale (= *D. catenatum*)

D. williamsonii

Guizhou Province

D aduncum

D aphyllum

D. denneanum

D. devonianum

D. fimbriatum

D. guangxiense

D. hancockii

D. henryi

D. lindleyi

D. loddigesii

D. moniliforme

D. nobile

Hainan Province

D. denneanum

(continued)

Table 10.1   (continued)

D. densiflorum

D. lindleyi

D. nobile

D. salacense

D. williamsonii

Henan Province

D. hancockii

D. moniliforme

Hubei Province

D. hancockii

D. moniliforme

D. nobile

Hunan Province

D. falconeri

D. hancockii

D. henryi

D. moniliforme

Jiangxi Province

D. moniliforme

Sichuan Province

D. hancockii

D. moniliforme

D. officinale (= *D. catenatum*)

Taiwan

D. denneanum

D. falconeri

D. miyakei

D. moniliforme

D. nobile

Yunnan Province

D. aduncum

D. aphyllum

D. bellatulum

D. bryamerianum

D. capillipes

D. cariniferum

D. christyanum

D. chrysanthum

D. chrysotoxum

D. crystallinum

D. denneanum

D. devonianum

D. falconeri

D. fimbriatum

D. finlayanum

D. gibsonii

D. gratiosissimum

D. guangxiense

D. hancockii

D. harveyanum

(continued)

Table 10.1 (continued)

D. henryi
D. heterocarpum
D. infundibulum
D. jenkinsii
D. lituiflorum
D. loddigesii
D. longicornu
D. minutiflorum
D. moniliforme
D. moschatum
D. nobile
D. officinale (= *D. catenatum*)
D. pendulum
D. primulinum
D. salacense
D. stuposum
D. sulcatum
D. thyrsiflorum
D. trigonopus
D. wardianum
D. williamsonii
Xizang Province (Tibet)
D. chrysanthum
D. densiflorum
D. devonianum
D. longicornu
D. moniliforme
D. nobile
D. salacense
Zhejiang Province
D. moniliforme
D. officinale (= *D. catenatum*)

(Table 10.2). A few species are listed under two categories in this list.

Commercial classification of *shihu* in Yunnan was observed in 1990 to be very complicated. Here, additional classifications (descriptions) included *Xian huangcao* (fresh *Dendrobium*), *Xi huangcao* (slender *Dendrobium*), *Cu huangcao* (thick *Dendrobium*), *Cha-huangcao* (branch *Dendrobium*) and *Xianggun huangcao* (stick *Dendrobium*) (Zheng 1990). *Shihu* plants belonging to the Section *Flickingeria* would thus be classified as *Cha-huangcao* (branch *Dendrobium*) by nature of its pseudobulbs being widely separated on a branching rhizome.

Table 10.2 Market Classification of *Shihu* (Hew et al. 1997; Bao et al. 2001)

***Jinchai shihu* (Golden Hairpin *Dendrobium*)**
D. linawianum Rchb. f.
D. nobile Lindl;

***Er Huan Shihu* (Earring *Dendrobium*)**
D. candidum Wall ex Lindl. (= *D. moniliforme*)
D. hercoglossum Rchb. f.
D. moniliforme (L.) Sw.

***Ma pien Shihu* (Horse Whip *Dendrobium*)**
D. chrysanthum Wall ex. Lindl
D. denneanum Kerr
D. fimbriatum Hook. var *oculatum*
D. hancockii Rolfe

***Huangcao Shihu* (Yellow Herb *Dendrobium*)**
D. aduncum Wall ex. Lindl.
D. crepidatum Lindl. ex Paxton
D. devonianum Paxton
D. houshanense Tang et Cheng (= *D. catenatum* Lindl.)
D. loddigesii Rolfe
D. lohohense Tang et Wang
D. tosaense Makino (= *D. catenatum* Lindl.)
D. wilsonii Rolfe

Yue ku Shihu (Melon *Dendrobium*)
D. bellatulum Rolfe
D. catenatum Lindl. (syn. *D. officinale* Kimura et Migo)
D. crispulum Kimura et Migo
D. chryseum Rolfe (syn. *D. flaviflorum* Hayata)
D. crumenatum Sw. (syn. *D. ceraia* Lindl.)
D. kwangtungense Makino (= *D. wilsonii* Rolfe)
D. lindleyi (syn. *D. aggregatum*)
D. pulchellum Roxb. ex Lindl.
D. thyrsiflorum B.S. Williams
Flickingeria fimbriata (Bl.) Hawkes (= *D. macraei* Lindl.)
Ephemerantha lonchophylla Hook. f. (= *D. xantholeucum* Rchb. f.).

Golden, pink and white flowers predominate among the medicinal *Dendrobium*, and an untrained observer may find difficulty in distinguishing some species from others. The confusion is compounded when the plant is not in bloom. To cope with the twin problems of substitution and contamination, it was earlier proposed that morphological details be supplemented with histological characteristics and TLC spectra to facilitate proper identification (Zhao et al. 1998). This has now been superceded by the various PCR procedures which allow for a far more precise identification (Lau et al. 2001; Xu et al. 2001, 2006; Qian et al. 2008). While PCR elicits correct species identification, by itself correct identification does not ensure the quality of the product. Factors like source of supply (Bai et al. 2007), host tree (Guo et al. 2014), age of the plant, season of harvest (Table 10.3, Ding et al. 1998), portion of the stem (Chen et al. 2001), processing (Wu et al. 2007), and storage all affect the final quality (Bai et al. 2007; Ding et al. 1998). Capillary electrophoretic fingerprinting has also been proposed as a means of identification (it has 15 characteristic peaks). *D. candidum* (= *D. moniliforme*) can be classified according to source and quality by this process (Zha et al. 2009).

Currently, between 32 and 40 species are accepted as *shihu* but there is no consistency as to which species constitute *shihu* in published texts, while many species do not occur as saxicolous plants. For instance, *A Coloured Atlas of Compendium of Materia Medica* (1998) leaves out *D. moniliforme* but includes *D. nobile* and *D. fimbriatum* var *oculatum* under *shihu* (Shen 1998). The *New Century Chinese-English Dictionary of Traditional Chinese Medicine* published by the People's Military Medical Press in Beijing lists only five species of *Dendrobum* under *shihu*, namely *D. nobile*, *D. loddigesii*, *D. candidum* (= *D. monililiforme*), *D. chrysanthum* and *D. fimbriatum* var. *oculatum* (Anonymous 2004). Presumbly, these are the commoner species used by the People's Liberation Army. *The Chinese Medical Encyclopaedia Vol 1 (Zhongyao Da Cidian)* published in Shanghai in 1986 lists 11 *Dendrobium* species under *shihu*, namely *D. nobile*, *D. linawianum*, *D. officinale* (= *D. catenatum*), *D. moniliforme*,

D. hercoglossum, *D. aduncum*, *D. wilsonii*, *D. hancockii*, *D. lohohense*, *D. loddigesii* and *D. bellatulum*. Molecular studies indicate that many of these species are closely related (Wongsawad et al. 2005). The list compiled by Wu Xiu Ren in Guangdong in 1994 is longer and includes *D. acinaciforme*, *D. aduncum*, *D. aphyllum*, *D. bellatulum*, *D. candidum* (= *D. moniliforme*), *D. cariniferum*, *D. chrysanthum*, *D. crepidatum*, *D. crystallinum* var. *hainanense*, *D. denneanum*, *D. densiflorum*, *D. devonianum*, *D. falconeri*, *D. faulhaberianum*, *D. fimbriatum*, *D. hancockii*, *D. henryi*, *D. hookerianum*, *D. jenkinsii*, *D. linawianum*, *D. loddigesii*, *D. lohohense*, *D. moniliforme*, *D. nobile*, *D. sinense*, *D. tosaense*, *D. wangii* (= *D. hercoglossum*) and *D. wilsonii* (altogether 28 species). A new species, *D. huoshanense*, which occurs in Anhui province and was newly discovered in 1984 was also regarded as *shihu* (Tang and Cheng 1984); it is now considered to be synonymous with *D. catenatum*. However, another saxicolous species, *D. guangxiense* S.J. Cheng et C.Z. Tang, which occurs in Guangxi, Guizhou and Yunnan (Chen et al. 1999b) was not incorporated into medicinal usage. In Vietnam, *D. dalhousieanum* (*D. pulchellum*; Vietnamese names: *Hoang thao; po len; co anh*) and *D. gratiosissimum* (Vietnamese names: *Hoa thao; Hoang thao*) are accepted as substitutes for *Den. nobile* (Doung 1993). An expert from the Shanghai University of Traditional Chinese Medicine stated that 32 species of orchids carry the name "*Huangcao Shihu*" in the herbal medicine market (Xu et al. 2006). *D. nobile* is the commonest *Dendrobium* species marketed as *shihu* in China and Vietnam (Doung 1993). Several *Dendrobium* species not previously considred as *shihu* are now offered as substitutes

Table 10.3   The time of harvest influences the efficacy of *shihu*

Therapeutic Index for newly planted *D. candidum*				
Therapeutic function	Year 1	Year 2	Year 3	Year 4
Curing throat/eye diseases	5.69	9.54	17.9	22.2
Boosting immunity	11.2	14.6	21.6	17.6
Total (average)	8.44	12.1	19.7	19.9

Data from Ding et al. (1998)

Conclusion: Taking both indices into consideration, the best time for harvest is in the autumn of the third year for *Huangcao shihu* in Xishuangbanna, Kunming Province. They were identified by rDNA ITS sequence analysis as *D. brymerianum*, *D. capillipes*, *D. ellipsophyllum*, *D. exile*, *D. salaccense* and *D. williamsonii*. There is concern that they may cause inconsistent therapeutic effects or even jeopardise the safety of *shihu* (Xu et al. 2005).

The various provinces in China enjoy different plant resources and sometimes each has its own classification of medicinal herbs. For instance, Yunnan Province which has the highest number of *Dendrobium* species proposed its own complicated classification of *shihu* which differed from the system used in Guangdong Province. The principal Yunnan commodities include (1) *Xifengduo* (spiral *Dendrobium*) which is divided into three classes; (i) *Huangcao* (yellow herb) which is subdivided into (a) *Xi-huangcao* (slender yellow herb), (b) *Cu-huangcao* (thick yellow herb) (c) *Bian huangcao* (goblet yellow herb) (d) *Xiaogua-huangcao* (small melon yellow herb) and (e) *Xian-huangcao* (fresh yellow herb). (2) Minor *Huangcao* consists of (i) *Cah-huangcao* (branch yellow herb), (ii) *Xianggun –huangcao* (stick yellow herb) and (iii) *Yougua huangcao* (melon yellow herb). A new commodity is (3) *Diaolan fengduo* (rough spiral *Dendrobium*) (Zheng 1990). In Sichuan Province, among the 11 species marketed as *shihu*, there are two new species whose medicinal effects were only described in 1995 (Li and Xiao 1995).

There are altogether 63 (Li et al. 2005) or 74 (Xu et al. 2006) species of *Dendrobium* in China, and with all this confusion, effort is now being directed towards correct identification of species using DNA probes and species-specific primers as a first step in quality control (Ding et al. 2002a, b). Diversity in DNA sequences among species ranged from 3.2 to 37.9 %, but variations within species were very low ranging from 0 to 3 % in ITS1 (internal transcribed spacer 1) and 0–4 % in ITS2 (Xu et al. 2006). Researchers at the Chinese University in Hong Kong found that inter-species variation of DNA sequences of ITS2 in 16 medicinal *Dendrobium* species was 12.4 % while intra-species variation

was only 1 % (Lau et al. 2001). They also concluded that ITS2 regions could be adopted as a molecular marker to differentiate among the different varieties of shihu, and also to detect non-orchids and other adulterants. DNA microarray managed to detect the presence of *D. nobile* in a Chinese herbal formulation containing nine herbal components (Zhang et al. 2003b).

A study of Random Amplified Polymorphic DNA (RAPD) markers in eight wild populations of *D. officinale* showed that there were distinct genetic differences and extensive genetic diversity among the wild populations, and that Primer S412 could be used to authenticate the eight wild populations completely (Ding et al. 2005). Allele-specific diagnostic PCR, using TP-JBO1S and TP-JBO1X for *D. officinale*, for instance, "is not only simpler and time-saving but also practical and effective" (Ding et al. 2004). Allele-specific diagnostic PCR for *D. thyrsiflorum* employing primers QH-JB1 and QH-JB2 also proved to be simple and fast (Ying et al. 2007). Use of trinucleotide microsatellite markers also help to identify the geographic origin of a species (Yuan et al. 2011b). However, each species needs its own primer for exclusive identification (Zhang et al. 2001b; Ding et al. 2002a,b,c, 2005, 2008, 2009; Li et al. 2005; Shen et al. 2006a; Xu et al. 2006; Qian et al. 2008). At Hangzhou Normal University, Wang and his associates has studied 31 *Dendrobium* species from Yunnan with ISSR markers. Altogether, 2369 bands were amplified by 17 ISSR primers, resulting in 278 ISSR loci with 100 % polymorphism at the genus level. The 31 species were unequivocally distinguished by ISSR fingerprinting even though species-specific markers were identified in only nine species (Wang et al. 2009a). A dot-blot hybridization assay which is based on species-specific amplified fragments derived from the ITS region of different *Dendrobium* species blotted as dots on a nylon membrane has recently been proposed as a superior process for identification of *Dendrobium* species in terms of speed, sensitivity and specificity (Xu et al. 2010). Besides being used to authenticate populations of the rarer *shihu* species like *D. catenatum* (Shen et al. 2006b), genetic fingerprinting is also useful in determining genetic diversity in various species of medicinal *Dendrobium* (Wang et al. 2007a) and for purity identification of germplasm (Xie et al. 2010), as well as analysis of genetic stability at various stages of development of a particular clone (Liu et al. 2007a).

Uses of *Shihu:* Chinese people in many parts of the world resort to traditional Chinese medicine (TCM) for cures and for tonics to improve their quality of life. *Shihu* is a popular tonic and it has a long history of use outside China (Hooper 1929). Of the over 300 Patent Chinese Herbal Medicines sold in the United States in 1999 by about a dozen companies, *Herba Dendrobii* (*shihu Dendrobium*) was included in seven and *Tianma* (*Gastrodia elata*) in five preparations. Most of the preparations were based on traditional formulae (Liu et al. 1999). Given the inordinate consumption of vitamin supplements by affluent, healthy Chinese with an adequate diet, it is not surprising that the natural existence of many such *Dendrobium* species is under threat. Hidden in the Chinese psyche is the need to invigorate oneself not solely through healthy lifestyle but additionally by the consumption of a tonic or supplement.

Tonics are substances which are thought to help in the recovery of those who have been ill, injured, stressed, or suffering from a degenerative condition. *Dendrobium* stems which are sweet to taste and regarded as "slightly cold", provide a *yin* tonic that nourishes the stomach (meridian), promotes the production of body fluids and removes "heat" (Anonymous 2004). They are used in combination with other herbs. The role of the tonic is to support the normal, whereas the function of the additional herbs is to eliminate the pathologic. Without the additional herbs, tonics may exacerbate rather than suppress an illness. This results in a lingering unease. Tonics are therefore not prescribed during the early stages of an illness. Chinese medical practitioners further caution that when a person is basically not deficient, tonics may introduce a range of problems that include indigestion, rash and irritability.

During the early 1970s, books on medicinal plants were distributed under Mao Zedong's directive to improve rural health care. In consequence, medicinal plant knowledge among the Bai minority living in western Yunnan came to be strongly influenced by mainstream Chinese traditional medicine, and this knowledge is much alive and practised today. Six species of orchids consisting account for 3 % of the medicinal plants used. *D. moniliforme* (and apparently this species alone) was used in decoctions as a kidney tonic by the Bai minority living in eastern Yunnan, and also to strengthen bones and tendons and to stave off diabetes. The *Dendrobium* was cultivated in gardens to provide a readily available, inexpensive and sustainable supply (Weckerle et al. 2009).

Shihu is used to quench thirst, increase salivation, improve eyesight, reduce fever and settle the stomach. It is administered to people suffering from fever, thirst, malaise and excessive perspiration. New scientific findings appear to justify this usage.

Aquaporins are endogenous membrane proteins which selectively transports water across cells. They regulate urine production by the kidneys and fluid secretion by other glands. Aquaporin-5 plays an important role in the production of tears, saliva and fluid in the lungs. *D. candidum* (= *D. moniliforme*) extract increases the expression of aquaporin-5 in labial glands from patients with Sjogren's syndrome (Xiao et al. 2011), a condition characterised by dry eyes and dry throat, and in which abnormal distribution of aquaporin-5 (loss of apical localization) has been demomstrated (Tsubota et al. 2001; Steinfield et al. 2001). The other classic *shihu*, *D. catenatum*, activates M3 muscarinic receptors and induces extracellular calcium influx which leads to translocation of aquaporin-5 to the apical membrane of human salivary cells, and thus it also promotes salivary secretion (Lin et al. 2015). *Shihu* may help patients with Sjogren syndrome; however, up-regulation of aquaporin-5 in patients with lung, breast, stomach, colo-rectal, and prostate cancer favours spread of the disease resulting in

Table 10.4 A sampling of prescriptions containing *Shihu*. Reference: (*Zhongyao Da Cidian* 1986)

1. For fever with sweats and dark coated tongue: Fresh *Shihu* 15 g, *Forsythia supensa* (莲翘) 15 g, *Trichosanthes kinlowii* (天花粉) 10 g, *Rehmannia glutinosa* (生地) 20 g, *Linope graminifoia* (麦冬) 20 g. Boil with water. (Source: *Treatise of Current Diseases*)

2. For "gastric fire" surging and causing depression and fear: *Shihu* 50 g, *Scrophularia oldhami* (玄参) 10 g. Boil with water. (Source: *Record of Evidence*)

3. For thirst after recovery from illness: fresh *Shihu* (*D. candidum*) 9 g, *Linope graminifolia* 9 g, *Schizandra chinensis* (五味子) 9 g. Boil with water and drink as tea.

4. For lung heat and dry coughs: fresh *Shihu* (*D. candidum*) 9 g, juice of *Enboti ya japonica* (枇杷) 9 g, *Glycyrrhiza globra* (甘草) 3 g, *Platycodon grandiflorum* (桔梗) 3 g. Boil with water. (Source: *Record of Herbs of Zhejiang Province*)

5. For poor night vision: *Shihu* 50 g, *Epimedium macranthum* (仙灵脾) 50 g, *Atractylodes japonium* (苍术) 25 g, ground to fine powder. Take 15 g, twice a day. (Source: *Sheng Ji General Record*)

6. For weak *yin qi*, backache, urinary frequency, spontaneous ejaculation, wet and pruritic scrotal skin, equal portions of *Shihu, Herpestis monniera* (巴戟天) *Tenodera sinensis* (桑螵蛸), *Euonymus chinensis* (杜仲). Make mixture in pill form. Take 10 pills with wine, twice a day. (Source: *Record of Tested Prescriptions*)

poor outcomes (Shen et al. 2010; Chae et al. 2008; Lee et al. 2014; Shan et al. 2014; Li et al. 2014c), whereas down-regulated aquaporin 5 inhibits proliferation and migration of human epithelial ovarian cancer 3AO cells in vitro and in vivo (Yan et al. 2014). Other aquaporins, AQP 1 and AQP 3, also facilitate tumour growth and migration (Verkman et al. 2008). Therefore, cancer patients should avoid consuming *shihu* which up-regulates aquaporin.

Table 10.4 features a sampling of TCM prescriptions that contain *shihu* and the indications for its use. This was translated from the *Zhongyao Da Cidian*. Note that the older prescriptions specify *D. candidum* (= *D. moniliforme*).

TCM practitioners maintain that the various species are not equal in their effects. Indeed, their chemical constituents are different. According to one source, "*D. candidum* (*tie pi shi hu*) is the most effective in enriching *yin,* generating fluids and eliminating heat. *D. nobile* (*jin chai shi hu*) is weaker although it contains six different classes of alkaloids in its leaves and stems. Earring *Dendrobium* (*er huan shi hu*; from the separated young stems of *D. chrysanthum* or *D. candidum*) generates fluids, but it is not cool in nature; it can be taken daily as a tea" (Bensky et al. 2004). *D. crispulum* Kim & Migo and *D. officinale* Kim & Migo were also grown specifically for use as tonics in Japan (Kimura and Migo 1936).

Chinese in Indochina regard *D. dalhousieanum* Wall, *D. gratiosissimum* Reichb.f., and *D. pulchellum* Roxb. as medicinal herbs or *shihu* (Petelot quoted by Perry and Metzger 1980). In Vietnam, *D. dalhousieanum, D. gratississimum* and *D. nobile* are used in a similar manner (Nguyen Van Duong 1993).

From a chemical standpoint, *D. nobile* and *D. moniliforme* contain the largest number of sesquiterpenoids. Cyclocopacamphane, cadinene, emmotin and muurolene are found mainly in *D. nobile*. The two related species, *D. densiflorum* and *D. thyrsiflorum*, are rich in coumarins. Polysaccharides constitute almost half the dry weight of *D. officinale* (Xu et al. 2013).

The source of the herb has traditionally had a bearing on its alleged potency. This has now been well demonstrated. Crude polysaccharides extracts of *D. officinale* (= *D. catenatum*) varied greatly in their ability to stimulate cytokine production in macrophages. At a concentration of 300 µg/ml, levels of IL-1a secreted was 108.2 pg/ml with the Yunnan product, 132.8 with the Anhui product and only 300 from the Zhejiang product. Corresponding levels for IL-6 were 5712 for Yunnan, 6-oxydendroxine 533.7 from Anhui and 58.4 from Zhejiang; for IL-10 were 858.7 from Yunnan, 457.3 from Anhui, and 117.9 from Zhejiang; and for TNF-alpha were

17,711 from Yunnan, 15,940.8 from Anhui and 16,377.8 from Zhejiang. Maximum phagocytosis was observed with crude polysaccharides of *D. officinale* (= *D.* catenatum) from Yunnan ((Meng et al. 2013). Species differences were also clearly evident. Crude polysaccharide of *D. nobile* from Yunnan at identical concentrations evoked minimal cytokine production in macrophages: 32.3 for IL-1 alpha, 8 for IL-6 and 7.5 for IL-10; minimal secretion of IL-1alpha but no IL-6 or IL-10 with *Dendrobrium chrysotoxum* polysaccharide; barely detectable cytokine production IL-10 secretion with polysaccharide of *D. fimbriatum* from Yunnan; and no significant cytokine production with polysaccharides of *Dandrobium chrysotoxum* and *D. fimbriatum* from Yunnan (Meng et al. 2013).

D. catenatum (syn. *D. officinale*), being one of two original shihu, has been selected as a gold standard for assessing the quality of *shihu*. The *Chinese Pharmacopoeia* (2010) states that the ratio of mannose to glucose in *D. officinale* should be 2.4:8.0. This standard is also met by *D. aphyllum* and *D. crystallinum*. Capillary electrophoresis and enzymatic fingerprints of polysaccharides were also used to identify *D. candidum* (= *D. moniliforme*) and *D. officinale* (= *D. catenatum*), respectively (Zha et al. 2009, 2012).

For the purpose of chemical identification, Chen and colleagues from the Chinese Academy of Medical Sciences and Peking Union Medical College recommend that, in addition to polysaccharide profiling, it is necessary to add the discriminatory power of narigenin, moscatilin, gigantol and 3,4'-dihydroxy-4'5-dimethoxy-bibenzyl (DBB-2). This would raise the quality control standard of *D. catenatum*. All compounds can be studied by HPLC analysis (Chen et al. 2012). Infrared spectroscopy was recently proposed as a simple and rapid method for discriminating between different species of *shihu* *Dendrobium* (Luo et al. 2013).

Alkaloids of *Dendrobium*
De Waldemann (1892) observed a high alkaloid presence in *D. nobile* when he undertook the first investigations on orchid alkaloids. But it was not until 40 years later that Suzuki and co-workers (1932) managed to obtain a crystalline alkaloid from '*Chin Shih Hu*', a herbal preparation derived from dried stems of *D. nobile*.

Additional alkaloids, dendroxine, 6-oxydendroxine and dendramine, were next isolated from *D. nobile* (Okamoto et al. 1966a, b). Subsequently, alkaloids were isolated from other species of *D.* (e.g. *D. linawianum, D. findlayanum, D. fredricksianum, D. hildebrandii*) (Leander and Luning 1968; Hedman et al. 1971; Eleander et al. 1973; Luning 1974a, b). Another 40 years passed before the correct structure of dendrobine was defined. Inubushi et al. (1972) and Yamada et al. (1972) independently succeeded in synthesising the alkaloid. Plant chemistry has since progressed by leaps and bounds, and much is known about alkaloids and other steroidal compounds produced by orchids. Much of the work still centres on *Dendrobium*. Since alkaloids are also present in hybrids of medicinal *Dendrobium*, the use of hybrid *Dendrobium* opens a vast resource for the preparation of *shihu* (Morita et al. 2000).

In a survey of the alkaloids of the *Orchidaceae*, Bjorn Luning found that 214 species in 64 genera had an alkaloid content of 0.1 % or greater, out of 2044 species in 281 genera tested (10.47 of the species tested positive; 29.36 % of the genera tested positive). Only 8.33of the *Dendrobium* species out of 384 species studied had alkaloid content amounting to 0.1 % or greater. Large amounts of alkaloid were found in numerous species referred to as *shihu* [*D. nobile, D. liniawanium, D. hildebrandii, D. fredericksianum, D. wardianum, D. crepidatum, D. pierardii* (= *D. aphyllum*), *D. chrysanthum, D. lohohense, D. primulum, D. parishii, D. anosmum*] and other species that belong to the Section *Dendrobium*, what Howard P. Wood (2006) refers to as the northern clade of the genus. Members of this section of *Dendrobium* require a distinct dry season to thrive and flower well, and, presumably, to also concentrate their alkaloids. But not every *Dendrobium* with high alkaloid content is used in TCM: for instance, *D. fredricksianum* and *D. hildebrandii*, possibly due to their slightly more southerly distribution outside China, are not used as *shihu*. *D. findlayanum* with uncharacteristic pseudobulbs is also not used as *shihu* but it contains three alkaloids, namely, dendromine, 2-hydroxydendrobine and nobilonine (Granelli et al. 1970), and a new picrotoxane type sesquiterpene, findlayanin (Qin et al. 2011). Only traces of alkaloid (0.001–0.01 % dry weight) is present in the two major *shihu* species, *D. moniliforme* (syn. *D. candidum*), *D. catenatum* and several other *shihu* species, namely, *D. acinaciforme*, *D. capillipes*, *D. chrysotoxum*, *D. dalhousianum* (=*D. pulchellum*), *D. devonianum*, *D. falconeri*, *D. hercoglossum*, *D. jenkinsii*, *D. thyrsiflorum*, *D. transparens*, *D. williamsonii*, and other non-*shihu* but medicinal species, namely *D. amoenum*, *D. crumenatum*, *D. cumulatum*, *D. densiflorum*, *D. heterocarpum*, *D. hookerianum* and *D. moschatum* (Luning 1964). Total and individual alklaoid concentrations between species, for example, *D. moniliforme* and *D. nobile*, are different (Chen et al. 2006).

Bjorn Luning listed 15 alkaloids isolated from *Dendrobium* before 1980 as follows: dendrobine (1), 2-hydroxy-dendrobine (2), nobilonine (3), dendrine (4), dendrowardine (5), dendroxine (6), dendroprimine (7), dendrochrysine (8), dendrocreptine (9), crepidine (10), crepidamine (11), hygrine (12), dendroparine (13), pierardine (14) and shihunine (15). Their structures are illustrated in Table D-3. Dendrobine, a sesquiterpenoid alkaloid, is the most abundant alkaloid from *D. nobile*, isolated in 1932 and subsequently thoroughly investigated. The 13 alkaloids and other miscellaneous alkaloids from *Dendrobium* are well reviewed by Slaytor (1977) in Arditti's *Orchid Biology, Reviews and Perspectives* Vol. 1. By 2003, 32 alkaloids had been isolated from 42 species of *Dendrobium* (Zhang et al. 2003a) and the work continues. More new alkaloids have since been isolated from *Dendrobium*: dendronobiline-A, a new dendrobine-type alkaloid from *D. nobile* (Liu and Zhao 2003); moniline from the leaves and stems of *D. moniliforme* (Liu et al. 2007b).

Several alkaloids present in *D. nobile* are also present in non-*shihu* species. Additionally, a few unique alkaloids have been isolated from *D. anosmum*, *D. findlayanum*, *D. friedricksianum*, *D. hildebrandii* and *D. wardianum*., species which are currently not used as *shihu* (Luning 1974a, b) (Fig. 10.52).

Alkaloids of *Dendrobium* can be classified into five types, namely sesquiterpenoids, indolizidine, pyrrolidines, phthalides and imidazoles, the last two being rare (Xu et al. 2013). The isolation of new picrotoxinin-type and dendrobine-type sesquiterpenoids from the hybrid *Dendrobium* Snowflake 'Red Star' opens a new direction to employ *Dendrobium* for the production of novel alkaloids, and possibly some of them might find a medicinal use. The new alkaloids are called flakinins A and B and mubironines A, B and C, respectively (Morita et al. 2000). *Dendrobium* Snowflake is a tetraploid hybrid that has 75 % *D. nobile* and 25 % *D. moniliforme* in its makeup. It was registered with the Royal Horticultural Society in 1904 and is not a new hybrid.

Alkaloids-enriched extract of *D. nobile* when fed to rats was found to be capable of attenuating tau protein hyperphosphorylation and apoptosis induced by direct injection of lipopolysaccharide into the animals' brain (Yang et al. 2014c). Whether this will usher in a new application for using *shihu* remains to be seen, but the findings will possibly boost the selling point of the herb.

Phylogenetic studies involving ITS and *matK* analyses revealed that *D. moniliforme* and *D. nobile* are closely related and were designated as Clade D1 by Wongsawad et al. (2005). It would be interesting to know whether hybrids between members of the Section *Callista* (e.g. *D. densiflorum*, *D. thyrsiflorum*) and members of other related Sections that have long-lasting flowers (*Cuthbertsonia*, *Pedilopnum* and *Oxyglossum*) are capable of producing higher yields of alkaloids in their canes. Tetraploidy may have contributed to *Dendrobium* Snowflakes' ability to synthesise more alkaloids than its diploid parent *D. moniliforme* (Karasawa and Shimai 2002). This phenomenon is

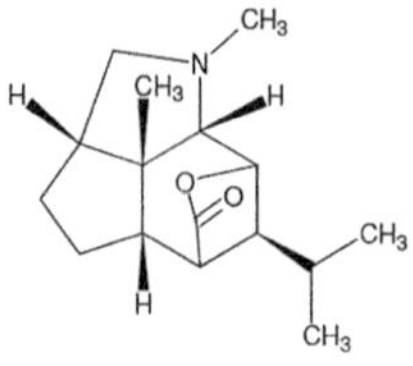

Fig. 10.52 Examples of alkaloids from *Dendrobium*

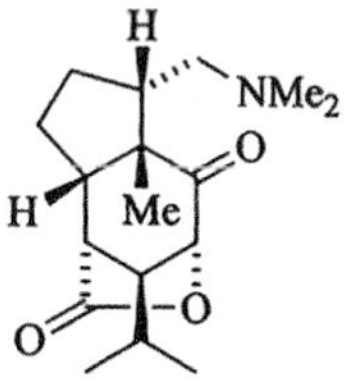

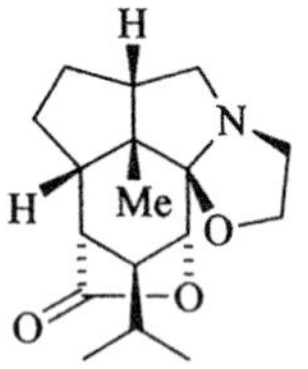

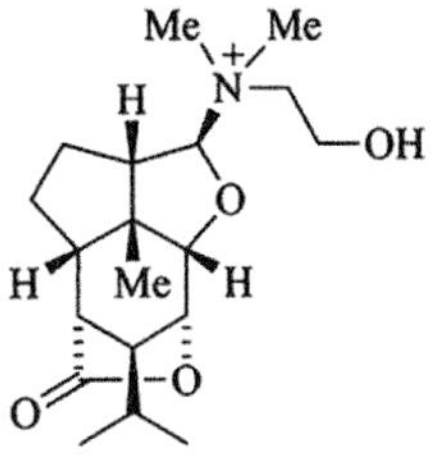

worthwhile examining because it would be a simple approach to producing *shihu* of stronger potency. Ultrasound can be used to optimise the yield of alkaloids during the extraction process (Liu and Dong 2013).

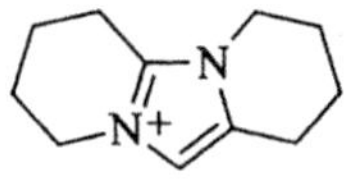

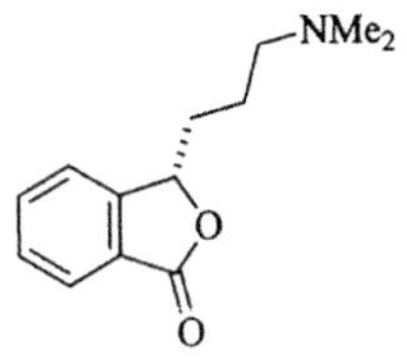

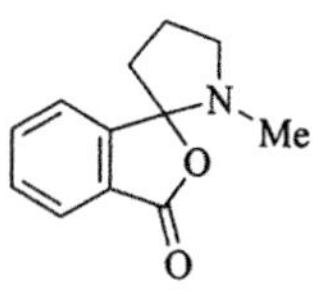

Genetic mapping has also been proposed as a way of identifying which *Dendrobium* species or their hybrids carry genes which enable them to produce medicinally useful compounds in quantity. A study in this direction was started by Lu and his colleagues at Nanjing Agricultural University who studied SSR and SRAP markers in 150 seedlings which they raised by breeding *D. officinale* with *D. aduncum* (Lu et al. 2012).

Chinese and Japanese doctors who studied *shihu* concluded that one of its components, dendrobine, has a feeble pain-relieving action but it had no effect on fever. (Here, it should be noted that the TCM term "heat" does not equate with fever.) Dendrobine did not appear to have any significant effect on the circulation, apart from a slight fall in blood pressure. It promoted the secretion of saliva and caused a rise in blood sugar. It provoked severe contraction of uterine muscles, and progressively paralysed intestinal muscles. In mice and rabbits, a large dose resulted in convulsions, paralysis and death (Chen and Chen 1935a, b).

Phenanthrenes and Bibenzyls

Denbinobin (5-hydroxy-3,7-dimethoxy-1,4-phenanthraquinone) was isolated from *D. nobile* in 1982 (Talapatra et al. 1982). Thirteen years later, Lee et al. (1995) reported that denbinobin exerted cytotoxic effects on A549 human lung cancer, SK-OV-3 human ovarian adenocarcinoma and HL-60 human promyelocyte leukaemia in vitro. However, animal studies were not forthcoming, and no clinical study was conducted. Meanwhile, a concise method was developed to synthesise the compound (Wang et al. 2005). With the availability of a pure denbinobin in quantity, it was then possible for scientists to conduct more pharmacological studies on the compound (Fig. 10.53).

Denbinobin was shown to suppress proliferation of colon cancer cells (COLO 205) in vitro in a dose-dependent manner. Denbinobin treatment

Fig. 10.53 Compounds with promising anti-tumour activity from *Dendrobium*

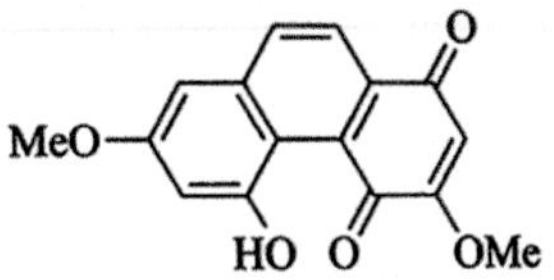

denbinobin

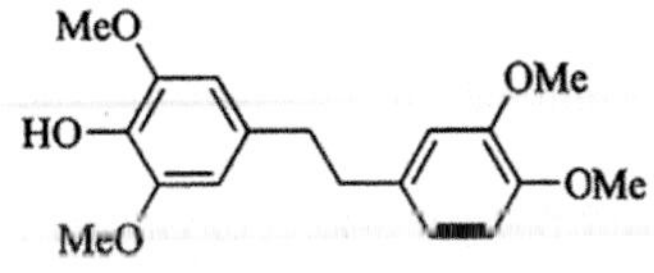

chrysotoxin

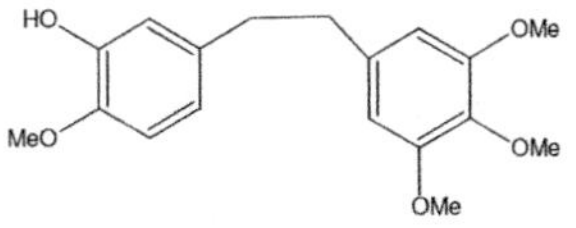

Erianin

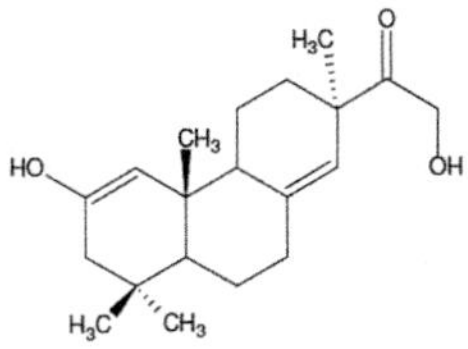

lonchophylloid A

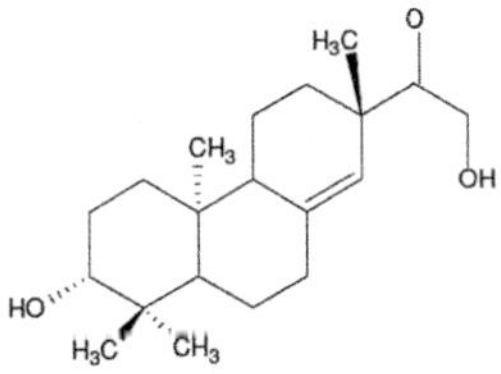

lonchophylloid B

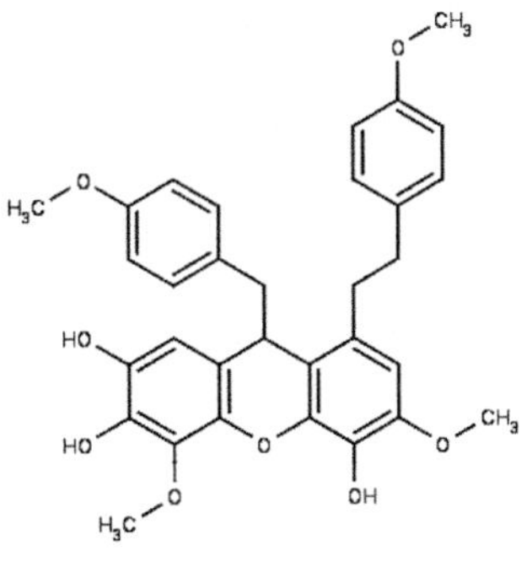

dendrofalconerol A

activated caspases 3, 8, 9 and Bid protein and translocation of apoptosis-inducing factor. In nude mice xenografts, regression of tumour of up to 68 % was observed (Yang et al. 2005a). Denbinobin-mediated cytotoxic effect on human leukemia cells (K562) was shown to be mediated by enhancing tubulin polymerisation and suppressing Bcr-Abl activity and phosphorylation of CrkL, which is a crucial adaptor protein in chronic myeloid leukaemia. It inhibited cancer cell viability in a dose-dependent manner with IC50 of 1.84 microM. Long-term treatment resulted in marked expression of CD11b (Huang et al. 2005).

Studying human colorectal cancer HCT-116 cells, Chen et al. (2008b) confirmed that denbinobin induced apoptosis by apoptosis-inducing factor release and DNA damage in the cancer cells. Denbinobin-induced apoptosis in human lung adenocarcinoma cells (A549) was shown to be mediated via caspase-3 activation, Akt inactivation, Bad activation, and mitochondrial dysfunction. It was effective at doses of 1–20 μM/L and it caused cancer cell death in a concentration-related manner (Kuo et al. 2008). In a xenograft model of antitumour and implant assays, denbinobin was shown to suppress angiogenesis and tumour growth in lung cancer. It suppressed iGF-1 receptors and its downstream pathways which are necessary for angiogenesis and tumour growth (Tsai et al. (2011). Apoptosis signal-regulating kinase 1 may also have a role in mediating apoptosis of human lung adenocarcinoma by denbinobin (Kuo et al. 2009).

Denbinobin exerts an inhibitory effect on the Src-mediated signalling pathways which are elevated in many human cancers and alleged to be involved in breast cancer migration and metastasis. Cell migration of human and mouse breast cancer cells is inhibited by denbinobin in vitro. This effect was removed by transfection of the breast cancer cells with a plasmid coding for Src. In a mouse metastatic model, denbinobin treatment resulted in a significant reduction in tumour volume, tumour metastasis and splenic enlargement (Chen et al. 2011).

Denbinobin inhibits the invasive phenotype of human gastric cancer cells (SNU-4484) by decreasing the expression of matrix metalloproteinases, MMP-2 and MMP-9. It induced programmed cell death by down-regulation of Bcl-2 and up-regulation of Bax (Song et al. 2012a).

The favoured site for metastatsis in prostate cancer, a common cancer affecting older men, is bone. CXCL12 preferentially expressed in bone activates Rac1, which in turn induces formation of lamellipodia by actin polymerisation. Prophorylated coractin, an actin-binding protein,

also plays a role in cell migration. Denbinobin was shown to inhibit Rac 1 activity and cortractin phosphorylation, thus preventing prostate cancer migration (Lu et al. 2014a).

With so many favourable Phase 1 results, denbinobin should be further developed and Phase 2 trials started. Meanwhile, it has been shown that denbinobin has anti-inflammatory properties and could be considered for development into an anti-inflammatory agent (Liu et al. (2011a). On osteoarthritis synoviocytes, denbinobin up-regulates miR-146a expression. Mi146a suppresses inflammatory responses by inhibiting nuclear factor NK-kB activity. (Yang et al. (2014a)

Subsequently, it was demonstrated, again in vitro, that dendronobin diminishes the levels of expression of the decoy receptor-3 and acts synergistically with Fas ligand to induce apotosis (programmed cell death) in a human pancreatic cancer cell line (Magwere 2009). The question that is being asked today is: can dendronobin be developed into an adjuvant to treat cancers that have become drug resistant? We are still a long way from being able to answer that question.

From *D. moniliforme*, Lin et al. (2001) isolated two phenanthraquinones, denbinobin and a second one which they named moniliformin (2,6-dimethoxy-1,4,5,8-phenanthradiquinone). Moniliformin is also formed in cereal by several *Fusarium* species of fungi. and it is toxic on heart muscle and causes enlargement of the ventricles. It is lethal to chicken and ducklings at low dosage (Rabie et al. 1982; Leoni and Soares 2003), and farmers are advised to be wary of discoloured pink or white corn that shows contamination by mould.

Zhao et al. (2003) isolated 8 sesquiterpenes, 4 stilbene derivatives and 2 lignens from the lipophilic fraction of the stems of *D. moniliforme*. In a separate publication the same year, another team from the same organisation, the Shanghai Institute of Materia Medica, announced the isolation of seven sesquiterpene glycosides with copacamophene, picrotoxane and alloromadendrane sesquiterpene aglycons together with three phenolic glycosides from the stems of *D. moniliforme*. The compounds are named dendromonilisides A–D for the four newly-discovered substances, and dendrosides A, C, and F, dendrominiliside E, vanilloloside, and acanthoside B for the known compounds. Dendromonilisides A and C promoted the proliferation of B cells and inhibited the proliferation of T-cells in vitro. Bioassay-guided fractionation of EtOAc-soluble extract of *D. moniliforme* yielded another new phenanthraquinone-type metabolite, 7-hydroxy-5,6-dimethoxy-1,4-phenanthenequinone, which inhibited VHR dual-specificity protein tyrosine phosphatase (DS-PTPase) (Bae et al. 2004). Lo et al. (2004) obtained antioxidant components from a methanolic extract from seedlings of *D. moniliforme* and *D. tosaense* Makino (*D. stricklandianum* Rchb. f.).

Denbinobin, fimbriol B and 2,3,5-trihydroxy-4,9-dimethoxyphenanthrene also exhibit antifibrotic activities possibly by the induction of selective cell death in hepatic stellate cells (HSCs) but not in normal hepatocytes. These compounds may be useful candidates for developing therapeutic agents to prevent the development of liver cirrhosis (Yang et al. 2011).

Two bibenzyls from *D. chrysotoxum* show promise as therapeutic agents. The first is erianin which inhibits angiogenesis resulting in extensive tumour necrosis, growth delay and vascular shutdown in hepatoma and melanoma models (Gong et al. 2004a), and causes programmed cell death of human leukaemia HL-60 cells in vitro (Li et al. 2001). This inspired the development of 23 analogues, one of which, isoerianin, strongly inhibits tubulin polymerisation (Massaoudi et al. 2011). ZJU-6, developed by another team of scientists, shows potent anti-tubulin polymerisation and anti-angiogenic activities (Lam et al. 2011). The second interesting bibenzyl is chrysotoxine which shows neuroprotective activity (Song et al. 2012b).

The other compound worth watching is moniliformediquinone which has been shown to destroy hormone-resistant prostate cancer cells in vitro. Prostate cancer ranks fifth for male cancer death. However, most prostate cancers are slow growing, hormone-sensitive, responsive to female hormone or anti-male hormone treatment, and are generally not fatal. Hormone-resistant and aggressive prostate cancers are a different matter, and it is here that moniliformediquinone may play a useful role in management (Hsu et al. 2014).

Two dimeric phenanthrenes and denthirsinin isolated from *D. thyrsiflorum* exhibit cytotoxicity against HeLa, K-562 and MCE cell lines (Zhang et al. 2005a). Moscatilin isolated from the stems of *D. loddgessi* induced G2 phase arrest in cancers of the placenta, stomach and colon but not in liver cancer (Ho and Chen 2003). It inhibited tubulin polymerisation and triggered the activation of the JNK and mitochondria-involved apoptosis pathway in HCT-116 colon cancer cell line (Chen et al. 2008b). Antitumour activity is also present in *D. formosum* which is not used medicinally (Prasad and Kock 2014).

The search for phenanthrenes and bibenzyl continues. Li et al. (2008) from Ningbo University isolated 6 phenanthrene compounds from *D. candidum* (= *D. moniliforme*), namely, 2,3,4,7-tetramethoxyphenanthrene, nakaharain, 2,5-dihydroxy-3,4-dimethoxyphenanthrene, confusarin, nudol and bulbophyllanthrin. Recently, Hwang et al. (2010) isolated one more new phenanthrene together with nine known phenanthrenes and three known bibenzyls from the stems of *D. nobile*. Five bibenzyl derivatives were isolated from *D. candidum* by a team led by Y. Li at the Institute of Medicinal Plant Development and PUMC. The five compounds have been given the names dendrocandin A–E, respectively (Li et al. 2008, 2009b). Four new bibenzyl derivatives, dendrocandin F-I have now been isolated (Li et al. 2009a). While *D. nobile* is often substituted for *D. candidum* in *shihu*, comparison of their chemical constituents showed that *D. candidum* has "a higher quality" for the desired medicinal effect; however, chemical studies showed that the amount of alkaloids in *Dendroobium nobile* far exceeds that of *D. candidum* (Chen 2006). Administration of *D. candidum* to mice at a dose of 400 mg/kg daily reduced the incidence of lung metastasis by 64.5 % when the animals were injected with 26-M3.1 colon cancer cells (Li et al. 2014b).

Meanwhile, other non-medicinal species of *Dendrobium* are also being investigated in the search for pharmacologically active compounds. Ten phenolic compounds were isolated from *D. ellipsophyllum* and four showed promising cytotoxic activity at non-toxic concentrations (Tanagornmeatar et al. 2014). Compounds with promising cyticToxic activity are shown in Figs. 10.53 and 10.54. Several compounds from *Dendrobium* exhibit anti-inflamatory and pain-relieving properties (Fig. 10.55), immunomodu-latory activity (Fig. 10.56) and anticlotting properties (Fig. 10.57).

Polysaccharides

Chinese centres have also made a considerable effort to study the polysaccharides in *shihu*. A 2-0-acetylglucomannan obtained from the dried stem of *D. catennatum* (syn. *D. officinale*) is composed of mannose, glucose and arabinose in a 40.2:8.4:1 molar ratio. Its structure has been elucidated by Hua et al. (2004). From the leaves and stems of *D. huoshanense*, Hsieh and co-workers isolated a unique mucilage polysaccharide which activated murine splenocytes to produce several cytokines (IFN-gamma, IL-10, IL-6, IL-alpha) and two haemopoietic growth factors, GM-CSF and G-CSF. However, the deacetylated mucilage obtained from alkaline treatment failed to induce cytokine production (Hsieh et al. 2008).

A polysaccharide with molecular weight of 150 kDa isolated from the stems of *D. chrysotoxum* lowered glucose levels in alloxan-induced diabetic mice, stimulated DNA synthesis and cytokine secretion in mouse lymphocytes, and prevented deoxyribose degra-dation by scavenging hydroxyl radicals and by chelating iron ions (Zhao et al. 2007). Four major polysaccharide fractions were obtained by hot water extraction of *D. nobile*, followed by chro-matography on DEAE-cellulose and Sephadex G-200. Their molecular weights vary consider-ably (11.4–136 kDa) and they are composed prin-cipally of mannose, glucose, and galactose with smaller amounts of rhamnose, arabinose and xylose. The polysaccharide with the lowest molecular weight showed promise as a potential anti-oxidant (Luo et al. 2009a, 2010). A water-soluble heteropolysaccharide DNP-W3, which is

Fig. 10.54 Additional cytotoxic phenanthrenes from *Dendrobium*. Chemical structures of moniliformidiquinone, denthyrsin, denthyrsinin denthyrsinol and denthyrsinone could not be located

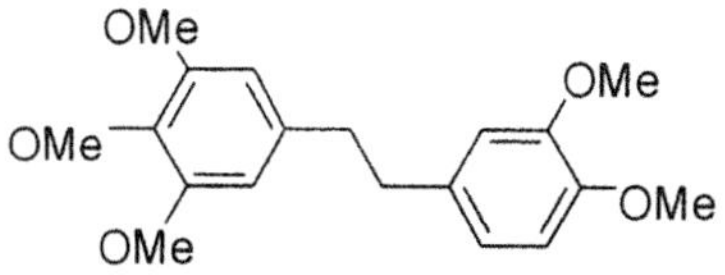

chrysotoxobibenzyl

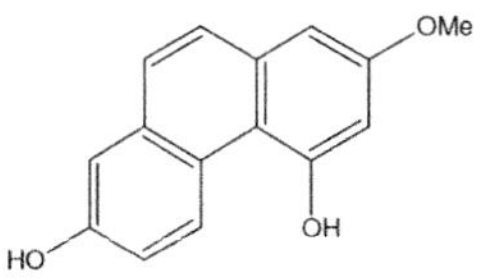

4,7-dihydroxy-2-methixt-9,10-dihydrophenanthrene

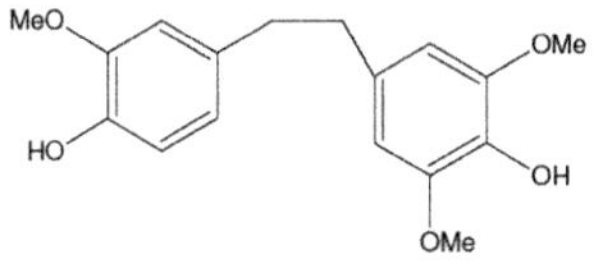

moscatilin

Fig. 10.55 Bibenzyl derivatives from *Dendrobium* with anti-inflammatory or pain-relieving properties

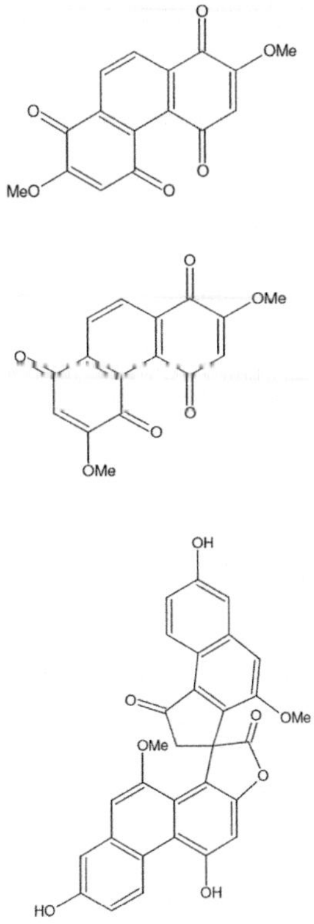

moniliformine

2,6-dimethoxy-1,4,5,8-phenanthradiquinone

dendrochrysanene

a rhamnoarabinogalactan with a molecular mass of 710 kDa isolated from *D. nobile*, stimulated Concanavalin-A (ConA) and lipopolysaccharide (LPS)-induced T- and B-lymphocyte proliferation (Wang et al. 2010a). The team also managed to isolate an acetylated galactomannoglucan from stems of *D. nobile* (Wang et al. 2010b). Although polysaccharides from different species of *Dendrobium* share several similar characteristics, there is sufficient variation for saccharide mapping to be used as a means of grading the quality of *shihu* (Xu et al. 2011, 2014b).

Sprague–Dawley rats injected subcutaneously with sodium slenite suffer serious liver injury leading to liver fibrosis, a condition similar to liver cirrhosis in the human. When these animals were fed *D. huoshanense* galactoglucomannan (polysaccharide) at a dose of 50–200 mg/kg body weight, the selenite-induced liver injury was reduced. This was reflected by decreased levels of liver enzymes used to detect liver injury, alanine aminotransferease (ALT), aspartate aminotransferase (AST), and latate

dehydrogenase (LDH). Reduced expression of transforming growth factor-beta1 and type 1 collagen and better appearance of the tissue on histology were additional evidence of liver injury protection offered by the galactoglucomannan (Pan et al. (2012). Another study showed that daily supplementation of *D. huoshanense* polysaccharide protecxted the liver of mice from alcoholic injury. Of interest in this study is the observation that polysaccharide feeding also reduced the level of low-density lipoprotein (LDL) (Wang et al. 2012).

Polysaccharide derived from *D. tosaense* is a galactoglucomannan which has immunomodulatory activity. Administered to mice, it increased their splenic natural killer cell population as well as cytotoxicity, macrophage phagocytosis and cytokine induction in spleen cells (Yang et al. 2014b).

Rats injected with streptozotocin became diabetic and many of them eventually developed diabetic cataract. The administration of polysaccharides from *D. huoshanense* at doses

dendroside D

dendroside D

dendroside E

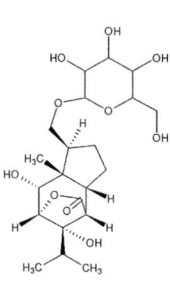

dendroside F

dendroside G

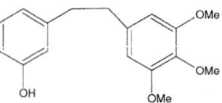

isoamoenylin

Fig. 10.56 Bibenzyl derivatives from *Dendrobium* with immunomodulatory activity

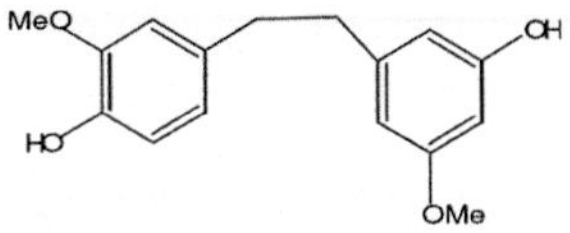

gigantol

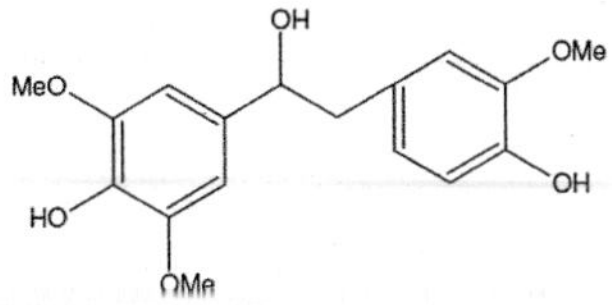

nobilin D

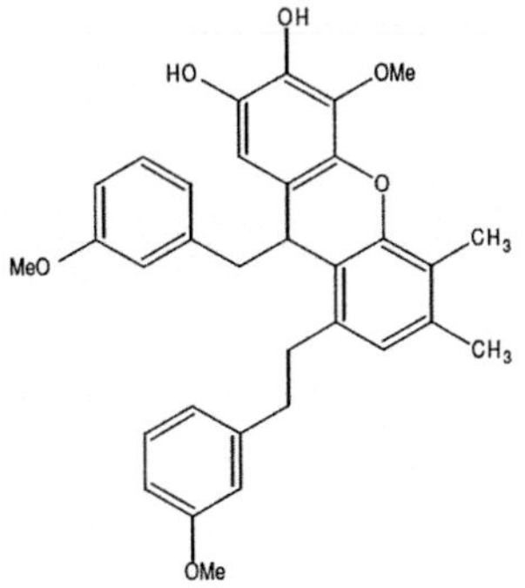

nobilin E

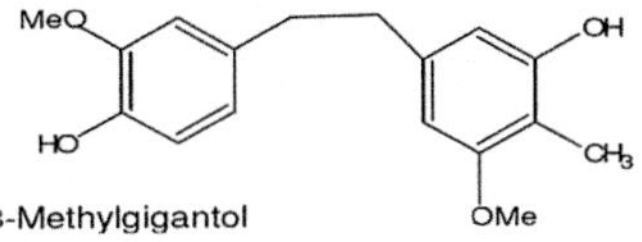

i-Methylgigantol

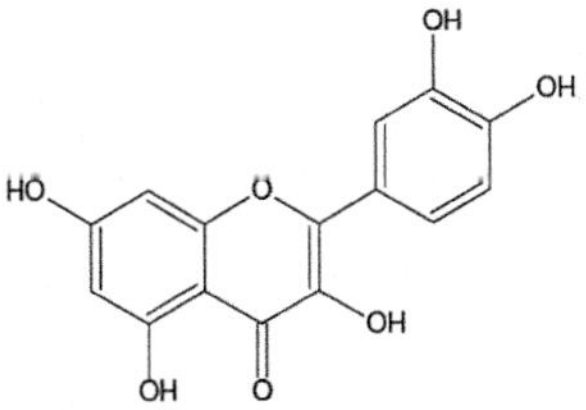

quercetin

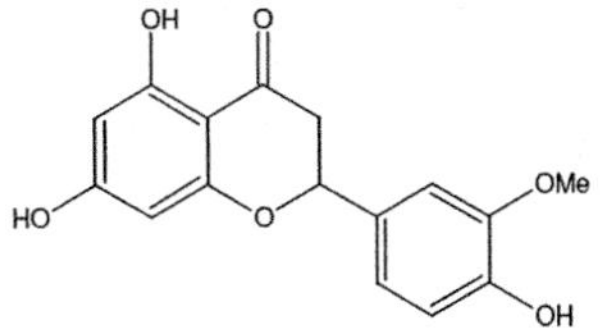

homoeriodictyol

Fig. 10.56 (continued)

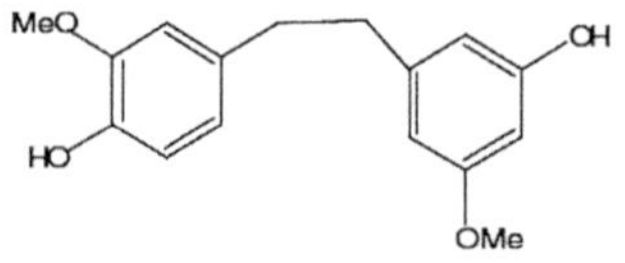

gigantol

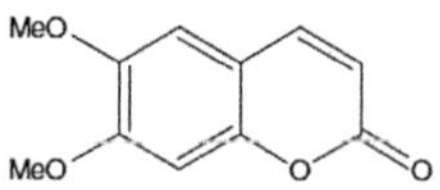

scoparone

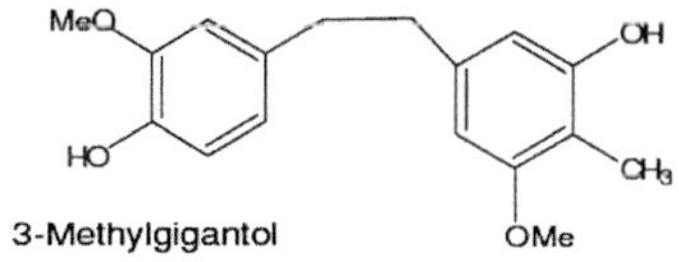

3-Methylgigantol

3 methyl gigantol

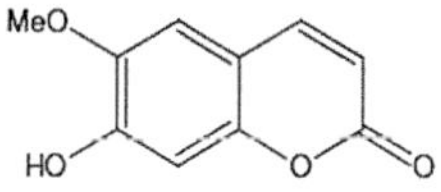

scopoletin

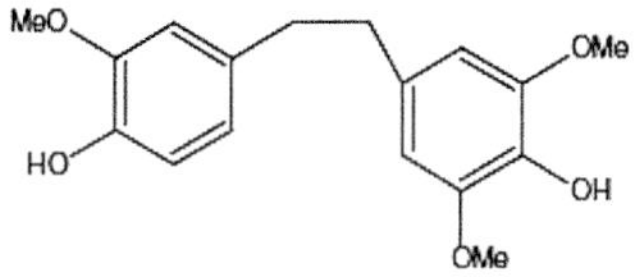

moscatilin

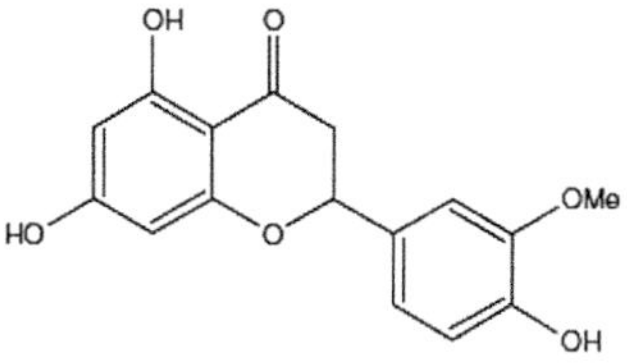

homoeriodictyol

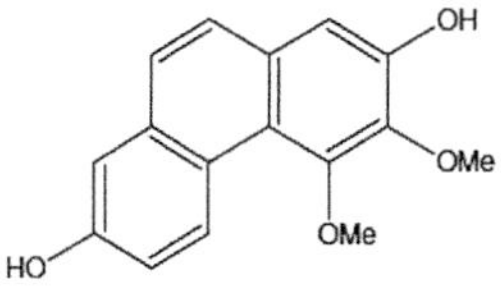

erianthridin

3,7-dihydroxy-2,4,methoxyphenanthrene

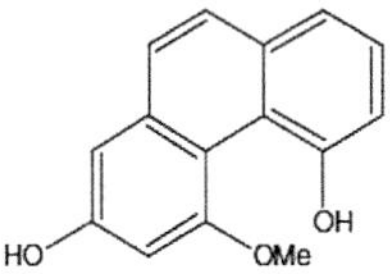

plicatol B

Fig. 10.57 Compounds from *Dendrobium* with antiplatelet or anticoagulant activity

of 50, 100 and 200 mg/kg/day to streptozotocin-induced diabeteic rats produced significant weight increases while simultaneously reducing blood sugar. Opacity of the lenses was reduced compared with rats receiving no treatment. RT-PCR analysis demonstrated inhibitory response of iNOSD gene expression with *D. huoshanense* polysaccharide treatment in a dose-dependent manner (Luo et al. 2008). However, this represents a very large dose. Even at 50 mg/kg, a person (animal) weighing 50 kg would need 2.5 g of polysaccharide or more than 5 g dry weight of *D. huoshanense* to benefit. This is much higher than the usual dose of *shihu* that is consumed. Sulphated modification enahanced the anti-glycation ability of polysaccharides from *D. houshanense* by 52.5 % (Qian et al. 2014; Li et al. 2014d).

The clinical evidence for the claims comes from the ability of mucilage to relieve coughs, improve digestion and sooth inflamed skin. Such a usage is not unique to TCM: indeed, another polysaccharide, wild cherry syrup, has, for a long time, been used in the famous western cough mixture, *Linctus Tussi Rubra*. Various species of *Dendrobium* are used to relief coughing in Papua New Guinea (Holdsworth 1974). However, the claim that *Dendrobium* polysaccharaides improves immunity is unique, if it is proven.

A few years ago, a new compound was isolated from *D. nobile* and given the code name Sg-168. In vitro experiments employing PC12 (brain cells) showed that Sg-168 protected the cells against hydrogen peroxide-induced oxidate stress and apoptosis (Yoon et al. 2011).

In order to bypass the depleting sources of wild *Dendrobium*, studies have also been conducted to recover polysaccharide from meristematic protocorm, plantlets and hybrids of various *Dendrobium* species. Many studies have demonstrated that cultivated *Dendrobium* produce equivalent or better yields of the desired medicinal constituents than wild specimens of *shihu*. Total alkaloid from *D. nobile* grown on sawdust media was higher than the content in specimens collected from the wild; the authors also found that the leaves had a higher content than the stems (Zhang et al. 2001a). Fan et al. (2005) found no observable difference in polysaccharide content between wild *D. candidum* and cultivated plants. However, Bai et al. (2007) reported that the polysaccharide content of *D. loddigesii* collected from different habitats showed remarkable differences in morphology and differences in their polysaccharide content. The polysaccharide content was 20.96 and 22.62 % higher in the plants collected from Anlong and Xingyi in Guizhou Province than elsewhere. The drying process and the method of extraction also affected recovery (Chen et al. 2001; Wu et al. 2007; Huang and Yu 2007).

The ability of protocorm-like bodies from stem explants of *D. huoshanense* cultured for 4 weeks to yield a water-soluble polysaccharide content of 3.75 % suggests a method for large-scale production of the desired polysaccharides. When long-term subcultures were performed every 30 days, the production of polysaccharide remained stable across 18 passages (Zha and Luo 2008). Furthermore, the polysaccharide retained the ability to stimulate interferon gamma release from splenocytes and tumour necrosis factor-alpha (TNF-alpha) from peritoneal macrophages (Luo et al. 2003, Zha and Luo 2008).

Sound wave stimulation (Li et al. 2006), addition of phosphate in two-stage cultivation (Wei et al. 2007a), addition of putrescine (Wei et al. 2007b), ground hyphae of *Mycena anoectochilus*, *Mycena dendrobii* and *Mycena orchidicola* and an ethyl acetate extract of *Mycena orchidicola* promoted protocorm multiplication of *D. candidum* (= *D. moniliforme*) (Gao and Guo 2001). Low concentrations of selenium (0.05 mg/l) promoted growth and antioxidative activity in protocorm-like bodies whereas high concentrations (0.2 mg/l) did the reverse (Xu et al. 2008).. Addition of putresin promoted the conversion of protocorm-like bodies of *D. catenatum* (syn. *D. officinale*) into shoots (Wei et al. 2011). Co-culture of *Sphingomonas paucimobilis* ZJSH1 with *D. catenatum* promoted growth of the orchid through phytohormone production and nitrogen fixation (Yang et al. 2014d). Micropropagation of *D. huoshanense*, so highly valued as a medicinal herb that it was commonly known as "thousand gold piece herb" or "soft gold" in Taiwan, can be initiated with root tip explants (Lee and Chen 2014).

Cryopreservation procedures have been successfully developed to store pollen, seeds and protocorms of *Dendrobium* species and hybrids (Zhang et al. 1999b; Vendrame et al. 2007, 2008; Vandrame and Faria 2011). Phloroglucinol enhances recovery and survival of cryopreserved *D. nobile* protocorms (Vandrame and Faria 2011).

Co-culture with endophytic fungi belonging to *Epulorhiza sp.* (Chen and Guo 2005), and hybridisation (Cai et al. 2005) separately and together both increase the polysaccharide yield in *D. hushanense*. In particular, co-culture with *Epulorhiza sp.* MF15 increased polysaccharide content by 153.4 % (Chen and Guo 2005). When inoculated with *Epulorhiza sp.* (GDB 181), the wet weight of seedlings of *D. officinale* nearly doubled, and this may be a useful approach for conserving the medicinal *Dendrobium* species which are threatened by overcollection and habitat deterioration (Jin et al. 2009). *D. linawianum, D. monilforme* and *D. tosaense* raised from seed by asymbiotic culture were tested for DPPH 1,1-diphenyl-2picrylhydrazyl) radical-scavenging ability and found to be potent (Lo et al. 2004b). Incidentally, *shihu* production has been supplemented by cultivation of the orchid in West Hubei and Sichuan at least as far back as 1929 (Hooper 1929).

Besides alkaloids, phenanthrenes and polysaccharides, a large number of bibenzyls, sesquiterpenes, coumarins and glycosides have also been isolated from *Dendrobium* species (Honda and Yamaki 2000; Ye and Zhao 2002; Yang et al. 2006b; Zhang et al. 2008a; Ou et al. 2009). Dihydroayapin is a novel coumarin isolated from *D. densiflorum* (Zheng et al. 2000). In *D. thyrsiflorum*, the coumarins are located mainly in the vascular bundles, especially in the wall of the outer fibre cells. The concentration was highest in the upper third of the stems, with a ninefold difference between the upper and lower third of the stem while the difference was fourfold between the upper and middle portion. Two-year-old stems are best for harvesting, and they should be harvested in February (Zheng et al. 2000).

Yang et al. (2006b) from the China Pharmaceutical University in Nanjing has now developed a new method for the simultaneous determination of phenols (bibenzyl, phenanthrene and flourenone) in *Dendrobium* using high-performance liquid chromatography with diode array detection that achieves satisfactory separation of the compounds within an hour. Applied to the detection of 11 phenols from 31 predominantly medicinal *Dendrobium* species, it yielded excellent results in terms of accuracy and reproducibility. Continuing their work, Zhang et al. (2008) reported the isolation of 9 bibenzyls and 2 benzylethanyl compounds from *D. aphyllum*, namely, moscatilin, gigantol, batatasin, tristin, 3,54′trihydroxybibenzyl, 3,5-dimethoxy-4,4′-dihydroxybibenzyl, moscatin, 2,4,7-trihydroxyl-9, 10-dihydro-phenanthrene, hircinol, 2-(4-hydrophenyl) ethyl-beta-D-glucopyrannoside, salidroside and *p*-hydroxybensylacetic acid. Working on *D. crystallinum*, they isolated 4,4′-dihydroxy-3,5-dimethoxybi-benzyl, *p*-hydroxybenzoic acid and seven other compounds which have previously been isolated from other *Dendrobium* species.

HPLC fingerprinting of flavonoids and phenols has been proposed as one way to establish quality control for *D. nobile*. In their hands, the method is simple and accurate, and it has good reproducibility (Ou et al. 2009).

Meanwhile, in very preliminary experiments, Devi et al. (2009) found that the aqueous extracts of the flowers and stem of *D. nobile* possessed antibacterial activity. They used five common bacteria in their experiments: *Escherichia coli, Bacillus subtilis, Proteus, Salmonella typhi* and *Staphylococcus aureus*. It is important to test for *Clostridium tetanii* which causes tetanus if the orchid is to be used for treating wounds.

Working on the Indian species, *D. amoenum* Wall ex Lind, Majumder and Bandyopadhyay (2010) have isolated a new bibenzyl derivative amoenylinin, besides the previously reported stilbenoids amoenylin [4-hydroxy-3,4′,5-trimethoxybibenzyl], isoamoenylin [3′-hydroxy-3,4,5-trimethoxybibenzyl], moscatilin [4,4′-dihydroxy-3,3′,5-trimethoxybibenzyl], batatasin-III [3,3′-dihydroxy-5-methoxybibenzyl], 3,4′-dihydroxy-5-methoxybibenzyl, the two phenanthrenes confusarin and 2,7-dihydroxy-3,4,6-trimethoxyphenanthrene, the two phenanthropyran derivatives imbricatin [2,7-dihydroxy-6-methoxy-9,10-dihydro-5*H*-phenanthro[4,5-*bcd*] pyran] and flaccidin [2,6-dihydroxy-7-methoxy-9,10-dihydro-5*H*-phenanthro[4,5-*bcd*]pyran], the two sesquiterpenoids amotin and amoenin, *p*-hydroxybenzaldehyde and *b*-sitosterol.

Amoenylinin was identified as 3,3′,4,4′,5-pentamethoxybibenzyl. *D. amoenum* is used as a tonic in Nepal (Pant and Raskoti 2013). Three bibenzyl derivatives, moscatilin, cumulatin and tristin, were isolated from *D. moschatum* (Majumder and Sen 1987a, b; Majumder and Pal 1993), and a phenanthrene derivative, moscatin, from the same species (Majumder and Sen 1987b).

The bright yellow coloration of *D. aggregatum* (= *D. lindleyi* or *D. jenkinsii*) is due to high concentrations of zeazanthin (15.5 mcg/g fresh weight) and antherxanthin (9.2 mcg/g fresh weight) and small amounts of neoxanthin (1.9 mcg/g). Lutein (8.1 mcg/g) is also present, but beta-carotene and chlorophylls are absent. *D. moschatum* has bright orange-yellow flowers that contain even higher levels of xanthins (zeathin 52.7 mcg/g; antheraxanthin 10.3 mcg/g; neoxanthin 2.4 mcg/g fresh weight) but no beta-carotene or chlorophylls (Thammasiri et al. 1986). The yellow flowers of several *shihu* species are now promoted as medicinal tea in Yunnan.

D. amplum and *D. fargesii* (formerly also known as *Epigenium amplum* and *E. fargesii*) are used in Chinese herbal medicine mainly to treat coughs and other symptoms associated with tuberculosis. *D. fargesii* is additionally used to treat gastritis and knife wounds. Their usage differs from those of *Dendrobium* species used as *shihu*. As there is no phytochemical publication on these two orchids, it is not possible to know whether there is any pharmacological basis from their usage which differs from *shihu*.

Dendrobium species are also used in Ayurvedic and other medicinal traditions. Five *Dendrobium* species are used in Thai native medicines: they are *D. cumulatum, D. draconis, D. indivisum, D.. leonis* and *D. trigonopus* (Chuakul 2002). We have allowed two spellings for the Thai word used to describe *Dendrobium* (*Uang* or *Ueang*) because the names are derived from different sources, and possibly both spellings are used in Thailand (Seidenfaden and Smitinand 1960; Vaddhanaphuti 2001). The same problem occurs in the transliteration of Indian names (e.g. *rasna* and *raasnna*).

In the Malay Archipelago, *D. crumenatum* is used to treat earache (Lawler 1986). *D. crumenatum* was the sole orchid listed among 194 species by the Administration Department of the Japanese Army in Singapore in their first compilation of useful plants produced in July 1944 *(Compilation of Medicinal Plants in the Malay District).* Burkill (1935) observed that the Malays do not distinguish between species in the manner of botanists, and they may use dissimilar species for the same medicinal purposes. For instance, whereas *D. crumenatum* was the common *Dendrobium* used for poulticing, sometimes in East Malaysia, *D. purpureum* was used instead; in West Malaysia, *D. planibulbe* might be used in its place. In Perak, *D. subulatum* was an acceptable substitute.

Several *shihu* species have become scarce in China and some may be facing extinction. Much work is being done to grow these plants in their natural environment to provide a sustainable source of *shihu* (Wu and Si 2010; Li et al. 2013b; Si et al. 2013a, b; Zhang et al. 2013c) instead of stripping plants from the forests of neighbouring countries like Laos, Myanmar and Thailand. Knowledge about mycorrhiza is important for the success of this project, and work is on-going in this area. Investigation of two habitats in Guangxi Province showed that *D. catenatum* associates with three to five fungi simultaneously whereas *D. fimbriatum* only associates with one fungal partner at a time, albeit this fungus differed between habitats. The two *Dendrobium* species did not share any fungal taxa. The common mycorrhiza of *Dendrobium* are members of Tulasnellaceae. In some *Dendrobium*, mycorrhiza belonging to Ceratobasidiaceae and Pluteaceae are present. Symbiotic seed germination of *D. draconis* which enjoys medicinal usage in Thailand has been described (Nontachalyapoom et al. 2011). The procedure might be useful for the propagation and conservation of *shihu Dendrobium* in China. The content of heavy metals and other elements in cultivated *D. catenatum* is being monitored and found to be safe, albeit one sample, cadmium, exceeded the allowable limit of 0.07 mg/kg (Zhu et al. 2011).

The polysaccharide content of *D. loddigesii* harvested from different habitats varies. A study from five regions found that samples from Xingyi and Anlong in Guizhou Province had the highest content, 22.62 and 20.96, respectively. The authors suggests that, for commercial purposes, *D. loddigesii* should be cultivated in these two districts of Guizhou (Bai et al. 2007). Twisting the stems of *D. loddigesii* which have been scaled by boiling water, followed by drying over fire, retains the highest amount of polysaccharide in the herb (Wu et al. 2007).

Host tree species also influence the polysaccharide content of *D. officinale* (= *D. catenatum*). This did not appear to be related to the quality of the bark and the scientists postulated that it was most probably due to the amount of light afforded by the tree species, light being a critical factor in photosynthesis (Guo et al. 2014). Neither study considered the role of endophytic mycorrhiza.

Concentrations of polysaccharides and alkaloids are highest in the upper parts of the stem, and vary in different parts of the plant (Chen et al. 2001). The drying process affects their ultimate concentration in the herbal preparation. Supplementing phosphorus and potassium up to an optimal concentration of 100 mg/l, and nitrogen, resulted in taller plants, more nodes per plant, more flowering nodes and increased total flower production in *D. nobile* (Bichsel et al. 2008). However, the effect of various fertiliser programmes on the medicinal content of *shihu* has not been studied (Table 10.5).

Genus: *Dienia*, Lindl.

There is much confusion in the identity of three genera *Dienia, Malaxis,* and *Microstylis* which some experts are now trying to solve (Margouska and Kowalkowska 2008). The generic name *Dienia* is derived from the Greek *dienos* (2-year-old). It was thought that the plant only flowers again after 2 years. Many species previously labelled as *Malaxis* are now classified under *Dienia* (e.g. *Malaxis latifolia* = *D. congesta*; *Malaxis cylindrostachya* = *D. cylindrostachya*). The problem should eventually be settled by DNA studies.

Dienia cylindrostachya Lindl.

syn. *Malaxis cylindrostachya* (Lindl.) Kuntze

Description: A terrestrial orchid distributed in the Himalayas (Xizang, Bhutan, Nepal and northeastern India), it also occurs in Pakistan, but there it is rare (Nasir and Ali 1972). Plants are 20–35 cm tall. Pseudobulb is conical, 1 cm tall, with a single, elliptical leaf, 4–9 by 3–4.5 cm. Inflorescence is 5–10 cm, densely many-flowered (Fig. 10.58). Flowers are yellowish-green, 2 mm across (Chen and Wood 2009a, b).

Usage: In Uttar Pradesh, it is sold as a nutrient and tonic for use in general debility (Trivedi et al. 1980); also in Uttarakhand (Jalal et al. 2008). Powdered pseudobulb is made into a tonic in Nepal (Subedi et al. 2013; Pant and Raskoti 2013).

Table 10.5 Trade in Shihu (in kilograms) in China, 1960 and 1980

	1960		1980	
	Fresh	Dry	Buying	Selling
All China	70,750	5800	600,000	550,000–600,000
Anhui	150	80,000		
Guangxi	5,000	2500	70,000	
Sichuan	22,500	2500	250,000	
Yunnan	25,000	750	40,000	
Guizhou	5000	100,000		
Others	18,100	50	50,000	

Data from Bao XS, Shun QS, and Chen LZ (eds.) The medicinal plants of Dendrobium (Shihu) in China. A coloured Atlas. Shanghai: Fudan Press, 2001

Fig. 10.58 *Dienia cylindrostachya* Lindl. [PHOTO: Bhaktar B. Raskoti]

Fig. 10.59 *Dienia ophrydis* (J. Koenig) Seidenf. [PHOTO: E.S. Teoh]

Dienia muscifera Lindl. [see **Malaxis monophyllos** (L.) Sw.]

Dienia ophrydis (J. Koenig) Seidenf.
syn. *Malaxis latifolia* Sm., *Anaphora lipaarioides* Gagnep.

Chinese names: *Kuoyezhao Lan* (broad-leaf mud orchid) *Huazhu Lan* (pillar flower orchid), *Guangyeruanye Lan* (floral pillar orchid), *Xiaozhu Lan* (small pillar orchid), *Guangyexiaozhu Lan* (tiny flower, small pillar orchid), *Suihuaxiaozhu* Lan (broad-leaved, small pillar orchid), *Ruanyezhao Lan* (soft-leaved mud or terrestial orchid)

Laotian names: Louang Prabang : *Van dong.* Vientiane: *Van nam*

Description: Plant is attractive Fig. 10.59). Stems are 10 cm tall (Indian variety 15 cm), stout, bearing 4–5 pleated leaves up to 20 cm long and 7 cm wide, with wavy edges (Figs. 10.60

Fig. 10.60 *Dienia ophrydis* (J. Koenig) Seidenf. close-up of flowers. [PHOTO: E.S. Teoh]

Fig. 10.61 *Dienia*
ophrydis (Koenig.)
Seidenf. [as *Gastroglottis*
montana Bl.]. From:
Reichenbach, H.G., Arnott,
G.A.W., *Xenia Orchidacea*
vol. 2: t. 129 (1900).
Courtesy of Missouri
Botanical Gardens,
St. Louis, USA

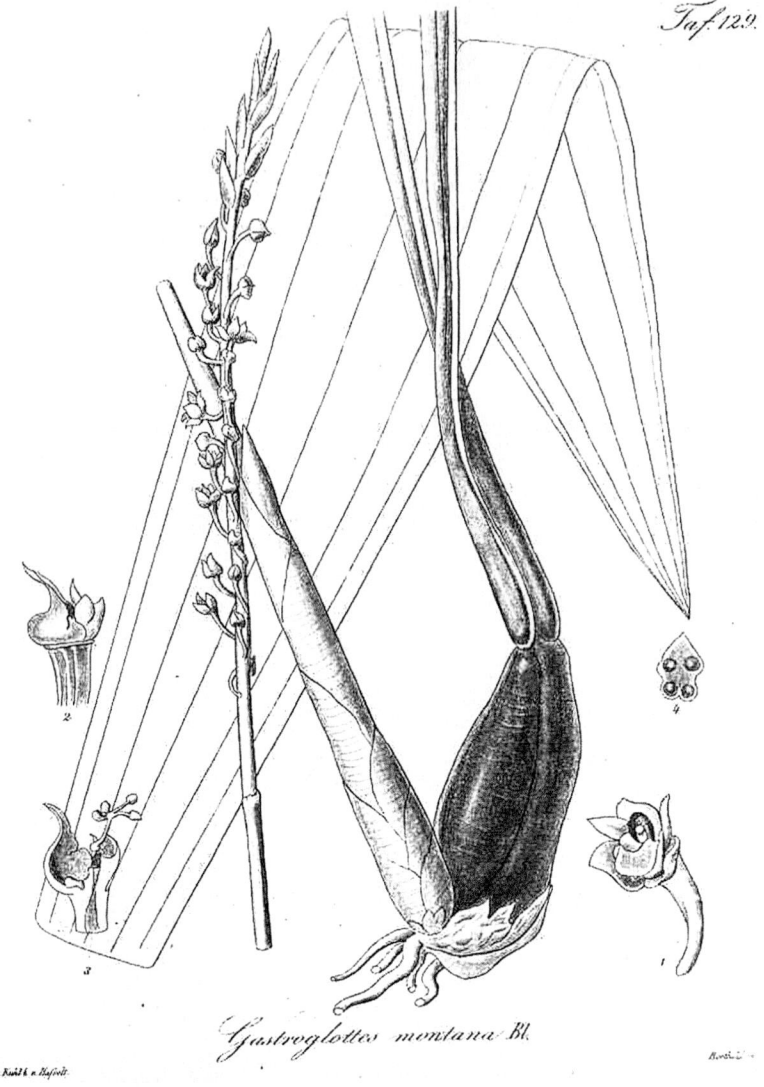

and 10.61). Rachis is erect, 5–20 cm tall, and carries numerous, tightly packed, small flowers which are usually yellow-green, yellow-green flushed with purple, deep purple throughout, or sometimes red (Fig. 10.60). It flowers from May to June in Taiwan (Lin 1977), August to December in most parts of the continent (Chen et al. 1999b), but May to June in Assam, India.

This shade-loving, terrestrial orchid has a wide distribution from the tropical Indian Himalayas and Sri Lanka eastwards to the Philippines, and from China and the Ryukyu Islands southwards across Indochina, Thailand, Malaysia, Singapore, Indonesia to New Guinea and Australia. It is a common herb in Taiwan below 1000 m, in forests throughout the island (Tang and Su 1978). It is widespread in lowland forests in Malaysia, including peat swamp forests where there is no constant flooding (Go and Hamzah 2008).

Phytochemistry: There is no pharmacological information on *Dienia*.

Usage: In Taiwan, the entire plant is used as an antipyretic, diuretic, detoxicant and to reduce swelling (Ou et al. 2003). Tubers are used to make a paste for application to burns in Vientianne, Laos (Vidal 1963).

Overview

D. ophrydis being rare in Indochina, commoner alternative herbs would be used for the treatment of the various conditions in the country. It is unlikely that the existence of *D. ophrydis* would be affected by its medicinal usage.

Genus: *Diploprora* Hook. f.

Chinese name: *Sheshe Lan*

The generic name is made up of two Greek words which mean 'double' (*diplous*) 'prow' (*prora*), referring to the conspicuous bifurcation of the lip. This is a small genus with only four members spread across Sri Lanka, India, Bangladesh, Myanmar, Thailand, Indochina and China. It belongs to the *Vanda* Tribe. While the flowering habit resembles *Luisia, Diploprora* has flat leaves whereas *Luisia* has terete leaves. It differs from the small-flowered *Cleisostoma* in not possessing a spur.

Diploprora championii (Lindl) Hook. f.

Chinese names: *Huangdiao Lan* (yellow hanging orchid), *Daodiao Lan* (hanging upside-down orchid), *Daochui Lan* (swaying orchid), *Niaolaidaochui Lan, Gaoshifodaochui Lan*
Taiwanese name: *Dao Diao Lan, Huang Diao Lan*

Description: *D. championii* is a small, pendulous, flat-leaved, vandaceous epiphyte with stems 5–37 cm long surrounded by persistent leaf sheathsand bending upwards terminally. Leaves are sessile, flat, linear, twisted, oblong and pointed at the apex, 8–12 by 2–2.4 cm, all facing one direction. Inflorescence is racemose, short, with 3–6 pale yellow flowers, 1.2–2 cm across, opening in succession. They are fragrant. Lip is spoon-shaped but the apex extends into a thin, bifid, tongue-ike projection which is forked; it is bordered in white and has four prominent

longitudinal yellow or red stripes. It does not have a spur (Fig. 10.62). Flowering season is February to September in China (Chen et al. 2009), summer in Taiwan (Su 1985), June or August in Thailand (Vaddhanaphuti 2001; Nanakorn and Watthana 2008), from June to November in India (Misra 2007), and February, June, August and December in Sri Lanka (Jayaweera 1981).

D. championii is distributed in tropical continental Asia (excluding Peninsular Malaysia, Laos and Cambodia), and in Taiwan at low altitudes, in montane forests (Schuiteman and de Vogel 2000). In Hong Kong, it is common on trees or rocks in forests and there it flowers from February to August (Wu et al. 2002).

Herbal Usage: The whole plant is used to treat traumatic injuries and fractures in Taiwan (Ou et al. 2003). It is also used for the treatment of physical injuries on the mainland (Wu 1994).

Phytochemistry: no data available

Overview

Pharmacological data on *Diploprora* are not available.

Genus: *Dipodium* R. Br.

The generic name *Dipodium* is derived from Greek *di* (two) and *podion* (foot) and alludes to the twin stipes holding up the pollinia. The genus consists of 12 species which are distributed in Southeast Asia and Australia. They are very different in vegetative appearance, some being leafless, achlorophylous terrestrials, others leafy terrestrials, and still others with climbing stems in the manner of the monopodial orchids but differing in being heavily clothed with leaves (Hawkes 1965). They are not commonly cultivated.

Dipodium pandanum F.M.Bailey

[syn. *Dipodium pictum* (Lindl.) Rchb. f.]

Description: Stem is long, erect, climbing, rooting at any point. Leaves are lanceolate, 25 cm long, folded along their axis, and tightly

Fig. 10.62 *Diploprora championii* (Lindl) Hook. f. From: Hooker, W.J., Hooker, J.D., *Icones Plantarum*, vol. 22: t.2120 (1894). [Drawing by M. Smith.] Courtesy of Missouri Botanical Gardens, St. Louis, USA

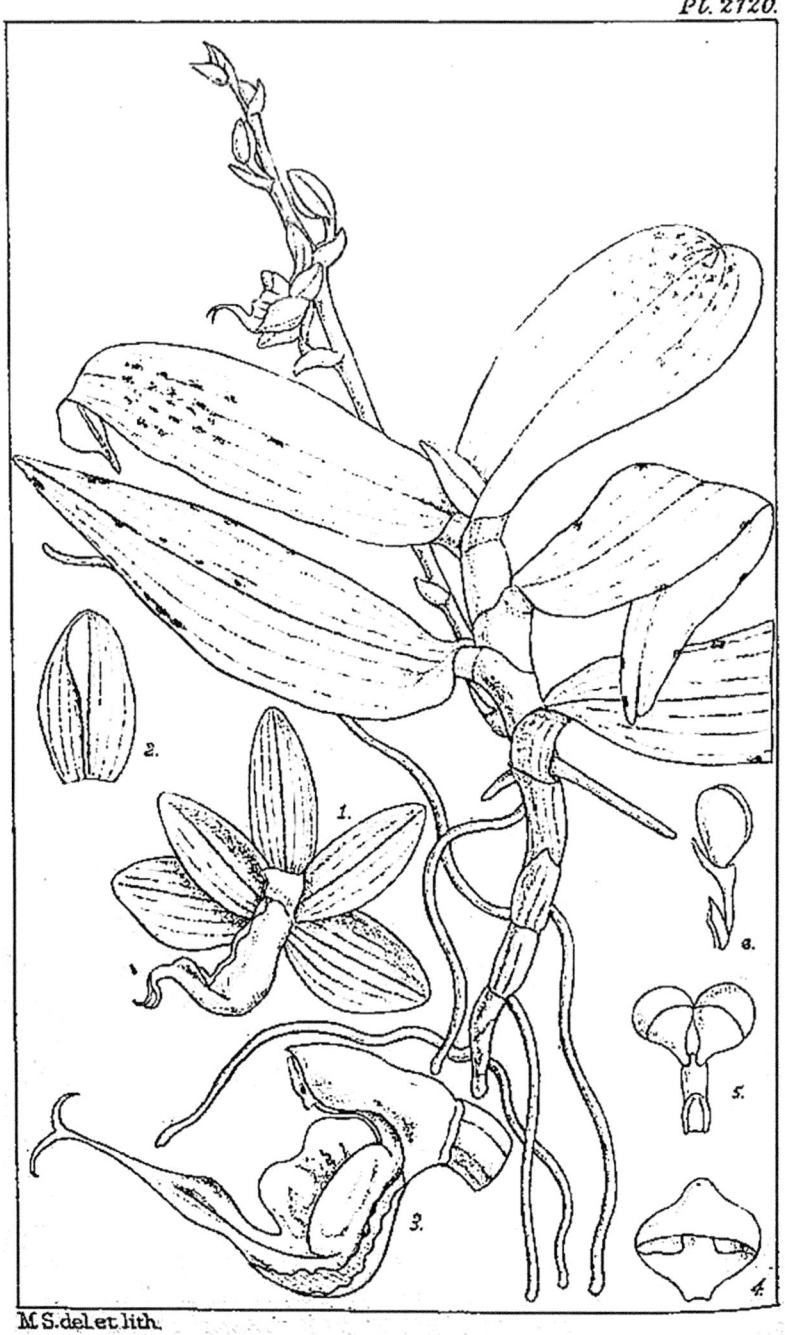

Pl. 2120.

M S.del.et lith.

ranked in two alternating rows. They are similar in appearance to those of *Pandanus* (which is not an orchid), thus earning the orchid its species name. Inflorescence is axillary, erect, 40 cm long with up to 12 pale yellow flowers with purplish-red blotches, loosely arranged. Tepals are narrow. Lip is striped with purple and bears long white hairs on its distal half (Millar 1978) (Fig. 10.63).

D. pandanum is a common epiphytic, terrestrial or saxicolous orchid in the lowland rain forests of

Fig. 10.63 *Dipodium pandanum* F.M.Bailey [PHOTO: Peter O'Byrne]

Papua New Guinea. The species is distributed eastwards and southwards from Borneo to the Philippines, Sulawesi, Java, Papua New Guinea, British Solomon Islands and Queensland.

Herbal Usage: An aqueous infusion of the leaves of *D. pandanum* is used to treat respiratory infections in Bougainville, Papua New Guinea (Lawler and Slaytor 1969; Lawler 1984).

Phytochemistry: Alkaloid was not detected in *D. pandanum* (Lawler and Slaytor 1969).

Overview

Currently, it has been very difficult to obtain information on the medicinal usage of orchids in Papua New Guinea. This is a rare instance of a Papuan orchid being mentioned in the ethnological review by Lawler (Lawler 1984). In this instance, I have retained the name, *D. pandanum* F.M. Bailey recorded by Lawler and not used the ? correct name provided in the Kew Monocot List, *D. scandens* (Blume) J.J. Sm because Peter O'Byrne who is an authority on the orchids of Papua New Guinea (O'Byrne 1994) informs me that *D. pandanum* is a common epiphytic, terrestrial or saxicolous orchid in the lowland rain forests of Papua New Guinea. The species is distributed eastwards and southwards from Borneo to the Philippines, Sulawesi, Java, Papua New Guinea, British Solomon Islands and Queensland, whereas the distribution of *D. scandens* is restricted to Java.

Genus: ***Doritis*** Lindl. (see ***Phalaenopsis***)
Doritis pulccherima Lindl. [see ***Phalaenopsis pulccherima*** (**Lindl.**) **J.J.Sm.**]

References

Alrich P, Higgins W (2008) The Marie Selby botanical gardens illustrated dictionary of orchid genera. Comstock Books, Carson City, NV

Anonymous (1989) Medicinal plants of China. Selection of 150 commonly used species. WHO Pacific Series No. 2. Manila: Regional Office of the Pacific.

Anonymous (2004) The new century Chinese-English Dictionary of traditional Chinese medicine. People's Military Medical Press, Beijing

Astuti IP, Hidayat S, Arinasa IBK (2000) Traditional plant usage in four villages of Baliaga. Botanic Gardens of Indonesia, Jakarta

Attundag E, Sevgi E, Kara O et al (2012) Comparative morphological, anatomical and habitat studies in Dactylorhiza romana (Schltr.) Soo, subsp. georgica (Klinge) Soo, Renze & Taub (Orchidaceae) in Turkey. Pak J Bot 44:143–152

Averjanova EA, Kharuta LG, Rybalko AE, Skipna KP (2014) Wild orchids of Colchis forests and save them as objects of eco-education and producers of medicinal substances. In: Haghi AK, Carvajal-Millan E (eds) Food composition analysis: methods and strategies. CRC Press, Boca Raton

Bae EY, Oh H, Oh WK et al (2004) A new VHR dual-specificity protein tyrosine phosphatase inhibitor from Dendrobium moniliforme. Planta Med 70 (9):869–70

Bai Y, Wang WQ, Bao YH, Sun ZR, Yan YN (2007) The comparative study on morphology and polysaccharides content of Dendrobium loddigesii from different habits. Zhong Yao Cai 30(2):130–2

Baker Margaret L, Baker Charles O (1996) Orchid species culture. Dendrobium. Timber Press, Portland

Bao X, Shun Q (1999) Investigation and identification of 'shihu' medicinal materials from Shanghai. Zhong Yao Cai 22(2):61–3

Bao XS, Shun QS, Chen LZ et al (2001) The medicinal plants of Dendrobium (Shi-Hu) in China. A coloured Atlas Shanghai. Fudan Press, Shanghai

Baral SR, Kurmi PP (2006) A compendium of medicinal plants in Nepal. Mrs Rachana Sharma and IUCN, Kathmandu

Bayarsukh N (2004): Inventory, documentation and status of medicinal plants research in Mongolia. In: Batugal PA, Kanniah J, Lee SY, Oliver JT (eds) Medicinal plants research in Asia (Vol. 1). The Framework and Project Workshops. Serdang: IPGRI-APO

Bensky D, Clavey S, Stoger E, Gambie A (2004) Herbal Medicine Materia Medica, 3rd edn. Eastland Press, Seattle WA

Bi ZM, Wang ZT, Xu LS, Xu GJ (2003) Studies on the chemical constituents of Dendrobium fimbriatum. Yao Xue Xue Bao 38(7):526–9

Bichsel RG, Starman TW, Wang YT (2008) Nitrogen, phosphorus and potassium requirements for optimizing growth and flowering of the nobile Dendrobium as a potted orchid. HortScience 43(2):328–332

Biomqvist L, Leander K, Luning B, Rosenblom J (1972) Studies on orchidaceae alkaloids. XXIX. The absolute configuration of dendroprimine, an alkaloid from Dendrobium primulinum Lindl. Acta Chem Scand 26 (8):3203–6

Bose TK, Bhattacharjee SK (1980) Orchids of India. Naya Prokash, Calcutta

Braun R (1888) China. Imperial Maritime Custums II. Special Series No. 8. List of Medicines exported from Hankow and the other Yangtze Ports. Shanghair Inspector-General of Customs, 1888

Burkhill IH, Haniff M (1930) Malay village medicine. Gardens Bull, Straits Settlements 6:165–321

Burkill IH (1920) Grammatophyllum flowering in January. Gardens Bull Straits and FMS 3:244

Burkill IH (1935) (1966 reprint, 2nd ed., with contributions by Birtwistle W, Foxworthy FW, Scrivenor JB, Watson IG). A Dictionary of economic products of the Malay Peninsula, Vol. II. London: Crown Agents for the Colonies. Kuala Lumpur:

Cai YP, Wang X, Lin Y, Li HS, Luo BS (2005) Studies on the growth, chemical components and physiological characteristics of F1 generation of Dendrobium huoshansense. Zhongguo Zhong Yao Za Zhi 30 (14):1064–8

Caius JF (1936) The medicinal and poisonous plants of India. J Bombay Natural Hist Soc 38(4):791–799

Chae YK, Woo J, Kim MJ et al (2008) Expression of aquaporin 5 (AQP5) promotes tumor invasion in human non small cell lung cancer. PLoS One 3(5), e2162

Chakrabarty M, Datta GK, Ghosh S, Debnath PK (2001) Induction of antioxidative enzyme by the ayurvedic herb Desmotrichum fimbriatum Bl. in mice. Indian J Exp Biol 39(5):485–6

Chang CC, Ku AF, Tseng YY et al (2010) 6,8-Di-C-glycosyl flavonoids from Dendrobium huoshanense. J Nat Prod 73(2):229–232

Chanvorachote P, Kowitdamrong A, Ruanghirun T et al (2013) Anti-metastatic activities of bibenzyls from Dendrobium pulchellum. Nat Prod Commun 8 (1):115–118

Charoenrungruang S, Chanvorachote P, Sritularak B, Pongrakhananon V (2014) Gigantol, a bibenzyl from Dendrobium draconis. Inhibits the migratory behavior of non-small cell lung cancer cells. J Nat Prod 77 (6):1359–1366

Chen KK, Chen AL (1935a) The alkaloid of Chin-shih-hu. J Biol Chem 111:653–658

Chen KK, Chen AL (1935b) The pharmacological action of dendrobine, the alkaloid of chin-shih-hu. J Pharamacol Exp Therap 55:319–325

Chen XM, Guo SX (2005) Effect of four species of endophytic fungi on the growth and polysaccharide and alkaloid contents of Dendrobium nobile. Zhongguo Zhong Yao Za Zhi 30(4):253–7

Chen KK, Rose CL (1936) Detoxification of dendrobine by "Sodium Amytal". Proc Soc Exp Biol Med 34:553–554

Chen SC, Tsi ZH (1998) The orchids of China. Wanhai Books, Beijing

Chen SC, Tang T (1982) A general review of the Orchid Flora of China. In: Arditti J (ed) Orchid biology reviews and perspectives II. Cornell University Press, Itcaca and London

Chen XQ, Wood JJ (2009a) Dienia Lindley. In: Chen XQ, Zj L, Zhu GH et al (eds) Flora of China—Orchidaceae. Science Press, Beijing

Chen XQ and Wood JJ (2009). Diplopropra J.D.Hooker. In: Chen XQ, Liu Zj, Zhu GH, et al. (eds.) Flora of China—Orchidaceae. Beijing: Science Press

Chen HC, Chu SH, Pei-Dawn LC (1991) Vasorelaxants from Chinese herbs, emodin and scoparone possess immunosuppressive properties. Eur J Pharmacol 198 (2–3):211–213

Chen CC, Wu LG, Ko FN, Teng CM (1994) Antiplatelet aggregation principles of Dendrobium loddigesii. J Nat Prod 57(9):1271–4

Chen S, Li Y, Wu Y, Zhou Z, Sun L (1995) Effect of Dendrobium nobile Lindl. on gastric acid secretion, serum gastrin and plasma somatostatin concentration. Zhongguo Zhong Yao Za Zhi 20(3):181–21

Chen HY, Hsiao MS, Huang YL, Shen CC, Lin YL, Kuo YH, Chen CC (1999a) Antioxidant principles from Ephemerantha lonchophylla. J Nat Prod 62 (9):1225–7

Chen SC, Tsi ZH, Luo YB (1999b) Native Orchids of China in colour. Science Press, Beijing

Chen CC, Huang YL, Teng CM (2000) Antiplatelet aggregation principles from Ephemerantha lonchophylla. Planta Med 66(4):372–4

Chen YL, Zhang M, Hua YF, He GQ (2001) Studies on polysaccharide, alkaloids and minerals from Dendrobium moniliforme (L.) Sw. Zhongguo Zhong Yao Za Zhi 26(10):709–10

Chen XM, Xiao SY, Guo SX (2006) Comparison of chemical composition between Dendrobioum candidum and Dendrobium nobile. Zhongguo Yi Xue Ke Xue Yuan Xue Bao 28(4):524–9

Chen TH, Pan SL, Guh JH, Chen CC, Huang YT, Pai HC, Teng CM (2008a) Denbinobin induces apoptosis by apoptosis-inducing factor releasing and DNA damage in human colorectal cancer HCT-116 cells. Naunyn Schmiedebergs Arch Pharmacol 378 (5):447–57

Chen TH, Pan SL, Guh JH, Liao CH, Huang DY, Chen CC, Teng CM (2008b) Moscatilin induces apoptosis in human colorectal cancer cells: a crucial role of c-Jun NH2—Terminal Protein Kinase Activation caused by tubulin depolymerization and DNA damage. Clin Cancer Res 14(13):4250–8

Chen YG, Liu Y, Jiang JH et al (2008c) Dendronone, a new phenanthraquinone from Dendrobium cariniferum. Food Chem 111(1):11–12

Chen XQ, Gale SW, Cribb PJ (2009) Dactylorhiza Necker ex Nevski. In: Chen XQ, Liu ZJ, Zhu GH et al (eds) Flora of China—Orchidaceae. Science Press, Beijing

Chen PH, Peng CY, Pai HC et al (2011) Denbinobin suppresses breast cancer metastasis through the inhibition of Src-mediated signaling pathways. J Nutr Biochem 22(8):732–740

Chen XM, Wang TT, Wang YQ et al (2012) Discrimination of the rare medicinal plant Dendrobium officinale based on narigenin, bibenzyl and polysaccharides. Sci China Life Sci 55(12):1092–109

Chen CA, Chen CC, Shen CC et al (2013a) Mocatilin induces apoptosis and mitotic catastrophe in human esophageal cancer cells. J Med Food 16(10):869–877

Chen XJ, Mei WL, Zuo WJ et al (2013b) A new antibacterial ohenanthrenequinone from Dendrobium sinense. J Asian Nat Prod Res 15(1):67–70

Chen JL, Zhong WJ, Tang GH et al (2014a) Norditerpenoids from Flickingeria fimbriata and their inhibitory activities on nitric oxide and tumour necrosis factor-alpha production in mouse macrophages. Molecules 19:5863–5875

Chen Y, Cai S, Den L et al (2014b) Separation and purification of 9,10-dihydrophenenthrene and bibenzyls from Pholidota chinensis by high speed countercurrent chromatography. J Sep Sci 38:453–9

Chopra RN, Chopra IC, Handa KL, Kapur LD (1958) Chopra's indigenous drugs of India. U.N. Dhur & Sons Pte, Calcutta

Chopra RN (1933): The indigenous drugs of India, Calcutta: The Art Press. Republished as Chopra's indigenous plants of India, 2nd edn. Academic Publishers, Kolkata (1986)

Chua TK, Koh HL (2006) Medicinal plants as potential sources of lead compounds with anti-platelet and anti-coagulant activities. Mini Rev Med Chem 6(6):611–24

Chuakul W (2002) Ethnomedical uses of Thai Orchidaceous plants. Mohidol Univ J Pharm Sci 29(3–4):41–5

Comber JB (2001) Orchids of Sumatra. Natural History Publications (Borneo), Kota Kinabalu

Cooray DA (1940) Orchids in oriental literature. Orchids Zelandica 7:73–80

Cribb PJ (1986) A revision of the Antelope and Latourea Dendrobiums. Kew Publications, London

Cribb P (2001) Dactylorhiza, uses. In: Pridgeon AM, Cribb PJ, Chase MWC, Rasmussen FN (eds) Genera Orchidacearum, vol 2, Orchidoideae (Part 1). University Press, Oxford, p 283

Dahmen J, Leander K (1978) Amotin and amoenin, two sesquiterpenes of the picrotoxane group from Dendrobium amoenum. Phytochemistry 17(11):1949–1952

Dahmen J, Rosenblom J, Leander K (1975) A new glycoside, 2-(beta-D- glucopyrannosyloxy)-4,5-dimethoxy-transcinnamic acid (Densifloroside) from Dendrobium densiflorum Wall: Studies on Orchidaceae glycosides.3. Acta chem Scand Ser B 29(5):627–628

Dalstrom S (2009) A tale of two Dendrobiums. Orchids 78(1):46–8

Das SP (2004) Indian orchids in indigenous medicine system. In: Britto SJ (ed) Orchids Diversity and Conservation. The Rapinat Herbarium, St John's College, Tiruchirappalli

Dash PK, Sahoo S, Bal S (2008) Ethnobotanical studies on Orchids of Niyamgiri Hill Ranges, Orissa, India. Ethnobot Leaflets 12:70–78

De Waldemann E (1892) Bull Soc/Belge Microscopie 18:101

Devi PU, Selvi S, Devipriya D et al (2009) Antitumor and antibacterial activities and inhibition of in-vitro lipid peroxidation by Dendrobium nobile. Afr J Biotechnol 8(10):2289–93

Ding Y, Wu Q, Yu L (1998) On the best time for harvesting Dendrobium candidum Wall. ex Lindl. Zhongguo Zhong Yao Za Zhi 23(8):458–60

Ding X, Xu L, Wang Z, Zhou K, Xu H, Wang Y (2002a) Authentication of stems of Dendrobium officinale by rDNA ITS region sequences. Planta Med 68(2):191–2

Ding XY, Wang ZT, Xu H, Xu LS, Zhou KY (2002b) Database establishment of the whole rDNA ITS region of Dendrobium species of "fengdou" and authentication by analysis of their sequences. Yao Xue Xue Bao 37(7):567–73

Ding XY, Xu LS, Wang ZT, Xu H, Zhou KY (2002c) Molecular authentication of Dendrobium chrysanthum from its allied species of Dendrobium. Zhongguo Zhong Yao Za Zhi 27(6):407–11

Ding X, Wang Z, Zhou K et al (2004) Alle-specific primers for diagnostic PCR authentication of Dendrobium officinale. Planta Med 69(6):587–8

Ding G, Ding XY, Shen J, Tang F, Liu DY, He J, Li XX, Chu DH (2005) Genetic diversity and molecular authentication of wild populations of Dendrobium officinale by RAPD. Yao Xue Xue Bao 40(11):1028–32

Ding G, Zhang D, Feng Z, Fan W, Ding X, Li X (2008) SNP, ARMS and SSH authentication of medicinal Dendrobium officinale KIMURA et MIGO and application for identification of Fengdou drugs. Biol Pharm Bull 31(4):553–7

Ding G, Li X, Ding X, Qian L (2009) Genetic diversity across natural populations of Dendrobium officinale, the endangered medicinal herb endemic to China, revealed by ISSR and RAPD markers. Genetika 45(3):3375–82

Doung NV (1993) Medicinal plants of Vietnam, Cambodia and Laos. Nguyen Van Duong, Vietnam

Duggal SC (1972) Orchids in human affairs (A Review). Pharm Biol 11(2):1727–1734

Elander M, Leander K, Rosenbiom J, Ruusa E (1973) Studies on orchidaceae alkaloids. XXXII. Crepidine, crepidamine and dendrocrepine, three alkaloids from Dendrobium crepidatum Lindl. Acta Chem Scand 27(6):1907–13

Fan C, Wang W, Wang Y, Qin G, Zhao W (2001) Chemical constituents from Dendrobium densiflorum. Phytochemistry 57(8):1255–8

Fan JA, Wang JS, Zhang Y, Ren LY, Qiu ZY, Xia YP (2005) A comparison of tissue formation and the content of polysaccharide between wild and cultured Dendrobium candidum. Zhongguo Zhong Yao Za Zhi 30(21):1648–1659

Fan YJ, He XJ, Zhou SD, Luo AX, He T, Chun Z (2009) Composition analysis and antioxidant activity of polysaccharide from Dendrobium denneanum. Int J Biol Macromol 45(2):169–173

Fan YJ, Chun Z, Luo A et al (2010) In vivo immunomodulatory activities of neutral polysaccharide (DPP1-1) from Dendrobium denneanum. Chin J Appl Environ Biol 16(3):376–379

Fang F, Liu G (2007) A novel cyclic squasmosamide analogue compound FLZ improves memory impairment in artificial senescence mice induced by chronic injection of D-galactose and NaNO$_2$. Basic Clin Pharmacol Toxicol 101(6):447–54

Fasihuddin BA, Hamsah R (1992) Medicinal plants of the Murut Community in Sabah. In: Ghazzaly I, Murtedza M, Omar S (eds) Proc intern conference on forest biology and conservation in Borneo, Centre for Borneo Studies Publication No. 2, Kota Kinabalu: Yayasan Sabah, pp 460–467

Gao WW, Guo SX (2001) Effects of endophytic fungi hyphae and their metabolites on the growth of Dendrobium candidum and Anoectochilus roxburghii. Zhongguo Yi Xue Ke Xue Yuan Xue Bao 23(6):556–9

Gao J, Jin R, Wu Y, Zhang H, Zhang D, Chang Y, Hu Z (2002) Comparative study of tissue cultured Dendrobium protocorm with natural Dendrobium candidum on immunological function. Zhong Yao cai 25(7):487–9

Gawell L, Leander K (1976) Phytochemistry 15 (12):1991–1992

Ghorbani A, Langenberger G, Feng L, Sauerborn J (2011) Ethnobotanical study of medicinal plants utilized by Hani ethnicity in Naban River Watershed National Nature Reserve, Yunnan, China. J Ethnopharmacol 134:651–67

Ghorbani A, Gravendeel B, Naghibi F, de Booer H (2014) Wild orchid tuber collection in Iran: a wake-up call for conservation. Biodivers Conserv 23. Doi: 10.1007/s10531-014-0746-y

Gimlette JD, Thomson HW (1939) A Dictionary of Malayan Medicine. Oxford University Press, London

Go R, Hamzah KA (2008) Orchids of Peat Swamp Forests in Peninsular Malaysia. Kuala Lumpur: Peat Swamp Forest Project, UNDP/GEF Funded (MAL/99/G321) Ministry of Natural Resources & Environment

Gong Y, Fan Y, Liu L et al (2004a) Erianin induces a JNK/SAPK-dependent metabolic inhibition in human umbilical vein endothelial cells. In Vivo 18(2):223–238

Gong YQ, Fan Y, Wu DZ et al (2004b) In vivo and in vitro evaluation of erianin, a novel anti-angiogenic agent. Eur J Cancer 40(10):1554–1565

Gong YQ, Yang H, Liu Y, Liang AQ, Wang ZT, Xu LS, Hu ZB (2006) Studies on the chemical constituents in stem of Dendrobium chrysotoxum. Zhongguo Zhong Yao Za Zhi 31(4):304–6

Gong CY, Yu ZY, Lu B et al (2014) Ethanol extract of Dendrobium chrysotoxum Lindl. ameliorates diabetic retinopathy and its mechanism. Vascul Pharamcol 62 (3):134–142

Granelli L, Leander K (1970) Studies on Orchidaceae alkaloids. XIX. Synthesis and absolute configuration of dendrine. Acta Chem Scand 24:1108–1109

Granelli I, Leander K, Luning B (1970) Studies on orchidaceae alkaloids. XVI. A new alkaloid, 2-hydroxydendrobine, from Dendrobium findlayanum Par. Ex Rchb. f. Acta Chem Scand 24(4):1209–12

Grant B (1895) The Orchids of Burma. Hanthawaddy Press, Rangoon

Guo SX, Fan L, Cao WQ, Chen XM (1999) Mycena dendrobii, a new mycorrhizal fungus. Mycosystema 18(2):141–144

Guo X, Li Y, Li C et al (2013) Analysis of the Dendrobium officinale transcriptome reveals putative alkaloid biosynthetic genes and genetic markers. Gene 15 527(1):131–138

Guo YY, Zhu Y, Si JP et al (2014) Effect of tree species on polysaccharide content of epiphytic Dendrobium officinale. Zhonggu Zhong Yao Za Zhi 39 (21):4222–4224

Gurong DB (2006) An illustrated guide to the Orchids of Bhutan. DSB Publications, Thimphu

Han BH et al (1981) Ann Rep Nat Prod Res Inst Seoul Nat Univ 1981(20):49

Handoyo F (2010) Orchids of Indonesia, vol 1. Indonesian Orchid Society, Jakarta

Hart R (1884) China Imperial Customs III. Misc. Series No. 17. List of Chinese medicines. Shanghai: Inspector Geeneral of Customs, 1884

Hawkes AD (1965) Encyclopaedia of cultivated orchids. Faber & Faber, London

He P, Song X, Luo Y, He M (2009) Reproductive biology of Dendrobium officinale (Orchidaceae) in Danxi landform. Zhongguo Zhong Yao Za Zhi 34(2):124–7

Hedman LK (1972) Studies on orchidaceae alkaloids. XXVII Qauternary salts of the Dendrobium type from Dendrobium nobile Lindl. Acta Chem Scand 26 (8):3177–80

Hedman K, Leander K, Luning B (1971) Studies on orchidaceae alkaloids. XXV. N-isopenttenyl derivaties of dendroxine and 6-hydroxydendroxine from Dendrobium fredricksianum Lindl. and Dendrobium hildebrandii Rolfe. Acta Chem Scand 25(3):1142–4

Hemsccheldt T, Spenser ID (1991) Biosynthesis of anosmine, an imidazole alkaloid of the orchid Dendrobium parishii. J Chem Soc, Chem Commun 1991(7):494–49

Hendren M, Nordstrom S (2009) Polymorphic populatios of Dactylorhza incarnata (Orchidaceae) on the Baltic Island of Gotland: morphology, habitat preference and genetic differentiation. Ann Bot 104:527–542

Hew CS, Arditti J, Lin WS (1997) Three orchids used as herbal medicines in China: an attempt to reconcile Chinese and Western pharmacology. In: Arditti J, Pridgeon AM (eds) Orchid biology reviews and perspectives VII. Kluwer Academic Publishers, Dordrecht, Boston, London

Ho CK, Chen CC (2003) Moscatilin from the orchid Dendrobium loddigesii is a potential anticancer agent. Cancer Invest 21(5):729–36

Holdsworth DK (1974) A phytochemical survey of Medicinal Plants in Papua New Guinea Part I. Science in New Guinea 2(2):142–154

Holttum RE (1949) Gregarious flowering of the terrestrial orchid, Bromheadia finlaysonniana. Gardens Bull Singapore 12(2):295–302

Honda C, Yamaki M (2000) Penanthrenes from Dendrobium plicatile. Phytochemistry 53(8):987–90

Hooper D (1929) On Chinese medicine: Drugs of Chinese pharmacies in Malaya. Gardens Bull Straits Settlements 6:1–163

Hsieh YS, Chien C, Liao SK et al (2008) Structure and bioactivity of the polysaccharides in medicinal plant Dendrobium huoshanense. Bioorg Med Chem 16 (11):6054–68

Hsu JL, Lee YJ, Leu WJ et al (2014) Moniliforme-diquinone induces in vitro and in vivo antitumor activity through glutathione involved DNA damage response and mitochondrial stress in human hormone refractory prostate cancer. J Urology 191(5):1429–1438

Hu SY (1970) Dendrobium in Chinese Medicine. Econ Bot 24:165–174

Hu SY (1971) Orchids in the life and culture of the Chinese People. Chuing Chi J 10:1–26

Hu JM, Chen JJ, Yu H et al (2008a) Two novel bibenzyls from Dendrobium trigonopus. J Asian Nat Prod Res 10(7–8):653–657

Hu JM, Chen JJ, Yu H, Zhao YX, Zhou J (2008b) Five new compounds from Dendrobium longicornu. Planta Med 74(5):535–9

Hu JM, Chen JJ, Yu H, Zhao YX, Zhou J (2008c) Two novel bibenzyls from Dendrobium trigonopus. J Asian Nat Prod Res 10(7):653–7

Hu JM, Zhou YX, Miao ZH, Zhou J (2009) Chemical components of Dendrobium polyanthum. Bull Korean Chem Soc 30(9):2098–2100

Hua YF, Zhang M, Fu CX, Chen ZH, Chan GY (2004) Structural characterization of a 2-0-acetylglucomannan from Dendrobium officinale stem. Carbohydr Res 339(13):2219–24

Huang JP, Yu L (2007) Studies on the extraction of active polysaccharide from Dendrobium hushansense with Box-Behnken method. Zhong Yao Cai 30(5):591–4

Huang HC, Lee CR, Weng YI et al (1992) Vasodiilator effect of scoparone (6,7-dimethoxycoumarin) from a Chinese herb. Eur J Pharmacol 218(1):123–128

Huang SY, Shi JG, Yang YC et al (2002a) Studies on the chemical constituents from Tibetan medicine wangle (rhizome of Ceologlossum viride (L.) Hartm. var, bracteatum). Zhongguo Zhong Yao Za Zhi 27 (2):118–120

Huang SY, Shi JG, Yang YC et al (2002b) Studies on the chemical constituents of Ceologlossum viride (L.) Hartm. var, bracteatum (Willd.) Richter. Yaoxue Xuebao 37(3):199–203

Huang SY, Li GQ, Shi JG, Mo SY (2004) Chemical constituents of the rhizomes of Coeloglossum viride var. bracteatum. J Asian Nat Prod Res 6(1):49–61

Huang YC, Guh JH, Teng CM (2005) Denbinobin-mediated anticancer effect in human K562 leukemia cells: role in tubulin polymerization and Bcr-Abl activity. J Biomed Sci 12(1):113–21

Hwang JS, Lee SA, Hong SS et al (2010) Phenanthrenes from Dendrobium nobile and their inhibition of the LpS-induced production of nitric oxide in macrophage RAW 264.7 cells. Bioorg Med Chem Lett 20 (13):3785–3787

Institute of Materia Medica Hanoi (1990) Medicinal plants in Viet Nam. WHO Regional Publications Western Pacific Series 3. World Health Organization, Manila

Inubishi Y, Sasaki Y, Tsuda Y et al (1964) Structue of dendrobine. Tetrahedrom 20:2007–2023

Inubushi Y, Tsuda Y, Konita T, Matsumoto S (1964) T and Shihunine. A new phthalide pyrrolidine alkaloid. Chem Pharm Bull 12:749–750

Inubishi Y, Nakano J (1965) Structure of dendrine. Getrahedron Lett 31:2723–2728

Inubushi Y, Kikuchi T, Ibuka T et al (1972) Total synthesis of the alkaloid dendrobine. J Chem Soc Chem Commun 22:1252–1253

Ishidoya T (1925) On the medicinal plants in Chosen, Korea. J Chosen Nat Hist Soc 3:1–10

Ito M, Matsuzaki K, Wang J et al (2010) New Phenanthrenes and Stilbenes from Dendrobium loddigesii. Chem Pharm Bull (Tokyo) 58(5):628–633

Jalal JS, Kumar P, Pangtey YPS (2008) Ethnomedicinal orchids of Uttarakhand, Western Himalayas. Ethnobot Leaft 12:1227–1230

Jayaweera DMA (1981) A revised handbook of the flora of Ceylon, vol II. A.A. Balkema, Rotterdam

Ji N, Li Y (2009) Effect of different strains of Amillaria mellea on the yield of Gastrodia elata f. glauca. J Fungal Res 6(4):231–233

Jiang T, Huang BK, Zhang QY, Han T, Zheng HC, Qin LP (2007) Effect of Liriope platyphylla total saponin on learning, memory and metabolites in aging mice induced by D-galactose. Zhong Xi Yi Jie He Xue Bao 5(6):670–4

Jin H, Xu ZX, Chen JH, Han SF, Ge S, Luo YB (2009) Interaction between tissue cultured seedlings of Dendrobium officinale and mycorrhizal fungus (Epulorrhiza sp.) during symbiotic culture. Chin J Plant Ecol 33(3):433–441

Jin J, Liang Y, Xie H, Zhang X, Yao X, Wang Z (2008) Dendiflorin retards the senescence of MRC-5 cells. Pharmazie 63(4):321–3

Jin YY, Fang CW, Yang QQ et al (2013) Investigation on wild resources of Dendrobium officinale distribution and ecological environment in Anhui. Zhongguo Zhong Yao Za Zhi 38(23):4024–4027

Jonathan KH, Raju AJS (2005) Terrestial and epiphytic orchids of Eastern Ghats. EPTRI-ENVIS Newsletter II 3:2–4

Kaul MK, Gaur RD (1995) Characteristics of ethnopharmacological resources in Kashmir Himalaya. In: Pushpangadan P, Nyman U, George V (eds) Glimpses of Indian ethnopharmacology. TBGRI, Thiruvananthapuram, pp 185–209

Karasawa K, Shimai H (2002) Horticultural History and Breeding of Dendrobium moniliforme in Japan. Proc. 17th World Orchid Conference, Shah Alam, pp 240–1

Karel CAJ, Kreutz K, Spensa J (2011) In the footsteps of Renz: Orhids of Iran

Kim JH, Oh SY, Han SB et al (2015) Anti-inflammatory effects of Dendrobium nobile derived phenanthrenes in LPS-stimulated murine macrophages. Arch Pharm Res 38:1117–26

Kimura K, Migo H (1936) New Species of Dendrobium from the Chinese Drug (Shih- hu). J Shanghai Science Institute Section III 3:121–124

Kimura T, But PPH, Guo JX, Sung CK (2002) International collation of traditional and folk medicine. Northeast Asia. Vol 1, 3 and 4.

Kizu H, Kaneko EI, Tomimori T (1999) Studies on Nepalese crude drugs. XXVI. Chemical constituents of Panch aunle, the roots of Dactylorhiza hatagirea. D Don Chem Pharm Bull 47(11):1618–25

Koopowitz H (2001) Orchids and their conservation. Timber Press, Portland

Koriba H, Watanabe K (1944) Illustrated useful plants in Malaya First compilation: Medicinal plants. Administrative Department, Japanese Army in Malay District, Singapore

Kovacs A, Vasas A, Hohmann J (2007) Natural phenanthrenes and their biological activity. Phytochemistry 69:1084–1110

Kowitdamrong A (2014) Effects of moscatilin on migration of non-small cell lung cancer cells. M SC Pharmacy Thesis. Chulalongkorn University, Bangkok. http://cuir.car.chula.ac.th/handle/123456789/42681

Kreis LM, Carreira EM (2012) Total synthesis of (−)-Dendrobine. Angewandte Chemie 51(14):3436–3439

Krohn K, Loock U, Paavilainen K et al (2001) Synthesis and electrochemistry of annoquinone-A, cypripedin methyl ether, denbinobin and related 1,4-phenanthrenequinones. ARKIVOC 2001(1):88–130

Kudo Y, Tanaka A, Yamada K (1983) Dendrobine, an antagonist of beta-alanine, taurine and of presynaptic inhibition in the frog's spinal cord. Br J Pharmacol 78 (4):709–715

Kunwar RM, Nepal BK, Kshhetri HB et al (2006) Ethnomedicine in Himalaya: a case study from Dolpa, Humla, Jumla and Mustang districts of Nepal. Ethnobiol Ethnomed 2:27

Kuo CT, Hsu MJ, Chen BC, Chen CC, Teng CM, Pan SL, Lin CH (2008) Denbinobin induces apoptosis in human lung adenocarcinoma cells via Akt inactivation, bad activation, and mitochondrial dysfunction. Toxicol Lett 177(1):48–58

Kuo CT, Chen BC, Yu CC et al (2009) Apoptosis signal-regulating kinase 1 mediates denbinobin-induced apoptosis in human lung adenocarcinoma cells. J Biomed Sci 16:43–57

Lam F, Bradshaw TD, Mao H et al (2011) ZJU-6, a novel derivative of Erianin, shows potent anti-tubulin polymerization and anti-angiogenic activities. Invest New Drugs 30:1899–907

Lau DT, Shaw PC, Wang J, But PP (2001) Authentication of medicinal Dendrobium species by the internal transcribed spacer of ribosomal DNA. Planta Med 67 (5):456–60

Lavarack B, Harris W, Stocker G (2000) Dendrobium and its allies. Timber Press, Portland

Lawler LJ (1984) Ethnobotany of the orchidaceae. In: Arditti J (ed) Orchid biology, reviews & perspectives 3. Cornell University Press, Ithaca

Lawler LJ (1986) Orchid ethnobotany in the Asean Area. In: Rao AN (ed) Proc 5th Asean Orchid Congress. Singapore: Parks & Recreation Department, Ministry of National Development (42–45)

Lawler LJ, Slaytor M (1969) The distribution of alkaloids in orchids from the Territory of Papua New Guinea. Proc Linnean Soc NSW 94:419–421

Lawler LJ, Slaytor M (1970) Uses of Australian Orchids by Aborigines and Early Settlers. Med J Australia 2:1259–1261

Lawler LJ (1984) Ethnobotany of the Orchidaceae. In: Arditti J (ed) Orchid biology reviews and perspectives, vol 3. Cornell University Press, Ithaca

Leander K, Luning B (1968) Studies on Orchidaceae alkaloids. 8. An imidazolium salt from Dendrobium anosmum Lindl. and Dendrobium parishii Rchb.f. Tetrahedron Lett 8:905–8

Lee PL, Chen JT (2014) Plant regeneration via callus culture and subsequent in vitro flowering of Dendrobium huoshanense. Acta Physiol Plant 36:2619–2625

Lee YH, Park JD, Baek NI, Kim SI, Ahn BZ (1995) In vitro and in vivo antitumoral phenanthrenes from the aerial parts of Dendrobium nobile. Planta Med 61 (2):178–80

Lee GH, Kim HR, Han SY et al (2012a) Gastrodia elata Blume and its pure compounds protect BV-2 microglial-derived cell lines against beta-amyloid: the involvement of GRP78 and CHOP. Biol Res 45:403–10

Lee W, Eom DW, Jung Y et al (2012b) Dendrobium moniliforme attenuates high fat induced renal damage in mice through regulation of lipid-induced oxidatice stress. Am J Chin Med 40(6):1217–1228

Lee SJ, Chae YS, Kim JG et al (2014) AQP5 expression predicts survival in patients with early breast cancer. Ann Surg Oncol 21(2):375–383

Leoni LAB, Soares LMV (2003) Survey of moniliformn in corn cultivated in the state of Sao Paulo and in corn products commercialized in the city of Campinas, SP. Braz J Microbiol 34(1):13–15

Li SZ (1578). Ben Cao Gung Mu

Li MF, Hirata Y, Xu GJ, Niwa M, Wu HM (1991) Studies on the chemical constituents of *Dendrobium loddigesii* Rolfe. Yao Xue Xue Bao 26(4):307–310

Li J, Xiao X (1995) Medicinal plant resources of Dendrobium in Sichuan Province. Zhongguo Zhong Yao Za Zhi 20(1):7–9

Li YM, Wang HY, Liu GQ (2001) Erianin induces apopstosis in human leukemia HL-60 cells. Acta Pharmacol Sin 22(11):1018–1012

Li T, Wang J, Lu Z (2005) Accurate identification of closely related Dendrobium species with multiple species specific gDNA probes. Biochem Biophys Methods 62(2):111–23

Li B, Wang BC, Liang YL, Tang K, Shu KK, Liu WQ (2006) Effect of polysaccharides content of tissue culturing seedlings on Dendrobium candidum under sound wave stimulation. Zhong Yao Cai 29:645–7

Li Y, Wang CL, Guo SX et al (2008) Two new compounds from Dendrobium candidum. Chem Pharm Bull (Tokyo) 56(10):1477–9

Li RS, Yang X, He P, Gan N (2009a) Studies on phenanthrene constituents from stems of Dendrobium candidum. Zhong Yao Cai 32(2):220–3

Li Y, Wang CL, Wang YJ et al (2009b) Three new bibenzyl derivatives from Dendrobium candidum. Chem Pharm Bull (Tokyo) 57(2):218–9

Li Y, Wang CL, Wang YJ et al (2009c) Four new bibenzyl derivatives from Dendrobium candidum. Chem Pharm Bull (Tokyo) 57(9):997–9

Li Y, Wang CL, Wang YJ et al (2010) Chemical constituents of Dendrium candidum. Zhongguo Zhong Yao Za Zhi 35(13):1715–1719

Li Y, Li F, Gong Q et al (2011) Inhibitory effect of Dendrobium alkaloids on memory impairment induced by lipopolysaccharide in rats. Planta Med 77(2):117–121

Li C, Ning LD, Si JP et al (2013a) Effects of post-harvest processing and extraction methods on polysaccharide content of Dendrobium officinale. Zhongguo Zhong Yao Za Zhi 38(4):524–527

Li G, Lu J, Chen X (2013b) Some worries about Dendrobium officinale industry. Zhongguo Zhong Yao Za Zhi 38(4):469–471

Li XH, Guo XH, Guo L et al (2014) Three new neolignan glucosides from the stem of *Dendrobium aurantiacum* var denneanum. Phytochem Lett 9:37–40

Li GJ, Sun P, Wang Q et al (2014a) Dendrobium candidum Wall. ex Lindl. attenuates CCl4-induced hepatic damage in imprinting control region mice. Exp Ther Med 8(3):1015–1021

Li GJ, Sun P, Zhou YL et al (2014b) Preventive effects of Dendrobium candidum Wall ex Lindl. on the formation of lung metastases in BALB/c mice injected with 26-M3.1 colon carcinoma. Oncology Lett 8(4):1879–1885

Li J, Wang Z, Chong T et al (2014c) Over-expression of a poor prognostic marker in prostate cancer: AQP5 promotes cell growth and local invasion. World J Surg Oncol 12(1):284

Li XL, Xiao JJ, Zha XQ et al (2014d) Structural denitification and sulfated modification of an antiglycation Dendrobium huoshanense polysaccharide. Carbohydr Polym 106:247–254

Lin TP (1977) Native Orchids of Taiwan, Vol 2, colour photo 98, pp.166–7

Lin TH, Chang SJ, Chen CC, Wang JP, Tsao LT (2001) Two phenanthraquinones from Dendrobium moniliforme. J Nat Prod 64(8):1084–6

Lin X, Shaw PC, Sze SC et al (2011) Dendrobium officinale polysaccharide ameliorate the abnormality of aquaporin 5, pro-inflammatory cytokines and inhibit apoptosis in the experimental Sjogren's syndrome mice. Int Immunopharmacol 11(12):2025–2032

Lin X, Liu JG, Chung WY et al (2015) Polysaccharides of Dendrobium officinale induce Aquaporin 5 translocation by activating M3 muscarinic receptors. Planta Med 81(2):130–137

Liu H, Dong HL (2013) Optimization of the extraction technology of total alkaloids from Dendrobium nobile by ultrasound assisted method

Liu TS, Su HJ (1978) Flora of Taiwan, Vol 5

Liu QF, Zhao WM (2003) A new Dendrobine-type alkaloid from Dendrobium nobile. Chinese Chem Lettrs 14(3):278–9

Liu CY, Deng Y, McIntyre A (1999) Encyclopedia of Chinese and U.S. Patent Herbal Medicines. Keats Publishing, Los Angeles

Liu SQ, Li XJ, Yu QB, Xie H, Zhuo GY (2007a) Analysis on genetic stability in different development stages of Dendrobium huoshanense by RAPD. Zhongguo Zhong Yao Za Zhi 32(10):902–5

Liu WH, Hua YF, Zhang ZJ (2007b) Moniline, a new alkaloid from Dendrobium moniliforme. J Chem Res 2007(6):317 8

Liu HS, Chang ASY, Teng CM et al (2011a) Shock 35(2):191–197

Liu XF, Zhu J, Ge SY et al (2011b) Orally administered Dendrobium officinale and its polysaccharides enhance immune functions in BALB/c mice. Nat Prod Commun 6(6):867–870

Lo SF, Mulabagal V, Chen CL et al (2004) Bioguided fractionation and isolation of free radical scavenging components from in-vitro-propagated Chinese medicinal plants Dendrobium tosaense Makino and Dendroboum moniliforme Sw. J Agric Food Chem 52(23):6916–6919

Lo SF, Mulabagal V, Chen CL, Kuo CL, Tsay HS (2004a) Bioguided fractionation and isolation of free radical scavenging components from in vitro propagated Chinese medicinal plants Dendrobium tosaense Makino and Dendrobium moniliforme SW. J Agric Food Chem 52(23):6916–9

Lo SF, Nalawade SM, Mulabagal V et al (2004b) In vitro propagation by asymbiotic seed germination and

1,1-diphenyl-2-picrylhydrazyl (DPPH) radical scavenging activity studies of tissue culture raised plants of three important species of Dendrobium. Biol Pharm Bull 27(5):731–5

Lu JJ, Wang S, Zhao HY et al (2012) Genetic linkage map of EST-SSR and SRAP markers in the endangered Chinese endemic herb Dendrobium (Orchidaceae). Genet Mol Res 11(4):4654–4667

Lu GY, Yan MQ, Chen SH (2013) Review of pharmacological activities of Dendrobium officinale based on traditional functions. Zhongguo Zhong Yao Za Zhi 38 (4):489–493

Lu KH, Han CK, Chang YS et al (2014a) Denbinobin, a phenanthrene from Dendrobium nobile, impairs prostate cancer migration by inhibiting Rac 1 activity. Am J Chin Med 42(6):1539–54

Lu Y, Kuang M, Hu GP et al (2014b) Loddigesinols G–J: alpha-gucosidase inhibitors from Dendrobium loddigesii. Molecules 19(6):8544–8555

Luning B (1964) Studies on the Orchidaceae alkaloids. I. Screening of species for alkaloids 1. Acta Chim Scand 18:1507–1516

Luning B, Leander K (1965) Studies on the Orchidaceae Alkaloids III. The alkaloids in *Dendrobium primulinum* Lindl. and *Dendrobium chrysanthum* Wall. Acta Chim Scand 19:1607–1611

Luning B (1974a) Studies on Orchidaceae alkaloids. IV. Screening of species for alkaloids 2. Phytochemistry 6:857–861

Luning B (1974b) Alkaloids of the Orchidaceae. In: Withner CL (ed) The Orchids. Scientific studies. John Wiley & Songs, New York

Luning B (1980) Alkaloids of the Orchidaceae. In: Sukshom Kashemsanta MR (ed) Proc. 9th World Orchid Conference, Bangkok

Luo JP, Zha XQ, Jiang ST (2003) Suspension culture of protocorm-like bodies from the endangered medicinal plant Dendrobium huoshanense. Zhongguo Zhong Yao Za Zhi 28(7):611–4

Luo JP, Deng YY, Zha XQ (2008) Mechanaism of polysaccharides from Dendrobium huoshanense on streptozotocin-induced diabetic cataract. Pharmaceutical Biol 46(4):243–9

Luo A, He X, Zhou S et al (2009a) Invitro antioxidant activities of a water-soluble polysaccharide derived from Dendrobium nobile Lindl. extracts. Int J Biol Macromol 45(1):359–363

Luo JP, Wawrosch C, Kopp B (2009b) Enhanced propogation of Dendrobium houshanense CZ Tang et SJ Cheng through protocorm-like bodies. The effects of cytokinins, carbohydrate sources and cold pretreatment. Scientia Hort (Amsterdam) 123(2):258–262

Luo AX, He XJ, Zhou SD et al (2010) Purification, composition analysis and antioxidant activity of the polysaccharides from Dendrobium nobile Lindl. Carbohydr Polym 79(4):1014–1019

Luo CP, He T, Chun Z (2013) Advances in infrared spectroscopy methods for discriminating Dendrobium. Chinese J Appl Environ Biol 19(3):537–541

Ma GX, Wang TS, Li Y, Yan P, Guo YL, Leblanc GA, Reinecke MG, Watson WH, Krawiec M (1998) Two pimarane diterpenoids from Ephemerantha lonchophylla and their evaluation as modulators of the multidrug resistance phenotype. J Nat Prod 61 (1):112–115

Ma B, Li M, Nong H et al (2008) Protective effects of extract of Coeloglossum viride var. bracteatum on ischaemia-induced neuronal death and cognitive impairment in rats. Behav Pharmacol 19 (4):325–333

Magwere T (2009) Escaping immune surveillance in cancer: is denbinobin the panacea? Br J Pharmacol 157 (7):1172–1174

Majumder PL, Bandyopadhyay D (2010) Stilbenoids and sequiterpene derivatives of the orchids Gastrochilum calceolaria and Dendrobium amoenum: applicationof 2D NMR spectroscopy in structural elucidation of complex natural products. J Indian Chem Soc 87:221–234

Majumder PL, Pal A (1993) Cumulatin and tristin, two bibbenzyl derivatives from the orchids Dendrobium cumulatum and Bulbophyllum triste. Phytochemistry 32:1561–1565

Majumder PL, Sen RC (1987a) Structure of moscatin—a new phenanthrene derivative from the orchid Dendrobium moscatum. Indian J Chem 26B:18–20

Majumder PL, Sen RC (1987b) Moscatilin, a bibenzyl derivative from the orchid Dendrobium moscatum. Phytochemistry 26(7):2121–2124

Majumder PL, Guha S, Sen S (1999) Bibezyl derivatives from the orchid Dendrobium amoenum. Phytochemistry 52(7):1365–1369

Manandhar NP, Manandhar S (2002) Plants and people of Nepal. Timber, Portland

Mao AA (2006) Notes on *Dendrobium denudans* D.Don – an endangered medicinal orchid of northeast India. J Orchid Soc India 20(1-2):53–55

Margouska HB, Kowalkowska A (2008) Taxonomic revision of Dienia. Ann Bot Fennici 45:97–104

Massaoudi S, Hamze A, Provot O et al (2011) Discovery of isoerianin analogues as promising anticancer agents. ChemMedChem 6(3):488–497

McKenna DJ (2014) Ayahuasca: an ethnopharmacologic history. In: Metzner R (ed) The Ayahuasca experience: a sourcebook on Sacred Vine of Spirits: Inner Traditions. Bear & Co., Rochester, Vermont

Meng LZ, Lv GP, Hu DJ et al (2013) Effects of polysaccharides from different species of Dendrobium (Shihu) on macrophage function. Molecules 18:5779–5791

Millar A (1978) Orchids of Papua New Guinea. Australian National University Press, Canberra

Misra S (2007) Orchids of India a glimpse. Bishen Singh Mahendra Pal Singh, Delhi

Miyazawa M, Shimammura H, Nakamura S, Sugiura W, Kosaka H, Kameoka N (1999) Moscatalin from Dendrobium nobile, a naturally occurring bibenzyl compound with potential antimutagenic activity. J Agric Food Chem 47(5):2163–7

Morita H, Fujiwara M, Yoshida N, Kobayashi J (2000) New picrotoxin-type and dendrobine-type sesquiterpenoids from Dendrobium Snowflake 'Red Star'. Tetrahedron 56(32):5801–5805

Nagananda GS, Satishchandra N (2013) Antimicrobial activity of cold and hot successive pseudobulb extracts of Flickingeria nodosa (Dalz.) Seidenf. Pak J Bio Sci 16(20):1189–1193

Neiland RM (2001) In: Pridgeon, Cribb, Chase and Rasmussen (eds) Genera Orchidaceaerum. Orchidoideae (part 1), vol 2. University Press, Oxford

Nanakorn W, Watthana S (2008) Queen Sirikit Botanic Garden (Thai Native Orchids 1 and 2). Wanida Press, Chiang mai

Nakamura S, Tokuda K, Omato A (1990) Japan prize fragrance competition. Am Orchid Soc Bull 59 (10):1031–1036

Vaddhanaphuti N (2001) A field guide to the wild orchids of Thailand, 3rd edn. Silkworm Books, Chiang Mai

Vaddhanaphuti N (2005) A field guide to the wild orchids of Thailand, Fourth and Expanded Edition. Silkworm Books, Chiang Mai

Nasir E, Ali SI (1972) Flora of West Pakistan. Pakistan Agricultural Research, Islamabad

Nath PC, Das DR (2013) Distribution of Dendrobium Swartz (Orchidaceae) in tropical evergreen forests of Upper Assam, India. Pleione 7(1):23–31

Ng GX, Wang TS, Yin L et al (1998) Two pimarane diterpenoids from Ephermerantha lonchophylla and their evaluation as modulators of the multidrug resistant phenotype. J Nat Prod 61:112–115

Ng TB, Liu F, Wang ZT (2000) Antioxidative activity of natural products from plants. Life Sci 66(8):709–723

Nielsen J, Chen S, Irwin SW et al (2006) Estrogen protects neuronal cells from amyloid-beta-induced apoptosis via regulation of mitochondrial proteins and function. BMC Neurosci 7:74

Nontachalyapoom S, Sasirat S, Manoch L (2011) Symbiotic seed germination of Grammatophyllum speciosum Blume and Dendrobium draconis Rchb. f., native orchids of Thailand. Scientia Hort 2130 (1):303–308

Nuesch J (1963) Defense reactions in orchid bulbs. Symp Soc Gen Microbiol 13:335–343

O'Byrne P (1994) Lowland Orchids of Papua New Guinea. National Parks Board, Singapore Botanic Gardens, Singapore

O'Byrne P (2001) A to Z of South East Asian orchid species. Orchid Society of South East Asia, Singapore

Okamoto T, Natsume M, Onaka T, Uchimaru F, Shimizo M (1966a) The structure of dendroxine. The third alkaloid from Dendrobium nobile. Chem Pharm Bull (Tokyo) 14(6):672–5

Okamoto T, Natsume M, Onaka T, Uchimaru F, Shimizo M (1966b) The structure of dendramine (6-oxydendrobine) and 6-oxydendroxine. The fourth and fifth alkaloid from Dendrobium nobile. Chem Pharm Bull (Tokyo) 14(6):676–80

Okamoto T, Natsume M, Onaka T et al (1972) Further 6466 studies on the alkaloidal constituents of Dendrobium nobile (Orchidaceae) – structure determination of 4-hydroxydendroxine and nobilomethylene. Chem Pharm Bull 20:418–421

Onaka TS, Kamata T, Maeda Y et al (1964) The structure of dendrobine. Chem Pharm Bull 12:506–512

Onaka TS, Kamata T, Maeda Y et al (1965) The structure of nobilonine. The second alkaloid from Dendrobium nobile. Chem Pharm Bull 13:745–747

Osugi M, Fan WZ, Hase K, Xiong QB (1999) Active-oxygen scavenging activity of traditional nourishing-tonic herbal medicines and active constituents of Rhodiola scara. J Ethnoparmacol 67(1):111–119

Ou JC, Hsieh WC, Lin IH, Chang YS, Chen IS (eds) (2003) The catalogue of medicinal plant resources in Taiwan. Department of Health, Executive Yuan, Taipei

Ou HJ, Chen JL, Li XX et al (2009) HPLC fingerprint of flavonoids and phenols of Dendrobium nobile. Zhong Yao Cai 32(6):871–4

Pai HC, Chang LH, Peng CY et al (2013) Moscatilin inhibits migration and metastasis of human breast cancer MDA-MB-231 cells through inhibition of Akt and Twist signaling pathway. J Mol Med 91 (3):347

Pan HP, Lu J, Luo JP et al (2012) Preventive effect of a glactoglucomannan (GGM) from Dendrobium huoshanense on selenium-induced liver injury and fibrosis in rats. Exp Toxicol Pathol 64(7–8):899–904

Pan HM, Chen B, Li F, Wang MK (2013a) Chemical constituents from the stems of Dendrobium denneanum (II). Chin J Appl Environ Biol 19(6):952–955

Pan HM, Chen B, Li F, Wang MK (2013b) Medicinal orchids and their uses: tissue culture a potential alternative for conservation. Afr J Plant Sci 7(10):448–467

Pant B, Raskoti BB (2013) Medicinal orchids of Nepal. Himalayan Map House (P) Ltd., Kathmandu

Pearce NR, Cribb PJ (2002) The Orchids of Bhutan Edinburgh. Royal Botanic Gardens and Thimpu: Royal Government of Bhutan

Pengpaeng P, Sritularak B, Chanvorachote P (2015a) Dendrofalconerol A suppresses migrating cancer cells via EMT and integrin proteins. Anticancer Res 35(1):201–205

Pengpaeng P, Sritularak B, Chanvorachote P (2015b) Dendrofalconerol A sensitizes anoikis and inhibits migration of lung cancer cells. J Nat Med 69 (2):178–190

Perner H, Luo Y (2007) Orchids of Huanglong. Huanglong National Park, Sichuan

Perry LM, Metzger J (1980) Medicinal plants of East and Southeast Asia: attributed properties and uses. MIT Press, Cambridge, MA

Petelot A (1953) Plantes medicinales du Cambodge, du Laos et du Vietnam, #18. Centre de Rech. Sci.et Tech.,

Arch. Des Rech. Agron. Au Camb., au Laos et Vietnam, Saigon (quoted by Perry, 1980)

Phechrmeekha T, Sritularak B, Likhitwitayawuid K (2012) New phenolic compounds from Dendrobium capillipes and Dendrobium secundum. J Asian Nat Prod Res 14(8):748–54.

Prasad R, Kock BP (2014) Antitumor activity of ethanolic extract of Dendrobium formosum in T-cell lymphoma: an in vitro and in vivo study. Biomed Res Int 2014:753451. doi:10.1155/2014/753451

Pridgeon AM, Cribb PJ, Chase MW, Rasmussen FN (eds) (2014) Genera Orchidacearum, vol 6, Epidendroideae (Part Three). Oxford University Press, Oxford

Purevsuren S, Tuya M (2013) Medicinal plants in Mongolia. WHO Western Pacific, Manila

Qian L, Ding G, Zhou Q et al (2008) Molecular authentication of Dendrobium loddgesii Rolfe by amplification refractory mutation system. Planta Med 74 (4):470–473

Qian XP, Zha XQ, Xiao JJ et al (2014) Sulfated modification can enhance antiglycation abilities of polysaccharides from Dendrobium houshanense. Carbohydr Polym 101:982–989

Qin XD, Qu Y, Ning L et al (2011) A new picrotoxane-type sesquiterpene from *Dendrobium findlayanum*. J Asian Nat Prod Res 13(11):1047–1050

Rabie CJ, Narasas WF, Thiel PG et al (1982) Moniliformin production and toxicity of different Fusarium species from Southern Africa. Appl Environ Microbiol 43(3):517–521

Rao RR (1981) Ethnobotany of Meghalaya. Medicinal plants used by Khasi and Garo tribes. Econ Bot 35:4–9

Rao AN (2004) Medicinal Orchid Wealth of Arunachal Pradesh. Newsletter of ENVIS NODE on Indian Medicinal Plants 1(2):1–7

Rao TA (2007) Ethno Botanical data on wild orchids of medicinal value as practised by tribals at Kudremukh National Park in Karnataka. Orchid Newslett 2(2):1–7

Rao TA, Sridhar S (2007) Wild Orchids in Karnataka. A pictorial compendium. Institute of Natural Resources Conservation, Education, Research and Training (INCERT), Bangalore

Raskoti BB (2009) The Orchids of Nepal. Bhakta Bahadur Raskoti and Rita Ale, Kathmandu

Rasolofonjatovo E, Provot O, Hamze A et al (2012) Conformationally restricted naphthalene derivatives type isocombretastatin and isoerianin analogues: synthesis, cytotoxicity and antitubulin activity. Eur J Med Chem 52:22–32

Reddy KN, Reddy CS, Raju VS (2005a) Ethno-orchidology of orchids of Eastern Ghats of Andhra Pradesh. EPRTI Newslett 11(3)

Reddy KN, Subha Raju GV, Sudhakar Reddy CH, Raju VS (2005b) Ethnobotany of Certain Orchids of Eastern Ghats of Andhra Pradesh. EPTRI-ENVIS Newslett II(3);5–9

Ridley HN (1906) Malay Drugs. Agricultural Bull. Straits Settlements & Fed Malay Staes 5, 245 and 277

Ridley HN (1907) Materials for a Flora of the Malay Peninsula, vol 1. Methodist Publishing House, Singapore

Rumphius GE (second half of 17th century, posthumus) Amboinsch Kruidboek (The Amboinese Herbal, Vols 1–6). Translated and annotated into English, with an introduction by E M Beekman (2011). Yale University Press, New Haven & London

Sanchez-Duffhues G, Calzado MA, de Venuesa AG et al (2009) Denbinobin inhibits nuclear factor-kB and induced apoptosis via reactive oxygen species generation in human leukemic cells. Biochem Pharm 77(8):1401

Sanchez-Duffhunes G, Calzado MA, de Venuesa AG et al (2008) Denbinobin, a naturally occurring 1,4-phenthrenequinone, inhibits HIV-1 replication through an NF-kappaB-dependent pathway. Biochem Pharmacol 76(10):1240–1250

Sandrasagaran UM, Subramaniam S, Murugaiyah V (2014) New perspective of Dendrobium crumenatum orchid for antimicrobial activity against selected pathogenic bacteria. Pak J Bot 46(2):719–724

Santapau H, Kapadia Z (1966) The orchids of Bombay. Government of India Press, Calcutta

Schuiteman A, de Vogel EF (2000) Cac Ci Ho Lan (Orchidaceae) Cua Thai Lan, Lao, Campuchia Va Viet Nam. Orchid Genera of Thailand Laos, Cambodia and Vietnam. (Vietnamese-English edition). National Herbarium Nederland, Leiden

Seidenfaden G, Smitinand T (1959) The orchids of Thailand: A preliminary list, Part I, vol 2. The Siam Society, Bangkok

Seidenfaden G, Smitinand T (1960) The orchids of Thailand. A preliminary list Part II (2). The Siam Society, Bangkok

Seidenfaden G (1985) Contributions to the orchid flora of Thailand. XII. Dendrobium Sw. Opera Bot 83 (Copenhagen)

Seidenfaden G, Wood JJ (1992) The orchids of peninsular Malaysia and Singapore. Olsen & Olsen, Fredensborg

Sezik E (1967) Turkiye'nin Salepgilleri Ticari Salep Cesitleri ve Ozellikle Mugla Salebi Uzerinde Arastirmalar. Doctoral Thesis. Istanbul Universitesi Eczacihk Fakultesinde (In Turkish. Summary in English)

Shan T, Cui X, Li W et al (2014) AQP5: a novel biomarker that predicts poor clinical outcome in colorectal cancer. Oncol Rep 32(4):1564–1570

Shao L, Huang WH, Zhang CF et al (2008) Study on chemical constituents from stem of Dendrobium aphyllum. Zhongguo Zhong Yao Za Zhi 33 (14):1693–5

Shen LS (chief ed.) (1998) Colored Atlas of Compedium of Materia Medica. Huaxia Publishing House, Beijing

Shen J, Ding XY, Ding G, Liu DY, Tang F, He J (2006a) Studies on population difference of Dendrobium officinale II. Establishment and optimization of the method of ISSR fingerprinting marker. Zhongguo Zhong Yao Za Zhi 31(4):291–4

Shen J, Ding X, Liu D, Ding G, He J, Li X, Tang F, Chu B (2006b) Intersimple sequence repeats (ISSR) molecular fingerprinting markers for authenticating populations of Dendrobium officinale Kimura aet Migo. Biol Pharm Bull 29(3):420–2

Shen L, Zhu Z, Huang Y et al (2010) Expression profile of multiple aquaporins in human gastric carcinoma and its clinical significance. Biomed Pharmacother 64 (5):313–318

Shin EJ, Whang WK, Kim SG et al (2010) Parishin C attenuates phencyclidine-induced schzophrenia in mice: involvements of 5-HT1A receptor. J Pharm Sc 113(4):404–408

Shrestha CB (2000) IUCN Nepal, 2000. National Register of Medicinal Plants. IUCN Nepal, Kathmandu

Shu Y, Zhang DM, Guo SX (2004) A new sesquiterpene glycoside from Dendrobium nobile Lindl. J Asian Nat Prod Res 6(4):311–314

Si JP, He BW, Yu QX (2013a) Progress and countermeasures of Dendrobium officinale breeding. Zhingguo Zhong Yao Za Zhi 38(4):475–480

Si JP, Yu QX, Song X, Shao WJ (2013b) Artificial cultivation modes for Dendrobium officinale. Zhongguo Zhong Yao Za Zhi 38(4):481–484

Singh A, Duggal S (2009) Medicinal orchids: An overview. Ethnobotanical Leaflets 13:351–63

Slaytor MB (1977) The distribution and chemistry of alkaloids in the Orchidaceae. In: Arditti JA (ed) Orchids biology reviews and perspectives, vol 1. Cornell University Press, Ithaca and London

Song JX, Shaw PC, Sze CQ et al (2010) Chrysotoxine, a novel bibenzyl compound, inhibits 6-hydroxydopamine apoptosis in SH-SY5Y cells via mitochondria protection and NF-kappa B modulation. Neurochem Int 57(6):676–689

Song JI, Kang YJ, Yong HY et al (2012a) Denbinobin, a phenanthrene from Dendrobium nobile, inhibits invasion and induces apoptosis in SNU-484 human gastric cancer cells. Oncol Rep 27(3):813–818

Song JX, Shaw PC, Wong NS et al (2012b) Chrysotoxine, a novel bibenzyl compound selectively antagpnizes MPP+, but not rotenone, neurotoxicity in dopaminergic SH-SY5Y cells. Neurosci Lett 521(1):76–81

Sood SK, Rana S, Lakhanpal TN (2005) Ethnic aphrodisiac plants. Scientific Publishers (India), Jodhpur

Sritularak B, Anuwat M, Likhitwitayawuid K (2011) A new phenanthrenequinone from Dendrobium draconis. J Asian Nat Prod Res 13(3):251–255

Steinfield S, Cogan E, King LS et al (2001) Abnormal distribution of aquaporin-5 water channel protein in salivary glands for Sjogren's syndrome patients. Lab Invest 81:143–148

Su HJ (1985) Native orchids of Taiwan, 3rd edn. Harvest Farm Magazine, Taipei

Subedi A, Kunwar B, Choy Y et al (2013) Collection and trade of wild harvested orchids in Nepal. J Ethnobiol Ethnomed 9:64–73

Subedi A, Kunwar B, Vermeulen JJ, Choi Y, Tao Y, van Andel TR, Chaudhary RP, Gravendeel B (2011) Medicinal use and trade of wild orchids in Nepal. In: Subedi A (ed) New species, pollinator interactions and pharmaceutical potential of Himalayan orchids. PhD Thesis, Leiden University, Leiden, pp 83–109

Suen K, Chung H, Ma H et al (2013) Dendrobium huoshanense can reduce hydrogen peroxide-induced toxicity in SH-SY5Y cells. Neuroscience 2013, An Diego, California, 9–13 Nov 2013. Conference paper, Society for Neuroscience

Sukphan P, Sritularak B, Mekboonsonglarp W et al (2014) Chemical constituents of Dendrobium venustum and their anti-malarial and anti-herpetic properties. Nat Prod Commun 9(6):825–827

Sun YD, Wang ZH, Ye QS (2013) Composition analysis and anti-proliferation activity of polysaccharides from Dendrobium chrysotoxum. Int J Biol Macromol 62:291–295

Sun J, Zhang F, Yang M et al (2014) Isolation of alpha-glucosidase inhibitors including a new flavonol glucoside from Dendrobium devonianum. Nat Prod Res 28 (21):1900–1905

Sung CK (2002) Dendrobium moniliforme. In: Guo JX, Kimura T, But PPH, Sung CK (eds) International Collation of Traditional and Folk Medicine Northeast Asia, Vol IV, p. 142

Suwal PN (1970) Medicinal plants of Nepal. H.M. Govt. of Nepal, Ministry of Forests and Soil Conservation, Department of Medicinal Plants, Thapathali, Kathmandu

Suzuki H, Keimatsu I, Ito M (1932) Alkaloid of the Chinese drug "Chin-Shih-Hu". II. Dendrobine. J Pharm Soc Jpn 52:1049–1060

Suzuki II, Keimatsu I, Ito M (1934) Alkaloids of the Chinese drug "Chin-Shih-Hu". III. Dendrobine. J Pharm Soc Jpn 54:802–819

Tadamasa O, Susumu K, Takashi M, Yutaka K, Mitsutaka N, Toshihiko O, Fumihiko U, Masao S (1965) The structure of nobilonine. The second Alakaloid from Dendrobium nobile. Chem Pharm Bull 13(6):745–7

Takamiya T, Wongsawad P, Tajima N et al (2011) Identification of Dendrobium species used for herbal medicine based on ribosomal DNA internal transcribed sequences. Biol Pharm Bull 34(5):778–782

Talapatra B, Mukhopadhyay P, Chaudhury P, Talapatra SK (1982) Denbinobin, a new phenanthraquinone from Dendrobium nobile Lindl (Orchidaceae). Indian J Chem 21B:386–387

Tanagornmeatar K, Chaotham C, Sritularak BC et al (2014) Cytotoxic and anti-metastatic activities of phenolic compounds from Dendrobium ellipsophyllum. Anticancer Res 34(11):6573–6579

Tanaka Y, Htun N, Yee TT (2003) Wild Orchids in Myanmar, vol 1–3. The Foundation of Agricultural Development and Education, Bangkok

Tang ZZ, Cheng SJ (1984) A study of the raw plants for the Chinese Traditional Medicine "Huoshan Shi-hu". Bull Bot Res 4(3):141–146

Tang SL, Su HJ (1978) Flora of Taiwan Vol. 5

Tezuka Y, Yoshida Y, Kikuchi T, Xu GJ (1993) Constituents of Ephemerantha fimbriata. Isolation and structure elucidation of two new phenanthenes, fimbriol A and fimbriol B, and a new dihydrophenanthrene, ephemeranthol-C. Chem Pharm Bull 41:1346–1349

Thakur M, Dixit VK (2007) Aphrodisiac activity of Dactylorgiza hatagirea (D. Don) Soo in Male Albino Rats. Evid Based Complement Alternat Med 4(Suppl 1):29–31

Thammasiri K, Tang CS, Yamamoto HY, Kamemoto H (1986) Carotenoids and chlorophylls in yellow-flowered Dendrobium species. Lindleyana 1 (3):215–218

Thangraj S, Tsao WS, Luo YW et al (2011) Total synthesis of moniliformediquinone and calanquinone A as potent inhibitors of breast cancer. Tetrahedron 67:6166

Tian CC, Zha XQ, Pan LH, Luo JP (2013) Structural characterization and antioxidant activity of a low molecular polysaccharide from Dendrobium huoshanense. Fitoterpia 91:247–255

Trivedi VP, Dixit RS, Lal VK (1980) Orchids in the drug markets of Bareilly, Kanpur and nearby districts. Nagarjun (Calcutta) 23(8):157–163

Tsai AC, Pan SL, Liao CH et al (2010) Moscatilin, a bibenzyl derivative from the India orchid Dendrobium loddgesii, suppresses tumour angiogenesis and growth in vitro and in vivo. Cancer Lett 292(2):163–170

Tsai AC, Pan SL, Lai CY et al (2011) The inhibition of angiogenesis and tumor growth by denbinobin is associated with the blocking of insulin-like growth factor-1 receptor signaling. J Nutr Biochem 22 (2011):625–633

Tsubota K, Hirai S, King LS (2001) Defective cellular trafficking of lacrimal gland aquaporin-5 in Sjogren's syndrome. Lancet 357:688–689

Urech J, Fechtig B, Nuesch J, Vischer E (1963) Hircinol, eine antifungisch wirksaame Substanz aus Knollen von Loroglossum hircinum (L.) Rich. Helv Chim Axcta 46:2758–2766

Van Rheede HA (1703) Hortus Malabaricus, vol 12. Dutch East India Company, Kerala

Vandrame WA, Faria RT (2011) Phloroglucinol enhances recovery and survival of cryopreserved Dendrobium nobile protocorms. Scientia Hort (Amsterdam) 128 (2):131–135

Veeraju P, Rao NSP, Rao LJ et al (1989) Amoenumin, a 9,10-dihydro-5H-phenanthro- (4,5-b.c.d.)-pyran from Dendrobium amoenum. Phytochemistry 28(3):950–951

Vendrame WA, Carvalho VS, Dias JMM (2007) In vitro germination and seedling development of cryopreserved Dendrobium hybrid mature seeds. Scientia Hort 114:188–193

Vendrame WA, Carvalho VS, Dias JMM, Maguire I (2008) Pollination of Dendrobium hybrids using cryopreserved pollen. Hort Sci 43(1):264–267

Venkateswarlu S, Raju MS, Subharaju GV (2002) Synthesis and biological activity of isoamoenylin, a metabolite of Dendrobium amoenum. Biosci Biotechnol Biochem 66(10):2236–8

Verkman AS, Hara-Chikuma M, Papdopoulos MC (2008) Aquaporins—new players in cancer biology. J Mol Med (Berl) 86(5):523–529

Vidal J (1963) Les plantes utiles Du Laos. Cryptograms—Gymnospermes—Monocotyledones. Museum National d'Historie Naturelle, Paris

Vij SP (1995) Orchid genetic diversity in India: conservation and commercialization. Proc. 5th Asia Pacific Orchid Conference & Show, Fukuoka, 20–39.

Wang T, Ma G, Yang G, Pan Y (1997) Isolation and structure elucidation of three dihydrophenanthrenes from Ephemerantha lonchophylla. Zhong Yao Cai 20 (7):353–5

Wang CN, Chen XM, Guo SX et al (2001) Studies in the pharmacological activity of Mycena dendrobii. Wei Shen wu xue Tong bao 28(2):73–76

Wang YC, Lin CH, Chen CM, Liou JP (2005) A concise synthesis of denbinobin. Tetrahedron Lett 46 (47):8103–8104

Wang HZ, Lu JJ, Shi NN, Zhao Y, Ying QC (2007a) Analysis of genetic diversity among 13 Chinese species of Dendrobium based on AFLP. Fan Zi Xi Bao Sheng Wu Xue Bao 40(3):205–10

Wang M, Zhang CF, Wang ZT, Zhang M (2007b) Studies on constituents of Dendrobium gratiosissimum. Zhongguo Zhong Yao Zha Zi 32(8):701–3

Wang L, Zhang CF, Wang ZT et al (2008) Studies on chemical constituents of Dendrobium crystallinum. Zhongguo Zhong Yao Za Zhi 33(15):1847–8

Wang HZ, Feng SG, Lu JJ et al (2009a) Phylogenetic study and molecular identification of 31 Dendrobium species using inter-simple sequence repeat (ISSR) markers. Scienta Hort (Amsterdam) 122(3):440–447

Wang L, Zhang CF, Wang ZT et al (2009b) Five new compounds from Dendrobium crystallinum. J Asian Nat Prod Res 11(11):903–11

Wang JH, Luo JP, Yang XF, Zha XQ (2010a) Structural analysis of a rhamnoarabinogalactan from the stems of Dendrobium nobile Lindl. Food Chem 122 (3):572–576

Wang JH, Zha XQ, Luo JP, Yang XF (2010b) An acetylated galactomannoglucan from the stems of Dendrobium nobile. Carbohydr Res 345(8):1023–1027

Wang Q, Gong Q, Wu Q, Shi J (2010c) Neuroprotective effects of Dendrobium alkaloids on rat cortical neurons injured by oxygen-glucose deprivation and reperfusion. Phytomedicine 17:108–115

Wang SW, Pan SL, Ou WC, et al. (2011) Moscatalin inhibits invasion by suppressing urokinase plasminogen activator expression through Akt inactivation in human hepatocellular varcinoma cells. Abstract 4236.

102nd Annual Meeting, Am Assoc Cancer Res, Orlando, Florida, USA

Wang XY, Luo JP, Chen R et al (2012) The effects of daily supplementation of Dendrobium huoshanense polysaccharide on ethanol-induced subacute liver injury in mice by proteomic analysis. Food Funct 5 (9):2020–2035

Wang CH, Fang JM, Yang WB (2015) Structure and bioactivity of the poysaccharides in medicinal plant Dendrobium husoshanense. Docket No 28A-970212, Department of Intellectual Property and Technology Transfer, Academia Sinica, Taipei, Taiwan

Wangchuk P (2009) High altitude medicinal plants of Bhutan. An illustrated guide for practical use. Pharmaceutical and Research Unit, Institute of Traditional Medicine Services, Ministry of Health, Thimphu

Warghat AR, Bajpai PK, Srivastava RB et al (2014) In vitro protocorm development and mass multiplication of an endangered orchid, Dactylorhiza hatagirea. Turk J Bot 38(4):737–746

Weckerle CS, Ineichen R, Huber FK, Yang YP (2009) Mao's heritage: Medicinal plant knowledge among the Bai in Shaxi, China at a crossroads between distinct local and common widespread practice. J Ethnopharmacol 123:213–228

Wei M, Jiang S, Luo JP (2007a) Study on the kinetics of two-stage cultivation of protocorm-like bodies from Dendrobium huoshanense for cell growth and synthesis of polysaccharides. Sheng Wu Gong Cheng Xue Bao 23(1):79–84

Wei M, Jiang S, Luo JP (2007b) Enhancement of growth and polysaccharide production in suspension cultures of protocorm-like bodies from Dendrobium huoshansense by the addition of putrescine. Biotechnol Lett 29(3):495–9

Wei M, Wei SH, Yang CY (2011) Effect of putrescine on the conversion of protocorm-like bodies of Dendrobium officinale to shoots. Plant Cell Tissue Organ Cult 102(2):145–151

Wongsawad P, Handa T, Yukawa T (2005) Molecular phylogeny of Dendrobium Callista-Dendrobium complex. In: Nair H, Arditti J (eds) Proc. 17th World Orchid Congress, Kuala Lumpur 2002. Kota Kinabalu: National History Publications (Borneo)

Wood HP (2006) The Dendrobiums. A.R.G. Ganter Verlag, Ruggell (Liechtenstein)

Wrigley TC (1960) Ayapin, scopoletin and 6.7-dimethoxycoumarin from Dendrobium thyrsiflorum (Reichb. f.). Nature 188:1108

Wu Y, Si J (2010) Present status and siustainable development of Dendrobium officinale industry. Zhongguo Zhong Yao Za Zhi 35(15):2033–2037

Wu TL, Hu QM, Xia NH, Lai PCC, Yip KL (2002) Check list of Hong Kong plants 2001. Hongkong, Agriculture, Fisheries and Conservation Department Bulletin 1

Wu XR (1994) A Concise Edition of Medicinal Plants in China. Guangdong Higher Education Publication House, Guangdong (in Chinese)

Wu KY, Wang WQ, Jin JX, Sun ZR, Long YL, Bai Y (2007) The comparison of different processing methods of Dendrobium loddigesii. Zhong Yao Cai 30(9):1067–9

Wu KG, Li TH, Chen CJ et al (2011) A pilot study evaluating the clinical and immunomodulatory effects of an orally administered extract of Dendrobium houshanense in children with moderate to severe recalcitrant atopic dermatitis. Int J Immunopathol Pharmacol 24(2):367–375

Wu CT, Huang KS, Yang CH et al (2013) Inhibitory effects of cultured Dendrobium tosaense on atopic dermatitis murine model. Int J Pharm 46(2):193–200

Xiang L, Sze SCW, Ng TB et al (2013) Polysaccharides of Dendrobium officinale inhibit TNF-alpha-induced apoptosis in A-253 cell line. Inflamm Res 62 (3):313–324

Xiao L, Ng TB, Feng YB et al (2011) Dendrobium candidum extract increases the expression of aquaporin-5 in labial glands from patients with Sjogren' syndrome. Phytomedicine 18(2–3):194–198

Xie MI, Hou BW, Han L et al (2010) Development of microsattelites of Dendrobium officinale and its application in purity identification of germplasm. Yao Xue Xue Bao 45(5):667–72

Xing YM, Chen J, Cui JL et al (2011) Antimicrobial activity and biodiversity of endophytic fungi in Dendrobium devonianum and Dendrobium thyrsiflorum from Vietnam. Curr Microbiol 62(4):1218–1224

Xu XH, Zhao TQ (2002) Effects of puerarin on D-galactose-induced memory deficits in mice. Acta Pharmacol Sin 23(7):587–90

Xu H, Li XB, Wang ZT, Ding XY, Xu LS, Zhou KY (2001) rDNA its sequencing of Herba Dendrobii (Huangcao). Yaoxue Xuebao 36(10):777–783

Xu J, Wang ZT, Ding XY, et al (2005): Differentiation of Dendrobium species used as "Huanghua Shihu" by rDNA ITS sequence analysis. Planta 2005-03-0232 Letter Neu −094-PDF1, 14.9.05/Druckerel Sommer

Xu H, Wang Z, Ding X, Zhou K, Xu L (2006) Differentiation of Dendrobium species used as "Huangcao Shihu" by rDNA ITS sequence analysis. Planta Med 72(1):89–92

Xu GH, Chang J, Mao SG et al (2008) Effects of sodium selenite on growth and anti- oxidative system in protocorm-like bodies of Dendrobium officinale in vitro. J Nat Sci Nanjing Normal Univ 12(1):62–66

Xu H, Ying Y, Wang ZT, Cheng KT (2010) Identification of Dendrobium Species by dot blot hybridization assay. Biol Pharm Bull 33(4):665–668

Xu J, Guan J, Chen XJ et al (2011) Comparison of polysaccharides from different Dendrobium using saccharide mapping. J Pharm Biomed Anal 55 (5):977–983

Xu J, Han QB, Li SL et al (2013) Chemistry, bioactivity and quality control of Dendrobium, a commonly use of tonic herb in traditional Chinese medicine. Phytochem Rev 12(2):341–367

Xu FQ, Xu FC, Hou B et al (2014a) Cytotoxic bibenzyl dimers from stems of Dendrobium fimbriatum Hook. Bioorg Med Chem Lett 24(22):5268–5273

Xu J, Li SL, Yue RQ et al (2014b) A novel and rapid HPGPC-based strategy for quality control of saccharide-dominant herbal materials: Dendrobium officinale, a case study. Anal Bioanal Chem 406 (25):6409–6417

Yamada K, Suzuki M, Hayakawa Y et al (1972) Total synthesis of denbinobine. J Am Chem Soc 94 (23):8278–8280

Yamamura S, Hirata Y (1964) Structures of nobiline and dendrobine. Tetrahedron Lett 79–87

Yan C, Zhu Y, Zhang X et al (2014) Down regulated aquaporin 5 inhiits proliferation and migration of human epithelial ovarian cancer 3AO cells. J Ovarian Res 7:78–87

Yang H, Chou GX, Wang ZT et al (2004) Two new fluorenones from Dendrobium chrysotoxum. J Asian Nat Prod Res 6(1):35–38

Yang KC, Uen YH, Suk FM et al (2005a) Molecular mechanisms of denbinobin-induced anti-tumourigenesis effect in colon cancer cells. World J Gastroenterol 11(20):3040–3045

Yang L, Zhang CF, Yang H et al (2005b) Two new alkaloids from Dendrobium chrysanthum. Heterocycles 65(3):633–636

Yang L, Qin LH, Bligh SW, Bashall A, Zhang CF, Zhang M, Wang ZT, Xu LS (2006a) A new phenanthrene with a spirolactone from Dendrobium chrysanthum and its anti-inflammatory activities. Bioorg Med Chem 14(10):3496–3501

Yang L, Wang Z, Xu L (2006b) Simultaneous determination of phenols (bibenzyl, phenanthrene, and fluorenone) in Dendrobium species by high-performance liquid chromatography with diode array detection. J Chromatogr A 1104(1–2):230–7

Yang H, Sung SH, Kim YC (2007a) Antifibrotic phenanthrenes of Dendrobium nobile stems. J Nat Prod 70(12):1925–9

Yang L, Han H, Nakamura N, Hattoori M, Wang Z, Xu L (2007b) Bio-guided isolation of antioxidants from the stems of Dendrobium aurantiacum var. denneanum. Phytother Res 21(7):696–8

Yang L, Wang Y, Zhang G et al (2007c) Simultaneous quantitative and qualitative analysis of bioactive phenols in Dendrobium aurantiacum var denneanum by high-performance liquid chromatography coupled with mass spectrometry and diode array detection. Biomed Chromatogr 21(7):687–694

Yang YJ, Lee HJ, Huang HS et al (2009) Effect of scoparone on dopamine biosynthesis and L-DOPA-induced cytotoxicity in PC12 cells. J Neurosci Res 87(8):1929–1937

Yang H, Lee PJ, Jeong EJ et al (2011) Selective apoptosis in hepatic stellate cells mediates the antifibrotic effect of phenanthrenes from Dendrobium nobile. Phytother Res 26(7):974–980

Yang CB, Shih KS, Liou JP, et al (2014) Denbinobin upregulates miR-146a expression and attenuates IL-1beta-induced upregulation of ICAM-1 and VCAM-1 expression in osteoarthritis fibroblast-like synoviocytes. J Mol Med

Yang LC, Lu TJ, Hsieh CC et al (2014b) Characterization and immunomodulatory activity of polysaccharides

derived from Dendrobium tosaense. Carbohydrate Polymers 111:856–863

Yang S, Gong QH, Wu Q et al (2014c) Alkaloids enriched extract from Dendrobium nobile Lindl. attenuates tau protein hyperphosphorylation and apoptosis induced by lipopolysaccharide in rat brain. Phytomedi (Jena) 21(5):712–716

Yang SJ, Zhang XH, Cao ZY et al (2014d) Growth promoting Sphingomonas paucimobilis ZJSH1 associated with Dendrobium officinale through phytohormone production and nitrogen fixation. Microbiol Biotechnol 7(6):611–620

Yang D, Liu LY, Cheng ZQ et al (2015a) Five new phenolic compounds from Dendrobium aphyllum. Fitoterapia 100:11–18

Yang L, Liu SJ, Luo HR et al (2015b) Two new dendrocandins with neurite-outgrowth-promoting activity from Dendrobium officinale. J Asian Nat Prod Res 17(2):125–131

Ye Q, Zhao W (2002) New alloaromadendrane, cardinene and cyclocopacamphene type sesquiterpene derivatives and bibenzyls from Dendrobium nobile. Planta Med 68(8):723–9

Ye Q, Qin G, Zhao W (2002) Immunomodulatory sesquiterpene glycosides from Dendrobium nobile. Phytochemistry 61(8):885–90

Ye QH, Zhao WM, Qin GW (2003) New fluorenone and phenanthrene derivatives from Dendrobium chrysanthum. Nat Prod Res 17(3):201–5

Ye QH, Zhao WM, Qin GW (2004) Lignans from Dendrobium chrysanthum. J Asian Nat Prod Res 6(1):39–43

Ying Y, Xu H, Wang ZT (2007) Allele-specific diagnostic PCR authentication of Dendrobium thyrsiflorum. Yao Xue Xue Bao 42(1):98–103

Yoon MY, Hwang JH, Park JH et al (2011) Neuroprotective effects of Sg-168 against oxidative stress-induced apoptosis in PC 12 cells. J Med Food 14(1–2):120–127

Yonzone R, Lama D, Bhujel RB, Rai S (2013) Present availability status, diversity resources and distribution of medicinal orchid species in Darjeeling Himalaya of West Bengal, India. Int J Pharm Nat Med 1(1):14–35

Yuan H, Bai Y, Si J et al (2011a) Variation of monosaccharide composition of polysaccharides in Dendrobium officinale by pre-column derivatization HPLC method. Zhongguo Zhong Yao Za Zhi 36 (18):2465–2470

Yuan YH, Hou BW, Xu HJ et al (2011b) Identification of the geographic origin of Dendrobium thyrsiflorum on Chinese herbal market using trinucleotide microsatellite markers. Bio Pharm Bull 34(12):1794–1800

Zha XQ, Luo JP (2008) Production stability of active polysaccharides of Dendrobium huoshansense using long term cultures of protocorm-like bodies. Planta Med 74(1):90–3

Zha XQ, Luo JP, Jiang ST (2007) Induction of immuno-modulatory cytokines by polysaccharide from Dendrobium huoshanense. Pharm Biol 45(1):71–76

Zha XQ, Luo JP, Wei P (2009) Identification and classification of Dendrobium candidum species by fingerprint technology with capillary electrophoresis. S Afr J Bot 75(2):276–282

Zha XQ, Pan LH, Luo JP et al (2012) Enzymatic fingerprints of polysaccharides of Dendrobium officinale and their application in identification of Dendrobium species. J Nat Med 66(3):525–534

Zhang D, Liao J (2005) Determination of polysaccharides of Dendrobium candidum in test tube culture. Zhong Yao Cai 28(6):450–1

Zhang D, Zhang JJ (2005a) Effect of Coeloglossum viride var bracteatum extract on oxidation injury in subacute senescent model mice. Zhongguo Yi Xue Ke Xue Yuan Xuc Bao 27(6):729–33

Zhang D, Zhang JJ (2005b) Effect of Coeloglossum viride var. bracteatum extract on oxidation injury in subacute senescent model mice. Zhongguo Yi CXue Ke Xue Yuan Xue Bao 27(6):729–733

Zhang EQ (1999) Chinese materia medica. Shanghai University Publishing, Shanghai

Zhang B, Feng G, He H, Zhang D, Li T, Li S, Wei Q, He Y, Huang P (1999a) Study on cultivation of Dendrobium flexicaule by habitat imitation. Zhong Yao Cai 22(5):217–20

Zhang Z, Lin H, Wang L, Zhang L (1999) Cryopreservation of seeds and mass propagation in vitro of Dendrobium candidum. Proc. 16th World Orchid Conference, p. 440

Zhang M, Chen S, Li Q et al (2001a) Comparison with the content of total alkaloid of Dendrobium nobile in different growing conditions. Zhong Yao Cai 24(10):707–8

Zhang M, Huang HR, Liao SM, Gao JY (2001b) Cluster analysis of Dendrobium by RAPD and design of specific primer for Dendrobium candidum. Zhongyao Yao Za Zhi 26(7):442–7

Zhang GN, Bi ZM, Wang ZT, Xu LS, Xu GJ (2003a) Advances in studies on chemical constituents from plants of Dendrobium Sw. Chin Trad Herbal Drugs 34:S5–S8 (Appendix)

Zhang YB, Wang J, Wang ZT et al (2003b) DNA microarray for identificationof the herb of Dendrobium species from Chinese medicinal formulations. Planta Med 69(12):1172–4

Zhang GN, Zhong LY, Bligh SW, Guo YL, Zhang CF, Zhang M, Wang ZT, Xu LS (2005a) Bi- cyclic and bi-tricyclic compounds from Dendrobium thyrsiflorum. Phytochemistry 66(10):1113–20

Zhang T, Xu LS, Wang ZT, Zhou KY, Zhang N, Shi YF (2005b) Molecular identification of medicinal plants: Dendrobium chrysanthum, Dendrobium fimbriatum and their morphologically allied species by PCR-RFLP analyses. Yao Xue Xue Bao 40 (8):728–33

Zhang CF, Shao L, Huang WH et al (2006a) Phenolic components from herbs of Dendrobium aphyllum. Zhongguo Zhong Yao Za Zhi 33(24):2922–5

Zhang G, Zhang F, Yang L et al (2006b) Simultaneous analysis of trans and cis isomers of 2-glucosyloxycinnamic acids and coumarin derivatives on Dendrobium thyrsiflorum by high performance liquid chromatography (HPLC) photodiode array detection (DAD)-electrospray ionization (ESI)-tandem mass spectroscopy (MS). Anal Chim Acta 57(1):17–24

Zhang D, Liu GT, Shi JG, Zhang JJ (2006c) Effects of Coeloglossum viride var bracteatum extract on memory deficits and pathological changes in senescent mice. Basic Clin Pharmacol Toxicol 98(1):55–60

Zhang D, Liu G, Shi J, Zhang J (2006d) Coeloglossum viride var bractetatum extract attenuates D-galactose and NaNO₂ induced memory impairment in mice. J Ethnopharmacol 104(1–2):250–6

Zhang D, Zhang Y, Liu G, Zhang J (2006e) Dactylorhin B reduces toxic effects of beta- amyloid fragment (25–35) on neuron cells and isolates of rat brain mitochondria. Naunyn Schmiedebergs Arch Pharmacol 374(2):117–125

Zhang X, Gao H, Wang NL, Yao XS (2006f) Three new bibenzyl derivatives from Dendrobium nobile. J Asian Nat Prod Res 8(1–2):113–8

Zhang SX, Li X, Yin JL, Chen LL, Zhang HQ (2007a) Effect of c21 teriodal glycoside from root of Cynanchum auriculatum on D-galactose induced aging model mice. Zhongguo Zhong Yao Za Zhi 32 (23):2511–4

Zhang X, Xu JK, Wang J, Wang NL, Kurihara H, Kitanaka S, Yao XS (2007b) Bioactive bibenzyl derivatives and flourenones from Dendrobium nobile. J Nat Prod 70(1):24–8

Zhang CF, Shao L, Huang WH et al (2008) Phenolic components from herbs of Dendrobium aphyllum. Zhongguo Zhong Yao Za Zhi 33(24):2922–2925

Zhang CF, Wang M, Wang L et al (2008a) Chemical constituents of Dendrobium gratiosissimum and their cytotoxic activities. Indian J Chem 47(6):952–956

Zhang X, Tu FJ, Yu HY, Wang NL, Wang Z, Yao XS (2008b) Copacamphane, picrotoxane and cyclocopacamphane sesquiterpenes from Dendrobium nobile. Chem Pharm Bull (Tokyo) 56(6):854–7

Zhang XL, An LJ, Bao YM, Wang JY, Jiang B (2008c) D-galactose administration induces memory loss and energy metabolism in mice: protective effects of catalpol. Food Chem Toxicol 46(8):288–294

Zhang AL, Yu M, Xu HH et al (2013a) Constituents of Devonianum devonianum and their antioxidant activity. Zhongguo Zhong Yao Za Zhi 38(6): 844–847

Zhang XL, Liu JJ, Wu L et al (2013b) Quantitative variation of polysaccharides and alcohol-soluble extracts in F1 generation of Dendrobium officinale. Zhongguo Zhong Yao Za Zhi 38(21):3687–3690

Zhang XL, DSi JP, Wu LS et al (2013c) Field experiment of F1 generation and superior families selection of DEndrobium officinale. Zgongguo Zhong Yao Za Zhi 38(22):3861–3865

Zhao Q, Dan J, Xu Y, Zhanghao (1998) Pharmocognostic identification on crude drug of Dendrobium denneanum. Zhong Yao Cai 21(6):282–4

Zhao W, Ye Q, Tan X, Jiang H, Li X, Chen K, Kinghorn AD (2001) Three new sesquiterpene glycosides from

Dendrobium nobile with immunomodulatory activity. J Nat Prod 64(9):1196–200

Zhao W, Ye Q, Dai Martin MT, Zhu (2003) allo-aromadendrene and pricrotoxane-type sesquiterpenes from Dendrobium moniliforme. Planta Med 69 (12):1136–40

Zhao Y, Song YO, Kim SS, Jang YS, Lee JC (2007) Antioxidant and anti-hyperglycaemic activity of poly-saccharide isolated from Dendrobium chrysotoxum Lindl. J Biochem Mol Biol 40(5):670–7

Zhao MM, Zhang G, Zhang DW, Guo SX (2013) Molecular cloning and characterization of S-adenosyl-L-methionine decarboxylase gene (soSAMDC1) in Dendrobium officinale. Yaoxue Xuebao 48(6): 946–952

Zhonghua Bencao (2000) Health Department and National Chinese Medical Management Office (ed). Science and Technology, Shanghai

Zheng B (1990) The status of Yunnan Dendrobium drugs and their botanical origin. Zhongguo Zhong Yao Za Zhi 15(1):9–12, 61–2

Zheng WP, Tang YP, Zhi F, Lou FC (2000) Dihydroayapin, a new coumarin compound from *Dendrobium densiflorum*. J Asian Nat Prod Res 2 (4):301–4

Zheng Y, Xu LS, Wang ZT, Zhang CY (2005a) Location and relative quantity of coumarins in the stem of Dendrobium thyrsiflorum. Yao Xue Xue Bao 40 (3):236–40

Zheng Y, Xu LS, Wang ZT, Zhang CY (2005b) Distribution and dynamic change of coumarins in leaf and root of Dendrobium thyrsiflorum. Zhongguo Zhong Yao Za Zhi 34(9):1071–1074

Zhongyao Da Cidian (1986) Jiangsu New Medical College (eds). Shanghai Science and Technology, Shanghai

Zhou GF, Chen SH, Lv GY, Yan MQ (2013) Determination of naringenin in Dendrobium officinale by HPLC. Zhongguo Zhong Yao Za Zhi 38(4):52–523

Zhu GH, Ji ZH, Wood JJ, Wood HP (2009) Dendrobium Swartz. In: Chen XQ, Liu ZJ, Zhu GH et al (eds) Orchidaceae, Flora of China 25. Science Press/Missouri Botanical Garden Press, Beijing/St Louis

Zhu Y, Yuan H, Li G et al (2011) Study of metal contents in Dendrobium officinale. Zhongguo Zhong Yao Za Zhi 36(3):356–360

Genus: *Epipactis* Zinn.

Chinese name: *Huoshao Lan* (flaming orchid)

Epipactis is a terrestrial herb with erect, leafy stems. Rhizomes are subterranean. Leaves are sessile, plicate, spirally arranged, convolute and persistent. Inflorescence is erect and carries large, prominent bracts that are longer than the flowers, and many droopy flowers which are yellow, green or purple.

The genus *Epipactis* comprises some 25 species of handsome terrestrial orchids which are mainly distributed through the temperate zone of Asia, Europe and North America, but Thailand, Laos and Vietnam each has one species (Schuiteman and de Vogel 2000). *E. helleborine,* one of the most widely distributed orchids, is found throughout the temperate areas of the northern hemisphere (Fig. 11.1). It has even invaded gardens and other urban habitats in Glasgow, Scotland. Introduced from the Old World into North America, it escaped into the wild and spread rapidly across the United States (Drew and Giles 1951). Such an aggressive orchid is uncommon in the temperate zone. It is unlikely that this species will be threatened.

Like all orchids, *E. helleborine* depends on fungus for its germination (fully myco-heterotrophic), but unlike other species, it remains partially myco-heterotrophic throughout its adult stage. A myco-heterotrophic species is an achlorophyllous plant (one devoid of chlorophyll) that obtains carbon from mycorrhizal fungi because it is unable to capture it from the air. Y. Ogura-Tsujita and T Yukawa (2008) reported that *E. helleborine* found in Japan thrive in locations where they are able to establish an association with ectomycorrhizal taxa of the Pezizales. In the coastal dunes, *E. helleborine* co-exists with *Wilcoxina* which is ectomycorrhizal on pine trees growing in the dunes. In Europe, *E. microphylla* is myco-heterotrophic and derives its sustenance from *Tuber species* of the *Pezizales* (Seiosse et al. 2004).

A plant used by the ancient Greeks and given the name *epipaktis* by Theophrastus shares the same name as this genus which was named by Swartz in 1800. His choice of the name was not explained.

Epipactis helleborine (L.) Crantz.

Chinese name: *Xiaohuahuoshao Lan* (small-flowered flaming orchid)
Chinese medicinal name: *Yezhulan*

Description: An alpine, terrestrial orchid with an erect stem carrying several green leaves and a terminal inflorescence with 3–40 droopy, greenish-white to green flowers with purple markings on the lip. Bracts are foliaceous,

© Springer International Publishing Switzerland 2016
E.S. Teoh, *Medicinal Orchids of Asia*, DOI 10.1007/978-3-319-24274-3_11

Fig. 11.1 *Epipactis*
helleborine (L.) Crantz [as
Helleborine latifolia (L.)
Druce]. From: Lindman,
C.A.M., *Bilder ur Nordens*
Flora, vol. 3: t. 648 (1922–
1926). Courtesy of
plantillustrations.org

648

BREDBLADET HULLÆBE, HELLEBORINE LATIFOLIA

linear-lanceolate and are larger than the flowers, the latter 8–13 mm long (Fig. 11.2). It flowers in July to August. *E. helleborine* is found in grassy locations and along ravines at 300–3600 m across the northern temperate region in Europe and Asia, and in northwest Africa.

E. helleborine has a wide pH tolerance. Although it is autotrophic, i.e. it has green leaves and is capable of supporting itself through photosynthesis, it may live underground in total dependence on its mycorrhiza for up to 3 years (Light and MacConaill 1991). There is also an albino form which is less robust (Rasmussen 1995).

There is some disagreement as to whether the green-coloured *E. tangutica* Schltr. found in

Fig. 11.2 *Epipactis helleborine* (L.) Crantz. [PHOTO: E.S. Teoh]

Minshan and Huanglong in Sichuan Province is a separate species or a variety of *E. helleborine* as proposed by Chen and Zhu in 2003 (Perner and Luo 2007).

Phytochemistry: *E. helleborine* produces the glucoside, loroglossin and quercetin O-glycosides (Williams 1979). Mannose-specific lectins with antiviral (Balzarini et al. 1992) and antifungal activities (Van Damme et al. 1994) has also been isolated from the orchid. Alkaloid is present (Luning 1964).

Herbal Usage: The medicinal plant is collected from Dongbei, Hubei, northwest Sichuan and Yunnan. It is used as a tonic. The root is employed in TCM to clear heat from the lungs and liver, stop coughs, clear phlegm, and improve breathing and blood circulation. It is used in the treatment of trauma, sore throat, toothache and painful eyes (*Zhongyao Da Cidian* 1986). It regulates the flow of vital energy (Chen and Tang 1982). The manual, *Kunming Commonly Used Folk Herbs*, advocates consuming a decoction made with 9 – 15 g of *Yezhulan* (*E. helleborine*) when one is afflicted with any of these conditions (*Zhongyao Da Cidian* 1986).

E. helleborine is employed to treat insanity, gout, headache and stomach disorders in Nepal (Baral and Kurmi 2006).

"Bastard Helleborine" is a common European name for *E. helleborine* which was an old folk remedy for gout.

Epipactis helleborine, var. *helleborine*

syn. *Epipactis teneii* Schltr.

Chinese name: *Huiyanhuoshao Lan* (grey rock flaming orchid), *Huoshao Lan* (flaming orchid)

Description: A terrestrial orchid with a short rhizome, numerous roots, and a tall (30–70 cm) stem with 5–8 ovate leaves, it is found in mountain thickets and grassy slopes at 1200–3200 m in Central and Southwestern China, from Shaanxi to Xizang. It is widespread throughout Europe, and also occurs in Pakistan and Nepal. It flowers in June and July. Between 10 and 20 nodding, well-distributed, yellowish to pale purple flowers with a dark red lip are carried on the terminal, erect inflorescence.

Herbal Usage: The medicinal plant is collected from Yunnan (Wu 1994), although the species and this variety is widely distributed throughout the temperate regions of the northern hemisphere. In TCM it is a substitute for *E. helleborine*, i.e. it is principally employed as a tonic. The entire plant is used to clear heat from the lungs and liver, and to stop coughs and clear phlegm, improve blood circulation, relieve toothache, diarrhoea, backache and to treat snake bites (*Zhongyao Da Cidian* 1986). Juice of the roots is administered to treat insanity and gout in Nepal (Vaidya et al. 2002; Das 2004; Subedi et al. 2013).

Epipactis latifolia var. *papillosa* (Franch. & Sav.) Maxim ex Kom. (see ***Epipactis papillosa*** Franch. & Sav.

Epipactis mairei Schltr.

Common name: *Dayehuoshao Lan* (big flower
   flaming orchid)
Other names (at Emei Shan): *Shan chu hua*
Chinese medicinal name: *Xiaozihanxiao*

Description: An attractive, terrestrial orchid with
a short rhizome, numerous roots and a tall
(30–70 cm) stem bearing 5–8 ovate leaves, it is
found in mountain thickets and grassy slopes at
1200–3200 m in central and southwestern China,
from Shaanxi to Xizang, and in the Indian
Himalayas. It is a common orchid at Huanglong
in Sichuan Province at 1700–2500 m, in exposed
locations such as scrub, woodland margins,
roadsides and open forests. It flowers in June
and July in China and until August in the
Himalayas. The newly emergent inflorescence
is a deep crimson bleaching to green when the
leaves develop. Between 10 and 20 nodding,
well-distributed, 2-cm-wide, yellowish to pale
purple flowers with a dark red lip are carried on
the terminal, erect, 40–60 cm tall inflorescence
(Chen et al. 1999; Perner and Luo 2007)
(Fig. 11.3).

   Usage: The plant is used as a tonic in Chinese
herbal medicine (*Zhonghua Da Cidian*, 1986). A
paste made from the vegetative portion is applied
to burns (Hu 1971). It dissolves extravasated
blood and improves the circulation (Chen and
Tang 1982). The medicinal plant is collected
from Gansu, Shanxi, Hunan, Hubei, Sichuan,
Yunnan and Xizang.

Epipactis papillosa Franch et Sav.

Chinese name: *Ximaohuoshao Lan* (fine fur
   flaming orchid)
Chinese medicinal names: *Jisuzihua* (cockerel
   crop flower); *Niushepian* (slice of a cow's
   tongue)

Description: A lowland, terrestrial orchid found
in the depths of temperate forests, its clustered
stems are erect, smooth, with about six oval,
plicate leaves and a tall, terminal, many-flowered
inflorescence. Flowers are droopy, star-shaped,

Fig. 11.3 *Epipactis mairei* Schltr. [PHOTO: Courtesy of
Plant Photo Bank of China]

cupped olive green with deep maroon on the
upper surface of the lip (*The Wild Orchids of
Japan* 1976, p 72). Flowering period is summer.
The species occurs in Japan, Korea, southern
Liaoning Province and the Russian Far East.

   Usage: In India, the stems and rhizomes were
employed to treat insanity (Duggal 1972). The
entire plant is used in Chinese herbal medicine to
strengthen the "middle burner" and replenish *qi*
after a spate of illness (i.e. as a tonic), and also to
treat cholera, testicular swelling and hernia. The
medicinal herb comes from Guizhou Province
(Wu 1994). Plants are collected in summer and
autumn (Anonymous 1970). Some examples of
Chinese herbal prescriptions employing
Epipactis are illustrated in Table 11.1.

Epipactis teneii Schltr. (see ***Epipactis
helleborine*, var. *helleborine*)**

Overview

E. royleana is known as *chhasakrunga*i to the
Chepangs of Nepal who regard it as food. The
orchid is found in pastureland at 1600–3500 m

Table 11.1 Herbal Remedies employing *Epipactis papillosa* and *E. helleborine*

1. Indication: weakness after an illness
   Cook *E. papillosa* 30 g with pork and consume

2. Indication: vomiting and diarrhea
   Boil *E. papillosa* 15 g and take three times a day

3. Indication: inguinal hernia
   Soak *E. papillosa* 30 g; with *Polygonum cuspidatum* 15 g; and *Akebia guinata* 15 g in wine and consume three times a day.

4. Indication: For swelling of male testicles (perhaps also hydrocoele)
   Add *Jisuzihua* (*E. papillosa* Franch. et Sav.) 30 g, *huzhang* (*Reynoutria japonica* Houtt.) 15 g, *xiaomutong* (*Clematis armandii*; evergreen Clematis) 15 g. Soak in wine.
   Take this 3 times a day, each dose 15 g

5. Indication: Chest pain; to remove stagnation of vital energy (*qi xie*): allergic purpura
   Use *Huoshaolan* (*E. helleborine*) 9 g, *hongmaoqi* (*Callophyllum robustum*) 9 g, *Sikuaiwa* [*Chloranthus holostegius* (Hand.-Mazz) Pei et Shan var. *trichoneurus* K.F.Wu; Henry chloranthus herb] (in chinese four pieces of baked clay) 9 g.
   Simmer in water.
   Add rice/millet wine to drink

References: *Zhongyao Da Cidian* (1986), *Zhonghua Bencao*, 2005

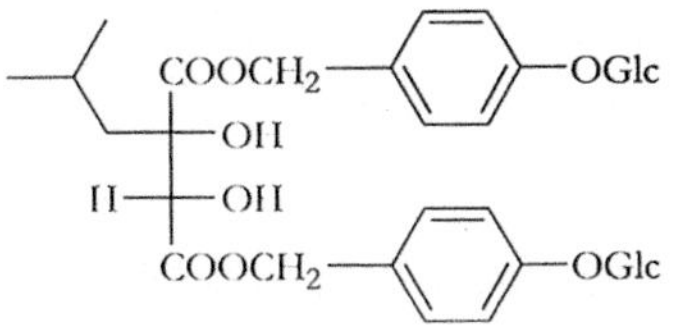

Fig. 11.4 Phytoalexins from *Epipactis*. Loroglossin is present in *E. helleborine* and *E. palustris*, whereas 1,7,-dihydroxy-5-methoxy-9,10-dihydrophenanthrene and 3,3′-dihydroxy-5,4′-dimethoxybibenzyl are present in *E. palustris*

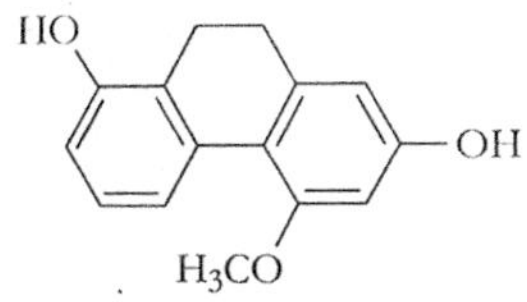

from Pakistan to northern India, Nepal, Bhutan and Tibet (Manandhar and Manandhar 2002).

Wu Xiu Ren (1994) reported that four species of *Epipactis*, namely *E. helleborine*, *E. latifolia*, *E. mairei* (syn. *E. yunnanensis*) and *E. tenii* (syn. *E. helleborine* var. *helleborine*) are employed as medicinal plants in China. So far, chemical studies have only been conducted on *E. helleborine* which botanically includes the so-called *E. tenii*. The structure of three phytoalexins isolated from *Epipactis* are illustrated in Fig. 11.4.

Fungi are generally a bane for orchid growers. *E. helleborine* produces a monomeric mannose-binding protein (a lectin) with in vitro antifungal properties (Van Damme et al. 1994) that enables it to cope with its permanent symbiont. The lectin is homologous with gastrodianin employed by

Gastrodia elata to control the fungus *Armillaria mellea* on which the orchid feeds (Wang et al. 2001). Gastrodin-like proteins (GLPs) are also present in other orchid species; there are seven such in *Listera ovata*. They are all related and some parts of their molecular sequences show a high degree of sequence homology (Van Damme et al. 1994).

Gastrodia elata and *E. mairei* are employed as tonics in Chinese herbal medicine. Their common denominator is a close association with mycorrhiza.

On the issue of medicinal usage, it should be noted that heavy metal localisation has been reported in mycorrhizas of *E. atrorubens* collected from mine tailings in Poland (Jurkiewicz et al. 2001). This should be a cause for concern, indicating that medicinal herbs must be regularly examined for heavy metal contamination.

Whereas Van Damme et al. (1994) found that this monomeric mannose-binding protein from *E. helleborine* tested negative for antiretroviral activity against immunodeficiency (AIDS) virus type 1 and type 2, the same team led by Balzarini et al. (1992) in Poland had earlier reported having found a mannose-specific lectin in *E. helleborine* (EHA) which was highly inhibitory to HIV virus type 1 and type 2 in MT-4 testing. It also showed a marked anti-human cytomegalovirus (CMV), respiratory syncytial virus (RSV) and influenza A virus activity in HEL, HeLa and MDCK cells, respectively. The 50 % effective concentration (EC50) was about three orders of magnitude below its toxicity threshold.

CMV is a silent but potentially serious infection if acquired during early pregnancy, and there is no antiviral drug for its treatment on the market. If a pregnant woman is freshly infected with CMV during her pregnancy, she has a 40 % chance that she will give birth to a child who is either mentally handicapped or who will develop deafness in childhood. While a current infection can be identified through sequential screening during pregnancy, tests so far are unable to distinguish between a normal foetus and an afflicted one. The problem with an antiviral for CMV is that it would need to be tested for potential teratogenicity before it can be administered to pregnant women. Presently, there is no reliable test that will guarantee absence of teratogenicity on humans.

Genus: *Eria* Lindl.

Chinese name: *Mao lan* (wool orchid)

Members of this large genus of epiphytic and terrestrial orchids look somewhat like star-shaped *Dendrobium*. However, *Eria* is immediately distinguishable from the latter by the presence of fine white hairs, rather like down, on the outer surfaces of the sepals and petals. They are among the commonest epiphytic orchids in Southeast Asia with the 375 species enjoying a distribution that extends to Tahiti in the east and Sri Lanka in the west. *Eria* are present in the lowlands and montane forests.

The generic name *Eria* derived from Greek, *erion* (wool), which describes the woolly perianth.

Recently, Chen and Wood (2009a, b) proposed a reclassification of *Eria* and, according to their proposal, *E. coneri* remains in the genus, but its correct name is *E. scabrilinguis*. *E. graminifolia* and *E. spicata* are moved to the genus *Pinalia*; *E. bambusifolia* is renamed *Callostylis babbusifolia* (Lindl.) Kuntz.; *E. pannea* is *Mycarantes pannea* (Lindl.) S.C. Chen and J.J. Wood; and *E. muscicola* is *Conchidium muscicola* (Lindl.) Rausch. These medicinal species are here located in their new respective genera.

Eria bambusifolia Lindl. [see **Callostylis bambusifolia (Lindl.) S.C.Chen & J.J.Wood**]

Eria bractescens Lindl.

Description: *E. bractescens* is a showy species (Figs. 11.5 and 11.6). Pseudobulbs are oblong, 6 by 2 cm, covered by large, striated sheaths. Leaves are 3–4, broad, 6–20 by 2 cm, blunted at the tip. Inflorescence is 12–20 cm long with numerous membraneous bracts and cream to yellow flowers. Flowers are up to 2 cm across, 1.4 cm long. Tepals are erect, lanceolate recurved at their tips. Lip is trilobed, bright yellow, and bears three keels marked with orange. The species is widely distributed from India to Southeast Asia and occurs in lowland forest as an epiphyte (Teoh 1980; Comber 2001).

Herbal Usage: In the Nicobar Islands, *E. bractescens* is employed to treat fever, malaria, or body and chest pain (Dagar and Dagar 2003).

Eria corneri Rchb. f. (see ***Eria scabrilinguis* Lindl.**)
Eria gramiifolia (Lindl.) Kuntz (see ***Pinalia graminifolia* Lindl.**)
Eria muscicola (Lindl.) Lindl. [see ***Conchidium muscicola* (Lindl.) Rausch.**]
Eria pannea Lindl. [see ***Mycaranthes pannea* (Lindl.) S.C. Chen & J.J. Wood**]

Fig. 11.5 Eria bractescens Lindl. From: Edwards Botanical Register, vol 30: t.29 (1844) [Original a colour drawing by S.A. Drake]. Courtesy of Missouri Botanical Gardens, St. Louis, USA

Fig. 11.6 *Eria bractescens* Lindl [PHOTO: Bhaktar B. Raskoti]

Eria scabrilinguis Lindl.

syn. *Eria corneri* Rchb. f.

Chinese name: *Banzhumao Lan* (half pillar woolly orchid), *Huangrong Lan* (yellow woolly orchid), *Darong Lan* (big woolly orchid, *Dayemaihu Lan* (big leaf wheat orchid), *Rong Lan* (wool orchid), *Suihuarong Lan* (spike woolly orchid), *Mengbi Lan* (grasshopper legs orchid), *Ganshimao Lan* (Gan family woolly orchid), *Fangguanmao Lan* (Fang and Guan woolly orchid). In Hong Kong: Corner's *Eria*, Four-corners *Eria*.

Chinese medicinal name: *Mengbi Lan*

Description: A beautiful, *Eria* worthy of horti-cultural consideration, this orchid has long, many-flowered inflorescences carrying small (1.5 cm in diameter) neatly spaced, yellow, star-shaped flowers with pointed sepals and petals. Fine hairs on the perianth (outer surface of the petals and sepals) are a prominent feature of its flowers. The species is distributed in south-ern China, from Taiwan to Fujian, Guangdong, Guangxi, Guizhou and Yunnan and Hainan. It is also \present in the Ryukyu Islands and in Vietnam

and Laos. It grows on rocks and trees in broad-leaved forests at low to medium altitudes (from 500 to 1500 m). In Taiwan, the flowering season is from September to November (Lin 1975), on mainland China, from August to October (Chen et al. 1999), in Hong Kong, from October to December (Wu et al. 2001).

Usage: The whole plant is antipyretic. It detoxifies, benefits the stomach, produces saliva and is used to treat hot illnesses, lack of saliva-tion, thirst, night sweats and running sores (Lin et al. 2003). Plants are harvested in summer or autumn, washed, steamed, then sun-dried. It is sweet in taste, and neutral in nature. Decoction is prepared with 6–15 g of the dry herb (*Zhongyao Da Cidian*, 1986).

Eria spicata (D. Don) Hand.- Mazz. [see *Pinalia spicata* (**D.Don**) **S.C. Chen & J.J Wood**]

Overview

Majumder's group working in Calcutta is the only team which has studied the chemical constituents in "*Eria*". However, of the six spe-cies that they worked on (Majumder and Banerjee 1988, 1990; Majumder and Kar 1987, 1989; Majumder and Rahaman 2006), four have now been assigned to the genus *Pinalia* (*E. acervata, E. confusa, E. stricta* and *E. convallarioides*). Of the two remaining within the genus *Eria*, only *E. carinata* is correctly named; for *E. flava* Lindl. the accepted name is *E. lasiopetala* (Willd.) Ormerod. Majumder and Rahaman (2006) did not give the full name of the *E. carinata*, but it is assumed that they were working on *E. carinata* Gibson and not on *E. carinata* (Blume) Rchb. f., because the former is an Indian species whereas the latter is distributed in Southeast Asia and Queensland.

E. lasiopetala (Willd.) Ormerod (syn. *E. flava*) contained a dimeric 9,10-dihydrophenanthrene derivative, flavanthrin (Majumder and Banerjee 1988), coelonin and two stilbenoids, flavathrindin and flavathinrinin (Majumder and Banerjee 1990). Nudol, a phen-anthrene originally isolated from *Eulophia nuda*, was found to be present in *E. carinata*, *Pinalia stricta* (LIndl.) Kuntz. (syn. *E. stricta*) (Bhandari et al. 1985) and *Pinalia spicata* (D.Don) Ormerod (syn. *E. covallarioides*) (Majumder and Kar 1989). Erianol, a 4-alpha-methylsterol, is present in *E. convallarioides* (= *Pinalia spicata*) (Majumder and Kar 1989) (see *Pinalia* for chemical structure of compounds.)

Eria and the related genera *Pinalia*, *Callostylis* and *Mycarantes* consist of hundreds of species which are widely distributed from the Himalayas to Southeast Asia, and more work needs to be done on their phytochemical analysis.

Genus: *Eulophia* R. Br.

Chinese name: *Meiguan Lan* (beautiful crown orchid)

Eulophia is derived from the Greek *eu* (well-developed) and *lophos* (crests) which describe the prominent calluses or crests on the lip of this terrestrial genus which thrives in sunny, exposed habitats. There are more than 300 species in the genus, which is distributed worldwide in the tropics. Pseudobulbs are angular. Leaves are long, narrow, lanceolate and leathery, appearing together with or after the flowers. An erect inflorescence arises from the base and carries many showy flowers which appear successively over a long period. Flowers may be white, yellow, greenish or purple.

Eulophia bicallosa (D. Don) P.F.Hunt & Summerh.

Chinese names: *Taiwan Meiguan Lan* (Taiwan beautiful crown orchid), *Lianchiyangersuan* (sickle wing goat ear garlic)
Japanese name: *Yukoku-ran* (deep valley orchid)

Description: This variable species is widely distributed from Japan (Kyushu and Okinawa) and Taiwan, Fujian, Guangdong, Guangxi, Hunan, Yunnan, Guzhou and Xizang provinces

to Nepal, Bangladesh, Myanmar, Thailand, Malaysia, Indonesia and northern Australia. Published descriptions vary. Plants in Japan are much smaller compared with those in Indonesia. In Japan, bulbs are green, 2 cm across, with 3–5 leaves, 7–15 by 1.5–4 cm, occasionally variegated. Inflorescence is erect, 27–85 cm tall, with 8–12 flowers, well spaced and facing all directions. Flowers are olive green, small. Petals and sepals narrow, lanceolate, pointed. Lip is white. By comparison (Fig. 11.7), *E. bicallosa* in Thailand are 60–80 cm tall, leaves are 60 by 1.5 cm, flowers are 3–3.3 cm across, olive green, and the lip is pale green with purple veins (Nanakorn and Watthana 2008). Both colour forms occur in Assam, India. It flowers in July in Japan, and in June in China and Assam, India (Chen et al. 2009a, b).

Herbal Usage: Chinese herbalists use the whole plant to enrich *qi*. It enriches the blood. It is employed as an antipyretic and an antidote for the treatment of tuberculosis, tuberculous lymphadenopathy, sores and ulcers, abdominal pain or distension and schistosomiasis (Wu 1994).

Eulophia campestris Wall ex Lindl. [see ***Eulophia dabia*** (D. Don) Hochr.]

Eulophia dabia (D. Don) Hochr.

syn. *Eulophia campestris* Wall ex Lindl.

Indian names: *Salibmisri, Sung Misrie, Charlemichhri* in Bengal, *Salum* (in Mumbai), *Salibmisri* (in Punjab); *Salu* (Gujerati dialect), *Salibmisri* (Hindi), *Salamisri* (Marathi), *Bongataini* (Santal), *Salabmisri* (Urdu), *Amrita, Amritobhava, Jiva, Jivani, Pranabhrita, Pranada, Sudhamuli, Virakanda* (Sanskrit)
Nepalese name: *Hattipaila*
Persian name: *Sungmisri*
Arabic name: *Kusyu-uth-thalab*

Description: *E. dabia* is distributed in a broad belt from Afghanistan, Baluchistan, Uzbekistan, the southern Himalayas, and south China

Fig. 11.7 *Eulophia bicallosa* (D.Don) P.F. Hunt & Summerh. Var. major [as *Eulophia bicarinata* (Lindl.) Hook. f. var. major. (From: Annals of the Royal Botanic Gardens, Calcutta, vol. 8(3): t.244A (1891) [drawn by R. Pantling]. Courtesy of Missouri Botanical Gardens, St. Louis, USA

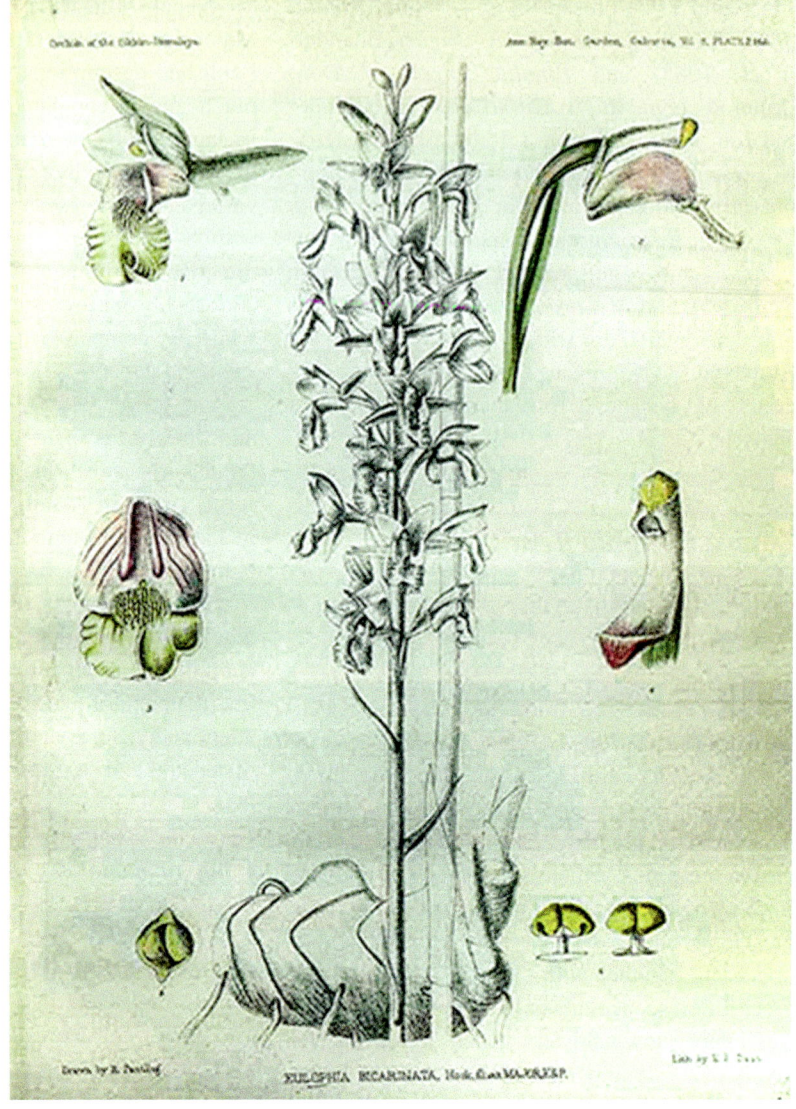

southwards across India and Upper Myanmar to the Andaman and Nicobar Islands. Pseudobulbs are well formed, bearing two linear leaves, 40 by 1 cm and a 45-cm-tall inflorescence which is many-flowered. Flowers are 2 cm across, pale pink with darker lines with narrow petals and sepals, the former not well extended, the latter curled backwards. Lip is narrow, as long as the sepals; purple on the mid-lobe (Manilal and Kumar 2004; Bose and Bhattacharjee 1980) (Fig. 11.8).

Phytochemistry: Tubers contain *n*-heacosyl alcohol and lupeol (Sood et al. 2005). Mucilage

of *E. campes*tris (= *E. dabia*) is eminently suitable for use as a binder in the preparation of tablets for the pharmaceutical trade. Such tablets display excellent physicochemical properties in terms of uniformity, hardness, friability, disintegration time and in vitro dissolution profiles. Ideal concentration of mucilage is 6–8 %. Using paracetamol as a model drug, the tablets employing *E. dadia* mucilage as binder released more than 85 % of the medication within 3 h (Ghule et al. 2006).

Herbal Usage: Tubers are eaten as tonic and aphrodisiac in India and Nepal. They are

Fig. 11.8 *Eulophia dabia* (D.Don) Hochr. (as *Eulophia hormusjii* Duthie). From: *Annals of the Royal Botanic Gardens, Calcutta,* vol. 9(2): t.109 (1906) [drawing by H. Hormusji]. Courtesy of Missouri Botanical Gardens, St. Louis, USA

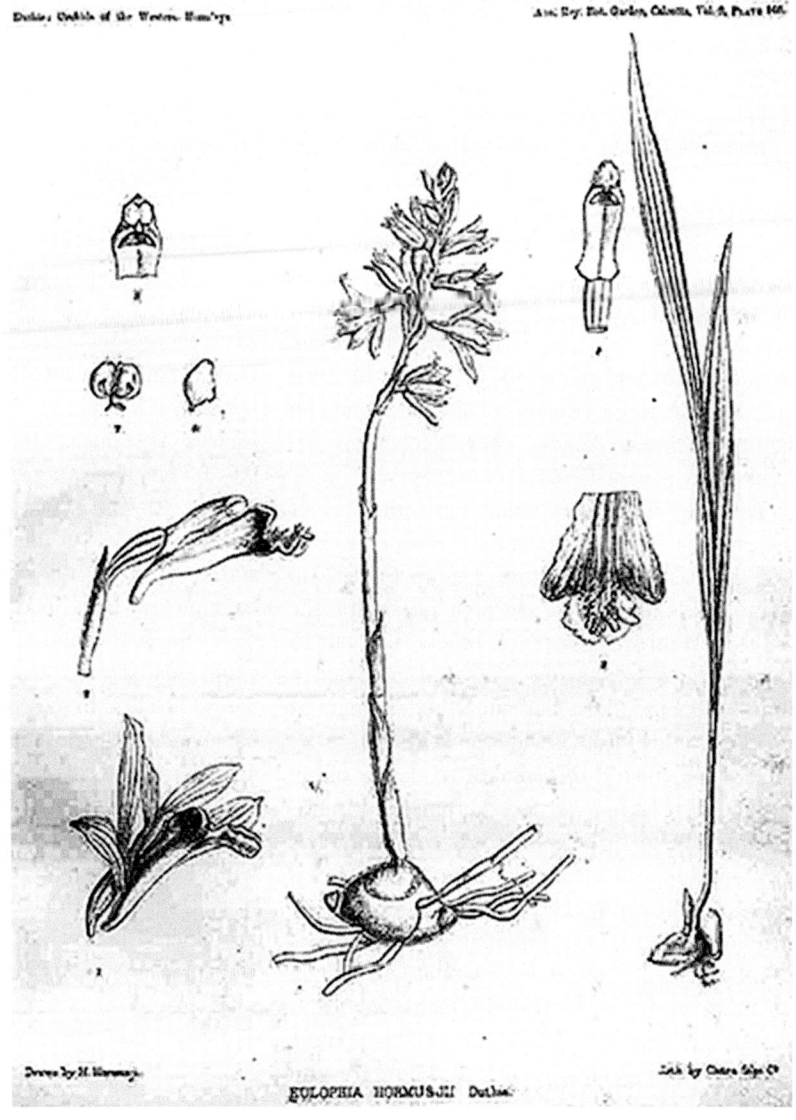

astringent and stimulate the appetite. Ayurveda practitioners prescribe it for stomach ache, poor appetite and to stimulate blood flow in patients suffering from heart disease, albeit its major role is as a tonic and aphrodisiac. For consumption, it is usually mixed with milk and flavoured with spices and sugar (Dymock 1885). Yunani practitioners recognise similar properties in pseudobulb powder and employ it to treat stomach ache, purulent coughs, and heart ailments and paralysis. They also use it as a tonic and aphrodisiac. Prepared as *salep*, it was once a prized tonic and aphrodisiac (Chopra 1933; Caius 1936; Chopra et al. 1958; Suwal 1970); Duggal 1972; Trivedi et al. 1980; Vaidya et al. 2002; Das 2004; Subedi et al. 2013). It is employed both internally and externally to treat scrofulous diseases of the neck, worms, poisoning (Nadkarni 1954), stomatitis, purulent coughs and heart ailments (Chopra et al. 1958). Tubers have similar usage in Nepal (Baral and Kurmi 2006). Tubers of *E. dabia* are also employed in Iranian medicine (Ghorbani et al. 2014).

Eulophia epidendraea (J.Koenig ex Retz) C.E.C. Fisch.

Indian Names: *Katou kaida maravara; Katou theka maravara; Segadomma gaddalu*

Description: Pseudobulbs are above-ground, green, leafy. Leaves are linear, grass-like, 30–60 cm long. Inflorescence is erect, simple or branched, and carries many pale green or yellow-green flowers with a white lip which is marked by pink veins and red crests. It flowers in April and May in India (Reddy et al. 2005), and in February, June, August and October in Sri Lanka (Jayaweera 1981).

E. epidendraea is an Indian terrestrial species distributed in southern India (Maharashtra, Kerala and the Eastern Ghats from Andhra Pradesh to Tamil Naidu), Sri Lanka and Bangladesh. It is the only terrestrial orchid found at sea level in southern India and grows among rocks in scrub, grassland or dry deciduous forests at an altitude range of 0–30 m (Sathish Kumar and Manilal 1994).

Phytochemistry: A dose of 200 mg of an extract of *E. epidendraea* administered intraperitoneally to diabetic rats reduced the levels of blood sugar in a glucose tolerance test, but it has only a fraction of the efficacy of the old antidiabetic agent, tolbutamide (Maridass et al. 2008). Tubers of *E. epidendraea* contain beta-sitosterol, beta-sitosterolglucoside, beta-amyrin and lupeol, whereas the leaves contain flavonoids, apigenin, luteolin, kaempferol and quercetin (Maridass and Ramesh 2010).

Herbal Usage: The tuber of *E. epidendraea* is sometimes used as vermifuge, demulcent and analgesic. It is applied externally for muscular pain (Trivedi et al. 1980). In the Anantapur district of southern India, tubers of the orchid with those of two non-orchidaceous plants, *Withania somnifer* (*Penneru gaddalu*) and *Curculigo orchioides* (*Nela Taadi*), in the ratio of 2:1:1 are crushed with a sufficient amount of pepper and garlic, and the extract is administered once a day for a week to restore appetite in anorexic subjects (Reddy et al. 2005).

Another remedy from the same district employs 100 g of the tubers, mixed with 50 g each of the fruit of *Terminalia bellirica,* *Terminalia chebula* and *Emblica officinalis* and a sufficient amount of pepper. These constituents are crushed and the mixture is administered orally once a day for 15 days to treat anthrax in domestic animals (Reddy et al. 2005).

Eulophia graminea Lindl.

Indian name: *Kattuvegaya*

Description: Pseudobulbs are above-ground, clustered, oval to long, 15 cm, with several internodes. Leaves are linear, thin, grass-like, 10–30 by 0.6–1.5 cm, appearing after anthesis. Inflorescence is erect, 40–75 cm, slender, paniculate, with 10–12 flowers spaced 1.5 cm apart. Flowers are 3 cm across, with lateral sepals that are horizontal, outstretched, lanceolate, brownish-green with reddish veins and a faint reddish flush. Dorsal sepal is slight smaller, erect and similarly coloured. Petals are borne close to the column, smaller, and curled backwards at the tip. Lip is 1.0 by 0.5 cm, trilobed and spurred at the base; side lobes are small, erect; mid-lobe, orbicular, crimson centrally, with a wide, white border, undulate margin and covered with numerous white tubercles over the crimson patch (Fig. 11.9). Flowering season is April to May in China (Chen et al. 1999) July to August in Bhutan (Gurung, 2006).

E. graminea is widely distributed from India and Sri Lanka to Southeast Asia, China and Japan (Ryukyu Islands), occurring in open spaces, grassland, wasteland and open forests at low elevations (Holttum 1964; Comber 2001; Chen et al. 2009a, b, c).

Herbal Usage: Eardrops made with juice of *E. graminea* are employed by the Paliyan tribes of Sirumalai Hills of southern India to treat earache (Karuppusamy 2007).

Eulophia herbacea Lindl.

Graphorkis herbacea (Lindl.), *Limodorum bicolor* Roxb., *Eulophia vera* Boyle, *E. albiflora* Edgew. ex Lindl., *E. brachypetala* Lindl., *Geodorum bicolor* (Roxb.) Voight, *Graphorkis bicolor* (Roxb.) Kuntz.

Fig. 11.9 *Eulophia graminea* Lindl. is a very robust, tropical species. [PHOTO: E.S. Teoh]

Chinese name: *Maochun Meiguan Lan*
Thai Name: *Wan Mangmum*

Description: *E. herbacea* is a terrestrial, herbaceous orchid occurring in lowlands in Indochina, Thailand, Bangladesh, throughout India from the western Himalayas to the Deccan (Karthikeyan et al. 1989), and in scrub or montane grasslands below 1500 m in southern Guangxi and southern Yunnan (Chen et al. 2009a, b). Plants are 25–50 cm tall. Pseudobulbs grow above the ground. They are ovoid, with a few thin roots and bear two or three leaves. Leaves are lanceolate, 15–30 by 2–5.5 cm, narrowing at the base to enclose a pseudostem of 15 cm height. Inflorescence is 22–50 cm tall with 6–10 flowers, loosely arranged. Flowers are 3–5 cm in diameter, sepals pale green to yellowish-green, petals and lip white flushed with yellowish-green at the base (Fig. 11.10). It flowers in May in Thailand (Vaddhanaphuti 2005) and June in China (Chen et al. 2009a, b).

Fig. 11.10 *Eulophia herbacea* Lindl. [From: Teoh Eng Soon: *Orchids in a Garden City*, Singapore: Marshall Cavendish, 2010]

Herbal Usage: *E. herbacea* tubers are sold in Indian bazaars as *salep misri*. It is used as a tonic and blood purifier for illnesses which are thought to be caused by impurities in the blood (Trivedi et al. 1980). In Thailand, the pseudobulbs are used to treat insect bites (Chuakul 2002).

Phytochemistry: 1-phenanthrene carboxylic acid was isolated from the tubers of *Eulophia herbacea* (Tatiya et al. 2014).

Eulophia hildebrandii Schltr. [see ***Eulophia spectabilis*** (Dennst.) Suresh]
Eulophia nuda Lindl. [see ***Eulophia spectabilis* (Dennst.) Suresh**]

Eulophia ochreata Lindl.

Indian names: *Amarkand, Singadyakand*

Description: *E. ochreata* is a pseudobulbous, terrestrial orchid, endemic in Peninsular India (Gujarat, Maharashtra, Karnataka, Andhra Pradesh, and Odisha) (Jalal and Jayanthi 2012). It is found on the humus-rich forest floor near watercourses at 900–1800 m (Misra 2007). A few lanceolate, plicate, undulate leaves arise from the base of a short stem completely ensheathing the latter. Flowers are a deep yellow with brown spots and crowded near the top of the scape. Lip is trilobed but the lateral lobes are indistinct. Flowering season is June to August (Rao 2007).

Phytochemistry: A phenanthrene, 9,10-dihydro-2,5 methoxyphenanthrene-1,7-diol isolated from the tubers of *E. ochracea* down-regulated NF-kB-regulated inflammatory cytokines and is a potential anti-inflammatory agent (Datla et al. 2010). It reduces inflammatory signalling by Toll-like receptors in mice. Pretreatment with the phenanthrene inhibited lipopolysaccharide (LPS)-stimulated NF-kB-activated inflammatory genes in vitro, and reduced LPS-induced TNF-alpha release and carrageenan-induced paw oedema in rats. Lipopolysaccharide (LPS)-stimulated mRNA expression of TNF-alpha, COX-2, intercellular-adhesion molecule-1, interleukin (IL)-8 and IL-1-beta were all reduced by treatment with the phenanthrene. What appeared to be more important was that, in the absence of provocation by LPS, the phenanthrene did not interfere with any cellular process (Datla et al. 2010). *E. orchreata* also contains 5,7-dimethyoxyphenanthrene-2,6-diol (Kshirsagar et al. 2010).

Usage: Tubers of *E. ochracea* are employed for a variety of conditions in Mumbai. The juice is applied externally to treat rheumatism. Tubers are eaten as a traditional vegetable. They have a high mineral content exceeding 2 % of dry weight (Aberoumard 2009). The author proposed cultivating this orchid and seven other plants with high mineral content as food supplements for the local people in southern India.

In Rajasthan, India, it is also reported that powdered pseudobulbs of *E. ochreata* are mixed with powdered tubers of *Chlorophytum borivilianum* in equal proportions and a teaspoon of the mix dissolved in milk is consumed daily for a month to boost the immune system. Another preparation consists of 100 g of *E. ochreata* pseudobulb, 100 g of *Sterculia urens* bark and 50 g of *Chlorophytum borivilianum,* ground into powder; half a teaspoon of the mix added to milk is consumed twice a day for 15–30 days to correct anaemia and eliminate fatigue. Tuber powder of *E. ochreata* is also administered to patients with leukaemia (Ketawa 2008).

In adjacent Maharasthra, tubers of *E. ochreata* are a speciality food, tonic, rejuvenating herb, aphrodisiac, and "blood purifier". They are employed to treat intestinal worms, coughs and cold and cardiac complaints. They are anti-inflammatory and anti-oxidant (Suresh et al. 2009). The Pawra tribe living in the Satpuda mountain ranges of Maharashtra employ the tubers to treat rheumatism and for rejuvenation and sexual dysfunction (Kshirsagar et al. 2010).

Eulophia pratensis Lindl.

syn. *Eulophia ramentacea* Wight

Local name: (Marathi): *Satavari*

Description: Caius (1936) mentioned that it was endemic in the Western Deccan, being found in pastureland. It is endemic in the Deccan, growing in exposed areas at 1400–2100 m. Flowers are bright yellow and appear in April (Matthew 1995), with the flowering stem, 35–50 cm tall, appearing before the leaves. Pseudobulbs are onion shaped, and covered with membraneous sheaths when young. Leaves are lanceolate, 25–30 cm long, with long petioles.

Herbal Usage: Tubers of *E. pratensis* are a main source of *salep misri* (Caius 1936). Fresh tubers were pounded and applied over the abdomen to kill worms [Greshoff 1900 quoted by Burkill (1935); Caius 1936]. They were also applied externally and taken orally to treat enlarged scrofulous glands in the neck (Chopra 1933; Singh et al. 1983). It was sold in the drug markets of Kanpur and the nearby districts in Uttar Pradesh in 1980 for such usage (Trivedi et al. 1980). In India, the dried roots were placed in small bags set around a building to drive away snakes (Lawler 1984).

Eulophia spectabilis (Dennst.) Suresh

Eulophia nuda Lindl., *E. squalida* Lindl., *E. bicolor* Dalz., *E. burkei* Rolfe ex Downie, *E. holochila* Collett & Hemsley, *E. macgregorii*

Ames, *Wolfia spectabilis* Denns, *Cyrtopera nuda* (Lindl.) Rchb., *Semiphajus chevalieri* Gagn.

Chinese names: *Meiguan Lan* (Beautiful crown orchid) *Zihuameiguan Lan* (purple flower beautiful crown orchid)

Indian names: *Bonga taini* in Orissa, *Amarcana, Manya* (Sanskrit), *Ambarkand, Goruma* (Hindi), *Ambarakand, Bhuikakali, Manakanda* (Marathi) *Rudbar* (in Bengal), *Mankand* (in Bombay); also *Balakanda, Granthidala, Kandalata, Malakanda, Panktikanda, Trishikhadala* (from Caius 1936: dialects/region not identified)

Nepalese name: *Amarkand*

Thai names: *Wan hua khru, Wan ung*

Description: Pseudobulbs are subterranean, subglobose, 3–4 cm in diameter with a few stout roots (Fig. 11.11). Leaves are 2 or 3, oblong-lanceolate, 20–40 by 2.5–6 cm (Chen et al. 2009a, b). Inflorescence is erect, 30–80 cm tall, and bears a dozen, well-spaced flowers which open about five at a time. Comb-like appendages are present on the lip (Fig. 11.12). It flowers February to April in southern India (Abraham and Vatsala 1981), April and May in Thailand (Vaddhanaphuti 2001), and up to June in China where it also grows at somewhat higher altitude of 1400–1500 m in mixed forests or grassy slopes (Chen et al. 1999). Thai–Malaysian flowers of this variable species have greenish petals and sepals with a purple lip (Go and Hamzah 2008), whereas the Chinese flowers are generally purplish-red (*E. nuda* var. *andersonii*) (Chen et al. 2009a, b). This purple variety is also extant in the Western Ghats of India (Abraham and Vatsala 1981). Floral and leaf shoots appear simultaneously and the leaves are still young when the plant is in bloom.

A widespread, terrestrial orchid distributed in tropical and subtropical Asia, it is found in the Western Ghats of India, tropical Himalayas, Myanmar and south China, Indochina, Malaysia, Indonesia, Philippines and the Pacific Islands. It thrives in the lowlands to 900 m, in the open or in lightly shaded areas.

Phytochemistry: *E. spectabilis* contains at least 9 phenanthrenes including coelonin, nudol, eulophiol, nudol, lupeol, 9,10 dihydro-2,5 dimethyl phenanthrene-1, and 9,10 dihydro-4-methoxy-phenanthrene-2,7-diol (Bhandari et al. 1985; Tuchinda et al. 1988, 1989). Nudol, a phenanthrene present in *E. spectabilis* (syn. *E. nuda*), is also present in *Eria carinata* and *Eria stricta* (Brandon et al. 1985; Bhandari et al. 1985; Tuchinda et al. 1988; Kovacs et al. 2007). It is 2,7-dihydroxy-3,4-dimethoxyphenethrene, and both nudol and its dimethyl ether was successfully synthesised in 1985 (Brandon et al. 1985).

The phenanthrene derivative isolated from *E. spectabilis* tubers, 9,10 dihydro-2,5-dimethoxyphenanthrene-1,7-diol, showed antiproliferative activity against human breast cancer cell lines MCF-7 (91 %) and MDA-MB-231 (85 %) at 1000 mcg/ml (Shriram et al. (2010). Tubers also contain alkaloids, saponins, cardiac glycosides, steroids and flavonoids (Ruchi et al. 2012). When tested in vitro for antiglycation effect, glycated products were decreased only with the highest concentration of Whitton root (*E. spectabilis*) extract (30 mg/ml) Lower concentrations did not produce an appreciable effect.

Herbal Usage: In Malaysia, a poultice made from the pseudobulbs is applied to the abdomen to kill intestinal worms, or to treat abscesses and infected wounds (Burkill 1935) This usage may be derived from Ayurvedic medicine because, in India, tubers were later reported to be used for tumours, scrofulous glands of the neck, bronchitis, and as vermifuge, blood purifier and an antidote for poisoning (Chopra et al. 1958; Duggal 1972; Trivedi et al. 1980; Rao and Sridhar 2007). Today, tribal dwellers at Kudremukh National Park in Karnataka prepare a decoction with the tubers (*amarcana*) to treat tumours and bronchitis or merely to be consumed as appetisers (Rao 2007; Rao and Sridhar 2007). The Dongria hill tribe in Orissa uses its leaves in decoction as vermifuge (Dash et al. 2008).

Fresh poultice made from a living plant was applied to boils and abscesses to get them to point and drain. Powder made from the tubers was given for intestinal worms (Caius 1936). Seeds are used to treat worms and scrofula (Nadkarni 1954).

Fig. 11.11 *Eulophia spectabilis* (Dennst.) Suresh (as *Eulophia nuda* Lindl.). Reproduced with permission from *Introductions to Orchids* by Abraham and Vatsala, Parlode, Thiruvananthapuram: Tropical Botanic Garden and Research Centre (TBGRI), 1981

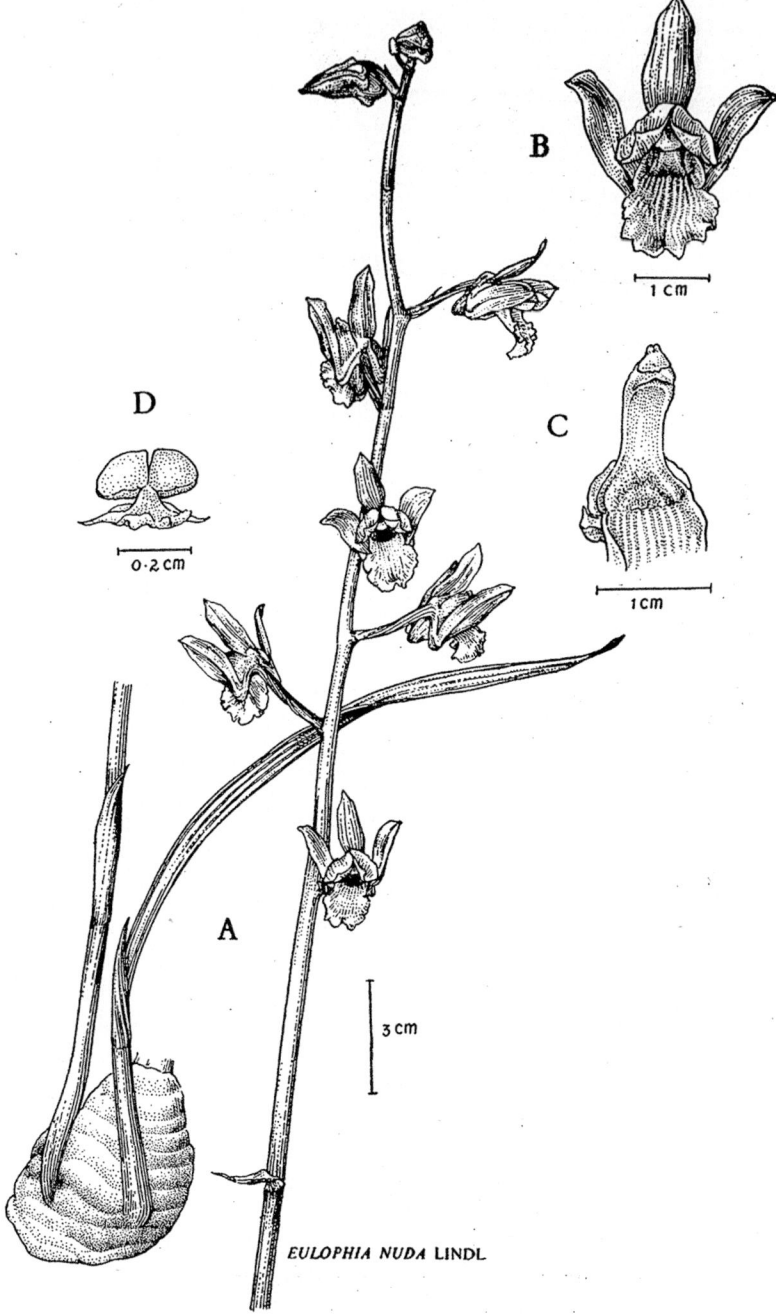

In India, the tubers are also employed as an aphrodisiac. Paradoxically, the Dongria Tribe in Orissa believes that a mixture of this orchid's dried pseudobulb when suitably combined with other herbs (10 g dried tuber, 5 g dried leaves of *Withonia somnifera*, 5 g dried leaves of *Curculigo* *orchioides* and 5 g black pepper, pounded and mixed with water, and consumed daily for 20 days) is effective as an anti-aphrodisiac (Dash et al. 2008). *E. spectabilis* is incorporated into many medicinal formulations. One out of six Nicobarese experts who had knowledge of plants

Fig. 11.12 *Eulophia spectabilis* (Dennst.) Suresh [From: Teoh Eng Soon: *Orchids in a Garden City,* Singapore: Marshall Cavendish, 2010]

in their traditional life on the islands in the Bay of Bengal volunteered that *E. spectabilis* could be used to treat stomach ache and related complaints (Dagar and Dagar 2003). Juice of the plant is used to treat snake bites (Ruchi et al. 2012).

Its usage in Nepal is similar to that in India. Powdered pseudobulb is employed to treat intestinal worms, scrofula, blood disorders, bronchitis and as an appetiser (Vaidya et al. 2002; Baral and Kurmi 2006; Subedi et al. 2013). In Chinese herbalists in Yunnan use the stem of *E. spectabilis* (syn. *E. hildebrandii*) to stop bleeding and pain from trauma or snake bite (Wu 1994).

Overview

Van Rheede (1693) was the first European to describe and illustrate *E. epidendraea* as a medicinal plant in his monumental *Hortus Indicus Malabaricus* in 1703. He referred to it by the Indian names, *Katou kaida maravara and Katou theka maravara*. Tubers of *E. campestris, E. epidendraea, E. herbaceae, E. pratensis and E. spectablis* are sold by herbalists in India as *salam-misri. Salep* is not mentioned in ancient

Sanskrit medical texts, and it appears to have been introduced via Arabic medicine. *Salep* of Bombay commerce is all imported from Iran, Afghanistan and northern India. In the later 19th century, three types of *salep* were available in the Bombay market: *Abushaheri* or *Lasaniya* which fetched Rs 15–25 per *maund* of 41 lbs, *Panjabi* which retailed for Rs 2.5–7 per lb., and *Panjah-i-s* (palmate or Persian *salep*) which fetched Rs. 10 per pound. Imitation *salep* prepared from potato and gum was sold to the unwary (Dymock 1885). *E. ochrrata* Lindl. which has a high mineral content is a vegetable in southern India (Aberoumard 2009). Luning (1974) tested 13 species of *Eulophia* for alkaloid and found that two had an alkaloid content exceeding 0.1 % of its tissues but his paper did not identify the species.

Of the 28 species that occur in India, a quarter are present in Maharashtra State and they are used medicinally. Roasted tubers are eaten to increase sperm count. Tuber extract is consumed to reduce liver swelling; powdered seed is added to sweetmeats to provide a tonic; and juice of the roots is an aphrodisiac as well as a remedy for snake bite and rheumatoid arthritis (Joshi et al. 2013). Trade in wild-harvested orchids for use as 'tonic' or aphrodisiac provides income for poor, isolated farmers but conservationists are concerned over its impact on biodiversity (Subedi 2005).

One encounters considerable confusion when studying the accounts of Indian usage of medicinal orchids arising from the mix-up with the names. One review of the ethnobotany of *Eulophia* describes an *E. ramenacea* Lind ex Wight (sic) which is used as an aphrodisiac, also to increase semen production and to treat gynaecological disorders. There is no species of this name. However, there is *E. ramentacea* (Roxb.) Lind. which is a synonym for *E. dabia* (D.Don) Hochr., and there is also *E. ramentacea* Wight which is a synonym for *E. pratensis* Lindl. Since the review goes on to mention that the orchid is employed by the Gujjar and Banjara tribes, one should deduce that the orchid referred to is *E. dabia* because the Gujjar tribe resides in northern India and Pakistan and *E. dabia* occurs in southern Himalayas, whereas *E. pratensis* is a south Indian species.

A recent Indian study reported that immature rats maintained on restricted feeding supplemented with a polyherbal aphrodisiac preparation which contained, among other things, flowers of *E. dabia* (syn. *E. campestris*), produced twice as many males in their litters when they were bred John et al. (2014)). However, the study did not prove that *E. dabia* flowers had any role to play in the result.

Many more species of *Eulophia* occur in Africa than in Asia, and in the former continent, a dozen species are employed medicinally, and some are used as charms. Aphrodisiac properties are attributed to roots of *E. reticulate*. There is an active trade in medicinal orchids (Chinsamy et al. 2011).

Five known phenanthrenes, four of which are present in *E. spectabilis*, and a mixture of phyosterols have been isolated from the roots of the African *E. petersii* (Blitzke et al. 2000). This species which thrives in the desert also occurs in Saudi Arabia; it is one of the local medicinal plants (but details are not available on its medicinal usage) and a methanolic extract showed cytotoxic activity (Almehdar et al. 2012). The structure of phenanthrenes isolated from *E. epidendraea* are shown in Fig. 11.13.

Genus: Flickingeria Hawkes

Desmotrichum Bl.; *Ephemerantha* Hunt & Summerh.

Chinese name: *Jin Shihu* (golden medicinal Dendrobium)

All member of the genus *Flickingeria* are now classified as *Dendrobium* Sw. The name *Flickingeria* was proposed by Alex Hawkes to distinguish a genus with some 65–70 epiphytic or lithophytic species which are characterised by a creeping, branching rhizome which bear a series of fusiform, ribbed, branching pseudobulbs of one inter-node and carrying a single leaf at intervals. The flowers are small, borne singly or occasionally in pairs, and barely last a day. The mentum is rather long and the lip is trilobed. *Flickingeria* is distributed from India across Southeast Asia to the Pacific Islands in lowland and montane forests. It is related to the genus *Dendrobium* and now included under *Dendrobium*

as a Section. Some species were previously listed under *Ephemerantha* (so-called because of the ephemeral nature of the flowers, thus unlike *Dendrobium*), and where the chemical research mentions this name we have retained it. In India, *Flickingeria fimbriata* was almost invariably referred to as *Dendrobium macraei*.

Overview

Flickingeria (*Ephemerantha*), being essentially an Asian tropical genus, one would expect to find many medicinal uses in Indian, Thai, Malay or Indonesian medicine. Its constituent species have changed names many times following various attempts to systematise the classification of the large genus *Dendrobium*. *Flickingeria* is probably here to stay. Its various species do not enjoy the range of medicinal uses accorded to say *Dendrobium* or *Bletillia* in China, or *Anoectochilus* in Taiwan, but, given its numerous Indian names, *F. macraei* appears to be widely regarded as an aphrodisiac in that subcontinent. There was a report that once 100 truckloads containing perhaps hundreds of thousands of *Dendrobium macraei* (*F. fimbriata*) were shipped from Nepal to India for medicinal use (Koopowitz 2001). *F. fimbriata* was shortlisted as one of two possible candidates for the role of *Sanjeevani*, the magical Indian herb mentioned in the *Ramayana* which revived the unconscious Lakshmana who had been hit by a poison arrow (Ganeshaiah et al. 2009). However, there has been no pharmacological investigation on the orchid in India.

There is disagreement on the proper identification of the *Flickingeria* species that occurs in northeast India and Nepal. Raskoti (2009) and Satish-Kumar (personal communication 2011) maintain that *F. fimbriata* does not occur in India, and the correct species should be *F. fugax* (Rchb. F.) Seidenf.

F. xantholeuca (the correct botanical name), frequently referred to as *Ephemerantha lonchophylla*, is lumped with numerous *Dendrobium* species employed for the production of the Chinese herbal remedy, *shihu*. TCM classifies *shihu* as an enhancer. It boosts the effects of a curative medicine when it is used in the recovery phase of an illness. On the other

beta-sitosterol 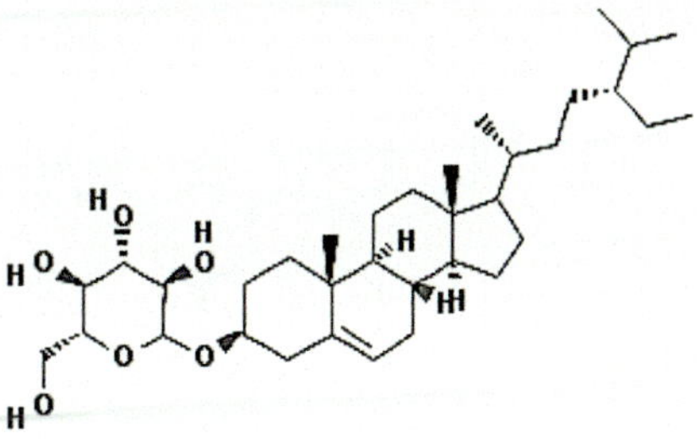   daucosterol, beta-sitosterol glucoside

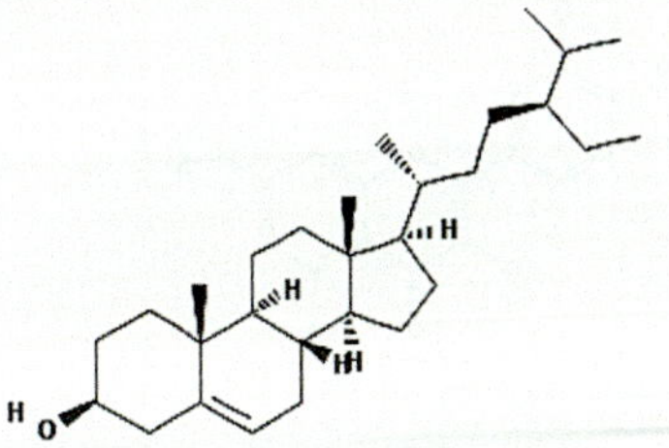

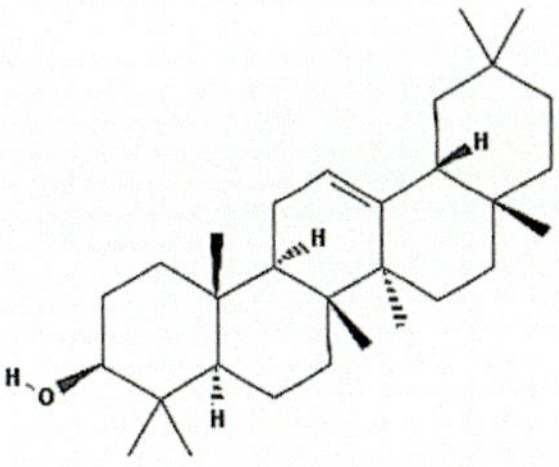

beta-amyrin 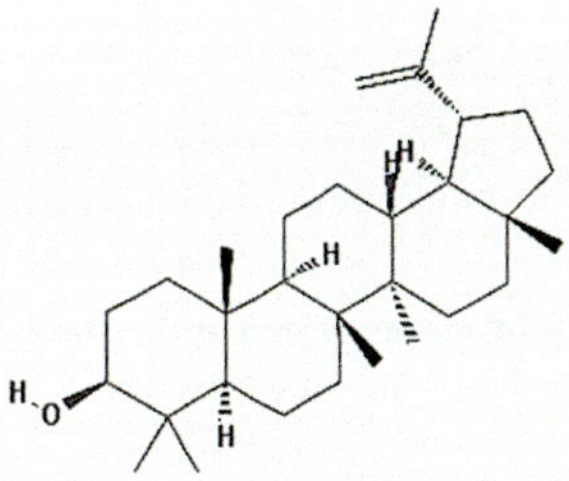   lupeol

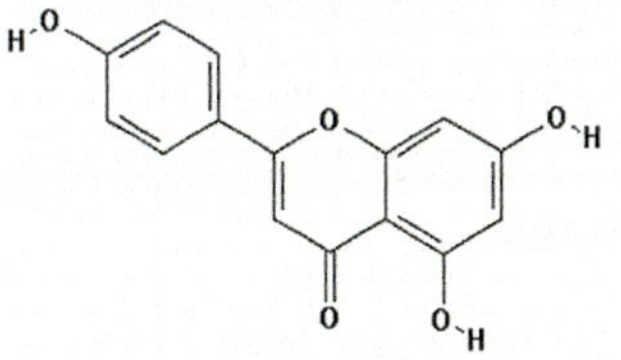

apigenin, chamomile 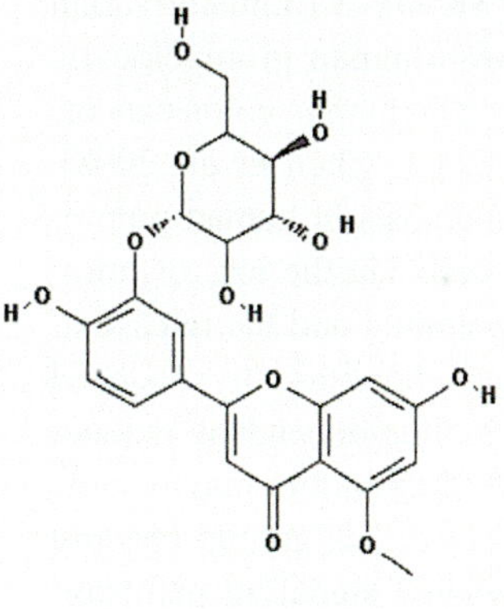   luteorin

Fig. 11.13 Phenanthrenes from *Eulophia epidendraea*

hand, if it is used prematurely, it may aggravate or prolong the illness.

Chinese scientists have studied *Flickingeria* for more than a decade. In 1997, Wang and his colleagues from the Nanjing University of Traditional Chinese Medicine isolated three dihydrophenanthrenes from the rhizomes and pseudobulbs of *F. xantholeuca* (*E. lonchophylla*), namely, lusianthridin, ephemeranthol-beta and erianthridin (Wang et al. 1997). At Taiwan's

National Research Institute of Chinese Medicine, Chen et al. (1999) managed to isolate three phenanthrene antioxidants from an ethanolic extract of *F. xantholeuca* (*E. lonchophylla*), but only one, ephenanthrene, was able to inhibit LDL oxidation in vitro. Extracts of *E. lonchophylla* also contained anti-platelet aggregation compounds (Chen et al. 2000; Chua and Koh 2006).

Denbinobin has been isolated from *F. xantholeuca* (*E. lonchophylla*) and

kaempferol 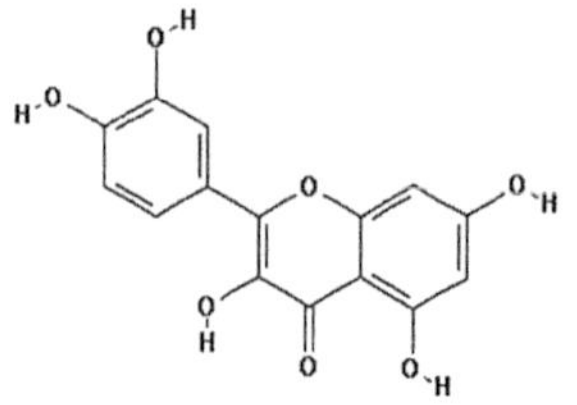   quercetin

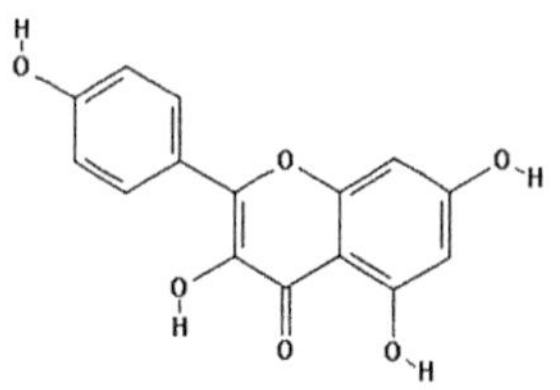

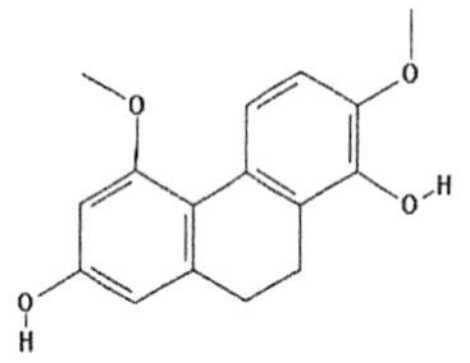

9,10-dihydro-4-methoxyphenanthrene-2,7-diol

Fig 11.13 (continued)

Dendrobium nobile at the Taipei Medical University. It is cytotoxic to the cell lines A549 (human lung cancer), SK-OV-3 (human ovarian adenocarcinoma) HL-60 (human promyelocytic leukaemia) and COLO 205 (human colon cancer) in vitro (Yang et al. 2005; Chen et al. 2008). Denbinobin induces apoptosis in human pulmonary adenocarcinoma cells via the loss of mitochondrial membrane potential and the release of mitochondrial apoptotic proteins. It produced AKT inactivation in a time-dependent manner (Kuo, Hsu et al. 2008). Denbinobin may be able to suppress breast cancer cell migration because it inhibits the Src-mediated signaling pathways that control cancer cell migration (Chen et al. 2011).

Of far greater interest is the isolation of two pimarane diterpenoids by researchers from Nanjing University of Traditional Chinese Medicine, their American collaborators in North Carolina State University in Raleigh and the Texas Christian University in Fort Worth. The two compounds, named lonchophylloids A and B, share an ability to overcome the resistance of myeloid leukaemia cells to the cytotoxic agent, doxyrubicin (Ma et al. 1998). However, there has been no follow-up publication in the last 15 years. Herbal medicinal usage of *Flickingeria* species in

China and India did not hint that there was any cytotoxic potential in the orchids.

Flickingeria fugax (Rchb. f.) Seidenf. (see ***Dendrobium fugax* Rchb.**)
Flickengeria fimbriata (Blume) A.D. Hawkes (see ***Dendrobium plicatile* Lindl.**)
Flickingeria nodosa (Dalzell) Seidenf, (see ***Dendrobium nodosum* Dalzell**)

References

Aberoumard A (2009) Preliminary assessment of nutritional value of plant based diets in relation to mineral nutrition. Int J Food Sci Nutr 60(Suppl 4):155–162

Abraham A, Vatsala P (1981) Introduction to Orchids, with illustrations and descriptions of 150 South Indian Orchids. TPGRI, Trivandrum

Almehdar H, Abdallah HM, Osman AM, Abdel-Sattar EA (2012) In vitro cytotoxic screening of selected Saudi medicinal plants. J Nat Med 66(2):406–412

Anonymous (1970) The wild orchids of Japan (Yasei ran jun yu). Seibundo Shinkoha, Tokyo

Balzarini J, Neyts J, Schols D, Hosoya M, Van Damme E, Peomans W, De Clerq E (1992) The mannose-specific plant lectins from Cymbidium hybrid and Epipactis helleborine and the (N-acetylglucosamine)n-specific plant lectin from Urtica dioica are potent and selective inhibitors of human immunodeficiency virus and cytomegalovirus replication in vitro. Antiviral Res 18 (2):191–207

Baral SR, Kurmi PP (2006) A compendium of medicinal plants in Nepal. Mrs Rachana Sharma and IUCN, Kathmandu

Bhandari SR, Kapadi AH, Mujumder PL, Joardar M, Shoolery JN (1985) Nudol, a phenanthrene of the orchids Eulophia nuda, Eria carinata and Eria stricta. Phytochemistry 24(4):801–804

Blitzke T, Masaoud M, Schmidt J (2000) Constituents of Eulophia petersii. Fitoterapia 71(5):593–594

Bose TK, Bhattacharjee SK (1980) Orchids of India. Naya Prokash, Calcutta

Brandon SR, Kapadi A, Majumder PL et al (1985) Nudol, a phenanthrene from the orchids, Eulophia nuda a, Eria carinata and Eria stricta. Phytochemistry 24 (4):801–804

Burkill IH (1935) (1966 reprint, 2nd edn, with contributions by Birtwistle W, Foxworthy FW, Scrivenor JB, Watson IG) A dictionary of economic products of the Malay Peninsula, Vol II. Crown Agents for the Colonies, Ministry of Agriculture & Co-operatives, London, Kuala Lumpur

Caius JF (1936) The medicinal and poisonous plants of India. J Bombay Nat Hist Soc 38(4):791–799

Chen SC, Tang T (1982) A general review of the Orchid Flora of China. In: Arditti J (ed) Orchid biology. Reviews and perspectives II. Cornell University Press, Itcaca

Chen SC, Tsi ZH, Luo YB (1999) Native orchids of China in colour. Science Press, Beijing

Chen CC, Huang YL, Teng CM (2000) Antiplatelet aggregation principles from *Ephemerantha lonchophylla*. Planta Med 66(4):372–374

Chen TH, Pan SL, Guh JH, Chen CC, Huang YT, Pai HC, Teng CM (2008) Denbinobin induces apoptosis by apoptosis-inducing factor releasing and DNA damage in human colorectal cancer HCT-116 cells. Naunyn Schmiedebergs Arch Pharmacol 378(5):447–457

Chen XQ, Cribb PJ, Gale SW (2009a) Erythroorchis Blume. In: Chen XQ, Liu Z, Zhu GH et al (eds) Flora of China—Orchidaceae. Science Press, Beijing

Chen XQ, Cribb PJ, Gale SW (2009b) Eulophia R. Brown. In: Chen XQ, Liu Z, Zhu GH et al (eds) Flora of China—Orchidaceae. Science Press, Beijing

Chen XQ, Luo YB, Cribb PJ, Gale SW (2009c) Epipactis Zinn. In: Chen XQ, Zj L, Zhu GH et al (eds) Flora of China—Orchidaceae. Science Press, Beijing

Chen PH, Peng CY, Pai HC et al (2011) Denbinobin suppresses breast cancer metastasis through the inhibition of Src-mediated signaling pathways. J Nutr Biochem 22(8):732–740

Chen XQ, Wood JJ (2009a) Epigeneium. In: Chen XQ, Liu ZJ, Zhu GH et al (eds) Flora of China—Orchidaceae. Science Press, Beijing

Chen XQ, Wood JJ (2009b) Flickingeria, A. D. Hawkes. In: Chen XQ, Zj L, Zhu GH et al (eds) Flora of China—Orchidaceae. Science Press, Beijing

Chinsamy M, Finnie JF, Van Staden J (2011) The ethnobotany of South African medicinal orchids. S Afr J Bot 77(1):2–9

Chopra RN (1933) The indigenous drugs of India. The Art Press, Calcutta. Republished as Chopra's Indigenous Plants of India, 2nd edn. Academic Publishers (1986)

Chopra RN, Chopra IC, Handa KL, Kapur LD (1958) Chopra's indigenous drugs of India. U.N. Dhur & Sons Pte Ltd., Calcutta

Chua TK, Koh HL (2006) Medicinal plants as potential sources of lead compounds with anti-platelet and anti-coagulant activities. Mini Rev Med Chem 6 (6):611–624

Chuakul W (2002) Ethnomedical uses of Thai Orchidaceous plants. Mahidol J Pharm Sci 29(3–4):41–45

Comber JB (2001) Orchids of Sumatra. Natural History Publications, Kota Kinabalu

Da Cidian Z (1986) Edited by Jiangsu New Medical College. Science and Technology Press, Shanghai

Dagar HS, Dagar JC (2003) Plants used in ethnomedicine by the Nicobarese of islands in Bay of Bengal, India. In: Singh V, Jain AP (eds) Ethnobotany and medicinal plants of India, vol 2. Scientific Publishers, Jodhpur, pp 773–775

Das SP (2004) Indian orchids in indigenous medicine system. In: Brirto SJ (ed) Orchids diversity and conservation. The Rapinat Herbarium, St John's College, Tiruchirappalli

Dash PK, Sahon S, Bal S (2008) Ethnobotanical studies on Orchids of Niyamgiri Hill Ranges, Orissa, India. Ethnobot Leafl 12:70–78

Datla P, Kalluri MD, Basha K et al (2010) 9,10-dihydro-2,5-dimethoxyphenanthrene-1,7-diol from Eulophia ochreata inhibits inflammatory signally mediated by Toll-like receptors. Br J Pharmacol 160(5):1158–1170

Drew WB, Giles RA (1951) Epipactis helleborine (L.) Crantz in Michigan, and its general range in North America. Rhodora 53:240–242

Duggal SC (1972) Orchids in human affairs (a review). Pharm Biol 11(2):1727–1734

Dymock W (1885) The vegetable Materia Medica of Western India. Education Society's Press, Bombay

Ganeshaiah KN, Vasudeva R, Shaanker RU (2009) In search of Sanjeevani. Curr Sci 97(4):484–489

Ghorbani A, Gravendeel B, Naghibi F, de Booer H (2014) Wild orchid tuber collection in Iran: a wake-up call for conservation. Biodivers Conserv 23(11):2749–2760. doi:10.1007/s10531-014-0746-y

Ghule BV, Darwhekar GD, Jain DK, Yeole PG (2006) Evaluation of binding properties of Eulophia campestris wall mucilage. Indian J Pharm Sci 68:566–9

Go R, Hamzah KA (2008) Orchids of peat swamp forests in Peninsular Malaysia. Kuala Lumpur, Peat swamp forest project, UNDP/GEF Funded (MAL/99/G321) Ministry of Natural Resources & Environment

Greshoff (1900) Greshoff 1900, quoted by Burkill, 1935

Gurung DB (2006) An illustrated guide to the orchids of Bhutan. DSB Publications, Thimphu

Holttum RE (1964) Orchids of Malaya, 3rd edn. Government Printers, Singapore

Hu SY (1971) Orchids in the life and culture of the Chinese people. Chung Chi J 10:1–26

Jalal JS, Jayanthi J (2012) Endemic orchids of peninsular India: a review. J Threat Taxa 4(15):3415–3425

Jayaweera DMA (1981) A revised handbook of the flora of Ceylon, vol II. A.A. Balkema, Rotterdam

John PK, Tripathi R, Johri R (2014) Effect of restricted feeding of normal food with Aphrodisiac Polyherbal preparation on prepubertal and adult male albino rats on fertility and sex ratio. J Exp Zool India 17(1):189–191

Joshi AR, Patil VN, Duokule SS (2013) Medicinal value of Eulophia (Orchidaceae) and its distribution in Maharashtra State, India: a review. Adv Plant Sci 26 (2):417–420

Jurkiewicz A, Turnau K, Mesjasz-Przybylowicz J, Przybylowicz W, Godzik B (2001) Heavy metal localization in Mycorrhizas of Epipactis atrorubens (Hoffm.) Besser (Orchidaceae) from zinc mine tailings. Protoplasma 218(3–4):117–124

Karthikeyan S, Jain SK, Nayar MP, Sanjappa M (1989) Florae Indicae Enumeration: monocotyledonae. Botanical Survey of India, Calcutta

Karuppusamy S (2007) Medicinal plants used by Paliyan tribes of Sirumalai hills of southern India. Indian J Nat Prod Res 6(5):436–442

Ketawa SS (2008) Indigenous people and forests: perspectives of an ethnobotanical study in Rajasthan (India). In: Raamawat KG (ed) Herbal Drugs: ethnomedicine to modern medicine. Springer, Berlin, pp 33–55

Koopowitz H (2001) Orchids and their conservation. Timber Press, Portland

Kovacs A, Vasas A, Hohmann J (2007) Natural phenanthrenes and their biological activity. Phytochemistry 69:1084–1110

Kuo CT, Hsu MJ, Chen BC, Chen CC, Teng CM, Pan SL, Lin CH (2008) Denbinobin induces apoptosis in human lung adenocarcinoma cells via Akt inactivation, Bad activation, and mitochondrial dysfunction. Toxicol Lett 177(1):48–58

Kshirsagar RD, Kanekar VB, Jagtap SD et al (2010) Phenanthrenes of Eulophia ochreata Lindl. Int J Green Pharm 2(2):76–78

Lawler LJ (1984) Ethnobotany of the Orchidaceae. In: Arditti J (ed) Orchid biology reviews and perspectives 3. Cornell University Press, Ithaca

Light MHS, MacConaill M (1991) Patterns of appearance in Epipactus helleborine. In: Wells TC, Willems JH (eds) Population ecology of terrestrial orchids. SPB Academic Publishing, The Hague

Lin IH, Chang YS, Chen IS, Hsieh WC (2003) The catalogue of medicinal plant resources in Taiwan, p 153

Lin TP (1975) Native orchids of Taiwan, vol 1. Southern Materials Center Inc, Taipei

Luning B (1964) Studies on the Orchidaceae alkaloids. I. Screening of species for alkaloids 1. Acta Chim Scand 18:1507–1516

Luning B (1974) Studies on *Orchidaceae* alkaloids. IV. Screening of species for alkaloids 2. Phytochemistry 6:857–861

Ma GX, Wang TS, Li Y, Yan P, Guo YL, Leblanc GA, Reinecke MG, Watson WH, Krawiec M (1998) Two pimarane diterpenoids from Ephemerantha lonchophylla and their evaluation as modulators of the multidrug resistance phenotype. J Nat Prod 61(1):112–115

Majumder PL, Banerjee S (1988) Structure of Flavanthrin, the first dimeric 9,10-dihydrophenanthrene derivative from the orchid, Eria flava. Tetrahedron 44(23):77303–77308

Majumder PL, Banerjee S (nee Bhattacharyya) (1990) Two stilbenoids from the orchid *Eria flava*. Phytochem 29(9): 3052–3055

Majumder PL, Kar A (1987) Confusarin and confusaridin, two phenanthrene derivatives of the orchid, Eria confusa. Phytochemistry 26(4):1127–1129

Majumder PL, Kar A (1989) Erianol, a 4alpha-methylsterol from the orchid Eria convallarioides. Phytochem 28(5):487–490

Majumder PL, Rahaman B (2006) Triterpenoids from the orchid Eria acervata. J Indian Chem Soc 83(1): 58–64

Manandhar NP, Manandhar S (2002) Plants and people of Nepal. Timber, Portland

Manilal KS, Kumar CS (2004) Orchid memories. A tribute to Gunnar Seidenfaden. Mentor Books, Calicut

Maridass M, Ramesh U (2010) Investigation of phytochemical constituents from Eulophia enpidendraea. Int J Biol Tech 1(1):1–7

Maridass M, Thangavel K, Raju G (2008) Antidiabetic activity of tuber extract of Eulophia epidendraea (Retz. Fisher (Orchidaceae) in alloxan diabetic rats. Pharmacologyonline 3:606–617

Matthew KM (1995) An excursion flora of central Tamilnadu, India. A.A. Balkaema, Rotterdam

Misra S (2007) Orchids of India A Glimpse. Bishen Singh Mahendra Pal Singh, Delhi

Nadkarni AK (1954) Dr. K.M. Nadkarni's Indian Materia Medica, vol 2, 3rd edn. Popular Book Depot, Bombay

Nanakorn W, Watthana S (2008) Queen Sirikit Botanic Garden Thai Native Orchids 1 and 2. Wanida Press, Chiang Mai

Ogura-Tsujita Y, Yukawa T (2008) Epipactis helleborine shows strong mycorrhizal preference towards ectomycorrhizal fungi with contrasting geographic distributions in Japan. Mycorrhiza 18(6-7):331–338

Perner H, Luo Y (2007) Orchids of Huanglong. Huanglong National Park, Sichuan Province, China

Rao TA (2007) Ethno Botanical data on wild orchids of medicinal value as practised by tribals at Kudremukh National Park in Karnataka. Orchid Newslett 2(2):1–7

Rao TA, Sridhar S (2007) Wild Orchids in Karnataka. A pictorial compendium. Institute of Natural Resources Conservation, Education, Research and Training (INCERT), Bangalore

Raskoti BB (2009) The orchids of Nepal. Bhakta Bahadur Raskoti and Rita Ale, Kathmandu

Rasmussen HN (1995) Terrestrial orchids from seed to mycotrophic plant. Cambridge University Press, Cambridge

Reddy KN, Reddy CS, Raju VS (2005) Ethno-orchidology of orchids of Eastern Ghats of Andhra Pradesh. EPRTI Newslett 11(3)

Ruchi KS, Ravi U, Sharad TU, Tiwari S (2012) Qualitative phytochemical analysis of Eulophia nuda Lind an endangered terrestrial orchid. Int J Pharm Res Bio-Sci 1(5):456–462

Satish Kumar C, Manilal KS (1994) A catalogue of Indian orchids. Bishen Singh Mahendra Pal Singh, Hehra Dun

Schuiteman A, de Vogel EF (2000) Cac Ci Ho Lan (Orchidaceae) Cua Thai Lan, Lao, Campuchia Va Viet Nam. Orchid Genera of Thailand Laos, Cambodia and Vietnam (Vietnamese-English edition). National Herbarium Nederland, Leiden

Seiosse MA, Faccio A, Scappaticci G, Bonfante P (2004) Chlorophyllous and achlorophyllous specimens of Epipactis microphylla, (Neottieae, Orchidaceae) are associated with ectomycorrhizal septomycetes, including truffles. Microb Ecol 47(4):416–426

Shriram V, Kumar V, Kishor PBK et al (2010) Cytotoxic activity of 9.10- dihydromethoxyphenanthrene-1,7-diol from Eulophia nuda against human cancer cells. J Ethnobotany 128(1):251–253

Singh U, Wadhwari AM, Johri BM (1983) Dictionary of economic plants in India. Indian Council of Agricultural Research (ICAR), New Delhi

Sood SK, Rana S, Lakhanpal TN (2005) Ethnic aphrodisiac plants. Scientific Publishers, Jodhpur

Suwal PN (1970) Medicinal plants of Nepal. H.M. Govt. of Nepal, Ministry of Forests and Soil Conservation, Department of Medicinal Plants, Thapathali, Kathmandu

Subedi A (2005) Orchids and sustainable livelihood: an initiative in Nepal Himalayas to manage globally threatened biodiversity. In: Raynal-Rogues A, Roguenant A, Prat D (eds) 18th world orchid conference proceedings, Djon, France. Naturalia Publications, Paris, pp 474–479

Subedi A, Kunwar B, Choi Y et al (2013) Collection and trade of wild-harvested orchids in Nepal. J Ethnobiol Ethnomed 9:64–72

Suresh J, Suhit G, Prashant B et al (2009) Validation of the potential of Eulophia ochreata L. tubers for its anti-inflammatory and antioxidant activity. Pharmacologyonline 2:307–316

Tatiya AU, Bari N, Surana SJ, Kalskar MG (2014) Effect of bioassay guided isolation of 1-Phenanthrene carboxylic acid from Eulophia herbacea Lindl., tubers on human cancer cell lines. Res J Phytochem 8(4):155–161

Teoh ES (1980) Asian Orchids. Times, Singapore

Trivedi VP, Dixit RS, Lal VK (1980) Orchids in the drug markets of Bareilly, Kanpur and nearby districts. Nagarjun 23(8):157–163

Tuchinda P, Udchachon J, Khumtaveeporn K et al (1988) Phenanthrenes of Eulophia nuda. Phytochemistry 27:3267–3271

Tuchinda P, Udchachon J, Khumtaveeporn K et al (1989) Benzylated phenanthrenes from Eulophia nuda. Phytochemistry 28(8):2463–2466

Vaddhanaphuti N (2001) A field guide to the wild orchids of Thailand, 3rd edn. Silkworm Books, Chiang Mai

Vaidya BN, Shrestha M, Joshee N (2002) Report of Nepalese orchid species with medicinal properties. In: Watanabe T, Tokano A, Bista MS, Saiju HK (eds) The Himalayan plants/can they have us? Proceedings of Nepal-Japan Joint Symposium on conservation and utilization of Himalayan medicinal resources. Japan Society for Conservation and Development of Himalayan Medicinal Resources

Van Damme EJ, Baklzarini J, Smeets K, Van Leuven F, Peumans WJ (1994) The monomeric and dimeric mannose-binding proteins from the Orchidaceae species Listera ovata and Epipactis helleborine: sequence homologies and differences in biological activities. Glycoconj J 11(4):321–332

Van Rheede HA (1693) Hortus Indicus Malabaricus, vol 12. Dutch East India Company, Kerala

Wang T, Ma G, Yang G, Pan Y (1997) Isolation and structure elucidation of three dihydrophenanthrenes from Ephemerantha lonchophylla. Zhong Yao Cai 20 (7):353–355

Wang X, Bauw G, van Damme EJ et al (2001) Gastrodianin-like mannose-binding proteins: a novel class of plant proteins with antifungal properties. Plant J 25(6):651–661

Williams CA (1979) The leaf flavonoids of the Orchidaceae. Phytochemistry 18(5):803–813

Wu XR (1994) A concise edition of medicinal plants in China. Guangdong Higher Education Publication House, Guangdong (in Chinese)

Wu TL, Hu QM, Xia NH, Lai PCC, Yip KL (2001) Check list of Hong Kong plants 2001. Agriculture, Fisheries and Conservation Department, Hong Kong, Bulletin 1

Yang KC, Uen YH, Suk FM et al (2005) Molecular mechanisms of denbinobin-induced antitumourigenesis effect in colon cancer cells. World J Gastroenterol 11(20):3040–3045

Zhonghua Bencao (2005) Edited by Health Department and National Chinese Medical Management Office. Science and Technology Press, Shanghai, 2000

Zhongyao Da Cidian (1986). Edited by Jiangsu New Medical College. Shanghai: Science and Technology Press, 1986

Genus: *Galeola* Lour.

Chinese name: *Shan Shan hu*
Japanese name: *Tsuchi Akebi*

Galeola is one of the several genera of parasitic, achlorophyllous, heterotrophic orchids that do not fix energy from sunlight but obtain it from complex organic substances which they absorb from the soil, substrate or fungi. They are leafless. There are around ten species in the genus (Fig. 12.1). The climbing, sprawling stems of the pan-Asian *G. nudifolia* may cover an area of 15 m², it is possibly the largest saprophyte in the world. Some species are spectacular when they are in bloom (Chen et al. 1999).

The generic name is derived from Latin *galea* (helmet) referring to the shape of the lip.

Galeola septentrionalis Reichb. [see *Cyrtosia septentrionalis* (**Rchb.**) **Garay**]

Galeola faberi Rolfe

Chinese names: *Shanshanhu* (Hill coral), *Shanhu Lan* (Coral orchid)
Chinese medicinal name: *Jinganyikehao*

Description: Rhizome is thick, creeping, 2 cm broad and 1–2 m tall, subterranean. Stems are reddish brown and woody towards the base.

Inflorescence is erect, paniculate. Raceme is 5–10 cm long with 4–7 yellow flowers, 3–3.5 cm across (Fig. 12.2). It is in bloom from May to July (Chen et al. 1999; Jin et al. 2009). Saprophytic *G. faberi* thrives in damp, humus-rich areas in sparse woods and bamboo forests at 1800–2300 m in Anhui, Sichuan, Guizhou and Yunnan provinces in China. It also occurs in Nepal, Vietnam and Sumatra.

Phytochemistry: Eight phenolic derivatives including gastrodin, *p*-hydroxybenzyl alcohol and hydroxybenzaldehyde, which are compounds originally isolated from *Gastrodia elata* and present in several species of achlorophyllous orchids, have been isolated from the rhizome of *G. faberi* but their actions were not evaluated. The five compounds are 4,4′-dihydroxy-diphenylmethane, 2,4-bis-(4-hydroxybenzyl)phenol, 5-methoxy-3-(2-phenyl-E-ethenyl)-2,4-bis(4-hydroxybenzyl phenol, bis-[4(beta-D-glucopyranosyloxy)-benzyl](S)*(-)-2-isopropylmalate and bis[4-(beta-D-glucopyranosyloxy) benzyl)] (S)-2-s-butylmalate (Li, Zhou and Hong 1993a). 4-(beta-D-glucopyranosyloxy) benzyl alcohol is also present. Bis[4(beta-D-glucopyranosyloxy)-benzyl](S)(-)-2-isopropylmalate was earlier isolated from *Cyrtosia septentrionalis* (*G. septentrionalis*) (see Fig. 12.3) (Yi et al. 1993).

Herbal Usage: Herb is obtained from Zhejiang, Yunnan, Guizhou and Sichuan. Entire plant and the large fruits are used as medicine for the treatment of gonorrhoea and scabies. It shares

Fig. 12.1 *Galeola falconeri* Hook. f. From: *Annals of the Royal Botanic Gardens, Calcutta*, vol. 8(3): t. 353 (1891). (Drawing by R. Pantling) Courtesy of Missouri Botanical Gardens, St. Louis, USA

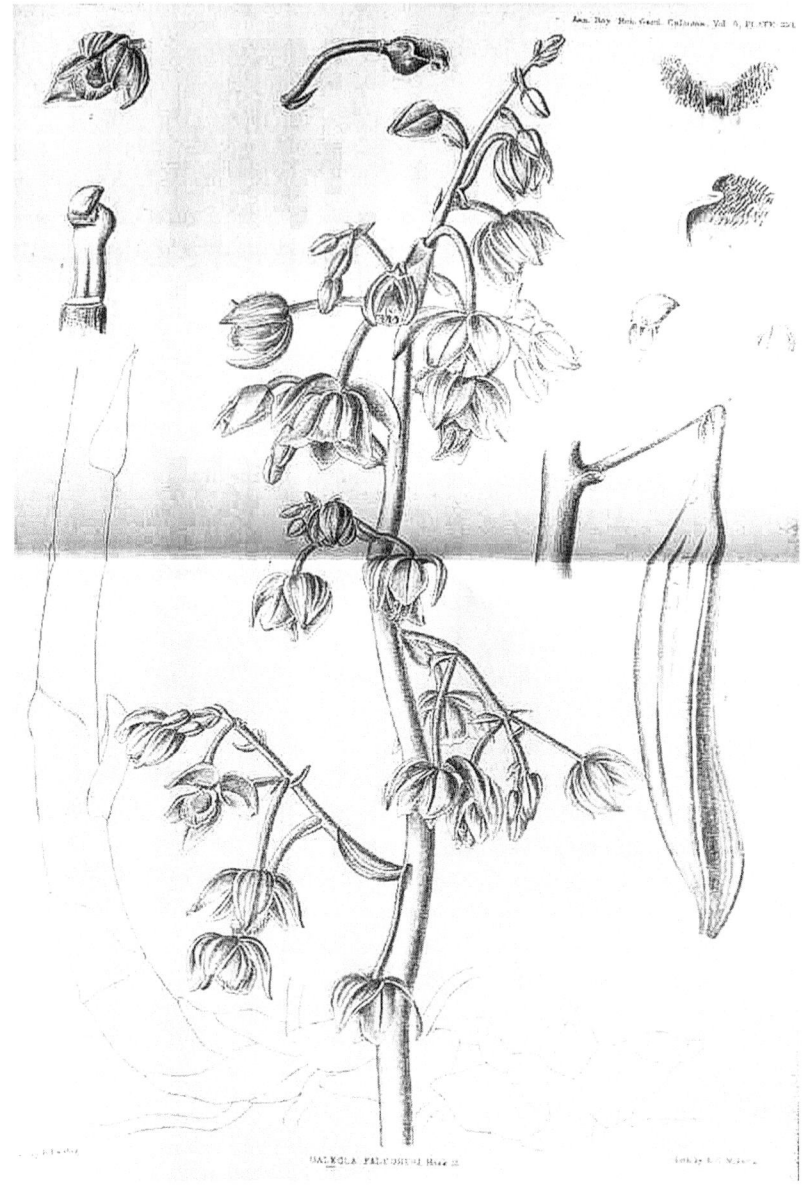

some similarity with *Gastrodia elata* (*Tienma*) in its constituents, but is rather weak in medicinal potency (Wu 1994). *Jinganyikehao* is also used to treat uterine prolapse (Chen and Tang 1982). Caution: **The plant is poisonous**.

Galeola lindleyana (Hook f. and Thomson) Rchb. f.

Chinese name*: Maoeshanhu Lan* (Hair stem coral orchid)

Taiwanese name: *Shan Hu* (mountain coral)

Description: This large saprophyte has stout, sub-terranean, horizontal, woody rhizomes 2–3 cm in thickness and 1–3 m tall giving rise to thick, erect, reddish-brown stems that are woody towards the base. Stems are furry, unbranched, with few, well-spaced bracts. Inflorescence is terminal, paniculate; racemes up to 10 cm long, with up to 15 flowers, the flowers 3.5 cm across, wide, yellowish, opening in succession from

Fig. 12.2 *Galeola faberi* Rolfe (Photo: Courtesy of Plant Photo Bank of China)

May to August. Seed-pods are large. Seeds have wings that are 1–1.3 mm wide.

G. lindleyana has a wide distribution in a crescent from Taiwan, across southern Shanxi, Henan, Anhui, Hunan, northern Guangdong, northern Guangxi, Guizhou, Yunnan and Xizang provinces in China to Vietnam, Peninsular Malaysia, Sumatra, Sikkim, Nepal and northeast India. It grows in rich, moist, rocky areas in sparse woods and thickets or along ravines at 1300–2200 m (Chen et al. 1999; Jin et al. 2009). In Bhutan, it flowers in June to July (Gurong 2006), while at Rishap Valley in Sikkim, it was observed to flower in July (Bose and Bhattacharjee 1980). In Malaysia, it is found in many localities in the lowlands up to 1200 m, in rather open places (Seidenfaden and Wood 1992). Plants grow in deep, moist organic matter but also thrive in clay or sandy soil in exposed places at Cameron Highlands in Peninsular Malaysia (Cheam 2009).

Herbal Usage: Herb is obtained from Shanxi, Hubei, Guangdong, Guangxi, Sichuan, Guizhou and Yunnan. Entire plant is used to combat heat. It promotes diuresis, stops bleeding, reduces swelling and is used in the herbal treatment of nephritis, haematuria (Wu 1994) and uterine

Fig. 12.3 Phenolic compounds from *Galeola* (Note: Glc = beta-D-glucopyranosyl)

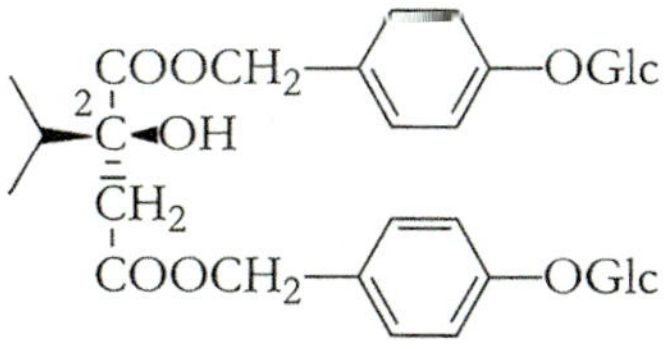

bis[4-(beta-D-glucopyranososyloxy)-benzyl](S)(-)-2-isopropropylmalate

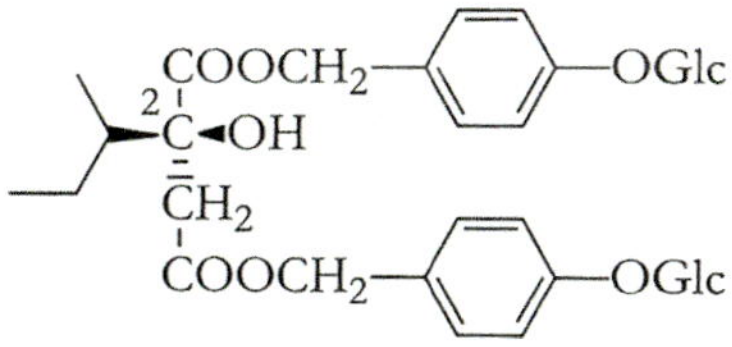

bis[4-(beta-D-glucopyranosyloxyloxy)benzyl](*S*)-(-)-2-sec-butylmalate.

prolapse (Chen and Tang 1982). Like the former species, *G. lindleyana* should also be regarded as poisonous.

Overview

Being an achlorophyllous orchids, all *Galeola* species should be investigated for potentially useful compounds. This has been done for *G. septentrionalis* Reichb. However, that species is now reclassified as *Cyrtosia septentrionalis* (Rchb.) Garay, and thus does not enter into the discussion here. Studies conducted at the Papua New Guinea University of Technology in Lae, found that leaves and stems of *G. foliata* (correct name: *Erythrorchis altissima* (Bl.) Bl.; Chinese name: *Daodiao Lan*) possessed broad spectrum bactericidal activity against 24 species of bacteria (Khan and Omoloso 2004). Pharmacological information is not available on the eight phenolic compounds that were isolated from *G. faberi* (Li, Zhou and Hong 1993a, b). There are no pharmacological data on *G. lindleyana*.

Genus: *Gastrochilus* D. Don.

Chinese name: *Penju Lan* (pot orchid)

A genus of epiphytic, monopodial orchids with close-set leaves and short, crowded racemes which bear 6–10 colourful flowers, *Gastrochilus* is a common orchid but it has not attracted much horticultural interest. Sepals and petals are well extended, and lip is saccular, the latter conferring the name to the genus—Greek *gastro* (stomach) *cheilos* (lip). There are about 40 species distributed from Sri Lanka through Southeast Asia to China and Japan.

Gastrochilus distichus (Lindl.) Kuntze

Chinese name: *Lieyepenju Lan* (leaves in a row, pot distance orchid)
Chinese medicinal name: *Fenghuangmao*

Description: Plant is 1.5 to 4.5 cm tall. Stems are slim, 2 mm in diameter, long, pendulous and branched. Leaves are lanceolate 1.5–3 by 0.4–0.6 cm. Inflorescence carries 2–4, greenish-yellow flowers with brown dots on all floral parts except the white circular extension of the prominent saccular lip. Its flowering season is May to June, the flowers lasting only a few days (Jin et al. 2009). The species occurs in southeast Tibet, western Yunnan, Sikkim, Bhutan and Nepal (Chen et al. 2009b). It is epiphytic or lithophytic at 1100–2700 m (Yang et al. 1998).

Phytochemistry: *G. distichus* (syn. *Saccolabium distichum* Lindl.) contains small amounts (around 0.001 % dry weight) of alkaloid (Luning 1964).

Herbal Usage: Herb is obtained from Yunnan and Tibet. The plant is used to treat mastitis (Wu 1994). Decoction is prepared with 3–9 g of the orchid plant (Hu et al. 2000).

Gastrochilus formosanus (Hayata) Hayata

Chinese names: *Taiwansong Lan* (Taiwan pine orchid), *Taiwannanchun Lan* (Taiwan pocket lip orchid), *Taiwan Pengju Lan* (Taiwan pot orchid), centipede orchid

Description: This small epiphytic orchid has a long flat stem that runs along tree branches, rooting at close intervals, somewhat in the fashion of *Vanilla*. Hence, its common name, Centipede orchid. Leaves are elliptic, pointed distally, 1.5–2.5 cm long and 0.4–0.6 cm wide. Inflorescence is short and carries 2–3 yellowish or greenish flowers which are minutely brown-spotted. Lip is shaped like a sac (*gaster*, Gr. belly; *cheilos*, lip) to which is attached a fan-shaped mid-lobe. The column is extremely short. It flowers erratically, mostly during winter. It occurs in Taiwan Fujian, Hubei and Shaanxi. In Taiwan, it is found on trees in broad-leaved forests at 1500–2000 m, in semi-shade (Lin 1975; Tang and Su 1978).

Herbal Usage: Whole plant is antipyretic, and is used for detoxification (Ou et al. 2003).

Gastrochilus obliquus (Lindl.) Kuntze

Chinese name: *Wujingpenju Lan*
Thai names: *Lin krabue noi; Sua luang; Chang rop kho*

Description. This is a widely distributed Thai epiphyte that bears a short inflorescence densely crowded with 20–26 yellow flowers in November. Stem is short barely 1–2 cm tall, bearing 3–5 oblong leaves measuring 15–20 by 4–6 cm. Inflorescences are axillary, multiple, with short peduncles, and densely many-flowered. Flowers are 2–2.7 cm across; some are fragrant. Sepals and petals are narrow and yellow with tiny red dots. The white lip is yellow at the base and purple over the sides (Fig. 12.4). Flowering season is October to January. The species is distributed from Nepal, Bhutan, Sikkim, southwest Sichuan and southern Yunnan to Myanmar, Thailand, Laos and Vietnam (Vaddhanaphuti 2001; Nanakorn and Watthana 2008; Chen et al. 2009a).

Fig. 12.4 *Gastrochilus obliquus* (Lindl.) Kuntze (Photo: E.S. Teoh)

Herbal Usage: Whole plant is used to treat body aches among village folk in Thailand (Chuakul 2002).

Overview

Medicinal information on *Gastrochilus* is scarce. Majumder and Bandyopadhyay (2010) isolated two new bibenzyl derivatives, designated gastrochilin and gastrochilinin, together with the known compounds, confusarin [2,7-dihydroxy-3,4,8-trimethoxyphenanthrene] and 2,7-dihydroxy-3,4,6-trimethoxyphenanthrene, from *G. calceolaria*, a species which is distributed from India and Nepal to southern China (Tibet, Yunnan and Hainan), Myanmar, Thailand and Malaysia. No medicinal usage has been reported for this species.

(* Note: The generic name *Gastrochilus* is also present in the Zingiberaceae (gingers) which are commonly used in herbal remedies, and this may cause confusion during a search.)

Genus: *Gastrodia* R. Br.

Chinese name: *Tianma*
Japanese name: *Oni No Yagara*

Among orchids that enjoy a medicinal reputation, *G. elata* (*Tianma*) ranks highest. Chinese tradition claims that 5000 years ago, around the dawn of its civilisation, Shen Nong taught people how to grow crops and use various herbs for healing and to promote health and longevity. This included the use of *Tianma* (*G. elata*), an orchidaceous herb that enjoys widespread usage today. Served as tea or in soup when cooked with chicken and other meats, there is a belief that it aids in tissue repair and boosts one's resistance to disease (Hu 2005).

Tianma has been extensively investigated in China and Japan. In a WHO publication of the medicinal plants of Korea, *G. elata* is the only orchid listed among 150 medicinal plants (Han et al. 1998). Several tertiary medical centres in the USA are exploring the possibility of using it to treat Parkinsonism and strokes. It is a unique plant in many ways.

Gastrodia is a genus of achlorophyllous, par-asitic orchids. Plant is without green leaves and lives off a mould on the forest floor. Tubers are irregular, somewhat ovoid-shaped, rather like a stomach (hence, *gastro*, Greek), and subterra-nean. Inflorescence rises above the forest floor. It bears several membranous sheaths, and many, a few, or single, small to medium-sized flowers which are not resupinate. Sepals and petals are joined together to form a tube (Fig. 12.4).

Over 20 species of *Gastrodia* are distributed from southern and tropical Africa (Cribb et al. 2010; Hsu and Kuo 2010) and India, east-wards through the rest of Asia to Australia, New Zealand and the Pacific Islands. There are 13 species in China (Chen et al. 2009b). They are not cultivated except for *G. elata* which is grown on a large commercial scale in China for medici-nal use (Fig. 12.5).

Gastrodia elata Blume

Local Name: *Tianma* (Heaven's fibre, Sky burlap)
Other Common Names: Chinese: *Ming Tianma* (Bright heaven's fibre); *Chi Jian* (Crimson arrow); *Chijiangen* (Red arrow root); *Ding Feng Cao* (Wind-calming herb); *Bai Long Pi* (White dragon skin); *Bailongcao* (White dragon grass)
Japanese name: *Tenma* (Heaven's fibre); *Oni-no-yagara* (Orge's arrow)
Korean name: *Cheon ma*

Description: Parasitic and leafless, *G. elata* lives underground and only becomes visible when it sends out its yellowish-red flower stalk or "Crim-son arrow" (*Chi Jian*). Rhizome is tuber-like, 8–12 cm long and 3–7 cm thick (Chen et al. 1999). Inflorescence is terminal, 5–30 cm tall with 30–50 white to orange flowers, loosely arranged almost in opposite-facing pairs over the distal third of the raceme (Chen et al. 1999) (Figs. 12.6 and 12.7). Flowering season is May to July all over China (Chen et al. 1999; Perner and Luo 2007; Jin et al. 2009).

Tuber of *G. elata* is one of the oldest drugs in Traditional Chinese Medicine (TCM), being

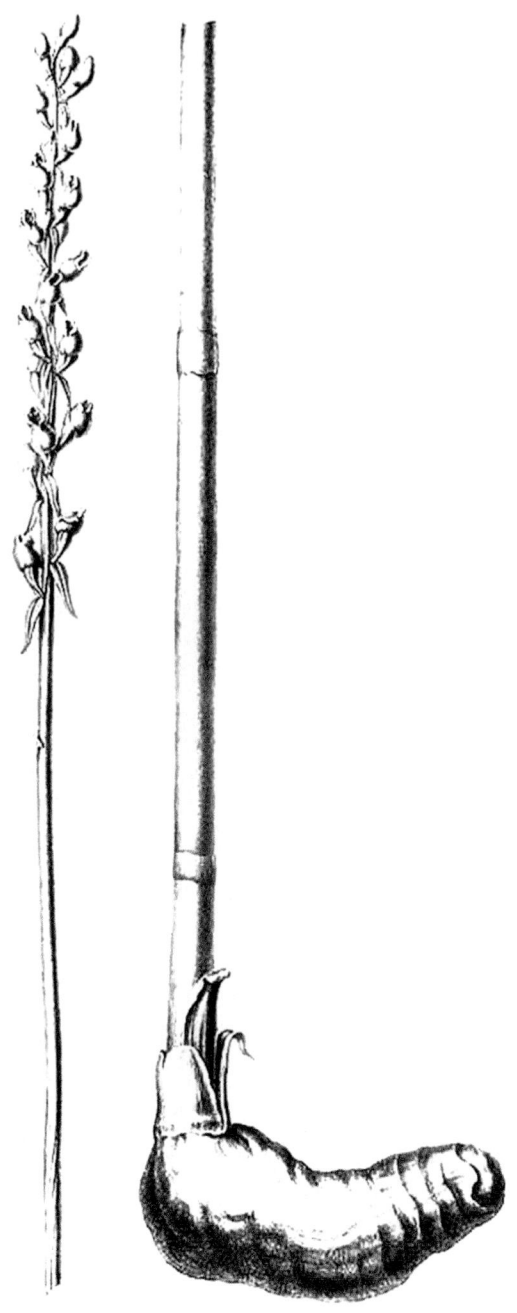

Fig. 12.5 *Gastrodia elata* Blume. From: Blume, C.L., *Collection des Orchidees les plus remarquables de l'archipel Indien et du Japon*, t. 53, Fig. 1 (1858). (Drawing by C.L. Blume)

listed in Shen Nong's *Pharmacopoeia* which is alleged to have been compiled at the dawn of Chinese civilisation, although existing records date only to the Han Dynasty (221 BC–220). Its

Fig. 12.6 *Gastrodia elata* Blume, inflorescence (Photo: qwert 1234 through Wikimedia Commons)

Fig. 12.7 *Gastrodia elata* Blume, seedpods (Photo: qwert 1234 through Wikimedia Commons)

original name was *Chi Jian* (Red arrow). The current name, *Tianma* (Heaven's fibre), was first mentioned during the Song Dynasty (960-1279) in *Kai Bao Bencao* published in 973 (Chen and Tang 1982). Likewise, the popular Japanese name *Oniniyagara* (Orge's arrow) refers to the inflorescence which resembles an arrow plunged into the ground. The scientific name is derived from ancient Greek, *gastrodes* (thick-bellied), possibly describing the tubers (Figs. 12.8 and 12.9), although some authors opined that it referred to the sepals that bulge laterally (Mayr 1998).

Although *G. elata* is widely distributed in China (occurring in northern Fujian, Gansu, Guangxi, Guizhou, Hebei, Henan, Hubei, Hunan, Jingsu, Jiangxi, Jilin, Liaoning, Nei Mongolia, Shaanxi, Shanxi, Shandong, Sichuan, Taiwan, Xizang, Yunnan and Zhejiang Chow and Chen 1983; Chen et al. 2009b), it is under threat because of overcollection for herbal usage during the mid-twentieth century. Listed under CITES Appendix II, *Tianma* requires government approval to be exported or imported. Today, marketed *Tianma* is mostly farmed, not collected from the wild. It is estimated that up to 80 % of *Tianma* consumed in China comes from farms. Korean scientists have also recently established in vitro production system of *G. elata* using symbiotic seed germination (Park et al. 2008).

For marketing purposes, *G. elata* is divided into seven varieties: forma (f.) *viridis* Mak., f. *glauca* S. Chow, f. *alba* S. Chow, f. *elata*, f. *flavida* S. Chow, f. *pilifera* Tayama (Chow and Chen 1983) and f. *obovata* Zhang (2010). Hybrids between different forms have been produced in the effort to enhance quality (Jiang, Wan, Wang 2001; Duan and Lu 2006).

It was well known that, when fungal hyphae grow into the cells of orchids, the latter digests them for their nourishment (Kusano 1911). Scientists from the Medicines Institute of the Chinese Academy of Medical Sciences developed a method of growing *G. elata* in the

Fig. 12.8 *Tianma* (processed tuber of *Gastrodia elata* Blume) from Quizhou Province, China (Photo: E.S. Teoh)

Fig. 12.9 *Tianma* from a Chinese apothecary in Yunnan Province: (*left*) large size tubers (*right*) small tubers(Photos. E.S. Teoh)

presence of a wood-decay fungus, *Armillaria mellea* (honey fungus) some 40 years ago. However, *Armillaria mellea* inhibits the germination of *Gastrodia* seeds (Xu 1990a). For that process, the orchid needs the help of a different fungus.

Mycena osmundicola Lange supplies the nutritional support for the germination of *G. elata* (Xu 1990a). Thriving on decaying tree leaves on the forest floor, its hyphae penetrate the orchid embryo through the suspensor cells. Meristem cells then divide, digest the hyphae, increase in size, and the embryo germinates (Xu 1990b). Following germination, the protocorm is invaded by *Armillaria mellea* (Xu and Guo 2000). Hyphae separate into layers before penetrating the cell wall of the "passage cell" of the orchid to spread in all directions and infect new cells. Their relationship is an ideal parasitic relationship resembling symbiosis, with the

orchid supplying minerals to the fungus and neither totally destroying the other. Fungus-free cells of protocorms will not grow unless they are allowed to be infected by *Armillaria mellea* (Xu 2001). Similarly, lateral tubers increase in size because they receive nutrients from the mother tubers, but if they do not acquire mycelia before separation and replanting they will perish unless they manage to establish a mycorrhizal relationship with fungi in the soil (Kusano 1911). In Hokkaido, Japan, a different species, *Armillaria nabsnona*, has been discovered in the tubers of *G. elata*. This fungal species was identified through the cultural morphology and features of its basidiomata, and molecular analysis of its ribosomal DNA (Sekizaki, Kuninaga, and Yamamoto 2008).

The relationship of *G. elata* to fungi is quite complex in the natural setting and not restricted to *Armillaria*. Sixty-two strains of endophytic fungi have been isolated from *G. elata* collected from different areas of Guizhou, Shanxi, Sichuan and Yunnan. They belong to 13 genera, 9 families, 7 orders, 5 classes and 3 phyla. Fungal species are dissimilar in different geographic regions and also in different parts of the orchid plant (Mo et al. 2009)

Protocorms become small, white *G. elata* within a year. After winter, they bud, blossom, and bear fruit, taking 3 years to complete the life cycle (Xu 1989).

Usage: After harvesting, stems and roots are removed, and mud is washed off the rhizomes before soaking in clean water. The rough outer coat is then removed and the rhizomes are steamed or boiled until they become totally translucent and devoid of white spots. They are left to dry and marketed in the form of hard, dry rhizomes. When needed, the rhizome is soaked in water until it is about 70 % wet. In this softened state, it is can be cut into slices which are allowed to dry. Sliced *Tianma* is fried in low to medium heat until both sides turn a light yellow. After transfer to paper, the slices are sprayed with water, then placed in a pot and cooked in low heat until the surface turns a dark yellow. As a decoction, the usual dosage is 3–9 g daily; as a powder 1–1.5 g per dose. *Tianma* is described as sweet in taste, neutral in property and it nourishes the liver meridian (Yang et al. 1999).

Sometimes, the herb is cooked with meat and served as soup. That is because it is "sweet, neutral and moist, with a nature that is slightly moistening and possessing a tonic effect." It "nourishes the *yin* fluid, calms the liver and extinguishes wind". It is used to treat all forms of *internal wind*, such as "wind" stroke, seizures, aphasia, slurred speech, blurred vision, numbness, tingling of the extremities, headache, vertigo and dizziness. In his *Grand Materia Medica* published in 1596, the celebrated Ming Dynasty herbalist, Li Shizhen, quoted past masters who maintained that without *Tianma* it was impossible to treat "darkening of vision with dizziness, a wind disorder". Thus, *Tianma* is also known as the wind-calming herb (*ding feng cao*). A number of prescriptions described in *Zhongyao Da*

Cidian (Chinese Medical Encyclopedia) are summarised in Table 12.1.

Gastrodin, vanillin and a few other compounds present in *Gastrodia* are present in other parasitic or terrestrial orchids like *G. faberi*, *Cremastra appendiculata* (proper name: *Cremastra wallichii*) (Liu et al. 2008) and *Coeloglossum viride* which enjoy a close relationship with fungi. *Coeloglossum viride* is the famous Tibetan drug *wangla* (Huang et al. 2002) which is purportedly used to treat dementia, but *G. faberi* and *Cremastra wallichii* do not have a similar usage.

Some publications reported that *Tianma* was used to treat internal haemorrhage. However, the *New Compilation of Materia Medica* warns that "It definitely should not be used lightly in those with *qi* and blood deficiency." Nor should it be used "whenever patients feel a lack of saliva with dry tongue and mouth, dry sore throat, constipation with dry stools, blazing fire causing dizziness, blood deficiency headaches without pathogenic wind" (Bensky, Clavey, Stoger, and Gambie 2004).

The inflorescence of *G. elata* was considered to be a tonic with aphrodisiac properties and it has its own distinctive medicinal name (*huant'ung-tzu*) [Stuart 1911; Li Shizhen 1596 quoted by Porter-Smith and Stuart 2003].

Tianma has very low toxicity and side effects are uncommon when it is consumed in the normal manner. However, there are reports of allergic reactions, vertigo, anaphylactic shock and acute renal failure when extract of *Tianma* is administered via injection. Administration of a very high dosage by whichever route (80 g in 3 h) resulted in poisoning, manifested by flushing, headache, dizziness, visual disturbance, general weakness, loss of muscle co-ordination and loss of consciousness. *Tianma* is present in varying amounts (2.8 to 27 %) in 9 % of patented herbal products manufactured in China. Three products are intended for use in infants whereas *Tianma* is contra-indicated during pregnancy (Fratkin 1997).

Commercial *G. elata* is most commonly harvested in winter or occasionally in spring. Wild herbs are known, respectively, as *Winter Gastrodia* (*Dong Tianma* or *Dong Ma*) and

Table 12.1 Prescriptions containing *Tianma* (*Gastrodia elata* Bl.) (Zhongyao Da Cidian (1986): *Zhongyao Da Cidian.* translated by Prof. Ong Siew Chey, MD, FACS]

---

1. For frightfulness of children: *Tianma* 25 g, *Buthus martensii* (全蝎) 50 g, *Ansaemi japonicum* (天南星) 25 g, *Bombyx batvyticatus* (白僵蚕)10 g, ground into fine powder and cooked with wine to make pills. Take 10–15 pills each time (Source: *Wei Family Formulae*)

---

2. For stroke: *Tianma*, "*tian zhu huang*" (天竺黄), *Ansaemi japonicum* (天南星), dry *Buthus martensii* (干蝎) in equal parts, mixed in powder form. Take 2.5 g each time (Source: *Sheng Ji General Records*)

---

3. For tetanus: *Tianma*, *Ansaemi japonicum* (天南星), *Siler divaricatum* (防风), *Notopterygium incisum* (蓋活), *Typhonium giganteum* (白附子) in equal parts, mixed in powder form. Take 10 g with a glass of wine each time (Source: *Mainstream Treatment of External Diseases*)

---

4. For migraine and other forms of headaches: *Tianma* 75 g, *Aconitum carmichaeli* (附子) 50 g, *Pineila ternate* (半夏) 50 g, *Schzonepeta tenouifolia* (荆芥) 25 g, *Inulara cemosa* (木香) 25 g, *Cinnamoum cassia* (桂) 0.5 g. "*Xiong Qiong*" (芎藭) 25 g, ground into powder and made into pills. Take 5 pills each time, increasing to 10 pills (Source: *Sheng Ji Records*)

---

5. For malaise, dizziness, sleepiness, headaches, etc.: *Tianma* 25 g, "*Xiong Qiong*" (芎藭) 50 g, ground into powder and made into pills. Take 1 pill after meal (Source: *Pu Ji Prescriptions*)

---

6. For excessive liver *yang*, surge of liver wind, headaches, dizziness, insomnia: *Tianma* 15 g, *Ourouparia rhyncopylla* (鉤藤) 20 g, *Cyathula officinalis* (川牛膝) 20 g, *Haliotis gigantea* (石决明) 30 g, *Gardenia florida* (山栀) 15 g, *Scutellara baicalensis* (黄芩) 15 g, *Euonymus ulmoides* (杜仲) 15 g, *Leonurus sibiricus* (益母草) 15 g, *Taxillus chinensis* (桑寄生) 15 g, *Polygonum multiflorum* (夜交藤)1 5 g, *Poria cocos* (朱伏神)15 g. Boil with water (Source: *New Treatise of Treatment of Miscellaneous Diseases*)

---

7. For high blood pressure: *Tianma* 5 g, *Euonymus ulmoides* (杜仲)10 g, *Chrysanthemum indicum* (野菊花) 10 g, *Conioselinum unvittatum* (川芎) 9 g. Boil with water (*Records of Natural Medicine of Qin and Ba Mountains*)

---

8. For postpartum blood circulation: *Tianma* 50 g, "*He Li Le*" (訶黎勒) 50 g, *Inula racemosa* (木香) 50 g, "*Yun Tai Zi*" (芸臺子) 25 g, mixed and strained. Take 10 g each time after boiling with a bowl of water to 70 % volume (*Sheng Ji General Records*)

---

9. For stroke with paralysis of extremities: *Tianma* 100 g, *Sanguisorba officinalis* (地榆) 50 g, *Commiphora molmol* (没药) 1.5 g, *Scrophulara oldhami* (玄参) 50 g, *Aconitum chinensis* (乌头) * 50 g, *Moschus moschiferus* (麝香) 0.5 g., ground into powder and made into pills. Take 20 pills with warm wine before dinner (Source: *Sheng Ji General Records*)

---

10. For rheumatism with stiffness of extremities in women: *Tianma*, *Achyranthes bidentata* (牛膝), *Aconitum carmichaeli* (俯子), *Euonymus ulmoides* (杜仲), 100 g each, Ground and placed in a silk sack and soaked in 5 *dou* (斗) of wine for 7 days. Take one cup warm each time (Source: *Shi Bian Good Prescriptions*)

---

11. For backache and pain of the lower extremities: *Tianma*, *Asarum sieboldi* (细辛), *Pineila ternate* (半夏), 150 g each. Placed in equal amounts in two silk sacks and cooked in hot water. Use the two sacks alternately to apply on site of pain. (*Traditional Effectual and Famous Prescriptions*)

---

12. For rheumatism, numbness and paralysis: *Tianma* 30 g, "*Niu Zi*" (扭子) 30 g, *Notopterygium incisum* (蓋活) 5 g, *Angelica grosserrata* (独活) 5 g, placed 500 ml of white wine for 7 days. Take appropriate amounts morning and night (Source: *Records of Natural Medicine of Qin and Ba Mountains*)

---

13. For lung "wind poison", pruritus and skin boils and ulcers: *Tianma* 50 g, *Cryptotympana atrata* (蝉壳) 50 g, *Glendischia horrida* (皂荚) 150 g, ground into powder, cooked with lean mutton and made into pills. Take 20 pills each time (Source: *Pu Ji Prescriptions*)

---

14. For generalized dermatitis: *Tianma* 37.5 g, *Schizonepeta tenouifolia* (荆芥) 17.5 g, *Mentha arvensis* (薄荷) 17.5 g, *Heydyotis diffusa* (白花蛇) 200 g, ground into powder, mixed with wine 4 *sheng* (升) and honey 200 g and cooked to a viscous consistency. Take 1 cup each time, three times a day (Source: *Yi Lei Yuan Rong*)

---

15. For "*Bai Lai Feng*" (白癞风): *Tianma* 800 g, "*Tian Liao Mu*" (天蓼木) 2.4 kg, boiled in 3 *dou* of water to 1.2 *dou*, cleared of the solid particles and cooked over slow fire to a soap form. Take ½ spoonful each time (Source: *Sheng Hui Prescriptions*)

Aconite is extremely poisonous. Together with other poisonous compounds, it was deleted from the Chinese Pharmacopoeia during the Yuan (Mongol) Dynasty and forbidden to be stored or prescribed because the rulers feared that they would be poisoned (Teoh)

Spring Gastrodia (*Chun Tianma*). The former is superior. *Tianma* from Sichuan is known as *Chuan Tianma*. Cultivated *Tianma* is only harvested in November, at the start of winter. The strain of co-cultured *Armillaria mellea* determines the yield of cultivated *G. elata*, Strain Av-4 providing the greatest yield in both wild cultivation and outdoor-box planting (Ji and Li 2009). Traditionally, wood of Chinese oak (*Quercus fabri* Hance) is used as a substrate for the culture of *G. elata*, but it has now been shown that wood of Chinese birch (*Betula luminifera*), Chinese aspen (a species of poplar, *Populus adenopoda*) and walnut (*Juglans regia*) also gave a satisfactory yield of *Tianma* tubers, the highest yield being obtained from *Betula luminifera* (Chinese birch, Chinese name: *Liang ye hua*) (Rong and Cai 2010). *Betula luminifera* occurs naturally at 100–1900 m in 14 provinces in China (Anhui, Fujian, Guangdong, Guangxi, Guizhou, Henan, Hubei, Hunan, Jiangsu, Jiangxi, S Shaanxi, Sichuan, Yunnan, Zhejiang), roughly overlapping much of the natural central and southern distribution of *G. elata*.

Excellent *Tianma* should have big, stout, compact rhizomes without holes at their core; creamy-white, without dark spots; and translucent and shiny when sectioned. *G. elata* put up for sale is sometimes adulterated by various tubers such as *Mirabilis jalapa*, *Dahlia pinnata*, *Canna edulis*, *Solanum tuberosum* or *Cacalia tangutica* (Bensky et al. 2004).

Phytochemistry: To date, 47 compounds, of which the majority were phenols, have been isolated from *G. elata* and identified by the group working with Zhou Jun at the Kunming Institute of Botany (Figs. 12.10, 12.11 and 12.12), viz.: (1) phenolic compounds—vanillin, vanillyl alcohol, *p*-hydroxybenzyl alcohol (gastrodigenin), *p*-hydroxybenzaldehyde, 3.4-dihydrobenzaldehyde, 4-4 dihydroxydibenzyl methane, *p*-hydroxybenzyl ethyl ether, 4-4 dihydroxydibenzyl ether, 4-ethyloxytolyl-4-hydroxybenzyl ether, 4-hydroxybenzyl methyl ether and gastrol; (2) glycosides: gastrodin, *p*-hydroxymethyl phenyl-beta-D-glucopyranoside, bis(4-hydroxy-benzyl) ether mono-beta-D-glucopyranoside, 4 (beta-D-glucopyranosyloxy) benzyl alcohol and parishins; and (3) other constituents such as

organic acids, *p*-hydroxy benzaldehyde, beta-sitosterol and daucosterol (Figs. 12.10, 12.11 and 12.12) (Zhou et al. 1979, 1980; Zhou et al. 1982; Hu and Zhou 2010).

Additional new compounds isolated from tubers of *Gastrodia elata* include trimethylcitryl-beta-D-galactopyranoside (a citryl glycoside) (Choi and Lee 2006); parishin F (1,3-di-[4-*O*-(β-D-glucopyranosyl) benzyl]-2-{4-*O*-[β-D-glucopyranosyl-(1 → 6)-β-D-glucopyranosyl] benzyl} citrate) and parishin G (2-[4-*O*-(β-D-glucopyranosyl)benzyl] citrate) (Wang et al. 2012); parishin J (2-[4-*O*-(β-D-glucopyranosyl) benzyl]-3-methyl-citrate) and parishin K 1(,2-di-[4-*O*-(β-D-glucopyranosyl)benzyl]-3-methyl-citrate) (Li et al. 2015); 4-hydroxy-3-(4-hydroxybenzyl) benzyl methyl ether and 4-(methoxymethyl) phenyl-1-*O*-beta-D-glucopyranoside (Wang et al. 2012); gastrodin B (4-(4′-hydroxybenzyl) phenyl glucoside and gastrol B (1′-hydroxymethyl-phenyl 4-hydroxy-3-(4″-hydroxybenzyl) benzyl ether) (Zhang et al. 2013); two benzofurans, 1-furan-2-yl-2-(4-hydroxyphenyl)-ethanone (1) and 5-(4-hydroxybenzyloxymethyl)-furan-2-carbaldehyde (Lee et al. 2007); seven 4-hydroxybenzyl-substituted amino acid derivatives (Guo et al. 2015); and two furaldehyes (5-[4′-(4″-hydroxybenzyl)-3′-hydroxybenzyloxymethyl]-furan-2-carbaldehyde) and 5-[4′-(4″-hydroxybenzyl)-3′-hydroxybenzyl]-furan-2-carbaldehyde (Huang et al. 2015).

Using column chromatography with silica gel and ODS, Xiao et al. (2002b) managed to isolate 6 compounds from *G. elata*, 3 of which were isolated for the first time. In the same year, the team found another new compound, alpha-acetylamino-phenylpropyl alpha-benzylamino-phenylpropionate together with a known compound, 4-hydroxybenzyl beta-sitosterol ether (Xiao et al. 2002a).

Parishin C is another major component of *G. elata* (Shin et al. 2010). The year 2007 witnessed the discovery of several new phenolic compounds and their derivatives in *Tianma* (Yang et al. 2007; Li et al. 2007; Wang et al. 2007c). Yang et al. (2007) described nine compounds, including parishin D and E which were new. Li, Wang, Chen, and Zhou (2007) also

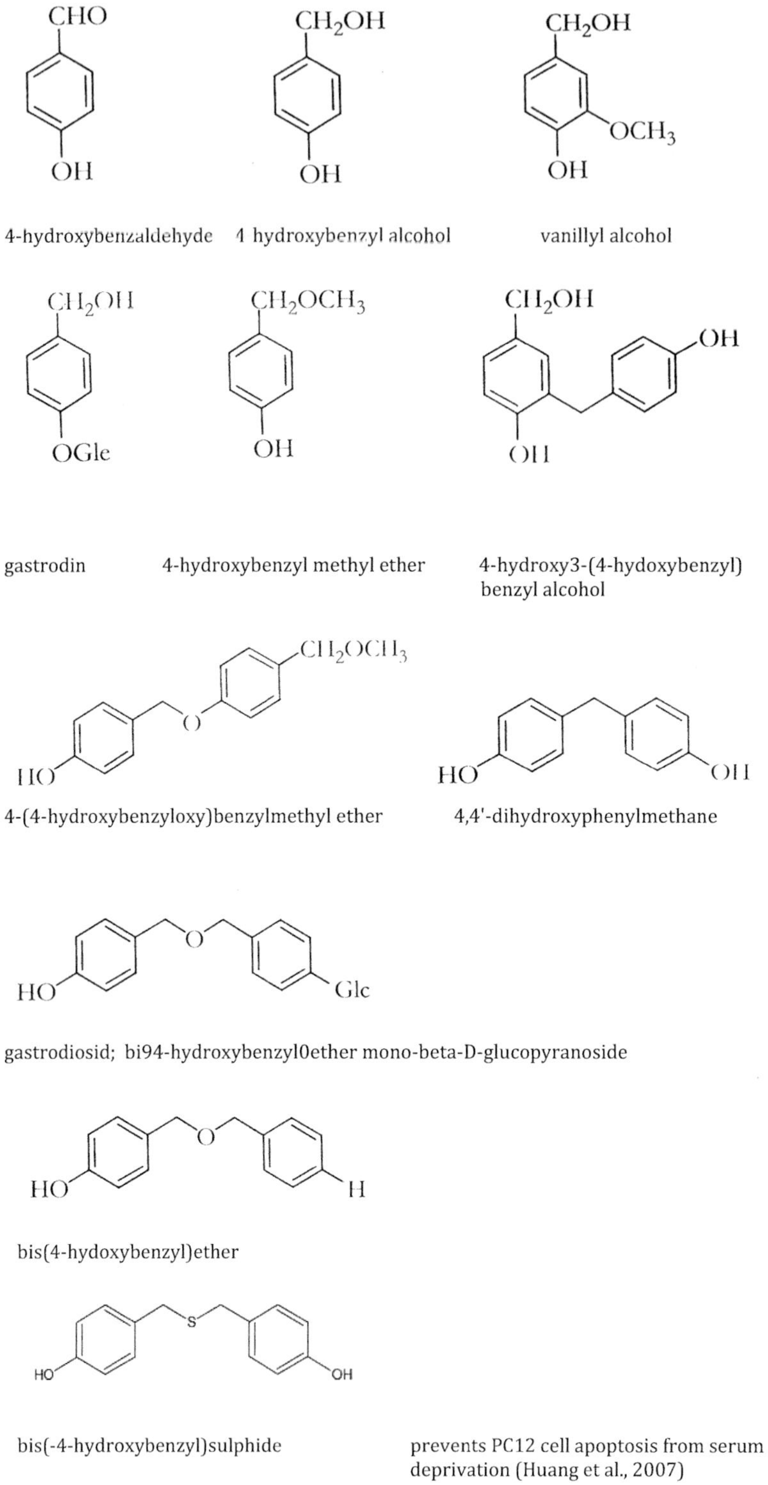

4-hydroxybenzaldehyde   4 hydroxybenzyl alcohol   vanillyl alcohol

gastrodin   4-hydroxybenzyl methyl ether   4-hydroxy3-(4-hydoxybenzyl)
                                                      benzyl alcohol

4-(4-hydroxybenzyloxy)benzylmethyl ether   4,4'-dihydroxyphenylmethane

gastrodiosid; bi94-hydroxybenzyl0ether mono-beta-D-glucopyranoside

bis(4-hydoxybenzyl)ether

bis(-4-hydroxybenzyl)sulphide   prevents PC12 cell apoptosis from serum
                                             deprivation (Huang et al., 2007)

Fig. 12.10   Some basic compounds isolated from *Gastrodia elata*

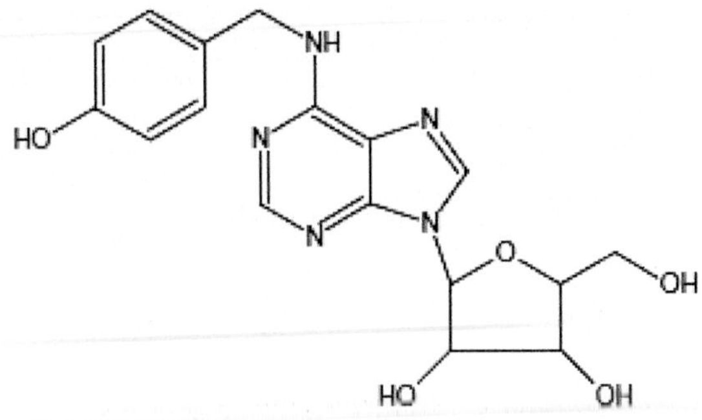

N6-(4-hydroxybenzyl)adenine riboside

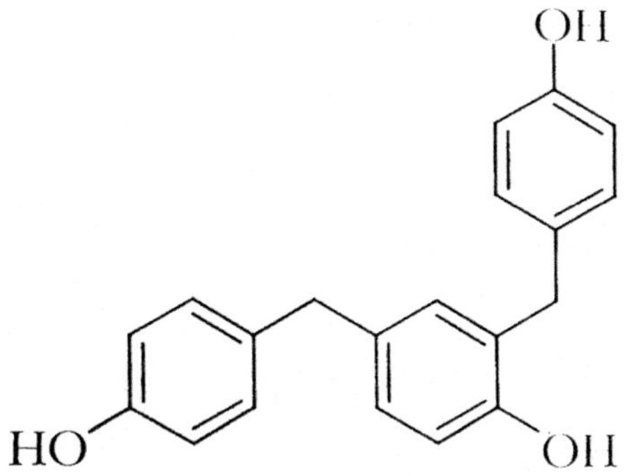

2,4=bis(4-hydroxybenzyl)phenol

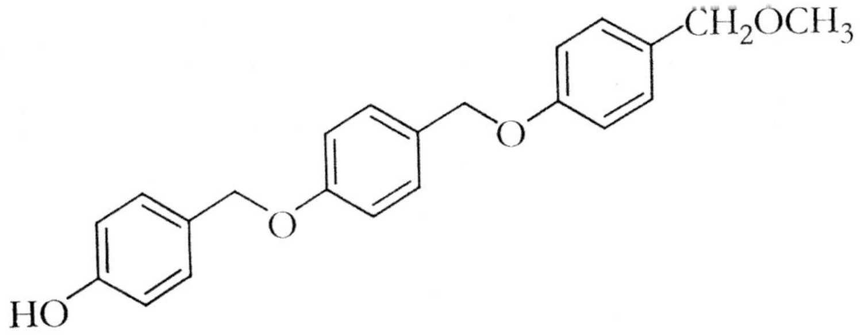

4-[4-(4-hydroxybenzyloxy}benzyloxy]benzyl alcohol

Fig. 12.10 (continued)

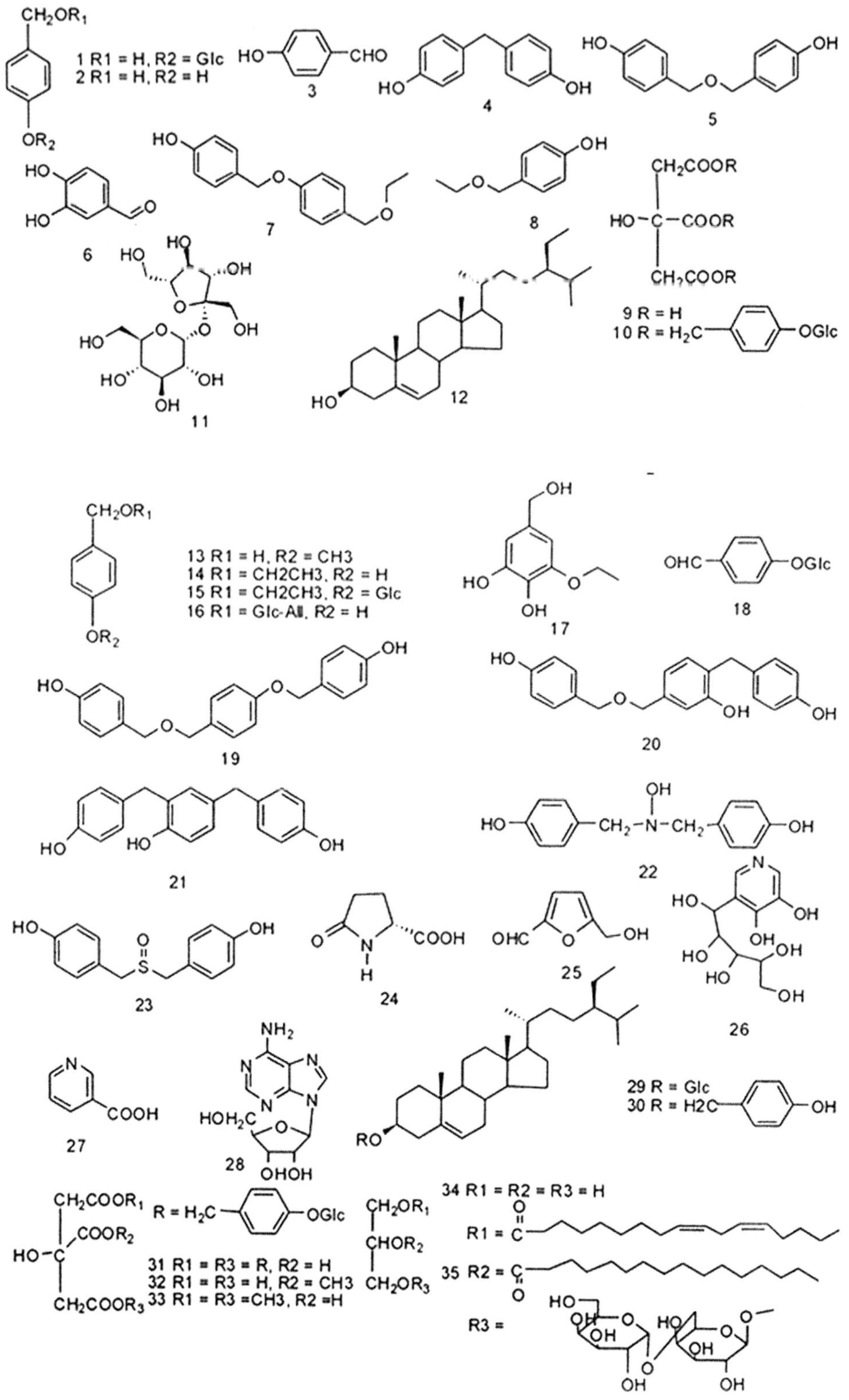

Fig. 12.11 Phytochemicals in *Gastrodia elata* identified by Jun Zhou and his team at the Kunming Institute of Botany. Between 1978 and 2010, 46 substances were identified

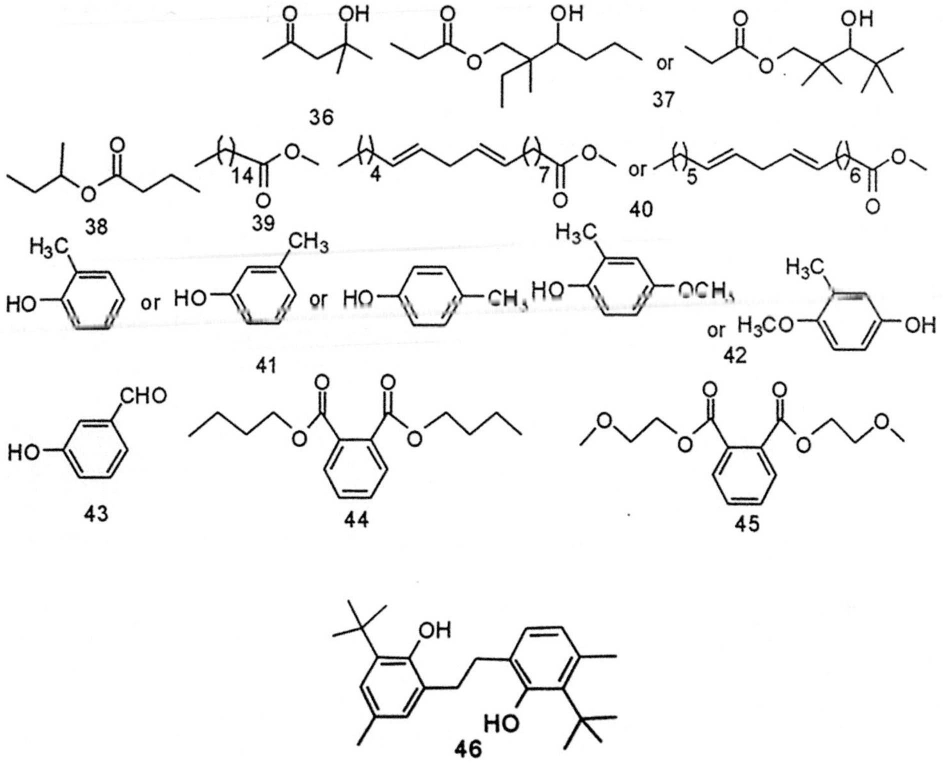

Fig. 12.12 Phytochemicals in *Gastrodia elata* identified by Jun Zhou and his team at the Kunming Institute of Botany (items 36–46) [Source: Poster presentation by Hu et al. (2010)]

described nine compounds but their list was different and contained two new compounds, gastrodin A and gastrol A. Wang et al, (2007c) identified 15 phenolics and 6 glycoside derivatives. Among the compounds that they described, 7 phenolics and 3 nucleoside derivatives had not been previously identified in *G. elata*.

Gas chromatography–mass spectroscopy revealed the presence of fructose (1.36 %), glucose (1.12 %) and sucrose (4.25 %) together with 4-hydroxybenzaldehyde (0.004 %), 4-hydroxybenzyl alcohol (0.03 %) and 4-(beta-D-glucopyranosyl oxy)-benzyl alcohol (GA) (1.97 %) (Li et al. 2001). This explains why the Chinese *Materia Medica* describes *Tianma* as "sweet".

Gastrodin is also present in the mycelia of *Armillaria mellea*, and it can be produced by the latter in the absence of the orchid (Muszyńska, Sułkowska-Ziaja, Wołkowska, Ekiert, 2011). For practical purposes, gastrodin obtained from batch fermentation of Armillaria is indistinguishable from the gastrodin present in the orchid. Gastrodin is a glycoside whose formula is known. It has been synthesised and developed into a drug which is available for both oral and parental use in China. Observing that chemical synthesis is a complicated procedure which is invariably associated with serious environmental pollution, a team of scientists at the Northwestern University in Xian developed a system for enzymatic synthesis which employs gastrodin biosynthesis enzyme (GBE) to biotransform p-hydroxy-benzylaldehyde into gastrodin. The enzyme itself, GBE, is obtained from *Rhizopus chinensis* SAITO AS3.1165 (Zhu et al. 2010). Gastrodin was also produced through biotransformation of p-hydroxybenzyl alcohol using hair root cultures of *Datura tatula* L. (Peng et al. 2007), *Aspergillus foetidus* and *Penicillium cyclopium*, the last giving the highest yield. *Penicillium cyclopium* AS 3.4513 has potential for use as an industrial biocatalyst

(Fan et al. 2013). Meanwhile, another new process has been described for the simple solid-phase extraction of gastrodin from *Gastrodia elata* (Ji et al. 2014).

TCM practitioners employ *Tianma* and pure gastrodin to treat hypertension, strokes and Parkinsonism. It is postulated that, in the human body, gastrodin is converted into methyl-dopa, ʟ-dopa, or a similar substance. For many decades, methyl-dopa was widely used to treat hypertension, particularly in pregnant women, but it is such a slow-acting drug with rather low potency that it cannot be relied upon in an emergency and is inadequate for the treatment of severe hypertension. Methyl-dopa has been retained in the modern scientific pharmacopoeia, but, like reserpine, it has generally been superseded by far more effective drugs even in third world countries. ʟ-dopa is used to treat Parkinsonism but patients on long-term treatment with ʟ-DOPA may develop dyskinesia (involuntary, sometimes painful, muscle movements). In a mouse model, *G. elata* extract attenuated ʟ-DOPA-induced dyskinesia (Kang et al. 2014). Work is underway to modify gastrodin in an effort to find more effective therapies.

Overview

With hundreds of tons gathered annually, *Tianma* was fast disappearing in its natural habitats in China during the 1960s. Fortunately, a breakthrough occurred during the 1970s when Chinese scientists found a way to cultivate *G. elata* Bl. that empowered them to overcome the dependence on wild sources of this parasitic orchid (Wen 1979). Numerous studies on the botany, breeding, commercial propagation, cultivation and processing of *Tianma* (*G. elata* Bl.) have since been published in Chinese medical journals (Luo 1985; Gan 1986; Ran and Xu 1988, 1990; Wang et al. 1989; Guo and Xu 1990; Wu et al. 1998; Jiang et al. 2001; Duan and Lu 2006; Zou et al. 2006; Tao et al. 2009).

Composition of nutrient media affects the content of gastrodin, free amino acids, and nutrient elements in *G. elata* grown on nutrient agar.

Wu et al. (1998) found that ½ MS media was the best for tissue culture.

Tianma sold on the market is now predominantly derived from *G. elata* cultivated on various species of wood. Tree species affect the rate of tuber growth and the chemical content of the tubers. *Tianma* is commonly grown on oak, but recent studies found that other types of wood gave better results. A Chinese study found that growing *Gastrodia* in wood of *Betula luminifera* produced the highest yield, and walnut and aspen were also satisfactory (Rong and Cai 2010).

In a study conducted in Korea, 30 tree species were successfully infected by *Armillaria mellea* but only 14 species supported growth of immature *G. elata* tubers. *Ulmus davidiana* (elm) produced the biggest tubers (7.8 g) that were almost nine times heavier than those recovered from *Abies holophylla* (Manchurian fir) wood. There was a negative correlation between tuber size and levels of ergothioneine (ERG), but gastrodin levels were similar in large and small tubers (Park and Lee 2013a). Gastrodin levels per tuber remained fairly constant across size tuber size and tree species. It is mainly affected by genetic constitution and place of origin (Tao et al. 2009).

Ergothioneine (ERG) is an amino acid produced by some plants, in particular by fungi, but not by the human body. It is sometimes used as a surrogate marker of fungal mass. Following ingestion, it is found in various human organs but its actions, if any, are unknown. However, there are claims that it is anabolic and anti-inflammatory, and protects against cataracts (Park and Lee 2013b).

From the commercial viewpoint, it would appear that cultivating *G. elata* in elm (*Ulmus davidiana*) would provide the largest and possibly the most saleable tubers because they meet the referred characteristics of "excellent *Tianma*" rhizomes, being big, stout and compact; however, these would not necessarily enhance the medicinal value of individual tubers. Indeed, if ergosterol is figured to be of additional value, smaller tubers appear to have an advantage.

In an earlier study, Wang et al. (1989) tested 14 different types of fungus-growing materials and they found that the properties and chemical composition (including gastrodin content) of their associated *Tianma* were similar. However, ergothioneine levels were not reported.

From the commercial viewpoint, it would appear that *Ulmus davidiana* is the best medium for cultivating *Tianma* because it improves tuber yield and produces more impressive tubers. However, it is important to determine how different ratios of gastrodin/ergothioneine or lower ERG affect the therapeutic actions of *Tianma* before attempts are made to grow *G. elata* in a large scale on *Ulmus davidiana* instead of traditional *Quercus* species (oak). Over 50 compounds are present in *Tianma*, but so far no proposal has been made as to which compounds are important and what are their appropriate ratios to one another.

Geographical distribution and growing environment are related to the genetic polymorphism of genomic DNA present in wild and cultivated *G. elata* (Zou et al. 2006). Tao and his associates (2009) studied eight populations of *G. elata* and they found that they clustered into three major groups which showed marked differences in their gastrodin content. The gastrodin content correlated to genetic makeup and place of origin. The ecological environment of Guizhou and Shanxi appears to be conducive to the evolution of the species and biosynthesis of gastrodin, with the two provinces being more suited for the cultivation of the herb (Tao et al. 2009). The market considers that the best *Tianma* comes from Guizhou.

Crossing *G. elata* with *G. glauca* resulted in a hybrid with improved zymogram when peroxidase (POX) isoenzyme content of the orchid was analysed (Jiang et al. 2001). Following upon this, the proposal has been made that hybrid combinations be developed to overcome the reduced output of *Tianma* brought on by drought or other less favourable environment (Duan and Lu 2006).

Pharmacology of *Tianma*

The *Ban Cao* describes *Tianma* as "sweet". Gas chromatography–mass spectroscopy revealed the presence of fructose (1.36 %), glucose (1.12 %) and sucrose (4.25 %), together with 4-hydroxybenzaldehyde (0.004 %), 4-hydroxybenzyl alcohol (0.03 %) and 4-(beta-D-glucopyranosyl oxy)-benzyl alcohol (GA) (1.97 %) (Li, Ding and Yu 2001).

TCM practitioners use *Tianma* to treat hypertension, dizziness, numbness, tingling of the extremities, strokes and parkinsonism. *Tianma* also enjoys a long classical usage in the treatment of a variety of other neurological disorders. Research has led to the discovery of numerous chemical compounds in *Tianma* which affect the central nervous system, in particular gastrodin, gastrodigenin (Baek et al. 1999; Cai et al. 2008a; Lu et al. 2009), 4 hydroxybenzaldehyde and its analogue 4 hydroxy- 3 methoxybenzyl-aldehyde (Ha et al. 2000) and their alcohols (Wu et al. 1996a), as well as the citryl glycoside, trimethylcitryl-beta-D-galactopyranosdie (Choi and Lee 2006).

Gastrodin is a phenol, the glucoside of the 4-hydroxybenzyl alcohol, gastrodigenin. Gastrodin administered into the femoral vein of anesthetised rats rapidly finds its way into the brain and bile of the animal, reaching peak concentrations in 15 min. It penetrates the blood–brain barrier and is excreted by the liver (Lin, Chen, Tsai, and Tsai 2007). The amount of gastrodin retained by the brain was maximum in the cerebellum, followed by frontal cortex, thalamus and hippocampus (ratios of AUC relative to plasma were 6.1, 3.3, 3.3 and 3.0, respectively) (Wang et al. 2008). Oral administration of gastrodin results in a peak concentration of gastrodin in 30 min and a third of this is available in the brain. Brain targeting and sustained residence in the tissue is improved by intranasal delivery of gastrodin in gel (Cai et al. 2008b).

It is postulated that gastrodin does not act directly on the central nervous system (Huang 1989). In vivo, it is first rapidly metabolised to gastrodigenin in the brain, blood and liver. Gastrodigenin binds to the benzyodiazepine (BZ) receptor on rat brain cell membrane but gastrodin does not (Guo et al. 1991a; You et al. 1994).

Compounds derived from other constituents of complex herbal remedies could influence the absorption rate and half-life of gastrodin. Pre-feeding with borneol for 20–40 min or its simultaneous administration results in an accelerated absorption of gastrodin (5–15 min vs. 30 min with gastrodin alone), followed by an increase in bioavailability of its active metabolite, gastrodigenin, in the brain to 108.8 %. However, the enhancing effect is attenuated if the dose of borneol is too high (600 mg/kg) or if pre-feeding occurred earlier than 40 min. Intra-gastric co-administration of gastrodin and puerarin (from *Radix Puerariae*) to rats reduced absorption rate and lowered the peak level of gastrodin in retinal veins but increased the bio-availability of both compounds (Jiang et al. 2013).

Extract of *G. elata* inhibits kainic acid binding to glutamate receptors in neurons (Anderson et al. 1995). It reduces the severity of convulsions and hippocampal neuronal damage following kainic acid administration to mice (Kim et al. 2001; Hsieh et al. 1999, 2005, 2007). Using serum deprivation-induced apoptosis of neuronal-like PC12 cells as an ischaemic/hypoxic model, two neuroprotective compounds from *G. elata* were identified as bis(4-hydroxybenzyl) sulfide and N6-(4-hydroxybenzyl)adenine riboside (Huang et al. 2007). *G. elata* extracts exhibited protective effects on gerbil brain homogenates subjected to lipid peroxidation induced by hydrogen peroxide or ferrous ammonium sulfate. Anti-oxidant potency was maximal with p-hydroxybenzyl alcohol followed by vanillyl alcohol, vanillin and hydroxybenzaldehyde in that order, the last being more potent than melatonin (Jung et al. 2007). Anti-oxidant actions of 4-hydroxybenzaldehyde and vanillin also present in *G. elata* were demonstrated at the cellular and molecular level by Liu and Mori (1992). Hydroxybenzyl-alcohol is present in natural *Vanilla* and its anti-oxidant activity from this source is documented (Shyamala et al. 2007). Strokes arising from cerebral haemorrhage results in extravasation of blood. Haemoglobin and the iron released from its breakdown generate reactive oxygen species (ROS) in rats. This results in seizures. Natural anti-oxidants from numerous sources, including *G. elata*, were shown to scavenge ROS and may serve as prophylaxis against seizures (Mori et al. 2004).

The Range of Bioactive Phenolic Compounds in *Tianma*

Zhou (1991) in a review of bioactive glycosides present in a few famous Chinese herbs mentioned that his group in Kunming discovered phenolic glycosides in *Tianma*. Some of these compounds are present in the achlorophyllous, parasitic orchid, *G. faberi*, namely p-hydroxybenzaldehyde, p-hydroxybenzyl alcohol and gastrodin (Li, Zhou and Hong 1993a). Gastrodin has also been isolated from *Coeloglossum viride* var. *bracteatum* (Huang et al. 2002; Huang et al. 2004).

Four new 4-hydroxybenzyl alcohol derivatives and numerous known compounds were recovered from the methanol extract of fresh *G. elata* tubers by two teams of scientists working in Korea (Table 12.2) (Li et al. 1993a; Yun-Choi et al. 1998). In the second study, the new 4-hydroxybenzyl alcohol derivatives were identified as 3-0-(4′hydroxybenzyl)-beta sitosterol and 4-[4′-(4″-hydroxybenzyloxy)benzyloxy]benzyl methyl ether (Yun-Choi et al. 1998). Other new compounds isolated from *G. elata* include alpha-acetylamino-phenylpropyl alpha-benzylamino-phenylpropionate (Xiao et al. 2002b), and parishin D and E (Yang et al. 2007). Parishin C is a major component of *G. elata* (Shin et al. 2010; Xiao et al. 2002a, b; Yang et al. 2007;

Table 12.2 *Tianma Guoteng Yin* (*Gastrodia and Uncaria Decoction*) from *Zubing Zhengzhi Xinyi* or New Significance of Patterns and Treatment in Miscellaneous Diseases, Qing Dynasty. This is similar to Item 6 in Table 12.1

Haliotis	18 g
Uncaria	12 g
Achyranthes	12 g
Gastrodia elata	9 g
Gardenia	9 g
Scute	9 g
Eucommia	9 g
Leonurus	9 g
Polygonium stem	9 g
Hoelen	9 g

Decocted and used to treat headache, vertigo, insomnia and symptoms of liver wind agitation and hyperactivity of liver *yang*

Li, Wang, Chen and Zhou 2007; Wang et al. 2007c)

Using TLC and HPLC to analyse gastrodin in various types of *G. elata*, Pan and Xu (1998) found similar peak numbers and retention times in the different types. Only peak heights and areas under the curve (AUC) were dissimilar. Thus, quality meant quantity, a good quality *Tianma* having more of the bio-active ingredients, but not different ingredients. The chemico-physical properties of the active components of *Tianma* have also been studied by capillary zone electrophoresis (Wang et al. 2002).

To date, 47 compounds, of which the majority are phenols, have been isolated from *G. elata* (Xiao et al. 2002a, b; Yang et al. 2007; Li et al. 2007; Wang et al. 2007c; Hu and Zhou 2010 Their molecular structures were identified by the group working with Zhou Jun at the Kunming Institute of Botany (Figs. 1–4) (Hu et al. 2010).

Their Source

Gastrodin and the related compounds isolated from *G. elata* can also be produced by pure cultures of *Armillaria mellea*. These bioactive compounds are recoverable from the nutrient media. N6-(5-hydroxy-2-pyridyl)-methyl-adenosine (HMPA), a novel N6-substituted adenosine analogue obtained from *Armillaria mellea*, decreased contractions of the vas deferens of the rat evoked by the addition of phenylephrine, norepinephrine and acetylcholine to the bath solution in a dose-dependent manner (Xiong and Huang 1998). At high dosage, HMPA even abolished the neurogenic twitch responses evoked by electric field stimulation. The experiments showed that fungal HMPA caused both pre-synaptic and post-synaptic depression.

Three taxa of endophytic fungi isolated from *G. elata* were found to produce gastrodin, but with different yields of 57, 89 and 184 mcg/g of mycelia. Gastrodin was present in the grounded hyphae but it was not present in the culture broth (Su et al. 2014). Thus, it is not the orchid but rather its fungus which is the source of gastrodin and the related compounds. Flavonoids and phenols are also produced by the fungus.

Aqueous and ethanolic extracts of *Armillaria mellea* exhibit potent free radical scavenging and antilipid peroxidation activities on brain homogenates from rats (Ng et al. 2007). Thus, one could dispense with the orchid and rely on laboratory cultures of *Armillaria mellea* to obtain pure chemical compounds. Indeed, over the last three decades, TCM practitioners are prescribing *Armillaria Fungus Tablets* to treat "deficiency of yin and flourishing yang", neurasthenia and hypertension. However, a PubMed Search on *Armillaria mellea* showed that extracts from the hyphae of the fungus contained an extremely wide range of compounds, many of which are not associated with *Tianma*.

Effects on the Central Nervous System Effects

Central nervous system control of muscle activity is elicited through the release of acetylcholine and tempered by dopamine. GABA (gamma-aminobutyric acid) is the chief neurotransmitter in the vertebrate central nervous system. Compounds that act on GABA-A receptors which are ligand-gated ion channels produce a fast response, whereas action on GABA-B receptors which are protein-coupled receptors result in a slow but sustained response. Medications like baclofen (beta-parachlorophenyl GABA) simulate the effect of GABA by binding to GABA-B receptors and reduce or prevent spasms of the muscles.

GABA is deactivated by GABA-transaminase. In keeping with the idea that traditional herbs act slowly and help the body to regain its balance, researchers discovered that hydroxybenzaldehyde (a constituent of *Tianma*) blocks the action of GABA transaminase to a greater extent than valproic acid, which is a proven anticonvulsant. It also reduces the extent of lipid peroxidation which occurs in the brain of rats treated with pentylenetetrazole (PTZ). It has been suggested that GABAergic neuro-modulation and anti-oxidation may be the pathways by which *Tianma* produces its anti-epileptic and anticonvulsive effects (Ha et al. 2000). Investigation of 10 analogues showed that the aldehyde group and the hydroxyl group at C-4 are necessary for the inhibitory effect on GABA-transaminase, and that 4-hydroxybenzaldehyde and 4-hydroxy-3-methylbenzylaldehyde are the most potent (Ha et al. 2001). Trimethylcitryl-beta-ᴅ-galactopyranoside recently isolated from *Tianma* by Choi and Lee (2006) has been shown to have an inhibitory effect on GABA transaminase (56.8 % at 10 mcg/ml).

Another GABA degradative enzyme is SSADH (succinic semialdehyde dehydrogenase). Gastrodin inactivates this brain enzyme, and this is another plausible explanation for the beneficial effect of gastrodin (Baek et al. 1999). In seizure-sensitive gerbils, gastrodin decreases immunoreactivities of gamma-aminobutyric acid shunt enzymes (GABA transaminase, succinic semialdehyde dehydrogenase and succinic semialdehyde reductase) in the hippocampus (An et al. 2003). Preconditioning with gastrodin reduces the ischaemia-induced elevation of glutamate and the glutamate/GABA ratios in the rat brain during ischaemia and reperfusion (Zeng et al. 2007).

Wu et al. (1996a) investigated the effect of administering p-hydroxybenzyl alcohol (HBA), an aglycone of gastrodin, to rats by assessing their ability to acquire an inhibitory avoidance response. Various drugs produced an impaired response: scopolamine, p-chloroamphetamine (a serotonin releaser) and apomorphine (a dopamine receptor agonist) all produced memory impairment when administered to rats. HBA reduced the impairment produced by p-chloroamphetamine and apomorphine but had no effect on the amnesia induced by scopolamine. This suggests that HBA (or *Tianma*) may act by suppressing the dopaminergic and serotonergic activities in the brain and thus improve learning. *G. elata* extracts, and their active constituents, gastrodin and p-hydroxybenzyl alcohol counteract scopolamine-induced amnesia in rats (Wu et al. 1996b). Gastrodin and p-hydroxybenzyl alcohol counteracts scopolamine-induced amnesia in rats (Wu et al. 1996b; Hsieh et al. 1997).

Phencyclidine-induced abnormal behaviour in mice was significantly reduced by parishin C, purified from *G. elata*. The effect was reversed by a 5-HT1A-receptor antagonist, WAY 100635.

Parishin C showed high affinities for 5-HT1A receptors and 5-HT1A agonist activity in an 8-OH-DPAT-simulated [35S]GTP-gamma S binding assay. The results suggest that the antipsychotic action of parishin C requires activation of the 5-HT1A receptor, an important therapeutic target in schizophrenia (Shin et al. 2010). An alcoholic extract of *G. elata* showed antidepressant-like activity in mice. An aqueous extract had a similar antidepressive effect on rats (Li et al. 2014).

Gastrodin is used in TCM to treat Tourette's syndrome (Lu et al. 2009), an inherited neuropsychiatric disease of childhood. The condition usually improves as the person gets older. Many patients become symptom-free spontaneously in adulthood. Tourette's syndrome is characterised by multiple motor tics and at least one vocal tic. The latter is embarrassing to the sufferer but amusing to bystanders because it usually involves the exclamation of obscene or socially inappropriate words or phrases; this is called coprolalia. The phenomenon often finds its way into Peranakan plays providing instant amusement when the matriarch of the family utters a shocking expletive. Tourette's syndrome is fairly common, its incidence being variously quoted as from 1:100 to 1:1000 children. Tics are caused by random discharges from the brain stem, but what triggers such discharges is not known. Afflicted individuals have normal intelligence and normal life expectancy.

Lu et al. (2009) simulated Tourette's syndrome in rats by injecting the animals intraperitoneally with apomorphine. Stereotyped behaviour of the rodents was reduced and levels of homovanillic acid (HVA) in the animals increased. Gastrodin counteracted the effect of apomorphine. Although gastrodin controls TS-like symptoms in rats by promoting dopamine metabolism and counteracting increased serum levels of HVA, this does not demonstrate that gastrodin will work on people with Tourette's syndrome.

A wide range of medications have been tried to manage patients with Tourette's syndrome. None works well because the cause is unknown. Treatment generally consists of explanation and reassurance, and medication is not recommended. In severe cases, and where there is risk of psychological trauma to the child, *Tianma* would be worth a trial, if it works, but that remains to be proven by a randomised clinical trial.

Tianma and *Uncaria lonchophylla* are two Chinese herbs that enjoy long usage in the treatment of epilepsy and involuntary muscle contraction from numerous causes (Shen and Chang 1963). Extracts of both herbs produced dose-dependent anti-oxidant and free radical scavenging effects in the rat brain following injections of ferric chloride into the lateral cortex to induce an epileptogenic focus (Liu and Mori 1992). 4-hydroxy-benzyl alcohol and vanillin, two of several compounds present in *Tianma* which possess anti-oxidant activity, are thought to play a role in preventing seizures (Liu and Mori 1993; Shyamala et al. 2007). Vanillin administered intraperitoneally 1 h before stimulation, or phenytoin at a non-toxic dose of 50 mg/kg ip, reduced stage 5 seizures produced by repeated low intensity electrical stimulation to the animal's baso-amygdala of the brain (Wu et al. 1989).

When Sprague-Dawley rats were administered a lateral ventricle injection of p-amyloid peptide (1–40) and intra-abdominally with D-galactose to simulate Alzheimer's disease, intra-gastric feeding with gastrodin appeared to improve their orientated learning and memory capacity (Fu et al. 2010). An ethyl ether fraction of *Gastrodia elata* was found to protect neural cells from damage caused by transient global ischaemia (Kim et al. 2003a): it also prevents beta-amyloid-induced cell death of IMR-32 neuroblastoma cells (Kim et al. 2003b). There are no data on the use of *Tianma* to treat human Alzheimer's disease.

Tianma is used in the TCM to treat hypertension and strokes (both to prevent and to treat strokes). In 2003, Kim, Lee and Moon decided to see what it will do for brain cells when these are denied their oxygen supply for a brief period. To do this, they treated rodents (gerbils) by oral dosing with the ether fraction of methanol extracts of *G. elata* at rates of 200 mg/kg or 500 mg/kg per day for 14 days. The animals were then subjected to transient global ischaemia. At 200 mg/kg, the GE extracts failed to attenuate the hippocampal neuronal damage in the CA1 region. However, the higher dosage reduced the hippocampal neuronal damage. Rats fed with 500 mg/kg of an ether fraction of methanolic extracts of *G. elata* for 14 days before being intraperitoneally injected with kainic acid also showed less hippocampal neuronal damage, and the severity of the convulsions following the toxic injection was reduced (Kim et al. 2001). Feeding rats *G. elata* 1 g/kg 30 min before the administration of kainic acid also reduced "wet dog shakes" (WDS), paw tremors and facial myoclonia, and delayed the onset of WDS in the animals. Serum luminol-CL and lucigenin-CI were lowered, and the level of lipid peroxidase in the rat brain was reduced. These findings showed that *G. elata* has an anticonvulsive effect, but whether this translates into clinical effectiveness in human remains to be investigated (Hsieh et al. 2001). Vanillin and p-hydroxybenzaldehyde which are constituent of *G. elata* inhibited intracellular Ca2+ increase and apoptosis of human neuroblastoma cells induced by glutamate (Lee, Ha, Yong, et al. 1999).

Another team of Korean scientists evaluated the effects of *G. elata* extracts and p-hydroxybenzyl alcohol (HBA) on brain damage and transcriptional levels of protein disulfide isomerase (PDI) and 1-Cys peroxiredoxin (1-Cys Prx) genes which play a role in anti-oxidant systems after transient focal ischaemia in the rat brain (Yu et al. 2005). To produce damage to the rat brain, the middle cerebral artery of the animals were occluded for 1 h. This was followed by 24 h of reperfusion before the animals were sacrificed. Brain infarct sizes (the area of dead tissue) of animals that had received *G. elata* extracts or HBA were smaller compared to untreated animals. Levels of PDI and 1-Cys Prx were significantly increased in both groups of treated animals when compared with controls. The authors concluded that their study demonstrated that *G. elata* extracts and p-hydroxybenzyl alcohol provided neuroprotection in the rat brain through the increased expression of genes encoding anti-oxidant proteins after transient cerebral ischaemia. Gastrodin protected neuronal cells in vitro from oxygen/glucose deprivation and glutamate-induced injury. At a high dosage (100 mg/kg), gastrodin also reduced infarct size and extent of brain oedema caused by experimental transient middle cerebral artery occlusion. Neurological function after the ischaemic episode was also better (Zeng et al. 2006). Elevation of glutamate in the hippocampus during the post-ischaemic period was less pronounced in treated animal's brain. Extracellular GABA was elevated (Zeng et al. 2007).

Intracarotid injection p-hydroxybenzyl alcohol (HBA), and HBA together with neural progenitor cells (NPCs) derived from mesenchymal stem cells in rats resulted in reduced infarct volume and improved neurological function 3 days after middle cerebral artery occlusion. HBA and HBA with NPCs induced expression of genes encoding anti-oxidant proteins like PDI, Nrf2, endogenous neurotrophic factor gene, brain-derived neurotrophic factor, NGF and VEGF which enhance angiogenesis in an ischaemic brain (Kaengkan et al. 2013).

Paradoxically, 4 hydroxybenzyl alcohol (HBA) has been shown to inhibit tumour angiogenesis in colon cancer cells (C126 WT) transplanted on dorsal skinfold chamber of mice without affecting the normal behaviour of the animals (Laschke, van Oigen, Koerbel, et al. 2013).

A wide range of n-methyl-D-aspartate receptors (NMDARs) which are glutamate-gated ion channels are preset in the central nervous system. They play a key role in excitatory synaptic transmission but their functional significance in normal subjects and in disease is yet to be determined (Cull-Candy et al. 2001; Paoletti and Neyton 2007). Muscarinic cholinergic receptors have an inhibitory effect on the brain. GPx (glutathione peroxidase) is an endogenous regulator of hydrogen peroxide and its content is reduced in scrapie (prion)-infected ewes (Gudmundsdottir et al. 2008). GPx was increased in the brain of rats exposed to hyperoxia for up to 72 h (Ismaeel bin-Jaliah 2008). Thus, a rise a GPx can be viewed as beneficial.

When mice prefed with *Tianma cuzhi keli* were subjected to repetitious cerebral ischaemia-reperfusion, glutamic acid levels in the cortex and hippocampus were higher than in controls. Aspartic acid in the hippocampus increased, while GABA content in the cortex decreased. *Tianmacuzhi* (TMC) granules reduced n-methyl-D-aspartate (NMDA) receptor activities and significantly increased M receptor activity and GPx activity of the cerebral cortex and hippocampus tissues. JH Fu and GY Du who conducted these studies concluded that *Tianma* might attenuate ischaemia-reperfusion injury of the brain by reducing glutamate neurotoxicity and increasing cholinergic function in the brain (Fu and Du 2001).

An expanded team next studied the effect of *tianma gouteng fang* on transmitter amino acids in hippocampal extracellular fluid in freely moving rats by subjecting them to 120 min of incomplete brain ischaemia, and then followed through with measurements after 2 h of reperfusion. During ischaemia, glutamic acid in the hippocampal extracellular fluid decreased by 31–38.64 % in treated rats compared to controls (100 %). The decrease dropped to 11.48–14.55 % after reperfusion. Tau rose by 12.86–13.99, while GABA rose 25.89–33.99 % during ischaemia. Tau was relatively unchanged while GABA improved only marginally on reperfusion. Cysteine dropped by 40.93–42.08 % while arginine rose by 108.96–116.95 % (Zhang et al. 2004). These studies showed that *Tianma* had an effect on the rat brain during ischaemia.

Zeng and colleagues in Hangzhou (2006) performed similar experiments to test the effect of gastrodin. A high dose of 100 mcg/kg markedly decreased infarct volume and cerebral oedema, and neurological recovery in the treated animals was better than in untreated controls. In separate experiments, they showed that cultured hippocampal neurons treated with 15 mcg/ml or 30 mcg/ml gastrodin had a lower neuronal cell death rate and reduced extracellular glutamate (an indication of nerve cell damage) when subjected to oxygen/glucose deprivation. Continuing their studies, they performed microdialysis sampling of the rat hippocampus during ischaemia and reperfusion. Glutamate and GABA in the dialysate were measured by HPLC (high-performance liquid chromatography). Administration of gastrodin (100 mg/kg) before ischaemia significantly reduced the post-ischaemic elevation of glutamate, increased extracellular GABA during the reperfusion period, and decreased glutamate/GABA ratios during both ischaemia and reperfusion (Zeng et al. 2007). The protective effect of gastrodin on the ischaemia-reperfusion injury in the rat striatum was been confirmed in a separate study reported by Bie et al. (2007). Experiments employing transient cerebral ischaemia in rats showed that the neuro-protective effect of *Tianma* was retained when it was combined with five herbs in a capsule marketed as *Nao-Shuan-Tong* (NST) in China. NST reduced cerebral infarct area, attenuated neurological deficits, and reduced neuronal apoptosis in the ischaemic cortex. It suppressed overexpression of Bax and activated caspases-3, -8 and -9, inhibited reduction of Bcl-2 expression, and markedly depressed Bax/Bcl-2 ratio (Xiang et al. 2010). *G. elata* activates the PI3K signaling against oxidative glutamate toxicity in HT22 hippocampal (brain) cells (Han et al. 2014).

Primary cultures of rat cortical neurons were significantly damaged when subjected to hypoxia for up to 24 h. Neuron survival improved when the cells were pretreated with 100 mcg/ml or 200 mcg/ml gastrodin, while extracellular glutamate decreased (Xu et al. 2007). Employing serum deprivation induction of neuronal-like PC12 apoptosis as an ischaemic/hypoxic model, Huang et al. (2007) screened compounds extracted from *G. elata* for neuroprotective capability. They found that bis(4-hydroxybenzyl) sulfide and N6-(4-hydroxybenzyl) adenine riboside potently prevented PC12 cell apoptosis.

Kim et al. (2006) studied plant extracts from various species for their ability to protect against CT-105-induced neuronal cell death, a phenomenon which is suspected of having a role in Alzheimer's disease. A dozen extracts of various plants were found to possess considerable protective effects, in particular *Uncaria ramulus et Uncus* (UREU), *G. elata, Evodia officianalis* and *Panax ginseng*, with the first two showing the strongest protective effect (Kim et al. 2006).

Employing an MTT assay, scientists at Singapore's Nanyang Technological University observed that neural cell proliferation peaked at 10 micromolar concentration of gastrodin. Treatment with 10 mcl of crude *Tianma* extract resulted greater cell numbers and extensive dendritic proliferation. The scientists postulated that the exuberant neuronal growth and dendritic formation may be due to unidentified compounds present in the extract, additional to gastrodin (Nah et al. 2010). Other active components of *Tianma* that exhibit a protective effect against transient global ischaemia in Mongolian gerbils are vanillin, 4-hydroxybenzyl aldehyde and HBA (Kim et al. 2007).

Lead poisoning damages the nervous system. In particular, it targets the hippocampus which is responsible for learning and memory. The effects are long lasting and usually irreversible. Childhood exposure to lead leads to cognitive and behavioural deficits. Young mammals absorb lead more readily and their brain can absorb eight times as much lead as adult brain. When iron is deficient in the diet, more lead is absorbed. Lead is ubiquitous in the environment and it requires only a small amount to cause damage in children, so there is really no safe level for lead (Neal and Guilarte 2013). Many health authorities routinely screen for lead contamination in herbal products because these are sometimes contaminated with unacceptable levels of lead. C-fos expression is a biological marker of nerve cell activity. *Rats* fed *G. elata* and E-gelatin (an ancient TCM item, which is donkey-hide gelatin that contains 18 amino acids) were protected against lead-induced down-regulation of c-fos expression when their brains were examined for c-fos protein (Hu et al. 2003a, b). Working with 22-day-old rats, Yong et al. (2009) found that gastrodin improved synaptic plasticity in the hippocampal CA1 region which had been impaired through lead exposure during the developmental period. Improvements occurred in input/output (I/O) function, paired pulse facilitation (PPF), and long-term potentiation of field excitatory postsynaptic potential (pEPSP) in the hippocampal CA 1 region. Accumulation of aluminum in the brain also impairs learning and memory. Experiments with adult rats showed that administration of *G. elata* by injection protected the animals against learning impairments and poor memory induced by aluminum (Niu et al. 2004; Shuchang et al. 2008).

Methamphetamine causes dopamine depletion and formation of 3-nitrotyrosine in the nigrostriatum of the brain that eventually leads to a spectrum of neurodegenerative disorders (Imam, el-Yazal, Newport, et al. 2001). After 4 intraperitoneal injections of methamphetamine, rats showed significant decreases in behavioural activity, dopamine levels, tyrosine hydroxylase activity and tyrosine hydroxylase protein expression, the last evaluated by immunochemistry. Methamphetamine caused significant increases in oxidative stress evidenced by increases in lipid peroxidation, protein oxidation and reactive oxygen species (ROS) formation. In these animals, treatment with *G. elata* attenuated the methamphetamine-induced dopaminergic toxicity by inhibiting oxidative burdens (Shin, Bach, Nguyen, et al. 2011). In the long term, methamphetamine abuse may lead to Parkinson's disease. Vanillyl alcohol possesses free-radical scavenging activity and inhibited "wet dog shakes" induced by intracortical injection of ferric chloride into rat brain (Hsieh et al. 2000). In TCM, gastrodin has been used in the prevention of Parkinson's disease, but its mode of action has not been defined. It is suggested that down-regulation of connexin 43 may be a mode of action (Wang, Wu, Liu, Fu 2013b). N6-(5-hydroxy-2-pyridyl)-methyladenosine (HPMA), a novel N6-substituted adenosine analogue isolated from *Armillaria mellea*, inhibited contractions of rat vas deferens induced by adrenalin, noradrenalin and acetylcholine in a dose-dependent manner. It also abolished the neurogenic twitch response to electrical stimulation. These experiments showed that the compound blocked nerve transmission as well as direct nerve stimulation of the muscle (Xiong and Huang 1998). HMPA is likely to be present in *Tianma*.

Another controversial area is the use of *Tianma* to prevent or treat Alzheimer's disease. To study this, a team from Chonbuk National University Medicinal School in Korea established a rat model of Alzheimer's disease by injecting A beta (25–35) into bilateral hippocampi (the area of the brain involved in learning and memory). Rats were fed 0.5–1 g/ kg *G. elata* for 52 days. These rats fared better at a water maze test for spatial memory. There were fewer amyloid deposits in their hippocampus, while choline acetyltransferase expression was significantly increased in the medial septum and hippocampus, and activity of acetylcholinesterase was decreased in the prefrontal cortex, medial septum and hippocampus (Huang et al. 2013).

The above results offer good evidence of neuro-protection by *Tianma* in animals subjected to sudden ischaemia or chronic poisoning during their developmental stage. In TCM circles, the studies are viewed as evidence which support the claim that *Tianma* is effective in preventing decrepitude and memory loss with advancing age (Sun et al. 2004). This opinion now has some supporters from the west (Ojemann et al. 2006).

The French team of Descamps et al. (2009) found that induction of protein disulfide isomerase (PDI) is the main mechanism by which 4-hydroxybenzyl alcohol (4-HBA) exerts its effect on reducing cerebral infraction in a murine model of focal brain ischemia/reperfusion. The effect is blocked by bacitracin, a known inhibitor of PDI. The scientists postulate that PDI is the key protein to target in order to control brain injury disorders and that 4-HBA, being devoid of neurotoxicity at 200 mg/kg in the rotarod test, can be promoted for neuroprotective usage.

Meanwhile, in New York, Lebesgue et al. (2009) reported that the very much maligned but universal natural compound, oestradiol, rescues neurons from global ischaemia-induced cell death. If proven by clinical studies, it would be a cheaper and simpler alternative than *Tianma* for women and it has additional benefits, albeit oestradiol also has numerous side effects and contra-indications. Unfortunately, in clinical trials, neurological protection by oestrogen remains unproven.

Phaeochromocytoma is a neuro-endocrine tumour which arises from the adrenal gland. It produces adrenalin and nor-adrenalin, and spurts of hormone production in patients with phaeochromocytoma usually result in sudden, transient, severe rises in blood pressure which may result in a stroke or worse. Huang et al. (2004) studied the effects of a methanol extract of *G. elata* on serum-deprived rat phaeochromocytoma and found that it prevented apoptosis of the tumour cells by activation of the serine/threonine kinase -dependent pathway and suppression of JNK activity. It increased cAMP formation, PKA activity and phosphorylation of CREB protein (Tsai et al. 2011). Subsequently, they also showed that *Gastrodia elata* reversed mutant Htt-induced protein aggregations and proteosome de-activation through A (2A) signalling in phaeochromocytoma (PC12) cells transfected with mutant Huntington aggregates (Huang et al. 2011). Alternative mechanisms have also been proposed (Jiang et al. 2014).

Using similar technology, different groups of scientists have isolated between 7 and 11 phenolic compounds from the tubers of *G. elata* (Hayashi et al. 2002; Pyo et al. 2004; Lee et al. 2006; Huang et al. 2006). Gastrol and 10 known phenolic compounds isolated by methanol extraction from *Tianma* were shown to possess smooth muscle relaxant activity of isolated guinea pig ileum (Hayashi et al. 2002).

The evidence that *Tianma* has neuro-protective properties appears to be sound, but one has to remember that they are all based on animal experiments. What is very lacking is proof from randomised controlled clinical trials on actual patients. Recently, Dr. SH Zhang advertised to recruit participants to be involved in a double-blind, randomised controlled study to test whether gastrodin could prevent cognitive decline related to cardiopulmonary bypass (Clinical Trials.gov Identifier NCT00297245) in 2006 and was able to publish his results in 2011; a remarkable achievement. Cardiopulmonary bypass involves an extra-corporeal circulation to supply oxygenated blood to the brain when the heart is stopped to permit it to be operated upon. The trail required 200 participants, all adults between 18 and 65 years who were scheduled for mitral valve replacement surgery. This was a Phase IV clinical trail that would provide valuable data (Zhang 2006; Zhang et al. 2011).

Patients received either gastrodin (40 mg/kg) or saline after induction of anaesthesia. They were evaluated for cognitive function with a battery of 5 neurocognitive tests before surgery, at discharge from hospital and 3 months after surgery. Cognitive decline was present in 9 % of patents who had received gastrodin and 42 % in the saline (control) group at discharge (p < 0.01). Cognitive testing at 3 months was available for 87 patients in the gastrodin group and 88 in the controls. The incidence of cognitive decline was 6 % in the gastrodin group and 31 % in the control group (p < 0.01). There was no difference in other adverse outcome in both groups (Zhang et al. 2011).

Well-designed clinical trials of herbal therapies in patients with epilepsy are scarce, and methodological issues prevent any conclusions of the efficacy or safety of herbal preparations in this population. The way to proceed is to conduct further preclinical evaluation with a view towards clinical development under the new US Food and Drug Administration guidelines (Schachter 2009).

Other Effects of *Tianma*

Using similar technology, different groups of scientists have isolated between 7 and 11 phenolic compounds from the tubers of *G. elata* (Hayashi et al. 2002; Pyo et al. 2004; Lee et al. 2006; Huang et al. 2006). Gastrol and 10 known phenolic compounds isolated by methanol extraction from *Tianma* were shown to possess smooth muscle relaxant activity of isolated guinea pig ileum (Ha et al. 2002; Hayashi et al. 2002).

Previous screening showed that such *G. elata* extracts also possess anti-inflammatory properties. Using paw oedema, arachidonic acid-induced ear oedema, and analgesic activity in acetic acid-induced writhing response in rats as in vivo tests, Lee et al. (2006) showed that eight phenolic compounds from *G. elata* possessed anti-inflammatory and analgesic activities. The most potent was 4-hydroxy-3-methoxyben-zaldehyde. Some of the compounds inhibited reactive oxygen species generation or DPPH radical scavenging activity while others inhibited the activity of COX I and II in vitro. The presence of anti-inflammatory and anti-angiogenic activities in *G. elata* extracts were confirmed by Ahn et al. (2007) using chick chorioallantoic membrane assay, the rat-pouch model, acetic acid induced abdominal writhing in mice, and in in vitro assays. *G. elata* extract significantly suppressed the TNF-alpha–induced increase in monocyte adhesion of HUVEC, inhibited TNF-alpha-induced intracellular reactive oxygen species (ROS) production and p65 NF-kappaB activation by preventing lkappaB-alpha phosphorylation (Ahn et al. 2007). Pretreatment with *G. elata* extracts attenuated the TNF-alpha-induced increase in expression levels of cell adhesion molecules in primary cultures of human umbilical vein endothelial cells (HUVEC). MRNA expression levels of intracellular adhesion molecule-1 (ICAM-1), vascular cell adhesion molecule-1 (VCAM-1) E-selectin and macrophage chemo-attractant protein-1 (MCP 1) and interleukin-8 (IL-8) all decreased (Hwang et al. 2009).

This anti-inflammatory activity translates into anti-asthmatic activity when tested in guinea pigs with igE-mediated asthma challenged with aerosolised ovalbumin. 4-Hydroxy-3-methoxybenzyl alcohol significantly reduced airway blockage in the immediate and late phases of the asthmatic response by 52 and 40 %, respectively, compared to controls. It suppressed the recruitment of leukocytes, in particular eosinophils and neutrophils, release of histamine (by 31 %), eosinophil peroxidase activity (by 21 %) and specific phospholipase activity (by 16.6 %) in the broncho-alveolar lavage fluids. Other phenolic compounds with C4 hydroxy or c3 methoxy radicals also demonstrated anti-asthmatic activities (Jang et al. 2009).

"Silky chicken *G. elata* Blume nutrient solution" fed to mice increased the thymus/body weight ratio, macrophage and NK cell activities. This observation was interpreted to suggest that the nutrient solution boosted the animal's immunity (Xiao et al. 2009). However, the action of the thymus in the human is generally regarded as inconsequential.

Pain Relief

From ancient times, pain relief has been stated as a property of *Tianma*, albeit it is not usually administered as a first-line treatment for pain but only forms part of a complex prescription. Gastrodin relieves the painful diabetic neuropathy which affects streptozotocin-induced diabetic rats. Abnormal hyper-excitability of the dorsal root ganglion was abolished by gastrodin administered intraperitoneally. Diabetes caused an enhancement of sodium current and a decrease in potassium currents in the dorsal root ganglia. These effects were reversed by gastrodin in a dose-dependent manner and normalised (Sun et al. 2012). Gastrodin effects pain relief by decreasing the excitability of nociceptive primary sensory neurons (Sun et al. 2012).

Gastrodin also attenuated vincristine-induced pain in mice. The effect is blocked by a selective serotonin receptor antagonist (Guo et al. 2014b). *G. elata* and *Lingusticum chuanxiong* are important components of a traditional TCM preparation, *DaChuanXiongFang*, which originated from the Jin Dynasty (1115–1234). Used to relieve migraine, it is believed that the analgesic effect is conferred by *Lingusticum chuanxiong* (Guo et al. 2014a), but it would seem that *G. elata* also has a role.

Anticoagulant Effects

Anticoagulant effects of Chinese herbs can be a boon or a bane depending on their appreciation and how the herb is used. A novel phenolic compound from *Tianma*, 4,4′-dihydroxybenzyl sulfone, suppressed platelet aggregation by U466 19; in this respect, compound 9 (4,4′dihydroxy-dibenzyl ether) was 2.5 to 20 times more potent than the first compound (Pyo et al. 2004). Gastrodin binds to fibrinogen causing fibrinogen depletion, and prolongation of coagulation time without affecting the kaolin partial thromboplastin time (KPTT) or prothrombin time (PT) in rats. It inhibits the formation of clots and the risk of thrombosis (Liu, Tang, Pei, et al. 2006). The effect appears to be also attributable to polysaccharide 2-1 from *G. elata* which prolongs both clotting and bleeding times and prevented platelet aggregation (Ding et al. 2007). There is interest in studying the use of *Tianma* as a possible antithrombotic agent, but here we must caution that *Tianma* should now be included in the list of herbs from which one should abstain for several weeks prior to any surgery or invasive intervention such as angioplasty.

Pharmacokinetics

Transnasal Administration

The Pali *Vinaya* records the earliest usage of a trans-nasal route to deliver drugs to the body. The method was used by the Buddha's physician, Jivaka Komarabhacca in the sixth century BC in northern India to treat King Pajjoto of Ujjeni. He used ghee as a carrier for the drug (Horner 1982). This intranasal route is suitable only for small molecules that are non-polar (Illum 2004). During the 1980s, salmon calcitonin, a small peptide which plays a crucial role in calcium metabolism, was presented in this way for the treatment of osteoporosis (Riis et al. 1986; Rigmister 1988). Even so, small amounts that entered the blood stream were sufficient to stimulate the production of antibodies (because the calcitonin was a fish peptide and not identical with the human hormone). After some time, the patient required higher dosages of the calcitonin and eventually it stopped working altogether. The same route is also used to deliver a gonadotrophin-releasing hormone (GnRH) agonist (Burserelin, GnRHa) which initially stimulates, and later suppresses the pituitary production of gonadotrophins that regulate ovulation. The method is not popular because injecting GnRHa under the skin works just as well. GnRH is a small molecule, a decapeptide.

Nevertheless, the blood–brain barrier is a formidable obstacle to the delivery of some essential drugs which are necessary for the treatment of many diseases. For instance, cytotoxic agents to treat metastatic cancer in the brain, and some antibiotics fail to cross the blood–brain barrier. Vyas et al. (2006) and Wu et al. (2008) discussed the various strategies that are being explored to resolve this problem.

Intranasal administration of gastrodin was reported by Wang, Chen and Zeng from Zhejiang University in Hangzhou in 2007. They found that, in rats, intranasal and intravenous administration of 50 mg/kg gastrodin achieved comparable concentrations in the cerebro-spinal fluid (CSF). However, with intranasal administration, the plasma concentration of gastrodin was very low. Expressed as AUC (area under the curve), the ratios of AUC values of intranasal and intravenous administrations were 8.85 % in plasma and 105.5 % in cerebro-spinal fluid, giving a drug targeting index (DTI) of 12.34. This meant that the intra-nasal route could be used to deliver gastrodin to the brain without incurring its generalised side effects, such as the increased bleeding tendency. Unfortunately, pharmacokinetic evidence in humans is not forthcoming (Merkus and van den Berg 2007).

Intravenous Administration

Gastrodin administered intravenously to rats reaches peak concentration in the brain and bile at 15 min (Lin, Chen, Tsai, and Tsai 2007). It disappears rapidly from the circulation (Wang et al. 2008); in rats, T-max is 15 min and half-life is 2.81 h (Jiang et al. 2013). Ratios of AUC (area under the curve) of brain/plasma were not high. The individual ratios of AUC in various parts of the brain and CSF were: for frontal cortex 3.3/1.2, hippocampus 3.0/0.7, thalamus 3.3/1.3, cerebellum 6.1/1.9 and cerebro-spinal fluid (CSF) 4.8/2.4. The AUC in the cerebellum was higher than in other parts of the brain. The hippocampus had the highest ratio (4.3) compared with the plasma level in the test animal.

Injectio G. elata evoked a sedative effect on the central nervous system in mice manifested by a decrease in motor activity and prolonged sleeping time. This was not due to gastrodin; rather, in the experiments, it was due to the gastrodin-free component in the injection (Huang 1989). Tritiated alpha-isobutylhydroxybenzyl alcohol (3H-G018) and 125-I G018 (radioactively labelled G018) bound to the benzodiazepine (BZ) receptor on the rat brain membrane (Guo et al. 1991b). Gastrodigenin and its derivatives showed competitive antagonist BZ receptor binding and inhibited the effect of 125-G018 on the BZ receptor, but no inhibition was produced by gastrodin. Guo and his colleagues (1991b) theorised that gastrodin converted to gastrodigenin in vivo and that was how it produced its effect on the central nervous system.

This was proven by the study of You et al. (1994). They examined the bio-distribution and metabolism by administering radioactively labelled 3H-gastrodin and 3H-gastrodigenin to mice. They found that gastrodin crossed the blood–brain barrier, and was rapidly converted into gastrodigenin. The latter was retained in the brain and mediated inhibitive effects on the central nervous system. Gastrodin and gastrodigenin were excreted mainly by the kidney but gastrodin also entered the hepato-biliary system and could be partially excreted via this route. In the study by Wang et al. (2007d) after intravenous administration of gastrodin, another metabolite, p-hydroxybenzyl alcohol, was only present in small amounts CSF and plasma.

Stress Reduction

Various studies have tried to show that *Tianma* is beneficial because it tampers with the body's response to stress. Oral administration of *G. elata* extract or intraperitoneal injections of either of its phenolic constituents, 4 hydroxybenzylalcohol and 4-hydroxybenzaldehyde, to mice prior to testing significantly increased the time that they spent in an elevated maze device used for testing anxiety. This anxiolytic-like effect of *G. elata* was blocked by two compounds, WAY 100635, a 5HT(1A) receptor antagonist, and also by flumazenil, a GABA(A) receptor antagonist. The anxiolytic effect of 4-hydroxybenzylalcohol was blocked by WAY 100635, whereas the action of 4-hydroxybenzaldehyde was antagonised by flumazenil. *Tianma* thus exert its anxiolytic effect on mice via two pathways: through the serotonic nervous system through 4-hydroxybenzyl alcohol, and via the GABAergic nervous system through 4-hydroxybenzaldehyde (Jung et al. 2006).

Pretreatment with *G. elata* extracts (GEE) protected mice against stress-induced gastric lesions (An et al. 2007). When rats were subjected to forced swimming, pre-feeding with GEE significantly increased dopamine concentration and decreased its turnover in their brain (corpus striatum) (Chen, Hsieh and Su 2008). Concentrations of serotonin (5HT) in the frontal cortex and dopamine in the striatum were significantly increased following administration of aqueous extracts to rats stressed by forced swimming compared to untreated rats, and their turnover was also lowered. Chen, Hsieh and Su et al. (2008, 2009) concluded that GEE probably possesses an antidepressant effect in rats. Hopefully, this might translate into the human situation.

In a different study, rats were gavaged with distilled water or increasing concentrations of *G. elata* extract (at 200 mg/kg, 500 mg/kg and 1000 mg/kg) and stressed by being subjected to forced swimming to study the effect of GE on exercise-induced fatigue recovery. GEE increased swimming time and led to lower levels of blood lactate (Wang and Yan 2010).

N-6-(3-methoxyl-4-hydroxybenzyl) adenine riboside (B2), an analog of N-6-(4-hydroxybenzyl) adenine riboside (NHBA) which was originally isolated from *G. elata*, has strong sedative and hypnotic effects and has a potential to be developed as a medication for handling sleep disorders (Shi et al. 2014). It significantly decreased spontaneous locomotor activity and potentiated the hypnotic effect of sodium pentobarbital in mice (Zhang et al. 2012b).

Anti-osteoporotic Effect

Aqueous extracts of *G. elata* enhanced proliferation, differentiation and bone nodule formation and inhibited osteoclastogenesis in vitro. When administered to oophorectomised rats, *G. elata* extracts prevented the decreases in trabecular bone volume and bone weight, decline of cortical bone thickness and thickness of bone at the epiphyseal regions in a dose-dependent manner (Seo et al. 2010).

On Skin Pigmentation

Korean, Chinese and other Asian scientists are screening medicinal plants for tyrosinase and DOPA auto-oxidation inhibitory activities to possibly employ them as skin whiteners in the cosmetic industry. Tyrosinase is the catalysing enzyme in the initial stages of melanin (the skin pigment) formation. Several plant extracts including one from *G. elata* inhibit mushroom tyrosinase activity, whereas others suppress DOPA auto-oxidation activity (Lee et al. 1997). Used together, they might inhibit melanogenesis. The tyrosinase inhibitor in *G. elata* is bis (4-hydroxybenzyl) sulfide (Chen, Tseng, Hsiao, et al. 2015).

Yan et al. (2010) screened 32 Chinese herbs and 18 TCM formulae which are used as folk skin whiteners in China and found that 10 out of 50 extracts demonstrated greater than 50 % inhibition of mushroom tyrosine activity. The main herbs possessing such activities were *Ampelopsis japonica, Lindera aggregata* and *Polygonatum odoratum* (none of these three are orchids). The single orchid that they tested, *Bletilla striata*, was inactive in this respect; they did not test *G. elata*.

On the Cardiovascular System

Human hypertension is frequently associated with a rise in Angiotensin II which is produced in the kidneys through the conversion of Angiotensin I by ACE (angiotensin-converting enzyme). A whole range of ACE-inhibitors have been developed by the pharmaceutical industry to control hypertension, and these drugs are effective but have the rather undesirable effect of provoking coughing in some people. Enalapril is an ACE-inhibitor. Enalapril also blocks the inactivation of bradykinin (a vasodilator) and it potentiates the prostaglandin system. These actions result in a lowering of the peripheral blood pressure, increased blood flow through the kidneys (which in turn reduces the production of Angiotensin I and II), reduced work load on the heart, increased cardiac output, and reduced ventricular hypertrophy.

To investigate the effect of *Tianma Gouteng* Recipe on left ventricular and aortic hypertrophy and Angiotensin II, Wang, Wang, Sun and their colleagues at the China Academy of Traditional Chinese Medicine devised the following experiment. Rats were randomly assigned to three groups: a control without treatment, a second

group on enalapril, and the third given *Tianma Gouteng* Recipe (TGR). The rats were rendered hypertensive by performing an operation to cause narrowing of their renal vessels. A week later, the blood pressure of the control animals increased by an average of 37.4 mmHg and another 6 weeks later their left ventricle and aorta were enlarged and circulating Angiotensin II was markedly elevated. In both, the enalapril and TGR groups, ventricular and aortic enlargement did not occur, and their levels of Angiotensin II were suppressed. The rise in blood pressure was lower. This experiment suggests that *Tianma* has an antihypertensive effect on rats (Wang et al. 2005).

It is now recognised that even isolated raised systolic blood pressure can be as damaging as, if not more damaging than, a raised diastolic blood pressure. Most of the current antihypertensive drugs effectively lower diastolic blood pressure, but they are not as effective in lowering systolic blood pressure. Zhang et al. (2008) who treated 63 elderly patients with refractory hypertension and 30 patients with essential hypertension with an intravenous infusion of 1000 mg gastrodin for 4 weeks found that the systolic blood pressure decreased significantly (but not stating by how much). A drug that selectively and effectively lowers systolic blood pressure would be a valuable addition to the therapeutic armamentarium. However, the present study is too small and the data would not qualify as being of high quality. Furthermore, maintenance would be impractical if the medication needs to be administered by intravenous infusion. More studies in this area would be helpful.

Protecting the carotid arteries which supply the brain with oxygenated blood and nutrients is an important aspect of stroke prevention. Wang et al. (2007e) showed that serum from rats fed a decoction of *Tiamma* and *Uncaria* inhibited vascular smooth muscle cell proliferation in vitro, and presumably this might lessen the risk of atherosclerotic plaque formation and stenosis. Pretreatment with *G. elata* extract reduced endothelial extracellular damage of human umbilical veins induced by decreased tumour necrosis factor (TNF)-alpha. This is reflected by suppression of the increase of matrix metalloproteinase (MMP)-2/-9 activities induced with TNF-alpha (Lee, Hwang, Kang, et al. 2009).

Tianma is a principal component of *Tianma Gouteng Yin* Formula (TGYF) which is being used to treat hypertension, and this intervention was assessed by a study group led by Zhang, Hong and Wei from the Chinese University of Hong Kong for the Cochrane Database. The formula which is widely used in East Asia contains 11 different herbs, and besides *Tianma*, it contains *Gouteng* (*Ramulus Uncariae Cum Unicis*) *Shjueming* (*Concha Haliotidis*), *Zhizi* (*Fructus Gardeniae*), *Huanqin* (*Radix Scutellriae*), *Chuanniuxi* (*Radix Cyathulae*), *Duzhong* (*Cortex Eucommiae*), *Yimucao* (*Herba Leonuri*), *Sangjisheng* (*Herba Taxilli*), *Yi Jiaoteng* (*Cauls Polugoni Multiflora*) and *Fushen* (*Poria*). Amounts of each component vary, being decided by the preference of individual physicians or the needs of the patient. TGYF is served as decoction, 300–400 ml, twice daily. Capsule form is also available. *Tianma Gouteng Yin* is one of top ten prescriptions for hypertension offered by TCM practitioners in Taiwan (Tsai et al. 2014).

To produce the meta-analysis, English databases including Cochrane Central (CCTR), MEDLINE, EMBASE, AMED, CINAHL, IPA and the Hypertension Group Specialised Register were searched up to July 2011. Chinese databases including Chinese Biomedical Database, TCMonline, Chinese Dissertation Database, CMAC and the Index to Chinese Periodical Literature were searched up to April 2010. No study was identified which met the inclusion criteria for this review. The authors concluded that, as no trials could be identified, no conclusions can be made about the role of TGYF in the treatment of primary hypertension. Well-designed randomised controlled studies need to be conducted and published (Zhang et al. 2012a).

A separate meta-analysis was conducted by Wang, Feng, Yang, and their team from Department of Cardiology, Guang'anmen Hospital, China Academy of Chinese Medical Sciences in Beijing. They examined the use of *Tianma gouteng yin* as adjunctive treatment for essential hypertension in patients already receiving conventional medication for hypertension and managed to identify 22 randomised controlled trials. Both systolic and diastolic pressure improvements were more marked when patients on ARB, ACE and CCB additionally received TGY, but it was not seen in the group treated with diuretics and TGY. When patients were assessed by TCM's zheng classification of syndromes, patients on combination therapy had the better improvement.

The zheng classification is a traditional diagnostic method to categorise patients into different syndromes determined by the use of four standard TCM procedures: observation, history, inquiry and pulse examination. All diagnostic and therapeutic methods in TCM are based on the differentiation of illness accorded by the *zheng* classification, a concept that has prevailed for millennia. TGY is used as adjunctive treatment for essential hypertension associated with liver *yang* hyperactivity syndrome (LYHS) and liver-kidney *yin* deficiency syndrome (LKYDS). LYHS is characterised by vertigo, tinnitus, headache, flushing, red eyes, irritability, insomnia, lassitude in loin and legs, bitter mouth, red tongue and wiry pulse. LKYDS is always characterised by dizziness, tinnitus, headache, low fever, flushing, burning sensation of five centres, hypochondriac pain, hypopsia, lassitude in loin and legs, red tongue with scanty coating and rapid wiry pulse. LYHS is usually caused by LKYDS. However, outside of TCM, essential hypertension is known to be a silent disease that affects a large proportion of people above the age of 50. It must be diagnosed early through routine, periodic, at least annual, examination of the blood pressure of individuals who are otherwise healthy.

Nevertheless, the authors concluded that the effectiveness and safety of combining TGY with standard antihypertensives is still uncertain because of the poor quality of the trials. Statistical methodology was poor as it was difficult to conduct blinded trials with herbal medicine due to appearance, taste and smell of the herbs, and only 8 out of the 22 studies documented side effects of the treatment that included rash, coughs, nausea, vomiting and oedema. There is some concern about possible herb–drug interactions. More rigorous trials are needed to confirm the results (Wang et al. 2013).

Another herbal combination, *Yinian Jiangya Yin*, which contains no *G. elata*, was found to be slightly superior to *Tianma Gouteng Yin* when both were administered singly to treat early primary hypertension. Average blood pressure in the 79 patients was 155/94 mmHg. Serum endothelin was lower and serum nitrous oxide higher in the group treated with *Yinian Jiangya Yin* (40 patients) than with *Tianma Guoteng Yin* (39 patients). There were no side effects in either group (Zhao, Liu, Guan and Liu 2010).

Tianma is not the only herb contributing an alkaloid content. Apart from gastrodin, TGYF contains rhynchophylline, leonurine, stachydrine, graveoline, baicalin, geniposide, quercetin (also isolated from in vitro-propagated *Dendrobium tosaense*) emodin and cyasterone (Zhang Tong, Zhou, et al. 2012). It was reported that TGYF lowers blood pressure and reduces total cholesterol. It enhances the effect of candesartan (an Angiotensin II Receptor Blocker) in reversing carotid vascular remodelling in patients with essential hypertension (Wang, Fend, Yang et al. 2013). Whether this translates into improved quality of life, reduction in the incidence of stroke and other complications of hypertension, and improved survival will be the objective of the meta-analysis.

G. elata is one of eight different herbs that constitute HMCO5, an aqueous extract based on a modification of Banhabackchulchunmatang (BCT), a Korean herbal medicine which has been in use for over a thousand years. HMCO5 is reported to have a vasorelaxant and anti-atherosclerotic action. The other herbs are *Pinellia ternate, Poria cocos, Atractylodes macrocephala, Sigesbeckia pubescens, Coptis japonica, Crataegus pinnatifida* and *Citrus unshiu*. HMCOS extract relaxed vascular strips prepared from rat aorta and precontracted with sodium fluoride, regardless of the endothelial integrity, and decreased GTP RhoA activated by the sodium fluoride (Seok et al. 2011). The contribution of *G. elata* to this relaxing effect of HMCO5 on vascular constriction remains to be elucidated.

Cytotoxic Properties

Going off on a different track, two teams from Korea have reported their investigations on the possible cytotoxic effect of *Gastrodia elata* on tumour cells. Ethyl ether extracts of *G. elata* Bl. (GEB) was found to increase GTP-Ras of tumour cells in vitro in a dose-dependent manner. It is speculated that this might have potential in alleviating tumourigenesis. GTP-Ras is an active form of G-coupled protein (Heo et al. 2007).

Two new benzofuran, 1-furan-2-yl-2-(4-hydroxyphenyl)-ethanone (1) and 5-(4-hydroxy-benzyloxymethyl)-furan-2-carbaldehyde (2) and five known compounds including gastrodin were isolated from the tubers of *G. elata* Bl. by Lee et al. (2007) in Korea. Although compound 1 showed potent inhibitory activity comparable to camptothecin and etoposide in DNA topoisomerase 1 and 11 inhibition assays, all the compounds showed weak or no cytotoxic activity against human colon, breast and liver cancer cell lines in vitro.

A water-soluble glycan, code-named WTMA, isolated from the rhizome of *G. elata* with a mean molecular weight of $7.0 \times 10(5)$ Da inhibited pancreatic cancer cells in vitro. This effect was absent when the polysaccharide was deprived of its side-chain following hydrolysis with isoamylase (Chen, Cao, Zhou, et al. 2011).

Antidengue Virus Bioactivity of 2 alpha-D Glucans from *G. elata*

In November 2007, Qiu, Tang, Tong (2007) and their colleagues from Shanghai reported the preparation of sulfated derivatives of two glucans WGEW and AGEW from *G. elata* which showed strong antidengue virus bio-activities. They found that the potency of antiviral activity was directly correlated with the degree of substitution (DS) (Qiu et al. 2007). The development of an effective antidengue medication will be an important contribution to medicine.

High Fructose-Induced Metabolic Syndrome

Fructose is sweeter than glucose or sucrose. It has a low glycaemic index and is present in many fruits and grain (corn). When sucrose is consumed, it is broken down into glucose and fructose in a 1:1 ratio. A high fructose intake results in raised triglycerides, elevated LDL and adiposity. High fructose items encourage adiposity because high fructose in the blood does not confer a sense of satiation in the brain. The end result is a pre-diabetic state or metabolic syndrome and an increased risk of cardiovascular disease. Administering *G. elata* at a rate of 100 mg/kg to rats on a high fructose diet suppressed the development of obesity, insulin resistance, dyslipidemia, hypertension and fatty liver (Kho et al. 2013)

Identification of Genuine *Tianma* and Quality Control

Fifteen species of *Gastrodia* occur in China (Chen et al. 2009a) and some of these may be sold as *Tianma* in place of *G. elata*. Other tubers (*Mirabilis jalapa, Dahlia pinnata, Canna edulis, Solanum tuberosum* or *Cacalia tangutica*) are sometimes present as adulterants (Bensky et al. 2004). *Tianma* powder has numerous counterfeits (Ji et al. 2008). Unfortunately, even employing PCR technology, it is not easy to identify genuine *Tianma*. Using an RAPD method, Zou et al. (2006) examined the genomic DNA of 15 samples from wild and cultivated *G. elata* Blume in Guizhou for genetic polymorphism. The incidence of polymorphic loci (PPB) was 70.97 %. Geological distribution and growing environment were significant factors for polymorphism. They did not identify any particular DNA sequence that characterised *G. elata*.

DNA-sequencing of nine populations of *G. elata* coupled with the determination of their gastrodin content by HPLC was performed by Tao and his associates at Changsha University of Science and Technology. The distribution of the DNA sequence in the five DNA fragments studied showed great variation. However, sequence 1 was common to all, and this would confirm that the *Tianma* was genuine. Sequence 2 correlated with higher gastrodin content. The authors concluded that DNA marker sequences can be used to identify better varieties of *G. elata* and optimise their selection for cultivation (Tao et al. 2006). Accurate authentication of many

Chinese herbs is possible using PCR-amplified ITS2 with specific primers, and many authors are now proposing that the widespread use of this method would ensure quality and raise the status of medicinal herbs (Chiou et al. 2007).

High-performance liquid chromatography is a simple, sensitive and efficient method for quality control of *Tianma*. An HPLC method for simultaneous quantification of thee principal components of *G. elata* (gastrodin, p-hydroxybenzyl alcohol and 4-hydroxybenzaldehyde which are responsible for most of the pharmacological effects of *Tianma*) was introduced in 2001. Its recovery rates ranged from 96.6 to 98.5 % (Li, Ding, Yu 2001).

HPLC has also been used to check the content of gastrodin in *zhennaonin* capsules. The procedure has a recovery of 98.8 %; with RSD (range of standard deviation) of 2.5 % (Su and Wang 1996). HPLC is a reliable way to evaluate the quality of *Tianma* based on the content of gastrodin (GAS) and p-hydroxybenzyl alcohol (HBA). This methodology was used by Zhang et al. (2007) to compare samples of *Tianma* from various sources. They found that the method of processing was the single most important factor which affected the final results. Samples from different habitats subjected to similar processing showed little difference in their quality. The procedure of freezing to dryness was superior to other methods (Zhang et al. 2007).

Calibration curves is linear for concentration ranges between 2.00 and 200.00 mcg/ml, with a correlation coefficient of 0.9997 when employing HPLC–electrospray ionisation mass spectrometric method for quantification of gastrodin (GAS) and p-hydroxybenzylalcohol (BHA) in rat plasma. In the laboratories of Peking Union Medical College, intra-day precision was 17.82 %, and inter-day precision was 10.21 %, which we take to mean that intra-assay variation was up to 17.82 %. Recoveries were in the range of 91.12–108.64 % (Zhang, Sheng and Zhang 2008). Maintenance of such records for laboratory tests over the long term is an essential quality-control procedure for laboratories.

Liu, Dong, Yu, Liu, et al. (2004) proposed the use of Fourier transform infrared spectroscopy as an alternative method to identify genuine from fake *Tianma* and grade its quality. It is easy to pick out the fakes by their infrared spectra which are different from the characteristic infrared spectrum of *G. elata*. Within this spectrum, it is possible to distinguish wild winter, wild spring and cultivated *Tianma* according to the differences in their spectral peaks and absorbance ratios. The method is rapid and non-destructive. This proposal has the support of Ji et al. (2008) who found that it could distinguish genuine *Tianma* powder from its five common counterfeits, *Polygonatum sibiricum* Red., *Colocasia esculenta* (L.) Schott., *Helianthus tuberous* L., *Solanum tuberosum* L. and *Ipomoea batatas* Lam.

Extraction Procedures

To optimise extraction of bioactive compounds from *Tianma*, scientists have used the following methods: microwave-assisted extraction (Zhang, Yu and Wu 2004); high-speed counter-current chromatography (Li and Chen 2004); pressurised solvent extraction (Li, Li, Fu, et al. 2004); and pressurised hot water and microwave extraction (Teo et al. 2008). The latest innovation is the employment of imprinted polymers with high selectivity and affinity to gastrodin in water to capture the compound after which the gastrodin is recovered with an eluting solvent. Pure gastrodin is obtained through the process. Recovery of gastrodin from *Tianma* is 76.6 % (Ji et al. 2014). To obtain polysaccharides from *G. elata* on an industrial basis, Zhu and Luo (2007) found that extraction at 120 °C for 3 h with a 1:40 ratio of orchid to water was optimal. Post-harvest handling is important in ensuring a good gastrodin content, and steaming has the greatest impact on gastrodin content. Shen-Hao et al. (2008) found that this single step was more important than the herbal varieties or the source of the *Tianma*.

An Important New Development: Possible Reduction in Reliance on Pesticides by Agricultural Crops

G. elata is able to co-exist with *Armillaria mellea* which is a dangerous root-rot fungus. What prevents the orchid from being destroyed by the fungus are its GAFPs (Gastrodia antifungal proteins) or gastrodianins which exist in four isoforms (Wang et al. 2001). Structurally, GAFP resembles insecticidal lectins. They encode two different mature proteins, and they bind mannose. Purification, crystallisation and X-ray diffraction analysis show that they are present in four independent isoforms in an asymmetrical unit (Liu, Hu, Wang et al. 2002; Liu, Yang, Ding, et al. 2005). GAFP-2 promoter controls the expression of the reporter gene beta-glucuronidase (GUS) in transgenic tobacco plants in a tissue-specific manner, expressing mainly in the vascular cells, with the highest GUS activity in the roots, followed by the stems. GAFP-2 in stably transformed tobacco was triggered by the fungus *Trichoderma viride* and by plant stress regulators, salicylic acid and jasmonic acid (Sa et al. 2003). GAFP-1 expressing lines of tobacco originating from *Nicotiana tabacum* cv. Wisconsin 38 had reduced symptom development and improved plant vigour compared with non-transformed plants and empty vector lines when they were challenged with two important crop-destroying fungi, *Rhizoctonia solani* and *Phytophthora nicotianae*. The transformed lines showed reduced root galling when challenged with *Meloidogyne incognita*, a roundworm which attacks the roots of plants, but against the water-borne bacterium, *Ralstonia* (*Pseudomonas*) *solanacea*, there was neither conferred resistance nor exacerbated disease (Cox et al. 2006).

These studies suggest that the development of GAFP-1 and GAFP-2 crops could help prevent fungal disease in agriculture and horticulture and reduce reliance on chemical fungicides. Gastrodianin being homologous to monomeric mannose-binding proteins in other orchids, particularly the achlorophyllous species, there is hope that gastrodianin-like proteins (GLIPs) may become a novel class of endogenous antifungal agents in genetically manipulated plants. Transgenic plum (*Prunus domestica*) lines expressing GAFP-1 possess resistance to the fungus, root fungus and nematodes, whereas the protein and its transcript are absent from scions of chimeric transfected plants (Nagel, Kalariya and Schnabel 2010). With the successful transfer of GAFP-gene to 280 plants of *Doritaenopsis* Tailin Angel using *Agrobacterium tumefaciens* for transfection Li, Kuang, Liu, et al. (2013) have opened up an avenue for making *Phalaenopsis* orchids resistant to bacterial rot to which they are currently especially prone. At least one other mannose-binding protein also from an orchid (Epipactis helleborine mannose-binding protein) has been shown to possess antifungal properties in vitro (Wang et al. 2001).

Wang et al. (2007a) determined the complete cDNA sequence of 2 types of cDNA clones in the subspecies *G. elata* Blume f. glauca matching the known gastrodianins and designated them as gastrodianin 4A and gastrodianin 4B. Gastrodianins appear to be abundantly present in fully opened flowers, in the outer cortical layers of secondary corms and in vascular cells of the orchid.

Comment

The numerous publications of well-designed and careful laboratory research mentioned in the foregoing discussion constitute substantial evidence for the neuroprotective effect of *Tianma*. Nevertheless, by themselves, they are not enough to convince medical science that the *Tianma* is effective in preventing memory loss and promote recovery from stroke, qualities for a medication which is sorely needed in modern medicine. Proof, in the modern setting, requires randomised controlled clinical trials, the more the better. Granted, there is a plethora of health supplements in the billion dollar market with even weaker evidence than *Tianma*, but that is not an valid excuse for a potentially valuable medicine to go untested. Randomised clinical trials should be performed for the advancement of TCM, perhaps beginning with *Tianma* and going on to other promising herbal remedies.

Unfortunately, in a recent review on the efficacy of TCM for stroke published in the refereed journal Stroke, the authors commented that, despite finding 11,234 articles on TCM stroke therapies, the review team from Italy and China were unable to conduct a meta-analysis. They found 34 randomised controlled trials, and all but one reported results in favour of TCM. To the reviewers, this suggests a strong publication bias. Wide variations in the studies prevented them from being pooled for analysis. Another meta-analysis conducted by the Stroke Clinical Research Unit of West China Hospital, Sichuan University in Chengdu also concluded, after studying data from 191 trials involving 19,338 patients, that there was insufficient good quality evidence on the effects of traditional Chinese patent medicine on the primary outcome in ischaemic stroke. This was not due to the TCM, which appeared to be potentially beneficial, but rather it was due to the design and conduct of the trials (Wu et al. 2007). A double-blind, placebo-controlled, randomised, multicenter study investigating the efficacy of a Chinese medicinal preparation, Neuroaid, on stroke recovery (CHIMES Study) found that it was no better than placebo. Neuroaid is constituted by nine herbal and five animal components but did not include any orchid among the herbs (Venketasubramaniam et al. 2009; Chen, Young, Gan, et al. 2013).

Here, it may be stressed that perhaps *Tianma* is better considered for primary and secondary stroke prevention, rather than for treatment.

High allozymic diversity was found to be present in 19 natural populations of *G. elata* occurring in central China (Chen et al. 2011). Overcollection at the rate of 100 t per annum during the 1960s has led to the depletion of natural sources, and *G. elata* is considered an endangered species in China. Autogamy and inbreeding within clone patches has resulted in a low level of genetic variation among wild populations. Thus, it is important to promote a conservation strategy which seeks to protect populations with the greatest genetic variation (Chen et al. 2014a). To alleviate heterozygote deficit in existing cultivated populations, it is suggested that competent research groups should conduct controlled hybridisation and have hybrid seedlings raised, evaluated and introduced into populations (Chen et al. 2014b).

G. elata Bl. f. *obovata* Y.J. Zhang is a new form (variety) from Jiuzhongjin in Lueyang

County, Shaanxi Province, China. It differs from the type in having spirally arranged flowers, obovate and basally oblique capsula and three nodding scapules (Zhang 2010). Crossing *G. elata* f. *elata* with *G. elata* f. *glauca* produced a hybrid with improved zymogram when the peroxide isoenzyme content of the orchids were analysed (Jiang et al. 2001). Variations in medicinal actions of different varieties of *G. elata*, if any, have yet to be determined. There is speculation that hybrids might better withstand unfavourable environments and weather conditions. They might, for instance, help overcome the reduced output of *Tianma* brought on by drought (Duan and Lu 2006).

Genus: *Geodorum* Jacks.

Chinese name: Dibao Lan

A genus of about ten species of terrestrial herbs, *Geodorum* is distributed from India and Sri Lanka through Southeast Asia and southern China to New Guinea, Australia and the Pacific Islands, in the lowland and montane forests. Rhizomes are short and pseudobulbs are subterranean. Leaves are large, elliptic, plicate, convolute, sheathing at the base and deciduous. Inflorescence arises from the side of the leafy pseudobulb. A nodding raceme crowded with medium-sized flowers is typical of the genus (Fig. 12.13).

The name *Geodorum* is derived from Greek *geo* (earth) and *darom* (present).

Geodorum attenuatum Griff.

Chinese name: *Dahuadibao Lan*
Thai Name: *Euong po*

Description: *G. attenuatum* is found in the edge of lowland forests in Hainan Island, Indochina, Thailand, Myanmar and Yunnan. Plants are 15–30 cm tall. Pseudobulbs are 2–3 by 1–1.5 cm, subterranean, and carry 3–4 oblong, lanceolate leaves, 9–22 by 2.5–4.2 cm, with long petioles 4–9 cm that form a pseudostem. Inflorescence arises from the base of leafy pseudobulbs and is shorter than the leaves. Raceme is nodding and crowded with 8–10 flowers of 2 cm diameter which are white with a yellow lip. The Yunnan variety has only 2–4 flowers which are rounder in form. It flowers in February in the south and April to May in the north of Thailand (Vaddhanaphuti 2001; Nanakorn and Watthana 2008) and May to June in China (Chen et al. 1999).

Herbal Usage: Pseudobulbs of *G. attenuatum* are used as a tonic in Thailand (Chuakul 2002).

Geodorum densiflorum (Lam) Schltr.

Bangladeshi name: *Kukurmuria* (Garo tribe)
Chinese name: *Dibao Lan* (precious ground orchid)
Indian name: *Kukurmuria* in Orissa

Description: *G. densiflorum* is a tuberous terrestrial species which appears above ground in India during the rainy season (Abraham and Vatsala 1981). Pseudobulbs are globular, subterranean, greenish, 3–3.5 cm across, and bear two or three stalked or sessile, plicate, coracious leaves, 15–20 cm long, 10–12 cm wide. Scape is 45 cm tall, tending to bend over, and the short rachis bears a few small, white or pink flowers, 2 cm across, which are generally not widely open. Lip is broad, concave, and marked with orange stripes over its basal half (Choo, van der Ent, Abdullah and Perumal 2009). Flowering season is from April to May in Assam, West Bengal and the Andaman Islands (Bose and Bhattacharjee 1980), March to May in the south (Abraham and Vatsala 1981), May to June at Parlode near Trivandrum (Nayar et al. 1986), June in Mumbai (Santapau and Kapadia 1966) and from June to July in China and the Philippines (Lin 1975; Davis and Steiner 1982; Chen et al. 2009c).

This terrestrial orchid is distributed in Sri Lanka, Myanmar, Thailand, Indochina, Malaysia, Indonesia, the Philippines, the Ryukyu Islands; and in Taiwan, Hainan, Guangdon,

Fig. 12.13 *Geodorum densiflorum* (Lam.) Schltr. Reproduced with permission from *Introductions to Orchids* by Abraham and Vatsala, Parlode, Thiruvananthapuram: Tropical Botanic Garden and Research Centre (TBGRI), 1981

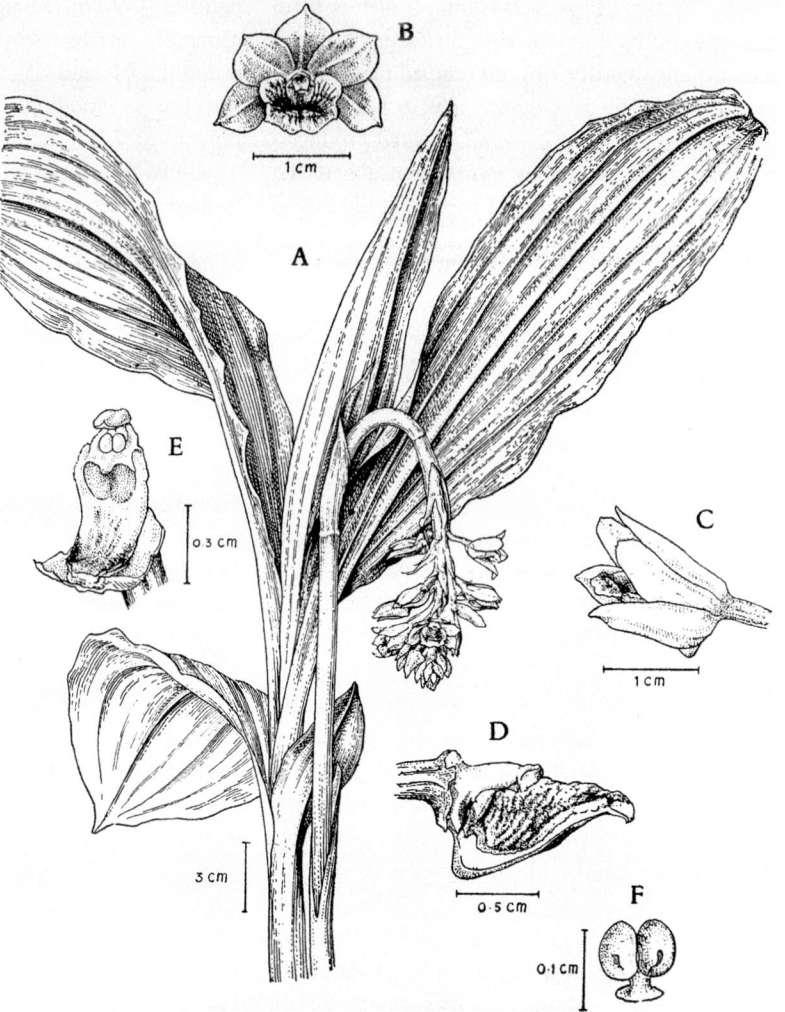

GEODORUM DENSIFLORUM (Lam.) Schltr.

Guangxi, Guizhou, Yunnan and Sichuan in China; also in Queensland (Australia) and Western Pacific, at 300–2400 m. It prefers sunny or slightly shaded locations and is found beside wooded streams in Hong Kong (Wu et al. 2002). *G. densiflorum* is widely distributed throughout the low elevations of the Western Ghats in India; at Nilgiris, Eastern Ghats, between 90 and 900 m (Joseph 1982; Rao and Sridhar 2007; Chen et al. 2009c) in open grasslands (Nayar et al. 1986), in the sun, on exposed slopes in the crevices of rocks, in shade, or on damp ground at 700–1400 m in the hills of Tamil Nadu (Matthew 1995).

Herbal Usage: Poultice made from the pseudobulbs of *G. densiflorum* is used in the Philippines (Gurrero 1922, quoted by Burkill 1935) and as a disinfectant in the province of Pangasinan of central Luzon, whereas further north in the provinces of Ilocos Norte and Ilocos Sur, a mucilage extracted from these tubers is used for gluing guitars and other musical instruments (Davis and Steiner 1982). Root paste mixed with ghee and honey is used by a hill tribe of Orissa to correct menstrual disorders (Dash et al. 2008). Further south in Andra Pradesh, the Kondareddis of Khammam and the Chenchus of Kurnool apply a root paste for insect bites and wounds. In the Khammam district of the same state, Koyas feed the tubers of the orchid, pepper and garlic to cattle struck down with intermittent fever (Reddy et al. 2005). It is fed to goats when the animals suffer from diarrhoea. Crushed rootstocks are rubbed on cattle to kill fleas in Uttar Pradesh (Trivedi et al. 1980).

A liniment made with the bulb of *G. purpureum* R. Br. (= *G. densiflorum*) and rice water was applied to heal skin inflammation, abscesses and other tumours in Northwestern India (Dalgado 1896, quoted by Lawler 1984). The roots are said to be insecticidal (Rao 2007). Herbal practitioners in Bangladesh use tubers and roots of *G. densiflorum* to treat unspecified diseases of women, swellings and skin infection (Musharof Hossain 2009 (line drawing: Abraham & Vatsala, vol. 1, p. 165)

Geodorum dilantum R. Br. (see **Geodorum recurvum (Roxb.) Alston.)**
Geodorum nutans (C. Presl.) Ames [see **Geodorum densiflorum (Lam.) Schltr.]**

Geodorum recurvum (Roxb.) Alston

Indian name: *Tajraj*

Description: A terrestrial herb, it is 25–30 cm tall and has elliptical pseudobulbs. Leaves are elliptic, plicate, undulate, measuring 17–20 by 5–7 cm, pointed at the tip and ensheathing the stem. Inflorescence reaches only up to the base of the unfolded portion of the leaves and carries 10–18 white flowers, 2–3 cm across. Lip is white and yellow, streaked with reddish-brown at the throat (Fig. 12.14). Flowering season is February to May in Thailand (Vaddhanaphuti 2005; Nanakorn and Watthana 2008) and April to June in China (Chen et al. 2009b). The species is distributed from Orissa in India to Myanmar, Thailand, Cambodia and Vietnam to south and southeast Yunnan, Guangdong and Hainan at 500–900 m (Chen et al. 2009b; Dash et al. 2008).

Herbal Usage: A hill tribe in Orissa uses *G. recurvum* to treat malaria. Decoction of 100 g of dried tuber, with 15–20 g black pepper,

Fig. 12.14 *Geodorum recurvum* (Roxb.) Alston (Photo: E.S. Teoh)

20–25 g garlic is drunk twice a day for 15 days as a remedy for malaria. Root paste is used to treat tumours (Dash et al. 2008).

Overview

Although all three species are present in China, there is no Chinese herbal application for these orchids. In India, *G. densiflorum* is used internally as an antidiarrhoeal and externally as a pesticide to treat goats and cattle in India. In the herbal markets of Kanpur, the orchids are offered separately as *G. dilantum*, *G. nutans* and *G. densiflorum* through distinctive herbal names (Trivedi et al. 1980). These herbal names are probably bestowed on variable forms of the species because all three names refer to a single species, namely *G. densiflorum*. In any case, their usage is similar. Local practitioners at Mount Popa in central Myanmar employ pseudobulbs of *Pecteilis hawkesiana*, and species of *Eulophia*, *Habenaria* and *Geodorum* as medicine. In the Shan state of Myanmar, tubers of *Geodorum* are eaten in the hope of increasing life expectancy. There are five species of *Geodorum* in Myanmar. The Burmese word for orchid is *thitkhwa* (Kurzweil and Lwin 2014).

Alkaloids are present in *G. citrinum* (Luning 1967). However, there is no pharmacologic publication on medicinal *Geodorum*.

Genus: *Goodyera* R. Br.

Chinese name: *Banye Lan*
Chinese medicinal name: *Banyelan*
Japanese name: Shusu Ran

A genus of terrestrial orchids with some 100 species, *Goodyera* grows in the dense shade in coniferous forests and alpine valleys across the northern hemisphere. Like other alpine orchids, they have adapted to the cold and can withstand being buried under layers of snow during the long winter months. *Goodyera* are characterised by creeping rhizomes terminating in short, succulent, erect stems with rosettes of evergreen leaves which are marked with white veins. Members of the genus are known as "rattlesnake plantain" in North America. A patch of hairs over the basal concavity of the lip distinguishes *Goodyera* from the plants of related genera such as *Anoectochilus, Ludisia* and *Zeuxine* which have similar foliage and are known as Jewel Orchids (Schuiteman and de Vogel 2000). The genus was named for an English botanist, John Goodyer (1592–1664) (Schultes and Pease 1963).

Living on the poorly lit forest floor amidst fallen leaves and decaying vegetation, *Goodyera* species have developed an elaborate leaf pigmentation involving an abaxial anthocyanin layer which enables them to capture energy from dim light (Benzing 1987). Even so, like many terrestrial orchids, *Goodyera* are also dependent on specific groups of fungi for their existence. This restricts their distribution. They become endangered whenever the environment is threatened (Wong and Sun 1999). *G. procera* is an endangered species in Hong Kong.

When McCormick and his colleagues studied the orchid–fungus relationship of *G. pubescens* in North America, they discovered that protocorms and adults associated with a single fungus species at a time, the relationship persisting for many years. Nevertheless, when stressed by drought, the orchid is capable of switching to a different fungal partner, albeit this is at a considerable cost in terms of mortality risk for the smaller individuals (McCormick, Whigham, Sloan, et al. 2006). Similarity-based large-scale distribution mapping twhichhat includes soil properties and the presence of associated species may assist in locating orchid species like *G. repens, Gymnadenia conopsea, Dactylorhiza incarnata* and *Dactylorhiza russowii* (Remm and Remm 2009). Such studies should also be able to determine suitable locations for in situ conservation of endangered or desirable medicinal species.

Nitrogen acquisition by the orchid was predominantly through a fungus-dependent pathway (Cameron, Johnsson , Leake, Read 2007). In contrast to mycorrhizal associations in other plants, orchid–mycorrhizal associations are often thought to benefit only the orchid (McCormick, Whigham, Sloan, et al. 2006). Not so. A study which used labelled carbon and nitrogen sources (C13 N15 glycine) demonstrated a bidirectional transfer of carbon between the green *G. repens* and its symbiotic fungus (Cameron et al. 2007). Initial transfer of carbon from fungus to orchid (up-flow) occurs, then after 8 days, there is a fivefold greater transfer of labelled carbon from the *G. repens* to the fungus, *Ceratobasidium cornigerum* (up-flow). The fungus also acquires inorganic phosphorus from the orchid (Cameron et al. 2007).

Goodyera biflora (Lindl.) Hook f.

Syn. *Goodyera pauciflora* Schltr.

Chinese name: *Danhuabanye Lan* (Big flower spotted leaf orchid) for *Goodyera pauciflora* from Sichuan and Yunnan: *Shaohuabanye Lan* (Few flowered spotted leaf orchid)

Fig. 12.15 *Goodyera biflora* (Lindl.) Hook. f. (Photo: Bhaktar B. Raskoti)

Description: This small, terrestrial orchid has a long creeping rhizome with an erect stem that is 4–7 cm tall. Leaves are 4–5 in number, elliptic, 2–4 by 1.2–2 cm, green marked by a reticulation of white veins above, pale green beneath, and supported by a 1–2 cm petioles. Inflorescence is almost sessile, pubescent, and two-flowered. Flowers barely open. Tepals are 2.5 cm long by 4 mm in width, cream-coloured, darkening to brown at the base (Fig. 12.15). *G. biflora* flowers in February or March in Taiwan (Lin 1975), February to July on the Chinese mainland (Chen et al. 2009c) and July to September in Bhutan (Gurong 2006). The species is distributed from Nepal, Bhutan and India across Xizang, Yunnan, Sichuan, Shanxi, Hubei, Hunan and Guangdong Provinces in China to Japan and Korea. In Bhutan, it is found in moist *Castanopsis* and *Quercus* (oak) forests at 1500–2400 m in the Thimphu district (Gurong 2006).

Herbal Usage: Herb is obtained from Shanxi, Hubei, Hunan, Guangdong, Sichuan, Yunnan and Tibet. In China, the whole plant is used to treat haemetemesis associated with tuberculosis, anorexia and neurosis. It enriches yin, and benefits the lungs (Wu 1994). Chinese herbalists use the whole plant of *Shaohuabanye Lan* (Few-flowered spotted leaf orchid; *G. pauciflora*) collected from Yunnan and Sichuan Provinces for detoxification and as an anti-inflammatory preparation to treat snake bites, sores and ulcers (Wu 1994). Interestingly, the leaves of *G. pubescens* were used in decoction and externally to treat scrofula in the Americas (Griffith 1847).

Goodyera brachysteia Hand. – Mazz

Chinese name: *Duanbaobanye Lan* (Short bud spotted leaf orchid)

Description: Plant 18–20 cm tall, bears 4–5 elliptic, green leaves, 2–2.7 by 0.8–1.5 cm supported by short petioles. Scape is slim, erect, hairy, a pale green, 15–22 cm in length. A short raceme carries many small, white flowers and lanceolate bracts, loosely arranged. The flowering season is May (Chen et al. 1999). *G. brachysteia* is an endemic, small, terrestrial herb found in forests at 1300–2000 m only in southwest Guizhou province and the adjacent northeast Yunnan.

Herbal Usage: Medicinal plants are obtained from Yunnan. In TCM, the entire plant is used to prepare the medicine. It is used to "strengthen the middle burner and replenish *qi* (internal energy or life force)". A decoction of the orchid plant is used to promote blood flow and to relax the muscles and tendons; it benefits the waist and kidneys, relieves backache, testicular swelling, dizziness, tinnitus and traumatic injuries (Wu 1994).

Goodyera foliosa (Lindl.) Benth ex C.B. Clarke

Chinese names: *Houchunbanye Lan* (Thick lipped, etched leaf orchid), *Gaolinbanye Lan* (High mountain ridge etched leaf orchid); *Duoyebanye Lan* (Multiple etched leaf orchid)

Fig. 12.16 *Goodyera foliosa* (Lindl.) Benth ex C.B. Clarke (Photo: E.S. Teoh)

Description: Stem is erect, 14–25 cm tall, commonly bearing 5 green, elliptical leaves which measure 2.7–4 cm by 1.5–2 cm, acute at the apex, petioled at the base, wavy at the edges, and marked by five longitudinal ribs, the central one being most prominent. Inflorescence is terminal, 5–7 cm in length, and many-flowered. The white flowers are shielded by prominent bracts. They do not open broadly. Ovary and dorsal sepal are pubescent (Fig. 12.16). Flowering period is July to September in China (Chen et al. 2009c), and August to November in Japan (Kanda 1977). It occurs at 300–1800 m, growing on the humus-rich floor of broad-leaved, evergreen forests from the Himalayan foothills of India, Bhutan, Nepal, Myanmar, across southern China (Xizang, Yunnan, Sichuan, Guangxi, Guangdong, Hong Kong, Fujian, Taiwan) to Japan and Korea. G. foliosa var. alba S.Y. Hu & Barretto is endemic to Hong Kong (Wu, Hu, Xia et al. 2001).

Herbal Usage: The whole plant is used in decoction. It is antipyretic, detoxifies, improves blood flow, reduces swellings and is used in the treatment of tuberculosis, hepatitis, weepy sores and snake bites (Ou et al. 2003).

Goodyera henryi Rolfe

Chinese name: *Guangebanye Lan* (Naked bud spotted leaf orchid)

Description: *G. henryi* has a long, creeping rhizome which terminates in a 10- to 15-cm-tall, erect stem that carries 4–6 ovate, pointed leaves, 2–5 by 1.6 cm. Inflorescence is short, smooth, and bears 3–9 white flowers. Blooms appear in August to September but they may sometimes be seen in October (Chen et al. 2009c). A montane Chinese species, it was recently encountered in the Hengshan Mountains in Sichuan at a height of 2130–2200 m

Herbal Usage: Herb is obtained from Guangdong and Guangxi. The whole plant is used to promote blood flow, and to treat snake bite, dysentery and lymphatic tuberculosis in Chinese herbal medicine (Wu 1994).

Goodyera kwangtungensis C.L. Tso

Chinese name: *Guangdongbanye Lan* (Guangdong province spotted leaf orchid), *Huayebanye Lan*

Description: Medicinal plants are harvested from Guangdong and Guangxi. Flowering plants have been found at the roadside at 1500–1800 m in

Taiwan. *G. kwangtungensis* is a lovely terrestrial, 18–30 cm tall, with 6–8 lanceolate leaves of jade green, broken by blotches of light green, most markedly at the midrib. Inflorescence is 25 cm long and carries 6–15 flowers over its distal half, all arranged in a single row. They barely open. Tepals are 8 mm in length. Flowering period is May to June (Chen et al. 2009c).

Herbal Usage: Herb is obtained from Guangdong and Guangxi. The entire plant is used to improve the condition of the lungs and to clear phlegm (Wu 1994).

Goodyera nantoensis Hayata [**see G. repens (L.) R.Br.**]
G. pauciflora Schltr. (see ***Goodyera biflora, (Lindl.) Hook. f.***]

Goodyera procera (Ker-Gawl.) Hook.

Chinese names: *Tushagen* (convex yarn root), *Gaobanyelan* (tall etched leaf orchid) *Zhengxijiao* (middle brook abaca/leaf), *Zhengxi Lan* (middle brook orchid)

Chinese medicinal name: *Shifengdan* (stone wind pellet), *Lanhuacao* (orchid flower herb)

Taiwanese names: *Peng Sha Gen* (borax root), *Zheng Xi Jiao* (straight stream leaf), *Sui Hua Ban Ye Lan* (spiking etched leaf orchid)

Description: Although an endangered species in India (Himalayas, Sikkim, Nilgiris, Western Ghats), *G. procera* enjoys a wide distribution, occurring in Nepal, Bhutan and China (including Hong Kong where attempts are being made at its conservation), Sri Lanka and Southeast Asia. It occurs in the Cameron Highlands in Malaysia (Cheam 2009). It is generally lithophytic, growing on rocks beside streams at different elevations. Plant is 15 cm tall, with many elliptic leaves that measure 11–8 by 1.5–3 cm

(Fig. 12.17). Inflorescence is 40 cm tall, with numerous, densely packed, minute (2 mm), white to cream-coloured flowers, which only partially open (Fig. 12.18). Flowering season is December to March in Thailand (Vaddhanaphuti 2005), February to April or May in China, depending on location (Chen et al. 1999, 2009c), April to May in southern India (Abraham and Vatsala 1981), June and November at Nilgiris in Tamil Nadu (Joseph 1982) but peaking in February to April in the Palni Hills which are an extension of the Western Ghats into Tamil Nadu (Matthew 1995). The species is deciduous (Abraham and Vatsala 1981; Lin 1975). Although it was once abundant in Hong Kong, here it has become rare due to overcollection (Wu, Hu, Xia et al. 2001).

Usage: TCM states that *G. procera* "dispels wind, eliminates dampness, nourishes blood, relaxes muscles and tendons, removes numbness, and promotes recovery from hemiplegia". Use of this orchid in pregnancy is contra-indicated (*Zhongyao Da Cidian* 1986). To promote recovery from hemiplegia, patients are fed the herb cooked with *Hong Hua Ma* and *Hong Niu Xi* (*Chengdu Chinese Herbs* quoted in *Zhongyao Da Cidian* 1986). Elsewhere, it is reported that the entire plant relieves rheumatism (Chen and Tang 1982), relaxes the muscles, enlivens the blood, smooth the lungs, suppresses coughs and stops bleeding. It is used to treat tuberculosis, weak kidneys, backache, jaundice, asthma and traumatic injuries (Wu 1994; Ou et al. 2003).

Plants are collected in winter. In the dim light at the winter forest floor, *G. procera* is unable to perform photosynthesis and it becomes mycotrophic, once again dependent on mycorrhiza for organic carbon and other nutrients (see Chap. 8). These mycorrhiza may therefore be the primary producers of the medicinal secondary metabolites isolated from *G. procera*.

Fig. 12.17 *Goodyera procera* Hook. f. Reproduced with permission from Introductions to Orchids by Abraham and Vatsala, Parlode, Thiruvananthapuram: Tropical Botanic Garden and Research Centre (TBGRI), 1981

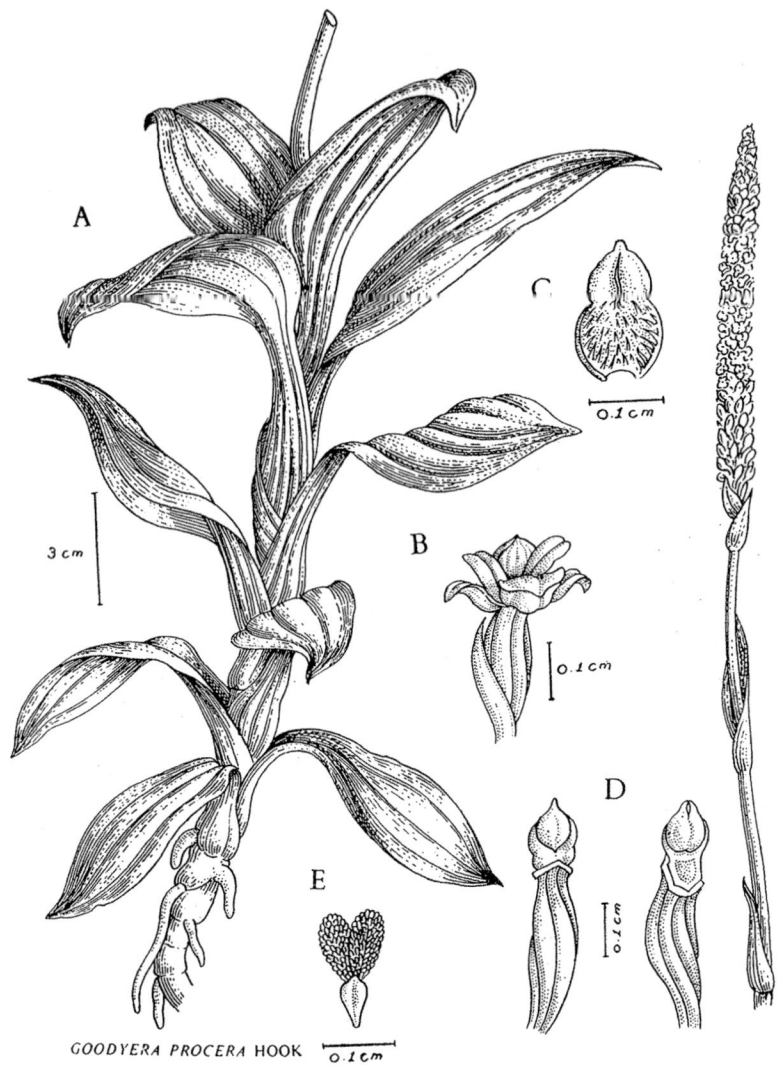

A. habit of the plant; B. flower; C. perianth part; D. column—front and top views; E. pollinia.

Phytochemistry: Only traces of alkaloid (0.001–0.01 % dry weight) is present in *G. procera* (Luning 1964).

Goodyera repens (L) R.Br.

Chinese names: *Nantoubanye Lan* (Pocket-sized, etched/reticulated leaf orchid), *Xiuzhenbanye Lan (Pocket-sized etched leaf orchid);*

Xiaobanye Lan (Small speckled leaf orchid); *Huasheyizhijian* (Floral snake single arrow.)

North American Names: Dwarf rattlesnake plantain; Lesser rattlesnake plantain

Pakistani common name: Creeping Ladies Tresses

Description: This is a small terrestrial herb with a tall, slim erect stem (10–25 cm in height), at its base bearing several dark green leaves with

Fig. 12.18 *Goodyera procera* (Ker-Gawl) Hook f. (Photo: E.S. Teoh)

Fig. 12.19 *Goodyera repens* (L.) R. Br. (Photo: Bhakter B. Raskoti)

heavy venation, measuring 1.5–3.5 by 1–1.5 cm. Most of the plant is an underground creeper with clusters of dark evergreen leaves which send out single spikes that bear small white flowers arranged in a spiral, in the manner of the genus *Spiranthes*. Flowers generally twist themselves to face the light, and appear to be arrayed on one side. They are flushed with pink on the back and pubescent over the dorsal sepal. Lateral sepals are well spread out. Petals form a hood with the dorsal sepal (Fig. 12.19). Flowering period is June to August in Taiwan (Lin 1975), and July to August in continental China (Chen et al. 1999), the United States and Canada (Christian 1975).

G. repens is distributed throughout the northern hemisphere from Scandinavia across Russia and China to Canada and northern USA. At the Huanglong Valley in Sichuan, only scattered communities occur in the lower and central valleys (Perner 2002). It grows in shady, leaf-strewn, humus-rich, moist ground in coniferous and deciduous forests or ravines and hill slopes at 2000–3800 m, in typically acid soil (pH 4.8–7.5). In North America, the orchid is found on the floor of isolated forests and bogs.

The life history of the orchid is interesting. Seed released in autumn stay dormant until spring when they germinate. During summer, the seedlings reach 1 mm in diameter but they acquire no roots or leaves until the next summer, their fourth growing season. Individual plants of *G. repens* may take up to 8 years to flower, after which the stem dies down, leaving a cluster of axillary shoots which then form a colony (Rasmussen 1995).

Phytochemistry: A small amount of alkaloid (0.01–0.1 % dry weight) is present in *G. repens* (Luning 1964). It also produces loroglossin (Aasen, Behr, and Leander 1975; Veitch and Grayer 2003).

Usage: In Chinese herbal medicine, the whole plant is used to nourish the lungs and kidneys, or to relieve pain. It is used to treat fever, weepy sores, tuberculosis, coughs, weak lungs, weak kidneys, asthma, dizziness, backache, nocturnal emission, impotence and snake bites (Wu 1994; Ou et al. 2003).

In India, this rare plant is used to treat illnesses of women, stomach and bladder diseases. The chewed leaves are applied to reptile bites. The mashed leaves are used to prevent infant rash. It is also used as salep (Pandey et al. 2003).

Goodyera schlectendaliana Rchb. f.

Chinese names: *Dabanye Lan* (Large speckled orchid); *Jinbian Lian* (Gold border lotus); *Yinherhuan* (Silver ear orchid); *Yinzong Lian* (Silver palm lotus); Mountain jewel orchid

Taiwanese names: *Da Wu Shan Ban Ye Lan* (Mountain dawn etched/speckled orchid), *Gao Shan Lian* (Mountain lotus)

Japanese name: *Miyama uzura* (Quail of the deep mountain)

Description: Plant is 5 cm tall. Leaves are heart-shaped, green with white spots or blotches on the top, pale green below (Fig. 12.20). Inflorescence is 9–15 cm tall, hairy, with 5–10 white flowers that are sometimes tinged with green. Flowers do not open fully, only the lower sepals are spread out. It flowers from August to October (Chen et al. 2009c). This dwarf species is native to China, Japan and Southeast Asia, occurring at 2000–2700 m. It was a favourite during the Edo Period in Japan because its beautiful dark green leaves with white snake-skin markings makes it shine like a gem. The white flowers are equally delicate.

Phytochemistry: *G. schlectendaliana* contains a unique, complex flavonol glucoside that was

Fig. 12.20 *Goodyera schlectendaliana* Rchb. f. (Photo: E.S. Teoh)

given the name Goodyerin (Du et al. 2000) (Fig. 12.12). It is 8-(4-hydroxy-3,5-dimethoxy-phenylmethyl)quercetin-3-O-rutinoside (Du, Sun and Shoyama 2000) and has sedative and anticonvulsant activities (Du et al. 2002a, 2008).

Usage: In TCM, the entire plant is used to clear the lungs, stop coughs, reduce swelling and pain, treat tuberculosis, coughs, phlegm, asthma and weak kidneys. A poultice of the leaves is applied for pain relief. Tincture of *Goodyera* (in rice wine) is consumed with rice as a tonic for internal injuries (Wu 1994; Ou et al. 2003).

Goodyera velutina Maxim

G. morrisonicola Hay.

Chinese names (in Taiwan): *Niaozui Lan* (Bird mouth orchid), *Rongyebanye Lan* (Velvet leaved etched leaf orchid)

Fig. 12.21 *Goodyera velutina* Maxim. (Photo: E.S. Teoh)

Description: *G. velutina* is distributed in mainland China and Taiwan, Korea and Japan. Leaves of this species are ovate, dark green marked with a white or brick-red streak at the centre, and 3–5 cm long (Fig. 12.21). They are also borne on the lower two-thirds of the flower spike. Unlike other species of *Goodyera*, in *G. velutina*, there are only a few (up to 7) partially open, hirsute, pinkish-white to reddish-brown flowers on a spike. The tip of the mid-lobe of the lip is white. In Taiwan the orchid is found in shady locations in primeval forests at 700–2500 m. It flowers in September and October in China and Taiwan (Chen et al. 2009c; Lin 1975).

Usage: TCM states that the whole plant is antipyretic and detoxifies, and that it enlivens the blood. It is also used as a painkiller (Ou et al. 2003).

Overview

Chinese herbalists are not partial to any particular species of *Goodyera* (*Banye Lan*) and would employ any available species already recorded in their pharmacopoeia, but usage varies among provinces. A decoction is usually prepared with 9–15 g of the herb, or for external

application 'a suitable amount'. In Hunan, it is used to "moisten the lungs", improve kidney qi and treat haematuria; in Guangxi, to improve circulation, reduce swelling, remove toxins, treat tuberculous coughs, bronchitis, dizziness, vertigo, fainting spells, fatigue, neurasthenia, erectile dysfunction, trauma, arthritic pain, sore throat, mastitis, tuberculous lymphadenitis, snake bites and skin diseases presenting with swelling and ulcers; in Fujian, to treat diabetes; in Hubei, to treat gastritis, tonsillitis and mastitis; in Zhejiang, to remove toxins and cause swellings to subside and for snake bites; and in Xizang (Tibet), to treat tuberculous lymphadenitis; in Guizhou, to treat predominantly neurasthenia or erectile dysfunction (Hu et al. 2000). Some prescriptions for their use are listed in Table 12.3.

A flavonol glycoside, goodyerin and 3 flavonoids (rutin, kaempferol-3-0-rutinoside and isorhamnetin-3-0-rutinoside) were isolated from *G. schlechtendaliana* by Du, Su and Shoyama in 2000. A few months later, they reported the isolation of a hepatoprotective aliphatic glycoside from three *Goodyera* species (*G. schlechtendaliana*, *G. matsumurana* and *G. discolor*). These glycosides protected primary cultured rat liver cells from carbon tetrachloride damage (Du, Sun and Shoyama 2000a; Du, Sun, Chen, et al. 2000a). *Goodyera* has been used as a substitute for the drug prepared from *Anoectochilus formosanus* because, like extracts of the latter orchid, Goodyerosides A and B blocked the elevation of the liver enzymes LDH, GOT and GPT in cultured rat hepatocytes following toxic exposure to carbon tetrachloride (Du et al. 2002). However, Goodyeroside A did not possess an 'antihyperliposis effect' unlike kinsenoside recovered from *Anoectochilus formosanus* which prevented increases in body and liver weights and deposition of uterine fat pads in rodents (Du et al. 2008). Goodyerin produced a significant, dose-dependent sedative and anticonvulsant effect on rats (Du et al. 2002, 2008).

These pharmacological observations do not shed any light on its traditional medicinal

Table 12.3 Prescriptions employing *Goodyera repens* or *G. schlechtendalia* (Jiangsu College of New Medicine 1986; *Zhongyao Da Cidian*)

1. Indication. General (dispel 'wind' and toxins, promote blood circulation and relieve pain, with the main usages being the treatment of bronchitis, arthritis, external injuries, tuberculous lymphadenitis, skin ulcers and carbuncles). Boil fresh plants 30–60 g for consumption. For external application: Grind the plants and apply to affected part.
2. Indication: Pulmonary tuberculosis Cook Banyelan (*G. repens* or *G. schlechtendalia*). Source: *Zhejiang Commonly Used Folk Herbs*
3. Indication: bronchitis Boil fresh plants 3–6 g for consumption Source: *Zhejiang Commonly Used Folk Herbs*
4. Indication: arthritis Fry grounded plants with wine and apply to affected joints once daily. Source: *Guizhou Folk Medicine*
5. Indication: Snake Bites Prepare poultice with fresh herb for external application Source: *Zhejiang Commonly Used Folk Herbs*

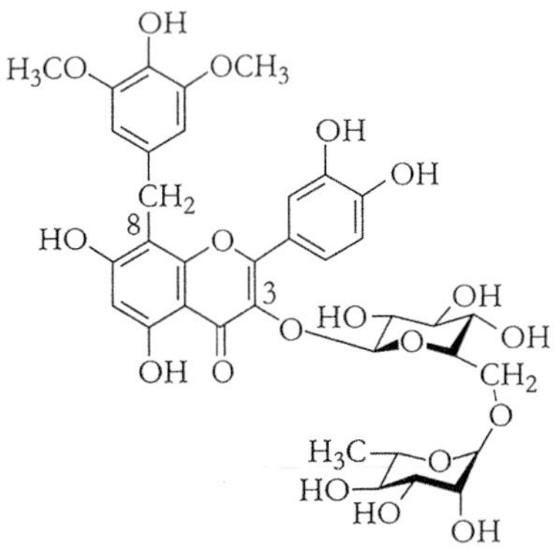

Fig. 12.22 Chemical structure of goodyerin from *Goodyera schlectendaliana*

applications, but in themselves they show promise for useful medicinal applications. *Goodyera* are small plants and their glycosides would need to be synthesised before anyone can perform extensive testing on them (Fig. 12.22).

Genus: *Grammatophyllum* Blume

The generic name, *Grammatophyllum*, is derived from the Greek *gramma* (letter) and *phyllon* (leaf) referring to the conspicuous marking on the petals and sepals that look rather like letters (Schultes and Pease 1963). *Grammatophyllum*s are large plants with cane-like or ovoid pseudobulbs that produce spectacular displays of numerous, arching inflorescences when in bloom. *G. speciosum* is the largest orchid in the world, in terms of plant mass (Sagarik and Guy 2011). *Grammatophyllums* are distributed in Southeast Asia, Papua New Guinea and the Solomon Islands.

In a bygone era, flowers of the green form of *G. scriptum* were regarded by the Moluccan natives as sacred, and only high-borne court ladies were entitled to sport them in their hair. The flowers were known as bunga putri, or "flower of princesses". Women on the island of Bali loved to wear flowers of *G. speciosum* in their hair because these flowers were such a rarity, the plant being found only in mountainous areas. The practice gave rise to a local Balinese proverb: *buka anggerek garingsinge, jabaning akatik di ja milike* (like the flowering of the Tiger Orchid, an entire family should not be judged by only a single member). *Garingsinge* is the local Balinese name for *G. speciosum* (Rifai 1975).

Grammatophyllum scriptum (L.) Blume

Common Names: Leopard orchid (adapted from *Gramm. leopardinum* Reichb. f.); tiger orchid (for var. *tigrinum* Lindl.)

Indonesian names provided by Rumphius (1627–1702): *Angrec calappa, Angrec lida* in Bali; *Angrek boki, Bonga boki, bonga putri* (Malay); *Saja baki, Saja ngawa, Ngawan, Saja ngawa* (Ternate), *Anggrek kringsing* (Bali)

Contemporary Indonesian names: *Bunga Bidadari, Anggerik Bidadari, Bunga Puteri* (Malaka); *Anggerik Puteri; Anggerik Matjan; Anggerik Harimau; Anggerik Boki* (Maluku); *Anggerik Garingsinge* (Bali); *Anggerik Tiwu Anggerik Susuru* (Sunda); *Saja Bake; Saja Ngawa; Saja Ngawan* (Ternate) *Tijgerorchidee* (Belanda).

Fig. 12.23 *Grammatophyllum scriptum* forma *alba* (L.) Blume (Photo: E.S. Teoh)

Description: This beautiful *Grammatophyllum* grows into a large plant when it is attached to trees, and it produces thousands of flowers on long arching inflorescences that are up to 1 m long. Pseudobulbs are cone-shaped, rounded at the base, green, with a few long thin leaves. Stiff white roots growing outwards and upwards all around the plant is typical of the genus *Grammatophyllum*, and are very striking in this species. Flowers of the common variety are 5 cm across, of a light brownish-green, marked with brown patches (appearing rather like tiger skin leading to the title "var. tigrinum" of one variety). Flowers of the alba variety are apple green and devoid of markings (Fig. 12.23). Main flowering season is April to May (Davis and Steiner 1982). *G. scriptum* is distributed from Borneo eastwards to the Pacific.

Usage: In his delightful accounts of the medicinal herbs of Maluku (the Moluccas), *Amboinscha Kruidbock* (The Amboinese Herbal), Geogius Everhardus Rumphius (1627–1702) reported that the pulp of *G. scriptum* pseudobulbs was pounded together with *Curcuma* (*kunyit*, turmeric, *Curcuma domestica*), a common spice in an Indonesian kitchen, and salt water to produce a paste for application on fingers afflicted by whitlow (bacterial infection of the nailbed) (Figs. 12.24 and

12.25). Rinsing one's mouth with sap of pseudobulbs was a remedy for thrush. A poultice prepared from mashed orchid pseudobulbs and ginger was applied to the abdomen as a vermifuge. It was believed to be capable of expelling all bad humours from the bowel, even shrinking a swollen spleen. (There are numerous causes of enlarged spleen but at that time, most swollen spleens would have been caused by malaria.) The same was applied on swollen legs to drain the fluid (Heyne 1927; Beekman 2002). Beri-beri was treated by rubbing the abdomen with the poultice (Rifai 1975).

Fruits were collected and seeds used for treating dysentery (Heyne 1927; Burkill 1935; Rifai 1975) or made into a love philter (Burkill 1935). The famous German taxonomist, H.G. Reichenbach, who was a leading authority on orchids during the second half of the nineteenth century, commenting on the work of Rumphius, noted that seeds of *G. scriptum* (*bonga boki, bonga putri*) was made into "a philtre which has a surprising effect upon the ladies who swallow it" (Hawkes 1953). Lawler was informed by Doinau that, in Bougainville, the seeds mixed with coconut milk were used to treat skin disease in children (Lawler and Slaytor 1969; Lawler 1984).

Fig. 12.24 *Grammatophyllum scriptum* (L.) Blume [as *Angrek scriptum* Rumphius]. From: Rumphius, G.E., *Amboinscha Kruidbock* (*The Amboinese Herbal*, 1747)

Fig. 12.26 *Grammatophyllum speciosum* Blume (Photo: E.S. Teoh)

Fig. 12.25 *Grammatophyllum scriptum* (L.) Blume (Photo: E.S. Teoh)

Grammatophyllum speciosum Blume

Common name: Tiger Orchid

Indonesian names; *Anggkrek tebu* (Malay) (sugar cane orchid); *Angkrek tiwu* (sugar cane orchid), *Kadaka susuru* (Sundanese)

Malay names: *Bunga puteri* (Princess flower); *Bunga bidadari* (nymph's flower)

Thai name: *Wan phetchahueng*

Myanmar name: *Kyar ba hone*

Philippine names: Giant Orchid, Queen Orchid, Sugar-cane Orchid

Description: *G. speciosum* is probably the largest orchid species in the world, producing individual epiphytic or terrestrial plants that weigh over two tons, each sometimes bearing hundreds of sprays with 5000–7000 flowers. Stems are 3 m long and 5 cm in diameter, covered by yellow membranous sheaths. Internodes are 4 cm long, and bear thin, duplicate, slender leaves 50–60 cm long and 3 cm wide, curving downwards to a pointed apex. Inflorescences arise from the base of the plant. They are erect or arching, 2 m long, with many large flowers, 10 cm across. The first two or three flowers at the bottom of the inflorescence are abnormal, with two sepals and two petals, without a lip, and they are widely spaced. All other flowers are not widely spaced but they do not overlap and are arranged uniformly on the raceme. They are normal, with all segments present. Petals and sepals are flat, yellow with blotches of dull yellowish-brown throughout. Lips are small, hirsute on the inner surface, yellow with red stripes (Figs. 12.26 and 12.27).

The species is distributed in Burma, Thailand, Laos, Malaysia, Singapore, Indonesia, the Philippine, and Papua New Guinea in the lowlands. Although it is epiphytic, owing to its enormous proportions, 13 m in diameter in both Penang and Singapore (Ridley 1894), it is generally grown on the ground. Plant sits on the ground but its roots grow upwards instead of penetrating into the earth.

In Singapore, it flowers from August to September. Ridley noted that plants had to reach a large size before they will flower but afterwards they would flower every year (Ridley 1894). Apparently, it requires high average precipitation, high relative humidity, and high temperatures to flower. These conditions are met in the province of Quezon (Tayabas) in eastern Luzon where the species occurs naturally. When the plants are brought to Luzon which experiences a long dry season from November to April, the plants do not flower (Davis and Steiner 1982). Plants grow very well in Penang

Fig. 12.27 *Grammatophyllum speciosum* Blume is a massive plant that can weigh over 2000 kg (Photo: E.S. Teoh)

in northern Malaysia which has a similar long dry season from November to February, but Michael Ooi also experienced difficulty in flowering his hundreds of large clumps of *G. speciosum* (Ooi, personal communication 2010). The flowers are pollinated by two species of carpenter bees, *Xylocopa latipes* Drury and *Xylocopa aestuans* Linn. (Ridley 1894).

Herbal Usage: The stem of *G. speciosum* is used to treat fever and anaemia in Thailand (Chuakul 2002). Roots of *Wan phet heung* (*G. speciosum*) are listed as an insect bite remedy in old Thai drug recipes (Daduang and Uawonggul 2008). Kelabit in Sarawak treat the orchid stems (*ubud aram*) as food, although it is slightly bitter when cooked (Tasn 2008).

Phytochemistry: Three new glucopyranosyloxybenzyl derivatives of (R)-2-benzylmalic acid and (R)-eucomic acid, grammatophyllosides A–C, a new phenolic glycoside, grammatophylloside D, along with cronupapine, vandateroside II, gastrodin, vanilloloside, orcinol glucoside and isovitexin, were isolated from pseudobulbs of *G. speciosum* (Sahakitpichan et al. 2013).

Overview

Plants of *G. scriptum* apparently differ in their ability to produce alkaloids. Whereas Luning (1967) found alkaloids to be present in moderate concentration (0.01–0.1 %) in *G. scriptum*, no alkaloid was detected in plants of the same species studied by Lawler and Slaytor (1970a, b). *G. speciosum* contains only traces of alkaloid (Luning 1967). *G. scriptum* and *G. papuanum* (= *G. speciosum*) did not show any antimicrobial activity against *Candida albicans*, *Staphylococcus aureus*, *Escherichia coli*, *Pseudomonas aeruginosa* and *Proteus mirabilis* (Lawler and Slaytor 1970a).

Obtaining sufficient material for chemical and pharmacological studies from this gigantic orchid should be simple; nevertheless, the sole report on the species is the above quoted study from Thailand (Sahakitpichan et al. 2013). Vitrification-based cryopreservation of *G. speciosum* protocorms for conservation has been described (Thammasiri 2010).

Genus: *Gymnadenia* R.Br.

Chinese name: *Shou shen* (hand ginseng).
Chinese medicinal name: *Shouzhangshen* (palm ginseng)
Japanese name: *Tegata Chidori*

This is a terrestrial herb with flattened tubers and an erect stem which is ensheathed by linear-lanceolate leaves. Inflorescence is terminal, racemose, cylindric or capitate and densely many-flowered. Flowers are small (Pearce and Cribb 2002).

The name *Gymadenia* is derived from Greek *gymnos* (naked) and *aden* (gland). This refers to the sticky disc of the pollinia which are free on either side of the rostellum. A small Eurasian temperate genus, its distribution extends to arctic North America, occurring in soil that is rich in lime (Rasmussen 1995).

Gymnadenia conopsea (L.) R.Br.

Chinese name: *shou shen* (hand ginseng), *Shouzhangshen* (Palm ginseng), *Foshoushen* (Buddha hand ginseng), *Zhangshen* (palm ginseng)

Chinese medicinal name: *Shouzhangshen* (Palm ginseng)

Japanese name: *Tegata-chidori*

Description: Plants are slender to robust, 20–60 cm tall with inflorescence. Tubers are ovoid with numerous short lobes. Stem is stout and carries 3–5 linear-lanceolate leaves, 5–11 by 1–2.5 cm sheathing the base. Inflorescence is terminal, slender, erect, 11–26 cm, with a densely flowered raceme 4–12 cm. Flowers are pale to purplish-pink, 7–10 mm across, fragrant, and arranged all round the rachis facing all directions. Flowering season is July and August.

Gymnadenia conopsea is found in open forests, rocky slopes, grasslands and water-logged meadows at 200–4700 m in Japan, Korea, Russia and Europe and in the temperate provinces of Heilongjiang, Jilin, Liaoning, Nei Mongolia, Shanxi, Shaanxi, and northern Sichuan, Yunnan and Xizang in China (Chen et al. 2009b). In the Czech Republic, tetraploid and octoploid cytotypes of *G. conopsea* occur in mixed ploidy populations (Jersakova et al. 2010). The ploidy of Chinese populations of *G. conopsea* is unknown.

Herbal Usage: Herb is obtained from Dongbei, Huabei, Sichuan and Tibet (Wu 1994). Shoushen (hand ginseng) enjoys a wide application in Chinese herbal medicine, almost similar to that of ginseng (*Panax ginseng*) which is not an orchid. The stem benefits the kidney and is used to replenish the vital essence, stop bleeding, or to alleviate lassitude caused by illness. It is used to treat coughs attributed to weak lungs, impotence and other forms of sexual dysfunction, discharge, traumatic injuries, thrombosis, chronic hepatitis and failure of lactation (*Zhonghua Bencao* 2000, 2005). It nourishes and strengthens a weakened body (Chen and Tang 1982).

Several prescriptions from various central Chinese provinces that employ *Gymnadenia conopsea* in a variety of conditions including malaise after an illness, coughs and dyspnoea, chronic bloody diarrhea, leucorrhoea, external injuries, hepatitis and poor lactation are collated from *Zhongyao Da Cidian* 1986) (see Table 12.4). Cooked with meat, it can also be served in a soup (Hu 2005).

Phytochemistry: Methanolic extract of *Gymnadenia conopsea* tubers produced an anti-allergic effect on cutaneous anaphylaxis reactions in mice. Following this demonstration, the team of Matsuda, Morikawa, Xia and Yoshikawa (2004) isolated three new dihydrophenanthrenes, gymconopins A–C, a new dihydrostilbene, gymconopin D, and 10 known compounds (phenanthrenes and stilbenes) from the methanolic extract of the orchid. Antigen-induced degranulation of RBL-2HC cells as measured by the release of beta-hexosaminidase was inhibited by 65.5 to 99.4 % at 100 µM by 11 of the isolated compounds, namely, 5 phenanthrenes and 6 stilbenes. Gymconopins A and C were not effective (Matsuda, Morikawa, Xia and Yoshikawa 2004). The group went on to isolate 7 glucosyloxybenzyl 2 isolbutylmalates termed gymnosides I–VII; three more gymnosides VIII–X, with radical scavenging activities from the methanolic extract; and 58 known constituents consisting of phenanthrenes, dihydrostilbenes, alpha-tocopherol and catechin (Morikawa et al. 2006a).

Alcoholic extract of *Gymnadenia conopsea* protected mice against silica-induced fibrosis of the lungs through its anti-oxidant action (Wang et al. 2007b). Ethanolic extraction of *Gymnadenia conopsea* tubers yielded four new minor constituents: two cyclodipeptides and two cyclodipeptide derivatives together with four known cyclodipeptides (Zi et al. 2010). A few months later, this team reported the isolation of 34 compounds from *Gymnadenia conopsea*, of which 22 were newly identified in the orchid species. Among these, three lignans, arctigenin, lappaol A and lappaol showed anti-oxidant activity (Yue et al. 2010).

Quantitative analysis of four active constituents, namely, dactylorhins A and B,

Table 12.4   Chinese herbal prescriptions employing *Gymnadenia conopsea* (Jiangsu College of New Medicine 1986; *Zhongyao Da Cidian*; *Bencao* 2000, 2005)

1. Indication: Weakness after an illness Prepare decoction with 9 g of the herb and consume (Primary source: *Handbook of Hebei Chinese Herbs*)
2. Indication: Coughs and Breathlessness Decoct: *Shouzhangshen* (*G. conopsea*, *G. crassinervis* or *G. orchidis*) 60 g with *Lilium brownie* 120 g and *Zizyphus vulgaris*[a] 120 g Take half a tea-cup of decoction each time (Primary Source: *Shaanxi Chinese Herbs*)
3. Indication: Chronic bloody diarrhoa; leucorrhoa Boil 9 g of *G. conopsea* and consume twice a day. (Primary source: *Handbook of Ningxia Chinese Herbs*)
4. Indication: External injuries Boil 9 g of *Shouzhangshen* (*G. conopsea*, *G. crassinervis* or *G. orchidis*) and consume (Primary source: *Handbook of Ningxia Chinese Herbs*)
5. Indication: Hepatitis: Appropriate amounts of *Foshoushen* (*G. conopsea*), *Polygonatum chinense* and *Fen Bao Ji* Prepare in syrup form. Take 10–15 ml each time, three times a day. (Primary source: *Quan Zhan Selected Chapters*)
6. Indication: Bu Lu's Disease (?) Equal amounts of *Foshoushen* (*G. conopsea*), *Sophora flavescens* and *Di Ding*. Boil 9 g each time and consume. (Primary source: *Selected Chapters of New Chinese Herbal Treatment of Inner Mongolia*)
7. Indication: Poor lactation Boil *Shouzhangshen* (*Gymnadenia conopsea*, *G. crassinervis* or *G. orchidis*) 9 g; with *Astragalus henryi*, 30 g; *Angelica sinensis* 15 g; *Stemmacantha uniflora* 6 g; and *Polygonatum thunbergii*, 9 g, with pig's trotters and consume. (Primary source: *Handbook of Highland Central China*)
8. Indication: Urinary frequency, spermatorrhoea and impotence Decoct flowers of *Gymnadenia* spp. 15 g; *Psoralea coryfolia* 9 g; *Epimedium* 15 pieces (?leaves); *Cynomorium songaricum* 12 g; yam 15 g; *Alpina cxyphylla* 4.5 g; *Achyranthes bidentata* 9 g

[a] *Zizyphus vulgaris* (syn. *Z. jujube*) fruit is often included in herbal prescriptions as a counter-poison. It lessens the toxicity of the principal herb, modifies the flavour of the decoction and lessens the effect of stimulants

loroglossin and militarine, by a method of extraction that gave a recovery of between 99.0 and 99.5 % was proposed as a method for gauging the quality of this Tibetan herb (Yang et al. 2009). Another team proposed the addition of one more compound, dactylorhin E. Their method gave a recovery of 97.7 to 101.0 %. The second method can also be applied to grade *Dactylorhiza viridis* (syn. *Coeloglossum viride*) (Li, Guo, Wang and Xiao 2009).

Alkaloid was not detectable in *G. conopsea* (Luning 1964).

Gymnadenia crassinervis **Finet**

Chinese name: *Duanjushoushen, Cumaishouzhangshen* (large venation hand palm ginseng), *Shou zhang zhen* (palm ginseng)

Chinese medicinal name: *Shouzhangshen* (Palm ginseng), *Cumaishouzhangshen* (large venation hand palm ginseng)

Description: Plant is stout or slender, 7–20 cm tall with inflorescence bearing lanceolate, pointed leaves, 3.5–7 by 1–2 cm, ensheathing the base. Inflorescence is 4–10 cm tall, erect, with a densely flowered rachis, 2–4 cm long. Flowers are pink, rarely tinged with white (Chen et al. 2009b); at Gaoligongshan in west Yunnan, light apple green flowers have been recorded (Jin et al. 2009). Flowering season is June to July (Chen et al. 2009b; Jin et al. 2009). The species is endemic to a small region in China. It is found in *Rhododendron* thickets, in crevices on rocky slopes at 2000–3800 m in west Sichuan, northwest Yunnan and eastern and southern Tibet (Chen et al. 2009b).

Usage: Herbs are obtained from Yunnan and Tibet (Wu 1994). *The Zhongyao Da Cidian* (Chinese Medicinal Encyclopedia) stated that "there are three types of palm ginseng or *Shou zhang shen*, namely, (1) *Gymnadenia conopsea* (*Shou Shen* in Chinese), (2) *Gymnadenia crassinervis* (*Shou Shen* with coarse veins) and (3), another herb in the "same family" called *Coeloglossum viride* "which has long, concave buds" All three "nourish wind and blood", quench thirst, treat weak lungs, coughs, dyspnoea, weakness from exertion and weight loss, neurasthenia, chronic diarrhea, blood loss, vaginal discharge, poor lactation, chronic hepatitis. Table 12.4 contains examples of Chinese herbal prescriptions employing *Shouzhangshen*, a preparation that may be derived from one of several species of *Gymnadenia*.

Gymnadenia orchidis Lindl.

Syn. *Gymnadenia conopsea* (L.) R.Br. var. *yunnanensis* Schltr.

Chinese names: *Shouzhangshen* (name also refers to *Gymnadenia crassinervis* Finet). *Xinanshoushen* (southwest hand ginseng), *Xinanshouzhangshen* (southwest hand palm ginseng)

Indian names: *Salam panja, Salam punja, Salep*

Nepali names: *Hati Jara; Panch aunle* ("five fingers", referring to the root) in Nepali; *Ongbu lakpa* (Sherpa)

Description: This terrestrial orchid is found in wet grasslands, or in thickets and forests along valleys at altitudes of 2400–4000 m from the southern parts of Shaanxi, Gansu and Qinghai across Hubei, western Sichuan, northwest Yunnan and Tibet in China to the Himalayan regions of Pakistan, Kashmir, Nepal, Bhutan and Sikkim. The 3-cm-long, paired tubers send off an erect plant 20–55 cm in height, with 3–5 elliptical leaves of diminishing size towards the apex (4–16 cm by 1.5 to 4 cm). Inflorescence is erect, densely flowered distally, clustered like a hyacinth, with pink to deep purple flowers,

0.6–1 cm across. Dorsal sepal forms a hood over the column and, with the outstretched lateral sepals, it forms an equilateral triangle. Petals are small, unextended and they lie just beneath the dorsal sepal. Lip is trilobed (Fig. 12.28). Flowering period is May to August in China (Chen et al. 2009b) or until September at Gaoligongshan (Jin et al. 2009). A period of chilling stimulates seed germination, and green pod culture (sowing 4–5 weeks post-pollination) works well in vitro (Rasmussen 1995).

In the *Iconographia Cormophtorum Sinicorum Tomus V* (1975), *G. conopsea* Linn. is differentiated from *G. orchidis* Lindl. by its narrow lanceolate leaves which are rounded at the apex, and broader flower parts. *G. orchidis* has broad elliptical leaves and narrower pointed petals and sepals and lip. The former species is found at 265–3500 m, whereas *G. orchidis* is found from 3100 to 4000 m (Lang and Tsi 1976).

Herbal Usage: In India, salep made from the tubers is consumed in the belief that it is a tonic with aphrodisiac properties (Rao 2004). Roots

Fig. 12.28 *Gymnadenia orchidis* Lindl. (Photo: Bhaktar B. Raskoti)

are used to heal wounds in Sikkim (Pearce and Cribb 2002). It was reported to be endangered in the Kumaum Himalayas (Jain 2003).

G. orchidis pseudobulb powder is applied to cuts and wounds in Nepal. A decoction of the pseudobulbs is taken three times a day, in moderation, to relieve stomach ache and liver or urinary disorders (Manandhar and Manandhar 2002; Baral and Kurmi 2006; Pant and Raskoti 2013). *G. conopsea* is used in Tibetan medicine, but we were unable to get more information on its usage.

Xinan Shoushen (hand ginseng) is collected from Tibet, Hubei, DShanxi, Gansu, Qinghai, Dichuan and Yunnan (Wu 1994). It enjoys a wide application in Chinese herbal medicine, almost similar to that of ginseng (*Panax ginseng*) which is not an orchid. The stem benefits the kidney and is used to replenish the vital essence, stop bleeding, and to overcome lassitude resulting from illness. It is used to treat coughs caused by weak lungs, cure impotence and other forms of sexual dysfunction, discharge, traumatic injuries, thrombosis, chronic hepatitis and failure of lactation (Anonymous 1989; Wu 1994).

Overview

Five species of *Gymnadenia* occur in China. Three species are used in herbal medicine. They have identical medicinal usage and are usually referred to by the same name, *Shouzhangzhen* (palm ginseng), although sometimes one or another could be identified by its origin, or by the appearance of the leaves. For instance, the specific Chinese name, *Xinan shou shen* (west and south palm ginseng), refers to *Gymnadenia orchidis* Lindl. whose distribution extends from the western and southwestern Chinese provinces to Sikkim, Bhutan, Nepal, Arunachal Pradesh and Kashmir: *Cumaishouzhangsheng* (large venation hand palm ginseng) describes the longitudinal veins on the leaves of *Gymnadenia crassinervis* Finet. Such labelling is commonly used in Chinese herbal terminology to define the source or variety of the herb, a factor that may influence its potency for a particular usage. Leaves of *Gymnadenia conopsea* contain the flavonols, kaempferol and quercetin. Flowers contain cyanidin 3,5-diglucoside or cyanin, and orchicyanin I and II. Chrysanthemin, seranin, ophrysanin and serapianin are also present in the plant (Strack et al. 1989). Floral scent in European strains of *G. conopsea* is due to benzyl acetate, benzyl benzoate, methyl eugenol, eugenol, elemicine, benzyl alcohol, cinnamic alcohol with trace amounts of phenylethylalcohol, phenylethylacetate and (Z)-3-hexenol, in varying proportions depending on the clone. Evaluation of four clones revealed that the scent in "spicy-floral" clones was attributable to eugenol and cinnamic alcohol, whereas aromatic floral clones had a dominance of benzyl acetate (Kaiser 1993).

Ordinarily, rats exposed to silica suffer from silica-induced pulmonary fibrosis as a consequence. In one experiment, a group of 60 rats was additionally treated with an alcoholic extract of *Gymnadenia conopsea* (GcAE) for up to 60 days. This produced a reduction in the lung/body weight ratio, reduction in types I and III collagen in the lungs, lowered lipid peroxidation, and an increase in the activities of superoxide dismutase (SOD) and glutathione peroxidase (GPx). These findings suggest that GcAE may have some protective activity against silica-induced fibrosis (Wang et al. 2007b).

Passive cutaneous anaphylaxis reactions produced on the ears of mice was attenuated by a methanolic extract *Gymnadenia conopsea* tubers, demonstrating that the latter possessed an anti-allergic principle (Matsuda, Morikawa, Xie and Yoshikawa 2004).

The methanol-eluted fraction of a methanolic extract of *Gymnadenia conopsea* exhibits radical-scavenging activities for DPPH and superoxide anion radicals. In this study, Morikawa et al. (2006a) isolated 3 new glucosyloxybenzyl 2-isobutylmalates, gymnosides VIII, IX and X, and an additional 58 known constituents which included the phenanthrenes and dihydrostilbenes responsible for the radical-scavenging activities. They had previously isolated 3 new dihydrophenanthrenes (gymconopins A–C), a new dihydrostilbene (gymconopin D) and 10 known phenanthrene

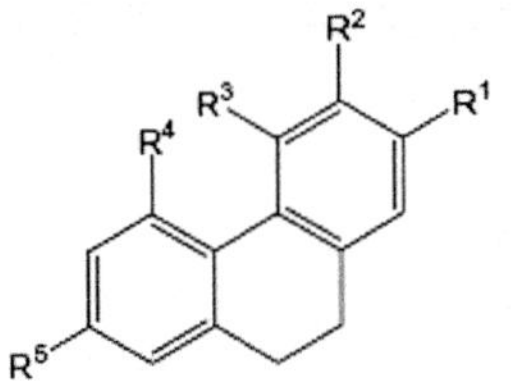

R1 = OMe, R2 = H, R3 = OH R4 = OH, R5 = H

R1 = OH, R2 = OMe, R3 = H, R4 = OMe, R5 = H

R1 = OMe R2 = OMe, R3 = H, R4 = OH, R5 = H

R1 = Ome, R2 – OMe, R3 = H, R4 = OMe, R5 = H

R1 – OH, R2 = OMe, R3 = H, R4 = H, R5 = OMe

R1 = OMe, R2 = OMe, R3 = H, R4 = H, R5 = OMe

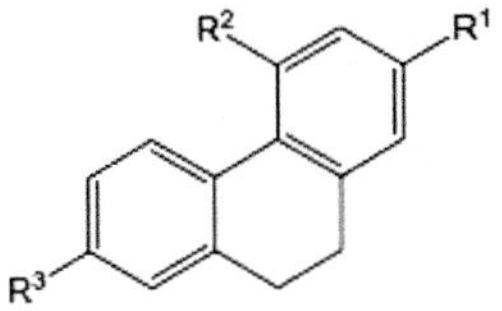

Coelonin: R1 = OH, R2 = OMe, R3 = OH

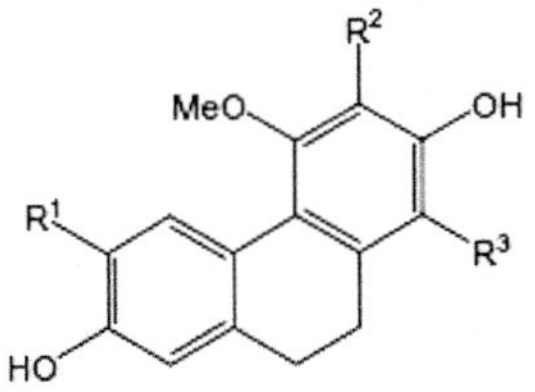

R1 = , R2 = hydroxybenzyl

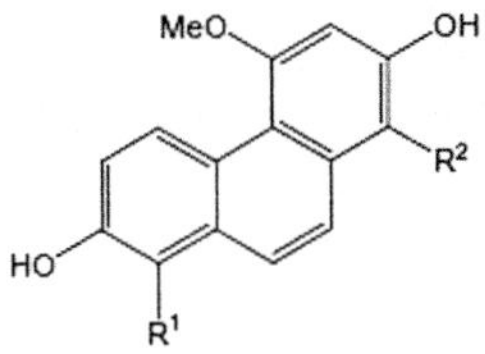

R1 = H, R2 = hydroxybenzyl

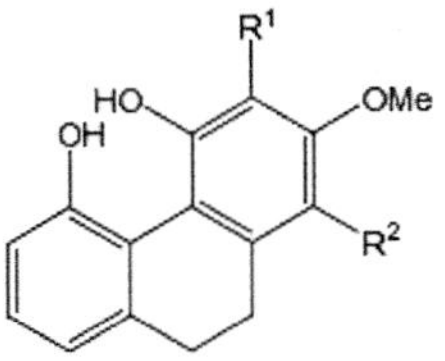

Gymconopin A: R1= H, R2 = hydroxybenzyl

Gymconopin B: R1 = hydroxybenzyl, R2 = H

Fig. 12.29 Phenanthrenes isolated from *Gymnadenia conopsea* by the Kyoto Pharmaceutical University group (Matsuda, Morikawa, Xie, Yoshikawa 2004; Morikawa et al. 2006b)

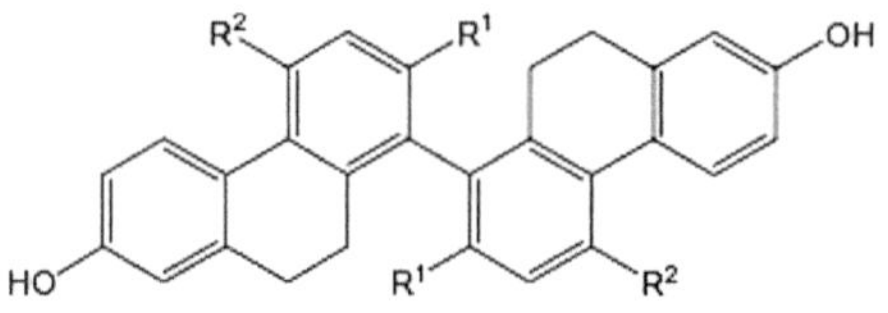

Blestriarene A, Flavantrin: R1 = OH, R2 = OMe

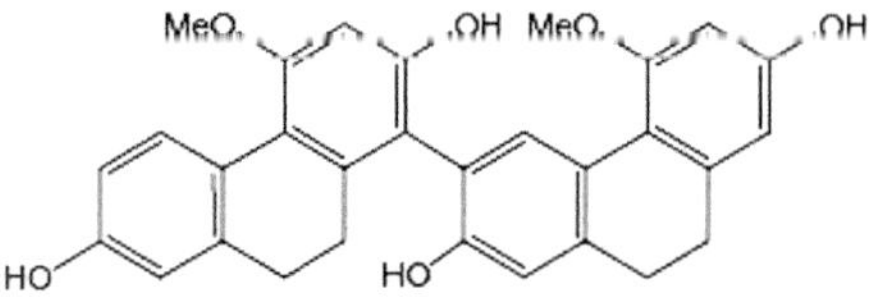

Gymconopin C

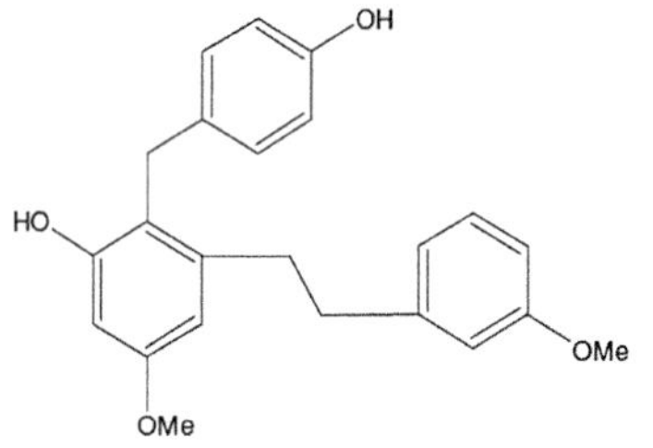

Gymconopin D

Fig. 12.29 (continued)

and stilbene compounds (Matsuda, Morikawa, Xie and Yoshikawa 2004). Subsequently, this active group from Kyoto Pharmaceutical University reported the isolation of 7 new gylcosyloxybenzyl 2 isobutylmalates which they named gymnosides I–VII (Fig. 12.29) (Morikawa et al. 2006a, b).

Five glycosyloxybenzyl 2-isobutylmalates (Dactylorhin B, loroglossin, dactylorhin A, dactylorhin E and militarine) were isolated from *G. conopsea* (Cai et al. 2006; Li, Guo Wang and Xiao 2009; Yang et al. 2009). Although their pharmacological actions have not been defined, it has been proposed that the quantitative analysis of these compounds by HPLC could be used as a yardstick for quality evaluation of this traditional Tibetan herb which is collected from Sichuan, Qinghai and Hebei provinces (Cai et al. 2006).

Average recoveries equal to or exceeding 99 % were described by Yang et al. (2009).

References

Aasen A, Behr D, Leander K (1975) Studies on Orchidaceae Gylcosides 2. The structure of loroglossin and militarine, two glycosides from Orchis militaris L. Acta Chem Scand 29b:1002–1004

Abraham A, Vatsala P (1981) Introduction to Orchids, with illustrations and descriptions of 150 South Indian Orchids. TPGRI, Trivandrum

Ahn EK, Jeon HJ, Lim EJ, Jung HJ, Park EH (2007) Anti-inflammatory and anti-angiogenic activities of Gastrodia elata Blume. J Ethnopharmacol 110 (3):476–482

An SJ, Park SK, Hwang IK et al (2003) Gastrodin decreases immunoreactivities of gamma-aminobutyric acid shunt enzymes in the hippocampus of seizure sensitive gerbils. J Neurosci Res 71(4):534–543

An SM, Park CH, Heo JC et al (2007) Gastrodia elata Blume protects against stress- induced gastric mucosal lesions in mice. Intern J Mol Med 20(2):209–215

Anderson M, Bergendorff O, Nielsen M et al (1995) Inhibition of kainic acid binding to glutamate receptors by extracts of Gastrodia. Phytochemistry 38(4):835–836

Anonymous (1989) Medicinal plants of China. Selection of 150 commonly used species. WHO pacific series no. 2. Regional Office of the Pacific, Manila

Baek NI, Choi SY, Park JK (1999) Isolation and identification of succinic semialdehyde dehydrogenase inhibitory compound from the rhizome of Gastrodia elata Blume. Arch Pharm Res 22(2):219–224

Baral SR, Kurmi PP (2006) A compendium of medicinal plants in Nepal. Rachana Sharma, Kathmandu, IUCN

Beekman EM (2002) (transl., ed) Rumphius Orchids. Orchid Texts from the Ambonese Herbal by Georgius Everhardus Rumphius. Yale University Press, New Haven

Bensky D, Clavey S, Stoger E, Gambie A (2004) Herbal medicine materia medica, 3rd edn. Eastland Press, Seattle, WA

Benzing DH (1987) Major patterns and processes in orchid evolution. In: Arditti J (ed) Orchid biology. Reviews & perspectives IV. Cornell University Press, Ithaca, NY

Bie X, Chen Y, Han J et al (2007) Effects of gastrodin on amino acids after cerebral ischemia-reperfusion injury in rat striatum. Asia Pac J Clin Nutr 16(Suppl 1):305–8

Bin-Jaliah I (2008) Comparison of glutathione peroxidase activity and free radicals production in the lungs and the brain of rats during graded hyperoxia. J Med Sci 8 (1):54–61

Bose TK, Bhattacharjee SK (1980) Orchids of India. Naya Prokash, Calcutta

Burkill IH (1935) (1966 reprint, 2nd edn, with contributions by Birtwistle W, Foxworthy FW, Scrivenor JB, Watson IG) A dictionary of economic products of the Malay Peninsula, vol II. Crown Agents for the Colonies, London. Ministry of Agriculture & Co-operatives, Kuala Lumpur

Cai M, Zhou Y, Gesang S et al (2006) Chemical fingerprint analysis of rhizomes of Gymnadenia conopsea by HPLC-DAD-MSn. J Chromatogr B Analyt Technol Biomed Life Sci 844(2):301–307

Cai Z, Hou S, Li Y et al (2008a) Effect of borneol on the distribution of gastrodin to the brain in mice via oral administration. J Drug Target 16(2):178–184

Cai Z, Hou SX, Yang ZX et al (2008b) Brain targeting of gastrodin nasal in situ gel. Sichuan Da Xue Xue Bao Yi Xue Ban 39(3):438–440

Cameron DD, Johnsson I, Leake JR, Read DJ (2007) Mycorrhizal acquisition of inorganic phosphorus by the green-leaved terrestrial orchid Goodyera repens. Ann Bot (Lond) 99(5):831–834

Chen WC, Tseng TS, Hsiao NW, et al (2015) Discovery of highly potent tyrosine inhibitor, T1 with significant anti-melanogenesis ability by zebra-fish in vivo assay

and computational molecular modeling. Sci Reports, 5 Article No: 7995

Cheam MC (ed) (2009) Wild orchids of Cameron Highlands. Regional Environmental Awareness Cameron Highlands, Cameron Highlands

Chen SC, Tang T (1982) A general review of the orchid flora of China. In: Arditti J (ed) Orchid biology. Reviews and perspectives II. Cornell University Press, Ithaca

Chen SC, Tsi ZH, Luo YB (1999) Native orchids of China in colour. Science Press, Beijing

Chen PJ, Hsieh CL, Su KP et al (2008) The antidepressant effect of Gastrodia elata on the forced swimming test in rats. Am J Chin Med 36:95–106

Chen PJ, Hsieh CL, Su KP et al (2009) Rhizomes of Gastrodia elata Bl. poosses antidepressant-like effect via Monoamine Modulation in Sub-chronic Animal Model. Am J Chin Med 36:1113–1124

Chen XQ, Lang KY, Gale SW et al (2009a) Goodyera R. Br. In: Chen XQ, Liu ZJ, Zhu GH et al (eds) Flora of China—Orchidaceae. Science Press, Beijing, pp 45–54

Chen XQ, Cribb PJ, Gale SW (2009b) Geodorum Jackson. In: Chen XQ, Zj L, Zhu GH et al (eds) Flora of China—Orchidaceae. Science Press, Beijing

Chen XQ, Gale SW, Cribb PJ (2009c) Gymadenia R. Br. In: Chen XQ, ZJ L, Zhu GH et al (eds) Flora of China—Orchidaceae. Science Press, Beijing

Chen YY, Bao ZX, Li ZZ (2011) High allozymic diversity in natural populations of mycoheterotrophic orchid Gastrodia elata, an endangered medicinal plant in China. Biochem Syst Ecol 39(4–6):526–535

Chen X, Cao DX, Zhou L et al (2011) Structure of a polysaccharide from Gastrodia elata Bl., and oligosaccharides prepared thereof with anti-pancreatic cancer cell growth activities. Carbohydr Polym 86 (3):1300–1305

Chen CL, Young SH, Gan HH et al (2013) Chinese medicine neuroaid efficacy on stroke recovery: a double blind, placebo-controlled randomized study. Stroke 44(8):2093–2100

Chen YY, Bao ZX, Qu Y, Li W (2014a) Genetic diversity and population structure of the medicinal orchid Gastrodia elata revealed by microsatellite analysis. Biochem Syst Ecol 54:182–189

Chen Y-Y, Bao Z-X, Qu Y, Li Z-Z (2014b) Genetic variation in cultivated populations of Gastrodia elata, a medicinal plant from central China analyzed by microsatellites. Genet Res Crop Evol 61 (8):1523–1532

Chiou SJ, Yen JH, Fang CL et al (2007) Authentication of medicinal herbs using PCR-amplified ITS2 with specific primers. Planta Med 73:1421–1426

Choi JH, Lee DU (2006) A new citryl glycoside from Gastrodia elata and its inhibitory activity on GABA transaminase. Chem Pharm Bull (Tokyo) 54:1720–1721

Choo CM, van der Ent A, Abdullah E, Perumal B (2009) Wild orchids of Cameron Highlands. Regional

Environmental Awareness Cameron Highlands, Cameron Highlands, Malaysia

Chow S, Chen SC (1983) Notes on Chinese Gastrodia. Acta Bot Yunnanica 5(4):361–368

Christian P (1975) Creeping ladies' tresses. Alpine Gard Soc Great Britain Quart Bull 43(4):322–324

Chuakul W (2002) Ethnomedical uses of Thai orchidaceous plants. Mahidol Univ J Pharm Sci 29 (3–4):41–45

Cox KD, Layne DR, Scorza R, Schnadbel G (2006) Gastrodia anti-fungal protein from the orchid Gastrodia elata confers disease resistance to root pathogens in transgenic tobacco. Planta 224(6):1373–1383

Cribb P, Fischer E, Killmann D (2010) A revision of Gastrodia (Orchidaceae: Epidendroideae, Gastrodieae) in tropical Africa. Kew Bull 65(2):315–321

Cull-Candy S, Brickley S, Farrant M (2001) NMDA receptor subunits: diversity, development and disease. Curr Opin Neurobiol 11(3):327–335

Daduang S, Uawonggul N (2008) Herbal therapies of snake and insect bites in Thailand. In: Watson RR, Preedy VR (eds) Botanical medicine in clinical practice. CAB International, Wallingford, pp 814–822

Dash PK, Sahon S, Bal S (2008) Ethnobotanical studies on orchids of Niyamgiri Hill Ranges, Orissa, India. Ethnobot Leaflets 12:70–78

Davis RS, Steiner ML (1982) Philippine Orchids. A detailed treatment of some 100 native species. M & L Licudine Enterprises (reprint), Manila

Descamps E, Petrault-Laprais M, Maurois P et al (2009) Experimental stroke protection induced by 4-hydroxybenzyl alcohol is cancelled by bacitracin. Neurosci Res 64(2):137–142

Ding CS, Shen YS, Li G et al (2007) Study of a glycoprotein from Gastrodia elata: its effects of anticoagulation and antithrombosis. Zhongguo Zhong Yao Za Zhi 32 (11):1060–1064

Du XM, Sun NY, Shoyama Y (2000) Flavonoids from Goodyera schlechtendaliana. Phytochemistry 53:997–1000

Du XM, Sun NY, Chen Y, Irino N, Shoyama Y (2000a) Hepatoprotective aliphatic glycosides from three Goodyera species. Biol Pharm Bull 23(6):731–734

Du XM, Sun NY, Takizawa N et al (2002) Sedative and anticonvulsant activities of goodyerin, a flavonol glycoside from Goodyera schlechtendaliana. Phytother Res 16(3):261–263

Du XM, Irino N, Furusho N, Hiyashi J, Shoyama Y (2008) Pharmacologically active compounds in the Anoectochilus and Goodyera species. Nat Med (Tokyo) 62(2):132–148

Duan N, Lu XQ (2006) Effects of drought on output of Gastrodia elata. Zhong Yao Cai 29(1):3–5

Fan LL, Dong YC, Xu TY et al (2013) Gastrodin production from p-2-hydroxybenzyl alcohol through biotransformation by cultured cells of Aspergillus foetidus and Penicillium cyclopium. Appl Biochem Biotechnol 170(1):138–148

Fratkin J (1997) Chinese herbal patent formulas: a practical guide. Pelanduk Publications, Selangor, Darul Ehsan

Fu JH, Du GY (2001) Effect of tianma cuzhi keli (TMCZKL) on contents of transmitter amino acids of brain tissue in the mouse repetitious cerebral ischemia reperfusion. Zhongguo Zhong Yao Za Zhi 26 (1):53–55

Fu XY, Li SM, Wang TT et al (2010) Antioxidation effects and mechanism of Gastrodin in AD rat models. Jiepou Xuebao 41(4):485–490

Gan ZJ (1986) Processing of Gastrodia elata where it is produced. Zhong Yao Tong Bao 11(1):27–28

Griffith RE (1847) Medical botany of descriptions of the more important plants used in medicine, with their history, properties and mode of administration. Lea and Blanchard, Philadelphia

Gudmundsdottir KB, Kristinsson J, Sigurdarson S et al (2008) Glutathione peroxidase (GPX) activity in blood of ewes on farms in different scrapie categories in Iceland. Acta Veterinaria Scand 50:23

Guo Q, Wang Y, Lin S et al (2015) (2015): 4-Hydroxybenzyl-substituted amino acid derivatives from Gastrodia elata. Acta Pharm Sin B 5(4):350–357

Guo S, Xu J (1990) Determination of primary metabolic products of fungi promoting seed germination of Gastrodia elata Bl. and other Orchidaceae medicinal plants. Zhongguo Zhong Yao Za Zhi 15(6):332–334

Guo Z, Tan T, Zhong Y, Wu C (1991a) Study of the mechanism of gastrodin and derivatives of gastrodigenin. Hua Xi Yi Ke Da Xue Xue Bao 22 (1):79–82

Guo Z, Tan T, Zhong Y, Wu C (1991b) Study of the mechanism of gastrodin and derivatives of gastrodigenin. Xi Yi Ke Da Xue Xue Bao 22(1):79–82

Guo JM, Pan WW, Qian DW et al (2014a) Analgesic activity of DaChuanXiongFang after intranasal administration and its potential active components in vivo. J Ethnopharmacol 150(2):649–654

Guo ZG, Jia XP, Su XJ et al (2014b) Gastrodin attenuates vincristine-induced mechanical hyperalgesia through serotonin 5-HTIA receptors. Bangladesh J Pharmacol 8(4):414–419

Gurong DB (2006) An Illustrated guide to the Orchids of Bhutan. DSB Publications, Thimphu

Ha JH, Lee DU, Lee JT et al (2000) 4-Hydroxybenzaldehyde from Gastrodia elata Bl. is active in the antioxidation and GABAergic neuromodulation of the rat brain. J Ethnopharmacol 73(1–2):329–333

Ha JH, Shin SM, Lee SK et al (2001) In vitro effects of hydroxybenzylaldehydes from Gastrodia elata and their analogues on GABAergic neurotransmission, and a structure-activity correlation. Planta Med 67 (9):877–880

Ha JH, Hayashi J, Sekine T, Deguchi S et al (2002) Phenolic compounds from Gastrodia rhizome and relaxant effects of related compounds on isolated smooth muscle preparation. Phytochemistry 59(5):513–519

Han BH, Suh Y, Chi HJ (1998) Medicinal plants of the Republic of Korea. WHO Regional Publications. Pacific Series No. 21. WHO, Manila

Han YJ, Je JH, Kim SY et al (2014) Gastrodia elata shows neuroprotective effects via activation of PI3K signaling against oxidative glutamate toxicity in HT22 cells. Am J Chin Med 42(4):1007–1019

Hawkes AD (1953) Snippet. Orchid J 1(3):105

Hayashi J, Sekine T, Deguchi S et al (2002) Phenolic compounds from Gastrodia rhizome and relaxant effects of related compounds on isolated smooth muscle preparation. Phytochemistry 59(5):513–519

Heo JC, Woo SU, Son MS et al (2007) Anti-tumor activity of Gastrodia elata Blume is closely associated with GTP-Ras-dependent pathway. Oncol Rep 18(4):849–853

Heyne K (1927) De nuttige planten van Nederlandsche Indie, vol 1. Uitgave van het Departement van Landbouw, Nijverheid & Handel in Nederlandsche-Indie, pp 508–513

Horner IB (ed) (1982) Book of discipline, vol I–IV. Pali Text Society, London

Hsieh MT, Wu CR, Chen CF (1997) Gastrodin and p-hydroxybenzyl alcohol facilitate memory consolidation and retrieval but not acquisition, on the passive avoidance task in rats. J Ethnopharmacol 56(1):45–54

Hsieh CL, Tang NY, Chiang SY et al (1999) Anticonvulsive and free radical scavenging actions of two herbs, Uncaria rhynchophylla (Miq) Jack and Gastrodia elata Bl., in kainic acid-treated rats. Life Sci 65(20):2071–2082

Hsieh CL, Chang CH, Chiang SY et al (2000) Anticonvulsive and free radical scavenging activities of vanillyl alcohol in ferric chloride-induced epileptic seizures in Sprague-Dawley rats. Life Sci 67(10):1185–1195

Hsieh CL, Chiang SY, Cheng KS et al (2001) Anticonvulsive and free radical scavenging activities of Gastrodia elata Bl in kainic acid treated rats. Am J Chin Med 29(2):331–341

Hsieh CL, Chen CL, Tang NY et al (2005) Gastrodia elata BL mediates suppression of nNOS and microglia activation t protect against neuronl damage in kainic-acid treated rats. Am J Chin Med 33(4):599–611

Hsieh C, Lin JJ, Chiang SY et al (2007) Gastrodia elata modulated activator protein 1 via c-Jun N-terminal kinase signalling pathway in kainic acid-induced epilepsy in rats. J Ethnopharmacol 109(2):241–247

Hsu TC, Kuo CM (2010) Supplements to the orchid flora of Taiwan (IV): four additions to the genus gastrodia. Taiwania 55(3):243–248

Hu SY (2005) Food plants of China. The Chinese University Press, Hong Kong

Hu JM, Zhou J (2010) Phytochemical progress at KIB of two species in orchidaceae—dendrobium and gastrodia. Recent Development in Chinese Herbal Medicine. Nanyang Technological University, p 90

Hu XM, Zhang WK, Zhu QZ et al (2000) Zhonghua Bencao, vol 8. Shanghai Science and Technology Publication, Shanghai

Hu JF, Li GZ, Li MJ (2003a) Protective effect of Gastrodia elata and E-gelatin on lead-induced damage to the structure and function of rat hippocampus. Zhonghua Lao Dong Wei Sheng Zhi Ye Bing Sa Zhi 21(2):124–127

Hu JF, Li GZ, Li MJ (2003b) The antagonistic action of Gastrodia elata combined with E-gelatin on lead-induced down regulation of c-fos expression in rat brain. Zhonghua Lao Dong Wei Sheng Zhi Ye Bing Sa Zhi 21(2):128–131

Hu JM, Zhao YX, Li N et al (2010) Pharmaceutical research of Gastrodia elata. Conference on Recent Development in Chinese Herbal Medicine. Nanyang Technological University, Singapore, 25–26 January 2010. Poster presentation, uncatalogued

Huang JH (1989) Comparison studies of pharmacological properties of injection Gastrodia elata, gastrodin-free fraction and gastrodin. Zhongguo Yi Xue Ke Xue Yuan Xue Bao 11:147–150

Huang SY, Shi JG, Yang YC, Hu SL (2002) Studies on chemical constituents from Tibetan medicine wangle (rhizome of Coeloglossum viride var. bracteatum). Zhongguo Zhong Yao Za Zhi 27(2):118–120

Huang NK, Lin YL, Cheng JJ, Lai WL (2004) Gastrodia elata prevents rat pheochromocytoma cells from serum-deprived apoptosis: the role of the MAPK family. Life Sci 75(13):1649–1657

Huang SY, Li GQ, Shi JG, Mo SY (2004) Chemical constituents of the rhizomes of Coeloglossum viride var. bracteatum. J Asian Nat Prod Res 6(1):49–61

Huang ZB, Wu Z, Chen FK, Zou LB (2006) The protective effects of phenolic constituents from Gastrodia elata on the cytoxicity induced by KCl and glutamate. Arch Pharm Res 29:963–968

Huang NK, Chern Y, Fang JM et al (2007) Neuroprotective principles from Gastrodia elata. J Nat Prod 70:571–574

Huang CL, Yang JM, Wang KC et al (2011) Gastrodia elata prevents Huntington aggregations through activation of the adenosine A(2A) receptor and ubiquitin proteasome system. J Ethnopharm 138(1):162–168

Huang GB, Zhao T, Muna SS et al (2013) Therapeutic potential of Gastrodia elata Blume for the treatment of Alzheimer's disease. Neural Regen Res 8(1):1061–1070

Huang LQ, Li ZF, Wang Q et al (2015) Two new furaldehyde compounds from the rhizomes of Gastrodia elata. J Asian Nat Prod Res 17(4):352–356

Hwang SM, Lee YJ, Kang DG, Lee HS (2009) Anti-inflammatory effect of Gastrodia elata rhizome in human umbilical vein endothelial cells. Am J Chin Med 37:395–406

Illum L (2004) Is nose-to-brain transport of drugs in man a reality? J Pharm Pharmacol 56:3–17

Imam SZ, El-Yazal J, Newport GD et al (2001) Methamphetamine-induced dopaminergic neurotoxicity: role of peroxynitrite and neuroprotective role of antioxidants and peroxynitrite decomposition catalysts. Ann N Y Acad Sci 938:366–380

Jain SP (2003) An inventory of threatened medicinal and aromatic plants of northwestern India. In: Singh V,

Jain AP (eds) Ethnoboany and medicinal Plants of India and Nepal. Scientific Publishers (India), Jodhpur, pp 908–913

Jang YW, Lee JY, Kim CJ (2009) Anti-asthmatic activity of phenolic compounds from the roots of Gastrodia elata Bl. Int Immunopharmacol 10(2):147–154

Jersakova J, Castro S, Sonk N et al (2010) Absence of pollinator-mediated barriers in mixed-ploidy populations of Gymnadenia conopsea s.l. (Orchidaceae). Evol Ecol 24(5):1199–1218

Ji N, Li Y (2009) Effect of different strains of Armillaria mellea on the yield of Gastrodia elata f. glauca. J Fungal Res 6(4):231–233

Ji XH, Li N, Wang JR, Zhang YJ (2008) Fingerprint identification between gastrodia elata Blume and its counterfeits with infrared spectroscopy. Xibei Zhiwu Xuebao 28:831–835

Ji W, Chen L, Ma X, et al (2014) Molecularly imprinted polymers with novel functional monomer for selective solid-phase extraction of gastrodin from the aqueous extract of Gastrodia elata

Jiang L, Wan S, Wang S, Yu C (2001) Isoenzyme analysis of Gastrodia elata f. elata and G. elata f. glaucca and their hybrid. Zhong Yao Cai 24:547–548

Jiang L, Dai J, Huang Z et al (2013) Simultaneous determination of gastrodin and puerarin in rat plasma by HPLC and the application to their interaction on pharmacokinetics. J Chromatogr B Analyt Technol Biomed Life Sci 915–916:8–12

Jiang GL, Wu HY, Hu YQ et al (2014) Gastrodin inhibits glutamate-induced apoptosis of PC12 cells via inhibition of CaMKII/ASK-1/p38 MAPK/p53 signalling cascade. Cell Mol Neurobiol 34(4):591–602

Jin XH, Zhao XD, Shi XC (2009) Native Orchids from Gaoligongshan Mountains, China. Science Press, Beijing

Joseph J (1982) Orchids of Nilgiris, vol XXII, Records of the botanical survey of India. Botanical Survey of India (Department of Environment), Howrah

Jung JW, Yoon BH, Oh HR et al (2006) Anxiolytic-like effects of Gastrodia elata and its phenolic constituents in mice. Biol Pharm Bull 29:261–265

Jung TY, Suh SI, Lee H et al (2007) Protective effects of several components of Gastrodia elata on lipid peroxidation in gerbil brain homogenates. Phytother Res 21 (10):960–964

Kaengkan P, Baek SE, Ji YK et al (2013) Administration of mesenchymal stem cells and ziprasidone enhance amelioration of ischemic brain damage in rats. Mol Cells 36(6):534–541

Kaiser R (1993) The scent of orchids: olfactory and chemical investigations. Editiones Roche, Basel

Kanda K (1977) The native orchids of Japan. Seibundo-Shinkosha, Tokyo

Kang OS, Doo AR, Kim SN, et al (2014) Gastrodia elata extract attenuates the L-Dopa-induced dyskinesia in Parkinson's disease mouse model. Society for Neuroscience Abstract Viewer and Itinerary Planner 43, 2013

Khan MR, Omoloso AD (2004) Antibacterial activity of Galeola foliata. Fitoterapia 75(5):494–496

Kho MC, Lee YJ, Ahn YM et al (2013) Effect of ethanol extract of Gastrodia elata Blume on high-fructose induced metabolic syndrome. FASEB J 27:1108

Kim HJ, Moon KD, Oh SY et al (2001) Ether fraction of methanol extracts of Gastrodia elata, a traditional medicinal herb, protects against kainic acid-induced neuronal damage in mouse hippocampus. Neurosci Lett 314:65–68

Kim HJ, Lee SR, Moon KD (2003a) Ether fraction of methanol extracts of Gastrodia elata, medicinal herb protects against neuronal cell damage after transient global ischemia in gerbils. Phytother Res 17:909–912

Kim HJ, Moon KD, Lee DS, Lee SH (2003b) Ethyl ether fraction of Gastrodia elata Blume protects amyloid beta peptide-induced cell death. J Ethnopharmacol 84:95–98

Kim ST, Kim JD, Lyu YS et al (2006) Neuroprotective effect of some plant extracts in cultured CT105-induced PC12 cells. Biol Pharm Bull 29:2021–2024

Kim HJ, Hwang IK, Won MH (2007) Vanillin, 4-hydroxybenzyl aldehyde and 4-hydroxybenzyl alcohol prevent hippocampal CA1 cell death following global ischemia. Brain Res 1181:130–141

Kurzweil H, Lwin S (2014) A Guide to Orchids of Myanmar. Natural History Publications (Borneo), Kota Kinabalu

Kusano S (1911) Gastrodia elata and its symbiotic association with Armillaria mellea, vol IV(I). College of Agriculture, Imperial University of Tokyo, Tokyo, pp 1–73

Lang KY, Tsi ZH (1976) Orchidaceae. Iconographia Cormophytorum Sinicorum V. Science Press, Beijing

Laschke MW, Vorsterman van Oijen AE, Scheuer C, Menger MD (2013) In-vitro and in-vivo evaluation of the anti-angiogenic actions of 4-hydroxybenzyl alcohol. Br J Pharmacol 163(4):835–844

Lawler LJ (1984) Ethnobotany of the orchidaceae. In: Arditti J (ed) Orchid biology reviews and perspectives, vol 3. Cornell University Press, Ithaca

Lawler LJ, Slaytor M (1969) The distribution of alkaloids in orchids from the Territory of Papua New Guinea. Proc Linnean Soc NSW 94:419–421

Lawler LJ, Slaytor M (1970a) Uses of Australian Orchids by Aborigines and Early Settlers. Med J Australia 2:1259–1261

Lawler LJ, Slaytor M (1970b) A simple method for screening plants for antibacterial activity. Aust J Pharmacy 51:609

Lebesgue D, Chevaleyre V, Zukin RS, Etgen AM (2009) Estradiol rescues neurons from global ischemia-induced cell death: multiple cellular pathways of neuroprotection. Steroids 74:555–761

Lee KT, Kim BJ, Kim JH et al (1997) Biological screening of 100 plant extracts for cosmetic use (i) inhibitory activities of tyrosinase and DOPA auto-oxidation. Int J Cosmet Sci 19(6):291–298

Lee YS, Ha JH, Yong CS et al (1999) Inhibitory effects of constituents of Gastrodia elata Bl. On glutamate-induced apoptosis in IMR-32 human neroblastoma cells. Arch Pharm Res 22:404–409

Lee JY, Jang YW, Kang HS et al (2006) Ant-inflammatory action of phenolic compounds from Gastrodia elata root. Arch Pharm Res 29:849–858

Lee YK, Woo MH, Kim CH et al (2007) Two new benzofurans from Gastrodia elata and their DNA topoisomerases I and II inhibitory activities. Planta Med 73(12):1287–1291

Lee YJ, Hwang SM, Kang DG et al (2009) Effect of Gastrodia elata on tumor necrosis factor-alpha-induced matrix metalloproteinase activity in endothelial cells. J Nat Med 63(4):463–467

Li YM, Zhou ZL, Hong YF (1993a) Studies on the phenolic derivatives from Galeola faberi Rolfe. Yao Xue Xue Bao 28(10): 766–771

Li HX, Ding MY, Yu JY (2001) Simultaneous determination of p- hydroxybenzyladehyde, p-hydroxybenzyl alcohol, 4(beta-D-glucopyranosyloxy)- benzyl alcohol, and sugars in Gastrodia elata Blume measured as their acetylated derivatives by GC-MS. J Chromatogr Sci 39:251–4

Li HB, Chen F (2004) Preparative isolation and purification of gastrodin from the Chinese medicinal plant Gastrodia elata by high speed counter-current chromatography. J Chromatogr A 1052:229–232

Li P, Li SP, Fu CM et al (2004) Pressurized solvent extraction in quality control of Chinese herb. Zhongguo Zhong Yao Za Zhi 29:723–726

Li N, Wang KJ, Chen JJ, Zhou J (2007) Phenolic compounds from the rhizomes of Gastrodia elata. J Asian Nat Prod Res 9:373–377

Li M, Guo SX, Wang CL, Xiao PG (2009) Quantitaive determination of five glycosyloxybenzyl 2-isobutylmalates in the tubers of Gymadenia coinopsea and Coeloglossum viride var. bracteatum by HPLC. J Chromatogr Sci 47(8):709–713

Li J, Kuang P, Liu RD et al (2013) Transfer of GAFP and NPI, two disease-resistant genes, into a Phalaenopsis by Agrobacterium tumefaciens. Pak J Bot 45 (5):1761–1766

Li SH, Chen WC, Lu KH et al (2014) Down-regulation of Slit-Robo pathway mediating neuronal cytoskeletal remodeling processes facilitates the antidepressive activity of Gastrodia elata Blume. J Agric Food Chem 63(43):10493–10503

Li Z, Wang Y, Ouyang H et al (2015) J Chromatogr B Analyt Technol Biomed Life Sci 988:45–52

Lin TP (1975) Native Orchids of Taiwan, vol 1. Southern Materials Center, Taipei

Lin LC, Chen YF, Tsai TR, Tsai TH (2007) Analysis of brain distribution and billiary excretion of a nutrient supplement, gastrodin, in rat. Anal Chim Acta 590:173–9

Liu J, Mori A (1992) Antioxidant and free radical scavenging activities of Gastrodia elata Bl. and Uncaria rhynchophylla (Miq.) Jacks. Neuropharmacology 31 (12):1287–1298

Liu J, Mori A (1993) Antioxidant and pro-oxidant activities of p-hydroxybenzyl alcohol and vanillin: effects on free radicals, brain peroxidation and

degradation of benzoate, deoxyribose, amino acids and DNA. Neuropharmacology 32:659–69

Liu W, Hu YL, Wang M et al (2002) Purification, crystallization and preliminary X-ray diffraction analysis of a novel mannose-binding lectin from Gastrodia elata with antifungal properties. Acta Crystallogr D Biol Crystallogr 58:1833–1835

Liu G, Dong Q, Yu F et al (2004) Identification of Gastrodia elata Blume by Fourier transform infrared spectroscopy. Guang Pu Xue Yu Guang Pu Fen Xi 24:308–310

Liu W, Yang N, Ding J et al (2005) Structural mechanism governing the quaternary organization of monocot mannose-binding lectin revealed by the novel monomeric structure of an orchid lectin. J Biol Chem 280:14865–14876

Liu Y, Tang X, Pei J et al (2006) Gastrodin interaction with human fibrinogen: anticoagulant effects and binding sites. Chemistry 12:7807–7815

Liu J, Yu ZB, Ye YH, Zhou YW (2008) Chemical constituents from the tubers of Cremastra appendiculata. Yao Xue Xue Bao 43(2):181–184

Lu H, Li AY, Liu FY et al (2009) Effects of gastrodin on the dopamine system of Tourette's syndrome rat models. Biosci Trends 3:58–62

Luning B (1964) Studies on the Orchidaceae alkaloids. I. Screening of species for alkaloids 1. Acta Chim Scand 18:1507–1516

Luning B (1967) Studies on Orchidaceae alkaloids IV. Screening of species for alkaloids 2. Phytochemistry 6:857–861

Luo HR (1985) Brief report on the planting of Gastrodia elata in the Dinghu Mountain area. Zhong Yao Tong Bao 10:8–10

Majumder PL, Bandyopadhyay D (2010) Stilbenoids and sequiterpene derivatives of the orchids Gastrochilum calceolaria and Dendrobium amoenum. application of 2D

Manandhar NP, Manandhar S (2002) Plants and people of Nepal. Timber, Portland, OR

Matsuda H, Morikawa T, Xie H, Yoshikawa M (2004) Antiallergic phenanthrenes and stilbenes from the tubers of Gymnadenia conopsea. Planta Med 70(9):847–855

Matthew KM (1995) An excursion flora of Central Tamilnadu, India. A.A. Balkaema, Rotterdam, The Netherlands

Mayr H (1998) Orchid names and their meanings. Gantner-Verlag K.G, Vaduz

McCormick MK, Whigham DF, Sloan D et al (2006) Orchid-fungus fidelity: a marriage meant to last? Ecology 87(4):903–911

Merkus FW, van den Berg MP (2007) Can nasal drug delivery bypass the blood-brain barrier? Questioning the direct transport theory. Drugs 8:133–144

Mo L, Kang JC, He J et al (2009) A preliminary Study of the Composition of Endophytic Fungi from Gastrodia elata. J Fungal Res 6(4):211–215

Mori A, Yokoi I, Noda Y, Willmore LJ (2004) Natural antioxidants may prevent posttraumatic epilepsy: a proposal based on experimental animal studies. Acta Med Okayama 58(3):111–118

Morikawa T, Xie H, Matsuda H et al (2006a) Bioactive constituents from Chinese natural medicines. XVII. Constituents with radical scavenging effect and new glucosyloxybenzyl 2-isobutylmalates from Gymnadenia conopsea. Chem Pharm Bull 54(4):506–513

Morikawa T, Xie H, Matsuda H, Yoshikawa M (2006b) Glucosyloxybenzyl 2-isobutylmalates from the tubers of Gymnadenia conopsea. J Nat Prod 69(6):881–886

Musharof Hossain M (2009) Traditional therapeutic uses of some indigenous orchids of Bangladesh. Med Aromat Plant Sci Biotechnol 3:100–106

Muszyńska B, Sułkowska-Ziaja K, Wołkowska M, Ekiert H (2011) Chemical, pharmacological, and biological characterization of the culinary-medicinal honey mushroom, Armillaria mellea (Vahl) P. Kumm. (Agaricomycetideae): a review. Int J Med Mushrooms 13(2):167–175

Nagel AK, Kalaariya H, Schnabel G (2010) Gastrodia Antifungal Protein (GAFP-1) and its transcript are absent from scions of chimeric transgenic-grafted plum. AGRIS 45(2):188–192

Nah ELQ, Ng XW, Chen KS (2010) An in-vitro study of the effect of gastrodin on neuronal cells. Recent development in Chinese herbal medicine. Nanyang Technological University, 2010. Programs and Abstracts, p 67

Nanakorn W, Watthana S (2008) Queen Sirikit Botanic Garden (Thai Native Orchids 1 and 2). Wanida Press, Chiang Mai

Nayar TS, Koshy KC, Sathish Kumar C, Mohanan N, Mukunthakumar S (1986) Flora of Tropical Botanic Garden, Palode. TBGRI, Thiruvananthapuram

Neal AP, Guilarte TR (2013) Mechanisms of lead and manganese toxicity. Toxicol Res (Camb) 2(2):99–114

Ng LT, Wu SJ, Tsai JY et al (2007) Antioxidant activities of cultured Armillaria mellea. Appl Biochem Microbiol 43(4):444–448

Niu Q, Niu P, He S (2004) Effect of Gastrodia elata on learning and memory impairment induced by aluminum in rats. Wei Sheng Yan Jiu 33:45–48

Ojemann LM, Nelson WL, Shin DS et al (2006) Tian ma, an ancient Chinese herb. offers new options for the treatment of epilepsy and other conditions. Epilepsy Behav 8(2):376–383

Ou JC, Hsieh WC, Lin IH, Chang YS, Chen IS (eds) (2003) The catalogue of medicinal plant resources in Taiwan. Department of Health, Executive Yuan, Taipei

Pan R, Xu J (1998) Analysis gastrodin in various types of Gastrodia elata Bl. Zhongguo Zhong Yao Za Zhi 23:336–337

Pandey NK, Joshi GC, Mudaiya RK et al (2003) Management and conservation of medicinal orchids of Kumaon and Garhwal Himalaya. In: Singh V, Jain AP (eds) Ethnoboany and medicinal plants of India and Nepal. Scientific Publishers (India), Jodhpur, India, pp 114–118

Pant B, Raskoti BB (2013) Medicinal orchids of Nepal. Himalayan Map House (P) Ltd, Kathmandu

Paoletti P, Neyton J (2007) NMDA receptor subunits: function and pharmacology. Curr Opin Pharmacol 7(1):38–47

Park EJ, Lee WY (2013a) Quantitative effects of various tree species on tuber growth and pharmacological compositions of Gastrodia elata. Hort Environ Biotechnol 54(4):357–363

Park EJ, Lee WY (2013b) In vitro symbiotic germination of myco-heterotrophic Gastrodia elata by Mycena species. Plant Biotechnol Rep 7(2):185–191

Park EJ, Ahn JK, Lee WY, Kim ST (2008) Establishment of in-vitro production system of Gastrodia elata immature tubers followed by symbiotic seed germination. Plant Biol (Rockville) 2008:172–173

Pearce NR, Cribb PJ (2002) The Orchids of Bhutan. Royal Botanic Garden, Edinburgh

Peng CX, Gong JS, Zhang XF et al (2007) Production of gastrodin through biotransformation of p-hydroxybenzyl alcohol using hair root cultures of Datura tatula L. Afr J Biotechnol 7:211–216

Perner H (2002) Orchids and eco-tourism: the world natural heritage and biosphere reserve, Huanglong. Proceedings of the 17th World Orchid Conference, Shah Alam, Malaysia, 158–164

Perner H, Luo Y (2007) Orchids of Huanglong. Sichuan Fine Arts Publishing House, China

Porter-Smith F, Stuart GA (2003) Chinese Medicinal Herbs: A Modern Edition of a Classic Sixteenth-Century Manual by Li Shih Chen. Dover Publications Inc., Mineola, NY

Pyo MK, Jin JL, Koo YK, Yun-Choi HS (2004) Phenolic and furan type compounds isolated from Gastrodia elata and their anti-platelet effects. Arch Pharm Res 27:381–385

Qiu H, Tang W, Tong X et al (2007) Structure elucidation and sulfated derivatives preparation of two alpha-D--glucans from Gastrodia elata Bl. and their anti-dengue virus bioactivities. Carbohydr Res 342:2230–2236

Ran YZ, Xu JT (1988) Studies on the inhibition of seed germination of Gastrodia elata Bl. by Armillaria mellea Qul. Zhong Yao Tong Bao 13:15–17

Ran Y, Xu J (1990) Selection of the germination strain of Mycena osmundicola Lange in Gastrodia elata Bl. seeds. Zhongguo Zhong Yao Za Zhi 15:271–274

Rao AN (2004) Medicinal Orchid Wealth of Arunachal Pradesh. Newsl ENVIS NODE Indian Med Plants 1(2):1–7

Rao TA (2007) ETHNO BOTANICAL DATA ON WILD ORCHIDS OF MEDICINAL VALUE AS PRACTISED BY TRIBALS AT KUDREMUKH NATIONAL PARK IN KARNATAKA. Orchid Newsl 2(2):1–7

Rao TA, Sridhar S (2007) Wild Orchids in Karnataka: a pictorial compendium. Institute of Natural Resources Conservation, Education, Research and Training (INCERT), Bangalore

Rasmussen HN (1995) Terrestrial orchids from seed to mycotrophic plant. Cambridge University Press, Cambridge, UK

Reddy KN, Reddy CS, Raju VS (2005) Ethno-orchidology of orchids of Eastern Ghats of Andhra Pradesh. EPRTI Newsl 11(3)

Remm K, Remm L (2009) Similarity-based large-scale distribution mapping of orchids. Biodivers Conserv 18(6):1629–1647

Ridley H (1894) The Orchidaceae and Apostasiaseae of the Malay Peninsula. J Linn Soc 32:335–338

Rifai MA (1975) Extraordinary uses of orchids in Indonesia. Report First ASEAN Orchid Congress, Bangkok

Riis BJ, Christensen C, Overgaard K (1986) The effect of nasal calcitonin on post-menopausal osteoporosis. In: Christensen C (ed) New horizons in Osteoporosis. Lancaster, Parthenon

Rigmister JY (1988) Intranasal calcitonin: a new horizon in prevention of early post-menopausal bone loss. In: Christensen C (ed) New horizons in Osteoporosis. Lancaster, Parthenon

Rong LH, Cai CT (2010) Effect of different woods on the yield of Gastrodia elata. Wuhan Zhiwuxue Yanjiu 28(6):761–766

Sa Q, Wang Y, Li W (2003) The promoter of an antifungal protein gene from Gastrodia elata confers tissue-specific and fungus-inducible expression patterns and responds to both salicylic acid and jasmonic acid. Plant Cell Rep 22:79–84

Sagarik R, Guy S (2011) Grammatophyllum speciosum. A gigantic specimen in flower and fruit. Orchids 80(12):730–733

Sahakitpichan P, Mhidol C, Disadee W et al (2013) Glucopyranosyloxybenzyl derivatives of (R)-2-benzylmalic acid and (R)-eucomic acid, and an aromatic glucoside from the pseudobulbs of Grammatophyllum speciosum. Tetrahedron 69(3):1031–1037

Santapau H, Kapadia Z (1966) The Orchids of Bombay. Government of India Press, Calcutta

Schachter SC (2009) Botanicals and Herbs: a traditional approach to treating epilepsy. Neurotherapeutics 6(2):415–420

Schuiteman A, de Vogel EF (2000) Cac Ci Ho Lan (Orchidaceae) Cua Thai Lan, Lao, Campuchia Va Viet Nam. Orchid Genera of Thailand Laos, Cambodia and Vietnam. (Vietnamese-English edition). National Herbarium Nederland, Leiden

Schultes RE, Pease AS (1963) Generic names of orchids. Their origin and meaning. Academic Press, New York

Seidenfaden G, Wood JJ (1992) The Orchids of Peninsular Malaysia and Singapore. Olsen & Olsen, Fredensborg

Sekizaki H, Kuninaga S, Yamamoto M (2008) Identification of Armillaria nabsnona in Gastrodia tubers. Biol Pharm Bull 31:1410–1414

Seo BI, Joo SJ, Park JH et al (2010) The anti-osteoporotic effect of aqueous extracts of Gastrodiae Rhizoma in vitro and in vivo. J Health Sci 56(4):422–424

Seok YM, Jin FX, Shin HM et al (2011) HMCO5 attenuates vascular contraction through inhibition of RhoA/Rho-kinase pathway. J Ethnopharmcol 133(2):484–489

Shen D, Chang H (1963) The anticonvulsant and analgesic activities of Tian-Ma (Gastrodia elata Bl.). Acta Pharm Sin 10:242–245

Shen-Hao Y, Wang D, Xiang-Lan M, Ying-Jun Z, Yang C R (2008) The influencing factors on gastrodin content in the herbal materials of Gastrodia elata (Orchidaceae). Acta Bot Yunnanica 30(1):110–114

Shi Y, Dong JW, Tang LN et al (2014) N-6-(3-methoxy-4-hydroxybenzyl) adenine riboside induces sedative and hypnotic effects via GAD enzyme activation in mice. Pharmcol Biochem Behav 126:146–151

Shin EJ, Whang WK, Kim SG et al (2010) Parishin C attenuates phencyclidine-induced schizophrenia-like psychosis in mice: involvements of 5-HT1A receptor. J Pharm Sci 113(4):404–408

Shin EJ, Bach JH, Nguyen TT et al (2011) Gastrodia elata attenuates methamphetamine-induced dopaminergic toxicity via inhibiting oxidative burdens. Curr Neuropharmacol 9(1):118–121

Shuchang H, Qiao N, Piye N et al (2008) Protective effects of gastrodia elata on aluminium-chloride-induced learning impairments and alterations of amino acid neurotransmitter release in adult rats. Restor Neurol Neurosci 26:467–473

Shyamala BN, Naidu MM et al (2007) Studies on the anti-oxidant activities of vanilla extract and its constituent compounds through in vitro models. J Agric Food Chem 55(19):7738–7743

Strack D, Busch E, Klein E (1989) Anthocyanin patterns in European orchids and their taxonomic and phylogenetic relevance. Phytochemistry 28:2127–2139

Stuart GA (1911) Chinese Materia Medica: Vegetable Kingdom. (A revision of a work by F. Porter Smith.) American Presbyterian Mission Press, Shanghai

Su H, Kang JC, Cao JJ et al (2014) Medicinal plant endophytes produce analogous bioactive compounds. Chiang Mai J Sci 41(1):1–13

Su J, Wang B (1996) Quantitative method for the determination of contents of gastrodin in zhennaonin capsules. Zhongguo Zhong Yao Za Zhi 21:284–285

Sun XF, Wang W, Wang DQ, Du GY (2004) Research progress of neuroprotective mechanisms of Gastrodia elata and its preparation. Zhongguo Zhong Yao Za Zhi 29:292–295

Sun W, Miao B, Wang XC et al (2012) Gastrodin inhibits allodynia and hyperalgesis in painful diabetic retinopathy rats by decreasing excitability of nociceptive primary sensory neurons. PLoS One 7(6):e39647

Tang SL, Su HJ (1978) Flora of Taiwan, vol 5. Taiwan National University Press, Taipei

Tao J, Fu TX, Luo Z et al (2006) Cloning of distinguishing DNA sequences of Gastrodia elata Bl. and application of them in identifying gastrodia tuber. Sheng Wu Gong Cheng Xue Bao 22:587–591

Tao J, Luo ZY, Msangi CI et al (2009) Relationships among genetic makeup, active ingredient content, and place of origin of the medicinal plant, gastrodia tuber. Biochem Genet 47:8–18

Tasn DGH (2008) Grammotophyllum speciosum Bl. http://orchidaceaemalaysiana2.blogspot.com/2008/07/grammatophyllum-speciosum-bl_21html

Teo CC, Tan SN, Yong JWH et al (2008) Evaluation of the extraction efficiency of thermally labile bioactive compounds in Gastrodia elata Blume by pressurized hot water extraction and microwave-assisted extraction. J Chromatogr 1182:34–40

Thammasiri K (2010) Vitrification-based cryopreservation of Grammatophyllum speciosum protocorms. Cryoletters 31(4):347–357

Trivedi VP, Dixit RS, Lal VK (1980) Orchids in the drug markets of Bareilly, Kanpur and nearby districts. Nagarjun (Calcutta) 23(8):157–163

Tsai CF, Huang CL, Lin YL et al (2011) The neuroprotective effects of an extract of Gastrodia elata. J Ethnopharm 138(1):119–125

Tsai DS, Chang YS, Li TC, Peng WH (2014) Prescription pattern of Chinese herbal products for hypertension in Taiwan: a population-based study. J Ethnopharm 155(3):1534–1540

Vaddhanaphuti N (2001) A field guide to the wild orchids of Thailand, 3rd edn. Silkworm Books, Chiang Mai

Vaddhanaphuti N (2005) A field guide to the wild orchids of Thailand, Fourth and Expanded Edition. Silkworm Books, Chiang Mai

Veitch N, Grayer R (2003) 143. Goodyera. Phytochemistry. In: Pridgeon AM, Cribb PJ, Chase MW, Rasmussen FN (eds) Genera Orchidacearum, vol 3, Orchidoideae (Part 2) Vanilloideae. Oxford University Press, Oxford, p 97

Venketasubramaniam N, Chen CL, Gan RN et al (2009) A double blind placebo-controlled, randomized, multicenter study to investigate Chinese Medicine Neuroaid Efficacy on Stroke recovery (CHIMES Study). Int J Stroke 4(1):54–60

Vyas TK, Tiwan SB, Amiji MM (2006) Formulation and physiological factors influencing CNS delivery upon intranasal administration. Crit Rev Ther Drug Carrier Syst 23:319–347

Wang ZB, Yan B (2010) Gastrodia elata Blume extract ameliorates exercise-induced fatigue. Afr J Biotechnol 9(36):5978–5982

Wang YZ, Li JX, Wu QA et al (1989) Studies on the quality of Gastrodia elata Bl. cultivated with various kinds of fungus-growing materials. Zhongguo Zhong Yao Za Zhi 14:15–18

Wang X, Bauw G, Van Damme EJ et al (2001) Gastrodianin-like mannose-binding proteins: a novel class of plant proteins with antifungal properties. Plant J 25:651–661

Wang D, Yang G, Li B et al (2002) Investigation of the chemico-physical characteristics of the active compounds in the Chinese herb Gastrodia elata Bl. by capillary zone electrophoresis. Anal Sci 18:409–412

Wang DQ, Wang W, Sun XF et al (2005) Effect of Tianma gouteng recipe on interfering LV and aortic hypertrophy in renovascular hypertension rats. Zhongguo Zhong Yao Za Zhi 30(8):606–609

Wang HX, Yang T, Zeng Y, Hu Z (2007a) Expression analysis of the gastrodianin gene ga4B in an achlorophyllous plant Gastrodia elata Bl. Plant Cell Rep 26:253–259

Wang J, Zeng JB, Zhao XF et al (2007b) Effects of Gymnadenia conopsea alcohol extract on collagen synthesis in rat lungs exposed to silica and its mechanism of antioxidative stress. Zhong Xi Yi Jie He Xue Bao 5(1):50–55

Wang L, Xiao H, Liang X, Wei L (2007c) Identification of phenolics and nucleoside derivatives in Gastrodia elata by HPLC-UV-MS. J Sep Sci 30:1488–1495

Wang Q, Chen G, Zeng S (2007d) Pharmacokinetics of Gastrodin in rat plasma and CSF after i.n. and i.v. Int J Pharm 341:20–25

Wang SJ, Chen Y, He DD et al (2007e) Inhibition of smooth muscle cell proliferation by serum from rats treated orally with gastrodia and Uncaria decoction, a traditional Chinese formulation. J Ethnopharmacol 114:458–462

Wang Q, Chen G, Zeng S (2008) Distribution and metabolism of gastrodin in rat brain. J Pharm Biomed Anal 46:399–404

Wang Y, Lin S, Chen M (2012a) Y. Zhongguo Zhong Yao Za Zhi 37(12):1775–1781

Wang L, Xiao HB, Yang L, Wang ZT (2012b) Two new phenolic glycosides from the rhizome of Gastrodia elata. J Asian Nat Prod Res 14(5):457–462

Wang Y, Wu Z, Liu X, Fu Q (2013a) Gastrodin ameliorates Parkinson's disease by downregulating connexin 43. Mol Med Rep 8(2):585–590

Wang J, Feng B, Yang XC et al (2013b) Tianma Gouteng Yin as adjunctive treatment for essential hypertension and a systemic review of randomized clinical trials. Evid Based Complement Alternat Med 2013:706125

Wen W (1979) China: a new medicine born of tradition. UNESCO Cour 7:25–27

Wong KC, Sun M (1999) Reproductive biology and conservation genetics of Goodyera procera (Orchidaceae). Am J Bot 86(10):1406–1413

Wu XR (1994) A concise edition of medicinal plants in China. Guangdong Higher Education Publication House, Guangdong (in Chinese)

Wu HQ, Xie L, Jin XN et al (1989) The effect of vanillin on the fully amygdala-kindled seizures in the rat. Yao Xue Xue Bao 24:482–486

Wu CR, Hsieh MT, Liao J (1996a) p-Hydroxybenzyl alcohol attenuates learning deficits in the inhibitory avoidance task: involvement of serotonergic and dopaminergic systems. Clin J Physiol 39:265–273

Wu CR, Hsieh MT, Huang SC et al (1996b) Effects of Gastrodia elata and its active constituents on scopolamine-induced amnesia in rats. Plant Med 62:317–321

Wu Y, Liu N, Long Q (1998) Effect of various nutrient solutions on quality of Gastrodia elata. Zhong Yao Cai 21:1–3

Wu TL, Hu QM, Xia NH, Lai PCC, Yip KL (2001) Check List of Hong Kong Plants 2001. Agriculture, Fisheries and Conservation Department Bulletin 1, Hongkong

Wu TL, Hu QM, Xia NH, Lai PCC, Yip KL (2002) Check List of Hong Kong Plants 2001. Agriculture, Fisheries and Conservation Department Bulletin 1 (Revised), Hongkong

Wu B, Liu M, Liu H (2007) Meta-analysis of traditional Chinese patent medicine for ischemic stroke. Stroke 38(6):1973–1979

Wu H, Hu K, Jiang X (2008) From nose to brain: understanding transport capacity and transport rate of drugs. Expert Opin Drug Deliv 5:1159–1168

Xiang J, Tang YP, Wu P et al (2010) Chinese medicine Nao-Shuan-Tong attenuates cerebral ischemic injury by inhibiting apoptosis in a rat model of stroke. J Ethnopharm 131(1):174–181

Xiao YQ, Li L, Yuo XL (2002a) Studies on chemical constituents of effective part of Gastrodia elata. Zhongguo Zhong Yao Za Zhi 27:35–36

Xiao YQ, Li L, You XL et al (2002b) A new compound from Gastrodia elata Blume. J Asian Nat Prod Res 4:73–79

Xiao R, Tian Y, Liu L, Xu K (2009) Action of silky chicken-Gastrodia elata Blume nutrient solution on immunoregulatory function in mice. Wei Sheng Yan Jiu 38:283–286

Xiong J, Huang J (1998) The A1- and non A1-effects of N6-(5-hydroxy-2-pyridyl)- methyl-adenosine on rat vas deferens. Yao Xue Xue Bao 33(3):175–179

Xu JT (1989) Studies on the life cycle of Gastrodia elata. Zhongguo Yi Xi Xue Ke Xue Yuan Xue Bao 11:237–241

Xu J (1990a) Cytological observation on hyphae invading Mycena osmundicola in the process of germination of Gastrodia elata Bl. Zhongguo Yi Xue Ke Xue Yuan Xue Bao 12:313–317

Xu J (1990b) Studies on nutrition source of seeds germination of Gastrodia elata Bl. Zhongguo Yi Xue Ke Xue Yuan Xue Bao 12:431–434

Xu JT (2001) The changes of cell structure in the courses of Armillaria mellea penetrating the nutritional stems of Gastrodia elata. Zhongguo Yi Xue Ke Xue Yuan Xue Bao 23:150–153

Xu J, Guo S (2000) Retrospect on the research of the cultivation of gastrodia elata Bl, a rare traditional Chinese medicine. Chin Med J 113:686–692

Xu X, Lu Y, Bie X (2007) Protective effects of gastrodin on hypoxia-induced toxicity in primary cultures of rat cortical neurons. Planta Med 73:650–654

Yan Y, Chou GX, Dan M et al (2010) Screening of Chinese herbal medicines for antityrosinase activity in a cell free system and B16 cells. J Ethnopharm 129:387–390

Yang ZH, Zhang QT, Feng ZZ, Lang KY, Li H. English edition translated by ZR Xiong (1998) Orchids. China Esperanto Press, Beijing

Yang XD, Zhu J, Yang R et al (2007) Phenolic constituents from the rhizomes of Gastrodia elata. Nat Prod Res 21:180–186

Yang B, Li S, Zhang R et al (2009) Quantitative analysis of four active constituents in Tibetan herb Gymadenia conopsea by high-performance liquid chromatography. Zhongguo Zhong Yao Za Zhi 34(14):1819–1822

Yang YL, Wu KJ, Lu G (eds) (1999) Traditional Chinese Materia Medica. Wuhan University Press, Wuhan

Yi ML, Zhou LZ, Yong FH (1993) New phenolic derivatives from Galeola faberi. Planta Med 59 (4):363–365

Yong W, Xing TR, Wang S et al (2009) Protective effects of gastrodin on lead-induced synaptic plasticity deficits in rat hippocampus. Planta Med 75:1112–1117

You J, Tan T, Kuang A et al (1994) Biodistribution and metabolism of 3H-gastrodigenin and 3H-gastrodin in mice. Hua Xi Yi Ke Da Xue Xue Bao 25:325–328

Yu SJ, Kim JR, Lee CK et al (2005) Gastrodia elata Blume and an active component, p-hydroxybenzyl alcohol reduce focal ischemic brain injury through antioxidant related gene expression. Biol Pharm Bull 28:1016–1020

Yue Z, Zi J, Zhu C et al (2010) Constituents of Gymnadenia conopsea. Zhongguo Zhong Yao Za Zhi 35(21):2852–2856

Yun-Choi HS, Pyo MK, Park KM (1998) Isolation of 3-0-(4'-hydroxybenzyl)-beta-sitosterol and 4-(4'-hydroxybenzyloxy)benzyloxy]benzyl methyl-ether from fresh tubers of Gastrodia elata. Arch Pharm Res 21:357–360

Zeng X, Zhang S, Zhang L et al (2006) A study of the neuroprotective effect of the phenolic glucoside gastrodin during cerebral ischemia in vivo and in vitro. Planta Med 72L:1359–1365

Zeng XH, Zhang Y, Zhang SM, Zheng XX (2007) A microdialysis study of effects of gastrodin on neurochemical changes in the ischemic/reperfused rat cerebral hippocampus. Biol Pharm Bull 30:801–804

Zhang SH (2006) Gastrodin prevents cognitive decline related to cardiopulmonary bypass. Invitation to participate in a clinical trial (Clinical Trials.gov Identifier NCT00297245)

Zhang CY, Du GY, Wang W et al (2004) Effects of tianma gouteng fang on transmitter amino acids in the hippocampus extracellular liquids in freely moving rats subjected to brain ischemia. Zhongguo Zhong Yao Za Zhi 29(11):1061–1065

Zhang W, Sheng YX, Zhang JL et al (2007) Evaluation of the quality of Gastrodia elata by HPLC-DAD/MS. Yao Xue Xue Bao 42:418–423

Zhang W, Sheng YX, Zhang JL (2008) Determination and pharmacokinetics of gastrodin and p-hydroxynebzyl alcohol after oral administration of Gastrodia elata Blume. Extract in rats by high performance liquid chromatography-electrospray ionization mass spectrometric method. Phytomedicine 15:844–850

Zhang Q, Yang YM, Yu YG (2008) Effects of gastrodin injection on blood pressure and vasoactive substances in treatment of old patients with refractory hypertension: a randomized controlled trial. Zhong Xi Yi Jie He Xue Bao 6:695–699

Zhang YJ (2010) Gastrodia elata var obovata Yue J Zhang. Acta Bot Boreal -Occid Sin 30:1277

Zhang Z, Ma P, Xu YN et al (2011) Preventive effect of gastrodin on cognitive decline after cardiac surgery with cardiopulmonary bypass: a double blind, randomized study. J Huazhong Univ Sci Tech (Med Sci) 31(1):120–127

Zhang HW, Tong J, Zhou G et al (2012) Tianma Gouteng Yin Formula in treating hypertension. Cochrane Data base Syst Rev 6:CD008166. doi:10.1002/14651858 CD008166.pub2

Zhang Y, Li M, Kang RX et al (2012b) NHBA isolated from Gastrodia elata exerts sedative and hypnotic effects in sodium barbital treated mice. Pharm Biochem Behav 102(3):450–457

Zhang ZC, Su G, Li J et al (2013) Two new neuroprotective phenolic compounds from Gastrodia elata. J Asian Nat Prod Res 15(6):619–623

Zhang Y, Yu ZY, Wu XQ (2004) A new technique of extracting effective components from Chinese herb and natural plan-microwave assisted extraction. MAE Zhongguo Zhong Yao Za Zhi 29:104–108

Zhao YH, Liu YD, Guan Y, Liu NW (2010) Effect of Yinian Jiangya Yin on primary hypertension in early stage—a clinical observation on 40 patients. J Tradit Chin Med 30(3):171–175

Zhonghua Bencao (2000) Edited by Health Department and National Chinese Medical Management Office. Science and Technology Press, Shanghai

Zhonghua Bencao (2005) Edited by Health Department and National Chinese Medical Management Office. Science and Technology Press, Shanghai

Zhongyao Da Cidian (1986) Edited by Jiangsu New Medical College. Science and Technology Press, Shanghai

Zhou J (1991) Bioactive glycosides from Chinese medicines. Mem Inst Oswaldo Cruz 86(Suppl 2):231–234

Zhou J, Yang YB, Yang CR (1979) The chemistry of Gastrodia elata Bl. 1. The isolation and identification of chemical compounds from Gastrodia elata Bl. Acta Chem Sin 37(3):183–189

Zhou J, Yang YB, Yang CR (1980) Phenolic constituents of fresh Gastrodia elata Blume. Acta BotanicaYunnanica 2(3):370

Zhou J, Pu XY, Yang YB (1982) Nine phenolic compounds of fresh Gastrodia elata Blume. Kexue Tongbao 27(2):179–181

Zhu XX, Luo XG (2007) Optimization of extraction parameters and de-protein process for Gastrodia elata. Polysaccharides. Zhong Yao Cai 30:724–726

Zhu HL, Dai PG, Zhang W et al (2010) Enzymic synthesis of gastrodin through microbial transformation and purification of gastrodin biosynthesis enzyme. Biol Pharm Bull 33(10):1680–1684

Zi JC, Lin S, Zhu CG et al (2010) Minor constituents from the tubers of Gymnadenia conopsea. J Asian Nat Prod Res 12(6):477–484

Zou JN, Song JX, Chang CR, Wang XL (2006) RAPD analysis on the germplasm resources of Gastrodia elata in Guizhou. Zhong Yao Cai 29:881–883

Genus: *Habenaria* Willd.

Chinese name: *Yufeng hua* (phoenix/heron flower)
Japanese name: *Mizu Tombo*
Sanskrit name: *Riddhi*

Habenaria is the largest genus of terrestrial orchids with 600 species inhabiting lowland and montane forests in the tropics and the subtropical region. Plants are short with a whorl of leaves and a subterranean tuber. Inflorescence is terminal with few to many flowers that are generally distinguished by a conspicuous, large expanse of lip that is flat and multi-lobed. Flowers are monochromatic, commonly green or white. Some species have brilliant yellow, orange or pink flowers. Lip and petals may be entire, bilobed or trilobed (Fig. 13.1). Classification of members in the genus is based on lip form and petal shape.

Habenaria thrives in areas with distinct wet and dry seasons. It needs a dormant phase to flower properly, or even just to survive. With the arrival of the first rain, the plant sends out an aerial shoot which grows rapidly during the 2 months of heaviest rainfall, flowers, and when the rainy season is over the aerial portion dies down leaving the underground tuber to await the next rainy season (Abraham and Vatsala 1981).

The generic name is derived from Latin *habena* (bridle, whip, strap, veins) which describes the thread-like fringe on the lip in some species, e.g. *H. ciliolaris*.

Habenaria acuminata (Twaites) Trimen [see **Platanthera edgeworthii (Hook f. ex Collett) R.K. Gupta**]

Habenaria aitchisonii H.G. Reich.

Syn. *Habenaria diceras* Schltr.

Chinese names: *Duiduisheng* (couple ginseng), *Luodijinqian*, *Shuangxianerye Lan* (two threads, two-leaves orchid)
Chinese medicinal name: *Shangicao* (name also applies to *H. ciliolaris*)

Description: Plants are 12–33 cm tall with fleshy, oblong to ellipsoid tubers, 1–2.5 long and 0.8–1.5 cm in diameter. Stem is erect, terete, pubescent, with paired leaves at its base, directly facing each other. Raceme is 5–15 cm tall, with several to numerous, small, yellowish-green to green flowers. In Bhutan and China, flowering season is July to September (Gurong 2006; Chen and Cribb 2009); The species occurs in forests, thickets and grassland at 2100–4300 m in Guizhou, Yunnan, Sichuan, Xizang, Qinghai and Bhutan, and the temperate Himalayan region of India, Kashmir, Pakistan and Afghanistan (Nasir

© Springer International Publishing Switzerland 2016
E.S. Teoh, *Medicinal Orchids of Asia*, DOI 10.1007/978-3-319-24274-3_13

Fig. 13.1 *Habenaria longicorniculata* Graham. Reproduced with permission from *Introductions to Orchids* by Abraham and Vatsala, Parlode, Thiruvananthapuram: Tropical Botanic Garden and Research Centre (TBGRI), 1981

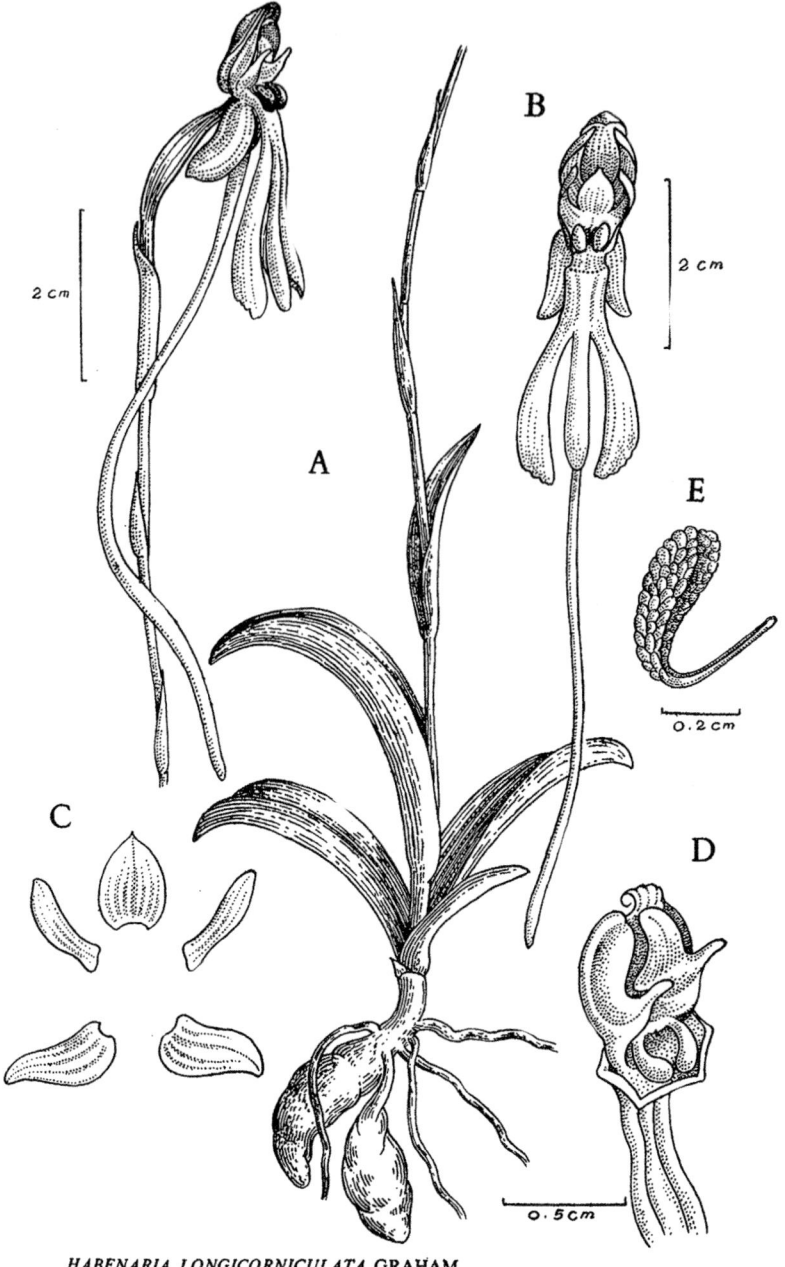

HABENARIA LONGICORNICULATA GRAHAM

and Ali 1972; Bose and Bhattacharjee 1980; Chen and Cribb 2009). Medicinal plants come from Yunnan, Sichuan and Tibet (Wu 1994).

Herbal Usage: Herbs are collected from Yunnan, Sichuan and Tibet (Wu 1994). Roots benefit the kidneys. They are used to treat nephritis and haematuria in Chinese Herbal Medicine (Wu 1994; Hu et al. 2000).

Habenaria arietina **Hook. f.**

Syn. *Habenaria intermedia* D. Don. var. *arietina* (Hook. f.) Finet

Common name: Reindeer Orchid
Chinese name: *Mao ban yu feng hua*

Fig. 13.2 *Habenaria arietina* Hook. f. (Photo: Bhaktar B. Raskoti)

Indian *Ayurvedic* names: *Riddhi, Vriddhi, Laksmi, Mangala, Rathanga, Risisrista, Saravajanpriya, Siddhi, Sukha, Vasu, Yuga.* (Note that the *Ayurvedic* names are identical with those of *H. edgeworthii.*) Other Indian name*: Safed musli* (in Garhwal), *Dakshinavarta, Himadrija.* (Note: Similar Ayurvedic names for *Habenaria edgeworthii.*)
Nepali name: *Thunma*
Pakistani common name: reindeer orchid

Description: Plant is robust, erect, 25–45 cm tall, with lanceolate leaves, 5.5–10 by 2–3 cm, and with a large, ovoid tuber at the base. *H. arietina* is a handsome species with densely clustered, fragrant, flowers, 6 cm across, of a creamy-white, flushed with green. Lip is large, ciliate, trilobed, the bifurcating lateral lobes each dividing into 10 filiform lobules which constitute a prominent feature of its attractive flowers (Fig. 13.2). It flowers from July to August in northeast India (Bose and Bhattacharjee 1980), autumn in Pakistan (Nasir and Ali 1972), August in China (Chen and Cribb 2009) and in September in

Thailand. The Thai variety has smaller flowers, 1.3 cm across (Vaddhanaphuti 2005).

H. arietina (syn. *H. intermedia*) is a variable species that is distributed in temperate Pakistan; northeast India (Ranikhet, Uttar Pradesh, Himachal Pradesh and Sikkim); Bhutan, Nepal, Bangladesh; on grassy hill slopes at 1500–2750 m; in southeast Xizang and Yunnan in China, also on grassy slopes but at 2300–2400 m (Nasir and Ali 1972; Chen and Cribb 2009); and in northern Thailand (Vaddhanaphuti 2005) and Vietnam. *H. intermedia, H. pectinata* and *H. arietina* are the same species, their differences merely reflecting physical extremes within a variable species (Pearce and Cribb 2002).

Herbal Usage: *H. arientana* (*H. intermedia*) is *Riddhi*, an *Ashtavarga* ingredient in *Chyawanprash*, a tonic used as a blood purifier and for rejuvenation (Puri 1970; Jalal et al. 2008). Rhizome of *H. arietina* (syn. *H. intermedia*) is used to produce an aphrodisiac in India (Sood et al. 2005) (also see *Malaxis acuminate*). Singh and Duggal (2009) stated that the leaves and roots of *H. intermedia*, sweet to taste, cooling, and spermatopoietic (or aphrodisiacal), are used to treat diseases of the blood. Crushed leaves of *H. pectinata* (locally known as *Safed musli* in Himachal Pradesh) are applied to snake bites: tubers are taken with condiments for arthritis. *Pueraria tuberosa* is sometimes used as a substitute for *H. arietina* (Singh and Duggal 2009). As noted above, *H. intermedia* and *H. pectinata* are synonyms for the same species.

H. arientana (*H. intermedia*) is *Riddhi*, an *Ashtavarga* ingredient in *Chyawanprash*, a tonic used as a blood purifier and for rejuvenation (Puri 1970; Jalal et al. 2008). The Ayurvedic prescription calls for 1 g of powdered *H. arientana* pseudobulb to be mixed with other components of *Ashtavarga* and consumed with milk in the morning (Dhayani et al. 2011).

In Nepal, the young leaves are cooked and tubers are boiled and eaten as a vegetable. It is claimed that this vegetable promotes vitality (Manandhar and Manandhar 2002). Tubers are used as emollient, aphrodisiac, appetiser and rejuvenating tonic. *Chyawanprash* is popular in Nepal. Tubers are used to treat thirst, fever, coughs, asthma, anorexia, haematemesis, worms, emaciation, general debility, skin diseases, leprosy, cataplexy and insanity. It is also claimed that they confer intelligence (Baral and Kurmi 2006).

Chinese herbalists employ *H. arientina* to treat people who suffer from weakness, insufficient *qi* or inflammation of the kidneys (Wu 1994).

Habenaria burchneroides Schltr. [see ***Habenaria densus* (Lindl.) Santapau & Kapadia**]

Habenaria ciliolaris Kraenzl.

Chinese names: *Yufeng Lan* (jade phoenix orchid), *Cugenyufeng Lan* (a bundle of roots jade phoenix orchid); *Maotingyufeng Lan* (Maoting jade phoenix orchid); bird's bill orchid

Description: A terrestrial herb reaching 25–60 cm in height from the ground to the tip of the terminal inflorescence, *H. ciliolaris* has cylindrical underground tubers and a stout stem with 5–6 leaves occupying the upper half of the stem, starting about 10 cm from the ground surface. Leaves are 12 by 3 cm, elliptic, pointed at the tip, and crenulate. Flowers are white to greenish, rarely tinged with pink, open, with very narrow sepals and petals except for the dorsal sepal which forms a hood over the column. Lip is trilobed, all very narrow, long and pointed, the central one directed downwards and curling a little forward. Lateral side lobes of the lip curl into a semi-circle and their tips point vertically upwards. It flowers in July and August in Taiwan (Lin 1975). and from July to September on mainland China (Chen and Cribb 2009).

H. ciliolaris is found in Hong Kong and Taiwan around 800 m in fairly exposed areas like forest paths. It additionally occurs in shaded locations in forests and along valleys at 100–1800 m on the mainland, in northern Fujian, southeast Gansu, Guangdong, Guangxi, Guizhou, Hainan, Hubei, Hunan, Jiangxi, Sichuan, Zhejiang and in Vietnam.

Herbal Usage: In TCM, it is stated that the stem improves weak kidneys, *yang* elements,

removes heat and detoxifies. Herb is used to treat nocturnal emissions, impotence, urinary problems, hernia, leucorrhoea, gonorrhoea, stomach ache, tuberculous cough, kidney infection and snake bites (Ou et al. 2003; Hu, Zhang, Zhu, et al. 2000). Tincture of *Habenaria* is taken as an alternative to tincture of *Goodyera schlectendaliana* or *Pholidota chinensis* in rice wine consumed with rice as a tonic for internal injuries (Wu 1994).

Habenaria commelinifolia (Roxb.) Wall ex Lindl.

Chinese name: *Fueyufeng hua*
Indian name: *Devsunda; Jadu, Jaitjadu* (Sadani); *Ridhi Vridhi*
Myanmar name: *Kadaw sut*

Description: Plants are 42–78 cm tall, with stout stems and 1–2 ellipsoid tubers, the latter 4–8 cm long and 1–5 cm in diameter. Leaves are lanceolate, distichous, 6–20 by 1.5–3.5 cm, becoming smaller and gradually merging into the inflorescence bracts; margins white or yellow and minutely papillate (Santapau and Kapadia 1966). Inflorescence carries 10–12 pure white flowers, 1.6–2 cm across, which open in succession. It flowers in August in China (Chen and Cribb 2009), August to September or October in the Deccan (Santapau and Kapadia 1966; Bose and Bhattacharjee 1980). *H. commelinifolia* occurs in Yunnan at 900–1200 m (Chen and Cribb 2009), Vietnam, Thailand, the Shan state of Myanmar, Nepal and throughout India (Karthikeyan et al. 1989)

Usage: The plant is eaten as a vegetable (Trivedi et al. 1980; Kumar 2003; Rao 2007) and alleged to be a blood purifier. It is used to cure blebs on the palm (Trivedi et al. 1980). Dried root of the orchid is used to treat spermatorrhoea. It is prepared in the following manner:

An equal quantity of dried roots of *H. commelinifolia* and *Saraca indica* (Ashoka tree) are boiled in 1000 ml of water until the volume is reduced to 100 ml. To cure spermatorrhoea, it is advised that 6–8 drops of the concentrated decoction be taken on an empty stomach for 10 days (Dash et al. 2008).

Saraca indica may have an oestrogenic (consequently, anti-androgenic) effect because, in India, its bark has the reputation of keeping women youthful and healthy and it is used to treat gynaecological conditions.

H. commelinifolia is also a source of Indian, Nepalese and Burmese *salep* (Caius 1936; Nair 1963; Puri 1970; Pandey et al. 2003; Baral and Kurmi 2006).

Habenaria crinifera Lindl.

Description: A terrestrial, sometimes epiphytic, leafy herb, *H. crinifera* is distributed in southern Deccan, the Western Ghats and in Sri Lanka (Abraham and Vatsala 1981; Jayaweera 1981). Stem is 5–15 cm long. Tubers are oblong or ovoid, 2 by 1 cm, with lateral roots above that develop into secondary tubers. Leaves, are 2–4, lanceolate, 5–13 by 1–2 cm, spreading, many-veined, ensheathing the stem. Scape is green, 7–12 cm tall, and raceme carries 2–4 beautiful, kite-shaped, white flowers. It flowers in August and September in southern India (Santapau and Kapadia 1966; Abraham and Vatsala 1981).

Santapau and Kapadia (1966) earlier commented that this species is rare: however, Abraham and Vatsala (1981) subsequently reported that it is rather common in the Western Ghats, occurring in great numbers and occupying large patches on the roadside. The species enjoys a distribution from Maharastra to Kerala.

Usage: Tubers of *H. crinifera* are used to treat headache by the tribes living in the Kudremukh National Park in Karnataka (Rao 2007).

Habenaria davidii Franch.

Syn. *Habenaria leucopecten* Schltr.

Chinese name: *Changjuyufenghua* (Long distance jade phoenix flower)
Chinese medicinal name; *Shuangshencao*

Description: Plants are 65–75 cm tall with fleshy, oblong tubers, 2–5 cm long and 0.8–1.5 in diameter. Stem is erect, stout, terete, slim bearing 5–7 ovate-lanceolate leaves 5–12 by 1.5–4.5 cm. Raceme is 4–21 cm tall, 4–15 flowered. Flowers are white, sometimes yellowish over the lip. Lip is trilobed, spurred at the base. Lobes are linear with ciliated margins. There are 7–10 filiform lobules on the margin of the lateral lobes, lobules usually branching. It flowers from June to August. This endemic species is found in forests, thickets, and grasslands in a crescent from Hubei, Hunnan and Guizhou to Sichuan, Yunnan and southern Xizang Yunnan and Tibet, at 600–3200 m (Chen and Cribb 2009).

Herbal Usage: Herb is collected in Yunnan and Tibet. Roots of *H. davidii* are used in Chinese herbal medicine to reduce swelling and to protect the kidneys. The herb is used to treat hernia and firm swellings of the lymph nodes (Wu 1994; Hu et al. 2000). Decoction is made with 9–15 g of the herb (Hu et al. 2000).

Habenaria delavayi Finet

Chinese name: *Houbanyufenghua* (thick petal jade phoenix flower)
Chinese medicinal names: *Jishenshen* (chicken kidney ginseng); *Duiduishen* (paired ginseng)

Description: Luo (2004) studied a plant collected from Lijiang, Yunnan, at 2700 m: size of the plant, with height varying from 9 to 47 cm. Tubers are oblong or ovoid, 1–2 cm long and 1.1.5 cm in diameter. Stem is erect, 3–5 mm in diameter with a dense basal rosette of 4–6 ovate leaves. It bears a few white flowers, usually 3–4, and rarely up to 6; 1.3–1.8 cm across. However, more flowers, 7–20, have been observed by Chen and Cribb (2009). Flowering season is May to August (Jin, Zhao and Shi 2009). *H. delavayi* is endemic in China occurring in Sichuan, Guizhou, Yunnan and Tibet.

Herbal Usage: *Jishenshen* is collected in autumn from Sichuan, Guizhou, Yunnan and Tibet. Tubers are either used fresh or dried for storage (Wu 1994). To treat renal insufficiency, waist pain and nephritis, Chinese herbalists recommend chicken soup containing 15–30 g *Jishenshen* (*H. delavayi*) (*Zhongyao Da Cidian* 1986; Hu, Zhang, Zhu, et al. 2010). Sometimes used as a tonic (Chen and Tang 1982), it is also suitable for lumbago, weakness of "kidney", dizziness, tinnitus, hernia and neurosis (Wu 1994). It is reputed to strengthen the body (Chen and Tang 1982).

Habenaria densa Wall (see **Platanthera clavigera Lindl.**)

Chinese medicinal names: *Jishencao* (chicken kidney grass); *Jishenzi* (chicken kidney); *Yaoshenzi* (waist and kidney); *Shuangren* (double kernels); *Shenjingcao* (kidney herb)

Fig. 13.3 *Habenaria dentata* (Sw.) Schltr. (Photo: E.S. Teoh)

Habenaria dentata (Sw.) Schltr.

Chinese names: *Emaoyufeng Hua* (feather jade phoenix flower), *Baifeng Lan* (white phoenix orchid), *Dalucao* (large heron grass), *Yufeng Lan* (jade phoenix), *Dongpuyufeng Lan* (Dongpu flaked teeth heron orchid); *Dongfubaifeng Lan* (Dongpu white phoenix/white heron/phoenix orchid); *Chipianlu Lan*; *Emaoyufenghua* (goose- feather jade blossom)
Chinese medicinal names: *Shuangshenzi* (*two kidney son*); *Baihuacao* (white flower herb); *Tianaebaodan* (swan carrying an egg); *Yufenghuagen* (jade-phoenix-flower root); *Duiduishen* (double ginseng)
Taiwanese name: *Bai Feng Lan* (White phoenix orchid)
Thai names: *Naang Oua Noi, Nang ua noi*

Description: Plant with inflorescence may reach a height of 80 cm. but is generally shorter. Flowers are pure white, 1.8–3 cm across and 8–10 flowers are loosely arranged near the tip of the erect rachis (Fig. 13.3). Flowering season is July to September (Lin 1975; Chen and Cribb 2009), and September to October in Thailand and at Gaoligongshan (Vaddhanaphuti 2005; Jin, Zhao and Shi 2009).

H. dentata enjoys a wide distribution that extends from the Ryukyu Islands of Japan to Taiwan and Hong Kong (flowers there from July to September) across southern in Fujian, Guangdong, Hong Kong, Guizhou and Yunnan to the Philippines, Indonesia, Indochina, Thailand, Myanmar, Nepal and Himalayan India at 200–2300 m. In Taiwan, it occurs on grassland below 1400 m in the central and southern parts of the island.

H. miersiana Champ described by Ohwi (1965) as a separate species is now considered to be synonymous with *H. dentata*. It is found in thickets from Ryukyu to Honshu and flowers from August to October (Ohwi 1965). Plants for medicinal use are collected in autumn.

Herbal Usage: Chinese herbals mention that stems "benefit the lungs and kidneys". They are diuretic, anti-inflammatory and detoxify. Stems are used to treat weak kidneys, impotence, stomach ache, orchitis, dysuria, swollen kidneys, carbuncles and coughs caused by tuberculosis (*Zhongyao Da Cidian* 1986; Hu et al. 2000; Ou et al. 2003). Prescriptions used in China are shown in Table 13.1. Lawler (1984), quoting

Table 13.1 Chinese herbal prescriptions containing *Habenaria dentata* (*Zhongyao Da Cidian*; Anonymous 1986)

1. Indication: Orchitis Prepare a drink or soup by boiling 15–30 g of the whole plant of *Habenaria dentata*. Optional: May add 2 small pig testes. (Source: *Jiangxi Herbs*)
2. Indication: Urethritis Boil 15 g of the stems of *Habenaria dentata* to make a drink. (Source: *Kunming Commonly Used Folk Herbs*)
3. Indication: Cough Boil roots of *Habenaria dentata* and add red sugar to prepare a drink (Source: *Kunming Commonly Used Folk Herbs*)
4. Indication: Carbuncle Grind fresh roots of *Habenaria dentata*, add sweet wine: for external application only (Source: *Jiangxi Herbs*)
5. Indication: Snake Bite Grind the roots and take internally. Also prepare grounded fresh roots for external application. (Source: *Jiangxi Herbs*)

various authors, reported that the pounded root of *H. miersiana* Champ ex Benth. was used by aboriginal mountain tribes in Taiwan to dress wounds and swellings. Chuakul (2002) found that Thai herbalists used the tubers for abscesses and bodily discomfort.

Habenaria diphylla (Nimmo) Dalzell

Syn. *Habenaria humistrata* Rolfe ex Downie

Thai name: *Tupmup mot lin*

Description: Plants are 7–25 cm tall, with solitary, ovoid, fleshy tubers 1 cm long. Stem is erect, terete, glabrous, with 2 opposing leaves at the base and bract-like leaves above. Leaves are heart or kidney shaped, 1.2–3.5 by 1–5 cm, with yellow or pale-coloured margins. Raceme carries one or a few flowers. Flowers are small, greenish-white. The species is found in damp locations and on rocks in forests or valleys in southern Yunnan at 1000–1400 m, Philippines, Thailand, Myanmar, Bangladesh, Nepal and northern India (Pearce and Cribb 2002; Chen and Cribb 2009), but at low elevations on the west coast of southern India (Bose and Bhattacharjee 1980). It flowers in June in China (Chen and Cribb 2009), July to September (Santapau and Kapadia 1966) or just August in India (Bose and Bhattacharjee 1980).

Herbal Usage: The whole plant is used for treating insect bites in Thailand (Chuakul 2002). Flowers of *H. diphylla* popularly known as *Jeevahi Purusharatna* are used to treat asthma in the Western Ghats (Rao 2004).

Habenaria diplonema Schltr.

Chinese name: *Xiaoqiaoyufeng Hua*
Chinese herbal name: *Shuangxianerye Lan* (two thread, two leafed orchid)

Description: Plants are 8–13 cm tall; tubers oblong, fleshy, 1 by 0.5 cm. Stem is erect, slender, pubescent, bearing 2 leaves at midpoint, the two leaves facing each other. Leaves are hairy, orbicular, with yellowish-white veins, 1.5–2 by 1.4–1.7 cm. Inflorescence carries 4–14 small, green flowers in August (Chen and Cribb 2009). An endemic species, plants are found on soil-covered rocks at 2800–4200 m in northern Fujian, southwest Yunnan and southwest Sichuan.

Herbal Usage: Herbs are obtained from Yunnan. Roots are used to improve kidney and liver function and to regulate menstruation (Wu 1994).

Habenaria disceras Schltr. (see ***Habenaria aitchisonii*** Rchb. f.)

Habenaria edgeworthii Hook f. ex Collett, [see **Platanthera edgeworthii (Hook f. ex Collett) R.K. Gupta**

Habenaria fordii Rolfe

Chinese name: *Changjukuorui Lan* (long distance broad pistil orchid), *Xianbanyufeng Hua*

Description: Plant is 30–60 cm tall. Tubers are 3–4 by 2–3 cm, fleshy. Stem is erect, stout, bearing 4 or 5 tufted leaves near the base and bract-like leaflets higher up. Leaf blades are oblong-lanceolate to elliptic. Inflorescence is many-flowered. Flowers are white and appear in July and August. *H. fordii* is an endemic species, distributed in *uah* Guangdong, Guangxi and Yunnan, in damp locations and on soil-covered rocks in forests and along valleys at 600–2200 m (Chen and Cribb 2009; Anonymous 1986).

Herbal Usage: Herbs are obtained from Guangxi and Yunnan. Root is used to treat indigestion in children (Wu 1994).

Habenaria furcifera Lindl.

Syn. *Habenaria ovalifolia*

Chinese name: *Mihuayufeng Hua*

Description: This is a terrestrial herb of 40–60 cm height with 1–2, ovate, ellipsoid tubers, 3 by 1.5 cm. Stems are slender or robust, up to 60 cm tall. Leaves are 3–5, oblong to oblong-elliptic leaves, sheathing at the base, 6–17 by 2.8–5.5 cm (in southern India up to 25 by 7 cm; Bose and Bhattacharjee 1980) and gradually passing into bracts of the scape. Inflorescence is 25–40 cam tall, stout, erect and laxly many-flowered (10–15), but the flowers are small (0.7–1 cm) and green or greenish-white. Lip is tripartite, lobes are linear, and the spur is as long as the ovary (Fig. 13.4). It occurs in deciduous forests along moist slopes besides streams in western Peninsular India, Pakistan, Nepal, Bhutan, Sikkim, Myanmar, China and Thailand and Laos at 1100–1200 m (Chen and Cribb 2009; Pearce and Cribb 2002). Flowering period is August to September in India (Santapau and Kapadia 1966), with a more extended period in some parts of July to October (Joseph 1982) or August to October (Matthew 1995); July to August in Bhutan (Gurong 2006) but September in neighbouring Nepal (Raskoti 2009); July to August in China (Chen and Cribb 2009); and October in Thailand (Vaddhanaphuti 2001).

Herbal Usage: Tubers of *H. furcifera* (syn. *H. ovalifolia*) are used in *Ayurveda* to treat wasting diseases, fever blood disorders, haemorrhage and fainting (Yoganarasimhan and Chelladurai 2000). A paste of this orchid is used

Fig. 13.4 *Habenaria furcifera* Lindl. (Photo: Bhaktar B. Raskoti)

for cuts, wounds and snake bites by the Chenchus in India (Reddy et al. 2005).

Habenaria hollandiana Santapau

Description: Plant is robust, erect, terrestrial, with stem 40–50 cm tall. Tubers are 2, oblong-ovoid to ellipsoid, 3 by 1–1.5 cm. Leaves are clustered around lowest quarter of the stem, obovate to lanceolate, with prominent longitudinal veins, 2–12 by 0.5–3.5 cm and minutely papillate along the margin. Inflorescence is up to 30 cm tall, with small, greenish-white flowers which appear in November (Santapau and Kapadia 1966). *H. hollandiana* is an Indian endemic species that is distributed in eastern Himalaya, Assam and the Eastern Ghats. Plants are found in cool, shady locations near waterfalls in the Eastern Ghats (Suryanrayana and Rao 2005).

Herbal Usage: Kondareddies and Valmikis of Andra Pradesh use a fresh paste of the plant to treat scorpion stings (Akarsh 2004). Tubers are also made into a paste to treat scorpion stings and maggot-infected sores in Bangladesh (Musharof Hossain 2009). Tubers of *H. hollandiana* play a role in the magic art practised by the Pawra tribe in Maharashtra (Jagtap et al. 2008).

Fig. 13.5 *Habenaria intermedia* (= *Habenaria arientina*. see Fig. 13.2.) (Photo: Bhaktar B. Raskoti)

Habenaria humistrata Rolfe ex Downie [see **Habenaria diphylla** (**Nimmo**) **Dalzell**]
Habenaria intermedia D. Don (see. **H. arietina Hook f.**) (Fig. 13.5)
Habenaria. leucopecten Schltr. (see **Habenaria davidii** Franch.)

Habenaria limprichtii Schltr.

Chinese name: *Kuanyaogeyufeng Hua*

Description: A robust terrestrial herb with stems 35–45 cm tall in Thailand (Nanakorn and Watthana 2008) but more variable in China, 18–60 cm tall, plants of *H. limprichtii* turn black when dried (Chen and Cribb 2009). Tubers are ovoid-ellipsoid or oblong, 1.5–3 by 1–1.5 cm. Stem is erect, terete, firm with 4–7 leaves of diminishing size towards the apex. Leaves are oblong-ovate, pointed, 8 by 3 cm, with three longitudinal veins. Raceme loosely 3- to 20-flowered, the flowers facing all directions, 4 cm across. Floral bracts are large, green, lanceolate. Sepals are light green. Dorsal sepal forms a hood and lateral sepals are horizontal. Petals are white, erect, abutting the dorsal sepal. Lip is white, trilobed, the lobes linear, of equal length. Lateral lobes bear 8–10 filiform lobules on their outer margin. Flowering season is June to August

in China (Chen and Cribb 2009), and July to August in Thailand. The species is found in thickets and grasslands at 1500–2000 m in northern Thailand (Nanakorn and Watthana 2008), and at 1900–3500 m in western Hubei, Sichuan and Yunnan. It also occurs in Vietnam.

Herbal Usage: In China, it is used for "feminine nourishment", and to treat nephritis and improve renal function (Wu 1994). Oestrogens made their first appearance on Earth in fungi and oestrogenic compounds are quite prevalent in plants. However, it has not been shown that phyto-oestrogens occur in *H. limprichtii*.

Habenaria linguella Lindl.

Chinese name and medicinal name: *Poshen*

Description: Plants are 20–50 cm tall with fleshy tubers 3–5 by 1–2 cm. Leaves are narrowly oblong-lanceolate, 5–12 (sometimes up to 27) by 1.2–2 cm. Raceme is densely many-flowered. Flowers are yellow, small 1 cm across. Dorsal sepal forms a hood with the petals. Lateral sepals are spread out, oblique to nearly horizontal, obovate, 6–7 by 4–4.5 mm. Lip is trilobed, side lobes tiny, thorn-like, mid-lobe shaped like a tongue or blade, pendulous, 1 cm long. It flowers from June to August (Chen and Cribb 2009). The species occurs in forests and grasslands at 500–2500 m in Guangdong, Guangxi, Guizhou, Yunnan and Hong Kong; also in Vietnam (Wu et al. 2002).

Herbal Usage: The plant is used to clear "heaty lungs" (Wu 1994). A decoction is made with 9–15 g of the herb (Hu et al. 2000).

Habenaria longecalcarata A. Rich. (see **Habenaria longicorniculata** **J Graham**)

Habenaria longicorniculata **J Graham**

Syn. *Habenaria longecalcarata* A. Rich.

Indian name: *Devasunda*. Tamil name: *Kozhikilangu*
Japanese name: *Oze-no-sawa-tombo*

Description: The Indian plants are large, 26–96 cm tall, with single or paired tubers, 2–3 by 1.3 cm. About 6 leaves are borne near the base. Leaves are elliptic 4–14 by 1.6–2.6 cm, ensheathing the base of the stem. Flowers are white, commonly 2 (1–3), 2 cm across. Lateral sepals are 10–14 by 4–7 mm, growing straight down, parallel to the lip. Petals are white 10–12 by 3.5–4 mm, with green at the base, and spathulate. Lip is 2.5 cm long, reflexed, trilobed, mid-lobe lingulate, lateral lobes much broader and fanning sideways. It flowers in July to September in the Western Ghats (Santapau and Kapadia 1966); August to November in Tamil Nadu (Seidenfaden 1999). It is found at 800–1900 m (Joseph 1982), extending to Orissa and Bihar in the north-east at 1400–2300 m (Santapau and Kapadia 1966; Matthew 1995) and in Sri Lanka. In Tamil Nadu, Matthew (1995) reported seeing it growing gregariously in full sun, on exposed slopes especially in a thin layer of soil by rocks, at 1200 to 1400 m.

Herbal Usage: Natti Vaidyas (folk practitioners) reported during a meeting at a Natti Vaidyas Sammelan convened in the Kolli Hills of Tamil Nadu in August 1997 that fresh tubers of *H. longicorniculata* were eaten to reduce scrotal enlargement (Subramani and Goyara 2003). All parts of the plant can be used to control pain and swelling (Rao and Sridhar 2007). A paste of crushed tuber is mixed with an equal volume of turmeric powder, and the resultant coloured paste is applied to the affected site to correct leucoderma (Dash et al. 2008).

Habenaria marginata Colebr.

Indian and Bangladeshi name: *Humari*
Common Name: Golden Yellow *Habenaria*

Description: Plants are 8–36 cm tall, with 1–2 ovoid or ellipsoid tubers. Leaves are 2–5, oblong, to ovate-lanceolate, fleshy, with whitish-yellow margins, clustered at the base. Inflorescence is 10–25 cm tall and laxly or densely many-flowered. Flowers are greenish-yellow. It flowers from July, and fruits ripen by November in India

(Santapau and Kapadia 1966). Flowering season is August to September in Bhutan (Pearce and Cribb 2002) and October to November in Nepal (Raskoti 2009).

H. marginata is a small terrestrial herb distributed over a large area that extends from Pakistan across northern India to Orissa, Nepal, Bhutan and Bangladesh to Myanmar and Thailand. It is common in the paddy fields at 500–2000 m. It occurs in open, sandy grassland at 1680–1770 at Tashigang District in central Bhutan (Pearce and Cribb 2002).

Herbal Usage: In the Niyamgiri Hills of Orissa in India, *H. marginata* is used to treat malignant ulcers. The prescription and orchid collected from *Jiniguda* calls for 250 g of the orchid tuber to be boiled in 1000 ml of water until the volume of the decoction is reduced to 250 ml. A teaspoon of honey is added and the decoction is drunk for 14 days (Dash et al. 2008). It is also reported as being used in Bangladesh to treat malignant ulcers (Musharof Hossain 2009). Tubers are cooked and eaten as a vegetable at the Sanjay National Park in Madhya Pradesh (Kumar 2003).

Habenaria miersiana Champ ex Benth. [see **Habenaria dentata (Sw.) Schltr.**]
Habenaria ovalifolia Wight. (see **Habenaria furcifera Lindl.**)

Habenaria pectinata D. Don

Chinese name: *Jianyeyufeng Hua*
Indian name: *Safed musli*

Description: Plants are terrestrial, 55–70 cm tall, with many distichous, linear-lanceolate, deeply-channelled leaves, 6–20 by 0.5–3.5 cm that turn black when dried. Tubers are fleshy, 2–3 by 1–1.5 cm. Raceme is many-flowered, the flowers green to greenish-white with a white lip. Dorsal sepal is disproportionately large, erect, concave, broadly lanceolate with an obtuse apex. Lip is trilobed; lateral lobes pectinate; mid-lobe linear and spurred (Fig. 13.6). Flowering season is August. The species occurs in forests around 1800 m in temperate northeast

Description: Plants are 35–60 cm tall with oblong tubers, 3–4 by 1–2 cm. Stem is erect, terete, with 5–6 leaves at mid-point and bract-like leaves further up. Leaves are elliptic, 3–15 by 2–4 cm. Inflorescence is loosely 3- to 12-flowered. Flowers are greenish-white, 2–3 by 2 cm. Lip is deeply trilobed, lobes linear. Flowering season is from July to September. The species is found in forests and along valleys in the southern provinces of China at 300–1600 m (Anhui, Zhejiang, Fujian, Guangdong, Jiangxi, Hunan, Guizhou, Guangxi, southeast Yunnan) and in Vietnam (Chen and Cribb 2009).

Herbal Usage: *H. petelotii* is used in China to treat renal insufficiency, coughs from "heat lungs", external injuries with bleeding, erectile dysfunction, hernia and nocturnal bed-wetting in children (Wu 1994). A decoction is made with 9–15 g of the herb (Hu et al. 2000).

Fig. 13.6 *Habenaria pectinata* D. Don (Photo: Bhaktar B. Raskoti)

Habenaria plantaginea Lindl.

Bangladeshi name: *Kusuma gadda*

India, Nepal and Yunnan (Chen and Cribb 2009). It was observed to inhabit shady banks on the edges of temperate forests at 2000–3000 m in Indian Himalayas (Pradhan and Pradhan 1997), and open grassy meadows or *Pinus wallichiana* forests at 1520–2900 m in Bumthang, Bhutan (Gurong 2006). In Pakistan, it occurs at 800–1100 m (Nasir and Ali 1972).

Herbal Usage: The leaves are crushed and used to treat snake bites in India. Mixed with condiments, the tubers provide an herbal remedy for arthritis (Singh and Duggal 2009). Its usage in China is different. Here, the whole plant is used to treat coughs arising from weakness, nephritis and pain at the waist (Wu 1994; Hu et al. 2000).

Habenaria petelotii Gagnep.

Chinese name: *Liebanyufeng Hua*
Chinese medicinal name: *Danshencao*
Taiwanese name: *You Mao Yu Feng Lan* (Little feather jade phoenix orchid)

Description: Plant is medium-sized, 10–40 cm tall, with small, widely spaced, lanceolate, many-veined, membranous leaves, 6–10 by 3 cm, prostate on the ground. Tuber is ovoid to globular, hairy, 1.5–2 cm by 1 cm in diameter. Inflorescence bears 5–9 fragrant, white flowers that are lightly tinged with green, 2 cm across and 3 cm tall. Lip is longer than the sepals with a linear mid-lobe and broad, oblique, rhomboid side lobes of equal length and finely serrated margins. It flowers in August to October in Bhutan (Pearce and Cribb 2002), and in September to November in India (Santapau and Kapadia 1966; Bose and Bhattacharjee 1980; Joseph 1982), March, April and June in Sri Lanka (Jayaweera 1981). In large groups, "the slender, elegant orchid among grass and herbs on rocky hill slopes . . .are pleasing with the (white) flower bunches standing out from the well appressed carpet of green leaves" (Joseph 1982).

This common *Habenaria* species is found in the under-storey in the dry zone forests in Sri Lanka (Jayaweera 1981), all over India below 900 m (Abraham and Vatsala) and in

Bhutan between 1000 and 2500 m (Pearce and Cribb 2002). It occurs in the plains, on the floor of scrub forests at the border of thickets that receive some direct sunlight (Matthew 1995). It also occurs in Bangladesh, Myanmar and the Lesser Sunda Islands.

Herbal Usage: Tubers of *H. plantaginea* are used to treat wasting diseases, fever, disorders of blood, haemorrhage and fainting (Yoganarasimhan and Chelladurai 2000). In the Eastern Ghats of Andra Pradesh, tubers of *H. plantaginea*, together with black pepper and garlic, are pounded into a paste and converted into tablets. One or two tablets are given to relieve chest pain and stomach ache (Rao and Henry 1995). In Bangladesh, tubers are also used to treat chest pain and stomach ache (Musharof Hossain 2009).

Habenaria platyphylla Spreng (see **Habenaria roxburghii,** Nicolson)

Habenaria purpureo-punctata K.Y. Lang (see **Hemipiliopsis purpureopunctata Y.B. Luo & S.C. Chen**)

Habenaria rariflora A. Rich.

Description: A small, saxicolous herb with white flowers on 12-cm-tall, long racemes, the endemic *H. rariflora* only occurs in southwest India and Nilgiris, at 1000–1900 m on bare, exposed slopes or in grasslands. The plant has one or two ovoid-oblong tubers, 1–2.5 by 0.5–1 cm. It bears 3–5 lanceolate leaves that radiate horizontally, 3–7 by 1–1.7 cm. Flowering is gregarious, the flowers 1–2 on a rachis; sepals green, petals and lip white (Seidenfaden 1999). Lip is tripartate, central lobe triangular, lateral lobes filiform, extending backwards into a curved spur, 2.5–4 cm long. Flowering season is July or August to September (Santapau and Kapadia 1966; Abraham and Vatsala 1981), or up to October in Tamil Nadu (Joseph 1982; Matthew 1995).

Herbal Usage: same as *H. furcifera* (Yoganarasimhan and Chelladurai 2000),

Habenaria rhodocheila Hance

Chinese name and medicinal name: *Chenghuangyufeng Hua*
Thai names: *Sanh hin, Lin mangkon, Pat daeng*

Description: Plants are 20 cm tall. Leaves are lanceolate, 15 by 5–6 cm, green mottled with light green and occasionally suffused with brown. Raceme is 2- to 15-flowered. The attractive part of the flower is the broad, 'four-lobed' lip, measuring 3 by 2.5 cm, which looks like an insect in flight. In fact the lip is trilobed, with the distal portion of the mid-lobe broadening to two symmetrical lobules that gives the lip an appearance of having four lobes. Colour ranges from light to dark pink, cinnabar to bright red, yellow to orange, and purple (Vaddhanaphuti 2005). Thai and Malaysian varieties bear more flowers, and their colour is more intense (Fig. 13.7). Flowering season is September to November. Stem and leaves dry out after flowering and remain dormant during the dry season. When the rains appear in May, vegetative growth begins (Kamemoto and Sagarik 1975).

This saxicolous or terrestrial *Habenaria* with large scarlet, light to dark pink, purple, yellow or orange flowers is popular among orchid growers. The species is widely distributed in Southeast Asia and from Hainan northwards to Guangdong, Hong Kong, Guangxi, Guizhou, Jiangxi and Fujian. It occurs at 300–1500 m in shaded places or soil-covered rocks in forests and along valleys in China (Chen and Cribb 2009), and in full sun on rocks along valleys in Penang, Malaysia, nearer the equator, at 1500 m.

Herbal Usage: In China, it is applied on finger ulcers to promote their healing (Wu 1994). Decoction prepared with 3–9 g of the herb is used to treat 'heatiness', swellings, traumatic injuries and for pain relief (Hu et al. 2000).

Fig. 13.7 *Habenaria rhodocheila* Hance (Photo: E.S. Teoh)

Fig. 13.8 *Habenaria stenopetala* Lindl. (Photo: Bhaktar B. Raskoti)

Habenaria roxburghii, Nicolson

Syn. *Habenaria platyphylla* Spreng

Description: *H. roxburghii* is a terrestrial herb, up to 40 cm tall with 2–3 sessile, elliptical or oval, coriaceous, slightly pubescent leaves that lie flat on the ground. Tubers are ovoid, one or two, 2.4 by 1.5–2.5 cm. Raceme is 5–6 cm tall, densely many-flowered. Flowers are white, 1 cm across, 3–9 cm long and fragrant. Flowering season is August to September. The species is endemic in Peninsular India (Abraham and Vatsala 1981) and is fairly common along the Coromandel Coast (Santapau and Kapadia 1966). It occurs in the plains and scrub jungle up to 800 m, often sheltered within thorny bushes and on exposed rocks (Matthew 1995).

Herbal Usage: In Tamil Nadu, *H. roxburghii* is used in a similar fashion as *H. rariflora*, i.e. Ayurveda practitioners use the tubers to treat wasting diseases, fever, disorders of the blood, haemorrhage and fainting (Yoganarasimhan and Chelladurai 2000). About 10–15 g of the tubers is crushed with 2–3 g of pepper and garlic, and the extract is taken orally for snake bites by the Konda reddis in Andra Pradesh (Reddy et al. 2005).

Habenaria stenopetala Lindl.

Chinese name: *Xiabanyufenghua*
Chinese medicinal name: *Jishencao*

Description: A terrestrial herb of variable height (35–45 cm in Thailand; to 60 cm in northern India; up to 90 cm in China), it has narrow oblong tubers 2.5 cm long by 1–2 cm in diameter. Stem is erect, stout, and terete. Leaves are 5–8, elliptic, undulate, 8–16 by 3–5.5 cm, whorled around the middle third of the stem, and several bract-like leaves are present above. Raceme is erect; densely many-flowered (up to 30). Flowers are 2 cm across, of greenish-white. Lip is trilobed; lobes linear (Fig. 13.8). It flowers in August in Thailand (Nanakorn and Watthana 2008), and to October in China (Chen and Cribb 2009), also August to October in India (Santapau and Kapadia 1966; Bose and Bhattacharjee 1980).

H. stenopetala is found in open areas or in dipterocarp forests in Pakistan, northern India, Southeast Asia and in Tibet, Guizhou, Taiwan and the Ryukyu Islands of Japan at 300–1800 m.

Herbal Usage: In Chinese herbal medicine, *Jishencao* (*H. stenopetala*) is used to treat erectile dysfunction and hernias. It enhances kidney and sexual functions. The medicine is prepared by boiling 3–9 g of the dried plant (Hu et al. 2000).

Habenaria stenostachya (Lindl. ex Benth.) Benth. [see ***Peristylis densus* (Lindl.) Santapau & Kapadia**]

Habenaria tentaculata Reichb. f. [see ***Peristylis tentaculatus* (Lindl.) J.J. Sm.**]

Overview

There are 54 species of *Habenaria* in China (Chen and Cribb 2009), of which 13 are used in Chinese herbal medicine (Wu 1994; Hu et al. 2000). An equal number of *Habenaria* species are used medicinally in India. The usages in the two large Asian nations do not overlap and many species which occur in both countries may be used in one but not in the other. The exception is *H. arietina* (syn. *H. intermedia*) which is used as a tonic in both countries (by saying that it is used to treat people with weak *qi* means that it is used as a tonic). *Habenaria* species are present in tonics used as blood purifiers, and in tonics used for treating lapses of consciousness by practitioners of Siddha and Unnani (Greek) medicine in India (Rao 2004).

When F Porter Smith edited GA Stuart's *Chinese Materia Medica* in 1911, he described *H. sagittifera* which the Chinese called *mao-yu-feng hua*. Today, *H. schindleri* Schltr. has replaced *H. sagittifera* as the proper name for the orchid species. Chinese herbalists apply the name *mao-yu-feng-hua* to numerous species of *Habenaria*, and Porter Smith was probably referring to *H. ciliolaris*, the species native in Zhejiang which then included Shanghai where he worked. However, he did not specify the medicinal value of the plant.

Other sources report that in China several species share a common usage: they "replenish kidney *yin*" and are used to treat sexual dysfunction (particularly male impotence), menstrual disorders, hernia, spermatorrhoea, haematuria, backache, tinnitus and nervousness (Anonymous 1986; Wu 1994; Hu et al. 2000). Thai and Taiwanese herbalists use *H. dentata* to treat infected wounds (Chuakul 2002; Ou et al. 2003).

There is much confusion with some Indian species, but *Habenaria* was one of the very few orchids mentioned in ancient Indian texts. This suggests that *Habenaria* probably had medicinal usage. It was given the Sanskrit names, *Riddhi* and

Vriddhi (Rao 2004; Yoganarasimhan and Chelladurai 2000). In India, *Habenaria* are considered to be *Siddhi* and *Vriddhi* tonics which are used to treat fainting spells, for de-worming and as a "blood purifier" (Rao 2007). *H. arietina* (syn. *H. intermedia*) is used to produce an aphrodisiac (Sood et al. 2005). However, the Indian name, *Riddhi*, is non-discriminatory between species of *Habenaria*: it is used collectively for *H. acuminata*, *H. arietina*, *H. edgeworthii*, *H. ovalifolia*, *H. rariflora*, *H. roxburghii* and other species, all species being used in *Ayurveda* for similar purposes (Yoganarasimhan and Chelladurai 2000; Rao 2004), i.e. as a revitalising tonic or aphrodisiac. It does not make much medicinal sense to distinguish among the various Indian species because their identification by the herbalists is uncertain (Rao 2004).

H. arietina has an undeserved reputation as a remedy for snake and scorpion bites (Caius 1936). *H. longicorniculata* is used to treat leucoderma. Tubers of *H. commelinifolia* and *H. marginata* are cooked and eaten as a vegetable in the Surguja District of Chhattisgarh in India (Kumar 2003). *H. multipartia* Blume ex Kraenzl. (*oowi oowi*) roots are consumed as food in Java, and *H. rumphii* (Brongn.) Lindl. tubers are made into a preserve (Tanaka 1976). Tubers of the largest and most showy native *Habenaria* in the Philippines, *H. malintana* (Blanco) Merr. are reported as edible (Davis and Steiner 1982).

There is no report on the chemistry of Chinese medicinal *Habenaria* species. The semi-aquatic, North American species, *H. repens* produces habenariol or bis-*p*-hydroxy-benzyl-2*S*-isobutylmalate to deter feeding by freshwater crayfish. Habenariol is an aglycone of militarine which has been isolated from *Orchis militaris* (Veitch and Grayer 2001).

Genus: *Hemipilia* Lindl.

Chinese name: *Shehui Lan* (beak tongue orchid), *Duyeyizhi Hua* (one leaf flower)

Hemipelia is a tuberous herb with an erect stem, several centimeters tall, terminating in an inflorescence which bears several white to purplish flowers that are loosely arranged. When approached from the front, the flowers resemble the little decorative angels one sees on a Christmas tree, their outstretched sepals streaked with green appearing like arms fully clothed, the large spreading lip resembling a skirt, while the petals and dorsal sepal are clustered together in the position of the head. The side view reveals the long trumpet shape of the spur. However, it is the single, large, heart-shaped leaf located near the bottom of the stem that is the most prominent part of the orchid (Fig. 13.9). There are about ten species distributed in China, India, Myanmar and Thailand (Chen et al. 2009a), with one species in northern Vietnam (Averyanov et al. 2003) and another in Taiwan (Liu and Su 1978).

The generic name *Hemipelia* is constituted from two Greek words, *hemi* (half) and *pilos* (felt), possibly referring to the sparsely hirsute lip of the type species.

Hemipilia cordifolia Lindl.

Syn. *Hemipelia yunnanensis* (Finet) Schltr.

Chinese names: *Dianshehui Lan* (Yunnan beak tongue orchid), *Xinyeshehui Lan*
Chinese medicinal name: *Niudanshen*
Taiwanese name: *Yu Shan Yi Ye Lan* (Jade mountain single leaf orchid)

Description: *Hemipilia cordifolia* is a dwarf, alpine, terrestrial with oblong, bluish tubers 1.5–4.5 cm long. Plant is 13–30 cm tall. It carries a solitary, small, cordate leaf, 2.5–4.5 by 3–6 cm, dark green with purple spots on its upper surface. Inflorescence is terminal, bearing 8–10 small, pink to purplish flowers, 1–2 cm across with a long cone-shaped spur, 1.2 cm long, which points horizontally backwards (Fig. 13.10). It flowers from June to August in China (Chen et al. 1999), August to September in Himachal Pradesh (Bose and Bhattacharjee 1980) and July to September in Bhutan (Gurong 2006).

The species is distributed in Nepal, Bhutan, northeast India, Myanmar, Xizang Yunnan, Sichuan at 1500–3500 m (Chen et al. 2009a) and in Taiwan. On the island, it is found in the

Central Mountain Range at 2500–3000 m on grassland (Liu and Su 1978). It occurs in forests, on grassy slopes and along the roadsides in Sichuan and Yunnan (Chen et al. 2009a) and in open *Pinus wallichiana* forests in western Bhutan at 2500–2900 m (Gurong 2006).

Herbal Usage: Chinese herbalists claim that the root benefits the kidney. It is diuretic, and is also used to treat hernia and kidney diseases (Wu 1994). A decoction is made with 6–15 g of the herb (Hu et al. 2000).

Hemipilia flabellata Bureau & Franch

Chinese names: *Duyeyizhi Hua* (one leaf flower), Meteor rain grass, *Shanchunshehui Lan*
Chinese medicinal name: *Duyeyizhi Hua* (solitary leaf flower)

Description: Stem is erect, 20–28 cm tall, arising from oblong-ellipsoidal tubers, near which it carries a single cordate, green leaf spotted with purple on its upper surface, and purplish on its undersurface. Inflorescence is terminal and carries 3–15 widely spaced, white to pink flowers, 1.5 cm across. Lip is deep purple with a central white streak. Flowering season extends from June to August in China (Chen et al. 1999; Jin, Zhao, Shi 2009). *Hemipilia flabellata* grows in damp, humus-rich soil, sometimes in mossy places on the ground or in rock crevices at the edge of limestone forests at 2500–3200 m in Sichuan, Yunnan and Tibet and in the Shan state of Myanmar.

Herbal Usage: Herb is obtained from Sichuan, Yunnan and Guizhou. The drug prepared from the orchid is used to "moisten the lungs", and is used to treat dry coughs, tuberculosis, trauma, excessive sweating and renal colic (Wu 1994). It is mainly used in Yunnan and Quizhou (Table 13.2).

Hemipilia yunnanensis (Finet) Schltr. (see *Hemipilia cordifolia* Lindl.)

Overview

Hemipilia are rare orchids that occur in the highlands, in an alpine zone. They have not been chemically investigated and pharmacological information is lacking.

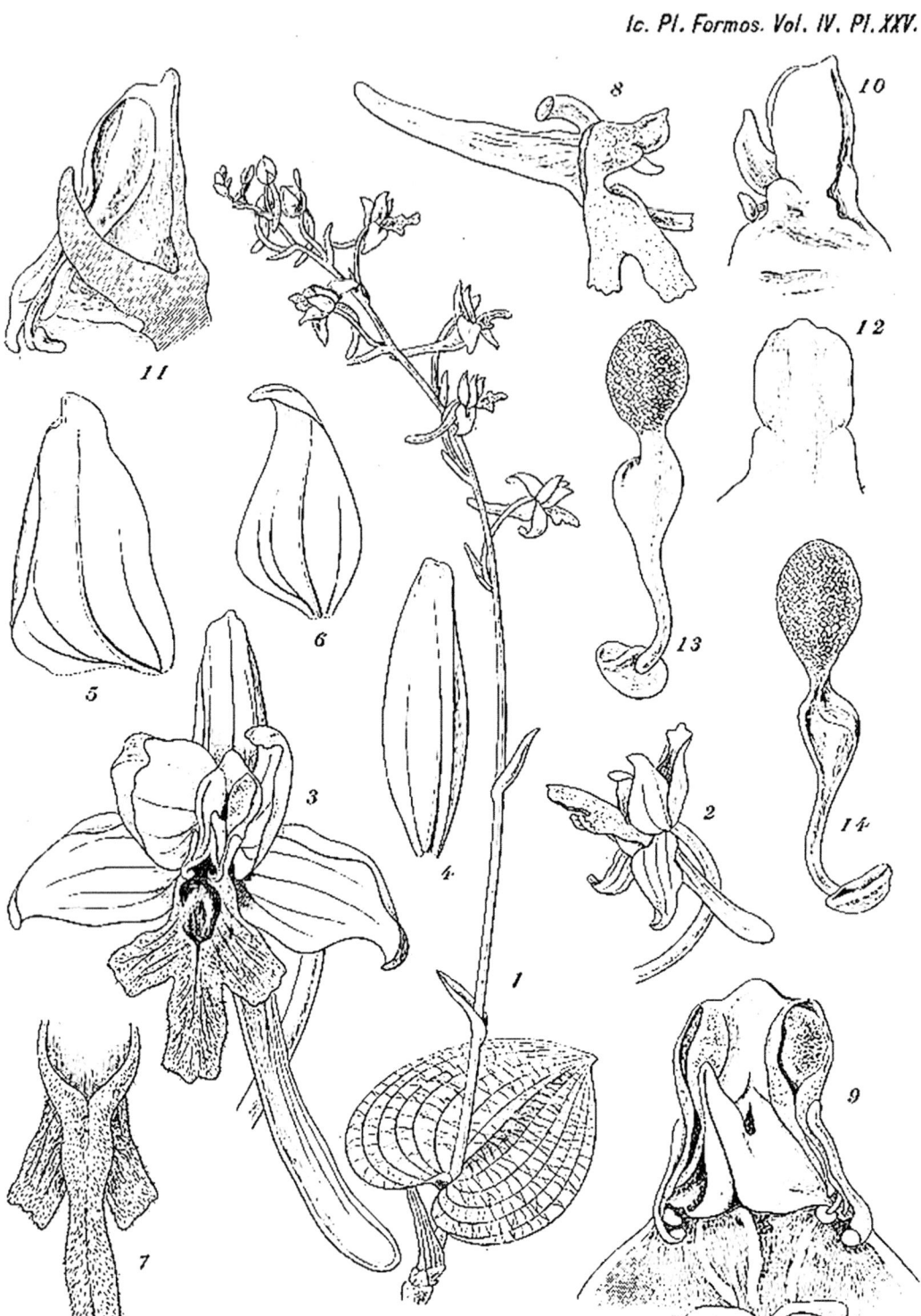

F. Hayami del.   T.Arai sculp.

Fig. 13.9 *Hemipelia cordifolia* Lindl. (as *Hemipilia formosana* Hayata) From: Hayata, B., Icones plantarum formosanarum, vol. 4: t. 25 (1914) Drawing by F. Hayami. Courtesy of University of California Libraries, USA

Fig. 13.10 *Hemipilia cordifolia* Lindl. (Photo: Liu Ming)

Table 13.2 Chinese herbal prescriptions advocating *Hemipilia flabellata* (*Zhongyao Da Cidian* 1986)

1. Indication: To augment "*yin*" and nourish the lungs; for treatment of weakness with low- grade fever and odourous productive cough. Decoct 15–30 g *Duyeyizhi Hua* (*Hemipilia flabellata*) for consumption.
2. Indication: graying hair: Grind with Amomum medium, dry, pulverize and make into tablet form. Take 3 g each time with 3 g of *Biota orientalis*. (Primary Source: *Southern Yunnan Herbs*)
3. Indication: foul productive cough Boil 15 g with 15 g each of *Verbena officinalia* and *Plantago major*, and consume. (Primary Source: *Quizhou Herbs*)

Genus: *Hemipiliopsis* Y.B. Luo & S.C. Chen

Chinese name: *Ziban Lan*

A genus with only one species which was formerly included in *Habenaria*, it differs from other members of that genus by its habit, structure of the stigma and lack of obvious anther canals (Luo and Chen 2009). Terrestrial herb with ellipsoid, fleshy tubers and filiform roots, its stem is erect, with purple spots, with one or rarely two leaves arising near the base. Leaves are elliptic to ovate-oblong, suberect, with short petiole, base ensheathing the stem. Inflorescence is terminal with a few or many well-spaced, pink to purple flowers.

Hemipiliopsis purpureopunctata (K.Y. Lang) Y.B. Luo & S.C. Chen

Syn. *Habenaria purpureopunctata* K.Y. Lang

Chinese name: *Ziban Lan*

Description: This terrestrial orchid has paired ellipsoid or subellipsoid fleshy tubers. Stem is erect with one or two elliptical leaves at the base, 5–15 by 2–5 cm. Plant is green, spotted with purple. Inflorescence is terminal, erect with 2–20 pale- pink flowers that are purple-spotted on the back of all sectors with flowers. Flowering season is June to July. It occurs in broad-leaved evergreen forest, alpine oak forest, grassy slopes and sandy river banks at 2100–3400 m in southeast Xizang (Tibet) and northeast India (Luo and Chen 2009; Chen et al. 2009a).

Herbal Usage: In China, the herb is used to relax tense muscles. The entire plant is used (Wu 1994) (Fig. 13.11).

Genus: *Herminium* L.

Chinese name: *Jiaopan Lan*
Japanese name: *Mukago So*

Herminium is a genus with some 35 members widespread in Asia and Europe, half found in China alone. One species, *Hermiunium lanceum*, is found as far south as Johore at the southern tip of Peninsular Malaysia (Seidenfaden and

Fig. 13.11 *Herminium lanceum* (Thunb. ex Sw.) Vuijk (as *Aceras angustifolium* Lindl.). From: Wight R., *Icones Plantarum Indiae Orientalis*, vol. 5 (1): t. 1691 (1846). Drawing by Govindoo

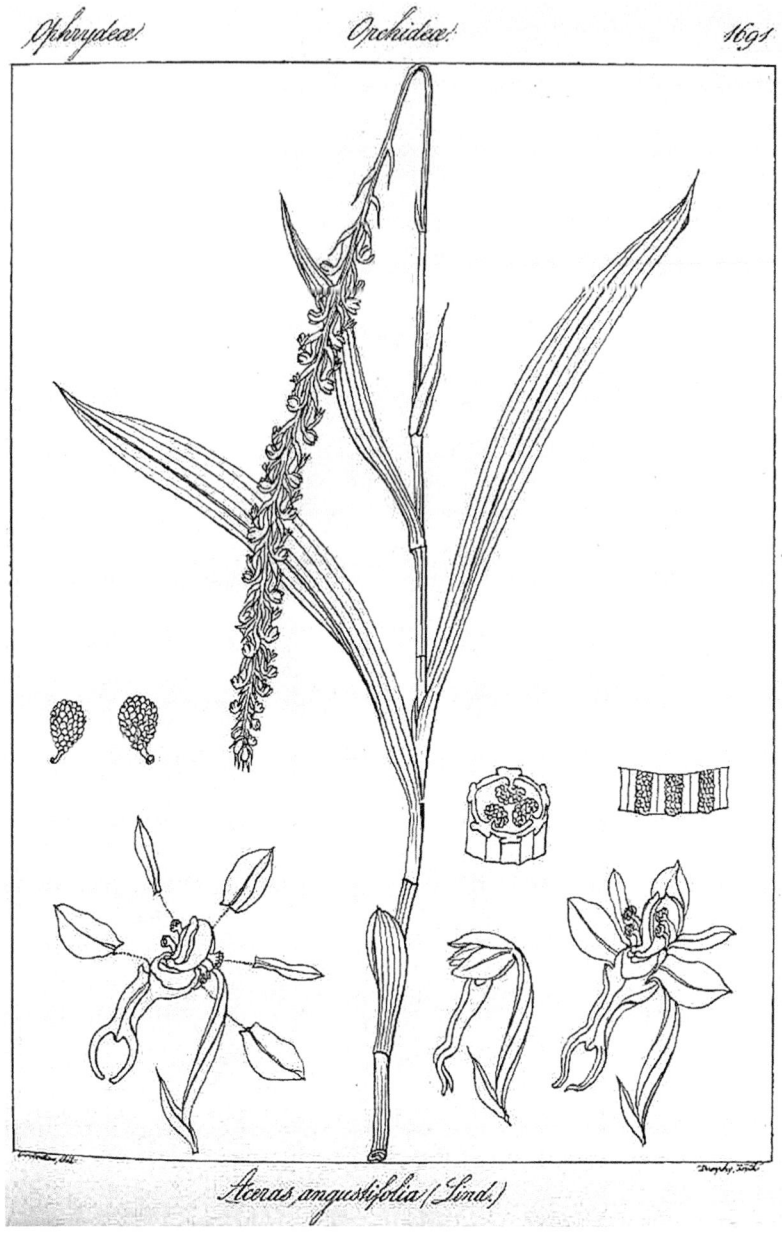

Wood 1992). Plant is grass-like. Stem is erect, sheathed by narrow, linear leaves. Roots are tuberous single or paired, and ovoid. Inflorescence is erect and many-flowered but the flowers are small, not spreading and greenish. *Herminium* is related to *Habenaria*.

There is some doubt regarding the intent of Robert Brown when he made up this generic name in 1813. On the one hand, the Greek word *hermin* (bed-post) might have been coupled to the Latin *(h)ermineus* (white like ermine), thus either referring to the stunted staminodia flanking the anther or to the resemblance of the inflorescence to a carved bed-post. Alternatively, it might have some relationship to Hermes (Mercury) (Schultes and Pease 1963).

Herminium bulleya (Rolfe) Tang & Wang [see **Peristylis bulleyi (Rolfe) K.Y. Lang**]

Herminium lanceum (Thunb. ex Sw.) Vuijk

Chinese names: *Shuangchunjiaopan Lan* (two lips, angle plate orchid), *Shuangshencao* (two kidney grass), *Chachunjiaopan Lan*
Chinese medicinal name: *Yaozicao*

Description: Plant is up to 50 cm tall with 3–4 linear leaves, up to 20 by 1.2 cm, sheathed at the base. Inflorescence is terminal, slim, densely many-flowered. Flowers are white to green, not spreading, 1 cm across. The distinctive feature is the trilobed lip with its linear, semi-circular base (Fig. 13.12).

H. lanceum is a terrestrial herb, thriving among grasses in forests, thickets and grassy slopes, and among rocks at 1100–3500 m. The species is widely distributed from Korea and Japan through China and throughout Southeast Asia to India, Bhutan, Nepal and Pakistan. In China, it is present over a large area covering most of the central and southern provinces from Shandong to Tibet, and from Shaanxi to Guangdong and Taiwan. It flowers from June to August or September in mainland China (Chen et al. 1999; Jin, Zhao and Shi 2009), April to September in Taiwan (Lin 1977), July to August in Sikkim (Bose and Bhattacharjee 1980), July to October in Bhutan (Pearce and Cribb 2002; Gurong 2006) and June to August in Nepal (Raskoti 2009).

Herbal Usage: Herb is obtained from Shandong to Tibet and from Dongbei (Northeast China, Manchuria) to Guangxi and Taiwan. The root is said to benefit the lung; it is used to treat tuberculosis. It also benefits the kidney, strengthens the muscles and bones and stops bleeding (Wu 1994). The decoction is prepared with 6–15 g of the herb (Hu et al. 2000). In India, it enjoys usage as a nutrient or a tonic (Trivedi et al. 1980; Pandey et al. 2003).

Phytochemistry: *H. laceum* tested negative for alkaloids (Luning 1974).

Herminium monorchis (L.) R. Br.

Chinese names: *Jiaopan Lan* (angle plate orchid), *Ren shen guo*

Fig. 13.12 *Herminium lanceum* (Thunb. ex Sw.) Vuijk (Photo: Bhaktar B. Raskoti)

Chinese medicinal name: *Rentouqi*

Description: Plant is 6–35 cm tall with an erect stem sheathed by 2–3 lanceolate leaves, 4–10 by 1–2.5 cm, near the base. Inflorescence is terminal and bears up to 20 small, yellowish-green, nodding flowers which are closely spaced. Flowering period is June to September (Chen et al. 1999). Two ovoid tubers are formed each season, the larger going on to produce leaves and flower the following season, the smaller detaching to start a new plant. Its principal form of reproduction is vegetative. Roots are short, thin and generally free of fungus, but the tip of the tuber usually carries some mycorrhiza (Fuchs and Ziegenspeck 1925, quoted by Rasmussen 1995).

H. monorchis enjoys a wide distribution across temperate Eurasia from Japan across Korea, northern China and Russia, and all of Europe. It also occurs in central Asia and the Himalayas. This terrestrial orchid is found in moist locations, in damp grassland or short turf, on non-calcareous soils, at 600–4300 m (Fig. 13.13).

Herbal Usage: Herb is obtained from the northern provinces and the Yangzi region (Wu 1994). Entire plant is used to enrich *yin*. It benefits the "kidney and stomach" and regulates menstruation. It is prescribed for a nervous breakdown, confusion, insomnia, thirst, anorexia, and precocious greying of hair (*Zhongyao Da Cidian* 1986; Wu 1994), or to strengthen and nourish a weak body (Chen and Tang 1982).

Overview

No alkaloid was detected in either species (Luning 1964, 1974). Glucomannans, hydrophilic carbohydrates of high viscosity, are present in the tubers of *H. augustifolium* (syn. *H. lanceum*) (Ohtsuki 1937). In a small, non-controlled, open trial of 93 patients suffering from chronic constipation, oral treatment with glucomannan 1 g three times a day resulted in improvement and there were no side effects (Passaretti et al. 1991). However, medicinal *Herminium* has not been used in this manner (Fig. 13.14).

Fig. 13.13 *Herminium monorchis* (L.) R. Br. (Photo: Courtesy of Plant Photo Bank of China)

The generic name, *Hetaeria*, is derived from Greek *hetaireia* (companionship), referring to the close relationship between members of this species and those of *Goodyera* and similar terrestrial orchids (Schultes and Pease 1963).

Genus: *Hetaeria* Blume

Chinese name: *Fanchun Lan*
Japanese name: *Hime No Yagara*

Hetaeria is a Malayan genus of terrestrial orchids with soft leaves and long erect inflorescences that carry many tiny flowers. There are about 13 or more species distributed from India to Fiji, at least 8 of which are endemic. It is similar to *Zeuxine* but its lip is non-resupinate (unlike most orchids, the lip is at the top of the flower). Lip is also pointed instead of flat and it bears a few glands or papillae at the base. The genus has no horticultural value.

Hetaeria obliqua Blume

Chinese name: *Xiebanfanchun* Lan
Malay name: *Pokok tumbak hutan*

Description: Plant is 25–40 cm tall, with 5–8 widely separated leaves at its lower third. Leaves are oval and pointed, 3–7 by 1.2–2.8 cm with their petioles sheathing the stem. Inflorescence is 25 cm long with several bracts at the base, and a many-flowered rachis. Reddish hairs cover the inflorescence up to the outer surface of the flowers which are 5 mm across, white, and do not open widely. Petals are oval (Comber 2001). The species is found in lowland forests in Sumatra, Thailand, Nicobar Islands, Peninsular

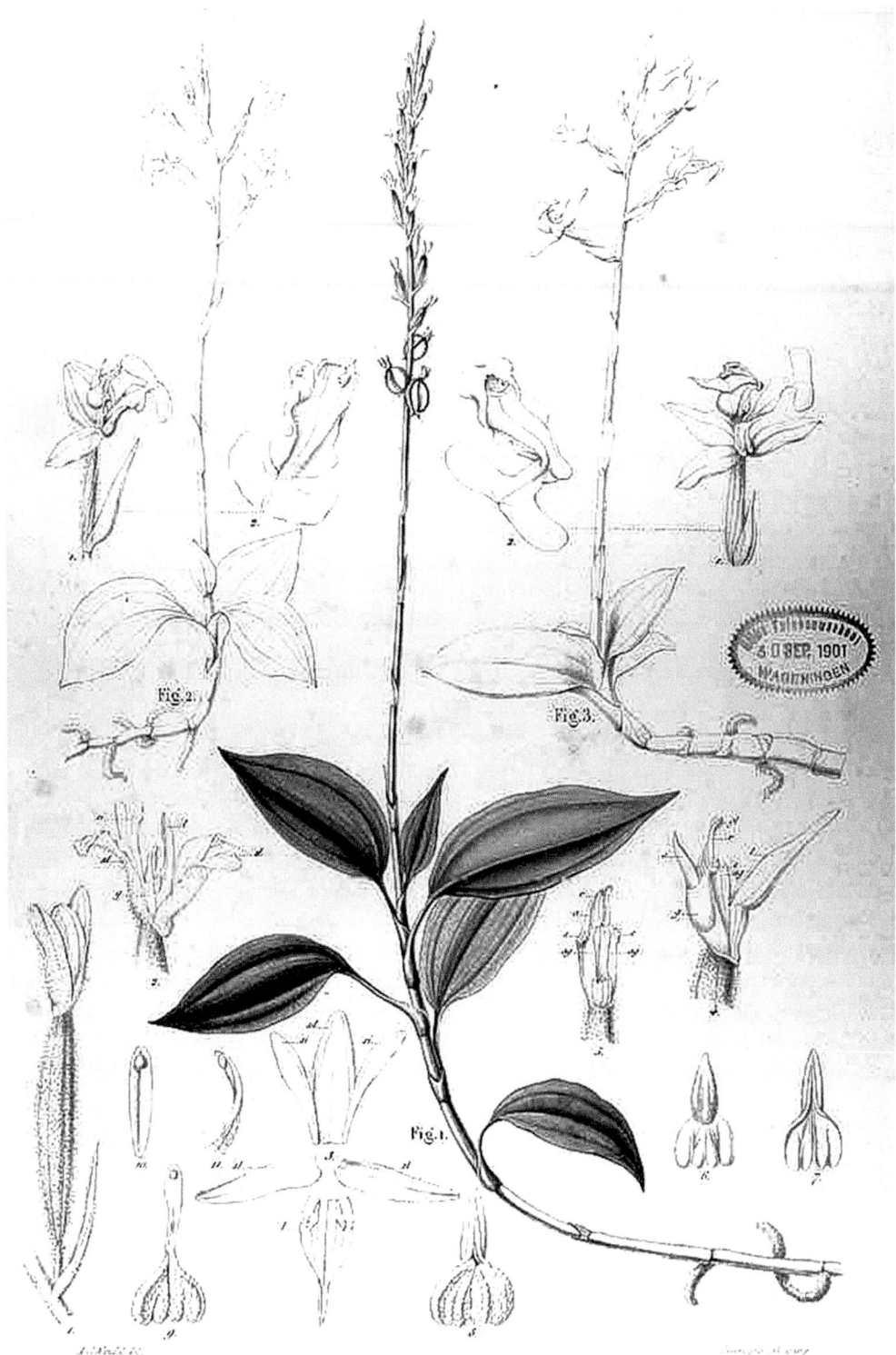

HETAERIA *(ETHELIS)* OBLIQUA, Fig... , LUHISIA ODORATA, Fig.2, L.FURETH, Fig.3.

Fig. 13.14 *Hetaeria obliqua* Blume. From: Blume C.L., *Collection des Orchidees les plus remarquables de l'archipel Indien et du Japon*, t. 34, Fig. 1 (1858). Drawing in colour and pencil by A.J. Wendel. Courtesy of plantillustrations.org

Malaysia and Borneo (Seidenfaden and Wood 1992). It also occurs in dense forests in Hainan where it flowers in March (Chen et al. 2009b).

Herbal Usage: Leaves of *H. oblique* were pounded to make a poultice which was used to treat sores and infected wounds in Peninsular Malaysia (Ridley 1907; Caius 1936). Leaves of a second orchid, *Thrixspermum pardale*, were added in making the poultice for treating ulcers of the nose by villagers in Malaya (Burkill 1935).

Overview

Five species of *Hetaeria* were investigated for the presence of alkaloids and none showed any promise (Luning 1974). The relief resulting from the application of a poultice of *H. oblique* on sores and infected wounds might be due to some anti-inflammatory constituents or to phytoalexins which are bacteriostatic. However, hard data are not available. *Haetaria* and *Goodyera* generally associate with *Ceratobasidium* (Shefferson et al. 2010).

Genus: *Hippeophyllum* Schltr.

Hippeophyllum is a small genus with five or six species distributed in the Malay Archipelago, Papua New Guinea and the Solomon Islands, with one species, *H. pumilum* T.P. Lin reported from Taiwan and southern Gansu (Chen 1995). It is vegetatively similar to *Oberonia* but the stems are more widely spaced, flowers are larger, column is slender, and the ovary is hairy. The generic name is derived from Greek, *hippeo* (horseman) and *phyllon* (leaf) referring to the equitant leaves (Schultes and Pease 1963).

Hippeophyllum scortechinii (Hook f.) Schltr.

Malay name: *Setawar baker perah*

Description: Plant has long, stout, creeping rhizomes that produce stems which are spaced 4 cm apart. Stems are short with 4 straight leaves, 20 by 1 cm, which are held close together.

Inflorescence is 30 cm long with a short scape carrying about 20 to well over 100 small (3 mm), non-resupinate, pale, greenish flowers with yellow lips (Seidenfaden and Wood 1992; Comber 2001).

This epiphytic species is distributed in Sumatra, the Malay Peninsula, Java, Borneo and Sulawesi. It is found in the lowlands of Pahang and Perak in shady locations, usually near rivers.

Herbal Usage: Gimlette and Thomson (1939) reported that Malays in Kuala Lipis, a town in Pahang, Peninsular Malaysia, squeezed the hot juice from heated leaves of *H. scortechinii* into the ear to relieve earache.

Overview

No experimental data on *Hippeophyllum* were found in the literature search (Fig. 13.15).

Genus: *Holcoglossum* Schltr. 1

Chinese name: *Caoshe Lan* (slot tongue orchid)

Holcoglossum is a small genus of epiphytic, monopodial orchids related to *Vanda* and *Papilionanthe*. Stems are short. Leaves are terete or triangular in section, grooved at the upper surface, dark green, arranged in two rows, sheathing at the base, and deciduous. Inflorescence is axillary, with few to many medium-sized, showy flowers in white, pink or yellow, resupinate with a large lip that is furrowed, devoid of keels and often in a contrasting colour. Petals and sepals are narrow, the lower sepals commonly bow-legged.

The name *Holcoglossum* is derived from Greek, *holkos* (furrowed) and *glossum* (tongue), referring to the lip.

Holcoglossum amesianum (Rchb. f.) Christenson

Chinese names: *Dagencaoshe Lan* (big slot tongue orchid), *Wanda Lan* (ten thousand generation orchid), *Diao lan* (hanging orchid), *Jiegucao* (bone setting herb)

Chinese medicinal name: *Jiuzhualong*

Fig. 13.15 *Hocoglossum amesianum* (Rchb.f.) Christenson (as *Vanda amesiana* Rchb. f.) From: *Lindenia, Iconographie des orchidees* by E. von Lindemann, Plates 673–720, vol. 15: t. 690 (1899). Original painting in colour by O. de Pannemaeker. Courtesy of the Smithsonian Institute, USA

Myanmar name: *Moe kadol*

Description: Stem is 2–5 cm long, rooting at the base and carries 4–7 semi-terete leaves, 1–3 by 0.5–1 cm which arise close to the base of the stem. Inflorescence is short, and carries 5–10 (up to 15 in Myanmar variety), small, thinly textured, white flowers with a light pink blush, 3-4 cm across, with a more intensely coloured lip. Some varieties are very fragrant (Fig. 13.16). It is not commonly seen because it does not do well in the lowland cities of Southeast Asia, but in its highland habitat it is free flowering. Flowering season is December to March in China (Chen et al. 1999), the flowers lasting several weeks on the plant. In northern Thailand, it flowers in October to January (Vaddhanaphuti 2001). Flowering season in the Shan state of Myanmar is December to January (Grant 1895; Tanaka and Yee 2007). This semi-terete, epiphytic, evergreen, small, montane orchid is found in southern China, Indochina, Thailand and Myanmar (the Shan state), at elevations of 1300–1700 m. It is epiphytic on trees growing on limestone rocks in the shade.

Herbal Usage: Chinese herbal medicine regards the whole plant as an antipyretic. It diminishes inflammation, improves blood flow, and removes gas and humidity. It is used to treat malaria, sore throat, mastitis, urinary infection, rheumatic pain, backache, irregular menses, traumatic bleeding and traumatic fractures (Wu 1994; Hu et al. 2000).

Fig. 13.16 *Holcoglossum amesianum* (Rchb.f .) Christenson (Photo: Peter O'Byrne)

Holcoglossum quasipinifolium (Hayata) Schltr.

Chinese names: *Songye Lan* (pine leaf orchid), (pine needle orchid), *Qiaochun Lan* (sledge lip orchid), *Caoshe Lan* (slot tongue orchid); *Yeludongqing* (green leaf pine).

Description: Plant has the habit of an *Ascocentrum*, but it is bigger, with a short stem and strap leaves, 10 cm by 3 cm. Inflorescence is axillary, 10 cm long, and carries 1–5 open flowers, 4 cm wide. Petals and sepals are narrow, with wavy edges, white and streaked in pink at the central vein. Lip is complex, trilobed, extending downwards into a narrow, pointed, deep spur, 1.8 cm in length (Liu and Su 1978; Lin 1975). It flowers from February to April in Taiwan (Su 1985) and from September to October in Yunnan and Sichuan (Yang et al. 1998). This vandaceous, semi-terete leaf orchid, once thought to be endemic in Taiwan, where it occurs in the Central Mountain Range at 2000–2500 m, often growing on the branches of *Quercus* spp. (Su 1985), was subsequently discovered occurring in Yunnan and Sichuan, in forests at 800–2200 m (Yang et al. 1998) (Fig. 13.17).

Herbal Usage: The whole plant is used to remove wind and dampness. It causes diuresis and is used for joint pains (Ou et al. 2003).

Overview

Both medicinal species in this four-member genus are used to treat joint pains, but *H. amesiana* has a broader range of medicinal usage being also used to treat varied infections (malaria, sore throat, mastitis and urinary tract infection) and bleeding disorders (irregular menstruation and traumatic bleeding).

Endophytic fungi are abundant in roots of *Holcoglossum* plants collected from Yunnan, Guangxi and Hainan in China (Tan et al. 2012). Nevertheless, a search of the English literature did not turn up any pharmacological data.

Genus: *Ipsea* Lindl.

Ipsea is a small genus of terrestrial, pseudobulbous, herbaceous orchids with only three species and confined to India and Sri Lanka. Plants have one or two narrow, lanceolate, plicate leaves with long petioles. Scape is tall, thin, erect, and few-flowered, but the flowers are large.

The derivation of the generic name is obscure. It is possibly derived from Greek, *ips* (woodworm) which could describe the appearance of the underground portion of the plant. The well-known Latin *ipse* (itself) is another possible derivation considering that the genus has only three members.

Ipsea speciosa Lindl.

Common name: Daffodil orchid
Sinhalese names: *Kiri Walla Kada; Naga-maru Ala* (tuber causing the sister's death)

Fig. 13.17 *Ipsea speciosa* Lindl. (*as Ipsia speciosa* LIndl.) From: Wight, R. *Icones Plantarum Indiae Orientalis* vol. 5 (1): t. 1663 (1846). Drawing by Govindoo. Courtesy of Missouri Botanical Garden, St. Louis, USA

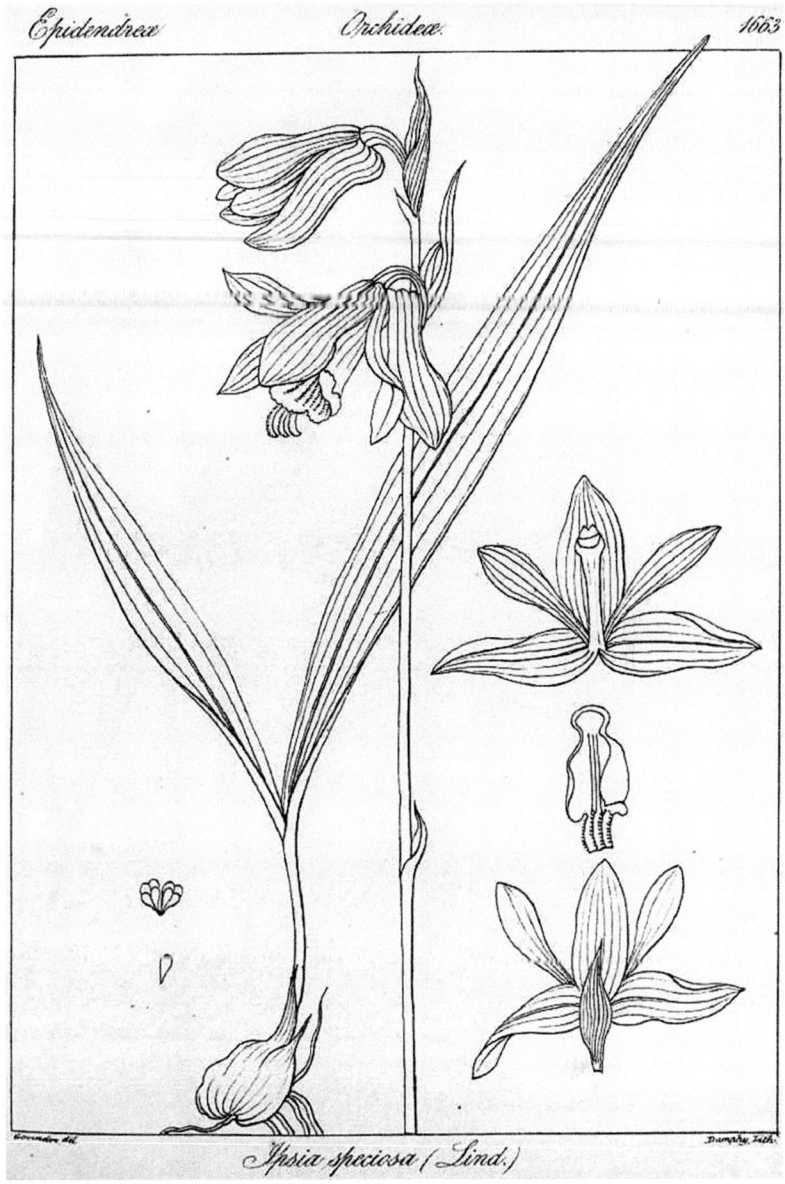

Description: Stem bears one or two long, upright leaves which are 15–25 by 0.5–2.2 cm, narrowly lanceolate, plicate, three-veined, and narrowing to a slender petiole at the lowest third of its length. Tubers are 2–3 cm long and 1.2–2.5 cm in diameter, clustered, and depressed. Long, filiform roots grow from their bases. Two or three large, bright yellow flowers, 5–6.6 cm across are carried near the apex of a slim, erect inflorescence (Jayaweera 1981) (Fig. 13.18). The species is in bloom in November and December in southern India (Bose and Bhattacharjee 1980), and from September to February in Sri Lanka (Jayaweera 1981). This yellow-flowered, terrestrial herb once thought to be endemic to Sri Lanka also occurs in southern India. It is rather common in *patana* lands in the montane zone at 1000–1800 m,

Fig. 13.18 *Ipsea speciosa*
Lindl. [From*: Curtis's*
Botanical Magazine, vol.
94 [ser. 3, vol. 24]: t.5701
(1868), Drawn by
W.H. Fitch. Illustration
through courtesy of
Missouri Botanical Garden,
St. Louis, USA

and is often found in association with grasses and two other orchid species, *Spiranthes sinensis* and *Satyrium nepalense*.

Herbal Usage: Tubers of *I. speciosa* are much sought after by sorcerers for making charms and love potions and by village quacks and medicine-men for use as an aphrodisiac (Jayaweera 1981). Cooray (1940) was possibly being euphemistic when he reported the claim that a decoction of

the tubers was capable of stimulating the nervous system and was much sought after for this purpose. The species is already rare, and may be rendered extinct through overcollection (Sumithraarachchi 1986).

Overview

This is one of the numerous terrestrial orchids with swollen, tuberous roots which are said to

have aphrodisiac properties. The popular Singhalese name *Naga-maru Ala* translates as "the tuber that caused a sister's death". It refers to a legend which spoke of a prince in ancient Sri Lanka who fell madly in love with his step-sister. When she rejected his advances, he killed her on the lonely moors. According to one version of the legend, the prince became lustful and crazy after he tasted the tubers of the orchid; in the second version, it was the blood of the princess which resulted in the appearance of the golden blossoms (Cooray 1940). Both legends engendered widespread belief in the efficacy of this so-called stimulant.

A search of the English literature did not turn up any pharmacological studies on the genus.

Genus: *Ischnogyne* Schltr.

There is only one species in this rare Chinese endemic orchid genus, *Ischnogyne*, which is saxicolous. Rhizome is short, with small pseudobulbs, each with a single oblong-elliptic, slightly undulate leaf. The short scape carries a single flower.

The generic name is derived from Greek *ischnos* (weak) and *gyne* (woman or pistil). It refers to the wiry column.

Ischnogyne manadarinanum (Kranzl.) Schltr.

Chinese name: *Shoufang Lan* (slim house orchid)

Description: Pseudobulb is curved, 1.5–3 cm long by 2.5–3.5 mm in thickness, and carries a single, apical, thick, oblong-elliptic leaf, 4–7 by 1.2–1.5 cm. A single, white flower is borne on both leafy and leafless pseudobulbs. Tepals are narrow. Lip is white, marked by a yellow patch in the centre. Flower is about 5 cm across and does not open widely. Flowering season is May to June. This saxicolous orchid grows on rocks in forest and ravines at 700–1500 m in Shaanxi, Gansu, Hubei, Sichuan and Guizhou in China (Chen et al. 1999).

Herbal Usage: Herb is obtained from Shanxi, Hubei and Sichuan. In China, the whole plant is sometimes used to treat tuberculous patients who cough, and bronchitis. The herb is obtained from Shaanxi, Hubei and Sichuan (Wu 1994).

Overview

No pharmacological research on this orchid has been reported in the available English literature.

Kingidium deliciosum (Rchb. f.) R.H. Sweet (see *Phalaenopsis deliciosa* Rchb.)

References

Abraham A, Vatsala P (1981) Introduction to Orchids, with illustrations and descriptions of 150 South Indian Orchids. TPGRI, Trivandrum

Akarsh (2004) Newsletter of ENVIS NODE on Indian Medicinal Plants 1(2), June 2004

Averyanov LV, Phan KL, Nguyen TH, Harder DK (2003) Phytogeographic review of Vietnam and adjacent areas of Eastern Indochina. Komarovia 3:1–83

Baral SR, Kurmi PP (2006) A Compendium of medicinal plants in Nepal. Mrs. Rachana Sharma & IUCN, Kathmandu

Bose TK, Bhattacharjee SK (1980) Orchids of India. Naya Prokash, Calcutta

Burkill IH (1935) (1966 reprint, 2nd ed., with contributions by Birtwistle W, Foxworthy FW, Scrivenor JB, Watson IG). A Dictionary of economic products of the Malay Peninsula, Vol. II. Crown Agents for the Colonies, London. Ministry of Agriculture & Co-operatives, Kuala Lumpur

Caius JF (1936) The medicinal and poisonous plants of India. J Bombay Nat Hist Soc 38(4):791–799

Chen SC (1995) Orchids and their conservation. Proceedings of the 5th Asia Pacific Orchid Conference and Show, Fukuoka, pp 15–18

Chen XQ, Cribb PJ (2009) Habenaria Willdenow. In: Chen XQ, Liu ZJ, Zhu GH et al (eds) Flora of China – Orchidaceae. Science Press, Beijing

Chen SC, Tang T (1982) A general review of the orchid flora of China. In: Arditti J (ed) Orchid biology. Reviews and perspectives II. Cornell University Press, Ithaca

Chen SC, Tsi ZH, Luo YB (1999) Native orchids of China in colour. Science Press, Beijing

Chen XQ, Gale SW, Cribb PJ (2009a) Hemipilia Lindl. In: Chen XQ, Liu ZJ, Zhu GH et al (eds) Flora of China 25 Orchidaceae. Science Press, Beijing, pp 98–100

Chen XQ, Gale SW, Cribb PJ, Ormerod P (2009b) Hetaeria blume. In: Chen XQ, Zj L, Zhu GH et al (eds) Flora of China - Orchidaceae. Science Press, Beijing

Chuakul W (2002) Ethnomedical uses of Thai orchidaceous plants. Mohidol Univ J Pharm Sci 29 (3–4):41–45

Comber JB (2001) Orchids of Sumatra. Natural History Publications (Borneo), Kota Kinabalu

Cooray DA (1940) Orchids in oriental literature. Orchids Zelandica 7:73–80

Dash PK, Sahon S, Bal S (2008) Ethnobotanical studies on orchids of Niyamgiri hill ranges, Orissa, India. Ethnobot Leaflets 12.70–70

Dassanayake MD, Fosberg FR (1981) A revised handbook of the flora of Ceylon, vol II. A.A. Balkema, Rotterdam

Davis RS, Steiner ML (1982) Philippine orchids. A detailed treatment of some 100 native species. : M & L Lucidine Enterprises, Manila (reprint)

Dhayani A, Nautiyal BP, Nautiyal MC (2011) Importance of Astavarga plants in traditional systems of medicine in Garhwal, Indian Himalaya. Int J Biodiv Sci Ecosyst Serv Manag 6(1-2):13–19

Gimlette JD, Thomson HW (1939) A dictionary of Malayan medicine. Oxford University Press, London

Grant B (1895) The orchids of Burma. Hanthawaddy Press, Rangoon

Gurong DB (2006) An illustrated guide to the orchids of Bhutan. DSB Publications, Thimphu

Hu XM, Zhang WK, Zhu QZ et al (2000) Zhonghua Bencao, vol 8. Shanghai Science and Technology Publication, Shanghai

Jagtap SD, Deokule SS, Bhosie SV (2008) Ethnobotanical uses of endemic and RET plants by Pawra tribe of Nandurbar district, Maharashtra. Indian J Tradit Knowl 7(2):311–315

Jalal JS, Kumar P, Pangtey YPS (2008) Ethnomedicinal orchids of Uttarakhand, Western Himalayas. Ethnobot Leaflets 12:1227–1230

Jin XH, Zhao XD, Shi XC (2009) Native orchids from Gaoligongshan Mountains, China. Science Press, Beijing

Joseph J (1982) Orchids of Nilgiris, vol XXII, Records of the botanical survey of India. Botanical Survey of India (Department of Environment), Howrah

Kamemoto H, Sagarik R (1975) Beautiful Thai orchid species. Orchid Society of Thailand, Bangkok

Karthikeyan S, Jain SK, Nayar MP, Sanjappa M (1989) Florae Indicae Enumeratio: Monocotyledonae. Botanical Survey of India, Calcutta

Kumar V (2003) Wild edible plants of Surguja District of Chhattisgarh State, India. In: Singh V, Jain AP (eds) Ethnobotany and medicinal plants of India, vol 1. Scientific Publications, Jodhpur

Lawler LJ (1984) Ethnobotany of the orchidaceae. In: Arditti J (ed) Orchid biology reviews and perspectives, vol 3. Cornell University Press, Ithaca

Lin TP (1975) Native orchids of Taiwan, vol 1. Southern Materials Center Inc., Taipei

Lin TP (1977) Native orchids of Taiwan, vol 2. pp. 166–7, colour photo 98. Southern Materials Center Inc., Taipei, pp 76 and 97–99

Liu TS, Su HJ (1978) Flora of Taiwan, vol 5. National University of Taiwan, Taipei

Luning B (1964) Studies in Orchidaceae alkaloids I. Screening of species for alkaloids. Acta Chim Scand 18(6):1507–1516

Luning B (1974) Alkaloids of the Orchidaceae. In: Withner CL (ed) The orchids. Scientific studies. Wiley, New York, 1974

Luo YB (2004) Cytological study on some representative species of the tribe Orchideae (Orchidaceae) from China. Bot J Linn Soc 145:231–238

Luo YB, Chen SC (2009) Hemipiliopsis, a new genus of Orchidaceae. Novon 13:450–453

Manandhar NP, Manandhar S (2002) Plants and people of Nepal. Timber, Portland

Matthew KM (1995) An Excursion flora of central Tamilnadu, India. A.A. Balkaema, Rotterdam

Musharof Hossain M (2009) Traditional therapeutic uses of some indigenous orchids of Bangladesh. Med Aromat Plant Sci Biotechnol 3:100–106

Nair DMN (1963) The families of Burmese flowering plants, vol 2. Rangoon University Press, Yangon

Nanakorn W, Watthana S (2008) Queen Sirikit botanic garden (Thai native orchids 1 and 2). Wanida Press, Chiang Mai

Nasir E, Ali SI (1972) Flora of West Pakistan. Pakistan Agricultural Research Council (Actual publisher not stated)

Ohtsuki T (1937) Untersuchungen uber das Bletillamannan, ein Mannan aus den Knollen von Bletilla striata. Acta Phytochim 10:29–41

Ohwi J (ed) (1965) Flora of Japan. English edition: FG Meyer, Walker EH (eds). Washington: Smithsonian Institution

Ou JC, Hsieh WC, Lin IH, Chang YS, Chen IS (eds) (2003) The catalogue of medicinal plant resources in Taiwan. Department of Health, Executive Yuan, Taipei

Pandey NK, Joshi GC, Mudaiya RK et al (2003) Management and conservation of medicinal orchids of Kumaon and Garhwal Himalaya. In: Jain AP, Singh V (eds) Ethnobotany and medicinal plants of India and Nepal. Scientific Publishers (India), Jodhpur, pp 114–118

Passaretti S, Franzoni M, Comin U et al (1991) Action of glucomannans on complaints in patients affected with chronic constipation: a multicentric clinical evaluation. Int J Gastroenterol 23(7):421–425

Pearce NR, Cribb PJ (2002) The orchids of Bhutan. Royal Botanic Gardens and Thimpu: Royal Government of Bhutan, Edinburgh

Pradhan UC, Pradhan SC (1997) 100 beautiful Himalayan orchids and how to grow them. Primulaceae Books, Kalimpong

Puri HS (1970) Indian medicinal plants used in elixirs and tonics. Quart J Crude Drug Res 10:1555–1566

Rao AN (2004) Medicinal Orchid Wealth of Arunachal Pradesh. Newsl ENVIS NODE on Indian Med Plants 1(2):1–7

Rao TA (2007) Ethno Botanical data on wild orchids of medicinal value as practised by tribals at Kudremukh National Park in Karnataka. Orchid Newsl 2(2):1–7

Rao TA, Sridhar S (2007) Wild orchids in Karnataka. A pictorial compendium. Institute of Natural Resources Conservation, Education, Research and Training (INCERT), Bangalore

Rao NR, Henry AN (1995) The Ethnobotany of Eastern Ghats in Andhra Pradesh, India. Botanical Survey of India. Govt. Press of India (quoted by Musharof Hossain, 2009)

Raskoti BB (2009) The orchids of Nepal. Bhakta Bahadur Raskoti and Rita Ale, Kathmandu

Rasmussen HN (1995) Terrestrial orchids from seed to mycotrophic plant. Cambridge University Press, Cambridge

Reddy KN, Reddy CS, Raju VS (2005) Ethno-orchidology of orchids of Eastern Ghats of Andhra Pradesh. EPRTI Newsl 11(3)

Ridley HN (1907) Materials for a flora of the Malay Peninsula, vol 1. Methodist Publishing House, Singapore

Santapau H, Kapadia Z (1966) The orchids of Bombay. Government of India Press, Calcutta

Schultes RE, Pease AS (1963) Generic names of orchids: Their origin and meaning. Academic, New York

Seidenfaden G (1999) 149. Orchidaceae. In Matthew KM (ed) The Flora of the Palni Hills, South India. Part 3. Tiruchirapalli: The Rapinat Herbarium. St. Joseph's College

Seidenfaden G, Wood JJ (1992) The orchids of Peninsular Malaysia and Singapore. Olsen & Olsen, Fredensborg

Shefferson RP, Cowden CC, McCormick MK et al (2010) Evolution of host breadth in broad interactions: mycorrhizal specificity in East Asian and North American rattlesnake plantains (Goodyera spp.) and their fungal hosts. Mol Ecol 19(14):3008–3017

Singh A, Duggal S (2009) Medicinal orchids: an overview. Ethnobot Leaflets 13:351–363

Sood SK, Rana S, Lakhanpal TN (2005) Ethnic aphrodisiac plants. Scientific Publishers (India), Jodhpur

Su HJ (1985) Native orchids of Taiwan, 3rd edn. Harvest Farm Magazine, Taipei

Subramani SP, Goyara GS (2003) Some folklore medicinal plants of Kolli Hills: record of a Natti Vaidyas Sammelan. In: Jain AP, Singh V (eds) Ethnobotany and medicinal plants of India and Nepal. Scientific Publishers (India), Jodhpur, pp 665–678

Sumithraarachchi DB (1986) Conservation of orchids in Sri Lanka. In: Rao AN (ed) Proceedings of 5th ASEAN orchid congress. Parks & Recreation Department, Ministry of National Development, Singapore, pp 140–144

Suryanrayana B, Rao AS (2005) Orchids and Epiphytes on veligonda hills—Eastern Ghats. EPTRI-ENVIS Newsl 11(4):1–3

Tan XM, Chen XM, Wang CL et al (2012) Isolation and identification of endophytic fungi in roots of nine Holcoglossum plants (Orchidaceae) collected from Yunnan, Guangxi and Hainan provinces of China. Curr Microbiol 64(2):140–147

Tanaka T (1976) Tanaka's cyclopedia of edible plants of the world. Keigaku Publishing, Tokyo (ed: Nakao S)

Tanaka Y, Yee TTA (2007) Wild orchids of Myanmar, vol 3. Orchid Press, Hong Kong

Trivedi VP, Dixit RS, Lal VK (1980) Orchids in the drug markets of Bareilly, Kanpur and Nearby Districts. Nagarjun (Calcutta) 23(8):157–163

Vaddhanaphuti N (2001) A field guide to the wild orchids of Thailand, 3rd edn. Silkworm Books, Chiang Mai

Vaddhanaphuti N (2005) A field guide to the wild orchids of Thailand, Fourth and Expanded Edition. Silkworm Books, Chiang Mai

Veitch NC, Grayer B (2001) Phytochemistry of habenaria and orchis. In: Pridgeon AM, Cribb PJ, Chase MW, Rasmussen FN (eds) Genera orchidacearum, vol 2, Orchidoideae (Part One). University Press, Oxford

Wu XR (1994) A concise edition of medicinal plants in China. Guangdong Higher Education Publication House, Guangdong (in Chinese)

Wu TL, Hu QM, Xia NH, Lai PCC, Yip KL (2002) Check list of Hong Kong plants 2001. Agriculture, Fisheries and Conservation Department Bulletin 1 (Revised), Hong Kong

Yang ZH, Zhang QT, Feng ZZ, Lang KY, Li H (1998) Orchids. English edition: ZR Xiong. China Esperanto Press, Beijing

Yoganarasimhan SN, Chelladurai V (2000) Medicinal plants of India: Tamil Nadu, vol 2. Regional Research Institute, Bangalore

Zhongyao Da Cidian (1986) Edited by Jiangsu New Medical College. Science and Technology Press, Shanghai

Limodorum spathulatum (L.) Willd. [see *Taprobanea spathulata* (L.) Christenson]

Genus: *Liparis,* Rich.

Chinese name: *Yanger suan* (goat ear garlic)
Japanese name: *Kumo kiri so*

An enormous genus with some 250 members, *Liparis* has several species that are now assigned to other genera and given different names. The *Index Kewensis* lists almost 400 taxa arising from duplication (Seidenfaden and Woods 1992). The genus is distributed in the Old and New Worlds with a concentration of species in Southeast Asia. It is absent in New Zealand. Some species are epiphytic, others terrestrial. *Liparis* and *Malaxis* burst into bloom in the western part of southern India when the great force of the changing monsoon abates and rain arrives, in August and September (Abraham and Vatsala 1981).

The generic name is derived from Greek, *liparos* (fat, greasy), referring to the shiny surface of the leaves in some species. Rhizomes of *Liparis* remain infected with mycorrhiza throughout their life-spans and, together with the leaf bases, they are the predominant sites of mycotrophy. Symbiotic seedlings have been successfully raised with *Rhizoctonia* species (Rasmussen 1995).

L. plicata Franch et Sav. is considered by many taxonomists to be synonymous with *L. bootanensis* Griff. (Chen et al. 2009c), but Wu Xiu Ren (1994) provided different Chinese common names for the species and indicated a wider distribution for *L. bootanensis* than for *L. plicata*. However, we are not able to clarify the basis for their separation into two distinct herbs. Incorrect identification or assignment of inappropriate names is commonplace in *Liparis*. In the present discussion, *L. plicata* is described and discussed under its accepted botanical name, *L. bootanensis* Griff.

Liparis is not commonly used for medicinal purposes except in China despite the presence of alkaloids in many species. Only a handful of sources mention its medicinal usage. The list produced by Wu Xiu Ren (1994) consisting of nine species is the longest, followed by *Zhonghua Bencao* (2000) with seven species. The only report of a *Liparis* being used in Indian medicine mentioned that *L. odorata* (or *L. nervosa*) is used to treat elephantiasis (Rao and Sridhar 2007).

Liparis was among the first orchid genera to be investigated for the presence of alkaloids. Boorsma (1902) reported the first discovery of alkaloids in *L. parviflora* over a century ago, but it was only in 1967 that Nishikawa and Hirata (1967a) in Japan and Luning and his colleagues in Sweden (Lindstrom et al. 1971, 1972) reported the earliest chemical studies reported the earliest chemical

© Springer International Publishing Switzerland 2016
E.S. Teoh, *Medicinal Orchids of Asia*, DOI 10.1007/978-3-319-24274-3_14

Fig. 14.1 *Liparis bootanensis* Griff. From: *Annals of the Royal Botanic Gardens, Calcutta*, vol. 8 (2): t. 40 (1891). Drawing by R. Pantling. Courtesy of Missouri Botanical Gardens, St. Louis, U.S.A

studies on the alkaloids of the related genera *Liparis* and *Malaxis*.

Liparis argentopunctata Aver. (see **Liparis cordifolia Hook f.**)
Liparis bicallosa (D. Don) Schltr. [see **Eulophia bicallosa (D. Don) P.F.Hunt & Summerh.**]

Liparis bootanensis Griff.

Syn. *Liparis plicata* Franch & Savat.

Chinese name*: Lianchiyangersuan* (sickle wing goat ear garlic). In Hong Kong and Bhutan *Twayblade*

Chinese medicinal name: *Jiuliandeng* (nine lotus lamps)
Taiwanese name: single leaf *Liparis*
Japanese name: *Chi-kei-ran* (bamboo grass Cymbidium)

Description: Herb is epiphytic or lithophytic, with small, clustered, ovoid pseudobulbs, 1.5 cm tall, each with a single, sessile, lanceolate leaf up to 24 by 2.3 cm (Fig. 14.1). Inflorescence is 18 cm long and carries a dozen light brown, green or yellowish-green, rarely nearly-white flowers with a greenish column. (*L. plicata*, the variety regarded by Chinese herbalists as a distinct species, has green flowers with a white column.) Petals and sepals are filiform, convex and curl backwards. Lip is broad, rectangular to ovate, with a deep central groove. Flowers open in succession and turn orange-yellow with age. The variety with a white column is striking (Fig. 14.2). It flowers from August to December at Gaoligongshan (Jin et al. 2009), but generally August to October in Mainland China (Chen et al. 1999) and from September to December in Taiwan. The species is abundant in Taiwan from the lowlands to 1200 m (Liu 1975, Chen et al. 2009b). Flowering in the Indian Himalayas occurs in August (Bose and Bhattacharjee 1980), and July to August in Bhutan (Gurong 2006). *L. bootanensis* is distributed from Nepal to southern China up to Hainan, Hong Kong, Taiwan, Japan (Ryukyu Islands, Kyushu), Korea, Indochina, Thailand, Philippines, Borneo, Java, the Langkawi Islands and Myanmar at 800–2300 m.

Herbal Usage: Botanically, *L. bootanensis* Griff and *L. plicata* Franch & Savat. are the same species, but in *A concise Edition of Medicinal Plants in China*, Wu Xiu Ren (1994) classifies them as separate herbs called *Lianchiyangersuan* and *Jiuliandeng* respectively. The former herb, *Lianchiyangersuan* is obtained from Taiwan, Fujian, Guangdong, Guangxi, Hunnan, Yunnan, Guizhou, Sichuan and Tibet; *Jiuliandeng* is obtained from Taiwan, Hongkong, Guangdong and Guangxi (Wu 1994). Subsequently, the *Zhonghua Bencao* (2000) recognised that the two species are similar and referred to them by their medicinal name, *Jiuliandeng*.

In Chinese herbal medicine, *Jiuliandeng*, the entire plant of *L. bootanensis*, is used to treat fever, enrich *qi* and blood, treat tuberculosis, lymph node enlargement in tuberculosis, sores and ulcers, abdominal pain, distension and schistosomiasis (Wu 1994). Cooked with pork, it provides a remedy for coughs and sore throat (Cheo, quoted by Perry and Metzger 1980; Wu 1994). Between 6 and 15 g of the herb is used to prepare the medicinal soup (*Zhonghua Bencao* 2000).

Fig. 14.2 *Liparis bootanensis* Griff. [PHOTO: Bhaktar B. Raskoti]

Liparis cathcartii Hook.f.

Chinese name: *Erzheyangersuan* (two fold sheep ear garlic

Description: A small terrestrial herb with ovoid pseudobulbs, 5–6 long and 4–5 mm wide, enclosed in white membraneous sheaths, it bears two ovate leaves 3.5–8 by 1.7–4 cm. Inflorescence is 7–25 cm with over 10 small pink, purple or green flowers (Fig. 14.3). Flowering season is June and July in China (Chen et al. 1999), and in August to September in Sikkim (Bose and Bhattacharjee 1980). It is distributed only in Sichuan and central and northwest Yunnan in China, but it also occurs in Laos, Myanmar, northeast India, Bhutan and Nepal. It is found in humid or grassy locations at 1900–2000 m in

Fig. 14.3 *Liparis
cathcartii* Hook.f. From:
Hooker, W. J. Hooker J. D.,
Icones Plantarum vol. 19:
t. 1808 (1889). Drawing by
M. Smith. Courtesy of
Missouri Botanical arden,
St Louis, U.S.A

China (Chen et al. 2009c), 2600–3000 m in Sikkim (Bose and Bhattacharjee 1980).

Herbal Usage: Chinese herbal medicine employs the whole plant to stop pain and to clear colds (Wu 1994).

Liparis cespitosa (Lam.) Lindl.

Chinese name: *Xiaohuayanger Suan* (small flowered goat ear garlic), *Shisuantou* (stone garlic), *Yeshengyangersuan* (goat ear garlic), *Congshen-gyangersuan*

Description: Herb is epiphytic or lithophytic. Inflorescence is borne on young pseudobulbs, 15 cm tall, with about 25 small white or light green flowers, closely arranged but not crowded (Figs. 14.4 and 14.5). It does not have a distinct flowering season (O'Byrne 2001), except in more northern climes where cold weather may deter the formation of new shoots. Flowering season is June to October at Gaoligongshan (Jin et al. 2009). In Taiwan, it flowers continuously, possibly with a peak in October to November (Lin 1975). It flowers in September, January and March in Sri Lanka (Jayaweera 1981). The species enjoys a wide distribution from East Africa across the Himalayan foothills, southern China and Southeast Asia to the Pacific Islands, generally from sea level to 1500 m (O'Byrne 2001); in China, up to 2400 m (Jin et al. 2009).

Herbal Usage: The whole plant is used in Taiwan to treat fever and remove toxins. In Chinese herbal terminology, it cools blood and stops bleeding (Ou et al. 2003).

Liparis condylobulbon Rchb. f.

Syn. *Liparis treubii* J.J. Smith

Indonesian name: *Anggrek gajang*

Description: *L. condylobulbon* (syn. *L. treubii*) is an epiphyte with flask-shaped pseudobulbs that carry two thin, narrow, lanceolate leaves at their apices. Inflorescence is 10–25 cm long and carries up to 50 closely placed, light green flowers, each 4 mm broad. Floral parts are tinged a light brown near their apices. Lip is strongly recurved, ciliate, oblong, white basally, light brown apically (Fig. 14.6). It is distributed from Myanmar, Thailand to Malaysia, Indonesia, the Philippines eastwards to Fiji, at 500–1700 m (Comber 2001).

Phytochemistry: Three new nervogenic acid glycosides, named condobulbosides A–C and an apigenin glycoside, schaftoside, are present in the leaves of *L. condylobulbon* (Slapetova et al. 2009).

Herbal Usage: Natives in Sulawesi (Indonesia) once believed that chewing on the young pseudobulbs and rubbing the heated leaves of the orchid on the abdomen facilitated bowel movements and relieved a distended stomach [Heyne (1927) and Burkill (1935) both quoting from Rumphius' *Herbal Amboinense* published in the late seventeenth century].

Liparis cordifolia Hook. f.

Syn. *Liparis argentopunctata* Aver., *Liparis keitaoensis* Hayata

Chinese name: *Xinyeyangersuan* (Silver cricket orchid)

Description: Plant is 6–9 cm tall. Pseudobulbs are obliquely ovoid, laterally flattened, green, 1.5–3 cm tall, sheathed by a single plicate, ovate, yellowish-green leaf spotted with white on its upper surface, 5–10 cm by 4–7 cm, and pointed at the tip. Inflorescence is terminal, erect, 7–15 cm tall with around 10 flowers of light green, 2 cm across, tepals linear, lip flat, rounded or teardrop, and marked by a median groove (Lin 1977; Tang and Su 1978). Flowering season is October to December (Lin 1977). This small species is endemic in Taiwan in forests at 1000 m. It grows in moist forest or bamboo stands in the company of other herbs (Fig. 14.7).

Phytochemistry: Two alkaloids, keitaonine and keitine were isolated from *L. keitaoensis* (=

Fig. 14.4 *Liparis cespitosa* (Lam.) Lindl. (as *Liparis pusila* Ridl.). From: Hooker, W. J, Hooker, J. D., *Icones Plantarum*, vol. 19: t. 1856 (1889). Drawing by M. Smith. Courtesy of Missouri Botanical Gardens

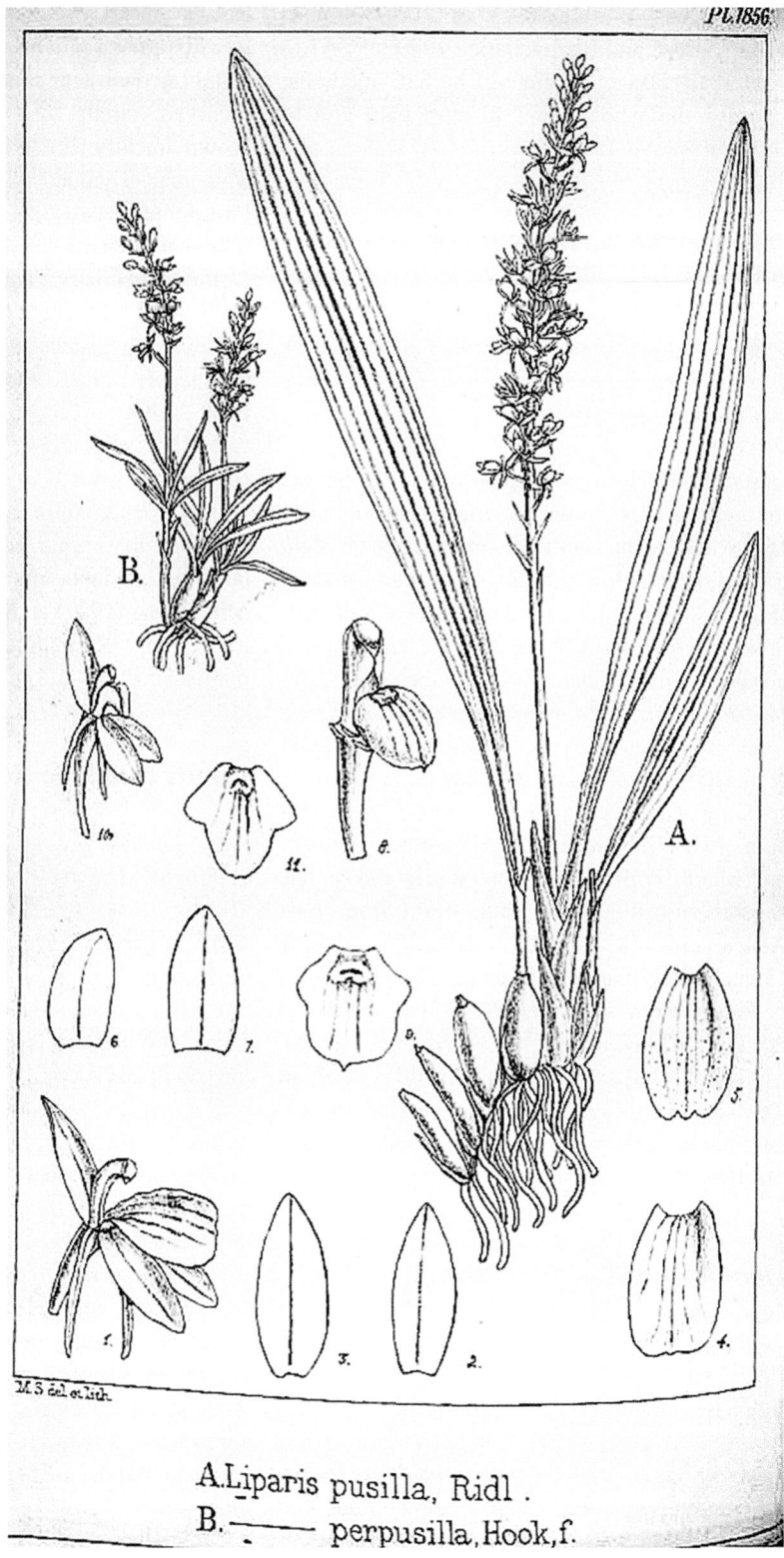

A.Liparis pusilla, Ridl .
B. ——— perpusilla, Hook, f.

Fig. 14.5 *Liparis cespitosa* (Lam.) Lindl. [PHOTO: Bhaktar B. Raskoti]

Fig. 14.6 *Liparis condylobulbon* Rchb. f. [PHOTO: Peter O'Byrne]

L. cordifolia). Keitaonine is a pyrrolizidine-based alkaloid, an ester of 3-methoxy-malaxinic acid and laburnine. Keitine is an aglycone of keitaoine but it is possible that it might not exist in the living plant (Lindstrom and Luning 1972).

Herbal Usage: A decoction of the root is a Taiwanese remedy for abdominal pain (Liu, quoted by Perry and Metzger 1980).

Liparis distans C.B. Clarke

Syn. *Liparis yunnanensis* Rolfe

Chinese names: *Yunnan yanger Lan* (Yunnan sheep ear orchid), *Dahuayan-gersuan* (big flower sheep ear garlic) *Yunnan Yangercao* (Yunnan sheep ear herb)
Chinese medicinal name: *Hushitou*

Description: *L. distans* is epiphytic, has yellow flowers and is allied to *L. plantaginea* Lindl. from which it differs by having longer leaves, narrower bracts and narrower sepals. Pseudobulb carries 2 leaves which are subcoriaceous, elongate-lanceolate, and measure 10–15 by 10–16 cm. Unopened buds are half as stout as those of *L. plantaginea* (Dunn 1903). Flowering period is October to February in China (Chen et al. 1999; Jin et al. 2009). *L. distans* was first discovered growing among rocks at 1600 m on Mengtze Mountains in Yunnan. It enjoys a wide distribution from Guangdong, Guangxi, Guizhou, Yunnan, Sichuan and Tibet in China to Assam, Thailand, Laos Cambodia, Vietnam and the Philippines (Chen et al. 2009c).

Herbal Usage: Herb is obtained from Guangdong, Guangxi, Yunnan, Sichuan and Tibet. Entire plant is used to treat pneumonia (Wu 1994). A decoction is made with 6–16 g of the whole plant (Hu et al. 2000).

Liparis dunnii Rolfe

Chinese names: *Dachunyangersuan* (big lip sheep ear garlic), *Fujianyangersuan* (Fujian Province goat ear garlic)
Chinese medicinal name: *Shuangyejinqiang*

Fig. 14.7 *Liparis cordifolia* Hook. f. (as *Liparis keitaoensis* Hayata). From: Hayata, B., *Icones plantarum formosanarum*, vol. 7: t. 13 (1918). Drawing by F. Hayami. Courtesy of University of California Libraries, U.S.A

Ic. Pl. Formos. Vol. VII. Pl. XIII.

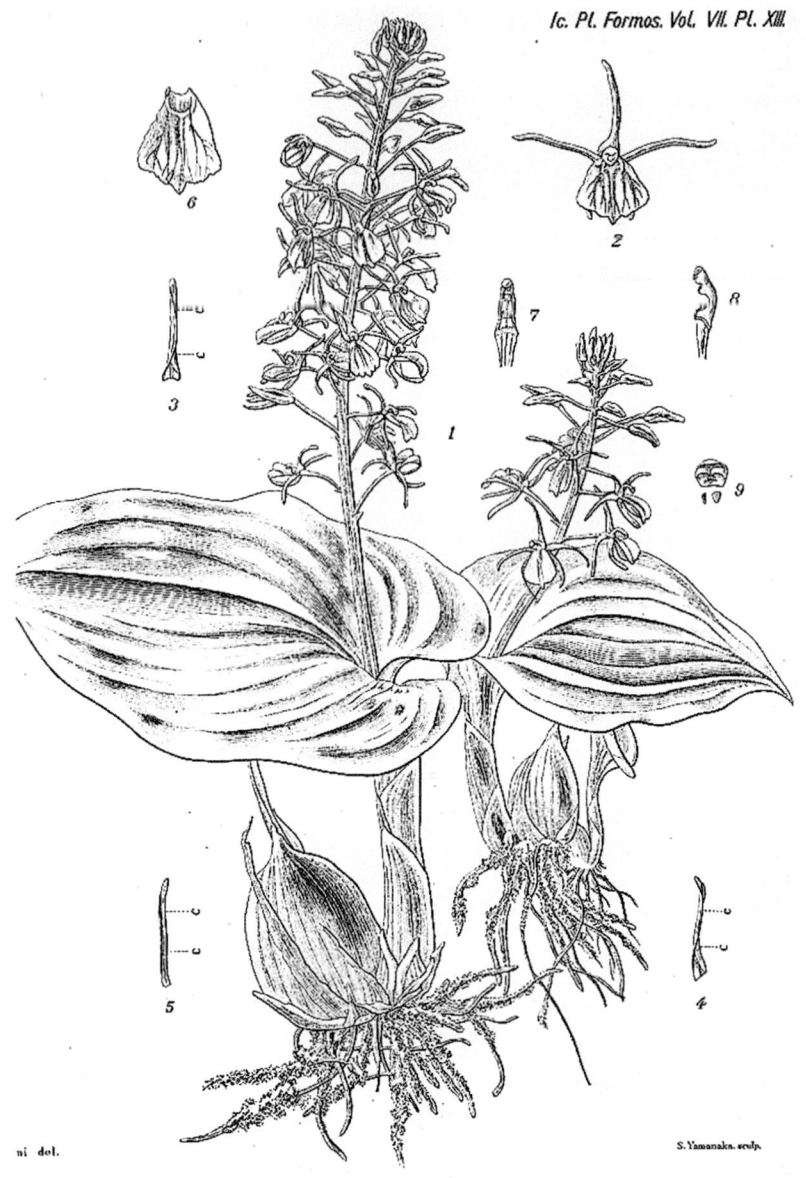

Description: Plant is 15 cm tall, bearing ovate, membranous leaves, 13 by 6 cm. Raceme is 15–18 cm tall with numerous flowers. Petals and sepals are filiform. Lip is shaped like a shield. Flowers are metallic black but the unusual colour is not retained when the flowers are dried. The species resembles *L. pauciflora* Rolfe in habit and foliage but it carries a much stouter, more densely flowered scape, and the flowers are twice as large (Dunn 1903).

This narrowly-endemic species was originally discovered during a botanical expedition in central Fujian province in 1903 growing on rocks at Tze Chuk Hang at 1000 m by Stephen Troyte Dunn (1868–1938) who was superintendent of the Botany and Forestry Department in the British colony of Hong Kong. The species only occurs in northern and western Fujian Province.

Herbal Usage: Herb is obtained from the Yangzi River and the southern regions of

China. The whole plant used to relieve pain and to stop bleeding. It is used in trauma and to treat intermenstrual bleeding (Wu 1994).

Zhonghua Bencao (2000) states that *L. cucullata* Chien is a synonym for this species (Hu et al. 2000). However, in *Flora of China* (2009), *L. cucullata* is synonymous with *L. pauliana* but it is not an alternative name for *L. dunnii*. The source of the medicinal herb should be the decisive factor as to whether the herbal name *Shuangyejinqiang* refers to one or two species of *Liparis*. On this matter, *L. dunnii* occurs only in Fujian province whereas *L. pauliana* (syn. *L. cucullata*) enjoys a wide distribution that covers Zhejiang, Hunan, Hubei, Jiangxi, Guangdong, Guangxi, Guizhou and Yunnan, but it does not occur in Fujian (Chen et al. 2009c). It was reported that the medicinal herb comes from the Yangzi and the southern areas (Wu 1994). Thus, the medicinal herb probably consists of two rather than a single species.

Liparis fargesii Finet

Chinese name: *Xiaoyangersuan* (small lip sheep ear garlic)
Chinese medicinal name: *Shimi*

Description: Pseudobulb is flat, 7–14 by 3 mm. Leaf is single, elliptic, 1–3 by 0.5–0.8 cm, with a 3–6 mm petiole. Inflorescence is 2–4 cm, with 2 or 3 flowers of pale green, 1 cm in diameter. Flowering season is September to October and May to June. A small, Chinese endemic herb, it often forms dense clusters on rocks or trees. It is found at 300–1700 m in Southern Gansu, Guizhou, Hubei, Hunan, Shaanxi, Sichuan and Yunnan (Chen et al. 2009c). The plant is cold-resistant (Chen 1995).

Herbal Usage: Herb is obtained from Shanxi, Sichuan and Gansu. Entire plant is used in Chinese herbal medicine. It benefits the lungs, reduces heat, stops coughs, and is prescribed for tuberculosis, heat coughs and whooping cough (Wu 1994). To prepare a decoction, 6–9 g of the dried herb is used at a time (Hu et al. 2000).

Liparis japonica (Miq) Maxim (see ***Malaxis monophyllos* var. *monophyllos***)
Liparis keitaoensis Hayata (see ***Liparis cordifolia* Hook. f.**)

Liparis kumokiri F. Maek.

Chinese name: *Xuesancao* (spread snow grass)
Japanese name: *Kumokiri so*

Description: Pseudobulbs are 1–1.2 cm long with two elliptic leaves, 5–12 by 2.5–5 cm. Raceme is 3–7 cm with 5–15 green to purplish flowers, about 1 cm across. Lip is cuneate, abruptly reflexed at mid-point (Ohwi 1965). The medicinal herb comes from Shanxi (Wu 1994). *L. kumokiri* is common terrestrial herb in the Russian Far East, Korea and in Japan between Hokkaido and Kyushu.

A natural hybrid between *L. kumokiri* and *L. makinoana* exists in Korea (Chung, Nason and Chung 2005). On the basis of DNA studies, it has been postulated that the epiphytic *L. fujisanensis* evolved from *L. kumokiri* (Tsutsumi, Yukawa, Lee, et al. 2007). No further information is available, and we were unable to obtain any mention of this species in *Flora of China* (Chen et al. 2009c)

Phytochemistry: *L. kumokiri* produces the pyrrolizidine alkaloid, kumokirine, an ester of kumameric acid and kumokiridine (*N*-methyl laburnine) (Slaytor 1977).

Herbal Usage: Roots and stem are used to strengthen the constitution, improve blood and air flow. It stops persistent intermenstrual bleeding and cures discharge (Wu 1994).

Liparis nakaharae Hayata

Chinese name: *Aochunyangersuan*

Description: Pseudobulb carries two petiolated, oblong-lanceolate leaves, 18–35 by 1.7–2 cm. Petiole is 10 cm long. Inflorescence is 18 cm, laxly flowered, with lanceolate bracts 7–10 mm.

Flowers are 1–2 cm across, dull cream lightly tinged with green. Sepals and petals are linear, the former spreading, the latter reflexed. Lip is ovate. Flowering season is winter, January. *L. nakaharae* is a medium-sized, epiphytic herb, endemic to Taiwan (Lin 1977).

Herbal Usage: It is used in Taiwanese folk medicine to treat cancer (Kuo et al. 2007).

Liparis nervosa (Thunb.) Lindl.

Syn. *Liparis bicallosa* (D. Don) Schltr.

Chinese names: *Honghuayanger Suan* (red flowered goat eating garlic), *Hei Lan* (black orchid), *Shixiagong* (stone shrimp), *Banbian Lan* (half sided orchid) *Roupangxie* (crab meat); *Lidihao* (well grounded); *Maocigu* (hairy kind aunt); *Yanyu* (stone yam); *Tiepashu* (Steel rake); *Daoyanti* (dangling on the cliff); *Zuozicao* (moving seed herb); *Roulongjian* (dragon meat arrow). In Hong Kong: nerved *twayblade*, purple star *Liparis*

Chinese medicinal name: *Jianxueqing*
Japanese name: *Koku ran*

Description: A small terrestrial herb no taller than 12 cm, stem is fleshy, erect, sheathed by two plicate, ovate to ovate-elliptical, pointed leaves, 5–8 cm by 3–4.5 cm. Inflorescence is erect, 15 cm tall, with 3–18 flowers, 1.5–1.8 cm across (Fig. 14.8). Tepals are narrow, spreading. Lip is cuneate-flabellate (Fig. 14.9). It flowers in February to July on the Chinese mainland (Hu et al. 2007; Chen et al. 2009a; Jin et al. 2009) and in June and July in Taiwan (Lin 1975). In Sri Lanka, it flowers in February to March, and again in June to September (Jayaweera 1981). Flowering season is July in Nilgiris, Tamil Nadu (Joseph 1982), August in Mumbai (Santapau and Kapadia 1966) and August to September in the Western Ghats (Abraham and Vatsala 1981).

L. nervosa is an extremely variable, pan-tropical species which enjoys a wide distribution throughout the humid tropical and sub-tropical regions of the Old and New World (Garay and Sweet 1974). It is an endangered species in Florida where it sometimes goes under the name of *L. elata*; the Floridan variety has inflorescences that may reach 60 cm in height (McCartney 2000). It is found at 1000–1500 m in Taiwan, and is related to or identical with *L. formosana* and *L. odorata*.

Some taxonomists maintain that *L. nervosa* and *L. bicallosa* (*Eulophia bicallosa*) are the same species (Chen et al. 2009c). However, chemical fingerprinting shows a difference between the two. The alkaloid in *L. nervosa* is nervosine, whereas the alkaloid in *Eulophia bicallosa* is malaxin. The chemical structures of the two alkaloids are different (Luning 1980).

Phytochemistry: Nervosin was isolated by Nishikawa and Hirata in 1967 from *L. nervosa*. It is an ester of lindelofidine and nervosinic acid, the latter an arabinosyl-glucosyl derivative of nervogenic acid which is a starting point for alkaloids of two other *L.* species, *L. kurameri* and *L. kumokiri* (Slaytor 1977). All alkaloids of *L. nervosa* inhibited each of the 12 bacterial and 4 fungal species tested with a paper diffusion method. Anti-oxidant effect ranged from 5 to 93.5 % at concentrations of 0.5 to 100 mcg/ml (Dong et al. 2010). Recently, six new pyrrolizidine alkaloids and two previously identified alkaloids were isolated from the whole herb of *L. nervosa*. They had no cytotoxic effect against A549, HepG2 and MCF-7 human cancer cell lines; however, they showed anti-oxidant activity on RAW264.7 macrophages (Huang et al. 2013). The scientists also isolated 10 new nervogenic acid derivatives from *L. nervosa* (Huang et al. 2013).

Liu et al. (2009) from the College of Life Sciences at Sichuan University demonstrated that *L. nervosa* lectin, a monocot mannose-binding lectin, exhibited haem-agglutinating activity and induced apoptosis in MCF-7 cells through a caspase-dependent pathway. *L. nervosa* lectin (LNL) was found to have a similar tertiary structure and three mannose-binding sites in common with *Polygonatum cyrtonema* lectin (PCL) and *Ophiopgon japonicus* lectin (OJL), and to exhibit similar inhibitory activity on the proliferation of

Fig. 14.8 *Liparis nervosa* (Thunb.) LIndl. (as *Katou ponnam maravara*) from van Rheede tot Drakestein, H. A., *Hortus Indicus Malabaricus*, vol. 12: t. 28 (1703-). Courtesy of Missouri Botanical Gardens, St. Louis, U.S.A

MCF-7 cells. There appears to be a close phylogenetic relationship among the three lectin producing orchid species. *L. nervosa* also produces a nervogenic acid glycoside which promotes blood clotting in vitro (Song et al. 2013).

Herbal Usage: The whole plant is used as an antipyretic, to cool blood, stop bleeding and reduce heat in the lungs. It is used for cramps in children, haemetemesis, coughs, and rheumatic pain. It is also used as an emollient for traumatic injuries, skin infection and snake bites (Wu 1994; Hu et al. 2000; Ou et al. 2003). It is reputed to reduce inflammation, dissolve extravasated blood and cause swellings to subside (Chen and

Fig. 14.9 *Liparis nervosa* (Thunb.) Lindl. [PHOTO: Bhaktar B. Raskoti]

Tang 1982). The herb is collected throughout the year. For decoction, 3–6 g of the fresh herb or 6–12 g of dried herb are used. For external use, the whole plant is pounded and soaked in wine (*Zhongyao Da Cidian* 1986).

Tubers are used to treat stomach disorders and a paste is applied on chronic ulcers in Nepal (Baral and Kurmi 2006).

Syn. *Liparis plicata* Franch & Savat. (see ***Liparis bootanensis* Griff.**)

Liparis odorata (Willd.) Lindl.

Syn. *Liparis paradoxa* (Lindl.) Rchb.f.

Chinese name: *Xianghuayangersuan*
Chinese medicinal name: *Erxiantao*

Description: *L. odorata* is a terrestrial herb, 30 cm tall, with sub-ovoid pseudobulbs, 1.3–2.2 by 1–1.5 cm, enclosed in membraneous sheaths. Leaves are 2–3, ovate to elliptic-lanceolate, 6–17 by 1.5–6 cm, membranous or herbaceous, conspicuously veined, petiole sheath-like. Inflorescence is 14–40 cm, erect, laxly several-flowered.

Fig. 14.10 *Liparis odorata* (Willd.) Lindl. [PHOTO: Bhaktar B. Raskoti]

Flowers are greenish-yellow to greenish-purple, or green with longitudinal purple band on either side of the lip, 0.6–1 cm across. Dorsal sepal and petals are linear, petals purplish. Lateral sepals are ovate-oblong and slightly oblique (Fig. 14.10). In India, it flowers from July to September (Bose and Bhattacharjee 1980), in Bhutan, from May to September (Gurong 2006), in Nepal, July (Raskoti 2009) and in China, April to July or August (Chen et al. 2009c; Jin et al. 2009).

Following the advice of Sathish Kumar and Manilal (1994), I have kept this species distinct from *L. nervosa* because the latter species does not occur in India. Nevertheless, some taxonomists consider this species to be identical with *L. nervosa* (Thunb.) Lindl. In the illustrations in *The Native Orchids of Japan*, *L. odorata* appears to prefer a more exposed habitat; its raceme is longer with numerous flowers whereas *L. nervosa* is few-flowered. Tepals are a light purple. Lip is a dull yellow. The species is distributed in forests and grassy slopes at

600–3100 m in Taiwan, Fujian, Guangdong, Hong Kong, Guangxi, Guizhou, Jiangxi, Hubei, Hunan, Zhejiang, Yunnan and Sichuan, and in India, Nepal, Bhutan, Myanmar, Thailand, Vietnam, Japan and the Pacific Islands. It occurs in *Pinus roxburghii* and broad-leaved forests in central and eastern Bhutan (Gurong 2006).

Herbal Usage: In his monumental *Hortus Indicus Malabaricus,* van Rheede (1703), the first European to record the medicinal uses of plants in India, reported that *L. paradoxa* was used to treat elephantiasis, but the fact went unnoticed by English-speaking orchidists because his work was rendered in Latin and comprised 12 volumes each of 500 pages! Subsequently, Rao and Sridhar (2007) mentioned that *L. odorata* is used in Malabar to treat elephantiasis. In this disease, the lower limbs of affected individuals take on a permanent resemblance to the legs of elephants because of obstruction to lymphatic flow by filarial worms transmitted through mosquito bites. Tribals in Karnataka offer the orchid pseudobulbs for sale as tribal medicine for this condition (Rao 2007). Juice extracted from the leaves was used to treat fever and oedema. Juice from the roots was used to treat burns, inflammation, gangrene and tumours (Dalgado 1896, 1898 both quoted by Lawler 1984), a practice still continuing in north-western India (Medhi and Chakrabarti 2009).

L. odorata appears to be a new addition to the Chinese pharmacopoeia as its usage is described only in *Zhonghua Bencao* (Hu et al. 2000). It is used to treat flu-like symptoms, peripheral neuritis, leucorrhoea, discomfort at the waist, ulcers and swellings. The herb removes "wind" and dispels "dampness". Decoction is prepared by boiling 6–15 g of sliced, dried pseudobulbs (Hu et al. 2000).

Liparis petiolata (D. Don) P.F. Hunt & Summerh.

Chinese name: *Bingyeyangersuan* (handle leaf sheep ear garlic)

Description: This is a terrestrial herb with ovoid pseudobulbs, 1.5–3 by 1–1.5 cm in diameter, spaced 2–4 cm apart on the rhizome. Pseudobulbs are covered with membranous sheaths and bear two ovate leaves, 5–11 by 3.5–8 cm. Inflorescence is 10–24 cm tall with 10 or more greenish-white flowers. Lips are purplish-green (Figs. 14.11 and 14.12). It flowers in May and June. The species is found near streams at 1000–2900 m in Hunan, Jiangxi, Guangxi, southern Yunnan and southeast Xizang; also in Vietnam, Thailand, Bhutan, Nepal and northeast India.

Herbal Usage: It is reported to benefit the lungs (Wu 1994).

Liparis plicata Franch et Savat. (see ***Liparis bootanensis* Griff.**)

Liparis rheedii Lindl.

Indian Name: *Simil*

Description: *L. rheedii* is a terrestrial herb with pseudobulbs close to one another on a 10- to 16-cm-long rhizome, each bearing 3–5 ovate, acute, plicate, green leaves, 15–25 cm by 4–12 cm, with ensheathing petioles of 3 cm length and wavy margins. A non-flowering plant resembles *Malaxis latifolia*. Inflorescence is 20–45 cm tall, erect, bearing numerous flowers that open gradually from the base. Flowers are green, red, or bicolored, usually changing from green to red as they age. They are 1 cm across. Petals and sepals are filiform. Lip is dark purple, reniform with a central groove, recurved at its mid-point, and carries a dentate border (Fig. 14.13).

The colour of the plant and its flowers is dependent on the amount of light that it receives. Plants growing in the shade are deep purple whereas those growing in the light are a pure green. Likewise on an inflorescence leaning towards light on the roadside, whereas the lowermost flowers still in shade are purple, the middle

Fig. 14.11 *Liparis petiolata* (D. Don) P. F. Hunt & Summerh. (as *Liparis pulchella* Hook. f.). From: Hooker, W. J., Hooker, J. D., *Icones Plantarum*, vol. 19: t. 1810 (1889). Drawing by M. Smith. Courtesy of Missouri Botanical Gardens, St. Louis, U.S.A

Fig. 14.12 *Liparis petiolata* (D. Don) P. F. Hunt & Summerh. [PHOTO: Bhaktar B. Raskoti]

forest floor in humus, in deep shade. It flowers in July and August at Karnataka and the Western Ghats (Santapau and Kapadia 1966), July at Nilgris (Joseph 1982), and from August to September in Tamil Nadu (Matthew 1995; Seidenfaden 1999).

Herbal Usage: It is used as a tonic in Karnataka (Rao 2004, 2007). It is one of the eight ingredients of the Ayurvedic drug known as *Ashtavarga*, another being the orchid, *Microtis muscifera* Ridl. (Rao 2004). The root is used by the hill tribes in Orissa to treat cholera. About 250 g of root is decocted in a litre of water until the volume is reduced to 333 ml. After cooling, 5 ml of the decoction is mixed with 2 ml of honey and orally administered twice a day on an empty stomach for 15–21 days as a remedy for cholera (Dash et al. 2008).

Liparis rostrata Rchb. f.

Chinese name: *Chitu yangersuan*

Description: Pseudobulb is small, ovoid and produces a stem from its base. Stem is 5–20 cm tall, with two opposite, elliptic-ovate, petioled leaves, 7–10 by 3–5 cm (Fig. 14.14). Flowers are tiny with linear yellowish-green sepals, dark purple, filiform petals and a green lip which is channelled (www.eFloras.org Flora of Pakistan; Chen et al. 2009c; Jin et al. 2009). The species is distributed from northern Pakistan to western Nepal, southern Xizang and at Gaoligongshan in Yunnan, on soil-covered rocks along valleys at 2000–2700 m (Jin et al. 2009).

Usage: Tubers of *L. rostrata* are used to treat stomach disorders in India (Duggal 1972) and as a tonic (Bhattacharjee 1998).

Fig. 14.13 *Liparis rheedii* Lindl. [PHOTO: Peter O'Byrne]

ones are yellow, and the uppermost flowers receiving full sunlight are green. On plants growing in deep shade on the forest floor, all flowers are of a deep purple (Santapau and Kapadia 1966).

The species occurs in mountain forests from southern India, across Thailand, Indochina, Malaysia and Indonesia to New Guinea at 450–1500 m (Comber 2001) growing at the

Liparis sootenzanensis Fukuyama

Syn. *Liparis nigra* Seidenf. var. *sootenzanensis* Fukuyama

Chinese name: *Chatianshanyangersuan; Zhihuayangersuan* (purple flower sheep ear garlic). In Hong Kong: giant purple *Liparis*

Fig. 14.14 *Liparis rostrata* Rchb. f. From: Hooker, W. J., Hooker, J. D., Icones Plantarum, vol. 19: t. 1813 (1889). Drawing by M. Smith. Courtesy of Missouri Botanical Gardens. U.S.A

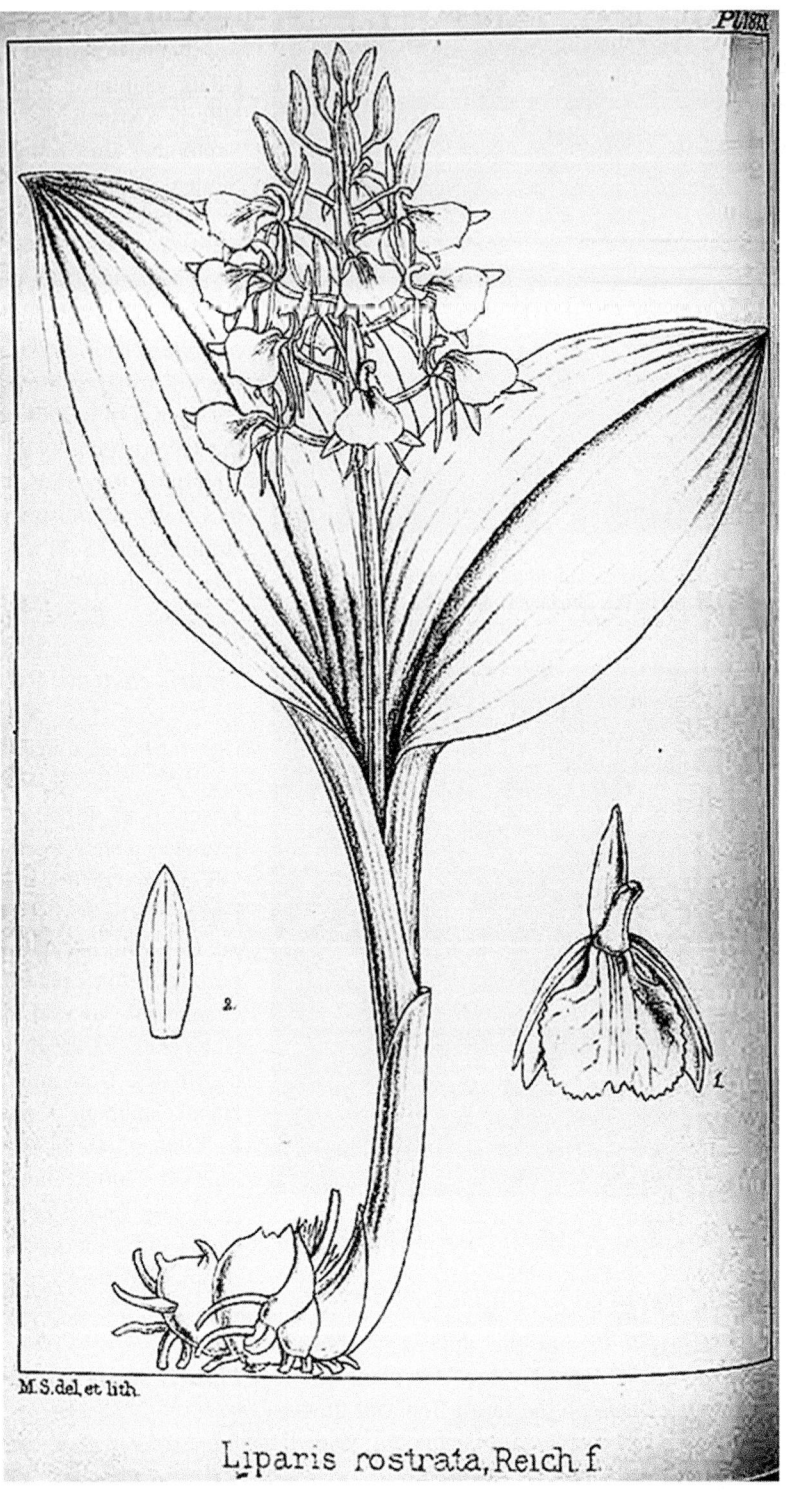

Liparis rostrata, Reich. f.

Description: This robust terrestrial herb is distributed from Yunnan to Guangxi, Guangdong, Hainan, Taiwan, Hong Kong and Lantau Islands on sheltered slopes in open forests from lowlands to 1500 m. Stem is cylindrical, 15–25 cm tall, 2–3 cm in diameter, with several nodes and completely ensheathed. Leaves are several, large, 15–25 by 6–12 cm, plicate, broadly elliptic. Inflorescence is stout, 20–30 cm, ridged, laxly many-flowered. Flowers are green to greenish-yellow, about 2 cm across. Plants in Taiwan are smaller than those on the mainland. It flowers from April to May in Taiwan (Lin 1977), February to May in Hongkong (Wu, Hu Xia et al. 2002) and April to May on mainland China (Chen et al. 2009c).

Herbal Usage: The whole herb is used to treat arthritis, numbness and skin disease in Vietnam (Hung 2014).

Liparis stricklandiana Rchb. f.

Chinese names: *Shanchunyangersuan* (fan lip sheep ear garlic), *Luhuayangersuan* (green flower sheep ear garlic)

Description: Plant is 21–32 cm tall; pseudobulbs ovoid, cylindrical and compressed. Inflorescence is densely many-flowered, the flowers 4 mm long and yellowish (Fig. 14.15). Flowers appear from September to December (Pearce and Cribb 2002; Jin et al. 2009). *L. stricklandiana* is an epiphytic species distributed from Hong Kong, Guangdong, Guangxi, Yunnan and Tibet in China to Vietnam, Sikkim, Bhutan and Nepal, at 1350–1800 m, in forests on hill slopes.

(pix Lin Gaoligonshan p. 320)

Herbal Usage: Herb is obtained from Guangdong, Guangxi and Yunnan. The whole plant used to treat sores, abscesses and ulcers (Wu 1994).

Liparis treubii J.J. Smith (see **Liparis condybulbon Rchb. f.**)

Fig. 14.15 *Liparis stricklandiana* Rchb. [PHOTO: Bhaktar B. Raskoti]

Liparis tschangii Schltr.

Chinese name: *Xizang Yangersuan* (Tibetan sheep ear garlic), *Zhebao Yangersuan*

Description: *L. tschangii* is a terrestrial herb with small, ovoid pseudobulbs 1–2 by 0.7–1.3 cm, nodded and enclosed by white, membranous sheaths. The plant carries two ovate leaves that spread horizontally and measure 5–13 by 2.5–7.3 cm. Inflorescence is 11–29 cm, and many-flowered. Flowers are green, tepals filiform, lip broad, ovate to broadly elliptic with a vertical band of dark colour at the base. It flowers from July to August (Chen et al. 2009c). *L. tschangii* is found in Tibet, south-west Sichuan, Yunnan, Laos, Thailand and Vietnam at 1100–1700 m.

Herbal Usage: Herb is obtained from Tibet and Yunnan. Stems are used in Chinese herbal

Fig. 14.16 *Liparis viridiflora* (Blume) Lindl. From: *Annals of the Royal Botanic Gardens, Calcutta*, vol. 8 (2): t. 47 (1891). Drawing by R. Pantling. Courtesy of Missouri Botanical Gardens, St. Louis, U.S.A

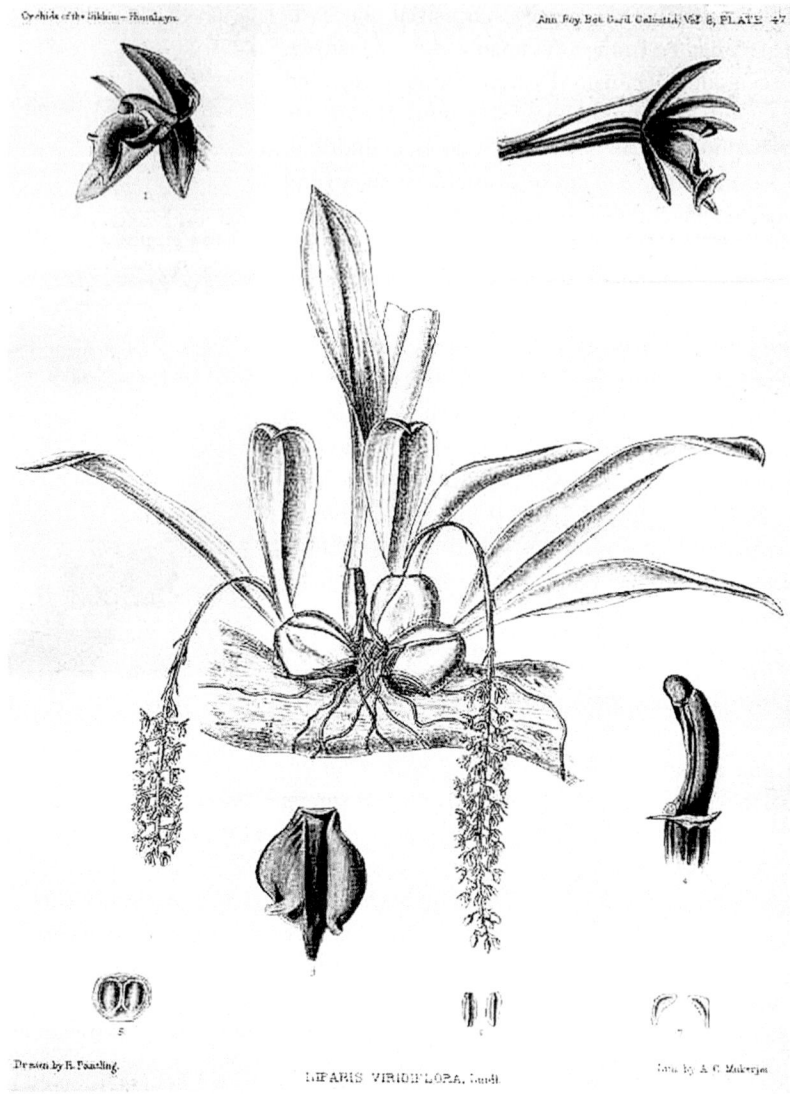

medicine to improve blood flow and stop bleeding, especially persistent vaginal bleeding (Wu 1994).

Liparis viridiflora (Blume) Lindl.

Chinese names: *Luhuayanger Lan* (green flower sheep ear orchid), *Changjing-yangersuan* (long stem sheep ear garlic)

Description: Plants are lithophytic or epiphytic, with cylindrical pseudobulbs, 3–9 cm long (in China, it may reach 18 cm), 1 cm in diameter. Leaves are linear, 8–25 cm long (Fig. 14.16). Inflorescence is arching, densely many-flowered. Flowers are small, non-resupinate, greenish-white, pale green or yellowish, with an orange-coloured lip. Tepals are 3 cm long and reflexed (Fig. 14.17). It flowers in September to December in China (Chen et al. 1999; Jin et al. 2009), November in Assam and Sikkim (Bose and Bhattacharjee 1980) and August to September in the Western Ghats Abraham and Vatsala 1981). Flowering period is also September in Thailand (Vaddhanaphuti 2001).

Fig. 14.17 *Liparis viridiflora* (Blume) Lindl. [PHOTO: E. S. Teoh]

L. viridiflora is an extremely widespread species distributed from Sri Lanka and the tropical Himalayas across southern China and Southeast Asia to the Pacific Islands, occurring at 200–2300 m in forests on hill slopes. It is not uncommon to find *L. viridiflora* growing together with other orchids and ferns forming large colonies that encircle tree trunks and branches (Yang et al. 1998).

Herbal Usage: Herb is obtained from Taiwan, Guangdong, Guangxi and Yunnan. Stems are used to treat coughs, poisoning, fever and fractures. Roots of *Juhuayangersuan* (gathering flower sheep ear garlic) identified by Wu (1994) as *L. pleistantha* Schltr. (now considered to be synonymous with *L. viridiflora*) are used to treat hernia (Wu 1994).

Liparis yunnanensis Rolfe (see **Liparis distans C.B. Clarke**)

Overview

Liparis is a genus with numerous alkaloid-containing species. When Boorsma tested the orchids at the Buitenzorg Botanic Gardens in Java for the presence of alkaloids, *L. parviflora* showed a positive alkaloid reaction. His paper, published in the *Bulletin of the Botanical Institute of Buitenzorg* in the Dutch East Indies, was among the first to demonstrate the presence of alkaloids in orchids (Boorsma 1902). Thirty-five

years passed before another species, *L. kurameri*, native to Japan, was reported to have tested positive for alkaloids. The discovery was announced in a Japanese journal by Eguchi and Wakasugi, so again the information failed to spark international interest until kumokirine and kuromerine were identified in 1967 and their structures elucidated (Nishikawa and Hirata 1967a; Nishikawa et al. 1967b; Nishikawa and Hirata 1968). After reporting on the medicinal usage of *L. treubii* (proper name: *L. condybulbon* Rchb. f.) in the Celebes, as described by Rumphius, Burkill (1935) commented that nothing of any economic interest is on record regarding the 17 Malaysian species.

Alkaloids were rediscovered in *Liparis* and *Malaxis* when Japanese and the Swedish teams of scientists independently commenced their search for alkaloids in orchids during the 1960s. Out of 67 species tested, 28 were found to have an alkaloid content equal to or exceeding 0.1 % of the plant's weight. Lindstrom and Luning found high alkaloid contents in the cool-growing *L. loeselli* (L.) L.C. Rich. and *L. keitaonsis* Hay. (Lindstrom and Luning 1971, 1972), but they did not immediately extend their work on the species because the plants were rare and they probably could not obtain enough material for detailed chemical studies at that time.

The Japanese team led by Nishikawa identified nervosine from *L. nervosa,* kuramerine from *L. krameri,* kumokirine from *L. kumokiri,* malaxin from *L. hacijoensis,* and auriculine from *L. auriculata* (Nishikawa and Hirata 1967a; Nishikawa and Hirata 1968).

Liparis and *Malaxis* are related genera, and one is often confused with the other. It turns out that not only do the two genera share similar physical characteristics but their alkaloids are also related. *Liparis* species invariably contain alkaloids which are glycosides of *p*-hydroxy-benzoates of pyrrolizidine carbinols (Luning 1980). Out of 67 species of *Liparis* tested for alkaloids, 28 (42 %) showed an alkaloid content exceeding 0.1 % (Luning 1974). Numerous *Liparis* species native to Japan (*L. nervosa* Lindl., *L. krameri* Fr. et Sav., *L. kumokiri* F. Maekawa, *L. bicallosa* Schltr.,

L. hachijonensis Kitamura, *L. makinoana* Schltr. and *L. japonica* Maxim.) each contain one principal alkaloid and sometimes one or two other alkaloids in minute amounts. Rf values and chemical properties of these alkaloids are similar. They are made up of three basic components: (1) aminoalcohol, (2) carboxylic acid and (3) sugar (Nishikawa et al. 1967b).

Malaxine isolated by Luning's team from *Malaxis congesta* has the skeletal form of nervosine. Grandiflorine from the New Guinean *Malaxis grandiflora* is also very similar to nervosine. These pyrrolizidine alkaloids are esters of *p*-hydroxybenzoic acid. By 1980, ten such alkaloids had been identified (Luning 1980), including malaxin from *L. bicallosa* Schltr. and *L. hachijoensis* Nakai, auriculine from *L. auriculata* and *L. loeselii* L.C. Rich, and keitaoine and keitine from *L. keitaoensis* Hay., earlier described by Luning (1974). The structures of some of these alkaloids are shown in Table 14.1 (Luning 1980, p. 202). In Fig. 14.18, Synthetic malaxine (dehydroartemisin) is used to treat uncomplicated falciparium malaria that is resistant to chloroquine.

Six pyrrolizidine alkaloids newly isolated from *L. nervosa* evaluated for cytotoxicity against several human cancer cell lines (A549, HepG2 and MCF-7) failed to show any effect, but they exhibited anti-oxidant activity on RAW264.7 macrophages (Huang et al. 2013).

Table 14.1 Chinese herbal prescriptions using *Luisia morsei* (*Zhongyao Da Cidian* 1986)

1. Indication: detoxification Wash one or two handfuls of leaves of *Chaizigu* (*L. morsei*) and squeeze the juice out to drink. Poisons can be eliminated by vomiting*
2. Indication: carbuncle Wash a handful of the leaves of *Chaizigu*, grind and apply. Change twice a day.
3. Indication: edema Cook *Chaizigu* 24–36 g with pig knuckle for 2 h. Eat before meal once a day.
4. Indication: Syphilis Boil fresh Chaizigu roots 60 g. Take before meal, twice a day.
5. Indication: Sore throat Use a gargle made with juice extracted from the whole plant, 30 g.

A new phenanthrene, 7-hydroxy-2,3,4-trimethoxyphenanthrene, and three new prenylated benzoic acids named liparacids A, B and C, together with muscatin, batatasin III and 2,5-dihydroxy-4-methoxy-9,10-dihydrophenanthrene were isolated from the rhizome of *L. nakaharai* which has been used in Taiwanese folk medicine for the treatment of cancer (Kuo et al. 2007). The new phenanthrene exhibited anti-inflammatory activity. Nudol and batatasin III have elsewhere been shown to be cytotoxic against several tumour cell lines. Unfortunately, the low yield of the liparacid acids precluded their testing for biological activity.

Three new bioactive phenolic glycosides were isolated from whole plant of *L. odorata* (willd.) Lindl., named liparisglycosides A–C, liparis glycoside A and C exhibited lipid-lowering effects in vitro (:I, Li et al. 2014). Three nervogenic acid glycosides (condobulbosides A–C) and an apigenin C-glycoside (schaftoside) are present in *L. condylobulbon* (Rchb. f.) (Slapetova et al. 2009). *L. nervosa* has yielded one nervogenic acid glycoside which promotes blood clotting in vitro. This compound is 3,5-bis (3-methyl-but-2-enyl)-4-*O*[beta-D-xylopyranosyl-(1->2)-beta-D-glucopyranosyl]benzoic acid (Song et al. 2013).

Rhizomes of *Liparis* species remain infected with mycorrhiza throughout their life-span (Rasmussen 1995), and since *L. stricklandiana* is used to treat abscesses, sores and ulcers, it might be worthwhile to investigate this species and other *Liparis* for antimicrobial properties. The recent discovery of apoptosis-inducing properties in *L. nervosa* (Liu et al. 2009) is interesting and should stimulate more work in this direction.

Six *Liparis* species (*L. caespitosa, L. dunii, L. japonica, L. kumokiri, L. nervosa* and *L. tschangii*) are used to stop bleeding. In four species (*L. dunii, L. japonica, L. kumokiri* and *L. tschangii*), the emphasis is that they stop persistent vaginal bleeding, in particular prolonged intermenstrual bleeding. There are a number of compounds that might achieve this effect; for instance, a sex steroid or an alkaloid with an action similar to that of ergometrine recovered from the fungus, *Claviceps purpurea*.

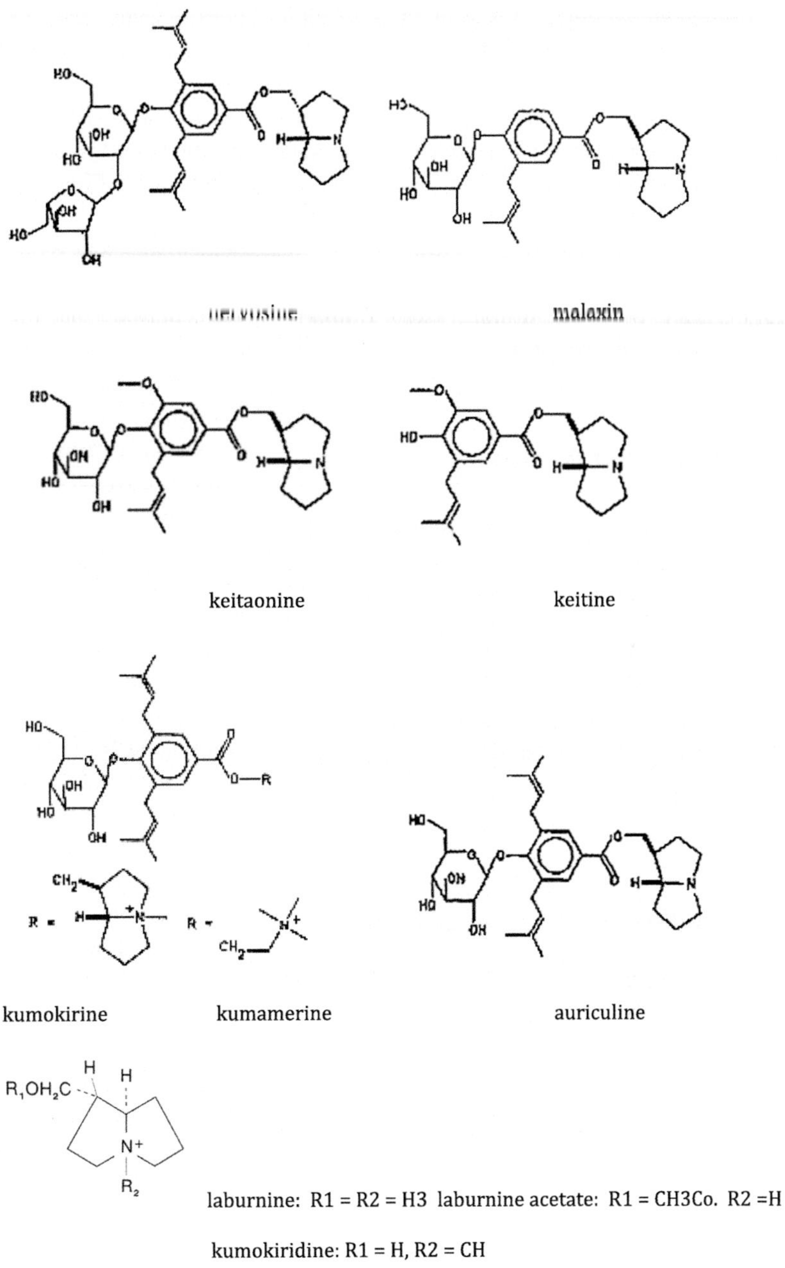

Fig. 14.18 Some alkaloids from *Liparis* species

nervosine

malaxin

keitaonine

keitine

kumokirine

kumamerine

auriculine

laburnine: R1 = R2 = H3 laburnine acetate: R1 = CH3Co. R2 =H

kumokiridine: R1 = H, R2 = CH

We are not aware of any work to investigate such compounds in *Liparis*.

Pyrrolizidine alkaloids are known to cause fatality in grazing cattle. Although there is concern that a single consumption of such alkaloids would eventually lead to liver cirrhosis and cancer, it was stated that pyrrolizidine alkaloids from orchids do not cause such problems (International Programme for Chemical Safety 1988).

The floral scent of *L. viridifolia* contains two isomeric compounds which are also insect pheromones, (E) and (Z) isomers of 7-methyl-1,6-dioxaspirol[4,5]decane which had been previously identified in the pentane extract of female

Paravespula vulgaris, the common Yellow Jacket, a wasp which feeds on aphids, flies, caterpillars and nectar (Kaiser 1993). In tropical South America, *Epidendrum paniculatum* Ruiz & Pav. is pollinated by butterflies that seek pyrrolizidine alkaloids which are necessary for the biosynthesis of male pheromones and mating success (De Vries and Stiles 1990).

Lissochilus arabicus Lindl. (see ***Eulophia streptopetala var. streptopetala***)

Genus: *Ludisia* A. Rich.

Chinese name: Xueye Lan (blood leaf orchid)

This mono-specific genus in the Jewel Orchids inhabits shady, moist places in forest valleys, often near streams at 200–1000 m. It is distributed from southern China through Indochina, Thailand and Malaysia to Indonesia. The brilliant leaves of the herb give the plant a jewelled appearance. Flowers are white. The origin of its generic name has not been explained.

Ludisia discolor (Ker.-Gawl.) A. Rich.

Chinese names: *Xueye Lan* (blood leaf orchid), *Yisexueye Lan* (special colour blood leaf orchid), *Shishangou* (lotus on the rock): In Hong Kong: rock silk-worm, twisty-flowered orchid
Chinese medicinal names: *Shishangou; Xueye Lan; Shicha*n (rock moth) *Zhenjincao* (genuine golden grass)
Thai name: *Wan nam thong*
Malay names: *Beledu merah* (red velvet plant), *Baldu merah* (red plant) (only applicable to some varieties)

Description: Plant has a creeping rhizome and the erect portion may be 10–20 cm tall, erect, carrying a whorl of 3–4 leaves which are elliptic-lanceolate, 3–7 by 2–3 cm, red above, usually marked with golden veins, and purplish beneath. Colour and shape of the leaves are variable in the species. Inflorescence is terminal, with brown lanceolate bracts and 6–10 unusual, white flowers. Lip is yellow (Figs. 14.19, 14.20, and 14.21). It flowers in July on the Chinese mainland, but in Hong Kong it flowers in February to April (Bechtel et al. 1980; Seidenfaden and Wood 1992; Chen et al. 1999). It flowers in March in Singapore.

L. discolor occurs in southernmost Chinese provinces of Fujian, Guangdong, Hong Kong, Guangxi, Yunnan and Hainan, extending to Myanmar, Thailand, Vietnam, Malaysia and Indonesia. It grows on the ground or on rocky surfaces covered by leaf mould in ravines.

Herbal Usage: Herb is obtained from Fujian, Guangdong, Guangxi and Yunnan (Wu 1994). The whole plant is used in traditional Chinese medicine and it may be collected throughout the year. It is used fresh or sun-dried. *L. discolor* is characterised as sweet, slightly astringent, and 'cool'. It nourishes the lungs, regulates body fluids, purifies the blood, and is anti-inflammatory. In Hong Kong, it is used to treat haemoptysis caused by pulmonary tuberculosis, neurasthenia, anorexia, employing 3–10 g in decoction (Li 1988). It relieves coughs (Chen and Tang 1982).

On the Chinese mainland, Wu (1994) reported that the entire plant is used in traditional Chinese herbal medicine to enrich *yin*; it benefits the lungs, cools the blood and stops bleeding. It is used to treat pulmonary tuberculosis, anorexia and neurosis. For decoction, 3–9 g of dried herb is used; the amount of the fresh herb is 9–15 g (Hu et al. 2000). Sometimes, the fresh herb is washed clean and chewed. Alternatively, juice is extracted from it to be consumed (*Zhongyao Da Cidian* 1986).

Rhizomes of *Wan num thong* (*L. discolor*) are listed as 1 of 57 herbal remedies for insect bites compiled from old Thai remedies by Daduang and Uawonggul (2008). *Grammatophyllum speciosum* is the other orchid featured in the list, and with the latter species, the roots are used.

Fig. 14.19 *Ludisia discolor* (Ker Gawl.) A. Rich. [as *Haemaria dawsoniana* (H. Low ex Rchb. f.) Hook.f.]. From: *Curtis Botanical Magazine*, vol. 122 [ser.3, vol.52]: t.7486 (1896) Coloured drawing by M. Smith. Courtesy of Missouri Botanical Gardens, St. Louis, U.S.A

Overview

L. discolor and *Anoectochilus* species look much alike when the plants are not in bloom. *L. discolor* is commonly offered as a fresh herb in herbal shops in Malaysia and Singapore. *L.a discolor* is not native to Taiwan and therefore it does not enjoy the same reputation as the King Medicine (*Anoectochilus*) of Taiwan.

There is no pharmacological information on *Ludisia*.

Fig. 14.20 *Ludisia discolor* (Ker.-Gawl.) A. Rich.
[PHOTO: E. S. Teoh]

Fig. 14.21 *Ludisia discolor* (Ker.-Gawl.) A. Rich.
[PHOTO: E. S. Teoh]

Genus: *Luisia* Gaudich.

Chinese name: *Chaizigu Lan* (hairpin strand
   orchid)

Luisia is a genus which is distributed from India
to southern China, Southeast Asia and Japan. It

was named for the nineteenth century Portuguese
botanist, Don Luis de Torres. With colourful lip
and widespread petals, its curious flowers resemble the Kathakali dancers of Kerala. *Luisia* is an
distinctive, unmistakable genus with terete leaves,
short inflorescence, and tight clusters of colourful
flowers (Fig. 14.22). A tough epiphyte, it is easily
cultivated. Nevertheless, it is not favoured by
growers because its flowers are small and not
showy. There are some 40 species in the genus
(Seidenfaden and Wood 1992), and some hybrids
have now been made by crossing with *Vanda* with
which it is inter-fertile. India has 14 species.

Luisia birchea Blume (see **Luisia tenuifolia
Blume**)

Luisia brachystachys (Lindl.) Bl.

Syn. *Luisia indivisa* King & Pantl.

Chinese name: *Xiaohuachaizigu*

Description: *L. brachystachys* is a tough, robust
epiphyte with terete stems and leaves. Inflores-
cence is short, usually with 2–3 (up to 6) small
flowers 8 mm across. Sepals and petals are light
green. Lip is crimson. It flowers in April in
Thailand and China (Vaddhanaphuti 2005;
Chen and Wood 2009), and September to
October in the Indian Himalayas (Bose and
Bhattacharjee 1980). It is distributed in southern
Yunnan, Thailand, Bhutan, India, Myanmar and
Vietnam, occurring at 600–1300 m.

Phytochemistry: Seven compounds were discovered in *L. brachystachys* by Majumder and
Lahiri (1989), namely, beta-sitosterol, betulinic
acid, *p*-hydroxybenzaldehyde, *p*-hydroxyphenyl-
propionic acid 3,4 = dihydroxy-3,4′-dimethoxy-
dibenzyl, together with two perfumery
constituents, methyl 2,4-dihydroxy-6-
methylbenzoate and methyl 2,4-dihydroxy-3,6-
dimethylbenzoate. Boorsma (1902) found traces
if alkaloids in *L. brachystachys*.

Herbal Usage: *L. brachystachys* is used as
a nutrient and emollient for rheumatic pain. Plant
is pounded to make a poultice for the treatment of
boils, abscesses and tumours (Trivedi et al. 1980).

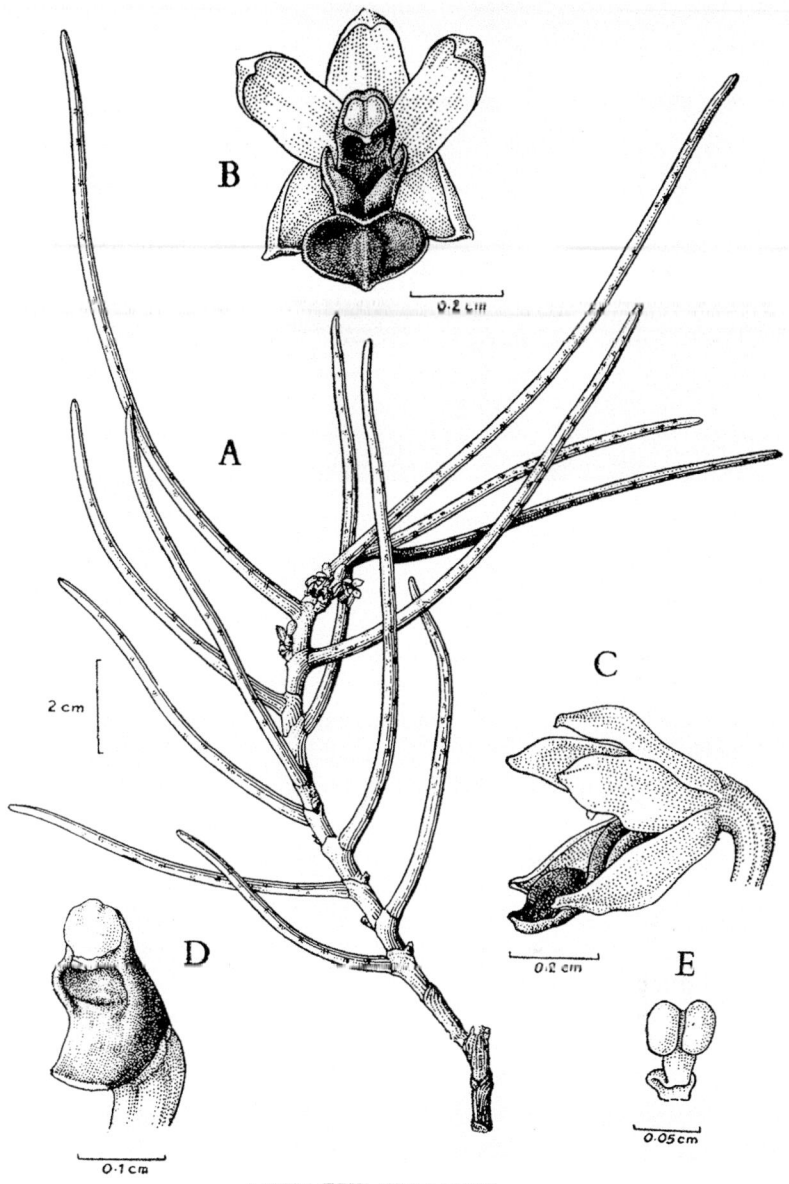

LUISIA ZEYLANICA LINDL.

Chopra (1933) and Nadkarni (1954) mentioned that there is an alkaloid in the species but neither author provided any details.

Luisia curtisii **Seidenf.**

Thai name: *Kho sing*

Description: Internodes are 1.5–2.5 cm and the terete leaves are 18 cm long. Inflorescence is 1.5–2 cm long, and carries several flowers, but only one or two are developed at any one time. Tepals are narrow, flared, a dirty yellow flushed with purple. The purple lip is divided into two parts: the proximal half is 1 cm wide and oblong, and separated from the distal half which is expanded and heart-shaped. Side lobes of the lip are tiny and erect (Fig. 14.23). *L. curtisii* is distributed in Thailand, Cambodia, Vietnam, Peninsular Malaysia, Borneo and the Philippines (Seidenfaden and Wood 1992). Plants in

Fig. 14.23 *Luisia curtisii* Seidenf. [PHOTO: Peter O'Byrne]

Peninsular Malaysia have larger flowers than those in East Malaysia. In Borneo, it occurs in lowland *dipterocarp* and lower montane forests on granite, sandstone and ultramafic substrates at 900–1500 m and it flowers in May and December (Chan et al. 1994).

Herbal Usage: The whole plant is considered to be diuretic in Thailand (Chuakul 2002).

Luisia hancockii Rolfe

Chinese name: *Xianyechaizigu, Qianyechaizigu* (slim leaf hairpin)

Description: Stems are rigid, terete, up to 20 cm long and 4 mm in diameter, with fleshy, terete leaves 5–9 cm long. Inflorescence is 1–1.5 cm arising 180° from the leaf axils, directly opposite the leaf and bears up to three flowers which are fleshy, 1.5 cm across, with narrow green sepals and petals and a broad, oblong lip slightly tapering distally to an undulate margin. Lip is splashed proximally with a broad patch of purplish-red. Flowering season is May to July (Chen and Wood 2009).

L. hancockii is an endemic, terete, monopodial orchid which occurs as an epiphyte on trees in sparse woods or as a lithophyte on rocks and cliffs along valleys from sea level to 1300 m (Chen et al. 1999), but commonly at 200–300 m (Chen and Wood 2009), in Zhejiang, Hubei and Fujian Provinces in China.

Herbal Usage: Entire plant is used to clear gas and phlegm, remove toxins and reduce swelling. The herb comes from Zhejiang Province (Wu 1994).

Luisia indivisa King & Pantl. [see ***Luisia brachystachys* (Lindl.) Bl.**]

Luisia morsei Rolfe

Chinese names: *Dayangjiao* (big goat horn), *Chaizigu* (hairpin strand), *Jinhuancao* (gold ring grass), *Shucong* (tree onion).

Chinese medicinal names: *Chaizigu* (hairpin strand); *Jinchagu* (gold hairpin section); *Sanshi Gen* (30 root); *Songjisheng* (pine parasite); *Chongjisheng* (worm parasite) *Haibanhu* (sea strip tiger) *Tanxiangxian* (sandlewood thread) *Longxucao* (dragon beard grass)

Description: Vegetatively, it is similar to *L. hancockii* but with a more open, spreading arrangement of the stem and leaves. Inflorescence carries 4–6 flowers, 1.5 cm across, with yellowish-green sepals and petals which are not well extended. Lip is maroon. Flowering period is April to May (Chen et al. 1999). *L. morsei* is an epiphyte in hilly forests at 500–700 m in southern China (Hainan, Guangxi, Guizhou and Yunnan), Japan, Vietnam, Laos and Thailand.

Herbal Usage: TCM states that the whole plant of *L. morsei* promotes movement of "wind", boosts *yang* elements, and stops vomiting. It detoxifies, removes gas and dampness and is used in Taiwan for the treatment of infantile paralysis, rheumatism, malaria, oedema, hypertension and for removal of poisons (Ou et al. 2003). It is also used to kill bugs (Ou et al. 2003). Four examples of its usage in Fujian are listed in Table 14.1 (*Zhongyao Da Cidian* 1986).

Luisia tenuifolia Blume

Syn. *Luisia birchea* Blume

Description: Plants varies considerably in size. Stems are woody, thick, slender, pendulous, 10–15 cm (or 17–75 cm) long, with 6–10 terete leaves, 8–9 cm long, slimmer than the stems, grooved on one side, and constricted at the tip to produce a nail-like beak.. Roots are vermiform. A short raceme bears a single flower, 2.5 cm tall, 1.1–1.5 cm across, with greenish sepals, faint purple petals, and a conspicuous lip, 2 cm long, white with deep purple markings, flaring into two narrow, blunted, divergent lobes at the tip. Flowering period is variously stated: March to April in Mumbai (Santapau and Kapadia 1966), May without specifying eastern or western part of India (Bose and Bhattacharjee 1980), May to June in Sri Lanka (Jayaweera), July to September or October in Tamil Nadu (Matthew 1995; Seidenfaden 1999) or March to April in the wild, but also in October to November when cultivated west of the Deccan (Abraham and Vatsala 1981).

This robust *Luisia* is distributed in southern India and prefers high elevations above 1500 m. It occurs in large clusters on tree trunks and is common in Peninsular India (Seidenfaden 1999). It is found in tropical dry mixed evergreen forests to tropical, wet, evergreen forests up to 900 m Sri Lanka but the species is rare (Jayaweera 1981). It also occurs in the Andaman and Nicobar Islands.

Herbal Usage: The whole plant is pounded for use as an emollient. It is also applied as a poultice to heal swellings such as boils, abscesses and tumours (Caius 1936; Singh et al. 1983; Jain and Defilipps 1991). After application of the poultice, abscesses can be painlessly drained. A powder prepared from the plant is mixed with vinegar to treat kidney disease, scalding, leucorrhoea and gonorrhoea (Dalgado 1896, 1898 quoted by Lawler 1984).

Luisia teres (Thunb.) Blume

Chinese names: *Jinchajiao* (gold hairpin), *Chaizigu* (bunched hairpins), *Yuanzhuchaizigu* (round pillar hairpin), *Yanggunzi* (goat stick), *Yangangdou* (stony long bean), *Xiaohuangcao* (small yellow grass), *Chachunchaizigu*

Description: This is an epiphyte with long (55 cm), thin (4–5 mm), terete stems that carry many leaves, 7–13 cm by 2–2.5 mm. Inflorescence is 1 cm long, arising directly opposite the leaf-axil, and bears up to 7 flowers despite its short length. Flowers are 1.5 cm across, yellowish-green with deep maroon marking on the prominent oblong lip which is bilobed at the apex. A Japanese variety has a solid maroon lip. Flowering season is March to May. It is found attached to the trunks of trees in exposed locations at 1200–1600 m from Sichuan to Guangxi and Taiwan to Japan and Korea (Chen et al. 1999; Chen and Wood 2009).

Herbal Usage: The entire plant is used to reduce swelling especially oedema caused by fractures. It is also used for treating tumours and as a counter-poison in China and Indochina (Lawler 1984). Early in the twentieth century, American missionary doctors in Shanghai reported that *L. teres* was "a much vaunted counter-poison, especially against infection. It (was) also prescribed for carcinoma, malaria and to counteract all sorts of medicinal poisons" (Stuart 1911; Porter-Smith and Stuart 2003). It is a specific for gout (Uphof 1968). Taiwanese herbalists use it to reduce swelling and heal fractures (Ou et al. 2003). Referred to as *Ch'ai Tzu Ku*, it was one of the few orchids (belonging to 4 genera) described as medicinal plants in Li Shizhen's *Bencao Gangmu* (*Pen Ts'ao Kung Mu*) of 1596 (Read 1936).

Luisia teretifolia Gaudich.[see **Luisia tristis (G. Forst.) Hook. f.**]

Luisia trichorrhiza (Hook.) Blume

Nepalese name: *Arjona*
Thai name: *Kluai nam thai*

Description: Stem is stout, terete, not branching, 10–25 cm long, sheathed and woody at the base. Leaves are terete, fleshy, slightly tapered at the tips, 10–17.5 cm long. Inflorescence is short, bearing 4–5 pale green flowers, 1 cm across. Lip is crimson to dark purple. It flowers in May in Nepal (Raskoti 2009). *L. trichorrhiza* is a monopodial epiphyte which is distributed in Nepal (at 1000–1400 m), Assam, Myanmar and Thailand.

Phytochemistry: Small amounts of alkaloid are present in *L. trichorrhiza* (Luning 1967; Lawler and Slaytor 1969).

Herbal Usage: In Thailand, the entire plant is used to treat liver dysfunction and diabetes mellitus (Chuakul 2002). In Nepal, plant paste is applied directly on painful muscles for relief (Singh et al. 1983; Baral and Kurmi 2006; Subedi et al. 2013). In Bangladesh, roots of *L. trichorrhiza* are used to treat jaundice, muscle pain and diarrhea (Musharof Hossain 2009).

Fig. 14.24 *Luisia tristis* (G. Forst.) Hook. f. [PHOTO: Bhaktar B. Raskoti]

Luisia tristis (G.Forst.) Hook. f.

Syn. *Luisia teretifolia* Gaudich., *Luisia zeylandica* Lindl.

Indian and Bangladeshi name: *Koira*
Nepalese names: *Bori jhaar, Kuwaa ko keraa*
Sri Lankan name: *Muwa kiriya* (Stems like deer horns), *soma valli* (in Sanskrit)
Thai name: *Kluai nam thai*
Chinese medicinal name: *Jinchaigu*

Description: This is the type species of *Luisia*. Stem is stout, erect, 10–15 cm tall, attached to tree trunks only at the base, and it carries green, purple-spotted, terete leaves 10–12 cm by 2 mm thickness. Inflorescence is axillary, short, with up to 5 flowers, 0.5 cm across. Petals and sepals are creamy-yellow or pale green tinged with purple, while the lip is deep purple (Figs. 14.24 and 14.25). It flowers twice a year, in March to June and again in December in Sri Lanka (Jayaweera 1981). In India, various flowering seasons are reported: May (Santapau and Kapadia 1966; Bose and Bhattacharjee 1980), August in Tamil Nadu (Matthew 1995) and June to July (Abraham and Vatsala 1981). In cultivation, it has flowered in January, "possibly a result of constant watering" (Santapau and Kapadia 1966). It forms large clumps on *Hora* (*Dipterocarpus scabridus*) in exposed positions in Sri Lanka (Cooray 1940).

L. tristis is distributed from Bhutan, the Sikkim Himalayas to south India and Sri Lanka, and across Bengal and Myanmar to the Andamans, Thailand, Malaysia, the Philippines and New Caledonia. It is found at low elevations.

Phytochemistry: A small amount of alkaloid is present in *L. tristis* (Luning 1967; Lawler and Slaytor 1969).

Herbal Usage: *L. tristis* is used in the preparation of medicinal oils used in the treatment of fractures by Sri Lankan *Ayurvedic* practitioners (Cooray 1940). A poultice prepared with crushed plants is applied to boils, abscesses and tumours in Uttar Pradesh (Trivedi et al. 1980) and Karnataka (Rao 2007). Emollient made from the plant is used for abscesses and burns in Nepal, and juice from the leaves is used for treating chronic wounds (Manandhar and Manandhar 2002; Raskoti 2009; Pant 2011). The juice is also used to get rid of worms (Manandhar and Manandhar 2002; Baral and Kurmi 2006). In Nepal, Bangladesh and Karnataka (India), stems are used to treat boils, burns and fractures (Baral and Kurmi 2006; Rao 2007; Musharof Hossain 2009). A paste

Fig. 14.25 *Luisia tristis* (G. Forst.) Hook. f. (as *Luisia occidentalis* Lindl.) Reichenbach, H. G., Arnott, G. A. W., *Xenia Orchidacea*, vol 3: t. 238 (1900). Courtesy of Missouri Botanical Garden, St. Louis, U.S.A.

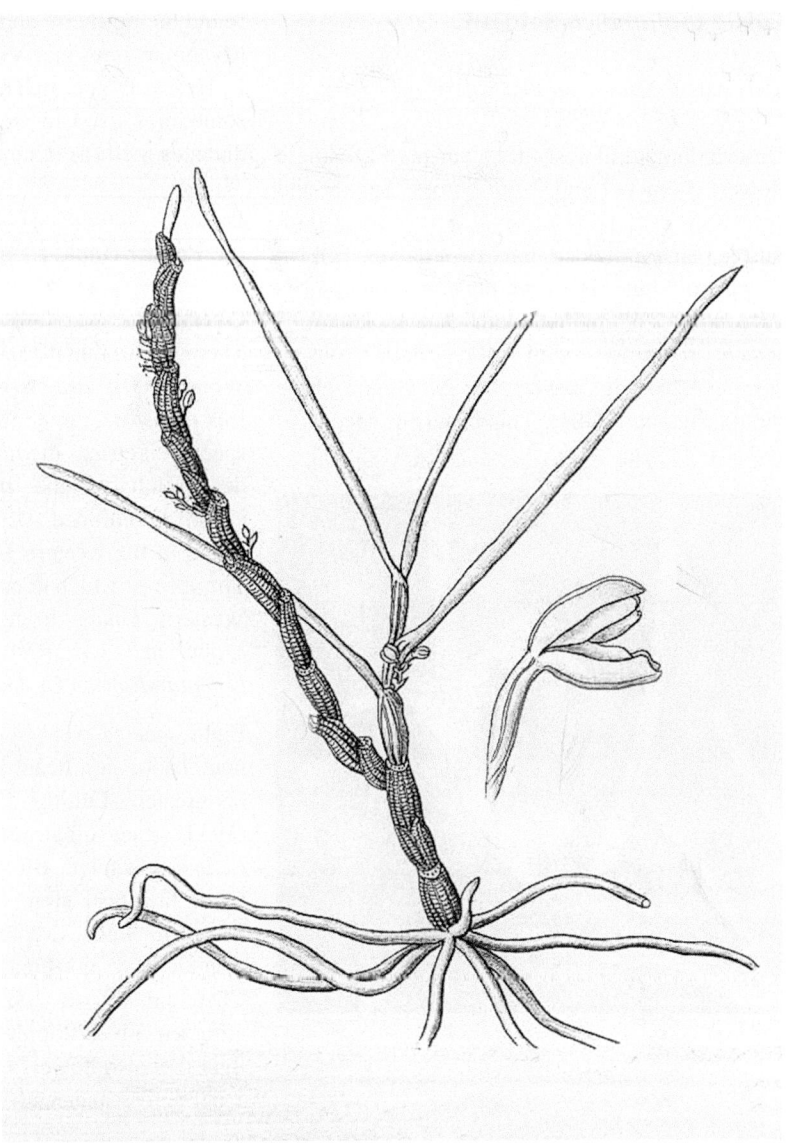

consisting of the dried plants of *L. tristis*, tumeric and ginger is taken three times a day for 10 days to treat jaundice. Root extract is used for myalgia, and to stop diarrhoea in cattle. In Bisamkatak Orissa, locals call it *koira* (Dash et al. 2008). Valmikis of Viskhatnam district in Andra Pradesh employ a paste made with the entire plant of *L. tristis*, egg white, turmeric and calcium to apply as a plaster which is then and bound with a bandage on fractured limbs to set fractures, this being retained for 2–3 weeks (Reddy et al. 2003).

L. zeylandica is listed as a herbal remedy, *Jinchaigu*, in *Zhongyao Bencao* (Hu et al. 2000). However, *L. zeylandica* (= *L. tristis*) does not occur in China, and the herb is imported from India and Myanmar. In Yunnan, it is used to treat 'heatiness', remove toxins, cure malaria, pruritus, sore throat, otitis media and food poisoning. Depending on the indication, the dose may be varied, using either 6–9 g or 15–30 g in decoction, the latter being considered a big dose. It is also used externally (Hu et al. 2000).

Luisia thailandica Seidenf.

Thai name: *Kluai nam thai yai*

Description: Inflorescence carries a single flower, 2 cm tall and 2 cm across. Sepals and petals are a clear greenish-yellow. Lip is dark purple, large, divided into two halves like in *L. curtisii*, but the distal portion (epichile) is very wide and marked with thin tessellations which contrast with the dark purple (Fig. 14.26). It flowers in May and June (Vaddhanaphuti 2001). This beautiful species is

Fig. 14.26 *Luisia thailandica* Seidenf. [PHOTO: E. S. Teoh]

found in northern and north-eastern Thailand, Myanmar, Laos and Vietnam.

Herbal Usage: In Thailand, the whole plant is sometimes used to treat liver dysfunction and diabetes mellitus (Chuakul 2002).

Luisia zeylandica Lindl. [see ***Luisia tristis* (G. Forst.) Hook. f.**]

Overview

There are around 40 species of *Luisia* (Chen and Wood 2009). Three species are used in Chinese folk medicine, three in Indian medicine. A single species is used medicinally in Nepal. In his description of the medicinal *Luisia*, Caius (1936) mentioned that the orchid plant was found in the western Deccan and in Sri Lanka. The species which occurs in Sri Lanka and in the Western Ghats from Konkan to Kerala is *L. birchea* (A. Rich. Bl.), also known as *L. tenuifolia* (L.) Bl. (Jayaweera 1981).

Eight species were screened for alkaloids and none had a significant alkaloid content of 0.1 % or greater (Luning 1974). However, Chopra (1933) stated that an alkaloid is present in *L. brachystachys* Blume, an epiphytic species found in forests along valleys at 600–1300 m in Vietnam, Laos, Myanmar, southern Yunnan, Bhutan and north-east India (Chopra 1933; Chen and Wood 2009). The species has not attracted any horticultural attention because its light greenish flowers with a purple lip are under 1 cm across and, with the very short stem, they

Fig 14.27 Phenanthrenes from *Luisia indivisa* (Ref.: Majumder and Lahari 1989)

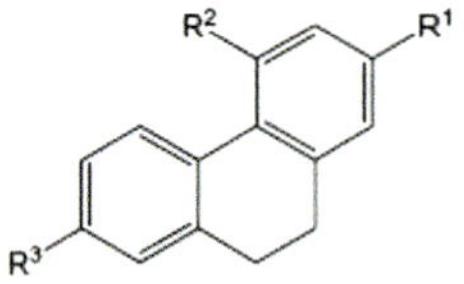

Lusianthridin: R1 = OMe, R2 = OH, R3 = OH

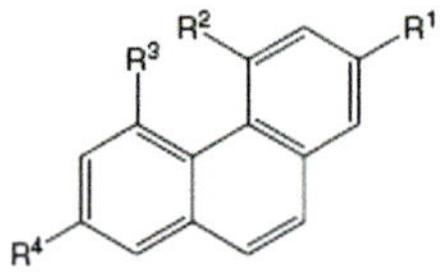

Lusianthrin: R1 = OMe, R2 = OH, R3 = H, R4 = OH

are not well displayed (Vaddhanaphuti 2005). There are no new pharmacological data on *L. brachystachys*.

The Himalayan species, *L. brachystachys* (syn. *L. indivisa*) was investigated by Majumder and Lahiri (1989, 1990) who discovered that it contained beta-sitosterol, betulinic acid, *p*-hydroxybenzaldehyde, *p*-hydroxyphenylpropionic acid, 3′,4-dihydroxy-3,5′-dimethoxydibenzyl and two stilbenoids which they named luslanthrin and lusiathridin (Fig. 14.27). They further isolated two perfumery constituents from this orchid: methyl 2,4-dihydroxy-6-methylbenzoate and methyl 2,4-dihydroxy-3,6-dimethylbenzoate. The pharmacological effects of these compounds have not been described and there are no clinical studies involving *Luisia*. Lignans were isolated from *L. volcris* (Majumder et al. 1994).

References

Abraham A, Vatsala P (1981) Introduction to orchids, with illustrations and descriptions of 150 South Indian Orchids. TPGRI, Trivandrum

Baral SR, Kurmi PP (2006) A compendium of medicinal plants in Nepal. Mrs. Rachana Sharma, Kathmandu, 281 Maujubahal, Chaabahil, Kathmandu, Nepal

Bechtel H, Cribb P, Launert E (1980) The manual of cultivated orchid species. Blandford, Poole

Bhattacharjee SK (1998) Handbook of medicinal plants. Pointer Publishers, Jaipur

Boorsma WG (1902) Pharmakologische Mitteilungen I. Bulletin de l'INstitut Botanique de Buitenzorg 14:1–39

Bose TK, Bhattacharjee SK (1980) Orchids of India. Naya Prokash, Calcutta

Burkill IH (1935) (1966 reprint, 2nd edn, with contributions by Birtwistle W, Foxworthy FW, Scrivenor JB, Watson IG). A dictionary of economic products of the Malay Peninsula, vol II. Crown Agents for the Colonies, London. Ministry of Agriculture & Co-operatives, Kuala Lumpur

Caius JF (1936) The medicinal and poisonous plants of India. J Bombay Nat Hist Soc 38(4):791–799

Chan CL, Lamb A, Shim PS, Wood JJ (1994) Orchids of Borneo. Vol. 1. Introduction and Selection of Species. The Sabah Society and Kew, Kota Kinabalu: Royal Botanic Gardens

Chen SC (1995) Orchids and their conservation. Proceedings of the 5th Asia pacific orchid conference & show, Fukuoka, pp 15–18

Chen SC, Tang T (1982) A general review of the orchid flora of China. In: Arditti J (ed) Orchid biology.

Reviews and perspectives II. Cornell University Press, Ithaca

Chen XQ, Wood JJ (2009) Luisia Gaudichaud. In: Chen XQ, Liu ZJ, Zhu GH et al (eds) Flora of China – Orchidaceae. Science Press, Beijing

Chen SC, Tsi ZH, Luo YB (1999) Native orchids of China in colour. Science Press, Beijing

Chen XQ, Gale SW, Cribb PJ (2009a) Ludisia A. Richard. In: Chen XQ, Zj L, Zhu GH et al (eds) Flora of China - Orchidaceae. Science Press, Beijing

Chen XQ, Liu ZJ, Zhu GH et al (eds) (2009b) Flora of China – Orchidaceae. Science Press, Beijing

Chen XQ, Ormerod P, Wood JJ (2009c) Liparis Richard. In: Chen XQ, Liu ZJ, Zhu GH et al (eds) Flora of China – Orchidaceae. Science Press, Beijing

Chopra RN (1933) The Indigenous Drugs of India. The Art Press, Calcutta. Republished as Chopra's indigenous plants of India, 2nd edn. Academic, Kolkata (1986)

Chuakul W (2002) Ethnomedical uses of Thai orchidaceous plants. Mohidol Univ J Pharm Sci 29 (3–4):41–45

Chung MY, Nason JD, Chung MG (2005) Patterns of hybridization and population genetic structure in the terrestrial orchids Liparis kumokiri and Liparis makinoana (Orchidaceae) in sympatric populations. Mol Ecol 14(14):4389–4402

Comber JB (2001) Orchids of sumatra. Natural History Publications (Borneo), Kota Kinabalu

Cooray DA (1940) Orchids in oriental literature. Orchids Zelandica 7:73–80

Daduang S, Uawonggul N (2008) Herbal therapies of snake and insect bites in Thailand. In: Watson RR, Preedy VR (eds) Botanical medicine in clinical practice. CAB International, Wallingford, pp 814–822

Dash PK, Sahoo S, Bal S (2008) Ethnobotanical studies on orchids of Niyamgiri Hill Ranges, Orissa, India Ethnobot Leaflets 12:70–78

Da Cidian Z (1986) Edited by Jiangsu New Medical College. Science and Technology Press, Shanghai, p 1986

De Vries PJ, Stiles FG (1990) Attraction of pyrrolizidine alkaloid seeking Lepidoptera to Epidendrum paniculatum orchid. Biotropica 22(3):290–297

Dong YF, Li WY, Ye RC, Wang L (2010) Antimicrobial and antioxidant activities of total alkaloids of Liparis nervosa (Thunb.) Lindl. Sichuan Daxue Xuebao (Ziran Kexueban) 47(3):669–673

Duggal SC (1972) Orchids in human affairs (a review). Pharm Biol 11(2):1727–1734

Dunn ST (1903) Expedition to central Fokien. J Linnean Soc XXXVIII, p 368

Garay LA, Sweet H (1974) Orchids of the southern Ryukyu Islands. Harvard University, Cambridge, MA

Gurong DB (2006) An illustrated guide to the orchids of Bhutan. DSB Publications, Thimphu

Heyne K (1927) De nuttige planten van Nederlandsch Indie, vol 1, pp 508–513. Uitgave van het Departement van Landbouw, Nijverheid & Handel in Nederlandsch-Indie

Hu XM, Zhang WK, Zhu QZ et al (2000) Zhonghua Bencao, vol 8. Shanghai Science and Technology Publication, Shanghai

Hu QM, Wu TL, Xing FW, et al. (eds) (2007) Flora de Macau, vol 3. South China Botanic Gardens, Chinese Academy of Sciences, Macau

Huang S, Zhou XL, Wang CJ et al (2013) Pyrrolizidine alkaloids from Liparis nervosa with inhibitory activities against LPS-induced NO production in RAW264.7 macrophage. Phytochemistry (Amster) 93:154–161

Hung T (2014) Lan Nhan diep, Lan tai de Liparis. Google (ucchau ndcinh com)

International Programme for Chemical Safety (1988) Pyrrolizidine alkaloids. Environmental healthy criteria 80. WHO, Geneva

Jain SK, Defilipps RA (1991) Medicinal plants of India, vol 2. Reference Publications, Algonac

Jayaweera DMA (1981) A revised handbook of the flora of Ceylon, vol II. A.A. Balkema, Rotterdam

Jin XH, Zhao XD, Shi XC (2009) Native orchids from Gaoligongshan Mountains, China. Science Press, Beijing

Joseph J (1982) Orchids of Nilgiris. Records of the botanical survey of India, vol XXII. Botanical Survey of India (Department of Environment), Howrah

Kaiser R (1993) The scent of orchids: olfactory and chemical investigations. Editiones Roche, Basel

Kuo WL, Huang YL, Shen CC et al (2007) Prenylated benzoic acids and phenanthrenes from Liparis nakaharai. J Chin Chem Soc 54:1359–1362

Lawler LJ (1984) Ethnobotany of the orchidaceae. In: Arditti J (ed) Orchid biology reviews and perspectives, vol 3. Cornell University Press, Ithaca

Lawler LJ, Slaytor M (1969) The distribution of alkaloids in New South Wales and Queensland Orchidaceae. Phytochemistry 8:1959–1962

Li NH (ed) (1988) Chinese medicinal herbs of Hong Kong, vol 2. Hong Kong Chinese Medicinal Research Institute, Hong Kong, p 198

Li L, Zhang D et al (2014) Three new bioactive phenolic glycosides from Liparis odorata. Nat Prod Res 28 (8):522–529

Lin TP (1975) Native orchids of Taiwan, vol 1. Southern Materials Center Inc., Taipei

Lin TP (1977) Native orchids of Taiwan, vol 2

Lindstrom B, Luning B (1971) Studies on orchidaceous alkaloids. 23. Alkaloids from Liparis loeselii (L.) L.C. Rich and Hammarbya paludosa (L.)).K. Acta Chem Scand 25(3):895–897

Lindstrom B, Luning B, Siiral-Hansen K (1971) Studies on orchidaceae alkaloids. XXVI. New glycosidic alkaloid from Malaxis grandiflora Schltr. Acta Chim Scand 25

Lindstrom B, Luning B (1972) Studies on the Orchidaceae alkaloids XXXV. Alkaloids from Hammarhya paludosa (L.) O.K. and Liparis keitaonsis Hay. Acta Chem Scand 25:895–897

Liu B, Eng H, Yao Q, Li J, Van Damme E, Balzarini J, Bao JK (2009) Bioinformatics analyses of the mannose-binding lectins from Polygonatum cyrtonema, Ophiopogon japonicus and Liparis nervosa with antiproliferative and apoptosis inducing activities. Phytomedicine (Jena) 16(6–7):601–608

Luning B (1967) Studies on Orchidaceae alkaloids IV. Screening of species for alkaloids. Phytochemistry 6:857–861

Lüning B (1974) Alkaloids of the Orchidaceae. In: Withner CL (ed) The orchids: scientific studies. Wiley, New York

Luning B (1980) Alkaloids of the Orchidaceae. In: Sukshom Kashemsanta MR (ed) Proceedings of the 9th World Orchid Conference, Bangkok

Luning B, Slaytor M (1969) The distribution of alkaloids in orchids from the Territory of Papua New Guinea. Proc Linn Soc NSW 94:419–421

Majumder PL, Lahiri S (1989) Chemical constituents of the orchid Lusia indivisa. Indian J Chem 28B:771–774

Majumder PL, Lahiri S (1990) Lusianthrin and Lusianthridin, two stilbenoids from the orchid Lusia indivisa. Phytochemistry 29(2):621–624

Majumder PL, Lahiri S, Pal S (1994) Occurrence of lignans in the Orchidaceae plants Luisia volucris and Bulbophyllum triste. J Indian Chem Soc 71:645–647

Manandhar NP, Manandhar S (2002) Plants and people of Nepal. Timber, Portland

Matthew KM (1995) An excursion flora of Central Tamil Nadu, India. A.A. Balkaema, Rotterdam

McCartney C (2000) African affiliates. Part 1: The surprising relationship of some of Florida's wild orchids. Orchids 69(2):130–139

Medhi RP, Chakrabarti S (2009) Traditional knowledge of NE people on conservation of wild orchids. Indian J Tradit Know 8(1):11–16

Musharof Hossain M (2009) Traditional therapeutic uses of some indigenous orchids of Bangladesh. Med Aromat Plant Sci Biotechnol 3:100–106

Nadkarni AK (1954) Dr. K.M. Nadkarni's Indian Materia Medica, vol 2, 3rd edn. Popular Book Depot, Bombay

Nishikawa K, Hirata Y (1968) Chemotaxonomical alkaloid studies III. Further studies on Liparis alkaloids. Tetrahedron Lett:6289–6291

Nishikawa K, Hirata Y (1967a) Chemotaxonomical alkaloid studies. I. Structure of nervosine. Tetrahedron Lett 27:2591–2596

Nishikawa K, Miyamura M, Hirata Y (1967b) Chemotaxonomical alkaloid studies II. Structures of kuramerine and kumokirine. Tetrahedron Lett 27:2597–2600

O'Byrne P (2001) A to Z of South East Asian orchid species. Orchid Society of South East Asia, Singapore

Ohwi J (ed) (1965) Flora of Japan. English edition: Meyer FG, Walker EH (eds). Smithsonian Inst, Washington

Ou JC, Hsieh WC, Lin IH, Chang YS, Chen IS (eds) (2003) The catalogue of medicinal plant resources in Taiwan. Department of Health, Executive Yuan, Taipei

Pant B (2011) Medicinal orchids of Nepal and their conservation by in-vitro technique, PowerPoint presentation. 20th World Orchid Conference. Singapore, Nov 14, 2011

Pearce NR, Cribb PJ (2002) The orchids of Bhutan. Royal Botanic Gardens and Royal Government of Bhutan, Edinburgh and Thimpu

Perry LM, Metzger J (1980) Medicinal plants of East and Southeast Asia: attributed properties and uses. MIT Press, Cambridge, MA

Porter-Smith F, Stuart GA (2003) Chinese medicinal herbs. A modern edition of a classic sixteenth-century manual by Li Shih Chen. Dover, Mineola, NY

Rao AN (2004) Medicinal orchid wealth of Arunachal Pradesh. Newslett ENVIS NODE Indian Med Plants 1(2):1–7

Rao TA (2007) Ethno botanical data on wild orchids of medicinal value as practised by tribals at Kudremukh National Park in Karnataka. Orchid Newslett 2(2):1–7

Rao TA, Sridhar S (2007) Wild orchids in Karnataka. A pictorial compendium. Institute of Natural Resources Conservation, Education, Research and Training (INCERT), Bangalore

Raskoti BB (2009) The orchids of Nepal. Bhakta Bahadur Raskoti and Rita Ale, Kathmandu

Rasmussen HN (1995) Terrestrial orchids from seed to mycotrophic plant. Cambridge University Press, Cambridge

Read BE (1936) Chinese medicinal plants from the Pen Ts'ao Kang Mu A.D. 1596, 3rd edn. Peking Nat Hist Bull, p 206

Reddy CS, Nagesh K, Reddy KN, Vatsavaya SR (2003) Plants used in ethnoveterinary practices by Gonds of Karimnagar District, Andhra Pradesh, India. In: Singh V, Jain AP (eds) Ethnobotany and medicinal plants of India and Nepal. Scientific Publishers (India), Jodhpur, pp 631–634

Santapau H, Kapadia Z (1966) The orchids of Bombay. Government of India Press, Calcutta

Sathish Kumar C, Manilal KS (1994) A catalogue of Indian orchids. Bishen Singh Mahendra Pal Singh, Hehra Dun

Seidenfaden G, Wood JJ (1992) The orchids of Peninsular Malaysia and Singapore. Olsen & Olsen, Fredensborg

Seidenfaden G (1999) 149. Orchidaceae. In: Matthew KM (ed) The flora of the Palni Hills South India, vol 3. The Rapinat Herbarium, St. Joseph's College, Tiruchirapalli

Singh U, Wadhwari AM, Johri BM (1983) Dictionary of economic plants in India. Indian Council of Agricultural Research (ICAR), New Delhi

Slapetova T, Smejkal K, Innocenti G et al (2009) Glycosylated nervogenic acid derivatives from Liparis condyle bulbon (Reichb. F.) leaves. Carbohydr Res 344(13):1770–1774

Slaytor MB (1977) The distribution and chemistry of alkaloids in the Orchidaceae. In: Arditti JA (ed) Orchids biology reviews and perspectives, vol 1. Cornell University Press, Ithaca

Song Q, Shou Q, Goun X et al (2013) A new nervogenic acid glycoside with pro-coagulant activity from Liparis nervosa. Nat Prod Commun 8(8):1115–1116

Stuart GA (1911) Chinese Materia Medica. Vegetable Kingdom. (A revision of a work by F. Porter Smith.) American Presbyterian Mission Press, Shanghai

Subedi A, Kunwar B, Choy Y et al (2013) Collection and trade of wild harvested orchids in Nepal. J Ethnobiol Ethnomed 9:64–73

Tang SL, Su HJ (1978) Flora of Taiwan, vol 5. Taiwan National University, Taipei

Trivedi VP, Dixit RS, Lal VK (1980) Orchids in the drug markets of Bareilly, Kanpur and Nearby Districts. Nagarjun (Calcutta) 23(8):157–163

Tsutsumi C, Yukawa T, Lee NS et al (2007) Phylogeny and comparative seed morphology of epiphytic and terrestrial species of Liparis (Orchidaceae) in Japan. J Plant Res 120(3):405–412

Uphof JCT (1968) Dictionary of economic plants. Verlag von J. Cramer, Lehre

Vaddhanaphuti N (2001) A field guide to the wild orchids of Thailand, 3rd edn. Silkworm Books, Chiang Mai

Vaddhanaphuti N (2005) A field guide to the wild orchids of Thailand, Fourth and Expanded Edition. Silkworm Books, Chiang Mai

van Rheede HA (1703) Hortus Indicus Malabaricus, vol 12. Dutch East India Company, Kerala

Wu XR (1994) A concise edition of medicinal plants in China. Guangdong Higher Education Publication House, Guangdong (in Chinese)

Wu TL, Hu QM, Xia NH, Lai PCC, Yip KL (2002) Check list of Hong Kong plants 2001. Agriculture, Fisheries and Conservation Department Bulletin 1 (Revised), Hong Kong

Yang ZH, Zhang QT, Feng ZZ, Lang KY, Li H. English edition translated by ZR Xiong (1998) Orchids. China Esperanto Press, Beijing

Zhonghua Bencao (2000) Edited by Health Department and National Chinese Medical Management Office. Science and Technology Press, Shanghai

Genus: *Macodes* Lindl.

Macodes was originally classified under *Neottia*, and only later assigned a separate genus status by Lindley. Plants are terrestrial, herbaceous, with creeping, rooting rhizomes which produce, at intervals, very short stems surmounted by a rosette of beautiful dark green leaves marked with a reticulum of golden veins. In some varieties, the leaves may be purplish-green (Fig. 15.1). Inflorescence is erect and carries numerous flowers on the rachis. Both the inflorescence and the superior surface of the sepals are finely hirsute. Flowers are small, non-resupinate, with a complex trilobed lip, and they face all directions. About a dozen species have been described. They are distributed from Southeast Asia to New Guinea and Vanuatu (Comber 2001). The name of the genus is derived from Greek, *macro* (large), referring to the large lip (Schultes and Pease 1963).

Macodes petola (Blume) Lindl.

Indonesian names: *Daun aksara* (calligraphic leaf), *Djaksara*, *Djukut sastra*, *Ki-aksara*, *Kidjaksara* (Sundanese) *Daun patola* (Javanese)

Description: A Southeast Asian Jewel Orchid, it is characterised by beautiful, cordate, dark green leaves, 6 by 5 cm, marked with three main and four subsidiary longitudinal veins crisscrossed by irregular, transverse veins, all of light green or sparkling orange and gold. Inflorescence is erect, up to 15 cm tall, succulent, finely hirsute, and olive in colour. Flowers, about 15 in number, are spirally arranged near the distal end, facing different directions, 6–8 flowers opening at a time. They are 1 cm across, olive to reddish-brown, with the complex white lips uppermost. Sepals are ovate; petals are irregularly linear (Fig. 15.2). Flowers appear after the rainy season and last a fortnight.

The species is distributed in Malaysia, Singapore and Indonesia, from sea level to 1600 m, in moist, mossy locations, on rocks, near streams. Young plants are commonly found scattered along the edge of streams or tiny rivulets.

Herbal Usage: Juice extracted from leaves of *M. petola* was dropped into the eyes to correct myopia in Indonesia (Lawler 1984).

Overview

Indonesian legend says that this magical plant first arose from the fragments of a beautiful scarf which belonged to a goddess (Teoh 1982). Rumphius in the later seventeenth century referred to an Amboinese jewel orchid as *Daun petola* (petola + leaf), a name which refers to its resemblance to an expensive silk fabric which is dyed in many colours (de Witt 1977), a rare

© Springer International Publishing Switzerland 2016
E.S. Teoh, *Medicinal Orchids of Asia*, DOI 10.1007/978-3-319-24274-3_15

Fig. 15.2 *Macodes petola* (Blume) Lindl. (Photo: E.S. Teoh)

Fig. 15.1 *Macodes petola* (Blume) Lindl. From: Reichenbach, H.G., Arnott, G.A.W., *Xenia Orchidacea*, vol. 1: t.96 (1900). Courtesy of Missouri Botanical Gardens, St. Louis, USA

silken batik. Inspired by the legend and following the notion of the Doctrine of Signatures, ancient Javanese who saw similarities between the golden reticulations on the lip and Javanese writing (they called the orchid "letter-leaf") believed that the orchid could improve the eyesight of children. Juice made from *Macodes* leaves, macerated in water collected from the axil of a young leaf sheath of the banana plant on a Friday morning (the day when Muslims attend prayers at their mosques), was dropped into the eyes of children to improve their recognition of the written language (Smith, 1930

quoted by Lawler 1986; Glicenstein 2010). It was also dropped into the eyes to improve near vision (myopia). People who wished to improve their calligraphy also made use of the eye-drops (Dakkus, 1935 quoted by Lawler 1984).

Genus: *Malaxis* Sol. ex Sw.

Chinese name: Yuanzhao Lan
Japanese name: Yachi ran

Malaxis is a large genus of broad-leaved, terrestrial orchids whose 230 members are distributed all over the world, with the greater number inhabiting tropical montane forests. They are mainly terrestrial herbs. Stems are cylindrical, fleshy, rooting at the base, and leaves are thin and pleated, sheathing the stem at the base, but some may be found as lithophytes or epiphytes. Inflorescence is erect, with long, naked peduncles crowded with tiny, greenish, maroon,

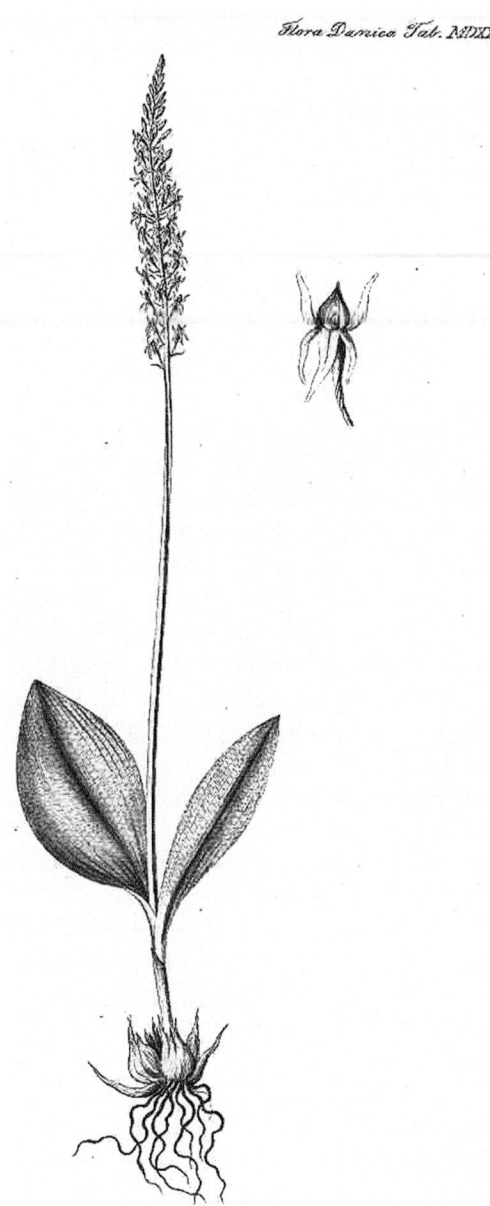

Flora Danica Tab. MDXXV.

Fig. 15.3 *Malaxis monophyllos* (Linn.) Sw. From: Oeder, G.C., et al., *Flora Danica*, fascicle 26: t. 1525, (1761–1853. Original in colour. Courtesy of the Royal Library, Copenhagen (Det Kongelige Biblotek) Denmark

Malaxis acuminata D. Don [see ***Crepidium acuminatum* (D. Don.) Szlach.**]

Malaxis cylindrostachya (Lindl.) Kuntze (see ***Dienia cylindrostachya* Lindl.**)

Malaxis latifolia Sm. [see ***Dienia ophrydis* (J. Koenig) Seidenf.**]

Malaxis monophyllos (Linn.) Sw. var. *monophyllos*

Syn. *Liparis japonica* (Miq.) Maxim

Chinese name and medicinal name: *Yangersuan* (sheep ear garlic)

Japanese name: *Seitakasuzumushiso*

Description: This is a small plant with a pair of opposing, elliptic, plicate, undulate, green leaves ensheathing the stem at the base. Inflorescence is short and carries 5–8 flowers opening in succession. Dorsal sepal is green. The filiform lateral sepals and petals and the shield-shaped lip are brown (Fig. 15.4). Flowering season is June to August. It occurs as a lithophyte on moss-laden rocks in shady locations in China (Dongbei, Shanxi, Hebei, Shaanxi, Gansu, Sichuan and Guizhou) and Japan (Jin et al. 2008).

Herbal Usage: Herb is obtained from Dongdei, Shanxi, Shaanxi, Gansu, Hebei, Sichuan and Guizhou (Wu 1994). The herbal properties of *M. monophyllos* var. *monophyllos* are: "stimulating blood circulation, regulating menstruation, haemostasis, relief of pain, strengthening the heart, tranquilising, stopping severe diarrhea, alleviating leucorrhea and relieving postpartum abdominal pain. It is also used as an emergency treatment for external injuries" (Zhongyao Da Cidian 1986). It is reputed to reduce inflammation, dissolve extravasated blood and cause swellings to subside (Chen and Tang 1982). The whole plant is used to stop pain and bleeding, especially in cases of trauma, continuous inter-menstrual bleeding and discharge (Wu 1994). Two prescriptions involving *M. monophyllos* var. *monophyllos* are listed in Table 15.1.

or white, non-resupinate flowers of complex form (Fig. 15.3).

The generic name is derived from Greek *malaxis* (softening). which refers to the texture of the leaves.

Table 15.1 Herbal remedies containing *Malaxis monophyllos* var. *monophyllos* (Zhongyao Da Cidian 1986)

1. Indication: severe diarrhea
Boil and drink 9 g of *M. monophyllos* var. *monophyllos* in water
2. Indication: postpartum abdominal pain
Prepare decoction of *M. monophyllos* var. *monophyllos* 9 g with Tao Nu 9 g
To be taken together with yellow wine

15–26 cm long and densely many-flowered (Fig. 15.5). Flowers are small, 4–6 mm across, of a pale green or yellowish-green, triangular in outline, with a large triangular, resupinate lip and small, linear petals and sepals all arranged in one plane (Fig. 15.6). They appear in June and July. Many capsules are formed after flowering (Lin 1975; Chen et al. 1999). Flowering season is June to August in the Thimphu and Bumthang districts of Bhutan (Gurong 2006); in Nepal, in the subalpine and alpine zone at 3000–4100 m (Raskoti 2009). In Alaska, Ossian observed that flowers and all plants parts are green to white in the typical form, with leaves darker than the blossoms. "These exceedingly inconspicuous plants are seldom seen except when in flower, and then only by experienced searchers" (Ossian 1984).

Fig. 15.4 *Malaxis monophyllos* (Linn.) Sw. (Photo: Courtesy of Plant Photo Bank of China)

Malaxis muscifera (Lindl.) Kuntze.

Ayurvedic names: *Jeevak, Jivaka, Chiranjivi, Dirghayu, Harsanga, Ksveda, Kurchasira, Pranda, Risvak, Sringaka, Svadu,* (in Gahwal) *Rishbhak*

Names: *Rishbhaka* and *Jivaka* are the medicinal names. Other Ayurvedic names: *Chiranjiva, Dirghayu, Harsanga, Ksveda, Kurchasira, Pranda, Sringaka, Svadu*

Additional Indian names: *Banndhura, Dheera, Durdhara, Gopati, Indraksa, Kakuda, Matrika, Vrisha* and *Vrishnaba.*

Nepalese name: *Jivaka*

Description: A short plant, the pseudobulbs are 1.5 cm long and bear two thin, oval to elliptic, faintly reticulated leaves, 5–10 cm by 3–5 cm, which are narrowed towards the base giving the impression of a stem. Inflorescence is erect,

This widespread terrestrial orchid is found in the subtropical and temperate regions of North America, Europe and Asia above 2000 m, but it is not common. In China, it occurs in Yunnan at Gaoligongshan (Jin et al. 2008), Xinjiang, Sichuan [e.g. Huanglong (Perner and Luo 2007)], Hubei, Hebei, Shanxi, Jinlin, Nei Mongolia and Taiwan (Chen et al. 1999). It grows among grass in exposed areas, or on forest humus close to streams in woodlands at 2000–4300 m. It is reported to be widespread in the Himalayan highlands (Pakistan, Kashmir, Sikkim and Arunachal Pradesh) (Hawkes 1965; Nasir and Ali 1972; Singh and Duggal 2009), but is thought to be rare and endangered (Chauhan et al. 2008). It occurs in mixed conifer forests in Bhutan (Gurong 2006). In Uttarakhand and Garwal Himalaya, the preferred habitats for *M. muscifera* are moss-laden moist slopes at 2500–3700 m (Chauhan et al. 2008).

Fig. 15.5 *Malaxis muscifera* (Lindl.) Kuntze. [as *Microstylis muscifera* (Lindl.) Ridley]. From: *Annals of the Royal Botanic Gardens, Calcutta*, vol. 8 (2): t. 25 (1891). Drawing by R. Pantling. Original has flowers in colour. Courtesy of Missouri Botanical Gardens, USA

In the western Himalayas, plants emerge a few days after snowmelt in April and complete one phenophase within 6 months. New plants commence their life by germinating from seed or from dormant apical buds on old pseudobulbs. The species is perennial and remains in vegetative phase during the first 2 years of growth. During the third or subsequent year, a flowering rachis emerges, completing flowering and fruiting in the brief summer. With the onset of winter in September, senescence of aerial parts occurs and underground tubers go into dormancy (Chauhan et al. 2008).

Usage: *Jeevak* is one component of an assemblage of eight vitalising herbs (*Jivaniya*) mentioned in the *Mahakashaya* of *Caraka Samhita*, the original *Materia Medica of Ayurveda*. These eight herbs are collectively referred to as *Ashtavarga*. *Rishbhaka* is said to be sweet in taste, and cold in potency. It pacifies *varta* and aggravates *kapha* and therefore it benefits hyperactive people but is unsuitable for overweight subjects. It is used to treat general debility and emaciation (Pandey et al. 2003; Chauhan et al. 2008). In Garhwal, which is located in the Indian Himalayas, the tonic is prepared by

Fig. 15.6 *Malaxis muscifera* (Lindl.) Kuntze (Photo: Bhaktar B. Raskoti)

dissolving dried, powdered pseudobulbs in boiled milk (Dhayani et al. 2011).

Ayurvedic practitioners claim that it exerts a cooling effect. It reduces fever, promotes spermatogenesis, and has the properties of an aphrodisiac (Vij 1995; Singh and Duggal 2009). In modern terminology, it is "anti-oxidant and anti-ageing". The indications for its use are bleeding diathesis, dysuria and phthisis (Singh 2006). A decoction of the tubers is used as a tonic to strengthen the kidneys (Rao 2004). Roots are used to promote milk production (Bhattacharjee 1998).

In Nepal, paste is made from pseudobulbs for use as an emollient to treat pruritus, fever and sores. It is also a tonic (Subedi et al. 2013).

Chinese herbal medicine maintains that the plant is antipyretic and detoxifies. It promotes regular menses and diuresis, It is used to treat sexual dysfunction, weak kidneys, coughs, discharges, menorrhagia and abdominal pain during the post-natal period (Ou et al. 2003). A decoction of the root was used in Sichuan Province of China for strengthening the kidneys (Cheo 1947 and Hu 1971, both quoted by Lawler 1984). Perry and Metzger (1980) reported that this tonic was prepared by cooking the roots of the orchid with pork.

Malaxis rheedei Blume [see **Liparis rheedii (Blume) Ldl.**]

Malaxis versicolor (Lindl.) Abeyw.

Syn. *Microstylis vesicolor* Lindl., *Seidenfia versicolor* Marg. and Szlach.

Description: This is a variable, terrestrial herb with stems 8–21 cm long, 1.5–2 cm in diameter, with pseudobulbs along its length supporting 2–3 sessile, thin, lanceolate, plicate leaves 6–10 by 3–5.5 cm, and 7-veined. Inflorescence is erect and carries numerous small flowers of greenish-yellow to orange or purple, 3.4 mm across. It continues to lengthen and produce new flowers which open successively over a long period. Flowers are non-resupinate. Lip is large, semi-circular, with dentate margin, the teeth long and pronounced in some varieties, and barely visible in others (Abraham and Vatsala 1981) (Fig. 15.7). The species occurs in southern India (Karnataka, Kerala and Tamil Nadu) and in Sri Lanka. Plants are found in shaded locations between 400 and 1800 m (Jayaweera 1981).

The colour of the plant and flowers are influenced by light intensity: pure green in bright light, deep purple in the shade, and yellowish inbetween. It is because of this variation in colour that Lindley gave the species the epithet *vesicolor* (Santapau and Kapadia 1966).

Herbal Usage: In the western part of the Indian peninsula, a potion made with the plant is used to treat fever, biliousness and infantile epilepsy (Delgardo, quoted by Lawler 1984).

Overview

M. muscifera or *Jeevak* is a popular India medicinal herb that is used as a tonic and aphrodisiac in addition to its application for diverse medical conditions like fever, dysentery, rheumatism and even minor complaints like insect bites. It

Fig. 15.7 *Malaxis vesicolor* (Lindl) Abeywickr. Reproduced with permission from *Introductions to Orchids* by Abraham and Vatsala, Parlode, Thiruvananthapuram: Tropical Botanic Garden and Research Centre (TBGRI), 1981

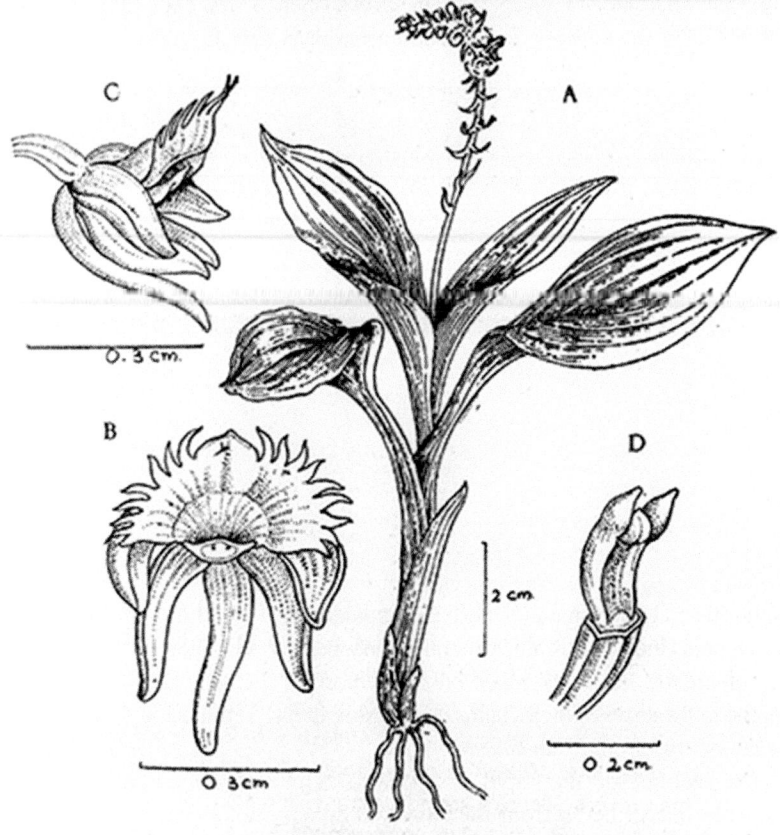

MALAXIS VERSICOLOR (LINDL.) ABEYWICKR. VAR. II

is alleged to improve semen output and boost fertility. In the herb market, it fetches Rs 100–120 kg^{-1}. Collection by pharmaceutical firms has severely threatened the existence of the herb, but this is only one reason why the plant is endangered.

When 10 natural populations of *M. muscifera* in the western Himalayas were studied from 2004 to 2006 by a team of scientists from HNB Garhwal University in Srinagar, two clusters of plant densities were observed. There were six low-density areas that had 0.6–2.8 plants/m^2 and four high-density areas with 7.2–10 plants/m^2. However, high density was not consistently related to species dominance. Maximum dominance of *M. muscifera* was recorded in a site with the lowest density of 0.6 plants/m^2. This suggests that habitat destruction was the principal cause of low *M. muscifera* density. It was a problem which also affected other plant species (Chauhan et al. 2008).

M. muscifera (*Jeevak*) and *Crepidium acuminatum* (*Rishbaka*) are two important constituents of *Astavarga*, an Indian "rejuvenating" tonic that substitutes as aphrodisiac and semen booster. *Asthavarga* is popular in Uttarakhand in the western Himalayas and many parts of the country (Jalal et al. 2008). Due to its popularity, most products selling on the market as *Jeevak* and *Rishbaka* are not genuine. Adulteration with tuberous roots of *Pueraria tuberosa*, *Ipomoea digitata* and *Centaurium roxburghii* (D. Don) Druce (local name: *Lal behmen*) is common (Chinmay et al. 2011; Balakrishna et al. 2013). Histological characteristics can easily differentiate between the species (Chinmay et al. 2011), but this requires some botanical training and laboratory facilities, and would be best conducted by regulatory

Fig. 15.8 Structures of
laburnine and
malaxinic acid

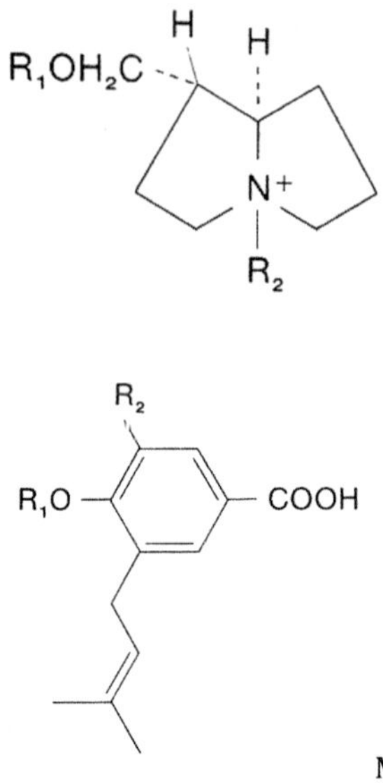

laburnine: R1 = R2 = H

Malaxinic acid: R1 = glu, R2 = H

authorities. Unfortunately, herbal preparations throughout the world are generally not being scrutinised by health or other public authorities for the truthfulness of their claims nor are suppliers penalised for offering substitutes.

On account of the rapid disappearance of valuable Indian herbs from their natural Indian habitats, Cheruvathur et al. (2010) undertook to propagate *Crepidium acuminatum* in tissue culture by inducing adventitious shoots in cultured internodal explants. Meanwhile, Deb and Temjensangba (2006) succeeded with in vitro immature seed germination of another threatened terrestrial Indian *Crepidium* species, *Crepidium khasianum* (Hook f,) Szlach. [syn. *M. khasiana* (Hook f.) Kuntz.], which is not medicinal. The plantlets showed 65 % survival under field conditions. Chemical constituents of *M. monophyllos* which feature in the Ayurvedic *asthavarga* should be studied but there is no publication on the topic.

There is some uncertainty regarding the herb labelled as *Liparis japonica* (Miq.) Maxim by Wu (1994) which could possibly be either *M. monophyllos* or a variant of *Liparis campylostalis* (H.G. Rchb.). Indeed, the herb could be a collection of both species which are vegetatively very similar and grossly indistinguishable in dry form. Shaanxi is listed within the distribution of *M. monophyllos*, but not

Guizhou; *Liparis campylostalis* occurs in Guizhou but not in Shaanxi (Chen et al. 2009c; Chen and Wood 2009). Both provinces are reported as sources for the herb (Wu 1994).

There is no phytochemical information on the medicinal species of *Malaxis* but two alkaloids have been isolated from other species (Slaytor 1977). Malaxine (dehydroartemisin) which is effective against chloroquine-resistant falciparium malaria is present in *M. congesta* (Lindl.) Deb. (Leander and Luning 1967), and several other related orchids (e.g. *LIparis bicallosa* Schltr. and *Liparis hachijoensis* Nakai). Grandiflorine is present in *M. grandiflora* (Schltr.) P.F. Hunt (Luning 1980) (Fig. 15.8).

Microstylis cylindrostachya (Lindl.) Rchb. [see **Malaxis cylindrostachya (Lindl.) Kuntz.**]
Microstylis muscifera Ridl. [see **Malaxis muscifera (Lindl.) Kuntz.**]
Microstylis vesicolor Lindl. [see **Malaxis versicolor (Lindl.) Abeyw.**]
Microstylis wallachii Lindl. [see **Crepidium acuminatum (D. Don). Szlach.**]

Overview

Microstylis is very similar to *Malaxis* in morphology, and, for that matter, also to *Liparis* and *Crepidium*. These orchids are used in a similar manner to prepare tonics and aphrodisiacs in India. There has been no pharmacological study on *M. muscifera*.

Genus: *Microtis* R. Br.

Chinese name: *Congye Lan* (onion leaf orchid)
Japanese name: *Niraba Ra*

Microtis is a small genus of terrestrial orchids with perhaps a dozen species distributed across eastern Asia from Japan and China through the Philippines and Indonesia to Australia, New Zealand and New Caledonia. From a small, globose subterranean tuber, an erect stem arises, bearing a single rod-shaped leaf and a terminal inflorescence which carries numerous, small, intricate, green or white flowers. Plants in the genus are perennial (Tang and Su 1978). *Microtis* is derived from two Greek words, *micros* (small) and *ous* or *otos* (ear), referring to the small membraneous auricle on the column (Schultes and Pease 1963).

Microtis unifolia (Forst.) Rchb. f.

Syn. *Microtis formosana* Schltr. ex Masam

Chinese names: *Jiuye Lan* (garlic leaf orchid), *Congye Lan* (onion leaf orchid),
Chinese medicinal names: *Shuangshencao* (two-kidney grass); *Zhuitaocao* (drop peach grass); *Yigencong* (single piece onion grass); *Chengtuocao* (weight grass)

Description: *Microtis* describes the tiny (micro-) flowers. This is a terrestrial herb related to *Liparis*. Root tuber is globular or ovoid, 1 cm long or smaller and carries a solitary, terete leaf 16 cm long. Without the inflorescence, it looks very much like a garlic plant, and hence its Chinese name. Inflorescence is as long as the leaves and densely many-flowered at its distal third. Flowers are small, 3.5 mm across, green, appearing in August to September (Fig. 15.9). The species is distributed in Japan, China, the Philippines, Java, Australia and Sri Lanka (Lin 1975; Tang and Su 1978). It is common in sunny grasslands at low altitudes, around 100–800 m (Chen et al. 2009a).

Usage: Plants intended for medicinal use are collected from the wild in April and May, washed and dried. The whole plant benefits the spleen and kidneys, and removes gas and humidity. It is used to treat weak spleen, anorexia, discharges, weak kidneys, backache and pain from hernia. To prepare a decoction, 30–60 g of the fresh plant or 15–30 g of the dried product is used (Zhongyao Da Cidian 1986; Hu et al. 2000).

Overview

There is no literature on the biochemistry of *Microtis*. The tiny tubers of *Microtis unifolia* (syn. *Microtis porrifolia* R. Br.; Maori name: *maikaika*) which also occurs in Australia and New Zealand are eaten by Maori children during periods of stress. They are sometimes roasted (Best 1903).

Genus: *Mycaranthes* Blume

Chinese name: *Nimao Lan*

A genus of 25 species distributed from the Himalayas to China and Southeast Asia, and established by Blume in 1825, its members were until recently classified as *Eria* where it was placed by H.G. Reichenbach. Plants are epiphytic or saxicolous, rarely terrestrial. Stem is slender, cylindrical, lacking pseudobulbs and carrying leaves along its entire length. Leaves are alternate or in a biseriate arrangement on the stem, conduplicate or terete. Inflorescence terminal or subterminal, covered with dense, white, woolly hairs. Flowers are small, widely open, cream-coloured, yellow or greenish-yellow. Lip is distinctly trilobed, with mealy ridge or callus (Comber 2001; Chen et al. 2009b).

Mycaranthes pannea (Lindl.) S. C. Chen and J. J. Wood

Syn.: *Eria pannea* Lindl.

Chinese name: *Zhiyemao Lan* (finger leaf hairy orchid), *Zhiyenimaolan*

Fig. 15.9 *Microtis unifolia* (Foster f.) Rchb. f. [as *Microtis porrifolia* (Sw.) R. Br.] From: Fitzgerald R.D., Australian Orchids, vol. 2: t. 74, Fig. 2 (1875–1882). Drawing by R.D. Fitzgerald

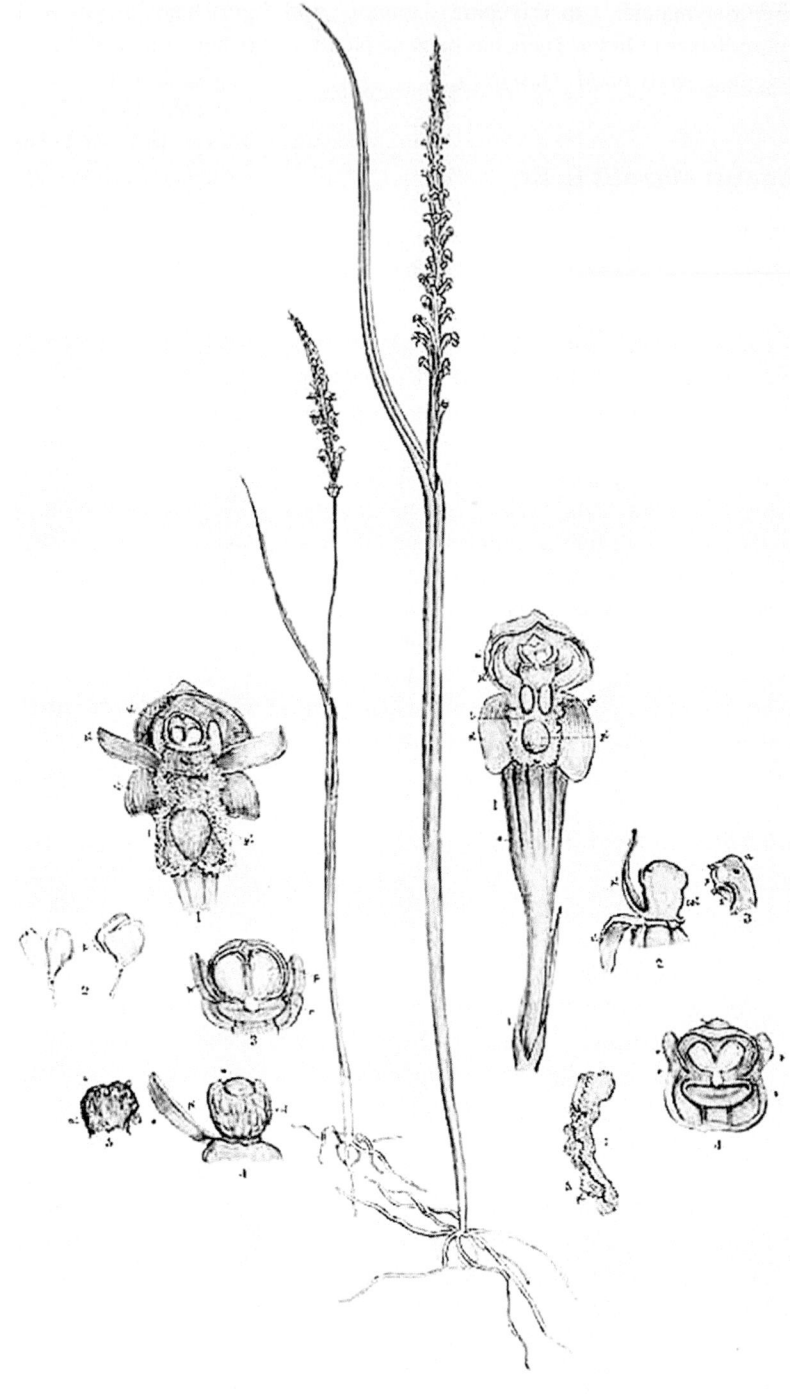

MICROTIS

Parviflora   Porrifolia

Chinese medicinal name: *Shucong*
Malay name: *Kura kubong*
Thai name: *Phrom hom mai*

Description: This attractive, vanilla-scented, small, creeping, epiphytic orchid is common in lowland forests and swamps from Arunachal Pradesh (India), Bhutan (at 1000–1330 m on sun-baked rocks in tropical valleys; Pearce and Cribb 2002), Myanmar, Thailand, Indochina and Malaysia to Sumatra and Kalimantan. It is also found in southern China (Hainan, Guangxi, southwest Guizhou, southwest Yunnan and southeast Xizang) in forests at elevations of 700–2200 m (Chen et al. 1999). Stem bears two or three terete, pendant leaves, 7–15 cm long, and between 1 and 3 flowers. Sepals and petals range from a pale yellow to yellowish-brown and they are covered with a white down on the dorsal surface (Fig. 15.10). Although its individual pseudobulb and leaves are small, the running rhizomes of *M. pannea* can stretch for several metres, completely covering the branches of a

Fig. 15.10 *Mycaranthes pannea* (Lindl.) S.C. Chen & J.J. Wood (Photo: E.S. Teoh)

big tree (Tanaka and Yee 2003) (Fig. 15.11). In Bhutan, it flowers from May to July (Pearce and Cribb 2002), and in Thailand from March to April (Vaddhanaphuti 2005).

Usage: *M. pannea* was used as a bath item when a person is afflicted with ague (malaria or an unremitting fever) in Peninsular Malaysia, Thailand and India (Ridley 1907; Chuakul 2002; Singh et al. 1983; Namsa et al. 2009). It is also an Indonesian custom to prepare a bath with a large collection of herbs for women following childbirth, but the inclusion of *M. pannea* in the puerperal bath is undocumented. In Peninsular Malaysia, the plant was boiled and the decoction used for bathing to relieve bone aches (Alvins quoted by Burkill 1935).

In a study of anti-inflammatory plants used by villagers consisting mainly of the Kampti tribe in the Lohit community of Arunachal Pradesh, India, Namsa and his colleagues came across only one orchid among 34 plants used for this purpose. The fresh leaves of *M. pannea* was made into a paste which was applied over dislocated joints to relieve pain and swelling. Of the 50 men aged 40–75 and 20 women aged 35–60 years that they interviewed, 60 % replied that they were familiar with the use of *M. pannea*. In fact, the orchid was cultivated near their homes to ensure a ready supply (Namsa et al. 2009).

A similar usage is also prevalent in southern China, where *M. pannea* (*Shucong*) is used to improve blood circulation and for detoxification, to treat external injuries, fractures, lower back or leg pain, carbuncles and burns. It has a further use as an antidote for poisoning caused by *Coriaria sinica* (=*C. nepalensis* Wall.), *Aconitum carmichaelii*, *Gelsemium elegans*, *Corydalis incisa* (*Duan Chang Cao*), phosphorus and zinc. The plants mentioned are all extremely poisonous: *Aconitum carmichaelii* (Chinese wolfbane) was used for making poison arrows; it was reported that a Chinese billionaire died after eating a stew containing *Gelsemium elegans*; and *Corydalis incisa* (*Chuan Duan Chang Cao*) which belongs to the opium family (Papaveraceae) contains 10 alkaloids and is a nerve poison (Huang 2010).

Fig. 15.11 *Mycaranthes pannea* (Lindl.) S.C. Chen and J.J. Wood. Reproduced with permission from *Introductions to Orchids* by Abraham and Vatsala, Parlode, Thiruvananthapuram: Tropical Botanic Garden and Research Centre (TBGRI), 1981

Table 15.2 Chinese herbal prescriptions involving the use of *Mycaranthes pannea* (*Shucong*)

1. Indication: To improve blood circulation, for detoxification, external injuries, fractures, carbuncle and burns
Boil 6–9 g of *Shucong* (*M. pannea*) for consumption. Grind the plant for external application
Contraindication: Pregnancy
(Primary Source: *Yunnan Selected Chinese Herbs*)
2. Indication: Fractures: Make a poultice of *M. pannea* with *Bletillia sinensis* and *Piper nigrum* for external application to fractured part
(Primary Source: *Yunnan Shi Mao Chinese Herbal Selections*)
3. Indication: Counter-poison. For poisoning caused by drugs such as *Coriaria sinica*, *Aconitum carmichaelii* (Chinese wolfbane), (Chuan Duan Chang Cao), *Gelsemium elegans*, *Corydalis incisa* (monkshood root), phosphorus and zinc compounds and for acute toxic rash
Boil 3–9 and consume twice a day
(Primary Source: *Wen Shan Chinese Herbs*

Source: Zhongyao Da Cidian (1986) and Zhonghua Bencao (2000)

Several prescriptions detailing the use of *M. pannea* are listed in Zhongyao Da Cidian (1986) and Zhonghua Bencao (2000). The Chinese *Materia Medica* state that the herb must not be used by pregnant women (Table 15.2).

References

Abraham A, Vatsala P (1981) Introduction to orchids, with illustrations and descriptions of 150 South Indian orchids. Tropical Botanic Garden and Research Institute (TBGRI), Trivandrum

Balakrishna A, Srivatava A, Mishra RK et al (2013) Astavarga plants—threatened medicinal herbs of the North-West Himalaya. Int J Med Arom Plants 2(4):661–676

Best E (1903) Food products of Tuhoeland: being notes on the food supplies of non-agricultural tribe of natives of New Zealand: together with some accounts of various customs, superstitions, etc. pertaining to food. Trans. New Zealand Inst 35:45–111

Bhattacharjee SK (1998) Handbook of medicinal plants. Pointer Publishers, Jaipur

Burkill IH (1935) A dictionary of economic products of the Malay Peninsula, vol 2. Crown Agents for the Colonies/Ministry of Agriculture & Co-operatives, London/Kuala Lumpur (1966 reprint, 2nd ed., with contributions by Birtwistle W, Foxworthy FW, Scrivenor JB, Watson IG)

Chauhan RS, Nautiyal MC, Prasad P, Purohit H (2008) Ecological features of an endangered medicinal orchid – Malaxis muscifera (Lindley) Kuntze in Western Himalaya. McAllen Int Orchid Soc J 9(6):8–12

Chen SC, Tang T (1982) A general review of the orchid flora of China. In: Arditti J (ed) Orchid biology. Reviews and perspectives, vol 2. Cornell University Press, Ithaca

Chen, Wood (2009) Malaxis Solander ex Swartz. In: Chen XQ, Liu ZJ, Zhu GH et al (eds) Flora of China – Orchidaceae. Science Press, Beijing

Chen SC, Tsi ZH, Luo YB (1999) Native orchids of China in colour. Science Press, Beijing

Chen, Gale, Cribb (2009a) Microtis R. Brown. In: Chen XQ, Liu ZJ, Zhu GH et al (eds) Flora of China – Orchidaceae. Science Press, Beijing

Chen, Luo, Wood (2009b) Mycaranthes Blume. In: Chen XQ, Liu ZJ, Zhu GH et al (eds) Flora of China – Orchidaceae. Science Press, Beijing

Chen, Ormerod, Wood (2009c) Liparis Richard. In: Chen XQ, Liu ZJ, Zhu GH et al (eds) Flora of China – Orchidaceae. Science Press, Beijing

Cheruvathur MK, Abraham J, Mani B, Thomas TD (2010) Adventitious shoot induction from cultured internodal explants of Malaxis acuminata D. Don, a valuable terrestrial medicinal orchid. Plant Cell Tiss Org Cult 101(2):163–170

Chinmay R, Kumari S, Bishnupriya D et al (2011) Phytocognostical studies of two endangered species of Malaxis (jeevak and rishibhak). Pharmacog J 3(26):77–85

Chuakul W (2002) Ethnomedical uses of Thai Orchidaceous plants. Mohidol Univ J Pharm Sci 29(3–4):41–45

Comber JB (2001) Orchids of Sumatra. Natural History Publications (Borneo), Kota Kinabalu

de Witt HCD (1977) Orchids in Rumphius' herbarium amboinense. In: Arditti J (ed) Orchid biology reviews and perspectives, vol 1. Cornell University Press, Ithaca

Deb CR, Temjensangba (2006) In vitro propagation of threatened terrestrial orchid, Malaxis khasiana Soland ex Swartz through immature seed culture. Indian J Exp Biol 44(9):762–766

Dhayani A, Nautiyal BP, Nautiyal MC (2011) Importance of Astavarga plants in traditional systems of medicine in Garhwal, Indian Himalaya. Int J Biodivers Sci Ecosyst Serv Manage 6(1–2):13–19, India J Ehtnopharm 125(2):234–235

Glicenstein L (2010) Macodes, dossinia and goodyera. Orchids 79:32–37

Gurong DB (2006) An illustrated guide to the orchids of Bhutan. DSB, Thimphu

Hawkes AD (1965) Encyclopaedia of cultivated orchids. Faber & Faber, London

Hu XM, Zhang WK, Zhu QZ et al (2000) Zhonghua Bencao, vol 8. Shanghai Science and Technology Publication, Shanghai

Huang YT (2010) Prospects and challenges of Chinese medicine. A Taiwanese perspective. Recent development in Chinese herbal medicine. In: Nanyang Tech Univ. Conference, 2010

Jalal JS, Kumar P, Pangtey YPS (2008) Ethnomedicinal orchids of Uttarakhand, Western Himalayas. Ethnobot Leaft 12:1227–1230

Jayaweera DMA (1981) A revised handbook of the flora of Ceylon, vol 2. A.A. Balkema, Rotterdam

Jin XH, Zhao XD, Shi XC (2008) Native orchids from Gaoligongshan Mountains, China. Science Pres, Beijing

Lawler LJ (1984) Ethnobotany of the Orchidaceae. In: Arditti J (ed) Orchid biology reviews and perspectives, vol 3. Cornell University Pres, Ithaca

Lawler LJ (1986) Orchid ethnobotany in the ASEAN Area. In Rao AN (ed) Proceedings of the 5th ASEAN orchid congress. Parks & Recreation Department, Ministry of National Development, Singapore, pp 42–45

Leander K, Luning B (1967) Studies on Orchidaceae alkaloids. VII. Structure of a glucosidic alkaloid from Malaxis congesta comb. No. (Rchb. f.). Tetraheddron Lett 36:3477–3488

Lin TP (1974) Alkaloids of the Orchidaceae. In: Withner CL (ed) The orchids. Scientific Studies. Wiley, New York

Lin TP (1975) Native orchids of Taiwan, vol 1. Southern Materials Center, Taipei

Luning B (1980) Alkaloids of the orchidaceae. In: Sukshom Kashemsanta MR (ed) Proceedings of the 9th world orchid conference, Bangkok

Namsa ND, Tag H, Mandal M, et al (2009) An ethnobotanical study of traditional anti-inflammatory plants used by the Lohit community of Arunachal Pradesh

Nasir E, Ali SI (1972) Flora of West Pakistan. Pakistan Agricultural Research Council (Publisher not stated)

Ossian CR (1984) Native orchids of Alaska. In: Tan KW (ed) Proceedings of the 11th world orchid conference, Miami, pp 267–277

Ou JC, Hsieh WC, Lin IH, Chang YS, Chen IS (eds) (2003) The catalogue of medicinal plant resources in Taiwan. Department of Health, Executive Yuan, Taipei

Pandey NK, Joshi GC, Mudaiya RK et al (2003) Management and conservation of medicinal orchids of Kumaon and Garhwal Himalaya. In: Jain AP, Singh V (eds) Ethnobotany and medicinal plants of India and Nepal. Scientific, Jodhpur, pp 114–118

Pearce NR, Cribb PJ (2002) The orchids of Bhutan. Royal Botanic Gardens/Royal Government of Bhutan, Edinburgh/Thimpu

Perner H, Luo Y (2007) Orchids of Huanglong. Huanglong National Park, Sichuan Province, Sichuan

Perry LM, Metzger J (1980) Medicinal plants of East and Southeast Asia: attributed properties and uses. MIT Press, Cambridge

Rao AN (2004) Medicinal orchid wealth of Arunachal Pradesh. Newsl ENVIS NODE Indian Med Plants 1 (2):1–7

Raskoti BB (2009) The orchids of Nepal. Bhakta Bahadur Raskoti and Rita Ale, Kathmandu

Ridley HN (1907) Materials for a flora of the Malay Peninsula, vol 1. Methodist Publishing House, Singapore

Santapau H, Kapadia Z (1966) The orchids of Bombay. Government of India Press, Calcutta

Schultes RE, Pease AS (1963) Generic names of orchids. Their origin and meaning. Academic Press, New York & London

Singh AP (2006) Asthavarga – rare medicinal plants. Ethnobot Leaft 10:104–108

Singh A, Duggal S (2009) Medicinal orchids: an overview. Ethnobot Leaft 13:351–363

Singh U, Wadhwari AM, Johri BM (1983) Dictionary of economic plants in India. Indian Council of Agricultural Research (ICAR), New Delhi

Slaytor MB (1977) The distribution and chemistry of alkaloids in the Orchidaceae. In: Arditti JA (ed) Orchids biology reviews and perspectives, vol 1. Cornell University Press, Ithaca

Subedi A, Kunwar B, Choy Y et al (2013) Collection and trade of wild harvested orchids in Nepal. J Ethnobiol Ethnomed 9:64 73

Tanaka H, Yee ATT (2003) Wild orchids in Myanmar, vol 1–3. The Foundation of Agricultural Development and Education, Bangkok

Tang SL, Su HJ (1978) Flora of Taiwan, vol 5. Department of Botany, Taiwan National University, Taipei

Teoh ES (1982) A joy forever. Vanda Miss Joaquim. Singapore's National Flower. Times, Singapore (reprinted 2008. Singapore: Marshall Cavendish)

Vaddhanaphuti N (2005) A field guide to the wild orchids of Thailand, 4th edn. Silkworm Books, Chiang Mai

Vij SP (1995) Orchid genetic diversity in India: conservation and commercialization. In: Proceedings of the 5th Asia Pacific orchid conference and show, Fukuoka, pp 20–39

Wu XR (1994) A concise edition of medicinal plants in China. Guangdong Higher Education Publication House, Guangdong (in Chinese)

Zhonghua Bencao (2000) Health Department and National Chinese Management Office (eds). Shanghai Science and Technology Press, Shanghai

Zhongyao Da Cidian (1986) Jiangsu New Medical College (eds). Shanghai Science and Technology Press, Shanghai

Genus: *Neottia* Guett.

Chinese name: *Niaochao Lan* (bird's nest orchid)
Japanese name: *Sakane Ran*

There are 70 species in *Neottia*, a genus of saprophytic orchids which are found in pine forests and grasslands in Europe and Asia. China is home to 35 species, 23 of which are endemic, i.e., they do not occur outside the country (Chen et al. 2009a). The name is derived from Greek, *neotteia* (bird's nest), referring to the large cluster of fleshy roots which surround the short rootstalk. Type species is the Eurasian, holomycotrophic *N. nidus-avis* (bird's nest orchid) which has no foliage leaves. Inflorescences grow towards light and generally appear above ground, but cleistogamous subterranean inflorescences have been reported. Vegetative reproduction is accomplished by the transformation of root tips into shoot meristems (Rasmussen 1995), a process that occasionally occurs in *Phalaenopsis* (Teoh 1980). Genome size and gene content are reduced in *N. nidus-avis* and the orchid lacks all genes encoding photosynthetic proteins and RNA polymerase subunits (Logacheva et al. 2011).

Neottia camtschatea (L.) Rchb. f.

Chinese names: *Kanchajianiaochao Lan* (Kamchatka bird's nest orchid), *Beifangniaochao Lan*

Description: *N. camtschatea* is found in the northern Chinese provinces of Xinjiang, Qinghai, Gansu, Shaanxi, Hebei and Nei Mongolia, and in Kazakhstan and eastern Siberia at 2000–2400 m in forests or forest margins, in humid locations. A holomycotrophic terrestrial, its stem is fleshy, 10–30 cm tall, with tiny sheathing scales, long fleshy roots and complex, resupinate, pale green to greenish-white or bluish flowers. Sheathing scales take the place of leaves which are absent in this saprophytic species. Flowers appear in July and August (Chen et al. 2009a).

Herbal Usage: Herb is obtained from Xinjiang, Shaanxi, Gansu, Qinghai and Hebei. The root-like stems of *N. camtschatea* are used by Chinese herbalists to remove heat and toxic materials (Wu 1994).

Overview

Alkaloids were not present in *Neottia* (Luning 1974). The chemical constitution of this holomycotrophic orchid should be investigated in detail because *N. camtschatea* like *Gastrodia elata* (*Tianma*), is associated with endophytic fungi throughout its entire life. Glycosides and lectins have been isolated from some species of *Neottia* (Veitch and Grayer 2010).

Genus: *Neottianthe* (Rchb. f.) Schltr.

Chinese name: *Doubei Lan* (pocket quilt orchid)

© Springer International Publishing Switzerland 2016
E.S. Teoh, *Medicinal Orchids of Asia*, DOI 10.1007/978-3-319-24274-3_16

Neottianthe is a genus of terrestrial or lithophytic orchids formerly considered to be a subgenus of *Habenaria*. There are seven or eight species distributed in the temperate region, from eastern Europe across Russia, China and Mongolia to Japan, and extending southwards to the Himalayas (Hawkes 1965). All species occur in China and five species are endemic (Chen et al. 2009b). The name is derived from Greek and means 'flowers resemble *Neottia*'.

Fig. 16.1 *Neottianthe cucullata* (L.) Schltr. [as *Gymadenia cucullata* (L.) Rich.] From: Schulze, M., *Die Orchidaceen Deutschlands, Deutsch-Oesterreichs und der Schweiz*, t. 45 (1894)

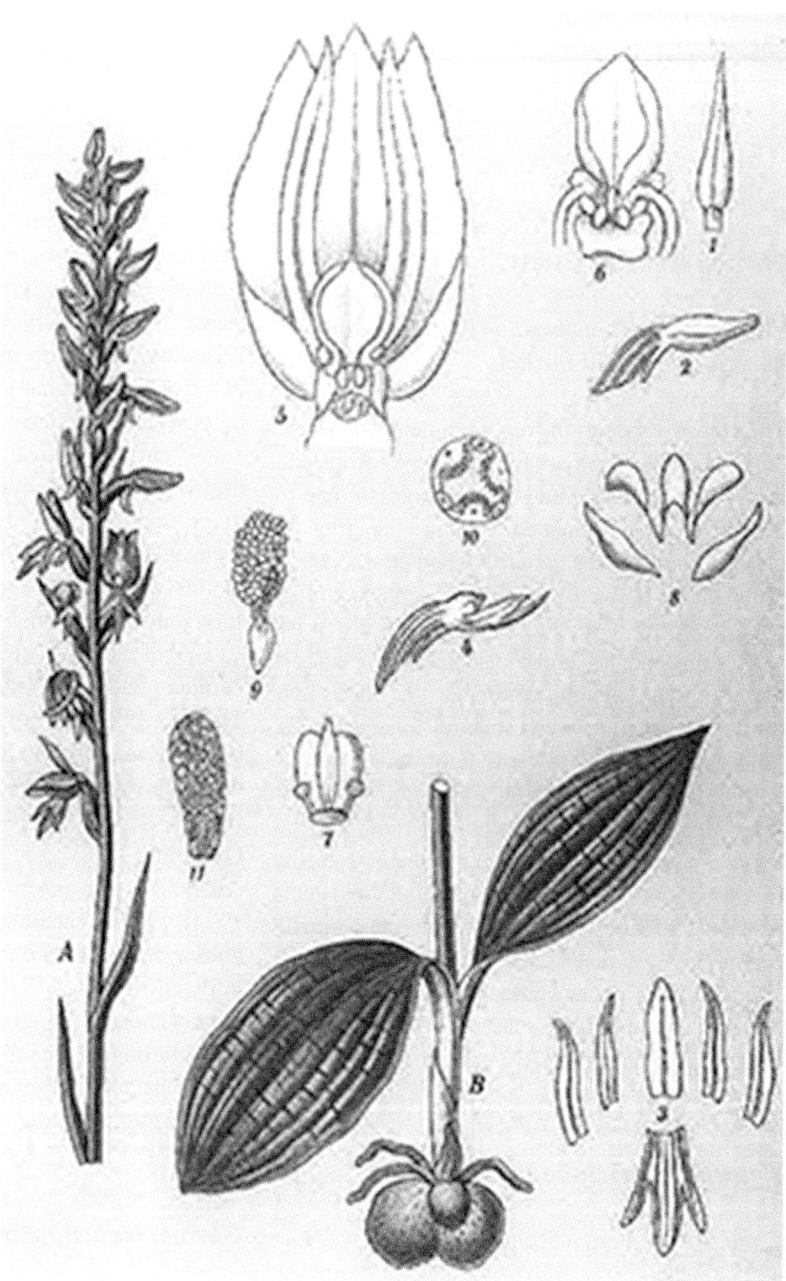

Neottianthe cucullata (L.) Schltr.

Chinese names: *Eryedoubei Lan* (two leaf pocket quilt orchid), *Doubei Lan* (pocket quilt orchid
Chinese medicinal name: *Baibuhuanyangdan*

Description: This small terrestrial orchid plant is 8–24 cm tall. Tubers are paired, ovoid, 1.5 cm long. Leaves also paired, elliptic, 2.2–9 by 1–3 cm, ensheathing at the base. Inflorescence is erect or curved and bears numerous rose-pink to purple flowers. Petals and sepals are narrowly lanceolate. Lip is narrow and shaped like a trident. It is found in forests, thickets and grassland at 400–4500 m in the subtropical or alpine forests of China, Nepal and the Eurasian steppes (Figs. 16.1, 16.2, and 16.3). It flowers from July to October in China (Chen et al. 2009b). The variety *calcicola* flowers in August in Nepal (Raskoti 2009) and August to September in Bhutan (Gurong 2006).

Herbal Usage: Entire plant is used in TCM as a cardiac stimulant, to improve blood flow, clear bruises and rejoin fractures. Its herbal usage extends to coma resulting from trauma (Wu 1994). The plant is obtained from Xinjiang, Shaanxi, Henan, Anhui, Sichuan and Yunnan. It is reputed to dissolve extravasated blood (Chen and Tang 1982). After the dried plant is ground, it can be used as an oral medication, 1.5–3 g per day; alternatively, a suitable amount may be used for external application (Hu et al. 2000).

Rhizome of *N. cucullata* (L.) Schltr. var. *calcicola* (W.W. Sm. Schltr. is used in Nepal to make a tonic (Pant and Raskoti 2013).

Fig. 16.2 *Neottianthe cucullata* (L.) Schltr. (Photo: Courtesy of Plant Photo Bank of China)

Fig. 16.3 *Neottianthe cucullata* (L.) Schltr. (Photo: Courtesy of Plant Photo Bank of China)

Overview

Cyanadin-3-*O*-beta-(6′-o-oxalyl) glucoside is responsible for the purple colour of *N. cucullata* flowers (Strack et al. 1989). There are no pharmacological data on *Neottianthe*. Cattle grazing poses a threat to *N. cucullata*.

Genus: *Nephelaphyllum* Blume

Chinese name: *Yunye Lan* (cloud leaf orchid)

Nephelaphyllum is a genus with a dozen species distributed in China, Japan and India through Southeast Asia to New Caledonia. Plants have short, creeping rhizomes bearing pedunculated, heart-shaped, marbled leaves (Fig. 16.4). Lip is held on top of the flower (Chen and Wood 2009). *Nephelaphyllum* is not a horticultural genus.

The name of the genus is derived from Greek *nephela* (cloud) and *phylum* (leaf), the description referring to the hazy cloudiness on the upper surface of the foliage.

Fig. 16.4 *Nephelaphyllum pulcrum* Blume. From: *Annals of the Royal Botanic Gardens, Calcutta*, vol 8 (3): t. 145 (1891). Original a colour painting by R. Pantling. Courtesy of Missouri Botanical Gardens, St Louis, USA

NEPHELAPHYLLUM PULCHRUM, Blume, var SIKKIMENSIS, Hook. fil.

Lith by A. C. Mukarjei

Nephelaphyllum pulchrum Blume

Description: *N. pulchrum* has a creeping rhizome, fusiform pseudobulbs and heart-shaped leaves, 4–10 by 2.5–6 cm, marbled with lighter and darker green on the upper surface and purple underneath. Inflorescence, 8 cm long, arises from the apex of new pseudobulbs. Flowers are up to 18, purplish, 0.9 mm across, and crowded together (Fig. 16.4). The species is not cultivated. It enjoys a wide distribution in India and Southeast Asia at 300–1000 m. In Peninsular Malaysia and Borneo, it is restricted to the mountains (Yong 1990; Comber 2001).

Phytochemistry: A small amount of alkaloid (0.01–0.1 % dry weight) is present in *N. pulchrum* (Luning 1964).

Herbal Usage: "Diuretic, its crushed rhizome vulnerary" (Cheo, quoted by Perry and Metzger 1980). It is used to treat injuries.

Overview

When Cheo (quoted by Perry and Metzger 1980) reported *Nephelaphyllum* being used as a medicinal herb he did not identify the species. The two species in Peninsular Malaysia are *N. pulchrum* Bl., which is widespread in the lowland and mid-mountain areas in Southeast Asia, and *N. tenuiflorum* Bl., which is an uncommon mountain plant. Thus, it is surmised that the medicinal plant mentioned by Cheo (Perry and Metzger 1980) is probably *N. pulchrum* Bl., the type species of the genus.

N. tenuiflorum Bl. is the only species listed in the recently published, authoritative *Flora of China* (2009), but there is no mention of a species occurring in Sichuan province (Chen et al. 2009c). However, it was reported that rhizomes of *N. chowii* (sic) cooked with pigs' feet was eaten to promote diuresis and a paste made with mashed rhizome was used for itching sores in Sichuan (Hu 1971). No Chinese pharmacopoeia published during the last three decades mentions *Nephelaphyllum* being used as a medicinal plant in China (Fig. 16.5).

Fig. 16.5 *Nephelaphyllum pulchrum* Blume (Photo: E.S. Teoh)

Three species of *Nephelaphyllum* tested negative when they were investigated for the presence of alkaloids and (Luning 1974).

Genus: *Nervilia* Comm. ex Gaudich.

Chinese name: *Yu Lan* (yam orchid)
Japanese name: *Mukago Saishin*

A genus with 100 constituent species, *Nervilia* is distributed from Africa across India, China and Southeast Asia to Australia. The unique feature of the genus is the single, large, heart-shaped leaf carried on a slender stalk. Prominent veins (*nervus* means 'vein' in Greek) on the leaves earned the name for this genus. Flowers appear before the leaf. Afterwards, underground runners, usually two, grow out from the base of the leaf and produce tubers at the tips. Young tubers produce their first leaves. When plant is mature, flowers appear before the leaves.

In India, *Nervilia* species are found in specific habitats. In cultivated land, they occur in shade under hedges, with rows of plants running parallel to the hedges. In forests, they are found in dense, rotting leaf litter. Flowers appear within a week of the arrival of monsoon rains, hence in the last week of May in Karnataka but not until the third week of June in Salsette Island (Santapau and Kapadia 1966).

Nervilia aragoana Gaud. [See ***Nervilia concolor* (Blume) Schltr.**]

Nervilia concolor (Blume) Schltr.

Syn. *Nervilia aragoana* Gaud.

Chinese names: *Dongyamaye Lan* (East Asian leaf orchid), *Yidianhong* (one dot of red); *Yidianguang* (one spot of brightness); *Qingtiankui* (azure sky fan palm); *Yu Lan* (taro orchid); *Yimianluo* (gong orchid), *Guangbu Yu Lan*
Chinese medicinal name: *Bailingzi*
Malay names: *Daun satu tahun* (one leaf a year), *Daun sahelai sa-tahun*
Thai names: *Phaendin yen, Bua sandot*

Fig. 16.6 *Nervilia concolor* (Blume) Schltr. (syn. *Nervilia aragoana* Gaud.). (Photo: Bhaktar B. Raskoti)

Description: *N. aragoana* is a terrestrial herb with a single petioled, heart-shaped, plicate, green, sometimes purple-blotched leaf, 7–15 by 10–17 cm, arising from a small, globular underground tuber. Inflorescence is erect, slim, 30 cm tall, and many-flowered. Flowers are yellow or pale green with a white lip, sometimes marked with purple veins on the lip and sepals (Fig. 16.6).

After a period of dormancy when only the underground tuber remains, the globose corm produces a bud from which two shoots arise, the first a floral shoot, followed later by the folia shoot which expands into a cordate, plicate, undulate leaf supported by a long petiole (Misra 2007). Leaf is apple-green with jagged patches of purple on the upper surface, and solitary. Flowering period is February to May in Thailand and Nepal (Nanakorn and Watthana 2008; Raskoti 2009) and May to July in India (Abraham and Vatsala 1981; Joseph 1982). Leaf appears in October. In China, it flowers from May to June (Chen and Gale 2009).

The species is distributed from the Deccan and the Himalayan foothills across Bangladesh, southern China (Xizang, Sichuan, Yunnan, Hubei), Southeast Asia to Papua New Guinea and Australia and to Taiwan and the Ryukyu Islands, at 500–1000 m. In Kerala, India, it occurs in both forested and non-forested regions of the state (Joseph and Mukkattu 2007). The orchid is often found under clumps of *Euphorbia neriifolia* Linn. in open fields and around bamboo clumps in forests (Santapau and Kapadia 1966). *N. aragoana* is quite variable (Seidenfaden and Woods 1992).

Phytochemistry: Chromatographic analysis showed that *N. aragoana* (=*N. concolor*) contained the same compounds as *N. purpurea* (=*N. plicata*). In that study, two triterpenes were identified, namely, cyclonervilol and cyclohomonervilol (Kikuchi et al. 1981a). *N. aragoana* (=*N. concolor*) tested negative for alkaloids (Luning 1967).

Herbal Usage: In China, the whole plant is considered to be antipyretic. It stops bleeding. The herb encourages diuresis and subsidence of swellings and is used in the treatment of discharges, menorrhagia and weak kidneys. Between 9 and 15 g goes into the preparation of a decoction. It kills bugs (Hu et al. 2000). In Peninsular Malaysia, where the orchid is found only in the northern half, Malays in Kuala Kangsar boiled the leaves and drank the decoction as *ubat meroyan* once considered an obligatory protective medicine after childbirth (Burkhill and Haniff 1930). Roots are chewed to allay thirst in Gaum (Safford, quoted by Burkill 1935). Pseudobulbs are used in the treatment of "disturbances of cardiac function" in Thailand (Chuakul 2002).

The plant of *N. aragoana* is bitter and acrid. Indian herbalists consider the plant to be cooling, and employ it to promote milk production and as a diuretic and tonic. It is used to treat urinary complaints, colic, mental instability, epilepsy, haemoptysis, diarrhoea, asthma, coughs, vomiting and "vitiated conditions of *pitta*" (Sala 1995). It has also been reported that it is used to ally thirst or as an antiseptic (Trivedi et al. 1980), and to treat eye infections in India (Rao 2004).

In some parts of Kerala, *Nervilla carinata* (Roxb.) Schltr. (correct name: *N. concolor*) is used as *ovilattamara* (Sala 1995), i.e. a lotus-like plant characterised by a single leaf and considered to fit the description of *padmacarini*, an Ayurvedic plant that is cooling, galactagogue, diuretic and tonic. There is considerable confusion over both Indian names. Kerala physicians consider *N. aragoana* (i.e. *N. concolor*), *N. plicata* and *N. prainiana* [correct name: *N. crociformis* (Zoll. and Mor.) Seidenf.] as *orilattamara* and *padmacarini*. The Korkus of Melghat Tiger Reserve employ *N. aragoana* (i.e. *N. concolor*) and *Habenaria grandifloriformis* (*Habenaria grandifloris*) as *padmacarini*. Both plants contain alkaloids, flavonoids, phenols and triterpenoids. *Hybanthus ennaespermus*, a non-orchidaceous plant, has also been identified as *padmacarini* even though its appearance does not fit the name (Bhogaonkar and Devakar 2007). (see *N. plicata*). Medicinal

usage in Nepal is similar to that in India (Baral and Kurmi 2006).

Nervilia biflora (Wight) Schltr. [See ***Nervilia plicata*** (Andrews) Schltr.]

Nervilia crociformis (Zoll. and Mor.) Seidenf.

Chinese name: *Baimaiyu* Lan
Thai name: *Bua sandot* (Note: this name is also used for *N. aragoana* Gaud.)

Description: Tubers are globose, whitish, 1.2 cm in diameter and subterranean. Leaf is 7 cm across, kidney-shaped to circular in outline with seven blunt angles, and is borne on a 1- to 1.5-cm-tall petiole. Inflorescence is 3.5 cm tall. It carries a single flower which is well expanded, 4 cm across, with light green, linear tepals and a white lip. In Karnataka, it flowers from May to June (Chen and Gale 2009), of just June (Santapau and Kapadia 1966).

The species is found in the tropical and subtropical Old World, in Australia and the Pacific in grasslands at 200–300 m, but it is not common in Java or Sumatra (Chen and Gale 2009; Comber 2001).

Herbal Usage: Pseudobulbs of *N. crociformis* are used to treat 'faintness' in Thailand (Chuakul 2002).

Nervilia fordii (Hance) Schultze

Chinese names: *Maochunyu Lan* (hairy lip yam orchid), *Qingtiankui* (blue sky sunflower), *Tiankui* (sky sunflower)
Chinese medicinal name: *Qingtiankui* (blue sky sunflower), *Tiankui* (sky sunflower)
Vietnamese names: *Cay moi la, Tran chau diep, Bau thooc, Thanh thien quy.*

Description: This is a perennial herb with a spherical tuber that gives rise to a vertical stem 10–20 cm high carrying usually only one, sometimes two, round, acuminate, cordate leaves, 6–7 by 8 cm wide, with an undulating margin.

Flowers are white and carried on erect spikes. It flowers in May. The species is present in Vietnam and in the Chinese provinces of Guangdong, Guangxi, Yunnan and Sichuan in shaded, damp locations in forests at 200–1000 m (Chen et al. 2009c).

Phytochemistry: Studies on the phytochemistry of *N. fordii* are recent and started with the isolation of five compounds from an ethyl acetate fraction using column chromatography with silica gel, namely, norleucine, 24(S/beta)-dihydrocycloeucalenol-(E)[p-hydroxy cinnamate, rhamnocitrin, rhamnazin and daucosterol (Chen et al. 2013a). Two years later, more isolates were reported. Five new 7-*O*-methylkaempferol and quercetin glycosides, named nevilifordins A–E, were obtained from entire plants of the orchid. Seven known flavonoids and one coumarin were also isolated, and one esculetin exhibited antimicrobial activity against Herpes Simples Type-1 virus (Tian et al. 2009). Simultaneously, another team reported isolating five different compounds, rhamnocitrin, rhamnazin, rhamnocitrin 3-*O*-beta-D-glucopyranoside, ramnocitrin 4'-*O*-beta-D-glucopyranoside and 4-hydroxybenzoic acid (Lu et al. 2009a). A rapid, accurate and highly reproducible HPLC method is now available for the simultaneous determination of seven flavonoids and perhaps this could serve as a method for quality control of the herb (Zhang et al. 2011). Meanwhile five flavonoid glycosides, named nervilifordins F–J, were isolated from aerial parts of *N. fordii,* and two of these compounds, nervilifordins G and J, revealed anti-inflammatory potential by inhibiting nitric oxide production of lipopolysaccharide-activated RAW264.7 macrophages (Qiu et al. 2013).

Petroleum ether and ethyl acetate extracts of *N. fordii* were found to have anticancer effects on two strains of mice (S180 and H22 mice) and prolonged the life of the H22 mice (Zhen et al. 2007b). However, the group did not trace this effect to any of the five compounds that had been isolated from an ethyl acetate extract of the orchid (Zhen et al. 2007a). The petroleum ether fraction of a 95 % ethanol extract of *N. fordii* shows significant antitumour activity in vitro.

Eight compounds were isolated in this study, namely, cycloeucalenol, stigmaterol, sitosterol, ursolic acid, aurantiamide, acetate, (20S,22E,24R)-ergosta-7,22-dien-3beta,5alpha 6beta-triol, 6-methoxy-cerevisterol and beta-daucosterol (Lu et al. 2009b). Six additional compounds were later isolated in the petroleum ether fraction, namely, (9R,10R,7E)-6-9,10-trihydroxyoctadec-7-enoic acid, stigmas-22-en-3beta,6beta,9beta-triol, ergosta-7,22-dien-3beta, 5alpha,6alpha-triol, 3beta-hydroxystigmasr-5, 22-dien-7-one, 3beta-hydroxystigmasr-5-eb-7-one and aurantiamide benzoate (Chen et al. 2013a); also, an acetyl flavonol, named 3-*O*-acetyl-7-*O*-methyl-kaemferol, with anti-oxidant properties (Zhou et al. 2009). Isolates from petroleum ether fractions appear to differ with different groups working on the same orchid species. The six compounds isolated by Chen et al. (2013a) were reported as cyclohomonenervilol, octacosanoic acid, stgmasterol, cyclohormonervilol-E)-*p*-hydroxy cinnamate, 24(R/alpha)-dihydrocycloeucalenol-(E)-*p*-hydroxy cinnamate, and docosanoic acid. One new flavonoid glycoside determined to be 7-*O*-beta-D-glucopyranosylapigenin-8-C-beta-D-glucopyrannosyl-(1->2)-beta-D-glucopyranoside, and a new triterpenoid cinnamate identified as 24(S/beta)-dihydrocycloeucalenol-3-(Z)-*p*-hydroxycinnamate, were recently isolated from *N. fordii* (Chen et al. 2013b).

Water-soluble constituents are more limited: norleucine, complanatuside, 5,7,4'-trihydroxy flavonoid-8-c-beta-D-glucosyl1->4-*O*-beta-D-glucoside and saponarin (Lu et al. 2010), and three new cycloartane glycosides, named nervisides A–C (Wei et al. 2012).

Herbal Usage: Leaves of *N. fordii* are used in Vietnamese folk medicine for managing tuberculosis and chest complaints, 10–15 g daily in decoction. Leaves are also pounded into a poultice to treat abscesses (Duong 1993).

Qingtiankui is bitter-sweet. The whole plant or stem is used to clear the lungs and stop coughs (Chen and Tang 1982). It benefits the stomach, relieves indigestion, relieves anxiety, stops pain, clears away heat and toxic materials, and removes toxins from swellings, and is also used to treat pulmonary tuberculosis, bronchitis, pneumonia, mental illness, sores and ulcers, traumatic injury, inflammation of the oral cavity and laryngitis (Wu 1994; Hu et al. 2000).

Qingtiankui is obtained from Guangdong, Guangxi, Yunnan and Sichuan. For use in traditional Chinese medicine, the herb may be collected throughout the year. It is used fresh or sun dried, pseudobulbs alone, or the entire plant. When used, a decoction is prepared using 10–15 g of the herb. Alternatively, the plant is steeped in alcohol for external use. For stomatitis, it is recommended to chew on fresh pseudobulbs. For infantile malnutrition, malabsorption or worms, 5–10 g (or 6–12 g) of the orchid is cooked with lean pork or hens' eggs and served as food in Guangxi Province (Li 1988; Hu et al. 2000). Mashed tubers are used for external use (Li 1988). Burkill reported that the leaves of *N. fordii* are a Chinese medicinal produce which is imported into Peninsular Malaysia, but he did not elaborate on its usage (Burkill 1935). Its principal usage appears to be "to improve and strengthen weak lungs and for the relief of coughs" (Chen and Tang 1982).

Nervilia gammleana (Hook.f.) Pfitzer.

Syn. *Pogonia gammiena* Hook f.

Indian name: *Shankhaluka*

Description: Plant is up to 20 cm tall with a subglobose tuber, 2 cm across. Stem is stout, green with 1–4 sheaths, and carries a single, terminal, glabrous, petiolate, cordate leaf with undulate margins, 10–15 by 10–15 cm, deep green above and light green underneath. Petiole is streaked with reddish-brown and sheathed along its lower half. Inflorescence bears 6–8 pink or pale lilac flowers, laxly arranged near the top. Flowers are 1.5–2.5 cm long, open minimally, and nodding; tepals lanceolate (Pearce and Cribb 2002). However, Raskoti (2009) illustrates an inflorescence with flowers which

are fully open and well arranged. Lip is trilobed, clawed, and pubescent along the veins. It flowers from April to June in Nepal (Pearce and Cribb 2002; Raskoti 2009).

N. gammieana is found in Pakistan, Nepal and Indian Himalayas, at 600–800 m in Garhwal (Bose and Bhattacharjee 1980) and in pine forest in north-east India.

Phytochemistry: Alkaloids are present in *N. gammieana* (Luning 1967).

Herbal Usage: The tubers are used as *salep* (Pandey et al. 2003).

Nervilia plicata (Andr.) Schltr.

Syn. *Nervilia purpurea* (Hay.) Schltr.

Chinese names: *Maoyeyu Lan* (hairy leaf yam orchid), *Qingtiankui* (blue sky sunflower)
Chinese medicinal name: *Qingtiankui* (blue sky sunflower) Note: same name for *N. fordii*
Indian names: *Padmacarini* (Sanskrit), *Oarilai thamarai* (in Tamil; vernacular name used by primitive Poliyar tribe)

Description: This is a small, herbaceous, variable, terrestrial, deciduous orchid. Soon after the start of the rainy season, the globular, 2-cm-diameter underground tuber sends up an inflorescence followed months later by a single leaf which is heart-shaped, 8–12 cm in diameter, brownish-purple, hirsute on both surfaces, and carried on a short stalk, lying flat on the ground (Figs. 16.7 and 16.8). Two pale olive-green to dull purplish flowers are carried on a short scape, 6–10 cm long. Lip is white and marked by a raised central yellow band and veins that are yellow, brown or purple. It is widely distributed throughout the Deccan where it grows in deep forests (Joseph and Mukkattu 2007), spreading eastwards across Bangladesh, Myanmar, Thailand, northern Peninsular Malaysia and Indochina to Indonesia, the Philippines and Australia (Seidenfaden and Wood 1992). It is also found in rocky, humus-rich soil in hilly forests at 600–1000 m in China, in southern Gansu, Sichuan, Yunnan, Xizang, Guangxi, Fujian,

Fig. 16.7 *Nervilia plicata*
(Andrews) Schltr.
(as *Pogonia*
pulchella Hook. f.) From:
Curtis Botanical Magazine,
vol. 111[ser. 3, vol. 41]:
t. 6851 (1885). Painting in
colour by M. Smith.
Courtesy of Missouri
Botanical Gardens,
St. Louis, USA

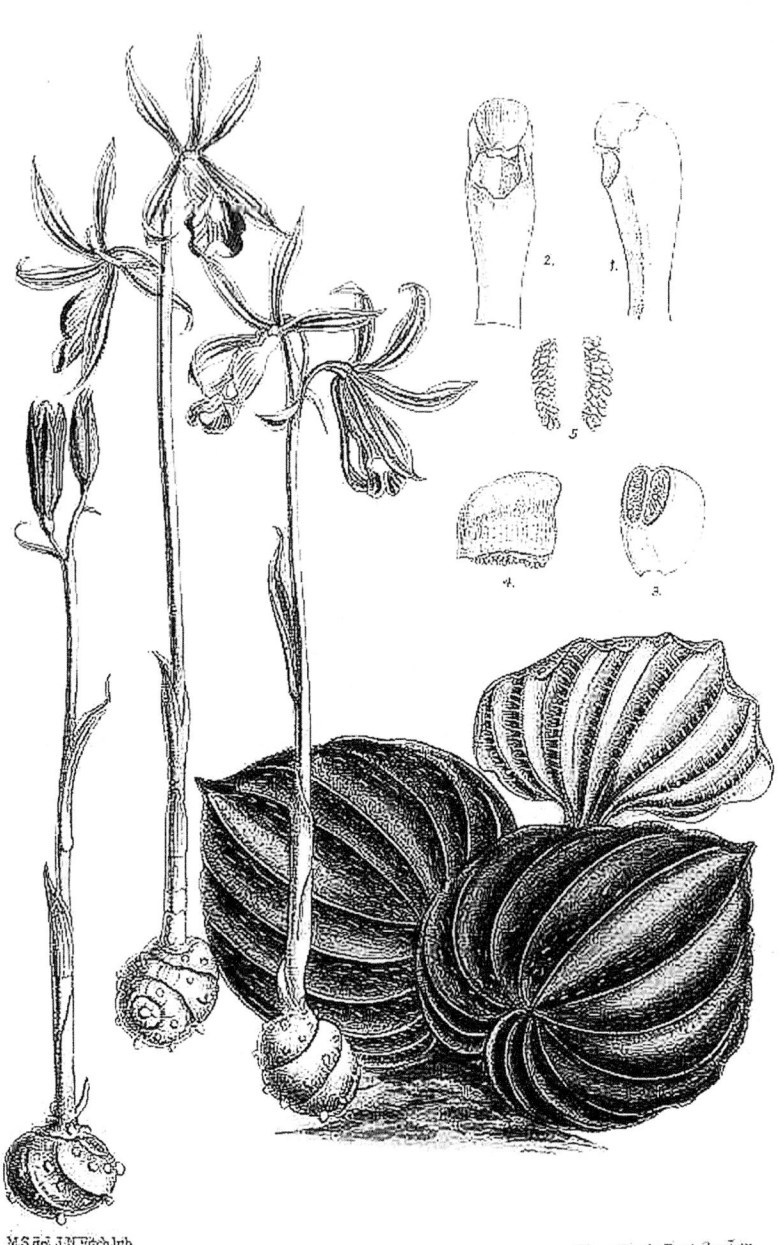

Guangdong and Hong Kong. It flowers in May in China, Thailand and Nepal (Chen et al. 1999; Nantiya 2001; Raskoti 2009), and May to June in Karnataka, Mumbai (Santapau and Kapadia 1966) and Tamil Nadu (Joseph 1982). Tetraploid forms ($2n = 108$) occur in India (Cheenaveeriah and Jorapur 1966). Leaves are present from June to November (Joseph 1982).

Plants growing in dense undergrowth with poor light intensity carry leaves which are deep purple

Fig. 16.8 *Nervilia plicata* (Andr.) Schltr. (Photo: E.S. Teoh)

to almost black in colour; in bright greenhouses, the leaves are green. The plants reported as *N. discolor* carry a yellow median band on the lip whereas in the plants labelled as *N. biflora*, the stripe is white to pale rose. However, the basic structure of the flowers is identical. Thus, they belong to the same species (Santapau 1958). Ancient Indian texts mention that this orchid appears suddenly, from nowhere as it were, and in profusion, just like the lotus. Hence, its common name, *Padmacarini* (resembling the pink lotus) (Kumar and Janardhana 2011).

Phytochemistry: *N. plicata* contains appreciable amounts of alkaloids (Luning 1967; Lawler and Slaytor 1970). Five new triterpenes, cyclonervilol, cyclohomonervilol, dihydrocyclo eucalenol, 24 (R/alpha)-dihydrocycloeucalenol and 24 (S/beta)-dihydrocycloeucalenol together with cyclofuntumienol and cycloeucalenol were isolated from *N. purpurea* Schltr. (=*N. plicata*) by a combination of silver nitrate-impregnated silica gel and reverse phase HPLC (Kikuchi et al. 1981a, b, 1984, 1985, 1986). Subsequently, isopropenylcholesterol was recovered (Kadota et al. 1987), followed by four new methylsterols, cyclonervilasterol, 24-epiccyclo nervilasterol, dihydrocyclonervilasterol and 24-epidihydronervilasterol, from the methylsterol fraction of the orchid plant (Kikuchi et al. 2001).

Herbal Usage: TCM uses the whole plant to clear the lungs and stop coughs, or to remove heat and toxic materials in China (Wu 1994). In Taiwan, the herb is known as *I-tian-hong*. There, it is used to treat bruises, weakness, pneumonia and hypertension. It exhibits analgesic and anti-inflammatory effects on mice (Hsieh et al. 1993). An aqueous extract of the leaves is drunk to facilitate childbirth, and chewed leaves are rubbed on the stomach for pain relief in the Philippines (Fox 1950).

N. plicata (or *padmacarini*) is an ingredient of *Priyagvadignana*, an Ayurvedic preparation mentioned in *Vagbhata* but it is not mentioned in the *Samhitas of Charaka and Susrutha*. Other preparations which include the orchid are *Vastyamayantaka ghratham* and *Satavari Ghrtam*. The mucilaginous extract of the pseudobulb is used, sometimes in combination with tumeric, as an infusion, to treat morbid *Kapha* (cough) and *Varta* (rheumatism), dysuria, urinary calculi, diabetes, diarrhoea, vomiting, jaundice, epilepsy and illnesses attributed to evil spirits. The drug is hot, bitter and astringent. In the Wayanad district of Kerala in Peninsular India, *N. plicata* is an old, traditional remedy for diabetes (Kumar and Janardhana 2011). Paste made with the leaves is administered orally to treat skin diseases by the Poliyers, one of the oldest tribal communities in the Anaimalai Hills of Tamil Nadu in southern India (Sivakumar et al. 2003). Best time for collecting the herb is just before the leaves die down, at which time the tubers are well developed (Comber 2001).

Overview

When *Nervilia* is in bloom, it has the appearance of a saprophytic orchid because its pseudobulbs are subterranean, plant is deciduous and flowers appear before the leaves. This may explain its selection for use in folk medicine. However, it is not a heterotrophic genus. *N. plicata* is used medicinally in the same manner as *Anoectochilus formosanus* in Taiwan. It was reported that *N. fordii* is so extensively collected for local use and export that the species is on the verge of extinction (Hu 2005). It is commonly used "to improve and strengthen weak lungs, stop coughs, and to reduce swellings and inflammation". There is now some experimental evidence that the herb may actually be effective. Pretreating rats with *N. fordii* before subjecting them to endotoxin challenge significantly reduced the swelling of their lungs and simultaneously increased aquaporin-1 and aquaporin-5 expression in the tissues. Thus, by promoting up-regulation of aquaporin 1 and 5 expression, *N. fordii* increases lung water transportation and clearance, preventing pulmonary oedema and damage. It promoted pulmonary function (Xu et al. 2010).

Successful propagation of *N. fordii* by green pod culture using 13- to 16-day-old capsules would be one way to provide a source for the herb (Lin and Yeh 2008). In Nepal, *Nervilia* species are under threat principally due to overgrazing, and secondarily from habitat destruction (Raskoti 2009).

N. plicata is an old traditional remedy used by native healers to treat diabetes in Wayanad, Kerala. There is some evidence to support its usage from animal experiments. Daily administration of 5 mg/kg of an alcoholic stem extract of the orchid to streptozoin-nicotinamide-induced diabetic rats showed 62.0 % and 76.3 % decrease of blood glucose levels on Days 0 and 30, to levels below those recorded in diabetic rats treated with the conventional antidiabetic agent, glibencamide, administered in the same dosage of 5 mg/kg. Two breakdown products of proteins, urea and creatinine, excreted by the kidneys show declines in serum levels of 61.5 % and 71.0 %, respectively, by Day 30. Histological examination of the kidneys provided further evidence of the protective effect of *N. plicata* on the kidneys of treated rats (Kumar and Janardhana 2011).

Numerous compounds have been isolated from *Nervilia* but whether any will join the modern therapeutic armamentarium must await the results of further studies. Two flavonoid glycosides, nervilifordins G and J, from *N. fordii* possess anti-inflammatory activity (Qiu et al. 2013).

Genus: *Neuwiedia* Bl.

Chinese name: *San rui lan*

Neuwiedia is a primitive genus of orchids that some botanists regard as not real members of Orchidaceae (the Orchid Family). The reason is that is it has three stalked anthers instead of the classical pollinia, and the column is very short. Sepals and petals are of equal size. Lip is not well developed, being slightly larger than the other petals and just distinguished by a raised midrib on its upper surface. *Neuwiedia* does not have pseudobulbs or rhizomes. The plant has a simple erect stem which bears a number of spirally arranged, lanceolate, plicate leaves sheathing at the base and a tall, erect inflorescence carrying many small flowers that barely open (Fig. 16.9). It belongs to the Subfamily Apostasioideae Reichb. f. There are eight species distributed in Malesia (Malay Archipelago and Thailand) (Seidenfaden and Wood 1992; Comber 2001). The genus was named after Prince Maximilian von Neuwied.

The primitive *Apostasioids* (*Apostasia* and *Neuwiedia*) are widespread but uncommon. They are usually found in specialised habitats, such as ultramafic substrate, e.g. on Bukit Ghemopuen in Sabah, Borneo, which are prone to destruction by fire, or in tropical heath (*kerangas*) forests (Kocyan 2010).

Neuwiedia singaporeana (Wallich ex Baker) Rolfe [see *Neuwiedia zollingeri* **Rchb. f. var. Singaporeana (Baker) de Vogel**].

Neuwiedia zollingeri Rchb. f. var. *Singaporeana* (Baker) de Vogel

Syn. *Neuwiedia singaporeana* (Wallich ex Baker) Rolfe

Chinese name: *Sanrui Lan*
Thai name: *Makphu makmia*

Description: Plant is 40–50 cm tall with a rhizome that is 10 cm long and 1–1.5 cm in diameter. It bears many lanceolate leaves 25–40 by 3–6 cm. Inflorescence is erect and carries 10–75 yellow, buff or white flowers which are glandular, pubescent on their outer surface. Flowers do not open widely. Lips only slightly larger than the other petals and has a raised midrib on its upper surface. Column is not well formed, and there are three anthers instead of the usual pollinia. Flowering season is May to June (Chen et al. 2009c).

Fig. 16.9 *Neuwideia zollingeri* Rchb. f. From: Reichenbach HG, *Xenia Orchidace*, vol. 2: tab 106, published by F.A. Brockhaus (1874). Courtesy of Biodiversity Heritage Library

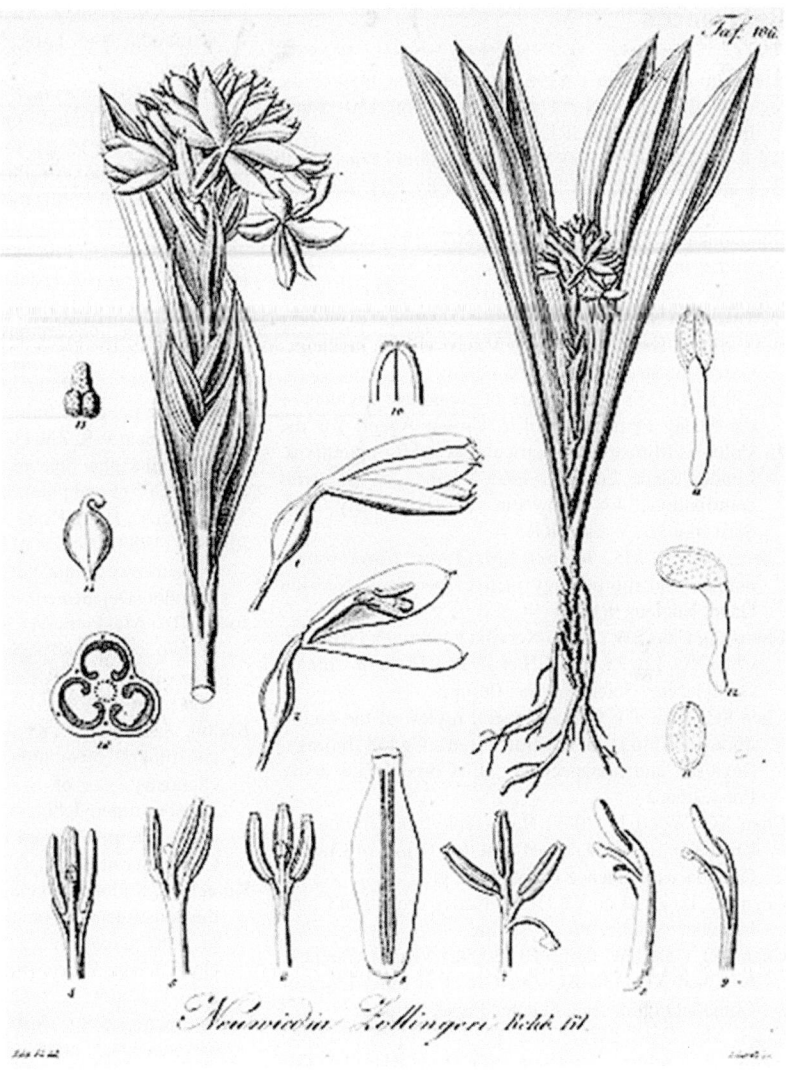

Herbal Usage: The stem and roots of *N. singaporeana* are used to treat furry tongue in Thailand (Chuakul 2002).

The species is distributed in Hong Kong, Hainan, southeast Yunnan, Vietnam, Thailand, Peninsular Malaysia, Borneo, Sumatra, the Lingga Archipelago and Bangka Island. It differs from the type (*N. zollingeri* Rchb. f. var. *zollingeri*) in that the plants are hairy, inflorescence is longer, sepals are also longer, lateral anthers are 5–5.5 mm long, and fruits are green when young, turning orange to red as they ripen (Comber 2001).

Overview

There is no pharmacological information on *Neuwiedia*. The primitive genus is worth investigating for its chemical constituents on account of its uniqueness. This may provide some insight into the origin and range of phytoalexins are developed by orchids for their symbiotic existence with fungi. Seedlings of *N. veratrifolia* collected from the wild in Sabah, east Malaysia, have typical orchidaceous mycotrophic protocorms (Kristensen et al. 2001).

References

Abraham A, Vatsala P (1981) Introduction to orchids, with illustrations and descriptions of 150 South Indian Orchids. TPGRI, Trivandrum

Baral SR, Kurmi PP (2006) A Compendium of medicinal plants in Nepal. IUCN. Published by Mrs. Rachana Sharma, Kathmandu

Bhogaonkar PY, Devakar VD (2007) Pharmaconostic studies on Padmacarini. Aryavaidyan XX 2:74–79

Bose TK, Bhattacharjee SK (1980) Orchids of India. Naya Prokash, Calcutta

Burkhill IH, Haniff M (1930) Malay village medicine. Gard Bull Straits Settl 6:165–321

Burkill IH (1935) A dictionary of economic products of the Malay Peninsula, vol 2. Crown Agents for the Colonies/Ministry of Agriculture & Co-operatives, London/Kuala Lumpur (1966 reprint, 2nd ed., with contributions by Birtwistle W, Foxworthy FW, Scrivenor JB, Watson IG)

Cheenaveeriah MS, Jorapur SM (1966) Chromosome number and morphology in five species of Nervilia Gaud. Nucleus 9(1):39–44

Chen XQ, Gale SW (2009) Nervilia Comm. ex Gaud. In: Chen XQ, Liu ZJ, Zhu GH et al (eds) Flora of China – Orchidaceae. Science Press, Beijing

Chen SC, Tang T (1982) A general review of the orchid flora of China. In: Arditti J (ed) Orchid biology. Reviews and perspectives II. Cornell University Press, Ithaca

Chen XQ, Wood JJ (2009) Nephelaphyllum, Blume. In: Chen XQ, Liu ZJ, Zhu GH et al (eds) Flora of China – Orchidaceae. Science Press, Beijing

Chen SC, Tsi ZH, Luo YB (1999) Native orchids of China in colour. Science Press, Beijing

Chen XQ, Gale SW, Cribb PJ (2009a) *Neottia* Guettard. In: Chen XQ, Liu ZJ, Zhu GH et al (eds) Flora of China – Orchidaceae. Science Press, Beijing

Chen XQ, Gale SW, Cribb PJ (2009b) Neottianthe (Rchb.) Schltr. In: Chen XQ, Liu ZJ, Zhu GH et al (eds) Flora of China – Orchidaceae. Science Press, Beijing

Chen XQ, Liu ZJ, Zhu GH et al (eds) (2009c) Flora of China – Orchidaceae. Science Press, Beijing

Chen JM, Wei LB, Zhang Y et al (2013a) Chemical constituents in petrol ether fraction of *Nervilia fordii*. J Jinan Univ (Nat Sci Med Ed) 03

Chen JM, Wei LB, Lu CL, Zhou GX (2013b) A flavonoid 8-C-glycoside and a triterpenoid cinnamate from *Nervilia fordii*. J Asian Nat Prod Res 15 (10):1088–1093

Chuakul W (2002) Ethnomedical uses of Thai orchidaceous plants. Mahidol Univ J Pharm Sci 29 (3–4):41–45

Comber JB (2001) Orchids of Sumatra. Natural History Publications (Borneo), Kota Kinabalu

Duong NV (1993) Medicinal plants of Vietnam, Cambodia and Laos. World Health Organization, Manila

Fox RB (1950) Notes on the orchids and people of northeast Polillo Island, Quezon Province, Philippines. Orchid Rev 3:16–21

Gurong DB (2006) An illustrated guide to the orchids of Bhutan. DSB Publications, Thimphu

Hawkes AD (1965) Encyclopaedia of cultivated orchids. Faber & Faber, London

Hsieh WT, Tsai HY, Hsieh MT, Cheh CC (1993) Analgesic and anti-inflammatory effects of *Nervilia purpurea* and its active components. Tradit Chin Med 4 (1):89–1107

Hu SY (1971) Orchids in the life and culture of the Chinese people. Chuing Chi J 10:1–26

Hu XM, Zhang WK, Zhu QZ et al (2000) Zhonghua Bencao, vol 8. Shanghai Science and Technology, Shanghai

Hu SY (2005) Food plants of China. The Chinese University Press, Hong Kong

Joseph J (1982) Orchids of Nilgiris. Records of the botanical survey of India, vol 22. Howrah, Botanical Survey of India (Department of Environment)

Joseph TS, Mukkattu M (2007) Description and ecology of two medicinally important species in the genus Nervilia Guad. In Kerala, India. J Econ Taxon Bot 31(4):996–999

Kadota S, Shima T, Kikuchi T (1987) Studies on the constituents of orchidaceous plants. VII. The stereochemistry of cyclohomonervilol and 24-isoppropenylcholesterol, non conventional side-chain triterpenes from *Nervilia purpurea* Schlechter. Chem Pharm Bull 35(1):200–210

Kikuchi T, Kadota S, Hanagaki S et al (1981a) Studies on the constituents of orchidaceous plants I. Constituents of *Nervilia purpurea* Schlechter and Nervilia aragoana Gaud. Chem Pharm Bull 29(7):2073–2078

Kikuchi T, Kadota S, Suehara H, Namba T (1981b) New triterpenes from *Nervilia purpurea* Schlechter, an orchidaceous plant. Structure of cyclonervolol, cyclohomonervilol and chemical correlation with cycloeucalenol. Tetrahedron Lett 22:465–468

Kikuchi T, Kadota S, Matsuda S, Suehara H (1984) Structure and C-13 signal assignments of new methylsterols from *Nervilia purpurea* by two dimensional NMR spectroscopy. Tetrahedron Lett 25 (24):2565–2568

Kikuchi T, Kadota S, Suehara H, Shima T (1985) Studies on the constituents of orchidaceous plants. II. Isolation, structure and stereochemistry of cyclonervilol, cyclohomonervilol and dihydroeucalenolC-24 epimers, new triterpenes from *Nervilia purpurea* Schlechter. Chem Pharm Bull 33:1914–1929

Kikuchi T, Kadota S, Matsuda, Suehara H (1986) Studies on the constituents of Orchidaceous plants, V: Isolation, structure, and C-13 signal assignments of novel methylsterols from *Nervilia purpurea*. Schlechter. Chem Pharm Bull 34(8):3183–3201

Kikuchi T, Kadota S, Matsuda S, Suehara H (2001) Structure and C-13 signal assignments of new methylsterols from Nervilia purpurea by two dimensional NMR spectroscopy. Tetrahedron Lett 25 (24):2565–2568

Kocyan A (2010) Apostasioideae – the least known orchid sub-family. Malesian Orchid J 5:125–138

Kristensen KA, Rasmussen FN, Rasmussen HN (2001) Seedlings of Neuwiedia (Orchidaceae subfamily Apostasiodeae) have typical orchidaceous mycotrophic protocorms. Am J Bot 88(5):9956–9959

Kumar EKD, Janardhana GR (2011) Antidiabetic activity of alcoholic stem extract of *Nervilia plicata* in streptozoin-nicotinamide induced type-2 diabetic rats. J Ethnopharm 133(2):480–483

Lawler LJ, Slaytor M (1970) The distribution of alkaloids in orchids from the territory of Papua and New Guinea. Proc Linnean Soc NSW 94:237–241

Li NH (ed) (1988) Chinese medicinal herbs of Hong Kong, vol 4, 200. Hong Kong Chinese Medicinal Research Institute, Hong Kong

Lin CC, Yeh MS (2008) Studies on seed development and germination of *Nervilia plicata* (Andr.) Schltr. Crop. Environ Bioinform 5:159–170

Logacheva MD, Schelkunov MI, Penin AA (2011) Sequencing and analysis of plastid genome in mycoheterotrophic orchid *Neottia nidus-avis*. Genome Biol Evol 3:1296–1303

Lu CL, Zhou GX, Wang HS (2009a) Studies on the chemical constituents of *Nervilia fordii*. Zhong Yao Cai 32(3):373–375

Lu CL, Wang H, Zhou GX et al (2009) Studies on chemical constituents of petroleum ether extract with antitumour activity from *Nervilia fordii*. J Jinan Univ (Nat Sci Med Ed) 05

Lu CL, Zhou GX, Wang HS et al (2010) Water-soluble constituents from *Nervilia fordii*. Lishizhen. Med Mater Med Res 12

Luning B (1964) Studies on the Orchidaceae alkaloids. I. Screening of species for alkaloids 1. Acta Chim Scand 18:1507–1516

Luning B (1967) Studies on the Orchidaceae alkaloids IV. Screening of species for alkaloids 1. Phytochemistry 6:857–861

Luning B (1974) Studies on Orchidaceae alkaloids. IV. Screening of species for alkaloids 2. Phytochemistry 6:857–861

Misra S (2007) Orchids of India: a glimpse. Bishen Singh Mahendra Pal Singh, Delhi

Nanakorn W, Watthana S (2008) Queen Sirikit botanic garden (Thai native orchids 1 and 2). Wanida Press, Chiang Mai

Nantiya V (2001) A field guide to the wild orchids of Thailand, 3rd edn. Silkworm Books, Chiang Mai

Pandey NK, Joshi GC, Mudaiya RK et al (2003) Management and conservation of medicinal orchids of Kumaon and Garhwal Himalaya. In: Singh V, Jain AP (eds) Ethnobotany and medicinal plants of India and Nepal. Scientific, Jodhpur, pp 114–118

Pant B, Raskoti BB (2013) Medicinal orchids of Nepal. Himalayan Map House, Kathmandu

Pearce NR, Cribb PJ (2002) The orchids of Bhutan. Royal Botanic Gardens/Royal Government of Bhutan, Edinburgh/Thimphu

Perry LM, Metzger J (1980) Medicinal plants of East and Southeast Asia: attributed properties and uses. MIT Press, Cambridge, MA

Qiu L, Jiao Y, Xie JZ et al (2013) Five new flavonoid glycosides from *Nervilia fordii*. J Asian Nat Prod Res 15(6):589–599

Rao AN (2004) Medicinal orchid wealth of Arunachal Pradesh. Newsl ENVIS NODE Indian Med Plant 1(2):1–7

Raskoti BB (2009) The orchids of Nepal. Bhakta Bahadur Raskoti and Rita Ale, Kathmandu

Rasmussen HN (1995) Terrestrial orchids from seed to mycotrophic plant. Cambridge University Press, Cambridge

Sala AV (1995) Indian medicinal plants. A compendium of 500 species, vol 4. Universities Press, Hyderabad

Santapau H (1958) Additions and corrections to the Indo-Nepalese Flora. Proc Natl Sci Inst India 24(3):133–139

Santapau H, Kapadia Z (1966) The orchids of Bombay. Government of India Press, Calcutta

Seidenfaden G, Wood JJ (1992) The orchids of peninsular Malaysia and Singapore. Olsen & Olsen, Fredensborg

Sivakumar A, Subramanian MS, Karunakaran M, Burkanudeen A (2003) Ethnobotany of Poliyars of Anailamai Hills, Tamil Nadu. In: Singh V, Jain AP (eds) Ethnoboany and medicinal plants of India and Nepal. Scientific, Jodhpur, pp 679–685

Strack D, Busch E, Klein E (1989) Anthocyanin patterns in European orchids and their taxonomic and phylogenetic relevance. Phytochemistry 28:2127–2139

Teoh ES (1980) Asian orchids. Times Books International, Singapore

Tian LW, Pei Y, Zhang YJ et al (2009) 7-O-methyl kaempferol and quercetin glycosides from the whole plant of *Nervilia fordii*. J Nat Prod 72(6):1057–1060

Trivedi VP, Dixit RS, Lal VK (1980) Orchids in the drug markets of Bareilly, Kanpur and nearby districts. Nagarjun (Calcutta) 23(8):157–163

Veitch NC, Grayer RJ (2010) Phytochemistry of Neottia. In: Pridgeon AM, Cribb PJ, Chase MC, Rassmussen FN (eds) Genera Orchidacearum, vol 4. Oxford University Press, Oxford

Wei LB, Chen JM, Ye WC et al (2012) Three new cycloartane glycosides from *Nervilia fordii*. J Asian Nat Prod Res 14(6):521–527

Wu XR (1994) A concise edition of medicinal plants in China. Guangdong Higher Education Publication House, Guangdong (in Chinese)

Xu YJ, Chen YB, Wang LL (2010) Effect of *Nervilia fordii* on lung aquaporin1 and 5 expression in endotoxin-induced acute lung injury in rats. Zhongguo Zhong Xi Yi Jie He Za Zhi 30(8):861–866

Yong HS (1990) Orchid portraits. Tropical Press Sdn Bhd, Kuala Lumpur

Zhang L, Zhu CC, Zhao ZX, Lin CZ (2011) Simultaneous determination of seven flavonoids in *Nervilia fordii* with HPLC. Yao Xue Xue Bao 46(10):1237–1240

Zhen HS, Zhou YY, Yuan YF et al (2007a) Studies on the chemical constituents of the ethyl acetate portion of *Nervilia fordii*. Zhong Yao Cai 30(8):942–945

Zhen HS, Zhou YY, Yuan YF et al (2007b) Study on the anticancer effect in vivo of active fraction from *Nervilia fordii*. Zhong Yao Cai 30(9):1095–1098

Zhou GX, Lu CL, Wang HS, Yao XS (2009) An acetyl flavonl from *Nervilia fordii* (Hance) Schltr. J Asian Nat Prod Res 11:498–502

Genus: *Oberonia*, Lindl.

Chinese name: *Yuanweilan* (eagle tail orchid)
Japanese name: *Yaraku Ran*

The genus *Oberonia* is fancifully named after Oberon, the Greek king of the fairies. It is a large genus of sympodial epiphytes that is widely distributed from East Africa across Southeast Asia to Australia in both lowland and montane forests. Stem is short or elongated, with leaves in two rows, sheathing at the base and laterally flattened, thus looking somewhat like a folding fan. Inflorescence is terminal with a many-flowered raceme covered by small to tiny flowers arranged in whorls, in green, white, yellow, orange and brown. When in bloom, it is easily recognisable. *Oberonia* is not a horticultural orchid (Fig. 17.1).

Oberonia caulescens Lindl. ex Wall.

Chinese names: *Erliechunebai Lan* (hare lip white orchid), *Lieyeyinluo Lan* (split leaf jade/pearl orchid), *Yedaiweilan* (slim leaf bird tail orchid); *Xiaxiaojinerhuan* (small gold earring), *Hucha* (curved hairpin), *Yancong* (rock spring onion) *Xiayeyuanwei Lan*
Description: Plant fits the description above for *O. anceps* but it is miniature with stems only 1–4.5 cm long, ensheathed by sword-shaped leaves that are laterally compressed and 1.5–5 cm long. Inflorescence is proportionately larger, up to 20 cm long and many-flowered, with pale yellow or pale green flowers that appear from July to October in China (Chen et al. 1999, 2009), July and August in Sikkim (Bose and Bhattacharjee 1980). It is epiphytic or lithophytic amidst moss in forests at 700–2600 m (in some places up to 3700 m), and is distributed across northern India, Sikkim, Bhutan, Nepal, Tibet, Yunnan, Sichuan, Hunan, Hubei, Guangdong, Taiwan and Vietnam (Chen et al. 2009) (Fig. 17.2).

Herbal Usage: The entire plant is used to stop bleeding and remove bruises. A paste made from the plant is applied directly on fractures and areas of traumatic bleeding (Wu 1994). In India, tubers of *O. caulescens* are used to treat biliary disorders. Tubers are rendered into a paste with black pepper in the ratio of 3:1. Cold water is added and the mixture is drunk before meals (Das et al. 2008).

Oberonia cavaleriei Finet.

Syn. *Oberonia myosurus* (Forster) Lindl.

Chinese name: *Bangyeyuanwei Lan* (stick leaf eagle tail orchid)
Chinese medicinal name: *Yancong*

© Springer International Publishing Switzerland 2016
E.S. Teoh, *Medicinal Orchids of Asia*, DOI 10.1007/978-3-319-24274-3_17

Fig. 17.1 *Oberonia*
falconeri Hook f. From:
Hooker W.J., Hooker, J.D.,
Icones Plantarum. vol.
18: t. 1780 (1888). Drawing
by M. Smith. Courtesy of
Missouri Botanical
Gardens, St. Louis, USA

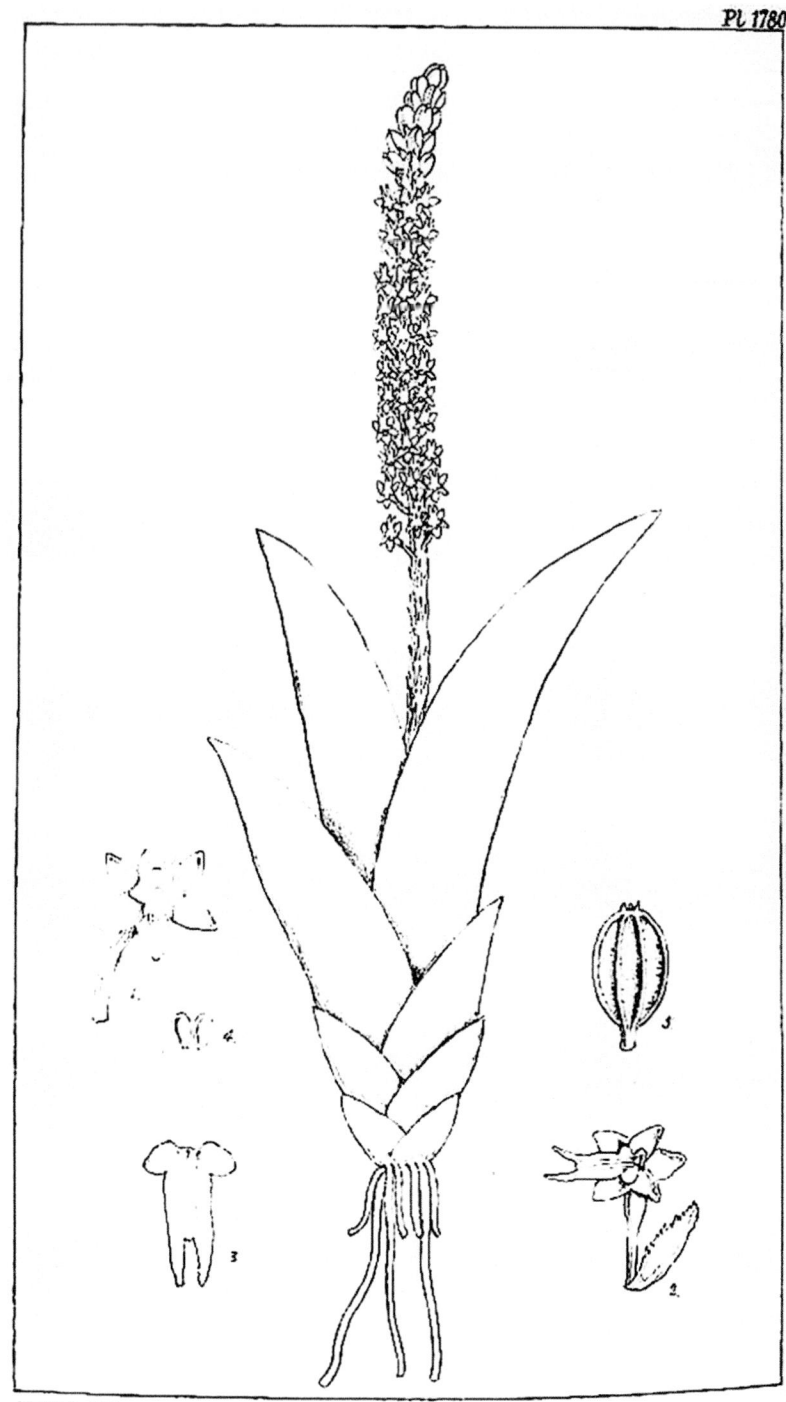

Oberonia Falconeri, Hk.f.

Fig. 17.2 *Oberonia caulescens* Lindl. ex Wall. (Photo: Bhaktear B. Raskoti)

Description: Plant is 6–15 mm tall. Inflorescences are as long as the leaves. Flowers are numerous, white or greenish-white, nearly microscopic and best examined with a hand lens. Sepals are triangular, broad; petals are linear. Lip is large with ciliated edges. It flowers from August to October (Chen et al. 2009). *O. cavaleriei* is found in at 1200–1500 m in China (Sichuan, Yunnan, Guizhou, Jiangxi, Guangxi), Vietnam, Thailand, Myanmar, Nepal (Chen et al. 2009) but in Nepal up to 2700 m (Raskoti 2009).

Herbal Usage: Herb is obtained from Yunnan, Guizhou and Jiangxi. It can be harvested the whole year round and used fresh or be divided into sections and dried. Herb is acrid in taste and slightly bitter. Its nature is cool. Chinese herbal medicine employs it to clear heat and urinary discharge, stop bleeding from superficial wounds and to remove bruises. It is used to treat traumatic injuries, fractures, and strangury resulting from urinary stones (Wu 1994). For the preparation of the decoction, 9–15 g of the orchid is used

(Zhonghua Bencao 2000). The following methods for using the herb are listed in the *Yunnan Record of Traditional Chinese Medicine* (Zhongyao Da Cidian 1986)

1. Ear infection: drip juice into the ear.
2. Bruises: Macerate the plant, add alcohol and fry when cool, place the mix over the wound.
3. Dog bite (!): Use fresh herb, add broad bean and make into a paste for application to wound.
4. Poisoning by other plants: 30 g of *Yancong*. Add 500 ml of water and decoct until the volume is reduced to 300 ml. Drink 100 ml every 4 h.

Tubers are used to treat liver problems in Nepal (Baral and Kurmi 2006).

Oberonia ensiformis (Sm.) Lindl.

Syn. *O. iridifolia* Lindl.

Chinese name: *Jianyeyuanweilan*
Chinese medicinal name: *Shubianzhu*

Description: *O. ensiformis* is a rather large plant with a short stem and subbasal, equitant leaves, laterally compressed, 15–50 by 1.2–2.3 cm, generally all curving in one direction. Inflorescence is stout, much shorter than the leaves (Fig. 17.3). Raceme is 10–25 cm long, bearing many green to yellowish-green flowers 2.5 mm across. Sepals and petals are recurved, petals erose-dentate or pubescent. Lip is erose-dentate, pubescent, trilobed, twice as long as the tepals and darker; upon drying, it turns black (Santapau and Kapadia 1966). The variety at Gaoligongshan has bright red flowers (Jin et al. 2008). Flowering season is September to November in China (Chen et al. 2009), October in Tamil Nadu (Matthew 1995), October to December at Palnis Hills (Seidenfaden 1999), or November in much of India (Sikkim, Orissa, Western Ghats) (Bose and Bhattacharjee 1980), the flowers lasting 2 months on the "rat's tail" flower-stalks (Davis and Steiner 1982). It is epiphytic on trees at 700–1600 m and is distributed in northern Guangxi and Yunnan

Fig. 17.3 *Oberonia ensiformis* (Sm.) Lindl. (Photo: E.S. Teoh)

Table 17.1 Chinese Herbal Prescriptions containing *Oberonia ensiformis* (*Zhongyao Da Cidian*, Jiangsu New Medical College, *1986*)

1. Indication: For diuresis; to improve blood circulation; to treat cystitis, urethritis, external injuries and fractured bones
Boil 9–15 g and consume. Grind for external application
2. Indication: Selected prescription for cystitis
Prepare a decoction of Shubianzhu (*O. ensiformis*), Coix lacrym-jobi and Houttuynia cordata and consume
(Primary Source: Yunnan Shi Mao Chinese Herbal Selections)

southwards to Vietnam, Laos, Thailand, Myanmar, Nepal and India (Chen et al. 2009). It was very common at the Palnis Hills, growing on stones or trees at 500–1200 m (Seidenfaden 1999).

Herbal Usage: As the orchid has a very narrow distribution in China, its usage is probably limited to the ethnic minorities living in Yunnan and Guangxi. It is used to encourage diuresis, improve blood circulation and to treat cystitis, urethritis, injuries and fractures. Two prescriptions listed in Table 17.1 shows how *O. ensiformis* is used medicinally (Zhongyao Da Cidian 1986).

Oberonia falconeri Hook f.

Indian name: *Kanchapra*

Description: Plant is epiphytic, 7–12 cm tall. Stem is short, inconspicuous. Leaves are 3–6, equitant, overlapping at the base, ensiform, laterally flattened, thick, 1.5–8 by 0.7–1.0 cm. Rachis is 6–13 cm, laxly many-flowered. Flowers are spirally arranged, 1–3 mm across, white or green to greenish-yellow. It flowers in August in Nepal and north-east India (Raskoti 2009), and August to September in China (Chen et al. 2009). *O. falconeri* is distributed from continental India and Yunnan (China) across Bangladesh, Myanmar, Thailand and Indochina into Peninsular Malaysia.

Herbal Usage: Plant is used to treat fractures in Indian Himalayas (Jalal et al. 2010).

Oberonia longibracteata Lindl.

Chinese name: *Changbaoyuanwei Lan*

Description: *O. longibracteata* is a tufted epiphyte with compressed, flexuous stems and fleshy, equidistant leaves, 3.7–5.5 by 0.3–0.5 cm, oblong to linear and acute at the apex. Leaf veins are not evident and the base is confluent with the stem. Inflorescence is terminal, dominated by bracts which outsize the tiny, yellowish-brown flowers. The very long floral bract is very characteristic and gives the name to the species. In Sri Lanka, where it is often found on *Palaquium rubiginosum* (Thw.) Engl., it flowers from September to March (Jayaweera 1981).

The species is distributed from Sri Lanka and the Indian Peninsula across Myanmar to Indochina and Hainan Island. In Vietnam, it occurs in dry lowland *Dipterocarp* forests as well as in the central Annamese floristic province which experiences high precipitation (Averyanov et al. 2003).

Herbal Usage: It is used for scorpion bites in Kampuchea (Uphof 1968; Lawler 1984).

Oberonia lycopodioides (J.Koenig) Ormerod

Syn. *Oberonia anceps* Lindl.

Malay name: *Sakat lidah buaya*

Description: Stem is 10–30 cm long, 2 cm wide and is covered with thick, short alternating leaves with blunted tips, 1.5–3 cm, arranged obliquely. Inflorescence is terminal and densely many-flowered, the flowers non-resupinate, orange, darkening at the throat and column, arranged in a spiral, facing all directions, opening simultaneously and lasting a long time (Millar 1978; Cheam et al. 2009).

O. *anceps* is a common, sun-loving, lowland, miniature species distributed in Indochina, Thailand, and from Sumatra eastwards to Malaysia, Kalimantan, the Philippines and Sulawesi.

Herbal Usage: The pulverised plant was used as a poultice to treat boils and infected wounds in Malaysia (Caius 1936).

Oberonia mucronata (D. Don) Omerod and Seidenf.

Chinese names: *Yuanweiyebai Lan* (eagle tail leaf, white orchid), *Yuanweiyebianzhu Lan* (eagle tail leaf, flat bamboo orchid); *Yuanwei Lan* (eagle tail orchid)

Description: Stem is short, inconspicuous. Leaves are 5–6, subbasal, distichous, equidistant, laterally compressed, thick, 6–16 by 0.6–1.5 cm, apex acuminate or obtuse. Peduncle is terete, 20–25 cm, densely many-flowered, flowers facing all directions. Flowers are reddish-brown and appear from August to December.

O. *mucronata* occurs in Yunnan, Laos, Philippines, Malaysia, Indonesia, Myanmar, Bangladesh, India, Nepal and Bhutan. In Yunnan it is epiphytic on trees in forests at 1300–1400 m (Chen et al. 2009).

Herbal Usage: The entire plant is used in TCM to relieve gas and help in digestion. It clears urinary discharge, stops coughs and pain.

It is also used to treat traumatic injuries, fractures and snake bites (Wu 1994). In the Pacific island of Fiji another species, *O. glandulosa* is used for pain relief (Pridgeon 2001).

Overview

Alkaloid contents equal to or exceeding 0.1 % were detected in 5 of 29 species of *Oberonia* (Lüning 1974) but only in small amounts (between 0.01 and 0.01 % dry weight) was present in *O. ensiformis* (Luning 1964). Despite this promising start, no further investigation has been reported. *Oberonia* does not pay an important role in Traditional Chinese Medicine as there is no mention of plants being saved in dry form.

Oberonia myosurus (Forst.) Lindl (see: *Oberonia cavaleriei* Finet.)

Genus: *Ophrys* L.

Ophrys is a large genus of perennial terrestrial herbaceous orchids which is distributed in central and western Europe from Scandinavia to the Mediterranean, with an extension to northern Africa, Asia Minor and northern Iran. There are 128 species. Tubers are oval and paired, the old tuber much smaller than the new. Stem is glabrous.

Leaves are basal, cauline, lanceolate, unspotted, ensheathing the stem. Inflorescence is erect, laxly several-flowered (Fig. 17.4). Flowers resemble bees or insects and are showy. Sepals and petals are free and spreading, sepals larger than the petals. Lip is entire or trilobed and brightly coloured.

Ophrys scolopax Cav.

Description: This is a small to medium-sized cold-growing, terrestrial herb. Tubers are oval, and paired. Stem is erect, 50 cm tall. Leaves are 2–6, lanceolate, upright, basal, ensheathing the bottom part of the stem. Inflorescence 10–50 cm, with 2–15 (commonly 3–10) flowers. Sepals are green to pink or violet, 0.7–1 by 0.3–1 cm. Petals 1.5–8 by 0.8–4 mm broad. Lip is trilobed, 0.6–1.6 cm

Fig. 17.4 *Ophrys scolopax* Cav. From: Cavanilles, A.J., *Icones et descriptions plantarum*, vol. 2: t. 161 (1793). Drawing by A.J. Cavanilles. Courtesy of Gottfried Wilhelm Leibniz Biblothek, Hannover, Germany

OPHRYS SCOLOPAX.   Tab. 161.

A J Cavanilles del.   M Gamborino Sculp.

long. It has a complex three-dimensional shape and intricate pattern, is brownish and very striking. Margin of central lobe is velvety; side lobes are pubescent (Fig. 17.4). Flowering season is March to June depending on locality. It is distributed from Hungary and the Mediterranean (including Algeria, Morroco and Canary Islands) to northern Iran.

Herbal usage: Tubers are used as *salep* (Ghorbani et al. 2014b).

Ophrys sphegodes Mill.

Common name: Early spider orchid

Description: Tubers are globular, subterranean. Plant is robust, slender, 40–50 cm tall, ensheathed at the base by a rosette of 4–5 small, ovate-lanceolate leaves which are green with darker longitudinal veins but no spotting. Inflorescence is 15–30 (up to 45) cm tall, laxly

Fig. 17.5 *Ophrys sphegodes* Mill. (Photo: Henry Oakley)

2- to 9-flowered, Flowers are small, with typical bee shape, but its common name attributes to the dark brown, oval lip which is likened to the body of a spider. Sepals or petals are yellow to brownish-green, white or pink, oblong-lanceolate. Petals are darker than sepals. Lip is oval, convex, crenate, hirsute at the base, glabrous along the yellow edges. Speculum at the centre is outlined by two parallel bluish to violet narrow bands (Fig. 17.5).

It is distributed from western and southern Europe to northern Iran in well-drained chalky soil in meadows, pastures, fallow and forest clearings; in Italy at 100–750 m (Italian Group for the Research on Wild Orchids 2013).

Phytochemistry: Loroglossin, a phenolic glycoside was isolated from *O. sphegodes* (Veitch and Grayer 2001). It is a phytoalexin and does not play any role in the alleged medicinal properties of *salep*.

Herbal usage: Tubers are used as *salep* (Ghorbani et al. 2014b).

Ophrys sphegodes subsp. *mammosa* (Desf.) Soo ex E.Nelson

Syn. *Ophrys mammosa* Desf.

Description: A handsome, robust, terrestrial species, its name refers to the two prominent lateral swellings below the base of the lip, *mammosa* meaning "full breasted". Stem is erect, cylindrical, usually 25–30 cm but may be up to 70 cm tall, with basal, lanceolate leaves arranged in a rosette and ensheathing the stem. Inflorescence is erect, with obtuse, inward curving bracts that are longer than ovaries, loosely several (2–15)-flowered. Flowers are vividly patterned and coloured. The subspecies is distinguished from the type by two protuberances that appear like breasts sticking out at the sides of the labellum (Fig. 17.6). Sepals are lanceolate, spread out and bicoloured; in the lateral sepals, the green is overlaid with various intensities of light red on the lower half. Petals are narrow, much smaller than the sepals, and monochromatic, either green or red to brown. Lip is entire, pendant, large, 1–2 cm, oval, convex, crenated, pointed at the tip, dark brown centrally fading to wine red at the rim. In some flowers, the yellow on the back of the lip extends to outline the front of the lip. The principal characteristic of this subspecies are the two conical, forward-pointing, velvety to hirsute protuberances at the base of the lip. Rest of the lip is velvety to subglabrous. The spur is edged by two narrow, parallel, grey or violet lines which give the impression of the letter H. Many intermediate forms occur, especially in Cyprus. These generally flower earlier and are the probably hybrids of *O. sphegodes* ssp. *mammosa* with other subspecies. Flowering season of typical *O. sphegodes* ssp. *mammosa* starts in early April in Cyprus (Kreutz et al. 2002).

It is distributed from south-eastern Europe to Iran, Iraq and Turkmenistan, in open woodland or light scrub in full sun or semi-shade.

Herbal usage: Tubers are used as *salep* (Ghorbani et al. 2014b).

Fig. 17.6 *Ophrys sphegodes* subsp. *mammosa* (Desf.) Soo ex E.Nelson (Photo: Henry Oakley)

Overview

In the Mediterranean where *Ophrys* predominate, many species occur in abundance but populations of *O. sphegodes* has been observed to decline in the British Isles due to poor seed set. *Ophrys* faces two principal threats, (1) from over-collection of its tubers for making *salep* in Turkey and Iran, and (2) farming and other forms of human encroachment. There is no merit in promoting *salep* because its taste is bland and all claims of its nutritious or aphrodisiac properties are fictitious.

Leaf rosettes of *Ophrys* appear in late autumn and plants remain green through the winter. Flowering occurs in early or late spring depending on species and environment. The aboveground portion of the plant senesces with seed set. *Ophrys* manage the summer heat by staying as tubers underground. Plants reach maturity rapidly and 70 % will flower in the first year of he appearance of the leaves. However, peak performance occurs at 4–7 years when 100 % of plants which are not dormant will

flower. However, few plants of *O. sphegodes* survive for more than 3 years (Hutchings 1987).

O. sphegodes are pollinated by sexually excited solitary bees, *Andrena nigroaenea*, that are lured by visual cues and volatile semiochemicals consisting of variable mixtures of alkanes and alkenes, especially the latter. Flower-specific odour variation between plants mimic sex pheromones of individual female bees preventing bees from revisiting flowers and effecting cross-breeding success (Ayasse et al. 2000; Schiestl 2000). After pollination, there is an increase in the amount of all trans-farnesyl hexanoate which signals pollinator bees not to visit the flower, in other words, to visit unpollinated flowers instead, thus maximising reproductive success (Schiestl and Ayasse 2001). In continental Europe, 45 % of seed capsules ripen, whereas in the British Isles, only 6–18 % do so. Pollination availability may be a reason for this difference. *O. sphegodes* is capable of spreading rapidly in favourable habitats. Conservation of *O. sphegodes* depends on the creation of conditions that are conducive to flowering, seed set and seedling establishment (Hutchings 1987).

Genus: *Orchis* Linn.

Common name: Dog stones, Satyrion
Chinese name: *Hongmenlan* (red door orchid)
Japanese name: *Hakusan Chidori*

The word 'orchid' is derived from the genus *Orchis*, but the distinctive character of *Orchis* is missing from epiphytic orchids. This special characteristic is the presence of pair of oval tubers in *Orchis* that look very much like testicles, hence *orchis*, Greek for 'testicles'. From ancient Greece to the eighteenth century, when the *Doctrine of Signatures* reigned supreme in the practice of European medicine, such orchid tubers were prized as aphrodisiacs.

Salep drinks were popular in the Ottoman Empire. Bars offering *salep* as a drink began to appear London and continental Europe during

Fig. 17.7 (*Left*) *Orchis militaris* L. (as *Orchis latifolia altera*) and (*Right*) *Orchis mascula* (L.) L. (as *Orchis V*). From: Clusius, C., *Rariorum plantarum historia*, vol. 1 Fasicle 2, p. 267; and P. 268, Fig. 2 (1601). Courtesy of Missouri Botanical Gardens, USA

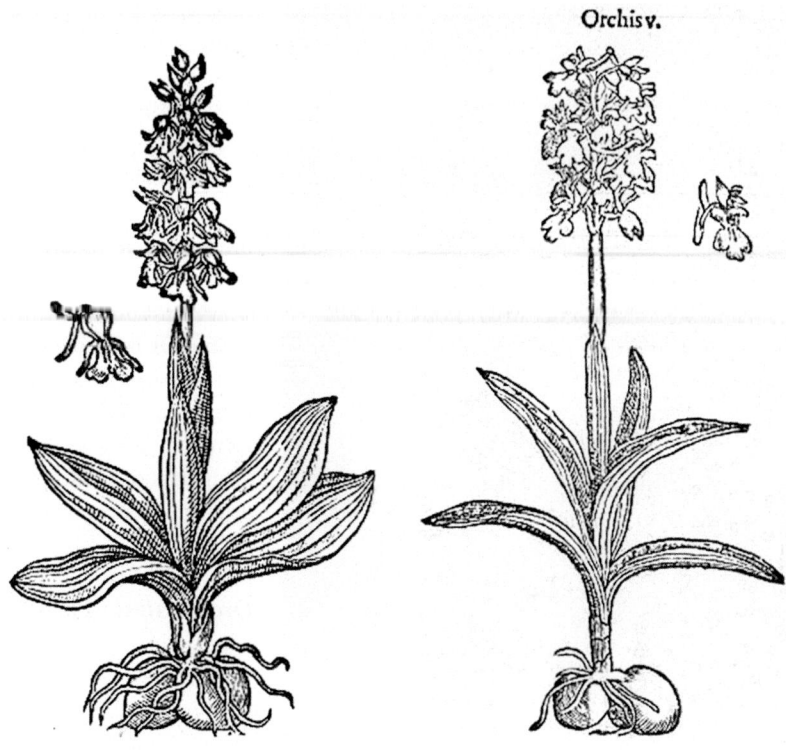

Orchis v.

the Middle Ages. The English writer, Charles Lamb (1775–1834), stated that a "Salopian shop" in Fleet Street and commented that the taste of *salep* had "a delicacy beyond the China luxury" to many people. Although the aphrodisiacal property of *Orchis* is mere fiction, *salep* is still sold today and it has become a component of Turkish ice cream. *O. latifolia* (proper name: *Dactylorhiza hatagirea*) still admired as an aphrodisiac in India, and is incorporated into its herbal preparations for "rejuvenation".

Mankind being so obsessed with aphrodisiacs, it is no surprise that an astounding number of orchidaceous plants (1924 in one count) have been described as distinctive species or varieties of *Orchis*. Today, only around 30 different species names are officially accepted. There are over 40 interspecific hybrids within the genus, and several natural inter-generic hybrids involving *Orchis* (Hawkes 1965). *Orchis* species are abundant in central and southern Europe, extending eastwards into the Himalayas and China. All species are terrestrial. Some are also found on rocks.

Orchis adenocheila Czerkiak

Description: *O. adenocheila* is a small, cool growing terrestrial herb. Stem is short, carries a whorl of 3–5 sessile, elliptic leaves with keels. Inflorescence is 20 cm tall with several to many flowers. Tepals are fused to form a dark green to dark brown helmet. Lip is trilobed. Lateral lobes are long, narrow, held at 45° to the horizontal: mid-lobes are large, diverging distally into two broad side lobules that are held parallel to the side lobes. Flowers are white or pale green, heavily spotted with red on the mid-lobe of the lip. Flowering season is spring. It is distributed from Trancaucasus to northern Iran at 1200 m (Fig. 17.7).

Herbal usage: Tubers are used as *salep* in Turkey and Iran (Ghorbani et al. 2014b).

Orchis latifolia Linn. (see *Dactylorhizia hatagirea*)
Orchis laxiflora Lam. (see *Anacamptis laxiflora*)

Fig. 17.8 *Orchis mascula* (L.) L. (Photo: Henry Oakley)

Orchis mascula (L.) L.

Common name: *The Salep Orchid*, *Early Purple*
Indian names: *Salab Misr*, *Salep Misri* (Hindi),
    Salum (Bombay)
Iranian names: *Punjah-i-salaba*; *Salab*

Description: Tubers of *O. mascula* are paired,
ovoid, 2 cm long. Stem is 20–60 cm tall sheathed
with lanceolate leaves, 7–15 cm long at the base.
Inflorescence is 8–12 cm tall, with many pinkish-
purple flowers (Fig. 17.8). Flowering season is
April to June. The species is distributed throughout
Europe, North Africa and temperate Asia to Siberia.

Usage: *Salep* is considered to be an
invigorating substance and an aphrodisiac. It
was boiled with milk and administered to people
suffering from phthisis, diabetes, chronic diarrhoea and dysentery in India (Chopra
et al. 1958). In Nepal, it is used as an expectorant,
astringent and nutrient (Suwal 1970). The tuber

is used as a demulcent, tonic and nutrient, especially when Iraqi infants suffer from diarrhoea
(Ali-Al-Rawi and Chakravarty 1964, quoted by
Lawler 1984). At one time, the *Pharmacopoeia
of Edinburgh* included root of *O. mascula* as a
demulcent. It was valued for its ability to expand
into a tremendous volume when water is added,
producing a mucilaginous or gelatinous substance referred to as *Bassorin* (Johnson and
Sowerby 1865) which had a sweetish taste and
a faint, if somewhat unpleasant, smell (Hooper
and Akerly 1829). Due to its high content of
mucilage, *salep* prepared with *O. mascula*
forms a thick jelly even when diluted with
40 parts of water. In the Peloponnesus, the
dried root is cooked and eaten (Hedrick 1919).

Orchis simia Lam.

Turkish   names: *Tavsantopugu*,   *topanbas*,
    solmazsoluk, *puskullu* (Sezik 1969)

Description: Plants are 20–40 cm tall, leaves 2–6,
ovate to lanceolate, 20 by 4.5 cm. Inflorescence
is short, flowers crowded into a dense cylinder,
3–9 cm long. Flowers are white, spotted with
lilac on the inner surface of the helmet and
mid-lobe of the lip. Sepals are not fused distally.
Lip is trilobed, 2 cm long, side lobes 1.1 cm by
1 mm. Filiform lobes are a deep purple. The
effect of the peculiarly shaped lip and helmet is
to produce a caricatured human figure which has
been likened to naked men (Fig. 17.9). Flowering
season is March to May.

It is distributed from Europe to Iran and north
Africa, occurring in meadows and sparse forests
in sunny locations on neutral soils or limestone
from lowlands to 1550 m.

Phytochemistry: *O. simia* contains a phenolic
glycoside, loroglossin, coumarin precursors,
anthocyanins, cyanin, orchicyanins I and II
(Ernst and Rodriguez 1984), mucilage 29.89 %,
starch 1.71 %, and sugars 1.97 % (Sezik 1967).

Herbal Usage: It is used as *salep* in Iran and
Turkey (Ghorbani et al. 2014b).

Fig. 17.9 *Orchis simia* Lam. (Photo: Henry Oakley)

Overview

European fascination with *Orchis* and *salep* vastly exaggerated the importance of this group of orchids in the old *Materia Medica*. Relying the *Doctrine of Signatures*, the ancient Greek physician, Pedanius Discorides (40–90 CE), alleged that the consumption of *Satyrion* not only stirred the fleshy lust, but also that, "if men ate the fat tubers they would beget male children, whereas if women ate the lesser, dry or barren root which was withered and shrivelled, they would bring forth girls". The Greek philosopher and botanist, Theophrastus (371–287 BC), wrote that on one occasion it enabled a man to have 70 consecutive acts of coitus (Wedeck 1961). Gaius Plinius Secundus (23–79 CE), or Pliny the Elder, described both aphrodisiacs and anti-aphrodisiacs in *The History of the World*. To *Orchis* and *Serapias* (now classified under *Epipactus*), he attributed the former property to the larger "or some say, the harder bulb of *Orchis*

when drunk in water". On the other hand, the lesser or softer bulb taken in goat's milk repressed the sexual appetite. Furthermore, "root of the former *Orchis* given to drink in the milk of an ewe bred up at home of a cade lamb, causeth a man's members to rise and stand; but the same taken in water, maketh it go down again and lie." Thus, it would seem that both the choice of orchid bulb and solvent had to be correct! The orchids were reported to equally effective when fed to goats, rams and stallions. Pliny offered mead or the juice of lettuce as an antidote when one became excessively lusty after consuming the orchids. The Roman historian also reported that the roots of *Orchis* healed mouth sores, and it was used to clear phlegm from the chest (Turner 1962).

Distinguished Arab physicians like Avicenna and al Rhazes subscribed to the *Materia Medica* of Discorides During the Ottoman Empire, the belief spread through northern Africa. In Algiers, of the nine orchid species which are locally used in herbal medicine, eight are used to treat male sexual dysfunction. They include *O. olbiensis* Reut. ex Gren. and *O. lactea* Poir. [=*Neotinea lactea* (Poir.) R.M.Bateman, Pridgeon and M.W. Chase] (Miara et al. 2013).

In Europe, these alleged aphrodisiac properties were revived and promoted following the Renaissance and the publication of Gerard's *Herbal* in 1633. Before the advent of scientific medicine, even a Cambridge-educated physician was misled by the ancient reputation of *Orchis* (Culpeper 1653). and the belief persisted into the nineteenth century (Parkins 1809). Shakespeare referred to the *salep* orchid in *Hamlet*:

> There with fantastic garlands did she come
>   Of crow flowers, nettles, daisies, and long purples
>   That liberal shepherds give a grosser name,
>   But our cold maids do dead men's fingers call them;
>   (*Hamlet* Act. 4, scene 7)

"Long purples" and "dead men's fingers" referred to *O. mascula*, *O. maculata* or *Dactylorhiza* (Lawler 1984; Bulpitt 2005).

Old English texts make fascinating reading. "A knobby root, somewhat long, two growing together, narrow, like an olive berry, the one above and the other beneath, and one of them is full and the other soft and full of wrinkles" is how the tubers were described (Turner 1551). Witches were alleged to use these tubers of *O. mascula* in the philters, the fresh tuber ensuring true love, and the withered one to check improper passions. When "bruised and applied to the place", they healed the King's Evil (Grieve 1971).

In the Roman play *Satyricon*, Petrionius (c. 54–68) featured the use of orchids by prostitutes. Among later writers, Bernard Shaw (1856–1950) described the character of a courtesan as "orchidaceous"; Arnold Bennett (1867–1931) introduced a new verb, *orchidise*, for the passionate behaviour of a courtesan (Lewis 1990).

Salep is a drink prepared from the starchy substance called *bassorin* which is extracted from the tuberous roots of many species of *Orchis* and other bulbous terrestrial orchids which are distributed in the Mediterranean, Middle East and Northern India. The term *salep* is an English corruption of the Arabic *sahlab*. It was sometimes spelt as *saloop* or *sahlop*. The drink originated in the Middle East, in the region of Asia Minor (Turkey), from whence it spread in all directions but no further east than India. However, Indian *salep* is not the same as the Mediterranean variety. *Salep misri* sold in the Indian bazaars are derived from various species of *Eulophia*, in particular *E. campestris* (*E. nuda*) and *E. herbacea* (Chopra 1933).

In Turkey, where *salep* is still widely enjoyed and sometimes made into ice cream, 30 species belonging to the genera *Orchis*, *Anacamptis*, *Himontoglossum*, *Orphys*, *Serapias* and *Dactylorhiza* (*Aceras*) are used to prepare *salep* (Sezik 1967, 1990; Tekinsen and Guner 2009). The composition of the tubers of the various species is shown in Table 17.2 (Sezik 1990). Mucilage (glucomannan) and starch are the major components, but the starch content of old tubers collected in autumn is low. *Salep* is sweet and has a faint, somewhat unpleasant smell.

Orchid species constituting *salep* varies from region to region, being dependent on which species are more prevalent in any one location.

O. anatolica Boiss is found throughout Turkey and is present in most *salep*. In north-western Turkey, *Orchis* and *Anacamptis* species are common, while in eastern Turkey *O. morio* thrives on the chalky soil. To meet the enormous appetite for *salep*, the various species of terrestrial orchids were stripped from their habitats by the millions and today they have been declared endangered in Turkey. It is illegal to export terrestrial orchid tubers from Turkey. *Salep* exporters deal in flour which has been artificially flavoured.

Lawler (1984) gave us some idea of what happened in the past. In 1892, Istanbul exported 19 t of *salep* and retained another 10 t as a reserve or for its own use. In 1879, Izmir exported 6.4 t, and from 1905 to 1908, 10.5 t. It was estimated that in the last quarter of the nineteenth century, 125 t of such orchid bulbs were dug up annually. Further afield in Nepal, about 5 t of *O. latifolia* (*Dactylorhiza hatagirea*) tubers are available for export each year. Going by the weight of *Dactylorhiza* tubers in Iran, 5 t of palmate *salep* would require over three million plants to be harvested (Ghorbani et al. 2014a). It is claimed that these orchids are replanted, but experts think that the following year's harvest probably comes from the smaller bulbs which were originally ignored. Lawler (1984) mentioned *O. laxiflora* and *O. longicornu* as two additional species which are used in western Asia, but the former is a west European species (it is the Jersey Orchid), whereas the second occurs in the Mediterranean countries and in Africa. They probably represent imported *salep*.

Anyone who has tasted *salep* knows that it is not an extraordinary drink, so why was there so much interest? The orchid powder from which the drink was prepared was believed to be an aphrodisiac. Additionally, it was a remedy for a wide range of debilities and illnesses, but already in 1829 it was noted by both British and American physicians that its nutritive powers were much overrated (Hooper and Akerly 1829). In the Middle East, the drug was called *Khus yatu's salab* (fox's testicle) or *Khus yaty'l klab* (dog's testicle). Some English names were mere translations of the Arabic (dogstones, foxstones),

Table 17.2 Orchid species used to prepare *Salep* (pooled data from: Hedrick 1919; Lawler 1982; Sezik 1969, 1990)

	Common names
Orchis coriophora Linn.	
O. longicruris Link.	
O. mascula	*Long Purples, Dead Men's Fingers* (in old English)
O. militaris	*Siliehongmenlan* (in China)
O. morio	*Gelincik, Dilicikik* (in Turkey)
O. maculata	*Long Purples, Dead Men's Fingers* (in old English)
O. indica	
O. tridentale	*Tavsantopugu*
O. simia	*Tavsantopugu, topanbas, solmazsoluk, puskullu* (in Turkey)
O. sancta	*Pirinc cicegi, purin cicegi* (in Turkey)
O. romana	*Camkoku* (in Turkey)
O. ustulata	
Platanthera bifolia	*Xijushechun Lan* (in China)
Eulophia campestris	*Salibmisri, Sung Misrie, Charle-michhri, Salum, Salib-misri, Salu Salibmisri, Salamisri, Bongataini, Salabmisri, Amrita, Amritobhava, Jiva, Jivani, Pranabhrita, Pranada, Sudhamuli, Virakanda* (in India); *Hattipaila* (in Nepal), *Sungmisri* (in Persia), *Kusyu-uth-thalab* (in Arabic)
E. herbacea	*Maochun Meiguan Lan* (in China); *Wan Mangmum* (in Thailand)
E. spectabilis (*E. nuda*)	*Zihuameiguan Lan* (in China); *Bonga taini, Amarcana, Manya, Goruma* (in India); *Wan hua khru, Wan ung* (in Thailand)
E. virens	*Satavari* (in India)
Habenaria commelinifolia	*Fueyufeng hua* (in China), *Devsunda* (in India), *Kadaw sut* (in Myanmar)
H. pectinata	*Jianyeyufeng Hua* (in China); *Safed musli* (in India)
Cymbidium aloifolium	*Wenban Lan, Yingyediao Lan, Chuihuadiao Lan, Diao Lan), Dabi Lan* (in China), *Ka Re Ka Ron* (*In Thailand*), *Kim bien* (*in Vietnam*); *Supurn* (in India), *Thit tet lin nay* (in Myanmar), *Harjor* (in Nepal)
Zeuxine strateumatica	*Shwethuli* (in India)
Anacamptis pyramidalis	*Peynir cicegi* (in Turkey) *Anacamptis laxiflora* (*O. laxiflora*) *Salap misri, shala misriri* (in India) *Himantoglossum longibracteatum ayikulagi, keskek cicegi, patpatan, Patpatinak* (in Turkey)
Serapias laxiflora	*Katirtirnagi, sigirkulagi* (in Turkey)
S. vomeracea	
Ophrys fucifera	*Kedigozu, pisipisi,tulekdokuyan* (in Turkey)
O. fusca	*Kedigozu, pisipisi, tulekdokuyan* (in Turkey)
Dactylorhiza hatagirea	*Marsh orchid* (in English); *Palma Christi* (Spain); *Kuanyehongmen Lan*, (*Orchis latifolia*) *Hongmen Lan, Mengguhongmen Lan, Zhanglie Lan* (in China); *Munjataka, Panja, Salampanja, Salep, Salap, Salap, Salimpanja* (in India), *Hathejadi, Lob, Panchaunle, Panchaunle, Ongu lakpa, dbang-po-la, Wonglak, Airalu* (in Nepal)

Note: The best *salep* is said to be constituted by the first four species and *Platanthera bifolia*

others locally invented (harestones, goatstones). Everywhere, fresh, plump tubers were preferred. Shrivelled tubers were discarded.

In his *Complete Herbal* published in 1653, an English physician, Nicholas Culpeper, attempted to make public all the directions for compounding medicines extracted from herbs so that physicians and apothecaries would not profit excessively by being exclusively privy to their preparation. For this reason, the organisation of apothecaries attempted to stop him from practising even though he was a Cambridge graduate. His description of the preparation of *salep* is as follows:

The best way to use it is to wash the new root in water; separate it from the brown skin which covers it, by dipping it in hot water, and rubbing it with a coarse linen cloth. When a sufficient number of roots have been thus cleaned, they are to be spread out on a tin plate, and placed in an oven heated to the usual degree, where they are to remain for five or six minutes, in which time they will have lost their milky whiteness, and acquired a transparency like horn without any diminution in bulk. When arrived at this state, they are to be removed in order to be dried and hardened in the air, which will require several days to effect; or by using a gentle heat, they may be finished in a few hours.

To prepare *salep*, the orchid tubers are collected in Turkey at the end of summer when they have the highest starch content. Washing and boiling for a short period removes their bitter taste. After removal of the outer covering, the tubers are dried in the sun or in the low heat of an oven until they turn from milky white to translucent. Further air-drying for a few days allows the tubers to harden. In this state, they can be stored for long periods (Ercisli and Esitken 2002). It takes 4–8 kg of wet tubers to produce 1 kg of dry *salep*. In the Golestan area, there are 304 (231–377) fresh tubers in 1 kg of fresh *salep* (Ghorbani et al. 2014a). Dried tubers are grounded before use (Ercisli and Esitken 2002). The ground powder has a slight yellow tinge like *baiji*. Despite the assaults on the tubers, pulverisation and prolonged storage, the population at large believes that *salep* still retains the aphrodisiacal property of the fresh *satyrion*. *Salep* orchids are present in pills sold as natural products that claim to promote male fertility and correct erectile dysfunction.

Although it is patently untrue, it was once believed that *salep* contained the greatest amount of nourishment in the smallest bulk and was thus useful in times of privation or famine (Culpeper 1653). A small amount of *salep* in a large volume of warm water converted into a jelly-like substance which was believed to be superior to rice. A basin of *salep* at three-halfpence, with a slice of bread was ideal breakfast for a chimney-sweep (Grieve 1971). To protect against famine at sea, it was proposed that *salep* should constitute part of a ship's provision at all times. *Salep* masked

the taste of salt water. It was also considered to have a role in treating scurvy, diarrhoea, dysentery and fever (Hooper and Akerly 1829). John Lindley, whose name is associated with numerous orchids, stated in his *Medical and Economical Botany* that *salep* prepared with *O. mascula* formed an agreeable diet which was nutritious and the powder could also be used as emollient and demulcent (Lindley 1858). It was seldom used in the United States, "except in the composition of Castillon powders, a nutritive and bland article of diet for invalids" (Griffith 1847). *Salep* was included in the 9th edition of the *State Pharmacopoeia of the Union of Soviet Socialist Republics* published in 1961. Russian *salep* was obtained from various species of *Orchis*, *Gymadenia*, *Anacamptis* and *Platanthera* (Puri 1970b).

Powdered *salep* is not readily miscible with water and Fernie (1914) recommended that it should first be stirred with wine. When the powder is well dissolved, water is added and the mixture brought to boil. Fernie's recipe called for 1 drachm (approximately 4 g) of *salep* powder to be dissolved in 1.5 fluid drachm (6 ml) of spirit, followed by 0.5 pint (284 ml) of water. Amber, cloves, cinnamon and ginger are added for flavour. Dr. Fernie, M.D., claimed that "*salep* is a most useful article of diet for those who suffer from chronic diarrhoea". *Salep* is still used as a Turkish folk remedy to treat infants and children when they suffer from diarrhoea.

Salep, not *khus yaty'l klab*, was the official name when it entered the standard pharmacopoeias of Europe, the United States, Japan and some South American nations in the nineteenth century. It was used as a restorative or tonic, an emollient, to treat stomach ache, heartburn, bilious colic, diarrhoea, dysentery, and other intestinal disorders. It was used for coughs, colds, tuberculosis and other respiratory disorders. It was used to treat infections of the bladder and kidneys, strangury, renal stones and acute and chronic fevers regardless of aetiology. It was used to stop haemorrhage, haemoptysis, improve fertility, prevent abortion, facilitate childbirth, expel the afterbirth and to cure venereal disease and sexual misconduct. It was

Table 17.3 Constituents of orchid tubers grown for s*alep* in Turkey (Sezik 1990)

	Mucilage (%)	Starch (%)	Reducing sugar (%)	N (%)	Moisture (%)	Ash (%)
Orchis maculata	35.95	3.37	1.38	0.91	10.62	2.76
Orchis indica	49.36	1.25	1.37	0.95	10.68	3.08
Orchis tridentate	24.50	36.04	1.35	0.80	11.14	4.26
Orchis mascula	50.11	0.69	1.33	0.90	11.70	1.37
Orchis simia	29.89	1.71	1.33	0.62	10.96	0.27
Orchis morio	32.11	25.04	2.67	0.57	10.96	3.14
Orchis sancta	15.70	10.64	1.72	0.49	8.65	2.27
Orchis romana	61.05	0.45	4.50	0.74	10.96	5.98
Anacamptis pyramidalis	44.72	5.94	2.81	0.92	9.76	1.05
Himontoglossum longibr	20.95	10.99	2.51	0.77	9.58	3.48
Serapias laxiflora	30.61	1.07	1.89	0.77	10.64	1.72
Serapias vomeracea	40.56	1.35	2.18	0.80	8.70	1.43
Ophrys fuciflera	9.60	18.78	1.05	0.64	6.40	0.49
Orphys fusca	6.82	12.77	1.01	0.72	8.60	

recommended for diabetes, scurvy, arthritis, scrofula, paralysis, nervous exhaustion. hoarseness, poisoning and to assist recovery from a prolonged illness. *Saloop* was a sovereign cure for drunkenness (Lawler 1984). In short, it was almost a panacea.

The constituents of *salep* are not remarkable (Table 17.3). The main constituents are mucilage, starch, cellulose, sugar (glucose, mannose, glucomannan), some proteins, a bit of fat, and traces of acetic acid, water and ash, the last consisting predominantly of chlorides and phosphates or potassium and calcium, and sometimes calcium oxalate. Different orchids and their source determine the actual amounts of the various components. Coumarin is present in its volatile oil. Alkaloids were not detected in the five species (*O. incarnata*, *O. mascula*, *O. militaris*, *O. morio and O. ustulata*) (Luning 1964).

In modern Turkey, *salep* is used as a stabiliser, particularly in Kahramanamaras-type ice cream. It is said to improve the taste and slows the melting of the ice cream. This property is attributable to glucomannan and to a lesser extent to starch. Tekinsen and Guner (2009) proposed that species which contain the highest amount of glucomannan and starch, like *O. italica*, *O. morio* and *O. anatolica*, should be more valuable and efforts should be made to turn them into commercial crops though tissue culture and

cultivation. Instead, it is reported that, despite a 1995 prohibition of export of *salep* in either tuber or powder form, 120 million wild orchid plants are damaged annually in Turkey alone (Tekinsen and Guner 2009). The reason: it takes 1000–4000 dried tubers of the terrestrial orchids to make 1 kg of *salep* and annual *salep* production in Turkey is around 45,000 kg (Krentz 2002). Efforts should be made to cultivate the *salep* orchids but the choice need to be based on further research and factors like ease of propagation and field tests. In vitro, asymbiotic germination of *O. mascula* has been achieved using max medium supplemented with benzyl-adenine and activated charcoal (Valetta et al. 2008).

Two phytoalexins, orchinol and p-hydroxybenzyl alcohol have been isolated from *O. mascula*, *O. militaris*, *O. morio*, *O. sambucina*, and in small amounts in *O. latifolia*, but the two compounds were not found to be present in *O. maculata* and *O. ustulata*. Orchinol was the first phytoalexin to be characterised (Nuesch 1963). The phenolic glycoside, loroglossin, is also present in *O. mascula* (Veitch and Grayer 2001): loroglossin is present together with militarine in *O. militaris* (Aasen et al. 1973). They possess antimicrobial properties and are predominantly fungistatic. Thus, they may play an import role in the ecology of *Orchis* but currently no medical application has been discovered. Nevertheless, a locally available, herbal capsule which claims to promote health and vitality for men is constituted of ten herbs among which is *O. latifolia*, purported to strengthen muscles, vitalise, and counteract premature aging by reducing the formation of free radicals.

Numerous tuberous orchids apart from *Orchis* have also been used as a nutrient in other parts of the world. As recently as 1987, Tim Low writing in *Australian Natural History*, confessed that, while conducting a long-term study of the traditional foods of Australian aborigines, he had dug up and tasted the tubers of 20 species of native orchids belonging to 12 genera. All were edible, although some were unpalatable. "A few were exceptionally tasty—especially the walnut-sized 'potatoes' of Brown Beaks (*Lyperanthus suaveolens*) and the fragrantly flavoured starch of the Horned Orchid (*Orthoceras strictum*). Most filling were the glutinous tubers of donkey orchids (*Diuris* spp.) and sun orchids (*Thelymitra* spp.) and I have no doubt that these were important aboriginal foods." Speaking to the Linnean Society of New South Wales in 1880, F. M. Bailey declared that, although he was unable to detect any medicinal value in any of the Australian orchids, "yet a wholesome food might be prepared from the thick starchy stems of several species" (Bailey 1881). In 1898, the Australian botanist Joseph Maiden commented that "There is hardly a country boy who has not eaten the so-called Yams, which are the tubers of numerous kinds of terrestrial or ground-growing orchids." (Low 1987). Tubers of *Geodorum pictum* (Aboriginal names *Yeenga*, *Uine*) are eaten by the aborigines living around Gladstone, Queensland, and *Cymbidium canaliculatum* (Aboriginal name: *Dampy-ampy*), grated and boiled, provided an arrowroot-type nutrient for weak children cut off from other supplies (Hedley 1888).

Collection of terrestrial orchid tubers for use as *salep* is illegal. Nevertheless, there is an extensive trade in Turkey. As these orchids have become scarce in Turkey, merchants started to source from Iran about 10 years ago. Presently, it is estimated that from the Golestan area alone, seven middlemen traded 24.5 t of *salep*. This required 7.4 (5.7–9.1) million individual orchid plants to be harvested. The western provinces of Iran are sourced for *Qolveh salep* which is made up by round or oval tubers that are smaller than the palmate tubers in *Panjehey salep*. One kilogram of *Qolveh salep* requires 1117 (881–1353) plants to be harvested (Ghorbani et al. 2014a). *Kastamonu salebi* regarded as one of three top varieties of *salep* in Turkey is made up of *O. mascula* and *O. purpurea*, species commonly found in the Kastamonu region: they have small tubers unlike tubers of *Dactylorhiza* and *Himantoglossum* which are large (Yaman 2013).

In the Tehran bazaar, six medicinal plant wholesalers traded 1920 kg of dried tubers between May and July 2013. Prices for tubers range from US$5–6 per kg for early tubers which are lighter and of poorer quality to US$22 per kg at the end of the season. Prices vary with the size of the tubers and availability in the market. Tubers collected after flowering are heaviest (Ghorbani et al. 2014a) and contain the highest concentration of mucilage.

Sustainable harvesting practices will be required to prevent overharvesting. Current harvesting practices are too destructive and unsustainable in the long run (Ghorbani et al. 2014a). In particular, early harvesting needs to be discouraged because this prevents seed set and dispersal; furthermore, it also makes sense to the gatherers because late-harvested tubers are three times more valuable, albeit they may be more difficult to find and harvest.

Genus: *Oreorchis* Lindl.

Chinese name: *Shan Lan* (mountain orchid)
Japanese name: *Kokei Ran*

The generic name is derived from Greek *oros* (mountain) and *orchis* (orchid), alluding to its habitat in Asian mountains. It is a genus of pseudobulbous, terrestrial orchids with 16–18 species distributed from Bhutan across northeast India, Myanmar, southwest and eastern China to eastern Russia, Korea and Japan. Of 11 species present in China, seven are endemic; however, the medicinal species are not. Some species are saprophytic. Plants are erect with corm-like pseudobulbs and one or two terminal, pleated leaves. Inflorescence is tall and bears few to numerous tiny, yellow, white, red or purple flowers arranged spirally around the rachis (Alrich and Higgins 2008; Chen et al. 2009).

Oreorchis foliosa (Lindl.) Lindl.

Chinese name: *Xiaoshan Lan* (small hill orchid), *Nangchunshan Lan*
Chinese medicinal name: *Duyeshan Lan*

Description: A terrestrial orchid, *O. foliosa* is found in Japan, Taiwan, Shanxi, Hubei, Sichuan, Yunnan and Tibet, Bhutan, Nepal, northeast India and Pakistan, in open shade, forests, scrubland and alpine meadows, at 2200–3400 m in China (Chen et al. 1999; Perner and Luo 2007), and in temperate Himalayas at 3000–4000 m. It flowers in September in China (Chen et al. 2009), and July in the Himalayas (Bose and Bhattacharjee 1980). Plant bears a single, terminal, lanceolate leaf, 10–14 cm long (up to 20 cm in the Himalayan variety) by 2.4 cm wide. Inflorescence is 25–35 cm tall, upright with 4–9 flowers, 15 mm across. Sepals are narrow pointed, the lower sepals bowed, a deep olive. Petals are not extended; they are flushed with crimson. Lip is white with crimson spots (Perner and Luo 2007).

Usage: Herb is obtained from Shanxi, Hubei, Sichuan, Yunnan and Tibet. The stem is used by Chinese herbalists as an antidote for snake bites and to treat tuberculosis of the lymph nodes, sores and ulcers (Wu 1994). For decoction, 3–9 g of the herb is used (Hu et al. 2000).

Oreorchis patens (Lindl.) Lindl.

Chinese name: *Shan Lan* (mountain orchid), *Shuangbanshan Lan* (double plate mountain orchid), (fine flower mountain orchid), (orchid herb), (gentle aunt of the mountain), (ice ball)

Description: A terrestrial orchid with a short rhizome linking subterranean, corm-like pseudobulbs which are 1–2 by 0.5–1.5 cm in diameter. It bears a solitary, terminal, narrowly lanceolate leaf, 13–30 by 1–2 cm with a petiole of moderate length. Scape is lateral, erect 20–50 cm tall, and loosely many-flowered (Fig. 17.10). Flowers are 1–1.5 cm across, dull yellow with pale veins and maroon margin, and not fully extended. Petals are narrow, pointed and spotted with maroon. Sepals are longer than the petals and they are not spotted. Lip is white and spotted with purple dots (Fig. 17.11). It flowers in June and July. *O. patens* is found in forests or at the edge of forests at 1000–3000 m from northern Yunnan, Jiangxi, Guizhou, Hunan and Taiwan, northwards into the northern provinces of China, and Korea, Japan and Siberia (Chen et al. 1999; Jin et al. 2009).

Usage: Herb is obtained from Gansu, Hubei, Sichuan, Yunnan and Xizang (Wu 1994; Hu et al. 2000), but some of the provinces are not mentioned as within the distribution of the orchid in *Flora of China*. The whole plant is used as a detoxicant but the herb itself is slightly poisonous. It enlivens the liver, clears phlegm, stops coughs and heals carbuncles and lumps in the skin (most probably referring to tuberculous lymphadenitis located at the neck). It is used to treat swollen lymph nodes, sores and snake bites. It is also used to kill bugs (Wu 1994; Hu et al. 2000; Ou et al. 2003). For decoction 3–9 g, for external use, pulverise and apply on the bites (*Xizang Commonly Used Herbs*, Hu et al. 2000).

Overview

No pharmacological information is available on the genus. Its entry into the Chinese *Materia Medica* is recent.

Genus: *Ornithochilus* (Lindl.) Wall ex Benth.

A genus with only one (or perhaps two) species, both rare and found only in the Himalayas and China, *Ornithochilus* is allied to *Phalaenopsis* and resembles the latter in vegetative form

Fig. 17.10 *Oreorchis patens* (Lindl.) Lindl. (as *Oreorchis gracilis* Franch. and Savatier). From: *Bulletin de la Societe botanique de France*, vol. 44: t. 3 (1897). Drawing by B. Herincq. Courtesy of Missouri Botanical Gardens, St. Louis, USA

Fig. 17.11 *Oreorchis patens* (Lindl.) Lindl. (Photo: Liu Ming)

(Fig. 17.12), but it is closer to *Aerides* in floral form (Hawkes 1965). Inflorescences are axillary and bear many small to medium-sized flowers. Although the lip extends into a downward and backward pointing spur, the rest of the floral structure makes it quite distinct from *Aerides*. The name is derived from Greek *ornis, -ithos* (bird) and *cheilos* (lip) (Fig. 17.13).

In the latest revision of the *Orchidaceae*, *Ornithochilus* is included under *Phalaenopsis*.

Ornithochilus difformis (Wall ex Lindl.) Schltr.

Syn. *Ornithochilus fuscus* Wall ex Lindl.

Chinese name: *Yixingxiachun Lan* (unique shape narrow lip orchid), *Yuchun lan* (feather lip orchid)

Thai name: *Soi Thong*

Description: *O. difformis* has the vegetative appearance of a small *Phalaenopsis* and may be epiphytic or lithophytic. Stem is very short and bears a few, broadly lanceolate, fleshy leaves, 10–15 by 2–4 cm. Inflorescence is long, axillary, often branching, with numerous (25 or more), widely-spaced flowers of a green or yellow brown base marked with crimson stripes and a crimson lip. Flowers are small, 6 mm wide, 10 mm long, resupinate and appear like tiny, colourful wasps in profile. This resemblance is accentuated by the long, thin ovaries and the wide spacing of the flowers which makes them seem like a swarm of insects in flight. Tepals are narrow and drawn back. Lip is large, complex with a prominent, cylindrical, curved and forward pointing, crimson spur. Lateral lobes white, with saw tooth borders stained with crimson at their base (Fig. 17.13). Flowering period is February to April in the Himalayas (Bose and Bhattacharjee 1980), May to July in China (Chen et al. 1999), July in Peninsular Malaysia (Seidenfaden and Wood 1992), and July and August in Thailand (Vaddhanaphuti 2001).

O. difformis is widely distributed from the tropical Himalayas (Garhwal to Sikkim and Assam) to southern China (Guangdong, Guangxi, Sichuan and Yunnan at 600–1800 m) and the highlands of western Malesia at 1800–1900 m (Chen et al. 1999; Seidenfaden and Wood 1992; Comber 2001; Vaddhanaphuti 2001). The Bornean *O. difformis* var. *kinabaluensis* J.J. Wood, A.L. Lamb and Shim, endemic in Sabah, differs from the Chinese var. *difformis* by having pure greenish-yellow tepals and a pure white lip with shorter fimbriate processes on the mid-lobe, more distinct, narrowly wing-like side lobes, and a shorter spur which is less convex (Chan et al. 1994; Chen et al. 2009). The Bornean plant flowers over a long period during the rainy season (Chan et al. 1994).

Herbal Usage: The whole plant is used to treat rheumatism and arthritis, sprains and soft tissue trauma in China. The herb is collected from Guangdong, Guangxi, Yunnan and Xizang Provinces (Wu 1994).

Fig. 17.12 *Ornithochilus difformis* (Wall. ex Lindl) Schltr. [as *Ornithochilus delavayi* Finet]. Bulletin de la Societe botanique de France vol 43:t11 (1896). Drawing by B. Henrincq. Courtesy of Missouri Botanical Gardens, St. Louis, USA

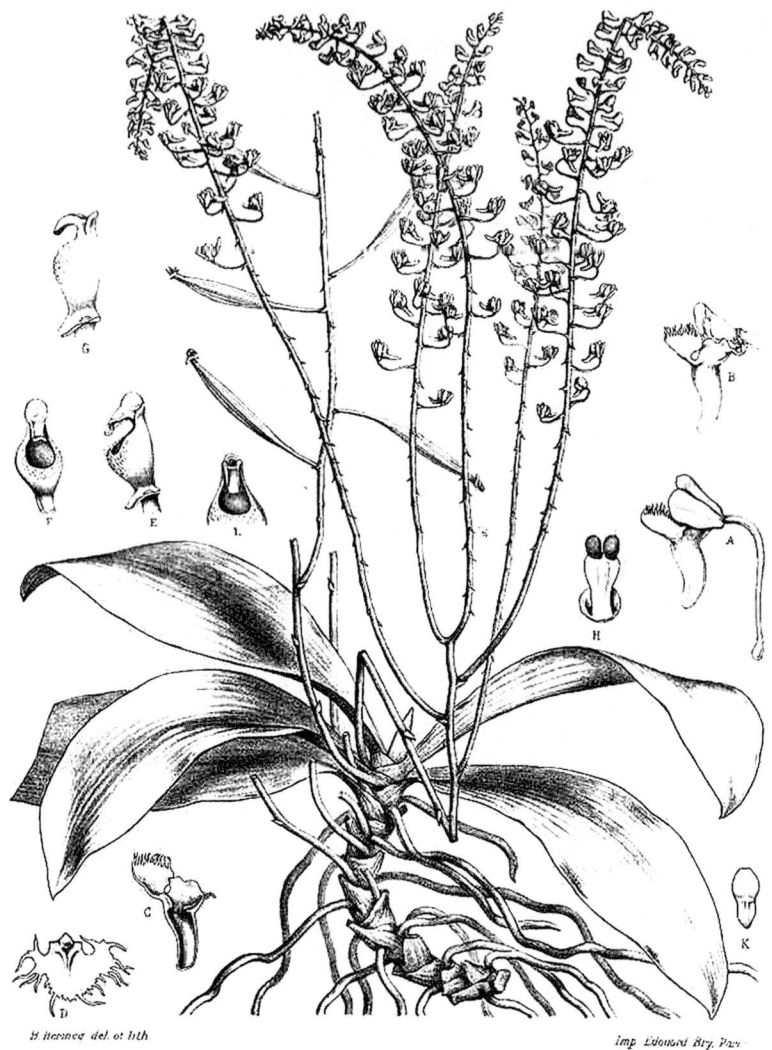

ORNITHOCHILUS DELAVAYI (*sp nov*)

Ornithochilus fuscus Wall ext Lindl., Schltr. (see *Ornithochilus difformis*)

Overview

Twenty-one species of *Sarcochilus* (under which *Ornithochilus* was formerly classified) were tested for alkaloids and two were found to have an alkaloid equivalent to or exceeding 0.1 % of the plant's weight (Lüning 1974). Luning did not state whether *Ornithochilus* was one of the four. Additional pharmacological data are not available.

In the latest taxonomic revision of Orchidaceae, *Ornithochilus* loses its generic status: it is included under *Phalaenopsis* (Pridgeon et al. 2014).

Genus: *Otochilus* Lindl.

Chinese name: *Erchun Lan*

Otochilus is a small epiphytic genus with only four species which are distributed in the southern

Fig. 17.13 *Ornithochilus* (=*Phalaenopsis*) *difformis* (Wall ex Lindl.) Schltr. (Photo: Peter O'Byrne)

Fig. 17.14 *Otochilus fuscus* Lindl. (Photo: Peter O'Byrne)

Himalayas, from Bhutan and Nepal to north-east India, Myanmar, China (south-east Xizang and Yunnan), Thailand and Indochina. *Otochilus porrectus* is sometimes saxicolous. The unique character of this genus is that the pseudobulbs are superposed, old psedobulbs giving rise to younger pseudobulbs at their apices. Flowers are small, white marked with brown.

The generic name is derived from Greek, otos (ear) and cheilos (lip), referring to the two small ear-shaped side lobes pf the lip which embrace the foot of the column.

Otochilus albus Lindl.

Chinese name: *Baihua Erchun Lan*

Description: An epiphyte with cylindrical, branching pseudobulbs, 2.5–11 by 0.7–1.3 mm, greenish-yellow when dry and sparsely wrinkled. Leaves are elliptic-oblong 6–16 by 1–2 cm. Inflorescence is pendulous, laxly several-flowered.

Bracts are prominent, brown. Flowers are white, tinged with yellow or marked with brown on the lip and anther cap. It flowers in June in Nepal (Raskoti 2009), and October to December in China (Chen et al. 2009). *O. albus* is distributed from Nepal to Indochina.

Usage: In Nepal, pseudoblubs are used to treat fractures (Pant and Raskoti 2013)

Otochilus fuscus Lindl.

Chinese name: *Xiaye Erchun Lan*

Description: A terrestrial herb in Nepal, pseudobulbs are cylindrical, tapering at both ends. Leaves are linear-lanceolate, 10–15 by 1–1.5 cm. Inflorescence is pendulous, many-flowered. Flowers are white with narrow, spreading tepals (Fig. 17.14). Flowering season is December to January in Nepal (Raskoti 2009),

Fig. 17.15 *Otochilus porrectus* Lindl. (as *Otochilus latifolius* Griff.). From: Griff, W., *Icones Plantarum Asiaticarum* 1851 vol. 3: p. index, t. 288 (1847–1854). Courtesy of Harvard University Botanical Libraries, USA

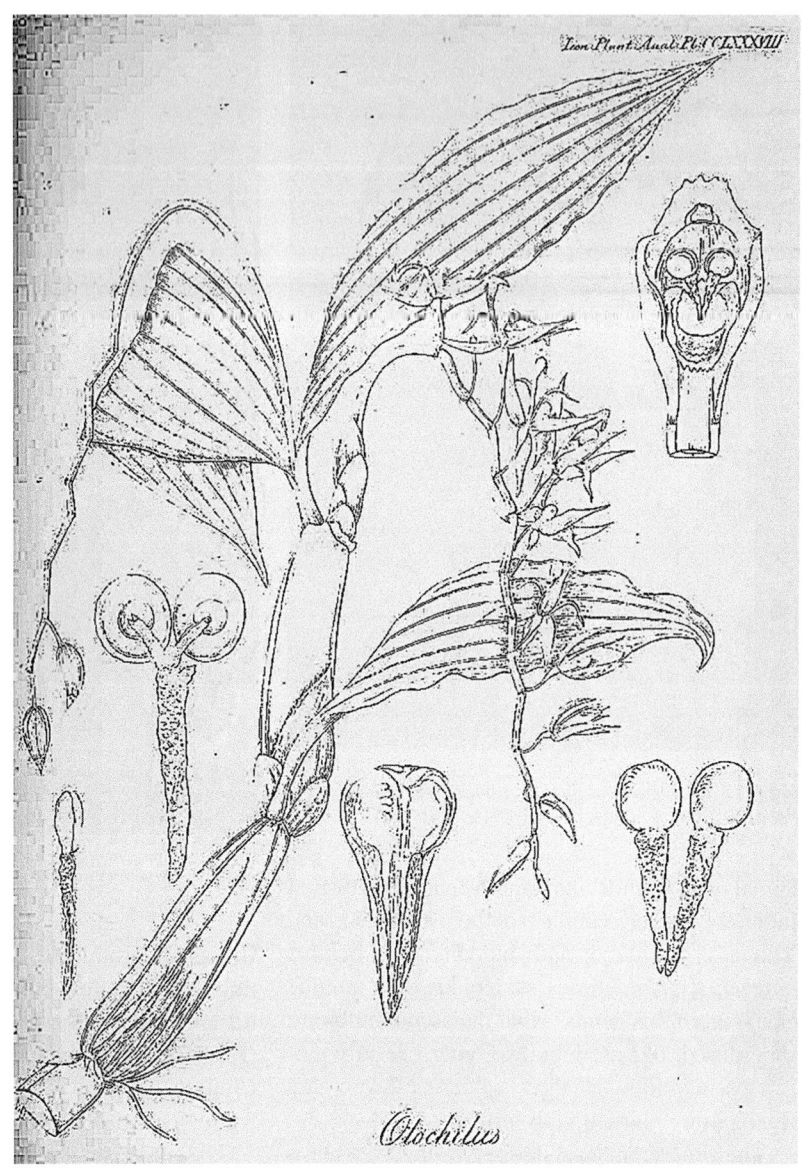

and November to January in Bhutan (Gurung 2006). It is distributed from Bhutan and Nepal to north-east India, Yunnan, Myanmar and Thailand to Indochina.

Phytochemistry: 14 chemical compounds, consisting of five bibenzyls, a bibenzyl with phenyl chromanol, one stilbene, two 9,10-dihydrophenanthreenes, two 9,10

phenanthropyrans, two lignans, one flavone and one steroidal ketone, have been isolated from *O. fuscus* (Wang et al. 2012).

Usage: Pseudobulbs are used to treat fractures (Pant and Raskoti 2013).

Otochilus porrectus Lindl.

Chinese name: *Erchun Lan*

Description: *O. porrectus* is epiphytic in Nepal, but sometimes also saxicolous in China. Pseudobulbs are cylindrical-fusiform, 2–10 by

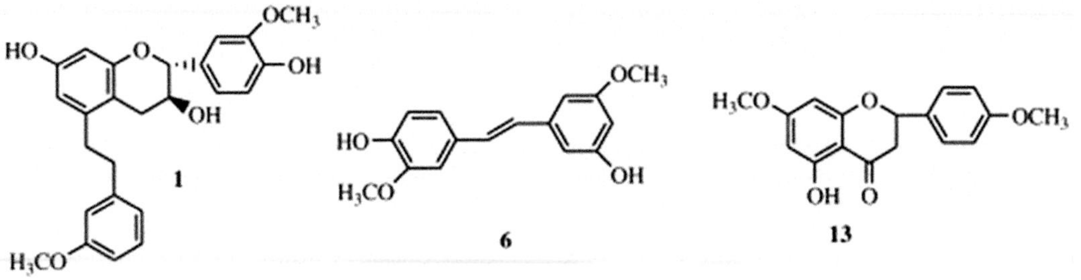

Fig. 17.16 Formulae of three compounds isolated from *Otochilus fuscus* by Wang et al. (2012); (1) bibenzyl with phenyl chromanol, (6) a stilbene and (13) a flavone

0.7–1.2 cm. Leaves are narrowly lanceolate, 10–18 by 2–4 cm. Inflorescence is pendulous, laxly several-flowered. Flowers are white, petals linear (Fig. 17.15). It flowers from October to December in Nepal. The species is distributed from Bhutan and Nepal to north-east India, Myanmar, China (Yunnan) to Thailand and Indochina.

Usage: In Nepal, pseudobulbs are used to treat sinusitis, rheumatism and as a tonic (Baral and Kurmi 2006; Pant and Raskoti 2013)

Overview

There are three species of *Otochilus* in Nepal, namely *O. albus* Lindl., *O. fuscus* Lindl. and *O. lancilabius* Seidenf. (Pant 2011); the first two are used to treat dislocations and fractures (Raskoti 2009); *O. porrectus* serves both as a tonic and also as a remedy for rheumatism and sinusitis. In Bhutan, a species of *Otochilus* (*O. porrectus*) features in traditional medicine. No medicinal usage has been recorded in China or the other countries where *Otochilus* is distributed (Fig. 17.16).

Fourteen chemical compounds have been isolated from *O. fuscus* which occurs in Yunnan. These consists of five bibenzyls, a bibenzyl with phenyl chromanol, one stilbene, two 9,10-dihydrophenanthreenes, two 9,10 phenanthropyrans, two lignans, one flavone, and one steroidal ketone (Wang et al. 2012).

References

Aasen A, Behr D, Leander K (1973) Studies on Orchidaceae glucosides. 2. The structures of loroglossine and militarine, two glucosides from *Orchis militaris*. L Acta Chi Scand B29:1002–1004

Alrich P, Higgins W (2008) The Marie Selby botanical gardens illustrated dictionary of orchid genera. Comstock Books, Carson City, NV

Averyanov LV, Phan KL, Nguyen TH, Harder DK (2003) Phytogeographic review of Vietnam and adjacent areas of Eastern Indochina. Komarovia 3:1–83

Ayasse M, Schiestl FP, Paulus HF et al (2000) Evolution of reproductive strategies in the sexually deceptive orchid, Ophrys sphegodes: how does flower specific variation in odor signals affect reproductive success? Evolution 54(6):1995–2006

Baral SR, Kurmi PP (2006) A compendium of medicinal plants in Nepal. Mrs. Rachana Sharma & IUCN, Kathmandu

Bose TK, Bhattacharjee SK (1980) Orchids of India. Naya Prokash, Calcutta

Bailey FM (1881) Medicinal plants of Queensland. Proc Linnean Soc New South Wales 5:1–29

Bulpitt CJ (2005) The uses and misuses of orchids in medicine. Quarterly J Med 98(9):625–631

Caius JF (1936) The medicinal and poisonous plants of India. J Bombay Nat Hist Soc 38(4):791–799

Chan CL, Lamb A, Shim PS, Wood JJ (1994) Orchids of Borneo, Volume 1: introduction and a selection of species. The Sabah Society and Kew Royal Botanic Gardens, Kota Kinabalu

Chen SC, Tsi ZH, Luo YB (1999) Native orchids of China in colour. Science Press, Beijing

Chen XQ, Ormerod P, Wood JJ (2009) *Oberonia* Lindley. In: Chen XQ, Liu ZJ, Lang KY et al (eds) Flora of China 25 (Orchidaceae). Science Press/Missouri Botanic Gardens, Beijing/St Louis

Cheam MC, van der Ent A, Abdullah E, Perumal B (2009) Wild orchids of Cameron Highlands. Regional Environmental Awareness Cameron Highlands (REACH), Bringchang

Chopra RN (1933) The indigenous drugs of India. The Art Press, Calcutta

Chopra RN, Chopra IC, Handa KL, Kapur LD (1958) Chopra's indigenous drugs of India. U.N. Dhur & Sons, Calcutta

Comber JB (2001) Orchids of Sumatra. Natural History Publications (Borneo), Kota Kinabalu

Culpeper N (1653) Complete herbal: consisting of a comprehensive description of nearly all herbs with their medicinal properties and directions for compounding the medicines extracted from them. W. Foulsham & Co., Ltd., London

Das PK, Sahoo S, Bal S (2008) Ethnobotanical studies on orchids of Niyamgiri Hill Ranges, Orissa, India. Ethnobot Leaft 12:70–78

Davis RS, Steiner ML (1982) Philippine orchids. A detailed treatment of some 100 native species. M & L Lucidine Enterprises, Manila (reprint)

Ercisli S, Esitken A (2002) Orchids (Salep) growing in Turkey. Proc. 12 WOC, Shah Alam, Malaysia, pp 242–244

Ernst R, Rodriguez E (1984) Carbohydrates of the Orchidaceae. In: Arditti J (ed) Orchid biology: reviews and perspectives, III. Cornell University Press, Ithaca, NY, pp 223–260

Fernie WT (1914) Herbal simples approved for modern uses of cure, 3rd edn. John Wright & Sons Ltd, Bristol

Ghorbani A, Gravendeel B, Zarre S, de Booer H (2014a) Illegal wild collection and international trade in CITES-listed terrestrial orchid tubers in Iran. Traffic Bull 26(2):52–58

Ghorbani A, Gravendeel B, Naghibi F, de Booer H (2014b) Wild orchid tuber collection in Iran: a wake-up call for conservation. Biodivers Conserv. doi:10.1007/s10531-014-0746-y

Grieve M (1971) A modern herbal. Vol. II. Hafner Publishing Co, New York, NY

Griffith RE (1847) Medical botany of descriptions of the more important plants used in medicine, with their history, properties and mode of administration. Lea and Blanchard, Philadelphia

Gurung DB (2006) An illustrated guide to the orchids of Bhutan. DSB Publications, Thimpu

Hawkes AD (1965) Encyclopaedia of cultivated orchids. Faber & Faber, London

Hedley C (1888) Uses of Queensland plants. Proc R Soc Queensland 5:10–13

Hedrick UP (1919) Sturtevant's notes on edible plants. State of New York: Dept of Agri, 27th Annual Report Vol. 2. and Albany, NY: J B Lyon

Hooper R, Akerly S (1829) Lexicon medicum or medical dictionary. 4th American edition. Collins and Hannay, New York, NY, p 137

Hu XM, Zhang WK, Zhu QZ et al (2000) Zhonghua Bencao, vol 8. Shanghai Science and Technology Publication, Shanghai

Hutchings MJ (1987) The population biology of the early spider orchid, Ophrys sphegoes Mill. II. Temporal patterns in behavior. J Ecol 75:729–742

Italian Group for the Research on Wild Orchids (2013) http://www.giros.it/Genera/ophrys_sphegodes.htm

Jalal JS, Kumar P, Tewari L, Pangtey YPS (2010) Orchids: uses in traditional medicine in India. In: National seminars on medicinal plants of Himalayas. Regional Research Institute, Himalaya, Tarikat

Jayaweera DMA (1981) A revised handbook of the flora of Ceylon, vol 2. A.A. Balkema, Rotterdam

Jin XH, Zhao XD, Shi XC (2008) Native orchids from Gaoligongshan Mountains, China. Science Press, Beijing

Jin SH, Zhao XD, Shi XC (2009) Native Orchids from Gaoligongshan Mountains, China. Science Press, Beijing

Johnson CP, Sowerby JE (1865) The useful plants of Great Britain. A treatise upon the principal native vegetables capable of application as food, medicine, or in the arts and manufactures. Robert Hardwicke, London

Krentz CAJ (2002) Turkiye'nin orkideleri salep, dondurma ve katliam. Yesil Atlas Degisi 5:99–109 (quoted by Tekinsen, 2009)

Kreutz K, Seger SM, Walraven H (2002) Contributions to the Ophrys mammosa group of Cyprus. Orchis alasiatica C.A.J. Kreutz, Segers & Walraven spec. nov. J Eur Orch 34(3):463–492

Lawler LJ (1984) Ethnobotany of the Orchidaceae. In: Arditti J (ed) Orchid biology reviews and perspectives, vol 3. Cornell University Press, Ithaca

Lewis MWH (1990) Power and passion: the orchid in literature. In: Arditti J (ed) Orchid biology reviews and perspectives. Timber Press, V. Portland

Lindley J (1858) Calanthe Dominii (Hybrida): gardeners' chronicle

Luning B (1964) Studies on Orchidaceae alkaloids. IV. Screening of species for alkaloids. Acta Chim Scand 18(6):1507–1516

Lüning B (1974) Studies on Orchidaceae alkaloids. I. Screening of species for alkaloids 2. Phytochemistry 6:857–861

Low T (1987) Australian wild foods. ground orchids: salute to saloop. Aust Nat Hist 22(5):202–203

Matthew KM (1995) An excursion flora of central Tamil Nadu, India. A.A. Balkaema, Rotterdam

Miara MD, Ait Hammou M, Hadjadi Aoul S (2013) Otherapie et taxonomie des plantes medicinales spontanees dans la region de Tiaret (Algerie). Phys Chem Chem Phys 11:206–218

Millar A (1978) Orchids of Papua New Guinea. Australian National University Press, Canberra

Nuesch J (1963) Defense reactions in orchid bulbs. Symp Soc Gen Microbiol 13:335–343

Ou JC, Hsieh WC, Lin IH, Chang YS, Chen IS (eds) (2003) The Catalogue of Medicinal Plant Resources in Taiwan. Department of Health, Executive Yuan, Taipei

Pant B (2011) Medicinal orchids of Nepaland their conservation by in-vitro technique powerpoint presentation. 20th World Orchid Conference, Singapore, 14 Nov 2011

Pant B, Raskoti BB (2013) Medicinal orchids of Nepal. Himalayan Map House, Kathmandu

Parkins (1809) The english physician, enlarged with 369 medicines made of English herbs not in any former impression of Culpeper's British herbal. Crosby & Co, London

Perner H, Luo Y (2007) Orchids of Huanglong. Huanglong National Park, Sichuan Province, China

Pridgeon AM, Cribb PJ, Chase MW, Rasmussen FN (eds) (2001) Genera Orchidacearum Vol. 2. Orchidoideae (Part 1). Oxford University Press, Oxford, pp 251–255

Pridgeon AM, Cribb PJ, Chase MW, Rassmussen FN (2014) Genera orchidacearum, vol 6, Epidendroideae (Part 3). University Press, Oxford

Puri HS (1970b) Salep: the drug from orchids. Am Orchid Soc Bull 39(1),723

Raskoti BB (2009) The orchids of Nepal. Bhaktar Bahadur Raskoti and Rita Ale, Kathmandu

Santapau H, Kapadia Z (1966) The orchids of Bombay. Government of India Press, Calcutta

Schiestl FP (2000) The evolution of floral scent and insect chemical communication. Ecol Lett 13(5):643–656

Schiestl FP, Ayasse M (2001) Post-pollination emission of a repellant compound in a sexually deceptive orchid: a new mechanism for maximizing reproductive success? Oecologia 126:531–534

Seidenfaden G (1999) 149. Orchidaceae. In: Matthew KM (ed) The flora of the Palni Hills, South India. Part 3. The Rapinat Herbarium, St. Joseph's College, Tiruchirapalli

Seidenfaden G, Wood JJ (1992) The orchids of peninsular Malaysia and Singapore. Olsen & Olsen, Fredensborg

Sezik E (1967) Turkiye'nin Salepgilleri Ticari Salep Cesitleri ve Ozellikle Mugla Salebi Uzerinde Arastirmalar. Doctoral Thesis, Istanbul Universitesi Eczacihk Fakultesinde (In Turkish. Summary in English)

Sezik E (1969) Mugla Civarinda Salep Elde Edilen Bitkilerin Mahalli Isimleri (The native names of salep producing plants around Mugla). Istanbul Ecz Fak Mec 5.77–79

Sezik E (1990) Turkiye'nin orkideleri. Bilim Teknik 269:5–8 (quoted by Ericisli, Esitken, 2002)

Suwal PN (1970) Medicinal plants of Nepal. H.M. Govt. of Nepal, Ministry of Forests and Soil Conservation, Department of Medicinal Plants, Kathmandu

Tekinsen KK, Guner A (2009) Chemical composition and physicochemical properties of tubera salep produced from some Orchidaceae species. Food Chem 121:468–471

Turner W (1551) A new herbal. London: S. Mierdman. Parts II and III edited by GTL Chapman, F McCombie and A Wesencraft. Cambridge University Press, Cambridge, 1995

Turner P (ed) (1962) Selections from The History of the World commonly called The Natural History of C. Plinus Secundus. Translated into English by P. Holland. Centaur Press Ltd, London

Uphof JCT (1968) Dictionary of economic plants. Verlag von J, Cramer, Lehre

Vaddhanaphuti N (2001) A field guide to the wild orchids of Thailand, 3rd edn. Silkworm Books, Chiang Mai

Valetta A, Attorre F, Bruno F, Pasqua G (2008) In-vitro asymbiotic germination of Orchis mascula L. Plant Biosystems 142(3):653–655

Veitch NC, Grayer B (2001) Phytochemistry of Habenaria, Himantoglossum, Ophrys, Orchis and Platanthera. In: Pridgeon AM, Cribb PJ, Chase MW, Rasmussen FN (eds) Genera Orchidacearum, Vol. 2. Orchidoideae (Part One). Oxford University Press, Oxford

Wang LQ, Wu MM, Tang ZG et al (2012) Chemical constituents of *Otochilus fuscus*. Biochem Syst Ecol 43:48–50

Wedeck HE (1961) Dictionary of aphrodisiacs. Philosophical Library, New York, p 216

Wu XR (1994) A concise edition of medicinal plants in China. Guangdong Higher Education Publication House, Guangdong (in Chinese)

Yaman K (2013) 1920' den Gunumuze T.C. Resmi Gazete Ar Invinde Salep ve Ticareti le Igili Yasal Duzenmeler (1920 to Present. The Official Gazette of the Republic of Turkey in Sahlep Archives and Trade Related Legislation/Regulations relating to the trade in Turkish Official Gazette Sahlep and its Archives from 1920 to the Present)

Zhonghua Bencao (2000) Edited by Health Department and National Chinese Medical Management Office. Science and Technology Press, Shanghai

Zhongyao Da Cidian (1986) Edited by Jiangsu New Medical College. Science and Technology Press, Shanghai

Genus: *Paphiopedilum* Pfitzer

Chinese name: *Dou Lan* (pouch orchid)

Paphiopedilum, the popular, beautiful Ladies Slipper Orchid, derives its name from Paphia, an alternative name for Aphrodite, the Greek goddess of beauty. Paphos is also the name of the town in Cyprus which once housed a famous temple to Aphrodite. An alternative popular name of the genus is Venus Slipper Orchid, Venus being the Roman equivalent of Aphrodite. *Pedilon* is Greek for 'sandal' or slipper. This refers to the helmet-shaped lip (Fig. 18.1).

When CITES enforcement was still lax, hundreds of thousands of plants were ripped from the wild and smuggled into Hong Kong from whence they found their way to the First World countries. *P. micranthum*, *P. armeniacum* and *P. malipoense* were the recent favourites. *Paphiopedilum* flowers lasts for a long time on the plant, so these orchids are ideal as pot plants, and a number of *Paphiopedilum* species and hybrids have made it to the supermarket. Flowers of *P. concolor* stay on the plant for 8 weeks (Yang et al. 1993). Following the rape of their habitats, CITES declared all *Paphiopedilum* species to be highly endangered, and they are listed in CITES I, which makes cross-border trade illegal in the absence of valid proof that the plant in question is raised from seed.

Paphiopediulun concolor (Lindl. ex Bateman) Pfitzer

Chinese name: *Tongseduo Lan* (uniform colour pouch orchid)
Chinese medicinal name: *Bazhangcao, Shizilixian*

Description: *P. concolor* is a member of the mottled-leaved *Paphiopedilum* that are members of the subgenus *Brachypetalum*. *P. concolor* is distinguished by the even yellow ground colour of its flowers. Plants are herbaceous, terrestrial or lithophytic. Leaves are 4–6 cm, oblong-elliptic, 7–15 by 3.5–4 cm, tessellated, pale and dark green on the upper surface, purple or densely purple-spotted underneath. Inflorescence may be one or two, 5–7 cm long, purple, fine white, pubescent, and each carries one or two flowers, 6–7 cm across. Flower is evenly yellow all over and finely purple-spotted (Fig. 18.2). Some varieties are white or ivory-coloured. Dorsal sepal is ovate-orbicular, calyx is ovate-elliptic. Petals are elliptic, rounded at the apex. Lip is ellipsoid-saccate. It flowers throughout the year with a peak in June to August in Thailand (Vaddhanaphuti 2001). Flowering period is April to June in Yunnan, Guangxi (Yang et al. 1993) and Vietnam (Averyanov et al. 2003).

The species is distributed in Guangxi, Guizhou, Yunnan, Indochina, Thailand and

© Springer International Publishing Switzerland 2016
E.S. Teoh, *Medicinal Orchids of Asia*, DOI 10.1007/978-3-319-24274-3_18

Fig. 18.1 *Paphiopedilum insigne* (Wall ex Lindl.) Pfitzer var. *chantini* [as *Cypripedium insigne* Wall ex Lindl var. *chantini*]. From: von *Lindemann E., Lindenia, Iconographie des orchidees, Plates 721 - 768*, vol. 16: t.738 (1900). Colour painting by G. Putzys. Courtesy of Smithsonian Institute, Washington, U.S.A

Myanmar on limestone at 300–1000 m, in the shade (Cribb 1987; Chen et al. 1999) or in full sun on vertical cliffs in Vietnam (Averyanov et al. 2003).

Herbal Usage: Herb collected from Yunnan and Guangxi Provinces. Nature of the herb is "hot and neutral". Entire plant is used in Chinese herbal medicine to relieve coughs and asthma, clear gas and for pain relief. It is used to treat pulmonary tuberculosis, bone and joint pains, rheumatoid arthritis and chronic gastroenteritis (Wu 1994). It is not used in isolation but together with other herbs. The amount of *P. concolor* used per decoction is 6–15 g of dried herb, or 30–60 g blended to extract the juice (Hu et al. 2000). The fresh plant is pounded to make a poultice for external application. The *Compilation of Guangxi Herbs* provides variations for the preparation of the poultice: (1) to treat soft tissue trauma, add alcohol to the mashed plant; (2) for sores, add red sugar; (3) for snake bites, mix pulverised whole plant of *P. concolor* 30–60 g with 30–60 ml of white wine. Drink the juice and apply the residue to the area of the bite (Bencao 2000).

Fig. 18.2 *Paphiopedilum concolor* (Lindl. ex Bateman) Pfitzer [PHOTO: E. S. Teoh]

Paphipedilum dianthum **Tang et Wang**

Paphiopedilum parishii (Rchb. f.) var. *dianthum* (Tang et Wang) Karasawa & Saito

Chinese name: *Changbanduo Lan, Shuanghuadou Lan*

Description: *P. dianthum* is a green-leaved epiphytic–lithophytic species. Plant carries 3–5 narrow leaves, 30–35 cm by 3.5–5 cm, which are slightly bilobed at the tips. Inflorescence is 30–80 cm long, horizontal, green with 2–4 greenish-brown flowers. Calyx is narrow, green at the base grading into white as it fans out towards the apex. The linear, tapering, twisted pendent petals are held about 40° from the perpendicular and are green with brownish-red stripes. Lip is helmet-shaped, narrow, 4–5 cm long, 2 cm wide, and of a greenish-brown. There are different reports on its flowering season in China: July to September (Chen et al. 1999) and September to November (Yang et al. 1993). It flowers from mid-September to end-November in Vietnam (Averyanov et al. 2003).

P. dianthum occurs in south-west Guangxi, Guizhou and southern Yunnan at 1000–2300 m

on trees and rocks in open forests. It was considered to be endemic in China until 2000, when Averyanov and his team discovered the species growing in northern Vietnam on small shady cliffs just below the summit of karst-limestone mountains and hills between 1000 and 1450 m. It sometimes occurs at lower elevations of 600–700 m. The plants were growing in deep shade under laurel, oak and conifers (Averyanov et al. 2003).

Herbal Usage: Herb is product of Guangxi and Yunnan. In China, it is used to treat swollen liver and spleen (Wu 1994), although an enlargement of these two organs can be due to numerous, unrelated causes.

Paphiopedilum insigne **(Wall. et Lindl.) Pfitzer**

Chinese name: *Naoquehua* (noisy sparrow orchid), *Bobandou Lan*

Description: This terrestrial, variable species has 5–6 narrow leaves, 15–40 by 2.5–3 cm, green above and a paler green underneath with purple spots towards the base on the under-surface. Scape is erect, 25–30 cm long, purple and pubescent, ending in a solitary bloom 7–9 cm across. Calyx is elliptic, broadest at the junction of its middle and outer third, 4–5 cm long by 2–2.5 cm broad, sometimes slightly undulate at the margin, of a pale green with a white border and heavily spotted with dark brown to maroon. Petals are narrow, spathulate, undulate at the margin, and bowed, 5–6 by 1.5–2 cm, of yellowish-brown and marked by linear red veins. Lip is helmet-shaped, 4.5–5 by 3 cm, yellow, marked with brownish-purple on the outside. Flowering season is October to December in China (Chen et al. 1999), and November to February at the Khasia Hills in India (Bose and Bhattacharjee 1980).

P. insigne is distributed in north-west Yunnan (at 1200–1600 m), Guangdong, Guangxi, Nepal and north-east India (Meghalaya) on grassy, rocky slopes. The species has been widely used in hybridisation.

Herbal Usage: Herb is product of Guangdong and Guangxi Provinces. Leaves are said to be poisonous. Boil the whole plant, or soak for 4–5 h with rice or wheat. They are used to kill sparrows (hence *Naoquehua*) and cockroaches which are regarded as pests and vectors of disease.

Paphiopedilum micranthum Tang et Wang

Chinese name: *Xiaohuadou Lan* (small flower pouch orchid)
Chinese medicinal name: *Huayezi*

Description: Plant is small, with 4–5 leaves, 5–12 by 5–2 cm, of pale green with darker tessellations above, and purple-spotted underneath. Lip is large, deeply inflated, ellipsoid-saccate, light to deep rose-pink and spotted on the inside, an attractive feature of the flowers. Unlike the older *Paphiopedilums*, the petals are ovate, 2–3 cm long and 2–3.2 cm wide, a pale greenish-yellow with crimson stripes. Calyx is small and more lightly coloured (Fig. 18.3). It flowers from mid-March to early May. The flowers are not scented and carry no nectar, and attract pollinators through food deception by floral mimicry (Averyanov et al. 2003).

P. micranthum is a newly discovered, unique, terrestrial–lithophytic species distributed Yunnan, Guizhou, Guangxi and Vietnam. In the last location, it only occurs in a district adjacent to the Chinese border (Averyanov et al. 2003). Immediately following its introduction, *P. micranthum* garnered numerous top awards from Orchid Societies in the Europe and USA and became a parent of many attractive hybrids. The species was so overcollected during the 1970s and 1980s for export to Europe and the USA when its discovery coincided with the opening up of China that it is now endangered and placed under CITES Appendix I. It is found in limestone regions at 1300–1700 m but occasionally as low as 600–800 m. in evergreen, broad-leaved forests. Plants grow on steep slopes and cliffs composed of crystalline marble-like limestone which are marked by vertical erosion.

Fig. 18.3 *Paphiopedilum micranthum* Tang & Wang [PHOTO: E. S. Teoh]

Herbal Usage: Product of Sichuan, Guizhou and Yunnan. It is bitter and cool. The whole plant is used to remove heat and toxins. It is said to benefit the brain and has the ability to calm the nerves. It is used to treat pneumonia, measles and neurosis (Wu 1994). For decoction, 10–15 g of the orchid is used (Hu et al. 2000). Considering that this is a small *Paphiopedilum*, that amounts to several plants being used at one time.

Paphiopedilum parishii (Rchb f.) Pfitzer

Chinese names: *Luoxuandoushe Lan* (spiral tongued orchid), *Duolan* (pocket orchid); *Banye Lan* (mottled leaf orchid), *Piaodaidou Lan*
Chinese medicinal name: *Qianlinglan*
Thai names: *Rongtao Naree Nuad Rue-Si, Rongtao Naree Chieng Dao, Rongtao Naree Muang Gaan.*

Description: Epiphtic or lithophytic, plant is large, with 5–8 leaves. Leaves are lingulate, up to 45 by 4.5–7 cm, thick and green. Inflorescence

Fig. 18.4 *Paphiopedilum parishii* (Rchb. f.) Pfitzer. From: Engler, H. G. A., *Das Pflanzenreich, Orchidaceae-Pleonandreae*, vol. 50: [Heft 12], p. 65, fig 30 (1903)

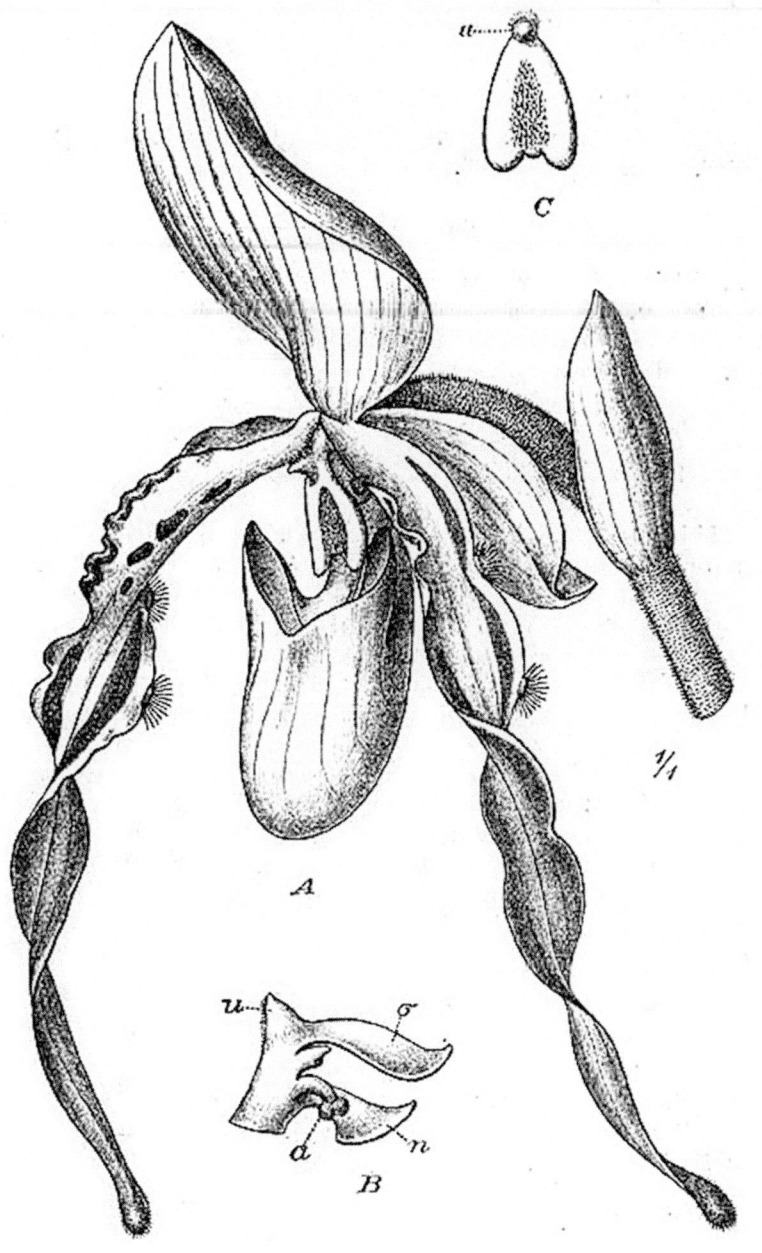

is 50–70 cm long, and it bears two to seven flowers which are each 7.5 cm across. Petals are long, twisted, greenish-yellow with dark purple spots proximally, and solid purple distally. Warts and hairs are present along the lower margin. The dorsal sepal is chartreuse with green stripes. Pouch is brownish-green (Figs. 18.4 and 18.5). It flowers from June to July in China (Liu et al. 2009b) and November to February in Assam (Bose and Bhattacharjee 1980).

P. parishii is found in northern and western Thailand, Laos, Myanmar Yunnan and Assam, in montane forests at 1200–2200 m, on trees or boulders in dense shade and high humidity.

Herbal Usage: Product of Yunnan. In China, *P. parishii* is used to dispel heat and for

Fig. 18.5 *Paphiopedilum parishii* (Rchb. f.) Pfitzer [PHOTO: E. S. Teoh]

detoxification. It is said to have a tranquilising effect. It is used to treat febrile rash, pneumonia and depression (Wu 1994). Character is bitter and cool. Entire plant is used as an antipyretic and detoxicant, to relieve unease of body or mind, skin rash, weak kidneys and dry cough. Between 6 and 15 g of the fresh plant is used to prepare a decoction (*Zhonghua Bencao*. 2000).

Overview

Boorsma (1902) found a trace of alkaloid in the leaves of *P. javanicum*, and he commented that saponin was also present in the roots and leaves.

If *Paphiopedilums* are to have a place in Chinese herbal medicine, they should be propagated by asymbiotic culture and cultivated. Simultaneously, the use of alternatives should be promoted. Nevertheless, damage caused by the use of a single or a few plants to treat someone in a remote village pales before the wanton stripping of entire populations by commercial collectors who supply the nurseries. In the conservation effort, attention is being paid to the isolation of mycorrhiza from roots of *Paphiopedilum*

species. *Epulorhiza calendunia*-like isolates are present in roots of *Paphiopedilum* species in northern Thailand together with *Epulorhiza repens*, the latter also being associated with *Dendrobium* and *Cymbidium* species (Nontachayappom et al. 2010). *Paphiopediulum* species in southwestern China associate with *Tulasnella* species (Yuan et al. 2010).

Genus: *Papilionanthe* Schltr.

Chinese name: *Fengdie Lan*

Formerly known as terete *Vanda,* members of the genus *Papilionanthe* are characterised by their rod-shaped leaves and stems (Fig. 18.6). They are epiphytic plants, which generally scramble over their neighbours, branching and forming dense clumps in open sunlight. Inflorescence is lateral, a raceme with a few medium-sized flowers which are resupinate with a lip that is spurred and devoid of callosities. The two medicinal species below are coincidentally the most popular species in cultivation. They are the parents of Singapore's national flower, the *Vanda* Miss Joaquim. Both species feature frequently in vandaceous hybrids as they impart floriferousness. *Papilionanthe* is related to the genus *Holcoglossum.*

The name of the genus is derived from Latin, *papilo* (butterfly) and Greek, *anthos,* (flower). "*Papilionanthe*" was coined by Rudolf Schlechter in 1915, but it was not popular with his contemporaries who preferred the term 'terete vanda'. Interbreeding between terete and strap leaf *Vanda* has produced hundreds of hybrids with intermediate-type leaves. Plants with such leaves are popular in the Southeast Asia where they were popularly known as semi-terete *Vanda*, quarter-terete *Vanda* and three-quarter terete *Vanda*, etc.

Papilionanthe hookeriana (Rchb.f.) Schltr.

Vanda hookeriana

Fig. 18.6 *Papilionanthe hookeriana* (Rchb. f.) Schltr. [as *Vanda hookeriana* Rchb. f.]. From: Warer, R, Williams B. S. The orchid album, vol. 883: t. 73 (1883). Colour painting by J. N. Fitch. Courtesy of Missouri Botanical Gardens, St. Louis, U.S.A.

VANDA HOOKERIANA

Malaysian names: Kinta weed, *pokok tulang* (bone plant)

Indonesian names: *Anggerik pensil* (pencil orchid), *Potloodorchidee* (in Dutch)

Description: Stems are long, terete, climbing, slender, with internodes of 4–5 cm. Leaves are 7–10 cm long, terete, 3 mm thick and straight. This is a smaller, slimmer plant than *P. teres*, and its leaves carry a constriction about 3 mm from the tips. Inflorescence is lateral, with two or three flowers, seldom up to five, 7 cm across. Petals and sepals are white with a tinge of pink to lavender, while the lip is large and dark purple (Fig. 18.7). Flowering is prolific when it receives sufficient sunlight.

A native of the mangrove swamps in the Kinta Valley, Peninsular Malaysia, this scrambling, terete *Vanda* occurs in swampy, open, sunny areas with high rainfall amidst thickets and shrubs in peninsular Thailand and Malaysia, Sumatra and Borneo and Indochina.

Fig. 18.7 *Papilionanthe hookeriana* (Rchb. f.) Schltr [PHOTO: E. S. Teoh]

Herbal Usage: A hot poultice of the orchid was used for treating painful joints in northern Peninsular Malaysia, the shape of its leaves giving rise to its usage as well as its vernacular name (Burkhill and Haniff 1930; Burkill 1935).

Fig. 18.8 *Papilionanthe teres* (Roxb.) Lindl. [PHOTO: E. S. Teoh]

Papilionanthe teres (Roxb.) Lindl.

Vanda teres Roxb.

Chinese names: *Banghua Lan* (stick flower orchid), *Jianyewandai Lan* (sharp leaved ten-thousand-generation/*Vanda* orchid, *Bangyewandai Lan* (terete leaf *Vanda* orchid); *Bangyeyu Lan* (terete leaf jade orchid), *Fengdie lan*

Indian names: *Chaitek Lei* in Manipuri, cylindrical *Vanda*

Indonesia: *Anggerik pensil*; (Belgian): *Potloodorchidee*

Nepalese names: *Harjor, Thurjo* in Nepali, *Harjor* (Gurung)

Description: Plant is ungainly and in its natural habitat it tends to scramble over rocks and trees in open spaces. It thrives in full sun. Stems are over 1 m long and may attain a great length, always flowering near the top. They are cylindrical, branching at the internodes, with pointed terete leaves that are equally fleshy and channelled along the axis. Inflorescence is lateral, with 2–5 pink flowers which are 6–8 cm across, usually 2–3 opening at a time. It flowers throughout the year, with a peak from March to July (Fig. 18.8). Under cultivation in the USA and Europe, peak flowering occurs in April to June (Hamilton 1990).

Formerly known as *Vanda teres*, this scrambling terete *Vanda* is the other parent of Singapore's national flower, the *Vanda* Miss Joaquim, or more correctly now, *Papillionanthe* Miss Joaquim (Teoh 1982). It is distributed from the foothills of the Himalayas to Upper Myanmar, southern Yunnan, Laos, Vietnam and Thailand at 600 m, and has been cultivated as a garden plant for centuries.

Phytochemistry: Eucomic acid [(2R)-2-(p-hydroxybenzyl)malic acid)] and vandateroside II isolated from stems of *P. teres* increased cytochrome c oxidase activity and/or expression without enhancing cellular mitochondrial content in a human immortalised keratinocyte cell line (HaCaT). Decline in mitochondrial functions occurs with age and may be an underlying cause of age-related changes in the body (Müller-Höcker 1992). Therefore, eucomic acid and one of its three glucopyranosyloxybenzyl eucomate derivatives (vandateroside II) are candidates as new natural ingredients for "anti-ageing" preparations to remedy age-related disorders such as skin aging (Simmler et al. 2011).

Herbal Usage: In TCM, the stems and leaves are used to improve blood flow and reduce swelling (Wu 1994). A paste made from the plant is used to treat dislocated bones in Nepal (Manandhar and Manandhar 2002; Subedi et al. 2013). In northeastern India, leaf paste is also applied to the forehead to reduce fever. Tribal peoples tie a length of the orchid stem to their loin to protect themselves from coughs and colds (Medhi and Chakrabarti 2009)!

Overview

The usage of the two *Papilionanthe* species in Malaysian and Nepalese traditional medicine appears to be an unconscious application of the *Doctrine of Signatures*. The discovery of potential antiskin ageing agents in *P. teres* is likely to lead to its incorporation in cosmetic preparations.

Genus: *Pecteilis* Raf.

Chinese name: *Baidiehua* (white butterfly flower)

Pecteilis is closely related to *Habenaria* from which it is separated by the characteristics of the column. There are fewer than ten species in the genus which is distributed over a large area from India to China and Japan and across Southeast Asia. Its name is derived from Latin *pecten* (comb), describing the lateral lobes of the lip

Pecteilis susannae (L.) Raf.

syn. *Habenaria susannae* (L.) R. Br.

Chinese names: *Longtou Lan* (dragon head orchid), *Baidiehua* (white butterfly flower), *Emaobaidiehua* (goose feather white butterfly flower), *Emaoyufenghua* (goose feather jade phoenix flower)

Chinese medicinal names: *Tu er cao* (rabbit ear herb); *Heqicao* (friendly herb); *Tuyuzhu* (mud jade bamboo); *Baidiehua* (white butterfly flower)

Indian medicinal name: *Riddhi Vriddhi;* also *Hukakanda* (Bihar), *Waghchoora* (Mumbai)

Description: This is a large, robust plant up to 120 cm tall, leafy throughout, the leaves 12 by 5 cm (Fig. 18.9). Tubers are ovoid-cylindric, 3–6 by 1–2.5 cm (Chen et al. 2009c, d). Inflorescence is 20 cm with 4–10 large, white flowers 9 cm across. Sepals are large and spreading, whereas the petals are small and narrow. Lip is 5 cm wide, the two side lobes forming a semicircle and fringed with long narrow teeth. The narrow mid-lobe extends like a tongue between the two side lobes (Fig. 18.10). It flowers in June and July in Hong Kong (Wu et al. 2001), July to September in mainland China (Chen et al. 2009; Jin et al. 2009), September in Thailand (Vaddhanaphuti 2001) and August to October in India (Santapau and Kapadia 1966; Bose and Bhattacharjee 1980).

P. susannae is distributed from Pakistan, the Deccan Peninsula and Assam, through Myanmar, southern China (Jiangxi, Sichuan, Guizhou, Yunnan, Guangxi, Guangdong, Hong Kong) and Indochina to Kedah in northern Peninsular Malaysia, Java, Sulawesi, Timor and Ambon. It is found at 540–2599 m in China. In Malaysia, it is only present in Kedah where it occurs in open grassy areas; the species is not present in Sumatra. The species was named after Susanna, the wife of the distinguished seventeenth century botanist G. Rumphius who authored *Herbarium Amboinense* (Beekman 2002).

Herbal Usage: Roots are collected in autumn from Jiangxi, Sichuan, Guizhou, Yunnan,

Fig. 18.9 *Pecteilis susannae* (L.) Raf. [as Flos Susannae] Rumphius, G. E., *Herbarium amboinense*, vol. 5: p.286, t. 99 (1747). Courtesy of Missouri Botanical Gardens, St. Louis, U.S.A

Guangxi, Guangdong Provinces and sun-dried for future use. They may also be used fresh. In Chinese herbal medicine, the root is said to benefit the "kidney"; it also strengthens *yang* and benefits the "spleen". The taste is sweet and 'slightly warm'. It is used in the treatment of low backache, chronic nephritis, impotence, nocturnal ejaculation, orchitis, hernia and indigestion.

Some Chinese herbal prescriptions employing *Tue Er Cao* (*Pecteilis susanne*) are listed in Table 18.1 (*Zhongyao Da Cidian*, 1986). Similar prescriptions are mentioned in *Zhonghua Bencao*, (Hu et al. 2000).

In India, pseudobulbs were used to treat blebs or bullae on the palm of the hand. Wild hogs sometimes make a meal of the pseudobulbs.

Fig. 18.10 *Pecteilis susannae* (L.) Raf. [PHOTO: Bhaktar B. Raskoti]

Table 18.1 Chinese herbal prescriptions employing *Tu Er Cao* (*Pecteilis susanne*)

1. Indication: renal insufficiency, waist pain, erectile dysfunction, nocturnal emission,
    (a) Mix *P. susanne* (*Tu Er Cao*) 15 g, *Ji Shen Shen* 15 g, *Epimedium macranthum* 6 g and grind to powder form. Steam with lard and red sugar and consume.
    (b) Steep in wine: *Tu Er Cao* 15 g, *Epimedium macranthum/ E. sagittatu*m 9 g and *Curculigo ensifolia* 9 g. Drink the wine.
    (c) Grind *Tu Er Cao* 15 g, *Psorales corylifolia* 9 g and *Lycium chin*ensis 15 g into powder and steam with pig kidney.

2. Indication: hernia
Boil *Tu Er Cao* 15 g, *Guoshangye* 9 g and *Suzhuguo* 9 g and consume with red sugar
(Source: *Kunming Commonly used Folk Herbs*)

Reference: *Zhongyao Da Cidian* (edited by Jiangsu College of New Medicine (1986)): and Wu (1994): *A Concise Edition of Medicinal Plants in China*

Pseudobulbs are relished by the jungle tribes and used to make *salep* (Pandey et al. 2003).

Overview

P. susannae is a large showy orchid in the highlands. Despite its alleged usefulness in Chinese herbal medicine, no effort has been directed towards elucidating its pharmacology. Researchers at the Queen Sirikit Botanic Gardens in Chiangmai managed to identify eight fungal

species associated with the roots of *P. susannae*. Seven species belonged to the genus *Eupulorhiza* and one to *Fusarium*. Germination and protocorm development were enhanced when seeds were cultured with *Eupuloriza* (Chutima et al. 2010 and 2011).

Genus: *Pelatantheria* Ridl.

Pelatantheria was split off from *Cleisostoma* with which it shares some similarities in floral form. The genus has five species distributed from tropical Himalayas to Sumatra, Korea and Japan. Thailand and China each have four species. Plants are epiphytic or saxicolous and monopodial. Stems are long, climbing, sometimes branched, with many leaves of the same size, which are oblong, fleshy, jointed, their bases sheathing the stem. Inflorescence is axillary, racemose, short and few-flowered. Flowers are small to medium size. Petals and sepals are marked with a deep purple line along their length (Comber 2001; Chen and Wood 2009).

Pelatantheria scolopedrifolia (Makino) Garay

syn. *Cleisostoma scolopedrifolium* (Makino) Garay

Chinese name: *Wugong Lan* (Centipede Orchid)

Description: This small epiphyte with a creeping stem 1–1.5 mm in thickness with short internodes and fleshy-coriaceous leaves, 5–8 mm by 1.5 mm does indeed look like a green centipede. Inflorescence is short, 1- to 2-flowered. Flower is small pink, 5–6 mm across, and appears in April. The species is distributed from Japan and South Korea to Shandong, Jiangsu, Anhui, Zhejiang, Fujian and Sichuan provinces in China (Chen et al. 1999).

Herbal Usage: Herbs are obtained from Guangdong and Zhejiang (Wu 1994). Plants may be harvested throughout the year and used fresh or sun-dried. Herb is slightly bitter and cool in nature. Its main functions are to clear internal

Table 18.2 Chinese Herbal Prescriptions employing *Wugong Lan* [*Pelatantheria scolopedrifolium* (Makino) Garay.]

1. Indication: Sore throat, bloody cough
Make a decoction with 15–30 g of *Wugong Lan* (*P. scolopedrifolium*)
Add rock sugar to taste and consume.
(*Zhejiang Common Folk Remedies*)

2. Indication: Sore throat, bloody cough
Prepare decoction with *Wugong Lan* 15 g, Baiji (*Bletilla striata*) 15 g, *Costus root* 15 g. hog fennel 30 g. and consume.
(*Zhejiang Common Folk Remedies*)
(CAUTION: Hogfennel has an antiplatelet action and must not be taken together with other blood thinners or prior to surgery. It may provoke haemorrhage.)

1. Indication: Chronic sinusitis
Decooction prepared with *Wugong Lan* 30 g and Yellow Millet Wine.
(*Zhejiang Common Folk Remedies*)

2. Indication: Acute Tonsillitis
Decoction prepared with 30 g *Wugong Lan*
(*Compilation of Chinese Herbal Medicine*)

3. Indication: Cholecystitis (Infection of the Gall Bladder)
Decoction prepared with *Wugong Lan* 30 g; Dried lychee 10 pieces; White sugar
(*Compilation of Chinese Herbal Medicine*)

4. Indication: Kidney Infection
Decoction with *Wugong Lan* 30 g.
(*Zhejiang Common Folk Herbal Meiicine*)

5. Indication: Infantile Convulsions
Decoction with *Wugong Lan* 15–30 g.
(*Zhejiang Common Folk Herbal Medicine*)

Reference: *Zhongyao Bencao* (2000)

heat, detoxify, sooth the lungs and arrest bleeding. It is used to treat sore throat, mouth ulcers, running nose, tonsillitis, blood in the sputum, infections of the gall bladder and kidneys, and convulsions in infants (Chen and Tang 1982; Wu 1994, *Zhongyao Bencao*, 2000). To treat sore throat and blood in the sputum, one consumes a drink prepared by boiling 15–30 g of *Wugong Lan* (*P. scolopedrifolium*) with rock sugar, the latter added to overcome its bitter taste (*Zhejiang Journal of Research into Medicinal Plants*, quoted by *Zhongyao Bencao*, 2000). A full list of prescriptions is listed in Table 18.2.

Platanthera stenostachya Lindl. [see ***Peristylus densus* (Lindl.) Santapau & Kapadia**]

Genus: *Peristylus* Blume

Chinese name: *Kuorui Lan* (broad pistil orchid)

Peristylis is a genus of montane, terrestrial orchids with 60–70 species distributed from the Himalayas eastwards to the Pacific. It is related to *Habenaria* but its flowers are much smaller. "*Peristylis*" is derived from Greek *peri* (around) and *stylos* (column) which alludes to the prominent glands around the gynostemium (Schultes and Pease 1963).

Peristylus affinis (D.Don) Seidenf.

syn. *Peristylus sampsonii* Hance

Chinese name: *Xiaohuakuorui Lan* (small flower broad pistil orchid)

Description: Plant slender, is 21–50 cm tall, with a vegetative form similar to *P. goodyeroides*. Tubers are oblong-ellipsoid, 1–2 by 0.4–1 cm. Stem is ensheathed at the base bearing 3–5 elliptic to ovate-lanceolate leaves, 2.5–9 by 1–3.5 cm. Inflorescence is terminal, 9–15 cm tall with 15–20 well-spaced, tiny flowers. Flowering season is June to September. It occurs in Hubei, Guangdong, Sichuan, Guizhou and Yunnan provinces in China at 1000–1700 m, in Nepal, northeast India, Myanmar, Thailand and Laos (Anonymous 1976; Chen et al. 2009c, d).

Herbal Usage: Herb is a product of Hubei, Guangdong, Sichuan, Guizhou and Yunnan. Entire plant is used in Chinese medicine to clear heat and remove toxin. It is used in the treatment of nephritis, weak kidneys and backache (Wu 1994).

Peristylus bulleyi (Rolfe) K. Y. Langherbal

Herminium bulleya (Rolfe) Tang & F.T. Wang

Chinese names: *Tiaoyekuorui Lan* (stripe leaf broad pistal orchid), *Tiaoyejiaopan Lan* (stripe leaf angle plate orchid),

Chinese medicinal name: *Zhulancaoz*

Description: Plants are slender, 15–35 cm tall with oblong tubers, 1–2 cm long and 0.5 cm in diameter. Leaves arise midway along the stem. They are linear, 4–10 cm by 2–6 mm. Inflorescence is erect, 13–30 cm, with several well-spaced, small, yellowish-green flowers. Ovaries are prominent, 5–8 mm, petals and sepals 3–4 mm long and 1.2 mm broad, erect or reflexed. Flowering season is July and August. This endemic Chinese species is found in pine forests and grassy slopes at 2500–3000 m, in southwest and west Sichuan and north and northwest Yunnan (Chen et al. 2009d; Jin et al. 2009). The medicinal herb comes from Hunan, Sichuan, Guizhou, Yunnan and Tibet (Wu 1994).

Herbal Usage: Herb is obtained from Hunan, Sichuan, Guizhou, Yunnan and Tibet. Chinese herbalists use the whole grass use to enrich kidney-*yang*. It invigorates the kidney, strengthens the loins, and is used for the treatment of lumbago and asthenia of kidney. It is especially well regarded as a remedy for impotence (Wu 1994).

Peristylis constrictus (Lindl.) Lindl.

Bangladeshi name: *Bhuinora* (Tanchinga tribe), *Cha-muinda* (Marma tribe)

Chinese name: *Dahuakuorui Lan*

Description: A robust plant 30–110 cm tall with 4–6 leaves clustered at the base. Leaves are broadly elliptic, 5–13 by 3.5–6.5 cm. Inflorescence is stout, 20–40 cm, densely many-flowered. Sepals are pale brown. Petals and lip are white (Fig. 18.11). Flowering season is June to August. *P. constrictus* is found on scrubby slopes at 1500–2800 m in Yunnan, Cambodia, Vietnam, Thailand, Myanmar, Bangladesh India, Nepal and Bhutan (Pearce and Cribb 2002; Chen et al. 2009d).

Herbal Usage: An extract of the roots is applied to boils in Bangladesh (Musharof Hossain 2009). The Marma tribe in the Chittagong Hill Tracts make pills from the

Fig. 18.11 *Peristylis constrictus* (Lindl.) Lindl. [PHOTO: Bhakter B. Raskoti]

whole plant of *P. constrictus* with leaves of *Plumbago indica* and employ them as contraceptive pills which are taken for 2–3 days after the period (Uddin and Yusuf 2011).

Peristylus densus (Lindl.) Santapau and Kapadia

syn. *Platanthera stenostachya* Lindl.

Chinese names: *Xiasuikuorui Lan, Xiasuilu Lan, Xiasuishechun Lan* (narrow tassel tongue lip orchid), *Xiasuilu Lan* (narrow tassel heron orchid), *Xiasuiyufenghua* (narrow tassel jade phoenix flower)

Description: *P. densus* is a slender to somewhat robust herb, 12–35 cm tall with a solitary, ovoid tuber, 0.8–2 by 0.6–1.5 cm. Stem is ensheathed at the base and bears 4–6 leaves, widely spaced, ovate to lanceolate, 2.5–10 by 0.6–2.5 cm, pointed at the tips, green or yellowish-green, the margins paler, minutely papillate (Santapau and Kapadia 1966; Chen et al. 2009c, d). Plants at Gaoligongshan are somewhat smaller (Jin et al. 2009). Inflorescence is 11–37 cm with many tiny, white to greenish-yellow, or pale green, sessile flowers well distributed around the rachis. Flowering season is May to October in most of China (Chen et al. 2009d), but shorter at Gaoligongshan (June to August) (Jin et al. 2009). It flowers from August to October in India (Santapau and Kapadia 1966; Bose and Bhattacharjee 1980).

P. densus is found in moist grasslands and forests at 300–2100 m from Zhejiang, northern Fujian and Guangdong, Hong Kong and Lantau Islands, in east Guizhou, Jiangxi and Yunnan. It is also found in Korea, Japan, Vietnam, Cambodia, Thailand, Myanmar Bangladesh and India, in Sikkim Himalaya and Western Ghats (Chen et al. 2009d).

Herbal Usage: Herbs are obtained from China's southwest (Wu 1994). The herbal prescription from the *Yunnan Shi Mao Chinese Herbal Selection* calls for 15–30 g of the *P. densa* to be cooked with meat. The preparation is used to treat physical weakness, fortify stomach and spleen, correct malnutrition and poor digestion in children, as well as for diarrhoea and rheumatism (*Zhonghua Bencao* 1986, *Zhonghua Da Cidian* 1986).

Peristylus goodyeroides (D. Don) Lindl.

Chinese names: *Luhuakuorui Lan* (green flower broad pistil orchid), *Yuanjukuorui Lan* (round distance broad pistil orchid), *Banyeyufeng Lan* (spotted leaf jade phoenix orchid), *Banyekuorui Lan* (spotted leaf broad pistil orchid)
Chinese medicinal name: *Shanshajiang*

Description: *P. goodyeroides* is a medium-sized herb with oblong-ovoid tubers, 1. 5–4 by 0.5–2 cm. Stem is erect, 30–75 cm tall, bearing three or four apple-green, 7-nerved, elliptic-ovate leaves, 10–15 by 4–6 cm, clustered around

and sheathing the lower third of the stem (Fig. 18.12). Inflorescence is terminal, long, and the 7- to 15-cm tall rachis is densely flowered, the flowers nodding and facing all directions. They are small with green sepals and greenish-white petals and a white lip (Fig. 18.13). Flowering period is May to June in Thailand (Nanakorn and Watthana 2008), July to August in China (Chen et al. 1999), August in India (Santapau and Kapadia 1966; Bose and Bhattacharjee 1980), but December is mentioned for Tamil Nadu (Matthew 1995).

P. goodyeroides is found in forests and grassy slopes at 500–2300 m from the Himalayas across Yunnan, Guizhou, Sichuan, Hunan, Hubei, Guangxi, Guangdong, Hong Kong and Lantau Islands, Taiwan and Southeast Asia to Papua New Guinea. The Chinese names are strange because the prominent part of the flower is the twisted ovary, but this is better described as large rather than broad. "Round distance" probably refers to the arrangement of the numerous (almost 100) flowers along the impressive length of the rachis.

Herbal Usage: Herb is obtained from Yunnan, Guizhou, Sichuan, Hunan, Hubei, Guangxi, Guangdong and Taiwan. The root is used to remove toxins from swellings (Wu 1994). For decoction, 6–15 g of the herb is used. Also used externally after pulverisation of fresh herb (Hu et al. 2000). The tubers are used as a tonic in India (Bhattacharjee 1998).

Peristylus sampsonii Hance [see **Peristylus affine (D.Don) Seidenf.**]

Peristylus tentaculatus (Lindl.) J.J.Sm.

Chinese name: *Chuxukuorui lan*

Description: *P. tentaculatus* is a terrestrial herb with globose to ovoid tubers, 1–2.2 by 0.5–1 cm. Stem is erect, 15–40 cm tall, ending in a terminal, racemose, laxly many-flowered raceme. Leaves are oblong, lanceolate, 4–6 by 1.5–2 cm, sheathing at the base, and carried very close to the

Fig. 18.12 *Peristylus goodyeroides* (D. Don) Lindl. Reproduced with permission from *Introductions to Orchids* by Abraham and Vatsala, Parlode, Thiruvananthapuram: Tropical Botanic Garden and Research Centre (TBGRI), 1981

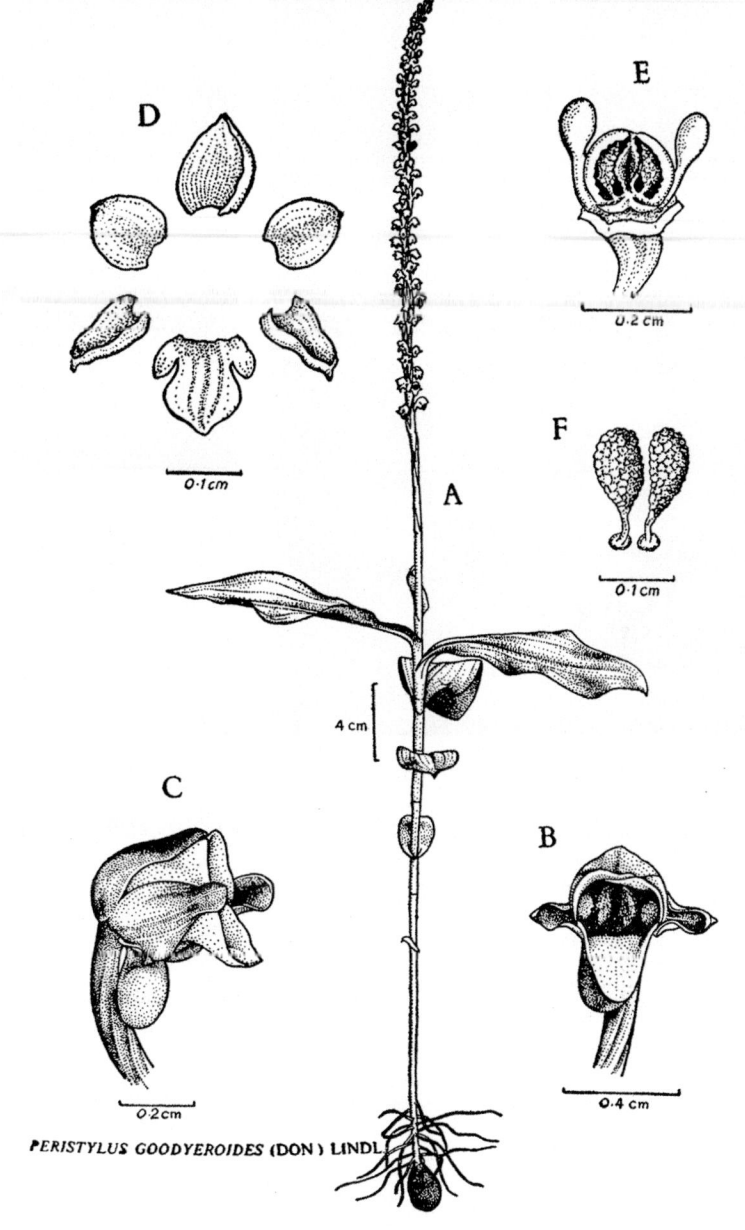

PERISTYLUS GOODYEROIDES (DON) LINDL

A. habit of the plant;   B & C.   flower—front and side views;   D. perianth parts;   E. column;   F. pollinia.

ground, leaving most of the stem bare. Flowers are green or yellowish-green with tiny petals and sepals 2–3 mm long and 1.5–2 mm broad. Lip is trilobed, and the two filiform side lobes measuring 1.5–2 cm in length are the most prominent feature of the flowers. Mid-lobe is only 3 mm long and slightly bifid at the tip (Fig. 18.14).

Flowering period is August in Thailand (Vaddhanaphuti 2001) and February to April in China (Hu et al. 2007; Chen et al. 2009e).

P. tentaculatus grows on rocky soil at the edge of forests at 500–1400 m in southern Yunnan, Guangdong, Hainan, Macau and Hong Kong; Indochina and Thailand (Chen et al. 1999). It is

Fig. 18.14 *Peristylis tentaculatus* (Lindl.) J.J.Sm. [PHOTO: Bhaktar B. Raskoti]

southern China through Southeast Asia to Australia and the Pacific. The name of the genus comes from Greek *phalos* (grey), the colour of its old leaves and the dorsal surface of the tepals.

Fig. 18.13 *Peristylus goodyeroides* (D. Don) Lindl. [PHOTO: Bhaktar B. Raskoti]

common in wet places or wasteland in Hong Kong (Wu et al. 2001).

Herbal Usage: The plant has many uses in China. A Chinese *Herbal* mentioned that *P. tentaculatus* is analgesic, anti-inflammatory and promotes healing, haemostasis, and circulation of *qi*, and improves kidney function (Wu 1994).

Overview

Peristylis has not been investigated pharmacologically. Its Chinese medicinal usage suggests a diuretic effect.

Genus: *Phaius* Lour.

Chinese name: *Heding Lan*
Japanese name: *Ganzeki ran*

Phaius is a genus of terrestrial orchids which are distributed from the Himalayan foothills and

Phaius callosus Lindl.

Indonesian: *Angkrek lemah* (Sundanese) (terrestrial orchid)

Description: *P. callosus* is a large terrestrial orchid with 6- to 12-cm-tall pseudobulbs which are hidden by leaf bases. Leaves are plicate, 10 by 25 cm with a 40-cm pedicle. Inflorescence is usually shorter than the leaves. It carries 10–20 flowers, 10 cm across which are white on their dorsal surface and brown to reddish in front. Tepals are lanceolate, blunted at the tip, fairly broad, and a little cupped. Lip is white with some red on the side lobes which curl over to form a tube around the column. Mid-lobe is large, broader than it is long, with undulate margins. It flowers in April (Comber 2001).

The species occurs in the Malay Peninsula, Sumatra, Java, Kalimantan and Sulawesi (Handoyo 2010).

Herbal Usage: Its roots are "tart and sternutatory" (Greshoff 1900, quoted by Burkill 1935). By that, Burkill probably meant that the roots caused sneezing if placed near the nose, and they might be used to rouse someone who had fainted.

Phaius flavus (Blume) Lindl.

Chinese names: *Huanghe Lan* (yellow crane orchid), *Huanghuaheding Lan* (yellow flower crane top orchid), *Banyeheding Lan* (spotted leaf crane top orchid), *Jiuzi Lan* (nine sons orchid), *Xiaohuaheding Lan* (small flowered crane top orchid)

Myanmar name: *Nay myo new thitkwa*

Description: *P. flavus* is a tall, terrestrial orchid with conical pseudobulbs, 10 cm long and 3.5 cm thick bearing 5–8 leaves, 45 by 11 cm (Fig. 18.15). Inflorescence emerges from the base, erect, tall (50–90 cm), and many-flowered (8–12). Flowers are lemon-yellow, 6 cm across. Lip is marked with orange (Fig. 18.16). Flowering occurs in July to September in the Philippines (Davis and Steiner 1982), August and September in Thailand (Vaddhanaphuti 2001; Nanakorn and Watthana 2008), April to October in China (Chen et al. 1999; Jin et al. 2009) and April in Sikkim and the Khasia Hills (Bose and Bhattacharjee 1980). It is found in shady, moist locations in forests at 300–2500 m in Japan, China, Taiwan, Vietnam, Laos, northern Thailand, Myanmar, Northeast India, Sri Lanka, Sumatra, Malaysia, the Philippines and New Guinea.

Herbal Usage: Taiwanese herbalists state that stems are antipyretic and have the capacity to detoxify. In Taiwan, they are used to treat sores and mouth ulcers (Ou et al. 2003).

Phaius tankervilleae (Banks) Blume

Chinese names: *Honghe Lan* (red crane orchid); *Guaiziye* (twisted leaf orchid); *Dabaiji* (large stone orchid)

Chinese medicinal name: *Hedging Lan* (crane top orchid)

Indian name: *Tipui*

Indonesian names: Indonesia name: *Angkrek apuj* (Sundanese) (fire orchid), *Anggerik Betul, Angkrek Bener*

Japanese name: *Kaku ran, kwaku ran, Kwa ran, Kakuchoran*

Myanmar name: *Zayti thitkhwa*

Papua New Guinea: *Kongimongo* (Ialibu tribe); common name—Kunai orchid

Description: *P. tankervilleae* is a giant sympodial orchid. Plants are 60–100 cm tall. Pseudobulbs are ovoid 6–8 by 3–6 cm. Leaves are 2–6 borne on the upper half of the pseudobulb, elliptic-lanceolate, 30–100 by 8–20 cm, glabrous, green. Inflorescence is large (up to 2 m tall), axillary or arising from the base of the pseudobulb, rising above the leaves, and carries 10–25 flowers, each 10 cm across which open in succession. Tepals are brown in front and white at the back, whereas the trumpet-shaped lip is maroon on the inside (Fig. 18.17). It flowers earlier in Thailand (January to March) and later in China (March to June) (Vaddhanaphuti 2001; Nanakorn and Watthana 2008; Chen et al. 1999; Jin et al. 2009) or Hong Kong (February to September) (Wu et al. 2001). Flowering season is March to April from the Khasia Hills to Sikkim and Himachal Pradesh (Bose and Bhattacharjee 1980). In Sri Lanka, it flowers in February, April to June and September to November (Jayaweera 1981).

P. tankervilleae is distributed in the lowlands (up to 1500 m) from Indonesia northwards to southern China and Taiwan and eastwards to Australia and the Pacific Islands (Liu and Su 1978). Like the former species, it is extremely widespread and enjoys the same distribution. It used to be a popular garden plant when orchid hybrids were still uncommon. In Papua New Guinea, *Phaius tankervilleae* is found in open grassland; hence the name, Kunai orchid, *kunai*, referring to tall grass (O'Byrne 1994).

Herbal Usage: Pseudobulbs are collected from cultivated plants in spring and summer in southern China and sun-dried for future use. The bulb is slightly acrid in taste and "warm'. It is mildly toxic. It is antitussive, promotes circulation and is haemostatic. A decoction is used to treat fever,

Fig. 18.15 *Phaius flavus*
(Blume) Lindl. From:
Blume, C. L. *Collection des
Orchidees les plus
remarquables de l'archipel
Indien et du Japon.* T. 3
(1858). Painting by Lateur.
Courtesy of
plantillustrations.org

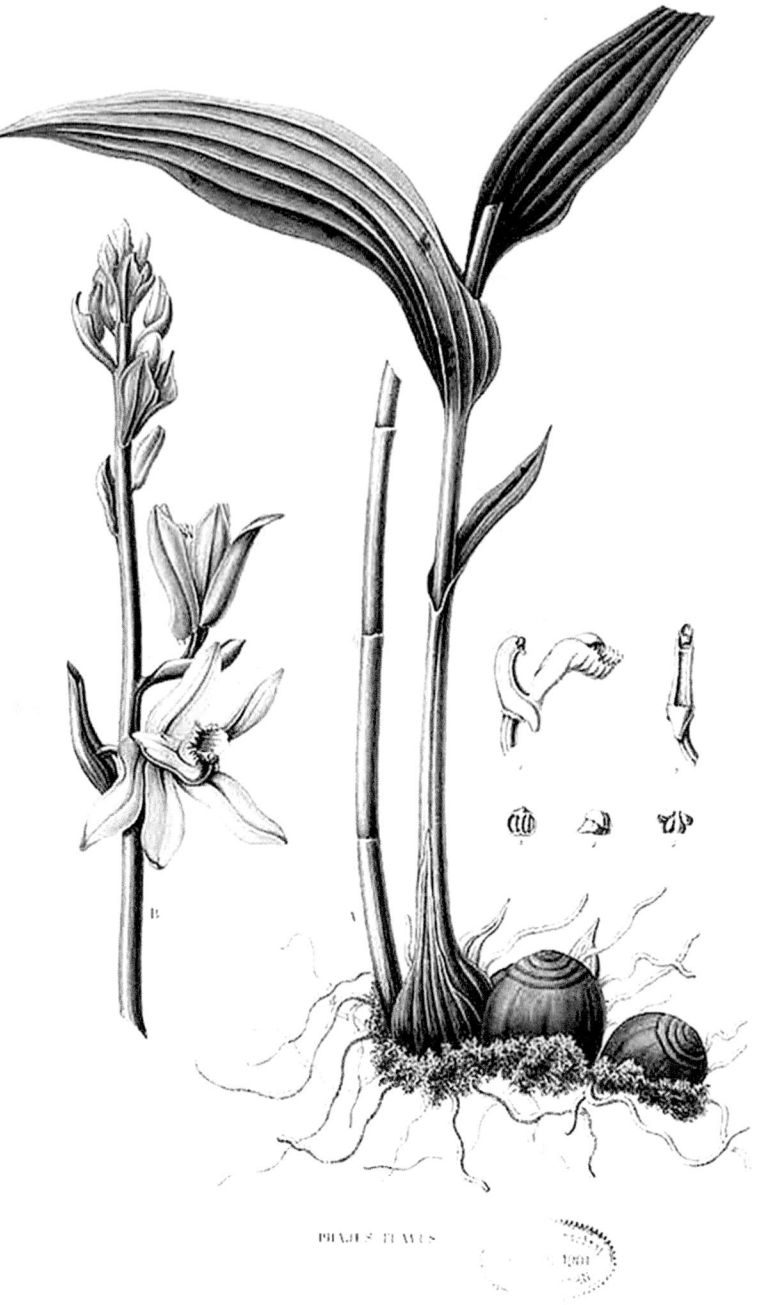

for detoxification and to prevent wet cough
(Wu 1994; *Zhonghua Bencao* 2000). The Hong
Kong Chinese Medical Institute warns that special precautions should be applied when it is used
in pregnancy. To arrest bleeding, the dried bulb
is ground and applied to the wound. To heal
infection of a lactating breast, a poultice is

prepared by mashing fresh pseudobulbs (Li and
Lau 1994).

In northeastern India, pseudobulbs are used to
treat fractures and dysentery (Medhi and
Chakrabarti 2009). A paste of the pseudobulbs
is also used to treat swellings of the hands and
legs, as vermifuge and for treating abdominal

Fig. 18.16 *Phaius flavus* (Blume) Lindl. [PHOTO: Bhaktar B. Raskoti]

Fig. 18.17 *Phaius takervillae* (Blume) Lindl. [PHOTO: E. S. Teoh]

disorders (Trivedi et al. 1980). Tubers provide a tonic in Nepal (Pant 2011). Poultice of *P. tankervilleae* is used to treat sores and infected wounds in Peninsular Malaysia, and to relieve the pain of abscesses in west Java (Rifai 1975).

The pale violet flower is heated in the smoke of a wood fire and eaten with any type of food by Ialibu women in Papua New Guinea who claim that this makes it easier for them to conceive (Holdsworth 1975). However, in an earlier publication, the author reported that *P. tankervilleae* was used as a contraceptive in the southern Highlands of Papua New Guinea (Holdsworth 1974). O'Byrne also reported that, in the

southern Highlands of Papua New Guinea, smoked flowers of *P. tankervilleae* are eaten for contraceptive purposes (O'Byrne 1994). Apparently in the hands of different native healers, a single plant, or different parts of it, might be used in a contradictory manner (also, see *Orchis*).

Phytochemistry: When bruised or in the process of dying, stem and leaves of *P. tankervilleae* become bluish due to the production of indigo. Alkaloids are present in this species (Lawler and Slaytor 1969.)

Overview

Given that two species of *Phaius* are used to treat mastitis, sores and infected wounds, it would be interesting to determine whether the phytoalexins of these orchids possess potent antimicrobial properties. Such studies have not been reported.

Using bioassay-guided chromatographic separation, Jao, Lin, Wu and Wu in 2008 managed to isolate eight indoloquinazolinones from a methyl alcohol extract of *P. mishmensis*. Six compounds were new, phaitanthrins A–E and methylisatoid. Phytanthrin A and tryptanthrin exhibited moderate cytotoxicity against several human cancer cell lines (Jao et al. 2008). However, *P. mishmensis* is not a medicinal species.

Genus: *Phalaenopsis* Blume

Chinese name: *Hudie Lan* (moth orchid)

Phalaenopsis occupies the top slot in the market of flowering orchid plants for home and commercial decoration with millions of mature flowering plants being produced in Taiwan, China, Europe and the United States. The potential is tremendous if *Phalaenopsis* can also be the source of a useful herbal medicine. Hybrids of the strikingly beautiful, long-lasting, white *P. amabilis* have now become so ubiquitous throughout the developed world that it hardly needs introducing. It is the national flower of Indonesia (Fig. 18.18).

The name is derived from the Greek *phalaena* (moth) and *opsis* (resembling). Only four species

phaitanthrin A   phaitanthrin B

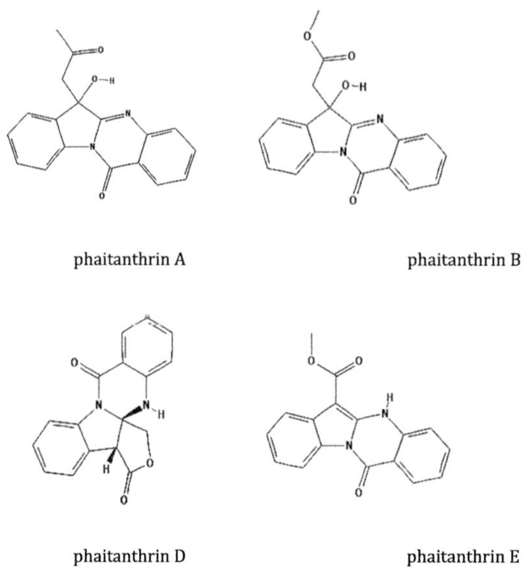

phaitanthrin D   phaitanthrin E

Fig. 18.18 Phaitanthrins A, B, D, E from *Phaius mishmensis* with the illustration of the four chemical formulae

of *Phalaenopsis* are reported to have medicinal usage.

Phalaenopsis amabilis (L.) Blume and Phalaenopsis aphrodite (L.) Blume

Common name: Moth orchid

Indonesian name: *Anggerik bulan* (moon orchid); *Tapak Djalak* in the Sunda Islands, *Anggrek poeti* (an old name recorded by Rumphius in 1701)

Description: *P. amabilis* is a robust epiphyte with a short stem and large, oblong leaves, up to 50 by 10 cm, thick, fleshy, of a deep green. Inflorescence is curved, sometimes branched, 40–60 cm long with numerous, white flowers that are well displayed. Flowers are showy, membraneous, 8 cm across, and lasting several months on the plant. Dorsal sepal is erect, elliptic-ovate; lateral sepals ovate-lanceolate. Petals are held horizontally and expand into a semicircle. Lip is trilobed with tendril-like appendages at the tip of the mid-lobe, and a shield-shaped callus between the side lobes. Lip is marked with yellow and reddish

streaks (Fig. 18.19). Flowering season is February to May. If not pollinated by carpenter bees (*Xylocopa* spp.), flowers remain on the inflorescence for up to 3 months. There are many famous cultivars and the variety *grandiflora* is prized for its reputation of producing more flowers than the other varieties. *P. amabilis* is native to East Malaysia, Indonesia, the Philippines, Papua New Guinea and Queensland from sea level to 1500 m.

P. aphrodite Rchb.f. is very similar to *P. amabilis* and generally only taxonomists and orchid experts can tell them apart. The difference lies in the shield-shaped callus seated at the top of the mid-lobe of the lip. In *P. amabilis*, its upper edge terminates in a single pair of divergent teeth; in *P. aphrodite*, there are two pairs of teeth along the upper border (Christensen 2001).

Phytochemistry: An alkaloid, Phalaenopsine La is present in *P. amabilis* and *P. aphrodite* (Luning et al. 1966). It also occurs in other *Phalaenopsis* species such as *P. fimbriata*, *P. hieroglyphica*, *P. lueddemanniana*, *P. schilleriana, P. sanderiana, P. violacea* and *Doritis* (=*Phalaenopsis*) *pulcherima* (Luning 1974a).

Fig. 18.19 *Phalaenopsis amabilis* (L.) Blume [as *Phalaenopsis grandiflora* Lindl.] From: de Puydt, P.E. *Les Orchidees*, t. 34 (1880). Courtesy of plantillustrations. org

PL.XXXIV.____PHALÆNOPSIS GRANDIFLORA (⅔ G.)

Herbal Usage: The two species *P. aphrodite* and *P. amabilis* being so similar, one species could well be substituted for the other in medicinal usage. *P. aphrodite* is used as a poultice for headaches, and as a plaster for backache and chest pain (Fox 1950). In India, it has been reported to be used as a shampoo to kill and remove leaches and as a poultice for insect bites (Trivedi et al. 1980).

Phalaenopsis deliciosa Rchb. f.

syn. *Kingidium deliciosum* (Rchb. f.)

Fig. 18.20 *Phalaenopsis amabilis* [PHOTO E. S. Teoh]

Chinese names: *Xiaoe Lan* (small moth orchid), *Dajiannanghudie Lan*

Description: This is a small flowered *Phalaenopsis* which stood out well enough among the *Phalaenopsis* species to earn a separate species status called *Kingidium* (Sweet 1969; Vaddhanaphuti 2001). Most taxonomists now agree with Christensen's reclassification of the genus to include it under *Phalaenopsis*.

Plants are short and form clumps by basal shoots. Leaves are oblong, dark green, 15.5 by 3.5 cm. Inflorescence is slim, upright or arching, bearing a few flowers that open in succession, 3–5 at any one time, almost continuously on mature plants. Flowers are fleshy, 1–1.5 cm across, with narrow, oblong sepals and petals, white or yellow, suffused with pink. Lip is white with pink to purple lines on the sidelobes and lip (Fig. 18.20).

This is the most widely distributed species in *Phalaenopsis* (Holttum 1964). It is distributed from Sri Lanka and India across Yunnan, Myanmar, Thailand, Indochina, Malaysia and Indonesia to the Philippines in riverine forests from sea level to 300 m; and on branches of trees overhanging streams, but in the rainshadow, drier foothills of the Crocker Range in Borneo (Chan et al. 1994). Thus, it would seem

that the plants need light and humidity but not water-logging. In China, it occurs in Yunnan and Hainan at 300–1600 m. Plants come into bloom in May or June in Thailand (Vaddhanaphuti 2001), and April and November in Borneo (Chan et al. 1994). Peak flowering in China is July (Chen et al. 1999).

Herbal Usage: The whole plant is used in Taiwan to dispel cold and wind, and clear dampness (Ou et al. 2003).

Phalaenopsis pulcherrima (Lindl.) J.J.Sm.

syn. *Doritis pulcherrima* Lindl.

Chinese name: *Wuchun Lan*
Thai name: *Ma wing*

Description: In habit, *P. pulcherrima* is unique, being terrestrial or lithophytic, growing in exposed localities. Sometimes, it is deciduous. Stem is short, stout and erect up to 15 cm tall with stiff pointed, lanceolate leaves 6–15 by 2.5–5 cm which are green, greyish-green or purple generally flushed with purple on the underside. Plant tends to develop numerous offshoots so that it soon becomes a community, and numerous roots are produced at the base (Fig. 18.21). Inflorescence is erect, up to a metre tall. Raceme carries 12–20 flowers which open in succession. Rarely, there is a panicle. The raceme keeps extending and the plant may continue flowering for up to a year. Flowers face all directions and are 1–3 cm across, varying from pure white through pink to deep magenta. Magenta flowers are smaller but rounder. Lip is trilobed and the base of the mid-lobe and the side lobes are marked with yellow stripes (Fig. 18.22). Lindley gave it the generic name, *Doritis*, because of the downward-pointing, trident lip which looked like a spear, the name being derived from Greek *doru* (spear).

Diploid and tetraploid populations occur in the wild. Peloric forms are now common. In these forms, the petals have the yellow stripe markings similar to those of the lip (Figs. 18.23 and 18.24). The tetraploid variety *buysonniana*

Fig. 18.21 *Phalaenopsis deliciosa* [PHOTO: E. S. Teoh]

are much larger with regards to both the size of the plant as well as the size of the flowers, but its flowers are considerably paler.

P. (syn. *Doritis*) *pulceherrima* is widely distributed in the lowlands throughout Myanmar, Thailand, Indochina, Hainan, northern Peninsular Malaysia and Sumatra. It is commonly found near the sea growing on sandy soil or on rocks. Some plants are deciduous. The variety *buysonniana* is present only in northeastern Thailand near the banks of the Mekong, and its distribution does not overlap that of the diploid variety. Triploid plants do not occur in nature. Flowering season is June to December for the diploid variety and June to August for the tetraploid (Teoh 2002).

Breeding with *Phalaenopsis* has resulted in heat-tolerant hybrids and backcrossing to the larger-flowered parent resulted in many-flowered *Doritaenopsis* with big flowers. These are currently the darlings of the parlour orchid flower industry.

Phytochemistry: An alkaloid, Phalaenopsine 1A, has been isolated from *Doritis pulcherrima* (Luning 1974a).

Herbal Usage: Leaves of *P. pulcherrima* are used to treat ear infections in Thailand (Chuakul 2002).

Phalaenopsis schilleriana Rchb. f.

Description: A beautiful, pink *Phalaenopsis* with an upright, monopodial habit and mottled leaves up to 40 cm by 8 cm. Inflorescence is upright, slightly arching, up to a metre long, sometimes branching, and many-flowered. Flowers are 5–7 cm in diameter, with broad, semicircular petals of powder pink. Lip is trilobed; lateral lobes upright and spotted with orange near their medial border; mid-lobe elliptical, narrowing towards the tip that carries two slender appendages and a pair of orange-spotted calli between the side lobes (Fig. 18.25). There is a white-flowered form. A natural hybrid with *P. aphrodite* is intermediate in appearance between its parents: it is known as *P. leucorrhoda*. Many pink *Phalaenopsis* hybrids have this endemic Philippine *P. schilleriana* in their parentage. It occurs in Luzon and Mindanao in the Philippines at 800–1200 m (Cootes 2001). Flowering period is February to May.

Phytochemistry: Two alkaloids, Phalanopsine La and Phalaenopsine T, have been isolated from *P. schilleriana* (Luning 1974a).

Herbal Usage: It is regarded as a counter-poison. A heated leaf of the orchid is placed over centipede bites in the Philippines. Decoction of the leaves is also prescribed for new cases of phthisis, and for stomach ache (Sulit 1934, quoted by Perry and Metzger 1980; Fox 1950; Lawler 1984a, b).

Phalaenopsis wilsonii Rolfe

Chinese names: *Die Lan* (butterfly orchid), *Huaxidie Lan* (Western China butterfly orchid), *Huaxihudie Lan* (Western China moth orchid)
Chinese medicinal name: *Die lan*

Description: This small Chinese species is found as an epiphyte on trees in sparse woods or as a lithophyte in shady spots along ravines at 800–2200 m in Yunnan, southeastern Xizang, western Guangxi, western Sichuan, southwestern Guizhou in China and in northern Vietnam. Whereas *Phalaenopsis* is a warm growing

Fig. 18.22 *Phalaenopsis pulcherrima* (Lindl.) J.J. Sm. [as *Phalaenopsis esmeralda* Rchb. f. From: Cogniaux A, Goossens A., *Dictionaire iconographique des orchidees. Phalaenopsis.* Vol. 15: t.5 (1896–1907). Painting by A. Goossens

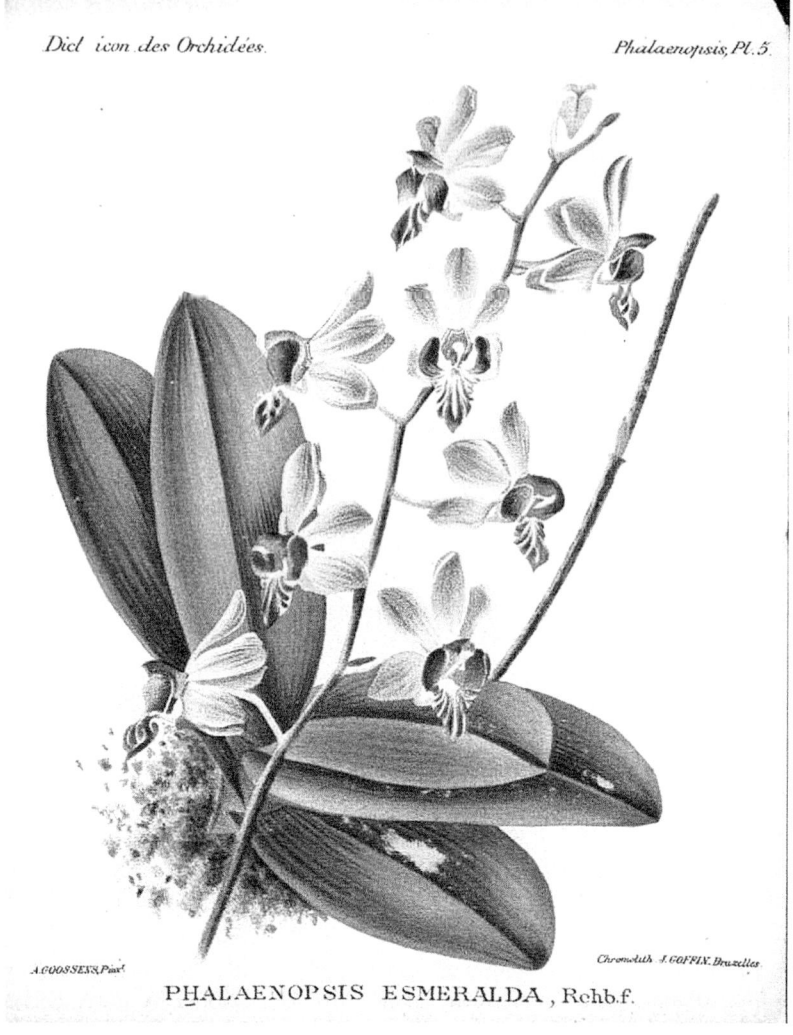

genus, *P. wilsonii* is distinguished by its cold resistance (Chen 1995). Stem is short and bears 4–5 elliptic leaves, 6.5–8 by 2.5–3 cm. Leaves are purplish on the under-surface when immature. Roots are numerous, flattened and green. Inflorescence is arching, 4–8 cm long, and it carries 2–5 pink flowers, 4 cm wide, with spreading, narrow petals and sepals and a purplish lip. Sepals are pointed at the tip, the petals rounded at the apex (Fig. 18.26). Flowering season is April to July (Chen et al. 1999; Jin et al. 2009), during which time the plant may shed its leaves (Yang et al. 1993).

Herbal Usage: Herb is the product of Guangxi, Yunnan, Sichuan and Tibet. Entire plant is used to treat headache, common cold and indigestion in children (Wu 1994).

Overview

The genus *Phalaenopsis* is promising as a source of secondary metabolites because out of 37 species investigated, 17 were found to contain alkaloids in excess of 0.1 % of total weight (Luning 1974a, b). Furthermore, the plants are large, easy to propagate and they mature rapidly. There are hundreds of hybrids in *Phalaenopsis* and most are large plants; this offers unlimited scope to study how hybridisation affects the production of secondary metabolites.

Fig. 18.23 *Phalaenopsis pulcherrima* (Lindl.) J. J. Sm. [PHOTO: E. S. Teoh]

Fig. 18.25 *Phalaenopsis schilleriana* Rchb. f. [PHOTO: E. S. Teoh]

Fig. 18.24 *Phalaenopsis pulcherrima* (Lindl.) J. J. Sm. var. *champorensis* [PHOTO: E.S.Teoh]

Fig. 18.26 *Phalaenopsis wilsonii* Rolfe [PHOTO: E. S. Teoh]

Alkaloids were discovered in *P. amabilis* and *P. leudemanniana* as early as 1900 (Greshoff, quoted by Burkill 1936). Several pyrrolizidine alkaloids have been isolated from *P. amabilis*, *P. manii* (Luning et al. 1966; Brandange and Luning 1969), *P. cornu cervi* (Brandange et al. 1971), *P. taenialis* (Brandange et al. 1970), other *Phalaenopsis* species (Slaytor 1977) and in some *Phalaenopsis* hybrids (Frolich et al. 2006).

Phalaenopsine is the major alkaloid accounting for more than 90 % of the alkaloid content in *Phalaenopsis*. It is present throughout the plant

Table 18.3 The distribution of phalaenopsine T, La, Is and cornucervine in various species of *Phalaenopsis*

Species	ph La	ph T	ph Is	cornu-cervine
Phalaenopsis amabilis		+		
P. amboinensis	+			
P. aphrodite		+		
P. cornu cervi				+
P. equestris			+	
P. fimbriata		+		
P. gigantean	–	–	–	–
P. hieroglyfica	+	or	+	
P. lindenii	–	–	–	–
P. lueddemanniana	+	or	+	
P. mannii	+			
P. sanderiana	+		+	
P. schilleriana	+			
P. stuartiana	+		+	
P. sumatrana	+			
P. taenialis	+			
P. violacea	+	or	+	
P. (Doritis) pulcherrima	+	or	+	

From Brandange et al. (1972): *Acta Chem Scand* 26: 2558–2560

with the highest concentrations located in young developing tissues such as root tips, immature leaves, flower stalks, buds and flowers. Alkaloid is produced by aerial roots and transported to other parts of the plant (Frolich et al. 2006). The alkaloids were named phalaenopsine T, phalaenopsine La and phalaenopsine Is, respectively (Brandange et al. 1972) (Fig. 18.27).

Cornucervine is present in *P. cornu cervi*, a species that is found in Thailand, Sumatra, peninsular Malaysia and Borneo. It is an ester of trachelanthamidine and monomethyl 2 isobutylmalate (Brandange et al. 1971). The distribution of the three phalaenopsines and cornucervine in *Phalaenopsis* is shown in the Table 18.3, adapted from Brandange et al. (1972)

Homospermdine synthase (HSS), the first pathway-specific enzyme of pyrolizidine alkaloid (PA) synthesis is located exclusively in the mitotically active cells of aerial root tips and young flower buds, but not in mature flowers

(Anke et al. 2008). Current thinking is that pyrolizidine alkaloids play a role in the plant's chemical defense against herbivores (Nurhayati et al. 2009). Within the flowers, the highest concentration is found in parts that play the dominant role in attracting insects such as the labellum with colourful crests, column and pollinia (Frolich et al. 2006).

Two phenanthropyran derivatives have been isolated from *P. equestris*. They are 3-methoxy-2,7-dihydroxy-5H-phenanthro(4,5-bcd) pyran and 2,3,7-trihydroxy-5H-phenanthro (4,5-bcd)pyran (Manako et al. 2001).

Geranyl diphosphate synthase is the precursor of monoterpenes, the major scent compound in strongly scented *P. bellina*. The scent biosynthesis pathway in *P. bellina* is controlled by geraniol and linalool metabolism (Tsai et al. 2008).

Pharmacological effects of the secondary metabolites of *Phalaenopsis* were not uncovered during the literature search. *Phalaenopsis* leaves can be eaten. Young leaves provide a salad in Java (Rifai 1975; Lawler 1984a, b). Snails and slugs often make a meal of the plants. Naturally occurring pyrolizidine alkaloids from plants are generally poisonous and are known to cause fatalities in grazing cattle, and liver cirrhosis and cancer in humans. However, the pyrolizidine alkaloids present in orchids are thought to be non-toxic (International Programme on Chemical Safety, WHO, 1988). Nevertheless, having cultivated *Phalaenopsis* for nearly 50 years, I noticed that *P. cornu-cervi* is one species that will thrive despite neglect. It does not get killed by bacterial rot, is immune to fungi, and it is not attacked by arthropods, snails and slugs, whereas other *Phalaenopsis* species and hybrids grown in the same vicinity are subject to the entire range of diseases and predators. Perhaps cornucervine, the prolizidine alkaloid of *P. cornu cervi*, is poisonous; or else the species may contain other poisonous compounds. *P. amabilis* and *P. wilsonii*, with entirely different geographic and national distributions, are both used to treat headache in separate countries.

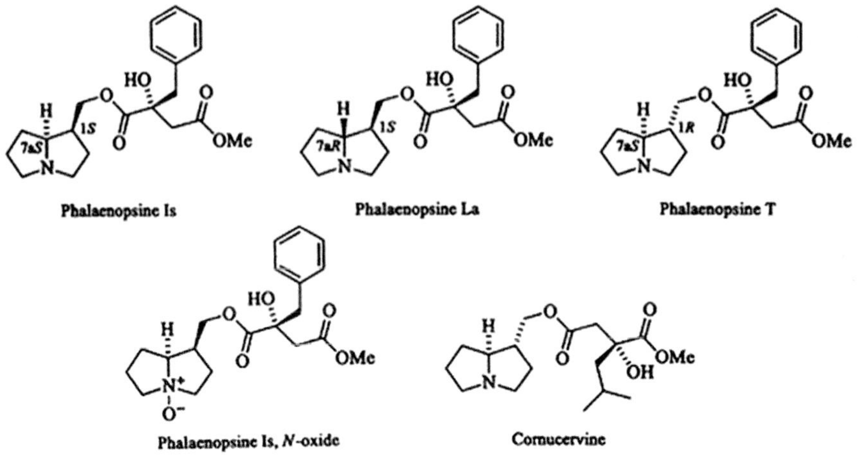

Fig. 18.27 Some Alkaloids from *Phalaenopsis*

Van den Brink (1937) recalled that *P. amabilis* was sometimes referred to as *Angkrek bintang* (star orchid) or *Tapak djalak*, the latter being a name given to the cross painted on the frontal aspect of the neck or above the door to ward off evil spirits and illnesses from people and cattle. In the mid-nineteenth century the white flowers of *P. amabilis* were likened to these charm crosses.

Genus: *Pholidota* Lindl.

Chinese name: *Shixiantao* (rock-living Immortal peach)
Chinese medicinal name: *Shilian* (stone lotus)

Pholidota is an Asian–Australian genus with some 40 mainly epiphytic, sometimes saxicolous, rarely terrestrial species that enjoy a distribution from India, southern China, Myanmar, Indochina, Malaysia, Indonesia, to Papua New Guinea and Australia. One or two linear to linear-lanceolate leaves arise from fusiform or cylindrical pseudobulbs. Prominent features of the genus are the arrangement of the flowers in two ranks, their globose shape, and prominent bracts. Its name is derived from Greek, *pholidotos* (scaly), referring to the large, imbricate bracts on the inflorescence.

In the Chinese herbal trade, *Pholidota* is commonly substituted for the desired *shihu* which should properly be one of several species of *Dendrobium*. Chinese medical establishments that value product purity have explored several approaches to distinguish between *Dendrobium* and *Pholidota*, such as gross morphology, cellular morphology, chromatographic analyses of their chemical constituents and DNA sequencing. The last is the most reliable.

Nevertheless, many species of *Pholidota* have their own place in traditional Chinese medicine (TCM), and *P. yunnanensis* has an important role as *Guoshangye, a* tribal remedy of the Miao and Tuja peoples in Yunnan.

Pholidota articulata Lindl.

Chinese name: *Jiejinshixiantao* (node stem, rock-living Immortal peach)
Chinese medicinal name: *Shilian* (stone lotus), *Shibangtui*
Indian name: *Harjojan; Jivanti*
Myanmar name: *Kwyet mee pan myo kywe*
Nepalese name: *Thurjo, Pathakera*

Description: *P. articulata* is found growing on trees or shaded rocks at 800–2500 m across a wide area extending from India, Nepal and

Fig. 18.28 *Pholidota articulata* Lindl [as *Pholidota griffithii*, Hook. f.] From: Hooker, W. J., Hooker, J. D., *Icones Plantarum*, vol. 19: t. 1881 (1889). Drawing by M. Smith. Courtesy of Missouri Botanical Gardens, U.S.A.

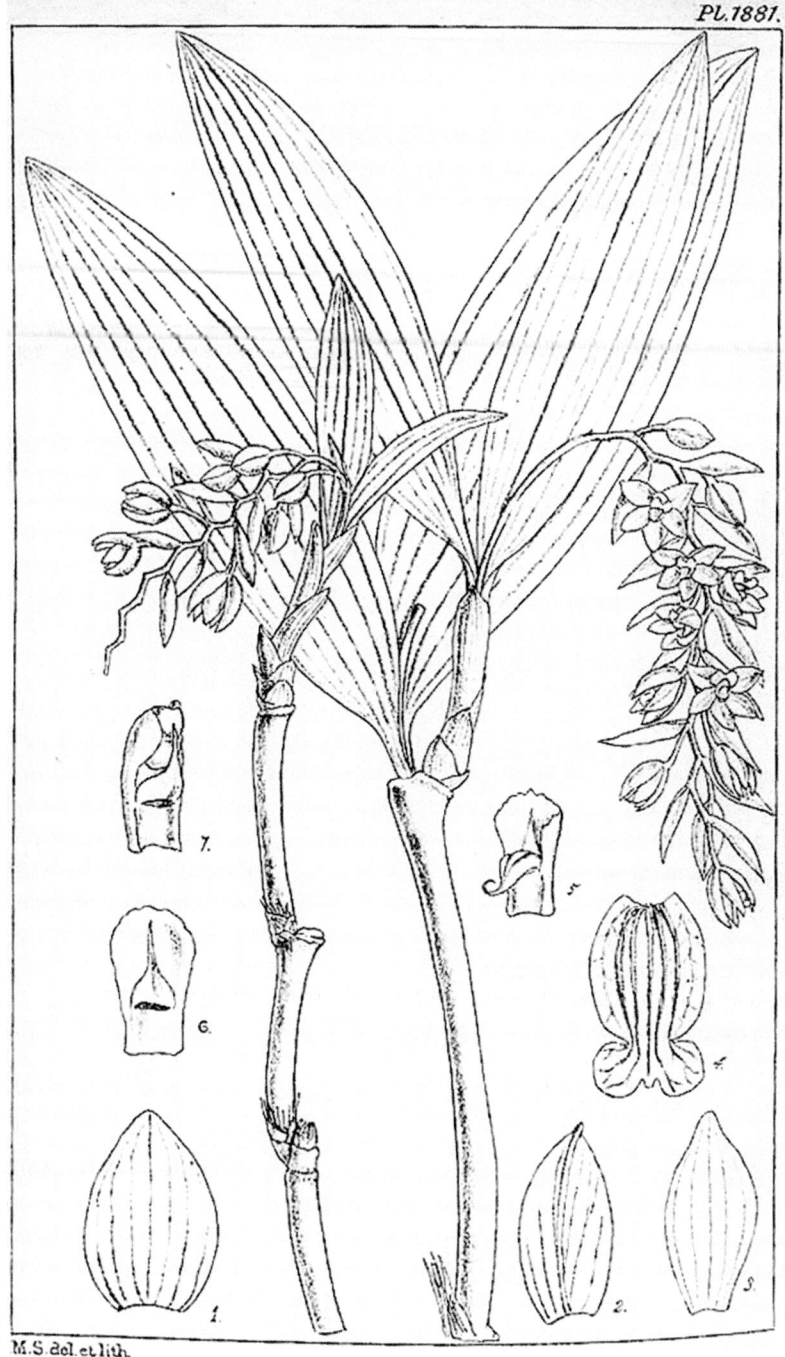

PL.1881.

M.S. del. et lith.

Pholidota Griffithii, Hk.f.

Bhutan, across the provinces of southeast Tibet, southwest Sichuan and Yunnan in China to Myanmar (Kachin, Kayin and Mon states), Thailand, Cambodia, Vietnam, Malaysia and Indonesia. Pseudobulbs are almost cylindrical, 4–12 by 1 cm, usually pendulous and carrying two elliptic, pointed leaves 6–13 by 1.5–5 cm with short stalks (Fig. 18.28). Inflorescence is droopy, with a zigzag rachis that carries 12 (up to 20), small,

Fig. 18.29 *Pholidota articulata* Lindl. [PHOTO: Bhaktar B. Raskoti]

Phytochemistry: Two 9,10-dihydrophenanthrenes, namely, isoflavidinin and iso-oxoflavidinin A and a novel 9,10-dihydrophenanthrene derivative named flavidin, were isolated from *P. articulata* (Majumder et al. 1982a,b). Flavidin is also present in two other Himalayan orchids, *Coelogyne flavida* and *Otochilus fusca*.

Herbal Usage: Herb is obtained from Yunnan and Tibet. The whole plant, 30–50 g is used in decoction (Hu et al. 2000); the pulverised plant is also used for external application. It enriches *yin*, removes gas and reduces swelling. It is used to treat coughs caused by body heat, headache, dizziness, traumatic injuries, sores and ulcers, irregular menses and uterine prolapse (Wu 1994; *Zhongyao Bencao* 2000).

A detailed prescription is provided by the *Yunnan Simao Compilation of Chinese Herbs.* For external use, prepare a poultice with the following: *Shibangtui* (*P. articulata*), *Guoshangye* (*Bulbophyllym odoratisssimum*), *Yeshanghua* [*Helwingia japonica* (Thunb.) Dietr.], *Citongpi* (bark of Oriental Varigated Coralbean or bark of Himalayan Coralbean) and *Zelan* (*Eupatorium japonicum* Thunb.). To this mix, add purple rice and blend. Apply to wounded part. (quoted by Hu et al. 2000).

In Uttarakhand in the western Himalayas, it is used to treat fractures (Jalal et al. 2008). It is also a stimulant, demulcent and tonic. *P. articulata* may become endangered in India because it is sometimes substituted for *Jivanti* in the popular tonic, *Asthavarga* (Pandey et al. 2003). In Nepal, *P.articulata* is regarded as a plant for making tonic. Paste made with pseudobulbs of the variety *griffithii* is applied on dislocated bones (Pant and Raskoti 2013). In this Himalayan country, powder prepared from the roots is used to treat cancer; whereas juice prepared from the capsule is used to treat skin eruptions and ulcers (Pant 2011).

Pseudomonas, Flavobacterium, Stenotrophomonas, Pantoea, Chryseobacterium, Bacillus, Agrobacterium, Erwinia, Burkholderia and *Paracoccus* strains of bacteria were found colonising roots of *P. articulata* but only three

pinkish to dull brownish flowers, 8–10 mm arranged in two rows. Tepals are widely spread and nearly equal 4–9 by 1.5–4.5 mm. Lip is bilobed at the base and is marked by 5 yellow ridges over its oval, hollowed part (Seidenfaden and Wood 1992) (Fig. 18.29). The species is variable and some Chinese varieties have greenish-white flowers. In China, the flowering period is June to August (Chen et al. 1999; Jin et al. 2009). It flowers from April to June in Nepal (Raskoti 2009), May to June in Myanmar (Tanaka and Yee 2003) and in June to August in northeast India (Bose and Bhattacharjee 1980).

Variety *griffithii* Hook. f. (syn. *P. griffithii*) distinguished by pure white flowers bloom from July to August, after the type form has finished flowering. This variety is found throughout Nepal and Northeast India at 800–1400 m.

genera (*Pseudomonas, Bacillus and Flavobacterium*) were endophytic. These bacteria produce indole-3-acetic acid (IAA), the highest concentration of auxin being produced during the stationary phase of the orchid. IAA stimulates root formation and growth (Tsavkelova et al. 2007). Although an ethanolic extract of *P.articulata* showed antibacterial activity against *Staphylococcus* and *Bacillus subtilis* at a concentration of 5 mg/ml (Marasini and Joshi 2012), some strains of bacteria from the root may infect human tissue, so they pose a risk when the orchid is used in the raw state to treat injured skin.

Paste made with pseudobulbs of *P. articulata* var. *griffithii* (Hook f.) King & Pantling [Nepali name (in Tamang): *Syabe lamda; Timpuno*] is applied to dislocated limbs in Nepal (Manandhar and Manandhar 2002). Paste of pseudobulb is used to relieve fever and powdered pseudobulb is used as a tonic (Subedi et al. 2013).

Pholidota cantonensis Rolfe

Chinese names: *Xiyeshixiantao* (fine leaf rock-living Immortal peach), *Yandou* (rock bean)
Chinese medicinal name: *Xiaoshixiantao* (small rock-living Immortal peach)

Description: This is a dwarf *Pholidota*. Pseudobulbs are ovoid 1–2 cm by 5–8 mm, covered with leathery sheaths and borne on a creeping rhizome, 1–3 cm apart. Leaf is linear to lanceolate, 2–8 cm by 5–7 mm, papery, and commonly recurved. Flowers are white to pale yellow, 4 mm in diameter, appearing in April. It is lithophytic on rocks in forests at 200–900 m in Zhejiang, Jiangxi, Fujian, Guangdong, Guangxi Hunan and Taiwan.

Phytochemistry: Four active compounds were isolated from *P. cantonensis*, namely, pholidonone (1), ephemeranthoquinone (2), orchinol (3) and batatasin III (4). Pholidonone is a new compound (Li et al. 2008). Subsequently, densiflorol, 3,5-dimethoxy-4-hydroxy-propiophenone, cinnamic acid, syringaaresinol, 24-methylenencycoartanol, ergosterol peroxide and beta-sitosterol were isolated from the whole plant of *P. cantonensis* (Li et al. 2014).

Herbal Usage: Herb is obtained from Zhejiang, Jiangxi, Fujian, Taaiwan, Guangdong, Guangxi and Hunan. Whole plant is used to treat high fever, eczema and haemorrhoids (Wu 1994). *P. cantonensis* is an occasional adulterant of *Herba Dendrobii (shihu)* but its occurrence can be easily discovered by applying the recommended criteria provided by Lin et al. (1998) from the Wuhan Institute for Drug Control for the identification of the adulterant.

Pholidota chinensis Lindl.

Chinese names *Datiao Lan* (large hanging orchid), *Foushihu* (floating *shihu*), *Shanxi xiantuo* (Shanxi's Immortal stone peach), *Chuanjiacao* (Sichuan's real grass), *Maliugen* (horse pomegranate root)
Chinese medicinal names: *Shixiantao* (Immortal stone peach); *Shishanglian* (lotus on the rock); *Shiganlan* (Rock olive); *Shichuanpan* (rock piercing plate) *Shiyurou* (rock dogwood); *Guoshangye* (leaf above the fruit); *Qiannianai* (thousand year short); Xiaokouzi Lan (Small button orchid); *Fu Shihu* (floating shihu); *Chuanjiacao* (Sichuan's real grass), *Maliugen* (horse pomegranate root)
Myanmar Name: *Kwyet mee pan myo kywe* (Note: The Myanmar name for *P. chinensis* and *P. articulata* are identical.)

Description: Pseudobulbs are well spaced, ovoid, 1.6–8 cm long and 0.5–2.3 cm in diameter, and each carries two thin, elliptical leaves 5–22 by 2–6 cm. Inflorescence is apical bearing double rows of 20–30, white or pale yellow flowers, 1.5–2 cm across and capped by large, prominent orange-brown bracts (Fig. 18.30). Flowering season is April to May (Chen et al. 1999; Hu et al. 2007; Jin et al. 2009), a month earlier in the Kachin state of Myanmar.

An epiphytic orchid with a creeping rhizome, *P. chinensis* grows on the trunk of medium-sized trees or on rocks in sparse forests and on the edge of forests from 1000 to 2500 m. Its distribution stretches in an arc from Zhejiang across Fujian, Guangdong, Guangxi and Guizhou to Yunnan and Tibet, and southwards to Hainan, Vietnam

Fig. 18.30 *Pholidota chinensis* Lindl. [PHOTO: Courtesy of Plant Photo Bank of China]

and Myanmar. It is common in Hong Kong (Wu et al. 2002).

Phytochemistry: Initial testing for alkaloids in 14 species of *Philodata* showed that none of them possessed an alkaloid content which was worth investigating (Luning 1974a, b). although Arthur and Cheung (1960) had earlier reported finding alkaloids in *P. chinensis*. Subsequently, Wen and colleagues at the Fujian Institute of Chinese Medicine in Fuzhou managed to isolate and determine the chemical structure of two new triterpenoids, cyclopholidonol (1) and cyclopholidone (2), from the crude drug *Shi Xian Tao* which was prepared with *P. chinensis* (Wen et al. 1986).

Wang, Matsuzaki and Kitanake demonstrated that an ethyl acetate extract of *P. chinensis* forcibly inhibited nitric oxide production in the murine macrophage-like cell line, RAW 264.7 subjected to lipopolysaccharide (LPS) and interferon-gamma (IFN-gamma). Further studies on the extract resulted in the isolation of two new stilbene derivatives, Pholidotol A and B, together with six known stilbene derivatives. Pholidotols A and B also inhibited nitric oxide (NO) production (Wang et al. 2006). Continuing their work on the species, they isolated two new stilbene derivatives from air-dried whole plants, namely, (e)-s'3,3'-trihydroxy-5-methoxystilbene, which they named pholidotol C, and (Z)-3,3'-hydroxy-5-methoxystilbene named pholidotol D, together with eight known dihydrophenanthrene derivatives, namely, lusianthridin (III), cannabidihydrophenenthrene, coelonin, hircinol, erianthridin, 4,5,-dihydroxy-2-methoxy-9,10-dihydrophenanthrene, eulophiol, a 2,4,7-trihydroxy-0,10-dihydrophenanthrene and a benzoxepin derivative, bulbophylol B. The last item bulbophylol B proved to be most potent in inhibiting nitric oxide production and in radical-scavenging activity. It reduced nitric oxide synthase mRNA expression (Wang et al. 2007).

Three more new stilbenoids have now been isolated at Zhejiang University in Hangzhou, their structures established, and a series of spin-labelled stilbene derivatives synthetised. These derivatives showed even stronger cytotoxicity than the original stilbenes (Wu et al. 2008). Coelonin, batatasin III and pholidotol D present in a dicholoromethane extract of stems and roots of *P. chinensis* enhanced GABA-induced chloride currents in *Xenopus laevis* oocytes (Rueda et al. 2014). GABA being an important inhibitory neurotransmitter in the brain, this finding suggests that *P. chinensis* may have a useful role in the treatment of neurological conditions.

Batatasin III having pain-relieving spasmolytic properties (Estrada et al. 1999; Morales-Sanchez et al. 2014) suggests that it may give patients some relief when they suffer from painful conditions like toothache or traumatic injuries, or when they have abdominal colic associated with an overactive bowel. Orchinol, batatasin III, 3'-*O*-methylbatatsin III, 1-(3,4,5-trimethoxyphenyl)-1',2'-ethanediol, coelonin, 3,4'-dihydroxy-5,5'-dimethoxybibenzyl and 2,7-dihydroxy-3,4,6-trimethoxy-9,10-dihydro-phenanthrene were recovered from *P. chinensis* by high-speed counter-current chromatography (Chen et al. 2015).

Herbal Usage: Plants are collected in the fall, boiled in water and then allowed to dry before storage. Fresh plants are also used as medicine. The character of the orchid is stated as cooling; it is pleasant and bland to taste. It nourishes the *yin* elements, moistens the lungs, cools the blood and promotes salivation (*Zhongyao Da Cidian* 1986). Various TCM practitioners use the whole plant to treat tuberculosis-associated haemoptysis, acute or chronic bronchitis, dry cough, pharyngitis, tonsillitis, toothache, peptic ulcer, gastroenteritis, dizziness, headache, post-concussion syndrome, neurasthenia, osteomyelitis and trauma (Anonymous 1974, *Zhongyao Da Cidian* 1986; Li 1988; Wu 1994). An aqueous extract of the pseudobulbs is prescribed for dysentery and cholera.

Two examples of prescriptions which specify *P. chinensis* are listed in the *Barefoot Doctors Manual* as follows:

(1) for neurasthenia: *P. chinensis* 30 g, *Polygonium multiflora* vine 30 g, use in decoction;
(2) for acute tonsillitis: fresh *P. chinensis* 30 g, fresh *Polygonium perfoliatum* 6 g, fresh *Solidago virgo aurea* 15 g, in decoction.

Table 18.3 shows the full range of prescriptions contained in *Zhongyao Da Cidian*, 1986.

In India, where the orchid is rare, an aqueous extract of the pseudobulbs is taken for scrofula, fever, stomach ache and toothache, while a tincture is used to arrest bleeding, and treat asthmatic coughs, tuberculosis and dysentery (Rao 2004)

Pholidota griffithii Hook, f., (see **P. articulata var. *griffithii* Lindl.**)

Pholidota imbricata Hook. f.

Chinese name: *Subaoshixiantao*
Sri Lankan name: Necklace Orchid
Nepal name: *Syalamba, Timyuno* (in Tamang)

Name in Papua New Guinea: Necklace orchid

Description: *P. imbricata* is a montane, epiphyte or lithophyte with creeping rhizome, tightly spaced, broadly ovoid-conical pseudobulbs, 2–6 by 1.5–3 cm, and large, solitary, leathery, erect, plicated, greyish-green leaves from the top of the pseudobulbs (Fig. 18.31). Pendent, 20- to 30-cm-long scapes arise from the tips of younger pseudobulbs and each carries two lateral rows of small, white to dirty pink, poorly expanded flowers (Fig. 18.32). Flowering period is June or July to September (Chen et al. 1999; Jin et al. 2009). *P. imbricata* is found on trees and karst formations at 1000–2700 m from southwest Sichuan, Yunnan and eastern Xizang in China through the Himalayas to Sri Lanka, Southeast Asia, New Guinea and Australia (Chen et al. 1999). In Sri Lanka, it is a common epiphyte on trees and rocks in moist areas up to 1200 m, and flowers in January, March and June to August (Jayaweera 1981), from May to August in India (Bose and Bhattacharjee 1980), but more specifically, June to July in Mumbai (Santapau and Kapadia 1966). It flowers in September and October in the Philippines (Davis and Steiner 1982).

The interesting life history of *P. imbricata* described by Santapau and Kapadia (1966) is illustrative of many pseudobulbous epiphytic orchids which we summarise as follows: "Towards the close of the dry season, leaves are given off from the rhizome in the axils of large bracts near the base of old pseudobulbs. The inflorescence emerges near the base of a leaf with the arrival of the rains; gradually as the season advances, the base of the leaf begins to swell forming a pseudobulb below the attachment of the rachis of the fruiting inflorescence. Young pseudobulbs are ovoid and smooth, but with advancing age grooves appear and mature pseudobulbs are conspicuously angulated."

Phytochemistry: A 9,10-dihydrophenanthre derivative named imbricatin was isolated from *P. imbricata* (Majumder and Sarkar 1982). Imbricatin is a phytoalexin which is bacteriostatic. Coelonin was also isolated from *P. imbricata* (Majumder and Sarkar 1982). Ethanolic extract of *P. imbricata* at a dosage of 5 mg/ml was

Fig. 18.31 *Pholidota imbricata* Hook. f. [as *Pholidota crotalina* Rchb. f.]. From: Engler, H.G.A., *Das Pflanzenreich, Orchidaceae—Monoandreae—Coelogyninae*, vol. 50 (II. B.7.): [Heft 32], p. 152, fig. 52 A,B (1907). Courtesy of University of Toronto Library, Canada

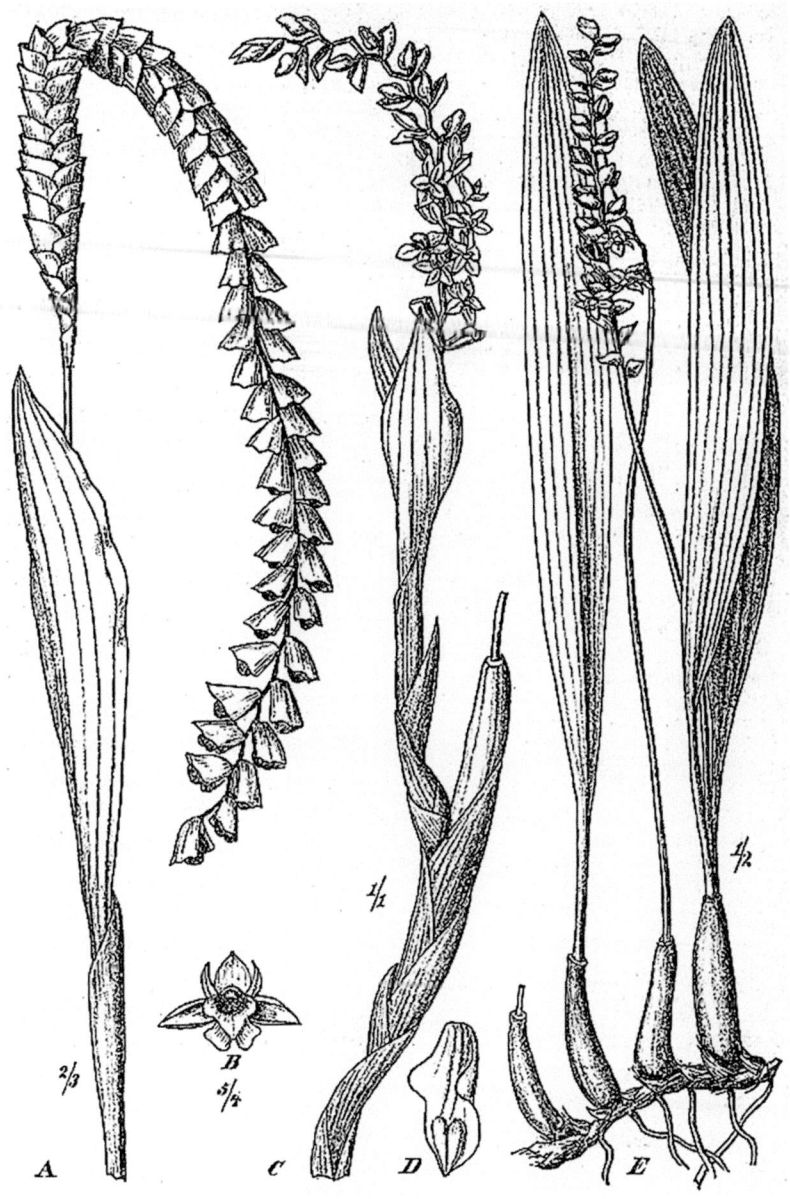

bactericidal against *Vibrio cholera* and *Staphylococcus aureus*. It had weaker activity against *Bacillus subtilis, Escherichia coli* and *Klebsiella pneumoniae* (Marasini and Joshi 2012).

One new phenanthrene derivative, phoimbrtol A, together with seven known compounds, loddigesiinol B, shanciol B, (-)-medioresinol, (-)-pinoresinol, quercetin 3-O-β-L-arabinofuranoside, luteolin 7-O-β-glucoside and platycaryanin D, were isolated from an ethyl acetate extract of the air-dried whole plant of *P. imbricata* Hook. Their inhibitory effects on nitric oxide (NO) production and 1,1-diphenyl-2-picrylhydrazil (DPPH) radical scavenging activity were examined. Platycaryanin D exhibited most potent activity at NO synthesis-inhibition and in the DPPH radical scavenging assay, exceeding those of the familiar anti-oxidative agents, quercetin and resveratrol (Wang et al. 2013a, c).

Herbal Usage: On the Malabar Coast of India, crushed roots were applied on the head, or the entire plant applied to the soles, to relieve fever.

Fig. 18.32 *Pholidota imbricata* Hook. f. [PHOTO: E.S. Teoh]

A poultice made from the entire plant was applied to the loins to facilitate childbirth, induce menstrual flow and diuresis. The fruit was used as a sedative. Both fruit and pseudobulbs were used to treat ulcers (van Rheede 1693; Lawler 1984a, b). The plant is also used to prepare a tonic

(Bhattacharjee 1998). Indian traditional healers make use of the pseudobulbs finely macerated in mustard oil to relieve rheumatic pain. The paste is applied to joints (Rao 2004). In Nepal, juice from pseudobulbs is applied to boils (Manandhar and Manandhar 2002), or over the navel to relieve pain at the navel, abdominal pain and rheumatic pain. Plant is also used to make a tonic (Baral and Kurmi 2006). Leaves and roots are made into a paste for treating fractures in Bangladesh (Musharof Hossain 2009).

Pholidota pallida Lindl.

Chinese name: *Eumaishixiantao*

Description: Pseubulbs are ovoid, 5–6 mm in diameter covered with scales and contiguous. Apex bears a single leaf, 10–20 by 4–6 cm (Fig. 18.33). Inflorescence arises apically from young pseudobulbs, 10–25 cm, rachis nodding, densely many-flowered. Flowers are white, lightly tinged with pale red (Fig. 18.34). *Pholidota pallida* is distributed from central Nepal, Bhutan, northeast India across south and southwest Yunnan to Myanmar, Thailand, Laos and Vietnam in forests at 800–2700 m in Yunnan and 500–2000 m in Nepal; in the latter country, it may be saxicolous. It flowers in June and July in Yunnan (Chen and Wood 2009), August in Nepal (Raskoti 2009).

Herbal Usage: Paste of root and pseudobulb is used to relieve fever, powder to induce sleep and relieve abdominal pain, and juice for abdominal pain at the navel (Subedi et al. 2013; Pant and Raskoti 2013). Pain that is caused by serious abdominal conditions like appendicitis, twisted bowel, a twisted ovarian cyst, infection of the fallopian tubes, infection of the gall bladder and pancreas, or internal hernia often starts by presenting as pain at the navel which later shifts to the affected site. These acute conditions are serious and usually need prompt surgical attention. For juice to relieve pain at the navel, the cause would need

Fig. 18.33 *Pholidota pallida* Lindl. Reproduced with permission from *Introductions to Orchids* by Abraham and Vatsala, Parlode, Thiruvananthapuram: Tropical Botanic Garden and Research Centre (TBGRI), 1981

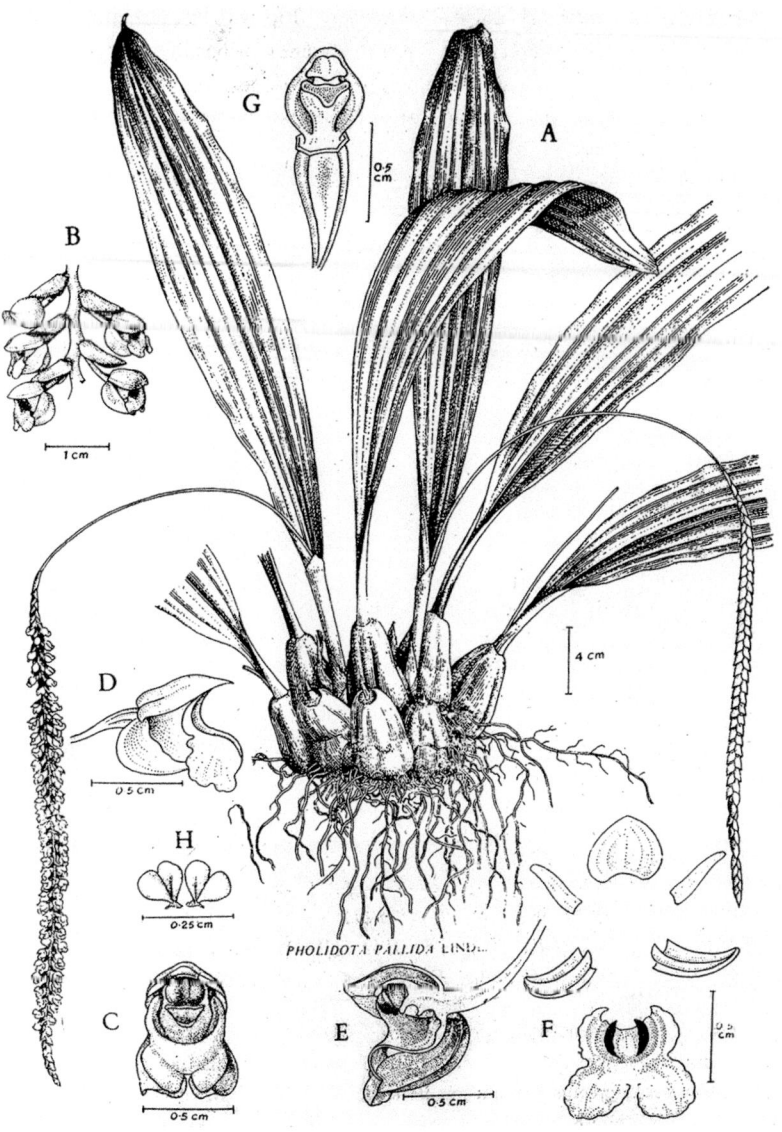

to be minor such as food poisoning or indigestion.

Pholidota yunnanensis Rolfe

Chinese medicinal name: *Shizaozi*

Description: Unlike the former species, the pseudobulbs on the creeping rhizome of *P. yunnanensis* are widely spaced, 1–3 cm apart. Plant is smallish, with cylindrical pseudobulbs, 2–5 cm in length, carrying narrow, lanceolate, coriaceous leaves 6–15 cm long. Scape arises from the younger pseudobulbs. It is 7–9 cm in length and bears two rows of small, white flowers (Fig. 18.35). Flowering period is May to June (Chen et al. 1999; Jin et al. 2009).

P. yunnanensis grows on karst formations and tree trunks in sparse woods at 1200–1700 m in southeast Yunnan, southwest Guizhou, Guangxi, Hubei and Hunan and in Vietnam (Chen et al. 1999).

Fig. 18.34 *Pholidota pallida* Lindl. in Bhutan [PHOTO: E. S. Teoh]

Fig. 18.35 *Pholidota yunnanensis* Rolfe [PHOTO: E. S. Teoh]

Phytochemistry: An early study on the chemical constituents of *P. yunnanensis* yielded seven compounds: *n*-nonacosane, cyclopholidone, *n*-dotriacontanoic acid, *n*-octacostyl ferulate, cyclopholidonol, cycloneolitsol and beta-sitosterol. While this was the first time that *n*-octacostyl ferulate and cycloneolitsol were isolated from the genus *Pholidota*, none of the compounds were new to science (Bi et al. 2004). Two years later, another group of researchers led by XY Guo at Shenyang Pharmaceutical University managed to isolate six new stilbenoids from air-dried, whole plant of the orchid. There were four stilbenes, phoyunbene A, B and C, a bibenzyldihydrophenanthrene ether designated phoyunbene D, and a bis(dihydrophenanthrene ether, phoyunbene E, all natural resveratrol analogues which possessed anti-oxidant properties. Nitric oxide production in the murine macrophage-like cell line RAW 264.7 activated by lipopolysaccharide and interferon-gamma was inhibited by the new stilbenoids (Guo et al. 2006). The team had earlier isolated three new 9,10-dihydrophenanthrene derivatives with unusual structures, named phoyunnanins A, B and C, together with six other 9,10-dihydrophenanthrenes among them lusianthrin, imbrictin and eulophiol, but this work was not reported until 2007. All eight compounds showed DPPH free radical-scavenging activity (Guo et al. 2007). Later, another team discovered an additional three new stilbenoids in *P. yunnanensis*, namely, 1-(4′-hydroxybenzyl)-imbricatin, (E)-4′-hydroxy-2′,3,3′,5-tetramethoxystilbene and (E)-3,4′-dihydroxy-2,6-bis (4-hydroxybenzyl)-2′3′5-trimethoxystilbene (Dong et al. 2013a).

Phuyunbene B prevented cell division (induces G2/M phase cell cycle arrest), caused programmed cell death (apoptosis) and inhibited invasion of liver cancer cells (Wang et al. 2013a, c). These are important properties if they can be translated into clinical efficacy.

Herbal Usage: Entire plant is cooked in pork porridge which is served to treat ordinary coughs and asthma (Hu 1971). Chinese herbalists recommend *Tincture of the Pholidota* (in distilled rice spirit) as an alternative to *Tincture of Goodyera schlectendaliana* or *Habenaria ciliolaris*. This is consumed with rice as a tonic for internal

Table 18.4 Chinese herbal prescriptions that contain *Shixiantao* (*Pholidota*)

1. Indication: Tuberculosis, chronic coughs, acute gastroenteritis and chronic gastritis
Fresh *Shixiantao* 30–60 g or desiccated preparation 9–15 g. Boil and consume.
(Primary Source: *Handbook of Commonly Used Chinese Herbs of Guangzhou Armed Forces*)

2. Indication: External injuries
Grind fresh *Shixiantao*, add wine and apply
(Primary Source: *Handbook of Commonly Used Chinese Herbs of Guangzhou Armed Forces*)

3. Indication: Coughs, difficulty of urination, edema, children's intestinal parasites
Shi Ganlun (*P. chinensis*) 9–15 g. Boil and consume
(Primary source: *Guangxi Chinese Herbs*)

4. Indication: Coughs, throat swelling and pain
Desiccated *Guoshangye* (*P. yunnanensis*) 15–30 g. Boil and take three times a day. (Primary Source: *Wen Shang Chinese Herbs*)

5. Indication: External injuries
Apply desiccated *Guoshangye* (*P. yunnanensis*) in powder form to wound
(Primary Source: *Wen Shang Chinese Herbs*)

6. Indication: Gastric and duodenal ulcers
Boil the whole plant of *Shixiantao* (*P. chinensis*) 15–30 g and consume. Reference: (Primary Source: *Records of Hunan Medicine*)

7. Indication: Numbness and waist pain
Boil the stems with pseudo-scales of fresh Shi Xian Tao 60–120 g. Take it with wine.
(Primary Source: *Fujian Chinese Herbs*)

8. Indication: Gonorrhea
Fresh *Shixiantao* (*Pholidota*) whole plant 30–60 g. Boil and consume.
(Primary Source: *Fujian Chinese Herbs*)

9. Indication: Nocturnal emission
Boil stems with pseudo-scales of fresh *Shixiantao* (*Pholidota*) 30 g, and fresh *Jin Shi Cao* 15 g and consume.
(Primary Source: *Fujian Chinese Herbs*)

10. Indication: Gastric heat, toothache and sore throat
Take decoction prepared with stems with pseudo-scales of fresh *Shixiantao* 30–60 g.
(Primary Source: *Fujian Chinese Herbs*)

11. Indication: Chronic inflammation of bone marrow
Grind fresh *Shishanglian* whole plant and apply externally
(Primary Source: *Quan Zhan Selected Chapters*)

Reference: *Zhongyao Da Cidian*, 1986

Table 18.5 Chinese herbal prescriptions employing *Pholidota pallida* Reference: *Zhongyao Bencao*, 2000). Translated by Jolene Tay

Caution: Pregnant women are advised not to use this herb.

For decoction, 15–30 g.

1. Indication: Tuberculosis
Prepare decoction with *P. pallida* 500 g whole plant; *Xiakucao* (*Prunella vulgaris*) 1000 g; Boil twice until concentrated. Add red sugar 180 g. Consume 15 ml of decoction two times a day.

2. Indication: chronic sore throat
Prepare decoction using whole plant of *P. pallida*, 15 g; with *Qianhu* [root of purple flowered *Peucedanum decursivum*; or *P. praeruptorum* Dunn.] 9 g; *Pipaye* (leaves of *Eriobotrya japonica* Thunb.) 15 g; *Tubeimu* (*Rhizoma Bolbostematis*) 15 g; *Aidicha* (Japanese Ardisia Herb or *Ardisia bicolor* Walker) 30 g; *Yuxingcao* (Heartleaf *Houttuynia* Herb or *Houttuynia cordata* Thunb.) 60 g.

3. Indication: For sores and swelling
Use fresh plant to prepare poultice for application to affected parts.
Primary Source: *Hunan Journal of Medicine*

and *B. kamgtimgemse* (Qu et al. 2006). Examples of prescriptions employing *P. pallida* are illustrated in Table 18.4.

P. yunnanensis is a common adulterant of "*Fengdou*" or *shihu*. Ding et al. (2002) found that the ITS sequence differences between *P. yunnanensis* and the *Dendrobium* species of "*Fengdou*" are so obvious as to make rDNA sequencing a reliable tool for distinguishing between the two. Prior to this, general morphological and histological characteristics served to identify *P. yunnanensis* (Gan and Zheng 1998).

Overview

Pholidota appears in Chinese herbs generally as a contaminant of or substitute for *shihu*. In the southern region of China, *P. yunnanensis* is one of the orchids used in the folk medicine of the Tuja and Miao tribes as *guoshangye*, which is used in a variety of apparently unrelated conditions such as fractures, tuberculosis and pain from hernia. The full range of medicinal usage or *Pholidota* species in general (*Shilian*) is set out in Table 18.4 (adapted from *Zhongyao Da Cidian* 1986). Table 18.5 lists the additional usage of *P. pallida* adapted from *Zhongyao*

injuries. *P. yunnanensis* is one of the principal plants of *Guoshangye*, the folk medicine of the Tuja and Miao tribes in the Yunnan-Guizhou region. Other orchids regarded as *Guoshangye* are *Bulbophyllum andersonii*, *B. odoratissimum*

Bencao (2000) which cautions that the herb is not to be taken by pregnant women.

Fourteen species of *Pholidota* were tested for the presence of alkaloids and all read negative with the Dragendorff spot test (Luning 1974a, b.) However, this did not deter scientists from looking for other types of physiologically active compounds because two species, *P. articulata* and *P. imbricata* are used as tonics in India (Bhattacharjee 1998). Majumder's famous team working in Calcutta reported the isolation of two 9,10-dihydrophenanthrenes, namely isoflavidinin and iso-oxoflavidinin from *P. articulata* (Majumder et al. 1982a) and two 9,10-dihydrophenanthrene derivatives from two Himalayan species of *Pholidota*: imbricatin from *P. imbricata* (Majumder and Sarkar 1982), and flavidin from *P. articulata* (Majumder et al. 1982). Pholidotin is a triterpene isolated from *P. rubra* and *Bulbophyllum* (*Cirrhopetalum*) *elatum* by Majumder et al. (1987). *P. rubra* also contained coelonin, imbricatin and 24-methlenecycloartanyl p-hydroxy-trans-cinnamate.

Pholidotal A and B, two new stilbene derivatives from *P. chinensis*, exhibit anti-inflammatory activity. They inhibit nitric oxide production from activated macrophages (Wang et al. 2007; Wang et al. 2013a, c). In inflammatory conditions like rheumatoid arthritis, excessive nitric oxide production by activated macrophages is thought to be related to worsening of the disease. GABA is the major inhibitory neurotransmitter in the brain. It acts through GABA-A receptors, ligand-gated chloride channels of which there are 19 different subunits that are linked together in diverse combinations. These subunits exhibit distinct pharmacological profiles and the search is on to discover compounds that are subtype-specific. Such a drug would be devoid of undesirable side effects which are produced by the currently non-specific nervous system depressants or relaxants like barbiturates, benzodiazepines (e.g. Valium), neuroactive steroids and anaesthetics. When Mathias Hamburger from the University of Basel in Switzerland and Diana C Rueda from the

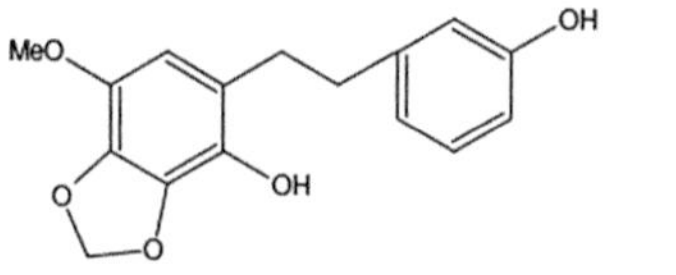

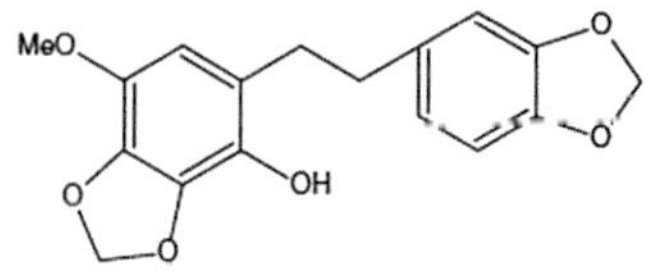

Fig. 18.36 Bibenzyl derivatives from *Pholidota*

University of Vienna in Austria screened a library of 880 plant extracts for activity on GABA-A receptor of the alpha1, beta2, gamma 25 subtype, they found that the dichloromethane extract of roots and stems of *P. chinensis* showed promising activity. Chinese Pharmacopoeia mention the use of *P. chinensis* to treat headache, dizziness and post-concussion syndrome (Anonymous 1974, 1986). Earlier studies reported sedative and anticonvulsant activity of the orchid species. Three stilbenes, coelonin, batatasin III and pholidotal D were the major compounds in the active fraction. Batatasin III displayed the strongest GABA-A receptor modulatory activity. It had a higher efficiency than any natural compound tested so far. Thus, it represents a new scaffold for GABA-A receptor modulators (Rueda et al. 2014).

Batatasin III is not confined to *P. chinense*. Batatasin III (3,3'-dihydroxy-5-methoxybibenzyl) It is fairly common in plants, for example root vegetables like the Chinese yam (*Dioscorea batatas*). It is the commonest bibezyl isolated from orchids. It is present in *P. cantonesis* (Li et al. 2008) and has been isolated from 13 species of *Dendrobium*: *D. aureum* (*D. heterocarpum*), *D. amplum*, *D. aphyllum*, *D. cariniferum*, *D. chrysotoxum*, *D. draconis*, *D. gratiosissimum*, *D. loddigesii*, *D. longicornu*, *D. nobile*, *D. plicatile*, *D. polyanthum D. rotundatum* and *D. venustum* (Sritularak et al. 2011; Xu et al. 2010; Sukphan et al. 2014), *Sunipia scariosa* (Yang et al. 2014), *Epidendrum rigidum* and *Cryptopodium macrobulbon*, the last two being orchids native to Mesoamerica (Hernandez-Romero et al. 2005;

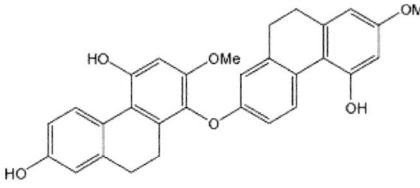

2,4,7-trihydroxy-9,10-dihydrophenanthrene

3,7-dihydroxy-2,4,8-trimethoxyphenanthrene

coelonin

3,7-dihydroxy-2,4-dimethoxyphenanthrene

Phoyunnanin D

Phoyunnanin D

Fig. 18.37 Phenanthrenes from *Pholidota yunnanensis*

Morales-Sanchez et al. 2014). It inhibits germination, photosynthesis and respirations in plants (Hernandez-Romero et al. 2005). Although it is a potent phytoalexin, it shows minimal cytotoxicity on animal cells (Wallstedt et al. 2001; Hernandez-Romero et al. 2005). It is spasmolytic on the guinea pig ileum (Estrada et al. 1999; Gonzales et al. 2014). It shows moderate activity against the malaria parasite and weak activity against herpes simplex virus (Sukphan et al. 2014). From the pharmaceutical viewpoint, Batatasin III is interesting because it has high efficiency in GABA A receptor modulation and could act as a new scaffold for drugs targeted at the brain. However, it lacks subunit specificity (Rueda et al. 2014).

On a different note, a method for the identification of *P. yunnanensis* based on morphological and histological features was described by Gan and Zheng (1998). Simple enough to perform, it may be a convenient alternative for correct identification when more sophisticated methodology is unavailable (Figs. 18.36 and 18.37, Table 18.5).

Genus *Pinalia*

Chinese name: *Ping Lan*

Pinalia is a large genus consisting of about 160 species enjoying a wide distribution from the Himalayas to China, Indochina, Malesia, Australia and the Pacific. Although it was first described by Lindley in 1826, many taxonomists down to the twenty-first century continued to classify these plants as *Eria*. They are epiphytic or terrestrial herbs. Stems are clustered, elliptic on transverse section, with several nodes, and covered with a semi-transparent leaf-sheath that give rise to an apparent venation on the stem. Leaves borne on the upper half of the stem are leathery, linear, terete or lanceolate and devoid of petiole. Inflorescence is axillary, with scaly brown hairs, leaving a pit on the stem when shed. Bracts are large and prominent. Flowers are small to medium-sized, of variable colour; sepals hirsute dorsally. Lip is trilobed, hinged at the base to column foot (Fig. 18.38). Papillose

keels are present in many species (Chen et al. 2009f).

Pinalia graminifolia (Lindl.) Kuntze

Syn. *Eria graminifolia* Lindl.

Chinese name: *Heyemao Lan* (grainy leaf wooly orchid), *Heyiping Lan*
Medicinal name: *Heyemohu* (grainy leaf black epiphyte)

Description: Plants are 8–17 cm tall with 2–6 lanceolate leaves. Inflorescence is axillary, short, and carries 3 white flowers with a yellow lip. Each flower is about 1 cm across (Fig. 18.39). Flowering season is June and July at Gaoligongshan in Yunnan province (Jin et al. 2009). A robust orchid which grows in large communities, this species is found in central and eastern Himalayas at 2000–3320 m (Pearce and Cribb 2002), Myanmar, Tibet and Yunnan.

Usage: The whole plant is used as a tonic. A decoction is prepared with 3–9 g of the orchid. It is used to nourish the stomach, promote the production of body fluids, and to treat hot flushes, fever, sweating, malaise and anorexia (*Zhongyao Da Cidian* 1986). It helps digestion (Chen and Tang 1982).

Pinalia spicata (D.Don) S.C. Chen & J.J Wood

syn. *Eria spicata* (D. Don) Hand.- Mazz.

Chinese name: *Mihuaping Lan*

Description: Epiphytic on tree trunks, pseudobulbs are flattened, bearing 4–6 elliptic-lanceolate leaves, 10–18 by 2.5–5 cm. Inflorescence is densely many-flowered, arising near apex of pseudobulb. Flowers are white, 6 mm across. Lip is yellow at the tip. It flowers in July and August in Nepal (Raskoti 2009), July to October in Yunnan (Chen et al. 2009e). It is distributed in central and eastern Nepal, northeast India, Myanmar, southern Yunnan (China), Thailand and northern Vietnam at 800–2800 m.

Fig. 18.38 *Pinalia graminifolia* (Lindl.) Kuntze. From: Hooker, W. J., Hooker, J. D., *Icones Plantarum*, vol. 19: t. 11847 (1889). Drawing by M. Smith. Courtesy of Missouri Botanical Gardens, St. Louis, U.S.A

Fig. 18.39 *Pinalia graminifolia* (Lindl.) Kuntze [PHOTO: Bhaktar B. Raskoti]

Phytochemistry: *P. spicata* (syn. *Eria convallarioides*) contains nudol, erianthridin, sitosterol, erianol and an uncharacterised fatty alcohol (Majumder and Kar 1989).

Herbal Usage: In Nepal, stems are rendered into powder and consumed to treat stomach ache, or are made into a paste and applied as a poultice for the relief of headache (Vaidya et al. 2002; Subedi et al. 2013; Pant and Raskoti 2013).

Overview

Majumder's group worked on four species of *Pinalia* and reported their findings using the generic name *Eria* which was appropriate at that point in time. They isolated two phenanthrene derivatives, confusarin and confusaridin, from *P.amica* (Rchb. f.) Kuntz. (syn. *Eria confusa*) (Majumder and Kar 1987). Two triterpenoids (acervatol and acervatone) were isolated from *P. acervata* (Lindl.) Kuntz. (syn. *Eria acervata*) (Majumder and Rahaman 2006). Nudol, a phenanthrene originally isolated from *Eulophia nuda*, was found to be present in *Eria carinata*, *P. stricta* (Lindl.) Kuntz. (syn. *Eria*

stricta) (Bhandari et al. 1985) and *P. spicata* (D.Don.) S.C.Chen & J.J.Wood (syn. *Eria covallariodes*) (Majumder and Kar 1989). The last species also yielded erianthridin, sitosterol, erianol and an uncharacterised fatty alcohol (Fig. 18.40).

Genus: *Platanthera* L.C. Rich.

Chinese name: *Shechun Lan*
Japanese name: *Tsure Sagi So*

Platanthera is a large group of cold-tolerant, terrestrial orchids with some 250 members occurring mainly of the northern temperate zones, or at very high elevations in the subtropical region. A single species, *P. angustata* Lindl. inhabits the mountains of Malaysia and Indonesia (Yong 1990). The generic name is rather esoteric as it describes the flat (*platy*, Greek for broad) anther which is hardly ever noticed by the usual orchid enthusiast who is not well versed in taxonomy. Plants have a tuberous root and stout scapes clothed with elliptic-lanceolate leaves at various levels and increasing in numbers towards the base. Inflorescence is terminal, erect, often bearing many flowers and it is quite impressive. They were once included among *Habenaria* by some taxonomists (Fig. 18.41).

Platanthera bifolia (L.) Rich.

Scandinavian name: *Nattviole* (night violet)
German name: *Waldhyazinthe* (wood hyacinth)
Chinese name: *Xijushechun Lan*

Description: A terrestrial herb with tuberous, ovoid rootstock, the erect stem carries two subbasal, elliptic leaves, 9–12 by 0.8–3.5 cm. Inflorescence is erect 9 up to 20 cm tall, many-flowered. Flowers are fragrant, white or green, 1.5 cm across. Dorsal sepal is erect, forming a hood with the petals. Lateral sepals are spreading, lanceolate (Figs. 18.42 and 18.43). Flowering season is August to

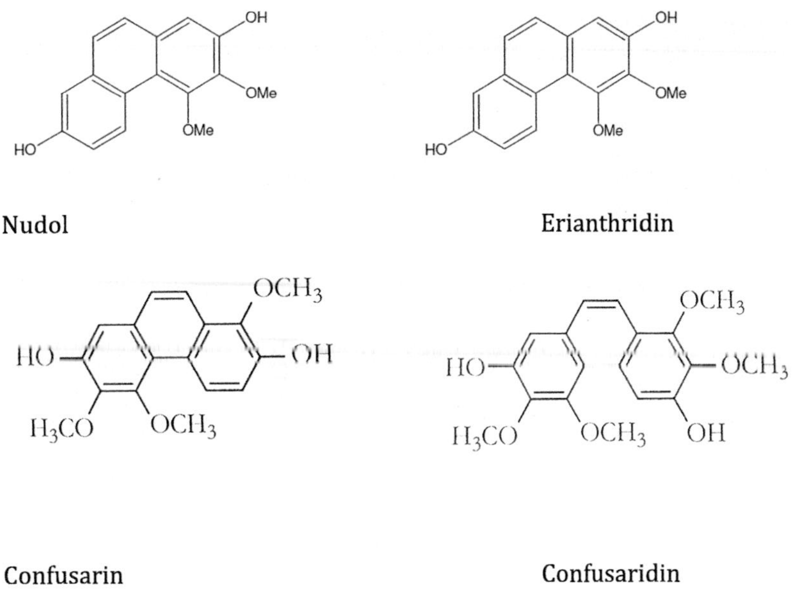

Fig. 18.40 Some phenanthrenes from *Pinalia*

Nudol

Erianthridin

Confusarin

Confusaridin

September. The species is widespread and is found in Europe, northern Africa, Japan, Korea, Mongolia and the northern Chinese provinces from Heilongjiang to Shandong and westwards to Gansu and Sichuan (Chen et al. 2009c, d).

Phytochemistry: Loroglossin, a phenolic glycoside, and phytoalexin are present in *P. bifolia*. It is one of the earliest phytoalexins to be identified. Leaves contain two flavonols, quercetin and kaempferol (Veitch and Grayer 2001). Floral fragrance which attracts the silver moth, *Autographa gamma*, for pollination in Sweden is mainly a blend of benzyl benzoate, benzyl salicylate, cinnamyl alcohol, lilac aldehydes, methyl benzoate and methyl salicylate, with the lilac aldehydes playing the key role in attracting the moth to the flowers (Plepys et al. 2002).

Herbal Usage: It is used as *salep* in Turkey (Sezik 1967) and Iran (Ghorbani et al. 2014).

Platanthera chlorantha Cust. ex Rchb.

Chinese name: *Eryeshechun Lan*
Chinese medicinal name: *Tubaiiji*

Description: *P. chlorantha* is a beautiful, white, terrestrial orchid with ovoid, underground tubers and an erect stem, 30–50 cm tall, with two broad, elliptic leaves at the base, 10–20 by 4–8 cm. Inflorescence is terminal, majestic, with green, lanceolate bracts and over a dozen white flowers, 1–1.2 cm across (Fig. 18.44). It inhabits temperate forests and grasslands at 2000–3500 m and is distributed in Japan, Korea, Siberia, Europe and China (in the provinces of Heilongjiang, Jilin, Inner Mongolia, Hebei, Shanxi, Shaanxi, Gansu, Qinghai, Xinjiang, Shandong, Anhui, Sichuan, Yunnan and Tibet). Flowering season is June to August, earlier or later depending on latitude (Chen et al. 1999; Perner and Luo 2007).

Herbal Usage: The herb is usually collected from August to October, after the flowering season from north and northeast China, Shaanxi, Gansu, Sichuan, Yunnan and Tibet (Wu 1994). It is commonly used in Xizang (Tibet); 3–9 g in decoction to "nourish the lungs, to treat people who cough up blood, vomit blood, or bleed from their nose. The plant is also ground and made into a poultice to treat lacerations, skin infections and burns (*Commonly Used Chinese Herbs in Tibet*, quoted by *Zhongyao Da Cidian* 1986).

Fig. 18.41 *Platanthera bifolia* (L.) Rich. From: Auerswald, B. A., Robmabler, E. A. *Botanische Unterhaltungen zum Verstandnib der helmathlichen Flora,* p. 143, t. 16 (1858). Courtesy of plantillustrations.org

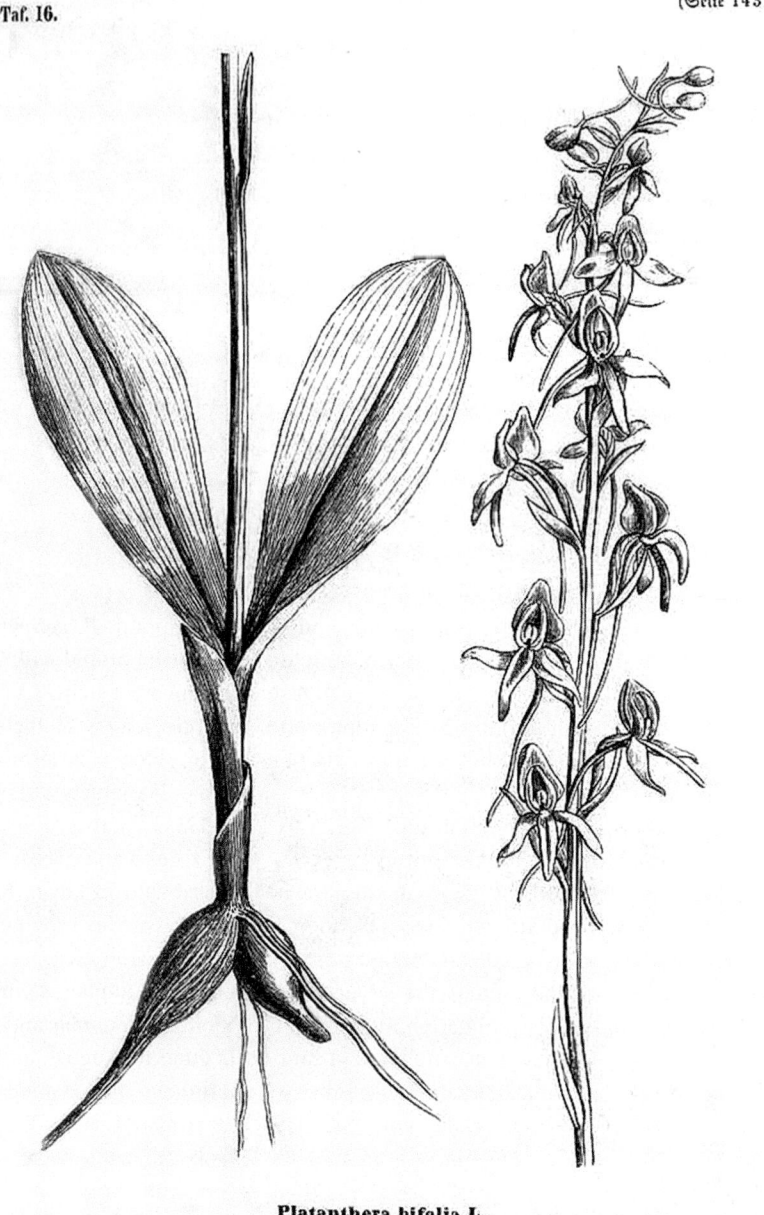

Taf. 16.

(Seite 143.)

Platanthera bifolia L.

Zweiblättrige Platanthere.

Platanthera clavigera Lindl.

syn. *Habenaria densa* Wall.

Chinese name: *Zangnanshechun Lan*

Chinese medicinal names: *Jishencao* (chicken kidney grass); *Jishenzi* (chicken kidney); *Yaoshenzi* (waist and kidney); *Shuangren* (double kernels); *Shenjingcao* (kidney herb)

Fig. 18.42 *Platanthera bifolia* (L.) Rich. [PHOTO: Henry Oakley]

Fig. 18.44 *Platanthera chlorantha* Cust. Ex. Rchb. [PHOTO: Courtesy of Plant Photo Bank of China]

Fig. 18.43 *Platanthera bifolia* in natural habitat [PHOTO: Henry Oakley]

Description: *P. clavigera* is a robust herb with tuberous, ovoid rootstock, one larger than the rest and a thick upright stem bearing 4–5 widely spaced, plicate, undulating, elliptic leaves, 3.5–10 by 1.5–3 cm. Inflorescence is terminal, 8–30 cm, many-flowered, with very small, yellowish-green, pubescent flowers, not densely arranged. It flowers in August and September in southern Tibet. *P. clavigera* also occurs in Kashmir, Nepal and Bhutan at 2300–3400 m (Chen et al. 2009e).

Usage: *Jishencao* is collected in August and September. The entire plant is used to strengthen the kidneys and loins, especially for the impotence or sexual dysfunction, hernia, and enuresis affecting children. The herb is reported to originate from Xizang and Hubei (*Zhongyao Da Cidian* 1986; Wu 1994), although the centre of

Table 18.6 Chinese herbal prescriptions that employ *Platanthera clavigera* (*Habenaria densa*) (*Jishencao*)

---

1. Indications: nourishing kidneys and masculinity, erectile dysfunction and hernia
Boil 6–12 g *Jishencao* (*Habenaria densa*) and drink the decoction
(*Tibet Commonly Used Herbs*)

---

2. Indication: Bed wetting by children
Boil *Ji Shen Cao* 6–12 g and consume (*Jingzhou Chinese Herbs*)

Ref: Jiangsu College of New Medicine (1986): *Zhongyao Da Cidian*

distribution of the orchid is the Himalayan region of Kashmir, northeast India, Nepal, Bhutan and southern Xizang (Chen et al. 2009c, d). It nourishes the lungs and kidneys (Table 18.6) (Chen and Tang 1982).

Plantanthera edgeworthii (Hook f. ex Collett) RK Gupta

syn. *Habenaria edgeworthii* Lindl., *Habenaria acuminata* (Twaites) Trimen

Indian *Ayurvedic* names: *Riddhi, Riddhi Vridhi, Laksmi, Mangala, Rathanga, Rissisrista, Saravajanpriya, Siddhi, Sukha, Vasa, Yuga, Kakoli.* (in Garhwal): also *Pranda, Talgranthisamakand, Vamavartal, Vrisya*

Description: This stout herb is 35–50 cm tall with 3–4 small lanceolate leaves, 6–8 by 2–4.5 cm. Tubers are small and slim, 1–2.4 long by 0.4–0.9 cm in diameter. Inflorescence is tall and many-flowered, but its yellowish-green flowers are small, 3–5 mm across, but in some places the flowers are 1–1.5 cm across. It flowers from June to September (Pearce and Cribb 2002). It is distributed from China to the northwest Himalayas, in scrub and open hillside at 1200–3600 m. *Habenaria acuminata* (= *H. edgeworthii*) also occurs at the Anamalai Hills of Tamil Nadu, and in Kerala at 1300–1700 m (Bose and Bhattacharjee 1980). However, Sathish Kumar and Manilal in Kerala report that this species is restricted to a narrower altitude range of 900–1500 m (Sathish Kumar and Manilal 1994).

Herbal Usage: *Riddhi* is an important medicine in the Ayurvedic tradition. Roots of *P. edgeworthii* (syn. *Habenaria acuminata, H. edgeworthii*) are eaten as food or a tonic (Duggal 1971; Pandey et al. 2003). It is one component of a tonic in Karnataka (Vij 1995; Rao 2004) and sometimes promoted as an aphrodisiac in India (Sood et al. 2005). Sweet in taste, it "pacifies *vata* and *pitta* but aggravates *kapha*", which translates as meaning that it is good for slim or driven people, but bad for those who are laid back or obese. It is cooling and promotes semen production. Leaves and roots are used for treating 'diseases of the blood'. In Garhwal, Indian Himalayas, the powdered orchid pseudobulb is added to *Ashtavarga* and *Swarnabhashma* (from gold calcinations) to prepare a potion to promote milk production (Dhayani et al. 2011). *Vriddhi* is used as brain tonic, general tonic, depurant, appetiser, "*rasayan*" and emollient. Oral dose is 2–3 g of powdered tuber. It features in *Astavarga churna, Chyavanprash rasayan* and *Mahamayura ghrita*. Substitutes are *Dactylorhiza hatagirea* (*salam panja*, an orchid also highly valued as tonic and aphrodisiac in Himalaya), *Tacca integrifolia* (local name: *Varachi kand*) and *Sida acuta* (local name: *Maha bala*) (Balakrishna et al. 2013). *Pueraria tuberosa* (Indian *kudzu*, a tuberous vine used as tonic, aphrodisiac, etc.) is also described as a substitute for *P. edgeworthii* (Chauhan 1999, quoted by Singh and Duggal 2009). Powder or paste made with leaves and rhizomes are used to treat blood disorders and for cooling in Nepal (Subedi et al. 2013).

Platanthera fuscescens (L.) Kraenz.]

syn. *Platanthera souliei* Kraenz., *Tulotis asiatica* Hara

Chinese names: *Qingting Lan* (Dragonfly orchid), *Zhuye Lan* (Bamboo leaf orchid), *Qingtingshechun lan*
Chinese medicinal name: *Qingting Lan*
Korean name: Broad leaf dragonfly orchid

Description: Plants are 20–60 cm tall. Stem is erect stout with 1–2 tubular sheaths and carries 1–3 cauline, obovate to elliptic leaves 6–15 by 3–7 cm. Rootstock is fleshy, slender, stoloniferous, with filiform roots. Inflorescence is erect and carries numerous small, greenish-white to yellowish-green, resupinate flowers. Dorsal sepal is pale green, oval, and it forms a hood over the column whereas the lateral sepals are narrow and clear white. Petals are linear and pointed, flushed a deep green towards the tip. The mid-lobe of the lip extends into a deep long spur. Flowering season is June to September (Lang and Tsi 1976; Chen et al. 2009e).

P. souliei occurs in Korea, Japan and Siberia in open grassland, but the orchid plants are well shaded by taller grass. In China, it is found along gullies at 400–4300 m in Qinghai, Sichuan, northwest Yunnan eastwards across Gansu, Nei Mongolia, Henan, Heibei and Shandong to Liaoning, Jilin and Heilongjiang.

Herbal Usage: Herb is obtained from northeastern and northwestern China and from Yunnan. The whole plant is used to treat burns (Wu 1994)

Platantheru japonica (Thunb.) Lindl.

Chinese names: *Changju Lan* (Long distance orchid); *Shuimaidong* (water winter wheat); *Shechun Lan* (tongue-lip orchid); *Qimashen* (horse riding ginseng); *Longzhuashen* (dragon claw ginseng)
Chinese medicinal name: *Guanyinzhu*

Description: Plant is 50–60 cm tall, with a few fleshy roots and an erect stem carrying 3–6 lanceolate leaves at the mid section, 8–16 cm by 3–5 cm. Inflorescence is terminal, 10–20 white flowers which possess a 2-cm-long, narrow, tapering, fleshy lip. It flowers from May to July in Japan and China (Anonymous 1976; Chen et al. 1999; Jin et al. 2009).

This terrestrial orchid is found in forests and on grassy slopes of hillsides at 700–2500 m in Japan, Korea, the Chinese provinces of Zhejiang,

Anhui, Hubei, Henan, Hunan, Shaanxi, Gansu, Sichuan, Yunnan and Guizhou and Nepal. At Huanglong, *Platanthera japonica* grows in open scrub at 2000 m. Plants are smaller, 30–50 cm tall, and carry on each inflorescence 10–25 striking white flowers, 2–2.5 cm across, which are distinguished by long, thin, curved, downward-pointing spurs of 3–6 cm length. Flowering period in this mountain Nature Reserve is June (Perner and Luo 2007).

Herbal Usage: Herb is obtained from Yunnan, Guangdong and the Yangzi River region. Plants are collected in summer, washed to remove the mud, then divided, dried and stored for future use. In western China, the whole plant is cooked by boiling over a slow fire, and eaten to improve the circulation and hasten recovery from injuries or sickness (Hu 1971). A decoction is advocated for this purpose by *E Mei Medicinal Plants*. Between 9 and 15 g of the herb is boiled to prepare the decoction. This decoction is also used in Shaanxi Province to nourish the lungs, relieve coughs and phlegm and breathlessness (*Zhongyao Da Cidian* 1986). The *Zhejiang Journal of Research into Medicinal Plants* states that it detoxifies, and heals toothache, discharge and snake bites. The herb may be used in decoction or as a poultice for application. A soup prepared by boiling the herb with meat is consumed for obtain relief from coughs or "heatiness in the lungs". For toothache, the root is steamed with 15–30 g of white sugar for consumption (*Zhongyao Bencao*, 2000).

Platanthera minor (Miq.) Rchb. f.

Chinese names: *Dayizhijian* (big arrow), *Luanchunfendie Lan* (soft lipped, pink butterfly orchid), *Xiaoshechun Lan* (small tongue lip orchid), *Guanyinzhun* (Guanyin bamboo), Chinese medicinal name: *Zhuliaoshen*

Description: Plant usually carries two tubers and two to three leaves. Inflorescence is 15–18 cm tall bearing many tiny white or green flowers 1 cm across. It flowers in early summer

(Perner and Luo 2007; Hu et al. 2007; Jin et al. 2009). *Platanthera minor* is a small terrestrial orchid found in China, Korea and Japan.

Herbal Usage: Its Chinese medicinal character is that the entire plant enlivens *yin* elements, benefits the lungs, improves air currents and promotes salivation. It is used to treat weak bodies, mental weakness, nocturnal emission, discharges and infants with hernia (Ou et al. 2003).

It helps to strengthen kidney and lungs, relieves asthmatic coughs, kidney weakness and lower backache, dizziness, people recuperating from long illness and nocturnal emission. For decoction, 15–60 g of the plant is used (*Sichuan Journal of Chinese Medicine, 1960*)

Platanthera sikkimensis (Hook f.) Kraenz.

Chinese name: *Changbanshechun Lan*

Description: Plants are terrestrial, occasionally epiphytic in Nepal, 17–21 cm tall. Rootstock is tuberous, cylindrical, fleshy, 0.5–1.5 by 0.3–0.5 cm. Stem is slender, erect, with 5–6 leaves and leaf-like bracts. Leaves are oblong-lanceolate, 4–5 by 1–1.5 cm, base clasping, apex pointed. Raceme is 4–7 cm, with 5–9 flowers, laxly arranged. Flowers are yellowish-green, appearing in July to August at Gaoligongshan and August to September in Nepal. *Platanthera sikkimensis* occurs in Nepal at 2600–2900 m, Sikkim and northwest Yunnan at 2000 m (Raskoti 2009; Chen et al. 2009d).

Herbal Usage: Tubers are used as a tonic in Nepal (Pant and Raskoti 2013).

Platanthera souliei Kraenz. [see **Platanthera fuscescens (L.) Kraenz.**]

Platanthera stenoglossa Hayata

Chinese names: *Xiabanfendie Lan* (slim petal pink butterfly orchid), *Xiachunfendie Lan* (slim lip pink butterfly orchid), *Xiasuishechun Lan* (slim spiked tongue, lip orchid)

Description: Pseudobulb of *P. stenoglossa* is fusiform, slim, producing a single ovate-elliptic leaf, 15 by 4 cm, which arises close to the ground. Inflorescence is 30 cm tall and many-flowered, the flowers yellow green to deep green, 8 mm across, facing in one direction. Flowering season is March to May. The species is found in Taiwan (only in Ilan) and the Ryukyu Islands of Japan (Liu and Su 1978; Lin 1987).

Usage: The entire plant is antipyretic and detoxifies, clears phlegm, stops coughs, improves blood flow and stops bleeding (Ou et al. 2003).

Platanthera ussuriensis (Regel) Maxim.

syn. *Tulotis ussuriensis* (Regel) Hara

Chinese names: *Xiaohuaqingting Lan* (small flower dragonfly orchid), *Fengchun Lan* (rich spring orchid), *Dongyashechun Lan*
Chinese medicinal names: *Banchunlian*
Japanese name: *Tombo-so*

Description: *P. ussuriensis* is a glabrous herb with fleshy, creeping root stock and filiform roots (Tang and Su 1978). Stem is erect, 20–55 cm tall with 2–3 lanceolate leaves, 7–12 by 2–3 cm. Raceme is 3–8 cm long with many small, pale green flowers. Dorsal sepal is hooded; the lateral sepals are wing-like, curling backwards; the small petals and side lobes of the lip are of equal size and curve forwards; while the mid-lobe of the lip is long and dips downwards, giving the entire flower the appearance of an insect in flight. It flowers in July and August (Tang and Su 1978).

The typical *P. ussuriensis* occurs in Japan at 1300 m and northeastern China but it is also found in other northern Chinese Provinces from Jilin to Xinjiang (Hebei, Shanxi, Jiangsu, Zhejiang, Jiangxi, Hunan, Hubei, Sichuan) at 400–2800 m, and in Korea and eastern Russia. The variety *transnokoensis* (Ohwi & Fukuyama) Liu & Su is found in an alpine meadow at 2500–3000 m in Taiwan.

Herbal Usage: Herb is obtained from Jilin, Hebei, Xinjiang, Shanxi, Jiangsu, Zhejiang, Jiangxi, Hunan, Hubei and Sichuan. The root is used for removing toxin, and in the treatment of swellings, abscesses and other inflammatory conditions, traumatic injury and thrush (Wu 1994; *Zhonghua Bencao*, 2000). *Liparis nervosa, L. japonica, Pleione bulbocodioides* and *Oberonia mysorus* are other orchids which are similarly used to treat swellings and inflammation and to remove extravasated blood (Chen and Tang 1982).

Overview

Usage of these orchids is new to Traditional Chinese Medicine. Three species, *P. chlorantha, P. japonica* and *P. ussuriensis* are used in treating injuries, albeit their other uses are different from one another. *P. clavigera* Lindl. is used in Chinese herbal medicine to treat impotence. Several species are used to treat swellings and injuries. It is remarkable that widely separated species of *Platanthera* are used, the species being spread over Tibet, Shaanxi and Taiwan.

Luning tested only one species of *Platanthera* for alkaloids and he did not find any (Luning 1974a, b). Loroglossin was discovered in *P. bifolia* (*Orchis bifolia*) by Delauney in 1921 (Ernst and Rodriguez 1984), but the search for orchinol in *P. bifolia* proved negative (Nuesch 1963). Glucomannans were present in two Japanese species, *P. florentii* and *P. ophrydioides* (Ohsuki 1937).

Leaves of *P. bifolia* contain flavonol and flavone C-glucosides, quercetin and kaempferol, respectively. Flowers of *P. bifolia* and *P. chlorantha* emit their scent at night. Floral scent of *P. bifolia* is attributed to methyl benzoate, methyl salicylate, linalool and E-ocimene and other minor components, whereas the scent of *P. chlorantha* consists principally of methyl benzoate, linalool and geraniol (Kaiser 1993; Veitch and Grayer 2001).

Nattviole (*P. bifolia*) which blooms in midsummer symbolises the mood of the season, its white flowers luminous in the light of the midnight sun. The breeze carries its floral fragrance, a mixture of linalool, methyl benzoate and methyl salicylate with small quantities of four isomers of lilac aldehyde, their corresponding alcohols and acetates, cinnamic aldehyde, geraniol, eugenol and vanilline (Kaiser 1993). Kaiser recommends that the best time to view this not-to-be-missed, eyecatching, fragrant orchid is at, or just after, dusk.

P. chlorantha is scentless during the day but it emits an intense white floral scent at night. This scent is based on linalool, methyl benzoate and geraniol (Kaiser 1993).

Genus: *Pleione* D. Don

Chinese name: *Dusuan Lan* (single bulb orchid)

Pleione is a genus of cool-growing orchids found in the foothills of the Himalayas, across southern China, Myanmar, Thailand and Indochina. They are medium-sized and may be epiphytic, lithophytic or terrestrial, some species being all three. They are dwarf plants with clustered pseudobulbs but proportionally long leaves which are deciduous. Flowers are large, resembling those of *Coelogyne* with an ornate lip which is frilled or dentate at the margin, and borne singly on leafless pseudobulbs (Chen et al. 2009a) (Fig. 18.45).

Pleione bulbocodioides (Franch.) Rolfe

Chinese names: *Taiwanyiye Lan* (Taiwan single leaf orchid), *Yiye Lan* (one leaf orchid), *Dusuan Lan* (single bulb orchid), *Bingqiuzi* (iceball)
Japanese name: *Sanjiko* (This name also applies to *Cremastra appendiculata*)
Vietnamese names: *Som tu co, Mao tu co*

Description: In Greek mythology, *Pleione* is the mother of the Pleiades, the small cluster of stars in the northern sky known in the East as the Seven Sisters. The genus was established by David Don who gave it the name to indicate the characteristic clustering of pseudobulbs. They are 1–2.5 by 1–1.5 cm in diameter. Leaf is solitary, oblong-lanceolate, borne at the apex of the

Fig. 18.45 *Pleione*
praecox (Sm.) D.Don.
From: *Orchis* vol. 8, t.2
(1907). Courtesy of
Missouri Botanical
Gardens, St. Louis, U.S.A

Tafel 2.

Orchis 1914.

Pleione praecox Don.

pseudobulb. Flowers appear in April and last until June. They appear at the same time as the leaves and are usually solitary, large (nearly 10 cm across), pink to dark purple, with stronger coloured markings on the lip which is papillose along its margin. There is much variation in the size, shape and colour of the flowers which resemble those of *Coelogyne*. *Pleione bulbodioides* is a fairly popular orchid among collectors in the USA and Europe: March is the peak flowering season for cultivated *P. bulbodioides* (Hamilton 1990).

P. bulbocodioides is distributed across central and southern China at 630–3600 m, growing in humus-rich or rocky soil or moss-laden rocks in sparse woods and shrubbery slopes (Anonymous 1976; Chen et al. 1999; Perner and Luo 2007). It is tolerant of cold, but requires its own microclimate (Chen 1995).

Herbal Usage: The whole plant is used for treatment of wet sores, sore throat and rabies. It clears phlegm is antipyretic and detoxifies. It was regarded as a counter-poison and was used for snake bites in China (Wilson, quoted by Perry and Metzger 1980). Pseudobulbs of *Pleione henryi* (now considered to be merely a form of *P. bulbocodiodes*; Chen, Cribb and Gale, 2009d), albeit it has smaller flowers, were used in treatment of tuberculosis and asthma (Wilson, quoted by Perry and Metzger 1980; Uphof 1968). The Japanese use it to treat boils and carbuncles (Guo 1996). It reduces inflammation, removes extravasated blood and causes swellings to subside (Chen and Tang 1982). Some compounds isolated from the pseudobulbs possess strong anti-inflammatory activity (Li et al. 2015).

China supplies Vietnam with the pseudobulbs which are used in that country to combat food poisoning and intoxication. It is also used on boils, and snake and insect bites (Duong 1993).

Phytochemistry: *P. bulbocodiodes* has been extensively studied by Bai and his team, who isolated dihydrophenathropyrans, stilbenoids, lignans, bichromans, bibenzyl glycosides and flavonoids from the orchid (Bai et al. 1996, 1997a, 1997b, 1998a, 1998b, 1998c, 1998d). The dihydrophenanthropyrans are shanciol, and bletilols A–C (Bai et al. 1996, 1998). Stilbenoids are shancidin, shancinlin, shanciguol and shanciol C and D (Bai et al. 1996, 1998). The lignans were named sanjidins A and B (Bai et al. 1997). Pleionin A was the bichroman isolated together with sanjidin A. Bibenzyl glycosides were 3′-hydroxy-5-methoxybibenzyl-3-*O*-beta-D-glucopyranoside, 3′,5′-dimethoxy-bibenzyl-3-*O*-beta-D-glucopyranoside, batatasin III and 3′-*O*-methylbatatasin III (Bai et al. 1997a). Flavonoids are shanciols A and B (Bai et al. 1998) and shanciol E and F (Bai et al. 1998a). Other known compounds such as lusianthidin, coelonin and 2,7-dihydroxy-1-(*p*-hydroxybenzl)-4-methoxy-

9,10-dihydrophenanthrene have also been isolated from the tubers of *P. bulbodidioides* (Bai et al. 1996a, b).

Two new stilbenoids, 9-(4′hydroxy-3′methoxyphenyl)-10-(hydroxymethyl)-11methoxy-5,6,9,10-tetrahydrophenanthro[2,3-*b*]furan-3-ol and 2-(4″-hydroxybenzyl)-3-(3′hydroxyphenethyl)-5-methoxy-cyclohexa-2,5-diene-1,4-dione; and three known stilbenoids were recently isolated from *P. bulbocodioides* by Liu et al. (2009a). The team also isolated two new phenanthrofurans, shanciol G and H, from the orchid (Liu et al. 2009). In their latest publication, they reported the isolation from rhizome of *P. bulbocodioides* of amentoflavone, kayaflavone, gymconopin D, methyl(4-OH) phenylacetate, *p*-hydroxybenzaldehyde, *p*-hydroxy benzoidc acid, 4-oxopentanoic acid, D-dihydroxybenzene and gastrodine (Yuan and Liu 2012). Amentoflavone, also present in the herb, St. John's Wort, and *Biophyllum sensitivum*, is a potent, caffeine-like ca^{2+} releaser (Suzuki et al. 1999). It inhibits COX-2 expression, is anti-inflammatory (Sakhivel and Guruvayoorappan 2013) and it has an effect on the brain similar to that of benzadiazepine (Hansen et al. 2005). Amentoflavone has significant antiviral activity against influenza A and B viruses and moderate antiHsV 1 and antiHSV-2 activities (Lin et al. 1999). It causes apoptosis in B16F-10 melanoma cells (Guruvayoorappan and Kuttan 2008a). Amentoflavone inhibits angiogenesis (Guruvayoorappan and Kuttan 2008b; Tarallo et al. 2011; Zhang et al. 2014) and induces apoptosis in melanoma and human breast cancer cells (Guruvayoorappan and Kuttan 2008a; Lee et al. 2009a), thus opening up a new range of angiogenesis compounds. Amentoflavone is present in several plants which do not belong to the orchid family, e.g. *Cycas rumphii, Trifolium alexandrinum* and *Selaginella tamariscina*, the last a traditional oriental medicine that has been used to treat cancer (Lee et al. 2009b). It inhibits angiogenesis of endothelial cells and stimulates apoptosis in hyperplastic scar fibroblasts. Perhaps it might have a role in the treatment of burns and scars (Zhang et al. 2014). Amentoflavone inhibits UVB-induced matrix metalloproteinase-1 expression of normal human skin fibroblasts. It

Fig. 18.46 *Pleione hookeriana* (Lindl.) Rollisson
[PHOTO: E. S. Teoh]

would be interesting to know whether amenotoflavone could suppress skin photo-aging (Lee et al. 2012).

Elution-extrusion counter-current chromatography was used to recover and separate five high polarity glucosides, namely, (E)-4-β-D-glucopyranosyloxycinnamic acid 9-(4-β-D-glucopyranosyloxybenzyl) ester, (Z)-2-(2-methylpropyl) butenedioic acid bis(4-β-D-glucopyranosyloxybenzyl) ester, gastrodin, dactylorhin A and militarine (Wang et al. 2013b). Meanwhile, four new pyrrolidone-substituted bibenzyls, dusuanlansins A–D, which contain nitrogen were isolated together with 19 other known compounds from the pseudobulbs. Several of the known compounds exhibited strong anti-inflammatory activity (Li et al. 2015).

Pleione henryi (Rolfe) Schlecht. [see *Pleione bulbocodiodes* (**Franch.**) **Rolfe**]

Pleione hookeriana (Lindl.) Rollisson

Chinese name: *Maochundusuan Lan* (hairy lip single garlic/bulb orchid)

Description: *P. hookeriana* is a cool-growing species found at 1800–3100 m from Nepal, Bhutan and Sikkim to southern China (Xizang, Yunnan, Guizhou, Guangxi and Guangdong) to Myanmar, Laos and Thailand. It is well adapted to its variable habitats, thriving equally as an epiphyte on trees, as a lithophyte on rocks, or as a terrestrial on mossy cliffs. Pseudobulbs are ovoid, 1–2 by 0.5–1 cm, green or purple with a single leaf, 6–10 (rarely 20) by 2–2.8 (rarely 4.6) cm at its apex. Flowers are borne singly, with white, rose, or purplish-pink petals and pointed sepals. Lip is white marked with blotches of orange (Fig. 18.46). Flowering season is April to June in China (Lang and Tsi 1976; Chen et al. 1999), and May in Sikkim and Manipur (Bose and Bhattacharjee 1980).

Herbal Usage: Herb is obtained from Hubei, Guangdong, Guangxi, Guizhou, Yunnan and Tibet. The pseudobulbs are used to remove heat and toxins. *Maochundusuan Lan* is used to treat 'heat', abscesses, snake bites and lymphatic tuberculosis (Wu 1994).

Pleione humilis (Sm.) D. Don,

Pleione diantha Schltr., *Epidendrrum humile* Smith; *Coelogyne humilis* (Smith) Lindl., *Cymbidium humile* (Smith) Lindl.

Nepalese name: *Hathi tauke* (meaning elephant head and referring to the shape of the flower), *Shaktigumba*

Description: This epiphytic or lithophytic orchid is distributed in central Nepal, Bhutan, Tibet, India and Myanmar at 2400–3000 m in shady, moist locations. Pseudobulbs are ovoid, conical 3.5 cm long with a single leaf up to 26 by 4.5 cm and a short inflorescence. Flowers are purplish and appear in September to November in Sikkim (Bose and Bhattacharjee 1980), and from February to March in Bhutan and Darjeeling (Pearce and Cribb 2002) (Fig. 18.47).

Fig. 18.47 *Pleione humilis* [PHOTO: Bhaktar B. Raskoti

Usage: A paste made from the pseudobulbs of *P. humilis* is applied to cuts and wounds in Nepal (Manandhar and Manandhar 2002). Powder is consumed as a tonic (Subedi et al. 2013).

Pleione maculata (Lindl.) Lindl. & Paxton

Chinese name: *Qiuhuadusuan Lan* (autumn flowering single bulb orchid)
Myanmar Name: *Phar la tet thitkhwa phyu*

Description: Pseudobulb is small, 2.5 cm, bottle-shaped, cylindrical in its lower two-thirds and tapering into a cone in the upper third. Leaf is lanceolate, 10–25 cm in length and 1.5–3.5 cm broad. Flowers appear on mature, deciduous bulbs in October to November. They are solitary or appear in pairs, 3–4.5 cm across, fragrant, and white, with a yellow splash and reddish or purple markings on the lip (Vaddhanaphuti 2001; Nanakorn and Watthana 2008).

P. maculata is an epiphyte on tree trunks in broad-leaved, evergreen forests in western Yunnan, Bhutan, Sikkim, Arunachal Pradesh and Assam, Myanmar and northern Thailand. Its flowering season is September to November in Thailand (Vaddhanaphuti 2001; Nanakorn and Watthana 2008), a month later in China (Chen et al. 1999), and October to November in northern and northeast India (Bose and Bhattacharjee 1980). It flowers in March and April in the Kachin state of Myanmar (Tanaka and Yee 2003).

Herbal Usage: The pseudobulbs are used to treat liver complaints and stomach ache in India and Nepal (Rao 2007; Pant and Raskoti 2013). The species is rare and endangered in India (Rao 2007).

Pleione pogoniodes Rolfe [see *Pleione bulbocodioides* (**Franch.**) **Rolfe**]

Pleione praecox (Sm.) D. Don

Epidendrum praecox Smith, *Coelogyne praecox* (Smith) Lindl., *Cymbidium praecox* (Smith) Lindl., *Coelogyne wallichiana* Lindl.

Nepali name: *Lasun pate, Shaktigumba*
Myanmar name: *Phar la tet thitkhwa*

Description: *P. praecox* is an epiphytic orchid with lightly clustered, bottle-shaped pseudobulbs, 1.3–4 cm long, green, heavily marked with maroon and covered with warty sheaths. It carries two narrow, elliptic, plicate leaves measuring 20 cm by 1.7–6.7 cm (Fig. 18.45). Inflorescence is terminal and bears one to three pink flowers, 10 cm across, in September to December. The upper two-thirds of the lip fold into a tube which completely hides the column, the lower third thus forming an opening which is surrounded by bristles. A yellow streak lined with fine bristles is present at the centre of the lip (Bose and Bhattacharjee 1980; Vaddhanaphuti 2001; Nanakorn and Watthana 2008; Jin et al. 2009) (Fig. 18.48).

P. praecox is found on tree trunks or on rocks at 1200–2500 m in Nepal, Bhutan, Northern India, Myanmar (Chin and Kachin states), Thailand and China (southern Yunnan and Xizang) (Fig. 18.49).

Fig. 18.48 *Pleione praecox* (Sm.) D.Don. [PHOTO: E. S. Teoh]

Fig. 18.49 *Pleione praecox* (Sm.) D.Don. habitat [PHOTO: E. S. Teoh]

Herbal Usage: A paste made from the pseudobulb is applied to cuts and wounds (Manandhar and Manandhar 2002). Powder is mixed with milk to form a tonic and energiser in Nepal (Subedi et al. 2013).

Pleione yunnanensis (Rolfe) Rolfe

Coelogyne yunnanensis Rolfe

Chinese name: *Yunnandusuan Lan* (Yunnan single garlic orchid, Yunnan single bulb orchid),

Duyebaiji (single leaf *Baiji*), *Dusuan Lan* (single garlic orchid)

Description: Pseudobulb is ovoid or conical, 1.5–3 cm long and it carries a long, slender, lanceolate leaf that measures 6.5–25 by 1–3.5 cm. Floral scape develops before the leaf and grows to 10–20 cm. It is single-flowered Flower is 6–7 across, with narrow, blunted petals and sepals of purplish-white or rose-pink and purple or red markings in the lip. Lip is oval, 3 cm broad, toothed at the margin and decorated by 3–5 high, crisp sharp, yellow lamellae in the centre. It flowers in April to May (Lang and Tsi 1976; Chen et al. 1999; Jin et al. 2009).

This pretty, pink *Pleione* is found on mossy rocks in light shade, or as a terrestrial on grassy slopes or amongst shrubbery at 1100–3500 m. It is endemic in southwestern China (southwest Sichuan, western Guizhou, southeast Xizang and Yunnan).

Phytochemistry: Four new bibenzyl derivatives, shancigusins A–D, and five known bibenzyls were isolated from the tubers of *Pleione yunnanensis* at the Institute of Medicinal Plant Development in Beijing. Shancigusin A and B were obtained as a yellow syrup, shancigusins C and D as a red syrup. Their structures were determined by spectroscopic analysis (Dong et al. 2009), but their pharmacological properties remain unreported. The team later reported successful isolation of three new dihydrophenanthrofurans, named pleionesins A–C (Dong et al. 2010), followed by five new glucosides, shancigusins E, F, G, H and I, together with 18 known compounds from tubers of the orchid (Dong et al. 2013b). A separate team managed to isolate 14 compounds from an ethanolic extract of *P, yunnanensis* (*shancigu*) that included 5 dihydrophenanthrenes, 4 bibenzyls, 2 triterpenoids and 3 phenylacrylic acids. These compunds were 4, 7-dihydroxy-2-methoxy-9,10-dihydrophenanthrene, 4, 7-dihydroxy-1-(*p*-hydroxybenzyl)-2-methoxy-9,10-dihydrophenanthrene, (2,3-trans)-2-(4-hydroxy-3-methoxyphenyl)-3-hydroxymethyl-10-methoxy-2,3,4,5-tetrahydro-phenanthro[2,1-b]furan-7-ol, pleionesin B, blestriarene A, batatasin III, 3,

3'-dihydroxy-2-(*p*-hydroxybenzyl)-5-methoxy-bibenzyl, 3',5-dihydroxy-2-(*p*-hydroxybenzyl)-3-methoxybibenzyl, 3,3'-dihydroxy-2,6-bis (4-hydroxybenzyl)-5-methoxybibenzyl, triphyllol, pholidotin, (E)-p-hydroxycinnamic acid, (E)-ferulic acid and (E)-ferulic acid hexacosyl ester (Wang et al. 2014).

Herbal Usage: Herb is the product of Yunnan, Guizhou and Sichuan. The stem is used to remove heat and toxin, stop coughs, clear phlegm, arrest bleeding and promote tissue healing (Wu 1994). It is said to nourish the lungs, and is used to treat coughs, pulmonary silicosis, tuberculosis and bronchitis. Additionally, it is prescribed for gastro-intestinal bleeding, carbuncle and external injuries (*Zhongyao Da Zidian* 1986).

The manual *Yunnan Chinese Herbs* recommends boiling 9–15 g for consumption, and a poultice prepared from ground tubers for external application (*Bencao*, 2005). Alternative usage and prescriptions are:

1. for coughs, tuberculosis or bronchitis: 3 g in powder form to be taken with honey.
2. for asthma: drink a decoction made with 9 g of *Pleione yunnanensis* tubers.
3. for suppurative bone marrow infection resulting from fracture: grind fresh roots and apply.
4. for carbuncle, pound the fresh roots and apply.

Overview

Four species of *Pleione* (*P. formosana, P. humilis, P.maculata and P. praecox*) tested for alkaloids did not contain any appreciable amount of the substance (Luning 1974a, b). However, they contain numerous phenenathrenes. *P. bulbocodioides* which has been most extensively studied contains more than 20 compound (Bai et al. 1996, 1997a, 1997b, 1998a, 1998b, 1998c, 1998d; Liu et al. 2009b) (Table 18.7). Measurement of dactylorhrhin A and militarine by HPLC is proposed as a simple method for quality control of the herb (Cui et al. 2013).

Shancigu (medicinal species of *Pleione*) are used in the treatment of tumours, burns and frostbite in Traditional Chinese Medicine, whereas paste made with *P. humilis* is used to treat cuts and wounds in Nepal, and powdered tuber is a tonic. Native medicine in India uses pseudobulbs of the rare and endangered *P. maculata* to treat liver disorders (Rao 2007; Pant and Raskoti 2013).

P. bulbocodioides is one ingredient in the Chinese herbal medicine, *Bushen Huoxue*, which was administered daily for 14 days per cycle to infertile women suffering from polycystic ovary which had not responded to other forms of treatment. *Bushen Huoxue* contains dodder seed 20 g, prepared *Rhemannia* root 10 g, mulberry mistletoe 20 g, *Epimedium* 15 g, *Psoralea* fruit 10 g, *Solomonseal* rhizome 10 g, Honeylocust thorn 15 g, peach kernel 10 g, *Pleione bulbocodioides* 10 g, red sage root 20 g and liquorice root 6 g.

In a randomised study, 20 women were given *Bushen Huoxue* in addition to receiving standard human menopausal gonadotrophin (HMG) injections for egg maturation, and 24 women randomly assigned served as controls. The study group received a total of 585 ($\pm$195) i.u. of HMG whereas the control group received 1470 $\pm$ 532.5 i.u. for reasons which were not clearly explained in the published abstract. Women receiving *Bushen Huoxue* had one mature follicle one day before ovulation whereas women in the control group (who received almost three times more HMG) had three mature follicles and one of them developed mild ovarian hyperstimulation syndrome (OHSS). Pregnancy rate was 44.4 % in the *Bushen Huoxue* group, and 31.8 % in the control group. The authors concluded that the use of this Chinese herbal medicine could reduce the usage of HMG and avoid ovarian hyperstimulation (Liang et al. 2008).

A systematic review of four randomised controlled trials involving 344 participants published in the *Cochrane Database Systematic Reviews* concluded that the methodology of the trials was not adequately reported (Zhang et al. 2010). This makes it difficult for readers to appreciate the significance of the findings. Nevertheless, there was no evidence that the herbal medicine improved ovulation rate when administered to women who also received clomiphene or underwent ovulatian drilling. Pregnancy rate was increased when herbal medicine was given together with clomiphene, but there was no report on live births rate (Zhang et al. 2010).

It is notoriously difficult to evaluate infertility studies because individual women are not comparable and there are many confounding differences resulting from the multiple factors that affect fertility, including differences in the quality of the semen (Fig. 18.50, Table 18.7).

Genus: *Plocoglottis* Bl.

Plocoglottis is a genus of terrestrial orchids with 40–45 species distributed in the Andaman Islands, Thailand, Malaysia, Indonesia and New Guinea. They are found in the shade of humid rainforest and are uncommon. Plants have a slender pseudobulb and a single, lanceolate, pleated leaf with prominent veins, except for *P. gigantea* Hook. f., the sole species possessing multiple leaves. A long rachis carries many flowers but they open one at a time (Fig. 18.51).

The generic name is derived from the Latin *ploke* (wicker work) and *glottis* (tongue). This describes the unusual lip which has a spring mechanism that ensures pollination. When the flower is newly open, the lip is slanted slightly downwards and when it is touched, it springs upwards towards the column.

Only one species has been reported to have any medicinal usage while another has a role in Malay village culture.

Plocoglottis javanica Blume

Pocoglottis fimbriata Teijsm & Binn.
Description: A terrestrial orchid with creeping rhizomes and terete pseudobulbs, about 4 cm

apart, 9–12 cm in length. Leaf is solitary, elliptic, 20–30 cm by 9–12 cm. Inflorescence arises from the base of the pseudobulb, reaching a height of up to 90 cm, and bearing several well-spaced flowers, 1.7 cm in diameter. Petals and sepals are yellow, speckled with red spots and blotches. Lip is smooth, initially white but turning yellow with age. The narrow petals point forwards and curve inwards to surround the lip (Fig. 18.52). *P. javanica* is found in wet locations between 100 and 1000 m from the Andaman Islands and peninsular Thailand through Malaysia to Singapore, Sumatra, Java and Kalimantan. In Java, Comber reported that it was usually found around 1000 m, but he had collected a specimen at 100 m in Sumatra (Comber 2001; Ang et al. 2011).

Herbal Usage: It was reported by Holmes (1892) in Meldrum's *List of Johore Medicines* that the juice of the fruit was dropped into the ear to treat earache (Burkill 1935). This was a common Malay remedy, but other orchids were also used in a similar manner (see *Acriopsis javanica*, *Bulbophyllum vaginatum*, *Dendrobium crumenatum* and *Hippeophyllum scortechnii*).

Plocoglottis lowii Rchb. f.

Plocoglottis porphyrophylla Ridl.

Local name: *Daun sepuleh dudok* (? also *Sepuleh dudor*)

Description: This purplish herb has large conical pseudobulbs, 7 by 2 cm. Leaf is solitary, lanceolate, plicate, glabrous, up to 28 cm long, but usually much shorter, with a 4-cm petiole. It is greenish-purple on its upper surface and deep purple on the under surface. Pseudobulb is olive and purple. Inflorescence is up to 70 cm long, erect, pubescent, with many flowers, with one flower opening at a time. Flower is 3 cm in diameter. Sepals and petals are narrow, pointed, of a pale yellow. There is a reddish patch on the medial aspect of the lateral sepals. Lip is square, fleshy and red. Column is yellow (Fig. 18.53). It is found in the lowland forests from the Andaman

Fig. 18.50 Stilbenoids from *Pleione bulbocodioides* (Bai et al. 1996)

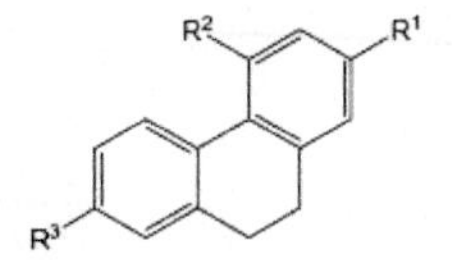

Coelonin: R1 = OH, R2 = OMe, R3 = OH

Lusianthridin: R1 = OMe, R2 = OH, R3 = OH

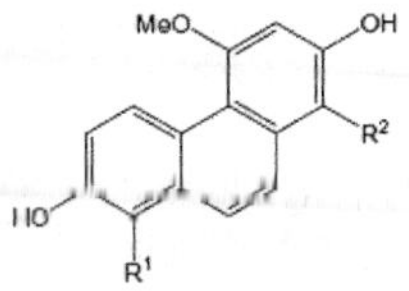

R1 = OMe, R2 = hydroxybenzyl

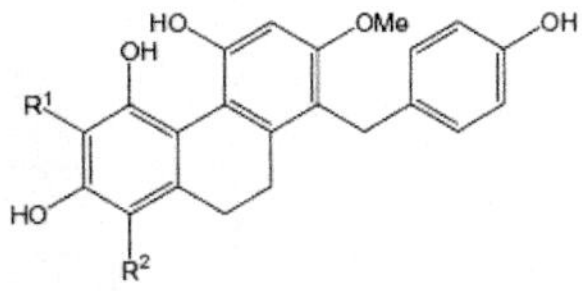

Shancidin: R1 = H, R2 = H

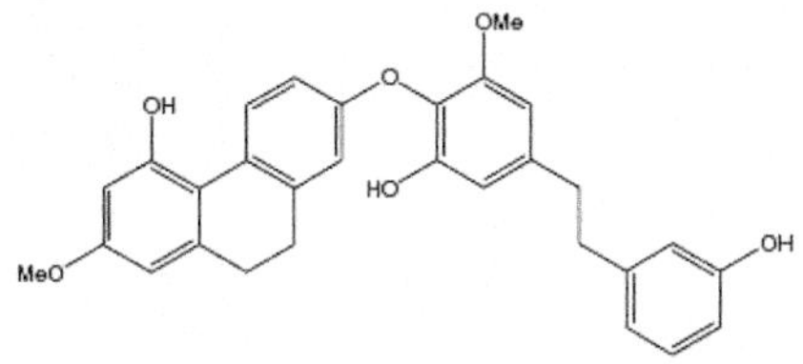

Shancilin

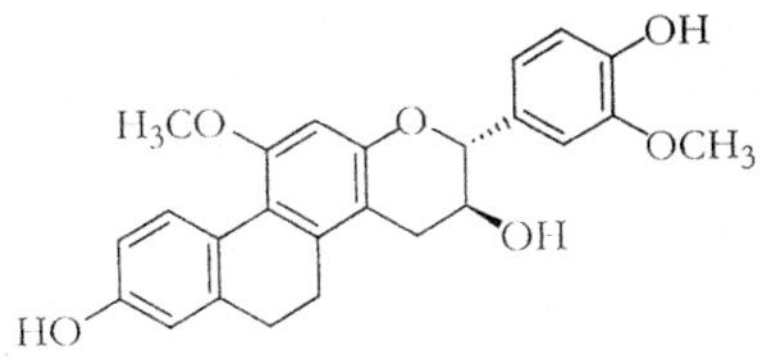

shancicol

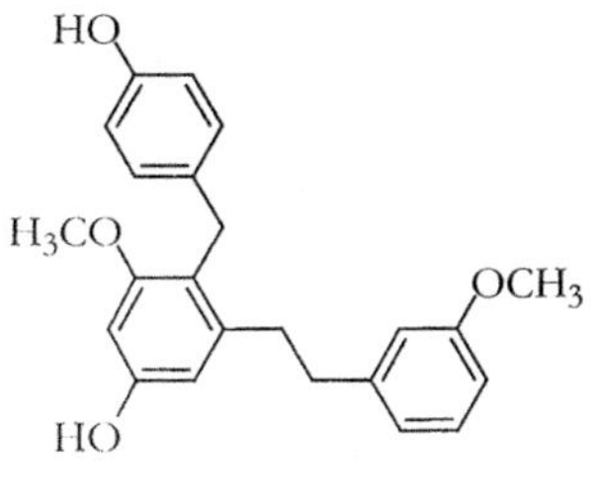

bulbocol

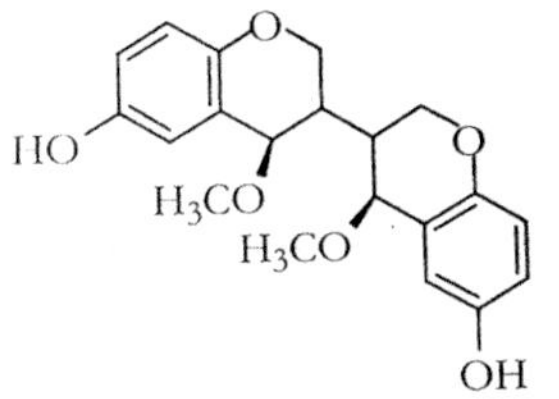

pleionin A

Islands of Myanmar to Thailand, Malaysia and Indonesia up to Amboin.

Table 18.7 Chemical constituents of *Pleione bulbocodioides*

9,10 dihydrophenanthrenes: coelonin; 7-dihydroxy-1-(p-hydroxybenzl)-4-methoxy-9,10-dihydrophenanthrene
Dihydrophenanthropyran: Shanciol, Bletilol A, Bletilol B, Bletilol C
Stilbenoids: Shancidin, Shancinlin, Shanciguol, shanciols C and D, 9-(4′hydroxy-3′methoxyphenyl)-10-(hydroxymethyl)-11methoxy-5,6,9,10-tetrahydrophenanthro[2,3-*b*]furan-3-ol; 2-(4″-hydroxybenzyl)-3-(3′hydroxyphenethyl)-5-methoxy-cyclohexa-2,5-diene-1,4-dione
Lignans: Sanjidins A and B
Bichromans: Pleionin A
Bibenzyl glycosides: 3′-hydroxy-5-methoxybibenzyl-3-*O*-beta-D-pyrannoside,
3′,5′-dimethoxybibvenzyl-3-*O*-beta-D-glucopyranoside, batattasin III, 3′O-methylbatatasin III
Other bibenzyl derivatives: shancigusins A–D
Flavonoids: Amentoflavone, Shanciols A, B, E, F
Glycosides: (E)-4-β-D-glucopyranosyloxycinnamic acid 9-(4-β-D-glucopyranosyloxy-benzyl) ester; (Z)-2-(2-methylpropyl)butenedioic acid bis(4-β-D-glucopyranosyloxybenzyl) ester; gastrodin; dactylorhin A; and militarine

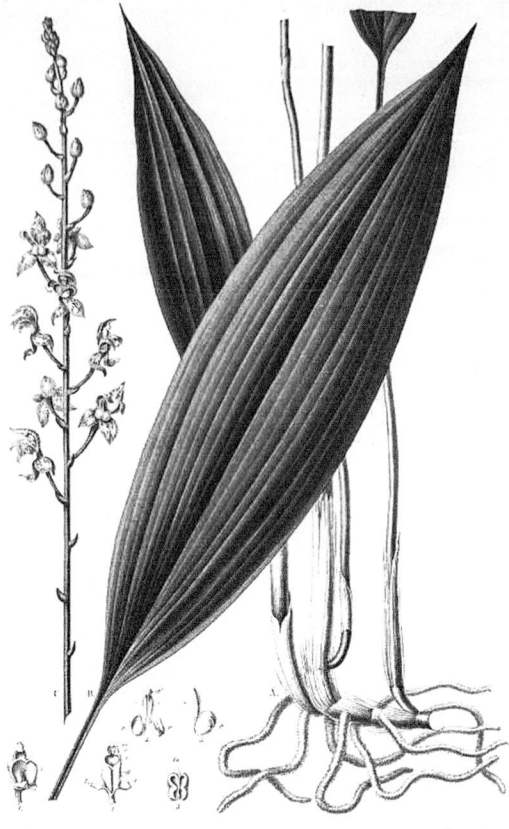

PLOCOGLOTTIS JAVANICA

Herbal Usage: Leaves of this orchid joined to other magical plants were made into a brush that was used to sprinkle consecrated rice gruel about the house to invite the return of benign spirits after a death (Burkill 1935). Earlier, Burkhill and Haniff (1930) mentioned that it was used to counter bewitchment.

Fig. 18.51 *Plocoglottis javanica* Blume. From Blume, C. L., *Flora Javae et insularum adjacenticum*, t. 15 (1858). Courtesy of Missouri Botanical Gardens, St. Louis, U.S.A.

Overview

Seven species were tested for alkaloid content and two were found to have an alkaloid content of 0.1 % or greater (Luning 1974a, b). There is no recent pharmacological research on *Plocoglottis*.

temperate regions of both the northern and southern hemispheres. There are three species in China. Stem is erect and carries a single lanceolate leaf. Inflorescence carries a single flower. The orchid is not grown for ornamental purposes.

Its generic name is derived from Greek, *pogonias* (bearded), which refers to the bearded crest of the lip typical of the genus.

Genus: *Pogonia* Juss.

Chinese name: *Zhu Lan* (red orchid)
Japanese name: *Toki so*

Pogonia is a small genus of small to medium-sized terrestrial orchids which are found in the

Pogonia japonica Rchb. f.

Chinese name and medicinal name: *Zhu Lan* (red orchid)
Japanese name: *Tokiso*

Fig. 18.52 *Plocoglottis javanica* Blume. flowering in a Singapore swamp [PHOTO: Ang Wee Foong]

Fig. 18.53 *Plocoglottis lowii* Rchb. f. [PHOTO: E. S. Teoh]

Description: Stems are slender, erect, 10–25 cm tall, bearing a single lanceolate leaf, 3.5–6 by 0.8–1.4 cm at its mid-section and a cluster of leaf-like bracts terminally. A solitary, white, or pink to purplish-red flower is borne at the end of the stem. Flower does not open fully. Petals and sepals are elliptical, the petals barely separate from each other while the sepals tend to curl

backwards. The inner surface of the petals and the filigree margin of the lip are marked with purplish stripes. Lip has a yellow patch on its central crest. It flowers in May to July (Chen et al. 1999; Chen et al. 2009c, d).

P. japonica is an alpine, terrestrial species, found in damp places on grassy slopes, sparse woods and ravines at 400–2000 m in northeastern China, the provinces immediately south of Yangzi river and in Korea and Japan.

Herbal Usage: The whole plant is used as an antidote for snake bite in China. Herb is from northeast China and the southern Yangzi region (Wu 1994).

Overview

Many plants, including orchids, have been used to treat snake bites in the various continents, but experiments by Caius (1936) using dogs and over three dozen Indian herbs offered as remedies for snake bite found that none of them had any therapeutic effect. The two anthocyanins, cyanidin 3-glucoxyloside and cyanidin monoglucoside, present in the perianth of *P. japonica* (Ueno et al. 1969) are also unlikely to have any medicinal significance. They are responsible for the purplish colour of the flowers.

Genus: *Polystachya* Hook.

Chinese name: *Duosui lan*

This is a large genus of sympodial epiphytic orchids with 150 members all but one limited to tropical Africa. The single Asian species, *Polystachya concreta*, enjoys a wide distribution from Sri Lanka to India, Thailand, Laos and Vietnam; curiously, it is also present in Florida, Guyana, Surinam, Brazil, the West Indies and Africa (Fig. 18.54).

The generic name is derived from Greek *poly* (many) and *stachys* (spike), referring to the numerous spikes that constitute the inflorescence in some species which rather resembles a spike of wheat (Schultes and Pease 1963).

Fig. 18.54 *Polystachya concreta* (Jacq.) Garay & H. R.Sweet. [as Polystachya luteola Hook. f.] From: Wright, R., Icones Plantarum Indiae Orientalis, vol. 5 (1): t. 1678 (1846). Drawing by Govindoo. Courtesy of Missouri Botanical Gardens, U.S.A.

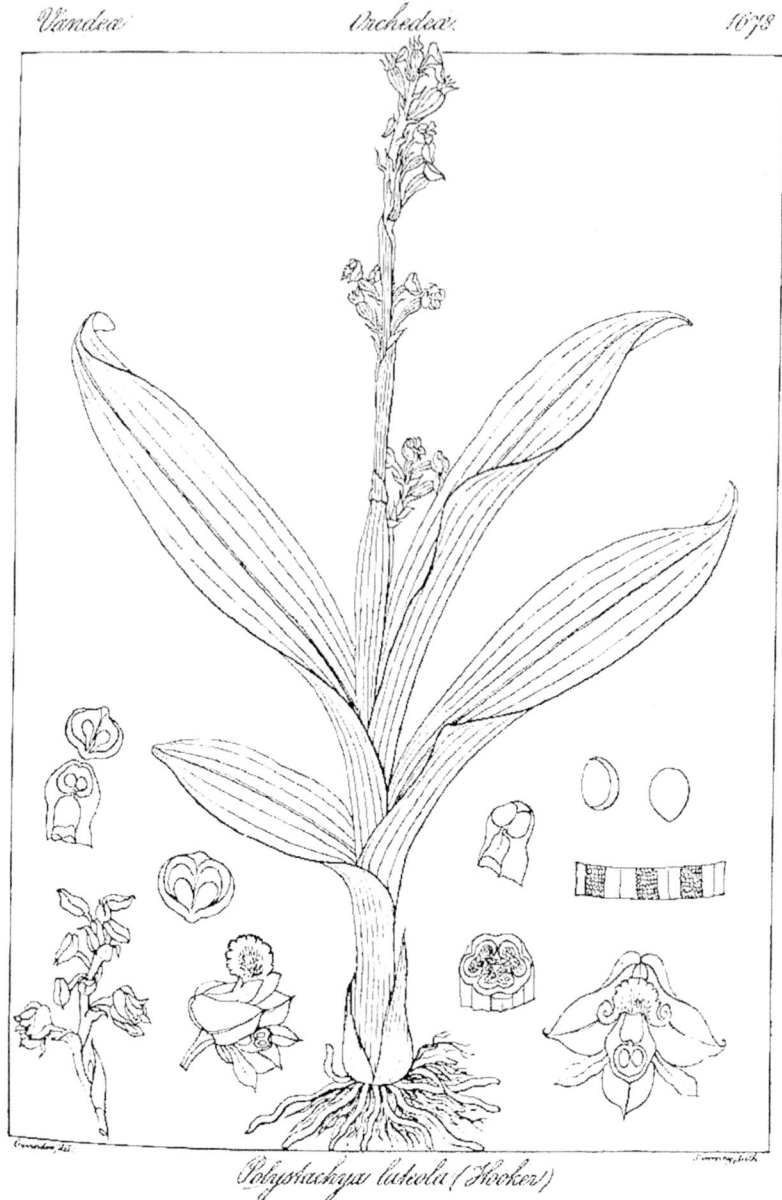

Polystachya concreta (Jacq.) Garay & H.R.Sweet

Polystachya zeylanica Lindl., *P. luteola* Lindl., *P. flavescens* (Bl.) Sm., *P. wightii* Rchb. f., *P. purpurea* Wt., *Dendrobium polystachyum* Thouars, *Epidendrum concretum* Jacq.

Chinese name: *Duosui lan*
Indian name: (Orissa): *Kakina*

Description: Pseudobulbs are short, with a few internodes and 3–4 leaves which are oblong-lanceolate, 8–20 by 1–2.5 cm, glabrous, duplicate, green, sometimes with a purplish tinge, many

Fig. 18.55 *Polystachya concreta* (Jacq.) Garay & H. R. Sweet [PHOTO: Courtesy of Plant Photo Bank of China]

veined with a distinct median vein, and deciduous. Inflorescence is erect, terminal and many-flowered. Flowers are small, 0. 5 cm across, yellow or greenish-yellow, non-resupinate, and they are arranged in several loose clusters. In this species, the inflorescence resembles a stalk of wheat (Fig. 18.55). Flowering season is February to March in the Western Ghats (Abraham and Vatsala 1981), August in Mumbai (Santapau and Kapadia 1966), March to April and July to October in Sri Lanka (Jayaweera 1981), April to September in the Philippines (Davis and Steiner 1982), and August to September in Thailand (Vaddhanaphuti 2001).

The orchid enjoys a pan-tropical distribution (Sathish Kumar and Manilal 1994) involving southern Yunnan in China, Indochina, Thailand, Malaysia, India, Sri Lanka, Africa and tropical and subtropical America (Chen and Wood 2009), and it has been assigned over 50 botanical names (Seidenfaden 1999). It is epiphytic on trees in dense forests at 600–1500 m and very common in the hills of Tamil Nadu above 1200 m (Matthew 1995). In Sri Lanka, it is found in montane forests and also on imported or domesticated trees like *Samanea saman* (rain tree), *Hevea brazilensis* (rubber), *Mangifera indica* (mango) and *Artocarpus heterophyllus* (jackfruit)

(Jayaweera 1981). This species was first discovered in Martinique in the West Indies by N J Joaquin and described in 1760 (Bechtel et al. 1980).

Herbal Usage: The tuber is used to treat arthritis by tribals of the Niyamgiri Hills in Orissa, India. Approximately 100 mg. of fresh tuber is boiled in 500 ml. water until the volume is reduced to 100 ml, and 3–4 ml of this decoction is taken with 7–8 drops of honey twice daily on an empty stomach as treatment for arthritis (Dash et al. 2008).

Overview

Fifteen species of *Polystachya* were tested for the presence of alkaloids and all the tests read negative (Luning 1974a, b). Although the orchid is present in China, it has no medicinal usage there.

Many species of *Polystachya* are scented. *P. fallax* smells of jasmine. Methyl anthranilate and indole conveys the scent of *sampaquita* (jasmine). *P. cultriformis* emits the scent of lime blossoms and lily of the valley whereas *P. campyloglossa* carries the fruity scent of bananas and strawberry. Lime blossom fragrance is produced by nerolidol, dihydrofarnesol and farnesol enhanced with a slew of minor constituents which includes *gamma*-decalactone, 2-amino benzyladehyde, *beta*-ionone and methyl jasmonate. The fruity fragrance of *P. camyloglossa* is conveyed by a combination of isoprenyl acetate, prenyl acetate with corresponding alcohols, small amounts of dihydro-beta-ionone, beta-ionone, linalool and over a dozen compounds present in trace amounts (Kaiser 1993).

Genus: *Pomatocalpa* Breda, Kuhl & Hasselt.

Pomatocalpa is a genus of small to medium-sized, monopodial, epiphytic orchids with long or short stems, flat leaves and branching inflorescences with numerous small flowers, closely arranged (Fig. 18.56). Its 40 member

Fig. 18.56 *Pomatocalpa spicatum* Breda, Kuhl & Hasselt. [as Cleisostoma uteriferum Hook. f.] From: *Annals of the Royal Botanic Gardens, Calcutta, Vol. 5 (1)*: t. 84 (1891) Drawing by G. C. Das and J. D. Hooker]. Courtesy of Missouri Botanical Gardens, St. Louis, U.S.A.

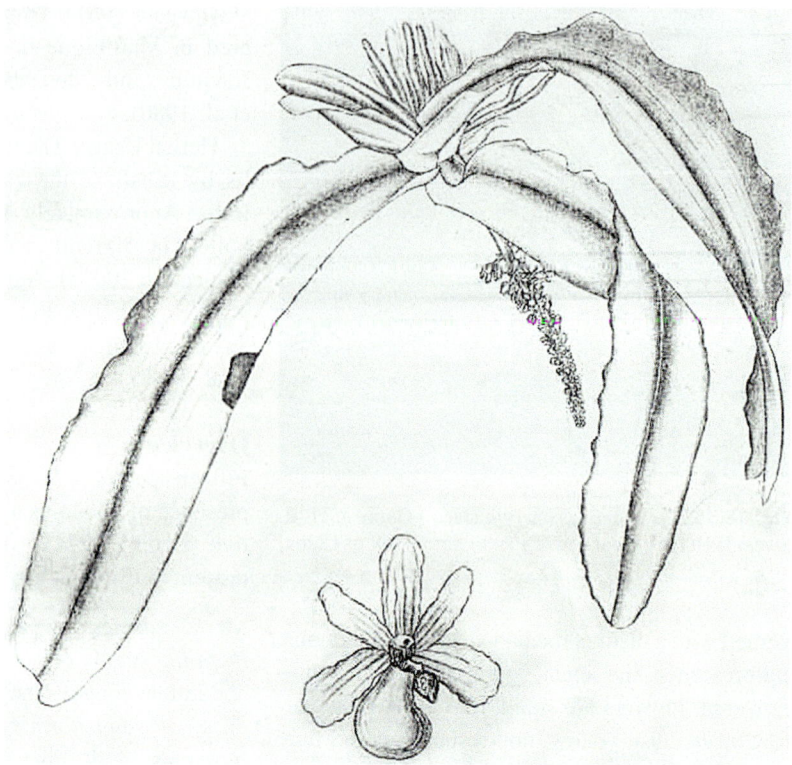

species are distributed from northeast India and China to Sri Lanka and Southeast Asia to Papua New Guinea and the Pacific Islands.

Pomatocalpa spicatum Breda, Kuhl & Hasselt.

syn. *Pomatocalpa wendlandorum* (Rchb.f.) J.J. Sm

Description: Plant is 13–25 cm tall with stout stems and coraceous, oblong leaves, unequally two-lobed 10–24 by 1.8–4 cm. Inflorescence is axillary, racemose or paniculate, densely many-flowered, several inflorescences appearing simultaneously. Flowers are yellow, marked with red at the base of the sepals and petals, 3–5 mm across (Fig. 18.57). Peak flowering occurs in May and June in Malenesia (O'Byrne 2001) but it is February to April in Assam (Bose and Bhattacharjee 1980), March to April in Thailand (Vaddhanaphuti 2001) and May to

Fig. 18.57 *Pomatocalpa spicatum* Breda, Kuhl & Hasselt. [PHOTO: E. S. Teoh]

June in Nepal and Sikkim (Pearce and Cribb 2002).

Widely distributed from central Bhutan and Sikkim to Myanmar, Thailand, Indochina, across Malaysia and Indonesia, the species occurs from the lowlands to 700 m (O'Byrne 2001).

Herbal Usage: The Nicobarese in the islands at the Bay of Bengal employ the orchid to treat fits (Dagar and Dagar 2003).

Overview

There is neither phytochemical nor pharmacological information on *Pomatocalpa*.

Genus: *Ponerorchis* Rchb. f.

Chinese name: *Huang hua xiao hong men lan*

Ponerorchis is a small to medium-sized, terrestrial, alpine orchid with globose cylindrical tubers and basal leaves. Stems are erect, terete, with tubular sheaths at the base and 1- to 5-leaved. Leaves are alternate, glabrous or pubescent, basally contracted into a sheath enclosing the stem. Inflorescence is terminal, erect, with few pink to purple (rarely yellow) flowers on the raceme. Dorsal sepal and petals overlap and form a hood over the column while the lateral sepals are spread out. Lip is large. The genus has 20 species distributed from Indian Himalaya to China, Siberia, Korea and Japan. Ten species are endemic in China.

All species formerly classified under the genus *Chusua* Nevski. are now moved to *Ponerorchis* Rchb. f (Chen et al. 2009b).

Ponerorchis chusua, (D. Don) Soo

syn. *Ponerorchis chusua* var. *nana* (King & Pantl.) R.C.Srivast.

Chinese name: *Guang bu xiao hong men lan*

Description: *P. chusua* is a small to a fairly large plant, 8–45 cm tall. Tubers are oblong or

Fig. 18.58 *Ponerorchis chusua* [PHOTO: E. S. Teoh]

globose, 1–2 by 1 cm. Stem is 5–23 cm tall, with tubular sheaths at the base and 2- to 5-leaved. Leaves are alternate, widely spaced, elliptic, green, 3–15 by 0.2–3 cm, becoming smaller above and merging into floral bracts. Inflorescence is erect, 2–20 cm, glabrous and few- or many-flowered (2–20). Flowers are pink, purplish-red to purple, 1.5–2 cm across (Fig. 18.58). Flowering season is July to August in China (Chen, et al. 2009a), and June to September in Bhutan (Pearce and Cribb 2002).

This lovely, alpine, terrestrial species is abundant in open scrub in the Huanglong area at 2800–3600 m in Sichuan Province (Perner and Luo 2007), and also reported as occurring in forests, *Rhododendron* scrub, grasslands, limestone outcrops and scree at 500–4500 m (depending on longitude) in Yunnan, Sichuan, Qinghai, Gansu, Ningxia, Shaanxi, Henan, Hebei, Nei Mongolia, Jilin, Heilongjiang provinces in China to Siberia, Korea and Japan in the north, and Myanmar, Indian Himalayas, Nepal and Bhutan (at 2890–4880 m) in the south (Pearce and Cribb 2002; Chen et al. 2009c,d).

Usage: The tubers are used in India to treat diarrhoea, dysentery and chronic fever (Rao 2007). There is no reported medical usage in China where the plant has its main distribution.

Overview

There is no report of any phytochemical and pharmacology study on *Ponerorchis*.

References

Abraham A, Vatsala P (1981) Introduction to Orchids, with illustrations and descriptions of 150 South Indian Orchids. TPGRI, Trivandrum

Ang WF, Lok AFSL, Yeo CK et al (2011) The status and distribution in Singapore of *Plocoglottis javanica* Blume (Orchidaceae). Nat Singapore 4:73–77

Anke S, Gonde D, Kaltenegger E et al (2008) Pyrrolizidine alkaloid biosynthesis in Phalaenopsis orchids: Developmental expression of alkaloid-specific homospermidine synthase in root tips and young flower buds. Plant Physiol 148:751–780

Anonymous (1974) Barefoot doctors manual. NIH, Bethesda

Anonymous (1976) The native orchids of Japan

Averyanov L, Cribb P, Phan KL, Nguyen TH (2003) Slipper orchids of Vietnam. With an introduction to the flora of Vietnam. Royal Botanic Gardens, Kew

Arthur HR, Cheung HT (1960) A phytochemical survey of the Hong Kong medicinal plants. 567–570

Bai L, Yamaki M, Tagaki S (1996) Stilbenoids from *Pleione bulbocodioides*. Phytochemistry 42(3):853–856

Bai L, Yamaki M, Yamagata Y, Tagaki S (1996) Novel dihydrophenanthropyran, shanciol isolated together with known compound bletilol B. Phytochemistry

Bai L, Masukawa N, Yamaki M, Tagaki S (1997a) Two bibezyl glycosides from Pleione bolbocodioides. Phytochemistry 44:1565–1567

Bai L, Yamaki M, Tagaki S (1997b) Lignans and bichroman from *Pleione bulbocodioides*. Phytochemistry 44:341–343

Bai L, Masukawa N, Yamaki M, Tagaki S (1998a) Four Stilbenoids from *Pleione bulbocodioides*. Phytochemistry 48(2):327–331

Bai L, Masukawa N, Yamaki M, Tagaki S (1998b) A polyphenol and two bibenzyls from *Pleione bulbocodioides*. Phytochemistry 47(8):1637–1640

Bai L, Yamaki M, Tagaki S (1998c) Flavan-3-ols and dihydrophenanthrenes from *Pleione bulbocodioides*. Phytochemistry 47(6):1125–1129

Bai L, Yamaki M, Tagaki S (1998d) Flavan-3-ols and dihydrophenanthropyrans from *Pleione bulbocodioides*. Phytochemistry 47:1125–1129

Baral SR, Kurmi PP (2006) A compendium of medicinal plants in Nepal. IUCN, Kathmandu, Published by Mrs. Rachana Sharma

Balakrishna A, Srivatava A, Mishra RK et al (2013) Astavarga plants—threatened medicinal herbs of the North-West Himalaya. Int J Med Arom Plats 2(4):661–67

Bechtel H, Cribb P, Launert E (1980) The manual of cultivated orchid species. Blandford, Poole

Beekman EM (2002) (transl., ed.) Rumphius Orchids. Orchid texts from the Ambonese Herbal by Georgius Everhardus Rumphius. New Haven, Yale University Press

Bencao Z (2000) Edited by Health Department and National Chinese Medical Management Office. Science and Technology Press, Shanghai

Bencao Z (2005) Edited by Health Department and National Chinese Medical Management Office. Science and Technology Press, Shanghai

Bhandari SR, Kapadi AH, Mujumder PL, Joardar M, Shoolery JN (1985) Nudol, a phenanthrene of the orchids Eulophia nuda, Eria carinata and Eria stricta. Phytochemistry 24(4):801–804

Bhattacharjee SK (1998) Handbook of medicinal plants. Pointer Publishers, Jaipur

Bi ZM, Wang ZT, Xu LS, Xu GJ (2004) Studies on the chemical constituents of *Pholidota yunnanensis*. Zhongguo Zhong Yao Za Zhi 29(1):47–9

Boorsma WG (1902) Pharmakologische Mitteilungen I. Bulletin de l'Institut Botanique de Buitenzorg 14:1–39

Bose TK, Bhattacharjee SK (1980) Orchids of India. Naya Prokash, Calcutta

Brandange S, Luning B (1969) Studies on orchid alkaloids. XII. Pyrrolizidine alkaloids from *Phalaenopsis amabilis* Bl. and Ph. Manii Rchb.f. Acta Chem Scand 23:1151–1154

Brandange S, Granelli I, Luning B (1970) Studies on Orchidaceae Alkaloids XVIII. Isolation of Phalaenopsin La from *Kingiella taenialis* (Lindl.) Rolfe. Acta Chem Scand 24(1):354

Brandange S, Luning B, Moberg C, Sjostrand E (1971) Studies on orchidaceae alkaloids. XXIV A pyrrolizidine alkaloid from Phalaenopsis corcu cervi Rchb f. Acta Chem Scand 25(1):349–50

Brandange S, Luning B, Moberg C, Sjostrand E (1972) Studies on orchidaceae alkaloids. XXX. Investigation of fourteen Phalaenopsis species. A new pyrrolizidine alkaloid from *Phalaenopsis equestris* Rchb f. Acta Chem Scand 26(6):2558–2560

Burkhill IH, Haniff M (1930) Malay village medicine. Gardens Bull Straits Settlements 6:165–321

Burkill IH (1935) (1966 reprint, 2nd ed., with contributions by Birtwistle W, Foxworthy FW, Scrivenor JB, Watson IG). A dictionary of economic products of the Malay Peninsula, Vol. II. London,

Crown Agents for the Colonies. Kuala Lumpur: Ministry of Agriculture & Co-operatives

Caius JF (1936) The medicinal and poisonous plants of India. J Bombay Nat Hist Soc 38(4):791–799

Chan CL, Lamb A, Shim PS, Wood JJ (1994) Orchids of Borneo. Introduction and a selection of species, vol 1. The Sabah Society and Kew, Royal Botanic Gardens, Kota Kinabalu

Chen SC (1995) Orchids and their conservation. Proc. 5th Asia Pacific Orchid Conference & Show, Fukuoka, 15–18

Chen SC, Tang T (1982) A general review of the orchid flora of China. In: Arditti J (ed) Orchid biology. Reviews and perspectives II. Cornell University Press, Itcaca and London

Chen XQ, Wood JJ (2009) Polystachya Hook. In: Chen XQ, Zj L, Zhu GH et al (eds) Flora of China—Orchidaceae. Science Press, Beijing

Chen SC, Tsi ZH, Luo YB (1999) Native orchids of China in colour. Science Press, Beijing

Chen XQ, Cribb PJ, Gale SW(2009a) Pleione D. Don. In: Chen XQ, Zj L, Zhu GH et al (eds) Flora of China—Orchidaceae. Science Press, Beijing

Chen XQ, Cribb PJ, Gale SW (2009b) Ponerorchis. In: Chen XQ, Zj L, Zhu GH et al (eds) Flora of China—Orchidaceae. Science Press, Beijing

Chen XQ, Gale SW, Cribb PJ (2009c) Platanthera Richard. In: Chen XQ, Zj L, Zhu GH et al (eds) Flora of China—Orchidaceae. Science Press, Beijing

Chen XQ, Gale SW, Cribb PJ (2009d) Peristylis Bl. In: Chen XQ, Zj L, Zhu GH et al (eds) Flora of China—Orchidaceae. Science Press, Beijing

Chen XQ, Gale SW, Cribb PJ (2009e) Peristylis BLume. In: Chen XQ, Zj L, Zhu GH et al (eds) Flora of China—Orchidaceae. Science Press, Beijing, p 141

Chen XQ, Luo YB, Wood JJ (2009f) Pinalia Lindley. In: Chen XQ, Zj L, Zhu GH et al (eds) Flora of China—Orchidaceae. Science Press, Beijing

Chen Y, Cai S, Deng L et al (2015) Separation and purification of 9,10- dihydrophenenthrenes and bibenzyls from Pholidota chinensis by high speed countercurrent chromatography. J Sep Sci 38(3):453–459

Christensen E (2001) Phalaenopsis. A monograph. Timber, Oregon

Chuakul W (2002) Ethnomedical uses of Thai orchidaceous plants. Mohidol Univ J Pharm Sci 29(3-4):41–5

Chutima R, Dell B, Vessabutr S et al (2010) Endophytic fungi from Pecteilis susannae (L.) Rafin (Orchidaceae), a threatened terrestrial orchid in Thailand. Mycorrhiza 21(3):221–229

Chutima R, Dell B, Lumyong S (2011) Effects of mycorrhizal fungi on symbiotic seed germination of Pecteilis susanne (L.) Rafin (Orchidaceae), a terrestrial orchid in Thailand. Symbiosis 53(3):149–156

Comber JB (2001) Orchids of Sumatra. Natural History Publications (Borneo), Kota Kinabalu

Cootes J (2001) Orchids of the Philippines. Marshall Cavendish, Singapore

Cribb PJ (1987) The Genus Paphiopedilum. Collingridge & Royal Botanic Gardens, Kew, Twickenham

Cui BS, Song J, Li S et al (2013) Determination of dactylorhin A and militarine in three varieties of Cremastrae Pseudobulbus/Pleiones Pseudobulbus by HPLC. Zhongguo Zhong Yay Za Zhi 38(24):4347–4350

Da Cidian Z (1986) Edited by Jiangsu New Medical College. Science and Technology Press, Shanghai

Dagar HS, Dagar JC (2003) Plants used in ethnomedicine by the Nicobarese of Islands in Bay of Bengal, India. In: Singh V, Jain AP (eds) Ethnobotany and medicinal Plants of India and Nepal. Scientific Publishers (India), . Jodhpur, pp 773–784

Dash PK, Sahoo S, Bal S (2008) Ethnobotanical studies on orchids of Niyamgiri Hill Ranges, Orissa, India. Ethnobotanical Leaflets 12:70–78

Davis RS, Steiner ML (1982) Philippine Orchids. A detailed treatment of some 100 native species. M & L Lucidine Enterprises, Manila

Dhayani A, Nautiyal BP, Nautiyal MC (2011) Importance of Astavarga plants in traditional systems of medicine in Garhwal, Indian Himalaya. Int J Biodiversity Sci Ecosyst Services Manag 6(1-2):13–19

Dong HL, Wang CL, Guo SX (2009) New Bibenzyl derivatives from the tubers of Pleione yunnanensis. Chemical and Pharmaceutical Bull (Tokyo) 57(5):513

Dong HL, Wang CL, Li Y et al (2010) Complete assignments of (1)H and (13)C NMR data of three new dihydrophenenthrofurans from Pleione yunnanensis. Magn Reson Chem 48(3):256–260

Dong FW, Fan WW, Xu FQ et al (2013a) Inhibitory activities on nitric oxide production of stilbenoids from Pholidota yunnanensis. J Asian Nat Prod Res 15(12):1256–1264

Dong HL, Liang HQ, Wang CL et al (2013b) Shancigusins E _ I, five new glucosides from the tubers of Pleione yunnanensis. Magn Reson Chem 51(6):371–377

Ding XY, Wang ZT, Xu H et al (2002) Database establishment of the whole rDNA ITS region of Dendrobium species of "fengdou" and authentication of their sequences. Yao Xua Xue bao 37(7):567–573

Duggal SC (1971) Orchids in human affairs (A review). Pharm Biol 11(2):1727–1734

Duong NV (1993) Medicinal plants of Vietnam, Cambodia and Laos. World Health Organization, Manila

Ernst R, Rodriguez E (1984) Carbohydrates of the Orchidaceae. In: Arditti J (ed) Orchid biology. Reviews and perspectives III. Cornell University Press, Ithca and London

Estrada S, Rojas A, Mathison Y et al (1999) Nitric oxide/cGNP mediates the spasmolytic action of 3,4'dihydroxy-5,5'-dimethoxybibenzyl from Scaphyglottis livida. Planta Med 65(2):109–114

Fox RB (1950) Notes on the orchids and people of northeast Polillo Island, Quezon Province, Philippines. Orchid Rev 3:16–21

Frolich C, Hartmann T, Ober D (2006) Tissue distribution and biosynthesis of 1,2- saturated pyrroliizidine alkaloids in Phalaenopsis hybrids (Orchidaceae). Phytochemistry 67(14):1493–502

Gan G, Zheng H (1998) Morphological and histological identification of *Pholidota yunnanensis*. Zhong Yao Cai 21(5):223–5

Ghorbani A, Gravendeel B, Naghibi F, de Booer H (2014) Wild orchid tuber collection in Iran: a wake-up call for conservation. Biodivers Conserv 23:2749. doi:10.1007/s10531-014-0746-y

Gonzalez VT, Junttila O, Lindgard B et al (2014) Batatasin III and the allopathic capacity of Empetrum nigrum. Nordic J Bot 000:001–007. doi:10.1111/njb.00559

Guo JX (1996) Bletilla striata (Thunb.) Reichb.f. In: Kimura (ed) et al. International collation of traditional and folk medicine. 1: 205

Guo XY, Wang J, Wang NL, Kitanaka S, Liu HW, Yao XS (2006) New stilbenoids from *Pholidota yunnanensis* and their inhibitory effect on nitric oxide production. Chem Pharm Bull (Tokyo) 54(1):21–5

Guo XY, Wang J, Wang NL, Kitanaka S, Yao XS (2007) 9,10-dihydrophenanthrene derivatives from *Pholidota yunnanensis* and scavenging activity on DPPH free radical. J Asian Nat Prod Res 9(2):165–74

Guruvayoorappan C, Kuttan G (2008a) Amentoflavone stimulates apoptosis in B16F-10 melanoma cells by regulating bci-2, p53 as well as caspase-3 genes and regulates the nitric oxide as well as proinflammatory cytokines production in B16F-10 melanoma cells, tumor associated macrophages and peritoneal macrophages. J Exp Ther Oncol 7(3):207–218

Guruvayoorappan C, Kuttan G (2008b) Inhibition of tumor specific angiogenesis by amentoflavoe. Biochemistry (Mosc) 73(2):209–218

Guruvayoorappan C, Kuttan G (2008c) Amentoflavone, a bioflavonoid from *Biophytum sensitivum* augments lymphocyte proliferation, natural killer cell and antibody dependent cellular toxicity through enhanced production of IL-2 and IFN-gamma and restrains serum sialic acid and gamma glutamyl transpeptidase production I tumor-bearing animals. J Exp Ther Oncol 6(4):285–295

Hamilton RM (1990) Flowering months of orchid species under cultivation. In: Arditti J (ed) Orchid biology reviews and perspectives, V. Timber Press, Portland

Handoyo F (2010) Orchids of Indonesia, vol 1. Indonesian Orchid Society, Jakarta

Hansen RS, Paulsen I, Davies M (2005) Determinants of amentoflavone interaction at the GABA(A) receptor. Eur J Pharmacol 519(3):189–207

Hernandez-Romero Y, Acevedo L, Sanchez ML et al (1953) Snippet. Orchid Journal 1(3):105

Hernandez-Romero Y, Acevedo L, Sanchez ML et al (2005) Phytotoxic activity of bibenzyl derivatives from the orchid Epidendrum rigidum. J Agric Food Chem 53(16):6276–6280

Holdsworth DK (1974) A phytochemical survey of medicinal plants in Papua New Guinea Part I. Sci New Guinea 2(2):142–154

Holdsworth DK (1975) Correspondence with J. Arditti

Holttum RE (1964) Orchids of Malaya, 3rd edn. Government Printers, Singapore

Holmes EM (1892) Malay material medica. Bull Pharm Detroit 6:108–117

Hu SY (1971) Orchids in the life and culture of the Chinese people, Chuing Chi J 10:1–26

Hu XM, Zhang WK, Zhu QZ et al (2000) Zhonghua Bencao, vol 8. Shanghai Science and Technology Publication, Shanghai

Hu QM, Wu TL, Xing FW et al (2007) Flora de Macau, vol 3. South China Botanic Gardens, Chinese Academy of Sciences, Macau

Jalal JS, Kumar P, Pangtey YPS (2008) Ethnomedicinal Orchids of Uttarakhand, Western Himalayas. Ethnobot Leaflets 12:1227–1230

Jayaweera DMA (1981) A revised handbook of the flora of Ceylon, vol II. A.A. Balkema, Rotterdam

Jin XH, Zhao XD, Shi XC (2009) Native orchids from Gaoligongshan Muntains, China. Science Press, Beijing

Kaiser R (1993) The scent of orchids: olfactory and chemical investigations. Editiones Roche, Basel

Lang KY, Tsi ZH (eds) (1976) Iconographia Cormophytorum Sinicorum 5:602–772. Orchidaceae. Beijing: Academia Sinca (Chinese Academy of Sciences, Istitute of Botany)

Lawler LJ (1984a) Ethnobotany of the orchidaceae. In: Arditti J (ed) Orchid biology reviews & perspectives 3. Cornell University Press, Ithaca

Ohsuki T (1937) Untersuchungen uber das Bletillamannan, ein Mannan aus den Knollen von Bletilla striata. Acta Phytochim 10:29–41

Ohsuki T (quoted by Lawler (1984)

Lawler LJ, Slaytor M (1969) The distribution of alkaloids in New South Wales and Queensland Orchidaceae. Phytochemistry 8:1959–1962

Lee JS, Lee MS, Oh WK, Sui JY (2009a) Fatty acid synthase inhibition by amentoflavone induces apoptosis and antiproliferation in human breast cancer cells. Biol Pharm Bull 32(8):1427–1432

Lee JS, Lee MS, Oh WK, Sui JY (2009b) Fatty acid synthase inhibition by amentoflavone induces apoptosis and antiproliferation in human breast cancer cells. Biol Pharm Bull 32(8):1427–1432

Lee CW, Na Y, Park NH et al (2012) Amentoflavone inhibits UVB-induced matrix- metalloproteinase-1 expression through modulation of AP-1 components in normal human fibroblasts. Appl Biochem Biotechnol 166(5):1137–1147

Li NH (1988) Chinese medicinal herbs of Hong Kong, vol 2. Hong Kong, Hong Kong Chinese Medical Research Institute, p 196

Li L, Lau KM (1994) Chinese medicinal herbs of Hong Kong, vol 6. Hong Kong, Hong Kong Chinese Medical Research Institute, p 198

Li JC, Feng L, Nohara T (2008) Chemical constituents from herb of *Pholidota cantonensis*. Zhongguo Zhong Yao Za Zhi 33(14):1691–3

Li B, Gao JY, Li J et al (2014) Chemical constituents of *Pholidota cantonensis*. Zhong Yao Cai 37(6):986–989

Li Y, Zhang F, Wu ZH et al (2015) Nitrogen-containing bibenzyls from Pleione bulbocodioides: absolute configurations and biological activities. Fitoterapia 102:120

Liang RN, Liu J, Lu J (2008) Treatment of refractory polycystic ovary syndrome by bushen huoxue method combined with ultrasound guided follicle aspiration. Zhongguo Zhong Xi Yi Jie He Za Zhi 28(4):314–317

Lin TP (1987) Native orchids of Taiwan, vol 3. Taipei, Southern Materials Center Inc, p 76, 97–99

Lin J, Zheng H, Huang J (1998) Pharmacognostical identification of *Pholidota cantonensis*—a confused variety of herba dendrobii. Zhong Yao Cai 21(7):340–342

Lin YM, Flavin M, Schure R et al (1999) Antiviral activities of bioflavonoids. Planta Med 66(2):120–125

Liu TS, Su HJ (1978) Flora of Taiwan, Vol. 5. 1978

Liu XQ, Yuan QY, Guo YQ (2009a) Two new stilbenoids from *Pleione bulbocodioides*. J Asian Nat Prod Res 11 (2):116–121

Liu XQ, Yuan QY, Gao WY et al (2009b) Two new phenanthrofurans from *Pleione bulboocodiodes*. J Asian Nat Prod Res 10:453–457

Liu ZJ, Chen XQ, Cribb PJ (2009c) *Paphiopedium* Pfitzer. In: Chen XQ, Liu ZJ, Zhu GH et al (eds) Flora of China, Orchidaceae, vol 25. Science Press, Beijing

Luning B (1974a) Studies on Orchidaceae alkaloids. IV. Screening of species for alkaloids 2. Phytochemistry 6:857–861

Luning B (1974b) Alkaloids of the Orchidaceae. In: Withner CL (ed) The Orchids. Scientific studies. John Wiley & Songs, New York

Luning B, Tranker H, Brandange S (1966) Studies on orchidaceae alkaloids V. A new alkaloid from *Phalaenopsis amabilis* Bl. Acta Chem Scand 20(7):2011

Majumder PL, Kar A (1987) Confusarin and confusaridin, two phenanthrene derivatives of the orchid, Eria confusa. Phytochemistry 26(4):1127–1129

Majumder PL, Kar A (1989) Erianol, a 4alpha-methylsterol from the orchid *Eria convallarioides*. Phytochemistry 28(5):1487–1490

Majumder PL, Sarkar AK (1982) Imbricatin, a new modified 9,10- dihydrophenanthrene derivative of the orchid *Pholidota imbricata*. Indian J Chem 21B:829–831

Majumder PL, Sarkar AK, Chakraborti J (1982a) Isoflavidinin and Iso-oxoflavidinin, two 9,10-dihydrophenanthres from the orchids *Pholidota articulata, Otochilus porecta* and *Otochilus fusca*. Phytochemistry 21(11):2713–2716

Majumder PL, Pal A, Lahiri S (1987) Structure of pholidotin, a new triterpene from orchids *Pholidota rubra* and *Cirrhopetalum elatum*. Indian J Chem 26B:297–300

Majumder Pl, Datta N, Sarkar AK, Chakraborti J (1982b) Flavidin, a novel 9,10dihydrophenanthrene derivative of the orchids Coelogyne flavida,

Pholidota articulate and Otochilus fusca. J Nat Prod 45(6):730–732

Manako Y, Wake H, Tanaka T et al (2001) Phenanthropyran derivatives from Phalaenopsis equestris. Phytochemistry 58(4):603–5

Manandhar NP, Manandhar S (2002) Plants and people of Nepal. Timber, Portland

Marasini R, Joshi S (2012) Antibacterial and antifungal activities of medicinal orchids growing in Nepal. J Nepal Chem Soc 29:104–109

Matthew KM (1995) An excursion Flora of Central Tamilnadu, India. A.A. Balkaema, Rotterdam

Majumder PL, Rahaman D (2006) Triterpenoids from the orchid Eria acervata. Chemiform 37 (45), doi 10.1002/chin.200645184

Medhi RP, Chakrabarti S (2009) Traditional knowledge of NE people on conservation of wild orchids. Ind J Tradit Know 8(1):11–16

Morales-Sanchez V, Rivero-Cruz I, Laguna Hernandez G et al (2014) Chemical composition, potential toxicity, and quality control procedures of the crude drug of *Cyrtopodium macrobulbon*. J Ethnoparmacol 154 (3):790–797

Müller-Höcker J (1992) Mitochondria and ageing. Brain Pathol 2(2):149–58

Musharof Hossain M (2009) Traditional therapeutic uses of some indigenous orchids of Bangladesh. Med Arom Plant Sci Biotechnol 3:100–106

Nanakorn W, Watthana S (2008) Queen Sirikit Botanic Garden (Thai Native Orchids 1 and 2). Wanida Press, Chiang Mai

Nontachayappom S, Sasirat S, Manoch L (2010) Isolation and identification of Rhizoctonia-like fungi from roots of three orchid genera, Paphiopedilum, Dendrobium and Cymbidium, collected in Chian Rai and Chiang Mai provinces of Thailand. Mycorrhiza 20(7):459–471

Nuesch J (1963) Defense reactions in orchid bulbs. Symp Soc Gen Microbiol 13:335–343

Nurhayati N, Gonde D, Ober D (2009) Evolution of pyrrolizidind alkaloids in Phalaenopsis orchids and other monocotyledons: identification of deoxyhypusine synthase, homospermidine synthase and related pseudogenes. Phytochemistry 70(4):508–516

O'Byrne P (1994) Lowland Orchids of Papua New Guinea. National Parks Board, Singapore Botanic Gardens, Singapore

O'Byrne P (2001) A to Z of South East Asian Orchid Species. Singapore, Orchid Society of South East Asia

Ou JC, Hsieh WC, Lin IH, Chang YS, Chen IS (eds) (2003) The catalogue of medicinal plant resources in Taiwan. Department of Health, Executive Yuan, Taipei

Pandey NK, Joshi GC, Mudaiya RK et al (2003) Management and conservation of medicinal orchids of Kumaon and Garhwal Himalaya. In: Singh V, Jain AP (eds) Ethnobotany and medicinal Plants of India and Nepal. Scientific Publishers (India), Jodhpur, pp 114–118

Pant B (2011) Medicinal Orchids of Nepal and their conservation by In-vitro Technique (Powerpoint Presentation). 20th World Orchid Conference. Singapore, Nov.14, 2011

Pant B, Raskoti BB (2013) Medicinal orchids of Nepal. Himalayan Map House (P) Ltd., Kathmandu

Perner H, Luo Y (2007) Orchids of Huanglong. Huanglong National Park, Sichuan Province, China

Perry LM, Metzger J (1980) Medicinal plants of East and Southeast Asia: attributed properties and uses. MIT Press, Cambridge, MA

Pearce NR, Cribb PJ (2002) The Orchids of Bhutan Edinburgh: Royal Botanic Gardens and Thimpu: Royal Government of Bhutan

Plepys D, Ubarra F, Lofstedt C (2002) Volatiles from flowers of Platanthera bifolia (Orchidaceae) attractive to the silver moth, Autographa gamma (LepidopteraL Noctuidae). Oikos 99:69–74

Qu XY, Qin SY, Yang DQ, Li QS, Peng FS (2006) Study on resource and varieties of Guoshangye. Zhongguo Zhong Yao Za Zhi 31(2):110–4

Rao AN (2004) Medicinal orchid wealth of Arunachal Pradesh. Newsletter of ENVIS NODE on Indian Medicinal Plants 1(2):1–7

Rao TA (2007) Ethno Botanical data on wild orchids of medicinal value as practised by tribals at Kudremukh National Park in Karnataka. Orchid Newslett 2(2):1–7

Raskoti BB (2009) The Orchids of Nepal. Bhakta Bahadur Raskoti and Rita Ale, Kathmandu

Rifai MA (1975) Extraordinary uses of orchids in Indonesia. Report First ASEAN Orchid Congress, Bangkok

Rueda DC, Schoffmann A, De Mieri M et al (2014) Identification of dihydrostilbenes in *Pholidota chinensis* as a new scaffold for GABA-A receptor modulators. Bioorg Med Chem 22:1276–1284

Sakhivel KM, Guruvayoorappan C (2013) Amenoflavone inhibits INOS, Cox-2 expression, and modulates cytokine profile, NK-kB signal transmission pathways in rats with ulcerative colitis. Int Immunopharmacol 17 (3):906–916

Santapau H, Kapadia Z (1966) The orchids of Bombay. Government of India Press, Calcutta

Sathish Kumar C, Manilal KS (1994) A catalogue of Indian Orchids. Bishen Singh Mahendra Pal Singh, Hehra Dun

Schultes RE, Pease AS (1963) Generic names of Orchids. Their origin and meaning. Academic, New York & London

Seidenfaden G (1999) 149. Orchidaceae. In: Matthew KM (ed) The Flora of the Palni Hills, South India, Pt 3. The Rapinat Herbarium. St. Joseph's College, Tiruchirapalli

Seidenfaden G, Wood JJ (1992) The Orchids of Peninsular Malaysia and Singapore. Olsen & Olsen, Fredensborg

Sezik E (1967) Turkiye'nin Salepgilleri Ticari Salep Cesitleri ve Ozellikle Mugla Salebi Uzerinde Arastirmalar. Doctoral Thesis. Istanbul Universitesi Eczacihk Fakultesinde

Simmler C, Antheaume C, André P et al (2011) Glucosyloxybenzyl eucomate derivatives from Vanda teres stimulate HaCaT cytochrome c oxidase. J Nat Prod 74(5):949–955

Singh A, Duggal S (2009) Medicinal orchids: an overview. Ethnobot Leaflet 13:351–63

Slaytor MB (1977) The distribution and chemistry of alkaloids in the Orchidaceae. In: Arditti JA (ed) Orchids biology reviews and perspectives, vol 1. Cornell University Press, Ithaca and London

Sood SK, Rana S, Lakhanpal TN (2005) Ethnic Aphrodisiac Plants. Scientific Publishers (India), Jodhpur

Sritularak BC, Anuwat M, Likhitwitayawuid K (2011) A new phenanthraquinone from *Dendrobium draconis*. J Asian Nat Prod Res 13(3):251–255

Subedi A, Kunwar B, Choi Y et al (2013) Collection and trade of wild-harvested orchids in Nepal. J Ethnobiol Ethnomed 9:64–73

Sukphan P, Sritularak BC, Mekboonsonglarp W et al (2014) Chemical constituents of *Dendrobium venustum* and their antimalarial and antiherpetic properties. Nat Prod Commun 9(6):625–627

Suzuki A, Matsunaga K, Mimaki Y et al (1999) Properties of amenoflavone, a potent caffeine-like Ca^{2+} releaser in skeletal muscle sarcoplasmic reticulum. Eur J Pharmacol 372(1):97–102

Sweet H (1969) A revision of the genus Phalaenopsis

Tanaka H, Yee ATT (2003) Wild Orchids in Myanmar, vol 1–3. The Foundation of Agricultural Development and Education, Bangkok

Tang SL, Su HJ (1978) Flora of Taiwan, vol 5. Department of Botany, Taiwan National University, Taipei

Tarallo V, Lepore L, Marcellini M et al (2011) The bioflavanoid amentoflavone inhibits neovascularization preventing the activity of proangiogenic vascular endothelial growth factors. J Biol Chem 286 (22):19641–19651

Teoh ES (1982) A joy forever. Vanda Miss Joaquim. Singapore's National Flower. Singapore: Times (reprinted 2008. Singapore: Marshall Cavendish)

Teoh ES (2002) Orchids of Asia, 2nd edn. Times Editions, Singapore

Trivedi VP, Dixit RS, Lal VK (1980) Orchids in the drug markets of Bareilly, Kanpur and nearby districts. Nagarjun (Calcutta) 23(8):157–163

Tsai WC, Hsiao YY, Pau ZJ et al (2008) Molecular biology of orchid flowers with emphasis on Phalaenopsis. In: Kader JC, Delseny M (eds) Advances in botanical research incorporating advances in plant pathology, vol 47. Elsevier, San Diego, pp 99–145

Tsavkelova EA, Cherdyntseva TA, Botina SG, Netrusov AI (2007) Bacteria associated with orchid roots and microbial production of auxins. Microbiol Res 162 (1):69–76

Uddin SB, Yusuf M (2011) Medicinal plants of Bangladesh. Ethnobotany Lab, Chittagong University, Chittagong

Ueno N, Takemura E, Hayashi K (1969) Additional data for the paper chromatographic survey of anthocyanins in the Flora of Japan (IV). Studies on anthocyanins, LXL. Bot Mag (Tokyo) 82:155–161

Uphof JCT (1968) Dictionary of economic plants. Verlag von J. Cramer, Lehre

Vaddhanaphuti N (2001) A field guide to the wild orchids of Thailand, 3rd edn. Silkworm Books, Chiang Mai

Vaidya BN, Shrestha M, Joshee N (2002) Report of Nepalese orchid species with medicinal properties. In: T Watanabe, A Tokano, MS Bista, HK Saiju (eds) *The Himalayan plant/Can they have us?* Proceedings of Nepal-Japan Joint Symposium on Conservation and Utilization of Himalayan Medicinal Resources. Japan Society for Conservation and Development of Himalayan Medicinal Resources

Van den Brink RCB (1937) Synopsis of the vernacular names and the economic uses of indigenous orchids of Java. Blumea 29(6):38–51

Van Rheede HA (1693) Hortus Indicus Malabaricus, vol 12. Dutch East India Company, Kerala

Veitch NC, Grayer B (2001) Phytochemistry of Habenaria and Orchis. In: Pridgeon AM, Cribb PJ, Chase MW, Rasmussen FN (eds) Genera Orchidacearum, Vol. 2. Orchidoideae (Part One) Orchidoideae. Oxford, Oxford University Press

Vij SP (1995) Orchid genetic diversity in India: conservation and commercialization. Proc. 5th Asia Pacific Orchid Conference & Show, Fukuoka, 20–39

Wallstedt A, Sommarin M, Nilsson MC et al (2001) The inhibition of ammonium uptake in excised birch (*Betula pendula*) roots by batatsin III. Physiol Plant 113(3):368–376

Wang SY, Kuo YH, Chang HN, Pl K, Tsay HS, Lin KF, Yang NS, Shyur LF (2002) Profiling and characterization antioxidant activities in Anoectochilus formosanus Hayata. J Agric Food Chem 50(7):1859–65

Wang J, Matzuzaki K, Kitanaka S (2006) Stilbene derivatives from *Pholidota chinensis* and their anti-inflammatory activity. Chem Pharm Bull (Tokyo) 54 (8):1216–9

Wang J, Wang LY, Kitanaka S (2007) Stilbene and dihydrophenanthrene derivatives from *Pholidota chinensis* and their nitric oxide inhibitory and radical scavenging activities. J Nat Med 61(4):381–6

Wang J, Wang T, Xie P et al (2013a) New phenanthrene derivatives with nitric oxide inhibitory and radical-scavenging activities from *Pholidota imbricata* Hook. Nat Prod Res (2013 Oct 15) [Epub ahead of print]

Wang Y, Guan SH, Feng RH et al (2013b) Elution-extrusion counter-current chromatography separation of two new benzyl ester glucosides and three other high-polarity compounds from the tubers of *Pleione bulbocodioides*. Phytochem Anal 24(6):671–676

Wang G, Guo X, Chen H et al (2013c) A resveratrol analog, phoyunbene B. induces G2/M cell cycle arrest and apoptosis in HepG2 liver cancer cells. Bioorg Med Chem Lett 22(5):2114–2118

Wang XJ, Cui BS, Wang C, Li S (2014) Chemical constituents of *Pleione yunnanensis*. Zhongguo Zhong Yao Za Zhi 39(5):851–856

Wen L, Weiming C, Zhi X, Xaotain L (1986) New triterpenoids of *Pholidota chinensis*. Planta Med 52 (1):4–6

Wu XR (1994) A concise edition of medicinal plants in China. Guangdong Higher Education Publication House, Guangdong (in Chinese)

Wu TL, Hu QM, Xia NH, Lai PCC, Yip KL (2002) Check list of Hong Kong Plants 2001. Hong Kong, Agriculture Fisheries and Conservation Department Bulletin 1 (Revised)

Wu B, Qu H, Cheng Y (2008) Cytotoxicity of new stibenoids from *Pholidota chinensis* and their spin-labelled derivatives. Chem Biodivers 5(9): 1803–10

Xu J, Zhao WM, Qian ZM et al (2010) Fast determination of five components of coumarin, alkaloids and bibenzyls in Dendrobium species using pressurized liquid extraction and ultra-performance liquid chromatography. J Sep Sci 33(11):1580–1586

Yang ZH, Zhang QT, Feng ZZ, Lang KY, Li H (1993) English edition translated by ZR Xiong (1993): Orchids. China Esperanto Press, Beijing

Yang JZ, Jiang JH, Huang GL et al (2014) Phenanthrenes from Eria stricta Lindl. Biochem Syst Ecol 54:333–336

Yong HS (1990) Orchid portraits. Tropical Press Sdn Bhd, Kuala Lumpur

Yuan QY, Liu XQ (2012) Chemical constituents from *Pleione bulbocodioides*. Zhong Yao Cai 35 (10):1602–1604

Yuan L, Yang ZL, Li SY et al (2010) Mycorrhizal specificity, preference and plasticity of six slipper orchids from South Western China. Mycorrhiza 20 (8):559–568

Zhang J, Li T, Zhou L, et al (2010) Chinese herbal medicine for subfertile women with polycystic ovary syndrome. Cochrane Database Syst Rev 20010 Sep 8, (9) CD007535

Genus: *Renanthera* Lour.

Chinese name: *Huoyan lan* (fire orchid), red coral

Renanthera are robust, spectacular, monopodial orchids with predominantly scarlet flowers borne on large, branching inflorescences (Fig. 19.1). They are the principal source of scarlet flowers in tropical, Asian monopodial orchids. There are 12–15 species in the genus distributed from the Himalayas across southern China and Southeast Asia to New Guinea. The generic name is derived from Greek, *renes* (kidney) and *anthera* (anther); it refers to the kidney-shaped pollinia (Schultes and Pease 1963).

Renanthera coccinea Lour.

Epidendrum renanthera Raecusch., *Gongora phillippica* Lianos.

Chinese name and medicinal name: *Huoyan lan* (fire orchid), red coral

Description: Stem is stout, up to 5 m tall, 1.5 cm thick with many oblong leaves that are rounded at the apex, 7–8 by 1.5–3.3 cm, carried in two alternating rows and sheathing the stem at the base. Several inflorescences may be borne simultaneously, opposite the leaves. They are racemose or paniculate, up to 1 m long, and loosely many-flowered (Fig. 19.2). Flowers are scarlet with yellow markings on the linear petals, 5–6 cm tall by 4–5 cm across. Dorsal sepal is also linear but the lateral sepals are oblong, clawed, with undulating margins and bluntly pointed at the tips. Lip is small, red on the mid-lobe, with a white blotch at the base, and yellow on the side lobes but streaked with red (Fig. 19.3). Peak flowering is February to April in Thailand (Nanakorn and Watthana 2008), April to June in southern China (Chen et al. 1999), but it can flower at any time of the year in a hot climate. This popular species is distributed from southern China (only southwest Guangxi, southeast Yunnan and Hainan) to Myanmar, Thailand and Indochina. It has a scrambling habit, sending off many stout roots from the leaf axils, and is found attached to tree trunks or rocks in open spaces, sparse woods, and along ravines from low elevations to 1400 m. There are two main cultivated strains, one diploid, the other hexaploid (Kamemoto and Sagarik 1975).

Usage: In Chinese herbal medicine it is used to remove gas and dampness, improve circulation, relieve rheumatic pain and to treat fractures Decoction is prepared with 9–15 g of the plant. The fresh plant is also rendered into a paste for application to affected parts of the body (Ou et al. 2003; Wu 1994; Zhong Yao Da Cidian 1986).

© Springer International Publishing Switzerland 2016

E.S. Teoh, *Medicinal Orchids of Asia*, DOI 10.1007/978-3-319-24274-3_19

Fig. 19.1 *Renanthera coccinea* Lour. From: *Curtis Botanical Magazine*, vol. 57 [ser. 2, vol. 4]: t. 2997 (1830). Colour painting by W.J. Hooker. Courtesy of Missouri Botanical Garden, St. Louis, USA

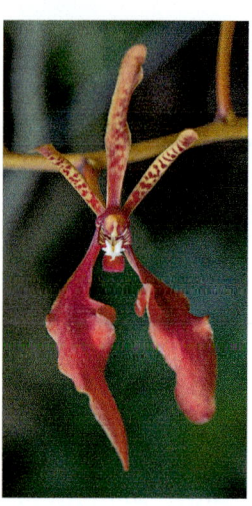

Fig. 19.3 *Renanthera coccinea* Lour. (Photo: E.S. Teoh)

Overview

Alkaloid is present in trace amounts (approximately 0.001 % dry weight) in *R. coccinea* (Luning 1964) but the level exceeds 0.1 % dry weight in *R. storiei* (Luning 1974). There is no additional pharmacological information on *R. coccinea*. Rumphius in the late seventeenth century reported that pickled young leaves of *R. moluccana* (Indonesian names: *Anggrek Merah*, *Boenga karang*) was considered a delicacy (Uphof 1968; Beekman 2002).

Genus: *Rhynchostylis* Blume

Chinese name: *Zuanhui Lan* (drill beak orchid)

This popular genus, widely known as the foxtail orchids, is much cultivated by orchid enthusiasts, and a large number of hybrids have been produced particularly in Thailand from the three common species which are all indigenous to that country. *Rhynchostylis* are robust epiphytes. Inflorescence is arching and carry numerous, closely well-arranged flowers and looking rather like a fox's tail.

The name of the genus is derived from Greek *rhyncos* (beak) and *stylis* (column), referring to the shape of the column in *R. retusa*, the first species to be described (Schultes and Pease 1963).

Fig. 19.2 *Renanthera coccinea* Lour. at The Botanic Gardens, Guangdong, China (Photo: E.S. Teoh)

Rhynchostylis retusa (L.) Blume

Common names: Fox Tail Orchid

Chinese name: *Zuanhui Lan* (drill beak orchid)

Indian names: *Dronpadi Mala* in Hindi, *Seetechi veni* (Gajaara), *Panas koli* (Konkani), *Pumam* (Orissa), *Rasna* (Arunachal Pradesh), *Banda, Rasna* (Uttarakahnd)

Bangladeshi names: *Sita pushpa, Pumam, Parada mura*

Thai name. *Alyarei, Hang Kraro*

Indonesian name: *Angkrek Lilin* in Sundanese and Malay (candle orchid)

Nepalese names: *Ghoge gava, Thur* in Nepali, *Gam* (Gurung) *Chadephuul*

Fig. 19.4 *Rhynchostylis retusa* (L.) Bl. (from Peninsular Malaysia; lip coloration and spots on the tepals are much paler) (Photo: E.S. Teoh)

Description: This is a popular, vandaceous orchid popular with growers of tropical orchids. It is a stout, tough, prolific epiphyte which thrives in deciduous and dry evergreen forests. Stem is short with thick, curved, elliptic leaves that appear bitten off (praemorse) at the tips, 15–50 by 1.6–5 cm. Inflorescence is pendulous, 30–40 cm with over 80 flowers, 1.2–1.5 cm across, and slightly fragrant. Flowers are white with tiny pink to purple spots, or pink; numerous, dense and beautifully arranged all round a stout but droopy inflorescence in the manner of a fox's tail. Lip is pink to bright purple (Fig. 19.4). Flower shape is variable, particularly in cultivated plants due to extensive line breeding in Thailand directed towards producing desirable horticultural varieties. Thirty-five years ago, flowering season was stated by leading Thai experts in Bangkok as April to May, the flowers lasting 2 weeks on the plants (Kamemoto and Sagarik 1975); today, in Chiang Mai, Thailand, it is May and June (Vaddhanaphuti 2001; Nanakorn and Watthana 2008). It also flowers in May to June in Mumbai (Santapau and Kapadia 1966), somewhat later, June to August, further south in the Western Ghats (Abraham and Vatsala 1981), while flowering seasons are in August and September in the Philippines (Davis and Steiner 1982) and in January, April, July and November in Sri Lanka (Jayaweera 1981). Plant grows well but does not flower if it does not have a dry period. Peak flowering of cultivated plants in the USA and Europe occurs from May to August, but some plants may be in bloom at any month of the year except December (Hamilton 1990), which rather suggests that they were imported from different countries. *R. retusa* has a wide distribution from Sri Lanka and India across Myanmar, south China, Thailand, Indochina and Malaysia to the Philippines. In Peninsular Malaysia, it occurs only in the northern states of Kelantan, Perlis, Kedah and Perak. It is common at low elevations in the Western Ghats (Abraham and Vatsala 1981). *R. retusa* occurs at sea level to 1200 m as an epiphyte in exposed locations in Thailand (Vaddhanaphuti 2001; Nanakorn and Watthana 2008). Plants from high elevations in Chiang Mai are large, with broad leaves, but these plants do not flower freely at low elevations (Kamemoto and Sagarik 1975). The fox-tail orchid forms massed displays on trees in the lowlands of Bhutan and Sikkim, even when the forests have been cleared (Pearce and Cribb 2002).

Phytochemistry: A small amount of alkaloid (between 0.001 and 0.01 % dry weight) is present in *R. retusa* (Luning 1964).

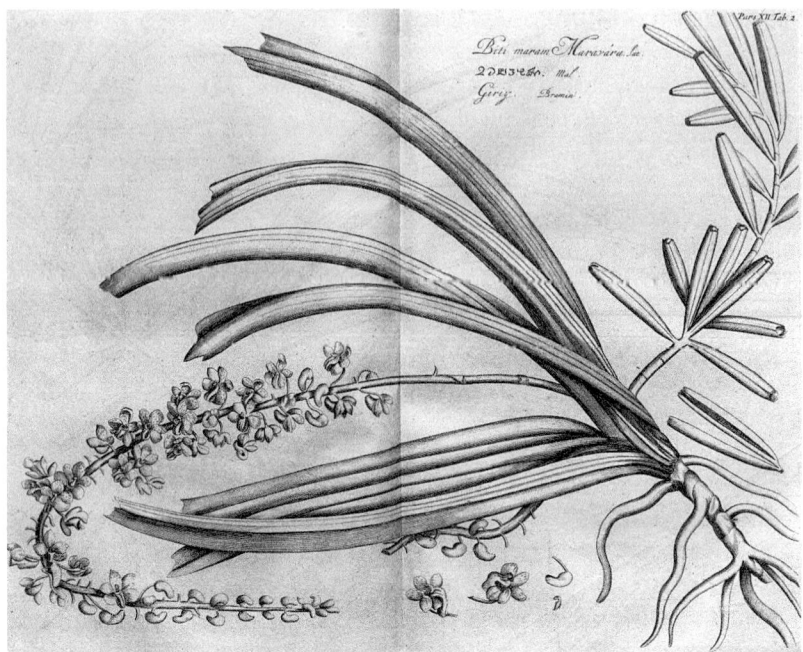

Fig. 19.5 *Rhynchostylis retusa* (L.) Blume. (as *Biti maram maravara*). From: van Rheede tot Drakenstein, H.A. *Hortus Indicus Malabaricus*, vol 22: t. 2 (1703–). Courtesy of Missouri Botanical Gardens, St. Louis, USA

Herbal Usage: In India, the fresh plant is used as an emollient (Dymock et al. 1893; Caius 1936; Suwal 1970; Jain and Defilipps 1991; Rao and Sridhar 2007). Leaves are used to treat rheumatism (Bhattacharjee 1998). Roots (known locally as *rasna* in Arunachal Pradesh) are used in a similar manner. Plant is used to manage asthma, tuberculosis, nervous twitching, cramps, infantile epilepsy, vertigo, palpitations, kidney stones and menstrual disorders (Lawler 1984; Rao 2004).

Root of *Pumam* (*R. retusa*) mixed with leafy shoots of *Pisum sativum* (pea) is made into a paste which forms a remedy for bloody diarrhoea among the primitive Dongria Kandha tribe of Orissa. A paste made with the leaves is used for wounds (Dash et al. 2008). Leaves of *R. retusa* are used to treat rheumatic disorders in Uttarakhand (Jalal et al. 2008). The plant is used to make an emollient in Kudremukh National Park in Karnataka, India (Rao 2007) and in Nepal (Baral and Kurmi 2006). In the latter country, juice from leaves (Pant 2011) or from root is applied to cuts and wounds (Manandhar and Manandhar 2002). Leaf powder is used for rheumatism and dried flowers are used as an insect repellent and to induce vomiting

(Subedi et al. 2013). Usage in Bangladesh is similar to the above (Musharof Hossain 2009).

Overview

Rhynchostylis has attracted much horticultural interest. *R. retusa* and other species of *Rhynchostylis* have been extensively line-bred in Thailand, but as far as we know there is no pharmacological work on the species. Investigation on the floral scent of *R. coelestis* revealed the presence of (*E*)-ocimene (47.0 %) and (E,E)-farnesal (34.4 %) with smaller amounts of their regioisomer and traces of anethole, benzyl, geranial, linalool and alpha-terpineol (Kaiser 1993). Although *Rhynchostylis* is abundant in Thailand, there is no medicinal usage for any of its species in this country, in contrast to the broad range of medicinal uses of *R. retusa* in India.

In van Rheede's *Hortus Indicus Malabaricus*, *R. retusa* was described twice, first as *Ansjeli maravara* (local name: *Ponoffou kely*), then as *Biti maram maravara* (local name: *Giriy*). Both descriptions were accompanied by line drawings of a beautiful vandaceous orchid with a long inflorescence carrying numerous spotted flowers (Fig. 19.5). The principal difference between the two plants was the spacing of the flowers. Van Rheede's description of the medicinal usage of *R. retusa* are the earliest records of its usage (van Rheede 1703). They do not differ from the contemporary usage described by Caius (1936), Lawler (1984) and later writers. Components of the scent of *R. coelestis* have been studied (Kaiser 1993) but the floral fragrance is not outstanding.

Genus: *Robiquetia* Guadich

Chinese name: *Jishu lan*

The genus is named after Pierre Robiquet, the French chemist who discovered morphine and caffeine. It consists of some 40 thick-leaved, epiphytic, monopodial orchids with pendent stems and densely-flowered inflorescences. *Robiquetia* is distributed from Myanmar, Thailand and Malaysia to Australia and the Pacific Islands.

Robiquetia succisa (Lindl.) Seidenf. and Garay

Chinese name: *Jishu lan*. In Hong Kong: precipice orchid, big ladder orchid
Chinese medicinal name: *Xiaoyejishu Lan*
Thai name: *Uang Man Pu*

Description: A small to medium-sized epiphyte, the plant is 15–25 cm tall. Leaves are oblong-elliptic, thick, measuring 1.5 by 1.2 cm. Inflorescence is pendulous, paniculate, 17–25 cm long, densely covered with 20–30 flowers which face all directions (Fig. 19.6). Flowers are 0.7 cm across, of heavy substance. Sepals and petals are yellow with tiny brown spots. Lip is white or pale yellow with red markings. Spur is a dirty yellow (Fig. 19.7). Flowering period is May to August in Thailand and northeast India (Vaddhanaphuti 2001; Pearce and Cribb 2002; Nanakorn and Watthana 2008). It flowers from June to September in China (Chen and Wood 2009).

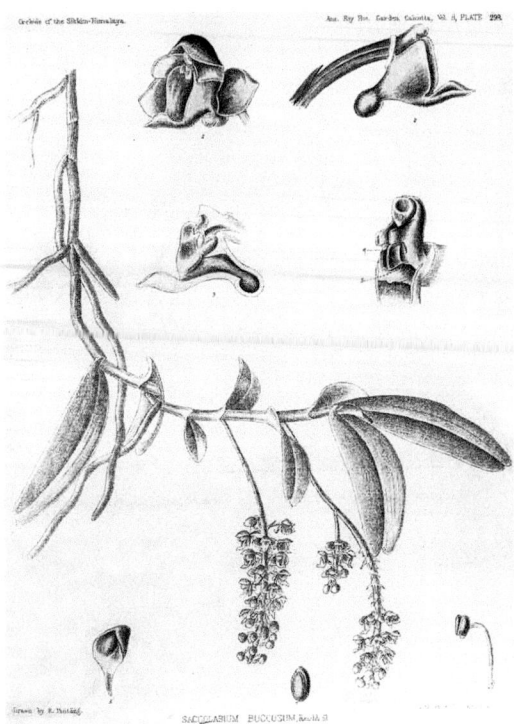

Fig. 19.6 *Robiquetia succisa* (Lindl.) Seidenf. and Garay (as *Saccolabium buccosum* Rchb. f.), From *Annals of the Royal Botanical Gardens, Calcutta*, vol. 8 (3): 298 (1891). Drawing by R. Pantling. Courtesy of Missouri Botanical Gardens, St. Louis, USA

R. succisa occurs in open forests and on cliffs at 500–1200 m. It is distributed from northeast India and Bhutan across Myanmar and Thailand to Laos and Vietnam, and across Yunnan, Guangxi, Fujian and Guangdong to Hainan (Chen and Wood 2009).

Phytochemistry: Alkaloid is present in small amounts in *R. succisa* (Lüning 1967).

Herbal Usage: The whole plant is used as a blood tonic (Chuakul 2002). It is also used in Chinese herbal medicine to treat "heaty coughs"; it soothes the lungs. After harvesting in spring or summer, the orchid plant is dried under the sun. To prepare the medicine, 9–15 g of the dried herb is decocted (Hu et al. 2000).

Overview

Pharmacological research on *Robequetia* has not been reported.

Fig. 19.7 *Robiquetia succisa* (Lindl.) Seidenf. and Garay (Photo: Courtesy of Plant Photo Bank of China)

References

Abraham A, Vatsala P (1981) Introduction to Orchids, with illustrations and descriptions of 150 South Indian Orchids. TPGRI, Trivandrum

Baral SR, Kurmi PP (2006) A compendium of medicinal plants in Nepal. IUCN. Published by Mrs. Rachana Sharma, Kathmandu

Beekman EM (2002) (trans, ed) Rumphius orchids. Orchid texts from the Ambonese Herbal by Georgius Everhardus Rumphius. Yale University Press, New Haven

Bhattacharjee SK (1998) Handbook of medicinal plants. Pointer, Jaipur

Caius JF (1936) The medicinal and poisonous plants of India. J Bombay Nat Hist Soc 38(4):791–799

Chen XQ, Wood JJ (2009) Robequetia Gaudichaud. In: Chen XQ, Zj L, Zhu GH et al (eds) Flora of China – Orchidaceae. Science Press, Beijing

Chen SC, Tsi ZH, Luo YB (1999) Native orchids of China in colour. Science Press, Beijing

Chuakul W (2002) Ethnomedical uses of Thai orchidaceous plants. Mohidol Univ J Pharm Sci 29 (3–4):41–45

Dash PK, Sahoo S, Bal S (2008) Ethnobotanical studies on orchids of Niyamgiri Hill Ranges, Orissa, India. Ethnobot Leaft 12:70–78

Davis RS, Steiner ML (1982) Philippine orchids. A detailed treatment of some 100 native species. M & L Lucidine Enterprises, Manila (reprint)

Dymock W, Warden CJH, Hooper D (1893) A history of the principal drugs of vegetable origin met with in British India. Education Society Press, Bombay

Hamilton RM (1990) Flowering months of orchid species under cultivation. In: Arditti J (ed) Orchid biology reviews and perspectives, vol V. Timber Press, Portland

Hu XM, Zhang WK, Zhu QZ et al (2000) Zhonghua Bencao, vol 8. Shanghai Science and Technology Publication, Shanghai

Jain SK, Defilipps RA (1991) Medicinal plants of India, vol 2. Reference Publications, Algonac

Jalal JS, Kumar P, Pangtey YPS (2008) Ethnomedicinal orchids of Uttarakhand, Western Himalayas. Ethnobot Leaft 12:1227–1230

Jayaweera DMA (1981) A revised handbook of the flora of Ceylon, vol 2. A.A. Balkema, Rotterdam

Kaiser R (1993) The scent of orchids: olfactory and chemical investigations. Editiones Roche, Basel

Kamemoto H, Sagarik R (1975) Beautiful Thai orchid species. Orchid Society of Thailand, Bangkok

Lawler LJ (1984) Ethnobotany of the Orchidaceae. In: Arditti J (ed) Orchid biology reviews and perspectives, vol 3. Cornell University Press, Ithaca

Luning B (1964) Studies on Orchidaceae alkaloids. Screening of species for alkaloids I. Acta Chem Scand 18(6):1507–1516

Luning B (1974) Alkaloids of the orchidaceae. In: Withner CL (ed) The orchids. Scientific studies. Wiley, New York

Lüning B (1967) Studies on Orchidaceae alkaloids. Screening of species for alkaloids IV. Phytochemistry 6:857–861

Manandhar NP, Manandhar S (2002) Plants and people of Nepal. Timber, Portland

Musharof Hossain M (2009) Traditional therapeutic uses of some indigenous orchids of Bangladesh. Med Arom Plant Sci Biotechnol 3:100–106

Nanakorn W, Watthana S (2008) Queen Sirikit botanic garden (Thai native orchids 2). Wanida Press, Chiang Mai

Ou JC, Hsieh WC, Lin IH, Chang YS, Chen IS (eds) (2003) The catalogue of medicinal plant resources in Taiwan. Department of Health, Executive Yuan, Taipei

Pant B (2011) Medicinal orchids of Nepal and their conservation by in-vitro technique PowerPoint presentation. In: 20th world orchid conference. Singapore, 14 Nov 2011

Pearce NR, Cribb PJ (2002) The orchids of Bhutan. Royal Botanic Gardens/Royal Government of Bhutan, Edinburgh/Thimpu

Rao AN (2004) Medicinal orchid wealth of Arunachal Pradesh. Newsl ENVIS NODE Indian Med Plants 1 (2):1–7

Rao TA (2007) Ethno botanical data on wild orchids of medicinal value as practised by tribals at Kudremukh National Park in Karnataka. Orchid Newsl 2(2):1–7

Rao TA, Sridhar S (2007) Wild orchids in Karnataka. A pictorial compendium. Institute of Natural Resources Conservation, Education, Research and Training (INCERT), Bangalore

Rheede V (1703) Hortus indicus malabaricus, vol 12. Dutch East India Company, Kerala

Santapau H, Kapadia Z (1966) The orchids of Bombay. Government of India Press, Calcutta

Schultes RE, Pease AS (1963) Generic names of orchids. Their Origin and Meaning. Academic, New York

Subedi A, Kunwar B, Choi Y et al (2013) Collection and trade of wild-harvested orchids in Nepal. J Ethnobiol Ethnomed 9:64–73

Suwal PN (1970) Medicinal plants of Nepal. H.M. Govt. of Nepal, Ministry of Forests and Soil Conservation, Department of Medicinal Plants, Thapathali, Kathmandu

Uphof JCT (1968) Dictionary of economic plants. Verlag von J. Cramer, Lehre

Vaddhanaphuti N (2001) A field guide to the wild orchids of Thailand, 3rd edn. Silkworm Books, Chiang Mai

Wu XR (1994) A concise edition of medicinal plants in China Guangdong Higher Education Publication House, Guangdong (in Chinese)

Zhong Yao Da Ci Dian (1986) The great dictionary of Chinese medicinals, Edited by Jiangsu New Medical College. Science and Technology Press, Shanghai

Genus: *Satyrium* to *Sunipia*

Genus: *Satyrium* Sw.

Chinese name: *Niaozu Lan*

Satyrium is a genus of terrestrial, cool-growing, herbaceous orchids with globular underground tubers. There are around a 100 species, mainly in South Africa and the Mediterranean, with two species in Asia. The name of the genus is derived from *satyrion*, the man-orchid of ancient Greek *Herbals* which alleged that the orchid had aphrodisiac properties. Satyrs were Greek demigods who were half man, half goat, and afflicted by insatiable lust. Such orchids were first described in the Greek *De Materia Medica* of Dioscorides (40–90), and drawings of the plants named *Satyrion* were copied by Kratevas in his *Codex ex Vindobonensis Greacus 1* sometime in the first century. However, many species that were called *Satyrion* up to the Middle Ages are now reassigned to other genera such as *Anacamptis*, *Dactylorhiza*, *Gymnadenia*, *Ophrys*, *Orchis* and *Platanthera* (Jacquet 2007). The inflorescence of *Satyrium* is terminal and carries few or many non-resupinate flowers in a range of colours (Fig. 20.1).

Satyrium nepalense var. *ciliatum* (Lindl.) Hook. f.

Satyrium ciliatum Lindl., *S. aceras* Schltr. ex Limpritch., *S. mairei* Schltr., *S. setchuenicum* Kraenz., *S. tenii* Schltr., *S. tschangii* Schltr.

Local and medicinal name: *Yuanmaoniaozu Lan* (hair edged bird feet orchid)

Description: Tubers are ellipsoid, 1–5 by 0.5–2 cm. Stem is erect, 14–32 cm tall. Leaves are ovate, pointed at the apex, 6–15 by 2–5 cm, sheathing the stem at the base. Inflorescence is terminal, 5–13 cm tall, and densely many-flowered. Flowers are pink, 1.3 cm across, with prominent, reflexed bracts. Lip and petals may be minimally denticulate at their apices (Fig. 20.2). Flowering period is August to October in China (Chen et al. 1999, 2009; Jin et al. 2008), and July to September in the Himalayas (Pearce and Cribb 2002).

This slender, terrestrial orchid is distributed in Hunnan, Guizhou, Sichuan, Yunnan and Tibet to Sikkim, Bhutan and Nepal. It is found on grassy slopes, sparse woods and alpine forests at 1800–4100 m.

Herbal Usage: Herb is a product of Tibet, Yunnan and Sichuan. Stem is used to strengthen the loins, invigorate the kidney, nourish blood and calm the mind. It is used to treat nephritis, weak kidneys and backache, swellings of the face and legs, and heart disease (Wu 1994); in decoction, 9–15 g (Zhonghua Bencao 2000).

Satyrium nepalense D. Don var. *nepalense*

Chinese name: *Dui dui shen*

© Springer International Publishing Switzerland 2016
E.S. Teoh, *Medicinal Orchids of Asia*, DOI 10.1007/978-3-319-24274-3_20

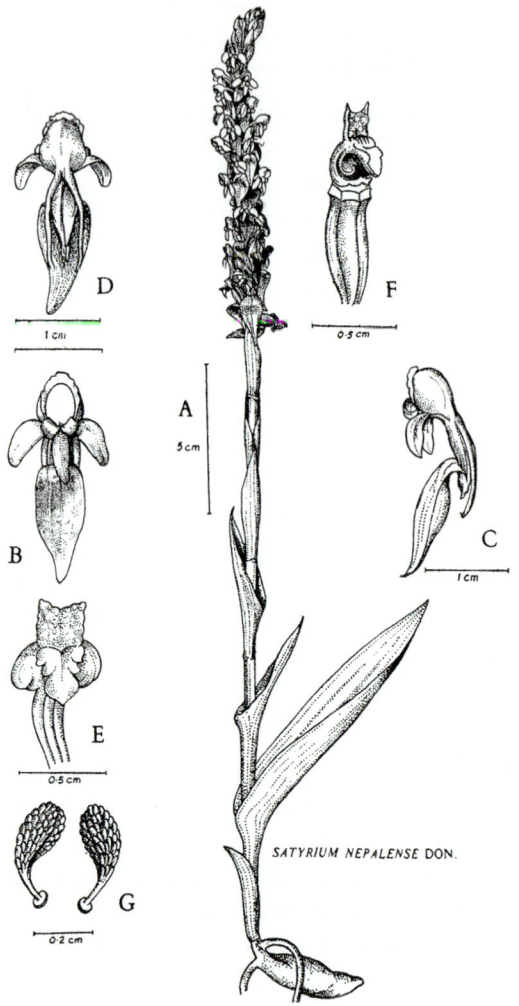

Fig. 20.2 *Satyrium nepalense* var. *ciliatum* Lindl.) Hook.
f. (Photo: Bhaktar B. Raskoti)

Fig. 20.1 *Satyrium nepalense* D. Don var. *nepalense*.
Reproduced with permission from *Introductions to Orchids*
by Abraham and Vatsala, Parlode, Thiruvananthapuram:
Tropical Botanic Garden and Research Centre (TBGRI), 1981

Chinese medicinal name: *Niaozu lan* (bird feet
orchid), *Dui dui shen*
Indian name: *Ezhtkwehhdr* in aboriginal Toda
(bullock's horns) referring to twin spurs of
flower; *salam misri*
Nepalese names: *Mishri*, *Thamni*

Description: This is a terrestrial, robust, leafy
herb with oblong, undivided tubers bearing 2–3
large lanceolate, plicate leaves, 4–25 by
1–10 cm, with long, broad petioles that ensheath

the stems. Flowers are densely clustered, white to
rose pink, fragrant, 0.7–1.5 cm across, borne on
an erect spike that varies from 3 to 15 cm tall.
Bracts are green in the white forms and green
with a pink flush in the pink forms. Lip is superior, hooded, trilobed (Fig. 20.3).

S. nepalense is a common, highly variable
plant found at 1300–3200 m in the Shan states
of Myanmar, Sikkim, Nepal, Bhutan, west
Yunnan, and in Pakistan at 1200–2400 m (Nasir
and Ali 1972); also in south India, from
Maharashtra to Kerala and Tamil Nadu above
1500 m on bare slopes (Seidenfaden 1999), and
in Sri Lanka where it is common among grass in
wet patina lands above 1220 m (Jayaweera
1981). Wherever the species is present, it occurs
in abundance (Abraham and Vatsala 1981). However, in the Palni Hills near Kodaikanal it has
become endangered because of overcollection

Fig. 20.3 *Satyrium nepalense* D. Don var. *nepalense* (Photo: Bhaktar B. Raskoti)

Table 20.1 A Chinese folk medicine prescription employing *Satyrium nepalense* (Zhongyao Da Cidian 1986)

1. Indications: Erectile dysfunction; chronic nephritis
Boil 15 g Satyrium nepalense with Plantago major 9 g and Huai Niu Xi (Achyranthes root) 6 g
(Source: Kuming Commonly Used Folk Herbs)
2. Indications: Low backache and weak kidneys
Consume a soup made with 10 pairs or 30 g Niaozu lan and pork tenderloin or chicken

dysentery (Rao 2007). A similar usage is reported from Uttarakhand in the Western Himalayas (Jalal et al. 2008). Likewise in Nepal, tubers are used to treat malaria, dysentery, or as a tonic (Baral and Kurmi 2006), and sometimes merely as an item of diet (Pant 2011). Dried tubers are consumed as a tonic or prophylaxis against dysentery, whereas juice is taken for fever and used on cuts and wounds (Subedi et al. 2013).

In Yunnan, herb is used to treat low backache, chronic nephritis, and weak kidneys, the last condition being possibly an euphemism for erectile dysfunction. Herb is harvested in autumn and sun-dried. Two typical Chinese folk herbal prescriptions are shown in Table 20.1. They are used to treat erectile dysfunction (Zhongyao Da Cidian 1986).

Overview

In ancient Rome, a drink called *satyrion* or *priapiscus* was prepared from the ground bulbs of terrestrial Mediterranean orchids (commonly *Orchis mascula* and *O. militaris* but also including species of *Satyrium*, *Anacamptis*, *Dactylorhiza*, *Gymnadenia*, *Ophrys*, *Orchis* and *Platanthera*) which were considered to be aphrodisiacs. Chinese folk or herbal usage of *S. nepalense* as an aphrodisiac for erectile dysfunction is probably borrowed from Arabic or Ayurvedic medicine introduced overland during the century of Mongol suzerain (1271–1368).

Belief in the legendary, if untruthful, aphrodisiac properties of *Satyrion* started in ancient Greece. Theophrastus (371–287 BC) stated that "once it occasioned 70 consecutive acts of coitus". Pliny the Elder (23–79 CE) was probably

(Seidenfaden 1999). Overgrazing in Nepal has seriously decimated its populations (Raskoti 2009). It flowers from September to January in Sri Lanka, August to September or October in India and Nepal (Jayaweera 1981; Abraham and Vatsala 1981; Pearce and Cribb 2002; Raskoti 2009), and September to December in China (Jin et al. 2008). The generic name uses the Greek word *satyr* to highlight the previously popular use of this orchid as an aphrodisiac.

Herbal Usage: Tubers of *S. nepalense* var. *nepalense* are eaten by the Monpa tribe (living predominantly in Arunachal Pradesh and Tibet) to treat malaria, dysentery and as an aphrodisiac. They are regarded as a tonic (Lawler 1984; Vij 1995; Rao and Sridhar 2007). Tribal people at the Kudremukh National Park in Karnataka combine the tubers of *S. nepalense* var. *nepalense* with tubers of other orchids to treat malaria and

echoing Theophrastus and Dioscorides when he declared that its power to arouse sexual excitement was common knowledge. The Roman novelist, Petronius wrote in *Satyrion*, "We saw in the chambers persons of both sexes acting in such a way that I concluded they must all have been drinking *satyrion*" (Wedeck 1961). Such was the power of belief and suggestion that this belief was carried well into the Middle Ages, and is sometimes alluded to even today. Orchids were invariably mentioned in *Herbals* and books written by scholar-physicians on the history of plants. Ten drawings were included the *Excellent Comments on Plant History*, originally written in Latin by Leonard Fuchs (1501–1566) and subsequently translated into French by Eloy Maignan. It carried very fanciful descriptions and names of the orchid species. From the drawings, the plant species were appropriately identified by Pierre Jacquet (2007) (Table 20.2). The French names are humorous: for instance, the first line reads "testicle of female dog, big" (in French *Couillon de chien femelle, grosse*). Similarly, another botanical work, *Historia Plantarum Eararum Earum Imagines, Qualities et Natale Solum, ex Dioscoride* translated by Geoffrey Linocier (c. 1550–1620), gave the name *testiculus canis femina* (testicle of a female

dog) for *Orchis morio* and *testiculus feminor minor* (small female testicle) for *Ophrys apifera*. This arose from the fact that shrivelled bulbs were considered to be female, the rounded bulbs male. Pierre Jacquet has recently identified the orchids in the illustrations (Jacquet 2007).

In the eighteenth century, Pierre de Garidel, a professor of medicine at the University of Aix, came out to refute the claims regarding the aphrodisiac properties of orchids. He published *Historie des Plantes qui Naissent aux Environs d'Aix* in 1715, in which he wrote, "It is true, and experience confirms it every day, that *Orchis*, whatever the species, has no such effect. All what Crollus had to say in his booklet *Signatura plantarum* and all that other chemists reiterated cannot support an opinion that daily experience contradicts. In a century so enlightened as ours, one must only find truth based on frequent experience. The great names of authors supported many stupidities in medicine" (translated by Jacquet 2007). Spoken like a scientist!

However, old beliefs hardly die, especially if there is money to be made! *S. nepalense* D. Don has been propagated by asymbiotic culture (Mahendra and Bai 2009). It is a matter of time before it reappears in herbal "anti-ageing" products. However, the predominant content is glucomannan, a polysaccharide which is totally devoid of aphrodisiac properties. Two species of *Satyrium* were screened for alkaloids and none were found to have a significant alkaloid content of 0.1 % or higher on screening (Luning 1974).

Four African species of *Satyrium* (*S. bicorne*, *S. candidum*, *S. carneum* and *S. erectum*) are used as a nutrient in the continent, in the manner of *salep* for their starch and mucilage content (Emboden 1974).

Table 20.2 Orchids mentioned by Eloy Maignan (c. eighteenth century) being used as aphrodisiacs in Europe (after Jacquet 2007, in Cameron, Arditti and Kull (eds): *Orchid Biology IX*. NY: New York Botanical Gardens)

Maignan's name in French	Current Latin name
Couillon de chien femelle, grosse	*Anacamptis pyrimidalis*
Satyrion royal, femell	*Dactylorhiza fuchsia*
Satyrion royal, masle	*Gymadenia conopsa*
Ophrys, autrement ellebore blanc	*Listera ovata*
Triple couillon de chien, femell	*Ophrys apifera*
Couillon de chien, masle, a feuilles etrites	*Orchis mascula*
Couillon de chien masle, a larges feuilles	*Orchis militaris*
Triple couillon de chien, masle	*Orchis morio*
Couillon de chien femelle, menus	*Orchis ustulata*
Satyrion a trios feuilles	*Platanthera bifolia*

Genus: *Sedirea* Garay and H. R. Sweet

Chinese name: *e ji lan*

Sedirea is a dwarf, monopodial epiphyte with short stems bearing several oblong to oblong-lanceolate leaves which are flat and fleshy. Inflorescence is axillary, racemose, laxly several-flowered.

Flowers open widely. Sepals and petals are free and similar. Lip is trilobed with a long narrow spur.

The 25generic name is an anagram (a mirror image) of *Aerides* (Chen and Wood 2009). It is a small genus with only two species distributed in China, Korea and Japan, and one endemic in China,

Sedirea subparishii (Z.H. Tsi) Christianson

Hygrochilus subparishii Z.H. Tsi

Chinese name: *Duanjingeji Lan*
Chinese medicinal name: *Zhijialan*

Description: This endemic Chinese species, *S. subparishii*, is a small to medium sized, epiphytic, monopodial orchid. It has a short stem and slightly flattened greenish roots. Leaves are several, oblong, 5.5–19 by 1.5–3.4 cm, sheathing at the base. Inflorescence is arching, 10 cm long, laxly several-flowered. Flowers are flat, star-shaped, fleshy, 4 cm across and scented. Tepals are pale yellow-green spotted with maroon. Lip is narrow, white to light yellow-green, with pink to purplish-red on the inner surface of the side lobes (Fig. 20.4). Flowering season is May. *S. subparishii* is found on

Fig. 20.4 *Sedirea subparishii* (Z.H. Tsi) Christianson (Photo: Courtesy of Plant Photo Bank of China)

tree trunks in wooded slopes at 700–1100 m in Zhejiang, Hubei, Hunan, Guangdong, Guizhou and Sichuan (Chen et al. 1999).

Herbal Usage: In Chinese herbal medicine, *Zhijialan* (*Sedirea subparishii*) is used to counter heat and wind. A decoction prepared with 30 g of the herb, sweetened with white sugar, is administered to children afflicted by acute illnesses involving the brain and nerves (possibly meningitis and encephalitis) in divided daily doses (Hu et al. 2000).

Overview
There are no pharmacological data on *Sedirea*.

Seidenfia versicolor Marg. and Szlach. [see **Malaxis vesicolor (Lindl.) Abeyw.**]

Genus: *Smitinandia* Holtt.

Chinese name: *Gai hou lan*

Smitinandia is a genus of small, monopodial, epiphytic orchids that very much resemble *Cleisostoma*. Stems are short, rigid, unbranched, and sheathed by bases of thick, coriaceous leaves and bearing numerous, flattened aerial roots. Inflorescence is axillary, horizontal, arching or pendent; simple or paniculate, and many-flowered. Flowers are flat, fleshy, with open petals and sepals. Lip is trilobed, with oblong mid-lobe and small, lateral lobes. There are three species, distributed from Himalaya, eastwards to Indochina, Thailand and northern Peninsular Malaysia (Fig. 20.5).

Smitinandia micrantha (Lindl.) Holtt.

Saccolabium micranthum Lindl., *Clerisostoma micranthum* (Lindl.) King & Prantl.

Chinese name: *Gai hou lan*
Thai names: *Khem Nu* (in Bangkok), *Kulap dong* (in Ubon Ratchathani)

Description: Stems are erect or pendant, 8–10 cm, enveloped by leaf sheaths. Leaves are

Fig. 20.5 *Smitinandia micrantha* (Lindl.) Holtt. (as Saccolabium micrantha Lindl.) From: *Annals of the Royal Botanic Gardens, Calcutta*, vol. 8 (3): t. 312 (1891). Drawing by R. Pantling. Courtesy of Missouri Botanical Gardens

Fig. 20.6 *Smitinandia micrantha* (Lindl.) Holtt. (Photo: E.S. Teoh)

fleshy, praemorse, oblong, 7.5–11 by 1.3–1.5 cm, Malaysian plants being larger. Inflorescence is 7 cm, and many-flowered. Flowers are small, pink, fleshy, fragrant and 1 cm across. The anther cap, and the mid-lobe and tips of the side lobes of the lip are dark pink (Fig. 20.6). A white-flowered form occurs in Langkawi (Go et al. 2010) and Nepal (Raskoti 2009). It flowers in May to July in Nepal (Raskoti 2009), April in China (Chen and Wood 2009), almost year round in Thailand [stated as June to February by Vaddhanaphuti (2001), or February to April by Nanakorn and Watthana (2008)], and April to May in Perlis in northern Peninsular Malaysia (Go et al. 2010). *S. micrantha* is distributed from Nepal and Sikkim Himalaya to Myanmar, Thailand, Laos, Cambodia and Vietnam south to Pulau Langkawi and Perlis in Peninsular Malaysia, occurring at 500–1400 m.

Herbal Usage: In Nepal, powdered plant is mixed with rice flour and butter, then baked, and served as tonic (Baral and Kurmi 2006). Leaves are used to treat rheumatism, and juice extracted from the roots is used to treat cuts and wounds (Pant 2011).

Overview

Pharmacological data on *Smitinandia* are not available.

Genus: *Spathoglottis* Blume

Chinese name: *Baoshe Lan* (bud tongue orchid)

Spathoglottis is a genus of 40 terrestrial orchids distributed from southern China and Indian Himalayas through Southeast Asia to New Guinea and Samoa. Pseudobulbs are ovoid, usually above–ground, and they bear several long-stemmed, thin, plicate, lanceolate leaves. Inflorescence arises from the base of the pseudobulb and bears several yellow, pink, purple or white flowers that face all directions, opening a few at a time, the raceme lasting several months. Flowers are star-shaped. Lip is trilobed and marked by a pair of callosities at the base of the mid-lobe (Fig. 20.7). Many species grow at the edge of lowland forests or on exposed hill slopes.

Spathoglottis is a popular garden flower in the tropics and numerous beautiful hybrids have been bred. The name of the genus is derived from Greek *spathe* (blade) and *glotta* (tongue). It describes the shape of the lip.

Fig. 20.8 *Spathoglottis affinis* de Vriese (Photo: E.S. Teoh)

Fig. 20.7 *Spathoglottis plicata* Blume (as *Bletia angustifolia* Gaudich). Gaudichaud-Beaupre, C., *Voyage autous du monde sur les corvettes de S.M. l'uranie et la Physicienne pendant les annees 1817–1820, publie par Louis de Freycinet, Atlas Botanique*, t. 32 (1826). Courtesy of Missouri Botanical Gardens, St. Louis, USA

Spathoglottis affinis de Vriese

Syn. *Spathoglottis lobii* Rchb.f.

Thai names: *Tan diao* (in the north), *Khao niao Hua khao nieo* (in Prachin Buri), *Luang Phitsamon* (in general); also *Luang Si Sa Ket*

Description: Pseudobulbs are small, flattened, bearing several narrow, plicate, lanceolate leaves, 30 by 2 cm. Floral scape is 20–30 cm tall, finely hirsute, with numerous well-spaced, yellow flowers, 2–3 cm across, which face all directions. The Thai variety has three linear longitudinal red markings across the lower half of the lateral sepals. About 3–5 flowers are open at any one time. Flower is usually star-shaped. Petals and sepals are equal in size and rounded at the tip. Lip is trilobed, the mid-lobe expanding from a narrow waist distally to a notched margin (Fig. 20.8). It flowers in May to September in north and northeast Thailand.

This small, yellow *Spathoglottis* distributed in Burma, Thailand, Peninsular Malaysia and Java. In Malaysia, it is found only in Gunong Jerai (Kedah Peak) in the north, and even there it is now quite rare. One variety, which is larger and deciduous, has 25 flowers opening simultaneously, and was previously described as a separate species, *S. lobbii* Rchb.f. but this is now considered to be just a different form of *S. affinis*.

Herbal Usage: Pseudobulbs of *S. affinis* are used to treat abscesses in Thailand (Chuakul 2002). In Indochina, *S. lobbii* (=*S. affinis*) was used to heal war wounds and infected wounds (Dournes 1955).

Spathoglottis eburnea Gagnep.

Thai name: *Ban duck, Sai pla kho* (in the north)

Description: Vegetatively, it is similar to *S. affinis* but plants are larger and more robust. Flowers are 2.5–4 cm across. It is commonly known as the ivory *Spathoglottis* or white

Fig. 20.9 *Spathoglottis eburnea* Gagnep (Photo: E.S. Teoh)

Fig. 20.10 *Spathoglottis plicata* Blume (Photo: E.S. Teoh)

Spathoglottis after the appearance of its flowers. The mid-lobe of the lip expands into paired circular lobes. Side lobes are yellow (Fig. 20.9).

S. eburnea is distributed in Thailand and Cambodia, Laos and Vietnam.

Herbal Usage: The whole plant is used as a tonic in Thailand (Chuakul 2002), whereas the pseudobulbs are food in Cambodia (Uphof 1968).

Spathoglottis lobii Rchb.f. (see ***Spathoglottis affinis* de Vriese**)

Spathoglottis plicata Blume

Common names: Pink *Spathoglottis*; Purple *Spathoglottis*

Chinese name: *Zihuabaoshe Lan*

Malay names: *Lumbah tikus* (little mouse); (in aboriginal Sakai): *Wah*

Indonesian names: *Angkrek Daun Tjongkok*; *Angkrek Tjongkok* (in the Sunda islands), *Antel-antelan, Djangkuawang* (Java), *Kupur* (Gajo), *Buluh Hutan* (Menado), *Daun korakora, Daun tana,* (Maluku), *Ahaan* (Amboin) *Lalagu* (Halmahera), *Kusuma raka* (Ternate), *Bure* (Seram utara)

Thai names: *Krathiam pa, Wan chuk* (in Trat), *Sapato, Ueang din* (in Bangkok)

Description: This is an extremely variable and widespread *Spathoglottis*. Pseudobulbs are corm-like, above-ground, 3 by 1.5–2 cm, ensheathed. Leaves are linear-lanceolate, plicate, 30–80 by 5–7 cm, ensheathing the pseudobulb at their bases. Axillary buds on the pseudobulb develop into inflorescences or into new pseudobulbs. Inflorescence is erect up to 60 cm tall, rachis 8–12 cm with up to 40 flowers, opening successively, usually 5 or 6 at a time. Flowers vary from dark to light purple, to white. It flowers throughout the year, with peaks after the change of monsoons, or during the dry seasons (Fig. 20.10). According to Issac Henry Burkill (1870–1965), Director of the Singapore Botanic Gardens from 1912 to 1925, who collected widely in Peninsular Malaysia, Javanese plants are much more attractive than the Malaysian variety, and it is the former that was being cultivated in the Malaysian-Singaporean gardens (Burkill 1935).

This is an extremely variable, widespread *Spathoglottis* which is found in grassy lowlands and foothills from India to the Philippines. It is a tough plant and was among the first re-colonisers of Krakatoa following the massive eruption of its volcano in the nineteenth century. In Singapore, it is found in open scrubland or *belukar* in the company of *Aurndina graminifolia, Nephentes* species, grasses and ferns. The Indonesian name

Daun korakora, describes the shape of the leaves which are likened to the Moluccan boat or *karakar* (Rumphius, quoted by de Wit 1977).

Herbal Usage: A decoction of the plant is used to treat rheumatism and as a hot foment in India. It is used to treat rheumatism in Bangladesh (Mollik, Hassan, Islam, et al. 2009). In the state of Perak in Peninsular Malaysia, aboriginal tribes also used a decoction as a foment to treat rheumatism. At the same time, they would drink a small amount of its decoction (Burkill and Haniff 1930; Burkill 1935). Its usage was confined to non-painful swelling of the limbs in Indonesia; if pain was present, one resorted to another orchid, *Calanthe veratrifolia* (Heyne 1927). In the Nicobar Islands, *S. plicata* is used to treat earache (Dagar and Dagar 2003). A yellow salve made from its powdery seeds is sometimes applied on children in place of *bedak* (jasmine-perfumed rice flower) to promote a fair countenance (van den Brink 1937; Rifai 1975).

Spathoglottis pubescens Lindl.

Spathoglottis fortunei Lindl., *S. plicata* Bl. var. *pubescens* (Lindl.) M. Hiroe

Chinese names: *Baoshe Lan* (bud tongue orchid), *Huanghuaxiaodusuan* (yellow flower small single garlic),
Chinese medicinal name: *Huanghuadusuan* (yellow flower single garlic)
Thai names: *Toe si re kho* (Karen and Mae Hong Son), *Ban Chuan* (Mae Hong Son), *Ueang din* (Lampang), *Ueang din lao* (in Chiang Mai), *Ueang nuan chan* (in the north)

Description: Plant is typical for *Spathoglottis*. Pseudobulb is 1–2.5 cm in diameter, covered with scale-like sheaths. Leaves are few, linear-lanceolate, 43 by 1–1.7 cm. Inflorescence is 6- to 8-flowered. Flowers are 2.5 cm across, of a very bright yellow, appearing from July to October. The species is common in Hong Kong (Wu et al. 2002). It is also present in the Khasia, Nagar and Manipur Hills of northeast India at 1000–2000 m (Bose and Bhattacharjee 1980). Flowering period is also July to October at Gaoligongshan (Jin et al. 2008). A yellow species from southern China and Myanmar,

S. pubescens grows in open spaces or sparse forests in the highlands at 400–1700 m, often in the company of *Arundina graminifolia*. It loves limestone.

Herbal Usage: The Chinese herb comes from the region south of the Yangzi. The stem is said to benefit the lungs. It stops coughs, promotes granulation, and heals sores (Wu 1994). To prepare a decoction, *Huanghuadusuan*, 9 g is used. The fresh herb is used in the preparation of a poultice (Hu et al. 2000).

Overview

Luning (1974) did not find any appreciable amount of alkaloid in the eight species of *Spathoglottis* that he tested. A search of the literature did not locate further investigation on the constituents of *Spathoglottis*.

S. plicata is one of the earliest colonisers of Krakatoa after it was shattered and minimized by a massive volcanic eruption in 1883. Its seeds germinate readily when sown around the parent plant (although fungicides must not be used when attempting to germinate the orchid). Fungal strains belonging to *Epulorhiza* isolated from several terrestrial orchids in Hong Kong appear to stimulate the germination of *S. pubescens* (Shan et al. 2002). Despite this apparent robustness, *Spathoglottis* is extremely prone to virus, and it is used as an indicator plant to identify the presence of virus in other plants. It would be interesting to know what is lacking in its genetic make-up vis-a-vis other orchids that makes it so susceptible to virus. Whether and how such susceptibility to virus applies to humans is another area that might be profitably investigated.

Genus: *Spiranthes* Rich.

Chinese name: *Shoucao* (tassel grass)
Japanese name: *Nejibana* (spiral flower)
Thai names: *Phak phai nam* (in Chiang Mai), *Phop plo mae* (Karen, Chiang Mai)

Spiranthes is a widespread terrestrial herb which is easily recognized by the spiral arrangement of the small, pink or white, partially-opened flowers on its long, terminal inflorescence. The generic name is derived from two Greek words, *speira* (coil) and *anthos* (flowers), referring to the unmistakable spiral arrangement of the flowers on the inflorescence. It is called "Tassel Grass" in China, or "Ladies' Tresses Orchid" in Australia and America. Leaves are slender, pointed and glabrous. They are arranged in a rosette from a short stem which is supported by numerous fleshy roots. The various species are found in montane grassland or in areas where the land has been disturbed.

S. sinensis is possibly the most widespread orchid species. It enjoys an extensive distribution throughout Asia (Japan, Korea, Russia, across China, Pakistan, India, and Southeast Asia) to Australia (including Tasmania), New Zealand and the Pacific Islands. In Peninsular Malaysia, it occurs in the highlands (Gunong Kali, Cameron Highlands and Genting Highlands (Seidenfaden and Wood 1992). Thirty related species extends the distribution of *Spiranthes* to North America (from Canada to the Caribbean), Europe and northern Africa. In the Far East, *S. sinensis* (Pers.) Ames is a medicinal plant; hence, the numerous fancy Chinese names which describe it as a dragon coiled around a post (*Qinglongchanzhu*, etc.). Some scholars believe that the first recorded orchid mentioned in the *Book of Odes* (circa sixth century BC) is tassel grass (Chen and Tang 1982), otherwise referred to as "wild grass" (Fig. 20.11).

Spiranthes autumnalis (Balb.) Rich. [see ***Spiranthes spiralis* (L.) Chevall**]

Spiranthes spiralis (L.) Chevall

syn. *Spiranthes autumnalis* (Balb.) Rich.

Description: *S. spiralis* is found in Afghanistan and Tibet, and in Asia Minor, Europe and Africa (Hawkes 1965). Gross appearance of the plant and its flowers is similar to that of *S. sinensis* and it takes a botanist to tell the difference. Recently, it was proposed that the white-flowered *S. hongkongensis* which hitherto has been found only in Hong Kong (Chen et al. 2009) is an allopolypoid probably derived from natural hybridisation between *S. sinensis* and *S. spiralis* (Sun 1996). If that is correct, *S. spiralis* must enjoy, or had enjoyed, a far wider distribution than what is described.

Herbal Usage: Its roots are used as an Indian aphrodisiac (Duggal 1971; Sood et al. 2005).

Fig. 20.11 *Spiranthe sinensis*. (Pers.) Ames var. *chinensis* (as *Neottia australis* R.Br. var *Chinensis*). From: *Botanical Register*, vol. 7: t: 602 (1822). Colour drawing by M. Hart. Courtesy of Missouri Botanical Garden, St. Louis, USA

Spiranthes sinensis (Pers.) Ames

Syn. *Spiranthes sinensis* (Pers.) Ames var. *amoena* (M. Bierberson) Hara

Chinese names: *Shoucao* (tassel grass), *Qinglongchanzhu, Jinlongpanshu, Panlongshen, Longbaozhu*, (dragon coiled around a post); *Qingmingcao* (bright spring herb)
Taiwanese name: *Chheng-thian-lion-thiau*
Japanese name: *Nejibana* (spiral flower), *Nezibana, Mojizuri*
Korean name: *Ta-rae-nan-cho*
Mongolian name: *Aolangheibu*.

Vietnamese name: *Ban long sam*

Medicinal names: *Panlongshen* in Chinese; *Poon lung sum* (Hong Kong); *Chheng thian liong thiau* (Taiwan); *Bamryongsam* (Korean)

Indonesian names: *Angkrek hindesan, Djukut hindesan* (Sundanese) (wool-comb orchid)

Description: *S. sinensis* is a small terrestrial orchid, a wisp of a plant, which grows almost as a weed in lowland fields, meadows and forests in both acidic and alkaline soils. It thrives even in disturbed areas and is to be found along roadside drains in Brisbane, preferring a moist to a dry environment. At Yercad in Tamilnadu, it occurs among grass in marshy ground fully exposed to the sun (Matthew 1995). Leaves are grass-like. They emerge from an underground tuber and surround the rachis when the flower spike appears. Flowering season is March to August in China (Chen et al. 1999, 2009), at Gaolingongshan only between July and August (Jin et al. 2008) and March to September in the Western Ghats (Santapau and Kapadia 1966), peaking in May to August at the adjacent Palni Hills (Seidenfaden 1999), July to August in Nepal (Raskoti 2009), and March to October in the Himalayas (Pearce and Cribb 2002; Gurong 2006). Up to 60 pink flowers with a white, translucent lip are arranged in a spiral on a long, slim rachis that may reach a length of up to 50 cm, but they are mostly shorter (Fig. 20.12). The variety at Palni Hills has white flowers (Seidenfaden 1999). Flowers emerge from large, persistent sheathing bracts. These flowers are only 6 mm across and they open in batches, in succession, along the rachis. Sepals are subequal with the dorsal sepal joining the smaller petals to form an erect hood. In Japan, Hirokazu Tsukaya (1994) identified two distinct varieties of *S. sinensis* var. *amoena* which flower in spring and autumn, respectively. Molecular studies showed wide heterogeneity among different populations.

After flowering, the plant enjoys a growth period of several months. Then it dies back to the bulb. Individual plants live for only 7 or 8 years, but they produce clusters of little bulbs before they die (Figs. 20.13, 20.14 and 20.15).

Fig. 20.12 *Spiranthes sinensis* (Pers.) Ames (Photo: E.S. Teoh)

Phytochemistry: The early work reported by Luning (1974) found no significant alkaloid content in seven species of *Spiranthes* that were screened. More recent research showed that *S. sinensis* contains a large range of interesting chemical compounds Spiranthols A and B, spirasineol A, orchinol, p-hydroxybesaldehyde, p-hydroxybenzyl alcohol, hydrocarbons, sterols and ferulates were isolated by Tezuka et al. (1989) from *S. sinensis* var. *amoena*. They gave the names Spiranthols A and B and spirasineol A to three of the seven new dihydrophenanthrenes which they isolated. Spiranthol A showed a weak cytotoxic activity on HeLa-S3 cells, with inhibition rates of 98.77 % at 25 mcg/ml and 7.03 % at 6.26 mcg/ml (Tezuka et al. 1989). The 9,10-dihydrophenanthrene derivatives showed bacteriostatic activity on Gram-positive bacteria (Tezuka et al. 1990).

Ten years later, YL Lin and his team in Taipeh reported the isolation of another 8 novel dihydrophenanthrene derivatives from *S. sinensis*, namely, sinensols A–H, a novel

Fig. 20.13 *Spiranthes sinensis var. alba* (Photo: E.S.Teoh)

homocyclotirucallane, sinetirucallol, together with other known compounds, spiranthesol, 2-(3′,4′-dihydroxyphenyl)-1,3-benzodioxole-5-aldehyde, ergosterol peroxide, *p*-hydroxybenzaldehyde, 3,4-dihydroxybenzylaldehyde, 3,4-dihydroxybenzyl alcohol, hydroquinone, 4-hydroxybenzyl ethylether, 4-hydroxybenzyl methyl ether and methyl 3-(4-hydroxyphenyl) propanoate (Lin et al. 2000, 2001). Another novel dimeric phenanthrene, 2,2′-dihydroxy-5,5′,7,7′-tetramethoxy-9,9′,10,10′-tetrahydro-3,3-′-biphenanthrene and a flavone, 5-hydroxy-3,7-dimethoxy-4′-(1-hydroxy-3-methylbut-3-en-2-yloxy)-flavone, were recently isolated along with three other known flavonoids from *S. sinensis* by Liu et al. (2013).

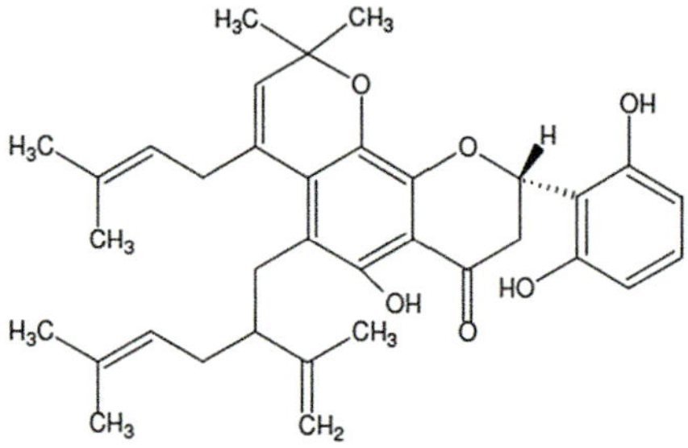

Fig. 20.14 A flavonoid from *Spiranthes sinensis*. It suppresses several human cancer cell lines in vitro but this is unrelated to the medicinal usage of the herb (Peng et al. 2007)

Two teams of scientists in northeastern China have also been working with *Spiranthes*. Dong et al. (2005) in Shenyang isolated a new

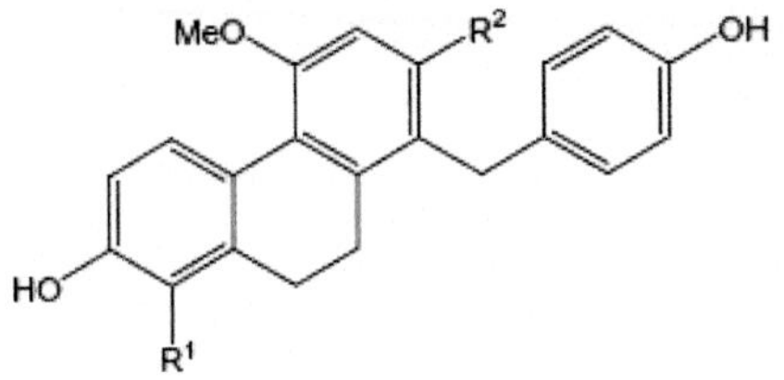

Sinesol A, arundinaol: R 1= H, R2 = OMe

Sinesol B: R1 = ‽ ‥ , R2 = OMe

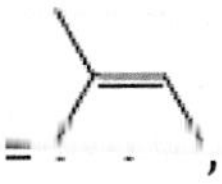

Sinesol C: R1 = ‽ ‥ , R Sinesol D

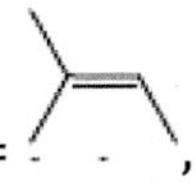

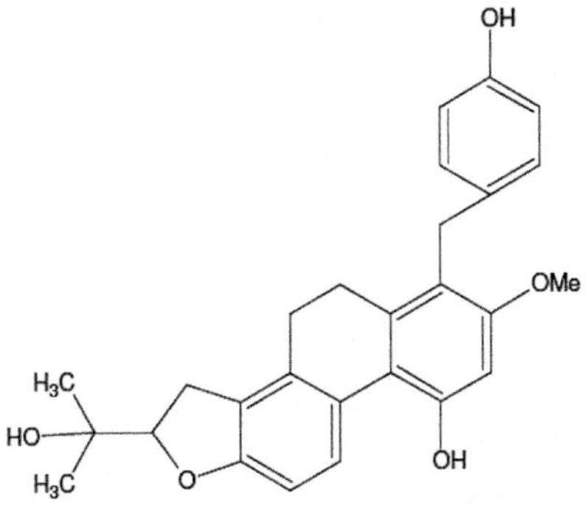

Sinesol E

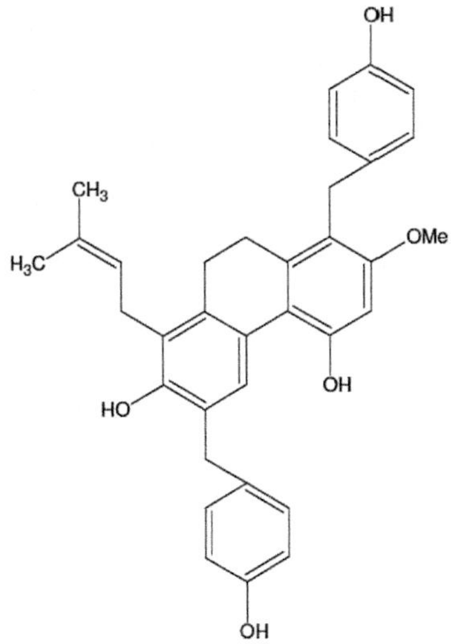

Sinesol F

Fig. 20.15 Phenanthrenes isolated from Spiranthes sinensis (Tezuka et al. 1990; Lin et al. 2000; Lin et al. 2001; Kovacs et al. 2008)

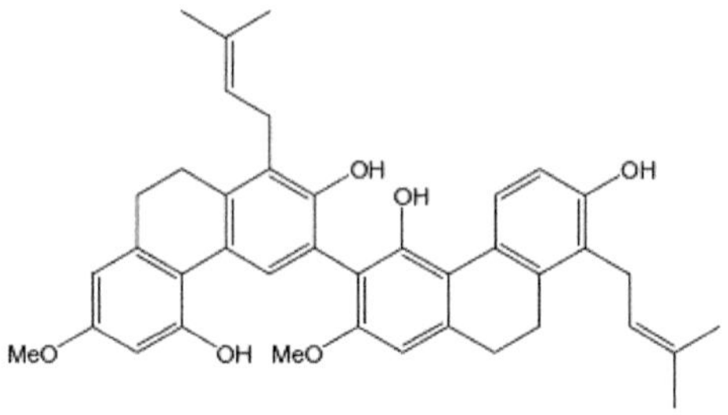

Spiranthesol

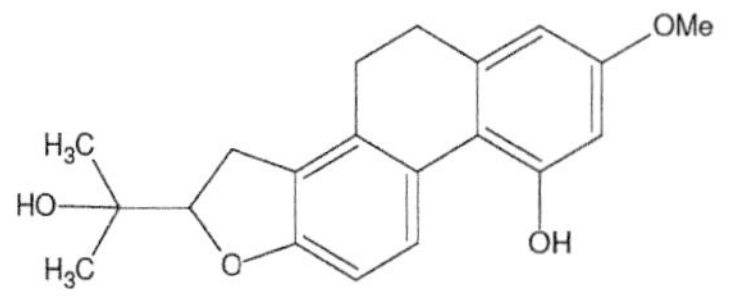

Spiranthol C

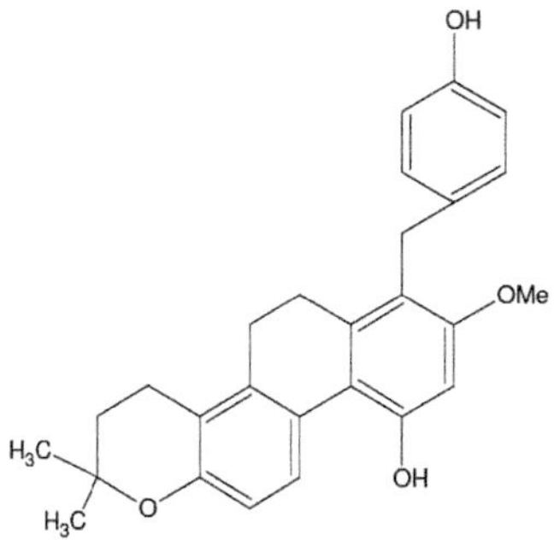

Spirasineol B

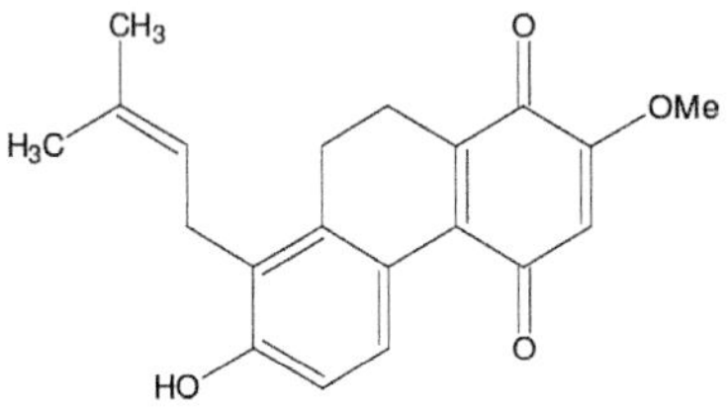

Spiranthoquinone

Fig. 20.15 (Continued)

flavonoid from the whole plant of *S. australis* [correct name: *S. sinensis* (Pers.) Ames]. Peng et al. (2007) in Dalian found that the new dihydroflavanoid which they isolated from a traditional Chinese medicine, constituted from the roots of *S. australis* (R. Brown) Lindl., inhibited cell growth in the human tumour cell lines A549, BEL-7402, SGC-79001, MCF-7, HT-29, K562 and A498 in vitro. However, the in vitro

cytotoxic activities of *S. sinensis* bear no relation to the traditional uses of *Spiranthes* in Chinese herbal medicine, nor to their uses in Korea, Vietnam, India and elsewhere. Kaempferol and quercetin are two flavonoids which were found in the leaves of *S. spiralis* (Williams 1979).

Two new prenylated coumarins were isolated from *S. sinensis* (Pers.) Ames by Peng et al. (2008) but the scientists did not report their pharmacological properties. It is uncertain whether they were working on two different species or varieties of *Spiranthes* in their two projects (Peng et al. 2007, 2008). Many taxonomists in China do not distinguish between *S. australis* and *S. sinensis*, and refer to all native *Spiranthes* as *S. sinensis* (Pers.) Ames (Table 20.3).

Herbal Usage: Plants are sun-dried for storage. In Traditional Chinese Medicine, the whole plant is decocted usually with pork and used to strengthen the "kidneys" (see Chap. 2) and "to cure spitting of blood" (Hu 1971). A similar preparation is used by the Bai minority in western Yunnan to treat kidney inflammation or to strengthen the kidneys (Weckerle et al. 2009). *Panlongshen* (*S. sinensis*) nourishes and strengthens a weakened body (Chen and Tang 1982). A Japanese team who studied the herb reported that it was a folk remedy in Taiwan for haemoptysis, epistaxis, headache, chronic dysentery and

meningitis, also being used as a tonic, whereas in China, it was used to treat fever, coughs, haemoptysis, vertigo, and low back pain (Tezuka et al. 1989). Other Chinese uses are leucorrhoea, diabetes and snake bites. In Korea, it is used to treat tuberculosis, haemoptysis, debility and coughs (Sung 2002).

The Hong Kong Chinese Medical Institute offers additional medicinal indications for its use: tuberculous coughs, tonsillitis and sore throat; debilitating neurasthenia; and summer fever in children. Either the roots or the entire plant may be used. The herb is collected in spring or summer when the plant is in bloom. Herb is sweet, bland and 'neutral'. It regulates body fluids, improves blood circulation, and is anti-inflammatory and bactericidal. *S. sinensis* is common in Hong Kong, growing on grassland (Wu et al. 2002).

A typical TCM prescription for sore throat is a decoction made by boiling together *S. sinensis* 15 g, *Empatorium chinensis* 15 g and *Ilex asprella* root 30 g. For tuberculous haemoptysis and dry coughs: *S. sinensis* 15 g, *Pholidota sinensis* 30 g and *Eclipta prostrata* 15 g in decoction. For summer fever in children, a different decoction: *S. sinensis* 6 g, and *Commelia communis* 15 g (Li 1988). A fuller list of herbal prescriptions is shown in Table 20.4.

Stems and roots of *S. sinensis* are used to treat sores by tribals in Kudremukh National Park in Karnataka on the western Deccan (Rao 2007). The entire plant is also used in Vietnamese folk medicine as a tonic to treat debility and haemoptysis (Duong 1993). Gerarde in his *Herbal* written in 1597 claimed that *S. spiralis* "provokes venery" (Lawler 1984) And Newaris in Nepal believe that it has aphrodisiacal properties (Balamai 2008). Another Nepali report mentions that decoction of the plant is administered to treat intermittent fever (possibly malaria?): tubers are used as a tonic (Pant and Raskoti 2013; Subedi et al. 2013), also to treat headache (Subedi et al. 2013). One method of treating sores in Nepal is to cover them with a paste made with roots and stem of *S. sinensis* (Baral and Kurmi 2006). Cherokee Indians in North America use *Spiranthes* to treat urinary disorders and they put it into infant baths to

Table 20.3 Chemical constituents of *Spiranthes sinensis*

Seven new dihydrophenanthrenes
Spiranthol A (weakly cytotoxic)
Spiranthol B,
Spirasineol A, B
Orchinol
p-hydroxybenzaldehyde
p-hydroxybenzyl alcohol
Hydrocarbons
Sterols
Ferulates (Tezuka et al. 1989)
Eight novel dihydrophenanthrene derivatives
Sinensols A–H
A novel homocyclotirucallane, sinetirucallol
Other known compounds (Lin et al. 2000)
Spiranthesol
2-(3′,4′-dihydroxyphenyl)-1,3-benzodioxole-5-aldehyde
Ergosterol peroxide
p-hydroxybenzaldehyde
3,4-dihydroxybenzylaldehyde
3,4-dihydroxybenzyl alcohol
Hydroquinone
4-hydroxybenzyl ethylether
4-hydroxybenzyl methyl ether
methyl 3-(4-hydroxyphenyl) propanoate (Lin et al. 2001)
Flavonoid (Dong et al. 2005)
Dihydroflavanoid (cytotoxic) (Peng et al. 2007)
Two new preylated coumarins (Peng et al. 2008)

Table 20.4 Chinese herbal prescriptions employing *Spiranthes sinensis* (Zhongyao Da Cidian 1986)

1. Indications: Augmenting yin and clearing heat, relieving coughs, treating weakness after an illness, haemoptysis, dizziness, waist aches, nocturnal emission, urethral discharge, skin ulcers and carbuncle
Drink a decoction made from 15 to 30 g of *S. sinensis*
Grind the herb for external application
(Primary Source: Records of Hunan Medicine)
2. Indication: Cough due to "empty heat"
Drink a decoction made from 9 to 15 g of *S. sinensis*
(Primary Source: Records of Hunan Medicine)
3. Indication: For nourishment after illness
Cook 30 g roots of "Jiang bean" with 250 g of pork or a small chicken
Eat this once every three days for three times
(Primary Source: Quizhou Folk Herbs)
4. Indication: Diabetes
Boil 30 g with a pig pancreas and Gingko biloba 30 g and consume
(Primary Source: Fujian Folk Herbs)
5. For gonorrhoea
Cook 30 g *S. sinensis* with a pig stomach. Add some salt and divide into two portions for consumption in morning and evening
(Primary Source: Fujian Folk Herbs)
6. Indication: Blood in the stools in old people
Cook 9–15 g *S. sinensis* with fresh ji fish 60 g. Add some sugar before consumption
(Primary Source: Records of Sichuan Chinese Herbs)
7. Indication: Gastric pain
Grind 6 g *S. sinensis* with Realgar 900 mg and Allium officinarum 2 pieces and consume
(Primary Source: Records of Hunan Medicine)
8. Indication: Carbuncle
Clean the roots of *S. sinensis* by washing, dry and seal in bottle with sesame oil. Remove for application when needed, once a day
(Primary Source: Jiangxi Folk Herbs)
9. Indication: Snake bites
Grind roots of *S. sinensis* and add in wine and some Realgar. Apply on wounds
(Primary Source: Jiangxi Folk Herbs)
10. Indication: Tonsillitis
Boil 9–15 g of *S. sinensis* and consume
(Primary Source: Handbook of Commonly Used Herbs of Guangzhou Armed Forces)
11. Herpes
Dry an appropriate amount of roots of *S. sinensis*, pulverize and add sesame oil for application
(Primary Source: Jiangxi Herbs)
12. Indication: Burns
Grind 30 g *S. sinensis* with five earthworms and add some sugar. Apply once a day
(Primary source: Shaanxi Chinese Herbs)

Note: Indications and the manner in which *S. sinensis* is used vary from one Chinese province to another

promote health and growth (Hamel and Chiltoskey 1975; Dalu and Aryesu 1985).

Overview

Bacteriostatic activity of 9,10 dihydrophenanthrene derivatives isolated from *S. sinensis* suggest that stems and roots of the orchid may actually help in the healing of sores.

Cryptosporiopsis ericae has been isolated from the root of *S. sinesis* collected from Tibet (Chen et al. 2010), a first record of the isolation of *C. ericae* from an orchid. *S. sinensis* is an extremely widespread species, and it is likely that other symbiotic fungi may also be associated with the orchid. The identities of mycorrhiza associated with *S. sinensis* in widespread habitats have not been studied; their role in ecology of the orchid and its medicinal applications are undetermined. Since the entire plant of *Spiranthes* is used in herbal medicine, not just the roots, the *C. ericae* present in Tibetan specimens of the orchid is more likely to have an ecological significance rather than a medicinal one. However, its role in the healing of sores cannot be ruled out without further investigation.

Genus: *Steveniella* Schltr.

A monotypic terrestrial genus distributed in Crimea, the Caucasus, Turkey and northern Iran.

Steveniella satyriodes Schltr.

Description: Plant is small to medium-sized, herbaceous, with paired oval tubers. Stem is erect, with two clasping sheaths at the base and a single, lingulate leaf. Leaf is usually bent backwards, dark olive green with brownish-purple stripes above and flushed with purple on the under-face. Inflorescence is loosely many-flowered. Flowers are green, flushed or spotted with purple. Dorsal sepal and petals form a large hood. Lip is trilobed, rounded at the tips and bent towards the ovary, olive green flushed with purple near the base (Wood 2001). *S. satyriodes* grows in mountain pastures up to 2000 m and woodlands in full sun or semi-shade on moss-covered slopes (Neiland 2001).

Herbal Usage: Tubers are regarded as both nutrient and aphrodisiac in the Middle East. Tubers are collected in northern Iran to supply the *salep* trade (Ghorbani et al. 2014a, b).

Genus: *Sunipia* Lindl.

Chinese name: *Dabao Lan*

Sunipia Buch. –Ham. Ex Sm. (*Ione* Lindl.) are sympodial, epiphytic or lithophytic orchids with distinct rhizomes and pseudobulbs that contain one internode and one leaf at the apex. The leaf does not sheath the base; it is glabrous and deciduous. Inflorescence arises from the base of the pseudobulb and carries one flower. *Sunipia* resembles *Bulbophyllum* but may be distinguished by the presence of four pollinia, arranged in two pairs each pair with a stipe (the Thailand, Laos and Vietnam. There are about 22 species, mostly epiphytes in montane forest at high altitudes (Schuiteman and de Vogel 2000). A single species is used in Thai native medicine.

Sunipia grandiflora (Rolfe) P.F. Hunt

Thai name: *Ma Tak Khok*

Description: *S. grandiflora* is an epiphytic, pseudobulbous orchid. Pseudobulbs are small, pyramidal, flattened and widened at the base, wrinkled, widely spaced on a creeping rhizome. Leaves 1–2, dark green, 5–6 by 0.8 cm. Inflorescence very short, arises from base of pseudobulb and bears a single flower, 2 cm across with narrow sepals and short, narrow petals of light purple, and a large, oval, dark purple lip which is hirsute and bears a median, longitudinal, convex band. Anther cap is white. It flowers in December (Vaddhanaphuti 2001). *S. grandiflora* is found in the north of Thailand, Yunnan, Myanmar, Laos and Vietnam at 1100 m.

Herbal Usage: the entire plant is used to prepare a tonic in Thailand (Chuakul 2002)

Overview

No additional information is available on the usage or pharmacology of the orchids in the genus.

References

Abraham A, Vatsala P (1981) Introduction to orchids, with illustrations and descriptions of 150 South Indian orchids. TPGRI, Trivandrum

Balamai NP (2008) Ethnomedicinal uses of plants among the Newari community of Pharping Village of Kathmandu District, Nepal. Source: Google

Baral SR, Kurmi PP (2006) A compendium of medicinal plants in Nepal. IUCN. Published by Mrs. Rachana Sharma, Kathmandu

Bose TK, Bhattacharjee SK (1980) Orchids of India. Naya Prokash, Calcutta

Burkill IH (1935) A dictionary of economic products of the Malay Peninsula, vol 2. Crown Agents for the Colonies/Ministry of Agriculture & Co-operatives, London/Kuala Lumpur (1966 reprint, 2nd ed. , with contributions by Birtwistle W, Foxworthy FW, Scrivenor JB, Watson IG)

Burkill IH, Haniff M (1930) Malay village medicine. Gard Bull Straits Settlements 6:165–321

Chen SC, Tang T (1982) A general review of the orchid flora of China. In: Arditti J (ed) Orchid biology. Reviews and perspectives, vol 2. Cornell University Press, Ithaca

Chen SC, Wood JJ (2009) Smitinandia. In: Chen XQ, Liu ZJ, Zhu GH et al (eds) Flora of China – Orchidaceae. Science Press, Beijing

Chen SC, Tsi ZH, Luo YB (1999) Native orchids of China in colour. Science Press, Beijing

Chen XQ, Gale SW, Cribb PJ (2009) Spiranthes. In: Chen XQ, Liu ZJ, Zhu GH et al (eds) Flora of China – Orchidaceae. Science Press, Beijing

Chen J, Dong HL, Meng ZX, Guo SX (2010) *Cadophora malorum* and *Cryptosporiopsis ericae* isolated from medicinal plants of Orchidaceae in China. Mycotaxon 112:457–461

Chuakul W (2002) Ethnomedical uses of Thai orchidaceous plants. Mohidol Univ J Pharm Sci 29(3–4):41–45

Dagar HS, Dagar JC (2003) Plants used in ethnomedicine by the Nicobarese of Islands in Bay of Bengal, India. In: Singh V, Jain AP (eds) Ethnobotany and medicinal plants of India and Nepal. Scientific, Jodhpur, pp 773–784

Dalu JA, Aryesu ES (1985) Medicinal plants of China, vol 2. Reference Publications, Algonac

de Wit HCD (1977) Orchids in Rumphius' herbarium amboinense. In: Arditti J (ed) Orchid biology reviews and perspectives, vol 1. Cornell University Press, Ithaca

Dong ML, Chen FK, Wu LJ, Gao HY (2005) A new flavanoid from the whole plant of *Spiranthes australis* (R. Brown) Lindl. J Asian Nat Prod Res 7(1):71–74

Dournes J (1955) Deuxieme contribution a ethnobotanique indochinoise. J Agric Trop Bot Appl 14:64–86 (quoted by LawlerLJ, 1984)

Duggal SC (1971) Orchids in human affairs (a review). Pharm Biol 11(2):1727–1734

Duong NV (1993) Medicinal plants of Vietnam, Cambodia and Laos. World Health Organization, Manila

Emboden WA (1974) Bizarre plants: magical, monstrous, mythical. Macmillan, New York

Ghorbani A, Gravendeel B, Zarre S, de Boer H (2014a) Illegal wild collection and international trade in CITES-listed terrestrial orchid tubers in Iran. Traffic Bull 26(2):52–58

Ghorbani A, Gravendeel B, Naghibi F, de Booer H (2014b) Wild orchid tuber collection in Iran: a wake-up call for conservation. Biodivers Conserv 23:2749–2760. doi:10.1007/s10531-014-0746-y

Go R, Yong WSY, Unggang J, Salleh R (2010) Orchids of Perlis. Jewels of the forest, Revised edn. Jabatan Perhutanan Negri/University Putra Malaysia, Kangar/Serdang

Gurong DB (2006) An illustrated guide to the orchids of Bhutan. DSB, Thimphu

Hamel PB, Chiltoskey MU (1975) Cherokee plants – their uses. A 400 year history. Herald Publishing, Sylva

Hawkes AD (1965) Encyclopaedia of cultivated orchids. Faber & Faber, London

Heyne K (1927) De nuttige planten van Nederlandsche Indie, vol. 1. Uitgave van het Departement van Landbouw, Nijverheid & Handel in Nederlandsche-Indie, pp 508–513

Hu SY (1971) Orchids in the life and culture of the Chinese people. Chung Chi J 10:1–26

Hu XM, Zhang WK, Zhu QZ et al (eds) (2000) Zhonghua Bencao, vol 8. Shanghai Science and Technology Publication, Shanghai

Jacquet P (2007) Orchid discoveries by French Botanists around the world. In: Cameron KM, Arditti J, Kull T (eds) Orchid biology, vol 9, Reviews and perspectives. New York Botanical Gardens, New York

Jalal JS, Kumar P, Pangtey YPS (2008) Ethnomedicinal orchids of Uttarakhand, Western Himalayas. Ethnobot Leaft 12:1227–1230

Jayaweera DMA (1981) A revised handbook of the flora of Ceylon, vol 2. A.A. Balkema, Rotterdam

Jin XH, Zhao XD, Shi XC (2008) Native Orchids from Gaoligongshan Mountains, China. Science Press, Beijing

Kovacs A, Vasas A, Hohmann J (2008) Natural phenanthrenes and their biological activity. Phytochemistry 69:1084–1110

Lawler LJ (1984) Ethnobotany of the orchidaceae. In: Arditti J (ed) Orchid biology reviews & perspectives, vol 3. Cornell University Press, Ithaca

Li NH (ed) (1988) Chinese medicinal herbs of Hong Kong, vol 4200. Hong Kong Chinese Medicinal Research Institute, Hong Kong

Lin YL, Huang RL, Don MJ, Kuo YH (2000) Dihydrophenanthrenes from *Spiranthes sinensis*. J Nat Prod 63:1608–1610

Lin YL, Wang WY, Kuo YH, Liu YH (2001) Homocyclotirucallane and two dihydophenanthrenes from *Spiranthes sinensis*. Chem Pharm Bull 49 (9):1098–1101

Liu J, Li C, Zhong YJ et al (2013) Chemical constituents of Spiranthes sinensis. Biochem Syst Ecol 47:108–110

Luning B (1974) Studies on Orchidaceae alkaloids. IV. Screening of species for alkaloids 2. Phytochemistry 6:857–861

Mahendra G, Bai VN (2009) Mass propagation of Satyrium nepalense D. Don—a medicinal orchid. Scientia Hort 119(2):203–207

Matthew KM (1995) An excursion flora of Central Tamil Nadu, India. A.A. Balkaema, Rotterdam

Mollik AH, Hassan S, Islam T et al (2009) Medicinal plants used against rheumatoid arthritis by traditional medical practitioners of Bangladash. Planta Med 75 (9):959

Nanakorn W, Watthana S (2008) Queen Sirikit botanic garden (Thai native orchids 1 and 2). Wanida Press, Chiang Mai

Nasir E, Ali SI (1972) Flora of West Pakistan. Pakistan Agricultural Research Council

Neiland MRM (2001) Stevenilla, ecology. In: Pridgeon AM, Cribb PJ, Chase MW, Rasmussen FN (eds) Genera Orchidacearum, vol 2, Orchidoideae. University Press, Oxford

Pant B (2011) Medicinal orchids of Nepal and their conservation by in-vitro technique Powerpoint presentation. In: 20th world orchid conference, Singapore, 14 Nov 2011

Pant B, Raskoti BB (2013) Medicinal orchids of Nepal. Himalayan Map House, Kathmandu

Pearce NR, Cribb PJ (2002) The orchids of Bhutan. Royal Botanic Gardens/Royal Government of Bhutan, Edinburgh/Thimpu

Peng J, Xu Q, Xu Y, Qi Y, Han X, Xu L (2007) A new anti-cancer dihydroflavanoid from the root of Spiranthes australis (R. Brown) Lindl. J Nat Prod Res 21(7):641–645

Peng JY, Han X, Xu LN, Qi Y, Xu YW, Xu QW (2008) Two new prenylated coumarins from Spiranthes sinensis (Pers.) Ames. J Asian Nat Prod Res 10 (3):256–259

Rao TA (2007) Ethno botanical data on wild orchids of medicinal value as practised by tribals at Kudremukh National Park in Karnataka. Orchid Newsl 2(2):1–7

Rao TA, Sridhar S (2007) Wild orchids in Karnataka. A pictorial compendium. Institute of Natural Resources Conservation, Education, Research and Training (INCERT), Bangalore

Raskoti BB (2009) The orchids of Nepal. Bhakta Bahadur Raskoti and Rita Ale, Kathmandu

Rifai MA (1975) Extraordinary uses of orchids in Indonesia. Report First ASEAN Orchid Congress, Bangkok

Santapau H, Kapadia Z (1966) The orchids of Bombay. Government of India Press, Calcutta

Schuiteman A, de Vogel EF (2000) Cac Ci Ho Lan (Orchidaceae) Cua Thai Lan, Lao, Campuchia Va Viet Nam. Orchid Genera of Thailand Laos, Cambodia and Vietnam, Vietnamese-English edn. National Herbarium Nederland, Leiden

Seidenfaden G, Wood JJ (1992) The orchids of peninsular Malaysia and Singapore. Olsen & Olsen, Fredensborg

Seidenfaden G (1999) 149. Orchidaceae. In: Matthew KM (ed) The flora of the Palni Hills, South India, Part 3. The Rapinat Herbarium, St. Joseph's College, Tiruchirapalli

Shan XC, Liew EC, Weatherhead MA, Hodgkiss IJ (2002) Characterization and taxonomic placement of Rhizoctonia-like endophytes from orchid roots. Mycologia 94(2):230–239

Sood SK, Rana S, Lakhanpal TN (2005) Ethnic aphrodisiac plants. Scientific, Jodhpur

Subedi A, Kunwar B, Choi Y et al (2013) Collection and trade of wild-harvested orchids in Nepal. J Ethnobiol Ethnomed 9:64–73

Sun M (1996) The allopolyploid origin of Spiranthes hongkongensis (Orchidaceae). Am J Bot 83 (2):252–260

Sung CK (2002) Dendrobium moniliforme. In: Guo JX, Kimura T, But PPH, Sung CK (eds) International collation of traditional and folk medicine Northeast Asia, vol 4. World Scientific Publishing Company, Singapore, p 142

Tezuka Y, Ueda M, Kikuchi T (1989) Studies on the constituents of orchidaceous plants, VIII. Constituents of Spiranthes sinensis (Pers.) Ames var. amoena (M. Bieberson) Hara. (1) Isolation and structure elucidation of Spiranthol-A, Spiranthol-B, Spirasineol-A, new isopentenyldihydrophenanthrenes. Chem Pharm Bull 37(12):3195–3199

Tezuka Y, Ji L, Hirano H et al (1990) Studies on the constituents of orchidaceous plants IX. Constituents of Spiranthes sinensis (Pers.) Ames var amoena (M. Bieberson) Hara. (2) Structures of spiranthesol, spiranthoquinone, spiranthol-C and spirasineol-B new isopentenyldihydrophenanthrene. Chem Pharm Bull 38:629–635

Tsukaya H (1994) Spiranthes sinensis var. amoena in Japan contains two seasonally differentiated groups. J Plant Res 107(2):187–190

Uphof JCT (1968) Dictionary of economic plants. Verlag von J. Cramer, Lehre

Vaddhanaphuti N (2001) A field guide to the wild orchids of Thailand, 3rd edn. Silkworm Books, Chiang Mai

Van den Brink ROB (1937) Synopsis of the vernacular names and the economic use of the indigenous orchids of Java. Blumea 1(Suppl 1):38–51

Vij SP (1995) Orchid genetic diversity in India: conservation and commercialization. In: Proceedings of the 5th Asia Pacific orchid conference and show, Fukuoka, pp 20–39

Weckerle CS, Ineichen R, Huber FK, Yang YP (2009) Mao's heritage: medicinal plant knowledge among the Bai in Shaxi, China at a crossroads between distinct local and common widespread practice. J Ethnopharmacol 123:213–228

Wedeck HE (1961) Dictionary of aphrodisiacs. Philosophical Library, New York, p 216

Williams CA (1979) The leaf flavonoids of the Orchidaceae. Phytochemistry 18(5):803–813

Wood J (2001) Steveniella. In: Pridgeon AM, Cribb PJ, Chase MW, Rasmussen FN (eds) Genera Orchidacearum, vol 2, Orchidoideae (Part one). University Press, Oxford

Wu XR (1994) A concise edition of medicinal plants in China. Guangdong Higher Education Publication House, Guangdong (in Chinese)

Wu TL, Hu QM, Xia NH, Lai PCC, Yip KL (2002) Check list of Hong Kong plants 2001. Agriculture, Fisheries and Conservation Department Bulletin 1, Hongkong (Revised)

Zhonghua Bencao (2000) Edited by Health Department and National Chinese Medical Management Office. Science and Technology Press, Shanghai

Zhongyao Da Cidian (1986) Edited by Jiangsu New Medical College. Science and Technology Press, Shanghai

Genus: *Taprobanea* Christenson

Taprobanea is an epiphytic, monopodial, climbing herb with long internodes and roots arising from the lower portion of the plant. Leaves are numerous, distichous, rigid with apices smoothly but unequally bilobed rather than praemorse. This is a monospecific genus and the rest of the description is given in *T. spathulata* (Fig. 21.1). It thrives in full sun and is cold-intolerant. It is distributed in Sri Lanka and South India. DNA analysis showed that *Taprobanea* falls outside *the Vanda* clade.

Taprobanea spathulata (L.) Christenson.

Syn. *Vanda spathulata* (L.) Spreng

Indian name: *Ponnamponmaraiva* in the Malayalam dialect

Description: Stems are long, 30–60 cm slim, with green, black-spotted internodes, 2.5–3 cm long, climbing rather like an *Arachnis* among bushes on rocks. Compared with the average *Vanda*, plants of *Trapobanea* are small. Leaves are stiff, flattened, red-speckled, up to 22 by 3 cm. Inflorescence is slim, up to 36 cm long, bearing 6–10 flat, small, uniformly chrome-yellow, sometimes fragrant, long-lasting flowers which open in succession, a few at a time. Tepals are narrow. Flower is only 3.5 cm across (Fig. 21.2). Polyploidy occurs naturally in *T. spathulata*: $n = 19$, and plants with 38, 76 and 114 chromosomes have been reported (Abraham and Vatsala 1981). The clone of *Vanda spathulata* formerly used for breeding in Singapore was hexaploid, thus dominating the characteristics of its hybrids. *T. spathulata* blooms freely throughout the year when it receives strong sunlight in the tropics (Jayaweera 1981), but in south (west) India, Abraham and Vatsala (1981) reported that its flowering season is October to December, whereas Matthew (1995) reported that it flowers from June to September in the southeastern state of Tamil Nadu. Peak flowering at mid-Palnis is August to September (Seidenfaden 1999). It is "almost ever-blooming" in Kerala (Bose and Bhattacharjee 1980).

This is a common species in Sri Lanka and southern India. However, it is very specific regarding habitat and altitudinal range like several other orchid species which are distributed only in south India and Sri Lanka (Sathish Kumar and Manilal 1994). It occurs in semi-arid, desert plains at low elevations in the Western Ghats (Abraham and Vatsala 1981) and from the foothills to 1000 m in the adjacent Palni Hills of Tamil Nadu (Seidenfaden 1999).

Herbal Usage: *T. spathulata* is used as a substitute for *Vanda tessellata* to treat diseases involving the nerves, rheumatism and scorpion

© Springer International Publishing Switzerland 2016
E.S. Teoh, *Medicinal Orchids of Asia*, DOI 10.1007/978-3-319-24274-3_21

Fig. 21.1 *Taprobanea spathulata* (L.) Sprengel (as *Ponnampu maravara*). From: van Rheede, tot Drakenstein, H.A., *Hortus Indicus Malabaricus*, vol. 12, t.3 (1703). Courtesy of Missouri Botanical Gardens, St. Louis, USA

Fig. 21.2 *Taprobanea spathulata* (L.) Christenson (Photo: Suranjan Fernando)

Duggal 2009). Crushed leaves and stems are made into an ointment for treating various skin lesions (Lawler 1984).

Overview

Usage of *T. spathulata* (syn. *Vanda spathulata*) as a medicinal herb was described for the European public by Hendrik Adriaan van Rheede tot Drakenstein (1636–1691) in 1703 in his monumental *Hortus Indicius Malabaricus* (Fig. 21.1), but no mention of it appeared in any European language until more than 300 years later when it appeared in Chopra's *The Indigenous Drugs of India*. Van Rheede referred to this orchid as *Ponnampu maravara*, which is very close to its current name in Malayalam dialect of Kerala, *Ponnamponmaraiva*. He reported that the entire plant was used. It was pounded, boiled with rice and coconut juice, then mixed with honey, and administered to stop diarrhoea and dysentery. It was also used to correct biliary disorders. Pulverised flowers were used to treat tuberculosis, asthma and mania (Van Rheede 1693). Later writers reported similar usages, but it is unclear whether they were merely quoting van Rheede or that the orchid had continued to be used in such ways in the Malabar region. For instance, it was reported that powder prepared from the dried flowers is administered for consumption, asthma, and psychiatric disorders (Dymock et al. 1893; Trivedi et al. 1980; Singh and Duggal 2009), whereas juice from the plant tempered bile and abated frenzy (Dymock et al. 1893; Caius 1936; Yaganarasimha and Chelladurai 2004), while leaves were used to treat consumption, asthma and mania (Nadkarni 1954; Duggal 1971).

stings (Chopra 1933). It is also used as a cholagogue and antispasmodic. Juice from the plant was reported being used "to temper bile and abate frenzy" (Dymock et al. 1893; Caius 1936; Yaganarasimha and Chelladurai 2004). Leaves and powder prepared from dried flowers were used to treat consumption, asthma and mania (Dymock et al. 1893; Nadkarni 1954; Duggal 1971; Trivedi et al. 1980; Singh and

Genus: *Thrixspermum* Lour.

Chinese name: *Baidian Lan*

Thrixspermum is a large genus of horticulturally unimportant, monopodial, epiphytic orchids which are distributed from India to the Pacific. The name is derived from the Greek, *thrix* (hair) and *sperma* (seed). The flowers are usually few and ephemeral, but the species flower gregariously, possibly in response to a sudden temperature change.

Two species have been used as medicinal herbs in quite separate countries.

Thrixspermum centipeda Lour.

Thai name: *Kratai hu dieo, Ta khap lueang, Tin ta khap, Ueang maeng mumkhao*

Description: *T. centipeda* is an epiphytic species. Stems are 15–20 cm, stout, terete, arching or pendulous, with thick, oblong leaves 6–24 by 1–2.5 cm. Inflorescence is borne directly opposite a leaf and carries several ephemeral, yellow flowers that emerge one or two at a time. Tepals are narrow, 3–5 cm long, pale to bright yellow and not spread out, barely opening. Lip is 0.5 cm wide, white with reddish-orange spots. The prominent bracts on the flattened inflorescence provide an impression of a green, segmented arthropod and inspired the species name *centipeda* (Fig. 21.3). Flowering period is June to July in China (Chen et al. 1999); July to September (Vaddhanaphuti 2001) or October to November (Nanakorn and Watthana 2008) in Thailand. *T. centipeda* is widely distributed from Bhutan eastwards to Hong Kong and southwards to Indonesia, occurring in hilly, broad-leaved forests at 700–1200 m.

Usage: The whole plant is used to treat asthma in Thailand (Chuakul 2002).

Thrixspermum pardale (Ridl.) Schlecht.

Sacrochilus pardalis, Ridl., *Dendrocolla pardalis* (Ridl) Ridl.

Description: Stem is 25 cm long with leaves 8 by 1 cm. Scape is thin, 25 cm long, ending in a 5-cm rachis bearing numerous small flowers, 1.4 cm in diameter, off-white to cream with brown spots in the Sumatran variety and purple-spotted in the Malaysian variety. Flowers open in succession, one to a few at a time, and smell of European Meadowsweet (*Spirea almaria*). It is

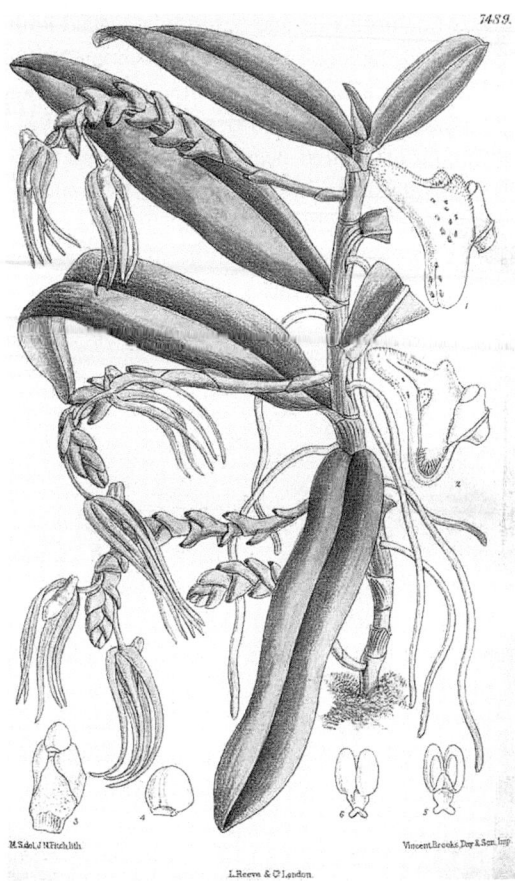

Fig. 21.3 *Thrixspermum centipeda* Lour. (as *Sarcochilus hainanensis* Rolfe). From: *Curtis Botanical Magazine*, vol. 122 [ser. 3, vol. 52]:t. 7489 (1896) (original drawing in colour by Matilda Smith). Courtesy of Missouri Botanical Gardens, St. Louis, USA

distributed in Sumatra, Peninsular Malaysia and Borneo, in the first locality at 1000 m (Comber 2001). In Peninsular Malaysia, it was once common on village trees in Perak and Pahang (Holttum 1964; Seidenfaden and Wood 1992).

Herbal Usage: Leaves of *T. pardale* were pounded with leaves of *Hetaeria oblique* to make a poultice for treating ulcers of the nose by villagers in Malaysia (Burkill 1935).

Overview

Luning (1974) investigated six species of *Thrixspermum* but did not find any significant amount of alkaloid in the plants.

Flowers of *T. centipeda* (*T. arachnites*) are strongly scented. Main components of the scent are *cis*-linalool oxide and nerolidol overlaid with (E,Z)-2,4-decadienal, ethyl(Z)-4-decenoate, ethyl (E,Z)-2,4-decadienoate with the corresponding (E,E)-isomers and *gamma*-decalactone. The typical aroma of Bartlett pears is produced by ethyl (E,Z)-2,4-decadienoate and the corresponding methyl ester. Ethyl(E,Z)-2,4-decadienoate is the "pear ester" which is used in the flavouring and perfume industries (Kaiser 1993).

Genus: *Thunia* Rchb. f.

Chinese name: *Sun Lan*

Thunia are large to medium-sized terrestrial orchids that are devoid of pseudobulbs. Stems are erect, clustered and covered with sheaths below and widely-spaced leaves above. Inflorescence is terminal, several-flowered, the flowers cymbidiform, showy but short-lived, white or amethyst with yellow on the lip. There are about six species distributed in India, China and Southeast Asia.

Thunia alba (Lindl.) Rchb. f.

Thunia marshalliana Rchb. f., *Phaius albus* Lindl.; *P. marshallianus* Rchb.f.; N.E. Brown

Chinese name: *Sun Lan* (bamboo orchid)
Chinese medicinal names: *Yan Sun* (rock bamboo); *Yanjiao* (rock horn); *Shizhuzi* (stone bamboo) *Jiegudan* (fracture union pill); *Shisun* (rock bamboo); *Yanzhu* (rock bamboo)
Thai names: *Chang Nga Dieo* (in general), *Phothuki* (in Karen Mae Hong Son), *Sawet sot si* (in Bangkok), *Ueang nga chang* (in Chiang Mai)
Nepalese names: *Goliano*

Description: Stems are 60 (30–100) cm tall, clustered, erect, covered with sheaths below and leafy above; grass-like. Leaves are deciduous, thin, elliptic, with a bluish, waxy surface, 10–20 by 2.5–5 cm. Inflorescence is terminal, 4–10 cm long with 2–7 pendulous flowers towards the

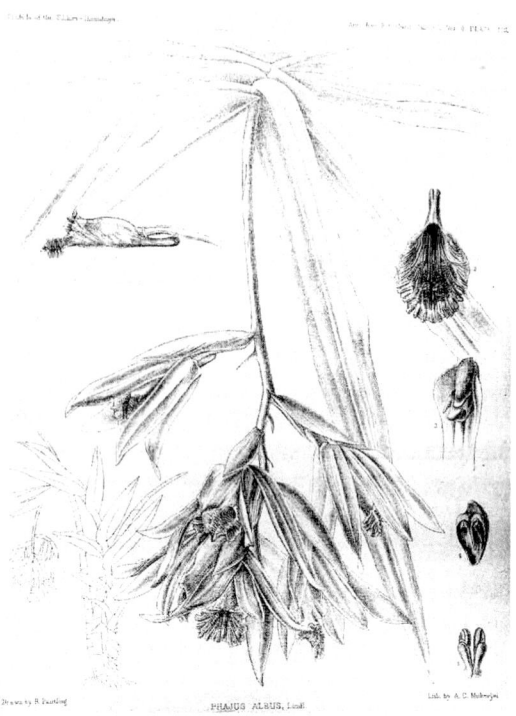

Fig. 21.4 *Thunia alba* (Lindl.) Rchb. f. (as *Phaius albus* Lindl.). From: *Annals of the Royal Botanical Gardens, Calcutta*, vol. 8 (3): t. 153 (1891) (drawing by R. Pantling). Courtesy of Missouri Botanical Gardens, St. Louis, USA

end, with large, persistent, pale green or white bracts (Fig. 21.4). Tepals are white, 6 cm long, narrow, pointed and not well extended. A large yellow lip encloses the column. It is crenated at the margins. The centre of the mid-lobe bears several keels or crests which are orange to maroon (Fig. 21.5). *T. alba* var. *bracteata* (Roxb.) Pearce and Cribb from the southern Himalayas lacks the yellow patch on the lip (Chen and Wood 2009). It flowers in June in China (Chen et al. 1999) and from March to August, sometimes to November, in Bhutan, Darjeeling and Sikkim (Pearce and Cribb 2002), but in January in Thailand (Vaddhanaphuti 2001).

T. alba is usually lithophytic, sometimes found in the hollows of large branches of trees growing in rocky soil. It enjoys a wide distribution from India, through Nepal, Bhutan, Myanmar, eastern Tibet, southwest Sichuan and southern Yunnan to Thailand and northern

Fig. 21.5 *Thunia alba* (Lindl.) Rchb. f. (Photo: Bhaktar B. Raskoti)

Table 21.1 Four Chinese herbal prescriptions that employ *Yan Sun* (*Thunia alba*) (*Zhongyao Da Cidian*, Anonymous, 1986)

1. Indication: cough and dyspnea, or asthma
Boil *Yan Sun* (*T. alba*) 30 g and consume
(*Quizhou Herbs*)
2. Indication: bone fracture
Fry fresh *Yan Sun* with wine and add egg white. Apply externally
(*Yunnan Chinese Herbs*)
3. Indication: closed bone fracture
Mix *Yan Sun*, *Shi Bang Dui* and *Shan Pi Cai* in wine and apply externally
(*Yunnan Si Mao Selected Chinese Herbs*)
4. Indication: external traumatic wounds, knife wounds
Boil *Yan Sun* 9–15 g and consume. Use fresh meshed *Yan Sun* for external application
(*Yunnan Chinese Herbs*)

Peninsular Malaysia (Kedah) (Seidenfaden and Wood 1992; Chen et al. 1999).

Herbal Usage: Herb is obtained from Yunnan, Guizhou, Sichuan and Tibet. It may be collected throughout the year. It is used fresh. For storage, it is first boiled, then dried. Chinese herbalists use the whole plant to enrich *yin*, benefit the lungs, clear phlegm and stop coughs, improve blood flow, remove bruises and assist the healing of fractures (Zhongyao Da Cidian 1986; Wu 1994). It dissolves extravasated blood and improves circulation (Chen and Tang 1982). Prescriptions for these conditions are listed in Table 21.1.

In Nepal, a paste made from the entire plant is used to help set fractures (Manandhar and Manandhar 2002; Subedi et al. 2013).

Phytochemistry: Thunalbene, a stilbene derivative, 3,3′-dihydroxy-5-methoxystilbene together with six stilbenoids, batatasin-III, lusianthridine, 3,7,-dihydroxy-2,4-dimethoxy-phenanthrene, 3,7-dihydroxy-2,4,8-trimethoxy-phenanthrene, cirrhopetalanthrin and flavanthrin were isolated from *T. alba* by Majumder et al. (1998). Their biological activities were not described. Luning (1974) did not find any alkaloid in the two species of *Thunia* that he tested.

Overview

Among the numerous Chinese *Materia Medica* that were referenced, *T. alba* was found in only three contemporary and extremely comprehensive lists (Anonymous, 1986, Zhongyao da Cidian 1986; Wu 1994; Zhonghua Bencao 2005). This supports the conclusion that its usage is provincial and until recently confined to Yunnan and Quizho

Genus: *Trias* Lindl.

The generic name *Trias* refers to the triangular shape of the open flower formed by the three large sepals (Fig. 21.6). Petals are minute. Lip is normal size and of variable shapes. Plants are small, pseudobulbous and epiphytic. *Trias* is very close to *Bulbophyllum* but differs from the latter in having a drawn-out appendage of variable shape on the anther cap (Seidenfaden and Smitinand 1960). The genus consists of fewer than ten species with six in Thailand, and the rest distributed in Vietnam, Myanmar and India.

***Trias disciflora* (Rolfe) Rolfe**

Bulbophyllum disciflorum Rolfe

Fig. 21.6 *Trias oblonga* Lindl., illustrating typical vegetative form and flower of *Trias* (Photo: E.S. Teoh)

Thai name: *Ma tak kok*

Description: This is a small epiphytic orchid. The short, ovoid pseudobulb carries a single, thick succulent leaf, 4–5 cm long, or longer. Inflorescence bears a single flower with the form of a narrow isosceles triangle, 2 cm across. Floral form is constituted by the large triangular sepals which are light greenish-yellow speckled with minute, reddish-brown spots all over. Lip is oval and covered with tiny purple-brown warts. It flowers in January (Vaddhanaphuti 2005). It occurs in the northeast of Thailand.

Usage: The whole plant is used as a tonic (Chuakul 2002).

Trias nasuta (Rchb. f.) Stapf.

Bulbophyllum nasutum Rchb. f.

Local Thai name: *Kratai khao, Kratai hu dieo, Ueang nok kra chip*

Description: A small epiphyte with ovoid pseudobulbs 1.5–2 cm tall and 1–1.5 cm in diameter, each bearing a single leaf that is elliptical, thick, pointed at the tip, and measuring 6–10 cm

long and 1.2–1.5 cm wide. Yellow flowers arise from the base of the pseudobulbs, singly or in pairs. The lip is yellow with red or purple on either side of the throat. Flowering season is November to December. It is found in evergreen forest in the eastern and southern Thailand, in Myanmar and Vietnam (Nanakorn and Watthana 2008).

Usage: The whole plant is used to treat asthma (Chuakul 2002).

Overview
Trias is not widely cultivated except by collectors of botanicals in Thailand. Not much is known about its member species, and there is no pharmacological information. *Trias* is closely related to *Bulbophyllum* and, therefore, it would be worthwhile to investigate its chemical constituents.

Genus: *Tropidia* Lindl.

Chinese name: *Zhujing Lan*
Japanese name: *Nettai Ran*

Tropidia is a genus of terrestrial orchids with erect, branching stems sheathed by bracts and leaves, and fibrous roots. It derives its name from the Greek *tropidios* (a ship's keel), which refers to the appearance of the lip is some species (Holttum 1964). The genus has some 30 member species, one in central America, and the rest scattered over the Indian subcontinent, southern China, Southeast Asia and Japan (Seidenfaden and Wood 1992). Only one species is used medicinally (Fig. 21.7).

Tropidia curculigoides Lindl.

Syn. *Tropidia graminea* Bl.; *T. formosana* Rolfe; *Schoenmorphus capitatus* Gagne.

Chinese name: *Duansuizhujing Lan*. In Hong Kong: bamboo stemmed orchid
Malay names: *Serugut, Ranchang hantu*

Fig. 21.7 *Tropidia curculigoides* Lindl. From: *Annals of the Royal Botanic Gardens, Calcutta*, vol. 8 (3): t. 365 (1891) (drawn by R. Pantling). Courtesy of Missouri Botanical Gardens, St. Louis, USA

Description: This is a very unusual terrestrial orchid which has the appearance of bamboo but with large, lanceolate, plicate leaves, 16 by 3 cm, which are widely spaced along the erect stem. Stems are up to 60 cm tall, thin with internodes 2–5 cm length. Inflorescences are numerous, axillary and terminal, very short, with a few white flowers tipped with green, tightly packed, about 2.5 cm across. *T. graminea* is a young version of *T. curculigoides* (Comber 2001). The species has no horticultural value. In Bhutan, it flowers in May (Pearce and Cribb 2002), in China, it flowers in June to August (Chen et al. 2009) and in northeast India, November (Bose and Bhattacharjee 1980). It is distributed from India, Bhutan and Bangladesh across Southeast Asia to southern China, Taiwan, Australia and Fiji. In most places, it is found in lowland forests (Tang and Su 1978). It occurs in Bhutan at 300–800 m in forests (Pearce and Cribb 2002).

Usage: A decoction of the plant is used to treat patients during the cold stage of malaria and decoction of the roots is used for diarrhoea in Peninsular Malaysia and India (Burkhill and Haniff 1930; Burkill 1935; Hawkes 1944; Rao 2004). *Ardisia* (not an orchid) is added to prepare the decoction which is administered to patients suffering from malaria (Jain and Defilipps 1991). It has a similar usage in Bangladesh (Musharof Hossain 2009). The species is endangered in India because of this medicinal usage for a common illness (Duggal 1972).

There is no medicinal usage for *T. curculigoides* in China.

Overview The antimalarial usage of *T. curculigoides* is interesting because fever in malaria is intermittent, occurring on alternate days, every third day (tertian malaria) or precisely every 72 h (quartan malaria), and lasting 12 h each time (Sandosham 1959). Thus, even as the disease is progressing, there are days without fever, and this could possibly explain why *Tropidia* "works". Nevertheless, *T. curculigoides* should be properly investigated before its medicinal value is dismissed. With respect to malaria, some strains are already developing resistance to artemisin so a new compound would be useful. Currently, pharmacological data on *Tropidia* are not available.

Tulotis asiatica Hara (see ***Platanthera souliei Kraenz.***)
Tulotis ussuriensis (Regel) Hara (see ***Platanthera ussuriensis***)

References

Abraham A, Vatsala P (1981) Introduction to orchids, with illustrations and descriptions of 150 South Indian orchids. TPGRI, Trivandrum

Bose TK, Bhattacharjee SK (1980) Orchids of India. Naya Prokash, Calcutta

Burkhill IH, Haniff M (1930) Malay village medicine. Gard Bull Straits Settlements 6:165–321

Burkill IH (1935) A dictionary of economic products of the Malay Peninsula, vol 2. Crown Agents for the Colonies/Ministry of Agriculture and Co-operatives,

London/Kuala Lumpur (1966 reprint, 2nd edn., with contributions by Birtwistle W, Foxworthy FW, Scrivenor JB, Watson IG)

Caius JF (1936) The medicinal and poisonous plants of India. J Bombay Nat Hist Soc 38(4):791–799

Chen SC, Tang T (1982) A general review of the orchid flora of China. In: Arditti J (ed) Orchid biology, vol 2, Reviews and perspectives. Cornell University Press, Ithaca

Chen SC, Tsi ZH, Luo YB (1999) Native orchids of China in colour. Science Press, Beijing

Chen XQ, Wood JJ (2009) Thunia H. G. Reichenbach. In: Chen XQ, Zj L, Zhu GH et al (eds) Flora of China—orchidaceae. Science Press, Beijing, p 315

Chen XQ, Gale SW, Cribb PJ (2009) Tropidia Lindley. In: Chen XQ, Zj L, Zhu GH et al (eds) Flora of China—orchidaceae. Science Press, Beijing, pp 195–197

Chopra RN (1933) The indigenous drugs of India. The Art Press, Calcutta

Chuakul W (2002) Ethnomedical uses of Thai orchidaceous plants. Mohidol Univ J Pharm Sci 29(3-4):41–45

Comber JB (2001) Orchids of Sumatra. Natural History Publications, Kota Kinabalu (Borneo)

Duggal SC (1971) Orchids in human affairs (a review). Pharm Biol 11(2):1727–1734

Duggal SC (1972) Orchids in human affairs (a review). Pharm Biol 11(2):1727–1734

Dymock W, Warden CJH, Hooper D (1893) A history of the principal drugs of vegetable origin met with in British India. Education Society Press, Bombay

Hawkes AD (1944) Orchids in human affairs (a review). Pharm Biol 11(2):1727–1734

Holttum RE (1964) Orchids of Malaya, 3rd edn. Government Printers, Singapore

Jain SK, Defilipps RA (1991) Medicinal plants of India, vol 2. Reference Publications Inc., Algonac

Jayaweera DMA (1981) A revised handbook of the flora of Ceylon, vol 2. A.A. Balkema, Rotterdam

Kaiser R (1993) The scent of orchids: olfactory and chemical investigations. Editiones Roche, Basel

Lawler LJ (1984) Ethnobotany of the Orchidaceae. In: Arditti J (ed) Orchid biology reviews and perspectives, vol 3. Cornell University Press, Ithaca

Luning B (1974) Alkaloids of the Orchidaceae. In: Withner CL (ed) The orchids. Scientific studies. Wiley, New York

Majumder PL, Roychowdhury M, Chakraborty S (1998) Thunalbene, a stilbene derivative from the orchid Thunia alba. Phytochemistry 49:2375–2378

Manandhar NP, Manandhar S (2002) Plants and people of Nepal. Timber, Portland

Matthew KM (1995) An excursion flora of Central Tamil Nadu, India. A.A. Balkaema, Rotterdam

Musharof Hossain M (2009) Traditional therapeutic uses of some indigenous orchids of Bangladesh. Med Aromat Plant Sci Biotechnol 3:100–106

Nadkarni AK (1954) Dr. K.M. Nadkarni's Indian Materia Medica, vol 2, 3rd edn. Popular Book Depot, Bombay

Nanakorn W, Watthana S (2008) Queen Sirikit botanic garden (Thai native orchids 1 and 2). Wanida Press, Chiang Mai

Pearce NR, Cribb PJ (2002) The orchids of Bhutan. Royal Botanic Gardens/Royal Government of Bhutan, Edinburgh/Thimpu

Rao AN (2004) Medicinal orchid wealth of Arunachal Pradesh. Newsl ENVIS NODE Indian Med Plants 1 (2):1–7

Sandosham AA (1959) Malariology, with special reference to Malaya. University of Malaya Press, Singapore

Sathish Kumar C, Manilal KS (1994) A catalogue of Indian orchids. Bishen Singh Mahendra Pal Singh, Hehra Dun

Seidenfaden G (1999) 149. Orchidaceae. In: Matthew KM (ed) The flora of the Palni Hills, South India, Part 3. The Rapinat Herbarium, St. Joseph's College, Tiruchirapalli

Seidenfaden G, Smitinand T (1960) The orchids of Thailand. A preliminary list, Part II. The Siam Society, Bangkok

Seidenfaden G, Wood JJ (1992) The orchids of peninsular Malaysia and Singapore. Olsen & Olsen, Fredensborg

Singh A, Duggal S (2009) Medicinal orchids: an overview. Ethnobot Leaft 13:351–363

Subedi A, Kunwar B, Choi Y et al (2013) Collection and trade of wild-harvested orchids in Nepal. J Ethnobiol Ethnomed 9:64–73

Tang SL, Su HJ (1978) Flora of Taiwan, vol 5. Department of Botany, Taiwan National University, Taipei

Trivedi VP, Dixit RS, Lal VK (1980) Orchids in the drug markets of Bareilly, Kanpur and Nearby Districts. Nagarjun (Calcutta) 23(8):157–163

Vaddhanaphuti N (2001) A field guide to the wild orchids of Thailand, 3rd edn. Silkworm Books, Chiang Mai

Vaddhanaphuti N (2005) A field guide to the wild orchids of Thailand, 4th and Expanded edn. Silkworm Books, Chiang Mai

Van Rheede HA (1693) Hortus indicus malabaricus, vol 12. Dutch East India Company, Kerala

Wu XR (1994) A concise edition of medicinal plants in China. Guangdong Higher Education Publication House, Guangdong (in Chinese)

Yaganarasimha SN, Chelladurai V (2004) Medicinal plants of India, vol 2. Regional Research Institute, Bangalore

Zhonghua Bencao (2005) Edited by Health Department and National Chinese Medical Management Office. Science and Technology Press, Shanghai, 2000

Zhonyao Da Cidian (1986)

Genus: *Vanda* R. Br.

Chinese name: *Wandai Lan*

Vanda is a Sanskrit name for *Vanda tessellata*, an outstanding Indian species with diverse forms. In Sanskrit *Vanda* should be pronounced as *Wanda*. This is a popular horticultural genus distributed from India through Southeast Asia and Yunnan. Its member species are generally inter-fertile with members of numerous genera in the *Vandeae* Tribe. Formerly, *Vanda* was divided into two groups according to the shape of their leaves, namely strap leaf *Vanda* and terete leaf *Vanda*. The latter is now recognised as a separate genus under the name *Papilionanthe* Schltr. *V. amesiana* has been reassigned as *Holcoglossum amesiana*, and *V. spathulata* is *Taprobanea spathulata*. Adopting this reclassification, they are not listed under *Vanda* but are listed separately in the present work.

Vanda are medium-sized to large epiphytic plants with branching erect stems sheathed by the bases of glabrous, flat, duplicate leaves which are arranged alternately in two rows in a single plane. Inflorescences are axillary, with several to numerous monochromatic or multi-coloured, sometimes spotted or tessellated flowers which are showy and resupinate.

Vanda coerulea Griff. ex Lindl.

Chinese name: *Dahuawandai Lan* (big flower ten thousand generation orchid, Large *Vanda*)
Thai name: *Fa Mui* (in the north), *Pho don ya*, *Pho thong* (Karen Mae Hong Son)
Myanmar Name: *Moe lone hmine*
Indian name (Tirap District, Arunachal Pradesh): *Rangpu*

Description: Stem is up to 50 cm tall, 1–1.5 cm in diameter. Leaves are distichous, rigid, oblong, curving laterally, 15–18 cm by 1.7–2 cm, apex slightly emarginated. Inflorescence is subapical, 20–35 cm long, sometimes branched. Flowers are pale blue or, rarely, pink, and commonly tessellated. Solid-coloured forms occur but they are extremely rare. Flowers are 7.5–10 cm across, round, usually 10–15 in number and well-spaced on the inflorescence, with the rachis borne above the uppermost leaves. Several line-bred strains produce flowers of excellent shape with overlapping sepals and petals (Figs. 22.1 and 22.2).

Much sought after and overcollected in the past, *V. coerulea* is an endangered epiphytic species, and it was briefly assigned to CITIES Appendix I. It is now back in CITES Appendix II, meaning that it is thriving in the wild, especially in Myanmar and northern Thailand.

© Springer International Publishing Switzerland 2016
E.S. Teoh, *Medicinal Orchids of Asia*, DOI 10.1007/978-3-319-24274-3_22

Fig. 22.1 *Vanda coerulea* Griff. Var. *rothschildii*. From: Warner, R., Williams,B.S., *The orchid album*, vol 1887: t. 517 (1887). Courtesy of Missouri Botanical Gardens, St. Louis, USA

Fig. 22.2 *Vanda coerulea* Griff. ex Lindl. (Photo: E.S. Teoh)

V. coerulea forms the backbone of most blue and red *Vanda* hybrids. A pure white variety occurs in Myanmar, but it is rare. *V. coerulea* is common in Northern Thailand and Myanmar (Khasia and Jyntea Mountains at 1300 m, growing on oak trees), but the species is uncommon in India and southern Yunnan. It grows on tree trunks in sparse forests near streams at 1000–1600 m. Flowering season is June or July to December with a peak in August in Thailand (Vaddhanaphuti 2001; Nanakorn and Watthana 2008) and Myanmar (Grant 1895; Tanaka and Yee 2003). It flowers from October to November in Yunnan (Chen et al. 1999). In the east Indian state of Manipur, *V. coerulea* is known as the "September Orchid" because the flowers first appear in September, the blooming season extending to January (Jojita Devi and Ghatak 1986). Flowers usually last for 6 weeks on the plant (Grant 1895). Dr. Yoshitaka Tanaka who has made a detailed study of the wild orchids in the Shan State of Myanmar reported coming across a *V. coerulea* with an inflorescence 1.5 m in length, bearing 300 blossoms. His photographs of *V. coerulea* growing in the wild are spectacular.

Repeated selfing of *V. coerulea* in Bangkok has greatly improved floral quality, and many clones flower when the plants are relatively young. Additionally, they may flower throughout the year. Under cultivation in the USA and Europe, peak flowering of *V. coerulea* occurs from August to November (Hamilton 1990). In Arunachal Pradesh, where some orchids are regarded as sacred, none is more venerated than the *Rangpu* (*V. coerulea*) which is associated with worship and the festivals of Vanchoo tribals in Tirap District (Bennet 1992). Unfortunately, such respect did not prevent the orchid from becoming endangered in the state.

Phytochemistry: Phytosterols, terpenoids, carbohydrates and stilbenoids were identified in extracts of *V. coerulea* stems. Further purification showed the principal stilbenoids to be imbricatin, methoxycoelonin and gigantol. These were shown to possess anti-oxidant properties and inhibited COX-2 production and activity in ultraviolet-irradiated skin cells, thus suggesting that they could protect skin against the harmful effects of sunlight and pollution (Simmler et al. 2009, 2010).

Herbal Usage: Juice prepared from the leaves is used to treat dysentery or diarrhoea, and applied for skin diseases (Nadkarni 1954; Rao 2004). It is also reported to be used extensively for (lesions of the) eye in northeastern India (Medhi and Chakrabarti 2009); however, the authors did not explain the indications and manner of its usage.

Vanda concolor Blume

Chinese name and medicinal name: *Qinchunwandai Lan*
Thai name: *Khao kae* (in Bangkok)

Description: Stems are 4–13 cm, 0.4–1 cm in diameter. Leaf is leathery, 15–30 by 1–3 cm, unequal at the tip. Inflorescence is axillary, arching, 11–17 cm long, sparsely 4- to 8-flowered. Flowers are heavy textured, well spread, 3–4 cm in diameter, fragrant, white, flushing to yellowish-brown, with yellow stripes and tessellations on the sepals and petals. Petals are subspatulate, with an undulating margin. Lip is trilobed, the same length as the sepals, white, spotted with purple. It flowers in April and May (Chen and Bell 2009). A small *Vanda* growing on tree trunks or rocks at forest margins, *V. concolor* is found only in Guangxi, southwest Guizhou, southern Yunnan and Vietnam.

Herbal Usage: Chinese herbal medicine maintains that *V. concolor* detoxifies and removes dampness. It is used to treat peripheral neuritis and ulcerative swellings such as carbuncles (Hu et al. 2000).

Vanda cristata Wall ex Lindl.

Syn. *Trudelia cristata* (Wall ex Lindl.) Senghas

Chinese name: *Chachunwandai Lan*
Nepali name: *Bhyagute phul* in Nepali dialect, *Vashgute phul*
Myanmar name: *Jyo koke thitkhwa*

Description: This is a common, epiphytic, vandaceous orchid, sometimes occurring on rocks in the Himalayas. Plant is 30 cm tall. leaves are oblong-linear, 7.5–12.5 by 0.6–1.8 cm (Fig. 22.3). Inflorescence is about the length of the leaves and carries up to 5 or 6 flowers with narrow, incurved sepals and petals which are yellow or green, and a white or golden lip that is striped with purple. Lip of the type flower is bifurcated at the tip (Fig. 22.4), whereas variety *multiflora* lacks the bifurcation, intermediate forms are common (Gurong 2006). It flowers in March to May in Nepal (Raskoti 2009), May and June in northern India, from July to August in the Shan state of Myanmar and May in Yunnan (Chen and Bell 2009). The Myanmar strain has fewer flowers, usually between one and three. *V. cristata* is distributed in Nepal, Bhutan, northern India, Tibet and southwest Yunnan at 1300–2000 m (Chen and Bell 2009). In Bhutan and northeast India, it commonly occurs on the trunks of *Rhododendron arboretum* and *Skimmia*

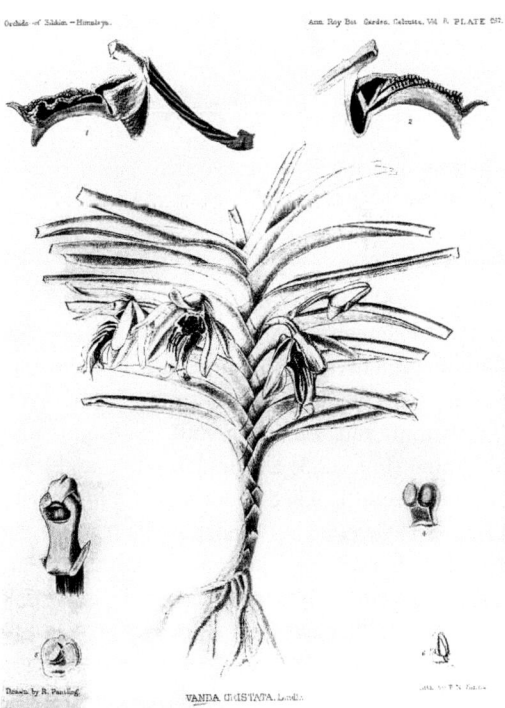

Fig. 22.3 *Vanda cristata* Wall. ex Lindl. From: *Annals of the Royal Botanic Gardens, Calcutta*, vol. 8 (3): t. 287 (1891). Drawing by R. Pantling. Courtesy of Missouri Botanical Gardens, St. Louis, USA

Fig. 22.4 *Vanda cristata* Wall ex Lindl. (Photo: E.S. Teoh)

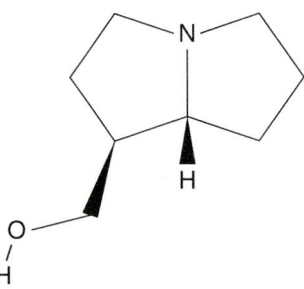

Fig. 22.5 Laburnine, an alkaloid from *Vanda cristata* with antimicrobial properties

spp. (Pearce and Cribb 2002). It likes bright light (Raskoti 2009).

Phytochemistry: The 1-hydroxymethylpyrrolizidine ester, laburnine acetate was isolated from *V. cristata* in 1969 (Lindstrom and Luning 1969). Subsequently, laburnine (Fig. 22.5) and lindelofidine and their acetates were isolated from other *Vanda* and *Vandopsis* species (Brandange and Granelli 1973). Laburnine is a poisonous alkaloid originally isolated from unripe seeds of *Laburnum anagyroides*. It possesses antimicrobial activity against fungi and *Shigella* (Li et al. 2011).

Herbal Usage: Plant is made into a paste to treat cuts and wounds in India. In Nepal, a paste made with the roots is used for treating boils and dislocated bones (Manandhar and Manandhar

2002). Leaf powder is used as an expectorant, whereas leaf paste is applied on cuts and wound (Subedi et al. 2013). Leaves are also used to make a tonic and expectorant in the northeastern region of India (Duggal 1971; Medhi and Chakrabarti 2009). It is used as a nutrient and tonic for general debility in Uttar Pradesh (Trivedi et al. 1980).

Vanda tessellata (Roxb.) Hook. ex G. Don

Vanda roxburghii R. Br.

Sri Lankan names: Anuradhapura Orchid, Grey Orchid, *Rat-tha, Arattha, Rasni*

Indian names: *Naguli* (Tamil), *Banki* (Orissa), *Nai* (Bengal), *Bandanike* (Canaarese), *Banda, Nai, Perasara, Persara, Vanda, San, Alisara* (Hindi), *Atirasa, Bhujangakshi, Chhatraki, Dronagandhika, Elaparni, Gandhanakuli, Muktarasa, Nakuleshta, Nakuli, Palankapa, Rasadhya, Rasana, Rasya, Sarpagandha, Shreyasi, Suggandha, Sugandhimula, Surasa, Suvaha, Vandaka, Vriksharuha, Yuktarasa* (Sanskrit), *Darebanki* (Santal), *Chittiveduri, Kanapabadanika, Mardaru, Vadanika* (Telugu), *Banda* (Urdu), *Knapachettu* (Madras Presidency in 1933), *Jarakindu, Japa* (Madhya Pradesh), *Ippa vajrnika* (Gonds of Andra Pradesh), *Vandekigidda* (Karnataka), *Maravazha* (Kerala)

Indian medicinal names: *Rasna* in Sanskrit, Hindi, Marathi, Bengal and Mumbai. However, *Rasna* refers more commonly to *Pluchea lanceolata* which is not an orchid. *Rasha* (Unani)

Nepalese names: *Parajiva, Rasna*

Arabic name: *Kharkittan*

Persian name: *Zanjabeel-e-shami*

Thai name: *Ueang sam poi India* (in Chiang Mai), *Ueang suea lek* (in Bangkok)

Description: *V. tessellata* is a robust, lowland, epiphytic, vandaceous orchid with stout stems and tough strap leaves. Stems are 30–60 cm tall, simple, its lower, leafless portion ensheathed by remnants of old leaf bases. Branching roots arise predominantly from the lower half of the stem. Leaves are distichous, coriaceous, thick, strapped and praemorse, 15–20 by 1.7–2.4 cm. Inflorescence is axillary, 15–20 cm long with 4–10 flowers, 5 cm across, expanded, commonly grey, greyish-blue, buff-yellow or red, some with light tessellations on the sepals and petals, waxy and of good substance. Margins of the sepals and petals are wavy (Fig. 22.6). Lip is funnel-shaped, trilobed. Mid-lobe is ridged, oval with a blunt, bifid apex and constricted below the bifid apex.

It occurs in Sri Lanka (once common in the forests surrounding Anuradhapura), India (from the peninsula to Bengal and Bihar), Nepal, and Myanmar (Tenasserim), close to human habitation and seldom in dense forests. It is abundant in the sacred groves of West Bengal (Basu 2010), forming massive clumps on tree trunks or rocks. Although confined to the dry zone in Sri Lanka, there it is a well-known species noted for its strongly scented flowers which present in a wide range of colour forms, with over 50 shades of colour, from red to pink, grey, yellow, green and blue (Jayaweera 1981). A popular form has petals and sepals of ash grey, tinged with blue, overlaid with tessellations of *café-au-lait*, and a contrasting, bright purple lip. The two best known varietal forms are the variety with bronze-pink to rose-pink flowers called var. *rufescens*, and the yellow variety known as var. *lutescens* (Figs 22.7, 22.8, 22.9 and 22.10). Leaves are recurved and channelled (Ekanayake 1975).

V. tessellata has a wide tolerance for sunlight, thriving in 27–93 % sunlight, high temperatures 26–31 °C and low relative humidity (31–44 %). Included among the numerous host plants in Sri Lanka are *Bauhinia tomentosa*, *Eugenia* sp., *Tamarindus indicus*, *Madhuca longifolia*, *Thespesia papulnea*, *Strychnos nux-vomica*, *Manilkara hexandra* and the non-native *Samanea saman* (Jayaweera 1981). Local names may allude to its habitat, for instance *Amara-vanda* refers to *V. tessellata* which is found growing on mango trees (*amara*) (Dymock et al. 1893).

PLATE No. 931

7

8

VANDA ROXBURGHII, *Br.*

Fig. 22.6 *Vanda tessellata* (Roxb.) Hook. ex G. Don (as *Vanda roxburghii*). From: Kirtika, K.R., Basu, B. D., *Indian Medicinal Plants*, *Plates*, vol. 5: t931 (1918). Courtesy of the Smithsonian Institute, Washington, USA

Fig. 22.7 *Vanda tessellata* (Roxb.) Hook. ex G. Don, popular cultivated form (Photo: E.S. Teoh)

Fig. 22.8 *Vanda tessellata* (Roxb.) Hook. ex G. Don, with a fuller flower (Photo: E.S. Teoh)

Fig. 22.10 *Vanda tessellata* (Roxb.) Hook. ex Don, yellow colour form (Photo: E.S. Teoh)

Fig. 22.9 *Vanda tessellata* (Roxb.) Hook. ex G. Don, another popular cultivated form (Photo: E.S. Teoh)

V. tessellata flowers in January, March to August and December in Sri Lanka (Jayaweera 1981), April to May in the Western Ghats (Abraham and Vatsala 1981), March to June at Mumbai (Santapau and Kapadia 1966), July to August in Tamil Nadu (Matthew 1995), with flowers peaking May to June at the Palni foothills

(Seidenfaden 1999). The orchid is disappearing in India due to heavy deforestation of their host trees (Pandey et al. 2003).

Phytochemistry: *V. tessellata* contains alkaloids (Dymock 1943), tannins, resin, saponin, beta and gamma sitosterols, fatty oil and colouring agents (Chopra et al. 1956). It had an anti-arthritic action in albino rats (Prasad and Achari 1966). In experimental animals, a glucoside from the plant, produces a transient initial rise in blood pressure which is then followed by a sustained fall: heart rate slows, cardiac output diminishes, and peripheral arterioles dilate. Atropine antagonises some of the effects, but does not entirely abolish them (Chopra et al. 1958). An extract of the entire plant produced an anti-acetylcholine effect on rats. Its LD50 exceeded 1000 mg/kg body weight when administered intraperitoneally (Bhakuni et al. 1969). Extract of *V. tessellata* assisted wound healing in rats (Nayak et al. 2005). *V. roxburghii* (=*V. tessellata*) contains two anti-inflammatory compounds, hepcosame (C_{27} H_{56}) and octacosonol (C_{28} H_{58} O).

Tessallatin, a phenanthropyran, was isolated from *V. tessallata* and found to be 3,-dihydroxy-2-methoxy-9,10-dihydrophenenthropyran (Anuradha and Prakasa Rao 1998a, b). Oxo-tessallatin was later isolated from *V. parishii* (Anuradha et al. 2008).

Roman Kaiser (1993), an authority on floral scents, explained that the strong floral scent of *V. tessellata* is based on linalool (23 %), methyl benzoate (61.5 %), cinnamic aldehyde (%.1 %) and methyl cinnamate (4.6 %), its ultimate character being derived from a variety of other fragrances that occur in traces, such as benzyl acetate, alpha-ionone, 3-phenylpropanal, p-cresol and indole. The scent of different plants varies, resulting from diverse combinations of scent skeletons. For example, two scent variants are based in one instance on (E)-ocimene, methyl salicylate and cinnamic aldehyde, and in the other, on (E)-ocimene, methylbenzyl alcohol, phenylethyl alcohol and the benzoates of the last two components. Fragrance of *V. tessellata* is transmitted to its F1 and F2 progeny in hybrids made with other species of *Vanda*. Roots of *V. tessellata* are fragrant (Chopra et al. 1958).

Herbal Usage: From the long list of local names cited from Caius (1936), it is evident that *V. tessellata* was much admired as *Rasna* (or *Raasnaa*), the ancient "anti-ageing" tonic. This orchid medication is mentioned by ancient Sanskrit writers under the names *Rasna* and *Gandhanakul* which are said to be bitter, aromatic and useful for rheumatism. It is taken orally and applied externally (Dymock 1885).

Rasna was used to treat nervous disorders, rheumatism and scorpion stings (Chopra 1933). A typical Sanskrit prescription, the *Rasnapanchaka*, is quoted in the nineteenth century Dutt's *Hindu Materia Medica* as follows:

> Take *rasna*, *Tinospora cordifolia*, wood of *Pinus longifolia*, ginger, root of *Ricinus communis*, each in equal parts, and prepare a decoction in the usual manner.

It was a popular prescription for rheumatism (Dymock et al. 1893). Several medicinal oils used in treating rheumatism, bone aches and venous disorders, such as *Mahamasha tada* and *Madhyama Narayana taila*, have *V. tessellata* among its ingredients (Dymock et al. 1893; Kapoor 1990). A liquid extract or syrup prepared from the orchid was used to treat dyspeptic conditions, bronchial affections and rheumatic

fever in India (Roberts 1931). Ear drops are prepared from the roots of *V. tessellata* (Caius 1936; Trivedi et al. 1980; Dwivedi 2003). The whole plant is used to treat fractures, and roots as a female contraceptive. Leaves are thought to be capable of preventing wounds from developing into purulent sores whereas the "bark" is mixed with the bark of *Emblica*, *Shorea robusta* and *Litsaea* for application on sores already present. Leaves are also used to treat tearing and earache, the latter either by itself or in combination with the aerial roots and neem oil (Chopra et al. 1956; Jain and Defilipps 1991). It was used by herbalists to treat secondary syphilis and for snake and scorpion bites (Chopra et al. 1958).

Rasna (root of *V. tessellata*) is an essential component of the much acclaimed *Asthavarga* (8-wonders remedy). Several native tribes in the eastern tropics use the juice of *V. tessellata* as a panacea for all ills (Duggal 1971). It was observed to act as an aphrodisiac when administered to mice (Kumar et al. 2000). The primitive Dongria Kandha tribe of the Niyamgiri Hills of southwest Orissa uses a decoction of the root in honey for the treatment of STDs (sexually transmitted diseases). First, 50 g of root is boiled in 250 ml of water until the volume is reduced to 100 ml. After cooling and filtration, 5 ml of the decoction is mixed with a teaspoon of honey and the sweetened mixture is taken orally on an empty stomach twice daily for one month (Dash et al. 2008).

V. tessellata was also reported to be used to treat secondary syphilis in India (Nadkarni 1954). However, secondary syphilis is asymptomatic and one wonders whether it is tertiary syphilis with its multi-organ damage that was being referred to. Since syphilis was introduced from the New World, this treatment could not have originated from ancient *Ayurveda*. If indeed *V. tessellata* was being used to treat syphilis, this would be an example of misguided faith in herbal remedies, when a curable disease is allowed to spread and lives are lost through an inappropriate herbal remedy. There is no place for using an orchid to treat syphilis because primary and secondary syphilis (the asymptomatic stage when the disease is recognised through positive serological tests) are easily treated and can be cured with penicillin. There is no satisfactory treatment for tertiary syphilis.

Gonds of Andhra Pradesh in the Deccan apply a root paste made with *V. tessellata* (*Ippa vajrnika*) once daily for 5–6 days for rheumatism (Reddy et al. 2003). An extract of root velamen is used to treat dysentery in Andra Pradesh Yoganarasimhan and Chelladurai (2000) reported that the roots are used in dyspepsia, bronchitis, rheumatism and fever. They are a constituent of medicated oils applied to relieve pain and swelling.

Dry roots of *V. tessellata* look superficially similar to the dry roots of *Pluchea lanceolata* and are commonly substituted for the latter in northern and eastern India for medicinal use. The latter fetches a very high price but dry roots of *V. tessellata* do not. On the other hand, it has been reported that native physicians in Bengal were not able to or did not bother to distinguish between the roots of *V. tessellata* and *Acampe papillosa*, both of which they regarded as *rasna* (Dymock 1885). Sarin (1996) in India stated that it is called *Raasnaa mool* or simply *Raasnaa*, while the drug indicated by the name *Raasnaa patra* is the root of *Pluchea lanceolata*.

To add to the confusion, an expensive *rasna* once sold in Bombay under the name *Khadaki-rasna* consisted of straight pieces of root and was identified as *Tylophora asthmatica* (Dymock et al. 1893). Ordinary bazaar *rasna* consisting of *Vanda* or *Acampe* roots were long and branched.

An ancient medicinal tome, the *Gandha-mula* states that *rasna* has strong smelling roots, a character absent from *V. tessellata* (Dymock et al. 1893). The *Sarasvathi Nighanduwa* (a medical dictionary) explains that there are three types of *rasna*, the first being the rare *mula rasna* which has the smelly root. The second variety, *pathra rasna* or *V. tessellata* is extensively used by Ayurvedic physicians in India (Cooray 1940). In the herb markets today, *V. tessellata* and *Acampe papillosa* are sold as *rasna*.

The issue is thus complex and unsettled. Cooray (1940) in Sri Lanka maintained that the *Mula Raasna*, *Pathra Raasna* and *Thurna*

Raasna mentioned in the *Sarasvathi Nighnduwa* are all varieties of *V. tessellata*. *Mula Raasna* is extremely rare so Ayurvedic physicians make extensive use of *Pathra Raasna* in their medications. In appearance, *Pathra Raasna* is cylindrical, bent and twisted, 3–4 mm in diameter and 20 cm long. The outer surface is brownish with longitudinal striations and transverse partial fractures. Odour is spicy, and taste slightly bitter.

Rasna guggulu is a *ghrita* composed of 8 parts of *rasna* and 10 of *bdellium* (an aromatic gum likened to myrrh) beaten into a uniform mass with clarified butter (*ghee*). This is given in drachm doses in sciatica. *Rasna* is also a component of medicated oils used for external application or massage in rheumatism and neuralgia, such as *Mahamasha taila*, *Madhyama Narayana taila*, etc. (Dymock et al. 1893).

Caius (1936) reported that, in Chota Nagpur, the leaves were pounded into a paste for application on the body during attacks of fever. Juice of leaves is instilled into the external ear to treat earache and other inflammatory conditions. In Madhya Pradesh, Upadhyay (2003) reported that juice made with the leaves is dropped into the eyes to treat sore eyes while leaf paste is applied on skin infections. Smelling the flower is advocated to relieve migraine. Bondo tribals of southern Orissa employ leaf paste to help heal fractures, and Kondhas in southeastern Orissa use it to relieve rheumatic pain. Paste made with the whole plant is used by Bhils of Rajasthan to remove acne (Parrotta 2001).

The plant is sometimes used for scorpion (Chopra 1933) and snake bites, especially in combination with other herbs. However, Caius and Mhaskar demonstrated that neither in isolation nor in combination with other herbs was it the least effective as an antidote to snake and scorpion venom (Caius 1936). Ayervedic healers employ *rasna* (or *raasnaa*) to treat diseases of the nervous system, sciatica, fistula, diseases of the ear and poisoning. The major preparations containing *Raasnaa mool* are *Raasna saptak kwaatha*, *Mahaamaash taila*, *Raasnaa panchak kwaatha*, *Kukuvaadi churna*, *Maha yograjgugglu* and *Sammirpananga* (Sarin 1996). The orchid was eaten with food by women who wanted to have sons (Hoernle early 1900s, quoted by Lawler 1984).

Yunani practitioners used it as a laxative, a tonic for the liver and the brain, and a remedy for bronchitis, piles, lumbago, toothache, boils on the scalp, inflammation and fractures (Caius 1936). In the Chhatarpur district of Madhya Pradesh, people living in remote villages use the stems of *V. tessellata* to treat fractures and sprains (Datt 1996).

In Nepal, roots of *V. tessellata* are used as antidote for scorpion stings, or to treat bronchitis and rheumatism. Leaf paste is used for fever (Subedi et al. 2013).

Vanda testacea (Lindl.) Rchb.f.

Syn. ***Vanda parviflora*** Lindl.

Indian name: *Malanga* in Orissa. Medicinal names: *Rasna, Banda*

Thai name: *Khem lueang* (in Bangkok)

Description: *V. testacea* is a small flowered *Vanda* with white or yellow tepals and a fringed lip that is marked with blue-ridged callosities. Leaves are coriaceous, channelled, erect or pendent, spreading, 3–15 by 0.4–0.7 cm, with 2 or 3 teeth at the apex. There are 10–20 flowers loosely arranged around an inflorescence of 12–20 cm length (Fig. 22.11). Petals and sepals are similar, narrow, of a light lemon-yellow (Fig. 22.12). *V. testacea* blooms in spring, or March to April (Abraham and Vatsala 1981), and in early summer, or May to June (Santapau and Kapadia 1966). It has a high tolerance for sunlight (37–90 %), temperatures of 30 °C and low humidity of 36–41 % (Jayaweera 1981; Dalstrom 2010).

V. testacea is found at the foothills on of the Himalayas, in Nepal, Bhutan and Sikkim, extending to northern Myanmar, Thailand and Sri Lanka at 800–2000 m, in dry, broad-leaved forests. It grows in clusters, abundantly on roadside trees under dappled shade, and sometimes under very arid conditions on the road from Inle Lake to Mandalay. In Bhutan and Sikkim, it

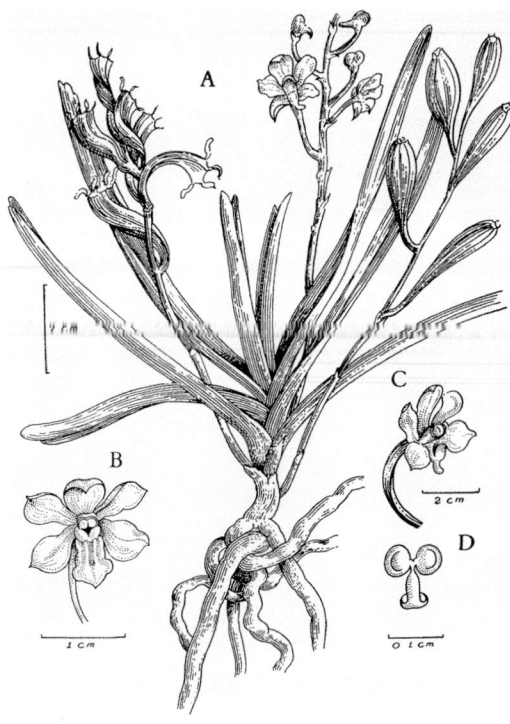

VANDA TESTACEA (LINDL.) REICH f.

Fig. 22.11 *Vanda testacea* (Lindl.) Rchb. Reproduced with permission from *Introductions to Orchids* by Abraham and Vatsala, Parlode, Thiruvananthapuram: Tropical Botanic Garden and Research Centre (TBGRI), 1981

occurs in broad-leaved forests, commonly on *Terminalia chebula* at 780–2000 m, and flowers in May to June (Pearce and Cribb 2002). However, it is a common epiphyte in moist low country extending up to medium elevations in Sri Lanka (Ekanayake 1975). It is also a very common epiphyte in the foothills to 1000 m in Tamil Nadu, where peak flowering is in April to May (Seidenfaden 1999). The species has become rare in the Punjab where it is now restricted to narrow pockets in its natural habitats, following extensive collection for medicinal usage and habitat destruction. With successful micropropagation from leaf-tip meristems, it is hoped that the medicinal plant will never be in short supply (Kaur and Bhutani 2009).

Phytochemistry: Parviflorin and tessalatin, two 9,10-phenanthropyran derivatives were isolated from whole plants of *V. testacea* (syn. *V. parviflora* Lindl.) (Anuradha and Prakasa Rao 1998a, b).

Fig. 22.12 *Vanda testacea* (Lindl.) Rchb.f. (Photo: Bhaktar B. Raskoti)

Herbal Usage: The orchid is used to treat rheumatism (Vij 1995). Crushed leaves are applied to cuts and wounds. Its decoction is used for earache (Yoganarasimhan and Chelladurai 2000). An Orissa hill tribe uses the plant to treat earache. Root is boiled with *Curculigo orchioides* to prepare a decoction that is taken twice daily to cure asthma.

The roots are used to treat dyspepsia, bronchitis, inflammation and coughs under the Unani system. They are also used to tone up the liver and brain, reduce inflammation and heal minor fractures and are sold as *rasna* in Ayurvedic shops (Rao and Sridhar 2007). Leaf paste mixed with mashed boiled rhizome of *Rhaphidophora glauca* in equal proportion is plastered over fractures before applying the bandage (Dash et al. 2008). Leaves and roots are also used to treat scorpion sting (Bhattacharjee 1998). Juice extracted from the leaves of *V. testacea* is dropped into the ear to relieve earache (Baral and Kurmi 2006).

Overview

Vanda is essentially a tropical Asian genus, so one would expect its usage to be prevalent in India.

Unani physicians employ *Vanda* roots to treat dyspepsia, bronchitis, inflammation and coughs. Roots of *V. testacea* and *V. tessellata* are sold as *rasna* in shops dealing with Ayurvedic medicine. Highly regarded as tonics for the liver and brain, they are also used to reduce inflammation and heal fractures (Rao 2007). Extracts of the plant show an anti-acetylcholine effect in rats. It lowers blood pressure, counteracts the contractile effect of acetylcholine on the isolated ileum and tempers electroshock seizures. With an LD50 exceeding 1000 mg/kg, it appears to be safer than coffee. The same Study Group found LD50 values to be 1000 for *Coffea arabica* Linn. and 500 for *C. bengalensis*, while only *C. khasiana* Hook f. had an equivalent value that exceeded 1000 mg/kg body weight (Bhakuni et al. 1969, 1971).

Healers among the Akha tribe in Northern Thailand recommend making poultices from *Vanda* or any of 17 other plant species which are not orchids to treat burns. Alternatively, the orchid may be made into an infusion to bathe the injury (Anderson 1993). *V. tessellata* is used by the Tharu community in Udham Singh Nagar, Uttarakhand, India to treat skin diseases (Sharma et al. 2014). It provided pain relief for mice injected intra-abdominally with acetic acid (Chowdhury et al. 2014).

A study conducted at the University of the West Indies in Trinidad reported that an extract of *V. roxburghii* (=*Vanda tessellata*) at a dose of 150 mg/day for 10 days produced a 60 % reduction in wound diameter in rats compared to controls. This improvement in wound healing was accompanied by measurable increases in wet and dry granulation tissue and their content of hydroxyproline (a component of collagen and bone) and hexosamine. The wounds healed by the 13th day in the treated group compared to the 20th day in the controls; wound size was smaller and the healing much faster (Nayak et al. 2005). Granted that orchids are hardly ever mentioned as possible sources of antimicrobials in any large review of medicinal plants (Cowan 1999), these findings are not surprising because all plants produce secondary metabolites to defend themselves against microbes and other predators. In the clinical setting, the issues concern efficacy, safety and convenience.

Surprisingly, evidence of widespread medicinal usage for *V. tesssellata* in Sri Lanka, the commonest source of this orchid is lacking. Numerous leading Ayurvedic physicians participated in the write-up of *Hortus Indicus Malabaricus* compiled by van Rheede and published between 1678 and 1703 (van Rheede 1703); they did not advise van Rheede on *V. tessellata*.

The strong floral scent of *V. tessellata* is due to the presence of various volatile oils, with linalool accounting for 23 % of the total (Kaiser 1993). Methyl benzoate, cinnamic aldehyde and methyl cinnamate are other major components but the final scent is derived from the addition of a variety of trace constituents which include benzyl acetate, alpha-ionone, 3-phenylpropanol, *p*-cresol and indole (Kaiser 1993). Linalool is a monoterpene compound which is present in many plants, including lavender. It is bacteriostatic (against *Escherichia coli* and *Staphylococcus aureus* which are common causes of skin, bladder and wound infections) and fungistatic (against *Candida albicans*), but it does not inhibit other bacteria. It is anti-inflammatory and relieves pain, enhances tissue permeability and assists in the delivery of topical medications. A sedative effect on mice is dose-dependent. Linalool is an anti-oxidant (Peana and Moretti 2008). However, there is no mention in the herbal literature that the flowers of *V. tessellate*, and extracts of the flowers in oil or alcohol have never been used medicinally, other than the observation that, in Madhya Pradesh, smelling the flower is an approach to handling migraine (Upadhyay 2003). It is unclear which clone should be used for this purpose.

V. tessellata is an extremely variable species, and different clones emit distinctive scents. In one instance, it is based on (E)-ocimene, methyl salicylate and cinnamic aldehyde. In another instance, it might be produced by (E)-ocimene, methyl benzoate, benzyl alcohol, phenylethyl alcohol and the benzoates of the last two compounds. Over the three scent skeletons, *V. tessellata* produces a variety of

scents by varying the minor components (Kaiser 1993). Other medicinal species, *V. coerulea*, *V. concolor*, *V. cristata* and *V. testacea* do not emit a scent. *V. dearei* and *V. tricolor* var. *sauvis* are scented, but not medicinal.

Of 22 species of *Vanda* that were screened for alkaloids, only three gave an alkaloid content of 0.1 % or greater (Lüning 1974). Laburnine acetate was isolated from *V. hindsii* Lindl. and *V. helvola*, with the latter also containing the alcohol, laburnine. An extract of *V. luxonica* Loher contained either laburnine or its enantiomer (Brandange and Granelli 1973). However, there has been no attempt to relate the usage of the four medicinal *Vanda* species to any pharmacologically active compound.

Stilbenoids with anti-inflammatory and antioxidant activities in-vitro have been isolated from *V. coerulea* by Simmler and her group at the University of Strasbourg, namely imbricatin, methoxycoelonin, gigantol, flavidin and coelonin (Simmler et al. 2009, 2010) (Fig. 22.13). Among the stilbenoids, the first three items were in the highest concentration in the crude hydro-alcoholic stem extract. This crude extract from *V. coerulea* demonstrated in-vitro inhibition of type 2 prostaglandin (PGE-2) release from UVB (60 mJ/cm^2) irradiated skin cells (HaCaT keratinocytes). Imbricatin and methoxycoelonin inhibited COX-2 activity in a dose-dependent manner, while gigantol inhibited PGE-2 production (Simmler et al. 2010). Replicative senescence of normal skin fibroblasts involves a reduction of cells in S-phase which correlates with decreases in cyclin E and cyclin dependent kinase 2, cdk2. Treatment with an ethanolic extract of *V. coerulea* stems titrated for the three stilbenoids, imbricatin, methoxycoelonin and gigantol, restores the percentage of skin fibroblasts to that of young cells together with restoration of cyclin E and cdk2 (Bonte et al. 2011). The studies suggest that stilbenoids of *V. coerulea* might have a potential use for skin protection from ultraviolet damage and ageing (Simmler et al. 2010; Bonte et al. 2011).

Eucomic acid [(2R)-2-(p-hydroxybenzyl) malic acid)] and vandateroside II isolated from stems of *V. teres* (=*Papilionanthes teres*)

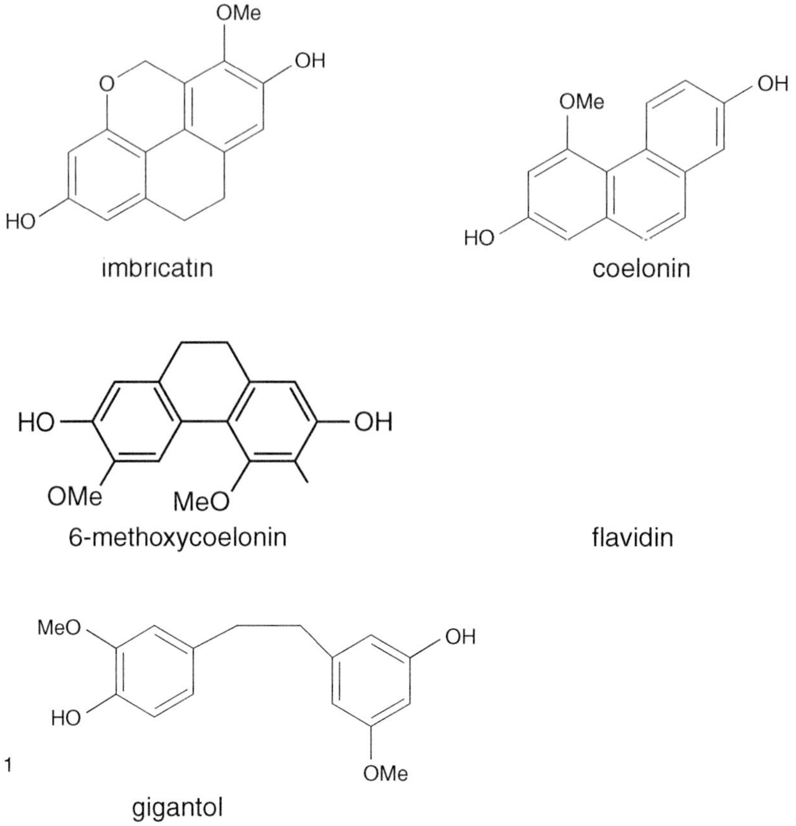

Fig. 22.13 Stilbenoids with anti-oxidant, anti-inflammatory properties from *Vanda coerulea*

imbricatin

coelonin

6-methoxycoelonin

flavidin

1

gigantol

increased cytochrome-c oxidase activity and/or expression without enhancing cellular mitochondrial content in a human immortalised keratinocyte cell line (HaCaT). Decline in mitochondrial functions occurs with age and may be an underlying cause of age-related changes in the body (Müller-Höcker 1992). Therefore, eucomic acid and one of its three glucopyranosyl-oxybenzyl eucomate derivatives (vandateroside II) are also candidates as new natural ingredients for "anti-ageing" preparations to remedy age-related disorders such as skin ageing (Simmler et al. 2011) (Fig. 22.14).

There are several problems about protecting skin. Humans need sunlight for their skin to produce vitamin D: otherwise they get osteoporosis. On the other hand, excessive sunlight leads to skin cancer, particularly rodent ulcers. It is unlikely that sun-blocks and other UV skin protectants can protect against excessive exposure to sunlight. Additionally, increasing mitotic activity in cells increases the risk

of mutations and carcinogenesis, this being evident from the data showing that progestogens increases breast cancer risk which is absent in women taking oestrogens alone (Teoh and Teoh 1991). Progestogens induces mitosis of breast cells. Oestrogens alone do not, but oestrogens set the stage for mitosis by increasing progestogen receptors in breast cells .

Two glucosides, namely tris[4-(beta-D-glucopyranosyloxy)benzyl] citrate or parishin and 4-(beta-D-pyranosyloxy)benzyl alcohol, were isolated from *V. parishii* which is not a medicinal orchid. Parishin is pharmocolgically inactive, but Parishin C which occurs in *Gastrodia elata* was shown in animal experiments to protect mice against phencyclidine-induced schizophrenic-like behaviour (Shin et al. 2010).

Meanwhile, an Indian team of scientists is attempting to develop an aphrodisiac from a new compound, 2,7,7-tri methyl bicyclo [2.2.1]

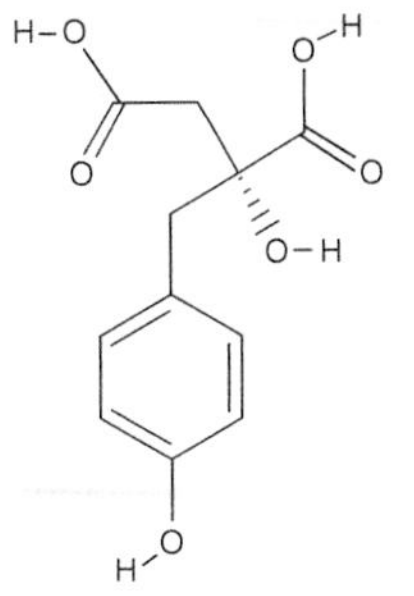

eucomic acid

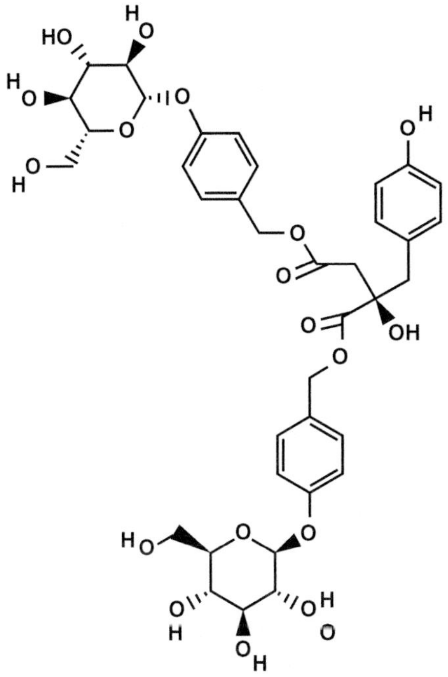

vandateroside II

Fig. 22.14 Two compounds from *Papilionanthe teres* that may have a role in protecting skin from changes associated with aging and ultraviolet damage. There are actually three vandaterosides, I, II, and III

heptane, which they have isolated from *V. tessellata* (Subramoniam et al. 2013).

Genus: *Vanilla* Plum. ex Mill.

Chinese name: *Xiangjia Lan*

Vanilla is a monopodial, terrestrial orchid with the climbing habit of an epiphyte. Stems grow in a zigzag manner producing a leaf and root at each node. Inflorescence arises from the leaf axils. It bears a few greenish or yellow flowers. Seedpods of *V. planifolia* are valued for their content of the fragrant flavouring agent, vanillin, which is widely used in condiments. This species is widely cultivated throughout the tropics (Fig. 22.12). Another aromatic species, *V. tahitensis*, was introduced from the Philippines into Tahiti in 1848 and it is still cultivated in several Pacific countries. There are between 90 and 110 species of *Vanilla*, with 52 in tropical America and 31 in Southeast Asia and New Guinea. Some 18–35 species are aromatic, but most do not provide a high yield of vanillin in their seedpods and they are economically insignificant (Bory et al. 2008). Three species have some medicinal usage.

Vanilla aphylla Blume

Thai Name: *Khot nok kut* (in Surat Thani), *Khruea ngu khieo* (in Nakhon Ratchasima), *Thau ngu khieo* (in Saraburi)

Description: Stems of *V. aphylla* are a dark green and flattened. They have taken over the work of the leaves which are reduced to small triangular green scales 7 mm long. Plants may attain a length of 200 cm with internodes measuring 6–8 cm (Seidenfaden and Wood 1992). Inflorescence is short, but it carries three 3-cm, light green flowers with a cream-coloured lip that is accentuated by numerous stiff hairs, 2 mm long on the circular disc located at the distal portion of the mid-lobe. Side lobes are erect with crisp, reflexed edges that overlap and cover the column (Fig. 22.13). Flowering period is October and November.

The species is distributed in the southern half of Thailand, Myanmar, Indochina, northern Peninsular Malaysia and Java. It is found among low bushes in open country in Perlis, the tiny northernmost state of Peninsular Malaysia.

Herbal Usage: The stem is used in the treatment of liver dysfunction in Thailand (Chuakul 2002).

Vanilla griffithii Rchb. f.

Indian name: *Telinga kerbaoo*

Description: *V. griffithii* is a creeping epiphyte, bearing leaves at 8–13 cm intervals. Leaves are ovate-elliptic, thick and fleshy, 9–16 by 3–8 cm, with short petioles of 5 mm (Comber 2001). Inflorescence is axillary and bears a single flower 5 cm across. Sepals and petals are greenish white. Lip is white-tipped with yellow, woolly hairs covering the centre of its mid-lobe (Fig. 22.14). It flowers in March to May in Thailand (Vaddhanaphuti 2005). Fruits are 7 by 1 cm at broadest diameter. *Vanilla griffithii* is distributed in India, southern Thailand, Sumatra, Peninsular Malaysia and Borneo at low elevations, rooting in the ground and climbing trees in shady locations.

Herbal Usage: It is eaten as a vegetable in India. The large fruit is said to be sweet and edible, tasting like small bananas (Burkill 1935; Rifai 1975; Tanaka 1976). An aqueous extract of its flowers is applied on the body to treat fever. Juice from the leaves is applied on the hair to encourage thick growth (Uphof 1968; Duggal 1971). In Indonesia, the milky sap of the stems is used as a hair-promoting shampoo (Rifai 1975).

Vanilla *pilifera* Holtt. (see **Vanilla borneensis** Rolfe.)

Vanilla planifolia Jacks. ex Andrews

Thai names: *Wanila* (in Bangkok), *Vanilla*

Description: The species has the typical *Vanilla* habit, with fleshy, elliptic leaves 20 by 5 cm. Pale green to yellowish-green flowers are produced in succession on the axillary inflorescences. They are hand-pollinated in the *Vanilla* farms. The exported products are the vanilla pods which are variable in quality and contents from source to source as well as from batch to batch (Gassenmeier et al. 2008).

V. planifolia is not native to Asia. It was introduced into cultivation in the French Reunion Island from Central America in 1817 by way of

Mauritius because of the high content of vanillin in its large, fleshy seedpods which have a length of up to 15 cm. Today, vanilla comes from many sources which include Central America, Mauritius, Reunion, Madagascar, Indonesia, the Pacific Islands, India (northeast India, Kerala, Karnataka, Tamil Naidu and Lakshadweep), the Andaman and Nicobar Islands, Thailand, Indochina and China. In its native Central America, it grows in damp forests up to 600 m.

Phytochemistry: The major volatile fragrance in vanilla pods is vanillin which accounts for about 2 % of the fermented fruit (range 1. 5–2.75 depending on provenance. However, vanillin is not present in newly harvested pods. It is formed during curing or fermentation when the precursor, non-volatile compound vanilloside (gluco-vanillin) undergoes beta-hydrolysis. Vanillin accounts for around 85 % of the volatile compounds. Another is p-hydroxybenzaldeyde (9 %), with p-hydroxybenzyl methyl ether

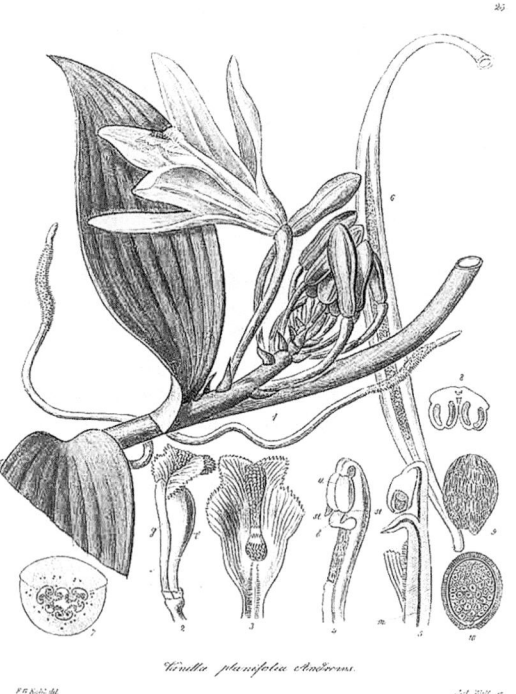

Fig. 22.15 *Vanilla planifolia* Jacks ex Andrews. From: Kohl, F.G., *Die officinellen Pflanzen der Pharmacopoea Germanica*, t. 25 (1891–1895) Painting by F.G. Kohl. Courtesy of the Universitats-und Landesbiblothek Dusseldorf, Germany

accounting for 1 % (Fig. 22.15). Two stereoiso-meric vitispiranes (2,10,10-trimethyl-1,6- and methylidene-1-oxaspiro(4,5)dec-7-ene), although only occurring in traces, also influence the aroma. Approximately 130 more compounds have been isolated from natural vanilla. These are made up of phenols, phenol ethers, alcohols, carbonyl compounds acids (e.g. vanillic acid, ferulic acid), esters, lactones, aliphatic and aromatic carbon hydrates and heterocyclic compounds. The actual flavour of any batch of natural vanilla is influenced by the plenitude of volatile compounds that are present in traces. Synthetic vanillin is commonly prepared from wood pulp. It does not contain the varied fragrances of the natural product (Figs. 22.16 and 22.17).

Herbal Usage: The whole plant is considered to have aphrodisiac value in India (Sood et al. 2005). The roots are used as a stimulant and to treat gonorrhoea and dysuria, possibly resulting from the sexually transmitted disease. They are mixed with onions, cumin, sugar and butter to prepare a confection. An extract of the root together with cumin and sugar is added to cold milk to provide a remedy for spermatorrhoea (Nadkarni 1954).

Fig. 22.17 *Vanilla griffithii* Rchb. f. (Photo: Peter O'Byrne)

Vanillin, better known as vanilla in common usage, is a popular flavour in ice cream, biscuits, cakes and other desserts. It has ceased to be considered a medicine but it continues to be used to flavour medicines for children. The orchid, *Vanilla*, does not have any significant amount of alkaloids (Lüning 1974).

Overview

Due to its important economic value, an enormous amount of literature on *Vanilla* exists, and no attempt will be made to summarise all of it in this volume. The Totonac were the first people to cultivate the plants, and referred to *Vanilla* as *xa'nat*. It was called *tlilxochitl* (black flower) by the Aztecs whose king, Montezuma offered chocolate (*xocoatl*) flavoured with vanilla to Cortez and his conquistadors. Hernandez, personal physician to Phillip II of Spain, sent to Mexico by the king in 1570 to study the natural history of the country (Oliver-Bever 1972), produced a *Thesaurus* in which he included a sketch and account of the plant which he named *Araco aromatico*. By 1721, it had appeared in the *London Pharmacopoeia*.

V. planifolia was introduced into Europe by the Spaniards in the sixteenth century, but nothing came out those introductions because the plant remained sterile in the absence of appropriate insect pollinators. In the first half of the nineteenth century, various individuals, in Belgium, France, Indonesia and Italy independently managed to hand-pollinate *Vanilla*. However, no commercial application resulted until Edmond Albius,

Fig. 22.16 *Vanilla aphylla* Blume (Photo: Chang Yoon Ching)

a black slave in Reunion Island, demonstrated a simple method for hand pollination and, in 1841, demonstrated the process to his employer who owned a plantation. Reunion grew the first crop of hand-pollinated *Vanilla* and remains a major producer of the flavouring agent.

In 1847, an American volume on *Medicinal Botany* by R.E. Griffith, MD, mentioned that vanilla acts "powerfully on the generative system as an aphrodisiac" and he advocated a dose as 8–10 grains (0.52–0.65 g). Vanilla was used as a remedy for hysteria, mild fever and impotence, but by 1880 Bentley and Trimen reported that its use was obsolete in England "although it is sometimes used in the continent and elsewhere. It is also frequently used for flavouring certain medicines, as lozenges and mixtures in the United States, etc." (Bentley and Trimen 1880). John Lindley wrote that in England *V. planifolia* was an aromatic stimulant which was also used to treat asthmatic fevers, rheumatism, hysteria, male impotence and infections (Lindley 1849).

During the nineteenth century, its reputation as an aphrodisiac became widespread. Cassanova and Marquis de Sade were reported to have used it regularly. One German physician, Bezaar Zimmermanm even claimed that "No fewer than 142 impotent men, by drinking vanilla decoctions, changed into astonishing lovers of at least as many women" (Ecott 2004). A recent study from India reported that vanillin at a dose of 100 mg/kg showed an antidepressant effect on rats tested by using forced swimming as a stress test, an effect comparable to that produced by fluoxetine (Shoeb et al. 2013).

Vanilla has been highly prized from ancient times by the Aztecs for the flavouring agent present in its seed pods. Mature green pods of *V. planifolia* accumulate 4-*O*-(3-methoxy-benzaldehyde)-beta-D-glucoside (glucovanillin or vanilloside) which yields vanillin upon hydrolysis by an endogenous beta-glycosidase. Both glucovanillin and beta-glucoside are predominantly located in the placentae (92 %) and marginally in the trichomes (7 %), with the latter storing massive amounts of a fluorescing oleoresin rich in alkenylmethyldihydro-lambda-

pyranones (44 %). The trichomes synthesise a mucilage constituted of a glucomannan and a pectic polysaccharide with monomeric arabinose and galactose side-chains (Odoux and Brillouet 2009).

Vanilla pods are harvested while they are still green, but they do not possess any flavour at this stage. The flavour and aroma only develops with curing which consists of four steps: scalding/killing, sunning/sweating, drying, and conditioning/ageing (Sinha et al. 2007). During this process, beta-glycosidases act on the glycosides to release the various compounds that make up the vanilla aroma. The quality of the aroma is dependent on species, source, climate and season, as well as the curing process. A 90 % conversion of glucovanillin to vanillin occurs in beans which are sweated continuously at 35 °C for 12 days, a rate which is much higher than the 70 % conversion for beans blanched at 67 °C and sweated at 45 °C for 4 days or at 35 °C for 5 days. In all instances, the beans had turned brown. The appearance of the beans which were blanched was more attractive but they lost out on aroma (van Dyk et al. 2010). Gamma radiation does not enhance the aroma of vanilla beans (Kumar et al. 2010).

Vanillin, the principal aromatic compound in vanilla, is now produced by enzymatic processes using wood or petrochemicals, and this preparation is widely used for food preparation and in the perfumery industry. Authenticity of vanilla extracts from *Vanilla* beans can be monitored by analysis of stable isotope ratios (C-13/C-12 ratio) of vanillin (Gassenmeier et al. 2013). Natural products are subject to variable quality depending on source and batch. One study found that almost half the beans supplied did not meet the desired ratio of vanillin, vanillic acid, *p*-hydroxybenzaldehyde and -*p*-hydroxybenzoic acid, and some do not even contain sufficient vanillin (Gassenmeier et al. 2008).

Bourbon vanilla is the most expensive product on the market. Aromatic compounds exclusive to bourbon vanilla are methyl vanillin (veratraldehyde), salicyaldehyde and guaiacol (Fig. 22.19) (Grayer and Veitch 2003). Tahitian vanilla contains piperonal (heliotropin, 3,4, dioxymethylenebenzaldehyde) and diacetyl (butandione) (Fig. 22.20).

The ability of vanilla to inhibit bacterial quorum sensing has been demonstrated in vitro by showing that the production of violacein by *Chromobacterium violaceum* CV026 was reduced by vanilla in a concentration-dependent manner (Choo et al. 2006). Vanilla extracts containing vanillic acid, 4-hydroxybenzyl alcohol, 4-hydroxy 3-methoxybenzyl alcohol, 4-hydroxybenzylaldehye and vanillin exhibited anti-oxidant activity in beta-carotene linoleate and DPPH models (Shyamala et al. 2007). These two experiments point to a possible use of vanillin as a food preservative.

Phenotypic and chromosomal variations have appeared following prolonged cultivation of *V. planifolia*. Several botanical institutions worldwide are making an effort to conserve the genetic resources of vanilla. Other species of *Vanilla* such as *V. abundiflora* J.J. Sm., *V. gardneri* Rolfe, *V. uianensis* Splitgerb., *V. phaeantha* and *V. pompona* are also used for flavouring in some countries, but none of them can rival *V. planifolia* in fragrance. At one time, *Selenipedium Chica* Rchb. f., a North American slipper orchid, referred to as *Vanilla Chica* was substituted for ordinary vanilla (Uphof 1968).

On an entirely different note, *V. claviculata* which is endemic in the Caribbean was once used as a folk remedy for wounds. The fruits were once believed to be a cure for syphilis (Griffith 1847; Duggal 1971). *V. decaryi* Perr. is an aboriginal aphrodisiac in southern Madagascar (Uphof 1968). In Indian the Irulas living at the Nilgiri Biosphere Reserve call *V. walkerie* Wight by the name *Kundu pirandi*, and they use its stem as a veterinary medicine (Balasubramaniam and Prasad 1996).

Genus: *Zeuxine* Lindl.

Chinese name: *Xianzhu Lan*
Japanese name: *Kinu Ran, Hosoba Ran*

Plants of *Zeuxine* have the appearance of forest herbs and the flowers carry granular pollinia, as do all other member genera of the Subfamily Neottioideae Lindl. There are some 50 species in this genus of terrestrial orchids, distributed from Africa through tropical Asia to Samoa.

The generic name is derived from Greek, *zeuxis* (yoking). It refers to the partial union of the lip and the column (Schultes and Pease 1963).

Zeuxine strateumatica (L.) Schltr.

Chinese name: *Xianzhu Lan*
Taiwanese name: *Cao Pu Lan* (bunched grass orchid)
Indian name: *Shwethuli* in Bengal
Nepalese name: *Kansjhar*
Bangladeshi name: *Swet huli, Shwet huli*

Description: A terrestrial herb, 5–20 cm tall with a soft, slender, purplish stem sheathed with several narrow, grass-like leaves, 4 cm long and a tenth as wide, with the edges turned backwards. A short rachis of 4 cm carries numerous small, white to pink flowers. The petals form a hood with the dorsal sepal to cover the column. However, flowers barely open and are quickly self-pollinating (Figs. 22.18 and 22.19). They appear in February to March in New Delhi in India (Maheshwari 1963), January to March in Mumbai (Santapau and Kapadia 1966), January to February in Sikkim (Pearce and Cribb 2002), December and January in Sri Lanka (Jayaweera 1981), January to April in Hong Kong (Wu et al. 2001) and March to July on the Chinese mainland (Chen et al. 2009). This species is naturalised in Florida (Hawkes 1965) and now Brazil (Neto et al. 2011).

This is the most widely distributed species in the genus. J. D. Hooker once called it the commonest Indian orchid (Santapau and Kapadia 1966; Seidenfaden and Wood 1992). It occurs abundantly in grassy locations in moist, swampy ground, even in running water, throughout India from the lowlands to 1600 m (Santapau and Kapadia 1966). However, few of the Indian regional orchid *Floras* published in the last 30 years carry a description of this orchid; with

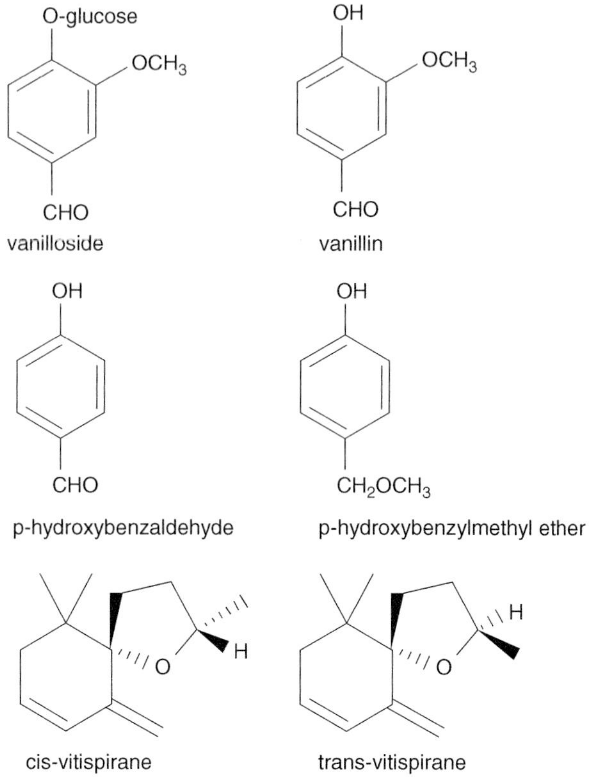

Fig. 22.18 Major volatile aromatic compounds in *Vanilla*: vanillin (85 %); *p*-hydroxy-benzaldehyde (almost 9 %); *p*-hydroxybenzylmethyl ether (1 %); vitispiranes—trace amounts

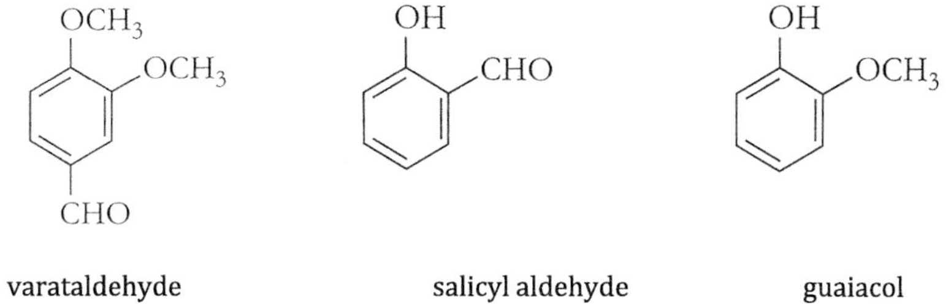

vanilloside

vanillin

p-hydroxybenzaldehyde

p-hydroxybenzylmethyl ether

cis-vitispirane

trans-vitispirane

Fig. 22.19 Volatile aromatic compounds present only in Bourbon vanilla (Grayer and Veitch 2003)

varataldehyde

salicyl aldehyde

guaiacol

17 synonyms listed by Jayaweera (1981), that is quite surprising. It is found throughout the warm regions of Asia, up to 1600 m on the Himalayan foothills. In Pakistan, it is often found growing amidst grass at the edge of watercourses (Nasir and Ali 1972). It is almost a weed of some padi at low elevations in Malaysia and Indonesia (Wood et al. 2011). Its subterranean rhizome confers a survival advantage in harsh environments (Ormerod and Cribb 2003) (Figs. 22.21 and 22.22).

Herbal Usage: Tubers of *Z. strateumatica* were used as *salep* (Caius 1936). They are combined with the roots of *Cymbidium aloifolium* (L.) Sw. to form a tonic in the Uttara Kannada district of Karnataka (Lawler 1984; Rao 2004). Considered to be a blood purifier and tonic, it is used externally for boils (Trivedi et al. 1980). *Z. strateumatica* is also used to treat infections of the eye (Rao 2004). It is still used as a tonic, or as a form of *salep*, in Bangladesh (Musharof Hossain 2009; Uddin and Yusuf 2011) and Nepal (Baral and Kurmi 2006; Subedi et al. 2013).

In Sri Lanka, Sumithraarachchi (1986) reported that *Z. strateumatica* (L) Schltr.,

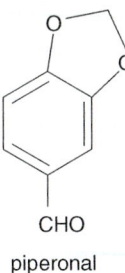

piperonal

Fig. 22.20 Piperonal (heliotropin) a unique aromatic constituent of Tahitian Vanilla, occurring in only trace amounts but imparting a unique fragrance

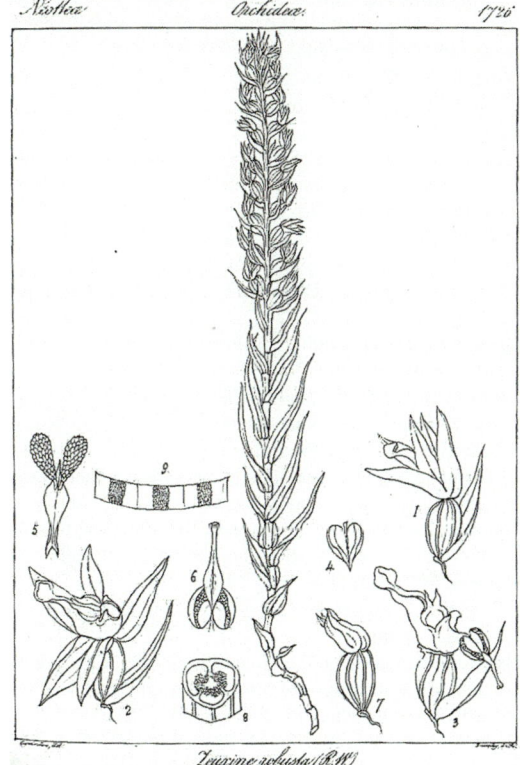

Fig. 22.21 *Zeuxine strateumatica* (L.) Schltr. [as *Zeuxine robusta* Wight]. From Wight, R., *Icones Plantarum Indiae Orientalis*, vol. 5 (1): t. 1726 (1846). Drawing by Govindoo. Courtesy of Missouri Botanical Gardens, St. Louis, USA

Z. longilabris (Lindl.) Trimen and *Z. flava* (Wall.) Trimen are all used in herbal medicine, but he did not elaborate on their usage. The last two species are rare in Sri Lanka, and *Z. flava* is vulnerable. Jayaweera (1981) also mentioned

Fig. 22.22 *Zeuxine* strateumatica (L.) Schltr. (Photo: Jagdeep Varma)

that *Z. regia* (Lindl.) Trimen was used medicinally but he did not specify its usage.

Overview

There is no pharmacological information on *Zeuxine*.

References

Abraham A, Vatsala P (1981) Introduction to orchids, with illustrations and descriptions of 150 South Indian orchids. TPGRI, Trivandrum

Anderson EF (1993) Plants and people of the golden triangle. Ethnobotany of the hill tribes of Northern Thailand. Dioscorides Press, Portland

Anuradha V, Prakasa Rao NS (1998a) Parviflorin a phenanthropyran from *Vanda parviflora*. Phytochemistry 46(1):181–182

Anuradha V, Prakasa Rao NS (1998b) Tessallatin, a phenanthropyran from *Vanda tessellata*. Phytochemistry 46(1):183–184

Anuradha V, Basaveswara Rao MV, Aswar AS (2008) Oxo-tessallatin, a novel phenethrapyrone isolated from *Vanda tessellata*. Orient J Chem 24 (3):1119–1122

Balasubramaniam P, Prasad SN (1996) Enthnobotany and conservation of medicinal plants by Irulas of Nilgiri biosphere reserve. In: Jain SK (ed) Ethnobiology in human welfare. Deep Publications, New Delhi, pp 271–273

Baral SR, Kurmi PP (2006) A compendium of medicinal plants in Nepal. Mrs. Rachana Sharma & IUCN, Kathmandu

Basu R (2010) Role of sacred groves in biodiversity conservation in Bankura District of West Bengal. J Econ Taxon Bot 34(1):12–17

Bennet SSR (1992) Venerated plants. Milton Book, Dehradun

Bentley R, Trimen H (1880) Medicinal plants being descriptions with original figures of the principal plants employed in medicine and an account of the characters, properties and uses of these plants and products of medicinal value. JA Churchill, London

Bhakuni DS, Dhar D et al (1969) Screening of Indian plants for biological activity: Part II. Indian J Exp Biol 7:250–262

Bhakuni DS, Dhar D et al (1971) Screening of Indian plants for biological activity: Part III. Indian J Exp Biol 9:91–102

Bhattacharjee SK (1998) Handbook of medicinal plants. Pointer, Jaipur

Bonte F, Simmler C, Lobstein A et al (2011) Action d'un extrait de Vanda coerulea sur la senescence de bibroblastes cutanes. Ann Pharm Fr 69(3):177–181

Bory S, Grisoni M, Duval MF, Besse P (2008) Biodiversity and preservation of vanilla: present state of knowledge. Genet Resour Crop Evol 55:551–571

Brandange S, Granelli I (1973) Studies on Orchidaceae alkaloids. XXXVI Alkaloids from some Vanda and Vandopsis species. Acta Chem Scand 73(3):1096–1097

Burkill IH (1935) A dictionary of economic products of the Malay Peninsula, vol 2. Crown Agents for the Colonies/Ministry of Agriculture & Co-operatives, London/Kuala Lumpur (1966 reprint, 2nd ed., with contributions by Birtwistle W, Foxworthy FW, Scrivenor JB, Watson IG)

Caius JF (1936) The medicinal and poisonous plants of India. J Bombay Nat Hist Soc 38(4):791–799

Chen XQ, Bell A (2009) Vanda Jones ex R. Brown. In: Chen XQ, Liu ZJ, Zhu GH et al (eds) Flora of China – Orchidaceae. Science Press, Beijing

Chen SC, Tsi ZH, Luo YB (1999) Native orchids of China in colour. Science Press, Beijing

Chen XQ, Gale SW, Cribb PJ, Ormerod P (2009) Zeuxine Lindley. In: Chen XQ, Liu ZJ, Zhu GH et al (eds) Flora of China – Orchidaceae. Science Press, Beijing

Choo JH, Rukavadi Y, Hwang JK (2006) Inhibition of bacterial sensing by vanilla extract. Lett Appl Microbiol 42(6):637–641

Chopra RN (1933) The indigenous drugs of India. The Art Press, Calcutta [Republished as Chopra's indigenous plants of India, 2nd ed. Kolkata: Academic (1986)]

Chopra RN, Nayar SL, Chopra IC (1956) Glossary of Indian medicinal plants. Council of Scientiific and Industrial Research (CSIR), New Delhi

Chopra RN, Chopra IC, Handa KL, Kapur LD (1958) Chopra's indigenous drugs of India. U.N. Dhur, Calcutta

Chowdhury MA, Rahman MM, Chowdhury MR et al (2014) Antinociceptive and cytotoxic activities of an epiphytic medicinal orchid: *Vanda tessellate* Roxb. BMC Complement Altern Med 14:464–471

Chuakul W (2002) Ethnomedical uses of Thai orchidaceous plants. Mohidol Univ J Pharm Sci 29 (3–4):41–45

Comber JB (2001) Orchids of Sumatra. Natural History Publications (Borneo), Kota Kinabalu

Cooray DA (1940) Orchids in oriental literature. Orchids Zelandica 7:73–80

Cowan MM (1999) Plant products as antimicrobial agents. Clin Microbiol Rev 12(4):564–582

Dalstrom S (2010) In the land of the Thunder Dragon. An introduction to the orchids of Bhutan. Orchids 79 (6):340–343

Dash PK, Sahon S, Bal S (2008) Ethnobotanical studies on orchids of Niyamgiri Hill Ranges, Orissa, India. Ethnobot Leaft 12:70–78

Datt B (1996) Ethnobotanical reserves of Chhatarpur District (Madhya Pradesh). In: Jain SK (ed) Ethnobiology in human welfare. Deep Publications, New Delhi, pp 400–402

Duggal SC (1971) Orchids in human affairs (a review). Pharm Biol 11(2):1727–1734

Dwivedi SN (2003) Ethnobotanicl studies and conservational strategies of wild and natural resources on Rewa District of Madhya Pradesh. In: Singh V, Jain AP (eds) Ethnobotany and medicinal plants of India and Nepal. Scientific, Jodhpur, pp 233–248

Dymock W (1885) The vegetable materia medica of Western India. Education Society's Press, Bombay

Dymock W (1943) Quoted by Chopra RN et al (1958) in Chopra's indigenous drugs of India. U.N. Dhur & Sons, Calcutta

Dymock W, Warden CJH, Hooper D (1893) A history of the principal drugs of vegetable origin met with in British India. Education Society Press, Bombay

Ecott T (2004) Vanilla. Travels in search of the Luscious Substance. Michael Joseph, The Penguin Group, London

Ekanayake DT (1975) Some wild orchid species of Sri Lanka (Ceylon). In: Senghas K (ed) Proceedings of the 8th world orchid conference, Frankfurt 10–17th Apr 1975, German Orchid Society, Frankfurt am Main, pp. 205–211

Gassenmeier K, Reisens B, Magyar B (2008) Commercial quality and analytical parameters of cured vanilla bean (*Vanilla planifolia*) from different regions from the 2006–2007 crop. Flavour Fragrance J 23(3):194–201

Gassenmeier K, Bingeli E, Kirsch T, Otiv S (2013) Modulation of 13C/12C ratio of vanillin from vanilla beans during curing. Flavour Fragrance J 28(1):25–29

Grant B (1895) The orchids of Burma. Hanthawaddy Press, Rangoon

Grayer RJ, Veitch NC (2003) Phytochemistry of vanilla. In: Pridgeon AM, Cribb PJ, Chase MW, Rasmussen FN (eds) Genera orchidacearum, vol 3, Orchidoideae (Part 2). Vanilloideae. University Press, Oxford

Griffith RE (1847) Medical botany of descriptions of the more important plants used in medicine, with their history, properties and mode of administration. Lea and Blanchard, Philadelphia

Gurong DB (2006) An illustrated guide to the orchids of Bhutan. DSB, Thimphu

Hamilton RM (1990) Flowering months of orchid species under cultivation. In: Arditti J (ed) Orchid biology reviews and perspectives, vol 5. Timber Press, Portland

Hawkes AD (1965) Encyclopaedia of cultivated orchids. Faber & Faber, London

Hu XM, Zhang WK, Zhu QZ et al (2000) Zhonghua Bencao, vol 8. Shanghai Science and Technology, Shanghai

Jain SK, Defilipps RA (1991) Medicinal plants of India, vol 2. Reference Publications, Algonac

Jayaweera DMA (1981) A revised handbook of the flora of Ceylon, vol 2. A.A. Balkema, Rotterdam

Jojita Devi RK, Ghatak (1986) A preliminary study of orchids in Manipur. In: Rao AN (ed) Proceedings of the 5th ASEAN orchid congress. Parks & Recreation Department, Ministry of National Development, Singapore, pp 72–78

Kaiser R (1993) The scent of orchids: olfactory and chemical investigations. Editiones Roche, Basel

Kapoor LD (1990) Ayurvedic medicinal plants. CRC Press, Boca Rotan

Kaur S, Bhutani KK (2009) In vitro propagation of *Vanda testacea* (Lindl.) Reichb. f. – a rare orchid of high medicinal value. Plant Tissue Cult Biotechnol 19 (1):1–7

Kumar S, Subramoniam A, Pushpangadan P (2000) Aphrodisiac activity of *Vanda tessellata* extract in mice. Indian J Pharmacol 32:300–304

Kumar KK, Kumar A, Arul A et al (2010) Effect of gamma radiation on major aroma compounds and vanillin glucoside of cured vanilla beans (*Vanilla planifolia*). Food Chem 122(3):841–845

Lawler LJ (1984) Ethnobotany of the Orchidaceae. In: Arditti J (ed) Orchid biology reviews and perspectives, vol 3. Cornell University Press, Ithaca

Li ZX, Pu XH, Zhou WW (2011) Thermodynamic properties of laburnine in saline and glucose solutions. J Therm Anal Calorim 107:1339, 1344

Lindley J (1849) Medicinal and oeconomical botany. Bradbury and Evans, London

Lindstrom B, Luning B (1969) Studies on Orchidaceae alkaloids XIII. A new alkaloid, laburnine acetate, from *Vanda cristata* Lind. Acta Chim Scand 23:3352–3354

Lüning B (1974) Alkaloids of the Orchidaceae. In: Withner CL (ed) The orchids: scientific studies. Wiley, New York

Maheshwari JK (1963) The flora of Delhi. Council of Scientific and Industrial Research, New Delhi

Manandhar NP, Manandhar S (2002) Plants and people of Nepal. Timber, Portland

Matthew KM (1995) An excursion flora of Central Tamil Nadu, India. A.A. Balkaema, Rotterdam

Medhi RP, Chakrabarti S (2009) Traditional knowledge of NE people on conservation of wild orchids. Indian J Trad Knowl 8(1):11–16

Müller-Höcker J (1992) Mitochondria and ageing. Brain Pathol 2(2):149–158

Musharof Hossain M (2009) Traditional therapeutic uses of some indigenous orchids of Bangladesh. Med Arom Plant Sci Biotechnol 3:100–106

Nadkarni AK (1954) Dr. K.M. Nadkarni's Indian materia medica, vol 2, 3rd edn. Popular Book Depot, Bombay

Nanakorn W, Watthana S (2008) Queen Sirikit botanic garden (Thai native orchids 1 and 2). Wanida Press, Chiang Mai

Nasir E, Ali SI (1972) Flora of West Pakistan. Pakistan Agricultural Research Council, Islamabad

Nayak BS, Suresh R, Rao AVC et al (2005) Evaluation of wound healing activity of *Vanda roxburghii* R. Br. (Orchidaceae): a preclinical study in a rat model. Int J Low Extrem Wounds 4(4):200–204

Neto M, Miranda MR, Cruz D (2011) *Zeuxine strateumatica* (Orchidaceae) goes south: a first record from Brazil. Kew Bull 66:1

Odoux E, Brillouet JM (2009) Anatomy, histochemistry and biochemistry of glucovanillin, oleoresin and mucilage accumulation sites in green mature vanilla pod (*Vanilla planifolia*; Orchidaceae): a comprehensive and critical reexamination. Fruits 64:221–241

Oliver-Bever B (1972) Drug plants in ancient and modern Mexico. Q J Crude Drug Res 12(4):1957–1972

Ormerod P, Cribb PJ (2003) Ecology of Zeuxine. In: Pridgeon AM, Cribb PJ, Chase MW, Rasmussen FN (eds) Genera orchidacearum, vol 3, Orchidoideae (Part 2) Vanilloideae. Oxford University Press, Oxford

Pandey NK, Joshi GC, Mudaiya RK et al (2003) Management and conservation of medicinal orchids of Kumaon and Garhwal Himalaya. In: Singh V, Jain AP (eds) Ethnoboany and medicinal plants of India and Nepal. Scientific, Jodhpur, pp 114–118

Parrotta JA (2001) Healing plants of Peninsular India. CABI, Wallingford

Peana AT, Moretti MDL (2008) Linalool in essential plant oils: pharmacological effects. In: Watson RR, Preedy VR (eds) Botanical medicine in clinical practice. CABI, Wallingford, pp 716–724

Pearce NR, Cribb PJ (2002) The orchids of Bhutan. Royal Botanic Gardens/Royal Government of Bhutan, Edinburgh/Thimphu

Prasad DN, Achari G (1966) A study of anti-arthritic action of *Vanda roxburghii* in albino rats. J Indian Med Assoc 46(5):234–237

Rao AN (2004) Medicinal orchid wealth of Arunachal Pradesh. Newsl ENVIS NODE Indian Med Plants 1 (2):1–7

Rao TA (2007) Ethno botanical data on wild orchids of medicinal value as practised by tribals at Kudremukh National Park in Karnataka. Orchid Newsl 2(2):1–7

Rao TA, Sridhar S (2007) Wild orchids in Karnataka. A pictorial compendium. Institute of Natural Resources

Conservation, Education, Research and Training (INCERT), Bangalore

Raskoti BB (2009) The orchids of Nepal. Bhakta Bahadur Raskoti and Rita Ale, Kathmandu

Reddy CS, Nagesh K, Reddy KN, Vatsavaya SR (2003) Plants used in ethnoveterinary practices by Gonds of Karimnagar District, Andhra Pradesh, India. In: Singh V, Jain AP (eds) Ethnobotany and medicinal plants of India and Nepal. Scientific, Jodhpur, pp 631–634

Rifai MA (1975) Extraordinary uses of orchids in Indonesia. Report First ASEAN Orchid Congress, Bangkok

Roberts E (1931) Vegetable materia medica of India and Ceylon. Platel, Colombo

Santapau H, Kapadia Z (1966) The orchids of Bombay. Govt. of India Press, Calcutta

Sarin YK (1996) Illustrated manual of herbal drugs used in Ayurveda. Indian Council of Medical Research, New Delhi

Schultes RE, Pease AS (1963) Generic names of orchids. Their origin and meaning. Academic, New York

Seidenfaden G, Wood JJ (1992) The orchids of Peninsular Malaysia and Singapore. Olsen & Olsen, Fredensborg

Seidenfaden G (1999) 149. Orchidaceae. In: Matthew KM (ed) The flora of the Palni Hills, South India. Rapinat Herbarium, Tiruchirapalli

Sharma J, Gairola S, Sharma YP, Gaur RD (2014) Ethnomedicinal plants used to treat skin diseases by Tharu community in district Udham Singh Nagar, Uttarakhand, India. J Ethnopharmacol 158(Pt A):140–206

Shin EJ, Whang WK, Kim SG et al (2010) Parishin C attenuates phencyclidine-induced schizophrenia in mice: involvements of 5-HT1A receptor. J Pharm Sci 113(4):404–408

Shoeb A, Chowta M, Pallempati G et al (2013) Evaluation of antidepressant activity of vanillin in mice. Indian J Pharm 45(2):141–144

Shyamala BN, Naidu MM, Sulochanamma G, Srinivas P (2007) Studies on the antioxidant activities of natural vanilla extracts and its constituent compounds through in vitro models. Agric Food Chem 55(19):7738–7743

Simmler C, Lobstein A, Antheaume C et al (2009) Isolation and structural identification of stilbenoids from *Vanda coerulea* (Orchidaceae). Planta Med 9:1020

Simmler C, Antheaume C, Lobstein A (2010) Antioxidant biomarkers from *Vanda coerulea* stems reduce irradiated HaCaT PGE-2 production as a result of COX-2 inhibition. PLoS One 5(10), e13713

Simmler C, Antheaume C, Andre P et al (2011) Glycosyloxybenzyl eucomate derivatives from *Vanda teres* stimulates HaCaT cytochrome c oxidase. J Nat Prod 74(5):949–955

Sinha AK, Sharma UK, Sharma N (2007) A comprehensive review on vanilla flavor: extraction, isolation and quantitation of vanillin and other constituents. Int J Food Sci Nutr 59(4):299–326

Sood SK, Rana S, Lakhanpal TN (2005) Ethnic aphrodisiac plants. Scientific, Jodhpur

Subedi A, Kunwar B, Choy Y et al (2013) Collection and trade of wild harvested orchids in Nepal. J Ethnobiol Ethnomed 9:64–73

Subramoniam A, Gangaprasad A, Sureshkumar PK, Radhika J, Arun BK (2013) A novel aphrodisiac compound from an orchid that activates nitric oxide synthases. Int J Impot Res 25(6):212–216

Sumithraarachchi DB (1986) Conservation of orchids in Sri Lanka. In: Rao AN (ed) Proceedings of the 5th ASEAN orchid congress. Parks & Recreation Department, Ministry of National Development, Singapore, pp 140–144

Tanaka T (1976) In: Nakao S (ed) Tanaka's cyclopedia of edible plants of the world. Keigaku, Tokyo

Tanaka H, Yee ATT (2003) Wild orchids in Myanmar, vol 1–3. The Foundation of Agricultural Development and Education, Bangkok

Teoh ES, Teoh K (1991) Over 45 feeling fabulous. Menopause and the hormone replacement controversy. Times Editions, Singapore

Trivedi VP, Dixit RS, Lal VK (1980) Orchids in the drug markets of Bareilly, Kanpur and Nearby Districts. Nagarjun (Calcutta) 23(8):157–163

Uddin SB, Yusuf M (2011) Medicinal plants of Bangladesh. Ethnobotany Lab, Chittagong University, Chittagong

Upadhyay R (2003) Ethnobotanical observations on some rare and endangered herbs of Singrauli in Madhya Pradesh. In: Singh V, Jain AP (eds) Ethnoboany and medicinal plants of India and Nepal. Scientific, Jodhpur, pp 989–992

Uphof JCT (1968) Dictionary of economic plants. Verlag von J. Cramer, Lehre

Vaddhanaphuti N (2001) A field guide to the wild orchids of Thailand, 3rd edn. Silkworm Books, Chiang Mai

Vaddhanaphuti N (2005) A field guide to the wild orchids of Thailand, 4th and Expanded edn. Silkworm Books, Chiang Mai

van Dyk S, MGlasson WB, Williams M, Gair C (2010) Influence of curing procedures on sensory quality of vanilla beans. Fruits 65(6):387–399

van Rheede HA (1703) Hortus indicus malabaricus, vol 12. Dutch East India Company, Kerala

Vij SP (1995) Orchid genetic diversity in India: conservation and commercialization. In: Proceedings of the 5th Asia Pacific orchid conference and show, Fukuoka, pp 20–39

Wood JJ, Beaman TE, Lamb A et al (2011) The orchids of Mount Kinabalu, vol 1. Natural History Publications (Borneo), Kota Kinabalu

Wu TL, Hu QM, Xia NH, Lai PCC, Yip KL (2001) Check list of Hong Kong plants 2001. Agriculture, Fisheries and Conservation Department Bulletin 1, Hong Kong

Yoganarasimhan SN, Chelladurai V (2000) Medicinal plants of India, vol 2, Tamil Nadu. Regional Research Institute, Bangalore

Part III

Future Directions

China has taken major steps to establish large forest reserves throughout the country. Unfortunately, in the process of playing catch-up, many third world countries are allowing their forests to be plundered or even destroyed. Studies are required to define forest regions that must be preserved, and governments need to be seriously involved in conservation rather than merely paying lip service.

Raising orchids require an understanding of its relationship with fungi, bacteria, habitats and climate. In addition to covering all these areas, this chapter also discusses micropropagation, germplasm preservation, seed banking, cryopreservation, vitrification, hybridisation, farming of medicinal species and good agricultural practice (GAP). Tables listing the medicinal orchid species found in a number of large national reserves are included.

Customs records from the nineteenth century revealed that tremendous quantities of *shihu* passed through ports where excise was collected. *Shihu* was the predominant medicinal orchid, followed by much lesser amounts of *Tianma* and *Baiji* (Hart 1884; Braun 1888). During the 1950s, when China was isolated from the rest of the world and the country had to rely heavily on herbs to treat disease, over 100,000 kg of *Tianma* (*Gastrodia elata*) was annually harvested entirely from natural sources. This nearly depleted wild populations of *Gastrodia elata*. In 1960, total trade in *shihu* was 70,750 kg of fresh pseudobulbs and 5800 kg of dried herb, with Sichuan and Yunnan Provinces supplying two-thirds of the total amount. In 1980, the quantity of herbs traded rose to 600,000 kg (Bao et al. 2011).

Trade in non-timber forest products (NTFP) is considered to be an important source of income for poor villagers in Laos. Unfortunately, most of this trade involves trading in wild orchid plants (Foppes Souvanpheng 2005). The collection of *Dendrobium* and *Anoectochius* species for sale to China as medicinal herbs is rapidly depleting wild sources in Laos (Lamxay 2007), and the same may soon happen in Myanmar. Illegal orchid trade in *Dendrobium catenatum* (recorded as *D. officinale*) from Laos in 2007 was estimated at 3500 kg. providing an income of US$52,500 or 16.6 % of the total value of illegal orchid trade in the country (Lovera and Laville 2009). To meet modern demand, and secondarily to protect such valuable plants from extinction through overcollecting and deforestation, these three major orchidaceous herbs, *Dendrobium*, *Bletillia*, *Gastrodia*, other orchids, and many herbs are now propagated and cultivated commercially (Lo et al. 2004; Li et al. 2006; Shen et al. 2010; Zhao et al. 2006).

Sources of Herbs Influence the Quality

Medicinal herbs with the same name collected from the wild differ in the proportion of their

© Springer International Publishing Switzerland 2016
E.S. Teoh, *Medicinal Orchids of Asia*, DOI 10.1007/978-3-319-24274-3_23

constituents and potency, these factors being dependent on correct species identification, the variety of the species, locality (province, altitude, orientation and soil conditions), maturity of the plant, timing of harvest and the preceding climactic conditions.

Over the centuries, herbalists have learnt to identify the optimum sources, and this information has been transmitted both orally and in their publications. In former times, the Chinese settlements grew fastest in Henan, the regions along the Huanghe and Yangzi Rivers from Sichuan to Zhejiang, and in Guangdong, a separate region in the south with its Pearl River. These four provinces are also the major sources of Chinese medicinal herbs. Classically, the source of many herbal products is designated by prefixing its name with the old names of these provinces. Thus, *Chuan Lian Zi*, *Chuan Bei Mu*, *Chuan Xiong* and *Chuan Wu* refer to herbs grown in Si-chuan, *Chuan* being the old name for Sichuan. They are considered to be superior to similar products obtained from other provinces. Similarly, there is *Zhe Bei Mu* and *Hang Ju Hwa*, both *Zhe* and *Hang* being old abbreviations for Zhejiang and its famous, classic capital, Hangzhou. *Huai* is an old name for *Henan* and its herbal products are thus designated; for example, Huai Ju Hua identifies *Chrysanthemum* as originating from *Henan*. A Guangdong source is indicated by *Guang*, for example *Guang Chen Pi* (Orange Peel from Guangdong).

The parasitic medicinal orchid, *Tianma* (*Gastrodia elata* Bl.), grows in shaded woods, in moist soil rich in organic matter at altitudes of 1200–1800 m. It is distributed in the northern and central states of Jilin, Liaoning, Hebei, Henan, Shanxi, Gansu, Anhui, Hubei, Sichuan, Guizhou and Xizang, but today the main sources are Guizhou, Sichuan, Yunnan and Xizang. Guizhou produces the best quality *Tianma*.

This came about because scientists at the Chinese Academy of Medicine led by Jintan Xu discovered a way to cultivate the orchid in vitro by switching the necessary symbiotic fungi, initially employing *Mycena osmundicola* for seed germination and then switching to *Armillaria mellea* for seedling development and maintenance of the mature plant. The discovery made it possible to cultivate the achlorophyllous orchid on a large commercial scale, rather than collecting it "from the wild". *Tianma* encompasses several varieties of *Gastrodia elata* and include such variants like *G. elata* Bl. f. var. *viridis* Makino, *G. elata* Bl. F. var. *glauca* S. Chow, *G. elata* Bl. F. var. *tuberculata* FY Lin et SC Chu, as well as *G. augusta* SC Chow et SC Chen.

Guizhou is an unsploit, subtropical, mountainous province with high rainfall and humidity and clear running streams. It is an ideal region for the cultivation of medicinal herbs. The province produces the best quality and the largest amount of *Baiji* (*Bletilla striata*) although *B. striata* is distributed in many provinces (Guizhou, Sichuan, Hunan, Hubei, Anhui, Henan, Zhejiang and Shaanxi), and the orchid is also cultivated in Yunnan, Jiangxi, Gansu, Jiangsu and Guangxi.

Two lithophytic species, *Dendrobium moniliforme* (syn. *D. candidum*) and *D. catenatum* (syn. *D. officinale*), which constitute the original *shihu*, were collected from the high mountains in central China, but these *Dendrobium* species also exist as epiphytes in woods. Acceptance of epiphytic *Dendrobium* as *shihu* eventually led to the inclusion of numerous epiphytic Chinese *Dendrobium* species in the medicinal produce. They come from numerous provinces (Table 23.1).

Dendrobium species are unequal in their potency, a point that may be reflected in their pricing. Earring *Dendrobium* retails for between Singapore $10 and $140 per 100 g from herbal outlets in Singapore. A comparison of the chemical composition of *D. candidum* and *D. nobile* was recently undertaken by Chen et al. (2006) from the Peking Union Medical College. They found great disparity in the total chromatographic peaks for alkaloids between the two species, 2.34 % for *D. moniliforme* (syn. *D. candidum*) and 41.87 % for *D. nobile*. However, *D. moniliforme* showed greater consistency in the composition of its alkaloids.

Table 23.1 Habitats of medicinal orchids and their host plants in Rolpa district, Nepal (data from: Pyakurei and Gurung 2008)

Epiphytic
1. Host tree: *Lyonia ovalifolia* (Angeri)
Bulbophyllum viridifolium, Chiloschista usneoides, Coelogyne corymbosa, C. cristata, C. ovalis, Cymbidium elegans, C. iridioides, Dendrobium aphyllum, D. bicameratum, D. chryseum, D. denudans, D. eriiflorum, D. heterocarpum, D. longicornu, Epigenium amplum, Gastrochilus calcoelaris, Kingidium taenialis, Oberonia acaulis, Pleione hookeriana, Vanda cristata, Cleisostema sp.
2. Host tree: *Benthamidia capitata* (Ban Litchi)
Den aphyllum, D. longicornu
3. Host tree: *Quercus leucotrichophora* (Banjh)
Bulb. viridifolium, Chiloschista usneoides, Coelogyne corymbooosa, C. cristata, Cym elegans, C. iridioides, Den aphyllum, D. bicameratum, D. chryseum, D. denudans, D. eriiflorum, D. heterocarpum, D. longicornu, Epigenium amplum, Gastrochilus calcoelaris, Pleione hookeriana, Vanda cristata, Cleisostema sp.
4. Host tree: *Diploknema butyracea* (Chiuri)
Aerides multiflora, A. odorata, Coelogyne flaccida, Coelogyne ovalis, Den aphyllum, D. heterocarpum, D. longicornu, Gastrochilus calcoelaris, Oberonia acaulis, O. sp., *Rhynchostylis retusa, Vanda cristata*
5. Host tree: *Berberis asiatica* (Chutro)
Chiloschista usneoides, Den denudans, D. eriflorum, Gastrochilus calceolaris, Kingidium taenialis
6. Host tree: *Myrica esculentia* (Kaphal)
Coelogyne corymbosa, C. flaccida, C. ovalis, Cym iridioides, Den aphyllum, D. bicameratum, D. denudans, D. eriiflorum, D. longicornu, Gastrochilus calcoelaris, Oberonia acaulis, Vanda cristata
7. Host tree: *Castanopsis indica* (Katush)
Bulb. viridifolium, Chiloschista usneoides, Coelogyne corymbosa, C. cristata, C. flaccida, C. ovalis, Den aphyllum, D. denudans, D. eriiflorum, D. heterocarpum, D. longicornu, Epigenium amplum, Oberonia acaulis, Vanda cristata
8. Host tree: *Persea odoratissima* (Kaulo)
Chiloschista usneoides, Coelogyne corymbosa, C. cristata, Cym elegans, C. iridioides, Den aphyllum, D. bicameratum, D. chryseum, D. denudans, D. eriiflorum, D. heterocarpum, D. longicornu, Epigenium amplum, Gastrochilus calcoelaris, Pleione hookeriana, Vanda cristata
9. Host tree: *Pinus roxburghii* (Khote salla)
Dendrobium aphyllum, D. denudans, D. longicornu, Rhyn retusa
10. Host tree: *Rhododendron aboreum* (Lali gurans)
Bulb. viridifolium, Chiloschista usneoides, Coelogyne corymbosa, C. cristata, C. flaccida, C. ovalis, Cym elegans, Cym iridioides, Den aphyllum, D. bicameratum, D. chryseum, D. denudans, D. eriiflorum, D. heterocarpum, D. longicornu, Epigenium amplum, Gastrochilus calcoelaris, Pleione hookeriana, Vanda cristata
11. Host tree: *Berberis aristata* (Lek chtro)
Chilochista usneoidoes, Den aphyllum, D. denudans, D. eriiflorum, D. longicornu, Kingidium taenialis, Vanda cristata
12. Host tree: *Engelhardia spicata* (Mauwa)
Aerides multiflora, A. odorata, Coelogyne flaccida, Coelogyne ovalis, Den aphyllum, D. bicameratum, D. eriiflorum, D. heterocarpum, D. longicornu, Gastrochilus calcoelaris, Oberonia acaulis, Rhynchostylis rertusa
13. Host tree: *Pyrus pashia* (Mayal)
Chiloschista usneoides, Den aphyllum, D. longicornu, Kingidium taenialis, Vanda cristata
14. Host tree: *Shorea robusta* (Sal)
Aerides multiflora, A. odorata, Coelogyne flaccida, Den aphyllum, D. eriiflorum, D. heterocarpum, D. longicornu, Gastrochilus calcoelaris, Oberonia acaulis, Rhynchostylis rertusa, Oberonia sp.
Lithophytic
Dendrobium denudans, Kingidium taenialis, Epigenium amplum
Terrestrial
Calanthe tricarinata, Coelogyne cristata, Coelogyne corymbosa, Cypripedium himalaicum, Dactylorhiza hatagirea, Herminium lanceum, Pleione hookeriana, Satyrium nepalense

One looks forward to seeing such data being collected in all countries

Identification of Source

Secondary metabolites are not the only thing that varies in orchids, depending on the source of the plant material. The composition of elements also differs with terrestrial species. Discriminant analysis using wavelength dispersive X-ray fluorescence (WDXRF) spectroscopy has been applied successfully to discriminate among 37 samples of *Vanilla planifolia* originating from Madagascar, Uganda, India and Indonesia and *V. tahitensis* from Papua New Guinea (Hondrogiannis et al. 2013). The process should be applied to Chinese medicinal herbs which are grown in the soil (Zhao et al. 2006), albeit it might be less helpful in the identification of epiphytic species.

Life Span of Above-Ground and Underground Organs of Orchids

Leaves, pseudobulbs, rhizomes, shoots and tubers of orchids are not immortal. They have a definite life span lasting from a few months to as much as 25 years (Tatarenko 2007). It follows that concentrations of secondary metabolites in such parts must reach their peak at some time during the orchid's life-span, and thereafter they decline. Medicinal orchids need to be harvested when their secondary metabolites are at their peak. The best time for harvesting an individual herb is as stated in the classic herb manuals, but it is often not strictly followed by herb gatherers.

Another conclusion that follows is that if such a back bulb of a sympodial orchid is not harvested, it simply goes to waste, albeit its component minerals are recycled. For instance, "*Gastrodia elata* produces a new underground stem tuber as an annual increment while the tuber produced during the previous year dies" (Tatarenko 2007). In the wild, it is impractical for the herb gatherer to remove only the older parts of each individual plant and to leave behind the juvenile portion. This process was advocated in Turkey to conserve the wild terrestrial orchids that were being collected for *salep*, but it did not work out. It is only practical in nursery settings where routine practice requires that old rhizomes be divided to encourage dormant buds on old pseudobulbs to develop into new plants before the old pseudobulbs perish. This is an additional argument for medicinal orchid farming.

Indispensable Mycorrhiza

Orchid seeds are present by the thousands in many orchid fruits. Wind dispersal is capable of distributing such seeds across long distances, so theoretically they should become established over vast areas, wherever suitable ecological conditions are present. Climate, altitude, soil conditions, host trees, pollinators, and the presence of suitable mycorrhiza determine whether suitable ecological conditions are met.

Understanding mycorrhiza is crucial for success with orchid conservation and there is now much research in this area because:

1. Mycorrhiza are essential for orchid seed germination. That knowledge is almost ancient (Bernard 1899; 1909; Burgeff 1909; 1934).
2. Following germination, many orchids switch to other symbiotic fungi which supply them with organic nutrients and minerals (Xu and Guo 2000; Gebauer and Meyer 2003; Bidartondo and Read 2008; Preiss et al. 2010). An orchid may associate with several species of mycorrhiza, and similarly the same fungus may be present in the roots of many orchid species and other plants (Hadley 1970; Chutima et al. 2010).
3. Second-wave mycorrhiza accelerate growth of protocorms by supplying the orchid with hormones and phytoalexins which enhance the orchid's defence against invasion by pathogenic organisms (Boller 1957; Chen et al. 2003; Liu et al. 2010b).
4. Survival rates for orchid seedlings raised by asymbiotic culture transferred to a natural habitat are abysmal, whereas seedlings germinated and raised with mycorrhiza have a better chance of surviving in the wild (Hollick 2004; Yam 2010). Hybrid *Cymbidium* plantlets achieve superior growth rates

and improved mineral uptake when inoculated with appropriate mycorrhiza derived from the roots of their parent (Zhao et al. 2014).

5. Orchids growing in dark forest floors, or species under stress, may need to switch from an autotrophic state to partial or even total temporary dependence on mycorrhiza (Preiss et al. 2010). In the northern temperate species, subterranean protocorms are devoid of chlorophyll, and the plants rely totally on mycorrhiza which have invaded their basal cells for their food supply (Hadley 1982).

6. Depending on their morphology, temperate terrestrial orchids differ in the extent of mycorrhizal infection in their adult stage. Apart from the mycotrophic orchids like *Gastrodia elata* and *Galeola septentrionalis*, orchids with long rhizomes, like *Epipactis* or thick storage roots like *Spiranthes*, are most heavily infected, while those with thin roots, like *Pogonia japonica* and *Liparis japonica*, are least infected (Tatarenko 2007). Ectomycorrhizal symbiosis is the rule in the temperate forests where hetereotrophic plants and trees share mycorrhizal fungi (Selosse et al. 2002). Tropical and subtropical heterotrophic orchids depend on saprophytic mycorrhiza.

7. The requirements of mycorrhiza should also be understood because their abundance is a prerequisite for the establishment of orchids in the natural habitat. Some mycorrhiza have special requirements for organic nitrogen. Some species flourish when supplied with vitamins; others do not (Stephen and Fung 1971; Hijner and Arditti 1973; Powell and Arditti 1975).

8. Mycorrhiza has an indispensable role in providing sugar for orchid seed germination in the wild. Recently, Cameron et al. (2007) showed that the relationship is symbiotic and not unilateral, and, at least in the case of *Goodyera repens*, the orchid supplies the fungus with nitrogen while the mycorrhiza continues to supply adult *G. repens* with phosphorus. Achlorophyllous orchids depend almost entirely on mycorrhiza for their carbon supply.

In the case of *Gastrodia elata*, a Japanese scientist demonstrated almost a 100 years ago that there was a symbiotic relationship between the orchid and *Armillaria mellea* (Kusano 1911). *G. elata* is widely distributed in Japan, and in the early twentieth century it was found growing in the Botanical Garden of the College of Agriculture, flourishing under *Sterculia platanifolia*. However, it generally grew under oaks (*Quercus serrata* and *Q. glandulifera*). Working on the tubers at the base of old inflorescences, Kusano established that *G. elata* was dependent on *A. mellea* for its carbon supply throughout its vegetative life (Fig. 23.1). Nourished by the fungus, the tuber reaches a length of 10–17 cm before flowering at the end of May. Meanwhile, it produces offshoots (Fig. 23.2) which on acquiring mycorrhiza develop and eventually flower. But many do not get infected with *Armillaria*, and these perish when they are still quite small or within a few generations (Kusano 1911). This knowledge did not lead to laboratory or open cultivation of commercially important *Tianma*

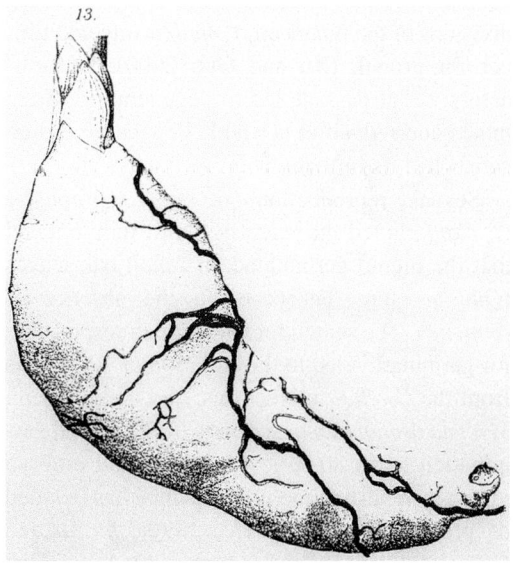

Fig. 23.1 Tuber of *Gastrodia elata* Bl. with *Amillaria mellea* growing on its surface. Histological examination of the tuber showed that the mycorrhiza also lived within the cells of the orchid (Adapted from the seminal paper by S. Kusano: *Gastrodia elata and its Symbiotic Association with Amillaria mellea*. J. Coll. Agri, Tokyo IV (1): 1–65, Fig. 13, 1911)

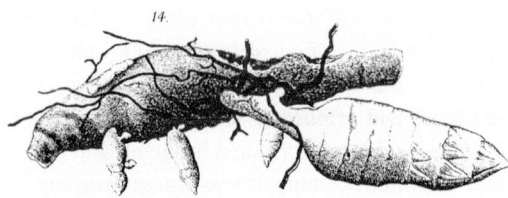

Fig. 23.2 Mycorrhizal invasion by *Amillaria mellea* of a withered *Gastrodia elata* tubercle laid under an oak tree in May promoted the formation and development of a offshoot tuber by September (Adapted from the seminal paper by S. Kusano: *Gastrodia elata and its Symbiotic Association with Amillaria mellea*. J. Coll. Agri, Tokyo IV (1): 1–65, part of Fig. 14, 1911)

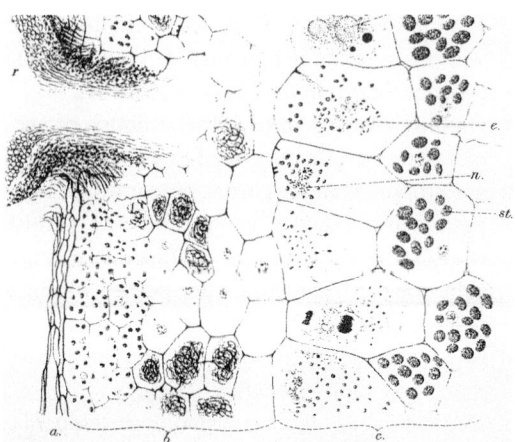

Fig. 23.3 Pelotons (intracellular coils of mycelia of *Amillaria mellea*) in the first layer cells of the tuber of *Gastrodia elata* (Adapted from the seminal paper by S. Kusano: *Gastrodia elata and its Symbiotic Association with Amillaria mellea*. J. Coll. Agri, Tokyo IV (1): 1–65, part of Fig. 17, 1911)

for another 50 years because the mycorrhiza inhibited germination of the orchid. A piece of the puzzle was missing.

A symbiotic relationship with *A. mellea* was essential for young tubers of *Gastrodia elata* to grow normally (Xu 1989). *A. mellea* is an edible mushroom (Kusano 1911), going by the name *Tianma mihuanjun* (*Gastrodia* honey mushroom) but it is not a common item in the menu. During the 1960s, Xu and Guo achieved the clonal propagation of *G. elata* by culturing budding divisions of the tubers on *A. mellea*-infected timber for growth (Xu and Guo 2000). Unfortunately, serial passage led to degeneration of the tubers and reduction in yield. Vegetative propagation had its limitations.

Sexual reproduction of the saprophytic *G. elata* eluded the scientists until they noticed that the orchid germinated on fallen oak leaves (*Quercus* spp., Fagaceae) in the absence of *A. mellea*. The search for mycorrhiza responsible for germination led to the isolation of 12 species from the *G. elata* protocorms, and one of them, *Mycena osmundicola*, supports *G. elata* seed germination to an 80 % level. The orchid embryo grows by digesting the fungus which has invaded its pro-embryonic cells (Xu 1990a, b; Xu and Guo 2000). Isolates from seeds which germinated carried through further passage provided strains of *Mycena osmundicola* which promoted better germination and protocorm development (Ran and Xu 1988). Subsequently, it was discovered that several mycorrhiza

isolated from other orchid species were also capable of supporting *G. elata* seed germination, among them *Mycena orchidicola* isolated from *Cymbidium sinense* (Fan et al. 1996), *M. anoectochila* from *Anoectochilus formosanus* (Guo et al. 1997) and *M. dendrobii* from *Dendrobium officinale* (Guo et al. 1999). *Epulorhiza albertaensis* from *Eria szetchuanica* also supported germination of *G. elata* (Fan and Guo 1998). On the other hand, the *A. mellea* on which adult *G. elata* is so dependent (Figs. 23.3 and 23.4) inhibits seed germination of the orchid (Xu and Guo 2000).

Liquid extract of *Mycena osmundicola* increased the percentage of seed germination of *Bletilla striata* in vitro, whereas a carbinol extract accelerated the differentiation and growth of the protocorm's cotyledon after initial incubation with the liquid extract. In nature, *B. striata* seeds germinate readily. *M. osmundicola* is the fungus which assists its germination (Guo and Xu 1992). Endophytic fungi in the roots of three terrestrial Thai orchids species (*Spathoglottis affinis*, *Paphiopedilum bellatulum* and *Phaius tankervilleae*) were found to produce high concentrations of an auxin, indole-acetic acid (IAA) (Chutima and Lumyong 2012), together

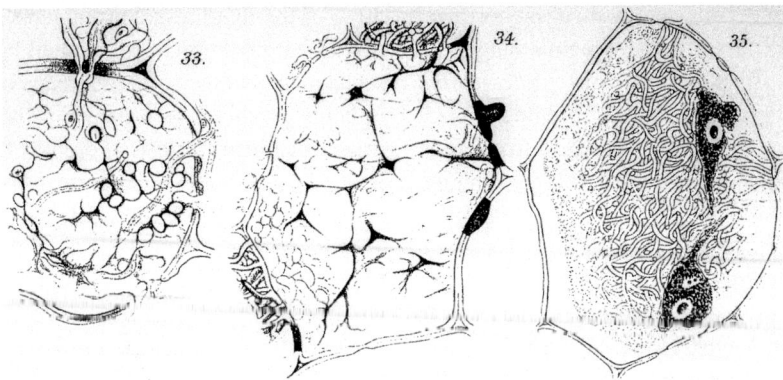

Fig. 23.4 Behaviour of *Amillaria mellea* in the tuber cells of *Gastrodia elata*. (33) "hyphae undergoing self-disorganization"; (34) nearly complete autolysis of the hyphae on which the orchid cell feeds; (35) in this cell, hyphae of *Amillaria mellea* persists even as the *Gastrodia elata* cell divides so both daughter cells will contain hyphae (Adapted from the seminal paper by S. Kusano: *Gastrodia elata and its Symbiotic Association with Amillaria mellea*. J. Coll. Agri, Tokyo IV (1): 1–65, part of Figs. 33–35, 1911)

with gibberellins (GA), dormin (ABA), zeatin (ZT) and zeatin riboside (ZR). Such compounds, referred to generically as fungal elicitors, promote seedling germination and differentiation, and overall they improve the growth of some medicinal orchids (Liu et al. 2010b), such as *Chiangniena amoena* (Yan et al. 2006), *Dendrobium candidum* (=*D. moniliforme*) (Chen et al. 2008) and *Cymbidium goeringii* (Dong et al. 2008). A 2.74-fold increase in the expression level of *S*-adenosyl-L-methionine decarboxylase was observed in protocorms of *D. officinale* (=*D. catenatum*) germinated via symbiotic culture with mycorrhiza. This key enzyme in the synthesis of polyamines is essential for numerous basic biochemical and physiological processes in the plant (Zhao et al. 2013). A growth promoting strain of *Sphingomonas paucimobilis* ZJSH1 associated with *D. officinale* (=*D. catenatum*) produces salicylic acid, indole-acetic acid, zeatin and abscisic acid. Seedlings inoculated with *S. paucimobilis* ZISHI contained higher levels of phyto-hormones and showed enhanced growth and greater polysaccharide levels (Yang et al. 2014).

Orchid growers are well aware that many orchid plants, especially the heavy feeders like *Phalaenopsis*, do better with organic than with chemical fertilisers (Hung Kuo Lian quoted by Raven-Riemann et al. 2010). The plants do even better with organic fertilisers fortified with amino acids. Photosynthesising green plants assimilate additional carbon and nitrogen sources provided by their mycorrhiza. This is proven in autotrophic and myco-heterotrophic orchids by the abundance of stable isotopic nitrogen 15 N and carbon 13C in their foliage and stems (Gebauer and Meyer 2003). *Cymbidium goeringii* switches its uptake of nitrogen from nitrates to ammonia when infected with mycorrhiza, and some fungal strains are capable of enhancing the orchid's uptake of glycine nitrogen (Wu et al. 2013).

Switching of mycorrhizal associates from typical *Rhizoctonia* species to ecto-mycorrhizal fungi that are simultaneously associated with trees widens the pool of organic nutrients available to the orchids (Bidartondo and Read 2008; Preiss et al. 2010). This is particularly important for myco-heterotrophic or partially heteromycotrophic species living on dark forest floors. Under low light conditions, two *Cephalanthera* species switch to a heterotrophic mode. When light intensity increases in late spring and summer, the orchids are fully autotrophic (Preiss et al. 2010). Adult orchids in open habitats in the Mediterranean and Macaronesia derived no carbon or nitrogen gains through myco-heterotrophy; indeed, isotope signatures showed instead orchid to fungus carbon transfer (Liebel et al. (2010). Nevertheless, even epiphytic orchids may need to switch their mycorrhizal associates at different developmental stages. In *Dendrobium chrysanthum*, one *Rhizoctonia* supports germination, another promotes plant growth, whereas the strain isolated from adult host plants did not promote seed germination and nor did it form associations with protocorms and plantlets (Hajong et al, 2013).

Armillaria causes root rot in many plants, so when *Gastrodia* is cultivated cognizance has to be taken of this factor. However, there seems to be no ecological danger, and it will probably remain this way provided there is no attempt to introduce genetic engineering. Gastrodin, vanillin and other compounds usually associated with the orchid are found in the culture media of pure cultures of *Armillaria*, and this is one way of obtaining the medicinal compounds. Gastrodin has also been synthesised. The synthetic material is available for oral and parental use in China.

Artificial propagation of *Anoectochilus roxburghii* which is a valued medicinal herb in China is difficult. Scientists at Peking Union Medical College have now discovered that eight strains of mycorhiza from various orchids encouraged survival and promoted growth of *A. roxburghii* plantlets in tissue culture medium. Of 42 strains studied, many killed the plantlets, but the following established a beneficial symbiotic system with the orchid plantlets: *Epulorhiza* sp. from *Dendrobium candidum* (=*D. moniliforme*) from Sichuan; two *Epulorhiza* spp. from *A. roxburghii* from Fujian; two *Gliocladium* spp. from *A. roxburghii* from Fujian; *Mycena anoectochilus* from *A. roxburghii* from Fujian; *Mycena dendrobii* from *D. candidum* (=*D. moniliforme*) from Sichuan; and *Moniliopsis* sp. from *D. nobile* obtained from Yunnan (Dan et al. 2011).

Epiphytic orchid species may be less particular than achlorophyllous or other terrestrial orchids about the mycorrhiza with which they associate to facilitate germination (Rasmussen 1995; Liu et al. 2010b). In a study involving mainly *Dendrobium* species, 12–100 % of the mycorrhiza tested were capable of promoting seed germination as compared with 0–33 % for terrestrial orchids (Liu et al. 2010b). More than

50 % of the fungi tested promoted seed germination of epiphytic orchids. "Fungal baiting" using seeds placed in the natural habitats of the species is being proposed as a means to obtain mycorrhiza of Chinese terrestrial orchids for investigation of their role in seed germination (Rasmussen and Whigham 1993; Liu et al. 2010b). On average, over 50 % of fungal species tested were able to promote germination of epiphytic orchids, whereas only 10 % of mycorhizza tested supported germination of terrestrial species (Liu et al. 2010b).

In some medicinal species, e.g. *Dendrobium catenatum* (syn. *D. officinale*) and *Cymbidium goeringii*, mycorrhiza prevented invasion by pathogens (Chen et al. 2003; Liu et al. 2010b). Seedlings raised by asymbiotic means rarely survive when transferred into natural habitats. For orchid conservation, it is better to use seedlings that have been germinated through the assistance of mycorrhiza (Hollick 2004). Studies in the field would help to promote the conservation of medicinal orchids in China and elsewhere. Plant size at transfer also determines survival rate (Yam 2010), and it is far better to transfer young, mature plants rather than seedlings. A total of 127 endophytic fungi were isolated from the protocorms and roots of *D. nobile* and *D. chrysanthum*, and there are plans to identify strains that would be useful for future large-scale cultivation of medicinal *Dendrobium* and to screen them for bioactive metabolites (Chen et al. 2008, 2011). Only 39 % of the fungi strains investigated supported seedling growth of terrestrial species, whereas 50 % supported seedling growth of terrestrial orchid species (Chen et al. 2012). Many terrestrial autotrophic orchids continue to obtain their carbon and nitrogen from mycorrhiza (Gebauer and Meyer 2003). Highly successful orchids which are extensively distributed probably owe their success to an ability to benefit from a wide spectrum of mycorrhizal partners; for instance, *Gymadenia conopsea* associates with typical orchid mycorrhizal families like Tulasnellaceae and Ceratobasidiaceae as well as generally non-orchidaceous mycorrhiza like the several taxa of Pezizales (Stark et al. 2009).

Bacteria

In horticultural circles, bacteria are generally considered as pests, *Pseudomonas* species being the cause of bacterial soft rot particularly in overwatered seedlings, but sometimes in adult plants, commonly in *Phalaenopsis*. Orchids produce phytoalexins to ward off bacterial invasion, but this does not prevent many bacteria from growing on the orchid stem and roots. Until recently, the importance of bacteria in the ecological relationship of orchids has largely been overlooked.

Twenty-seven strains of bacteria were recently isolated from the roots of the medicinal orchid *Cymbidium faberi* by Chu et al. (2010) at Wuhan Botanic Gardens in China. They belonged to 14 species and 8 genera, namely *Burkholderia phytofirmans*, *Pseudomonas jessenii*, *P. cedrina*, *P. mohnii*, *P. fragi*, *P. koreensi*, *Bacillus megaterium*, *B. atrophaeus*, *B. subtilis*, *Leifsonia shinshuensis*, *Variovora paradoxus*, *Erwinia rhapontici*, *Duganella zoogloeoides* and *Acinebacter woffi*. Their occurrence might be beneficial to the orchid. Some species of soil bacteria fix nitrogen; some synthesise vitamins. They also produce auxoins, in particular indole-3-acetic acid (IAA) (Tsavkaelova et al. 2007; Tong et al. 2014).

Scientists in South America support this view. It was suggested that bacteria such as *Bacillus megaterium*, *Pseudomonas koreensis* and *Actinobacter* spp. should be regarded as members of the beneficial microorganism population that exists in the rhizosphere of *Vanilla planifolia* and perhaps they could be used as biofertilisers to improve plant nutrition and growth of the valuable orchid (Alvarez-Lopez et al. 2013).

Bacterial associates of a terrestrial and epiphytic orchid species may differ. From the terrestrial orchid *Paphiopedilum appletonianum*, the following bacteria were isolated: *Streptomyces*, *Bacillus*, *Pseudomonas*, *Burkhoderia*, *Erwinia* and *Norcardia*. The following bacteria genera were found in epiphytic *Pholidota articulata*: *Pseudomonas*, *Flavobacterium*, *Stenotrophomonas*, *Pantoea*, *Chryseobacterium*, *Bacillus*, *Agrobacterium*, *Erwinia*, *Burkhoderia* and *Paracoccus* (Tsavkaelova et al. 2007).

In three Chinese provinces, *Dendrobium loddigesii* harbours 67 endophytic bacterial strains belonging to 12 genera, namely *Pseudomonas*, *Microbacterium*, *Eneterobacter*, *Bacillus*, *Spingomonas*, *Staphylococcus*, *Psychrobacter*, *Brevundimonas*, *Nesterenkonia*. *Paracoccus*, *Pantoea* and *Serratia*, with stems carrying the most strains (32). A total of 42 strains were present on plants from Guangxi Province which had the highest abundance of bacteria. Thirty strains solubilised organic and inorganic phosphorus, 22 delivered potassium and 24 produced the auxin, IAA. Only eight strains possessed all three orchid growth-promoting properties (Tong et al. 2014).

Light and Moisture

Trees that provide dappled sunlight are necessary for epiphytic orchids (Fig. 23.5), and some terrestrial species only thrive in the deep shade of forests (Figs. 23.6 and 23.7). Many terrestrials that bask in strong sunlight at high altitudes require open spaces devoid of trees and shrubs,

Fig. 23.5 *Dendrobium loddigesii* Rolfe is a small, delicate-looking species with branching stems. It forms dense clumps on trees or rocks in southern China and Laos (Photo: E.S. Teoh)

Fig. 23.6 *Malaxis acuminata* D. Don in the company of *Acanthephippium striatum* Lindl. growing in dense shade of broad-leaved forest at 1406 m. at Rimchu, Punakha, Bhutan (Photo: E.S. Teoh)

Fig. 23.7 *Cypripedium himalaicum* Rolfe growing in the open but amidst grasses and fern on the bank of a currently-dry, shallow stream at Chendebi, Bhutan (Photo: E.S. Teoh)

Fig. 23.8 *Spiranthesis sinensis* (Pers.) Ames in a bog which is exposed to full sun on a hill slope at Taktsang, Bhutan (Photo: E.S. Teoh)

but they need the company of long grasses to provide partial shade and humidity or, in the case of some *Cypripediums*, short grasses to obtain more light and afford necessary moisture retention around the roots (Figs. 23.8, 23.9, 23.10 and 23.11). In Japan, it is observed that favourable vegetation for the establishment of its seedlings are narrow-leaved, medium-sized grasses, sedges, herbs, mosses and prostate mat-forming shrubs which offer the orchids a stable moisture and temperature environment and the right amount of light (Kosaka et al. 2014).

Terrestrial and saprophytic orchids are especially susceptible to decreases in soil moisture, and potential climate change worries conservationists at the Yachang Reserve, a wild orchid hotspot in southwestern China where half its orchid population (68 species) are terrestrial or saprophytic. To make matters worse is the fact that the vast majority of species occur on limestone mountain tops, which narrows the scope for them to move higher to cooler climes in the event of global warming (Liu et al. 2010a).

Protecting the Sources: Conservation

Ignorance, expedience and the search for a quick profit have led to the destruction of natural orchid habitats by some local people. That is not quite the same as the vandalising of habitats

Fig. 23.9 *Ponerorchis chusua* (D. Don) Soo and *Aorchis spathulata* (Lindl.) Vermeulen on temperate cattle grazed meadow at 4000 m in Bhutan. Low shrubs and frequent mists ensures constant moisture supply during their growth and flowering periods (Photo: E.S. Teoh)

Fig. 23.10 *Anacamptis pyramidalis* (L.) Rich. also thrives in full sun but tall grass and tight communities prevent drying of its roots (Photo: Henry Oakley, taken in July 2012)

to prevent other people from finding the same orchid, and thus boost the value of their collection (Vij 1995). At the height of orchid mania during the nineteenth century, several famous collectors engaged in both practices, overcollection and the destruction of residual orchid communities (Swinson 1970; Koopowitz 2001).

In several Asian countries, there is widespread usage and high trade demand for medicinal orchids. Demand for *Dactylorhiza hatagirea*, which is valued as an aphrodisiac in India and Nepal, is estimated to be 5,000,000 kg per year, with 90–100 mature plants being required to constitute 1 kg of dried bulb, each kilogram fetching 100–200 rupees. Consequently, *D. hatagirea* is overcollected from its unique habitats by ignorant herb gatherers whose livelihood is ironically so dependent on the availability of these plants. The situation is worsened by cattle grazing (Pant and Rinchen 2012; Bhattarai et al. 2014), and the fact that the wild Impeyan pheasant (*Lophophorus impejanus*) digs up its

tubers for food (Giri et al. 2008). Some attempts are being made to produce plantlets for replanting and conservation either by vegetative subdivision of the digitate tubers (Rajasekaran et al. 2009), *Ceratobasidium*-assisted, symbiotic seed germination (Aggarwal and Zettler 2010) or asymbiotic seed germination (Warghat et al. 2014). However, one would be hard put to keep up with the rapacious harvest of mature plants by herb gatherers (Ghorbani et al. 2014).

Prof. Chen Sing Chi of the Chinese Academy of Sciences in Beijing, a leading authority on orchids in China, highlighted the overcollection of *Dendrobium* from the prefecture of Xin Yi in southwestern Guizhou. Formerly, this prefecture was so well known for its *Dendrobium* that it was referred to as *Huang Cao Ba* (*Dendrobium* Forest), but "recently hundreds of tons of

Fig. 23.11 *Arundina graminifolia* (D. Don) Hochr. and *Spathoglottis plicata* Blume growing wild at the edge of secondary forest in Singapore. They were two of the three first orchid species to re-establish on land laid waste by the massive volcanic eruption of Krakatoa in 1883. Their needs are full sunlight with shading for the roots, and tropical rains (Photo: E.S. Teoh)

Dendrobium have been gathered there for medicinal uses and this kind of orchid has become very rare now" (Chen 1995). Public education which emphasises the importance of conserving the natural environment, and regulations which provide some leeway for rural folk to benefit from their natural wealth without being destructive to natural habitats, should have a role in the conservation of nature reserves (Heinen 2010).

Overcollection is not the only threat faced by orchids. By far the most serious cause of the disappearance of wild orchids today is forest destruction caused by harvesting of timber and clearance of forests for commercial use and housing. Annually, 1.7 % of forests in Nepal is being destroyed. In eastern and central Nepal, where orchid biodiversity is highest, the annual loss rate is even higher, at 2.3 %. Now, the Annapurna Conservation project is trying to conserve the habitats where the pressure on wild orchid existence is most severe (Subedi 2005). Climate change, particularly global warming and extreme climatic fluctuations, encroachment of human populations which become barriers for

migration of plant populations, especially migration towards the poles, the lack of potential for migration to higher altitudes for orchids growing in limestone habitats already close to the mountain tops, and inadequate size of nature reserves, destabilise the ecological balance over time. Such factors all threaten the survival of orchid species (Liu et al. 2010a).

In a balanced ecosystem, orchid species survive on their own without human support, i.e. they do not need supplemental nutrients and pesticides. For orchid species to endure in the wild, such natural ecosystems need to be preserved through the maintenance of large forest preserves. Many countries have set aside land for nature reserves. However. Koopowitz (2001) pointed out that, generally, forest preservation consists mainly of words on paper. Economic goals, predatory businessmen and greedy politicians threaten the natural environment. Jungle clearing for commercial purposes has always been the greatest threat to wild orchids (Bacon 1975). *Paphiopedilum dianthum*, once thought to be endemic in China, was discovered in Vietnam in 2000, but logging in that country soon led to the disappearance of the species in many of its natural habitats (Averyanov et al. 2003). *Vanda spathulata* was previously very common at Veli near Trivandrum in Kerala State, but after the area was cleared and occupied by the Vikram Sarabhai Space Centre, the orchid disappeared from that locality (Abraham and Vatsala 1981). In many developing countries, agriculture and horticulture take back seats and conservation is a dirty word among influential politicians and their sponsors. In such a context, orchid conservation is not of real concern to the powers that be.

Conservation in China

Regarding medicinal orchids, efforts are being made in China and Thailand to ensure that they continue to survive in nature. Such efforts are laudable, but excessive human attention may tend to favour some species at the expense of others, and interfere with the larger picture of evolutionary radiation which orchids are undergoing. Land is set aside as nature reserves generally to protect water catchments, and secondarily to protect animal and tree species, and perhaps, but only in China, to preserve medicinal herbs which are crucial to affordable health-care. Orchids are a minor consideration. Nevertheless, many orchids have multiple precise requirements, such as the presence of specific mycorrhiza, insect pollinators, tree species for anchorage, mosses and lichens to protect the seeds and seedlings, proximity to streams, specific soil types, amount of sunlight, ambient temperature and seasonal rainfall, etc. Epiphytic orchids do better in specific host trees (Jayaweera 1981; Pearce and Cribb 2002; Harshani et al. 2014). Terrestrial orchids also have their preferences: some grow in dense shade, others in open scrub. For instance, favourable vegetation for *Cypripediums* consists of narrow-leaved, medium-sized grasses, sedges, short herbs, mosses and prostate mat-forming shrubs, as such ground cover plants provide sufficient light, moisture, temperature and ground surface for seedlings to germinate and plants to flourish (Kosaka et al. 2014).

Some orchid pollinators have specific dietary requirements and additionally they may need to collect various plant chemicals for defence and reproduction. Loss of critical biotic resources or relationships may compromise the abundance of orchid pollinators or their effectiveness and bring about a decline in the population size and genetic variation of orchid species. To ensure success in the conservation of vital pollinator resources, it is necessary to provide large areas of pesticide-free orchid habitats (Pemberton 2010). The larger the reserve, the greater is the likelihood that such requirements will be met.

China has an estimated 1200 species of orchids in approximately 170 genera, and because of the country's enormous size, a broad range of ecosystems. It has a wealth of geophytic (terrestrial) and numerous lithophytic species. With the growth of scientific awareness and progress in the nation, its government is now paying great attention to the conservation and reforestation. By the end of 1993, China had set up 766 nature reserves covering an area of over 66 million hectares. Many of the important orchid habitats are included, such as Xishuangbanna Nature Reserve in southern

Yunnan, Gaoligongshan Nature Reserve in western Yunnan, Dinghushan Nature Reserve in Guangdong, Jianfengling Nature Reserve in Hainan and Wuyishan Nature Reserve in Fujian (Chen 1995). Two-thirds of the orchid genera present in China occur in Yunnan (Chen and Tang 1982). Huanglong World Natural Heritage Site has 10 species of medicinal orchids in the warm temperate zone and 4 in the alpine zone (Perner 2002, 2007). At Gaoligongshan, of the 401 species documented by Jin et al. (2009), 110 are medicinal. At the Yachang Orchid Nature Reserve in Guangxi, there are 71 medicinal species (Liu 2010). Tables 23.1 and 23.2 list the medicinal orchids that are protected at the Gaoligongshan Nature Reserve in Yunnan and the Yachang Nature Reserve in Guangxi, respectively.

According to Jia Jiansheng, the Deputy Director of the Department of Wildlife Conservation, State Forestry Administration of China, orchids are an important component in the long-term project launched by the Chinese government to preserve wildlife. The project is tapping both local and international expertise. By 2007, it had designated 2349 nature reserves which cover a total of 150 million hectares which is more than 15 % of the territory. Of particular importance are the Yachang Orchid Nature Reserve in Guangxi province established in 2005, an ex-situ conservation centre in Shengzheng, Guangdong Province, and germplasm and seed banks focussing particularly on the resources of the southwestern region (Jia 2007). It is home to 36 terrestrial medicinal species (Table 23.3). Vast territories for conservation are needed because fragmented plant populations eventually lose genetic diversity. Diminishing allelic richness was observed when populations of *Cymbidium goeringii* became fragmented in Korea (Chung et al. 2014).

Korea

Although forests covers 64 % of the land in the Korean peninsula, 29 orchid species are listed as critically endangered in the country and 11 of these are medicinal: *Calanthe discolor* for. *sieboldii* (Decne.) Ohwi, *Cymbidium kanran*

Table 23.2 Medicinal orchid species at Gaoligongshan
Nature Reserve, Yunnan, China

Acampe rigida
Amitostigma simplex
Anoectochilus roxburghii
Anthogonium gracile
Arundina graminifolia
Bletilla formosanus
B. orchracea
Bulbophyllu. cylindraceum?
B. kwantungense
B. odoratissimum?
B. pectinetum
B. reptens
B. retusiusculum
B. umbellatum
Calanthe alismaefolia
C. alpina
C. brevicornu (lamellosa)
C. davidii
C. densiflora
C. puberula (similis)
C. tricarinata
C. triplicata
Cephalanthera falcata
C. longifolia
Cleisostoma williamsonii
Coeloglossum viride
Coelogyne barbata
C. corymbosa
C. elata (C. stricta)
C. flaccida
C. leucantha
C. nitida
C. occultata
C. ovalis
C. prolifera (35)
Cremastra appendiculata
Cymbidium faberi
C. floribundum
C. goeringii
C. hookeranum
C. kanran
C. lancifolium (42)
Cypripedium elegans
C. guttatum
C. henryi
C. tibeticum (Perner)
D. aphyllum
D. chrysanthum
D. falconeri

(continued)

Table 23.2 (continued)

D. fimbriatum (50)
D. hookerianum
D. longicornu
D. moniliforme (crispulum)
D. nobile
D. thyrsiflorum
Epigeneium amplum
Epipactus helleborine
Eria bambusifolia
E. graminifolia (60)
Eulophia spectabilis
Galeola lindleyara (Chandra Pradhan MOR 29/95 p 68)
Gastrochilus distichus
Gastrodia elata
Goodyera biflora (pauciflora)
G. foliosa
G. henryi
G. procera
G. repens (nantoensis)
G. schlectendaliana (70)
Gymnadenia orchidis (conopsea)
Habenaria delavayi
H. dentata
Hemipelia flabellata
Herminium lanceum
Liparis bootanensis
L. caespitosa
L. cathecartii
L. distans
L. japonica (80)
L. nervosa
L. odorata
L. rostrata
L. stricklandiana
L. viridiflora
Malaxis acuminata (Crepidium acuminatum)
M. monophyllos
Nervilia aragona
Oberonia caulescens
O. ensiformis (iridifolia) (90)
Oreorchis patens
Pecteilis susannae
Peristylis bulleyi (=Herminium bulleya)
P. densus (Habenaria stenostachya)
Phaius tankervilleae
P. wilsonii
Pholidota articulata
P. chinensis
P. imbricata
P. yunnanensis (100)

(continued)

Table 23.2 (continued)

| Platanthera japonica |
| P. minor |
| Pleione praecox |
| P. yunnanensis |
| Ponerorchis chrysea |
| Satyrium nepalense var. ciliatum |
| Satyrium nepalense var. nepalense |
| Spathoglottis plicata |
| S. pubescens |
| Spiranthes sinensis (110) |

Total Orchid Species: 401. Medicinal species from list: 110
Reference: Jin XH, Zhao XD, Shi XC (2009): *Native Orchids from Gaoligongshan Muntains*, China. Beijing: Science Press

Table 23.3 Medicinal orchid species found in the Yachang Orchids Nature Reserve in Guangxi Province, China

	Epiphytic (E)	Saxicolus (S)	Terrestrial (T)
Anoectochilus roxburghii			T
Bletilla formosana			T
B. ochracea			T
B. striata			T
Bulbophyllum ambrosia	E		
B. kwantungense	E		
B. odoratissimum	E		
Calanthe davidii			T
C. reflexa			T
C. sylvatica			T
C. triplicata			T
Cephalanthera longifolia			T
Cleisostoma paniculata	E		
C. williamsonii	E		
Coelogyne fimbriata	E		
C. flaccida	E		
Cremastra appendiculata			T
Cymbidium aloifolium	E		
C. bicolor			T
C. ensifolium			T
C. faberi			T
C. floribundum	E		
C. goeringii			T

(continued)

Table 23.3 (continued)

	Epiphytic (E)	Saxicolus (S)	Terrestrial (T)
C. goeringii var. serratum			T
C. kanran			T
C. lancifolium		S	T
C. simense			T
Cypripedium henryi			T
Dendrobium aduncum	E		
D. aphyllum	E		
D. chrysanthum	E		
D. denneanum	E		
D. devonianum	E		
D. fimbriatum	E		
D. hancockii	E		
D. henryi	E		
D. hercoglossum	E		
D. lindleyi	E		
D. loddigesii	E		
D. lohohense	E		
D. nobile	E		
D. officinale	E		
D. williamsonii	E		
Epipactus helleborine			T
Eria corneri	E		
Galeola lindleyana			T
Gastrodia elata			T
Geodorum densiflorum			T
G. recurvum			T
Goodyera schlechtendaliana			T
Habenaria ciliolaris			T
H. davidii			T
H. dentata			T
H. fordii			T
H. petelotii			T
Herminium bulleyi			T
H. lanceum			T
Liparis bootanensis	E		
L. japonica	E		
L. keitaoensis (L. cordifolia)			T
L. nervosa	E		
L. stricklandiana	E		

(continued)

Table 23.3 (continued)

	Epiphytic (E)	Saxicolus (S)	Terrestrial (T)
L. viridiflora	E		
L. yunnanensis (L. distans)	E		
Luisia teres	E		
Malaxis latifolia			T
M. monophyllos			T
Nervilia fordii			T
Oberonia myosuurus	E		
Paphiopedilum dianthum		S	
Zeuxine strateumatica			T

Terrestrial species 36; epiphytic species 33; saxicolus species 2; saxicolus and terrestrial species 1
This list is based on the List of Orchids at Yachang compiled by Liu et al. (2010a, b)

Makino, *C. lancifolium* Hook., *guttatum* var. *koreanum* Nakai, *C. japonicum* Thinb., *C. macranthos* Sw., *Dendrobium moniliforme* (L.) Sw., *Galeola septentrionalis* Rchb. f., *Goodyera repens* (L.) R. Br., *Gymadenia conopsea* (L.) R. Br., and *Liparis nervosa* (Thunb.) Lindl. Habitats for some species are scarce (in the case of *Dendrobium moniliforme* limited to one or two) and the numbers of individuals are few. *Bulbophyllum inconspicuum* occurs in small populations in fewer than 10 places surveyed, and the species is also considered as endangered, albeit not critically so. Illegal collection has decimated the ornamental species. Careless overcollection of *Cremastra appendiculata* resulted in habitat destruction and a rapid decline in plant population (Lee 2009). However, several of the medicinal species enjoy a widespread distribution either in Asia or even throughout the world, and, from a global viewpoint, they may not be critically endangered.

India

India is estimated to have about the same number (1100–1220) of orchid species as China, with 750 species in the northeast, and the rest spread over the country. The state of Arunachal Pradesh has 550 species, with 370 epiphytes, 160 autotrophic terrestrials and 20 saprophytes. Of the 550 species, 314 are endemic, 300 are graded as rare, 215 are endangered, 2 are placed under CITES Category I (*Renanthera imshootiana* and *Vanda oerulea*, the latter a medicinal species) and 14 are nearly or possibly extinct (Hegde 1997; Rao 2004). Only 37 are known to have a medicinal value (Rao and Henry 1995; Rao 2004), a very low percentage when compared to the situation in China. A dozen of the medicinal orchids are simultaneously of ornamental interest. Commercial exploitation of the forests has led to the disappearance of many species, and Rao observed that in Arunachal Pradesh the orchids are struggling to survive. Medicinal orchids are harvested indiscriminately, although trade in orchids is regulated under Schedule VI of the Wildlife Protection Act (1972). At least four endemic or rare species of *Calanthe* which were first collected from India have since not been encountered for the last 100 years, including *C. alpina* which is a medicinal species (Rathore 1983). Koopowitz (2001) was appalled by a report that 100 truckloads of *Dendrobium macraei* which must contain some 500,000 wild plants were being shipped from Nepal to India for medicinal use. How could nature repeatedly endure such reckless harvest?

This is surprising, given that archeological remains attest to an ancient love and respect for plants in that subcontinent (Teoh 2003). Gardens devoted to the cultivation of medicinal herbs under the supervision of experts were promoted during the Buddhist regimes, and Asoka the Great provided large state grants for their establishment (Chopra 1933). The Asoka Edict at Kalinga (now Orissa State) inscribed during the Mauryan Period in the third century BC called for the establishment of Gardens for Medicinal Herbs in all corners of the Emperor's domain to provide for the care of the sick and wounded (see photo).

Unfortunately, modern commerce has no sentimentality; it values only immediate returns. In the first quarter of the twentieth century, the government provided 15,000 acres for the cultivation of cinchona which is the source of the antimalarial drug, quinine. Actual cultivation was under 6000 acres. The major portion of the remaining land was occupied by narcotics such as hemp, tobacco and opium (Chopra 1933).

The giant shipment of *Flickingeria fimbriata* (=*Dendrobium plicatile* Lindl.) recalls an incident in the *Ramayana*, a legendary epic whose origin dates to the first millennium BC.

> During a battle waged by Rama against Ravana in Sri Lanka, the hero Lakshmana was mortally wounded by a poison arrow. He was saved by the asura physician, Susena, who asked for Hanuman to bring him the Sangeevani, a rare herb found only in the Himalayas. The plant could bring a dead person back to life. (This herb is now tentatively identified to be either Dendrobium plicatile or a fern.) When he arrived at the mountains, Hanuman could not decide whether he had the correct plant in hand. To settle his predicament, he ripped off a part of the Himalayas and took it back to Sri Lanka whereupon the physician picked the herb that he needed and Lakshmana was resurrected. Hanuman was no ordinary monkey: he was a god empowered with unlimited power and transformations and thus equal to the task (Fig. 23.12).
>
> *F. fimbriata* is now distributed from the Himalayas to Sri Lanka.

The story carries a simple lesson. When orchids are removed from the wild, distributed, and properly cultivated, ensuring their survival in a new, albeit man-made environment, it is not necessarily a bad thing. In third world countries, many governments commit forests to destruction in their effort to improve the national economy. When forest trees are felled, it is far better if their orchids are collected than left to burn or perish through neglect. Currently, CITES regulations do not permit the collection of orchids from fallen trees. There is much argument over the soundness of this prohibition (Hansen 2001; Koopowitz 2001).

Left to their own means, simple villagers take pains to cultivate some medicinal plants in their own gardens for easy access when they are required. In their study of anti-inflammatory plants used by the Lohit community in Arunachal Pradesh, Namsa and his colleagues discovered that *Eria pannea* (*Mycaranthes pannea*) was the only orchid among the 34 plants used for this purpose. That is probably because orchids as a rule are not as easy to come by as plants which also serve as vegetables. To have a ready supply of *Mycaranthes pannea*, whose fresh leaves are required to make a paste for application on dislocated joints and traumatic swellings, the orchid was cultivated near their homes. It was also an indication that the villagers have a respect for nature and that they appreciate the importance of conservation and sustainable utilisation of ethnomedicinally important plant species (Namsa et al. 2009).

India has embarked on conservation. The Niyamgiri Hills of Orissa are home to 31 species of orchids (19 epiphytes and 12 terrestrials), of which 20 are being used by the Dongaria Kandha tribes as medicinal remedies for the treatment of snake and scorpion bites, infections, alimentary disorders, bone aches, trauma, pregnancy and sexual dysfunction. Tribals consider the hills to be sacred. The region is noted for its biodiversity, but the entire ecosystem of the hill areas would be under threat if the proposal by Vedanta Resources to mine 72 million tons of bauxite from the area had succeeded (Dash et al. 2008). Following strong protests by the tribals and their international supporters, the Indian government banned mining in the Niyamgiri Hills of Orissa (Chaturvedi 2014).

On the other hand, it was observed that the intrusion of motorised traffic in the Western

Fig. 23.12 Hanuman carrying a piece of Himalaya and lotus to Sri Lanka. Bas relief at Jaina Cave Temple in Udayagiri, Odisha, India, first century BC (Photo: E.S. Teoh)

Ghats has encouraged the proliferation of epiphytic orchids along the highway which Rao and Sridhar attribute to the carbon emissions from the vehicles (Rao and Sridhar 2007). Light probably also plays a significant role.

Two orchid conservation sanctuaries have been established in the West Kameng district of Arunachal Pradesh. The sanctuary at Sessa is home to 400 species of orchids which include several saprophytes. An Orchid Research and Development Centre was set up in West Kameng, and it has successfully established a few hundred thousand orchid plants collected from various terrains in the state. Some of these will be returned to forests which are considered safe and harbouring suitable habitats. Seed- and micropropagation are being pursued (Hegde 1984, 1997).

Sacred Groves in India have traditionally been left undisturbed by development because these spots are considered the property of gods. Trees are not felled and both animal and botanical inhabitants are left untouched. There are thousands of these groves in India, with 65 in Arunachal Pradesh, 40 in Assam, 365 in Manipur, 83 in Medhalaya and 56 in Sikkim (Medhi and Chakrabarti 2009). They have been presented as suitable sites for orchid conservation. However, even though orchids are present in many of these sacred groves, unfortunately most of such patches of protected land are too small to be considered reliable as self-replicating communities for orchid plant life.

Ancient tribals living in the southeastern part of the Indian Peninsula may have retained some knowledge of Siddha medicine, but they have scarce knowledge of the medicinal usage of orchids. Three tribal communities in Tamil Nadu used a single species of orchid to treat disease (Sivakumar et al. 2003; Ganesan and Kesavan 2003), whereas many tribal communities in the same region have no medicinal usage for orchids (Muthu et al. 2006; Sandhya et al. 2006; Kottaimuthu 2008; Darairaj and Kanaraj 2013; Senthikumar et al. 2013; Sivasankari et al. 2013).

However, over 20 % (approximately 250) native orchid species occur in Peninsular India

with 5 perhaps already extinct and 25 under threat. Thirteen peninsular species are endemic. There is, therefore, also a need to establish orchid sanctuaries in Peninsular India. Currently, serious conservation work is being undertaken at the Tropical Botanic Garden and Research Institute at Pacha Parlode in Kerala, and at the Botany Department, Chandigarh and the Indian Institute of Horticultural Research at Hesarghatta, Karnataka. Laboratories for seed germination and tissue culture have been set up in West Bengal, Orissa and Andra Pradesh to propagate rare, endangered and ornamental orchids (Hegde 1997).

Ex situ conservation appears to be the answer for the numerous endemic species, because usage of land for agriculture makes more sense for most people. In the Shevaroy and Kolli Hills of Tamil Nadu, habitats of orchids are under severe pressure. In Shevaroy, forests have been converted into coffee plantations, and for pineapple cultivation.in the Kolli Hills Forest fires are deliberately set to produce and collect charcoal, and also to convert forests into grasslands for cattle grazing (Ansari et al. 1995). It does not appear that such activities are stoppable.

Sri Lanka

There are 170 orchid species belonging to 68 genera in Sri Lanka. Of the 69 species which are endemic to the island nation, the status of *Ipsea speciosa* is reported as vulnerable and the remainder are rare (Sumithraarachchi 1986). Two of the four biosphere reserves, namely Sinharaja Forest Reserve (88.64 km^2) and the Kanneliy-Dediyagala-Nakiyadeniya (10,139 ha), are rich in endemic flora. An additional 31,600 ha are designated as strict nature reserves. A further 44,000 ha are also designated as nature reserves where control is not so stringent, and here local residents are permitted to continue with traditional human activities (possibly meaning farming and herb collection) There is an obvious urgency to protect these endemic species, but conservation programmes in the country appear to be more concerned with protecting its fauna, in particular elephants, leopards and turtles (Wikimedia).

The main threat to native orchids in Sri Lanka is habitat destruction due to logging, collection of firewood, conversion to agriculture, monoculture, housing, urbanisation and construction of dams and roads. Collection of orchids from their habitats as novelties and medicine also contributes to the disappearance of species, but is of lesser importance. At the Royal Botanic Gardens in Peradeniya, rare plants are collected, cultivated, mass-propagated and returned to their natural habitats. Collection of wild plants by amateurs and plant collectors is discouraged. An exchange programme allows enthusiasts, research organisations and botanic gardens access to rare plants. Public education is on-going (Sumithraarachchi 1986).

Examples

Attempts to re-introduce five vigorous, epiphytic, native species (*Grammatophyllum speciosum*, *Bulbophyllum vaginatum*, *B. membranaceum*, *Cymbidium finlaysonianum* and *C. bicolor*) in Singapore produced survival percentages that varied from 10 to 95 %, 8 years after re-introduction (Yam and Thame 2005; Yam 2010). Seedling size at re-introduction, choice of host trees, and relative humidity appeared to be significant factors influencing the success rate. Mature plants are least likely to perish as a result of unexpected inclement weather. Much of the primary forest has disappeared in Singapore and many roadside trees are introduced species. Among the trees tested for re-introduction of the orchids, the Rain Tree (*Samanea samaan*) was the best host.

Large colonies of *Dendrobium crumenatum* and *Bulbophyllum vaginatum* are found on old Rain Trees (*Samanea samaan*) in Singapore (Fig. 23.13) and Peninsular Malaysia. Other native orchids are also found on Rain Trees in Sri Lanka. This suggests that Rain Trees would be the best host trees for a variety of native epiphytic orchids to re-establish themselves, provided that such species are first

Fig. 23.13 Gregarious flowering of *Bulbophyllum vaginatum* (Lindl.) Rchb. f. growing on *Albizia saman* in Singapore (Photo: E.S. Teoh)

re-introduced in sufficient numbers to establish self-sustaining reproductive colonies and that suitable pollinators are also present. Pyakurei and Gurung (2008) have carefully identified native tree species which are the favourite perch for epiphytic medicinal orchids in Nepal (Table 23.1).

Unlike the situation in the equatorial belt, in the Yachang Orchid Nature Reserve in southwestern China, 50 % of the orchids numbering 68 species are either terrestrial or saprophytic. The large numbers of such orchids highlight the importance of maintaining this unique Orchid Nature Reserve. Atmospheric temperature, evaporation rates, soil moisture and suitable mycorrhiza populations are important determinants for the survival of these orchids (Liu et al. 2010b).

Seventy-one out of the total of 139 orchid species in the Yachang Reserve have a usage in

Chinese herbal medicine (Table 23.2). Using a score of 1–6, 1 being the species at the lowest risk of extinction and 6 being the ones with the highest risk, six medicinal orchids have a score of 5 and 15 have a score of 4. The six with the high score are *Bletilla formosana*, *Cymbidium faberi*, *C. goeringii*, *C goeringii* var. *serratum*, *C. henryi* and *Herminium bulleyi*. These species need protection if they are to continue to exist in Yachang. However, the Yachang Orchid Nature Reserve represents the southernmost extant of their distribution, and is not the ideal place to focus on their protection for the purpose of protecting the species per se. Attention has been given to establishing nature reserves further north in China where habitats are more ideal for the survival of the threatened species.

Conservationists need to examine the global distribution of species which they wish to protect. An attempt to protect a species at the extreme limits of its distribution is not going to do much good at preserving the species. For instance, protecting *Vanda coerulea* which is listed under CITES Category I (i.e. it is considered to be under serious threat), in southwestern China or northeastern India will not ensure survival of the species. To accomplish that goal, it is necessary to protect *V. coerulea* in Myanmar and Thailand where this orchid is abundant. During a financial crisis, one is not concerned about small bank accounts, one focusses immediately on saving one's large accounts. That is not to say that it is not important to protect *V. coerulea* in China and India, but this is done for a different reason, the varieties in these areas may possess valuable genetic material which are not present in the plants at the centre of their distribution.

It is not absolutely necessary to rely on their natural habitats as essential sources for the supply of medicinal orchids. American ginseng cultivated in Canada possibly outsells Chinese ginseng (*Panax ginseng*). The Eurasian, alpine, myco-heterotrophic *Epipactis helleborine*, finding suitable mycorrhizal symbionts in North America, has become naturalised in a large swath of its temperate regions. Chinese *Bletilla striata* (*Baiqi*) grows wild in the Pensacola area of Florida's westernmost Panhandle; *Phaius*

tankervilleae, the terrestrial "Nun's Orchid" from Southeast Asia has naturalized in two disparate Central Florida counties, Hernando and Hardee, producing many seedlings; and the Asian *Zeuxine strateumatica* occurs in 46 of Florida's 67 counties as well as in coastal parts of Georgia, Mississippi, Louisiana and Texas, and perhaps also Hawaii, Puerto Rico, the Bahamas and Cuba (McCartney 2010). If it comes to the crunch, the vast African continent can easily become a production base, given that southern part of the continent is already home to an enormous wealth of terrestrial orchids.

Medicinal Orchids from the Cool Temperate Regions

Conservation work on orchids is mostly focused on the tropical, subtropical and warm temperate regions, but many Asian medicinal orchid genera occur in the cool temperate regions. The orchid genera of Russia and the adjacent countries which contain medicinal species include *Amitostigma*, *Anacamptis*, *Cephalanthera*, *Coeloglossum*, *Cremastra*, *Cypripedium*, *Dactylorhiza*, *Epipactis*, *Eulophia*, *Gymadenia*, *Habenaria*, *Herminium*, *Limodorium*, *Liparis*, *Malaxis*, *Neottia*, *Neottianthe*, *Oreorchis*, *Orchis* and *Spiranthes* (Vakhrameeva 2008). Many species growing in the dark on the forest floor during winter are mixotrophic. Although they possess green leaves, they are unable to photosynthesise given insufficient light and rely on mycorrhiza for carbon sources and energy. Research is now focussing on achieving consistently productive germination of some of these orchids (Valetta et al. 2008).

In the mixed broad-leaved alpine forests of the World Natural Heritage and Biosphere Reserve of Huanglong in Sichuan Province, Holger Perner in 2002 reported finding 24 species of orchids in the high alpine regions above 3800 m. There were only four medicinal species: *Cypripedium tibeticum* was numerous whereas communities of *Goodyera repens*, *Malaxis monophyllos* and *Ponerorchis chusua* were scattered (Perner 2002).

Artificial Pollination

Apart from those species which self-pollinate, the presence of insect pollinators is critical to the success of a programme to reintroduce extinct species to its former environment. When a species is overcollected, there is danger that its pollinator might disappear and prevent successful re-introduction. Another concern is that reduction in pollinator population would suppress the chances for out-crossing and reduce genetic variability in the population (Wong and Sun 1999). Some species which have managed to survive in pollinator-scarce environments have managed to survive by acquiring the means of autogamy (self-pollination) (Kenji 2013).

Artificial pollination is unnatural but it has a role in ensuring the survival of a species when natural pollinators are scarce (Sun et al. 2003, 2005). In this event, native pollen should be used, and, if scent is present, the scented individuals might receive preference because scent is an attractant for insect pollinators (although, one must agree, that such thinking itself smacks of human interference!). Genes introduced from a distant population often generate hybrid vigour, but, on the other hand, this may drive evolution away from its options as happened with North American *Cypripediums*. The process is known as out-breeding depression (Koopowitz 2001). This makes the population less fit for its immediate environment.

Micropropagation and Improved Culture Techniques

Assisted seed germination and micropropagation of selected clones with desirable characteristics, and continuous improvement of culture techniques, are essential for success in mass production of medicinal herbs and in any conservation effort that attempts to restore depleted plant communities (Pant and Raskoti 2013). Micropropagation of desirable clones should improve standardisation of herbal products and quality control.

Anoectochilus formosanus (Pearl orchid) is regarded as a King Medicine in Taiwan where it is used to treat hypertension, diabetes and disorders of the liver. Once prevalent throughout the mountainous areas of the island, it is now uncommon due to overcollection and development overtaking its habitats. The problem has been addressed by commercial cultivation employing multiple approaches: tissue culture of shoot tips and node segments, better control of growth, flowering, timing of pollination (2–5 days after full bloom), timing of seed pod harvest and improvement of germination media (Tsay 2002). *A. formosanus* is a typical shade plant. Increasing light intensity supplied to *A. formosanus* from 10 to 60 μmol m^{-2} s^{-1} or L-60 increased superoxide dismutase activity and total flavonoid content in the plants. Further increase in light intensity beyond this level stressed the plants and led to a reduction in flavonoid content which was found to be lowest at L-90 (Ma et al. 2010). The issues of clonal selection and mono-clonal propagation have not been addressed.

Extracts obtained from cultured *Dendrobium tosaense* produced by tissue culture were capable of protecting mice from skin lesions resembling atopic dermatitis, when the skin of these mice was inoculated with ovalbumin and 2,4,6, trinitro-1-chlorobenzene. Protection was reflected in the cytokine profiles and lowered anti-OVA IgE levels, and demonstrable on histology (Wu et al. 2014). Being the major component of the extracts, quercetin could be used for standardisation of potency and quality control, thus providing a means for standardising this particular *shihu* which has a role in TCM for the treatment of atopic dermatitis. This level of standardisation is difficult to achieve with plants obtained from the wild, and perhaps even with those propagated by seed.

Nevertheless, public preference for orchids harvested from the wild persists, both for horticultural specimens and for orchidaceous herbs. Many studies have shown that cultivated plants are either similar to or indeed superior to the wild-type. it is not possible to differentiate between wild and cultivated *Anoectochilus formosanus*, and indeed there is no difference in the range or concentration of their constituents. *Habenaria edgeworthii*, an important component herb in *Ashtavarga*, has been propagated in vitro from seeds, and phytochemical analysis of the plants revealed higher phenolic contents than the wild strains (Giri et al. 2012). The problem lies in public education and elimination of ancient bias.

Cryopreservation and Vitrification

Orchid seeds which have been stored in liquid nitrogen at −196 °C for several months or a few years have been shown to retain their ability to germinate (Vendrame et al. 2007; Thammasiri 2008). This works for the seeds of some terrestrial orchids (*Eulophia*, *Orchis*, *Disa*) (Pritchard 1984), but not of other orchids. Adequate desiccation of the seeds prior to super-cooling is possibly the key factor. Contact time with the plant vitrification solution (PVS) is a critical determinant on post-thaw seed germination; for wild mature seeds of *Dendrobium chrysanthum*, 45 min is optimal, as after 60 min there is a sharp decline (He et al. 2010). Zhang and co-workers in Hangzhou found that when they dehydrated the seeds of *Dendrobium candidum* and got its water content down to 8–18 % before storage in liquid nitrogen, 92–95 % of the thawed seeds germinated (Zhang et al. 1999). During cryopreservation, crystallisation of residual water in the seeds causes a sudden rise in the osmotic/oncotic pressure of the liquid plasm which is instantaneously lethal to cells. Vitrification which employs high concentrations of cryo-protectants and snap-freezing has been developed to overcome this problem. It has been shown to work where earlier cryopreservation procedures failed in the storage of the seeds of four Thai tropical species [*Dendrobium chrysotoxum*, *D. draconis*, *Doritis* (*Phalaenopsis*) *pulcherrima* and *Rhynchostylis coelestis*]. It had previously been successfully used to store seeds of *Bletilla striata* and immature seeds of *Ponerorchis graminifolia* in Japan and *Dendrobium candidum* (*D. moniliforme*) in China (Thammasiri 2002; Hirano et al. 2005).

Vitrification of pollen and seeds can play an important role in the hybridisation of species that flower at different seasons, and in the maintenance of a gene pool for endangered species. Vitrification of seeds and pollen have not been described for *Anoectochilus formosanus*. Vertrification of germinating seeds might also be another area to explore.

Good post-thaw germination was achieved following simultaneous cryopreservation of seeds of *Anacamptis morio* with its fungal symbiont has resulted in survival at −196 °C (Wood et al. 2000). This is a practical way to conserve valuable species of terrestrial orchids.

Protocorms of hybrid *Cymbidium* Twilight Moon "Day Light" encapsulated in alginate beads prior to vitrification were able to survive super-freezing for a year. Vitrified protocorms formed neo-pseudobulbs when thawed in a 2 % sucrose solution (da Silva 2013).

Longer-term studies on seeds, pollen and protocorms are necessary to ensure that the vitrification process can preserve viable plant tissues indefinitely.

Orchid Seed Banking

Orchid seed banking seeks to conserve maximum genetic diversity in orchid species for future use or to restore a species to its natural habitat when it is threatened with extinction by climate changes or human activity. The Darwin Inititative Project, Orchid Seed Stores for Sustainable Use (OSSSU), was initiated by the Royal Botanic Gardens at Kew, UK, and has started a collection of orchid seeds from countries with high orchid biodiversity in Asia and Latin America, with emphasis on China (Seaton et al. 2010).

At this stage, countries with vested interest in medicinal plants like China, Japan, Korea and India should undertake to establish cryopreserved seed banks for all their medicinal orchid species, regardless of their conservation status, as well as for all orchid species under threat, regardless of whether they are medicinal. The process is simple and banks are inexpensive to maintain. Thailand, with established local expertise in vitrification (Thammasiri 2000, 2002, 2008, 2010), should be well placed to maintain a cryopreserved seed bank with vitrified seeds obtained from throughout continental Southeast Asia using their ASEAN connection.

Genetic Diversity

DNA analysis has greatly advanced the study of genetic diversity, the preservation of which is important to species survival. Conservation is crucial because several Chinese medicinal orchids are endemic, i.e. they do not exist outside China, and some have very restricted distributions like being confined to a single mountain. *Dendrobium* has been extensively studied by Chinese scientists using rDNA ITS region sequencing, species-specific hybridisation arrays, re-amplified fragment length polymorphism (AFLP), inter-simple sequence repeat (ISSR) and random amplified polymorphic (RAPD) markers (Lau et al. 2001; Liu et al. 2001; Li et al. 2005, 2008; Shen et al. 2006; Wang et al. 2007; Ding et al. 2009). Comparable studies are also being conducted elsewhere (Begum et al. 2009). The technology has also been used for authenticating other important herbs, like *ginseng*, and found to be reliable. Cui et al. (2003) using SS-rRNA spacer domain and RAPD analysis was able to distinguish the different varieties of *ginseng* and separate them from their adulterants which, incidentally, included *Bletillia striata*.

There are 11 *Dendrobium* species in Sichuan province, and all are used as *shihu* (Li and Xiao 1995). There are over 60 (variously quoted as 63 or 74) species of *Dendrobium* in China and 32 of them are included in the medicinal classification of *Huangcao Shihu* on the herbal market, making it unlikely for the consumer to know what he is taking. The concise versions of the *Chinese Materia Medica* mention only a handful of *Dendrobium* species. The Beijing *Colour Atlas of Compedium of Materia Medica* lists only *D. officinale* Kimura et Migo

(=*D. catenaum*), *D. nobile* Lindl. and *D. fimbriatum* Hook. var. *oculatum* Hook as *shihu* the *Coloured Illustrations of Chinese Traditional and Herbal Drugs* lists *D. officinale* Kimura et Migo [=*D. catenaum* (*Shihu, Tiepi Lan* or Iron Skin Orchid)], *D. chrysanthum* Wall ex Lindl. (Golden Flower *Dendrobium*, *Shuhua Shihu* or a Bundle of Flowers *Dendrobium*), *D. loddigesii* Rolfe (*Huancao Shihu, Meihua Shihu*), *D. fimbriatum* Hook (*Mabian* or Horse Whip *Shihu*) and *D. nobile* Lindl. (*Jinchai* or Golden Hairpin *Shihu*), whereas Foozhou's *Color Illustrations of Chinese Materia Medica* mentions *Mabian Shihu* to be *D. loddigesii* Rolfe, *D. chrysanthum* Wall ex Lindl., *D. officinale* Kimura et Migo (=*D. catenaum*). The People's Military Medical Press has a 2004 edition of *The New Century Chinese-English Dictionary of Chinese Traditional Medicine* which also lists only five *Dendrobium* species under *Shihu*: *D. nobile* Lindl., *D. loddigesii* Rolfe, *D. candidum* Wall ex Lindl. (=*D. moniliforme*), *D. chrysanthum* Wall ex Lindl. and *D. fimbriatum* Hook var. *oculatum* Hook.

However, more comprehensive lists exists elsewhere. Shanghai's Chinese Medicine Big Dictionary *Shihu* sources have 11 species of *Dendrobium*. From the lists already mentioned, it has *D. nobile*, *D. officinale* (=*D. catenaum*) and *D. lodigessi*. It omits *D. chrysantum* and *D. fimbriatum* var. *oculatum* but it includes *D. linawianum* Rchb. f., *D. moniliforme* (L.) Sw., *D. hercoglossum* Rchb. f., *D. aduncum* Wall et Lindl., *D. wilsonii* Rolfe, *D. hancockii* Rolfe, *D. lohohense* Tang et Wang and *D. bellatulum* Rolfe. *Shihu* in Taiwan consists of *D. aduncum*, *D. chrysotoxum* Lindl., *D. falconeri* Hook., *D. linawianum* Rchb. f., *D. loddigesii* Rolfe, *D. moniliforme* (L.) Sw., *D. nobile* Lindl., *D. parishii* Rchb. and *D. primulinum* Lindl. (correct name: *D. polyanthum* Wall. ex Lindl.). Some members from the mainland Chinese lists are excluded and some new species have been added to the list. Five species (*D. aduncum*, *D. chrysotoxum*, *D. loddigesii*, *D. parishii* and *D. polyanthum*) are not native to Taiwan.

In the past, major herbal establishments relied on the physical appearances to distinguish among species. The first scientific advancement relied on chromatography. It would appear that the leading pharmaceutical establishments now rely on DNA sequencing, although at the village level, one must still rely on physical appearance and the reliability of the apothecary.

Xu and co-workers (2006) from the Shanghai University of Traditional Chinese Medicine proposed using rDNA ITS sequence analysis to distinguish among the species. Working on 18 *Dendrobium* species listed as *Huangcao Shihu*, they found that interspecific sequence divergence ranged from 3.2 to 37.9 % in ITS 1 and 5–26.6 % in ITS 2, while intra-specific variation ranged from 0 to 3 % in ITS 1 and 0–4 % in ITS 2. Thus, the species could be distinguished at their DNA level. They were also able to design five pairs of species-specific primers for rapid PCR identification of 5 *Dendrobium* species listed in the Chinese Pharmacopoeia (Xu et al. 2006).

In the case of the critically endangered *D. catenatum* (syn. *D. officinale*), Ding et al. (2009) found that the most variation occurred within populations rather than between populations, a phenomenon which they associate with small genetic drift caused by excessive human exploitation and habitat destruction during the last 60 years. They suggest that conservation should be aimed at protecting those populations which have the highest genetic diversity and, secondarily, populations with low genetic diversity but representing a novel evolutionary trend. Ex situ conservation with well-designed integrated germ-plasm banks is a complementary approach.

Finally, it might be mentioned that Myanmar is an untapped gold-mine for medicinal orchid species. Their ecology should now be extensively studied before Myanmar opens its doors to the rest of the world and its orchid wealth gets plundered. Currently, CITES lists 58 Myanmar orchids in the CITES Appendix 1, but this list does not include many medicinal orchids (Saw Lwin 2002). *D. gibsonii* whose golden flowers are a favourite Buddhist offering in Myanmar is offered for sale with flowers on pseudobulbs, shorn of roots. Here, tea farmers cultivate the orchid as an additional cash crop (Tanaka and Yee 2007).

Medicinal Orchids and Cites

CITES lists all orchids under Appendix 1 or II, implying that all species of orchids are threatened with extinction, to an extent that is beyond the average risk faced by all living organisms on earth. Those orchids placed under its Appendix I are thought to be already at risk of disappearing in the wild, but there is much disagreement on the real risk faced by the listed species, whilst many others that have almost entirely disappeared from the wild are not listed. Nevertheless, this shows a genuine worldwide concern about the conservation of orchid species and wild life in general, which a very good thing.

Thus, it might well be asked: is medicinal usage of an orchid species threatening its occurrence in its natural habitats? The answer is a qualified no. Turkey reports the annual destruction of 120 million wild orchid plants which are collected for the production of 45,000 t of *salep* which is now recognised as merely an imaginary aphrodisiac but appreciated as a beverage and food additive, while it is used as a stabiliser for the local Kahramananmaras-type ice cream (Tekinsen and Guner 2009). The export of orchid tubers intended for *salep* from Turkey is now banned. In East Asia, great efforts have been made to conserve orchid habitats, and to hand-pollinate and propagate orchids in laboratories and farms. There are still many problems to overcome, the most serious being the greed of the entrepreneur and the fact that developed nations still plunder the botanical wealth of developing nations. The great awareness among the scientists and conservationists has still to reach the villagers. Expertise and funding are in short supply. There is a problem of priorities, and the tendency has been to protect the attractive species. *Dactylorhiza osmanica* (Kl.) Soo var. *osmanica* End., endemic in Turkey, obviously needs to be protected (Bulut and Yilmaz 2010).

It is true that some orchids have been collected to extinction, but medicinal orchids have been collected in China for well over 2000 years and most species still exist in abundance, notwithstanding their former large-scale collection (Hart 1884; Chen 1995). The reason is that there is a fundamental difference in the attitude of the modern collectors, and their methods of collecting account for the extinction of a species on the one hand or their continued survival on the other. In the case of the early orchid hunters who worked for commercial orchid enterprises, their intent on exclusivity turned them into destructive creatures. They collected en masse, and what they could not collect they destroyed to prevent anyone else from obtaining the same item. During the heyday of orchid collecting in the late nineteenth century, Joseph Hooker complained that mountain roads of Assam were "become stripped like the Penang jungles, and for miles it sometimes looks as if a gale had strewed the road with rotten branches and Orchidae. Falconer's men sent down 1000 baskets the other day." A Swiss botanist wrote that "not satisfied with taking 300 or 500 specimens of a fine orchid, collectors must scour the whole country and leave nothing for many miles around" (quoted from Orleans 1999). Not so long ago, an entire population of the small-flowered, white *Phalaenopsis micholitzii*, already uncommon in nature and confined in its distribution to Mindanao, was stripped bare in its natural habitat, right down to its tiny seedlings (Koopowitz 2001). It was not as if the flowers of that orchid were showy. The main merits of *P. micholitzii* was its floriferousness and a tolerance for cold which was being exploited by a very few American and Taiwanese hybridisers. But only an extremely small number of orchid growers would offer the orchid a second glance, or ever think of owning a plant. Numerous orchid oddities have met the same fate in the recent decades, for instance, *P. gigantea* and *P. javanica*. Only after tens of thousands of the newly-discovered Vietnamese/southern Chinese *Paphiopedilums* had passed through Singapore en route to Europe and the USA in the 1980s did CITES step in to ban the international

distribution of *Paphiopedilum* altogether. In 1985, over 6000 plants of *Paphiopedilum micranthum* and *P. armeniacum* were taken from China to Hong Kong and from thence exported to other countries. In the 1990s, it was estimated that, annually, between 100,000 and 200,000 Chinese *Cypripediums* met the same fate (Chen 1995). The move gave additional benefit to those nurseries who had commissioned the stripping of those *Paphiopedilums* and *Cypripediums* from the Asian forests in the first place; they now have nursery-cultivated *Paphiopedilums* and *Cypripediums* and are well placed to monopolise *Paphiopedilum* and *Cypripedium* trade in their respective countries.

Paphiopedilum rothschildianum was once thought to be extinct in the wild, collected to extinction, yet within three short years following the rediscovery of its habitat in Sabah by Shim and Lamb in the early 1980s, over a hundred plants of this rare orchid were taken out of Mount Kinabalu National Park by people who ought to know better. "Orchidists from all four corners of the world came to Sabah and the Park to look for orchids. Some in the name of science, some for commercial interests and some just to collect and to gather plants from their natural habitat to wrap up and take home until the survival of the species is threatened. I plead guilty myself, having taken six of this species, not only to grow them in my orchid house but also to send to breeders and growers in Kenya, Germany, France and the USA." (quoted from Bacon 1986).

In his paper discussing the problems pertaining to conservation of *Paphiopedilums* of Sabah, Bacon (1986) concluded that the problem was really with orchidists who must look to themselves. They must show that they possess the discipline not to participate in the extinction of orchids in their natural habitats, the principal plant at risk being the treasured *P. rothschildianum*.

Arguing for the establishment of orchid centres, such as the one at Tenom in Sabah, to collect, cultivate, propagate, multiply and distribute desirable orchid species as a practical method of ensuring their survival, Lamb (1986) observed that. in Malaysia, for instance, it would be difficult "to conserve species in their natural habitat until education and greater awareness of their botanical heritage that needs conservation becomes accepted politically as well as generally by the public, and that planning of natural resources finally defines areas to be fully protected." Orchidologists in Sri Lanka apparently share his view. Here, rare plants are collected for cultivation in the Botanic Gardens, propagated in large numbers, then reintroduced into their natural habitats or into ecologically suitable forest reserves if their original home is being destroyed by progress. Rare plant collectors and amateur botanists are warned against depleting populations (Sumithraarachchi 1986).

Collectors of medicinal orchids on the other hand would be mindful of maintaining their source of supply, lest they destroy the very source of their livelihood. Despite population growth and possible increasing demand, thus far, the populations of medicinal orchids have been maintained, and already steps are being taken to increase the supply, simplify the "collection", and ensure the maintenance of uniformly high standards by mass cultivation of desirable species in nurseries. This is happening in China, Thailand, Myanmar and elsewhere. However, overcollection of *Dendrobium nobile* to meet medicinal demand from China is threatening the existence of the species in the wild in Myanmar (Tanaka and Yee 2007) and in Laos (personal communication).

Nevertheless, there are exceptions. In 1985, Bailes reported that a 1000 trucks of 8-t capacity transported *Flickingeria fagax* from Nepal to India for incorporation into Ayurvedic medicine (Bailes 1985). Many orchids from the high altitudes are highly valued as medicinal herbs, and poor rural women collect medicinal and ornamental orchids from the forests for sale in the local markets to supplement their tiny incomes. Now the Local Initiative for Biodiversity, Research and Development (LI-BIRD) in Pokhara, Nepal, and two women's groups have set up community nurseries to grow medicinal orchids to obviate the need to collect plants from the wild. Existing private nurseries were motivated to join the effort (Subedi 2005). Hopefully, medicinal orchids will not become extinct in the wild.

Several orchid species which find usage in the local herbal practice in Sri Lanka, namely *Ephemerantha macraei* (*Flickingeria fimbriata*), *Ipsea seciosa*, *Zeuxine longilabris, and Z. flava* are rare in Sri Lanka and are vulnerable; *Anoectochilus setaceus* and *Z. strateumatica* although not rare are also vulnerable (Sumithraarachchi 1986). However, there is no concerted effort to preserve these orchids and their natural habitats.

Jin Xiaohua of the Kunming Institute of Botany is concerned about the possible endangered status of *Dendrobium nobile*, and he urges more field studies to be performed on its status in nature. Its popularity as herbal medicine over the past 500 years, coupled with greater demand arising from population growth and prosperity, has led to overcollection from the wild and a reduction in the density of the wild populations. Although *Dendrobium nobile* is widely distributed across several countries, surveys show that in many places the wild plant is no longer common, habitat quality has declined, and human activity has caused its populations to be fragmented. Thus, Jin argues that the population of *D. nobile* is threatened. However, this is a beautiful orchid and many selected varieties have been mericloned in Thailand. They are readily available to growers at very reasonable prices and there is no real need for hobbyists to grow wild plants. Farming of *D. nobile* as a medicinal crop is also underway. However, *D. nobile* remains in Appendix II (www.cites. org).

The status of individual plants and animals is periodically reviewed and movement from one category to another is allowed. The beautiful *Vanda coerulea* which is the major source of blue colour in *Vanda* hybrids was originally listed under Appendix II in 1975. It was transferred to Appendix I in 1979, and back again to Appendix II on 12 January 2005 (www.cites. org). Despite the small Indian populations, a study that used two different single primer amplification reactions (SPAR) found high genetic diversity within populations (Manners et al. 2013).

Commercial interests should not dominate in conservation, but unfortunately they will. If one looks at *Vanilla*, that is such a valuable crop that many gene banks are being set up in various parts of the world to ensure the existence of a comprehensive *Vanilla* germ plasm. No other orchid receives this sort of attention. The various species of medicinal orchids are unequal from a medicinal or commercial point of view, and it is likely that the species which are currently important will receive attention, while the others may not. However, orchids as a whole should be seen as an indicator of the health of a natural environment. Their numbers and diversity in an environment are the best clue to its ecological health.

Fig. 23.14 *Dendrobium catenatum* Lindl., the original and a highly prized *shihu* is now cultivated on an extensive scale in China. Picture shows a small collection in Fujian Province (Photo: E.S. Teoh)

Farming of Medicinal Orchids

During the twentieth century, isolation and population growth rapidly exhausted traditional sources of medicinal herbs and new sources had to be developed. With regards to herbs collected from the wild, even secondary sources could not meet the demand. Inner Mongolia is the source of wild *Radix Astralagus* (not an orchid); the yield was 2000 t during the 1960s, now it is below 100 t. Another herb, *Radix glycorrhiza* (again not an orchid), also comes from Inner Mongolia and its annual yield has declined by 40 %. The situation is far worse for the achlorophyllous orchid, *Gastrodia elata* (*Tianma*), which was so overcollected from the wild that it is now an endangered species. Demand for *shihu* threatened the existence of the original lithophytic *Dendrobium* species, *D. catenatum* and *D. moniliforme*, centuries ago, forcing an accommodation that led to the admission of numerous epiphytic *Dendrobium* species which are now officially recognized as *shihu*. But still demand continues to outstrip supply. Fortunately, it is not difficult to cultivate *Dendrobium* in farms and this was one of the first medicinal orchids to be cultivated (Figs. 23.14, 23.15, 23.16, 23.17 and 23.18). *Gastrodia elata* is now

Fig. 23.15 Flowering plant of *Dendrobium catenatum* from the nursery above (Photo: E.S. Teoh)

farmed in several provinces, most notably Guizhou. *Bletilla striata* is farmed in Guizhou and in northern and central Vietnam near Hue. *Anoectochilus formosanus* which is not native on the Asian mainland is cultivated in Fujian Province. The medicinal herb is now sourced from this province and Taiwan (Wu 1994; Zhonghua Bencao 2000).

Between 1999 and 2006, 600 medicinal plantations were established in China. Simultaneously, orchid nurseries in neighbouring

Fig. 23.16 *Phalaenopsis* production at a nursery near Xiamen, Fujian demonstrates that China can produce enough medicinal orchids to meet demand within a very short period, provided sales returns justify the effort (Photo: E.S. Teoh)

Fig. 23.18 Old pseudobulbs of *Dendrobium* will generally produce plantlets at the nodes if they enjoy enough humidity but are not allowed to rot (Photo: E.S. Teoh)

Fig. 23.17 *Dendrobium chrysotoxum* Lindl. seedlings at a research nursery in Chiang Mai, Thailand. Most *Dendrobium* species can be easily raised from seed (Photo: E.S. Teoh)

countries started growing *shihu Dendrobium*. It is estimated that over a million families are involved in the cultivation of medicinal herbs in China, predominantly as smallholders. About 2000 Chinese herb dealers have set up their own plantations, amounting to around 5 % of the nation's total. (They are registered and qualify for GAP evaluations whereas the family farmer, being unregistered, do not qualify.)

Tissue culture holds the most promise for orchids. Only minor genetic variation resulted when *D. huoshanense* was subjected to serial passage in tissue culture medium (genetic similarity coefficient varied from 85.4 to 98.4 %), as expected occurring more with protocorm than in bilfoliated seedlings or inflorescences (Liu et al. 2007).

Farming of medicinal orchids like *shihu* which are propagated by meristematic tissue culture will go a long way to ensure that there is consistency in the product. Adulteration can then be easily detected using SSR technique for confirmatory identification of the germplasm (Xie et al. 2010). Since the public often show a preference for "wild orchids" as opposed to cultivated specimens, when meristemmed plants have reached a stage advanced enough for them to survive in nature, they should be planted on appropriate host trees or on the ground, returned to the wild as it were, and later harvested from such environments. As there is also a preference for wild ornamental orchids (Phelps et al. 2014), this recommendation would also apply to them. Replanting *D. catenatum* in *danxia* landform will be a challenge since this orchid reproduces poorly in that environment (He et al. 2009).

Numerous conservationists in India, Nepal and Bhutan report that *Dactylorhiza hatagirea* is under threat because its tubers are regarded as aphrodisiac. They fetch the best prices in the herbal market. The orchid occurs between 2500 and 4500 m in Central Himalaya. Formerly only collected from the wild, *D. hatagirea* is now cultivated in buffer zone villages in the Nanda Devi Biosphere Reserve (NDBR), in Uttarakhand, India. This Biosphere Reserve designated in 2004 has a core area of 514,857 ha and a buffer zone of 514,857 ha containing 55 villages which are home to 15,000 people who subsist on farming and herb collection.

Plants are propagated through seed and tuber division. *D. hatagirea* grows well in shaded locations in humus-rich, porous soil. Manuring twice during its growth period and frequent weeding are essential for optimum tuber yield. Maximum yield is reached after 5 years of growth, but often the tubers are harvested after 2–3 years (Nautiyal et al. 2003). This was probably during a transition period when loss of villagers' income enforcement of nature protection demanded short-term returns from cultivated crops. In order for conservation goals to be achieved, economic needs of tribal people must be achieved (Maikhuri et al. 2001). This should involve improvement in quicker, massive propagation of medicinal plants, and training in farming techniques for growing new non-traditional crops like medicinal orchids. Other precious, endangered medicinal orchids, e.g. *Malaxis muscifera*, whose pseudobulbs fetched Rs 100–120 per kg in 2008 should also be farmed, but currently there is no effective management plan for these species (Chauhan et al. 2008).

Nevertheless, not every species of medicinal orchids can be cultivated. *Herminium lanceum* and *H. monphyllum* [=*Androcorys monophylla* (D. Don) Agrawala and H.J. Chowdhery], whose tubers are valued as *salep* in India, are alpine orchids that grow at higher altitudes than any of the other Indian orchids. Proper cultivation is not practicable. Such orchids require natural protection in designated forest reserves (Pandey et al. 2003).

Hybridization for Improvement

Intra-specific hybridisation of *Dendrobium houshanense* (*D. catenatum*) yielded hybrids that exceeded the natural species in chlorophyll content, photosynthetic rate and the accumulation of medicinal components (Wang et al. 2006). Protocorm has been shown to have higher contents of pharmacologically active compounds than adult tissue, and such production can be further enhanced by the application of appropriate fungal elicitors. These studies confirm that the rapid propagation and cultivation of medicinal orchids is feasible, and, with care, more productive plants with stronger medicinal content may be obtained. Furthermore, hybrids are much easier to grow than species plants collected from the wild due to the phenomenon of hybrid vigour (Holttum 1978; Teoh 1980). The creation of medicinal hybrid orchids is not only a good way to improve yield but it will spare the species from overcollection.

It should be noted that hybrids may produce phytochemicals which are qualitatively or quantitatively dissimilar to compounds produced by the parental species. For instance, volatiles emitted by the flowers of the hybrid, *Vanda* Mimi Palmer (Fig. 23.19), are dissimilar to those of its dominant, fragrant parent *V. tessellata* (Mohd-Hairul et al. (2010). Nevertheless, this last study did not undertake an analysis of the actual *V. tessellata* clone used in the making of hybrid *Vanda* Mimi Palmer but relied on data from Kaiser (1993). There are approximately 30 different natural forms of *V. tesssellata*, so the spectra of their volatiles could possibly show differences among the different forms. Second-generation hybrids of *Vanda tessellata* are also fragrant (Fig. 23.20). This opens possibilities for medicinal orchids to be hybridised through several generations.

Genetic Manipulation

Production of transgenic plants that display enhanced protection against fungal and bacterial diseases will create healthier plants that do not

Fig. 23.19 *Vanda* Mimi Palmer (*V. tessellata* × *V.* Tan Chay Yan) looks like *Vanda tessellata*, smells Like *Vanda tessellata*. Intra-specific hybridization will provide hybrid vigour to the species; inter-specific hybridization may assist in preserving some precious genetic material of the species or impart it to a more vigorous plant that is easier to cultivate (Photo: E.S. Teoh)

Fig. 23.20 *Vanda* Overseas Union Bank is an F2 hybrid of *Vanda tessellata* derived from *Vanda* Mimi Palmer. It has larger flowers and retains the fragrance of *Vanda tessellata* (From: Teoh Eng Soon, *Orchids in a Garden City*. Singapore: Marshall Cavendish 2010)

require treatment with fungicides. Making transgenic orchids focusses on creating plants that possess novel colour, fragrance and better shape (da Silva et al. 2011), but Taiwanese scientists are simultaneously working on creating plants that are resistant to disease. The *pflp* gene was successfully tranfected into *Oncidium* and *Phalaenopsis* plants which made them resistant to *Erwinia carotovora*, a cause of soft rot disease in orchids (Chai and Yu 2007). Tomato plants transfected with *Arabidopsis* NPR1 gene displayed enhanced resistance to a spectrum of fungal and bacterial diseases (Lin et al. 2004), and this might be a useful gene to incorporate into medicinal plants. However, the author is not aware that there is any such study on medicinal orchids.

Good Agricultural Practice

With such a large number of producers, great variation in the quality of any herb can be expected. This applies equally to common agricultural products like fruit and vegetables. China Good Agricultural Practice (China GAP) was a response to the European Retail Merchant Association's Agricultural Product Work Team's recommendation that there should be a global uniform standard for agricultural produce. The European version is known as EUREGAP, but in the event the Chinese were the first to draw up the guidelines. Chinese herbal medicine was not left out from this working standard. CHMGAP enlists additional procedures to ensure that cultivated herbal products meet the required standards. They would probably apply only to *Gastrodia elata*, *shihu Dendrobium*, *Bletilla striata* and *Pholidota sinensis* because other orchid species are not required in massive quantities and farmers have yet to learn to cultivate the crops on an extensive–intensive scale. Experience with *Vanilla* (Gassenmeier et al. 2008) illustrates the difficulty of obtaining supplies that meet with set standards. With Chinese medicinal orchids, their active ingredients must be first agreed upon, their concentrations and their ratios defined. Species and varieties need confirmation by DNA fingerprinting and active ingredients determined by chromatography. However, if there is determination, one will find a way.

GAP takes into consideration:

1. The environment which must be free from pollution. Standards have been set at CB 3095-82 for air quality; CB 5084-92 for water quality and CB 15618-1995 for soil quality. Contamination with heavy metals has always been a major concern for medicinal herbs (Braithwaite 2008).
2. The germplasm, its correct identification, potential and conservation.
3. Cultivation methods, soil treatment, watering, use of fertilisers and control of disease and pests using Standard Operating Procedures Principles which are defined.
4. Harvesting, timing and procedure, and correct processing and drying procedures as set out for each specific medicinal product.
5. Packaging, transport and storage with attention to storage containers and facilities paying attention to light, temperature and humidity. Every batch of plant materials should be correctly labelled and proper records must be kept.
6. Quality control to ensure that the characteristics of the material are met. Foreign matter, water content, ash content are defined.
7. Training of operators and maintenance of equipment.
8. Documentation.

ChinaGAP announced several objectives: protection of the consumers, modernisation of agriculture, intensive farming, sustainable development, ecological balance and environmental protection, improving the income of farmers, promoting local economic development, standardised production system for Chinese herbal products, improving the quality of the products and upgrading competitiveness (Chonghua Certification Company Ltd. Declaration 2010). The task is daunting and many social issues are involved, such as the livelihood of the farmers working nearby or upstream who may have need of chemical fertilisers and pesticides for their crops (Liu 2010).

The industrial output of traditional Chinese medicine had already reached 177.2 billion yuan (US$25.9 billion) in 2007, accounting for 26.53 % of the total pharmaceutical output (Chonghua Certification Co., Ltd. Statement, 2008). While the European attitude towards GAP is concerned primarily with obtaining reliable starting materials for drug manufacture, China's focus is to obtain uniformly high-quality herbs for the use of its own population, so that it can continue to keep health care costs well below the world's average per capita expenditure while maintaining a high standard of health and extending longevity. When annual export sales of herbal medicine worldwide totalled US$15 billion, China's share was only US$600 million (Leung 2006).

Intercropping

Intercropping with *Vanilla* and areca nut has been used to increase the economic yield in India (Sujatha and Bhat (2010), and this seems like a good means to increase the supply of tropical orchids. Initial attempts with mericloned *Phalaenopsis* during the 1980s did not succeed in Malaysia, possible due to the decision to grow *Phalaenopsis* hybrids rather than better-adapted local species. It would work well provided the orchid species selected for cultivation match their habitats. Terrestrial orchids lend themselves well to intercropping.

Conclusion

Overcollection of orchids for medicinal usage is currently restricted to those *Dendrobium* species which are used as *shihu*, the few species which are used in Indian herbal medicine as aphrodisiacs or tonics, and orchids of the Middle East which are used for making *salep*. Although *Vanilla*, *Gastrodia elata* and *Bletilla striata* are consumed on an extensive scale, they are grown in farms and wild communities do not face a threat from extensive stripping. This proves that all desirable orchids can be mass-propagated, and every step should be taken by the countries supplying such threatened orchids to provide cultivated specimens for the medicinal market. At the same time, in situ and ex situ conservation approaches should be applied to save all orchids, and not only the known medicinal species.

References

Abraham A, Vatsala P (1981) Introduction to orchids, with illustrations and descriptions of 150 South Indian Orchids. TPGRI, Trivandrum

Aggarwal S, Zettler LW (2010) Reintroduction of an endangered terrestrial orchid, *Dactylorhiza hatagirea* (D.Don) Soo, assisted by symbiotic seed germination. First report from the Indian subcontinent. Nat Sci 8 (1):139–145

Alvarez-Lopez CL, Osorio V, Nelson W, Marin-Montoyo M (2013) Idenificacion molecular de microoganismos asociados a la Rizofera de plantas de Vanilla en Colombia. Acta Biol Colombiana 18(2):293–305

Ansari AA, Dwarakan P, Diwakar PG (1995) Conservayion of orchids – a review on few species of Shevaroy and Kolli Hills. J Econ Tax Bot Add Ser 11:73–75

Averyanov L, Cribb P, Phan KL, Nguyen TH (2003) Slipper orchids of Vietnam. With an introduction to the Flora of Vietnam. Royal Botanic Gardens, Kew

Bacon AJ (1975) The wild orchids of Sabah. Malayan Orchid Rev 12(2):42–43

Bacon AJ (1986) The Paphiopedilums of Sabah, and the problems of conservation. In: Rao AN (ed) Proceedings of the 5th ASEAN orchid congress. Parks & Recreation Department, Ministry of National Development, Singapore, pp 132–133

Bailes CP (1985) Orchids in Nepal. Conservation and development of a national resource. Advisory report recommendations. Royal Botanic Gardens, Kew, Richmond

Bao XS, Shun QS, Chen LZ et al (2011) The medicinal plants of Dendrobium (Shi-Hu) in China. A coloured Atlas. Fudan Press, Shanghai

Begum R, Alam SS, Menzel G, Schmidt T (2009) Comparative molecular cytogenetics of major repetitive sequence families of three dendrobium species (Orchidaceae) from Bangladesh. Ann Bot 104 (5):863–872

Bencao Z (2000) Edited by Health Department and National Chinese Medical Management Office. Science and Technology Press, Shanghai, p 2000

Bernard N (1899) Sur la germination du Neottia nidus-avis. C R Acad Sci Paris 128:1253–1255

Bernard N (1909) L'evolution dans la symbiose, les orchidees et leurs champignons commensaux. Ann Sci Nat Bot Ser 9(9):1–196

Bhattarai P, Oandey B, Gautam RK, Chhetri R (2014) Ecology and conservation status of threatened orchid *Dactylorhiza hatagirea* (D.Don) Soo in Manaslu Conservation Area, Central Nepal. Am J Bot Sci 5 (23):3483–3491

Bidartondo MI, Read DJ (2008) Fungal specificity bottlenecks during orchid germination and development. Mol Ecol 17:3707–3716

Boller AH, Corrodi F, Gaumann E et al (1957) Uber induzierte Abwehrstoffe bei Orchideen Pt. 1. Helv Chim Acta 40:1062–1066

Braithwaite R (2008) Heavy metals and herbal medicine. In: Watson RR, Preedy VR (eds) Botanical medicine in clinical practice. CABI, Cambridge

Braun R (1888) China. Imperial Maritime Custums II. Special Series No. 8. List of Medicines exported from Hankow and the other Yangtze Ports. Inspector-General of Customs, Shanghai

Bulut Z, Yilmaz H (2010) The current situation of threatened endemic flora in Turkey: Kemaliye (Erzican) case. Pak J Bot 42(2):711–719

Burgeff H (1909) Dir wurzelpilze der orchideen ihre kultur und ihre leben in der pflanze. Verlag von Gustav Fischer, Iena

Burgeff H (1934) Pflanzliche Avitaminose und ihre Behebung durch Vitamin=zufuhr. Ber Deutsch Bot Ges 52:384–390

Cameron DD, Johnsson I, Leake JR, Read DJ (2007) Mycorrhizal acquisition of inorganic phosphorus by the green-leaved terrestrial orchid Goodyera repens. Ann Bot (Lond) 99(5):831–834

Chai D, Yu H (2007) Recent advances in transgenic orchid production. Orchid Sci Biotech 1(2):34–39

Chaturvedi S (2014) India rejects plan to mine bauxite in Niyamgiri Hills. Setback for Vedanta resources and Anil Agarwal. Wall Street J, 12 Jan 2014

Chauhan RS, Nautiyal MC, Prasad P, Purohit H (2008) Ecological features of an endangered medicinal Orchid – *Malaxis muscifera* (Lindley) Kuntze in Western Himalaya. MIOS J 9(6):8–12

Chen SC (1995) Orchids and their conservation. In: Proceedings of the 5th Asia Pacific orchid conference & show, Fukuoka, pp 15–18

Chen SC, Tang T (1982) A general review of the orchid flora of China. In: Arditti J (ed) Orchid biology. Reviews and perspectives II. Cornell University Press, Ithaca

Chen RR, Lin XG, Shi YQ (2003) Research advances on orchid mycorrhiza. Chin J Appl Environ Biol 9 (1):97–101

Chen XM, Xiao SY, Guo SX (2006) Comparison of chemical composition between *Dendrobioum candidum* and *Dendrobium nobile*. Zhongguo Yi Xue Ke Xue Yuan Xue Bao 28(4):524–529

Chen XM, Guo SX, Meng ZX (2008) Effect of fungal elicitors on the growth of dendrobium candidum protocorms. Chin Trad Herb Drugs 39(3):423–426

Chen J, Hu KX, Hou XQ, Guo SX (2011) Endophytic fungi assemblages from 10 dendrobium plants (Orchidaceae). World J Microbiol Biotechnol 27 (5):1009–1016

Chen J, Wang H, Guo SX (2012) Isolation and identification of endophytic and mycorrhizal fungi from the seeds and roots of dendrobium. Mycorrhiza 22(4):297–307

Chopra RN (1933) The indigenous drugs of India, Calcutta: the art press. Republished as Chopra's Indigenous Plants of India, 2nd edn. Academic, Kolkata (1986)

Chu XL, Yang B, Gao L et al (2010) Species diversity of cultivable bacteria isolated from the roots of *Cymbidium faberi* Rolfe. Wuhan Zhiwuxue Yanjiu 28(2):199–205

Chung MY, Nason JD, Lopez Pujol J et al (2014) Genetic consequences of fragmentation on populations of terrestrial orchid *Cymbidium goeringii*. Biol Conserv 170:222–231

Chutima R, Dell B, Vessabutr S et al (2010) Endophytic fungi from *Pecteilis susannae* (L.) Rafin (Orchidaceae), a threatened terrestrial orchid in Thailand. Mycorrhiza 21(3):221–229

Chutima R, Lumyong S (2012) Production of indole-3-acetic acid by Thai native orchid-associated fungi. Symbiosis 56(1):35–44

Cui XM, Lo CK, Yip KL, Dong TT, Tsim KW (2003) Authentication of Panax notoginseng by SS-rRNA spacer domain and random amplified polymorphic DNA (RAPD) analysis. Planta Med 69(6):584–6

da Silva JAT (2013) Ammonium to nitrate ratio affects prootocorm like bodies PLB formation in vitro of hybrid cymbidium. J Ornam Plants 3(3):155–160

da Silva JAT, Chin DP, Van PT, Mii M (2011) Transgenic orchids. Sci Hortic 130(4):673–680

Dan Y, Yu XM, Guo SX, Meng ZX (2011) Effects of forty-two strains of orchid mycorrhizal fungi on growth of plantlets of *Anoectochilus roxburghii*. Afr J Microbiol Res 6(7):1411–1416

Darairaj P, Kanaraj M (2013) Traditional medicinal plant resources of Southern Pachchmalais in Trichirapalli of Tamil Nadu, India. Implication of traditional knowledge in health care systems. Int J Res Hum Arts Lit 1(6):39–46

Dash PK, Sahoo S, Bal S (2008) Ethnobotanical studies on orchids of Niyamgiri hill ranges, Orissa, India. Ethnobot Leaft 12:70–78

Ding G, Li X, Ding X, Qian L (2009) Genetic diversity across natural populations of *Dendrobium officinale*, the endangered medicinal herb endemic to China, revealed by ISSR and RAPD markers. Genetika 45(3):3375–3382

Dong F, Zhao JN, Liu HX (2008) Effects of fungal elicitors on the growth of the tissue culture of *Cymbidium goeringii*. North Hortic 5:194–196

Fan L, Guo SX (1998) The development of orchid fungi research. Microbiology 25:227–230

Fan L, Guo SX, Cao WQ et al (1996) Identification and biological activity of *Mycena orchidocola* sp. Nov. in *Cymbidium sinense* (Orchidaceae). Acta Mycol Sin 15:251–255

Foppes J, Souvanpheng P (2005) Experiences with market development of non-tomber Forest products in Lao PDR. In: International Workshop on "Market Development for Improving Upland Poor's Livelihood Security" organized by the Centre for Community Development Studies, Kunming, China. http://search4dev.nl/download/284094/116258.pdf

Ganesan S, Kesavan L (2003) Ethnomedicinal plants used by the ethnic group Valaiyans of Vellimalai hills (Reserved Forest), Tamil Nadu, India. J Econ Taxon Bot 27(3):754–760

Gassenmeier K, Reisens B, Magyar B (2008) Commercial quality and analytical parameters of cured vanilla bean (*Vanilla planifolia*) from different regions from the 2006–2007 crop. Flavour Fragrance J 23(3):194–201

Gebauer G, Meyer M (2003) 15N and 13C natural abundance of autotrophic and myco-heterotrophic orchids provide insight into nitrogen and carbon gain from fungal association. New Phytol 160(1):209–233

Ghorbani A, Gravendeel B, Naghibi F, de Booer H (2014) Wild orchid tuber collection in Iran: a wake-up call for conservation. Biodivers Conserv 23:2749–2760. doi:10.1007/s10531-014-0746-y

Giri D, Arya D, Tamta S, Tewari LM (2008) Dwindling of an endangered orchid *Dactylorhiza hatagirea* (D Don) Soo: a case study from Tungnath Alpine meadows of Garhwal Himalaya, India. Natl Sci 6(3):1545–1543

Giri L, Jugran A, Bhatt ID et al (2012) Promoting conservation and sustainable use of rare species of Himalayan orchid *Habenaria edgeworthii* through in-vitro propagation. In Vitro Cell Biol 48(Suppl 1):62

Guo S, Xu J (1992) The relationship between seed germination and seedling development of *Bletilla striata* and *Mycena osmundicola* fungi. Zhongguo Yi Xue Ke Xue Yuan Xue Bao 14(1):51–54

Guo SX, Fan L, Cao WQ et al (1997) *Mycena anoectochila* nov. isolated from mycorrhizal roots of *Anoectochilus roxburghii* from Xishuangbanna. Mycologia 89(6):952–954

Guo SX, Fan L, Cao WQ, Chen XM (1999) *Mycena dendrobii*, a new mycorrhizal fungus. Mycosystema 18(2):141–144

Hadley G (1970) Non-specificity of symbiotic infection in orchid mycorrhiza. New Phytol 69:1015

Hadley G (1982) Orchid mycorrhiza. In: Arditti J (ed) Orchid biology: reviews and perspectives II. Cornell University Press, Ithaca, pp 83–118

Hajong S, Kumaria S, Tandon P (2013) Compatible fungi, suitable medium, and appropriate developmental stage essential for stable association of *Dendrobium chrysanthum*. J Basic Microbiol 53(Sp Iss S1):1025–1033

Hansen E (2001) Orchid fever. Methuen, New York

Harshani HBC, Senanayake SP, Sandamali H (2014) Host tree specificity and seed germination of *Dendrobium aphyllum* (Roxb.) CEC Fisch in Sri Lanka. J Natl Sci Found Sri Lanka 42(1):71–86

Hart R (1884) China imperial customs III. Misc. Series No. 17. List of Chinese medicines. Inspector General of Customs, Shanghai

He P, Song X, Luo Y, He M (2009) Reproductive biology of *Dendrobium officinale* (Orchidaceae) in Danxia landform. Zhongguo Zhong Yao Za Zhi 34(2):124–127

He MG, Wang RX, Song XQ et al (2010) Study on cryopreservation of *Dendrobium chrysanthum* (Orchidaceae) Seeds. Acta Bot Yunnanica 32 (4):334–338

Hegde SN (1984) Review of orchid research in India and development of orchid sanctuary in Arunachal Pradesh. In: Rao AN (ed) Proceeding of the 5th ASEAN orchid congress. Parks & Recreation Department, Ministry of National Development, Singapore, pp 60 69

Hegde SN (1997) Orchid wealth of India. Proc Indian Natl Sci Acad B63(3):229–244

Hcinen JT (2010) The importance of a social science research agenda in the management of protected natural areas, with selected examples. Bot Rev 76 (2):140–164

Hijner JA, Arditti J (1973) Orchid mycorrhiza. Vitamin production and requirements by the symbionts. Am J Bot 60:829–835

Hirano T, Ishikawa K, Mii M (2005) Cryopreservation of immature seeds of *Ponerorchis graminifolia* var. suzukiana by vitrification. Cryo Letters 26 (3):139–146

Hollick SH (2004) Mycorrhizal specificity in endemic Western Australia terrestrial orchids (Tribe Diurideae): implications for conservation. PhD Dissertation, University of Murdock, Australia

Holttum RE (1978) The early days. In: Teoh ES (ed) Orchids. Orchid Society of Southeast Asia, Singapore

Hondrogiannis E, Rotta K, Zapf CM (2013) The use of wavelength dispersive X-ray fluorescence in the identification of the elemental composition of Vanilla samples and the determination of the geographic origin by discriminant function analysis. J Food Sci 77 (Epub 2013 Feb 8)

Jayaweera DMA (1981) A revised handbook of the flora of Ceylon, vol 2. A.A. Balkema, Rotterdam

Jia JS (2007) The status of orchid conservation in China. Lankesteriana 7(1–2):48

Jin XH, Zhao XD, Shi XC (2009) Native orchids from Gaoligongshan Mountains China. Science Press, Beijing

Kaiser R (1993) The scent of orchids: olfactory and chemical investigations. Editiones Roche, Basel

Kenji S (2013) Autogamous fruit set in a mycoheterotrophic orchid *Cyrtosia septentrionalis*. Plant Syst Evol 299(3):481–486

Koopowitz H (2001) Orchids and their conservation. Timber Press, Portland

Kosaka N, Kawahara T, Takahashi H (2014) Vegetation factors influencing the establishment and growth of the endangered Japanese orchid, *Cypripedium macranthos* var rubenense. Ecol Res 29(5):1003–1023

Kottaimuthu R (2008) Ethnobotany of the Valaiyans of Karandamalai, Dindigul District, Tamil Nadu, India. Ethnobot Leaft 12:195–203

Kusano E (1911) Gastrodia elata and its symbiotic association with *Armillaria mellea*. J Coll Agric Tokyo 4:1–66

Lamb A (1986) Progress in collecting lowland species of orchids of horticultural and botanical interest for the tenom orchid centre, Sabah, and the role of the centre with respect to a programme for conservation. In Rao AN (ed) Proceedings of the 5th ASEAN orchid congress. Parks & Recreation Department, Ministry of National Development, Singapore, pp 134–139

Lamxay V (2007) Case study on orchid exports from Lao PDR: recommendations for using the convention on international trade in endangered species of wild fauna and flora to increase sustainable orchid trade. Traffic SEA. IUCN, NUoL

Lau DT, Shaw PC, Wang J, But PP (2001) Authentication of medicinal dendrobium species by the internal transcribed spacer of ribosomal DNA. Planta Med 67 (5):456–460

Lee BC (ed) (2009) Rare plants data book of Korea. Korea National Arboretum, Gyeonggi-Do Korea

Leung PC (2006) Good agricultural practice – GAP does it ensure a perfect supply of medicinal herbs for research and drug development. In: Leung PC, Fong H, Xue CC (eds) Annals of traditional Chinese medicine, vol 2, Current review of Chinese medicine. Quality control of herbs and herbal medicine. World Scientific, Singapore, pp 27–57

Li J, Xiao X (1995) Medicinal plant resources of dendrobium in Sichuan Province. Zhongguo Zhong Yao Za Zhi 20(1):7–9

Li T, Wang J, Lu Z (2005) Accurate identification of closely related dendrobium species with multiple species specific gDNA probes. Biochem Biophys Methods 62(2):111–123

Li B, Wang BC, Liang YL et al (2006) Effect of polysaccharides content of tissue culturing seedlings on Dendrobium candidum under sound wave stimulation. Colloid Surf B Biointerfaces 63(2):269–275

Li X, Ding X, Chu B, Zhou Q, Ding G, Gu S (2008) Genetic diversity analysis and conservation of the endangered Chinese endemic herb *Dendrobium officinale* Kimura et Migo (Orchidaceae) based on AFLP. Genetica 133(2):159–166

Liebel HT, Bidartondo MI, Preiss K et al (2010) C and N stable isotope signatures reveal constraints to nutritional modes in orchids from the Mediterranean and Macaronesia. Am J Bot 97(6):903–912

Lin TH, Chang SJ, Chen CC, Wang JP, Tsao LT (2001) Two phenanthraquinones from *Dendrobium moniliforme*. J Nat Prod 64(8):1084–1086

Lin WC, Lu CF, Wu JW et al (2004) Transgenic tomato plants expressing the Arabidopsis NPR1 gene display enhanced resistance to a spectrum of fungal and bacterial diseases. Transgenic Res 13(6):567–581

Liu H (2010) Geodorum eulophioides. Challenges for conserving one of the rarest Chinese orchids. Orchids 79(3):161–162

Liu YP, Cao H, Komatsu K, But PP (2001) Quality control for Chinese herbal drugs using DNA probe technology. Yao Xue Xue Bao 36(6):475–480

Liu SQ, Li XJ, Yu QB, Xie H, Zhuo GY (2007) Analysis on genetic stability in different development stages of *Dendrobium huoshanense* by RAPD. Zhongguo Zhong Yao Za Zhi 32(10):902–905

Liu H, Feng CL, Luo YB et al (2010a) Potential challenges of climate change to orchid conservation in a wild orchid hotspot in Southwestern China. Bot Rev 76(2):174–192

Liu HX, Luo YB, Liu H (2010b) Studies of mycorrhizal fungi of Chinese orchids and their role in orchid conservation in China – a review. Bot Rev 76(2):241–262

Lo SF, Nalawade SM, Mulabagal V et al (2004) In vitro propagation by asymbiotic seed germination and 1,1-diphenyl-2-picrylhydrazyl (DPPH) radical scavenging activity studies of tissue culture raised plants of three important species of dendrobium. Biol Pharm Bull 27(5):731–735

Lovera P, Laville B (2009) Orchids trade study Laos 2009. Lad.nafri.org.la

Ma ZQ, Li SS, Zhang MJ et al (2010) Light Intensity affects growth, photosynthetic capability, and total flavonoid accumulation of Anoectochilus plants. Hortscience 45(6):863–867

Maikhuri RK, Nautiyal S, Rao KS, Saxena KG (2001) Conservation policy-people conflicts: a case-study from Nanda Devi Biosphere Reserve (a World Heritage Site), India. For Policy Econ 2(3 and 4):355–365

Manners V, Kumaria S, Tandon P (2013) SPAR methods revealed high genetic diversity within populations and high gene flow of *Vanda coerulea* Griff ex Lindl (Blue Vanda), an endangered orchid species. Gene 519 (1):91–97

McCartney C (2010) Aliens among us. Foreign orchids go wild in South Florida. Orchids 79(10):576–585

Medhi RP, Chakrabarti S (2009) Traditional knowledge of NE people on conservation of wild orchids. Indian J Tradit Knowl 8(1):11–16

Mohd-Hairul AR, Namasivayam P, Ee GCL, Ong-Abdullah J (2010) Terpenoid, benzenoid and phenylpropanoid compounds in the floral scent of Vanda Mimi Palmer. J Plant Biol 53(5):358–366

Muthu C, Ayyanar M, Raja N, Ignacimuthu S (2006) Medicinal plants used by traditional healers in Kanchee Puran District of Tamil Nadu, India. J Ethnobiol Ethnomed 2:43

Namsa ND, Tag H, Mandal M et al (2009) An ethnobotanical study of traditional anti-inflammatory plants used by the Lohit community of Arunachal Pradesh, India. J Ehtnopharmacol 125(2):234–235

Nautiyal S, Maikhuri RK, Rao KS, Saxena KG (2003) Ethnobotany of the Tolchha Bhotiya tribe of the buffer zone villages in Nanda Devi Biosphere Reserve, India. In: Singh V, Jain AP (eds) Ethnobotany and medicinal plants of India, vol 1, 2. Scientific, Jodhpur

Orleans S (1999) The orchid thief. Heinemann, London

Pandey NK, Joshi GC, Mudaiya RK et al (2003) Management and conservation of medicinal orchids of Kumaon and Garhwal Himalaya. In: Singh V, Jain AP (eds) Ethnobotany and medicinal plants of India and Nepal. Scientific, Jodhpur, pp 114–118

Pandey NK, Joshi GC, Mudaiya RK et al (2013) Medicinal orchids and their uses: tissue culture a potential alternative for conservation. Afr J Plant Sci 7 (10):448–467

Pant S, Rinchen T (2012) Dactylorhiza hatagirea: a high value medicinal orchid. J Med Plants Res 6 (19):3522–3524

Pant B, Raskoti BB (2013) Medicinal orchids of Nepal. Himalayan Map House (P) Ltd, Kathmandu

Pearce NR, Cribb PJ (2002) The orchids of Bhutan. Royal Botanic Gardens/Royal Government of Bhutan, Edinburgh/Thimpu

Pemberton RW (2010) Biotic resource needs of specialist orchid pollinators. Bot Rev 76(2):275–292

Perner H (2002) Orchids and eco-tourism: the world natural heritage and biosphere reserve, Huanglong. In: Proceedings of the 17th world orchid conference Shah Alam, Malaysia, pp 158–164

Perner H (2007) Cypripediums in China Part IV. Orchids 76(4):291–292

Phelps J, Carrasco LR, Webb EL (2014) A framework for assessing supply-side wildlife conservation. Conserv Biol 28(1):244–277

Powell KB, Arditti J (1975) Growth requirements of *Rhizoctonia repens* M32. Mycopathologia 55:163

Preiss K, Adam IKU, Gabauer G (2010) Irradiance governs exploitation of fungi: fine-tuning of carbon gain by two partially myco-heterotrophic orchids. Proc R Soc Biol Sci Ser B 277(1686):1333–1336

Pritchard H (1984) Terrestrial orchid seed storage. In: Tan K (ed) Proceedings of the 11th world orchid conference, Miami, p 290

Pyakurei D, Gurung K (2008) Assessment of orchids and estimation of current stock of traded orchids in Rolpa District, Nepal. District Forest Office, Rolpa

Rajasekaran C, Maikhuri IRK, Kuseum C et al (2009) Multiplication and conservation of *Dactylorhiza hatagirea* (D. Don) Soo – endangered medicinal orchid of higher Himalaya. MIOS J 10(1):7–16

Ran YZ, Xu JT (1988) Studies on the inhibition of seed germination of Gastrodia elata Bl. by *Armillaria mellea* Qul. Zhong Yao Tong Bao 13:15–17

Rao AN (2004) Medicinal orchid wealth of Arunachal Pradesh. Newsl ENVIS NODE Indian Med Plant 1 (2):1–7

Rao NR, Henry AN (1995) The ethnobotany of Eastern Ghats in Andra Pradesh, India. Botanical Survey of India. Government Press of India (quoted by Musharof Hossain, 2009)

Rao TA, Sridhar S (2007) Wild orchids in Karnataka. A pictorial compendium. Institute of Natural Resources

Conservation, Education, Research and Training (INCERT), Bangalore

Rasmussen HN (1995) Terrestrial orchids from seed to mycotrophic plant. Cambridge University Press, Cambridge

Rasmussen HN, Whigham DF (1993) Seed ecology of dust seeds in situ. A new study technique and its application in terrestrial orchids. Am J Bot 80:1374–1378

Rathore SR (1983) Endemic and rare species of calanthe R. Br. (Orchidaceeae) in India. In: Jian SK, Rao RR (eds) An assessment of threatened plants of India. Botanical Survey of India, Delhi

Raven-Riemann C, Fifghetti C, Ku M (2010) A bright new talent emerges in Taiwan's phalaenopsis hybridizing. Orchids 79(5):253–263

Sandhya B, Thomas S, Isabel W, Shenbagarathai R (2006) Ethnomedicinal plants used by the Valaiyan community of Piranmalai Hills (Forest Reserve), Tamil Nadu. A pilot study. Afr J Trad CAM 3(1):101–114

Saw Lwin (2002) Myanmar native orchids listed in CITES Appendix. In: Proceedings of the 12th world orchid conference, Kuala Lumpur, pp 207–210

Seaton PT, Hu H, Perner H, Pritchard HW (2010) Ex-situ conservation of orchids in a warming world. Bot Rev 76(2):193–203

Selosse MA, Weiss M, Jany JL, Tillier A (2002) Communities and populations of sebacinoid basidiomycetes associated with the achlorophyllous orchid Neottia nidus-avis (L.). L.C.M. Rich. and neighbouring tree ectomycorrhiza. Mol Ecol 11:1831–1844

Senthikumar K, Aravindan V, Rajendran A (2013) Ethnobotanical survey of medicinal plants used by Malayali tribes in Yercaud Hills of Eastern Ghats. J Nat Res 13(2):118–132

Shen J, Ding X, Liu D, Ding G, He J, Li X, Tang F, Chu B (2006) Intersimple sequence repeats (ISSR) molecular fingerprinting markers for authenticating populations of Dendrobium officinale Kimura aet Migo. Biol Pharm Bull 29(3):420–422

Shen J, Xu GH, Chang J et al (2010) Effects of sodium selenite on growth of protocorm-like bodies of Dendrobium officinale in vitro. J Nat Sci Nanjing Norm Univ 12(1):62–66

Sivakumar A, Subramanian MS, Karunakaran M, Burkanudeen A (2003) Ethnobotany of Poliyars of Anaimalai Hills, Tamil Nadu. J Econ Taxon Bot 27 (3):679–685

Sivasankari B, Pitchaimani S, Anandharaj M (2013) A study of traditional medicinal plants of Uthapuram, Madurai District, Tamil Nadu, South India. Asian Pac J Trop Biomed 3(12):975–979

Stark C, Babik W, Durka W (2009) Fungi from the roots of the common terrestrial orchid Gymadenia conopsea. Mycol Res 113(9): 952–959

Stephen RC, Fung KK (1971) Vitamin requirements of the fungal endophytes of Arundina chinensis. Can J Bot 49:411–415

Subedi A (2005) Orchids and sustainable livelihood: an initiative in Nepal Himalayas to manage globally threatened biodiversity. In: Raynal-Rogues A, Roguenant A, and Prat D (eds) Proceedings of the 18th world orchid conference, Dijon, France. Naturalia Publications, Paris, pp 474–477

Sujatha S, Bhat R (2010) Response of vanilla (Vanilla planifolia A.) intercropped in arecanut to irrigation and nutrition in humid tropical India. Agric Water Manag 97(7):988–994

Sumithraarachchi DB (1986) Conservation of orchids in Sri Lanka. In: Rao AN (ed) Proceedings of the 5th ASEAN orchid congress. Parks & Recreation Department, Ministry of National Development, Singapore, pp 140–144

Sun HQ, Liu YB, Song GE (2003) A preliminary on pollination of an endangered orchid, Changnienia amoena in Shennongjia. Acta Bot Sin 45 (9):1019–1023

Sun HQ, Li A, Ban W, Zheng XM, Ge S (2005) Morphological variation and its adaptive significance for Changnienia amoena, an endangered orchid. Biodivers Sci 13(05):376–386

Swinson A (1970) Frederick Sander, the orchid king. Hodder and Stoughton, London

Tanaka H, Yee ATT (2007) Wild Orchids in Myanmar, vol 3. The Foundation of Agricultural Development and Education, Bangkok

Tatarenko IV (2007) Growth habits of temperate terrestrial orchids. In: Cameron KM, Arditti J, Kull T (eds) Orchid biology. Reviews and perspectives, vol 9. The New York Botanical Garden Press, New York, pp 91–161

Tekinsen KK, Guner A (2009) Chemical composition and physicochemical properties of tubera salep produced from some Orchidaceae species. Food Chem 121:468–471

Teoh ES (1980) Asian orchids. Times, Singapore

Thammasiri K (2000) Cryopreservation of seeds of a Thai orchid (Doritis pulcherrima, Lindl.) by vitrification. CryoLetters 21:237–244

Thammasiri K (2002) Preservation of seeds of some Thai orchid species by vitrification. In: Proceedings of the 12th WOC, Kuala Lumpur, pp 248–249

Thammasiri K (2008) Cryopreservation of some Thai orchid species. Acta Hortic 788:53–62

Thammasiri K (2010) Vitrification-based cryopreservation of Grammatophyllum speciosum protocorms. CryoLetters 31(4):347–357

Teoh ES (2003) The lotus in the Buddhist Art of India. Singapore, doctorteoh@gmail.com

Tong WJ, Zhang L, Xue QY et al (2014) Isolation of endophytic bacteria in Dendroboiun loddigesii collected from different locations and comparison on their plant-growth-promoting potential. J Plant Resources Environ 23(1):16–23

Tsavkaelova EA, Cherdyntseva TA, Botina SG, Netrusov AI (2007) Bacteria associated with orchid roots and microbial production of auxins. Microbiol Res 162 (1):69–76

Tsay HS (2002) Tissue culture of *Anoectochilus formosanus*. In: Proceedings of the 16th World orchid conference, Kuala Lumpur, p 438

Vakhrameeva MG (2008) Orchids of Russia and adjacent countries. ARG Gantner-verlag KG, Ruggell

Valetta A, Attorre F, Bruno F, Pasqua G (2008) In-vitro asymbiotic germination of *Orchis mascula* L. Plant Biosyst 142(3):653–655

Vendrame WA, Carvalho VS, Dias JMM (2007) In vitro germination and seedling development of cryopreserved dendrobium hybrid mature seeds. Sci Hortic 114:188–193

Vij SP (1995) Orchid genetic diversity in India: conservation and commercialization. In: Proceedings of the 5th Asia Pacific orchid conference and show, Fukuoka, pp 20–39

Wang S, Lin Y, Cai YP, Zhan SH, Ma SJ (2006) Comparison on growth, physiology and medicinal components of *Dendrobium huoshanense* hybrid and its parents. Zhongguo Zhong Yao Za Zhi 31(9170):1401–1404

Wang HZ, Lu JJ, Shi NN, Zhao Y, Ying QC (2007) Analysis of genetic diversity among 13 Chinese species of dendrobium based on AFLP. Fen Zi Xi Bao Sheng Wu Xue Bao 40(3):205–210

Warghat AR, Bajpai PK, Srivastatva RB, Chaurasia OP (2014) In vitro protocorm development and mass multiplication of an endangered orchid, *Dactylorhiza hatagirea*. Turk J Bot 38(4):737–746

Wong KC, Sun M (1999) Reproductive biology and conservation genetics of *Goodyera procera* (Orchidaceae). Am J Bot 86(10):1406–1413

Wood CB, Pritchard HW, Millar AP (2000) Simultaneous preservation of orchid seed and its fungal symbiont using encapsulation-dehydration is dependent on moisture content and storage temperature. Cryo Letters 21(2):126–136

Wu XR (1994) A concise edition of medicinal plants in China. Guangdong Higher Education Publication House, Guangdong (in Chinese)

Wu J, Ma H, Xu X et al (2013) Mychorrhizas alter nitrogen acquisition by the terrestrial orchid *Cymbidium goeringii*. Ann Bot 111(6):1181–1187

Wu CT, Huang KS, Yang CH et al (2014) Inhibitory effects of cultured *Dendrobium tosaense* on atopic dermatitis murine model. Int J Pharm 463:193–20

Xie MI, Hou BW, Han L et al (2010) Development of microsatellites of *Dendrobium officinale* and its application in purity identification of germplasm. Yao Xue Xue Bao 45(5):667–672

Xu JT (1989) Studies on the life cycle of *Gastrodia elata*. Zhongguo Yi Xi Xue Ke Xue Yuan Xue Bao 11:237–241

Xu J (1990a) Cytological observation on hyphae invading *Mycena osmundiloca* in the process of germination of *Gastrodia elata* Bl. Zhongguo Yi Xue Ke Xue Yuan Xue Bao 12:313–317

Xu J (1990b) Studies on nutrition source of seeds germination of *Gastrodia elata* Bl. Zhongguo Yi Xue Ke Xue Yuan Xue Bao 12:431–434

Xu J, Guo S (2000) Retrospect on the research of the cultivation of *Gastrodia elata* Bl., a rare traditional Chinese medicine. Chin Med J 113:686–692

Xu H, Wang Z, Ding X, Zhou K, Xu L (2006) Differentiation of dendrobium species used as "Huangcao Shihu" by rDNA ITS sequence analysis. Planta Med 72(1):89–92

Yam TW (2010) Conservation of the native orchids through seedling culture and re-introduction: a Singapore experience. Bot Rev 76(2):263–274

Yam T, Thame A (2005) Conservation and reintroduction of the native orchids in Singapore. Selbyana 26(1/2):75–80

Yan RH, Liu HX, Cai HF, Ge S (2006) A preliminary study of *Changnienia officinale* mycorrhizal fungi. J Beijing Forest Univ 28(2):112–117

Yang SJ, Zhang XH, Cao ZY et al (2014) Growth promoting *Sphingomonas paucimobilis* ZJSH1 associated with dendrobium officinale through phytohormone production and nitrogen fixation. Microb Biotechnol 7(6 Sp. Iss.S1):611–620

Zhang Z, Lin H, Wang L, Zhang L (1999) Cryopreservation of seeds and mass propagation in vitro of *Dendrobium candidum*. In: Proceedings of the 16th world orchid conference, p 440

Zhao ZH, Yan ZJ, Iida O (2006) Supply and cultivation of medicinal plants in Japan. In: Leung PC, Fong H, Xue CC (eds) Annals of traditional Chinese medicine, vol 2, Current review of Chinese medicine. Quality control of herbs and herbal medicine. World Scientific, Singapore, pp 59–73

Zhao MM, Zhang G, Zhang DW, Guo SX (2013) Molecular cloning and characterization of S-adenosyl-L-methionine decarboxylase gene (soSAMDC1) in *Dendrobium officinale*. Yao Xue Xue Bao 48 (6):946–952

Zhao XL, Yang JZ, Liu S et al (2014) The colonization patterns of different fungi on roots of *Cymbidium hybridum* plantlets and their respective inoculation effects on growth and nutrient uptake of orchid plantlets. World J Microbiol Biotechnol 30 (7):1993–2003

The Randomized Clinical Trial and the Documentation of Side Effects

This chapter explains the nature of a RCT, and the meaning of probability as applied to medical data. The need for continuous collection of information for side effects from voluntary reporting is emphasised.

Contemporary medical thinking insists that randomised controlled trials (RCTs) of sufficient size are absolutely necessary to prove the effectiveness of any medical treatment. In the case of Western medicine, few entrenched treatments have been re-evaluated by RCTs, but nearly all new remedies are being studied in this way. Complementary and alternative medicine (CAM) have seldom been properly subjected to well-designed RCTs, and Professor R. Barker Bausell, who is Research Director of an NIH-funded CAM Specialized Research Center at the University of Maryland, USA, cautions that many of their claims can be explained by the common placebo effect (Bausell 2007).

Doctors frequently prescribe placebos to reassure their patients that something is being done for them. In fact, the word placebo is derived from the Latin and means "to please." A placebo is an irrelevant substance that shows a therapeutic effect when administered to a person who believes that he or she is receiving an effective treatment. The substance may be inert like cellulose, starch or sugar, or it may be irrelevant, like administering vitamin C for headaches. Since some treatments are physical and do not make use of drugs, the wider definition of placebo may be Bausell's suggestion of "a make believe therapy". Comparison with placebos is necessary to prove that a treatment is more than just placebo.

When treatments are being compared (placebo vs. agent under investigation), it is essential that both the person administering the treatment and the patients are unaware as to whether the latter is receiving the agent under test or the placebo. The selection is random, giving the agent or the placebo to alternate subjects, or by random pick. Such an investigation then qualifies for the definition of a "randomised double-blind, placebo-controlled clinical trial".

If the differences are expected to be dramatic, 50 patients assigned to each group should suffice. If the differences are slight, then a larger group may be required. At the end of the trial, the differences between effects produced in the test group and the placebo group are calculated for extent; for instance, a 70 % response in test group and a 35 % response in the placebo group means that the treatment is twice as effective as the placebo, which probably means there is a real difference. The test for statistical significance demonstrates whether the difference is likely to be a true difference or a chance effect. Both statisticians and scientists agree that, if the likelihood of a chance occurrence is under 5 %, then the difference can be considered to be statistically significant. Medical conclusions are based on the demonstration of such statistically significant differences in outcome.

© Springer International Publishing Switzerland 2016
E.S. Teoh, *Medicinal Orchids of Asia*, DOI 10.1007/978-3-319-24274-3_24

Professor Bausell provides a graphic explanation of statistical difference in his book, *Snake Oil Science*. He stated that if one flips a coin five times and it comes out heads five times consecutively, the likelihood of that coin being different from other coins is statistically significant. It is calculated like this. Each time the coin is flipped, there is a 50 % chance that it will turn up heads. So, for five times in a row, the likelihood is $0.5 \times 0.5 \times 0.5 \times 0.5 \times 0.5 = 0.03125$, i.e. 3.125 %. On the other hand, if the coin turns up heads four times out of five, that is not a scientifically acceptable significant difference because the likelihood of that occurring is 6.25 %.

Obviously, scientists and laymen do not see eye to eye, especially if the laymen are gamblers.

The problem of clinical testing does not stop here. When promising results are obtained, they need to appear in peer-reviewed journals of a high international stature. There are many such journals world-wide, but in order to receive prompt attention, it is better to publish in journals that use the English language. Several notable discoveries in endocrinology (at one time, the author's favourite field of research) published in Japanese have not been widely acknowledged.

Next, the findings have to be confirmed by other independent teams. Authors have to be prepared to be challenged, and they must be prepared to defend their methodology and statistics, and to be able to offer explanations of the mode of action.

Mode of action is an area in which scientific medicine and complementary-alternative medicine (CAM) often find no common ground. The concept of the constancy of the internal environment proposed by the French physiologist, Claude Bernard, appears to have some similarity of *yin–yang* balance entrenched in Traditional Chinese Medicine (TCM). But Bernard was talking about constant temperature, a narrow range of electrolyte concentrations, pH, osmotic pressure and so on, which do not have true equivalents in TCM. It would seem that the absence of scientifically acceptable explanations is the greatest hurdle for CAM to overcome, because credibility and plausibility rests on scientific evidence and explanations. Scientists working in the field of herbal medicine appreciate this fact, and are devoting an enormous amount of time and effort in performing biochemical, pharmacological and physiological studies in their respective fields.

Side Effects

Efficacy does not promise safety. Some herbal remedies are effective for the relief of pain. Everyone is familiar with opium the active principle of which is morphine. Doctors still rely on morphine for terminal pain, and its shorter acting derivative, pethidine (Demerol) for pain relief in trauma, surgery and childbirth. Less well known is the fact that salicylates, which reduce fever and relieve pain, are naturally present in a variety of herbal preparations, fruits and vegetables (for instance, berries, dry fruits, tea, liquorice, peppermint, rosemary, thyme, chilli and curry powder). In fact, willow bark which has a high content of salicylic acid was used by Edward Stone way back in 1763 to treat fever. Salicylic acid was isolated and identified in the early 19th century. Aspirin (acetylsalicylic acid) was synthesised by Gerhardt in 1852 and, some 30 40 years later, it was promoted for pain relief in rheumatoid arthritis. Many later trials showed that salicylates are not effective for pain relief in this condition. On the other hand, the Chinese herb, *lei gong teng* (*Tripterygium wilfordii*), is efficacious, but there is a serious safety issue. When mice were fed a decoction of *lei gong teng* at doses of 5 and 10 mg/25 g per day for 4 consecutive days, they suffered significant liver and kidney damage (Al-Achi 2008).

There is therefore a need to document the side effects of every treatment, regardless of whether it works as a remedy or is worthless. Nowhere is this more important than in the field of cardiovascular medicine, because herbal products consumed by a large proportion of patients with heart conditions often contain substances which influences blood clotting, and they compound the risk of anticoagulants used for the prevention of strokes or a recurrence of coronary thrombosis (Miller et al. 2004; Elmer et al. 2007). They may also cause serious bleeding during surgery. Nevertheless, the American College of Cardiology recognises the need to integrate complementary medicine into cardiovascular medicine because that is the wish of a significant proportion of their patients (Vogel et al. 2005). In the United States, about half the patients who are using complementary medicine do not wish to discuss the subject and do not make their usage known to their doctors (Conway 2009). Thus, the call is not for prohibition, rather the message is understanding by both physicians and public of the potential for dangerous interactions, while the emphasis is on the need for well-designed and carefully conducted randomised clinical trials with adequate power to detect the effects of herbal and other complementary therapies, coupled with adequate short- and long-term monitoring of side effects. The National Heart, Lung and Blood Institute and the National Center for CAM of the National Institutes of Health in the United Sates further emphasise the need for more data on the uptake, bioavailability, pharmacodynamics and mechanisms by which herbal remedies might exert their effects (Lin et al. 2001).

With some of the medicinal orchids, it might be argued that they have been used for over 2000 years, which suggests that they are safe. That is not absolutely true. In the *Shen Nong Ban Cao Jing*, the safe herbs are those classified as Superior Herbs, an example being *shihu* (*Dendrobium moniliforme* and *D. catenatum*); the rest have their side effects, and the Inferior Herbs can be dangerous if improperly used. The vast majority of orchids listed in the present publication were not in the ancient pharmacopoeia and their safety is not documented. Ancient or new, for orchids to be safely integrated into any holistic therapy, safety monitoring for side effects is essential.

Stroke and Traditional Chinese Patent Medicines

Stroke, defined as a sudden, severe attack of paralysis, usually followed by coma, and due to an acute vascular lesion of the brain, such as haemorrhage, thrombosis or embolism, is a major health problem and a catastrophe both for the patient and the family. There is no medicine that will improve the condition, but there are numerous claims that TCM improves the outcome. *Tianma* (*Gastrodia elata*), ginseng and *Ginkgo biloba* are some of the items used in the treatment (Gong 2002; Kim 2005; Zeng et al. 2005; Wu et al. 2007; Li et al. 2009). Several meta-analyses of the trials for a variety of TCM, namely *Dan Shen* (Wu et al. 2004), *Ginkgo biloba* (Zeng et al. 2005), *mailuoning* (Yang et al. 2009), *acanthopanax* (Li et al. 2009) and 22 traditional Chinese patent medicines (Wu et al. 2007), all concluded that, while the poor-quality trials showed a definite benefit, producing an "attractive forest plot of meta-analysis", this carried the risk of bias. The good-quality, placebo-controlled trials failed to show any improvement. These meta-analyses were prepared by staff of the Neurology Department of West China Hospital, Sichuan University, one of the leading medical schools in China, for the respected Cochrane Database Systematic Review (Wu et al. 2004; Zeng et al. 2005; Yang et al. 2009; Li et al. 2009), or for the journal *Stroke* (Wu et al. 2007). The authors found no convincing evidence of adequate methodological quality in the trials, and they urge that large-scale RCTs be performed to confirm efficacy, because the agents appear to be non-toxic, and they would therefore be an important contribution to medicine if they work.

A double-blind, placebo-controlled, randomised, multicentre study to investigate Chinese Medicine NeuroAID Efficacy on stroke recovery (CHIMES Study) was initiated at the Division of Neurology, National University Hospital in Singapore (Venketasubramanian et al. 2009). It enrolled 1100 individuals and employed a modified Rankin Scale (mRS) to study the primary efficacy endpoint at 3 months. Safety outcomes were monitored. The study was undertaken because, in animal stroke models, NeuroAID had been shown to induce neuroplasticity, reduce cell proliferation and stimulate dense zonal and dendritic networks

(Chen et al. 2013a). When the study group published their results in 2003, they concluded that serious and non-serious adverse events occurred with the same frequency in the treatment and placebo groups, but, at 3 months, the Chinese medicine was no better than placebo in improving outcome. However, in subgroup analysis, there was a trend for a benefit for patients who received treatment beyond 48 h of stroke onset. Longer durations of treatment will need to be studied (Chen et al. 2013b).

TCM for Subfertility

Four randomised control trials involving 344 participants comparing standard clomiphene treatment with standard treatment plus the addition of Chinese herbal medicine (CHM) were studied by meta-analysis. The reviewers found that, whereas there was no difference in ovulation rate, there was a difference in pregnancy rate favouring the addition of CHM (odds ratio 2.97; 95 % confidence limits 1.71–5.17, i.e. statistically significant) and improved clinical pregnancy outcome. However, the methodology was not adequately reported and there was a risk of methodological bias (Zhang et al. 2010).

Standards of Excellence in an RCT

The above comments are not specifically targeted at TCM. Similar criticisms are carried by Western medical journals whenever an important trial is reported. The quality of reporting in "randomised" surgical trials published in high-quality surgical journals in the West have also been graded as poor (Sinha et al. 2009). The findings and conclusions of a trial will stand up to criticism only if the following criteria are met: (1) the trial is well designed; (2) it is appropriately conducted; (3) there are sufficient numbers to show statistically valid differences; and (4) reasonable conclusions are drawn from it.

References

Al-Achi A (2008) An introduction to botanical medicines. Praeger, Westport

Bausell RB (2007) Snake oil science. The truth about complementary and alternative medicine. Oxford University Press, Oxford

Chen CL, Ikram K, Anqi Q et al (2013a) The NeuroAID II (ML901) in vascular cognitive improvement study (NEURITES). Cerebrovasc Dis 35(Suppl 1):23–29

Chen CL, Young SH, Gan HH et al (2013b) Chinese medicine neuroaid efficacy on stroke recovery: a double-blind, placebo-controlled, randomized study. Stroke 44(8):2083–2100

Conway C (2009) Use of complementary and alternative medicine before surgery poses risk to patient safety (internet release prior to publication: cc328@columbia.edu)

Elmer GW, Lafferty WE, Tyree PT, Lind BK (2007) Potential interactions between complementary/alternative products and conventional medicines in a medicare population. Ann Pharmacother 41(10):1617–1624

Gong X (2002) Stroke therapy in traditional Chinese medicine (TCM): prospects for drug discovery and development. Phytomedicine 9(5):475–484

Kim H (2005) Neuroprotective herbs for stroke therapy in traditional eastern medicine. Neurol Res 27(3):287–301

Li W, Liu M, Feng S et al (2009) Acanthopanax for acute ischaemic stroke. Cochrane Database Syst Rev 8(3):CD007032

Lin MC, Nathan R, Gershwin ME et al (2001) State of complementary and alternative medicine in cardiovascular, lung, and blood research. Executive summary of a workshop. Circulation 103:2038–2041

Miller KL, Liebowitz RS, Newby LK (2004) Complementary and alternative medicine in cardiovascular disease: a review of biologically based approaches. Am Heart J 147(3):401–411

Sinha S, Sinha S, Ashby E et al (2009) Quality of reporting in randomized trials published in high quality surgical journals. J A Coll Surg 209(5):565–571

Venketasubramanian N, Chen CL, Gan RN et al (2009) A double-blind, placebo- controlled, randomized, multicenter study to investigate Chinese medicine neuroaid efficacy on stroke recovery (CHIMES Study). Int J Stroke 4(1):54–60

Vogel JHK, Bolling SF, Costello RB et al (2005) Integrating complementary medicine into cardiovascular medicine: a report of the American College of Cardiology Foundation Task Force on Clinical Expert Consenses Documents (Writing committee to develop an expert consensus document on complementary and integrative medicine). J Am Coll Cardiol 2005 (46):184–221

Wu B, Liu M, Zhang S (2004) Dan Shen agents for acute ischaemic stroke. Cochrane Database Syst Rev 4:CD004295

Wu B, Liu M, Liu H (2007) Meta-analysis of traditional Chinese patent medicine for ischaemic stroke. Stroke 38(6):1973–1979

Yang W, Hao Z, Zhang S et al (2009) Mailuoning for acute ischaemic stroke. Cochrane Database Syst Rev 2:CDC007028

Zeng X, Liu M, Yang Y et al (2005) Ginkgo biloba for acute ischaemic stroke. Cochrane Database Syst Rev 4:CD003691

Zhang J, Li T, Zhou L (2010) Chinese herbal medicine for subfertile women with polycystic ovary syndrome. Cochrane Database Syst Rev 9:CD007535

Glossary

Abaxial Surface of an organ that is directed away from the main axis, e.g. under-surface of a leaf, or outer surface of petals and sepals

Achlorophyllous Lacking chlorophyll

Acuminate Tapering to a point

Adenomatous Pertaining to glandular change

Adenocarcinoma Cancer arising from a gland

Adrenalin A compound produced by the adrenal gland in response to stress

Aetiology The total knowledge of the cause of a disease

Aglycone The non-carbohydrate group of a glycoside molecule

Ague Malarial fever

Alba White. In plants, it would additionally refer to colour forms lacking in anthocyanins but not chlorophyll, i.e. the flowers may be green and not white

Albino Lacking colour pigments, chlorophyll and anthocyanins

Alkaloid Organic compound containing nitrogen and showing basic (alkaline) properties

Allogamy Outcrossing

Allopolyploid With more than two complete genomes (sets of chromosomes) that belong to more than a single species

Alzheimer's disease Presenile dementia

Amblyopia Dimness of vision without apparent organic cause

Analgesic Pain-relieving

Analogue A compound with the same action but of a different molecular structure from the original

Angiogenesis Formation of new blood vessels

Angiotensin Compound produced by the kidneys which contracts arteries and raises the blood pressure

Anodyne Medicine that relieves pain, like morphine or codeine

Anoikis Apoptosis induced by inadequate or inappropriate cell–matrix interactions

Anthesis The period when flowers are fully open and at their prime. The functional period of a flower

Anthocyanin Colour pigment in plants which imparts red to blue colour ranges in the flowers

Anti-oxidant Compound that slows oxidation and work by scavenging free radicals

Antipyretic Relieving or reducing fever

Antisense Mirror image of a genetic sequence. An antisense compound would theoretically inactive a genetic sequence

Aneuploidy Not possessing the proper number of chromosomes for a specific genome, i.e. lacking or having additional, odd chromosomes

Anoikis A programmed cell death that occurs when a cell is detached from its normal surrounding cells

Anorexia Loss of appetite

Antiplatelet In effect, anticoagulant

Apoptosis Programmed cell death, in contrast to cell death caused by injury or infection

Apotropaics Charms for averting evil influences that may cause illness. They were also employed in Greek medicine, and are commonplace in the Asian tradition

© Springer International Publishing Switzerland 2016
E.S. Teoh, *Medicinal Orchids of Asia*, DOI 10.1007/978-3-319-24274-3

Aquaporin A natural compound involved in the flow of water across cell membranes

Arcuate Arched

Arthralgia Painful joint

Astringent Causing contraction, or an ability to stop discharges

Axil Upper angle where a leaf joins the stem

Axillary Arising from the axil

Baclofen A drug which relaxes muscles

Bactericidal Causing death of bacteria

Bacteriostatic Suppressing multiplication of bacteria but not causing their demise

Besom **(Malay)** A brush made from small twigs, employed in ceremonies for dispelling harmful influences

Beta-amyloid Deposits in the brain associated with Alzheimer's disease

Bibenzyl An aromatic organic compound considered a derivative of ethane in which one phenyl group is attached to each carbon compound

Bifid Y-shaped

Calcareous Soil containing predominantly calcium carbonate, limestone

Calli Plural of callus

cAMP Cyclic adenosine monophosphate. It is a second messenger that conveys messages into the cell from hormones that are unable to cross the cell membrane

Capitate Carrying a head of flowers

Carbon tetrachloride A potent liver poison. Carbon tetrachloride is a liquid compound employed in dry cleaning

Carcinoma Cancer

Carrageenan A linear sulfated polysaccharide present in seaweed which is widely employed as a food additive because it binds to proteins and gels the preparation

Caruncle Small, fleshy eminence

Caudate Like a tail

Cauline Occurring on a stem

Cleistogamous Self-pollinating

Caespitose Growing in dense clusters

Catalase Enzyme present in all cells (except anaerobes) which catalyses the dissociation of hydrogen peroxide

Cerebellum Part of the brain which controls motor activity

Cerebral Referring to the brain

Cercaria Final, free-swimming, larval stage of the liver fluke

c-fos protein A proto-oncogene which is overexpressed in some cancers

Ciliate Fringed with thin processes

Cholangiocarcinoma Cancer of the gall bladder

Cholangiofibrosis Chronic inflammation and scarring of the gall bladder

Chromosome x-shaped, rod-like structures in the nucleus constituted by DNA

Chromatography A laboratory technique for separating compounds in a mixture by exploiting the rate at which they migrate through a stationary phase

Caespitose Matted or in clusters

Clade Monophyletic group

Coriaceous Leathery, usually referring to leaves

Corm Subterranean, condensed stem

CREB protein cyclicAMP response element-binding protein, a cellular transcription factor

Crenate Outlined in rounded scallops

Cucullate Hooded

Cytokines Substances which promote cell multiplication

Cytotoxic Inducing cell death

Deacetylated Removing the acetyl group from an organic compound

Decoction A medicine prepared by boiling

Demulcent A soothing mucilaginous or oily medication

Dendrite Branching projections from a nerve cell that receives messages

Dentate Toothed

Depolomerisation Break-up

Depurant A cleansing or purifying agent or drug

Dihydrophenanthrene Phenanthrene with two hydroxyl groups

Diaphoretic Medication which promotes sweating

Diploid With two sets of chromosomes; a normal genotype as compared with polyploids (triploid, tetraploid, pentaploid, hexaploid, etc.)

Dipterocarp Members of the tree family Dipterocarpaceae. Giant trees of tropical, lowland, primary rainforests in Southeast Asia. They grow to an enormous height, have straight trunks, often buttressed, huge crowns, thick leaves, and symmetrical flowers

Distichous In two ranks on opposite sides of an axis

Dopamine A neurotransmitter

Dropsy Abnormal accumulation of body fluid

Dysentery Illness(es) characterised by frequent stools containing mucus and blood, generally caused by viruses, bacteria, protozoa, worms or rarely by chemicals

Dyskinesia Involuntary, sometimes painful muscle movements

Dyspepsia Indigestion. Term is sometimes employed to describe excessive stomach acidity

Eczematous Referring to inflammatory skin lesions

Electrophoresis A process for separating large molecules

Emetic A substance that induces vomiting

Emollient An agent which softens or soothes the skin

Endemic Restricted in distribution to one country

Endothelial Pertaining to the endothelium

Endothelium Lining of the blood vessels and the heart

Endophytic Pertaining to a plant which grows inside another plant

Ensiform Long and narrow, shaped like a sword blade

Enuresis Involuntary discharge of urine at night; urinary incontinence

Eosinophils White blood cell which stains red with eosin. Their numbers increase in the presence of an allergen

Ephemeral Short-lived. In flowers, lasting less than a day

Epilepsy Disease characterised by temporary loss of consciousness, involuntary muscle movements, and psychic or sensory disturbances

Epiphyte Plant growing on trees

Equitant Fan-shaped

Erose Irregularly notched

Ethnobotany Study of the relationship of plants and humans

Ethnomedicine Study of folk medicine

Expectorant A compound which promotes the expulsion of mucus from the lungs. It promotes coughing

Falcate Sickle-shaped

Fas-ligand A protein that has an important role in the immune response to cancer; binding to its receptor induces cell death

Fibronectin A high molecular weight glycoprotein which binds collagen and fibrin

Filopedia Slender cytoplasmic projections in migrating cells

Fimbriate Fringed

Flavone A colourless compound that is the precursor of yellow dyestuffs

Flavonoid Yellow plant compound

Flexous Wavy

Fomentation Treatment with warm moist applications, e.g. a heated poultice

Free radicals Oxygen molecules missing electrons; they take them from normal, healthy cells thereby damaging the latter. The process is called oxidation

Frontal cortex The front part of the brain which is involved in conscious thought, memory, decisions and motor activity

Fusiform Spindle-shaped

GABA Gamma amino-butyric acid. It is a suppressive neurotransmitter

Gerbil Rodent

Glabrous Smooth

Glutathione A tripeptide present in plant and animal tissues; it carries oxygen

Glycoside Plant compound in which the sugar constituent is glucose

Glucosidase The enzyme which splits glucoside

GPx An enzyme; short form of glutathione peroxidase

Haemetemesis Vomiting out blood

Haemoptysis Coughing out blood

Haemolysis Disintegration of red blood cells

Haemopoeitic Stimulating the formation of blood

Haemoptysis Coughing out blood

Hepatoma Liver cancer

Hippocampal Pertaining to the hippocampus

Hippocampus The portion of the brain that deals with memory

Hippocampal Pertaining to the hippocampus

Hirsute Covered with rough hairs or bristles

Histochemical Microscopic examination of tissue after appropriate staining to display the compound under study

Holomycotrophic Referring to plants that are entirely dependent on fungi for complex carbon compounds

Homologous Identical in structure and origin

Hyperplasia Excessive growth of normal cells

Hypoxia Oxygen deprivation

Ileum Distal part of the small intestine

Imbricate Overlapping in an orderly manner, like scales on an armadillo

Infarct An area of necrosis (cell death) caused by interruption to its blood supply

Interferon A protein the body produces in response to infection

Intracarotid Into the carotid arteries at the neck

In vitro In a test tube

In vivo In a living animal

Ischaemia Being deprived of its blood supply

Iso-enzyme Enzyme with similar activity

kapha An Ayurvedic term referring to heavyweight or obese physique

kepialu **(Malay)** Prolonged or severe fever

Kinase A compound which activates an enzyme

Ketone Carbon compound formed by incomplete oxidisation and breakdown of sugar in the body

Lignin Compound derived from wood

Lingulate Resembling a tongue

Lithophyte A plant that grows on rocks

Lorate Strap

Lumbago Low backache

Lymphocyte White blood cell

Macrophage A large, mononuclear cell in the blood involved in the body's defence matrix

Metalloproteinase Enzyme involved in degradation of extracellular matrix

Melanoma Black-pigmented tumour of the skin

Meta-analysis Bringing together and analysing the results of several studies

Myrmecochores Plants which house an ant colony

Mesangial cells Specialised smooth muscle cells in renal capillaries that regulate blood flow through the kidneys

Metacercaria Encystic cercaria present in water, aquatic vegetation, fish and crabs, constituting the infective source for man and animals

MIT assay A cell proliferation assay

Mitochondria Rod-shaped structures in cytoplasm

Mononuclear Possessing one nucleus. Mononuclear cells are large white blood cells

Monopodial Growth habit in which new leaves arise from the same meristem as all previous leaves

Mutagenic Stimulating mutation

Mutation A permanent, transmissible change in the characteristics of an offspring which sets it apart from its parents

Myasthenia Muscle weakness

Myoclonia Involuntary, irregular contractions of muscle groups

Mycoheterotrophic Referring to plants that are partially or totally dependent on fungi for complex carbon compounds (their energy source)

Mycorrhiza Symbiotic fungi dwelling in the roots of higher plants

Mycotrophic Referring to plants that are totally dependent on fungi for complex carbon compounds

Myrmecochores Plants which employ ants to assist in seed dispersal (e.g. *Acriopsis*)

Necrosis Death of a cell or a group of cells that are in contact with living cells

Neovascularization Formation of new blood vessels

Neurasthenia Nervous exhaustion and abnormal fatigue

Neuroblastoma A form of brain cancer

Neuron Nerve cell

Neurotoxin Poison affecting the nervous system

Nocturesis Nocturnal urinary frequency

Obovate Between oblong and oval

Obtuse With a blunt or rounded tip

Osteoclast Cells which break down bone and deplete bone of calcium

Osteoclastogenesis Generation of osteoclasts

Ovalbumin Egg white

Ovate With an oval outline

Ovoid Egg-shaped

Paniculate Branched, resembling an inflorescence

Parkinsonism Chronic condition characterized by muscular rigidity but without resting tremor

Patana Scrub

Pectinate Resembling a comb

Pedicellate Resembling the stalk to which the flower is joined

Peloric Referring to a floral malformation such as petals taking on the appearance of a lip

Peloton Cell containing coils of fungal hyphae

Peranakan A descendant of early Chinese immigrants to Southeast Asia

Perianth The whorl of sepals and petals of a flower

Peroxidase An enzyme which promotes oxidation of substances

Pharmaceutical Pertaining to the development of drugs

Pharamacokinetics Metabolism of a drug within the body

Phenanthrene Polycyclic aromatic hydrocarbon composed of three fused benzene rings

Phenanthropyran A phenanthrene derivative with a five-carbon ring

Phenol Organic compound with molecular formula C6H5OH; carbolic acid

Pheromone A chemical substance which serves as a sexual attractant for individuals of the same species

Phytoalexin Secondary product of plants that ward off predators

Phytohaemagglutinin A compound present in red beans which causes agglutination of human red blood cells. It stimulates mitosis of white blood cells and it is employed for chromosome studies

PICA activity Primary inventory control activity

Platelet Small, non-nucleated particle in blood that plays an important role in clot formation

Pleurodynia A form of muscular rheumatism characterised by paroxysmal pain affecting the intercostal muscles

Pleurisy Inflammation of the covering of the lungs with formation of exudate (fluid)

Praemorse Appearing as though bitten off at the tip

Prostaglandin A ubiquitous compound found in plants and animal tissue with numerous physiological actions

Proteasome A multi-enzyme complex that regulates proteins involved in cell division and apoptosis. It is a target for anticancer therapy

Pseudobulb The swollen stem of a sympodial orchid plant

Pollinum Coherent mass of pollen grains

Poultice Herbs made into a paste

Pubescent Covered with short hairs

Raceme Inflorescence with flowers arranged on a single axis

Racemose Having flowers arranged in a single axis on an inflorescence

Reflexed Bent backwards

Resupinate Referring to an orchid flower in which the lip occupies the lowest position, lower than the sepals and petals. This is achieved by twisting of the pedicle or arching of the inflorescence

Retuse Possessing a rounded apex with a shallow notch

Rheumatism Inflammation of the connective tissue in muscle and joints with pain in these parts

Rhizome The horizontal portions of a sympodial plant which give rise to new shoots and flowers

Rhizosphere The narrow region of the soil which is directly influenced by root secretions and is associated with soil micro-organisms

Rugose Wrinkled

Sagitate Shaped like an arrow

sampaquita Jasmine (Tagalog)

Saprophyte Plant which derives its nourishment from dead organic matter. Achlorophyllous orchids are not saprophytic: they are parasitic on fungi

Saxicolous Growing on rocks, lithophytic

Saxitoxin A nerve poison

Schistosomiasis Disease of the liver caused by infestation of a parasite, the liver fluke,

Schistosoma. It leads to cirrhosis, liver abscess and liver cancer

Schizophrenia A serious mental disease characterised by breakdown of mental functions, hallucination, catatonia and, in severe forms, violence

Scrapie A form of dysentery in sheep once thought to be caused by bacteria; can also be caused by prions

Scrofula Tuberculosis of the lymph nodes and bones

Semiochemicals Small organic compound or mixtures of compounds that carry a signal for another organism. If it affects individuals of the same species, it is termed a pheromone, and of a different species, an allelochemical

Senescence Process of ageing

Senescent Ageing, old

Serotonin Neurotransmitter; it contracts blood vessels and regulates intestinal activity

Sessile Devoid of a stalk

shola Patches of broad-leaved evergreen forest scattered amidst montane grasslands in the Deccan

Spatulate Resembling a spatula, narrow at the base and expanding to a broad, rounded apex

Spectroscopy A process for analysing a compound that yields information about its molecular weight , functional groups and counts of atoms in different parts of the molecule. Originally, it referred to the study of visible coloured light dispersed by a prism

Spermiopoeitic An Anglo-Indian word which refers to sperm (or semen) production

Spur Saccate of tubular extension of he lip. It usually contains nectar

Stilbenoid Hydroxylated derivatives of stilbene. They are secondary products of plants that function as phytoalexins

Strangury Slow, painful discharge of urine

Sternutatory An agent that causes sneezing

Stipe Pollinium stalk

Styptic Stops bleeding

Subglobose Almost but not entirely globular

Sympodial Growth habit in which new shoots arise from axillary buds on the rhizome

Synapse The junction between nerve cells where nerve signals are transmitted

Tau protein Normally, tau protein stabilises microtubules in nerve cells. Defective tau protein forms neurofibrillary tangles and is associated with Alzheimer's disease

Taxonomy Science of identification, classification and naming of organisms

Tepals Sepals and petals which look alike

terai **areas** Mountain scrub

Terrestrial Growing on the ground

Terete Pencil-like

Tetramerous With four parts in a whorl, referring to floral parts

Tetraploid Possessing four sets of chromosomes

Thalamus A central area in the brain which serves as a relay station for sensory messages

Tinnitus Buzzing in the ears

Tonic Medicinal preparation alleged to have the ability to restore normal tone to tissues, i.e. restore health

Trabecular bone Spongy bone vertebrae, ribs, sternum

Transgenic Genetically modified by the introduction of a foreign gene

Transfection Transfer of genetic material by making use of an infective organism

Transcription First step in gene expression

Tubulin A vital cell protein that has an important role in cell division

Tuber Thickened, underground, storage root or stem

Ultramafic A special soil type which is usually inhospitable to plants but which hosts certain species of orchids, e.g. in Sabah, East Malaysia. It is high in iron and nickel but low in calcium, potassium and phosphorus

Vascular Pertaining to the vessels, blood vessels in animals, phloem and xylem in plants

Vata An Ayurvedic term which classifies personalities into three types: vital, energetic, and hyperactive

Vernalization Subjected to cold temperature for a length of time, it stimulates floral bud initiation

Vulnerary An agent that promotes wound healing

Vulnerary Wound healing

'Wind' In in traditional Chinese medicine, 'wind' refers to stroke, seizures, aphasia, slurred speech, blurred vision, numbness, tightness of extremities, headache, vertigo, dizziness

Xenograft A graft from a foreign species

Xylose A pentose sugar from wood

yang **element** A concept in traditional Chinese medicine. *Yang* must always be balanced with *yin*. Excessive *yang* results in "heat", headache, sore throat, red eyes, restlessness

yin **element** A concept in traditional Chinese medicine. Excessive *yin* results in ennui, cold limbs, weakness

Zymogram Output from a fermentation process

Index[1]

---

[1] **NOTE**: Names in bold are the accepted names of medicinal Asian orchid species

© Springer International Publishing Switzerland 2016
E.S. Teoh, *Medicinal Orchids of Asia*, DOI 10.1007/978-3-319-24274-3

Printed by Printforce, the Netherlands